AF333704

# EXPERIMENTAL MODELS OF MULTIPLE SCLEROSIS

# EXPERIMENTAL MODELS OF MULTIPLE SCLEROSIS


Edited by

Ehud Lavi
Weill Medical College of Cornell University, New York

Cris S. Constantinescu
University Hospital, Queen's Medical Centre, Nottingham, UK

Library of Congress Cataloging-in-Publication Data

Experimental models of multiple sclerosis / edited by Ehud Lavi, Cris S. Constantinescu.
     p. cm.
   Includes bibliographical references and index.
   ISBN-10: 0-387-25517-6 (alk. paper)
   ISBN-10: 0-387- 25518-4 (e-book)
   ISBN-13: 978-0387-25517-0
   ISBN-13: 978-0387-25518-7
   1. Multiple sclerosis.  2. Multiple sclerosis—Animal models.  I. Lavi, Ehud. II.
Constantinescu, Cris.S.

   RC377.E93 2005
   616.8'34—dc22

                                                                    2005043435

# Contents

# Dedication

The book is dedicated with love to our families:
Dara, Judy, Karen, Irene, Simona and Phillip.

# Preface

## Experimental Models of Multiple Sclerosis

Multiple Sclerosis (MS) is an enigmatic immune-mediated disease of the central nervous system that affects around 350,000 individuals in the United States. The pathologic hallmarks of this disease are inflammation and demyelination (the destruction of the myelin layer that insulates nerve cell processes, impeding the conduction of neuronal impulses within the brain and spinal cord). The mechanism of demyelination in MS is not completely understood. Immune modulating therapy can delay the progression of MS but there is still no cure for the disease. Thus the development of experimental animal models for MS is among the most powerful tools for studying its pathogenesis and for testing the safety and efficacy of experimental therapies before using these drugs in clinical trials on human patients.

The main experimental models for MS are divided into immune-mediated models or virus-induced models. In immune-mediated models an autoimmune reaction against myelin is induced by the injection of myelin molecules (or portions of it), in combination with immune boosters (Freund's adjuvant). The prototype of this category is EAE, or experimental allergic (or autoimmune) encephalomyelitis. This model uses a variety of experimental animals including mice, rats, rabbits and even primates. The viral models include several ubiquitous animal viruses such as coronaviruses (mouse hepatitis virus) and enteroviruses (Theiler's virus) in mice, visna virus in sheep, and distemper virus in dogs. The virus-induced model systems provide both a substrate for studying the pathogenesis of the disease and a hypothetical mechanism for virus-induced autoimmunity in humans.

In this book we have assembled the different experimental models for MS (both immune-mediated and viral). We have asked MS experts to examine aspects of these experimental models highlighting differences and similarities. Each chapter in this book deals with an element of an MS model: from individual cellular and molecular CNS factors, to cellular and molecular elements of the immune system. Thus the reader is able to read parallel chapters that deal with the same factor in different model systems. The reader of the book will be able to summarize and integrate the different disease components and highlight the common features and the differences that exist in the different model systems. We realize that a book of such magnitude cannot keep pace with all the new knowledge of immunology (in particular new cytokines being discovered), but it offers a conceptual framework. We hope that this book will improve our understanding of the mechanism of demyelination and the pathogenesis of MS. The analysis of common features in the various experimental models of demyelination may also stimulate the development of new ideas about prevention of demyelination and ultimately the treatment of MS.

Ehud Lavi and Cris S. Constantinescu<br>Editors

Part A

# EXPERIMENTAL ALLERGIC (AUTOIMMUNE) ENCEPHALOMYELITIS (EAE)

Chapter A1

# EAE: HISTORY, CLINICAL SIGNS, AND DISEASE COURSE

J. William Lindsey, MD

*Department of Neurology, University of Texas Health Science Center at Houston, Houston, Texas, USA*

**Abstract:**    EAE is a useful animal model of autoimmune disease of the central nervous system. In this chapter we discuss the history of EAE, including the original description of the model and the subsequent major developments, and then describe the usual clinical signs and disease course.

**Key words:**    experimental autoimmune encephalomyelitis, encephalitogen, adoptive transfer, active EAE, history, relapses, transgenic models

## 1.    INTRODUCTION

Experimental allergic encephalomyelitis (EAE) is an animal model of autoimmune disease of the central nervous system (CNS). It resembles multiple sclerosis (MS) in many respects, and is used in the laboratory investigation of MS. EAE also serves as a model for the study of organ specific autoimmune diseases in general. EAE was first described over 50 years ago, and it has been a popular and frequently used model. A Medline search of the term identifies more than 5000 citations, with more than 2500 references since 1990. In this chapter, we will review the history of EAE and describe the general features of the model, including the usual clinical signs and disease course.

## 2.     HISTORY

As mentioned in the introduction, there has been an enormous amount of work done in EAE over the last 50 years. In this section, we will briefly summarize the major developments in the field. These include the early studies that led to the establishment of the model, the extensive work done on various antigens, the development of adoptive transfer disease, the development of relapsing EAE, and recent work with EAE in transgenic mice.

## 2.1     Early Studies

The earliest attempts at producing EAE were efforts to understand the pathogenesis of post rabies vaccination encephalomyelitis. Pasteur's rabies vaccine consisted of a suspension of dessicated spinal cords from rabbits infected with rabies. After a series of injections, occasional patients developed an encephalomyelitis which was distinct from rabies. Early investigators were able to induce a similar encephalomyelitis in rabbits or monkeys by administering repeated injections of neural tissue, thus demonstrating that post rabies vaccine encephalomyelitis probably resulted from the unintentional induction of an autoimmune response against neural antigens [1-3].

In these early experiments the incidence of disease was low, and induction of disease required multiple injections over a period of many weeks. The introduction of complete Freund's adjuvant (CFA) made induction of EAE much simpler and more reliable. CFA consists of a mixture of mineral oil and *Mycobacterium* made into an emulsion with the antigen, and injecting antigen in CFA usually induces a vigorous and prolonged immune response against the antigen. Several investigators were quick to appreciate the usefulness of this technique. In 1946, Kabat et al published a preliminary report of the induction of EAE in 3 of 4 monkeys using three weekly injections of rabbit brain emulsion in Freund's adjuvant [4]. This was followed rapidly by more definitive reports of similar results in monkeys, rabbits, and guinea pigs [5-8]. In each case the majority of animals developed neurologic deficits after one or a few injections of neural tissue in Freund's adjuvant, and the onset of disease occurred three to four weeks after the first injection. EAE was subsequently induced in many other species, including dogs, cats, rats, mice, sheep, goats, pigs, chickens, and pigeons.

Attempts to induce EAE in mice with antigen in CFA were initially unsuccessful. Olitsky and Yager reported on EAE in mice two years later

[9]. The majority of mice developed neurologic symptoms after three to five injections of brain tissue in CFA. The symptoms began about three weeks after the first injection. They noted that their two strains of mice differed substantially in susceptibility to EAE, and future studies verified that susceptibility to EAE varies widely between strains. Subsequently, Lee and Olitsky demonstrated that pertussis vaccine given as a separate injection in addition to antigen in CFA increased the incidence of disease [10]. Munoz and coworkers found that the active ingredient in pertussis vaccine was the toxin, and purified pertussis toxin is now routinely used as an additional adjuvant to induce EAE in mice [11].

## 2.2 Identification of encephalitogens

The experiments described above were done with homogenate or extract of CNS tissue as the antigen. An early focus of EAE research was identifying the encephalitogenic component of the homogenate. The encephalitogen was present in white matter or myelin rather than gray matter, and appeared to be protein rather than lipid. At the time, there was much controversy over whether brain homogenate contained one or multiple encephalitogens, and over which of the partially purified mixtures prepared with the limited methods available was the relevant encephalitogen. In retrospect, it appears that both of major structural proteins of myelin, myelin basic protein (MBP) and proteolipid protein (PLP), were demonstrated to be encephalitogenic at an early date [12-14].

Subsequent work with MBP demonstrated that the intact protein was not required to induce EAE. Specific fragments of MBP generated by pepsin or trypsin digestion of the whole molecule were encephalitogenic. And in 1970 Eylar et al. induced EAE with a 9 amino acid synthetic peptide corresponding to part of the sequence of MBP [15]. This peptide was as active on a molar basis as the intact molecule. Since then, multiple T cell epitopes for MBP, PLP, and other encephalitogenic proteins have been reported, and synthetic peptides comprising encephalitogenic T cell epitopes are now a standard reagent for inducing EAE.

Recently many other myelin proteins or peptides based on myelin proteins have been demonstrated to be encephalitogenic. Other encephalitogenic myelin proteins include myelin oligodendrocyte glycoprotein (MOG), myelin-associated oligodendrocytic basic protein, and oligodendrocyte specific protein [16-20]. MOG is particularly interesting since it induces relapsing EAE with extensive demyelination. Non-myelin proteins, such as S100$\beta$ and glial fibrillary acidic protein (GFAP) also cause EAE, but the distribution of the lesions is different than that seen with myelin proteins [21, 22]. Not all CNS proteins will induce EAE. The myelin protein 2',3'-cyclic nucleotide 3'-phosphodiesterase and the stress protein alphaB crystallin did not induce EAE [19, 23].

## 2.3      Adoptive or passive EAE

The results discussed above have all been in active EAE, where the disease is induced by injection of CNS antigen in adjuvant. In this model, the induction phase and the effector phase of EAE occur in the same animal. In the induction phase, autoreactive cells are activated and multiply in peripheral lymph nodes in response to the injection of antigen and adjuvant. In the effector phase, the activated autoreactive cells migrate to the CNS and cause autoimmune damage and clinical signs. Another model of EAE is passive or transferred or adoptive disease. In passive EAE, the autoreactive cells specific for CNS antigens are generated in one group of animals, and disease is induced in a second group of animals by transfer of these autoreactive cells. In this way, one can study the effector phase in isolation from the induction phase.

Transfer of disease from EAE induced animals to naïve animals was first achieved by Paterson in 1960 using an elegant experimental design to overcome the difficulty of using outbred experimental animals [24]. Subsequent investigators refined this model by using inbred animals to avoid histocompatibility problems and by stimulating the donor cells *in vitro* with either mitogen or antigen to increase their encephalitogenic activity [25-27]. CNS antigen-specific T cell lines or clones maintained for prolonged periods *in vitro* can be used instead of freshly harvested lymph node cells [28]. Passive or transferred EAE is a very reproducible model of EAE since the variability of the induction phase is eliminated.

## 2.4      Relapsing EAE

Active or passive EAE is usually an acute, monophasic process similar to the human diseases acute disseminated encephalomyelitis or acute transverse myelitis. Investigators interested in multiple sclerosis would prefer a model with relapsing or progressive symptoms to more closely mimic the course of MS, and methods to produce such a model have been described. The induction of relapsing instead of acute EAE depends on the strain of animal, the antigen used, and the induction regimen. A relapsing course has been described in both active and passive EAE, and different regimens have been reported to reliably produce relapsing rather than acute disease [29, 30]. The tendency to relapse varies between strains of mice [31].

## 2.5      **Transgenic models of EAE**

The EAE model continues to develop as new methods and techniques become available. One recent development is the use of transgenic mice. Two types of transgenic mice have been constructed for the study of EAE. The first type is T cell receptor (TCR) transgenic mice. These mice express a receptor specific for MBP on all or most of their T lymphocytes [32, 33]. These mice may develop EAE following injection with adjuvant alone, and mice not maintained in sterile pathogen-free environments may develop spontaneous cases of EAE. A second type of transgenic mice was engineered to express a viral protein in oligodendrocytes [34]. Subsequent systemic infection with the corresponding virus resulted in autoimmune CNS damage.

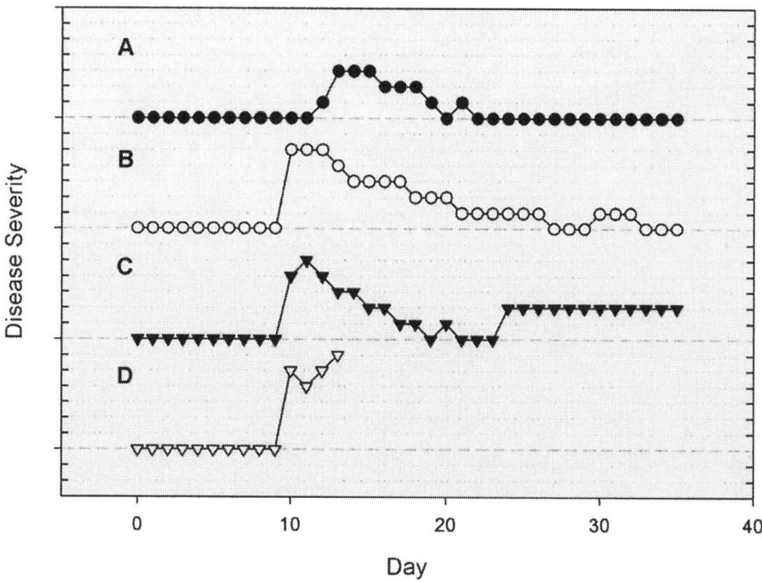



*Figure A1-1.* Clinical course of EAE. Representative examples of EAE in SJL mice injected with spinal cord homogenate in CFA and pertussis (see reference 35 for detailed methods). Mice were rated daily for severity of EAE using the scale in Table 1-1. Time in days is on the abscissa, with the day of injection as day 0. Severity of disease is on the ordinate. Mouse A developed monophasic EAE starting on day 12 and resolving by day 23. Mouse B had more severe disease. Mouse C recovered from the initial bout of disease, and then relapsed and had a chronic deficit. Mouse D had severe EAE with rapid onset and died on day 13.

# 3.    CLINICAL SIGNS AND DISEASE COURSE

The clinical signs and disease course of EAE vary depending on the strain and species of animal, the inciting neural antigen, the adjuvants, and the timing and dose of antigen and adjuvant. We will describe EAE induced in SJL mice by spinal cord homogenate with mycobacteria and pertussis toxin as adjuvants as a typical example [35]. The first sign is usually ruffled fur and weight loss. One or two days later the animal develops an ascending paralysis. This begins with loss of tail tone and progresses to hind limb weakness and sometimes forelimb weakness. Severely affected mice develop quadriparesis and labored respiration and may die from the disease. The weakness starts about 14 days after injection and worsens over 1 to 4 days. In mice which recover, the symptoms last about 7 days, and relapses after recovery are rare. The clinical course of EAE in representative individual mice is depicted in Figure 1-1.

Severity of disease is rated on an ordinal scale. The scale we use is given in Table 1-1 [31, 35]; other investigators use similar scales with different number of stages or different definitions of stages. We titrate the dose of antigen to minimize the number of mice with stage 5 or 6 disease. Proper veterinary care of severely affected mice is essential.

*Table A1-1.* Clinical rating scale for EAE

| Stage | Clinical deficit |
|---|---|
| 0 | no deficits |
| 1 | limp tail or slowed righting response |
| 2 | limp tail and slowed righting response |
| 3 | hind limb weakness with abnormal gait |
| 4 | hind limb paralysis, mobile using forelimbs |
| 5 | hind limb paralysis, forelimb weakness, mobility impaired |
| 6 | moribund |

The ascending paralysis described above is typical in mice, rats, and guinea pigs. Other species have different clinical courses. Rabbits may have ataxia and gait difficulties, while monkeys may have visual loss, cranial nerve deficits, and ataxia in addition to paralysis.

The acute, monophasic clinical course described above is typical for EAE in rodents. Other disease courses are seen, including a hyperacute form and relapsing or chronic forms of disease. The hyperacute form of disease is characterized by a short latent period between injection and symptom onset and a fulminant course with a high mortality. Chronic or relapsing disease has been described with particular strains of animals and with particular induction regimens.

## ACKNOWLEDGEMENTS

Work in our laboratory has been supported in part by the Clayton Foundation for Research and the National Multiple Sclerosis Society.

## REFERENCES

1. Koritschoner, R. and F. Schweinburg, Klinische und experimentelle Beobachtungen über Lähmungen nach Wutschutzimpfung. *Zeitschrift für Immunitätsforschung und experimentelle Therapie*, 1925. **42**:217-283.
2. Rivers, T.M., D.H. Sprunt, and G.P. Berry, Observations on attempts to produce acute disseminated encephalomyelitis in monkeys. *Journal of Experimental Medicine*, 1933. **58**:39-53.
3. Rivers, T.M. and F.F. Schwentker, Encephalomyelitis accompanied by myelin destruction experimentally produced in monkeys. *Journal of Experimental Medicine*, 1935. **61**:689-702.
4. Kabat, E.A., A. Wolf, and A.E. Bezer, Rapid production of acute disseminated encephalomyelitis in rhesus monkeys by injection of brain tissue with adjuvants. *Science*, 1946. **104**:362-363.
5. Kabat, E.A., A. Wolf, and A.E. Bezer, The rapid production of acute disseminated encephalomyelitis in rhesus monkeys by injection of heterologous and homologous brain tissue with adjuvants. *Journal of Experimental Medicine*, 1947. **85**:117-129.
6. Morgan, I.M., Allergic encephalomyelitis in monkeys in response to injection of normal monkey nervous tissue. *Journal of Experimental Medicine*, 1947. **85**:1331-140.
7. Morrison, L.R., Disseminated encephalomyelitis experimentally produced by the use of homologous antigen. *Archives of Neurology and Psychiatry*, 1947. **58**:391-416.
8. Freund, J., E.R. Stern, and T.M. Pisani, Isoallergic encephalomyelitis and radiculitis in Guinea pigs after one injection of brain and mycobacteria in water-in-oil emulsion. *Journal of Experimental Medicine*, 1947. **57**:179-194.
9. Olitsky, P.K. and R.H. Yager, Experimental disseminated encephalomyelitis in white mice. *Journal of Experimental Medicine*, 1949. **90**:213-223.
10. Lee, J.M. and P.K. Olitsky, Simple method for enhancing development of acute disseminated encephalomyelitis in mice. *Proceedings of the Society for Experimental Biology and Medicine*, 1955. **89**:263-266.
11. Munoz, J.J., C.C.A. Bernard, and I.R. Mackay, Elicitation of experimental allergic encephalomyelitis (EAE) in mice with the aid of pertussigen. *Cellular Immunology*, 1984. **83**:92-100.
12. Olitsky, P.K. and C. Tal, Acute disseminated encephalomyelitis produced in mice by brain proteolipide (Folch-Lees). *Proceedings of the Society for Experimental Biology and Medicine*, 1952. **79**:50-53.
13. Kies, M.W. and E.C. Alvord, *Encephalitogenic activity in guinea pigs of water-soluble protein fractions of nervous tissue.*, in *"Allergic" encephalomyelitis*, M.W. Kies and E.C. Alvord, Editors. 1959, Charles C Thomas: Springfield, IL. p. 293-299.
14. Kies, M.W., J.B. Murphy, and E.C. Alvord, Fractionation of guinea pig brain proteins with encepahlitogenic activity. *Federation Proceedings*, 1960. **19**:207.
15. Eylar, E.H., et al., Experimental allergic encephalomyelitis: Synthesis of disease-inducing site of the basic protein. *Science*, 1970. **168**:1220-1223.

16. Linington, C., et al., T cells specific for the myelin oligodendrocyte glycoprotein mediate an unusual autoimmune inflammatory response in the central nervous system. *European Journal of Immunology*, 1993. **23**:1364-1372.

17. Amor, S., et al., Identification of epitopes of myelin oligodendrocyte glycoprotein for the induction of experimental allergic encephalomyelitis in SJL and Biozzi AB/H mice. *Journal of Immunology*, 1994. **153**:4349-4356.

18. Johns, T.G., et al., Myelin oligodendrocyte glycoprotein induces a demyelinating encephalomyelitis resembling multiple sclerosis. *Journal of Immunology*, 1995. **154**:5536-5541.

19. Maeaettae, M.A., et al., Encephalitogenicity of myelin-associated oligodendrocytic basic protein and 2',3'-cyclic nucleeotide 3'-phosphodiesterase for BALB/c and SJL mice. *Immunology*, 1998. **95**:383-388.

20. Stevens, D.B., et al., Oligodendrocyte-specific protein peptides induce experimental autoimmune encephalomyelitis in SJL/J mice. *Journal of Immunology*, 1999. **162**:7501-7509.

21. Kojima, K., H. Wekerle, and C. Linington, Experimental autoimmune panencephalitis and uveoretinitis transferred to the Lewis rat by T lymphocytes specific for the S100Beta molecule, a calcium binding protein of astroglia. *Journal of Experimental Medicine*, 1994. **180**:817-830.

22. Berger, T., et al., Experimental autoimmune encephalomyelitis: The antigen specificity of T lymphocytes determines the topography of lesions in the central and peripheral nervous system. *Laboratory Investigation*, 1997. **76**.

23. Thoua, N.M., et al., Encephalitogenic and immunogenic potential of the stress protein alphaB-crystallin in Biozzi ABH (H-2Ag7) mice. *Journal of Neuroimmunology*, 2000. **104**:47-57.

24. Paterson, P.Y., Transfer of allergic encephalomyelitis in rats by means of lymph node cells. *Journal of Experimental Medicine*, 1960. **111**:119-135.

25. Stone, S.H., Transfer of allergic encephalomyelitis by lymph node cells in inbred guinea pigs. *Science*, 1961. **134**:619-620.

26. Panitch, H.S. and D.E. McFarlin, Experimental allergic encephalomyelitis: Enhancement of cell-mediated transfer by concanavalin A. *Journal of Immunology*, 1977. **119**:1134-1137.

27. Richert, J.R., et al., Adoptive transfer of experimental allergic encephalomyelitis: Incubation of rat spleen cells with specific antigen. *Journal of Immunology*, 1979. **122**:494-496.

28. Ben-Nun, A., H. Wekerle, and I.R. Cohen, The rapid isolation of clonable antigen-specific T lymphocyte lines capable of mediating autoimmune encephalomyelitis. *European Journal of Immunology*, 1981. **11**:195-199.

29. Brown, A.M. and D.E. McFarlin, Relapsing experimental allergic encephalomyelitis in the SJL/J mouse. *Laboratory Investigation*, 1981. **45**:278-284.

30. Mokhtarian, F., D.E. McFarlin, and C.S. Raine, Adoptive transfer of myelin basic protein-sensitized T cells produces chronic relapsing demyelinating disease in mice. *Nature*, 1984. **309**:356-358.

31. Lindsey, J.W., Characteristics of initial and reinduced experimental autoimmune encephalomyelitis. *Immunogenetics*, 1996. **44**:292-297.

32. Goverman, J., et al., Transgenic mice that express a myelin basic protein-specific T cell receptor develop spontaneous autoimmunity. *Cell*, 1993. **72**:551-560.

33. Lafaille, J.J., et al., High incidence of spontaneous autoimmune encephalomyelitis in immunodeficient anti-myelin basic protein T cell receptor transgenic mice. *Cell*, 1994. **78**:399-408.

34. Evans, C.F., et al., Viral infection of transgenic mice expressing a viral protein in oligodendrocytes leads to chronic central nervous system autoimmune disease. *Journal of Experimental Medicine*, 1996. **184**:2371-2384.
35. Lindsey, J.W., M. Pappolla, and L. Steinman, Reinduction of experimental autoimmune encephalomyelitis in mice. *Cellular Immunology*, 1995. **162**:235-240.

Chapter A2

# INDUCTION OF EAE

Eli Ben-Chetrit and Stefan Brocke
*Department of Pathology, Hebrew University Hadassah Medical School, P.O.Box 12272, 91120 Jerusalem, Israel*

**Abstract**:  Experimental autoimmune encephalomyelitis (EAE) can be induced by two principal approaches. The first is based on direct immunization with autoantigen in combination with different types of adjuvant. The second approach of inducing EAE consists in adoptive transfer of activated T cells that are specific for myelin-associated autoantigens of the central nervous system (CNS). This chapter presents an overview of some useful strategies and protocols by which EAE can be induced, and discusses the pros and cons of various approaches.

**Key words**:  Experimental autoimmune encephalomyelitis, active immunization, adoptive transfer, adjuvant, myelin antigens

## 1. INTRODUCTION

Methods of inducing experimental autoimmune encephalomyelitis (EAE) have come a long way since initial observations that postvaccinal encephalomyelitis is caused by spinal cord contaminants of rabies vaccine preparations rather than by the inactivated virus itself (1, 2). There are many species susceptible to EAE induction, including mice (3), rats (4), guinea pigs (5), pigs (6) and non-human primates (7). In addition, a human version of accidental EAE, 'human EAE', has been described (8, 9).

In 1949, Olitsky and Yager described the induction of experimental disseminated encephalomyelitis in white mice, thus establishing mouse EAE as a model for autoimmune inflammatory diseases of the central nervous system (CNS) (3). EAE in the mouse shares some clinical features with the human disease multiple sclerosis (MS), most prominently paresis/paralysis and ataxia. Depending on the model chosen, EAE can take an acute, a chronic, and/or relapsing-remitting course. Inadvertently induced 'human

EAE' took a course with striking resemblance to a rare form of MS, acute MS (8, 9). Sensitization with myelin antigens often results in perivenous encephalomyelitis and in some cases, depending on the animal model, in histopathologic changes similar to those seen in MS (9).

There is ample evidence that susceptibility of different animal strains to EAE is linked to MHC class II determinants (10, 11), although non-MHC genes also influence disease expression. Thus, murine EAE can be considered a useful model for some aspects of MS (12).

There are several strategies to induce EAE in animals. EAE can be induced by immunization of animals with mouse spinal cord homogenate (MSCH) or isolated myelin proteins and peptides. Alternatively, EAE is generated by adoptive transfer of lymphocytes that have been activated for myelin antigens *in vivo* and/or *in vitro* (11). Both immunized animals as well as mice expressing a transgenic T cell receptor specific for myelin basic protein (MBP) or other myelin antigens can function as a source of T cells suitable for adoptive transfer (13, 14, 15).

Several immunodominant epitopes of myelin components have been characterized, and peptides representing epitopes of MBP, proteolipid protein (PLP), or myelin oligodendrocyte glycoprotein (MOG) are widely used for the induction of EAE. (10, 16-20). As mentioned, the encephalitogenicity of myelin-derived peptides is dependent on the expression of particular MHC class II antigens in the appropriate mouse strain. Thus, induction of disease using PLP peptide $PLP_{139-151}$, MBP peptide $MBP_{87-99}$ or MOG peptide $MOG_{92-106}$ is suitable in SJL ($I-A^s$) mice. The N-terminal peptide $MBP_{Ac1-11}$ or $MBP_{87-99}$ are suitable in (PL X SJL)$F_1$ mice. $MOG_{35-55}$ is routinely used to induce EAE in C57BL/6 mice.

Of note, the human MOG-derived peptide $MOG_{97-108}$, which has a high binding affinity for the (human) HLA-DR4 MHC class II allele, induces severe EAE in HLA-DR4 transgenic mice (21).

In a recent study, inoculation of myelin-associated oligodendrocytic basic protein and one of its derived peptides, $MOBP_{37-60}$, were reported to produce severe clinical and histopathologic signs of EAE in SJL mice (22). Moreover, the encephalitogenic potential of synthetic peptides of myelin-associated glycoprotein (peptide $MAG_{97-112}$) and oligodendrocyte-specific glycoprotein (peptide $OSP_{57-72}$) has recently been demonstrated in ABH (H-$2A^{g7}$) and SJL mice, respectively, (23).

EAE in the rat is typically an acute paralytic disease from which most animals spontaneously recover. However, also chronic models have recently been described (24). Dominant MBP epitopes in Lewis rats are $MBP_{68-86}$ (25, 26), $MBP_{73-86}$ and $MBP_{87-99}$ derived from guinea pig myelin as well as rat $MBP_{73-86}$ (27). MAG has also been used in the induction of EAE in Lewis rats. In DA rats, $MBP_{63-81}$, $MBP_{79-99}$, $MBP_{101-120}$ or $MBP_{142-167}$ can be used for induction of EAE (28). DA rats are also susceptible to PLP-induced EAE.

Recently, EAE was induced in the swine by minipig spinal cord homogenate inoculation. The minipigs presented a monophasic course of disease with spontaneous improvement (6).

Weir et al. demonstrated induction of EAE by transferring bone-marrow-derived dendritic cells presenting $MOG_{35-35}$ into naïve C57BL/6 mice (29).

Finally, rabies vaccination in humans has been reported to result in sensitization against neural antigens and lead to acute perivenous encephalomyelitis clinically and histopatholigically indistinguishable from some types of EAE in animals or even MS-like lesions (8, 9). Such complications, however, have been very rare.

## 2. EAE INDUCED BY IMMUNIZATION (ACTIVE EAE) OR BY ADOPTIVE T CELL TRANSFER (PASSIVE EAE, AT-EAE)

## 2.1 Overview

Active EAE enables the study of both immunization and effector phases in experimental demyelinating disease. Typical immunization protocols require the emulsification of encephalitogens in complete Freund's adjuvant (CFA) containing inactivated mycobacteria such as *Mycobacterium tuberculosis* strain H37RA. Both reagents are obtainable from suppliers such as BD-DIFCO. Antigen concentrations and protocols used for immunizations vary widely between groups and should be adjusted to each laboratory's own conditions.

Direct immunization may complicate the interpretation of experimental data on EAE induction and expression. Due to the severe long-term inflammation caused by the use of CFA, some of the observed effects could be attributed to the mode of immunization rather than to the pathogenic process of inflammation and demyelination in the CNS. The induction of active EAE in some rat and mouse models is significantly enhanced by the administration of heat-killed whole organisms of *Bordetella pertussis* (30, 31) or their derived toxin, pertussis toxin (PTX) (32).

EAE can be induced by transferring sensitized lymph node cells (LNCs) from mice or rats that have been immunized with MSCH, MBP, PLP or other CNS antigens (33-36). Induction of EAE by cell transfer allows the separate study of the effector phase of the disease without the influence of immunization and inflammation outside the CNS. More importantly, the prime mediators of this type of disease, myelin antigen-specific T cells, can be studied in detail *in vitro* by standard immunological techniques such as proliferation assays for the determination of antigen specificity, surface marker analysis by flow cytometry, cyto- and chemokine measurements, and T cell receptor sequencing. *In vitro* analysis of enriched or cloned antigen-specific autoreactive T cells is essential for our understanding of cellular and molecular properties associated with demyelinating disease. In transfer protocols, cells can either be directly transferred after dissection of draining lymph nodes of immunized donor animals or be cultured for various time periods with the appropriate antigen to enrich autoreactive cells.

We describe methods that are routinely used in the laboratory to induce a chronic, often relapsing and remitting paralysis by immunization of mice with $PLP_{139-151}$ or MBP or by adoptive transfer of lymphocytes sensitized to $PLP_{139-151}$ or whole MBP (19, 36, 37).

The method for establishing long-term antigen-specific T cell lines and clones was developed by Kimoto and Fathman in 1980 (38). This technique formed the basis for the induction of autoimmune disease by transferring enriched autoreactive T cell populations. Utilizing this technique, it is possible to establish MBP-speccific T cell lines and clones transferring EAE in rats (39) and mice (40). T cell lines specific for MBP or MBP peptides can be established as described (39). The following method for the generation of T cell lines and clones specific for immunodominant and encephalitogenic MBP peptides can be applied to a variety of antigens and *in vivo* systems (41). Our protocol describes T cell cloning for the N-terminal peptide $MBP_{Ac1-11}$ in PL or (PL x SJL)F1 mice and MBP peptide $MBP_{87-99}$ in (PL x SJL)F1 and SJL mice. In general, this protocol will lead to the isolation of CD4-positive T cell clones with chemotactic activity *in vivo* (41-43). The optimal concentration for each antigen, e.g. protein or peptide, must be determined both for the immunization of mice and for *in vitro* primary cultures of LNCs.


## 2.2    Antigenic preparations


As mentioned before, a variety of antigens and antigenic peptides have been described as inducing EAE in the appropriate mouse strains (10, 16-18, 40, 44-47). Most antigens used for the induction of active EAE are suitable for the restimulation of T cells *in vitro*. Thus, MBP and many other myelin antigens and their derived peptides are capable of inducing T cells that can be cultured *in vitro* to adoptively transfer EAE.

Synthetic peptides are prepared by continuous flow solid-phase synthesis according to the desired sequences. Peptide purity is examined by high-performance liquid chromatography, and peptide identity is confirmed by amino acid composition analysis. Depending on the sequence, the peptide solution may require some pH adjustments before being dissolved in distilled water or phophate-buffered saline (PBS).

After synthesis and before use, peptides should be lyophilized several times to remove volatile organic compounds. Lyophilized peptides should be stored in the cold in a desiccator until use.

MBP for active immunization and for *in vitro* cell culture can be prepared from guinea pig spinal cords according to the protocol from Deibler *et al.* (48).

## 2.3    Active EAE

Mice of the appropriate strains at 6 to 12 weeks of age are immunized by s.c. inoculation with 20 to 400 µg of $PLP_{139-151}$, 400 µg of MBP, or 200 µg of MBP peptide in 0.1 or 0.2 ml of an emulsion of equal volumes of PBS and CFA. To enhance immunization with antigen, killed *Mycobacterium tuberculosis* strain H37RA is added to the emulsion at a concentration of 0.3 to 4 mg/ml. Injection sites are the regions above the shoulder and the flanks (25 to 50 µl at each injection site). In addition, 200 ng of PTX  is injected i.p. or i.v. on the day of immunization and 48 h post immunization. The onset of disease lies usually between 7 and 12 days post immunization. Figure 1 demonstrates a typical EAE course when disease is induced according to the active immunization protocol.

Some older protocols calling for s.c. immunization into the footpads or in the tail base in order to induce active EAE are obsolete due to the fact that the local inflammation can interfere with the gait of animals and hence with EAE scoring.

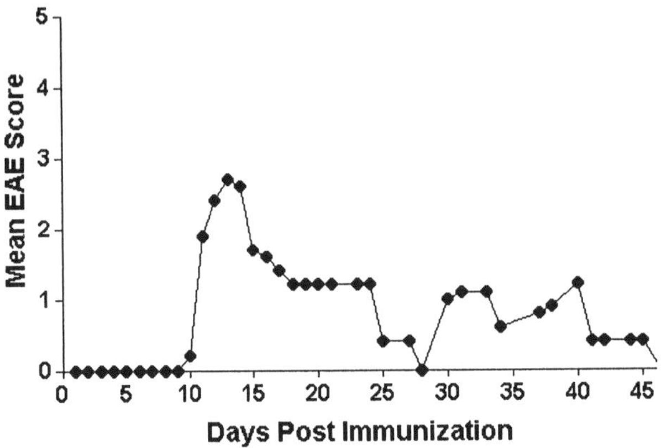

*Figure 1.* Active EAE. SJL female mice were immunized with $PLP_{139\text{-}151}$ in CFA followed by PTX induction and scored daily for clinical signs of disease.


## 2.4     Generation of PLP- or MBP-specific LNCs

Ten days after immunization, draining lymph nodes are harvested under sterile conditions. The lymph nodes are gently processed and washed through steel mesh until a single-cell suspension is obtained. Cells are then cultured at a concentration of $4 \times 10^{6}$ cells/ml for 4 days in RPMI 1640 supplemented with 2 mM glutamine, 100 U/ml penicillin, 100 µg/ml streptomycin, 10% fetal calf serum, and $5\text{--}15 \times 10^{-6}$ M $PLP_{139\text{-}151}$. Cultures are set in complete medium as described (36, 37) at 2 ml per well of a 24-well tissue culture plate. The generation of MBP-specific LNCs is performed according to the same technique, except that cells are stimulated with 25 µg/ml MBP.


## Isolation and characterization of autoreactive T cell lines and clones

Seven to twelve days after immunization with peptide antigen in an emulsion of PBS and CFA, mice are euthanized, and draining lymph nodes are harvested under sterile conditions. The lymph nodes are gently processed and washed through steel mesh until a single-cell suspension is obtained. Cells ($30 \times 10^{6}$) are cultured in 5 ml RPMI 1640 supplemented with 2 mM glutamine, 100 U/ml penicillin, 100 mg/ml streptomycin, $5 \times 10^{-5}$ M 2-mercaptoethanol, 10% fetal calf serum, and antigen ($5\text{--}15 \times 10^{-6}$ M). After 4 days of incubation, cells are washed and resuspended in 5 ml of the enriched medium as described above without antigen. Depending on cell growth, the cell culture medium can be supplemented with interleukin-2 (IL-2) or 10% (v/v) concanavalin A (Con A) supernatant as a source of IL-2 (and other growth factors) between stimulations. Con A supernatant is prepared from splenocytes of BALB/c mice by incubating $5 \times 10^6$ cells/ml with 2–5 µg/ml Con A for 24 h. The remaining Con A in the supernatant is removed by stirring it with 2 mg/ml Sephadex G-50 for 1 h. The resulting Con A supernatant is filtered sterile and stored at -70°C until use. T cells are kept in Con A supernatant enriched medium at a concentration of $1 \times 10^6$ cells/ml and restimulated every 14 days using antigen ($5–15 \ 10^{-6}$ M) presented on irradiated (3000 rad) syngeneic spleen cells at a ratio of 1/5 to 1/50 T cells versus antigen-presenting cells (APCs). T cell clones from the MBP-induced T cell lines are derived by the limiting dilution technique. Cells are diluted in medium and distributed in wells of flat-bottom microtiter plates at a minimal concentration of 0.2 cell/200 µl on $5 \times 10^5$ irradiated syngeneic splenocytes as APCs. Resulting clones are maintained by techniques described for the lines. T cell lines and clones generated by this protocol are usually characterized by the expression of CD4, VLA-4, and CD44 cell surface markers (Table 1; 49). Cells proliferate well according to their specific antigen and are often quite heterogeneous with regard to their encephalitogenicity and cytokine production (42, 43)

*Table 1*

**Properties of encephalitogenic T cells**

CD4 positive

Diverse TCR $\alpha/\beta$ chain gene usage
High expression of VLA-4
High expression of LFA-1
High expression of CD44
High expression of CD25 (IL-2 receptor $\alpha$-chain)
Very low or no expression of L-selectin
Production of TNF-$\alpha/\beta$
Possible production of IL-4

## 2.5    Lymphocyte proliferation

Specific recognition of myelin antigens, by sensitized T cells is tested using lymphocyte proliferation assays. Primary proliferative responses of

MBP-specific LNCs are assessed by incubating $2 \times 10^5$ cells with 100, 50, 25, and 12.5 µg/ml MBP in 200 µl of complete medium per well of a 96-well tissue culture plate. Cultures with no antigen and with 4 µg/ml Con A serve as negative and positive controls, respectively. After 80 h, cultures are pulsed with 1 µCi of [³H]thymidine per well. After 96 h, cells are harvested on glass filters, and radioactivity is determined in a β-counter. All cultures are set up at least in triplicate.

The best time period to examine the proliferation of long-term T cell lines and clones is when the cells are in a resting phase. Thus, 7 to 10 days after antigenic stimulation, cells of T cell lines or clones are tested for their specific proliferative responses to MBP peptides. Ten thousand T cells are cultured in 200 µl in a well of a 96-well tissue culture treated flat-bottom microtiter plate together with $0.5 \times 10^6$ irradiated (3000 rad) syngeneic spleen cells in the presence of various concentrations of the antigens. A typical concentration range for testing peptide antigens in a proliferation assay includes 0.1, 1, 10, 50, and $100 \times 10^{-6}$ M. At the end of a 48-h incubation period, 1 µCi of [³H]thymidine is added. Twelve hours later, cells are harvested and radioactivity is determined in a β-counter. All cultures are set up at least in triplicate.


## 2.6      EAE induced by adoptive T cell transfer (passive EAE, AT-EAE)

For AT-EAE, $15 \times 10^6$ to $50 \times 10^6$ activated autoreactive LNCs are washed in PBS and injected i.v. or i.p. in 200 to 500 µl PBS into naive (PL X SJL)F₁ or SJL mice 6 to12 weeks of age. The severity of EAE is dependent on the number of MBP-specific cells transferred and can be adjusted accordingly, but precise control of clinical EAE scores is very difficult to achieve.

Additional injections of PTX on the day of transfer and 48 h later have been recommended but may precipitate an exacerbative course of disease or even prevent optimal disease induction. In some cases, whole-body irradiation of mice with 350 rad facilitates the transfer of EAE by weakly encephalitogenic T cell clones.

Figure 2 demonstrates EAE induced by adoptive transfer as described above.

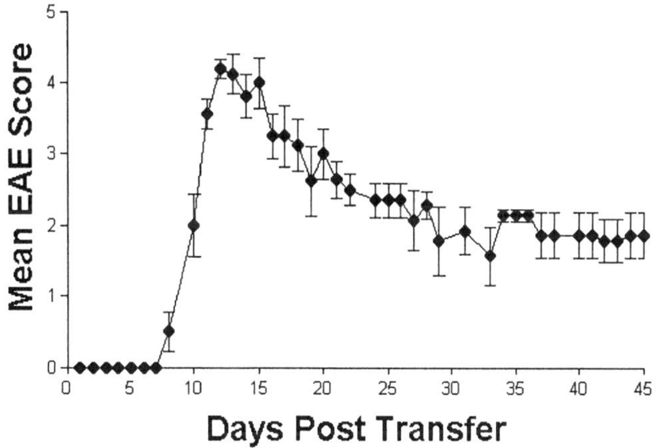

*Figure 2*. AT-EAE. SJL mice were injected i.p. with 36 X $10^6$ LNCs/mouse and clinical disease scores were recorded daily.

# 3. CLINICAL MANIFESTATION AND EVALUATION OF DISEASE

The clinical and histopathologic signs of EAE can be quantified according to various schemes. We evaluate all animals for clinical signs of EAE using a clinical scoring scale ranging from 0 to 5 presented in table 2. Scoring is performed by an examiner blinded to the experimental protocol and by the same observer throughout the experiment. Mean clinical scores at separate days are calculated by adding the scores of individual mice and dividing the sum by the number of mice in each group.

*Table 2*

| EAE score | Clinical disease |
| --- | --- |
| 0 | No clinical disease |
| 1 | Tail weakness |
| 2 | Hemi- or paraparesis (incomplete paralysis of one or two hindlimbs) |
| 3 | Hemi- or paraplegia (complete paralysis of one or two hindlimbs) |
| 4 | Paraplegia with forelimb weakness or paralysis |
| 5 | Moribund or dead animals |

The onset of actively-induced EAE lies usually between 7 and 12 days post immunization. Maximal clinical severity is often reached between day 9 and day 14, followed by a remitting disease course, although relapses can occur as shown in figure 1. In contrast, transfer of myelin antigen-specific T lymphocytes in the mouse rarely results in a full recovery and tends to leave long-lasting neurological deficits. More severe forms of the disease are associated with an earlier onset, a higher disease incidence, higher disease scores, a slower recovery, and a more synchronized course in the different animals. A significant drop in body weight usually precedes other clinical signs of EAE for about 1 to 2 days. The onset of clinical disease depends on the number and encephalitogenic capacity of the transferred cells.

Disease starts 6 to 12 days after transfer of encephalitogenic cells. Typically, maximal clinical severity is reached between day 8 and day 14 after transfer. Disease course and duration vary according to the animal strain used and the selection cycle of encephalitogenic cells. In primary T cell transfers, a relapsing-remitting disease course over a long time period can be observed, whereas T cell-line- or clone-mediated EAE is characterized by a monophasic disease course with lasting neurological deficits.

## 3.      DISCUSSION

EAE in the mouse is characterized by perivascular inflammatory demyelinating lesions in the CNS and an immunogenetically determined susceptibility. Clinical expression of the disease is obvious by neurological signs and can be easily quantified. The immune system of the mouse is extensively studied, and immune reagents are readily available, making mouse EAE an animal model for the study of some aspects of inflammatory demyelinating disorders of the CNS.

The preparation of T cell lines or clones is obviously much more labor-intensive than mixing myelin antigens in CFA. Why then use passive EAE when active EAE is so much easier? Our laboratories are using routinely passive EAE for several reasons. Passive EAE allows the separate study of immunization and effector events in the pathogenesis of the disease. This procedure avoids local and systemic side effects of adjuvant-based immunizations. Treatment protocols could interfere with the efficacy of autoantigen immunization rather than reflect the effects on the disease process. Finally, cells that mediate EAE *in vivo* are accessible to characterization and treatment *in vitro*. The disadvantage of the approach of inducing EAE by adoptive transfer is that every culturing process selects for certain properties and deselects others. The outcome of the described process cannot always be well controlled. Culture methods select cells that proliferate efficiently. Experience in EAE shows that T cells that proliferate best are not always the most encephalitogenic cell populations. Thus, any conclusions drawn from experiments that include cell culture procedures bear the potential risk of overgeneralization. This becomes evident when comparing diseases that are induced based on the different methods of culturing cells presented in this paper. However, all EAE models, both active and passive, display a wide variety of different disease courses. A careful matching of the method of choice to induce EAE with the objective of the study under investigation should help to navigate through the various options available to the investigator.


## ACKNOWLEDGMENTS

The authors acknowledge the support of the Chief Scientist's Office, MH (grant No. 5063) and BMBF/MOS (grant No. 1809).

## References

1.   Remlinger J (1905) *Ann. Inst. Pasteur* **19:**625–646.

2.   Rivers TM, Sprunt DH, and Berry GP (1933) *J. Exp. Med.* **58:**39–53.

3.   Olitsky PK, and Yager RH (1949) *J.Exp. Med.* **90:**213–223.

4.   Paterson PY, Drobish DG, Hanson MA, and Jacobs AF (1970) *Int. Arch. Allergy Appl. Immunol.* 37:26-40.

5.   Wisniewski HM, and Madrid RE (1983) *J. Neuropathol. Exp. Neurol.* **42**:243-55.

6.   Singer BA, Tresser NJ, Frank JA, McFarland HF, and Biddison WE (2000) *J. Neuroimmunol.* **105**:7-19.

7.   Rauch HC, and Einstein ER (1974) *J. Neurol. Sci.* **23**:99-116.

8.   Uchimura I, and Shiraki H (1957) *J. Neuropathol. Exp. Neurol.* **16** : 139-208.

9.   Prineas JW, McDonald WI, and Franklin RJM (2002) Demyelinating diseases. In: Greenfield's Neuropathology, 7th ed., Graham DI, and Lantos PL, (Eds.) Arnold, London, Vol 2, pp 471-550.

10.  Zamvil SS, and Steinman L (1990) *Annu. Rev. Immunol.* **8:**579–621.

11.  Brocke S, Gijbels K, and Steinman L (1994) In: Autoimmune Disease Models: A Guidebook, Cohen IR, and Miller A (Eds.), Academic Press, San Diego, pp 1–14.

12.  Steinman L, Schwartz G, Waldor M, O'Hearn M, Lim M, and Sriram S (1984) In: EAE: A Good Model for MS, Alvord EC, Kies MW, and Suckling A (Eds.), Liss, New York, pp 393–397.

13.  Goverman J (1999) *Immunol. Rev.* **169**:147-159.

14.  Kuchroo VK, Sobel RA, Yamamura T, Greenfield E, Dorf ME, and Lees MB (1991) *Pathology* **59**:305-312.

15.  Brocke S., Quigley L, McFarland HF, and Steinman L (1996) *Methods* **9**:458-462.

16.  Tuohy VK, Lu Z, Sobel RA, Laursen RA, and Lees MB (1988) *J. Immunol.* **141**:1126–1130.

17.  Tuohy VK, Lu Z, Sobel RA, Laursen RA, and Lees MB (1989) *J. Immunol.* **142**:1523–1527.

18.  Amor S, Groome N, Linington C, Morris MM, Dornmair K, Gardinier MV, Mattieu JM, and Baker D (1994) *J. Immunol.* **153:**4349–4356.

19.  Vanderlugt CL, Neville KL, Nikcevich KM, Eager TN, Bluestone JA, and Miller SD (2000) *J. Immunol.* **164**:670-678.

20.  Kuchroo VK, Anderson AC, Waldner H, Munder M, Bettelli E, and Nicholson LB (2002) *Annu. Rev. Immunol.* **20**:101-23.

21.  Forsthuber TG, Shive CL, Wienhold W, de Graaf K, Spack EG, Sublett R, Melms A, Kort J, Racke MK, and Weissert R (2001) *J. Immunol* **167**:7119-7125.

22.  Holz A, Bielekova B, Martin R, and Oldstone MBA (2000) *J. Immunol.* **164**:1103-1109.

23.  Morris-Downes MM, McCormack K, Baker D, Sivaprasad D, Natkunarajah J, and Amor S (2002) *J. Neuroimmunol.* **122**:20-33.

24. Dahlman I, Jacobson L, Glaser A, Lorentzen JC, Anderson M, Luthman H, and Olsson T (1999) *J. Immunol.* 162: 2581–2588.

25. Willenborg DO (1979) *J. Immunol.* **123**:1145-1150.

26. Mannie MD, Paterson PY, U'Prichard DC, and Flouret G (1985) *Proc. Natl. Acad. Sci. USA* **82**:5515-5519.

27. Swanborg RH (2001) *Immunol. Rev.* **181**:129-135.

28. Smeltz RB, Wolf NA, and Swanborg RH (1998) *J. Neuroimmunol.* **87**:43-48.

29. Weir CR, Nicolson K, and Backstrom BT (2002) *Immunol. and Cell Biol.* **80**:14-20.

30. Lee JM, and Olitsky PK (1955) *Proc. Exp. Biol. Med.* **89:**263.

31. Bernard CCA, and Carnegie PR (1975) *J. Immunol.* **114:**1537–1540.

32. Munoz JJ, Bernard CCA., and Mackay IR (1984) *Cell. Immunol.* **83:** 92–100.

33. Paterson PY (1960) *J. Exp. Med.* **111:**119–133.

34. Bernard CCA., Leydon J, and Mackay IR (1976) *Eur. J.Immunol.* **6**:655–660.

35. Paterson PY, Richardson WP, and Drobish DG. (1975) *Cell. Immunol.* **16**:48–59.

36. Pettinelli CB, and McFarlin DE (1981) *J. Immunol.* **127**:1420–1423.

37. Mokhtarian F, McFarlin DE, and Raine CS (1984) *Nature* **309**:356–358.

38. Kimoto M, and Fathman CG (1980) *J. Exp. Med.* **152**:759–770.

39. Ben-Nun A, Wekerle H, and Cohen IR (1981) *Eur. J. Immunol.* **11**:195–199.

40. Zamvil SS, Mitchell DJ, Moore AC, Kitamura K,Steinman L, and Rothbard J (1986) *Nature* **324**:258–260.

41. Brocke S, and Hahn H (1991) *Infect. Immun.* **59**:4531–4539.

42. Brocke S, Gaur A, Piercy C, Gautam A, Gijbels K, Fathman CG, and Steinman L (1993) *Nature* **365**:642–644.

43.  Brocke S, Gijbels K, Allegretta M, *et al.* (1996) *Nature* **379**:343–346.

44.  Zamvil S, Nelson P, Mitchell D, Knobler R, Fritz R, and Steinman L (1985) *J. Exp. Med.* **162**:2107–2124.

45.  Sakai K, Zamvil SS, Mitchell DJ, Lim M, Rothbard JB, and Steinman L (1988) *J. Neuroimmunol.* **19**:21–32.

46.  Sobel RA, Tuohy VK, Lu Z, Laursen RA, and Lees MB (1990) *J. Neuropath. Exp. Neurol.* **49**:468–479.

47.  Whitham RH, Jones RE, Hashim GA, Hoy CM, Wang RY, Vandenbark A, and Offner H (1991) *J. Immunol.* **147**:3803–3808.

48.  Deibler GE, Martenson RE, and Kies MW (1972) *Prep. Biochem.* **2**:139-165.

49.  Brocke S., Veromaa T, Weissman IL, Gijbels K, and Steinman L (1994) *Trends Microbiol.* **2**:250–254.

Chapter  A3

# HISTOPATHOLOGY OF EAE

Michael J. Day
*Department of Pathology & Microbiology, School of Veterinary Science, University of Bristol,
Langford BS40 5DU, UK*

**Abstract:**     This chapter reviews the histological structure of the normal central nervous
system and the basic pathology of inflammation and demyelination of this
tissue.  An outline of the pathogenesis of experimental autoimmune
encephalomyelitis (EAE) is given and pathology of multiple sclerosis is
reviewed.  The limitations of EAE as a model for multiple sclerosis are
discussed.  The approach to histopathological assessment of the lesions of
EAE is outlined, including collection and processing of appropriate samples
and a review of grading systems that have been used to quantify the
histopathological changes.  The specific histopathological features of EAE are
described with reference to selected model systems including the SJL/J,
C57Bl/6 and NOD mouse, the Lewis rat, rhesus monkey and marmoset.

**Key words:**    EAE, multiple sclerosis, histopathology, inflammation, demyelination

## 1. BASIC MICROANATOMY OF THE CENTRAL NERVOUS SYSTEM

The central nervous system (CNS) comprises the brain, spinal cord, optic
nerve and retina; the latter structures considered embryological extensions of
the brain.  Within the CNS, the broadest structural division is into the
myelinated axons of the white matter tracts, and the collections of neuronal
bodies, glial cells and surrounding neuropil that comprise the gray matter.

The basic cellular unit of the CNS is the neuron.  The neuronal cell
body typically receives input from the dendrites that emanate from the cell
body, and transmits output along a single axon through the axonal

telodendritic branches to the dendrites of other neurons. The terminal bulbs of the telodendritic branches store neurotransmitter substances that are released into the synapse. Most synapses are axo-dendritic or axo-somatic, but a range of other possible interneuronal contacts may be made. A single multipolar neuron may receive many thousands of synaptic inputs.

The smaller neuroglial cells produce structural and functional support to the neuroaxonal units, and are the predominant population in CNS tissue. The largest glial cells are the astrocytes that have numerous processes containing glial fibrils. The length and degree of branching of these processes is greater for the fibrous astrocytes of the white matter, than for the protoplasmic astrocytes of the gray matter. Adjacent astrocytes are linked to each other by gap junctions, and astrocyte processes terminate in end feet that are in close association with vascular endothelium and may induce the formation of inter-endothelial tight junctions, forming the basis of the 'blood-brain-barrier' (Couraud 1998). Astrocytes have a range of functions including provision of structural support and insulation for synapses, acting as an energy (glucose) source and acting to inhibit synaptic activity by taking up neurotransmitters. Astrocytes may be induced to express MHC class II and other co-stimulatory molecules (B7 and CD40) and are capable of antigen presentation to T lymphocytes *in vitro* (Dong and Benveniste 2001). Astrocytes may also have a role in down-regulation of CNS immune responses.

By contrast, oligodendrocytes are less branched cells that do not have gap junctions. White matter oligodendrocytes form the multilayered myelin sheath that surrounds and insulates the axon, and one oligodendrocyte may provide myelination for multiple axons (Baumann and Pham-Dinh, 2001). There are areas of unmyelinated axon (nodes of Ranvier) that form at the junction of myelin sheath produced by adjacent oligodendrocytes (internodes). The electrical insulation provided by myelin means that action potentials 'jump' between nodes which increases the speed of signal conduction relative to that which would occur in an unmyelinated axon. The function of gray matter oligodendrocytes is not defined, but they are arranged as 'satellite' cells to neurons and may be involved in neuronal homeostasis.

Microglia are relatively small glial cells with more sparse distribution in the neuropil. During CNS inflammation, microglia become phagocytic and capable of antigen presentation, and therefore act in similar fashion to infiltrating blood-derived macrophages to which they are likely related (Stoll and Jander 1999; Smith 2001). In demyelinating diseases these phagocytic cells accumulate intracellular lipid, appearing as 'gitter cells' with a foamy cytoplasm.

The ciliated ependymal cells form the lining to the ventricular spaces and modified ependyma line the choroid plexus and produce the cerebrospinal fluid (CSF) of the brain.

# 2.  BASIC CNS HISTOPATHOLOGY

The pathogenesis of the lesions of experimental autoimmune encephalomyelitis (EAE) is complex but involves two basic processes of inflammation and demyelination.  The generic features of these changes will be described in this section, and the specific pathology that characterises EAE is discussed subsequently.

## 2.1  Inflammation

The characteristic CNS inflammatory lesion in diseases such as EAE is the 'perivascular cuff' – an accumulation of leukocytes in the perivascular space, involving vessels of the neuropil, meninges or choroid plexus.  This accumulation of leukocytes involves the processes of leukocyte margination, molecular interactions with endothelia, breakdown of the blood-brain barrier and active egress of leukocytes from the bloodstream to the perivascular space in response to chemokine signalling (Ransohoff 1999; Boztug et al., 2002).  The role of matrix metalloproteinases in disruption of the basal lamina of the blood-brain-barrier has recently been appreciated (Rosenberg 2002).  These processes must be distinguished from primary vascular pathology (vasculitis).  The nature of the leukocytes within the cuff may vary, and the inflammation may therefore range from neutrophilic to pyogranulomatous to lymphoplasmacytic.  Additionally, resident pericytes and local microglial cells may form part of the cuff population.  Perivascular cuffing may be accompanied by gliosis of adjacent parenchyma – an increased prominence of glial cells (astrocytes, oligodendrocytes or microglia) due to proliferation or cellular hypertrophy.  In some models of EAE there may be resolution of the inflammatory lesion associated with a period of clinical remission.  In the resolution of T lymphocyte-mediated inflammatory lesions, it has been suggested that there is Fas-Fas ligand-mediated apoptosis of the infiltrating lymphoid cells and macrophages, with subsequent phagocytosis of the apoptotic cells by microglia (Ouallet et al., 1999; Zipp 2000; Chan et al., 2001).

## 2.2  Demyelination

The destruction of a normal myelin sheath resulting in an unsheathed ('naked') axon is primary demyelination, and characteristic of multiple sclerosis (MS) and some models of EAE.  By contrast, secondary demyelination involves primary axonal degeneration with resulting loss of the myelin sheath that cannot be maintained in the absence of the axon.  In primary demyelination, there may however be eventual axonal damage following a prolonged period of demyelination (Pivneva et al., 1999).

In MS and EAE the target of the disease process is therefore the oligodendrocyte-myelin unit.  Demyelination is unlikely to be directly mediated by lymphocytes, but may involve a range of effectors including lymphocyte and macrophage-derived cytokines, oxygen radicals, nitric oxide, matrix metalloproteinases, antibody, complement and phagocytic cells including macrophages and microglia (Smith 1999; Kieseier et al., 1999).  Proliferating glial cells will infiltrate the area of demyelination.  In the absence of effective remyelination, axonal degeneration will occur and this forms the basis for the progressive clinical nature of these diseases.  Failure of remyelination reflects oligodendrocyte apoptosis and failure to renew oligodendrocytes from progenitor cells (Nait-Oumesmar et al., 2000).


## 3.    EXPERIMENTAL AUTOIMMUNE ENCEPHALOMYELITIS


EAE is an experimentally induced, T cell-mediated inflammatory disease of the CNS that has been most widely studied in primates, mice, rats, rabbits and Guinea pigs as a model of organ-specific autoimmune disease.  The disease is characterised by a mixed inflammatory infiltration of the CNS and varying degrees of accompanying demyelination, which underlie the typical clinical signs.  The clinical course of EAE may be acute and monophasic or polyphasic with a chronic, relapsing and remitting time course.  The clinical course and specific pathology are dependent on the genetic strain, age and sex of animal used, and the method of disease induction.

EAE can be actively induced by immunisation with crude brain or spinal cord homogenates, whole myelin, isolated myelin antigens such as myelin basic protein (MBP), proteolipid protein (PLP) or myelin oligodendrocyte glycoprotein (MOG) or by synthetic peptides constituting defined T cell epitopes of these antigens. These immunogens are administered with adjuvant; for example most murine models of EAE use a combination of complete Freund's adjuvant (CFA) and *Bordetella pertussis* (entire organisms or, more recently, toxin).  EAE may also be passively induced by the adoptive transfer of various cell populations including lymph node or spleen cells from donors with actively induce EAE.  These donor cells are generally activated by *in vitro* stimulation with encephalitogen before transfer.  Disease may be similarly induced by encephalitogen-specific T cell lines or clones, or peptide-loaded antigen presenting cells.

EAE has also been studied in genetically modified experimental animals, for example those with targeted gene disruption, or mice transgenic for an encephalitogen-specific T cell receptor. Several studies have shown that cross-reactive viral peptides are capable of inducing clinical EAE with characteristic CNS lesions in susceptible animals (Ufret-Vincenty et al., 1998).

T cells, in particular $CD4^+$ T lymphocytes, play a pivotal role in the induction of EAE as demonstrated by various adoptive transfer and blocking studies. Later in the disease process other cell types are considered to play an important role in the full expression of the disease. It is thought that the autoreactive T cells are activated in the periphery, proliferate in peripheral lymphoid tissue and then migrate into the CNS. They gain entry having up-regulated various cell surface adhesion molecules such as LFA-1 and VLA-4; the ligands of these molecules, ICAM-1 and VCAM-1, show increased expression on CNS endothelial cells as inflammation develops. Local stimulation of T cells by microglia or astrocytes is likely to occur (Aloisi et al., 2000). The autoreactive T cells involved in EAE pathogenesis are most likely to be of the IFNγ producing Th1 subset, although it has been shown that Th2 cells may mediate pathology when transferred to immunodeficient recipients. A role for B lymphocytes and antibody in some models is also proposed (Lyons et al., 2002).

A number of EAE models show a monophasic disease course, suggesting that immune regulation may mediate recovery from disease. Numerous mechanisms have been investigated to account for this observation, but current emphasis is placed on regulatory populations of T lymphocyte, particularly $CD4^+CD25^+$ T cells, and on the regulatory cytokine IL-10 that is produced by these cells.

## 3.1    EAE as a model of MS

Multiple Sclerosis is a chronic inflammatory and demyelinating disease of the CNS, the histopathology of which shows marked heterogeneity between individuals (Hickey 1999). There is no clear correlation between lesion histopathology and clinical presentation, which may be broadly categorised as chronic relapsing-remitting or primary progressive in nature. Most studies of the histopathology of MS have been performed with chronic lesional material, but there is also heterogeneity in acute lesions, suggesting that MS may have multiple aetiologies and pathogenic mechanisms (Luchinetti et al., 2000). In classical MS, several broad 'stages' of the plaque lesion have been described. In an acute plaque there is a marked infiltration of T lymphocytes and macrophages, accompanied by endothelial activation, oedema, myelin swelling and oligodendrocyte loss. In chronic-active plaques lymphocytes are located both perivascularly and at (or just beyond) the advancing edge of the zone of demyelination. There is active demyelination with prominent foamy macrophages, and some axonal damage. Oligodendrocyte loss and astrocytosis are additional features. The chronic-inactive plaque has reduced inflammatory infiltration with persistent perivascular cuffs of lymphocytes, macrophages and plasma cells. There is sharp demarcation at the margin of the lesion with normal myelinated tissue. Within the plaque there is gliosis, reduced oligodendrocytes and axonal loss (Hickey 1999). An important feature of at least some forms of acute MS lesions is remyelination, however as disease progresses the ability to remyelinate clearly fails and this likely relates to loss of oligodendrocytes by apoptosis with failure of proliferation of oligodendrocyte progenitors (Luchinetti et al., 2000; Ransohoff et al., 2002). Additionally, a role for primary axonal loss in the pathogenesis of MS has been proposed (Anthony et al., 2000).

The exact aetiology of MS is unknown, although a role for T-cells is suggested by histopathological analysis of MS lesions, and this has been more specifically suggested to reflect a role for $CD4^+$ T cells by extrapolation from EAE. Again, by analogy with EAE, and in view of the localisation of MS lesions to the white matter, it has been assumed that the lymphocyte populations involved in MS respond to myelin antigens. In this context, T lymphocyte responses to MBP, PLP, MOG and myelin associated glycoprotein (MAG) have been demonstrated. A role for B lymphocytes and antibody in MS pathogenesis is suggested by the finding of increased CNS and cerebrospinal fluid (CSF) levels of immunoglobulin, and intralesional deposition of IgG in MS patients.

Although EAE has been widely employed as an experimental model of MS, it remains imperfect model system. One major limitation is that EAE is artificially induced rather than spontaneously occurring, hence it has limited use in determining the primary triggers in MS pathogenesis. It has been proposed that MS may be triggered by exposure to microbial (particularly viral) antigens cross-reactive to myelin (Weiner and Selkoe, 2002). An additional limitation is that the majority of EAE models show very little, if any, demyelination within the CNS and have an acute, monophasic course with spontaneous recovery. This contrasts with the chronic, progressive nature of MS, and the demyelination that is a feature of the human disease. Several EAE models have been developed in which the CNS lesions more closely approximate those of MS. For example, EAE induced in C57Bl/6 mice with $MOG_{35-55}$ (see below) or that generated in Biozzi AB/H mice with spinal cord homogenate (Baker et al., 1990) is characterised by demyelination. The later model also displays a relapsing-remitting, chronic time course.

# 4.    HISTOPATHOLOGY OF EAE

It is beyond the scope of the present review to describe the histopathological changes that occur in every reported model of EAE, however the pathology described in selected models will be summarised below.  The general approach to routine histopathological assessment of EAE involves the collection of brain and spinal cord into 10% neutral buffered formalin, although the use of other fixative solutions (e.g. 75% ethanol, 25% glacial acetic acid) has been reported.  In larger animal species, dissection of brain and spinal cord is less complex than for mice. For murine brain samples, the entire skull may be fixed and hemisectioned, allowing two halves of the brain to be freed from the cranial cavity.  Murine spinal cord is best removed from the vertebral canal by applying pressure with a saline-filled 10ml syringe and blunt ended 20-gauge needle. Alternatively, a variety of protocols for perfusion of entire animals with fixative under anaesthesia have been described.

Fixed tissues are embedded in paraffin wax.  Brain tissue obtained from mice as described above is best embedded for longitudinal sectioning; spinal cord may be embedded for either (or both) longitudinal or transverse sectioning.  Serial sections may be stained with haematoxylin and eosin for assessment of inflammation and inflammatory cell populations, Luxol fast blue to assess demyelination and Bielschowksy silver staining to assess axonal pathology (Figure 3-1).

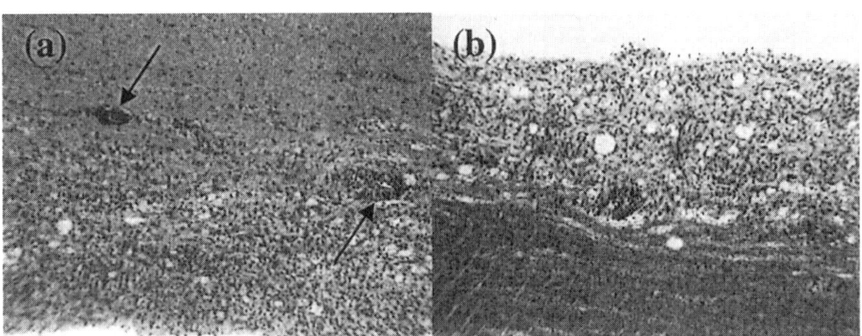

*Figure 3-1.*  Serial longitudinal sections of spinal cord from a C57Bl/6 mouse with MOG induced EAE. (a) In the haematoxylin-eosin-stained section there is an intense mononuclear inflammatory infiltration of the peripheral white matter, with two perivascular cuffs at the margin of the affected area (arrows).  (b) In the Luxol fast blue stained section the extent of the demyelination of this area of peripheral spinal cord white matter can be seen relative to the normally myelinated deeper tissue.

Immunohistochemical methodology has been widely applied to investigate the expression of specific molecules associated with EAE lesions.    Some immunohistochemical techniques may be applied to formalin-fixed tissue, but most will require frozen sections (Figure 3-2). Transmission electron microscopic studies of EAE lesions have also been reported, and such investigations will generally require collection of samples into glutaraldehyde for specialised embedding and processing protocols.    Thin sections (one micron) stained with toluidine blue may be prepared from tissue samples post-fixed in 1% osmium tetroxide and epoxy resin-embedded for electron microscopy.    Such thin sections are often used to evaluate the pathological features of EAE.

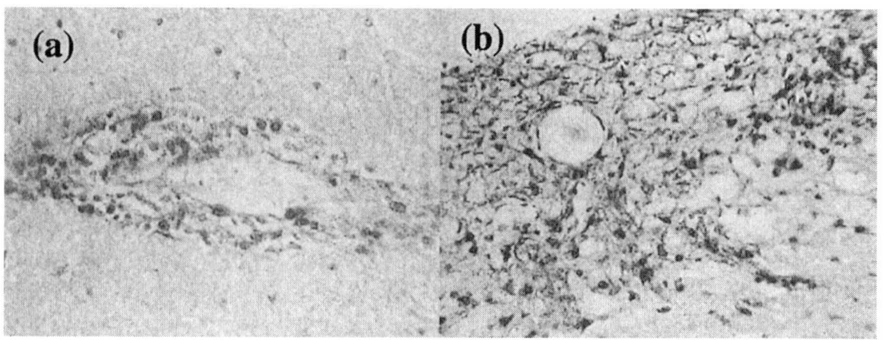


*Figure 3-2.* Immunohistochemical labelling of sections of spinal cord from a C57Bl/6 mouse with MOG induced EAE. These are formalin fixed, paraffin wax-embedded tissues that have been labelled for expression of CD3 using a cross-reactive polyclonal rabbit anti-human CD3 antiserum in an avidin-biotin immunohistochemical technique. (a) Shows T lymphocytes within a perivascular cuff in affected spinal cord white matter. (b) Shows an area of peripheral spinal cord white matter demyelination, throughout which are scattered individual T lymphocytes.

In general, some form of scoring of the histopathological lesions will be performed, but as the pathology in each model system varies, there is no consensus grading system.    Inflammation and demyelination may be scored separately.    An example of one scoring system applied to EAE lesions is given in Table 3-1.

*Table 3-1.* Example of scoring system for EAE histopathology (disease induced in SJL/J mice by adoptive transfer of encephalitogenic T cells)

**Inflammation**

| | |
|---|---|
| 0 | no inflammatory cells |
| 1 | a few scattered inflammatory cells |
| 2 | organisation of inflammatory infiltrates into perivascular cuffs |
| 3 | extensive perivascular cuffing with extension into adjacent subarachnoid space and CNS parenchyma |
| 4 | extensive perivascular cuffing with increasing subarachnoid and parenchymal inflammation |

**Demyelination**

| | |
|---|---|
| 0 | no demyelination |
| 1 | a few, scattered naked axons |
| 2 | small groups of naked axons |
| 3 | large groups of naked axons |
| 4 | confluent foci of demyelination |
| 5 | widespread demyelination |

Racke et al., 1995

Other means of quantifying lesions are to determine the mean number of discrete lesions within serial longitudinal sections of segments of spinal cord (O'Brien et al., 1999) or specific area of tissue in $mm^2$ (Furlan et al., 1999). Similarly, demyelination may be measured in a standard region of CNS tissue (Linginton et al., 1988), or expressed as the percentage of a defined area in $mm^2$ that is demyelinated (Furlan et al., 1999). In a further approach to measuring demyelination, sections of spinal cord were immunolabelled using monoclonal antibody to PLP, and captured digital images were used to assess the proportion of demyelinated tissue using image analysis software (Tuohy et al., 1999).

## 4.1 EAE in SJL/J mice

An early study reported detailed features of the kinetics of the histological and ultrastructural pathology induced in eight week old, female SJL/J mice following two subcutaneous injections of spinal cord homogenate in CFA (but without *Bordetella pertussis*) (Brown et al., 1982). In this model mice displayed clinical signs of EAE between days 14 and 18 post induction (PI), and had up to three episodes of disease relapse when monitored to 6 months PI. Lesions were reported in the optic nerve, medulla or pons, spinal cord, cerebellum and cerebral cortex, with the former three areas most severely affected.

The initial period of disease was characterised histologically by a mixed neutrophilic and lymphocytic meningitis and perivascular infiltration of these cells into the peripheral white matter. The inflammatory response was accompanied by local petechial haemorrhage, demyelination and macrophage phagocytosis of myelin, and the presence of naked axons. These latter changes only occurred in areas with inflammatory change. Mice sampled at the end of the period of initial disease (day 20 PI) had a reduced neutrophilic component to the inflammation, fewer haemorrhages and evidence of astrocyte proliferation.

By day 25 PI the inflammatory infiltrate had diminished but white matter damage was more extensive. There was gliosis, prominent phagocytic cells, naked axons and myelin debris with early evidence of remyelination. Plasma cells appeared within the meningeal lesions. Similar changes were observed at day 30 PI, however in addition to these chronic lesions, these mice had small acute lesions (as described before day 20) despite the fact that they were in clinical remission.

At the time of clinical relapse, mice also displayed a mixture of acute and chronic lesions with the latter dominating. The chronic lesions had minimal inflammation, but gliosis and naked axons were more prominent with some evidence of focal remyelination. The CNS of mice sampled after multiple episodes of clinical relapse had similar, but more extensive chronic white matter lesions with small acute lesions. There was no greater degree of remyelination in the chronic lesions at this time.

Disease may also be induced in the SJL/J mouse by administration of other encephalitogens (e.g. $PLP_{139-151}$, $MBP_{84-104}$), or by the passive transfer of encephalitogenic T cells. For example, transfer of a $CD4^+$ T cell line specific for peptide 139-151 of PLP induces a chronic relapsing disease with multifocal areas of lymphocyte and macrophage inflammation and associated demyelination in spinal cord peripheral white matter. When the cell line was preactivated in the presence of TGFβ, extensive areas of demyelination extending to the central gray matter were observed at day 70 after adoptive transfer (Weinberg et al., 1992). An example of a histological grading system applied to $PLP_{139-151}$ induced EAE in SJL/J mice is given in Table 3-2. F1 mice derived from crosses of a number of other strains with the SJL/J mouse have also been used in studies of EAE.

*Table 3-2.* Histological Grading of Inflammation in $PLP_{139-151}$ induced EAE in SJL/J mice

| 0 | absence of infiltrates |
|---|---|
| 1 | small, rare perivascular infiltrates |
| 2 | small, numerous perivascular infiltrates |
| 3 | numerous perivascular lesions and parenchymal infiltration |
| 4 | severe confluent lesions |

Langer-Gould et al., 2002


## 4.2    EAE in C57Bl/6 mice

Monophasic EAE has been induced in mice of the C57Bl/6 strain by immunisation with MOG peptide 35-55 in CFA, followed by intraperitoneal injections of pertussis toxin on days 0 and 2 of the induction protocol. The peak of clinical disease occurs on day 14 after induction. Histologically, this form of EAE is characterised by multifocal, coalescing areas of mononuclear inflammatory infiltration of peripheral white matter of spinal cord (sometimes involving perivascular cuffs in gray matter), with associated local demyelination. The infiltrates include macrophages, B lymphocytes and both αβ and γδ T lymphocytes (Spahn et al., 1999). Small clusters of neutrophils may occasionally be present. When inflammation and demyelination are graded, the peak of the inflammatory response occurs two days after the peak of clinical disease, and peak demyelination occurs several days after the maximal inflammatory response (Figure 3-3).

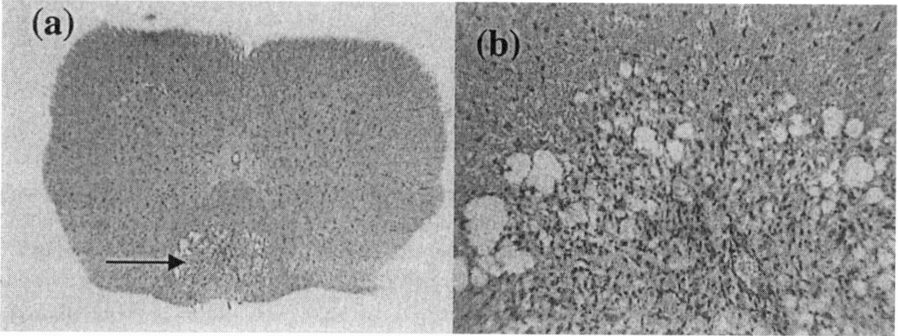


*Figure 3-3.* Sections of spinal cord from a C57Bl/6 mouse with MOG induced EAE-stained by haematoxylin-eosin. (a) There is an intense inflammatory and demyelinating lesion within the ventral white matter of the cord (arrowed). (b) Higher power view of the margin of this lesion with unaffected white matter, shows extensive macrophage-dominated inflammation.

Pathology also occurs in the brain, where there is evidence of meningitis, and perivascular cuffing with mixed mononuclear cells that occurs predominantly in the white matter of the cerebellum, and in the underlying hind brain area (Figure 3-4). The inflammation and demyelination in this model resolve, and when mice recover from clinical disease there is virtually complete resolution of the histopathological changes (Massey et al., 2002).

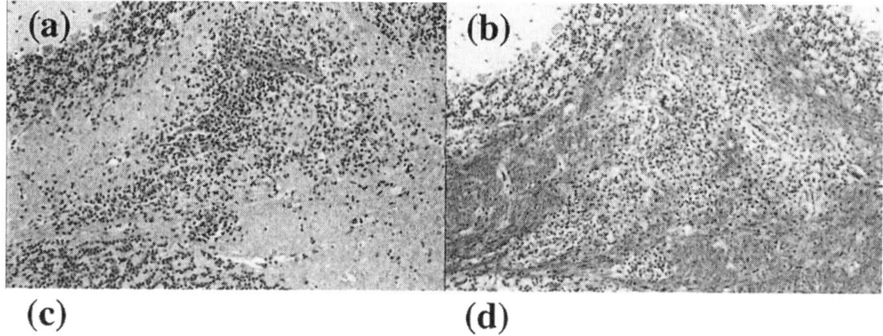

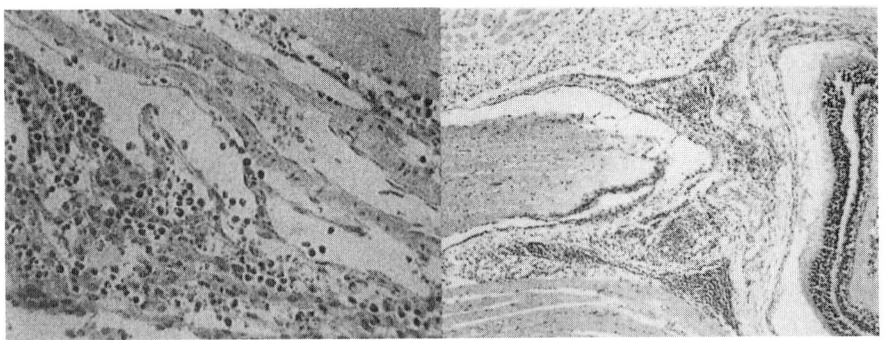

*Figure 3-4.* Tissues from C57Bl/6 mice with MOG induced EAE stained by haematoxylin and eosin (a, c and d) and Luxol fast blue (b). (a) High power view of cerebellar white matter showing a prominent area of perivascular cuffing with mononuclear inflammatory cells, with extension of the lesion into the parenchyma. (b) This serial section shows that demyelination is restricted to the area occupied by the inflammatory infiltrate, with uninflamed parenchyma showing normal staining reaction. (c) Section through the meninges, showing a mixed inflammatory infiltration throughout the meningeal layers. (d) Section through the retina, optic disc and optic nerve. There is an intense inflammatory infiltration surrounding the optic nerve (perineuritis). This comprises chiefly neutrophils and there are scattered neutrophils within the optic nerve tissue.

A chronic, sustained form of disease may be induced by subcutaneous injection with $MOG_{33-55}$ in CFA on day 0, followed by intravenous pertussis toxin on days 0 and 2, and a booster injection of $MOG_{33-55}$ in CFA on day 7. The pathology of this form of disease is similar to that described above, with extensive diffuse infiltration by mixed macrophages, T and B lymphocytes into the CNS white matter (extending to gray matter) accompanied by demyelination (Suen et al., 1997). A more severe clinical form of EAE is induced in $IL-10^{-/-}$ C57Bl/6 mice immunised with $MOG_{35-55}$ in CFA and pertussis toxin on days 0 and 2. These $IL-10^{-/-}$ C57Bl/6 mice also develop EAE and severe inflammatory lesions when pertussis is not administered (wild-type mice develop minimal disease) (Bettelli et al., 1998).

The C57Bl/6 MOG EAE model has recently been used to demonstrate the effect of $CD4^+CD25^+$ regulatory cells in this disease. When these T regulatory cells were harvested from the lymph nodes of naïve C57bl/6 mice and adoptively transferred to naïve recipients three days before induction of EAE, there was significant protection from both disease and pathological change in CNS (Kohm et al., 2002). The transferred T reg cells appeared not to enter the CNS, but to localise to the spleen and lymph nodes of the recipient mice.

## 4.3    EAE in the NOD mouse

A recently reported model of EAE has been used to investigate the clinical relationship between type I diabetes mellitus and multiple sclerosis (Winer et al., 2001). When diabetes-prone NOD (nonobese diabetic) mice over eight weeks of age are simply injected twice with pertussis toxin (in the absence of encephalitogen or CFA), they develop an acute onset malaise within 7 days. Most animals recover from this within 4-7 days, however a proportion die or develop clinical signs of EAE (primary, progressive monophasic disease). A proportion of recovered female mice will develop relapsing disease between days 15-90, but the severity and outcome of this is variable.

The brains of mice with acute onset malaise show small, multifocal, mononuclear cell perivascular cuffs in cortex and brainstem, with inflammation of the margin of the fourth ventricle. The lesions of mice with acute, progressive monophasic disease were more extensive and involved hippocampus, thalamus, cerebellum, brain stem, meninges and ventricles. This form of disease was demyelinating, with plaques present in cerebellar and brain stem white matter. Female mice with relapsing disease had similar pathology to that described for the phase of acute malaise, although the perivascular cuffing was more prominent.

## 4.4     EAE in the Lewis rat

EAE may be induced in Lewis strain rats by subcutaneous administration of Guinea pig MBP, or encephalitogenic peptide (amino acids 70-86) derived from this protein, in CFA (actively induced disease). These rats develop an acute monophasic clinical disease that resolves spontaneously within 5 days of onset. Recovered animals do not relapse and become refractory to further attempts to induce the disease. The disease is T cell dependent, and may also be induced in naïve recipients by adoptive transfer of spleen or lymph node cells from affected animals, or by administration of MBP-specific T cell lines or clones (passively induced disease). Spontaneous recovery from disease and subsequent refractoriness to disease reinduction has been associated with elevation in the concentration of serum antibody specific for MBP, endogenous corticosterone production, elevated circulating reactive nitrogen intermediates and a range of immunoregulatory events. Adoptive transfer of lymphocytes from recovered rats is reported to protect recipients from the active induction of EAE by immunisation with MBP in CFA, but cannot prevent EAE induced by adoptive transfer of effector cells. Relapsing disease can be induced by a number of protocols including administration of immunosuppressive agents such as cyclophosphamide (Minagawa et al., 1987) or cyclosporin (Polman et al., 1988).

In Lewis rats sacrificed 13 days after induction of disease by injection of MBP in CFA (at the peak of disease), there are perivascular infiltrations of inflammatory cells within both white and gray matter. There is however, no evidence of demyelination of CNS axons although there may be demyelination within the spinal roots. Spinal cords immunostained for expression of MBP display loss of staining only in the areas occupied by inflammatory cells (Sternberger et al., 1989). Demyelinating disease has been induced in Lewis rats by intraperitoneal injection of monoclonal antibody specific for MOG, given eight days after immunisation with MBP in CFA (Schluesener et al., 1988), or at the time of onset of passively induced disease (Linington et al., 1988), providing further evidence for synergy between T cell and antibody-mediated mechanisms in demyelinating disease.

A grading system for the histological lesions observed at the peak of actively induced disease has been described (Table 3-3). In Lewis rats recovered from monophasic disease and sacrificed on day 50 PI, there are residual small perivascular aggregates of T cells and macrophages within brain white matter (Polman et al., 1988).

*Table 3-3.* Histological grading system for lesions of actively induced EAE in Lewis Rats

| | |
|---|---|
| 0 | normal |
| 1 | meningeal hypercellularity or one perivascular cuff with a non-invasive margin per high power field |
| 2 | meningeal hypercellularity and one perivascular cuff per high power field, or two to three perivascular cuffs per field without meningeal hypercellularity |
| 3 | inflammatory cells extending from cuffs into CNS parenchyma |
| 4 | diffuse inflammation in either white or gray matter |
| 5 | inflammation extending throughout entire tissue section with or without primary demyelination |

(Lyman et al., 1989)

Pathological changes have also been assessed in the spinal cord of Lewis rats, five days after adoptive transfer of splenocytes obtained from MBP primed rats and restimulated *in vitro* with MBP before transfer. Such rats develop clinical disease, and have oedema and marked perivascular infiltration of spinal cord white and gray matter and meninges by macrophages and CD4$^+$ and CD8$^+$ T lymphocytes (Sedgwick et al., 1987). These central lesions are non-demyelinating although lymphocytes have been observed within the myelin sheaths (Blakemore et al., 1989). The infiltrates were much reduced when rats were treated with anti-CD4 antibody by daily intravenous injection following adoptive transfer. Studies in this model have suggested that the majority of infiltrating leukocytes are unrelated to the disease *per se* and may be present as a secondary response to tissue damage (Sedgwick et al., 1987; Sedgwick 1988). Following spontaneous recovery from passively induced disease the inflammatory infiltrates regress, and spinal cords are largely normal by 10 days post transfer (Blakemore et al., 1989). The kinetics of infiltration of adoptively transferred lymphocytes versus recipient cells has been examined in a chimeric model. Adoptively transferred cells dominate the preclinical inflammatory lesions, but during clinical disease large numbers of recipient T cells and macrophages infiltrate the parenchyma. During recovery, the residual T cells are primarily of donor origin (Matsumoto and Fujiwara, 1988). Donor T cells isolated from the recipient spinal cord produce higher levels of Th1 cytokines than recipient cells recruited to the same site (Weinberg et al., 1994).

## 4.4    EAE in rhesus monkeys and marmosets

EAE has been induced in a range of primate species including Cynomolgus monkeys (*Macaca fascicularis*), rhesus monkeys (*Macaca mulatta*), owl monkeys (*Aotus spp.*), squirrel monkeys (*Saimiri spp.*) and the common marmoset (*Callithrix jacchus*).

In rhesus monkeys an hyperacute form of EAE develops within weeks of immunisation with the recombinant human N-terminal extracellular Ig-like domain of MOG in CFA ('t Hart et al., 2000). This disease has a short clinical course, with death occurring in days, however when monkeys are given immunosuppressive therapy the disease becomes relapsing and remitting in nature. The pathological features of the hyperacute disease are extensive areas of necrosis with a neutrophilic inflammation and resultant loss of myelin and axonal structure. Although the number of infiltrating T lymphocytes is low, myelin-specific T cells are involved in the pathogenesis of this disease.

In marmosets, relapsing-remitting or primary progressive EAE has been induced by immunisation with myelin or myelin derivatives (particularly MOG) in CFA ('t Hart et al., 2000). EAE may also be induced by co-transfer of myelin-specific $CD4^+$ T lymphocytes and anti-myelin antibody (but not by T lymphocytes alone), providing good evidence in this model for a pathogenic role for B lymphocytes. Lesions develop in white matter with varying degrees of inflammation (with numerous macrophages and T lymphocytes), demyelination and remyelination, and axonal damage ('t Hart et al., 1998). Areas of demyelination are associated with local deposition of MOG-specific IgG (Raine et al., 1999). Relative to the pathology in rhesus monkeys however, there is preservation of much axonal structure. Small perivascular foci will eventually coalesce to form large demyelinating lesions. The extent of the inflammatory component of the lesions is exacerbated when disease is induced by concurrent intravenous injection of *Bordetella pertussis* with the myelin/CFA protocol.

## ACKNOWLEDGEMENTS

The histopathological images used in this chapter were prepared by the author whilst grading tissues provided from experiments performed in the laboratory of Professor D.C. Wraith, Department of Pathology and Microbiology, University of Bristol.

## REFERENCES

Aloisi F, Serafini B and Adorini L, 2000.  Glia-T cell dialogue.  Journal of Neuroimmunology 107: 111-117.

Anthony DC, Hughes P and Perry VH, 2000.  The evidence for primary axonal loss in multiple sclerosis.  Revista de Neurologia 30: 1203-1208.

Baker D, O'Neill JK, Gschmeissner SE, Wilcox CE, Butter C and Turk JL, 1990.  Induction of chronic relapsing experimental allergic encephalomyelitis in Biozzi mice.  Journal of Neuroimmunology 28: 261-270.

Baumann N and Pham-Dinh D, 2001.  Biology of oligodendrocytes and myelin in the mammalian central nervous system.  Physiological Reviews 81: 671-927.

Bettelli E, Das MP, Howard ED, Weiner HL, Sobel RA and Kuchroo VJ,  1998.  IL-10 is critical in the regulation of autoimmune encephalomyelitis as demonstrated by studies of IL-10- and IL-4-deficient and transgenic mice.  Journal of Immunology 161: 3299-3306.

Blakemore WF, Summers BA and Sedgwick J, 1989.  Lymphocyte-myelin sheath interactions in acute experimental allergic encephalomyelitis.  Journal of Neuroimmunology 23: 19-24.

Boztug K, Carson MJ, Pham-Mitchell N, Asensio VC, DeMartino J and Campbell IL, 2002.  Leukocyte infiltration, but not neurodegeneration, in the CNS of transgenic mice with astrocyte production of the CXC chemokine ligand 10.  Journal of Immunology 169: 1505-1515.

Brown A, Dale MD, McFarlin MD and Raine CS,  1982.  Chronic neuropathology of relapsing experimental allergic encephalomyelitis in the mouse.  Laboratory Investigation 46: 171-185.

Chan A, Magnus T and Gold R, 2001.  Phagocytosis of apoptotic inflammatory cells by microglia and modulation by different cytokines.  Mechanism for removal of apoptotic cells in the inflamed nervous system.  Glia 33: 87-95.

Couraud PO, 1998.  Infiltration of inflammatory cells through brain endothelium.  Pathologie Biologie 46: 176-180.

Dong YS and Benveniste EN, 2001.  Immune function of astrocytes.  Glia 36: 180-190.

Furlan R, Martino G, Galbiati F, Poliani PL, Smiroldo S, Bergami A, Desina G, Comi G, Flavell R, Su MS and Adorini L, 1999. Caspase-1 regulates the inflammatory process leading to autoimmune demyelination. Journal of Immunology 163: 2403-2409.

't Hart BA et al., 1998. Histological characterization of magnetic resonance imaging-detectable brain white matter lesions in a primate model of multiple sclerosis. American Journal of Pathology 153: 649-663.

't Hart BA, van Meurs M, Brok HPM, Massacesi L, Bauer J, Boon L, Bontrop RE and Laman JD, 2000. A new primate model for multiple sclerosis in the common marmoset. Immunology Today 21: 290-297.

Hickey WF, 1999. The pathology of multiple sclerosis: a historical perspective. Journal of Neuroimmunology 98:37-44.

Kieseier BC, Storch MK, Archelos JJ, Martino G and Hartung HP, 1999. Effector pathways in immune mediated central nervous system demyelination. Current Opinion in Neurology 12: 323-336.

Langer-Gould A, Garren H, Slansky A, Ruiz PJ and Steinman L, 2002. Late pregnancy suppresses relapses in experimental autoimmune encephalomyelitis: evidence for a suppressive pregnancy-related serum factor. Journal of Immunology 169: 1084-1091.

Linington C, Bradl M, Lassmann H, Brunner C and Vass K, 1988. Augmentation of demyelination in rat acute allergic encephalomyelitis by circulating mouse monoclonal antibodies directed against a myelin/oligodendrocyte glycoprotein. American Journal of Pathology 130: 443-454.

Luchinetti C, Bruck W, Parisi J et al., 2000. Heterogeneity of multiple sclerosis lesions: implications for the pathogenesis of demyelination. Annals of Neurology 47: 707-717.

Lyman WD, Sonett JR, Brosnan CF, Elkin R and Bornstein MB, 1989. $\Delta^9$-Tetrahydrocannabinol: a novel treatment for experimental autoimmune encephalomyelitis. Journal of Neuroimmunology 23: 73-81.

Lyons JA, Ramsbottom MJ and Cross AH, 2002. Critical role of antigen-specific antibody in experimental autoimmune encephalomyelitis induced by recombinant myelin oligodendrocyte glycoprotein. European Journal of Immunology 32: 1905-1913.

Massey EJ, Sundstedt A, Day MJ, Corfield G, Anderton S and Wraith DC, 2002. Intranasal peptide-induced peripheral tolerance: the role of IL-10 in regulatory T cell function within the context of experimental autoimmune encephalomyelitis. Veterinary Immunology and Immunopathology 87: 357-372.

Matsumoto Y and Fujiwara M, 1988. Adoptively transferred experimental allergic encephalomyelitis in chimeric rats: identification of transferred cells in the lesions of the central nervous system. Immunology 65: 23-29.

Minagawa H, Takenaka A, Itoyama Y and Mori R, 1987. Experimental allergic encephalomyelitis in the Lewis rat. A model of predictable relapse by cyclophosphamide. Journal of the Neurological Sciences 78: 225-235.

Nait-Oumesmar B, Lachapelle F, Decker L and Baron-Van Evercooren A, 2000. Do central nervous system axons remyelinate? Pathologie Biologie 48: 70-79.

O'Brien NC, Charlton B, Cowden WB and Willenborg DO, 1999. Nitric oxide plays a critical role in the recovery of Lewis rats from experimental autoimmune encephalomyelitis and the maintenance of resistance to reinduction. Journal of Immunology 163: 6841-6847.

Ouallet JC, Baumann N, Marie Y and Villarroya H, 1999. Fas system up-regulation in experimental autoimmune encephalomyelitis. Journal of the Neurological Sciences 170: 96-104.

Pivneva TA, Kolotoushkina EV and Mel'nik NA, 1999. Mechanisms of the demyelination process and its modelling. Neurophysiology 31: 403-412.

Polman CH, Matthaei I, de Groot CJA, Koetsier JC, Sminia T and Dijkstra CD, 1988. Low-dose cyclosporin A induces relapsing remitting experimental allergic encephalomyelitis in the Lewis rat. Journal of Neuroimmunology 17: 209-216.

Racke MK, Burnett D, Pak S-H, Albert PS, Cannella B, Raine CS, McFarlin DE and Scott DE, 1995. Retinoid treatment of experimental allergic encephalomyelitis. Journal of Immunology 154: 450-458.

Raine CS, Cannella B, Hauser SL and Genain CP, 1999. Demyelination in primate autoimmune encephalomyelitis and acute multiple sclerosis lesions: A case for antigen-specific antibody mediation. Annals of Neurology 46: 144-160.

Ransohoff RM, 1999. Mechanisms of inflammation in MS tissue: adhesion molecules and chemokines. Journal of Neuroimmunology 98: 57-68.

Ransohoff RM, Howe CL and Rodriguez M, 2002. Growth factor treatment of demyelinating disease: at last, a leap into the light. Trends in Immunology 23: 512-516.

Rosenberg GA, 2002. Matrix metalloproteinases and neuroinflammation in multiple sclerosis. Neuroscientist 8: 586-587.

Schluesener HJ, Sobel RA and Weiner HL, 1988. Demyelinating experimental allergic encephalomyelitis (EAE) in the rat: treatment with a monoclonal antibody against activated T cells. Journal of Neuroimmunology 18: 341-351.

Sedgwick J, Brostoff S and Mason D, 1987. Experimental allergic encephalomyelitis in the absence of a classical delayed type hypersensitivity reaction. Journal of Experimental Medicine 165: 1058-1075.

Sedgwick JD, 1988. Long-term depletion of $CD8^+$ T cells *in vivo* in the rat: no observed role for $CD8^+$ (cytotoxic/suppressor) cells in the immunoregulation of experimental allergic encephalomyelitis. European Journal of Immunology 18: 495-502.

Smith ME, 1999. Phagocytosis of myelin in demyelinative disease: A review. Neurochemical Research 24: 261-268.

Smith ME, 2001. Phagocytic properties of microglia in vitro: implications for a role in multiple sclerosis in EAE. Microscopy Research and Technique 54: 87-94.

Spahn TW, Issazadah S, Salvin AJ and Weiner HL, 1999. Decreased severity of myelin oligodendrocyte glycoprotein peptide 33-55-induced experimental autoimmune encephalomyelitis in mice with a disrupted TCR δ chain gene. European Journal of Immunology 29: 4060-4071.

Sternberger NH, Sternberger LA, Kies MW and Shear CR, 1989. Cell surface endothelial proteins altered in experimental allergic encephalomyelitis. Journal of Neuroimmunology 21: 241-248.

Stoll G and Jander S, 1999. The role of microglia and macrophages in the pathophysiology of the CNS. Progress in Neurobiology 58: 233-247.

Suen WE, Bergman CM, Hjelmstrom P and Ruddle NH, 1997. A critical role for lymphotoxin in experimental allergic encephalomyelitis. Journal of Experimental Medicine 186: 1233-1240.

Tuohy VK, Yu M, Yin L, Kawczak JA and Kinkel RP, 1999. Spontaneous regression of primary autoreactivity during chronic progression of experimental autoimmune encephalomyelitis and multiple sclerosis. Journal of Experimental Medicine 189: 1033-1042.

Ufret-Vincenty RL, Quigley L, Tresser N, Pak SH, Gado A, Hausmann S, Wucherpfenning KW and Brocke S, 1998. In vivo survival of viral antigen-specific T cells that induce experimental autoimmune encephalomyelitis. Journal of Experimental Medicine 188: 1725-1738.

Weinberg AD, Whitham R, Swain SL, Morrison WJ, Wyrick G, Hoy C, Vandenbark AA and Offner H, 1992. Transforming growth factor β enhances the in vivo effector function and memory phenotype of antigen-specific T helper cells in experimental autoimmune encephalomyelitis. Journal of Immunology 148: 2109-2117.

Weinberg AD, Wallin JJ, Jones RE, Sullivan TJ, Bourdette DN, Vandenbark AA and Offner H, 1994. Target organ-specific up-regulation of the MRC Ox-40 marker and selective production of Th1 lymphokine mRNA by encephalitogenic T helper cells isolated from the spinal cord of rats with experimental autoimmune encephalomyelitis. Journal of Immunology 152: 4712-4721.

Weiner HL and Selkoe DJ, 2002. Inflammation and therapeutic vaccination in CNS diseases. Nature 420: 879-884.

Winer S, Astsaturov I, Cheung RK, Gunaratnam L, Kubiak V, Cortez MA, Moscarello M, O'Connor PW, McKerlie C, Becker DJ and Dosch H-M, 2001. Type I diabetes and multiple sclerosis patients target islet plus central nervous system autoantigens; nonimmunized nonobese diabetic mice can develop autoimmune encephalitis. Journal of Immunology 166: 2831-2841.

Zipp F, 2000. Apoptosis in multiple sclerosis. Cell and Tissue Research 301: 163-171.

Chapter A4

# NEUROANTIGENS IN EAE
*Myelin genes, proteins and non-protein antigens*

James Garbern
*Department of Neurology and Center for Molecular Medicine and Genetics, Wayne State University School of Medicine*

**Abstract:** This chapter reviews briefly the biology of myelin and describes the major CNS myelin proteins, the potential neuroantigens in EAE. The genetics of the principal myelin proteins are also discussed. Myelin glycolipids with potential role in CNS demyelinating disease and the enzymes involved in their synthesis are also reviewed.

**Key words:** Proteolipid protein (PLP, PLP1); myelin basic protein (MBP); 2',3'-cyclic nucleotide 3' phosphodiesterase (CNPase); oligodendrocyte specific protein (OSP); myelin associated glycoprotein (MAG); myelin-oligodendrocyte glycoprotein (MOG); ceramide galactosyltransferase (CGT); Oligodendrocyte myelin glycoprotein (OMgp); Myelin-associated oligodendrocyte basic protein (MOBP); Nogo; Nogo receptor; ganglioside; sulfatide; cerebroside

## 1.    INTRODUCTION

Efficient conduction of neural signals along the axons of higher vertebrates occurs by the saltatory propagation of action potentials between narrow specializations of axonal membrane called nodes of Ranvier. Voltage regulated ion channels are clustered at the nodes and mediate the ionic currents that constitute the action potential. Between adjacent nodes the axon is ensheathed by spirally wound membranous sheets that define the myelinated internode. Myelin is composed primarily of lipids and proteins [for excellent comprehensive review see 1]. The complex composition of myelin offers many known and potential antigens that can induce autoimmune responses that will be reviewed in other chapters. In this chapter I review the composition and properties of central nervous system (CNS) myelin.


## 2.    THE CELL BIOLOGY OF MYELIN

The composition of myelin is best appreciated in a cellular context. CNS myelin is formed by oligodendrocytes, which elaborate multiple processes that ravel around the axon. At the point of axonal contact the glial membrane loops around the axon, forming the inner mesaxon. As the membranous sheet winds around the axon, most of the cytoplasm is extruded, allowing tight apposition of the adjacent membrane faces to form compact myelin. At regular intervals around successive wraps of myelin the compaction is interrupted by zones of non-compact myelin that form the incisures (or clefts) of Schmidt-Lanterman. These incisures may serve as conduits for small molecules to pass radially across the myelin sheath rather than having to diffuse circumferentially around the cytoplasmic space of the myelin leaflet [2].

The successive edges of each loop of the myelin sheet each contact the axon and are known as paranodal loops. The zone where these paranodal loops contact the axon is called the paranode, and marks a zone of separation of the voltage-gated sodium channels at the node from the voltage gated potassium channel enriched region known as the juxtaparanode. The paranodal loops are bridged to the axon by specialized, electron dense adhesions called septate-like junctions that form diffusion barriers between the periaxonal space and the interlamellar and periaxonal spaces of the internode [4, 5]. These junctions and a network of juxtamembranous filamentous proteins help to maintain the precise partitioning of the various ionic channels of the nodal region [reviewed in 6].

Although the classic view is that myelin and the subjacent internodal membrane are electically inert, myelin provides a relatively low electrical resistance pathway through which current can pass (probably through the septate-like junctions and Schmidt-Lanterman clefts) between the internodal axolemma and extracellular space that is essential for setting the nodal resting potential and also facilitates repolarization of the nodal membrane through dispersal of the charge carried into the axon by nodal sodium channels along the internodal axolemma and generating the depolarizing afterpotential [7, 8].

# 3.  THE BIOCHEMISTRY OF MYELIN

The analysis of myelin composition was facilitated by development of methods to isolate myelin-enriched preparations. Myelin was isolated from the brain using a series of centrifugation steps, including a sucrose gradient flotation that resulted in great enrichment of myelin fragments that could be identified electron microscopically. The composition of myelin prepared in this way reveals that it makes up 40 to 50 % of the dry weight of the brain, and is about two thirds lipid and one third protein. With the advent of newer, especially genetic methodologies, proteins and lipids have been characterized through their association with specific hereditary diseases or cellular properties and localization.

Myelin is essentially highly enriched cell membrane, with a dry weight composition of about two thirds lipid and one third protein [9]. While the conventional view of myelin as a passive electrical insulator is still prevalent, it is now becoming clear that there is a more active role for myelin, and/or other oligodendroglia functions, in the development and maintenance of axonal properties. The presence of oligodendrocytes, and quite possibly axonal contact with them is critical for the formation and spacing of the nodes of Ranvier. As discussed in more detail below, oligodendroglia are necessary for maintaining the long term integrity of axons, demonstrating the existence of important axo-glial signaling pathways.

## 3.1  Myelin lipids and their synthetic enzymes

### 3.1.1  Myelin lipids

The composition of myelin is similar with that of cellular membranes with respect to its enrichment in lipids [10]. While there are no known myelin-specific lipids, the distribution of lipids in myelin is distinct from that of other cell membranes. The major lipid in myelin is free cholesterol, which comprises about 30 % of total lipid by dry weight. Cholesterol esters are not found in normal mature myelin. Phospholipids are collectively the most abundant lipids in CNS myelin, constituting about 40 – 45 % of total lipid by dry weight. Most brain phospholipids are derivatives of phosphatidic acid (diacylated glycerol-3-phosphate). Plasmalogens are lipids where an aliphatic alcohol is ether-linked rather than acyl-linked to the glycerol backbone.

Glycerophospholipid classes are defined by the nature of the compound linked at the third position of the glycerol backbone. These compounds are short-chain, polar alcohols phosphodiester-linked to phosphatidic acid. The most abundant of these glycerophospholipids in adult human brain are phosphatidyl ethanolamine, phosphatidylcholine (lecithin), phosphatidylserine and phosphatidylinositides. The phosphoinositides include phosphatidylinositol and its phosphorylated derivatives that are quantitatively minor phospholipids but are important role in signal transduction.

Sphingolipids are distinguished by the substitution of sphingosine, a long chain aminodiol, for the glycerol backbone used by glycerophospholipids. Ceramides are sphingolipids where the amino group of sphingosine is N-acylated with long-chain fatty acid. The first carbon of ceramide is linked to different head groups to form various subclasses of sphingolipids. Sphingomyelin is the phosphodiester of ceramide and choline. Most glycolipids in brain consist of ceramide linked at the first carbon with different mono- or polysaccharides. The major glycolipid of mammalian CNS myelin is galactocerebroside (Gal-C), in which galactose is β-glycosidically linked to ceramide. Gal-C constitutes about 16% of total adult human brain lipid. Sulfatide constitutes about 6% of brain lipid and is formed from galactocerebroside esterified to sulfate at the 3' position of galactose. Other sugars including glucose, (to form glucocerebroside; Cer-Glc), N-acetylglucosamine, N-acetylgalactosamine, fucose and others, are conjugated to cerebroside to form other brain glycolipids. Sialic acid, N-acetyl-neuraminic acid (NANA), is an important N-acylated, nine-carbon amino sugar that is linked to glycolipids to form gangliosides. Gangliosides are acidic in nature because of the presence of a free carboxyl group. Gangliosides are present in a variety of brain regions as well as in other tissues, and are usually classified according to the number of sialic acid residues in the molecule and their relative migration rates on thin-layer chromatograms.

Although they make up a very small proportion of total myelin lipid, gangliosides play a critical role in the nervous system. Mice that are deficient in GM2/GD2 synthase, a key enzyme needed for synthesis of complex gangliosides, can only express the simple gangliosides GM3 and GD3 and have decreased central myelination and develop Wallerian degeneration [11]. In addition, complex gangliosides, such as GD1a and GT1b, are inhibitors of neural regeneration probably through their properties as ligands for myelin associated glycoprotein (see below) [12].

### 3.1.2    Myelin lipid synthetic enzymes

It is likely that the program of myelination coordinates not only the expression of myelin proteins, but also the synthesis of myelin lipids. Expression of enzymes key for the synthesis of these lipids is maximal during the peak of myelination. For example, expression of UDP-galactose:ceramide galactosyltransferase (CGT), which is essential for the synthesis of galactocerebroside and sulfatide, follows the same pattern as do structural proteins in myelin. [13-15]. There is evidence that some lipids are selectively complexed with myelin proteins for processing and ultimate insertion into the plasmalemma [16-18].

Knockout mouse technology has significantly added to our understanding and appreciation of the roles of myelin lipids in overall function and structure of the CNS. CGT knockout mice are completely deficient in galactolipids, and develop tremors and weakness before dying at about 3 months of age [19-21]. These mice have abnormal paranodal ultrastructure, most notably poor or absent apposition of the paranodal loops with the axon. Septate-like junctions fail to form, resulting in loss of compartmentation separating the voltage gated sodium channels at the node from the potassium channels normally restricted to the juxtaparanodal regions. In addition, while sodium channels remain confined to the nodes, paranodal proteins such as Caspr, and the potassium channels normally restricted to the juxtaparanode are more diffusely spread along the axonal membrane. Similarly, mice deficient for sulfatide synthesis due to knockout of the ceramide sulfotransferase (CST) gene also develop similar abnormalities of the paranodal apparatus as do mice deficient for both sulfatide and galactocerebroside [22, 23]. Thus, the lipid component of myelin plays a critical role in higher order structure of not only the myelin sheath, but also of more complex structures and molecular organization of the critical nodal and surrounding regions. Loss of this segregation these channels causes impaired conduction of action potentials.

Emerging evidence supports a role for aspartoacylase (N-acetylaspartate amidohydrolase; aminoacylase II) (Figure 1) in the synthesis of myelin lipids. This enzyme catalyzes the hydrolysis of N-acetylaspartate (NAA) into acetate and aspartate [24, 25]. Although NAA is one of the most abundant compounds in the CNS, its function has not been conclusively established. However, several observations suggest that it does play an important role in myelination. These include: 1) the expression of aspartoacylase, NAA and of aspartate N-acetyltransferase (ANAT: the synthetic enzyme for NAA) all rise in concert with the classical myelin proteins, 2) apparent deficiency of NAA results in leukodystrophy [26], 3) mutations affecting the aspartoacylase gene cause the fatal leukodystrophy Canavan disease [27], 4) NAA injected into rodent brains is incorporated into myelin lipids [28], 5) aspartoacylase mRNA is expressed specifically in oligodendrocytes [29, 30], 6) aspartoacylase itself is expressed in oligodendrocytes [31] and figure 1, and 7) NAA injected into the eye can be transported through the optic nerve and then incorporated into myelin [32]. The interpretation of NAA levels measured by magnetic resonance spectroscopy will be greatly impacted by elucidation of the role of NAA in nervous system function [33].

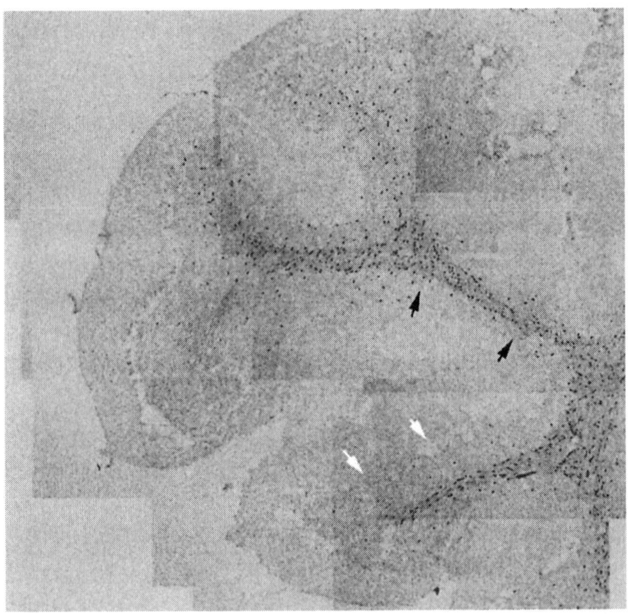

Figure 1. Immunocytochemistry of aspartoacylase in 1-month mouse cerebellum. Note the staining in the white matter tracts (black arrows), but not in Purkinje cells (white arrowheads). Distinct cellular staining is confined to the white matter.

## 3.2     Myelin proteins

Sodium dodecyl sulfate polyacrylamide gel electrophoretic analysis of myelin proteins reveals a complex composition with several prominent bands that have now been characterized.

### 3.2.1     Proteolipid protein 1 (PLP1; lipophilin) and DM20

The major protein in CNS myelin is a highly lipophilic protein now designated proteolipid protein 1, to distinguish it from the structurally similar gut-specific protein, proteolipid protein 2 [34]. PLP1 constitutes half the protein in CNS myelin, or roughly one sixth the dry weight of myelin [35-37]. It is one of several brain antigens that can elicit EAE. PLP1 has the unusual property of being soluble in chloroform/methanol, a solvent mixture used to extract lipids from other cellular compounds. Its unusually strong lipid association led to its alternate designation of lipophilin [reviewed in 38]. PLP1 is a transmembrane protein with four membrane-spanning domains [39-41]. Synthesized as a preprotein with 277 residues, the initiation methionine is cleaved generating the mature protein of 29,869

# MAJOR CNS MYELIN PROTEINS

| Protein | Abundance | Localization | Gene | Chromosome | Disease(s) | Mouse mutant(s) | Null phenotype | Proposed function(s) |
|---|---|---|---|---|---|---|---|---|
| Proteolipid protein | 50% | Compact myelin | PLP1 | Xq22.2 | PMD/SPG2 | jimpy; jp | Late progressive weakness, axonal degeneration | Intralamellar adhesion |
| Myelin basic protein | ~30% | Major dense line | MBP | 18q22-ter | None reported | shiverer; shi | Dysmyelination, tremors, seizures, abnormal axo-glial junctions | Myelin compaction |
| Myelin-associated oligodendrocyte basic protein | ~20% * | Compact myelin | MOBP | 3p22-21 | None reported | | Normal phenotype | |
| Oligodendrocyte specific protein | 7 % | Intralamellar tight junctions | CLDN11 | 3q26.2-26.3 | None reported | | Late hindlimb weakness; normal myelination; absent interlamellar tight junctions | Interlamellar tight junction component |
| 2',3'-cyclic nueleotide phosphodiesterase | 4% | Major dense line | CNP1 | 17q21 | None reported | | Late progressive weakness, ataxia and seizures; Axonal degeneration | Axo-glial signaling |
| Myelin-associated glycoprotein | 1% | Inner mesaxon | MAG | 19q13.1 | None reported | | Late onset weakness; redundant myelin; oligodendrogliopathy; peripheral axonopathy | Axo-glial signaling; inhibits axon regeneration |
| Myelin oligodendrocyte glycoprotein | <1% | Outer myelin sheath | MOG | 6p21.3 | None reported | | | |
| Connexin 32 | <1% | Non-compact myelin | GJB1 | Xq13.1 | CMTX1 | | Peripheral neuropathy; mild CNS dysmyelination | |
| Oligodendrocyte myelin glycoprotein | <1% | Paranodal myelin | OMgp | 17q11.2 | None reported | | | Inhibits axonal regeneration |

Refer to text for references.

*Estimate based on mRNA expression relative to other myelin protein genes.

daltons with 276 residues, highly enriched in hydrophobic amino acids. Both the amino and carboxyl termini lie within the cytoplasm, as does an intracellular loop that is enriched in charged residues.

The primary structure of PLP1 is remarkably highly conserved, and is identical in humans, mice and rats. As described in more detail subsequently, the *PLP1* gene undergoes alternative splicing, generating an additional product, DM20, that lacks 35 residues (the PLP1-specific domain) that lie within the cytoplasmic loop of PLP1. Although DM20 appears during embryonic development and is about equivalent in level to that of PLP1 in Schwann cells, PLP1 predominates in mature CNS myelin, where it is about 10 fold more abundant than DM20 [42]. During development, PLP1 expression peaks in concert with other major myelin proteins, and depends upon contact with axons [43].

PLP1 is covalently modified after its synthesis. and several cysteine residues are acylated by fatty acids that probably anchor the protein to the plasma membrane [44-46](figure 2). Proteolytic cleavage may also occur in a physiologically regulated manner to release a peptide with mitogenic properties on astrocytes as well as on oligodendrocyte precursors [47].

54

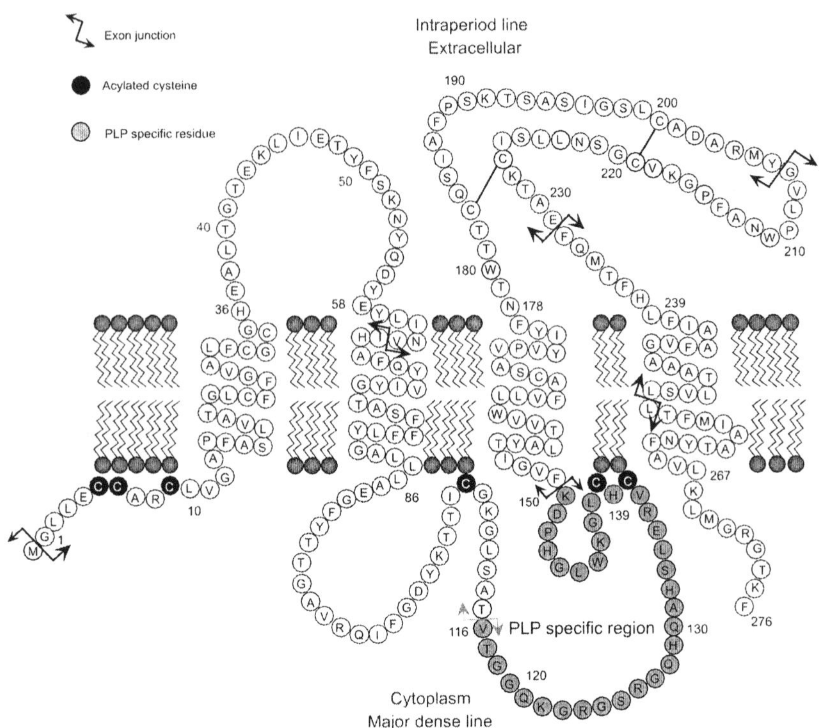



*Figure 2*. Proteolipid protein 1 structure. PLP1 is predicted to have four transmembrane domains, and both the amino and carboxy-termini in the cytoplamic compartment. A 35-amino acid long domain shaded in blue represents the PLP1-specific region of the protein. Cysteine residues that are predicted to be acylated are shaded in black.

Although studied for over 40 years, the function of PLP1 remains poorly understood. The topology of PLP1 and its abundance in myelin lead to speculation that it must act as an adhesion protein, locking together adjacent membrane leaflets in compact myelin. With the advent of knockout technology, it has been possible to generate *Plp1*-deficient mice. Surprisingly, compact myelin forms in these animals and only subtle ultrastructural abnormalities were seen [48, 49]. Thus, oligodendrocyte development doesn not require PLP1 or DM20, and the process of myelin compaction must depend upon other myelin components. The slight 'loosening' of myelin lamellae does support an adhesive role for PLP1, but the presence of compact myelin in these knockout animals suggests that

---

[1] By convention, nomenclature for a murine gene or protein requires only the first letter be capitalized, whereas that for the human gene or protein abbreviation requires all letters be capitalized. In addition, the protein name is indicated by normal font, whereas the gene is indicated by italicized characters.

other factors are also important for adhesion of myelin leaflets. The expression of other myelin proteins, such as MBP, is not altered when PLP1 is absent. This indicates that PLP1 is not required for the expression of other myelin proteins, and that there is not an obvious compensatory change in other myelin proteins that accounts for the surprisingly mild null phenotype.

Other PLP1 functions have been proposed, including ion channel and glial mitogenic factor [reviewed in 50]. The *Plp1* knockout studies, however, provide conclusive proof that this protein is necessary for axonal survival. Griffiths and colleagues made the surprising observation that Plp1 deficient mice, which are behaviorally normal until about 15 months of age, clinically deteriorate due to the degeneration of axons [51]. Similar findings were noted in human patients that have null mutations of the *PLP1* gene [52]. Thus, the integrity of the nervous system depends upon the synergistic relationships between axons and glia. PLP1 is also necessary for peripheral nerve function since PLP1-deficient patients develop demyelinating peripheral neuropathy, even though PLP1 and DM20 constitute only a very small fraction of peripheral nervous system myelin [53].

### 3.2.1.1    Proteolipid protein 1 genetics

The *PLP1* gene lies on the X chromosome band Xq22.2 [54]. It is approximately 20 Kbp in length and contains seven exons [55]. The longest open reading frame of the *PLP1* mRNA has 277 codons. Because the initiation methionine is cleaved during processing of the protein, by convention the numbering of amino acid residues as well as corresponding codons begins with the glycine that immediately follows the initiation methionine.

Alternative usage of a splice site in exon three of the *PLP1* gene generates a shorter transcript that encodes DM20, also a tetraspan protein, but that lacks 35 amino acids from the cytoplasmic loop [56]. DM20 is similar in sequence to the M6A and M6B proteins, which are expressed in neurons or in both neurons and oligodendrocytes, respectively [57].

Many mutations affecting the *PLP1* gene have been discovered that cause clinical syndromes ranging from 'pure' spastic paraplegia (SPG2) to severe Pelizaeus-Merzbacher disease [reviewed in 58]. Mutations occur in all regions of the protein, unfortunately providing little insight into the biological functions of this still enigmatic protein. The most common mutation causing Pelizaeus-Merzbacher disease is complete duplication of the *PLP1* gene, suggesting that precise regulation of PLP1 expression is important for myelination and oligodendrocyte health.

### 3.2.2    Myelin basic protein (MBP)

The second most abundant protein in CNS myelin, and the first identified EAE inducing antigen, is myelin basic protein (MBP) [59]. Although a myelin protein, constituting about 30 % of myelin protein mass, MBP is not a transmembrane protein. MBP is a hydrophilic protein with a highly basic isoelectric point, and is probably tightly associated with lipids and/or other myelin proteins, where it lies in the thin cytoplasmic space between cell membrane leaflets that constitutes the ultrastructural feature known as the major dense line of compact myelin [60].

MBP is actually a family of related proteins. The *MBP* gene undergoes a complex pattern of alternative splicing, generating 4 major products with molecular masses of 21.5 (encoded by the seven main exons), 20.2 (lacking exon 5), 18.5 (lacking exon 2) and 17.2 (lacking exons 2 and 5) kDa. In mice there are five MBP variants, with molecular masses of 21.5 (all 7 exons), 18.5 (lacking exon 2), 17.2 (lacking exons 2 and 5), 17.0 (lacking exon 6), and 14.0 (lacking exons 2 and 6) kDa [61, 62]. Other splice variants occur but are present at very low levels. These different forms are developmentally regulated, and in the adult brain the 18.5 and 17.2 kDa forms predominate, constituting about 95 % of the MBP. Why the different MBP forms exist is not clear, however, the exon 2 containing transcripts are expressed early during development and the encoded MBP variants can be transported into the nucleus, suggesting a transcription regulatory role for MBP [63]. All these isoforms are not needed for myelin compaction and maintenance, per se, since mice that express only the 14 kDa form can form normal myelin [64].

Campagnoni and colleagues discovered that the *MBP* gene is part of a much larger gene which they named *Golli-mbp* (gene expressed in the oligodendrocyte lineage) [65]. The *Golli-mbp* gene includes three additional exons spanning 73 Kbp that lie 5' of the classical *MBP* gene, and contains a transcription unit of 179 Kbp in humans, compared to 45 Kbp for the classical *MBP* gene. The expression of the *Golli-mbp* gene is complex and includes several alternatively spliced transcripts, including some that are found outside of the nervous system, including thymus, where it could potentially play a role as an autoantigen [65].

Although studied for many years, the role of MBP in myelin formation is not well understood. It is thought to be involved in compaction of myelin and is clearly part of the major dense line of compact myelin. Post-translational modification, such as methylation and citrullination, of MBP may facilitate compaction and interactions with other proteins or oligodendryocyte components [66-69].

### 3.2.2.1 *MBP* genetics

Mutations of the human *MBP* gene have not been identified as yet. There are individuals that have deletion of the long arm of chromosome 18, where the *MBP* gene lies. Patients with the 18q- syndrome have mild to severe cognitive and behavioral impairment and MRI abnormalities that suggest this gene is haploinsufficient in humans (i.e. only one copy of the gene is not enough for normal neurologic function) [70-72].

Although there are no known pathologically significant *MBP* mutations in man, there are at least three rodent *Mbp* gene mutations that cause dysmyelination. The best characterized of these is the autosomal recessive *shiverer* (shi) mutation in mice that results from deletion of part of the *Mbp* gene containing exons 3 − 7 [73-76]. Heterozygous (shi/+) mice are normal and have morphologically normal myelin. *shi* mice have a severe neurologic syndrome characterized by ataxia and tremor that begins at about 2 weeks of age followed by development of tonic seizures at about 2 to 3 months of age and they usually die by 4 to 5 months of age. Morphologically, shi myelin sheaths are thinner than normal myelin sheaths and lack the major dense line. Similar phenotypes occur in the myelin deficient (*shi$^{mld}$*) and Long Evans shaker rats, caused by inverted partial duplication of the *Mbp* gene [77] and by transposon insertion into intron 3 [78], respectively. Although Mbp is present in relatively large amounts in peripheral nerve myelin, peripheral nerve function as well as myelin ultrastructure, including the major dense line, is normal [79]. It has been speculated that the major protein of PNS myelin, myelin protein zero (MPZ) not only acts as an interlamellar adhesion protein, but also acts, through its intracellular domain, to promote myelin compaction and participate in the formation of the major dense line.

*Mbp* mutations most likely act as functional nulls. No Mbp protein is detectable in *shi* myelin, whereas other myelin proteins are present, although with alteration in some of the myelin protein levels. For example, Plp levels are reduced, but the level of 2',3'-cyclic nucleotide phosphodiesterase is not affected. The level of some lipids, such as gangliosides, is altered in *shi* mice [80]. Abnormal paranodal junctions and localization of ion channels occurs in *shi* mice, demonstrating that lack of MBP results in significant disruption of axo-glial organization as well as myelin and probably contribute to the clinical phenotype [81]. Brady et al. observed that *shi* mice have abnormal axonal cytoskeletal organization and axonal transport demonstrating the importance of normal glial interactions for establishing and maintaining axonal structure and function [82].

Interestingly, the phenotype of the *shi/shi*;jp/Y mutation (i.e. mice deficient for Mbp and with a severe mutation of the *Plp* gene) is less severe than that of either single mutation and suggests that Plp and Mbp functionally interact consistent with in vitro studies [80, 83].

### 3.2.3    2',3'-cyclic nucleotide phosphodiesterase (CNPase)

CNPase constitutes about 5 % of CNS myelin protein [84-86]. Although *in vitro* this protein catalyzes the hydrolysis of 2',3'-cyclic nucleotides to 2'-nucleotides, whether this enzymatic activity is active *in vivo* is not known. CNPase is one of the so-called Wolfgram proteins, and on SDS gels consists of two proteins of about 46 and 48 kDa that result from differential utilization of translation initiation sites. Both proteins are isoprenylated at their carboxy termini and are acylated, facilitating their association with membranes, although they are cytosolic proteins. The larger isoform especially is phosphorylated. Immunohistochemical studies have demonstrated that this protein is confined to non-compacted regions of myelin, such as the clefts of Schmidt-Lanterman, the paranodal loops and the inner mesaxon [87]. Expression of the two protein forms is differentially regulated during development, with only the larger protein expressed in oligodendrocyte precursors [88]. As with structural myelin proteins such as MBP and PLP, expression of CNPase by oligodendrocytes is dependent upon contact with axons [43].

The precise role of CNPase in myelin is unclear. CNPase knockout mice have ultrastructurally normal myelin, including compact myelin [89]. These mice have normal motor performance and behavior until about 4 months of age, but subsequently develop progressive motor weakness, ataxia and seizures, followed by death between about 7 to 11 months. Pathologically the most prominent abnormality is widespread axonal degeneration, characterized by axonal swellings filled with vesicles, mitochondria and cytoskeletal proteins, and reduction in axon numbers. These findings are similar to those seen in both mice and humans with *PLP1* null mutations. Thus, CNPase as well as PLP1 are essential for maintaining axonal integrity, further demonstrating the importance of the dynamic interaction between axons and their ensheathing oligodendrocytes.

## 3.3    Myelin associated glycoprotein (MAG)

Myelin contains several glycoproteins, the best studied of which is myelin associated glycoprotein [90, 91]. MAG comprises about 1% of CNS myelin and is actually comprised of two related proteins of 72 (large or L-MAG) and 67 (small or S-MAG) kDa molecular mass that are both products of alternative splicing of the MAG gene. Both are heavily N-glycosylated, with approximately 30 % of their mature mass constituted of carbohydrate. MAG is a membrane protein, with a single transmembrane domain, and five immunoglobulin (Ig) domains, the first of which has an RGD (Arginine-glycine-aspartate) motif characteristic of integrin-mediated cell adhesion molecules. The oligosaccharides in MAG are heterogeneous and complex, most being negatively charged due to sialic acid or sulfate groups, and include the L2/HNK-1 carbohydrate epitope.

The structure and localization of MAG is therefore consistent with it playing a role in axo-glial signaling. MAG is located in the inner mesaxon at the point of axo-glial contact. In the CNS, MAG is found only in the periaxonal oligodendroglial membrane, however, in the PNS it is also localized at other non-compacted areas of myelin, i.e. paranodal loops, Schmidt-Lanterman clefts and the outer mesaxons. There are no known mutations in the human MAG gene, but experimentally created MAG-null mice provide clear evidence that MAG mediates important signaling functions between axons and myelinating glia, both in the CNS and PNS [92, 93]. Mice deficient in both S-MAG and L-MAG have a delay in myelination, and later develop cognitive and locomotor deficits. Mice have a normal to near normal lifespan. Pathologically the most striking abnormalities in the CNS are aberrant myelin loops and redundant myelin sheaths, and later develop dying back oligodendrogliopathy [94]. This implicates MAG as a mediator of an axonal signal that exerts a trophic effect on oligodendrocytes.

The strong evidence that a significant cause of disability in patients with multiple sclerosis results from axonal degeneration has placed great importance on studies of neural regeneration [reviewed in 95]. MAG has been shown to have axonal growth-inhibitory effects in postnatal animals after injury [96]. The current evidence favors a complex set of interactions with the oligodendrocyte proteins MAG, Nogo66, and Oligodendrocyte-Myelin glycoprotein (OMgp, see below) interacting with the axonal Nogo receptor (NgR) and p75 neurotrophin receptor on lipid rafts containing complex gangliosides to activate the small GTPase Rho, which then leads to axonal growth inhibition. Nogo, originally identified by the Schwab group as an inhibitor of axonal regrowth, encodes three splice variants, Nogo-A, -B, and –C. Of these evidence from selective knockout of splice variants supports Nogo-A in particular as important in preventing axonal regrowth. However, not all lines of Nogo – deficient mice show the predicted improvement in neuronal regeneration [97-101]. Therefore, understanding the roles of other factors that inhibit regeneration, compensatory and genetic background factors will still be needed before we can develop effective approaches for neural regeneration.

# 3.4     Oligodendrocyte specific protein (OSP; aka Claudin-11)

OSP was originally found through a molecular strategy to isolate novel oligodendrocyte specific genes [102]. It is a 22 kDa protein that is a member of the tetraspanin and claudin families of transmembrane proteins and comprises about 7 % of CNS myelin protein, making it the third most abundant myelin protein. Mice that are deficient in OSP develop tremors at about two weeks of age but that later improves, followed by adult onset hindlimb paresis and male sterility. There is not a readily apparent deficiency of myelin in these animals, however with specific antibodies, and ultrastructural studies it has been found that OSP forms intracellular tight junctions at sites of apposition of noncompact myelin regions. In addition, it is also necessary for formation of tight junctions in Sertoli cells [103].

# 3.5     Myelin oligodendrocyte protein (MOG)

MOG is a minor component of CNS myelin that was originally identified through its ability to act as a brain antigen [104]. It is a glycoprotein of 26 to 28 kDa that is like MAG a member of the Ig superfamily of proteins [105, 106]. The *MAG* gene lies within the HLA gene complex on chromosome 6 [107]. It contains complex oligosaccharides, including the L2/HNK-1 epitope. It has a single Ig domain and probably has a single transmembrane domain. Its expression is restricted to the outer surface of myelin sheaths and oligodendrocytes, suggesting that it may function in mediating signals between the oligodendrocytes and the extracellular space [108]. Both T and B cell-mediated immune responses can be experimentally generated by MOG [109, 110]. Thus far, no naturally occurring or experimentally-generated mutations of MOG have been reported in any species.

# 3.6     Connexin 32 (Cx32; Gap junction β 1)

The gap junction-forming protein connexin 32 is a quantitatively minor component of CNS myelin, where it forms heterologous gap junctions between oligodendrocytes and astrocytes (Cx32 on oligodendrocyte side, and Cx26, Cx30 and Cx43 [111, 112]. The function of these gap junctions was uncertain since patients with mutations in Cx32 classically develop only peripheral neuropathy. In addition, Cx32 null mice only manifest a peripheral nervous system phenotype [113]. Recent reports demonstrate that these gap junctions do have functional significance since some patients with X-linked Charcot-Marie-Tooth disease develop deafness and transient CNS white matter abnormalities associated with bulbar palsy and quadriparesis [114-117].

## 3.7 The paranodal junction proteins (septate-like junctions)

Paranodal axo-glial junctions or septate-like junctions are specializations of critical importance to the normal electrophysiological function of myelinated axons [reviewed in 6, 118, 119, 120]. The proteins at these junctions not only interact to form intercellular junctions linking the axonal paranodal membrane to the paranodal myelin loops, but also interact with intracellular proteins to anchor the junctional complex to the cytoskeleton and to segregate ion channels and other proteins to the appropriate spatial domains. The expression of these junction proteins during development is critical to the formation of the nodal region. The best characterized components of septate-like junctions are axonal proteins, and include contactin, Caspr (contactin-associated protein; aka paranodin), and Caspr2, which are collectively also referred to as NCP proteins (neurexin, caspr and paranodin). Recently at least one of the glial receptors for the Caspr/paranodin complex was found to be neurofascin 155 (NF155) [5]. NF155 is encoded by the neurofascin gene, which also encodes additional proteins whose transcripts result from alternative splicing of the gene [121, 122]. NF155 is an ankyrin-bound transmembrane protein of the Ig superfamily related closely to the L1 cell adhesion molecule.

## 3.8 Other proteins

Myelin-associated oligodendrocyte basic protein (MOBP) is a small, basic protein present in compact myelin, where they probably associate with MBP [123, 124]. At least three splice variants exist, generating polypeptides of 8.2, 9.7 and 11.7 kDa. There are no human MOBP gene mutations reported, however, recently the murine gene has been knocked out. Mobp-deficient mice have no neurologic phenotype and have morphologically normal myelin [125].

Oligodendrocyte-myelin glycoprotein (OMgp) is a glycoprotein of 120 kDa localized to the paranodal areas. It contains the L2/HNK-1 epitope. This protein is implicated as an inhibitor of axonal regeneration, probably acting as a co-ligand for the Nogo receptor (see discussion in section 3.3).

Oligodendrocytes, as do other cells, express a variety of ion channels and ion transporters, but these have not been studied directly with respect to their possible roles in the etiology of EAE or MS. Interestingly, knockout of one potassium channel gene, the weak inwardly rectifying channel (Kir4.1) results in a severe syndrome characterized by tremor by 2 weeks and quadriparesis with death by about 3 weeks of age [126]. Pathologically the white matter is altered by spongiform change and axonal swellings, reminiscent of Canavan disease pathology.

## 4.        CONCLUSIONS

It is now clear that in addition to their likely roles in the formation, compaction and maintenance of myelin, myelin proteins and lipids are also critical for maintaining the ionic environment of the axon, and for the long term stability of axonal structure. The understanding of disease pathogenesis and development of more effective therapeutic strategies for multiple sclerosis must take into account the basic properties and functions of these oligodendrocyte components, as well as their potential as autoantigens in initiating EAE, and possible MS.

**ACKNOWLEGDMENT**
The author acknowledges support from the US National Multiple Sclerosis Society, research grant RG 3204A1/1

References

1.      Baumann, N. and D. Pham-Dinh, *Biology of oligodendrocyte and myelin in the mammalian central nervous system.* Physiol Rev, 2001. **81**(2): p. 871-927.
2.      Balice-Gordon, R.J., L.J. Bone, and S.S. Scherer, *Functional gap junctions in the schwann cell myelin sheath.* J Cell Biol, 1998. **142**(4): p. 1095-104.
3.      Wood, P. and R.P. Bunge, *The biology of the oligodendrocyte*, in *Oligodendroglia: Advances in Neurochemistry*, W.T. Norton, Editor. 1984, Plenum Press: New York. p. 1-46.
4.      Bhat, M.A., J.C. Rios, Y. Lu, G.P. Garcia-Fresco, W. Ching, M. St Martin, J. Li, S. Einheber, M. Chesler, J. Rosenbluth, J.L. Salzer, and H.J. Bellen, *Axon-glia interactions and the domain organization of myelinated axons requires neurexin IV/Caspr/Paranodin.* Neuron, 2001. **30**(2): p. 369-83.
5.      Charles, P., S. Tait, C. Faivre-Sarrailh, G. Barbin, F. Gunn-Moore, N. Denisenko-Nehrbass, A.M. Guennoc, J.A. Girault, P.J. Brophy, and C. Lubetzki, *Neurofascin is a glial receptor for the paranodin/Caspr-contactin axonal complex at the axoglial junction.* Curr Biol, 2002. **12**(3): p. 217-20.

6.      Scherer, S.S. and E.J. Arroyo, *Recent progress on the molecular organization of myelinated axons.* J Peripher Nerv Syst, 2002. **7**(1): p. 1-12.

7.      Baker, M.D., *Axonal flip-flops and oscillators.* Trends Neurosci, 2000. **23**(11): p. 514-9.

8.      Baker, M.D., *Electrophysiology of mammalian Schwann cells.* Prog Biophys Mol Biol, 2002. **78**(2-3): p. 83-103.

9.      Rumsby, M.G., *Organization and structure in central-nerve myelin.* Biochem Soc Trans, 1978. **6**(2): p. 448-62.

10.     Norton, W.T. and W. Cammer, *Isolation and characterization of myelin*, in *Myelin*, P. Morrell, Editor. 1984, Plenum Press: New York. p. 531.

11.     Sheikh, K.A., J. Sun, Y. Liu, H. Kawai, T.O. Crawford, R.L. Proia, J.W. Griffin, and R.L. Schnaar, *Mice lacking complex gangliosides develop Wallerian degeneration and myelination defects.* Proc Natl Acad Sci U S A, 1999. **96**(13): p. 7532-7.

12.     Vyas, A.A., H.V. Patel, S.E. Fromholt, M. Heffer-Lauc, K.A. Vyas, J. Dang, M. Schachner, and R.L. Schnaar, *Gangliosides are functional nerve cell ligands for myelin-associated glycoprotein (MAG), an inhibitor of nerve regeneration.* Proc Natl Acad Sci U S A, 2002. **99**(12): p. 8412-7.

13.     Benjamins, J.A., K. Miller, and G.M. McKhann, *Myelin subfractions in developing rat brain: characterization and sulphatide metabolism.* J Neurochem, 1973. **20**(6): p. 1589-603.

14.     Shimomura, K., S. Yahara, Y. Kishimoto, and J.A. Benjamins, *Metabolism of cerebrosides and sulfatides in subcellular fractions of developing rat brain.* Biochim Biophys Acta, 1984. **795**(2): p. 265-70.

15.     Stahl, N., H. Jurevics, P. Morell, K. Suzuki, and B. Popko, *Isolation, characterization, and expression of cDNA clones that encode rat UDP-galactose: ceramide galactosyltransferase.* J Neurosci Res, 1994. **38**(2): p. 234-42.

16.     Burkart, T., L. Caimi, H.P. Siegrist, N.N. Herschkowitz, and U.N. Wiesmann, *Vesicular transport of sulfatide in the myelinating mouse brain. Functional association with lysosomes?* J Biol Chem, 1982. **257**(6): p. 3151-6.

17.     Simons, M., E.M. Kramer, C. Thiele, W. Stoffel, and J. Trotter, *Assembly of Myelin by Association of Proteolipid Protein with Cholesterol- and Galactosylceramide-rich Membrane Domains.* J Cell Biol, 2000. **151**(1): p. 143-154.

18.     Lee, A.G., *Myelin: Delivery by raft.* Curr Biol, 2001. **11**(2): p. R60-2.

19.     Coetzee, T., N. Fujita, J. Dupree, R. Shi, A. Blight, K. Suzuki, and B. Popko, *Myelination in the absence of galactocerebroside and*

      *sulfatide: normal structure with abnormal function and regional instability.* Cell, 1996. **86**(2): p. 209-19.
20.   Dupree, J.L., T. Coetzee, K. Suzuki, and B. Popko, *Myelin abnormalities in mice deficient in galactocerebroside and sulfatide.* J Neurocytol, 1998. **27**(9): p. 649-59.
21.   Marcus, J. and B. Popko, *Galactolipids are molecular determinants of myelin development and axo-glial organization.* Biochim Biophys Acta, 2002. **1573**(3): p. 406-13.
22.   Ishibashi, T., J.L. Dupree, K. Ikenaka, Y. Hirahara, K. Honke, E. Peles, B. Popko, K. Suzuki, H. Nishino, and H. Baba, *A myelin galactolipid, sulfatide, is essential for maintenance of ion channels on myelinated axon but not essential for initial cluster formation.* J Neurosci, 2002. **22**(15): p. 6507-6514.
23.   Honke, K., Y. Hirahara, J. Dupree, K. Suzuki, B. Popko, K. Fukushima, J. Fukushima, T. Nagasawa, N. Yoshida, Y. Wada, and N. Taniguchi, *Paranodal junction formation and spermatogenesis require sulfoglycolipids.* Proc Nat Acad Sci Usa, 2002. **99**(7): p. 4227-4232.
24.   D'Adamo, A.F., Jr., J.C. Smith, and C. Woiler, *The occurrence of N-acetylaspartate amidohydrolase (aminoacylase II) in the developing rat.* J Neurochem, 1973. **20**(4): p. 1275-8.
25.   Hagenfeldt, L., I. Bollgren, and N. Venizelos, *N-acetylaspartic aciduria due to aspartoacylase deficiency--a new aetiology of childhood leukodystrophy.* J Inherit Metab Dis, 1987: p. 135-41.
26.   Martin, E., A. Capone, J. Schneider, J. Hennig, and T. Thiel, *Absence of N-acetylaspartate in the human brain: impact on neurospectroscopy?* Ann Neurol, 2001. **49**(4): p. 518-21.
27.   Kaul, R., G.P. Gao, K. Balamurugan, and R. Matalon, *Cloning of the human aspartoacylase cDNA and a common missense mutation in Canavan disease.* Nat Genet, 1993. **5**(2): p. 118-23.
28.   Burri, R., C. Steffen, and N. Herschkowitz, *N-acetyl-L-aspartate is a major source of acetyl groups for lipid synthesis during rat brain development.* Dev Neurosci, 1991. **13**(6): p. 403-11.
29.   Kirmani, B.F., D.M. Jacobowitz, A.T. Kallarakal, and M.A. Namboodiri, *Aspartoacylase is restricted primarily to myelin synthesizing cells in the CNS: therapeutic implications for Canavan disease.* Brain Res Mol Brain Res, 2002. **107**(2): p. 176-82.
30.   Kirmani, B.F., D.M. Jacobowitz, and M.A. Namboodiri, *Developmental increase of aspartoacylase in oligodendrocytes parallels CNS myelination.* Brain Res Dev Brain Res, 2003. **140**(1): p. 105-15.
31.   Garbern, J., K. Jjajj, G. Jayasundera, and R. Matalon. *Aspartoacylase, the Hydrolytic Enzyme for N-Acetylaspartate, Is an Oligodendrocyte Protein: Implications for Myelination.* in *Neurology.* 2003. Honolulu, HI.

32. Chakraborty, G., P. Mekala, D. Yahya, G. Wu, and R.W. Ledeen, *Intraneuronal N-acetylaspartate supplies acetyl groups for myelin lipid synthesis: evidence for myelin-associated aspartoacylase.* J Neurochem, 2001. **78**(4): p. 736-45.

33. Barker, P.B., *N-Acetyl Aspartate-A neuronal marker?* Ann Neurol, 2001. **49**: p. 423-24.

34. Milde, S., C. Viebahn, and C. Kirchner, *Proteolipid protein 2 mRNA is expressed in the rabbit embryo during gastrulation.* Mech Dev, 2001. **106**(1-2): p. 129-32.

35. Tenenbaum, D. and J. Folch-Pi, *The preparation and characterization of water-soluble proteolipid protein from bovine brain white matter.* Biochimica et Biophysica Acta, 1966. **115**(1): p. 141-7.

36. Griffiths, I.R., P. Montague, and P. Dickinson, *The proteolipid protein gene.* Neuropathology & Applied Neurobiology, 1995. **21**(2): p. 85-96.

37. Greer, J.M. and M.B. Lees, *Myelin proteolipid protein--the first 50 years.* Int J Biochem Cell Biol, 2002. **34**(3): p. 211-5.

38. Gow, A., *Redefining the lipophilin family of proteolipid proteins.* J Neurosci Res, 1997. **50**(5): p. 659-64.

39. Laursen, R.A., M. Samiullah, and M.B. Lees, *The structure of bovine brain myelin proteolipid and its organization in myelin.* Proc Natl Acad Sci U S A, 1984. **81**(9): p. 2912-6.

40. Popot, J.-L., D. Pham-Dinh, and A. Dautigny, *Major myelin proteolipid: the 4-$\alpha$-helix topology.* Journal of Membrane Biology, 1991. **120**: p. 233-246.

41. Inouye, H. and D.A. Kirschner, *Membrane topology of PLP in CNS myelin: evaluation of models.* Neurochemical Research, 1994. **19**(8): p. 975-81.

42. Pham-Dinh, D., M.C. Birling, G. Roussel, A. Dautigny, and J.L. Nussbaum, *Proteolipid DM-20 predominates over PLP in peripheral nervous system.* Neuroreport, 1991. **2**(2): p. 89-92.

43. Scherer, S.S., H.H. Vogelbacker, and J. Kamholz, *Axons modulate the expression of proteolipid protein in the CNS.* J Neurosci Res, 1992. **32**(2): p. 138-48.

44. Townsend, L.E., D. Agrawal, J.A. Benjamins, and H.C. Agrawal, *In vitro acylation of rat brain myelin proteolipid protein.* Journal of Biological Chemistry, 1982. **257**(16): p. 9745-50.

45. Agrawal, H.C., C.L. Randle, and D. Agrawal, *In vivo acylation of rat brain myelin proteolipid protein.* Journal of Biological Chemistry, 1982. **257**(8): p. 4588-92.

46. Bizzozero, O.A., S.P. Malkoski, C. Mobarak, H.A. Bixler, and J.E. Evans, *Mass-spectrometric analysis of myelin proteolipids reveals*

*new features of this family of palmitoylated membrane proteins.* J Neurochem, 2002. **81**(3): p. 636-45.

47.    Yamada, M., A. Ivanova, Y. Yamaguchi, M.B. Lees, and K. Ikenaka, *Proteolipid protein gene product can be secreted and exhibit biological activity during early development.* J Neurosci, 1999. **19**(6): p. 2143-51.

48.    Boison, D. and W. Stoffel, *Disruption of the compacted myelin sheath of axons of the central nervous system in proteolipid protein-deficient mice.* Proceedings of the National Academy of Sciences of the United States of America, 1994. **91**(24): p. 11709-13.

49.    Klugmann, M., M.H. Schwab, A. Pühlhofer, A. Schneider, F. Zimmermann, I.R. Griffiths, and K.-A. Nave, *Assembly of CNS myelin in the absence of proteolipid protein.* Neuron, 1997. **18**(1): p. 59-70.

50.    Knapp, P.E., *Proteolipid Protein: Is it more than just a structural component of myelin?* Developmental Neuroscience, 1996. **18**(4): p. 297-308.

51.    Griffiths, I., M. Klugmann, T. Anderson, D. Yool, C. Thomson, M.H. Schwab, A. Schneider, F. Zimmermann, M. McCulloch, N. Nadon, and K.-A. Nave, *Axonal swellings and degeneration in mice lacking the major proteolipid of myelin.* Science, 1998. **280**: p. 1610 - 1613.

52.    Garbern, J.Y., D.A. Yool, G.J. Moore, I.B. Wilds, M.W. Faulk, M. Klugmann, K.A. Nave, E.A. Sistermans, M.S. van der Knaap, T.D. Bird, M.E. Shy, J.A. Kamholz, and I.R. Griffiths, *Patients lacking the major CNS myelin protein, proteolipid protein 1, develop length-dependent axonal degeneration in the absence of demyelination and inflammation.* Brain, 2002. **125**(Pt 3): p. 551-61.

53.    Garbern, J.Y., F. Cambi, X.M. Tang, A.A. Sima, J.M. Vallat, E.P. Bosch, R. Lewis, M. Shy, J. Sohi, G. Kraft, K.L. Chen, I. Joshi, D.G. Leonard, W. Johnson, W. Raskind, S.R. Dlouhy, V. Pratt, M.E. Hodes, T. Bird, and J. Kamholz, *Proteolipid protein is necessary in peripheral as well as central myelin.* Neuron, 1997. **19**(1): p. 205-18.

54.    Willard, H.F. and J.R. Riordan, *Assignment of the gene for myelin proteolipid protein to the X chromosome: implications for X-linked myelin disorders.* Science, 1985. **230**(4728): p. 940-2.

55.    Milner, R.J., C. Lai, K.A. Nave, D. Lenoir, J. Ogata, and J.G. Sutcliffe, *Nucleotide sequences of two mRNAs for rat brain myelin proteolipid protein.* Cell, 1985. **42**(3): p. 931-9.

56.    Nave, K.A., C. Lai, F.E. Bloom, and R.J. Milner, *Splice site selection in the proteolipid protein (PLP) gene transcript and primary structure of the DM-20 protein of central nervous system myelin.* Proc Natl Acad Sci U S A, 1987. **84**(16): p. 5665-9.

57.  Yan, Y., C. Lagenaur, and V. Narayanan, *Molecular cloning of M6: identification of a PLP/DM20 gene family.* Neuron, 1993. **11**(3): p. 423-31.

58.  Garbern, J., *PLP related diseases*, in *Geneclinics*, R. Pagon, Editor. 2002.

59.  Laatch, R.H., M.W. Kies, S. Gordon, and E.C. Alvord, Jr., *The encephalitogenic activity of myein isolated by ultracentrifugation.* J Exp Med, 1962. **115**: p. 777.

60.  Kies, M.W., R.E. Martenson, and G.E. Deibler, *Myelin basic proteins.* Adv Exp Med Biol, 1972. **32**(0): p. 201-14.

61.  de Ferra, F., H. Engh, L. Hudson, J. Kamholz, C. Puckett, S. Molineaux, and R.A. Lazzarini, *Alternative splicing accounts for the four forms of myelin basic protein.* Cell, 1985. **43**(3 Pt 2): p. 721-7.

62.  Kamholz, J., F. de Ferra, C. Puckett, and R. Lazzarini, *Identification of three forms of human myelin basic protein by cDNA cloning.* Proc Natl Acad Sci U S A, 1986. **83**(13): p. 4962-6.

63.  Reyes, S.D. and A.T. Campagnoni, *Two separate domains in the golli myelin basic proteins are responsible for nuclear targeting and process extension in transfected cells.* J Neurosci Res, 2002. **69**(5): p. 587-96.

64.  Kimura, M., M. Sato, A. Akatsuka, S. Nozawa-Kimura, R. Takahashi, M. Yokoyama, T. Nomura, and M. Katsuki, *Restoration of myelin formation by a single type of myelin basic protein in transgenic shiverer mice.* Proc Natl Acad Sci U S A, 1989. **86**(14): p. 5661-5.

65.  Pribyl, T.M., C.W. Campagnoni, K. Kampf, T. Kashima, V.W. Handley, J. McMahon, and A.T. Campagnoni, *The human myelin basic protein gene is included within a 179-kilobase transcription unit: expression in the immune and central nervous systems.* Proc Natl Acad Sci U S A, 1993. **90**(22): p. 10695-9.

66.  Ulmer, J.B. and P.E. Braun, *In vivo phosphorylation of myelin basic proteins in developing mouse brain: evidence that phosphorylation is an early event in myelin formation.* Dev Neurosci, 1983. **6**(6): p. 345-55.

67.  Yamamori, C., M. Terashima, H. Ishino, and M. Shimoyama, *ADP-ribosylation of myelin basic protein and inhibition of phospholipid vesicle aggregation.* Enzyme Protein, 1994. **48**(4): p. 202-12.

68.  Boggs, J.M., P.M. Yip, G. Rangaraj, and E. Jo, *Effect of posttranslational modifications to myelin basic protein on its ability to aggregate acidic lipid vesicles.* Biochemistry, 1997. **36**(16): p. 5065-71.

69.  Boggs, J.M., G. Rangaraj, K.M. Koshy, C. Ackerley, D.D. Wood, and M.A. Moscarello, *Highly deiminated isoform of myelin basic*

*protein from multiple sclerosis brain causes fragmentation of lipid vesicles.* J Neurosci Res, 1999. **57**(4): p. 529-35.

70. Kline, A.D., M.E. White, R. Wapner, K. Rojas, L.G. Biesecker, J. Kamholz, E.H. Zackai, M. Muenke, C.I. Scott, Jr., and J. Overhauser, *Molecular analysis of the 18q- syndrome--and correlation with phenotype.* Am J Hum Genet, 1993. **52**(5): p. 895-906.

71. Loevner, L.A., R.M. Shapiro, R.I. Grossman, J. Overhauser, and J. Kamholz, *White matter changes associated with deletions of the long arm of chromosome 18 (18q- syndrome): a dysmyelinating disorder?* AJNR Am J Neuroradiol, 1996. **17**(10): p. 1843-8.

72. Mahr, R.N., P.J. Moberg, J. Overhauser, G. Strathdee, J. Kamholz, L.A. Loevner, H. Campbell, E.H. Zackai, M.E. Reber, D.P. Mozley, L. Brown, B.I. Turetsky, and R.M. Shapiro, *Neuropsychiatry of 18q-syndrome.* Am J Med Genet, 1996. **67**(2): p. 172-8.

73. Bird, T.D., D.F. Farrell, and S.M. Sumi, *Brain lipid composition of the shiverer mouse: (genetic defect in myelin development).* J Neurochem, 1978. **31**(1): p. 387-91.

74. Rosenbluth, J., *Central myelin in the mouse mutant shiverer.* J Comp Neurol, 1980. **194**(3): p. 639-48.

75. Roach, A., N. Takahashi, D. Pravtcheva, F. Ruddle, and L. Hood, *Chromosomal mapping of mouse myelin basic protein gene and structure and transcription of the partially deleted gene in shiverer mutant mice.* Cell, 1985. **42**(1): p. 149-55.

76. Kimura, M., H. Inoko, M. Katsuki, A. Ando, T. Sato, T. Hirose, H. Takashima, S. Inayama, H. Okano, K. Takamatsu, and et al., *Molecular genetic analysis of myelin-deficient mice: shiverer mutant mice show deletion in gene(s) coding for myelin basic protein.* J Neurochem, 1985. **44**(3): p. 692-6.

77. Akowitz, A.A., E. Barbarese, K. Scheld, and J.H. Carson, *Structure and expression of myelin basic protein gene sequences in the mld mutant mouse: reiteration and rearrangement of the MBP gene.* Genetics, 1987. **116**(3): p. 447-64.

78. O'Connor, L.T., B.D. Goetz, J.M. Kwiecien, K.H. Delaney, A.L. Fletch, and I.D. Duncan, *Insertion of a retrotransposon in Mbp disrupts mRNA splicing and myelination in a new mutant rat.* J Neurosci, 1999. **19**(9): p. 3404-13.

79. Inouye, H., A.L. Ganser, and D.A. Kirschner, *Shiverer and normal peripheral myelin compared: basic protein localization, membrane interactions, and lipid composition.* J Neurochem, 1985. **45**(6): p. 1911-22.

80. Uschkureit, T., O. Sporkel, J. Stracke, H. Bussow, and W. Stoffel, *Early onset of axonal degeneration in double (plp-/-mag-/-) and hypomyelinosis in triple (plp-/-mbp-/-mag-/-) mutant mice.* J Neurosci, 2000. **20**(14): p. 5225-33.

81.    Rosenbluth, J., *Axoglial junctions in the mouse mutant Shiverer.* Brain Res, 1981. **208**(2): p. 283-97.

82.    Brady, S.T., A.S. Witt, L.L. Kirkpatrick, S.M. de Waegh, C. Readhead, P.H. Tu, and V.M. Lee, *Formation of compact myelin is required for maturation of the axonal cytoskeleton.* J Neurosci, 1999. **19**(17): p. 7278-88.

83.    Wood, D.D., G.J. Vella, and M.A. Moscarello, *Interaction between human myelin basic protein and lipophilin.* Neurochem Res, 1984. **9**(10): p. 1523-31.

84.    Wells, M.R. and T.J. Sprinkle, *Purification of rat 2',3'-cyclic nucleotide 3'-phosphodiesterase.* J Neurochem, 1981. **36**(2): p. 633-9.

85.    Sprinkle, T.J., M.R. Wells, F.A. Garver, and D.B. Smith, *Studies on the Wolfgram high molecular weight CNS myelin proteins: relationship to 2',3'-cyclic nucleotide 3'-phosphodiesterase.* J Neurochem, 1980. **35**(5): p. 1200-8.

86.    Sprinkle, T.J., H.J. Sheedlo, T.B. Buxton, and J.P. Rissing, *Immunochemical identification of 2', 3'-cyclic nucleotide 3'-phosphodiesterase in central and peripheral nervous system myelin, the Wolfgram protein fraction, and bovine oligodendrocytes.* J Neurochem, 1983. **41**(6): p. 1664-71.

87.    Braun, P.E., F. Sandillon, A. Edwards, J.M. Matthieu, and A. Privat, *Immunocytochemical localization by electron microscopy of 2'3'-cyclic nucleotide 3'-phosphodiesterase in developing oligodendrocytes of normal and mutant brain.* J Neurosci, 1988. **8**(8): p. 3057-66.

88.    Scherer, S.S., P.E. Braun, J. Grinspan, E. Collarini, D.Y. Wang, and J. Kamholz, *Differential regulation of the 2',3'-cyclic nucleotide 3'-phosphodiesterase gene during oligodendrocyte development.* Neuron, 1994. **12**(6): p. 1363-75.

89.    Lappe-Siefke, C., S. Goebbels, M. Gravel, E. Nicksch, J. Lee, P.E. Braun, I.R. Griffiths, and K.A. Nave, *Disruption of Cnp1 uncouples oligodendroglial functions in axonal support and myelination.* Nat Genet, 2003. **33**(3): p. 366-74.

90.    Quarles, R.H., *Glycoproteins of myelin sheaths.* J Mol Neurosci, 1997. **8**(1): p. 1-12.

91.    Quarles, R.H., *Myelin sheaths: glycoproteins involved in their formation, maintenance and degeneration.* Cell Mol Life Sci, 2002. **59**(11): p. 1851-71.

92.    Montag, D., K.P. Giese, U. Bartsch, R. Martini, Y. Lang, H. Bluthmann, J. Karthigasan, D.A. Kirschner, E.S. Wintergerst, K.A. Nave, and et al., *Mice deficient for the myelin-associated glycoprotein show subtle abnormalities in myelin.* Neuron, 1994. **13**(1): p. 229-46.

93.     Fruttiger, M., D. Montag, M. Schachner, and R. Martini, *Crucial role for the myelin-associated glycoprotein in the maintenance of axon-myelin integrity.* European Journal of Neuroscience, 1995. **7**(3): p. 511-5.

94.     Lassmann, H., U. Bartsch, D. Montag, and M. Schachner, *Dying-Back Oligodendrogliopathy: a Late Sequel Of Myelin-Associated Glycoprotein Deficiency.* Glia, 1997. **19**(2): p. 104-110.

95.     Bjartmar, C. and B.D. Trapp, *Axonal and neuronal degeneration in multiple sclerosis: mechanisms and functional consequences.* Curr Opin Neurol, 2001. **14**(3): p. 271-8.

96.     Mukhopadhyay, G., P. Doherty, F.S. Walsh, P.R. Crocker, and M.T. Filbin, *A novel role for myelin-associated glycoprotein as an inhibitor of axonal regeneration.* Neuron, 1994. **13**(3): p. 757-67.

97.     McKerracher, L. and M.J. Winton, *Nogo on the go.* Neuron, 2002. **36**(3): p. 345-8.

98.     Simonen, M., V. Pedersen, O. Weinmann, L. Schnell, A. Buss, B. Ledermann, F. Christ, G. Sansig, H. van der Putten, and M.E. Schwab, *Systemic deletion of the myelin-associated outgrowth inhibitor nogo-a improves regenerative and plastic responses after spinal cord injury.* Neuron, 2003. **38**(2): p. 201-11.

99.     Woolf, C.J., *No nogo. Now where to go?* Neuron, 2003. **38**(2): p. 153-6.

100.    Zheng, B., C. Ho, S. Li, H. Keirstead, O. Steward, and M. Tessier-Lavigne, *Lack of enhanced spinal regeneration in nogo-deficient mice.* Neuron, 2003. **38**(2): p. 213-24.

101.    Kim, J.E., S. Li, T. GrandPre, D. Qiu, and S.M. Strittmatter, *Axon regeneration in young adult mice lacking nogo-a/b.* Neuron, 2003. **38**(2): p. 187-99.

102.    Bronstein, J.M., P. Popper, P.E. Micevych, and D.B. Farber, *Isolation and characterization of a novel oligodendrocyte-specific protein.* Neurology, 1996. **47**(3): p. 772-8.

103.    Gow, A., C.M. Southwood, J.S. Li, M. Pariali, G.P. Riordan, S.E. Brodie, J. Danias, J.M. Bronstein, B. Kachar, and R.A. Lazzarini, *CNS myelin and sertoli cell tight junction strands are absent in Osp/claudin-11 null mice.* Cell, 1999. **99**(6): p. 649-59.

104.    Linington, C., M. Webb, and P. Woodhams, *A novel myelin associated glycoprotein defined by a mouse monoclonal antibody.* J Neuroimmunol, 1984. **6**: p. 387-396.

105.    Gardinier, M.V., P. Amiguet, C. Linington, and J.M. Matthieu, *Myelin/oligodendrocyte glycoprotein is a unique member of the immunoglobulin superfamily.* J Neurosci Res, 1992. **33**(1): p. 177-87.

106.    Johns, T.G. and C.C. Bernard, *The structure and function of myelin oligodendrocyte glycoprotein.* J Neurochem, 1999. **72**(1): p. 1-9.

107.    Pham-Dinh, D., M.G. Mattei, J.L. Nussbaum, G. Roussel, P. Pontarotti, N. Roeckel, I.H. Mather, K. Artzt, K.F. Lindahl, and A.

Dautigny, *Myelin/oligodendrocyte glycoprotein is a member of a subset of the immunoglobulin superfamily encoded within the major histocompatibility complex.* Proc Natl Acad Sci U S A, 1993. **90**(17): p. 7990-4.

108.    Matthieu, J.M. and P. Amiguet, *Myelin/oligodendrocyte glycoprotein expression during development in normal and myelin-deficient mice.* Dev Neurosci, 1990. **12**(4-5): p. 293-302.

109.    Abo, S., C.C. Bernard, M. Webb, T.G. Johns, A. Alafaci, L.D. Ward, R.J. Simpson, and N. Kerlero de Rosbo, *Preparation of highly purified human myelin oligodendrocyte glycoprotein in quantities sufficient for encephalitogenicity and immunogenicity studies.* Biochemistry & Molecular Biology International, 1993. **30**(5): p. 945-58.

110.    de Rosbo, N.K. and A. Ben-Nun, *T-cell responses to myelin antigens in multiple sclerosis; relevance of the predominant autoimmune reactivity to myelin oligodendrocyte glycoprotein.* J Autoimmun, 1998. **11**(4): p. 287-99.

111.    Rash, J.E., T. Yasumura, K.G. Davidson, C.S. Furman, F.E. Dudek, and J.I. Nagy, *Identification of cells expressing Cx43, Cx30, Cx26, Cx32 and Cx36 in gap junctions of rat brain and spinal cord.* Cell Commun Adhes, 2001. **8**(4-6): p. 315-20.

112.    Rash, J.E., T. Yasumura, F.E. Dudek, and J.I. Nagy, *Cell-specific expression of connexins and evidence of restricted gap junctional coupling between glial cells and between neurons.* J Neurosci, 2001. **21**(6): p. 1983-2000.

113.    Scherer, S.S., Y.T. Xu, E. Nelles, K. Fischbeck, K. Willecke, and L.J. Bone, *Connexin32-null mice develop demyelinating peripheral neuropathy.* Glia, 1998. **24**(1): p. 8-20.

114.    Stojkovic, T., P. Latour, A. Vandenberghe, J.F. Hurtevent, and P. Vermersch, *Sensorineural deafness in X-linked Charcot-Marie-Tooth disease with connexin 32 mutation (R142Q).* Neurology, 1999. **52**(5): p. 1010-4.

115.    Paulson, H.L., J.Y. Garbern, T.F. Hoban, K.M. Krajewski, R.A. Lewis, K.H. Fischbeck, R.I. Grossman, R. Lenkinski, J.A. Kamholz, and M.E. Shy, *Transient central nervous system white matter abnormality in X-linked Charcot-Marie-Tooth disease.* Ann Neurol, 2002. **52**(4): p. 429-34.

116.    Schelhaas, H.J., B.G. Van Engelen, A.A. Gabreels-Festen, G. Hageman, J.H. Vliegen, M.S. Van Der Knaap, and M.J. Zwarts, *Transient cerebral white matter lesions in a patient with connexin 32 missense mutation.* Neurology, 2002. **59**(12): p. 2007-8.

117.    Takashima, H., M. Nakagawa, F. Umehara, K. Hirata, M. Suehara, H. Mayumi, K. Yoshishige, W. Matsuyama, M. Saito, M. Jonosono, K. Arimura, and M. Osame, *Gap junction protein beta 1 (GJB1)*

*mutations and central nervous system symptoms in X-linked Charcot-Marie-Tooth disease.* Acta Neurol Scand, 2003. **107**(1): p. 31-7.

118.   Bellen, H.J., Y. Lu, R. Beckstead, and M.A. Bhat, *Neurexin IV, caspr and paranodin--novel members of the neurexin family: encounters of axons and glia.* Trends Neurosci, 1998. **21**(10): p. 444-9.

119.   Trapp, B.D. and G.J. Kidd, *Axo-glial septate junctions. The maestro of nodal formation and myelination? [comment].* J Cell Biol, 2000. **150**(3): p. F97-F100.

120.   Girault, J.A. and E. Peles, *Development of nodes of Ranvier.* Curr Opin Neurobiol, 2002. **12**(5): p. 476-485.

121.   Collinson, J.M., D. Marshall, C.S. Gillespie, and P.J. Brophy, *Transient expression of neurofascin by oligodendrocytes at the onset of myelinogenesis: implications for mechanisms of axon-glial interaction.* Glia, 1998. **23**(1): p. 11-23.

122.   Tait, S., F. Gunn-Moore, J.M. Collinson, J. Huang, C. Lubetzki, L. Pedraza, D.L. Sherman, D.R. Colman, and P.J. Brophy, *An oligodendrocyte cell adhesion molecule at the site of assembly of the paranodal axo-glial junction [see comments].* J Cell Biol, 2000. **150**(3): p. 657-66.

123.   Yamamoto, Y., R. Mizuno, T. Nishimura, Y. Ogawa, H. Yoshikawa, H. Fujimura, E. Adachi, T. Kishimoto, T. Yanagihara, and S. Sakoda, *Cloning and expression of myelin-associated oligodendrocytic basic protein. A novel basic protein constituting the central nervous system myelin.* Journal of Biological Chemistry, 1994. **269**(50): p. 31725-30.

124.   Holz, A., N. Schaeren-Wiemers, C. Schaefer, U. Pott, R.J. Colello, and M.E. Schwab, *Molecular and developmental characterization of novel cDNAs of the myelin-associated/oligodendrocytic basic protein.* Journal of Neuroscience, 1996. **16**(2): p. 467-77.

125.   Yool, D., P. Montague, M. McLaughlin, M.C. McCulloch, J.M. Edgar, K.A. Nave, R.W. Davies, I.R. Griffiths, and A.S. McCallion, *Phenotypic analysis of mice deficient in the major myelin protein MOBP, and evidence for a novel Mobp isoform.* Glia, 2002. **39**(3): p. 256-67.

126.   Neusch, C., N. Rozengurt, R.E. Jacobs, H.A. Lester, and P. Kofuji, *Kir4.1 potassium channel subunit is crucial for oligodendrocyte development and in vivo myelination.* J Neurosci, 2001. **21**(15): p. 5429-38.

# Chapter A5

# ADJUVANTS IN EAE


Cris S Constantinescu and Brendan A Hilliard
*Division of Clinical Neurology, University of Nottingham, UK and Depertment of Molecular and Celluar Engineering, University of Pennsylvania, Philadelphia, USA*



**Abstract:** Adjuvants enhance autoantigens' immunogenicity and play a major role in the induction of EAE. Complete Freund's adjuvant (CFA) is the key adjuvant for cell-mediated immunity, including EAE. An additional adjuvant, pertussis toxin, is necessary for active EAE in mice. The mechanisms of action of adjuvants are multiple, but important actions are stimulation of proinflammatory cytokines and costimulatory molecules.



**Key words:**

adjuvant, experimental autoimmune encephalomyelitis, complete/incomplete Freund's adjuvant, pertussis toxin, immunopotentiation


The term *(immunological) adjuvant* refers to an agent that can potentiate an immune reaction (immunopotentiation). *Adjuvanticity* is the efficacy of an immunological adjuvant to enhance immunogenicity of an antigen. The final mode of presentation of the immunogen and adjuvant combination is called adjuvant formulation.

Adjuvants have been extensively investigated for vaccine development. Currently numerous synthetic adjuvants are being tested for vaccines and for effective immunization protocols for generation of antibodies. However, most of the traditional adjuvants are of bacterial origin. It is not surprising that bacteria and bacterial products have high adjuvanticity as infectious agents are strong stimuli for immunological defense. Freund's discovery that killed bacteria suspended in mineral oil enhanced antibody production [1] led to the use of antigen-adjuvant combination emulsified in oil. The oil may play the dual role of having adjuvant activity and acting as a depot for the antigen.

After this important discovery, Kabat [2] successfully and more reliably induced EAE in rhesus monkeys using the complete Freund's adjuvant.

Subsequently in efforts to induce EAE more reliably in mice, the ancillary adjuvant, *Bordetella pertussis* vaccine [3] was used. Subsequent studies have identified pertussis toxin as the component responsible for the adjuvanticity and currently many protocols use intravenous or intraperitoneal injection of killed *B. pertussis*, *pertussis* toxin, or the *pertussis* vaccine.

A large number of synthetic materials or bacterial products have been shown to have adjuvant properties.

| Adjuvants | Examples | Immunological effects |
|---|---|---|
| Depot forming | Water/oil emulsions, e.g., CFA | Antigen persistence, stronger memory T cell response |
| Macrophage activating | Muramyl dipeptide (MDP), LPS, CpG bacterial DNA, CFA *C. parvum*, etc. | Enhanced cytokine production by macrophages, Costimulation, MHC expression |
| Inert materials | Alum, latex, acrylic | Antigen persistence and aggregation |
| T/B cell activating | PPD, polyA:polyU, LPS | Provide T cell help/enhance B cell function |
| Surface active | Saponins, lysolecithin, Quil A, liposomes | Antigen aggregation, facilitate antigen presentation |

The general mechanism of action of adjuvants is an enhancement of antigen presenting cell (APC) function including increased MHC and costimulatory molecule expression and enhanced expression of relevant cytokines such as IL-1 and IL-12 for adjuvants of cell mediated immune responses and IL-4 for humoral immune responses. Another quality of adjuvants should be that of ensuring persistence of an antigen at the relevant sites. This is accomplished by the use of antigen depots such as oils.

**Complete Freund's Adjuvant** has all of these properties. It has been widely used in EAE, where its roles are multiple. CFA is an adjuvant preparation of killed mycobacteria, most commonly *M. tuberculosis*, in a water and oil emulsion. The emulsion without mycobacteria is called incomplete Freund's adjuvant (IFA). CFA has been shown to induce preferentially a cell mediated immune reaction, although it is also often used as an adjuvant for generation of antibodies and for humoral immune reactions. Through *Mycobacterium*, a strong inducer of IL-12, it is preferentially inducing Th1 responses [4]. The ability to induce Th1 responses to myelin antigen is considered to be a major correlate of the adjuvanticity of CFA and it is due to the mycobacterial component, immunization with antigen in IFA fails to induce EAE and appears to be tolerogenic [5]. The tolerogenicity of antigen in IFA is not due to the lack of an immune response, but to the development of a Th2-type response, in contrast to the CFA-associated Th1 response. This phenomenon occurs in a variety of cell mediated experimental autoimmune diseases including EAE [6] and across a spectrum of MHC backgrounds, indicating that the Th1/Th2 dichotomy is more likely a function of the adjuvant (infectious v. non-infectious) rather than of the animal's genetic background [7]. Moreover, IFA was shown to have a direct inhibitory effect on EAE: when administered intraperitoneally it inhibited EAE in various strains of mice. The effect was due to a down-regulation of antigen and mitogen-specific T cell proliferation. The CNS infiltrating cells expressed reduced cytokines, chemokines and chemokine receptors [8].

In addition to its role as an immunization adjuvant enhancing Th1 responses, activating the macrophages, ensuring antigen persistence and facilitating recruitment of memory T cells, *Mycobacterium* in CFA preparation may also have an important role in increasing the blood brain barrier permeability [9]. This process may be due to antibodies to a mannan component of the bacteria [10]. The CFA-induced increased permeability to various proteins of the blood brain barrier by itself does not result in reactive gliosis or microglia activation as shown by Rabchevsky et al [11].

The presence of Mycobacteria in an emulsion form has been considered important for the adjuvanticity of CFA and of many other adjuvants. Finer emulsions have stronger adjuvant effect as shown by Maata et al, who succeded in inducing EAE in otherwise resistant mice by ultrasound emulsification of the antigen-adjuvant mixture. However, Levine (S Levine et al) have shown that, in susceptible rats, immunization in the absence of an adjuvant water-oil emulsion still results in EAE.

**Pertussis**

Induction of active EAE in mice is greatly facilitated by Pertussis toxin. Pertussis toxin is an important virulence factor for Bordetella pertussis, the agent causing whooping cough. The toxin is a hexamer and its $\alpha$ subunit is a adenosine diphosphate (ADP)-ribosyl transferase which interferes with signal transduction by ribosylating the $\alpha$ subunit of trimeric $G_i$ proteins, while the $\beta$ subunit binds to cell surface receptors. The mechanisms of facilitation of EAE induction by pertussis toxin are complex. A histamine-sensitizing factor has been extracted from B. pertussis. The susceptibility to EAE in mice was shown to be in large part dependent of the MHC and the histamine sensitization genes. The adjuvant effects of pertussis toxin have been attributed to an increased vascular permeability of the brain and spinal cord. This is mediated by vasoactive amines and is controlled by an autosomal locus, *Bhps* (Bordetella pertussis-induced histamine sensitization). Recently, this locus has been identified as the histamine $H_1$ receptor.

Other mechanisms are also likely to be implicated in the ancillary adjuvant effects of pertussis toxin in EAE. One such mechanism is the stimulation of Th1 responses. Pertussis induces IL-12, which in turn activates Th1 responses [12]. This is due to a stimulatory effect on both macrophages and dendritic cells [13]. However, IL-12 alone is probably not sufficient as IL-12 as an adjuvant, although effective in vaccination for infectious disease models [14], was insufficient for modification of resistance (although it can increase disease severity) in several autoimmune disease models in mice (Constantinescu et al, unpublished observations); [15-17].

The adjuvant effect of pertussis, however, appears to involve both Th1 and Th2 immune responses [18] which are both enhanced by it. However, in the context of EAE, it appears, as expected, that the Th1 effect is more important. In studies showing that co-administered pertussis toxin overcomes the tolerogenic effect of IFA in proteolipid protein (PLP)139-151-induced EAE, both Th1 and Th2 PLP139-151-reactive Th1 and Th2 cells were generated but Th2 cells were neither protective nor pathogenic [19].

The expansion and stimulation of CD4+ T cells may be mediated through effects on macrophages and other APC and on T cells.

There is evidence that both the vascular and the immunologic effects of pertussis toxin (the increased vascular permeability and stimulation of encephalitogenic cells) are mediated through the histamine $H_1$ receptor. Indeed Pedotti et all have recently shown that encephalitogenic T cells express more H1 receptor and less H2 receptor than Th2 cells. Moreover H1 antagonists suppress EAE [20].

Like CFA, pertussis can overcome genetic resistance to EAE [21]. The adjuvant effects of pertussis toxin have have been shown in other autoimmune diseases, where pertussis overcomes resistance to disease. These include experimental autoimmune orchitis [22] uveitis [13], and neuritis [15]. Although generally known as an ancillary adjuvant for mouse autoimmune disease, pertussis has had adjuvant effects in rat EAE. These effects were shown to be associated with increased costimulation, TNF, and nitric oxide synthase. Pertussis can induce a hyperacute EAE in susceptible Lewis rats [23]and can overcome resistance in typically resistant rat strains [24, 25]. Recently, we have also shown that pertussis toxin as an ancillary adjuvant with CFA induced reproducible experimental autoimmune myositis in rats after immunization with skeletal muscle myosin [26].

**Protective effect of adjuvant pre-treatment**

Although pertussis toxin and CFA are used as adjuvants, pre-treatment prior to immunization for EAE results in protection from EAE. The mechanisms are unclear and it has been suggested that they may be different between CFA and pertussis. [27]. For the latter, it requires native, not genetically inactivated toxin [28]. The protective effect may be through interfering with TCR signaling. The effect was not present on preformed encephalitogenic T cells and it has been suggested that the inhibition is through affecting the development of encephalitogenic cells. For CFA, it has been shown that the effect is due to the PPD component of M. tuberculosis.

Recently it has been shown that cholera toxin, like pertussis toxin a member of the bacterial ADP-ribosylating exotoxins, also has potent adjuvant activity for mucosal immune responses, a property that may be useful in vaccine design. It preferentially induces a Th2-type response. Mucosal (e.g., nasal) administration of cholera toxin results in suppression of EAE via inhibition of IL-12 and IFN-$\gamma$ production in the CNS [29-31]. Coupling of cholera toxin with the autoantigen followed by mucosal administration also inhibits EAE, by reducing the autoantigen (myelin basic protein, proteolipid protein peptide)-specific delayed-type hypersensitivity and inflammatory infiltration in the spinal cord. It has been shown that the adjuvanticity of cholera toxin is due to the A subunit and is dependent on ADP ribosylation [32]. On the other hand, the tolerizing effect is induced through the B subunit, possibly via induction of IL-10. The mechanism may be similar for pertussis toxin as well. These findings show that bacterial toxins adjuvants can have, in certain circumstances, immunomodulatory effects.

**Other adjuvants**

**Muramyl dipeptide** (N-acetyl-muramyl L-alanyl D-isoglutamine) is a bacteria-derived substance widely used as immunostimulator/adjuvant. has been used in an experimental model of MS in guinea pigs displaying acute demyelination and inflammatory infiltrates similar to those of "classical" EAE [33, 34].

**Lysolecithin** not only can induce demyelination by itself but it can also act as an adjuvant and enhance myelin-induced EAE [35].

Lipid-bound neuroantigen (MBP, extracted using a non-ionic detergent that preserves its binding to myelin lipids also has an adjuvant effect in Lewis rats, where it induced more demyelination than lipid-free MBP.

Intense emulsification by **sonication** in adjuvant emulsion prior to immunization induces EAE in reputedly resistant mice (BALB/c) indicating that it may be the physical state of the neuroantigen that has the major influence on the development of the autoimmune response [36].

**Alum** adjuvant is thought to favor Th2 responses. This property has been used in experiments in which animals with EAE were immunized with soluble MHC class II molecules. The immunization elicited an antibody response thus blocking the TCR-MHC interaction. A phase I clinical trial using alum as an adjuvant for immunization with peptides derived from HLA-DR resulted in increased production of anti-HLA-DR antibodies and was well tolerated.

Alum was also used as an adjuvant and compared with IFA in an attempt to generate antibodies against myelin-associated inhibitors of axonal regeneration after experimental spinal cord injury. Although the SJL/J mice used are susceptible to EAE, the mice did not get EAE when immunized with alum. The implications are important, because alum can be used in humans. Interestingly, however, aluminium (tin) has been shown also to have an adjuvant effect in EAE along with a number of other particulate materials (see below) [37].

**Schiff base-forming drugs** are potent adjuvants for cell-mediated immune responses. Their immunopotentiating ability is mediated through enhancement of costimulation through B7.1 and B7.2 molecules and by preferentially inducing Th1 responses [38, 39]. To date no studies using these adjuvants in EAE are known, but the mechanisms of actions of these drugs are very relevant for the pathogenesis of EAE and are potentially useful adjuvants for the future.

**Particulate materials** (carbonyl iron, silicon and silica powders etc) inoculated either at the same time or (for some, like silicon and silica powder) prior to the immunization with adjuvants. The mechanism appears to be an increase in antigen persistence [37, 40]. Moreover, carbonyl iron, when used as adjuvant together with rat spinal cord tissue, is capable of inducing EAE in reputedly EAE-resistant Brown Norway (BN) rats. The augmentation of EAE was attributed to an increase in the absorption of the antigen [41].

**Live attenuated vaccines** also enhance the severity of EAE, having an adjuvant effect. Caspary has shown that the distemper vaccine, the BCG and the measles vaccines all enhanced EAE [42].

The search for alternatives to CFA, an adjuvant that causes pain, is important not only for EAE but also for generation of antibodies. Several additional adjuvants have been tested for their ability to induce monoclonal antibodies in mice with some, such as GERBU adjuvant producing titers of antibodies similar to Freund's adjuvant and poly(A).poly(U) producing lower titers but showing higher fusion efficiency [43]. These findings, however, have not been extrapolated to EAE, where CFA remains the strongest adjuvant. It is hoped, however, that less pain-producing adjuvants will become available.

In conclusion, a variety of adjuvants have been used for induction of EAE. The adjuvant effect appears to be in significant part through cytokine immunity induction, in particular Th1 cytokines associated with cell mediated immunity and also through stimulation of costimulatory molecules.

## ACKNOWLEDGEMENTS

Dr. Constantinescu's research has been supported in part by the NIH, University of Nottingham, The Stanhope Trust and the Multiple Sclerosis Society of the United Kingdom and Northern Ireland.

## REFERENCES

1.　Freund, J. and K. McDermott, *Sensitization to horse serum by means of adjuvants.* Proc Soc Exp Biol Med, 1942. **49**: p. 548-553.

2.　Kabat, E.A., A. Wolf, and A.E. Bezer, *The rapid production of acute disseminated encephalomyelitis in rhesus monkeys by injection of heterologous and homologous brain tissue with adjuvants.* J Exp Med, 1947. **85**: p. 117-129.

3.　Lee, J.M. and P.K. Olitsky, *Simple method for enhencing development of acute disseminated encephalomyelitis in mice.* Proc Soc Exp Biol Med, 1995. **89**: p. 263-266.

4.　Lucey, D.R., M. Clerici, and G.M. Shearer, *Type 1 and type 2 cytokine dysregulation in human infectious, neoplastic, and inflammatory diseases.* Clin Microbiol Rev, 1996. **9**(4): p. 532-62.

5.　Forsthuber, T., H.C. Yip, and P.V. Lehmann, *Induction of TH1 and TH2 immunity in neonatal mice.* Science, 1996. **271**(5256): p. 1728-30.

6.　Heeger, P.S., T. Forsthuber, C. Shive, E. Biekert, C. Genain, H.H. Hofstetter, A. Karulin, and P.V. Lehmann, *Revisiting tolerance induced by autoantigen in incomplete Freund's adjuvant.* J Immunol, 2000. **164**(11): p. 5771-81.

7.　Yip, H.C., A.Y. Karulin, M. Tary-Lehmann, M.D. Hesse, H. Radeke, P.S. Heeger, R.P. Trezza, F.P. Heinzel, T. Forsthuber, and P.V. Lehmann, *Adjuvant-guided type-1 and type-2 immunity: infectious/noninfectious dichotomy defines the class of response.* J Immunol, 1999. **162**(7): p. 3942-9.

8. Zamora, A., A. Matejuk, M. Silverman, A.A. Vandenbark, and H. Offner, *Inhibitory effects of incomplete Freund's adjuvant on experimental autoimmune encephalomyelitis.* Autoimmunity, 2002. **35**(1): p. 21-8.

9. Namer, I.J., J. Steibel, P. Poulet, Y. Mauss, M. Mohr, and J. Chambron, *The role of Mycobacterium tuberculosis in experimental allergic encephalomyelitis.* Eur Neurol, 1994. **34**(4): p. 224-7.

10. Namer, I.J. and J. Steibel, *Antibody directed against mannan of the Mycobacterium tuberculosis cell envelope provokes blood-brain barrier breakdown.* J Neuroimmunol, 2000. **103**(1): p. 63-8.

11. Rabchevsky, A.G., J.D. Degos, and P.A. Dreyfus, *Peripheral injections of Freund's adjuvant in mice provoke leakage of serum proteins through the blood-brain barrier without inducing reactive gliosis.* Brain Res, 1999. **832**(1-2): p. 84-96.

12. Mahon, B.P., M.S. Ryan, F. Griffin, and K.H. Mills, *Interleukin-12 is produced by macrophages in response to live or killed bordetella pertussis and enhances the efficacy of an acellular pertussis vaccine by promoting induction of Th1 cells.* Infect Immun, 1996. **64**(5295-5301).

13. Hou, W., Y. Wu, S. Sun, M. Shi, Y. Sun, C. Yang, G. Pei, Y. Gu, C. Zhong, and B. Sun, *Pertussis toxin enhances Th1 responses by stimulation of dendritic cells.* J Immunol, 2003. **170**(4): p. 1728-36.

14. Afonso, L.C., T.M. Scharton, L.Q. Vieira, M. Wysocka, G. Trinchieri, and P. Scott, *The adjuvant effect of interleukin-12 in a vaccine against Leishmania major.* Science, 1994. **263**(5144): p. 235-7.

15. Calida, D.M., S.G. Kremlev, T. Fujioka, B. Hilliard, E. Ventura, C.S. Constantinescu, E. Lavi, and A. Rostami, *Experimental allergic neuritis in the SJL/J mouse: induction of severe and reproducible disease with bovine peripheral nerve myelin and pertussis toxin with or without interleukin-12.* J Neuroimmunol, 2000. **107**(1): p. 1-7.

16. Germann, T., J. Szeliga, H. Hess, S. Storkel, F.J. Podlaski, M.K. Gately, E. Schmitt, and E. Rude, *Administration of interleukin 12 in combination with*

*type II collagen induces severe arthritis in DBA/1 mice.* Proc Natl Acad Sci U S A, 1995. **92**(11): p. 4823-7.

17.    Germann, T., H. Hess, J. Szeliga, and E. Rude, *Characterization of the adjuvant effect of IL-12 and efficacy of IL-12 inhibitors in type II collagen-induced arthritis.* Ann N Y Acad Sci, 1996. **795**: p. 227-40.

18.    Ryan, M., L. McCarthy, R. Rappuoli, B.P. Mahon, and K.H. Mills, *Pertussis toxin potentiates Th1 and Th2 responses to co-injected antigen: adjuvant action is associated with enhanced regulatory cytokine production and expression of the co-stimulatory molecules B7-1, B7-2 and CD28.* Int Immunol, 1998. **10**(5): p. 651-62.

19.    Hofstetter, H.H., C.L. Shive, and T.G. Forsthuber, *Pertussis toxin modulates the immune response to neuroantigens injected in incomplete Freund's adjuvant: induction of Th1 cells and experimental autoimmune encephalomyelitis in the presence of high frequencies of Th2 cells.* J Immunol, 2002. **169**(1): p. 117-25.

20.    Pedotti, R., J.J. DeVoss, S. Youssef, D. Mitchell, J. Wedemeyer, R. Madanat, H. Garren, P. Fontoura, M. Tsai, S.J. Galli, R.A. Sobel, and L. Steinman, *Multiple elements of the allergic arm of the immune response modulate autoimmune demyelination.* Proc Natl Acad Sci U S A, 2003. **100**(4): p. 1867-72.

21.    Blankenhorn, E.P., R.J. Butterfield, R. Rigby, L. Cort, D. Giambrone, P. McDermott, K. McEntee, N. Solowski, N.D. Meeker, J.F. Zachary, R.W. Doerge, and C. Teuscher, *Genetic analysis of the influence of pertussis toxin on experimental allergic encephalomyelitis susceptibility: an environmental agent can override genetic checkpoints.* J Immunol, 2000. **164**(6): p. 3420-5.

22.    Teuscher, C., W.F. Hickey, C.M. Grafer, and K.S. Tung, *A common immunoregulatory locus controls susceptibility to actively induced experimental allergic encephalomyelitis and experimental allergic orchitis in BALB/c mice.* J Immunol, 1998. **160**(6): p. 2751-6.

23.    Ahn, M., J. Kang, Y. Lee, K. Riu, Y. Kim, Y. Jee, Y. Matsumoto, and T. Shin, *Pertussis toxin-induced hyperacute autoimmune encephalomyelitis in*

*Lewis rats is correlated with increased expression of inducible nitric oxide synthase and tumor necrosis factor alpha.* Neurosci Lett, 2001. **308**(1): p. 41-4.

24.     Arimoto, H., N. Tanuma, Y. Jee, T. Miyazawa, K. Shima, and Y. Matsumoto, *Analysis of experimental autoimmune encephalomyelitis induced in F344 rats by pertussis toxin administration.* J Neuroimmunol, 2000. **104**(1): p. 15-21.

25.     Zhou, J., J.S. Zhang, B.L. Ma, and M.J. Mamula, *Experimental autoimmune encephalomyelitis in the Wistar rat: dependence of MBP-specific T cell responsiveness on B7 costimulation.* Autoimmunity, 2002. **35**(3): p. 191-9.

26.     Nemoto, H., M.K. Bhopale, C.S. Constantinescu, D. Schotland, and A. Rostami, *Skeletal muscle myosin is the autoantigen for experimental autoimmune myositis.* Exp Mol Pathol, 2003. **74**(3): p. 238-43.

27.     Ben-Nun, A., S. Yossefi, and D. Lehmann, *Protection against autoimmune disease by bacterial agents. II. PPD and pertussis toxin as proteins active in protecting mice against experimental autoimmune encephalomyelitis.* Eur J Immunol, 1993. **23**(3): p. 689-96.

28.     Robbinson, D., S. Cockle, B. Singh, and G.H. Strejan, *Native, but not genetically inactivated, pertussis toxin protects mice against experimental allergic encephalomyelitis.* Cell Immunol, 1996. **168**(2): p. 165-73.

29.     Sun, J.B., C. Rask, T. Olsson, J. Holmgren, and C. Czerkinsky, *Treatment of experimental autoimmune encephalomyelitis by feeding myelin basic protein conjugated to cholera toxin B subunit.* Proc Natl Acad Sci U S A, 1996. **93**(14): p. 7196-201.

30.     Czerkinsky, C., J.B. Sun, M. Lebens, B.L. Li, C. Rask, M. Lindblad, and J. Holmgren, *Cholera toxin B subunit as transmucosal carrier-delivery and immunomodulating system for induction of antiinfectious and antipathological immunity.* Ann N Y Acad Sci, 1996. **778**: p. 185-93.

31.     Yuki, Y., Y. Byun, M. Fujita, W. Izutani, T. Suzuki, S. Udaka, K. Fujihashi, J.R. McGhee, and H. Kiyono, *Production of a recombinant hybrid molecule of cholera toxin-B-subunit and proteolipid-protein-peptide*

*for the treatment of experimental encephalomyelitis.* Biotechnol Bioeng, 2001. **74**(1): p. 62-9.

32. Douce, G., C. Turcotte, I. Cropley, M. Roberts, M. Pizza, M. Domenghini, R. Rappuoli, and G. Dougan, *Mutants of Escherichia coli heat-labile toxin lacking ADP-ribosyltransferase activity act as nontoxic, mucosal adjuvants.* Proc Natl Acad Sci U S A, 1995. **92**(5): p. 1644-8.

33. Colover, J., *Acute demyelination in EAE after pretreatment with foreign protein and muramyl dipeptide (MDP).* Prog Clin Biol Res, 1984. **146**: p. 37-42.

34. Colover, J., *A new pattern of spinal-cord demyelination in guinea pigs with acute experimental encephalomyelitis mimicking multiple sclerosis.* Br J Exp Pathol, 1980. **61**: p. 390-400.

35. Gregson, N.A., M.C. Kennedy, and S. Leibowitz, *Immunological reactions with lysolecithin-solubilized myelin.* Immunology, 1971. **20**(4): p. 501-12.

36. Maatta, J.A., J.P. Eralinna, M. Roytta, A.A. Salmi, and A.E. Hinkkanen, *Physical state of the neuroantigen in adjuvant emulsions determines encephalitogenic status in the BALB/c mouse.* J Immunol Methods, 1996. **190**(1): p. 133-41.

37. Levine, S. and R. Sowinski, *Enhancement of allergic encephalomyelitis by particulate adjuvants inoculated long before antigen.* Am J Pathol, 1980. **99**(2): p. 291-303.

38. Rhodes, J., H. Chen, S.R. Hall, J.E. Beesley, D.C. Jenkins, P. Collins, and B. Zheng, *Therapeutic potentiation of the immune system by costimulatory Schiff-base-forming drugs.* Nature, 1995. **377**(6544): p. 71-5.

39. Chen, H. and J. Rhodes, *Schiff base forming drugs: mechanisms of immune potentiation and therapeutic potential.* J Mol Med, 1996. **74**(9): p. 497-504.

40. Levine, S. and R. Sowinski, *Carbonyl iron: a new adjuvant for experimental autoimmune diseases.* J Immunol, 1970. **105**(6): p. 1530-5.

41. Levine, S. and A. Saltzman, *The role of antigen absorption in the resistance of brown-Norway rats to experimental allergic encephalomyelitis.* Immunol Lett, 1988. **19**(2): p. 103-8.

42.    Caspary, E.A., *Effect of live attenuated vaccines on the course of experimental allergic encephalomyelitis. A pilot study.* Eur Neurol, 1977. **16**(1-6): p. 176-80.

**43**.    Ferber, P.C., P. Ossent, F.R. Homberger, and R.W. Fischer, *The generation of monoclonal antibodies in mice: influence of adjuvants on the immune response, fusion efficiency and distress.* Lab Anim, 1999. **33**(4): p. 334-50.

Chapter A6

# THE ROLE OF ASTROCYTES IN AUTOIMMUNE DISEASE OF THE CENTRAL NERVOUS SYSTEM

Olaf Stüve and Scott S. Zamvil
*Department of Neurology, University of California, San Francisco, CA 94143, USA*

**Abstract:** Astrocytes are the most abundant glial cell population in the central nervous system (CNS). In the healthy brain and spinal cord, the major function of astrocytes includes the formation and maintenance of the blood brain barrier (BBB), and the supply of structural support and nourishment to neurons. This article will discuss the role of astrocytes in multiple sclerosis (MS) and other inflammatory autoimmune diseases of the CNS. We will address the capacity of astrocytes to serve as immunocompetent cells, their role in major histocompatibility complex (MHC) class II restricted antigen (Ag) presentation, and their ability to express costimulatory molecules. We will also discuss astrocytes as the major CNS producers of several chemokines and cytokines, and their relevance to neurological disease.

**Key words:**
    astrocytes, antigen presentation, major histocompatibility complex, co-stimulatory molecules

Multiple sclerosis (MS) is a chronic inflammatory demyelinating disease of the CNS of unknown etiology, causing relapsing and progressive neurological disabilities (1-3). It is the most common and clinically important demyelinating disease in humans. Worldwide, it is estimated that at least 1.1 million people have physician-diagnosed MS (3, 4). The disease course can be separated into two-stages (5-8). While MS often starts with a relapsing-remitting (RR) disease course, an estimated 50% of patients RR-MS will convert to a secondary progressive (SP) clinical course within 10 years of disease onset. Pathologically, inflammation is a characteristic feature accompanying demyelination in early MS. Remyelination can occur during the early phase when initial damage may be reversible. Activated CD4[+] T cells, which recognize CNS autoantigen (Ag) in association with major histocompatibility complex (MHC) class II molecules, have been implicated in the inflammatory stages of MS.

The four types of glia in the central nervous system (CNS) can be divided into microglia and macroglia, which include astrocytes, ependymal cells, and oligodendrocytes. Astrocytes are the most abundant glial cell population in the brain and spinal cord. In fact, they outnumber neurons by 10-50 : 1. Using gold chloride- mercury bichloride staining technique at the end of the 19th century, Drs. Santiago Ramon y Cajal and Camillo Golgi first demonstrated the contributions of astrocytes to degeneration and regeneration of the CNS, a discovery that was awarded with the Nobel prize in 1906. The importance of astrocytes for homeostasis and immune function of the central nervous system (CNS) has relatively recently become a topic of scientific interest.

Astrocytes are easily identified by their star-shaped appearance. In the fully developed adult brain, astrocytes are intimately associated with blood vessels by wrapping their fine processes around the abluminal vascular surface. The interaction between astrocytes and endothelial cells occurs through soluble factors, and seems bidirectional: The induction of tight junctions between endothelial cells is thought be triggered by astocytes, whereas endothelial cells may play a crucial role in regulating the maturation of astrocytes (9). Thus, astrocytes play a crucial role in forming and maintaining the blood brain barrier (BBB), and shielding the CNS from blood-derived pathogens (10). Some data indicate that astrocytes express MHC class II molecules (11-14) and can serve as antigen presenting cells (APC) to CD4$^+$ T cells during the initial inflammatory phase of MS. However, astrocytes may also have a neuroprotective role: Not only do they communicate with neurons through the release of the amino acid glutamate, more importantly they prevent neuronal death from excess release of glutamate during ischemia or inflammation (15). Astrocytes also regulate the extracellular pH and potassium concentration. Physical support for neurons and other CNS cells is provided by the astrocytes through the formation of extracellular matrix (ECM), which also serves to isolate synapses and limits the dispersion of neurotransmitters (15). Astrocytes have an important role in CNS scar formation. After CNS injury or cell death, astrocytes quickly respond with upregulation of glial fibrillary acidic protein (GFAP), which is associated with astrogliosis (16, 17). Increased tissue repair and turnover is then triggered through the astocytic production of ECM, proteases, protease inhibitors, trophic factors, and cytokines (18-23). Around sites of mechanical injury, astrocytes become hypertrophic, and extend numerous fine processes around each other. The scar consists of tightly packed, hyperfilamentous astrocytes, bound together by tight and gap junctions.

The role of astrocytes in the formation of the BBB, in glutamate homeostasis, and scar formation in the CNS make these glial cells important participants and potential targets of therapeutic interventions in inflammatory diseases of the brain and the spinal cord, including MS. Here, we review investigations that address the role of astrocytes as antigen presenting cells (APC), and specifically their role in major histocompatibility complex (MHC) class II restricted antigen (Ag) presentation. Various costimulatory pathways and the release of inflammatory mediators by astrocytes during CNS autoimmune disease will also be discussed.

**The role of astrocytes in MHC class II restricted antigen presentation**

Experimental autoimmune encephalomyelitis (EAE) is a CNS inflammatory demyelinating disease that serves as a model for multiple sclerosis (MS) and other organ-specific autoimmune diseases (1, 24). EAE is primarily mediated by CD4$^+$ T lymphocytes that recognize CNS self-Ag's in the context of MHC class II molecules expressed on the surface of APC (1, 24). Ag presentation may be required at different stages of EAE pathogenesis. Peripheral (outside the CNS) activation facilitates T cell entry into the CNS (25, 26). Recent investigations further indicate that Ag processing by resident CNS APC is required for MHC class II-restricted Ag presentation within the CNS (27, 28). While the healthy brain and spinal cord are almost devoid of MHC class II molecules, it has been proposed that both, astrocytes and microglia, may participate in Ag presentation and CD4$^+$ T cell activation through IFNγ-inducible upregulation of MHC class II during inflammatory CNS conditions like MS (29-33). Specifically, interferon (IFN)γ a proinflammatory cytokine that is thought to have a key role in the pathogenesis of MS (34), is required for induction of class II MHC molecules on so called "nonprofessional APC". While the role of microglia as APC for MHC class II-restricted Ag presentation in CNS inflammatory disease was established (32, 33), the role of astrocytes remains controversial.(33, 35)  MHC class II molecules have been detected on astrocytes within inflammatory lesions of MS (11, 14) and EAE (12, 13), suggesting that these nonprofessional APC may participate in Ag presentation to CD4$^+$ T cells in vivo (12, 13). IFNγ-activated astrocytes upregulate MHC class II and MHC class I molecules in vitro, and can present antigen to CD4$^+$ T cells or CD8$^+$ T cells, respectively (30, 36-38).

T cells that express the accessory surface molecule CD4 are capable of recognizing and binding linear peptide antigen (Ag) in the context of the MHC class II (*Figure 1*). This high-affinity cell-cell interaction is termed the "immunological synapse" (39-41). MHC class II expression is regulated mainly on a transcriptional level (42-45). Three *cis*-acting elements in the promoter region of the MHC class II gene, namely W, X, and Y boxes, constitute the major transcriptional control elements. Numerous DNA binding proteins bind to these cis-acting sequences, which are essential but not sufficient to initiate transcription. In contrast to MHC class I, the expression of MHC class II is not ubiquitous. The MHC class II transactivator (CIITA), a transcriptional co-activator, is the essential molecule that directs IFNγ-inducible MHC class II expression in nonprofessional APC and constitutive MHC class II expression in professional APC (*Figure 2*) (46, 47).

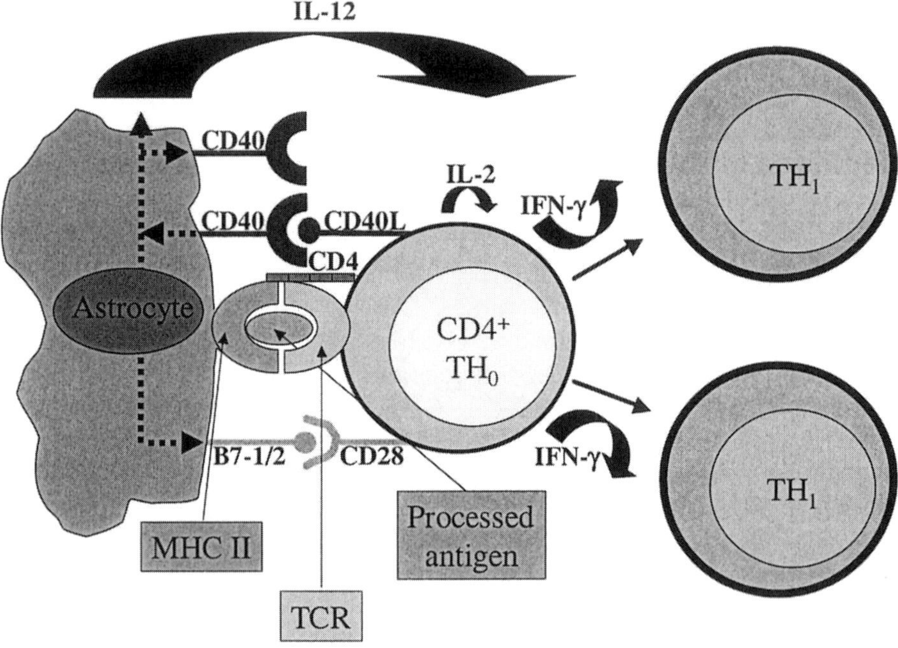


*Figure 1:* In order for antigen (Ag) presentation in the context of major histocompatibility complex (MHC) class II to occur, some absolute requirements have to be met. An MHC class II heterodimer on the cell surface of an antigen presenting cell (APC) contain and antigen binding groove, which can load a linear peptides that are typically 13-20 amino acids in length. A T cell can bind to Ag within the MHC class II binding groove with its T cell receptor. Only T cells that express the accessory surface molecule CD4 are capable of recognizing and binding linear peptide antigen (Ag) in the context of the MHC class II. Ag-activated T cells express CD40 ligand (CD40L), which engages to the costimulatory molecule CD40 on the cell surface of APC. Cross linking of CD40 and CD40L enhances expression of other costimulatory molecules on APC's, namely B7-1 (CD80), and B7-2 (CD86). Furthermore, engagement of CD40 leads to secretion of the cytokine IL-12, which induces the differentiation of naïve Th0 cells to effector CD4⁺ Th1 cells. These Th1 cells secrete increased amounts of proinflammatory cytokines like interferon (IFN)-γ and IL-2. INF- γ is required for induction of class II MHC molecules on nonprofessional antigen presenting cells (APC), including astrocytes. Some studies suggest that IFN-γ also causes astrocytes to upregulate expression of B7 (B7-1 and B7-2) molecules, which are required for CD28 T cell costimulation. Thus, in cell-mediated immunity, cross-activation occurs between APC

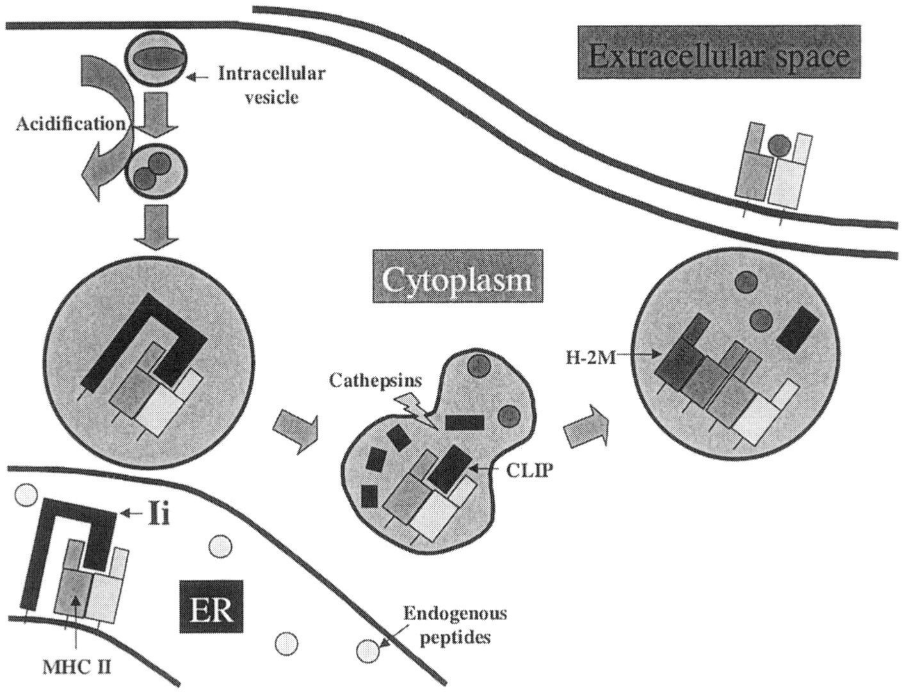


*Figure 2:* The endocytic compartment. Exogenous protein antigen (Ag) is taken up into an antigen presenting cell (APC) by endocytosis and broken down into peptide through acidification and proteolysis. The major histocompatibility complex (MHC) class II heterodimer is assembled in the endoplasmatic reticulum (ER) and then stabilized by the invariant chain (Ii). Within the ER, the Ii initially is a trimer that can bind three MHC class two heterodimers (not shown). The Ii blocks the MHC class II Ag binding groove, in order to prevent endogenous peptides in the ER from binding. Subsequently, the Ii targets the delivery of MHC class II dimers to the endosomal compartment, where it modulates proteolytic enzymes. Cathepsins are a group of cystein and aspartyl proteases that play a key role in the endosomal degradation of Ag and Ii. The last component of the Ii that remains bound to the MHC class II binding groove is named CLIP. The MHC class II like molecule (H-2M) catalyzes the release of CLIP and low stability peptides from the MHC class II Ag binding groove, and stabilizes empty MHC class II until high affinity peptide can be loaded. H-2M cannot bind Ag itself, and it is not expressed on the cell surface. Peptide loaded into the MHC class II binding groove can now be transported to the cell surface and presented to CD4$^+$ T cells.

Despite the fact that the genes encoding CIITA and MHC class II are located on different chromosomes in humans (chromosome 16 and 6, respectively), these molecules are developmentally coexpressed. CIITA interacts with transcription factors that are bound to the MHC class II promoter. In addition to regulation of MHC class II, CIITA also directs expression of invariant chain (Ii) and the MHC class II like molecule, HLA-DM (47), two molecules involved in MHC class II maturation and Ag presentation through the endocytic processing pathway (*Figure 3*) (48). Although considered the "master regulator" of the endocytic pathway (49), RNA message for MHC class II has been demonstrated in CIITA-deficient mice in secondary lymphoid tissue (50), and in some strains there was even residual expression of MHC class II protein in the thymus (51, 52), spleen (52), and lymph node cells (LNC) (52).

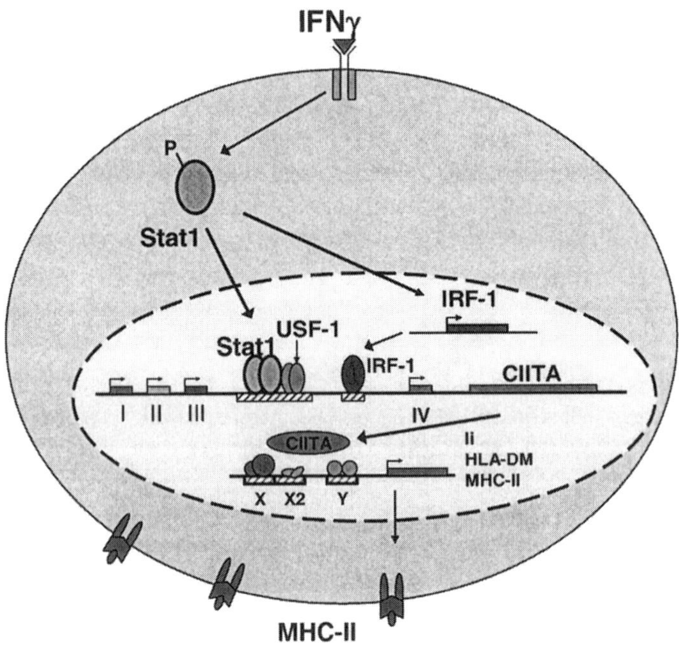


*Figure 3:* The major histocompatibility complex (MHC) class II transactivator (CIITA), a transcriptional co-activator, is the key molecule that directs interferon-gamma (IFNγ)-inducible MHC class II expression in nonprofessional antigen presenting cells (APC) and constitutive MHC class II expression in professional APC. CIITA expression is controlled at the level of transcription by differential activation of multiple nonhomologous promoters. CIITA promoter, pIV, which contains a GAS site, E-box, and IRF element, directs IFNγ-inducible CIITA expression in non-professional APC. It has been demonstrated that statins decrease IFNγ inducible MHC class II surface expression through inhibition of CIITA pIV.

CIITA expression is controlled at the level of transcription in a tissue-specific manner by differential activation of multiple nonhomologous promoters (53). Originally, it was observed that two CIITA promoter elements, promoter (p) I and pIII, directed constitutive CIITA expression in dendritic cells and B cells, respectively (53). Another CIITA promoter, pIV, which contains a GAS site, E-box, and IRF element (53, 54), directs IFN□-inducible CIITA, Ii, and H-2M expression in non-professional APC, including IFN-□-activated astrocytes (*Figure 3*) (35, 55, 56). While the impact of CIITA on processing and presentation of native Ag's appears cell-type dependent (57), it was demonstrated that activated astrocytes are capable of processing and presenting native Ag to MHC class II-restricted CD4$^+$ T cells in vitro (35).

Two recent studies reported that CIITA-deficient mice are resistant to clinical or histologic EAE induced by active immunization or by adoptive transfer of wild-type encephalitogenic class II-restricted CD4$^+$ T cells, and that CNS class II expression is not detectable in these animals (28, 58). Thus, MHC class II-restricted CNS Ag presentation is clearly CIITA-dependent. To further investigate the role of MHC class II-restricted Ag presentation by astrocytes in CNS autoimmunity, one group of investigators used the GFAP promoter to direct CIITA expression in astrocytes (58). Astrocytes transfected with CIITA constitutively upregulated cell surface MHC class II molecules, Ii and H-2M. The in vivo relevance of constitutive MHC class II expression on astrocytes was examined by generating GFAP-CIITA transgenic (Tg) mice using the same human GFAP promoter construct (59) used for transfection of astrocytes (58). While astrocytes in GFAP-CIITA Tg mice expressed cell surface class II molecules, these animals did not develop spontaneous CNS inflammation and, in comparison to control mice, there was no significant difference in EAE onset or clinical severity (58). These findings may partly be explained by the fact that CIITA-transfected astrocytes did not upregulate cathepsin (Cat) S, a cystein protease that may be required for processing of myelin basic protein (MBP) (60). Significant differences in histologic EAE were also not observed, although some astrocytes within EAE lesions did express MHC class II molecules in the GFAP-CIITA transgenic animals. Thus, by taking it to the extreme, these investigators demonstrated that constitutive overexpression of MHC class II molecules in astrocytes did not promote EAE, indicating that astrocytes may not play a significant role in CNS MHC class II restricted Ag presentation. Furthermore, like others (61), they did not detect MHC class II molecules on astrocytes in wild-type mice with CNS EAE.

## **Role of astrocytes in costimulation**

While binding of the TCR to the MHC class II molecule is an absolute requirement for class II restricted Ag presentation, and commonly referred to as "signal one", a second signal is required for the activation of naïve, unactivated T cells (39, 40) (*Figure 1*). Ag-activated T cells express CD40 ligand (CD40L), which engages to the costimulatory molecule CD40 on APC cell surface. Cross linking of CD40 and CD40L enhances expression of other costimulatory molecules on APC's, namely B7-1 (CD80), and B7-2 (CD86), which are required for CD28-mediated T cell costimulation. Furthermore, engagement of CD40 leads to secretion of the cytokine IL-12, which induces the differentiation of naïve Th0 cells to effector $CD4^+$ Th1 cells. These Th1 cells secrete increased amounts of proinflammatory cytokines like interferon (IFN)-$\gamma$ and IL-2.

CD 40 has not been demonstrated on astrocytes in vivo in adult human CNS tissue (62). Fetal human astrocytes express CD40 mRNA and protein constitutively. CD 40 expression can further be was upregulated by various inflammatory mediators, including LPS, IFN-$\gamma$, TNF-$\alpha$, or IL-1$\beta$ (63). The reports regarding CD 40 expression on murine astrocytes are controversial: While CD40 expression on astrocytes was detected by Tan et al. (64), this observation could not be confirmed by three other groups of investigators (33, 65, 66).

The evidence regarding B7 expression on astrocytes is no less confusing. No constitutive or IFN-$\gamma$-inducible expression of B7-1 or B7-2 was detected on human astrocytes in vitro (67) or in MS CNS tissue (68, 69). Reports regarding the expression of B7-1 and B7-2 on murine astrocytes have been conflicting: While Aloisi et al. could detect neither one of these costimulatory molecules on astrocytes after IFN-$\gamma$ stimulation in vitro (33), expression of B7-1 and B7-2 molecules on astrocytes was demonstrated by Nikcevich et al. (70). However, Soos et al. showed that B7-2, but not B7-1, was upregulated in response to IFN-$\gamma$ stimulation on primary and transformed murine astrocytes in short-term cultures (seven days), whereas in long-term cultures, both B7-1 and B7-2 were upregulated (65). Interestingly, short-term cultures of astrocytes that expressed B7-2, but not B7-1, promoted Th2 differentiation of naive Th0 cells (65). There has also been controversy with regard to determining B-7 in EAE tissue: Cross et al. did not detect either B7-1 or B7-2 in EAE lesions of different mouse strains (71), whereas Issazedeh et al. detected B7-2 expression on astrocytes during acute EAE, while B7-1 was detected on astrocytes during disease remission (72).

T cells also engage APC through binding of the lymphocyte function-associated antigen (LFA)-1 to the very late activation adhesion molecules of the immunoglobulin superfamily, intercellular adhesion molecule (ICAM)-1.  LFA-1 is considered one of the most important integrins for lymphocyte activation, and the activation of naïve T lymphocytes can be completely abrogated with antibodies to this molecule (73).  Interestingly, LFA-1 has been identified on the surface of inflammatory cells in perivascular MS lesions (74-78).  Furthermore, it has been demonstrated that the expression of its ligand, ICAM-1, on the cell surface of human astrocytes can be enhanced by proinflammatory cytokines (79), and that ICAM-1 is present within MS lesions and surrounding adjacent white matter beyond the lesion edge (76).  Constitutive ICAM-1 expression is strongly enhanced by TNF-$\alpha$ and IL-1$\beta$, and less so by IFN-$\beta$ (80, 81).  There is accumulating data which demonstrate that activated astrocytes can express a number of other adhesion molecules, including VCAM-1, E-selectin, and LFA-3, which is a ligand for LFA-2 on leukocytes, and known to participate in T cell activation (76, 82).  It is thus reasonable to propose that glial cell expression of adhesion molecules in an inflammatory environment may contribute to T cell activation and antigen presentation (78, 82-84).

Conflicting observations regarding the capability of astrocytes to express costimulatory molecules certainly need further clarification.  The absence of "signal two" could render astrocytes as incompetent APC, which may promote anergy and apoptosis rather than activation of naïve T cells.  The requirement for costimulation in CNS inflammation was recently demonstrated in GFAP-CIITA transgenic mice that constitutively expressed MHC class II surface molecules, Ii, and H-2M, but did not upregulate costimulatory molecules or increase EAE susceptibility (58).

## Expression of inflammatory mediators by astrocytes

The extravasation of leukocytes represents one of the pathological hallmarks of several inflammatory CNS diseases, including MS (85). There is ample evidence that the progression of this demyelinating disease is associated with an inflammatory reaction that involves activated lymphocytes, macrophages, and endogenous glial cells (85).

The migration of T cell from the blood into the CNS is a complex, multi-step event (86, 87). Chemoactive cytokines, also known as "chemokines", facilitate the migration of leukocytes into the CNS. Chemokines are produced by a wide spectrum of cell types relevant in EAE and MS, including astrocytes. Their secretion can be induced exogenously through cell-cell contact or bacterial cell wall components like lipopolysaccharide (LPS), the HIV-1 Tat protein, and by proinflammatory cytokines such as TNF-α, IFNγ, TGF-β, and the binding of Fas ligand to its receptor (88-94). Chemokines are currently categorized into four subfamilies, based on their three-dimensional structure (92). The β-chemokines contain two adjacent cysteine moieties (CC). Monocyte chemotactic protein (MCP)-1 is one member of the CC chemokine family that is mainly produced by astrocytes during CNS inflammation, and that binds exclusively to the chemokine receptor CCR2, which is expressed on activated T cells and macrophages (95). Chemokines from several subfamilies have recently been identified that selectively recruit T cell and monocytes into the CNS, and their secretion has been associated with clinical disease activity in EAE (96). MCP-1 appears to play an important, differential role in the pathogenesis of this archetypical CNS autoimmune disease (97), and in one study, monoclonal antibodies directed against MCP-1 significantly reduced the severity of the relapsing form of EAE (98). LNC from primed CCR-2-deficient mice showed decreased Ag-specific proliferative responses and secretion of IFN-γ (99, 100). The importance of MCP-1 in EAE pathogenesis was further underscored by the observation that both, MCP-1-deficient mice (101), and CCR2-deficient mice (99, 102) are resistant to EAE. These results underscore the key role of astrocytes in the recruitment of activated T cells into the brain and spinal cord during CNS autoimmune disease.

Astrocytes respond to and produce proinflammatory cytokines. Th1 CD4$^+$ T cells produce proinflammatory cytokines including IFNγ and TNFα, which lead to astrocyte activation. In EAE, Th1 T cells are strongly associated with clinical disease (103). Specifically, it has been demonstrated in several species that Th1 cytokines are upregulated during acute disease and relapses, but not during clinical remission (104, 105). In plaques from MS patients, immunohistochemical in situ analysis showed the presence of the Th1 cytokine TNF-α (106, 107). Furthermore, in another study TNFα was detected in chronic MS lesions, but not in other neurological diseases

(107). The major sources of TNF-α in the CNS are astrocytes and activated macrophages (108), and in an autocrine loop, activated macrophages in turn increase their cell surface expression of TNF receptors (109). A variety of cytokines, including IFN-γ and TNFα induce the production of nitric oxide (NO) and oxygen radicals by astrocytes (110, 111). NO is a vascular and neuronal messenger abundantly present in the CNS, which is produced by inducible NO synthase (iNOS). In acute MS lesions, intense reactivity for iNOS mRNA and protein was detected in reactive astrocytes throughout the lesion and in adjacent normal appearing white matter (112). Staining of macrophages and other inflammatory cell infiltrates detected iNOS also in acute MS lesions. In chronic MS lesions, reactive astrocytes were positive for iNOS at the lesion edge (112).

However, there is also substantial evidence indicating that astrocytes may serve to downregulate proinflammatory CNS responses (113): Astrocytes produce a number of immunosuppressive cytokines, including TGF-β (29), IL-10 (114, 115), IFNα (116), and IFNβ (117). One study demonstrated that astrocyte supernatants suppressed Ag-specific T cell proliferation (113). Furthermore, it has been demonstrated that astrocytes can promote naïve Th0 cells to differentiate into Th2 cells (65). Thus, while some data support the notion of astrocytes as producers of inflammatory mediators in CNS autoimmune disease, these observations are contradicted by reports of their beneficial effects during CNS inflammation. A better understanding of astrocyte function, specifically of the temporal expression of proinflammatory and anti-inflammatory mediators, is required for therapeutic interventions to reach their full benefit.

**Conclusion**

The role of astrocytes as APC in CNS autoimmune disease, and particular their role in MHC class II restricted Ag presentation, has been controversial. In vitro, IFNγ-activated astrocytes upregulate MHC class II molecules and can present antigen to $CD4^+$ T cells. While MHC class II molecules have been detected on astrocytes within inflammatory lesions of MS and EAE in some studies, these findings have not consistently been reproduced. A recent report demonstrated that while CIITA-directed MHC class II expression was required for EAE induction, CIITA-directed constitutive overexpression of MHC class II molecules on astrocytes did not increase EAE susceptibility. These results indicate that astrocytes do not have a prominent role as APC in MHC class II-restricted Ag presentation during acute EAE.

Astrocytes express a number of costimulatory cell surface molecules and ligands for costimulatory molecules on leukocytes. Their biological significance in CNS autoimmune diseases remains uncertain and requires further investigation, which should involve more recently discovered costimulatory pathways. Similarly, the contribution of several inflammatory mediators secreted by astrocytes, including that of chemokines, proinflammatory cytokines, and nitric oxide, in the initiation and continuance of CNS inflammation needs to be determined. In comparison to microglia, current data indicate that astrocytes have a more significant role in downregulating CNS inflammatory responses.

# ACKNOWLEDGEMENTS

Support to SSZ was provided by the Alexander M. and June L. Maisin Foundation (grant no. 98-416), the National Institutes of Health (grant no. K02 NS02207) and the National Multiple Sclerosis Society (NMSS) (RG 3206-A-3). SSZ is a 2002 recipient of a research grant from the Wadsworth Foundation. OS is supported by an advanced fellowship from the NMSS and is a Boehringer-Ingelheim Fonds Scholar.

# REFERENCES

1. Martin, R., H. F. McFarland, and D. E. McFarlin. 1992. Immunological aspects of demyelinating diseases. *Annu.Rev.Immunol.* 10:153-187.

2.   Steinman, L. 1996. Multiple sclerosis: a coordinated immunological attack against myelin in the central nervous system. *Cell* 85:299-302.

3.   Stuve, O. and S. S. Zamvil. 2001. Neurologic diseases. In *Medical Immunology*. T. G. Parslow, D. P. Stites, A. I. Terr, and J. B. Imboden, eds. McGraw Hill, San Francisco, pp. 510-526.

4.   Anderson, D. W., J. H. Ellenberg, C. M. Leventhal, S. C. Reingold, M. Rodriguez, and D. H. Silberberg. 1992. Revised estimate of the prevalence of multiple sclerosis in the United States. *Ann.Neurol.* 31:333-336.

5.   Koopmans, R. A., D. K. Li, J. J. Oger, J. Mayo, and D. W. Paty. 1989. The lesion of multiple sclerosis: imaging of acute and chronic stages. *Neurology* 39:959-963.

6.   Stone, L. A., P. S. Albert, M. E. Smith, C. DeCarli, M. R. Armstrong, D. E. McFarlin, J. A. Frank, and H. F. McFarland. 1995. Changes in the amount of diseased white matter over time in patients with relapsing-remitting multiple sclerosis. *Neurology* 45:1808-1814.

7.   Paty, D. W. and H. McFarland. 1998. Magnetic resonance techniques to monitor the long term evolution of multiple sclerosis pathology and to monitor definitive clinical trials. *J.Neurol.Neurosurg.Psychiatry* 64 Suppl 1:S47-S51.

8.   Steinman, L. 2001. Multiple sclerosis: a two-stage disease. *Nat.Immunol.* 2:762-764.

9.   Mi, H., H. Haeberle, and B. A. Barres. Induction of astrocyte differentiation by endothelial cells. 2001. *J Neurosci* 21:1538-1547.

10.   Wolburg, H. and W. Rissau. 1995. Formation of the blood brain barrier. In *Neuroglia*. B. R. Ransom, Kettenmann, and H, eds. Oxford University Press, New York, pp. 763-776.

11.   Traugott, U., L. C. Scheinberg, and C. S. Raine. 1985. On the presence of Ia-positive endothelial cells and astrocytes in multiple sclerosis lesions and its relevance to antigen presentation. *J.Neuroimmunol.* 8:1-14.

12.   Sakai, K., T. Tabira, M. Endoh, and L. Steinman. 1986. Ia expression in chronic relapsing experimental allergic encephalomyelitis induced by long-term cultured T cell lines in mice. *Lab Invest* 54:345-352.

13.  Traugott, U. 1989. Detailed analysis of early immunopathologic events during lesion formation in acute experimental autoimmune encephalomyelitis. *Cell Immunol.* 119:114-129.

14.  Krogsgaard, M., K. W. Wucherpfennig, B. Canella, B. E. Hansen, A. Svejgaard, J. Pyrdol, H. Ditzel, C. Raine, J. Engberg, and L. Fugger. 2000. Visualization of myelin basic protein (MBP) T cell epitopes in multiple sclerosis lesions using a monoclonal antibody specific for the human histocompatibility leukocyte antigen (HLA)-DR2-MBP 85-99 complex. *J.Exp.Med.* 191:1395-1412.

15.  Benveniste, E. N., Shrikant, P., Patton, H. K., and Benos, D. J. Neuroimmunologic mechanisms for disease in AIDS: The role of the astrocyte. Gendelman, H. E., Lipton, S. A., Epstein, L., and Swindells, S. 130-146. 2002. New York, Chapman & Hall. The neurology of AIDS. Ref Type: Serial (Book,Monograph)

16.  Abnet, K., J. W. Fawcett, and S. B. Dunnett. 1991. Interactions between meningeal cells and astrocytes in vivo and in vitro. *Brain Res.Dev.Brain Res.* 59:187-196.

17.  Eng, L. F., P. J. Reier, and J. D. Houle. 1987. Astrocyte activation and fibrous gliosis: glial fibrillary acidic protein immunostaining of astrocytes following intraspinal cord grafting of fetal CNS tissue. *Prog.Brain Res.* 71:439-455.

18.  Dusart, I., S. Marty, and M. Peschanski. 1991. Glial changes following an excitotoxic lesion in the CNS--II. Astrocytes. *Neuroscience* 45:541-549.

19.  Eddleston, M. and L. Mucke. 1993. Molecular profile of reactive astrocytes--implications for their role in neurologic disease. *Neuroscience* 54:15-36.

20.  Merrill, J. E. and E. N. Benveniste. 1996. Cytokines in inflammatory brain lesions: helpful and harmful. *Trends Neurosci.* 19:331-338.

21.  Norton, W. T., D. A. Aquino, I. Hozumi, F. C. Chiu, and C. F. Brosnan. 1992. Quantitative aspects of reactive gliosis: a review. *Neurochem.Res.* 17:877-885.

22.  Ridet, J. L., S. K. Malhotra, A. Privat, and F. H. Gage. 1997. Reactive astrocytes: cellular and molecular cues to biological function. *Trends Neurosci.* 20:570-577.

23. Wu, V. W. and J. P. Schwartz. 1998. Cell culture models for reactive gliosis: new perspectives. *J.Neurosci.Res.* 51:675-681.

24. Zamvil, S. S. and L. Steinman. 1990. The T lymphocyte in experimental allergic encephalomyelitis. *Annu.Rev.Immunol.* 8:579-621.

25. Wekerle, H., C. Linington, H. Lassmann, and R. Meyermann. 1986. Cellular immune reactivity within the CNS. *Trends Neurosci.* 9:271-277.

26. Hickey, W. F., B. L. Hsu, and H. Kimura. 1991. T-lymphocyte entry into the central nervous system. *J.Neurosci.Res.* 28:254-260.

27. Slavin, A. J., J. M. Soos, O. Stuve, J. C. Patarroyo, H. L. Weiner, A. Fontana, E. K. Bikoff, and S. S. Zamvil. 2001. Requirement for endocytic antigen processing and influence of invariant chain and H-2M deficiencies in CNS autoimmunity. *J.Clin.Invest* 108:1133-1139.

28. Tompkins, S. M., J. Padilla, M. C. Dal Canto, J. P. Ting, L. Van Kaer, and S. D. Miller. 2002. De novo central nervous system processing of myelin antigen is required for the initiation of experimental autoimmune encephalomyelitis. *J.Immunol.* 168:4173-4183

29. Shrikant, P. and E. N. Benveniste. 1996. The central nervous system as an immunocompetent organ: role of glial cells in antigen presentation. *J.Immunol.* 157:1819-1822.

30. Sedgwick, J. D., R. Mossner, S. Schwender, and M. ter, V. 1991. Major histocompatibility complex-expressing nonhematopoietic astroglial cells prime only CD8+ T lymphocytes: astroglial cells as perpetuators but not initiators of CD4+ T cell responses in the central nervous system. *J.Exp.Med.* 173:1235-1246.

31. Williams, K., Jr., E. Ulvestad, L. Cragg, M. Blain, and J. P. Antel. 1993. Induction of primary T cell responses by human glial cells. *J.Neurosci.Res.* 36:382-390.

32. Weber, F., E. Meinl, F. Aloisi, C. Nevinny-Stickel, E. Albert, H. Wekerle, and R. Hohlfeld. 1994. Human astrocytes are only partially competent antigen presenting cells. Possible implications for lesion development in multiple sclerosis. *Brain* 117 ( Pt 1):59-69.

33. Aloisi, F., F. Ria, G. Penna, and L. Adorini. 1998. Microglia are more efficient than astrocytes in antigen processing and in Th1 but not Th2 cell activation. *J.Immunol.* 160:4671-4680.

34. Panitch, H. S., R. L. Hirsch, A. S. Haley, and K. P. Johnson. 1987. Exacerbations of multiple sclerosis in patients treated with gamma interferon. *Lancet* 1:893-895.

35. Soos, J. M., J. Morrow, T. A. Ashley, B. E. Szente, E. K. Bikoff, and S. S. Zamvil. 1998. Astrocytes express elements of the class II endocytic pathway and process central nervous system autoantigen for presentation to encephalitogenic T cells. *J.Immunol.* 161:5959-5966.

36. Fontana, A., W. Fierz, and H. Wekerle. 1984. Astrocytes present myelin basic protein to encephalitogenic T-cell lines. *Nature* 307:273-276.

37. Takiguchi, M. and J. A. Frelinger. 1986. Induction of antigen presentation ability in purified cultures of astroglia by interferon-gamma. *J.Mol.Cell Immunol.* 2:269-280.

38. Frei, K., H. Lins, C. Schwerdel, and A. Fontana. 1994. Antigen presentation in the central nervous system. The inhibitory effect of IL-10 on MHC class II expression and production of cytokines depends on the inducing signals and the type of cell analyzed. *J.Immunol.* 152:2720-2728.

39. Dustin, M. L. and A. S. Shaw. 1999. Costimulation: building an immunological synapse. *Science* 283:649-650.

40. Grakoui, A., S. K. Bromley, C. Sumen, M. M. Davis, A. S. Shaw, P. M. Allen, and M. L. Dustin. 1999. The immunological synapse: a molecular machine controlling T cell activation. *Science* 285:221-227.

41. Delon, J. 2000. The immunological synapse. *Curr.Biol.* 10:R214.

42. Ting, J. P. and X. S. Zhu. 1999. Class II MHC genes: a model gene regulatory system with great biologic consequences. *Microbes.Infect.* 1:855-861.

43. Collawn, J. F. and E. N. Benveniste. 1999. Regulation of MHC class II expression in the central nervous system. *Microbes.Infect.* 1:893-902.

44. DeSandro, A., U. M. Nagarajan, and J. M. Boss. 1999. The bare lymphocyte syndrome: molecular clues to the transcriptional regulation of major histocompatibility complex class II genes. *Am.J.Hum.Genet.* 65:279-286.

45.  Boss, J. M. 1997. Regulation of transcription of MHC class II genes. *Curr.Opin.Immunol.* 9:107-113.

46.  Mach, B., V. Steimle, E. Martinez-Soria, and W. Reith. 1996. Regulation of MHC class II genes: lessons from a disease. *Annu.Rev.Immunol.* 14:301-331.

47.  Chang, C. H. and R. A. Flavell. 1995. Class II transactivator regulates the expression of multiple genes involved in antigen presentation. *J.Exp.Med.* 181:765-767.

48.  Wolf, P. R. and H. L. Ploegh. 1995. How MHC class II molecules acquire peptide cargo: biosynthesis and trafficking through the endocytic pathway. *Annu.Rev.Cell Dev.Biol.* 11:267-306.

49.  Harton, J. A. and J. P. Ting. 2000. Class II transactivator: mastering the art of major histocompatibility complex expression. *Mol.Cell Biol.* 20:6185-6194.

50.  Itoh-Lindstrom, Y., J. F. Piskurich, N. J. Felix, Y. Wang, W. J. Brickey, J. L. Platt, B. H. Koller, and J. P. Ting. 1999. Reduced IL-4-, lipopolysaccharide-, and IFN-gamma-induced MHC class II expression in mice lacking class II transactivator due to targeted deletion of the GTP-binding domain. *J.Immunol.* 163:2425-2431.

51.  Williams, G. S., M. Malin, D. Vremec, C. H. Chang, R. Boyd, C. Benoist, and D. Mathis. 1998. Mice lacking the transcription factor CIITA--a second look. *Int.Immunol.* 10:1957-1967.

52.  Chang, C. H., S. Guerder, S. C. Hong, W. van Ewijk, and R. A. Flavell. 1996. Mice lacking the MHC class II transactivator (CIITA) show tissue- specific impairment of MHC class II expression. *Immunity.* 4:167-178.

53.  Muhlethaler-Mottet, A., L. A. Otten, V. Steimle, and B. Mach. 1997. Expression of MHC class II molecules in different cellular and functional compartments is controlled by differential usage of multiple promoters of the transactivator CIITA. *EMBO J.* 16:2851-2860.

54.  Piskurich, J. F., Y. Wang, M. W. Linhoff, L. C. White, and J. P. Ting. 1998. Identification of distinct regions of 5' flanking DNA that mediate constitutive, IFN-gamma, STAT1, and TGF-beta-regulated expression of the class II transactivator gene. *J.Immunol.* 160:233-240.

55.  Dong, Y., W. M. Rohn, and E. N. Benveniste. 1999. IFN-gamma regulation of the type IV class II transactivator promoter in astrocytes. *J.Immunol.* 162:4731-4739.

56.  Nikcevich, K. M., J. F. Piskurich, R. P. Hellendall, Y. Wang, and J. P. Ting. 1999. Differential selectivity of CIITA promoter activation by IFN-gamma and IRF-1 in astrocytes and macrophages: CIITA promoter activation is not affected by TNF-alpha. *J.Neuroimmunol.* 99:195-204.

57.  Siegrist, C. A., E. Martinez-Soria, I. Kern, and B. Mach. 1995. A novel antigen-processing-defective phenotype in major histocompatibility complex class II-positive CIITA transfectants is corrected by interferon-gamma. *J.Exp.Med.* 182:1793-1799.

58.  Stuve, O., S. Youssef, A. J. Slavin, C. L. King, J. C. Patarroyo, D. L. Hirschberg, W. J. Brickey, J. M. Soos, J. F. Piskurich, H. A. Chapman, and S. S. Zamvil. 2002. The Role of the MHC Class II Transactivator in Class II Expression and Antigen Presentation by Astrocytes and in Susceptibility to Central Nervous System Autoimmune Disease. *J.Immunol.* 169:6720-6732.

59.  Brenner, M., W. C. Kisseberth, Y. Su, F. Besnard, and A. Messing. 1994. GFAP promoter directs astrocyte-specific expression in transgenic mice. *J.Neurosci.* 14:1030-1037.

60.  Beck, H., G. Schwarz, C. J. Schroter, M. Deeg, D. Baier, S. Stevanovic, E. Weber, C. Driessen, and H. Kalbacher. 2001. Cathepsin S and an asparagine-specific endoprotease dominate the proteolytic processing of human myelin basic protein in vitro. *Eur.J.Immunol.* 31:3726-3736.

61.  Butter, C., J. K. O'Neill, D. Baker, S. E. Gschmeissner, and J. L. Turk. 1991. An immunoelectron microscopical study of the expression of class II major histocompatibility complex during chronic relapsing experimental allergic encephalomyelitis in Biozzi AB/H mice. *J.Neuroimmunol.* 33:37-42.

62.  Togo, T., H. Akiyama, H. Kondo, K. Ikeda, M. Kato, E. Iseki, and K. Kosaka. 2000. Expression of CD40 in the brain of Alzheimer's disease and other neurological diseases. *Brain Res.* 885:117-121.

63.  Abdel-Haq, N., H. N. Hao, and W. D. Lyman. 1999. Cytokine regulation of CD40 expression in fetal human astrocyte cultures. *J Neuroimmunol.* 101:7-14.

64. Tan, L., K. B. Gordon, J. P. Mueller, L. A. Matis, and S. D. Miller. 1998. Presentation of proteolipid protein epitopes and B7-1-dependent activation of encephalitogenic T cells by IFN-gamma-activated SJL/J astrocytes. *J.Immunol.* 160:4271-4279.

65. Soos, J. M., T. A. Ashley, J. Morrow, J. C. Patarroyo, B. E. Szente, and S. S. Zamvil. 1999. Differential expression of B7 co-stimulatory molecules by astrocytes correlates with T cell activation and cytokine production. *Int.Immunol.* 11:1169-1179.

66. Nguyen, V. T. and E. N. Benveniste. 2000. Involvement of STAT-1 and ets family members in interferon-gamma induction of CD40 transcription in microglia/macrophages. *J Biol Chem* 271:23674-23684.

67. Satoh, J., Y. B. Lee, and S. U. Kim. 1995. T-cell costimulatory molecules B7-1 (CD80) and B7-2 (CD86) are expressed in human microglia but not in astrocytes in culture. *Brain Res.* 704:92-96.

68. Williams, K., E. Ulvestad, and J. P. Antel. 1994. B7/BB-1 antigen expression on adult human microglia studied in vitro and in situ. *Eur.J.Immunol.* 24:3031-3037.

69. Windhagen, A., J. Newcombe, F. Dangond, C. Strand, M. N. Woodroofe, M. L. Cuzner, and D. A. Hafler. 1995. Expression of costimulatory molecules B7-1 (CD80), B7-2 (CD86), and interleukin 12 cytokine in multiple sclerosis lesions. *J Exp Med* 1985-1996.

70. Nikcevich, K. M., K. B. Gordon, L. Tan, S. D. Hurst, J. F. Kroepfl, M. Gardiner, T. A. Barrett, and S. D. Miller. 1997. IFN-gamma activated primary murine astrocytes express B7 costimulatory molecules and prime naive antigen-specific T cells. *J.Immunol.* 158:614-621.

71. Cross, A. H. and G. Ku. 2000. Astrocytes and central nervous system endothelial cells do not express B7-1 (CD80) or B7-2 (CD86) immunoreactivity during experimental autoimmune encephalomyelitis. *J Neuroimmunol.* 110:76-82.

72. Issazadeh, S., V. Navikas, M. Schaub, M. Sayegh, and S. Khoury. 1998. Kinetics of expression of costimulatory molecules and their ligands in murine relapsing experimental autoimmune encephalomyelitis in vivo. *J Immunol* 161:1104-1112.

73. Hogg, N. and R. C. Landis. 1993. Adhesion molecules in cell interactions. *Curr.Opin.Immunol.* 5:383-390.

74. Sobel, R. A., M. E. Mitchell, and G. Fondren. 1990. Intercellular adhesion molecule-1 (ICAM-1) in cellular immune reactions in the human central nervous system. *Am.J.Pathol.* 136:1309-1316.

75. Washington, R., J. Burton, R. F. Todd, III, W. Newman, L. Dragovic, and P. Dore-Duffy. 1994. Expression of immunologically relevant endothelial cell activation antigens on isolated central nervous system microvessels from patients with multiple sclerosis. *Ann.Neurol.* 35:89-97.

76. Brosnan, C. F., B. Cannella, L. Battistini, and C. S. Raine. 1995. Cytokine localization in multiple sclerosis lesions: correlation with adhesion molecule expression and reactive nitrogen species. *Neurology* 45:S16-S21.

77. Cannella, B. and C. S. Raine. 1995. The adhesion molecule and cytokine profile of multiple sclerosis lesions. *Ann.Neurol.* 37:424-435.

78. Bo, L., J. W. Peterson, S. Mork, P. A. Hoffman, W. M. Gallatin, R. M. Ransohoff, and B. D. Trapp. 1996. Distribution of immunoglobulin superfamily members ICAM-1, -2, -3, and the beta 2 integrin LFA-1 in multiple sclerosis lesions. *J.Neuropathol.Exp.Neurol.* 55:1060-1072.

79. Frohman, E. M., T. C. Frohman, M. L. Dustin, B. Vayuvegula, B. Choi, A. Gupta, N. S. van den, and S. Gupta. 1989. The induction of intercellular adhesion molecule 1 (ICAM-1) expression on human fetal astrocytes by interferon-gamma, tumor necrosis factor alpha, lymphotoxin, and interleukin-1: relevance to intracerebral antigen presentation. *J.Neuroimmunol.* 23:117-124.

80. Shrikant, P., I. Y. Chung, M. E. Ballestas, and E. N. Benveniste. 1994. Regulation of intercellular adhesion molecule-1 gene expression by tumor necrosis factor-alpha, interleukin-1 beta, and interferon-gamma in astrocytes. *J.Neuroimmunol.*209-220.

81. Selmaj, K. W., M. Farooq, W. T. Norton, C. S. Raine, and C. F. Brosnan. 1990. Proliferation of astrocytes in vitro in response to cytokines. A primary role for tumor necrosis factor. *J.Immunol.* 144:129-135.

82. Moingeon, P., H. C. Chang, B. P. Wallner, C. Stebbins, A. Z. Frey, and E. L. Reinherz. 1989. CD2-mediated adhesion facilitates T lymphocyte antigen recognition function. *Nature* 339:312-314.

83. Damle, N. K., K. Klussman, G. Leytze, A. Aruffo, P. S. Linsley, and J. A. Ledbetter. 1993. Costimulation with integrin ligands intercellular adhesion molecule-1 or vascular cell adhesion molecule-1 augments activation-induced death of antigen-specific CD4+ T lymphocytes. *J.Immunol.* 151:2368-2379.

84. Damle, N. K. and A. Aruffo. 1991. Vascular cell adhesion molecule 1 induces T-cell antigen receptor- dependent activation of CD4+T lymphocytes. *Proc.Natl.Acad.Sci.U.S.A* 88:6403-6407.

85. Raine, C. S. 1994. The Dale E. McFarlin Memorial Lecture: the immunology of the multiple sclerosis lesion. *Ann.Neurol.* 36 Suppl:S61-S72.

86. Springer, T. A. 1994. Traffic signals for lymphocyte recirculation and leukocyte emigration: the multistep paradigm. *Cell* 76:301-314.

87. Springer, T. A. 1995. Traffic signals on endothelium for lymphocyte recirculation and leukocyte emigration. *Annu.Rev.Physiol* 57:827-872.

88. Hurwitz, A. A., W. D. Lyman, and J. W. Berman. 1995. Tumor necrosis factor alpha and transforming growth factor beta upregulate astrocyte expression of monocyte chemoattractant protein-1. *J Neuroimmunol.* 57:193-198.

89. Zhou, Z. H., P. Chaturvedi, Y. L. Han, S. Aras, Y. S. Li, P. E. Kolattukudy, D. Ping, J. M. Boss, and R. M. Ransohoff. 1998. IFN-gamma induction of the human monocyte chemoattractant protein (hMCP)-1 gene in astrocytoma cells: functional interaction between an IFN-gamma-activated site and a GC-rich element. *J.Immunol.* 160:3908-3916.

90. Oh, J. W., L. M. Schwiebert, and E. N. Benveniste. 1999. Cytokine regulation of CC and CXC chemokine expression by human astrocytes. *J.Neurovirol.* 5:82-94.

91. Kutsch, O., J. Oh, A. Nath, and E. N. Benveniste. 2000. Induction of the chemokines interleukin-8 and IP-10 by human immunodeficiency virus type 1 tat in astrocytes. *J Virol.* 74:9214-9221.

92. Godessart, N. and S. L. Kunkel. 2001. Chemokines in autoimmune disease. *Curr.Opin.Immunol* 13:670-675.

93. Choi, C., X. Xu, J. W. Oh, S. J. Lee, G. Y. Gillespie, H. Park, H. Jo, and E. N. Benveniste. 2001. Fas-induced expression of chemokines in human glioma

cells: involvement of extracellular signal-regulated kinase 1/2 and p38 mitogen-activated protein kinase. *Cancer Res.* 61:3084-3091.

94. Nelson, P. J. and A. M. Krensky. 2001. Chemokines, chemokine receptors, and allograft rejection. *Immunity.* 14:377-386.

95. Gerard, C. and B. J. Rollins. 2001. Chemokines and disease. *Nat.Immunol.* 2:108-115.

96. Karpus, W. J. and R. M. Ransohoff. 1998. Chemokine regulation of experimental autoimmune encephalomyelitis: temporal and spatial expression patterns govern disease pathogenesis. *J.Immunol.* 161:2667-2671.

97. Karpus, W. J., N. W. Lukacs, B. L. McRae, R. M. Strieter, S. L. Kunkel, and S. D. Miller. 1995. An important role for the chemokine macrophage inflammatory protein-1 alpha in the pathogenesis of the T cell-mediated autoimmune disease, experimental autoimmune encephalomyelitis. *J.Immunol.* 155:5003-5010.

98. Karpus, W. J. and K. J. Kennedy. 1997. MIP-1alpha and MCP-1 differentially regulate acute and relapsing autoimmune encephalomyelitis as well as Th1/Th2 lymphocyte differentiation. *J.Leukoc.Biol.* 62:681-687.

99. Izikson, L., R. S. Klein, I. F. Charo, H. L. Weiner, and A. D. Luster. 2000. Resistance to experimental autoimmune encephalomyelitis in mice lacking the CC chemokine receptor (CCR)2. *Immunol Today* 15:321-331.

100. Luther, S. A. and J. G. Cyster. 2001. Chemokines as regulators of T cell differentiation. *Nat.Immunol* 2:102-107.

101. Huang, D. R., J. Wang, P. Kivisakk, B. J. Rollins, and R. M. Ransohoff. Absence of monocyte chemoattractant protein 1 in mice leads to decreased local macrophage recruitment and antigen-specific T helper cell type 1 immune response in experimental autoimmune encephalomyelitis.

102. Fife, B. T., G. B. Huffnagle, W. A. Kuziel, and W. J. Karpus. 2000. CC chemokine receptor 2 is critical for induction of experimental autoimmune encephalomyelitis. *J Exp Med* 192:899-905.

103. Windhagen, A., L. B. Nicholson, H. L. Weiner, V. K. Kuchroo, and D. A. Hafler. 1996. Role of Th1 and Th2 cells in neurologic disorders. *Chem.Immunol* 63:171-186.

104.  Khoury, S. J., W. W. Hancock, and H. L. Weiner. 1992. Oral tolerance to myelin basic protein and natural recovery from experimental autoimmune encephalomyelitis are associated with downregulation of inflammatory cytokines and differential upregulation of transforming growth factor beta, interleukin 4, and prostaglandin E expression in the brain. *J.Exp.Med.* 176:1355-1364.

105.  Begolka, W. S., C. L. Vanderlugt, S. M. Rahbe, and S. D. Miller. 1998. Differential expression of inflammatory cytokines parallels progression of central nervous system pathology in two clinically distinct models of multiple sclerosis. *J.Immunol.* 161:4437-4446.

106.  Hofman, F. M., D. R. Hinton, K. Johnson, and J. E. Merrill. 1989. Tumor necrosis factor identified in multiple sclerosis brain. *J.Exp.Med.* 170:607-612.

107.  Selmaj, K., C. S. Raine, B. Cannella, and C. F. Brosnan. 1991. Identification of lymphotoxin and tumor necrosis factor in multiple sclerosis lesions. *J.Clin.Invest* 87:949-954.

108.  Segal, B. M., B. K. Dwyer, and E. M. Shevach. 1998. An interleukin (IL)-10/IL-12 immunoregulatory circuit controls susceptibility to autoimmune disease. *J.Exp.Med.* 187:537-546.

109.  Paulnock, D. M. 1992. Macrophage activation by T cells. *Curr.Opin.Immunol.* 4:344-349.

110.  Kawahara, K., T. Gotoh, S. Oyadomari, M. Kajizono, A. Kuniyasu, K. Ohsawa, Y. Imai, S. Kohsaka, H. Nakayama, and M. Mori. 2001. Co-induction of argininosuccinate synthetase, cationic amino acid transporter-2, and nitric oxide synthase in activated murine microglial cells. *Brain Res.Mol.Brain Res.* 90:165-173.

111.  Nomura, Y. 2001. NF-kappaB activation and IkappaB alpha dynamism involved in iNOS and chemokine induction in astroglial cells. *Life Sci.* 68:1695-1701.

112.  Liu, J. S., M. L. Zhao, C. F. Brosnan, and S. C. Lee. 2001. Expression of inducible nitric oxide synthase and nitrotyrosine in multiple sclerosis lesions. *Am.J.Pathol.* 158:2057-2066.

113. Meinl, E., F. Aloisi, B. Ertl, F. Weber, M. R. de Waal, H. Wekerle, and R. Hohlfeld. 1994. Multiple sclerosis. Immunomodulatory effects of human astrocytes on T cells. *Brain* 117 ( Pt 6):1323-1332.

114. Hulshof, S., L. Montagne, C. J. De Groot, and D. Van, V. 2002. Cellular localization and expression patterns of interleukin-10, interleukin-4, and their receptors in multiple sclerosis lesions. *Glia* 38:24-35.

115. Ledeboer, A., J. J. Breve, A. Wierinckx, J. S. van der, A. F. Bristow, J. E. Leysen, F. J. Tilders, and A. M. Van Dam. 2002. Expression and regulation of interleukin-10 and interleukin-10 receptor in rat astroglial and microglial cells. *Eur.J.Neurosci.* 16:1175-1185.

116. Rho, M. B., S. Wesselingh, J. D. Glass, J. C. McArthur, S. Choi, J. Griffin, and W. R. Tyor. 1995. A potential role for interferon-alpha in the pathogenesis of HIV- associated dementia. *Brain Behav.Immun.* 9:366-377.

117. Benveniste, E. N. 1997. Cytokines: Influence on glial cell gene expression and function. In *Neuroimmunoendocrinology*. J. E. Blalock, ed. S Karger, Basel, pp. 31-75.

# Chapter A7

# ROLE OF MICROGLIA AND MACROPHAGES IN EAE

Gennadij Raivich and Richard Banati
*Perinatal Brain Repair Centre, Dept Obstetrics and Gynaecology, Dept Anatomy, University College London, UK and Molecular Neuropsychiatry Laboratory, Department of Neuropathology, Charing Cross Hospital, Imperial College School of Medicine, London, UK*

**Abstract:** Microglia and macrophages are related cell types that play an important role in the pathogenesis of MS and EAE. This chapters reviews the role of these cells in the normal brain and their contribution to inflammatory demyelinating disease, including their role in antigen presentation, co-stimulation, and production of cytokines and other inflammatory mediators

**Key words:** experimental autoimmune encephalomyelitis, macrophages, microglia, inflammation, antigen presentation, perivascular, parenchymal, meningeal

The microglia, a brain-resident, non-neuronal cell and the blood-derived or haematogenous macrophage represent two related cell types involved in key events in the development of pathology in Multiple Sclerosis and its autoimmune animal model, the Experimental Allergic Encephalomyelitis. Microglia and macrophages fulfil a variety of different functions, but they are also recognized for their ability to act as a particularly fine sensor of brain pathology. Both cell types are rapidly activated and recruited to sites of infection, neurodegeneration, stroke, autoimmune inflammatory models such as EAE and its presumptive human counterpart, multiple sclerosis. Microglia and macrophages are stimulated by a variety of cytokines, neurotransmitters, modulators and putative neurotoxins, extracellular matrix molecules and proteases present in the inflamed central nervous system. Moreover, both cell types are plastic in their morphology and cellular identity. The presence of dying cells and cell debris will cause a transformation of phagocytic microglia into a detached, rounded and migratory or amoeba-like (amoeboid) macrophage (Streit et al., 1988; Bohatschek et al., 2001). This also works in reverse: surrounded by a CNS environment, non-phagocytic macrophages freshly recruited from the blood stream will gradually develop branched processes and transform into ramified microglia (Flugel et al., 2001, Bohatschek et al., 2001).

Activated microglia and macrophages synthesise a cornucopia of different cytokines, trophic factors, ECM components and neurotransmitter-like molecules that could exert a positive or damaging effect on the adjacent cells. They also interact with other cells of the immune system, particularly T-cells, which are recruited to the sites of CNS inflammation. Both in vitro and in vivo evidence also suggests that they may act as competent presenters of antigen, inducing and regulating the intensity of T-cell mediated inflammation and tissue injury. The aim of the current chapter will be to provide an overview on the different, related types of microglia and macrophages in the normal brain, describe their cellular and molecular response in EAE and multiple sclerosis and finally focus on their direct contribution to neuropathology in autoimmune demyelinating disease.

## 1.    MICROGLIA AND MACROPHAGES IN THE NORMAL BRAIN

The normal central nervous system consists of several different non-neuronal cell populations that are related to monocytes and macrophages in the bone marrow and peripheral tissues, based on the presence of specific cellular differentiation markers such as the aMb2 integrin (CD11b/CD18), IgG receptors (CD16/CD32), IBA1 and so on. The brain microglia comprise the largest component, located inside the neural parenchyma. In the normal resting state, they are highly ramified cells, with extensive branches that can cover spaces of 30-50 μm in diameter. These resting microglia are territorial, in that their cell bodies or branches are rarely seen to adhere to one another, unlike white matter oligodendrocytes contacting one another like pearls on a string (Suzuki and Raisman, 1992) or protoplasmic astrocytes with extensive cell process to process contacts at astrocyte boundaries which allow the spread of intracellular ions and other small molecules from one astrocyte to the next (Nedergaard, 1994; Bushong, 2002).

The perivascular macrophages are located in between the blood vessel endothelia, occasional perithelial cells and the basal membrane that separates the blood vessel from the surrounding neural parenchyma, the Robin-Virchow space. Perivascular macrophages are typically slender elongated cells (elongated in the direction of the blood vessel axis) with broad but short processes that sometimes go around the blood vessel. Most perivascular macrophages are located around small to moderate blood vessels inside the central nervous system. Unlike the microglia, they do not show the elaborate ramified structure typical of resting microglia, which could be due to spatial constraints, but also to a molecular and cellular micro-environment, different from that of neural parenchyma. These differences also extend to molecular markers, such as MHC2, cyclo-oxygenase or scavenger receptors, found on normal perivascular macrophages but not in resting microglia (Linnehan et al., 1999), and the paucity of the aMb2 integrin (Angelov et al., 1992).

The meningeal macrophages, a third group, are large and rounded cells located between meningeal epithelial cells and the basal membranes surrounding glia limitans, the astroglial lining encasing neural parenchyma. Immunohistochemically, meningeal macrophages are more closely related to perivascular macrophages, macrophages in chorio-epithelial and ventricular epithelial tissue and less to the highly ramified, resting microglia inside neural parenchyma. On the whole, the basal membranes of vessels meninges, ventricular and chorioepithelium, mark an anatomical border between two brain macrophage subpopulations: the ramified microglia inside the neural parenchyma that lacks intrinsic basal membranes, and perivascular, meningeal, ventricular or chorioepthelial cells sitting on the external side of the surrounding basal membranes.

Some publications also use the term "perivascular microglia", but the term is confusing, controversially defined, and frequently misleading. It is often used it as a synonym for rounded or process-poor perivascular macrophages (Linnehan et al., 1999; Stoll and Jander, 1999), even though these cells look very different from the ramified microglia. In others, it is used to denote microglia in the neural parenchyma with processes that contact blood vessels from the inside (Owens et al., 1998). Since each microglial cell covers a relatively large territory of well vascularized tissue, up to 50-70 μm in diameter, some contact is probably unavoidable, and labelling a microglial cell "perivascular", like that shown in figure 1C can simply reflect a particularly prominent process attached to a vessel co-stained by the same molecular marker.

Despite this clear anatomical partition, between microglia and macrophages, there is clearly at least some exchange between and plasticity in the individual compartments. Perivascular macrophages are gradually replenished by a pool of circulating monocytes, with a half-life of 1-2 months. A small population of macrophages migrate through basal membrane into neural parenchyma, to differentiate into ramified microglia, a process enhanced in different forms of neuropathology (Streit et al., 1989; Priller et al., 2001), including EAE (Fluegel et al., 2001). Resident microglia in the adult brain are themselves descendents of 2 waves of macrophage infiltration into neural parenchyma – a very early one, from the surrounding mesodermal tissue (Navascues et al., 1995; Cossmann et al., 1997; Kurz and Christ, 1998), then as a second wave, as "fountains of microglia" in CNS white matter during axonal pruning in the late fetus/newborn (Rio-Hortega cf Brockhaus et al., 1996), but this is followed by rapid differentiation, arborisation and quiescence, turning into the resting, ramified phenotype. However, the same resident microglia can be rapidly activated by a host of different pathologies, lose branching and transform into amoeboid macrophages. Some debris-laden transformed microglia also appear to migrate from neural parenchyma into the perivascular, Robin-Virchow space, where they can stay for a very long time (Kosel et al., 1997), which could set the scene for an interaction with T-cells homing onto perivascular macrophages (Walter et al., 2001).

## 2.        CELLULAR AND MOLECULAR RESPONSE IN EAE AND MS

The rapid recruitment of blood-borne monocytes, the activation of resident microglia and perivascular macrophages, together with the recruitment of T-cells, are among the most consistent changes observed in multiple sclerosis and its autoimmune animal models of experimental allergic encephalomyelitis (McCombe et al., 1994; Bruck et al., 1995; Ford et al., 1995; Li et al., 1996). Microglia display strong proliferative activity, particularly at the early active sites of demyelination (Matsumoto et al., 1992, Schonrock et al., 1998), and avid upregulation in their mitogen receptors, that is tuned down in later stages of the disease (Hulkower et al., 1993; Werner et al., 2002). Compared to neighbouring T-cells, microglia/macrophages show relatively little apoptosis (Nguyen et al., 1994; Smith et al., 1996; Bonetti et al., 1997) and much more proliferative activity (Ogmori et al., 1992). Interestingly, almost all of our knowledge of changes affecting brain microglia/macrophages in the human disease come from the post mortem analysis of the terminal stage, unlike the EAE models which allow to explore pathology at different phases of the disease - from the early preclinical stage, to onset of neurological symptoms, paralysis and remission, including the second and following bouts of the disease process in the relapsing forms of the EAE.

Recent increase in the use of diagnostic brain biopsies (Bruck et al., 1995; Lucchinetti et al., 2000), but particularly the introduction of the positron-emitting [$^{11}$C]-PK11195 in combination with positron emission tomography (PET) scanning has begun to change this situation. In brain tissue, the PK1195 binding site is highly selective for microglia and macrophages, it is rapidly activated in even in moderate forms of brain pathology (Stephenson et al., 1995; Banati et al., 1997) and can be used map the spatial pattern of microglial activation in multiple sclerosis (Banati et al., 1999; Cagnin et al., 2001). Compared with magnetic resonance imaging with or without Gadolinium, the PET-based technique shows a higher sensitivity with respect to identifying white mater regions at risk and provides a good correlation with disease process and appearance of neurological deficit (Banati et al., 2000). These recent human data underscore the importance of brain macrophage activation as a diagnostic tool, to identify the localization and disease activity in multiple sclerosis.

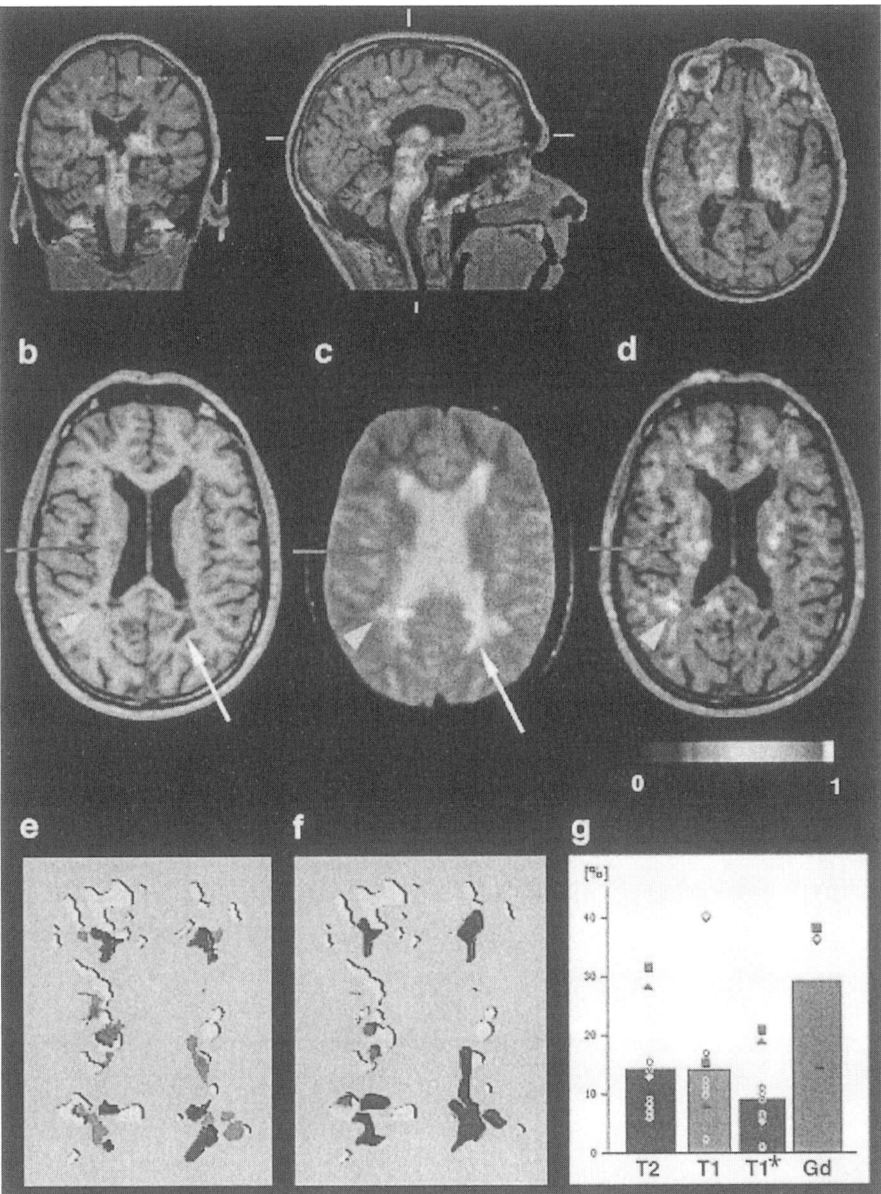

**Figure 1.** Detection of microglial activation in human patients with muliple sclerosis using positron emission tomography (PET) with [11C](R)-PK11195, combined with nuclear magnetic resonance imaging (MRI). All images follow the radiological convention, i.e. the left side of the image corresponds to subject's right side. (A) Three orthogonal views of

[11C](R)-PK11195 images co-registered and overlaid on the MRI of Patient 9, showing spinothalamic tract-associated [11C](R)-PK11195 signals extending through the brainstem and pons into the thalamus. (B-D) T1-weighted (B) and T2-weighted (C) MRI and [11C](R)-PK11195 PET (overlaid onto T1-weighted MRI) (D) of Patient 9 show lesions in all different spin-echo MRI sequences that partially overlap with areas of significantly increased [11C](R)-PK11195 binding (red arrow). The white arrow points to a `black hole' in an area that appears strongly hypointense in the T1-weighted MRI and has little binding of [11C](R)-PK11195. Note, however, that a similar black hole (yellow arrowhead) adjacent to the right occipital horn of the lateral ventricle shows significant [11C](R)-PK11195 binding. (E-F) Demonstration of the definition of the MRI lesion load masks in Patient 9 (purple, T1-weighted MRI lesions excluding black holes; blue, black hole only; green, gadolinium-enhancing areas; dark grey (in F), T2-weighted MRI lesions; red, areas of overlap between significantly increased [11C](R)-PK11195 binding and MRI-defined areas of pathology); yellow, areas of increased [11C](R)-PK11195 binding and no overlap with any MRI-defined pathology. (G) Average percentage volume of the MRI-defined lesions overlapping with increased [11C](R)-PK11195 binding. The red square represents Patient 8 and the red triangle Patient 6, who were both in relapse at the time of the scans. The yellow diamond represents Patient 9, who had secondary progressive multiple sclerosis. T1*, black holes. Reproduced from Banati et al., Brain. 123:2321-37, 2000.


On the biochemical level, macrophage and microglial activity in MS and EAE is associated with a strong upregulation of molecules involved in antigen presentation, myelin and tissue breakdown, production of reactive oxygen substances. They also synthesize components of the complement cascade, cytokines, growth factors and neurotrophins, chemotactic molecules, excitotoxins and apoptosis-inducing substances, and their receptors. These molecules, reviewed in the following paragraphs, as well as in more detail in the preceding and following chapters of this book, are involved in inducing and regulating the level of macrophage activation, interaction with encephalitogenic lymphocytes, mediating damage to myelin, axons and oligodendrocytes, as well as inducing the repair of the injured white matter.


## ANTIGEN PRESENTATION

***Major Histocompatibility Complex.*** T cells are known to recognize their specific antigen when associated to the class I or class II molecules of the major histocompatibility complex, abbreviated as MHC1 and MHC2 (for a review see Zinkernagel and Doherty, 1997). This recognition is aided by the binding of T-cell accessory molecules CD4 and CD8, expressed by the T-helper (mainly CD4+) and the T-suppressor/cytotoxic (mainly CD8+) lymphocytes, to their respective MHC2 or MHC1 ligands (Fleury et al., 1991; Miceli and Parnes, 1991). Despite the strong MHC class-selectiveness in the presentation of specific antigen – MHC1 for the endogenous (cytoplasmic) and MHC2 for the exogenous (phagocytosed) antigen, recent studies point to the existence of alternative and highly effective pathways for the presentation of exogenous antigen via MHC1 (Reimann et al., 1994;

Larsson et al., 2001). In addition, non-classical MHC1-like molecules such as CD1 can also present antigen, particularly glycolipids, to the CD8+ as well as to the CD4-CD8- (double null) T-lymphocytes (Sugita et al., 1998).

The rapid upregulation in the major histocompatibility complex antigens were amongst the first set of molecular changes described in multiple sclerosis as well as in different forms of experimental allergic encephalomyelitis using immunohistochemical techniques. Here, the upregulation of MHC2 was clearly restricted to reactive microglia, macrophages and blood-borne leukocytes, neighbouring GFAP+ astrocytes were MHC2 negative (Konno et al., 1989; Boyle and McGeer, 1990). Unlike MHC2, CD1 expression was not found on parenchymal macrophages or microglia, but rather on perivascular leukocytes and particularly, on the GMCSF+, hypertrophic astrocytes surrounding MS plaques (Battistini et al., 1996). Interestingly, the absence of overlap between MHC2 and CD1b suggests a high level of cell type selectivity in the presentation of MHC2 and CD1b-dependent antigens (Cipriani et al., 2003).

Up to now, most reports on antigen presentation and EAE have concentrated on MHC2, leading to the common assumption that only CD4+ lymphocytes are encephalitogenic. However, studies using beta2-microglobulin-deficient mice do show that MHC1 is involved in mediating EAE elicited by adoptively transferred, encephalitogenic CD8+ T-cells (Sun et al., 2001). This effect is direct and not mediated by some indirectly stimulated host CD4+ lymphocytes, a point demonstrated using the RAG1-/, immunodeficient mice (Sun et al., 2001). In the majority of cases, antigen presentation by CD1+ cells appears to inhibit EAE (Lider et al., 2001; Miyamoto et al., 2001), although this depends on the cytokine requirements of the specific form of EAE (Jahng et al., 2001). In the case of MHC2, most initial reports tended to emphasize that microglial expression of MHC2 may present antigen to encephalitogenic T-cells, needed to initiate or promote the inflammatory and demyelinating process (Hayes et al., 1987; McGeer et al., 1988). More recent studies have focused on their counter-regulatory and immunosuppressive properties. EAE-inducing effects are now attributed to perivascular macrophages.

Studies by Hickey and Kimura using bone marrow chimaeras clearly show that MHC2 expression by perivascular macrophages is sufficient for antigen presentation and onset of severe EAE, following the transfer of encephalitogenic CD4+ T-cells (Hickey and Kimura, 1988). On the other hand, high level of microglial MHC2 corresponds with resistance to EAE in different strains of rats (Sedgwick et al., 1993; Klyushnenkova et al., 1997). Exposure to microglial MHC2 also appears to induce T cell apoptosis, unlike the perivascular macrophages that promote T-cell survival (Ford et al., 1996; Klyushenkova et al., 1997). Presentation of antigen to non-encephalitogenic T-cells appears to play an important part in preventing excessive autoimmunity. Thus, adoptive transfer of encephalitogenic T-cells in mice with severe combined autoimmunity (scid) leads to a much more severe and recurrent form of EAE compared with immunocompetent mice (Jones et al., 1999). The particularly strong and profuse upregulation of MHC2 on parenchymal microglia during early remission (Konno et al., 1989; McCombe et al., 1992) thus appears to make an important contribution in strengthening the immunosuppressive action of microglial MHC2.


***Accessory Molecules.*** In addition to processed antigen embedded in MHC, effective antigen presentation requires the presence of co-stimulatory or accessory molecules on the surface of the antigen presenting cell, engaging their receptor counterparts on the T-lymphocyte. These accessory molecules belong to several different families of cell surface glycoproteins including B7, CD40, ICAM1-3 and the aXb2 integrin. aXb2 is a cell type-specific marker of professional, antigen-presenting cells (APC) also known as dendritic cells (Brocker et al., 1997; Suter et al., 2000).

MS is associated with a strong upregulation of CD40 (Gerritse et al., 1996; Laman et al., 1998; Weinberg et al., 1999) and B7.1 on the perivascular macrophages and microglia (Williams et al., 1994; De Simone et al., 1995). Inhibition or neutralization of CD40 and B7.1 prevented induction of EAE (Kuchroo et al., 1995; Gerritse et al., 1995; Weinberg et al., 1999; Becher et al., 2001). In the case of B7.1 this effect depended on the presence of IL4 (Kuchroo et al., 1995). Resting microglia already express moderate levels of B7.2 (Dangond et al., 1997), and its inhibition may enhance the severity of EAE (Kuchroo et al., 1995). Interestingly, combined inactivation of B7.1 and B7.2 strongly reduced the pathology and severity of neurological symptoms in adoptively transferred EAE (Chang et al., 1999). Both systems, CD40 and B7, appear to complement each other. Combined inactivation of the B7 receptor CD28, and inhibition of CD40 leads to a particularly strong resistance to the induction of EAE (Grivin et al., 2002).

Activated and phagocytic brain-resident microglia do express co-stimulatory molecules such as ICAM1-3, aXb2 or B7.2 in a variety of pathological conditions (Bo et al., 1996; Werner et al., 1998, Bohatschek et al., 1999; Kloss et al., 1999). However, most studies concur that the majority of the accessory molecule-positive cells in MS and EAE that present antigen to T-lymphocytes are hematogenous in origin and concentrated in the perivascular infiltrates (Williams et al., 1994; De Simone et al., 1995; Gerritse et al., 1996; Laman et al., 1998; Weinberg et al., 1999). Adoptive transfer of encephalitogenic T-cells strongly enhances the influx of bone marrow-precursors of dendritic cells to the site of CNS inflammation. Moreover, these newly recruited dendritic cells, with the appropriate MHC molecules, are fully sufficient to induce inflammation and myelin destruction in mice following adoptive transfer of rat bone marrow and rat encephalitogenic T-cells (Subramanian et al., 2001). This point is also underscored by recent study using bone marrow chimaeras between the CD40 wild type (CD40+/+) and CD40 null (CD40-/-) animals, with the latter normally resistant to EAE (Becher et al., 2001). Replacement of the CD40+/+ bone marrow with that from a CD40-/- animal, turning perivascular macrophages and newly recruited dendritic cells to CD40-/-, while retaining CD40+/+ microglia, prevents the appearance of EAE in almost all animals. On the other hand, the transfer of CD40+/+ bone marrow to the irradiated, CD40-/- host, causes an almost complete recovery of EAE susceptibility in the host, with just a minor delay (+20%) and reduction in the maximal severity of clinical symptoms (-25%), compared to normal, CD40+/+ animals receiving a CD40+/+ bone marrow transplant (Becher et al., 2001).

## CYTOKINES

Perivascular cuff macrophages, parenchymal macrophages and microglia show a strong upregulation for a long list of inflammation-associated, soluble cytokines, including interleukin-1/IL1 (Bauer et al., 1993), interleukin-10/IL10 (Jander et al., 1998; Hulshoff et al., 2002), transforming growth factor beta-1 (Kiefer et al., 1998; De Groot et al., 1999), macrophage-colony stimulating factor/MCSF (Hulkower et al., 1993; Werner et al., 2002), granulocyte-macrophage colony-stimulating factor (GMCSF), interleukin-12 (IL12) and tumour necrosis factor-alpha/TNFa (Hulkower et al., 1993; Renno et al., 1995; Bitsch et al., 1998; Laman et al., 1998; Fischer and Reichmann, 2001). There is also an upregulation of related cell surface molecules such as FAS and FAS-ligand (FasL), which are members of the TNF superfamily (Ouallet et al., 1999). Interestingly, expression is frequently focused to specific and different subpopulations of macrophages and microglia (Bitsch et al., 2000; Juedes et al., 2000). For example, the aMb2-positive microglia/macrophages expressing the dendritic cell marker aXb2 integrin secrete high amounts of IL12, while those negative for aXb2 produce GMCSF and TNFa (Fischer and Reichmann, 2001).

A subpopulation of these macrophage-produced cytokines and related molecules, such as IL10, TGFb1, or FasL has been shown to inhibit or prevent EAE (Rott et al., 19994; Stevens et al., 1994; Zhu et al., 2002; see also Wyss-Coray et al., 1997). However, a majority have a strong disease-promoting activity (Waldburger et al., 1996; Taupin et al., 1997; Marusic et l., 2002). Neutralization of TNF-alpha (Selmaj et al., 1991; Korner et al., 1997), IL1 (Jacobs et al., 1991), IL12 (Leonard et al., 1995) with antibodies or soluble receptors suppresses EAE. Genetic deletion of IL1R1, GMCSF, TNFa or IL12p40 has a similar effect, conferring resistance to EAE (Schiffenbauer et al., 2000; McQualter et al., 2001; Matejuk et al., 2002; Murphy et al., 2002; Gran et al., 2002; see however Liu et al., 1998).

Moreover, studies using bone marrow chimaeras between TNFa+/+ and TNFa-/- mice show that it is the TNFa which is produced by blood-borne leukocytes that plays a decisive role in the onset and severity of EAE (Murphy et al., 2002).

Importantly, moderate overexpression of TNFa or IL12, using the astrocyte GFAP promoter (Stalder et al., 1998; Pagenstecher et al., 2000) has been shown to lead to CNS inflammation and neurological disease. In the case of TNF, it causes overt demyelination and axonal damage very similar to that observed in EAE and MS. The fact that this demyelination is also observed in TNFa-overexpressing scid mice lacking T and B-cells (Stalder et al., 1998) strongly suggests that these cytokines not only promote the initial immune response, but also appear to play a crucial in the final steps following antigen recognition, that actually cause the brain pathology and neurological dysfunction.


CHEMOKINES

The activated macrophages and microglia produce a variety of chemotactic molecules including members of the chemokine family, but also many other chemoattractant factors such as secretoneurin (Storch et al., 1996), leukocyte chemotactic factor/LCF (Schluesener et al., 1996), endothelial-monocyte-activating polypeptide II (Schluesener et al., 1997) and chemotactic peptide-10/CP10 (Deininger et al., 1999). Amongst the chemokines, there is a strong increase in the macrophage chemotactic protein/MCP1 (Hulkower et al., 1993; Simpson et al., 1998, Jee et al., 2001), monocyte inflammatory protein 1 alpha/MIP1a (Balashov et al., 1999), neurotactin or fractalkine/CX3CL1 (Pan et al., 1997), TCA3 (Murphy et al., 2002), CCL19 and the macrophage-derived chemokine/CCL22 (Columba-Cabezas et al., 2002, 2003), which act as chemotactic ligands for macrophages, T-lymphocytes or both. There is also an upregulation for a list of macrophage chemokine receptors, such as CCR1, the receptor for MIP1a (Rottman et al., 2000; Trebst et al., 2001), CCR2, CCR3 and CCR5 (Simpson et al., 2000), CCR8 (Trebst et al., 2003), CXCR4 and CX3CR1 (Jiang et al., 1998), and the receptors for IL8 and N-formyl-Met-Leu-Phe/FMLP, particularly on foamy, phagocytic macrophages (Muller-Ladner et al., 1996).

Many of these macrophage-derived chemokines are produced by hematogenous macrophages, invading the inflamed CNS (Miyagishi et al., 1997; Sorensen et al., 1999; Matejuk et al., 2002) and are controlled by inflammation–associated cytokines such as TNF (Matejuk et al., 2002; Murphy et al., 2002). Studies using genetically deficient animals and/or neutralizing antibodies also show that the chemoattractive molecules play an important role in the pathology of EAE. Transgenic deletion of CCR1, the receptor for MIP1a, CCR2, the receptor for MCP1, or CCR8, the receptor for TCA3, strongly reduce the susceptibility, onset and severity of EAE (Rottman et al., 2000; Huang et al., 2001; Murphy et al., 2002). A similar inhibitory effect is also observed with the antibody inactivation of MCP1 (Karpus et al., 1995). That these molecules are produced in the brain has suggested that brain-resident cells, particularly microglia, may also play a role in the induction of EAE (Murphy et al., 2002). However, this point is contentious and needs to be confirmed, using bone marrow chimaeras. The fact that many of these chemokines and chemokine receptors are located on blood-borne macrophages, particularly in perivascular cuffs, could argue against a major contribution by microglia.

## REACTIVE OXYGEN SPECIES AND SIGNALLING ENZYMES

Acute inflammatory diseases such as EAE and multiple sclerosis are associated with strongly augmented production of reactive oxygen species (ROS), particularly by activated brain macrophages (Ruuls et al., 1995). Brain macrophages show increased deposition of iron (LeVine, 1997), myeloperoxidase (Nagra et al., 1997) and inducible NO synthase/iNOS in multiple sclerosis (Bagasra et al., 1995; De Groot et al., 1997) and in EAE (Van Dam et al., 1995; Tran et al., 1997). Similar upregulation is also observed in viral models of CNS demyelination, e.g. with Theiler's murine encephalomyelitis virus (Oleszak et al., 1997). In combination with peroxide radicals the synthesis of NO will lead to the formation of peroxynitrite, which is toxic for oligodendrocytes (Mitrosic et al., 1996). In vitro, oxidative stress causes macrophages to become toxic (Bartnik et al., 2000), and scavengers of ROS or their precursors such as uric acid or catalase are known to inhibit EAE (Ruuls et al., 1995; Kean et al., 2000).

Nonetheless, brain macrophages also produce a string of molecules that reduce oxidative stress. Perivascular macrophages, and to lesser extent microglia, show high levels manganese superoxide dismutase in EAE (Qi et al., 1997). Expression of metallothioneins 1&2/MT1&2 (Espejo et al., 2001) reduces the high susceptibility to EAE, shown in the MT1&2-/- mice (Penkowa et al., 2001). Multiple sclerosis and EAE also cause increased macrophage synthesis of stress protein heme oxygenase-1 (HO1) that produces CO (Emerson and LeVine, 2000; Schluesener and Seid, 2000), the inducer of cGMP-synthesizing enzyme guanyl cyclase (Brune and Ullrich, 1987). This increased HO1 degrades the pro-oxidant heme groups, but also reduces the availability of the NADPH cytochrome P450 reductase that is needed for the production of superoxide (Emerson and LeVine, 2000). Inducers of HO1 such as hemin, reduce the severity, and HO1 inhibitors enhance the severity of EAE (Liu et al., 2001).

Macrophage iNOS and NO synthesis and their effects in EAE have been a focus or particular attention. In most cases, early pharmacological inhibition of nitric oxide synthase has been shown to reduce EAE (Cross et al., 1994; Brenner et al., see however Ruuls et al., 1996). Late application interferes with the recovery process and enhances relapsing activity (Okuda et al., 1998; O'Brien et al., 2001). Interestingly, NO production by brain microglia and macrophages strongly inhibits T-cell proliferation (Juedes and Ruddle, 2001).

The importance of this point was illustrated by studies in the interferon gamma receptor deficient (IFNgR-/-) mice, that are normally unable to recover following the induction of EAE, and die from severe demyelinating illness (Willenborg et al., 1999). In vitro analysis showed that supernatants from IFNgR+/+ macrophages inhibit the proliferation of encephalitogenic T-cells while IFNgR-/- macrophages lack this ability. Moreover, the inhibitory effects of IFNgR+/+ macrophages could be suppressed by an inhibitor of iNOS, underscoring the importance of NO in regulating the T-cell response. Interestingly, bone marrow chimaeras between IFNgR+/+ and IFNgR-/- show that the presence of IFNg receptors on just one, blood-borne or brain resident component, was sufficient to prevent the normally lethal outcome following the induction of EAE in IFNgR-/- mice (Willenborg et al., 1999).


FUNCTIONAL ROLE

Both brain–derived microglia and blood borne macrophages are crucially involved in many consecutive stages of autoimmune demyelination in experimental allergic encephalomyelitis and multiple sclerosis. They play an important role as antigen-presenting cells in the initial demonstration of antigen (Hickey and Kimura, 1988; Jones et al., 1999), and secondary recruitment of T-cells, granulocytes and macrophages (Huitinga et al., 1995). They also produce a long list of potentially damaging substances, including reactive oxygen species, NO and peroxynitrite, TNFa, interleukin-1b and excitotoxins (see above). As a case in point, interleukin 1b has been shown to activate mixed glial cell cultures to produce glutamate agonist neurotoxins that cause oligodendroglial cell death (Takahashi et al., 2003). MS-associated inflammation leads to a strong increase in glutamate receptors on axons in the centre of CNS lesions and on neighboring, reactive astrocytes (Geurts et al., 2003), which could predispose them to damage by the glial-derived excitotoxins. Axonal damage is particularly intense at early disease stages of MS (Kuhlmann et al., 2002), which could correspond to particularly intense microglial activation (Bitsch et al., 2002). Microglia and macrophages are also the chief debris-removing cells that eliminate damaged myelin, resulting in the widespread loss of axonal covering the CNS white matter. Last, but not least, microglia and macrophages appear to play a decisive role in the induction of remission, as well as resistance to the induction of the disease (Konno et al., 1989; McCombe et al., 1992; Willenborg et al., 1999).

A key question raised with the respect to microglia and macrophages is the relative contribution of blood-derived versus brain resident cell populations to the overall pathology during the process of autoimmune destruction of myelin. This question is particularly appropriate, since activated, and particularly phagocytic microglia share most of molecular markers with the blood-borne macrophages that enter the damaged brain (Streit et al., 1988; Bruck et al., 1995; Raivich et al., 1999). Most studies concur that blood-derived macrophages are crucial in the induction of EAE. Removal of bone marrow monocyte precursors and circulating macrophages with clodronate reduces the parenchymal influx of new macrophages, interferes with lymphocyte recruitment and microglial proliferation (Bauer et al., 1995; Plofriet et al., 2002). The same macrophage depleting treatment also blocks adoptively transferred EAE (Huitinga et al., 1995; Tran et al., 1998). Similar, disease-abolishing results were also obtained in bone marrow chimaeras, when bone marrow-derived macrophages (but not microglia) lacked the appropriate MHC antigens, accessory molecules (CD40) or cytokines such as TNFa that promote demyelinating disease (Hickey and Kimura, 1988; Stalder et al., 1998; Subramanian et al., 2001; Becher et al., 2001; Murphy et al., 2002), underscoring the significance of blood derived macrophages.

Nonetheless, there are several lines of evidence that begin to shed light on the importance of brain resident microglia at different stages of the demyelinating disease. The initiation of the autoimmune response is an important case in point. For example, the acute transfer of rat encephalitogenic T-lymphocytes and appropriate antigen presenting cells (APC) to scid mice causes a delay phase of approximately 8 days before the onset of EAE (Subramanian et al., 2001), pointing to the importance of local APC in the early initial stages. This point is underscored by the very rapid microglial response to the adoptive transfer of EAE, in this case the induction of microglial amyloid precursor protein, within 24 hours after the infusion of encephalitogenic T-cells in animals with the appropriate MHC antigens (Banati et al., 1995). Studies using bone marrow chimaeras also show that microglia are much more effective removers of myelin than blood-derived macrophages (Rinner et al., 1995). Finally, brain-resident microglia also play a decisive role in limiting the extent of demyelination, preventing lethal outcome and inducing disease remission (Konno et al., 1989; McCombe et al., 1992; Willenborg et al., 1999). Here, new insights into the function and molecular signals of macrophages and microglia, their interaction with each other, as well as with T-lymphocytes, axons and myelin-producing oligodendroglia, could pave the way to introducing new and more effective therapies to the human demyelinating disease.

# REFERENCES


Angelov DN, Neiss WF, Streppel M, Walther M, Guntinas-Lichius O, Stennert E (1996) ED2-positive perivascular cells act as neuronophages during delayed neuronal loss in the facial nucleus of the rat. Glia 16:129-39

Bagasra O, Michaels FH, Zheng YM, Bobroski LE, Spitsin SV, Fu ZF, Tawadros R, Koprowski H. 1995 Activation of the inducible form of nitric oxide synthase in the brains of patients with multiple sclerosis. Proc Natl Acad Sci USA 92:12041-5

Balashov KE, Rottman JB, Weiner HL, Hancock WW. 1999 CCR5(+) and CXCR3(+) T cells are increased in multiple sclerosis and their ligands MIP-1alpha and IP-10 are expressed in demyelinating brain lesions. Proc Natl Acad Sci USA. 96:6873-8

Banati RB, Gehrmann J, Lannes-Vieira J, Wekerle H, Kreutzberg GW. (1995) Inflammatory reaction in experimental autoimmune encephalomyelitis (EAE) is accompanied by a microglial expression of the beta A4-amyloid precursor protein (APP). Glia 14:209-15

Banati RB, Myers R, Kreutzberg GW (1997) PK ('peripheral benzodiazepine')--binding sites in the CNS indicate early and discrete brain lesions: microautoradiographic detection of [3H]PK11195 binding to activated microglia. J Neurocytol 26:77-82

Banati RB, Goerres GW, Myers R, Gunn RN, Turkheimer FE, Kreutzberg GW, Brooks DJ, Jones T, Duncan JS (1999) [11C](R)-PK11195 positron emission tomography imaging of activated microglia in vivo in Rasmussen's encephalitis. Neurology. 53:2199-203

Banati RB, Newcombe J, Gunn RN, Cagnin A, Turkheimer F, Heppner F, Price G, Wegner F, Giovannoni G, Miller DH, Perkin GD, Smith T, Hewson AK, Bydder G, Kreutzberg GW, Jones T, Cuzner ML, Myers R (2000) The peripheral benzodiazepine binding site in the brain in multiple sclerosis: quantitative in vivo imaging of microglia as a measure of disease activity. Brain 123:2321-37

Bartnik BL, Juurlink BH, Devon RM. 2000 Macrophages: their myelinotrophic or neurotoxic actions depend upon tissue oxidative stress. Mult Scler. Feb;6(1):37-42

Battistini L, Fischer FR, Raine CS, Brosnan CF. 1996 CD1b is expressed in multiple sclerosis lesions. J Neuroimmunol. 67(2):145-51.

Bauer J, Berkenbosch F, Van Dam AM, Dijkstra CD. 1993 Demonstration of interleukin-1 beta in Lewis rat brain during experimental allergic encephalomyelitis by immunocytochemistry at the light and ultrastructural level. J Neuroimmunol 48:13-21

Bauer J, Huitinga I, Zhao W, Lassmann H, Hickey WF, Dijkstra CD (1995) The role of macrophages, perivascular cells, and microglial cells in the pathogenesis of experimental autoimmune encephalomyelitis. Glia 15:437-46

Becher B, Durell BG, Miga AV, Hickey WF, Noelle RJ. 2001 The clinical course of experimental autoimmune encephalomyelitis and inflammation is controlled by the expression of CD40 within the central nervous system. J Exp Med. 193:967-74

Bitsch A, da Costa C, Bunkowski S, Weber F, Rieckmann P, Bruck W 1998 Identification of macrophage populations expressing tumor necrosis factor-alpha mRNA in acute multiple sclerosis. Acta Neuropathol (Berl). 95:373-7

Bitsch A, Kuhlmann T, Da Costa C, Bunkowski S, Polak T, Bruck W (2000) Tumour necrosis factor alpha mRNA expression in early multiple sclerosis lesions: correlation with demyelinating activity and oligodendrocyte pathology. Glia. 29:366-75

Bo L, Peterson JW, Mork S, Hoffman PA, Gallatin WM, Ransohoff RM, Trapp BD (1996) Distribution of immunoglobulin superfamily members ICAM-1, -2, -3, and the beta 2 integrin LFA-1 in multiple sclerosis lesions. J Neuropathol Exp Neurol. 55:1060-72.

Bonetti B, Pohl J, Gao YL, Raine CS 1997 Cell death during autoimmune demyelination: effector but not target cells are eliminated by apoptosis. J Immunol. 159:5733-41

Boyle EA, McGeer PL. Cellular immune response in multiple sclerosis plaques. Am J Pathol 1990 Sep 137:575-84

Brenner T, Brocke S, Szafer F, Sobel RA, Parkinson JF, Perez DH, Steinman L. 1997 Inhibition of nitric oxide synthase for treatment of experimental autoimmune encephalomyelitis. J Immunol. 158:2940-6

Brocker T, Riedinger M, Karjalainen K (1997) Targeted expression of major histocompatibility complex (MHC) class II molecules demonstrates that dendritic cells can induce negative but not positive selection of thymocytes in vivo. J Exp.Med. 185 541-550.

Brockhaus J, Moller T, Kettenmann H (1996) Phagocytozing ameboid microglial cells studied in a mouse corpus callosum slice preparation. Glia 16:81-90

Bruck W, Porada P, Poser S, Rieckmann P, Hanefeld F, Kretzschmar HA, Lassmann H. 1995 Monocyte/macrophage differentiation in early multiple sclerosis lesions. Ann Neurol. 38:788-96

Bruck W, Porada P, Poser S, Rieckmann P, Hanefeld F, Kretzschmar HA, Lassmann H. 1995 Monocyte/macrophage differentiation in early multiple sclerosis lesions. Ann Neurol. 38(5):788-96

Brune B, Ullrich V (1987) Inhibition of platelet aggregation by carbon monoxide is mediated by activation of guanylate cyclase. Mol Pharmacol. 32:497-504

Bushong EA, Martone ME, Jones YZ, Ellisman MH. 2002 Protoplasmic astrocytes in CA1 stratum radiatum occupy separate anatomical domains. J Neurosci 22:183-92

Cagnin A, Brooks DJ, Kennedy AM, Gunn RN, Myers R, Turkheimer FE, Jones T, Banati RB (2001) In-vivo measurement of activated microglia in dementia. Lancet. 358:461-7

Chang TT, Jabs C, Sobel RA, Kuchroo VK, Sharpe AH. 1999 Studies in B7-deficient mice reveal a critical role for B7 costimulation in both induction and effector phases of experimental autoimmune encephalomyelitis. J Exp Med. 190:733-40

Cipriani B, Chen L, Hiromatsu K, Knowles H, Raine CS, Battistini L, Porcelli SA, Brosnan CF. 2003 Upregulation of group 1 CD1 antigen presenting molecules in guinea pigs with experimental autoimmune encephalomyelitis: an immunohistochemical study. Brain Pathol. Jan;13(1):1-9.

Columba-Cabezas S, Serafini B, Ambrosini E, Aloisi F. 2003 Lymphoid chemokines CCL19 and CCL21 are expressed in the central nervous system during experimental autoimmune encephalomyelitis: implications for the maintenance of chronic neuroinflammation. Brain Pathol 13:38-51

Columba-Cabezas S, Serafini B, Ambrosini E, Sanchez M, Penna G, Adorini L, Aloisi F. 2002 Induction of macrophage-derived chemokine/CCL22 expression in experimental autoimmune encephalomyelitis and cultured microglia: implications for disease regulation. J Neuroimmunol. Sep;130(1-2):10-21.

Cossmann PH, Eggli PS, Christ B, Kurz H (1997) Mesoderm-derived cells proliferate in the embryonic central nervous system: confocal microscopy and three-dimensional visualization. Histochem Cell Biol. 107:205-13.

Cross AH, Misko TP, Lin RF, Hickey WF, Trotter JL, Tilton RG. 1994 Aminoguanidine, an inhibitor of inducible nitric oxide synthase, ameliorates experimental autoimmune encephalomyelitis in SJL mice. J Clin Invest. 93:2684-90.

Dangond F, Windhagen A, Groves CJ, Hafler DA 1997 Constitutive expression of costimulatory molecules by human microglia and its relevance to CNS autoimmunity. J Neuroimmunol. Jun;76(1-2):132-8

De Groot CJ, Ruuls SR, Theeuwes JW, Dijkstra CD, Van der Valk P 1997 Immunocytochemical characterization of the expression of inducible and constitutive isoforms of nitric oxide synthase in demyelinating multiple sclerosis lesions. J Neuropathol Exp Neurol. 56:10-20

De Simone R, Giampaolo A, Giometto B, Gallo P, Levi G, Peschle C, Aloisi F. The costimulatory molecule B7 is expressed on human microglia in culture and in multiple sclerosis acute lesions.J Neuropathol Exp Neurol. 1995 Mar;54(2):175-87

Deininger MH, Zhao Y, Schluesener HJ. 1999 CP-10, a chemotactic peptide, is expressed in lesions of experimental autoimmune encephalomyelitis, neuritis, uveitis and in C6 gliomas. J Neuroimmunol. 93(1-2):156-63

Emerson MR, LeVine SM 2000 Heme oxygenase-1 and NADPH cytochrome P450 reductase expression in experimental allergic encephalomyelitis: an expanded view of the stress response. J Neurochem. 75:2555-62

Espejo C, Carrasco J, Hidalgo J, Penkowa M, Garcia A, Saez-Torres I, Martinez-Caceres EM. 2001 Differential expression of metallothioneins in the CNS of mice with experimental autoimmune encephalomyelitis. Neuroscience. 105:1055-65

Fischer HG, Reichmann G. 2001 Brain dendritic cells and macrophages/microglia in central nervous system inflammation. J Immunol. 166:2717-26

Fleury SG, Croteau G, Sekaly RP (1991) CD4 and CD8 recognition of class II and class I molecules of the major histocompatibility complex. Semin Immunol 3:177-85

Flugel A, Bradl M, Kreutzberg GW, Graeber MB. 2001 Transformation of donor-derived bone marrow precursors into host microglia during autoimmune CNS inflammation and during the retrograde response to axotomy. J Neurosci Res. Oct 1;66(1):74-82.

Ford AL, Foulcher E, Lemckert FA, Sedgwick JD 1996 Microglia induce CD4 T lymphocyte final effector function and death. J Exp Med. 184(5):1737-45.

Ford AL, Goodsall AL, Hickey WF, Sedgwick JD. 1995 Normal adult ramified microglia separated from other central nervous system macrophages by flow cytometric sorting. Phenotypic differences defined and direct ex vivo antigen presentation to myelin basic protein-reactive CD4+ T cells compared. J Immunol May 1 154(9):4309-21

Gerritse K, Laman JD, Noelle RJ, Aruffo A, Ledbetter JA, Boersma WJ, Claassen E 1996 CD40-CD40 ligand interactions in experimental allergic encephalomyelitis and multiple sclerosis. Proc Natl Acad Sci U S A. Mar 19;93(6):2499-504

Geurts JJ, Wolswijk G, Bo L, Van Der Valk P, Polman CH, Troost D, Aronica E. 2003 Altered expression patterns of group I and II metabotropic glutamate receptors in multiple sclerosis. Brain. Jun 4 [Epub ahead of print].

Girvin AM, Dal Canto MC, Miller SD (2002) CD40/CD40L interaction is essential for the induction of EAE in the absence of CD28-mediated co-stimulation. J Autoimmun. 18(2):83-94

Gran B, Zhang GX, Yu S, Li J, Chen XH, Ventura ES, Kamoun M, Rostami A. 2002 IL-12p35-deficient mice are susceptible to experimental autoimmune encephalomyelitis: evidence for redundancy in the IL-12 system in the induction of central nervous system autoimmune demyelination. J Immunol. 169(12):7104-10

Hayes GM, Woodroofe MN, Cuzner ML 1987 Microglia are the major cell type expressing MHC class II in human white matter. J Neurol Sci. 80:25-37

Hickey WF, Kimura H. 1988 Perivascular microglial cells of the CNS are bone marrow-derived and present antigen in vivo. Science. 239:290-2

Huang DR, Wang J, Kivisakk P, Rollins BJ, Ransohoff RM (2001) Absence of monocyte chemoattractant protein 1 in mice leads to decreased local macrophage recruitment and antigen-specific T helper cell type 1 immune response in experimental autoimmune encephalomyelitis. J Exp Med. 193:713-26

Huitinga I, Ruuls SR, Jung S, Van Rooijen N, Hartung HP, Dijkstra CD. 1995 Macrophages in T cell line-mediated, demyelinating, and chronic relapsing experimental autoimmune encephalomyelitis in Lewis rats. Clin Exp Immunol 100:344-51

Hulkower K, Brosnan CF, Aquino DA, Cammer W, Kulshrestha S, Guida MP, Rapoport DA, Berman JW. 1993 Expression of CSF-1, c-fms, and MCP-1 in the central nervous system of rats with experimental allergic encephalomyelitis. J Immunol. 150:2525-33.

Jacobs CA, Baker PE, Roux ER, Picha KS, Toivola B, Waugh S, Kennedy MK. 1991 Experimental autoimmune encephalomyelitis is exacerbated by IL-1 alpha and suppressed by soluble IL-1 receptor. J Immunol. 146:2983-9.

Jahng AW, Maricic I, Pedersen B, Burdin N, Naidenko O, Kronenberg M, Koezuka Y, Kumar V. 2001 Activation of natural killer T cells potentiates or prevents experimental autoimmune encephalomyelitis. J Exp Med. 194:1789-99

Jander S, Pohl J, D'Urso D, Gillen C, Stoll G. 1998 Time course and cellular localization of interleukin-10 mRNA and protein expression in autoimmune inflammation of the rat central nervous system. Am J Pathol. 152:975-82

Jee Y, Yoon WK, Okura Y, Tanuma N, Matsumoto Y. 2002 Upregulation of monocyte chemotactic protein-1 and CC chemokine receptor 2 in the central nervous system is closely associated with relapse of autoimmune encephalomyelitis in Lewis rats. J Neuroimmunol. Jul;128(1-2):49-57

Jiang Y, Salafranca MN, Adhikari S, Xia Y, Feng L, Sonntag MK, deFiebre CM, Pennell NA, Streit WJ, Harrison JK. 1998 Chemokine receptor expression in cultured glia and rat experimental allergic encephalomyelitis. J Neuroimmunol. 86:1-12

Jones RE, Mass M, Bourdette DN. 1999 Myelin basic protein-specific T lymphocytes induce chronic relapsing experimental autoimmune encephalomyelitis in lymphocyte-deficient (SCID) mice. J Neuroimmunol. Jan 1;93(1-2):92-101

Juedes AE, Hjelmstrom P, Bergman CM, Neild AL, Ruddle NH. 2000 Kinetics and cellular origin of cytokines in the central nervous system: insight into mechanisms of myelin oligodendrocyte glycoprotein-induced experimental autoimmune encephalomyelitis. J Immunol. 164:419-26

Juedes AE, Ruddle NH. 2001 Resident and infiltrating central nervous system APCs regulate the emergence and resolution of experimental autoimmune encephalomyelitis. J Immunol. Apr 15;166(8):5168-75.

Karpus WJ, Lukacs NW, McRae BL, Strieter RM, Kunkel SL, Miller SD. 1995 An important role for the chemokine macrophage inflammatory protein-1 alpha in the pathogenesis of the T cell-mediated autoimmune disease, experimental autoimmune encephalomyelitis. J Immunol. 155:5003-10

Kean RB, Spitsin SV, Mikheeva T, Scott GS, Hooper DC 2000 The peroxynitrite scavenger uric acid prevents inflammatory cell invasion into the central nervous system in experimental allergic encephalomyelitis through maintenance of blood-central nervous system barrier integrity. J Immunol. 165:6511-8.

Kiefer R, Schweitzer T, Jung S, Toyka KV, Hartung HP. 1998 Sequential expression of transforming growth factor-beta1 by T-cells, macrophages, and microglia in rat spinal cord during autoimmune inflammation. J Neuropathol Exp Neurol 57:385-95.

Klyushnenkova EN, Vanguri P 1997 Ia expression and antigen presentation by glia: strain and cell type-specific differences among rat astrocytes and microglia. J Neuroimmunol. 79(2):190-201

Konno H, Yamamoto T, Iwasaki Y, Saitoh T, Suzuki H, Terunuma H. 1989 Ia-expressing microglial cells in experimental allergic encephalomyelitis in rats. Acta Neuropathol (Berl) 77(5):472-9

Korner H, Lemckert FA, Chaudhri G, Etteldorf S, Sedgwick JD. 1997 Tumor necrosis factor blockade in actively induced experimental autoimmune encephalomyelitis prevents clinical disease despite activated T cell infiltration to the central nervous system. Eur J Immunol. 27:1973-81

Kosel S, Egensperger R, Bise K, Arbogast S, Mehraein P, Graeber MB (1997) Long-lasting perivascular accumulation of major histocompatibility complex class II-positive lipophages in the spinal cord of stroke patients: possible relevance for the immune privilege of the brain. Acta Neuropathol (Berl). 94:532-8

Kuchroo VK, Das MP, Brown JA, Ranger AM, Zamvil SS, Sobel RA, Weiner HL, Nabavi N, Glimcher LH. 1995 B7-1 and B7-2 costimulatory molecules activate differentially the Th1/Th2 developmental pathways: application to autoimmune disease therapy. Cell. 80:707-18

Kuhlmann T, Lingfeld G, Bitsch A, Schuchardt J, Bruck W. 2002 Acute axonal damage in multiple sclerosis is most extensive in early disease stages and decreases over time. Brain. 125:2202-12

Kurz H, Christ B (1998) Embryonic CNS macrophages and microglia do not stem from circulating, but from extravascular precursors. Glia 22:98-102

Laman JD, van Meurs M, Schellekens MM, de Boer M, Melchers B, Massacesi L, Lassmann H, Claassen E, Hart BA. 1998 Expression of accessory molecules and cytokines in acute EAE in marmoset monkeys (Callithrix jacchus). J Neuroimmunol. Jun 1;86(1):30-45

Larsson M, Fonteneau JF, Bhardwaj N (2001) Dendritic cells resurrect antigens from dead cells. Trends Immunol 22:141-8

Leonard JP, Waldburger KE, Goldman SJ. 1995 Prevention of experimental autoimmune encephalomyelitis by antibodies against interleukin 12. J Exp Med. 181:381-6

LeVine SM. 1997 Iron deposits in multiple sclerosis and Alzheimer's disease brains. Brain Res. Jun 20;760(1-2):298-303

Li H, Cuzner ML, Newcombe J 1996 Microglia-derived macrophages in early multiple sclerosis plaques. Neuropathol Appl Neurobiol. 22(3):207-15

Lider O, Miller A, Miron S, Hershkoviz R, Weiner HL, Zhang XM, Heber-Katz E (1991) Nonencephalitogenic CD4-CD8- V alpha 2V beta 8.2+ anti-myelin basic protein rat T lymphocytes inhibit disease induction. J Immunol. 1991 Aug 15;147(4):1208-13.

Linehan SA, Martinez-Pomares L, Stahl PD, Gordon S (1999) Mannose receptor and its putative ligands in normal murine lymphoid and nonlymphoid organs: In situ expression of mannose receptor by selected macrophages, endothelial cells, perivascular microglia, and mesangial cells, but not dendritic cells. J Exp Med 189:1961-72

Liu J, Marino MW, Wong G, Grail D, Dunn A, Bettadapura J, Slavin AJ, Old L, Bernard CC. 1998 TNF is a potent anti-inflammatory cytokine in autoimmune-mediated demyelination. Nat Med. 4:78-83

Liu Y, Zhu B, Luo L, Li P, Paty DW, Cynader MS. 2001 Heme oxygenase-1 plays an important protective role in experimental autoimmune encephalomyelitis. Neuroreport. 12:1841-5

Lucchinetti C, Bruck W, Parisi J, Scheithauer B, Rodriguez M, Lassmann H. 2000 Heterogeneity of multiple sclerosis lesions: implications for the pathogenesis of demyelination. Ann Neurol. 47:707-17

Marusic S, Miyashiro JS, Douhan J 3rd, Konz RF, Xuan D, Pelker JW, Ling V, Leonard JP, Jacobs KA. 2002 Local delivery of granulocyte macrophage colony-stimulating factor by retrovirally transduced antigen-specific T cells leads to severe, chronic experimental autoimmune encephalomyelitis in mice. Neurosci Lett. 332:185-9.

Matejuk A, Dwyer J, Ito A, Bruender Z, Vandenbark AA, Offner H. 2002 Effects of cytokine deficiency on chemokine expression in CNS of mice with EAE. J Neurosci Res. 67(5):680-8

Matsumoto Y, Ohmori K, Fujiwara M. 1992 Microglial and astroglial reactions to inflammatory lesions of experimental autoimmune encephalomyelitis in the rat central nervous system. J Neuroimmunol Mar 37(1-2):23-33

McCombe PA, de Jersey J, Pender MP. 1994 Inflammatory cells, microglia and MHC class II antigen-positive cells in the spinal cord of Lewis rats with acute and chronic relapsing experimental autoimmune encephalomyelitis. J Neuroimmunol 51(2):153-67

McCombe PA, Fordyce BW, de Jersey J, Yoong G, Pender MP. 1992 Expression of CD45RC and Ia antigen in the spinal cord in acute experimental allergic encephalomyelitis: an immunocytochemical and flow cytometric study. J Neurol Sci Dec 113(2):177-86

McGeer PL, Itagaki S, McGeer EG. 1988 Expression of the histocompatibility glycoprotein HLA-DR in neurological disease. Acta Neuropathol (Berl).;76(6):550-7

McQualter JL, Darwiche R, Ewing C, Onuki M, Kay TW, Hamilton JA, Reid HH, Bernard CC. 2001 Granulocyte macrophage colony-stimulating factor: a new putative therapeutic target in multiple sclerosis. J Exp Med. 194:873-82

Miceli MC, Parnes JR (1991) The roles of CD4 and CD8 in T cell activation. Semin Immunol 3:133-41

Mitrovic B, Parkinson J, Merrill JE 1996 An in Vitro Model of Oligodendrocyte Destruction by Nitric Oxide and Its Relevance to Multiple Sclerosis. Methods 10:501-13

Miyagishi R, Kikuchi S, Takayama C, Inoue Y, Tashiro K. 1997 Identification of cell types producing RANTES, MIP-1 alpha and MIP-1 beta in rat experimental autoimmune encephalomyelitis by in situ hybridization. J Neuroimmunol. 77:17-26

Miyamoto K, Miyake S, Yamamura T. 2001 A synthetic glycolipid prevents autoimmune encephalomyelitis by inducing TH2 bias of natural killer T cells. Nature. 413:531-4

Muller-Ladner U, Jones JL, Wetsel RA, Gay S, Raine CS, Barnum SR 1996 Enhanced expression of chemotactic receptors in multiple sclerosis lesions. J Neurol Sci. Dec;144(1-2):135-41

Murphy CA, Hoek RM, Wiekowski MT, Lira SA, Sedgwick JD. 2002. Interactions between hematopoiectically derived TNF and central nervous system-resident glial chemokines underlie initiation of autoimmune inflammation in the brain. J. Immunol 169:7054-62

Nagra RM, Becher B, Tourtellotte WW, Antel JP, Gold D, Paladino T, Smith RA, Nelson JR, Reynolds WF. 1997 Immunohistochemical and genetic evidence of myeloperoxidase involvement in multiple sclerosis. J Neuroimmunol. 78:97-107

Navascues J, Moujahid A, Almendros A, Marin-Teva JL, Cuadros MA (1995) Origin of microglia in the quail retina: central-to-peripheral and vitreal-to-scleral migration of microglial precursors during development. J Comp Neurol. 354:209-28.

Nedergaard M (1994) Direct signaling from astrocytes to neurons in cultures of mammalian brain cells. Science Mar 25;263(5154):1768-71

Nguyen KB, McCombe PA, Pender MP. 1994 Macrophage apoptosis in the central nervous system in experimental autoimmune encephalomyelitis. J Autoimmun 7(2):145-52

O'Brien NC, Charlton B, Cowden WB, Willenborg DO. 2001 Inhibition of nitric oxide synthase initiates relapsing remitting experimental autoimmune encephalomyelitis in rats, yet nitric oxide appears to be essential for clinical expression of disease. J Immunol. 167:5904-12

Ohmori K, Hong Y, Fujiwara M, Matsumoto Y. 1992 In situ demonstration of proliferating cells in the rat central nervous system during experimental autoimmune encephalomyelitis. Evidence suggesting that most infiltrating T cells do not proliferate in the target organ. Lab Invest Jan 66(1):54-62

Okuda Y, Sakoda S, Fujimura H, Yanagihara T. 1998 Aminoguanidine, a selective inhibitor of the inducible nitric oxide synthase, has different effects on experimental allergic encephalomyelitis in the induction and progression phase. : J Neuroimmunol. 81:201-10

Oleszak EL, Katsetos CD, Kuzmak J, Varadhachary A. 1997 Inducible nitric oxide synthase in Theiler's murine encephalomyelitis virus infection. J Virol. 71:3228-35

Ouallet J, Baumann N, Marie Y, Villarroya H. 1999 Fas system up-regulation in experimental autoimmune encephalomyelitis. J Neurol Sci. 170:96-104.

Owens T, Tran E, Hassan-Zahraee M, Krakowski M (1998) Immune cell entry to the CNS--a focus for immunoregulation of EAE. Res Immunol 149:781-9

Pagenstecher A, Lassmann S, Carson MJ, Kincaid CL, Stalder AK, Campbell IL. 2000 Astrocyte-targeted expression of IL-12 induces active cellular immune responses in the central nervous system and modulates experimental allergic encephalomyelitis. J Immunol. 164(9):4481-92.

Pan Y, Lloyd C, Zhou H, Dolich S, Deeds J, Gonzalo JA, Vath J, Gosselin M, Ma J, Dussault B, Woolf E, Alperin G, Culpepper J, Gutierrez-Ramos JC, Gearing D 1997 Neurotactin, a membrane-anchored chemokine upregulated in brain inflammation. Nature 387:611-7

Penkowa M, Espejo C, Martinez-Caceres EM, Poulsen CB, Montalban X, Hidalgo J. 2001 Altered inflammatory response and increased neurodegeneration in metallothionein I+II deficient mice during experimental autoimmune encephalomyelitis. J Neuroimmunol. 119:248-60

Polfliet MM, van de Veerdonk F, Dopp EA, van Kesteren-Hendrikx EM, van Rooijen N, Dijkstra CD, van den Berg TK 2002 The role of perivascular and meningeal macrophages in experimental allergic encephalomyelitis. J Neuroimmunol 122:1-8

Priller J, Flugel A, Wehner T, Boentert M, Haas CA, Prinz M, Fernandez-Klett F, Prass K, Bechmann I, de Boer BA, Frotscher M, Kreutzberg GW, Persons DA, Dirnagl U. (2001) Targeting gene-modified hematopoietic cells to the central nervous system: use of green fluorescent protein uncovers microglial engraftment. Nat Med 7:1356-61

Qi X, Guy J, Nick H, Valentine J, Rao N. 1997 Increase of manganese superoxide dismutase, but not of Cu/Zn-SOD, in experimental optic neuritis. Invest Ophthalmol Vis Sci. 38:1203-12

Raivich G, Bohatschek M, Kloss CU, Werner A, Jones LL, Kreutzberg GW. 1999. Neuroglial activation repertoire in the injured brain: graded response, molecular mechanisms and cues to physiological function. Brain Res Rev 30:77-105

Reimann J, Bohm W, Schirmbeck R (1994) Alternative processing pathways for MHC class I-restricted epitope presentation to CD8+ cytotoxic T lymphocytes. Biol Chem Hoppe Seyler 375:731-6

Renno T, Krakowski M, Piccirillo C, Lin JY, Owens T. 1995 TNF-alpha expression by resident microglia and infiltrating leukocytes in the central nervous system of mice with experimental allergic encephalomyelitis. Regulation by Th1 cytokines. J Immunol Jan 15 154(2):944-53

Rinner WA, Bauer J, Schmidts M, Lassmann H, Hickey WF. 1995 Resident microglia and hematogenous macrophages as phagocytes in adoptively transferred experimental autoimmune encephalomyelitis: an investigation using rat radiation bone marrow chimeras. Glia 14:257-66

Rott O, Fleischer B, Cash E (1994) Interleukin-10 prevents experimental allergic encephalomyelitis in rats. Eur J Immunol. 24:1434-40

Rottman JB, Slavin AJ, Silva R, Weiner HL, Gerard CG, Hancock WW. 2000 Leukocyte recruitment during onset of experimental allergic encephalomyelitis is CCR1 dependent. Eur J Immunol. 30:2372-7

Ruuls SR, Bauer J, Sontrop K, Huitinga I, 't Hart BA, Dijkstra CD. 1995 Reactive oxygen species are involved in the pathogenesis of experimental allergic encephalomyelitis in Lewis rats. J Neuroimmunol 56:207-17

Ruuls SR, Van Der Linden S, Sontrop K, Huitinga I, Dijkstra CD 1996 Aggravation of experimental allergic encephalomyelitis (EAE) by administration of nitric oxide (NO) synthase inhibitors. Clin Exp Immunol. 103:467-74

Schiffenbauer J, Streit WJ, Butfiloski E, LaBow M, Edwards C 3rd, Moldawer LL. 2000 The induction of EAE is only partially dependent on TNF receptor signaling but requires the IL-1 type I receptor. Clin Immunol. 95:117-23

Schluesener HJ, Seid K, Kretzschmar J, Meyermann R 1996 Leukocyte chemotactic factor, a natural ligand to CD4, is expressed by lymphocytes and microglial cells of the MS plaque. J Neurosci Res 44(6):606-11

Schluesener HJ, Seid K, Zhao Y, Meyermann R. 1997 Localization of endothelial-monocyte-activating polypeptide II (EMAP II), a novel proinflammatory cytokine, to lesions of experimental autoimmune encephalomyelitis, neuritis and uveitis: expression by monocytes and activated microglial cells. Glia. Aug;20(4):365-72

Schluesener HJ, Seid K. 2000 Heme oxygenase-1 in lesions of rat experimental autoimmune encephalomyelitis and neuritis. J Neuroimmunol. 110:114-20

Schonrock LM, Kuhlmann T, Adler S, Bitsch A, Bruck W. 1998 Identification of glial cell proliferation in early multiple sclerosis lesions. Neuropathol Appl Neurobiol. 24(4):320-30

Sedgwick JD, Schwender S, Gregersen R, Dorries R, ter Meulen V 1993 Resident macrophages (ramified microglia) of the adult brown Norway rat central nervous system are constitutively major histocompatibility complex class II positive. J Exp Med. 177(4):1145-52.

Selmaj K, Raine CS, Cross AH. 1991 Anti-tumor necrosis factor therapy abrogates autoimmune demyelination. Ann Neurol. 30:694-700

Simpson J, Rezaie P, Newcombe J, Cuzner ML, Male D, Woodroofe MN. 2000 Expression of the beta-chemokine receptors CCR2, CCR3 and CCR5 in multiple sclerosis central nervous system tissue. J Neuroimmunol. 108(1-2):192-200

Smith T, Schmied M, Hewson AK, Lassmann H, Cuzner ML 1996 Apoptosis of T cells and macrophages in the central nervous system of intact and adrenalectomized Lewis rats during experimental allergic encephalomyelitis. J Autoimmun. Apr;9(2):167-74

Sorensen TL, Tani M, Jensen J, Pierce V, Lucchinetti C, Folcik VA, Qin S, Rottman J, Sellebjerg F, Strieter RM, Frederiksen JL, Ransohoff RM. 1999 Expression of specific chemokines and chemokine receptors in the central nervous system of multiple sclerosis patients. J Clin Invest. 103:807-15

Stalder AK, Carson MJ, Pagenstecher A, Asensio VC, Kincaid C, Benedict M, Powell HC, Masliah E, Campbell IL. 1998 Late-onset chronic inflammatory encephalopathy in immune-competent and severe combined immune-deficient (SCID) mice with astrocyte-targeted expression of tumor necrosis factor. Am J Pathol 153:767-83

Stephenson DT, Schober DA, Smalstig EB, Mincy RE, Gehlert DR, Clemens JA (1995) Peripheral benzodiazepine receptors are colocalized with activated microglia following transient global forebrain ischemia in the rat. J Neurosci. 15:5263-74.

Stevens DB, Gould KE, Swanborg RH. 1994 Transforming growth factor-beta 1 inhibits tumor necrosis factor-alpha/lymphotoxin production and adoptive transfer of disease by effector cells of autoimmune encephalomyelitis. J Neuroimmunol. 51:77-83

Stoll G, Jander S (1999) The role of microglia and macrophages in the pathophysiology of the CNS. Prog Neurobiol. 58:233-47

Storch MK, Fischer-Colbrie R, Smith T, Rinner WA, Hickey WF, Cuzner ML, Winkler H, Lassmann H 1996 Co-localization of secretoneurin immunoreactivity and macrophage infiltration in the lesions of experimental autoimmune encephalomyelitis. Neuroscience. Apr;71(3):885-93

Streit WJ, Graeber MB, Kreutzberg GW (1989) Expression of Ia antigen on perivascular and microglial cells after sublethal and lethal motor neuron injury. Exp Neurol 105:115-26

Subramanian S, Bourdette DN, Corless C, Vandenbark AA, Offner H, Jones RE. 2001 T lymphocytes promote the development of bone marrow-derived APC in the central nervous system. J Immunol. 166(1):370-6

Sugita M, Moody DB, Jackman RM, Grant EP, Rosat JP, Behar SM, Peters PJ, Porcelli SA, Brenner MB. 1998 CD1--a new paradigm for antigen presentation and T cell activation. Clin Immunol Immunopathol. 87:8-14

Sun D, Whitaker JN, Huang Z, Liu D, Coleclough C, Wekerle H, Raine CS. Myelin antigen-specific CD8+ T cells are encephalitogenic and produce severe disease in C57BL/6 mice. J Immunol. 2001 Jun 15;166(12):7579-87

Suter T, Malipiero U, Otten L, Ludewig B, Muelethaler-Mottet A, Mach B, Reith W, Fontana A. 2000 Dendritic cells and differential usage of the MHC class II transactivator promoters in the central nervous system in experimental autoimmune encephalitis. Eur J Immunol 30(3):794-802

Suzuki M, Raisman G (1992) The glial framework of central white matter tracts: segmented rows of contiguous interfascicular oligodendrocytes and solitary astrocytes give rise to a continuous meshwork of transverse and longitudinal processes in the adult rat fimbria. Glia 6:222-35

Takahashi JL, Giuliani F, Power C, Imai Y, Yong VW. 2003 Interleukin-1beta promotes oligodendrocyte death through glutamate excitotoxicity. Ann Neurol. 53:588-95

Taupin V, Renno T, Bourbonniere L, Peterson AC, Rodriguez M, Owens T 1997 Increased severity of experimental autoimmune encephalomyelitis, chronic macrophage/microglial reactivity, and demyelination in transgenic mice producing tumor necrosis factor-alpha in the central nervous system. Eur J Immunol. 27:905-13

Tran EH, Hardin-Pouzet H, Verge G, Owens T. 1997 Astrocytes and microglia express inducible nitric oxide synthase in mice with experimental allergic encephalomyelitis. J Neuroimmunol. 74:121-9

Tran EH, Hoekstra K, van Rooijen N, Dijkstra CD, Owens T. 1998 Immune invasion of the central nervous system parenchyma and experimental allergic encephalomyelitis, but not leukocyte extravasation from blood, are prevented in macrophage-depleted mice. J Immunol 161:3767-75

Trebst C, Sorensen TL, Kivisakk P, Cathcart MK, Hesselgesser J, Horuk R, Sellebjerg F, Lassmann H, Ransohoff RM. 2001 CCR1+/CCR5+ mononuclear phagocytes accumulate in the central nervous system of patients with multiple sclerosis. Am J Pathol 159:1701-10

Trebst C, Staugaitis SM, Kivisakk P, Mahad D, Cathcart MK, Tucky B, Wei T, Rani MR, Horuk R, Aldape KD, Pardo CA, Lucchinetti CF, Lassmann H, Ransohoff RM. 2003 CC chemokine receptor 8 in the central nervous system is associated with phagocytic macrophages. Am J Pathol. 162:427-38

Van Dam AM, Bauer J, Man-A-Hing WK, Marquette C, Tilders FJ, Berkenbosch F: 1995 Appearance of inducible nitric oxide synthase in the rat central nervous system after rabies virus infection and during experimental allergic encephalomyelitis but not after peripheral administration of endotoxin. J Neurosci Res. Feb 1;40(2):251-60

Waldburger KE, Hastings RC, Schaub RG, Goldman SJ, Leonard JP 1996 Adoptive transfer of experimental allergic encephalomyelitis after in vitro treatment with recombinant murine interleukin-12. Preferential expansion of interferon-gamma-producing cells and increased expression of macrophage-associated inducible nitric oxide synthase as immunomodulatory mechanisms. Am J Pathol. Feb;148(2):375-82

Walther M, Popratiloff A, Lachnit N, Hofmann N, Streppel M, Guntinas-Lichius O, Neiss WF, Angelov DN. (2001) Exogenous antigen containing perivascular phagocytes induce a non-encephalitogenic extravasation of primed lymphocytes. J Neuroimmunol 117:30-42

Weinberg AD, Wegmann KW, Funatake C, Whitham RH. 1999 Blocking OX-40/OX-40 ligand interaction in vitro and in vivo leads to decreased T cell function and amelioration of experimental allergic encephalomyelitis. J Immunol 162(3):1818-26

Werner K, Bitsch A, Bunkowski S, Hemmerlein B, Bruck W. 2002 The relative number of macrophages/microglia expressing macrophage colony-stimulating factor and its receptor decreases in multiple sclerosis lesions. Glia 40:121-9

Willenborg DO, Fordham SA, Staykova MA, Ramshaw IA, Cowden WB. 1999 IFN-gamma is critical to the control of murine autoimmune encephalomyelitis and regulates both in the periphery and in the target tissue: a possible role for nitric oxide. J Immunol. 163:5278-86

Williams K, Ulvestad E, Antel JP. B7/BB-1 antigen expression on adult human microglia studied in vitro and in situ. Eur J Immunol. 1994 Dec;24(12):3031-7

Xue S, Sun N, Van Rooijen N, Perlman S. 1999 Depletion of blood-borne macrophages does not reduce demyelination in mice infected with a neurotropic coronavirus. J Virol 73:6327-34

Zhu B, Luo L, Chen Y, Paty DW, Cynader MS. 2002 Intrathecal Fas ligand infusion strengthens immunoprivilege of central nervous system and suppresses experimental autoimmune encephalomyelitis. J Immunol. 169:1561-9

Zinkernagel RM, Doherty PC (1997) The discovery of MHC restriction. Immunol Today 18:14-7

Chapter A8

# THE NEURON AND AXON IN EXPERIMENTAL AUTOIMMUNE ENCEPHALOMYELITIS


Nikos Evangelou and Cris S Constantinescu

*Division of Clinical Neurology, University of Nottingham, UK*



**Abstract:** Neuronal and axonal loss and damage are increasingly recognized major pathological features of MS that correlate with disability. Correspondingly, in EAE, these processes are being investigated in great detail. This chapters discusses the present knowledge of axonal and neuronal pathology in MS and details the relevant studies in EAE, that may form the basis to future neuroprotective/regenerative approaches to MS.



**Key words:** EAE, MS, axon, neuron, magnetic resonance imaging, atrophy


The importance of axonal dysfunction and axonal loss in Multiple Sclerosis (MS) has been at the centre of the pathological debate over the last decade. Since 1995 the number of publications listed in Medline related to "MS AND myelin" has increased by 30%, whereas articles on "MS AND axon*" increased by 300%. Equally, it is not surprising that more importance is also given to the extent of axonal involvement in the EAE models.

The first part of this review will examine the evidence of axonal involvement in MS, and the second part will focus to studies of axons in the EAE models.

## 1.1 History of axonal loss

Axonal injury has been recognized since the first descriptions of MS pathology. Neuropathologists usually refer to Charcot when describing the hallmarks of MS pathology, e.g. inflammation, demyelination, and relative preservation of axons. In the historical review on axonal loss in MS, Kornek and Lassmann [1] suggest that it is Marburg in 1906, who emphasized the term "relative" as he found more axons were lost than generally believed. Charcot clearly described demyelinated axons showing increased diameters, large axonal swellings and admitted that, due to methodic limitations, the presence of axonal degeneration could not be definitively excluded. Fromann provided a detailed description of axonal pathologic changes in the lesions of MS. He noticed that lesion formation not only led to the disruption of the myelin sheath, but also to the appearance of axonal transections. Quantifying the degree of axonal loss came several decades later. Putnam in 1936 found an "unmistakable decrease in number" of axons in 13 out of 35 plaques and in 18 they were "absent or the number was so far decreased that some high power fields were devoid of them". Greenfield and King produced conflicting results: in a series of 125 plaques from 13 cases they found that less than 10 percent showed "severe destruction," meaning a reduction in number to one-fifth or one-seventh.


## 2.        AXONAL LOSS IN MS


### 2.1        In demyelinated lesions in MS

MS is primarily a demyelinating disease of the central nervous system. Evidence of axonal loss derives indirectly from imaging studies and directly from pathological specimens. Magnetisation transfer imaging (MTI) is a MRI technique that can measure water content in its relation to macromolecules [2]. The MTI signal reflects protons bound to tissue (particularly myelin) and can therefore provide an index of tissue destruction. It is described as a ratio – the lower the ratio, the more destructive the lesion. A moderately strong correlation coefficient has been demonstrated between the MT ratio and disability[3]. Hypointense T1 lesions with decreased MTR are also associated with tissue destruction and axonal loss and are frequently found in MS patients [4, 5].

A number of studies have used Magnetic Resonance Spectroscopy (MRS) techniques in the last decade. MRS techniques permit inferences about chemical changes in tissues. N-acetylaspartate (NAA) is found exclusively or perdominantly in neurons and the NAA peak was found to be low both in lesions [6] and in the NAWM [7] in MS. Brain proton MR spectroscopy in patients with MS has demonstrated substantially reduced NAA/Cr ratios within both chronic [8, 9] and acute [6, 10] lesions.

Pathological studies have found that axonal damage was more evidently found with acute lesions associated with inflammatory activity [11, 12]. On average axonal loss of 59-82% was observed in demyelinated plaques [13].

## 2.2     *Normal appearing white matter in MS*

The term Normal Appearing White Matter (NAWM) is a technical MRI definition and was initially used to describe the white matter volume on T2-weighted images outside of areas of abnormally high white matter signal intensity. An increasing amount of evidence suggests that the white matter outside demyelinated lesions is abnormal mainly due to loss of axons.

### 2.2.1     Atrophy

Whereas late stages of chronic MS have long been recognised to be associated with brain atrophy, more recent quantitative MRI studies have demonstrated that large white mater volume changes occur even early in the progression of the disease. Losseff and colleagues have shown that the relative atrophy of the spinal cord at the level of C3 correlates with the degree of disability [14]. In addition, they described a measurable progressive change in the cross-sectional area of the spinal cord at C3 with time.   Just as for the spinal cord, it was found that MS patients show measurable atrophy of brain white matter with time. Although lesion formation resulting in tissue destruction certainly can contribute to volume loss, the degree of atrophy is out of proportion to the relatively small lesion volumes suggesting pathological processes outside lesions being involved.

### 2.2.2     Magnetic Resonance Spectroscopy shows evidence for diffuse axonal injury in NAWM in MS

Proton magnetic resonance spectroscopy (MRS) provides a unique tool for *in vivo* evaluation of biochemical changes associated with brain pathology. In MRS, information on the resonance frequencies of protons observed in the tissue is preserved.   This allows analysis of the relative spatial distributions and concentrations of small molecules. Water-suppressed localised $^1$H spectroscopy yields spectra dominated by resonances of N-acetyl aspartate (NAA), Creatine (Cr)/Phosphocreatine (PCr) and Choline (Cho). NAA is exclusively or, at least, predominantly localized within neurons in the mature brain [15, 16] and is therefore used as a marker of axonal density and integrity. Conventionally in brain MR spectroscopy NAA has been quantified in relation to the magnitude of the Cr, the concentration of which is believed to vary little.

Irreversible decreases in the relative NAA concentrations in lesions must reflect axonal volume loss (i.e., either a decrease in the number of axons or in their aggregate cross sectional area).  Reversible changes are likely to signify axonal dysfunction [17] The possibility that significant biochemical changes occur outside of areas of focal inflammation (i.e. within the NAWM), was highlighted by the finding that reductions in the NAA/Cr ratio from a large volume of interest within the brain were relatively independent of T2-hyperintense lesion volumes and could be found with even very large voxels that included relatively small lesions volumes [18].  Later studies directly demonstrated reduced NAA/Cr adjacent to [6] or even distant from lesions in regions of the brain containing only NAWM [19]. The reduction in the NAA/Cr ratio within the NAWM was shown to be more marked in patients with secondary progressive than relapsing-remitting MS [20].

The reduced NAA/Cr ratio or reduction in absolute NAA signal within the NAWM could arise from reduced axonal density, reduction in axonal NAA concentration with axons or altered relaxation times.  Early studies suggested that relaxation time changes are unlikely to be great enough to have a large effect on the NAA/Cr ratio [9]. Decreases in neuronal NAA concentration occur in response to an acute inflammatory insult and can be reversible [6, 17]. Hence, a reduction in the NAA signal within large areas of the macroscopically NAWM in patients who have not suffered a recent relapse most likely reflects chronic changes in axonal morphology or density.

### 2.2.3    Axonal loss in NAWM

Few studies have investigated the NAWM with pathological methods.  In an acute MS case, axonal changes were reported. By studying the corpus callosum we have shown substantially reduced axonal density and loss of volume in the NAWM. Considering both axonal density and volume measurements in the corpus callosum there is a mean estimated loss of 53 % of the total number of axons in the NAWM of MS brains in our samples [21]. We also demonstrated that the axonal loss in the NAWM is the result of Wallerian and post-synaptic degeneration of axons caused by axonal transection at the site of MS lesions.

## 2.3 Axonal damage is the likely morphological substrate of clinical disability

As mentioned, the association between axonal pathology and clinical disability was first discussed by Charcot. The volume of hypointense lesions (black holes) [22] and the extent of brain atrophy, both considered a marker of axonal loss, correlate with the degree of disability [23].

The significance of those MR abnormalities is strengthened by studies that have found strong correlations between the NAWM changes and clinical disability. Possibly, these MRI abnormalities of the NAWM may reflect the cumulative damage from MS lesions, which appears to occur both early in the disease, as well as later, in the secondary progressive phase.

The earliest MR spectroscopic studies reported that large voxel MRS measurements of brain NAA were lower in patients with higher disability scores than those with lower disability scores [8]. Fu and colleagues later co-registered MRSI with conventional images in which the lesions had been segmented in order to define independently the relationship between axonal injury in lesions or NAWM and progression of disability [7]. They found a correlation between decreases in NAA in the NAWM and the progression of disability. It is notable that a similar correlation was <u>not</u> found between NAA decreases in lesions and disability. While work of Davie *et al* had noted the correlation between overall axonal injury in lesions and disability [24], measurement of NAA in the aggregate lesion load is difficult and may not be as sensitive to change over time as changes in the NAWM. Longitudinal follow-up of the same RR and SP MS patient cohort followed by Fu *et al* demonstrated a strong correlation between the progression of disability and decreases in relative NAA concentrations in a large central volume of WM.

## 2.4 Pathological methods for detecting axonal loss

MR techniques are only indirectly reflecting axonal dysfunction and are unable to differentiate between acute and longstanding axonal damage.

One of the reasons the axonal element of MS has been given little attention until recently is possibly the ease of identifying demyelinated plaques using conventional pathological stains compared to the less striking pictures of axonal stains.

A number of techniques can be used to study axonal damage. While with the use of electron microscopy the number and size of axons can be measured accurately, this method is complicated by technical difficulties. The best images are acquired when the brain tissues are frozen immediately following the death of the patient, and as most MS patients do not die in hospital, pathological material is difficult to obtain. The use of the EM and the fixation techniques also require considerable expertise for good quality results. The availability of EM remains another limiting factor.

Silver-based methods (Bodian, Fink-Heimer, Nauta) are the choice of most neuropathology laboratories. Silver-based stains require good fixation of tissue. Modifications of those stains take advantage of the higher affinity of degenerating axoplasm for binding silver.

Most immunocytochemical techniques use antibodies to differentiate neurofilament components (NF70, NF150 and NF200) present in axons. Both silver-based and neurofilament methods visualise mainly the largest of the axons. Quantitative comparative studies examining what percentage of myelinated or unmyelinated axons are visualised using those techniques have not been conducted.

Another immunocytochemical technique involves the use of Amyloid Precursor Protein (APP). APP is a membrane glycoprotein and a normal constituent of neuronal cells. Normally APP is transported in an anterograde direction and is not visualised in healthy axons. If the transport is interrupted either due to axonal transection or possibly due to local dysfunction, APP accumulates in the proximal axonal ends and can be visualised by standard immunohistochemistry in formalin-fixed tissue as APP-positive spheroids. Therefore, beta APP has been suggested to be the immunocytochemical marker of choice for the detection of injured axons. This method was first used in traumatic brain injuries and found to be more sensitive than silver stains. As it has been demonstrated in fatal head injuries, APP stains damaged axons within 2 hours after injury and remains detectable in axons and bulbs for 10 to 14 days. With a survival time longer than 2 weeks, beta APP reactivity disappears from the injured axons as well as from bulbs. Beta APP therefore may serve as a marker for early axonal damage, while more advanced stages of axonal degeneration may not be detected by this method. In MS, APP has been found to be significantly expressed throughout acute lesions and at the border of active chronic lesions.

## 3.        AXONAL LOSS IN EAE

### 3.1      *Axonal dysfunction vs. axonal loss*

Using magnetic resonance spectroscopy, larger reductions of NAA have been demonstrated than could have been expected by plain axonal loss. The main proposed explanation relies on the existence of significant axonal dysfunction without leading ultimately to neuronal death. Pathologically proven axonal loss probably represents the extreme part of the spectrum of axonal dysfunction. In MS autopsies a reduced number of cortical synapses were found in paralysed compared to ambulatory patients [25].

There is supporting EAE evidence derived from a study of the anatomy and function of the optic nerve in guinea pigs. The axonal density was analysed, but also orthograde transport was studied following tritiated leucine injection into the vitreous by autoradiographic analysis of silver grain counts. The extent of reduction in radioactivity in acute EAE (74%) was significantly more than the axonal loss (16%) [26].

### 3.2      *Axonal damage in different EAE models*

EAE offers the possibility of providing detailed pathological correlation for the MRI technique used to assess axonal loss and dysfunction [27].

A number of EAE models have been to study axons. These include murine EAE induced by MBP or PLP- specific cell lines and clones [28-32]. Guinea pigs [26], and marmoset [33]. Amongst the different models MOG-induced EAE shows that, the highest incidence of acute axonal injury during active demyelination lead to a pronounced reduction of axonal density in all stages of demyelination [31]. The SWXJ mice, which have been used to study axonal loss in EAE, when immunized with PLP develop relapsing remitting chronic EAE that is very similar in its clinical course to MS. The mice have initial relapses and a late plateau of sustained functional disability [34, 35]. In most of those models a CNS antigen, usually myelin or a myelin protein peptide, was used to induce specific immune responses mimicking MS. A different approach was employed by Newman who used heat-killed bacillus Calmette-Guerin in a rodent mode in a two-staged approach. Initially a brain injection was given. At a second stage, after the blood brain barrier damage had healed, a subcutaneous injection of the bacilli induced CNS immune activation that resulted in axonal damage [36]. The mechanism of axonal damage by the attracted macrophages is not yet clear, although release of cytokines or matrix mettaloproteinases have been implicated.

## 3.3    *Axonal selectivity*

Recent studies have suggested that smaller axons may be preferentially injured in MS [37, 38]  We have also shown size-specific neuronal changes in the visual pathways [39]. Size-dependent differences in susceptibility could be a consequence of differences in surface area-to-volume ratio (e.g., if a non-specific, diffusible toxin such as NO is responsible) or other factors, such as the particular metabolic or antigenic characteristics of the different classes of cells. Confirming size-dependent correlations with injury throughout the CNS, for neurons and axons, would help to suggest candidate toxic factors.

Examining spinal cord axons immunostained for tyrosine hydroxylase, serotonin and substance P, differences were found in the degree of axonal damage in the paraplegic phase in guinea pig EAE.  The small diameter thinly myelinated or unmyelinated axons that course the spinal cord for long distances are more vulnerable to damage. Substance P containing fibres in the dorsal horns did not exhibit any changes, possibly because of their small intraspinal course [40].

## 3.4    *Timing of abnormalities*

The axonal and neuronal abnormalities seem to correlate with the degree of demyelination and the functional recovery following the induction of EAE.  As in MS, actively demyelinating lesions have many swollen and distorted axons, seen with Bielschowsky silver stain and immunohistochemistry for APP protein, whereas in inactive lesions and shadow plaques axonal spheroids are found only occasionally [31]. Early active lesions demonstrate higher levels of bAPP [31] as well as non-phosphorylated neurofilaments (SMI-32 positive) [33] compared to late active lesions.  In the paraplegic phase of EAE in rats, grossly distorted axons are seen with accompanying grey matter abnormalities. Even by the first day of recovery, those changes are less prominent.  Within the next 2 weeks, both the grey matter and the axons became increasing more normal in appearance [41].

In clinical relapses in EAE Lewis rats, the somatosensory evoked potentials was reduced in amplitude and prolonged in latency, presumably due to both demyelination and axonal dysfunction. In well-established remission, there was residual conduction failure, which can be mainly accounted for by axonal degeneration [42]. Most MS remyelinating lesions (shadow plaques) do not show significant demyelinating activity, and equally show very little acute axonal dysfunction by measuring betaAPP levels. This is in contrast to EAE where recurrent demyelination in the plaque is common, which may explain the high incidence of axonal bAPP reactivity [31].

## 3.5     *Molecular changes within neurons*

Molecular changes have been observed in CNS neurons both in EAE and in post mortem MS material. In a chronic relapsing EAE mouse model that consistently displays cerebellar lesions, Sensory Neuron-Specific (SNS) sodium channels were detected, which are not present in the cerebellum of control mice. Similarly, SNS expression was detected in Purkinje cells in MS tissues from patients that exhibited cerebellar deficits, but not in the controls. The up regulation of some sodium channels could be due to neurotrophic factors, altered electrical activity, or demyelination *per se* [43].

Abnormal localization of N-type calcium channel has been implicated as a mechanism of axonal degeneration. Prominent axonal transport of a pore-forming subunit of N-type voltage-gated calcium channel has been observed in actively demyelinated axons in MS and MOG induced EAE. The subunit is thought to then integrate in the axonal membrane of the injured axon, possibly causing influx of calcium with activation of proteases and degradation of the cytoskeleton [44]. Calpain is a proteinase, found to be expressed at clinical peaks of EAE in spinal cord microglia of Lewis rats, possibly being one of the causes of degeneration of the axonal cytoskeleton, as its activity is increased by free intracellular [45].

## 3.6     *Cytotoxic cells and axonal damage*

What causes this profound axonal damage is unclear. Acute axonal pathology correlates with the presence of macrophages in a Lewis rat EAE model [31]. Direct attack of the neurons by cytotoxic cells is a possibility. Cytotoxic cell precursors recognize antigens on the surface of cells in association with MHC class I molecules. Both neurons and axons in MS can express MHC I rendering them vulnerable to CD8 –mediated cytotoxicity.

Myelin specific CD8 T cells induced clinically and histopathologically severe EAE and these cells can directly cause neuronal and axonal damage [46, 47].

## 3.7    *Agents associated with axonal damage*

The severity of inflammation and the degree of axonal damage are positively correlated in both in MS and EAE.    Several mediators of inflammation were investigated as primary causes of axonal damage.

**Nitric oxide** is produced in high concentration by macrophages and possibly by other CNS cells at sites of inflammation. The inducible form of NOS can produce much larger quantities of NO for protracted periods. Inducible NOS has been detected in MS plaques [48] and in the CNS of EAE [49]. The NO levels in the CNS of animals with EAE are reported to be up to 30 times greater than in the control animals. In an in vivo study, Smith has shown that axons operating at physiological frequencies are prone to nitric oxide mediated damage [50].

**Other inflammatory mediators and cytokines** may also act via induction or enhancement of NO.  When relapses of the EAE were induced by intraperitoneal administration of IL-12 in Lewis rats, axonal death was observed in association with iNOS and tPA immunostaining throughout the gray and white matter. These agents might act either directly or indirectly by activating microglia and macrophages leading to axonal damage [51].

**Perforin** may be the principal mediator of axonal damage caused by cytotoxic T cells [47].

**Glutamate** neurotoxicity may be implicated in axonal damage and its blockade may be beneficial in EAE [52].

**Trophic factors** are important for the normal functioning of the complex glial-axon interaction.    Ciliary neurotrophic factor (CNTF) has been demonstrated to be an important surviving factor for both neuronal and oligodendroglial populations. Disruption of the CNTF gene results in motor neuron degeneration. A CNTF mutation was detected in a group of MS patients with earlier onset and more severe course than those with at least one functional allele [53]. In EAE induced by MOG CNTF- deficient mice caused histologically a severe oligodendrogliopathy with prominent secondary axonal dystrophy and transection. This resulted clinically in earlier onset disease, with a significantly prolonged recovery phase [54]. In conclusion, lack of trophic factors seems to be associated with more severe axonal damage.    Whether pharmacologically providing an excess of neurotrophic stimulation will lead to less severe disease remains to be tested. Ruffini [55] delivered basic fibroblastic growth factor in mouse by genetically modified herpes virus vector that led to improvement of the clinical course of chronic EAE.

## 3.8    *Correlation with disability*

Although the degree of axonal abnormalities observed depends on the different model of EAE, the extent of the axonal involvement in a given EAE model seems to correlate with the degree of clinical severity of the disease. Examining the spinal cord of paraplegic EAE guinea pigs showed abnormalities in axons that were more pronounced, compared to rats that eventually recovered from the illness [41].

With quantitative immunohistochemical methods examining the spinal cord in chronic EAE neurological disability has been correlated with inflammation and axonal loss. Whereas inflammation was the principal determinant of reversible neurological disability during the acute EAE, the fixed neurological deficits correlated with axon loss, that in turn correlated with the number of symptomatic attacks [35].

## 3.9    *Therapeutic interventions*

A number of agents that modulate inflammation in EAE can lead to protection of axons. Prazosin an alpha 1- adrenergic antagonist prevented monoaminergic axonal damage in the spinal cord of EAE rats [56]. When recombinant HSV expressing IL-4 a Th2 cytokine, was injected intracranially in the BALB/c model of mice EAE it prevented massive leucocyte infiltration resulting in protection of both the myelin and the axons [57].

Memantine, an NMDA receptor antagonist ameliorates neurological deficits in Lewis rats with EAE. As this was not found to be due to dampened CNS inflammation a direct effect on NMDA receptors had been postulated, suggesting that this might be an easy neuronal target for modulation during neuroinflammation [58, 59].

Similarly, the phosphodiesterase IV inhibitor roliparm was proven an effective anti-inflammatory agent in EAE, possibly because glutamate-induced cytotoxicity can be prevented by cyclic AMP elevating agents [60]. Studies in progressive MS with pentoxifylline, another phosphodiesterase inhibitor, did not show a significant effect [61]. As relapse frequency and MRI activity were the end point of the study, one wonders if the anti-inflammatory effect of those agents is less important than the effect on glutamate toxicity.

Cannabinoids are currently under investigation in MS especially for control of fatigue, tremor, spasticity and bladder dysfunction. Cannabinoids have bee successful in the control of spasticity in EAE [62]. Cannabinoid receptors are expressed in the CNS, and in vitro studies cannabinoids suppress nitric oxide production in neurons [63] and so possibly protect neurons from NMDA-induced neurotoxicity.  There are suggestions that cannabinoids may have a neuroprotective role [64].

## 3.10    *Immune modulation and protective autoimmunity*

The degree of neuronal survival following CNS insult is influenced by immune mediated mechanisms. Kipnis *et al* [65] showed that the survival of retinal ganglion cells in adult mice, after a variety of insults to the optic nerve or the vitreous, is influenced by the susceptibility status to EAE of the animal. T cells reactive against neural antigens but nonencephalitogenic may enhance recovery. Animals with EAE resistance had 60 % more surviving neurons than normal animals. Furthermore, an unrelated injury to the spinal cord prior to the retinal ganglion cell insult led to increased survival of retinal neurons. Survival was dependent upon the generation of a protective T cell response, suggesting T-cell dependent protection mechanisms as potential target for neuroprotective therapy.

**ACKNOWLEDGMENTS**
Dr. Constantinescu's research has been supported in part by the NIH, University of Nottingham, The Stanhope Trust and the Multiple Sclerosis Society of the United Kingdom and Northern Ireland.  Dr Evangelou's research is supported by the Multiple Sclerosis Society of the United Kingdom and Northern Ireland.

# REFERENCES

1.      Kornek, B. and H. Lassmann, Axonal pathology in multiple sclerosis. A historical note. Brain Pathol, 1999: p. 651-6.

2.      Grossman, R.I., Magnetization transfer in multiple sclerosis. Ann Neurol, 1994. **36(suppl)**: p. S97-9.

3.      Gass, A., G.J. Barker, D. Kidd, J.W. Thorpe, D. MacManus, A. Brennan, P.S. Tofts, A.J. Thompson, W.I. McDonald, and D.H. Miller, Correlation of magnetization transfer ratio with clinical disability in multiple sclerosis. Ann Neurol, 1994. **36**: p. 62-7.

4.      Bruck, W., A. Bitsch, H. Kolenda, Y. Bruck, M. Stiefel, and H. Lassmann, Inflammatory central nervous system demyelination: correlation of magnetic resonance imaging findings with lesion pathology. Ann Neurol, 1997. **42**: p. 783-93.

5.      van Walderveen, M.A., W. Kamphorst, P. Scheltens, J.H. van Waesberghe, R. Ravid, J. Valk, C.H. Polman, and F. Barkhof, Histopathologic correlate of hypointense lesions on T1-weighted spin-echo MRI in multiple sclerosis. Neurology, 1998. **50**: p. 1282-8.

6.      Arnold, D.L., P.M. Matthews, G.S. Francis, J. O'Connor, and J.P. Antel, Proton magnetic resonance spectroscopic imaging for metabolic characterization of demyelinating plaques. Ann Neurol, 1992. **31**: p. 235-41.

7.      Fu, L., P.M. Matthews, N. De Stefano, K.J. Worsley, S. Narayanan, G.S. Francis, J.P. Antel, C. Wolfson, and D.L. Arnold, Imaging axonal damage of normal-appearing white matter in multiple sclerosis. Brain, 1998. **121**: p. 103-13.

8.      Arnold, D.L., P.M. Matthews, G. Francis, and J. Antel, Proton magnetic resonance spectroscopy of human brain in vivo in the evaluation of multiple sclerosis: assessment of the load of disease. Magn Reson Med, 1990: p. 154-9.

9.      Matthews, P.M., G. Francis, J. Antel, and D.L. Arnold, Proton magnetic resonance spectroscopy for metabolic characterization of plaques in multiple sclerosis. Neurology, 1991. **41**: p. 1251-6.

10.     Davie, C.A., C.P. Hawkins, G.J. Barker, A. Brennan, P.S. Tofts, D.H. Miller, and W.I. McDonald, Serial proton magnetic resonance spectroscopy in acute multiple sclerosis lesions. Brain, 1994. **117**: p. 49-58.

11.     Ferguson, B., M.K. Matyszak, M.M. Esiri, and V.H. Perry, Axonal damage in acute multiple sclerosis lesions. Brain, 1997. **120**: p. 393-9.

12.     Trapp, B.D., J. Peterson, R.M. Ransohoff, R. Rudick, S. Mork, and L. Bo, Axonal transection in the lesions of multiple sclerosis. N Engl J Med, 1998. **338**: p. 278-85.

13.     Mews, I., M. Bergmann, S. Bunkowski, F. Gullotta, and W. Bruck, Oligodendrocyte and axon pathology in clinically silent multiple sclerosis lesions. Mult Scler, 1998. **4**: p. 55-62.

14.     Losseff, N.A., S.L. Webb, J.I. O'Riordan, R. Page, L. Wang, G.J. Barker, P.S. Tofts, W.I. McDonald, D.H. Miller, and A.J. Thompson, Brain 1996 Jun;119 ( Pt 3):701-8. Spinal cord atrophy and disability in multiple sclerosis. A new reproducible and sensitive MRI method with potential to monitor disease progression. Brain, 1996. **119**: p. 701-8.

15.     Simmons, M.L., C.G. Frondoza, and J.T. Coyle, Immunocytochemical localization of N-acetyl-aspartate with monoclonal antibodies. Neuroscience, 1991. **45**: p. 37-45.

16.     Bjartmar, C., J. Battistuta, N. Terada, E. Dupree, and B.D. Trapp, N-acetylaspartate is an axon-specific marker of mature white matter in vivo: a biochemical and immunohistochemical study on the rat optic nerve. Ann Neurol, 2002. **51**: p. 51-8.

17.     De Stefano, N., P.M. Matthews, J.P. Antel, M. Preul, G. Francis, and D.L. Arnold, Chemical pathology of acute demyelinating lesions and its correlation with disability. Ann Neurol, 1995. **38**: p. 901-9.

18.     Van Hecke, P., G. Marchal, K. Johannik, P. Demaerel, G. Wilms, H. Carton, and A.L. Baert, Human brain proton localized NMR spectroscopy in multiple sclerosis. Magn Reson Med, 1991. **18**: p. 199-206.

19.     Husted, C.A., D.S. Goodin, J.W. Hugg, A.A. Maudsley, J.S. Tsuruda, S.H. de Bie, G. Fein, G.B. Matson, and M.W. Weiner, Biochemical alterations in multiple sclerosis lesions and normal-appearing white matter detected by in vivo 31P and 1H spectroscopic imaging. Ann Neurol, 1994. **36**: p. 157-65.

20.     Matthews, P.M., E. Pioro, S. Narayanan, N. De Stefano, L. Fu, G. Francis, J. Antel, C. Wolfson, and D.L. Arnold, Assessment of lesion pathology in multiple sclerosis using quantitative MRI morphometry and magnetic resonance spectroscopy. Brain, 1996. **119**: p. 715-22.

21.     Evangelou, N., M.M. Esiri, S. Smith, J. Palace, and P.M. Matthews, Quantitative pathological evidence for axonal loss in normal appearing white matter in multiple sclerosis. Ann Neurol, 2000. **47**: p. 391-5.

22.     Grimaud, J., G.J. Barker, L. Wang, M. Lai, D.G. MacManus, S.L. Webb, A.J. Thompson, W.I. McDonald, P.S. Tofts, and D.H. Miller, Correlation of magnetic resonance imaging parameters with clinical disability in multiple sclerosis: a preliminary study. J Neurol, 1999. **246**: p. 961-7.

23.     Fisher, E., R.A. Rudick, G. Cutter, M. Baier, D. Miller, B. Weinstock-Guttman, M.K. Mass, D.S. Dougherty, and N.A. Simonian, Relationship between brain atrophy and disability: an 8-year follow-up study of multiple sclerosis patients. Mult Scler, 2000. **6**: p. 373-7.

24.     Davie, C.A., G.J. Barker, A.J. Thompson, P.S. Tofts, W.I. McDonald, and D.H. Miller, 1H magnetic resonance spectroscopy of chronic cerebral white matter lesions and normal appearing white matter in multiple sclerosis. J Neurol Neurosurg Psychiatry, 1997. **63**: p. 736-42.

25.     Peterson, J.W., L. Bo, S. Mork, A. Chang, and B.D. Trapp, Transected neurites, apoptotic neurons, and reduced inflammation in cortical multiple sclerosis lesions. Ann Neurol, 2001. **50**: p. 389-400.

26.     Guy, J., E.A. Ellis, E.F.r. Tark, G.M. Hope, and N.A. Rao, Axonal transport reductions in acute experimental allergic encephalomyelitis: qualitative analysis of the optic nerve. Curr Eye Res, 1989. **8**: p. 261-9.

27.     Dousset, V., R.I. Grossman, K.N. Ramer, M.D. Schnall, L.H. Young, F. Gonzalez-Scarano, E. Lavi, and J.A. Cohen, Experimental allergic encephalomyelitis and multiple sclerosis: lesion characterization with magnetization transfer imaging. Radiology, 1992. **182**: p. 483-91.

28.     Tabira, T. and K. Sakai, Demyelination induced by T cell lines and clones specific for myelin basic protein in mice. Lab Invest, 1987. **56**: p. 518-25.

29.     Raine, C.S. and A.H. Cross, Axonal dystrophy as a consequence of long-term demyelination. Lab Invest, 1989. **60**: p. 714-25.

30.     Storch, M.K., A. Stefferl, U. Brehm, R. Weissert, E. Wallstrom, M. Kerschensteiner, T. Olsson, C. Linington, and H. Lassmann, Brain Pathol 1998 Oct;8(4):681-94Autoimmunity to myelin oligodendrocyte glycoprotein in rats mimics the spectrum of multiple sclerosis pathology. Brain Pathol, 1998. **8**: p. 681-94.

31.     Kornek, B., M.K. Storch, R. Weissert, E. Wallstroem, A. Stefferl, T. Olsson, C. Linington, M. Schmidbauer, and H. Lassmann,

Multiple sclerosis and chronic autoimmune encephalomyelitis: a comparative quantitative study of axonal injury in active, inactive, and remyelinated lesions. Am J Pathol, 2000. **157**: p. 267-76.

32. McGavern, D.B., L. Zoecklein, S. Sathornsumetee, and M. Rodriguez, Assessment of hindlimb gait as a powerful indicator of axonal loss in a murine model of progressive CNS demyelination. Brain Res, 2000. **877**: p. 396-400.

33. Mancardi, G., B. Hart, L. Roccatagliata, H. Brok, D. Giunti, R. Bontrop, L. Massacesi, E. Capello, and A. Uccelli, Demyelination and axonal damage in a non-human primate model of multiple sclerosis. J Neurol Sci, 2001. **184**(1): p. 41-9.

34. Yu, M., A. Nishiyama, B.D. Trapp, and V.K. Tuohy, Interferon-beta inhibits progression of relapsing-remitting experimental autoimmune encephalomyelitis. J Neuroimmunol, 1996. **64**(1): p. 91-100.

35. Wujek, J.R., C. Bjartmar, E. Richer, R.M. Ransohoff, M. Yu, V.K. Tuohy, and B.D. Trapp, Axon loss in the spinal cord determines permanent neurological disability in an animal model of multiple sclerosis. J Neuropathol Exp Neurol, 2002. **61**(1): p. 23-32.

36. Newman, T.A., S.T. Woolley, P.M. Hughes, N.R. Sibson, D.C. Anthony, and V.H. Perry, T-cell- and macrophage-mediated axon damage in the absence of a CNS-specific immune response: involvement of metalloproteinases. Brain, 2001. **124**(Pt 11): p. 2203-14.

37. Ganter, P., C. Prince, and M.M. Esiri, Spinal cord axonal loss in multiple sclerosis: a post-mortem study. Neuropathol Appl Neurobiol, 1999. **25**(6): p. 459-67.

38. Lovas, G., N. Szilagyi, K. Majtenyi, M. Palkovits, and S. Komoly, Axonal changes in chronic demyelinated cervical spinal cord plaques. Brain, 2000. **123** ( **Pt 2**): p. 308-17.

39. Evangelou, N., D. Konz, M.M. Esiri, S. Smith, J. Palace, and P.M. Matthews, Size-selective neuronal changes in the anterior optic pathways suggest a differential susceptibility to injury in multiple sclerosis. Brain, 2001. **124**(Pt 9): p. 1813-20.

40. Vyas, D., D. Bieger, and S.R. White, Intraspinal nerve terminals immunoreactive for tyrosine hydroxylase, serotonin and substance P in guinea-pigs with acute experimental allergic encephalomyelitis. Neuroscience, 1988. **26**(1): p. 253-9.

41. White, S.R., D. Vyas, and D. Bieger, Serotonin immunoreactivity in spinal cord axons and terminals of rodents with experimental allergic encephalomyelitis. Neuroscience, 1985. **16**(3): p. 701-9.

42. Stanley, G.P. and M.P. Pender, The pathophysiology of chronic relapsing experimental allergic encephalomyelitis in the Lewis rat. Brain, 1991. **114**: p. 1827-53.

43. Black, J.A., S. Dib-Hajj, D. Baker, J. Newcombe, M.L. Cuzner, and S.G. Waxman, Sensory neuron-specific sodium channel SNS is abnormally expressed in the brains of mice with experimental allergic encephalomyelitis and humans with multiple sclerosis. Proc Natl Acad Sci U S A, 2000. **97**(21): p. 11598-602.

44. Kornek, B., M.K. Storch, J. Bauer, A. Djamshidian, R. Weissert, E. Wallstroem, A. Stefferl, F. Zimprich, T. Olsson, C. Linington, M. Schmidbauer, and H. Lassmann, Distribution of a calcium channel subunit in dystrophic axons in multiple sclerosis and experimental autoimmune encephalomyelitis. Brain, 2001. **124**(Pt 6): p. 1114-24.

45. Schaecher, K.E., D.C. Shields, and N.L. Banik, Mechanism of myelin breakdown in experimental demyelination: a putative role for calpain. Neurochem Res, 2001. **26**(6): p. 731-7.

46. Huseby, E.S., D. Liggitt, T. Brabb, B. Schnabel, C. Ohlen, and J. Goverman, A pathogenic role for myelin-specific CD8(+) T cells in a model for multiple sclerosis. J Exp Med, 2001. **194**(5): p. 669-76.

47.    Medana, I., Z. Li, A. Flugel, J. Tschopp, H. Wekerle, and H. Neumann, Fas ligand (CD95L) protects neurons against perforin-mediated T lymphocyte cytotoxicity. J Immunol, 2001. **167**(2): p. 674-81.

48.    Bo, L., T.M. Dawson, S. Wesselingh, S. Mork, S. Choi, P.A. Kong, D. Hanley, and B.D. Trapp, Induction of nitric oxide synthase in demyelinating regions of multiple sclerosis brains. Ann Neurol, 1994. **36**(5): p. 778-86.

49.    Koprowski, H., Y.M. Zheng, E. Heber-Katz, N. Fraser, L. Rorke, Z.F. Fu, C. Hanlon, and B. Dietzschold, In vivo expression of inducible nitric oxide synthase in experimentally induced neurologic diseases. Proc Natl Acad Sci U S A, 1993. **90**(7): p. 3024-7.

50.    Smith, K.J., R. Kapoor, S.M. Hall, and M. Davies, Electrically active axons degenerate when exposed to nitric oxide. Ann Neurol, 2001. **49**(4): p. 470-6.

51.    Ahmed, Z., D. Gveric, G. Pryce, D. Baker, J.P. Leonard, M.L. Cuzner, and L.T. Diemel, Myelin/axonal pathology in interleukin-12 induced serial relapses of experimental allergic encephalomyelitis in the Lewis rat. Am J Pathol, 2001. **158**(6): p. 2127-38.

52.    Pitt, D., P. Werner, and C.S. Raine, Glutamate excitotoxicity in a model of multiple sclerosis. Nat Med, 2000. **6**(1): p. 67-70.

53.    Giess, R., M. Maurer, R. Linker, R. Gold, M. Warmuth-Metz, K.V. Toyka, M. Sendtner, and P. Rieckmann, Association of a null mutation in the CNTF gene with early onset of multiple sclerosis. Arch Neurol, 2002. **59**(3): p. 407-9.

54.    Linker, R.A., M. Maurer, S. Gaupp, R. Martini, B. Holtmann, R. Giess, P. Rieckmann, H. Lassmann, K.V. Toyka, M. Sendtner, and R. Gold, CNTF is a major protective factor in demyelinating CNS disease: a neurotrophic cytokine as modulator in neuroinflammation. Nat Med, 2002. **8**(6): p. 620-4.

55.    Ruffini, F., R. Furlan, P.L. Poliani, E. Brambilla, P.C. Marconi, A. Bergami, G. Desina, J.C. Glorioso, G. Comi, and G. Martino, Fibroblast growth factor-II gene therapy reverts the clinical course and the pathological signs of chronic experimental autoimmune encephalomyelitis in C57BL/6 mice. Gene Ther, 2001. **8**(16): p. 1207-13.

56.    White, S.R., P.C. Black, G.K. Samathanam, and K.C. Paros, Prazosin suppresses development of axonal damage in rats inoculated for experimental allergic encephalomyelitis. J Neuroimmunol, 1992. **39**(3): p. 211-8.

57.    Broberg, E., N. Setala, M. Roytta, A. Salmi, J.P. Eralinna, B. He, B. Roizman, and V. Hukkanen, Expression of interleukin-4 but not of interleukin-10 from a replicative herpes simplex virus type 1 viral vector precludes experimental allergic encephalomyelitis. Gene Ther, 2001. **8**(10): p. 769-77.

58.    Wallstrom, E., P. Diener, A. Ljungdahl, M. Khademi, C.G. Nilsson, and T. Olsson, Memantine abrogates neurological deficits, but not CNS inflammation, in Lewis rat experimental autoimmune encephalomyelitis. J Neurol Sci, 1996. **137**(2): p. 89-96.

59.    Paul, C. and C. Bolton, Modulation of blood-brain barrier dysfunction and neurological deficits during acute experimental allergic encephalomyelitis by the N-methyl-D-aspartate receptor antagonist memantine. J Pharmacol Exp Ther, 2002. **302**(1): p. 50-7.

60.    Sommer, N., P.A. Loschmann, G.H. Northoff, M. Weller, A. Steinbrecher, J.P. Steinbach, R. Lichtenfels, R. Meyermann, A. Riethmuller, A. Fontana, and et al., The antidepressant rolipram suppresses cytokine production and prevents autoimmune encephalomyelitis. Nat Med, 1995. **1**(3): p. 244-8.

61.    Myers, L.W., G.W. Ellison, J.E. Merrill, A. El Hajjar, B. St Pierre, M. Hijazin, B.D. Leake, J.R. Bentson, M.R. Nuwer, W.W. Tourtellotte, P. Davis, D. Granger, and J.L. Fahey, Pentoxifylline is not a promising treatment for multiple sclerosis in progression phase. Neurology, 1998. **51**(5): p. 1483-6.

62.     Baker, D., G. Pryce, J.L. Croxford, P. Brown, R.G. Pertwee, J.W. Huffman, and L. Layward, Cannabinoids control spasticity and tremor in a multiple sclerosis model. Nature, 2000. **404**(6773): p. 84-7.

63.     Hillard, C.J., S. Muthian, and C.S. Kearn, Effects of CB(1) cannabinoid receptor activation on cerebellar granule cell nitric oxide synthase activity. FEBS Lett, 1999. **459**(2): p. 277-81.

64.     Baker, D., G. Pryce, G. Giovannoni, and A.J. Thompson, The therapeutic potential of cannabis. Lancet Neurol, 2003. **2**: p. 291-298.

65.     Kipnis, J., E. Yoles, H. Schori, E. Hauben, I. Shaked, and M. Schwartz, Neuronal survival after CNS insult is determined by a genetically encoded autoimmune response. J Neurosci, 2001. **21**(13): p. 4564-71.

Chapter  A9

# ENDOTHELIAL CELLS AND ADHESION MOLECULES IN EXPERIMENTAL AUTOIMMUNE ENCEPHALOMYELITIS


Jeri-Anne Lyons and Anne H. Cross
*Department of Neurology and Neurosurgery, Washington University, Saint Louis, MO USA*



**Abstract**:     Under normal circumstances, entry of immune system cells into the central nervous system (CNS) is restricted by the blood-CNS barrier. However, with activation, cells of the immune system undergo changes that allow for an immune response within the CNS. The animal model, experimental autoimmune encephalomyelitis (EAE), is an important tool by which to investigate these processes. Using EAE, it has been shown that the establishment of an immune response in the CNS involves activation not only of the infiltrating inflammatory cells, but also of the CNS endothelial cells. Receptor-ligand interactions between CNS-EC and invading immune cells involved in EAE are multi-factorial. Particularly critical interactions occur between cell surface selectins on leukocytes and addressins on CNS-EC, and between integrins on leukocytes and their ligands on CNS-EC and other CNS cells. These interactions are good targets for potential therapies of MS, the human disease for which EAE is a model.



**Key words**:     EAE, Endothelial Cells, Adhesion Molecules


## 1.      INTRODUCTION

Experimental Autoimmune Encephalomyelitis (EAE) is an autoimmune demyelinating disease of the central nervous system (CNS) that results from the infiltration of neuroantigen-specific T cells into the CNS. These T cells in turn initiate a cell- and humoral-mediated inflammatory response that leads to the destruction of the myelin sheath, resulting in histopathology and clinical signs similar to those observed in multiple sclerosis.

Under normal circumstances, entry of immune system cells (e.g. T cells, B cells, monocytes, polymorphonuclear cells--PMNs) into the CNS is restricted by the blood-CNS or blood-brain barrier (BBB; discussed elsewhere in this volume). However, upon cell activation, T cells of any antigen specificity are able to cross the BBB and enter the CNS. During immune surveillance, only those T cells specific for neuroantigen remain for an extended period of time (1). Although unproven, it is presumed that cellular activation enhances migration across the BBB by cell types other than T cells as well.

With immune activation, blood vessel endothelial cells (EC) and antigen-specific T cells undergo morphologic and phenotypic changes, allowing for the migration of T cells from the peripheral blood into tissues at the site of an immune response. These processes of cell activation and migration have been the subjects of much investigation. In many ways, migration of T cells across the BBB into the CNS is similar to T cell migration into other anatomic sites. However, there are differences unique to the CNS. Here we will discuss the role of CNS-EC and the adhesion molecules involved in the immunopathogenesis of EAE.


## 2.        INFLAMMATORY CELL MIGRATION IN EAE

EAE lesions are comprised of T cells, monocytes/macrophages, occasional B cells and plasma cells, and reactive glial cells. During early clinical EAE, PMNs are also observed. Migration of inflammatory cells from the peripheral circulation into the CNS is normally highly restricted.

### 2.1      T Cell migration

Migration of inflammatory cells from blood into tissues is a multi-step process (2; Figure 1). While this model was first derived from studies of monocyte and neutrophil migration, most of the current data, especially as it pertains to migration of cells across the BBB, comes from studies of T cell migration. Activated lymphocytes are far better able to adhere to EC and to migrate across them than are resting T cells. Upon interaction with antigen (Ag), T cells express active forms of a variety of cell surface molecules including chemokine receptors and adhesion molecules. Similarly, chemokine and adhesion molecule expression is induced in EC at the site of an inflammatory response. Expression is induced by cytokines produced by inflammatory cells. Invading inflammatory cells and intrinsic CNS cells at the site of the immune response produce chemokines, small negatively charged chemoattractant molecules (reviewed elsewhere in this volume), which bind to receptors expressed by inflammatory cells. Chemokines also interact with extracellular matrix via electrostactic interactions to form a

chemoattractant gradient to attract inflammatory cells to the site. CNS-EC
may also express chemokine receptors, to aid in formation of a
chemoattractant gradient (3). This gradient attracts cells bearing specific
chemokine receptors to the site. As T cells approach the target site, they
loosely adhere to the EC in an interaction referred to as "tethering", typically
mediated by interaction of selectin molecules on T cells with their addressin
receptors on EC. This interaction slows the velocity of the T cells and
induces the "rolling" stage of cell migration, which is mediated by
interaction of selectin and integrin molecules on T cells with their addressin
and immunoglobulin superfamily receptors on endothelial cells. "Rolling" is
followed by "firm adhesion" (integrin-mediated), and subsequent
transmigration of the T cell across the EC barrier. The secretion of matrix
metalloproteinases by invading inflammatory cells aids transmigration
across the EC layer and into the extracellular matrix.

## 2.2    B cells and polymorphonuclear cells

Although less is known regarding the molecular mechanisms used by B
cells and PMNs to access the CNS, it is widely believed that they are similar
to mechanisms used by T cells (4). However, evidence suggests that for
PMNs different receptor-ligand interactions are involved in rolling and firm
adhesion. (5-6; section 4.2.3)

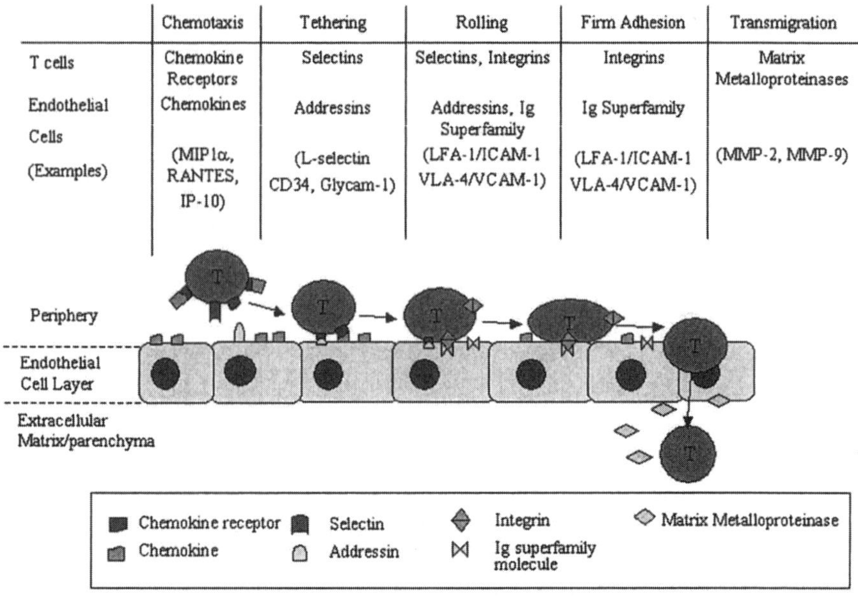

| | Chemotaxis | Tethering | Rolling | Firm Adhesion | Transmigration |
|---|---|---|---|---|---|
| T cells | Chemokine Receptors | Selectins | Selectins, Integrins | Integrins | Matrix Metalloproteinases |
| Endothelial Cells | Chemokines | Addressins | Addressins, Ig Superfamily | Ig Superfamily | |
| (Examples) | (MIP1α, RANTES, IP-10) | (L-selectin CD34, Glycam-1) | (LFA-1/ICAM-1 VLA-4/VCAM-1) | (LFA-1/ICAM-1 VLA-4/VCAM-1) | (MMP-2, MMP-9) |



*Figure A9-1.* T cell migration across an endothelial cell layer

# 3.       ROLE OF CNS ENDOTHELIAL CELLS IN EAE

## 3.1       Properties of CNS endothelial cells (CNS-EC)

The primary blood-CNS barrier is due to specialized properties of CNS-EC. Studies using tracer molecules point to the endothelial tight junctions, which involve cell membrane fusion between adjacent EC, as the primary barrier (7). CNS-EC are unique among EC in that they display little to no pinocytosis, a paucity of caveolae, reduced vesicle content, high mitochondrial content and a high electrical resistance. The luminal surface of CNS-EC is negatively charged (8), typical of EC throughout the body. The negative charge is believed secondary to localization of sialic groups and sulfated glycosaminoglycans at the luminal EC membrane (9).

Leukocyte recruitment into the CNS is not typical of that of any other body tissues, where recruitment of cells to a site of injury is rapid and often extensive. Nonetheless, during EAE, cells of the immune system locate the target CNS and penetrate the blood-brain barrier. Several studies have indicated that lymphocytes penetrate CNS-EC directly, during which time these same EC continue to display intact tight junctions with the adjacent EC

(8, 10). This phenomenon has also been reported in non-CNS tissues (11). Non-lymphocytic cells such as PMNs and monocytes do appear to invade the CNS by moving between the EC.

BBB breakdown is an early event in EAE, during which the blood-CNS barrier becomes opened at the level of both capillaries and venules at the sites of inflammation (8, 12). Gaps between EC are observed. Increased vesicular transport occurs (12). Such vascular leakage can be induced by leukotrienes, vasoactive amines and cytokines. For example, IL-1$\beta$ injected into the rabbit vitreous induces both acute and chronic inflammatory changes associated with altered blood-retinal barrier that occur within 6hr of injection and persist for 2 weeks or more (13).

Under normal circumstances and in contrast to systemic EC, lymphocytes do not constitutively adhere to CNS-EC in culture (14). Preincubation of rodent and human CNS-EC with Interferon-gamma (IFN$\gamma$) and Tumor Necrosis Factor (TNF) $\alpha$, and in some cases IL-1, increases the retention of lymphocytes on cultured CNS-EC monolayers (15). Direct injection of either TNF$\alpha$ or IFN$\gamma$ into rodent spinal cords leads to accumulation of mononuclear cells in perivenular cuffs, presumably due largely to the effects of these cytokines upon the CNS-EC (see section 3.2, below; 16).

Effects of cytokines on EC integrity are dramatically demonstrated by the vascular leak syndrome induced by systemic IL-2 administration (17). This syndrome is associated with the upregulation of E-selectin, ICAM-1 and HLA-DQ on peripheral postcapillary venular endothelium. In animal models injection of IL-2, as well as other cytokines such as IL-6, into the circulation increases BBB permeability (18-19). IL-2 also activates lymphocytes to display enhanced motility, adherence to, and migration across EC *in vitro* (20). Elevated serum IL-2 levels are found in MS patients (21), and are reported to correlate with disease activity and increased BBB permeability (22). Thus, circulating cytokines such as IL-2 may act directly or indirectly on cells of the BBB to increase permeability and enhance CNS inflammation.

EAE is the primary animal model for MS. As in EAE, MS lesions are perivenular and the microvascular EC display upregulated expression of adhesion related molecules (23-25). Several studies of MS patients have shown that BBB breakdown as detected by gadolinium enhancement on MRI, is an initial event in the development of both permanent and transient MS lesions (26).

## 3.2    Morphologic changes in CNS-EC that support cellular transmigration

Clearly, endothelial cells are not passive bystanders in the process of transmigration of cells across the blood CNS barrier. We and others have reported that CNS-EC associated with early EAE lesions display an altered, "activated" morphology at both light and electron microscopic levels (8, 27). These "activated" CNS-EC appear plumper, possess increased content of mitochondria and rough endoplasmic reticulum, and display frond-like appendages, which can be observed to contact adherent cells, at times appearing to "capture" circulating cells (14; Figure 2). This appearance is distinct from the smooth, flat appearance of CNS-ECs of normal mice.

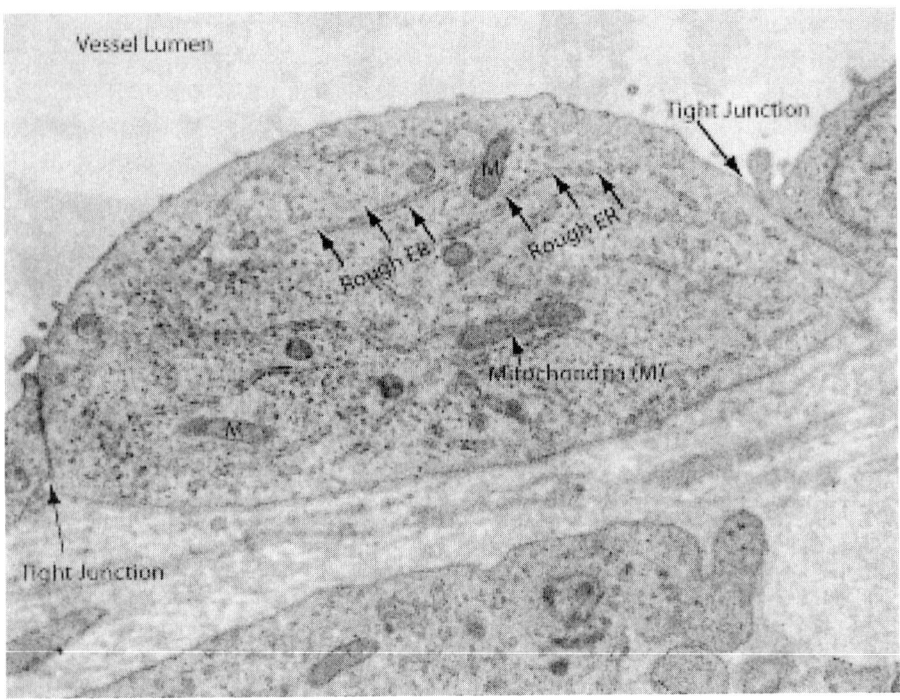


*Figure A9-2.* Electron micrograph of CNS-EC during acute EAE induced by passive transfer of myelin basic protein sensitized lymph node cells in SJL mouse. Typical of CNS-EC observed at sites of inflammation during acute EAE, this cell is plumper than normal with increased amounts of rough endoplasmic reticulum (rough ER) and mitochondria (M). The EC nucleus is out of the plane of section. Tight junctions with the adjacent EC, characteristic of the BBB, are noted at either side (long arrows). X 20,000. [See Ref. 10 for methods.]

Exposure to cytokines such as TNFα and IFNγ induces the expression of ICAM-1 and other adhesion related molecules by CNS-EC, as well as LFA-

1 by leukocytes. The interaction of ICAM-1 on CNS-EC with LFA-1 expressed by leukocytes leads to signal transduction events that alter EC morphology and function. (28). Upon ligation by LFA-1, the cytoplasmic domain of ICAM-1 interacts with α-actinin (29) and with the actin-binding protein and p60Src substrate, cortactin, leading to its tyrosine phosphorylation (28).The activation of actin-associated proteins and the subsequent cytoskeletal rearrangements induced with ICAM-1 ligation suggest a role for these interactions in the observed morphologic changes of CNS-EC (30).

## 3.3    Inflammatory mediators expressed by CNS-EC

### 3.3.1    Chemokine production by CNS-EC

CNS-EC elaborate a number of inflammatory mediators upon activation, including chemokines, cytokines, and vasoactive amines. At sites of EAE lesions, CNS-EC produce chemokines, such as MCP-1 (31). Cultured human brain ECs are reported to produce several chemokines, including IL-8, MCP-1, IP-10, MIP1β, and RANTES (32-33). Expression of these chemokines can be induced by cytokines such as TNFα, IFNγ, IL-1β (32-33). Chemokines are concentrated at the luminal and basal surfaces of CNS-EC via binding to glycosaminoglycans and chemokine receptors (33-34).

### 3.3.2    Cytokine production by CNS-EC

Both pro-inflammatory (e.g., IL-1α, IL-6, TNFα, GM-CSF) and regulatory cytokines (e.g., IFNβ and TGFβ1 & 2) are among the molecules reported to be secreted by CNS-EC. Production of cytokines by CNS-EC is regulated by glial cell-derived soluble cytokines, including IL-1β, IL-12, TNFα, IL-10 and TGFβ (35).

### 3.3.3    MHC & costimulatory molecule expression

The cell responsible for initial presentation of antigen to encephalitogenic T cells in EAE (and possibly MS) is unknown. CNS-EC remain a contender, as do astrocytes, but most evidence favors perivascular microglia. Based upon what is known regarding antigen presentation to CD4 T cells, the CNS antigen presenting cells must express MHC II and costimulatory molecules (B7-1, B7-2).

Although CNS-EC do not constitutively express MHC II, they can be induced to do so, and IFNγ is especially potent in this role. Early on, Sobel et al. showed expression of MHC on CNS-EC during EAE (36). Human and mouse CNS-EC have also been shown to express B7 costimulatory molecules when stimulated with IFNγ *in vitro* (37-38), but this has been difficult to demonstrate on EAE-affected CNS tissues immediately ex vivo (39). Several independent groups have each observed that CNS-ECs do not support primary antigen-induced T cell proliferation, even upon stimulation with cytokines such as IFNγ (40-42). Furthermore, mouse CNS-EC produce 6-KPGF1α, a stable product of prostaglandin $GI_2$ ($PGI_2$) that directly inhibits lymphocyte proliferation to antigen. Exogenous macrophages or recombinant IL-1 induce murine CNS-EC to produce $PGI_2$ in a dose dependent manner. Thus, CNS-EC may actively inhibit the response to antigen by T cells at the BBB by production of 6-KPGF1α (43).

# 4.      ROLE OF ADHESION MOLECULES IN EAE

## 4.1      Selectins and Addressins

The initial step of transmigration of inflammatory cells across an endothelial cell barrier is "tethering" of activated cells to activated endothelial cells via the interaction of selectin molecules with addressin molecules (44). These molecules can also mediate the subsequent rolling of T cells along the endothelial cell surface. This adhesion of cells to vascular endothelium is a long-recognized process termed "pavementing" by Cohnheim in the late 1800's (45; Figure 3).

The three members of the selectin family include L-selectin (CD62L, found on all leukocytes), P-Selectin (CD62P, found on platelets and endothelial cells), and E-Selectin (CD62E, found on endothelial cells). Selectins are transmembrane glycoproteins with intra- and extra-cellular domains that bind carbohydrate ligands via a calcium-dependent lectin domain. The intracellular domains of the selectins are structurally distinct from each other, but highly conserved between species for a given selectin, suggesting specific effector functions following the ligation of the particular selectin molecules. The extracellular domain is composed of a calcium-dependent lectin domain, an epidermal growth factor (EGF)-like domain, and a series of short consensus repeats (SCR) with homology to complement-binding proteins (reviewed in 46). The extracellular domains of the selectins differ primarily in the number of SCR present. The EGF-like and lectin-binding domains of L-, P-, and E-selectin are quite similar, suggesting that the molecules interact with similar, if not overlapping, ligands (47).

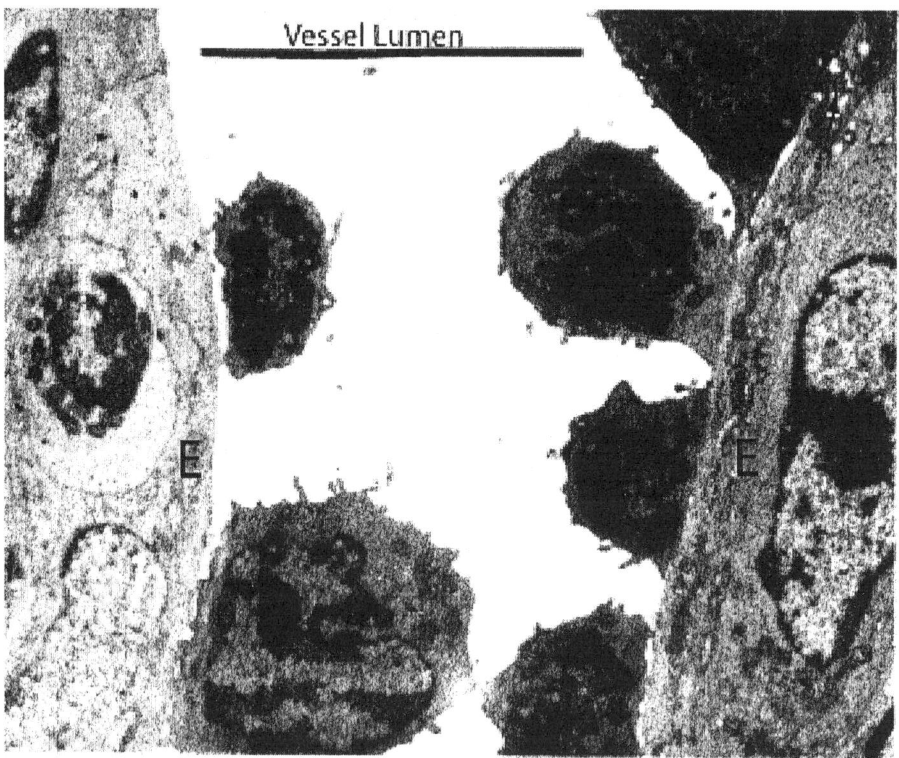


*Figure A9-3.* "Pavementing" is demonstrating in this electron micrograph taken from a well-perfused SJL mouse less than 24 hr following clinical onset of passively transferred EAE. Mononuclear cells line the luminal aspect of this small venule. Note the tiny finger-like projections of the lymphocytes contacting the endothelium (arrow upper right). At higher magnifications, clathrin coated pits are observed at these contact points. X 10,000

Early reports suggested a role for selectin molecules and their addressin receptors in EAE. Naparstek et al. were perhaps the first to recognize the importance of these molecules in the development of EAE (48). These investigators observed strong binding of neuroantigen-specific T cells to the lectin, peanut agglutinin (PNA) upon activation, suggesting that T cell activation induced expression of lectin receptors. Furthermore, about 75% of activated neuroantigen-specific T cells acquired the lectin binding, and this correlated strongly with ability to transfer EAE to naïve recipients.

Selectins are known to interact with sialylated mucin-like receptors on EC and other leukocytes, and these sialyl moieties are required for receptor-ligand interaction (discussed below). In a second study, pretreatment of CD4+ lymphocytes with sialidase prior to injection into MBP-primed Lewis rats led to a significant reduction in the accumulation of these cells in the

CNS of animals, compared to injection of untreated cells. Circulating levels of treated and untreated cells were similar. Daily injections of sialidase delayed the onset of clinical and histologic EAE in a passive transfer model (49).

### 4.1.1    L-selectin

The role of L-selectin in the initial stages of lymphocyte migration across most endothelial barriers is accepted (reviewed in 50). However, a function for L-selectin in migration across the BBB remains controversial. It has been suggested that the adhesion molecule phenotype of T cells in the inflamed CNS differs from the phenotype of T cells at other inflammatory sites, including a lack of L-selectin expression by CNS inflammatory T cells by flow cytometric methods (51). Separate studies in a mouse model of EAE did identify L-selectin[pos] cells in the CNS by histologic observation. These L-selectin[pos] cells were noted first prior to clinical onset, beginning 8d following immunization with CNS homogenate (52). L-selectin[pos] cell numbers increased through d18 and then declined with remission of signs. However, injection of anti-L-selectin monoclonal antibodies (mAb) had no effect on disease course or severity in this model or in a peptide-induced mouse EAE model (53). Moreover, L-selectin was not detected on T cell lines and clones used for adoptive transfer. Taken together these observations suggest that L-selectin is not necessary for migration of T cells into the CNS.

In contrast, in a model of MBP-induced EAE in the Lewis rat treatment with either a mAb to L-selectin or F(ab'$_2$) fragments of the mAb suppressed actively induced disease. Passively transferred disease was also mildly suppressed. Histological examination demonstrated a marked reduction in CNS inflammatory cell infiltration. T cell numbers were decreased in the draining lymph nodes and spleens in treated animals (54). Thus, suppression of EAE may have been due to insufficient trafficking of T cells to peripheral lymphoid organs leading to insufficient priming of encephalitogenic T cells, rather than defects in migration into the CNS.

The conflicting results on the effect of anti-L-selectin mAb on EAE induction could be due to differences in the EAE models employed. Differences in the detection of L-selectin[pos] cells between systems could be a reflection of the different methods of tissue isolation and detection, as it has been shown that L-selectin expression can be altered by various isolation techniques (55-56). Alternatively, it is known that L-selectin is shed readily upon cell activation (reviewed in 50). Thus, differences in the detection of cell surface L-selectin may be related to time-points chosen for observation.

Recent data has suggested a novel role for L-selection expression by CNS-infiltrating macrophages in EAE pathogenesis (57). Mice expressing an MBP-specific T cell receptor (MBP-TCR) transgene were made deficient for L-selectin to study the role of the latter molecule in EAE pathogenesis. L-selectin[neg] MBP-TCR mice were not susceptible to clinical EAE following immunization with the encephalitogenic MBP$_{Ac1-11}$ peptide. Histologic examination of CNS tissue revealed the presence of typical inflammatory infiltrates but a lack of demyelination in L-selectin[neg] mice. Moreover, while inflammatory cells were present, they were localized to the perivascular space and did not enter the CNS parenchyma as was observed in L-selectin[pos] mice. Passive transfer of L-selectin[pos], but not L-selectin[neg], macrophages resulted in myelin destruction and clinical EAE in L-selectin[neg] mice, indicating a critical role for L-selectin expression by macrophages in the effector phase of disease. Interestingly, while L-selectin[pos] macrophages did enter the parenchyma of L-selectin[neg] mice with EAE, the endogenous L-selectin[neg] T cells remained in the perivascular space, suggesting a role of L-selectin in the final step of transmigration of cells across the parenchymal basement membrane into the CNS tissue.

### 4.1.2    P-selectin

The role of P-selectin in EAE remains controversial. P-selectin can be induced on mouse CNS-EC by inflammatory cytokines (TNFα, IL-1β, IFNγ) *in vitro* (58). However, P-selectin could not be detected on CNS-EC *in vivo* by immunohistologic methods, although expression was detected on EC outside the CNS. In addition, administration of anti-P-selectin Ab had no effect on immune cell infiltration of the CNS or actively induced EAE in a mouse model of disease. Interestingly, induction of P- (and E-) selectin on primary CNS-EC *in vitro* was inhibited by astrocyte culture-derived supernatants, suggesting that active inhibition by the CNS microenvironment of adhesion molecule expression may play a role in maintenance of the BBB.

More recent studies have supported a role for P-selectin in EAE induction (59-62). Utilizing a passive transfer model and looking at early time-points (2h and 24h) post transfer (p.t.), Carrithers et al. (59) demonstrated a role for P-selectin in the earliest recruitment of T cells into the CNS. Mice received MBP-specific T cells labeled with a fluorescent dye, and frozen sections of CNS tissue were prepared at various times p.t. Ag-specific T cells were detected in the CNS 2h p.t., and these numbers were increased at 24h. Characterization of adhesion molecules in the CNS revealed expression of ICAM and VCAM (discussed below) beginning at 24h. P-selectin expression was detected 2h p.t. in the choroid plexus and meninges, but was undetectable at 48h. Thus, expression of P-selectin in the CNS coincided with Ag-specific T cell infiltration of the CNS at early time-points post adoptive transfer. Furthermore, the early migration of T cells into the CNS was inhibited by treatment with Ab recognizing P-selectin, but not $\alpha 4$ integrin (VLA-4, discussed below). These studies suggest a role for P-selectin in the migration of "pioneer" T cells (i.e., those T cells that will initiate the EAE-inducing inflammatory response) into the CNS.

Data from a rat model of EAE further support a role for P-selectin in the inflammatory infiltration of the CNS (61). Immunization of PVG rats deficient for complement component 6 (C6) with neuroantigen resulted in less severe clinical EAE than in control (C6$^{pos}$) animals. T cell and macrophage infiltration in the spinal cords of C6$^{neg}$ rats was also less than that observed in C6$^{pos}$ animals. Immunohistochemistry of CNS tissues revealed a decrease in ICAM-1 expression by EC and infiltrating cells and a lack of P-selectin on vascular endothelium in C6$^{neg}$ rats, suggesting a role for these adhesion molecules in immune cell infiltration of the CNS. Cytokine-induced activation of EC renders EC susceptible to complement activation (63). Thus, an effect of complement activation on activated EC at the site of inflammation may upregulate adhesion molecule expression and further compromise the BBB, leading to amplification of the immune infiltrate into the CNS.

Two independent groups employing intravital microscopy techniques have further defined a role for P-selectin in lymphocyte recruitment across the BBB (60, 62). Kerfoot and Kubes (62) utilized intravital microscopy to study leukocyte rolling and adhesion during the course of actively induced EAE in mice. Both rolling and adhesion were observed 2d prior to onset of clinical signs and remained elevated for 2 weeks before returning to normal levels by 5 weeks post onset of clinical signs. Expression of P-selectin in EAE-primed mice closely followed this profile. Furthermore, while anti-$\alpha 4$ integrin (discussed below) Ab were able to partially block the rolling and adhesion of leukocytes, especially at later time-points, co-treatment with anti-$\alpha 4$ and anti-P-selectin blocked all remaining rolling at all time-points. A role for P-selectin-mediated adhesion in regulation of $\alpha 4$ integrin-mediated adhesions was suggested by the finding that anti-P-selectin treatment lead to a complete block of leukocyte rolling and a 70% reduction in adhesion. In a separate study employing a similar technique, Piccio et al. (60) demonstrated that treatment with Ab recognizing P-selectin was able to block almost 100% of tethering and rolling of neuroantigen-specific T cells in their model. Co-treatment with antibodies recognizing P- and E-selectin had a similar effect, and treatment with anti-E-selectin alone blocked rolling by 77%, suggesting that both P- and E-selectin are important to primary adhesion events in the

CNS. Interestingly, treatment of mice with anti-L-selectin had no effect on Ag-activated lymphocyte rolling (see Section 4.1.1).

Although neutrophils are not a component of the chronic EAE lesion, these cells can be found in the CNS at the first signs of clinical disease. Recently, the role of selectins in the migration of neutrophils across the BBB was investigated in a rat system (64). It is known that injection of IL1-β into the CNS of juvenile rats leads to the accumulation of neutrophils at the site of the injection. This accumulation leads to further breakdown of the BBB (65). Bernardes-Silva et al. (64) demonstrated that the stereotactic injection of IL-1β into brains of juvenile rats resulted in the expression of E-selectin and P-selectin on CNS vessels by immunohistochemistry. Depletion of neutrophils prior to IL-1β treatment had little effect on E-selectin expression, but resulted in significantly lower expression of P-selectin, indicating that neutrophil recruitment may play an important role in the upregulation of P-selectin in this model. The functional importance of P-selectin to neutrophil migration across the BBB was demonstrated by the inhibitory effect of anti-P-selectin Ab, but not anti-E-selectin, on the accumulation of neutrophils in the CNS of treated rats. Thus, P-selectin may function in the migration of neutrophils into the CNS, which may in turn be important to the further compromise of the BBB and recruitment of inflammatory cells into the CNS.

### 4.1.3    E-selectin

The role of E-selectin in EAE has received little attention. E-selectin has been observed on CNS-EC in MS lesions (26). Induction of expression on cultured human astrocytes by TNFα and on mouse CNS-EC by TNFα, IL-1β, or IFNγ *in vitro* has been reported (58, 66). However, E-selectin expression by CNS-EC could not be detected *in vivo*, and administration of anti-E-selectin did not affect infiltration of inflammatory cells across the BBB or the development of EAE (58). As discussed above, more recent studies have suggested a role for E-selectin in immune cell migration across the BBB (60).

### 4.1.4    Ligands of selectin molecules ("Addressins")

The selectin molecules interact with their ligands through the N-terminal lectin domain. Consistent with this, most of the identified ligands of selectins are mucin-like molecules with carbohydrate modifications, including sialylation, fucosylation, and carbohydrate sulfation, necessary for receptor interaction (reviewed in 67). Such ligands, originally termed "Peripheral Node Addressins--PNAd", include GlyCAM-1, CD34, MadCAM-1, sialyl

Lewis x (sLe$^x$), and PSGL-1 (46, 68-71). More recently, other non-mucin ligands for selectins, including heparin, have been identified (72).

### 4.1.4.1    Addressins in EAE

Early studies by Cannella et al. noted expression of high endothelial cell (HEV) addressins localized to the CNS during periods of active inflammation in a mouse model of EAE (73). More recently, encephalitogenic T cells mediating EAE and inflammatory cells present in the CNS during active disease were shown to express ligands for P-selectin and, to a lesser extent, E-selectin (51, 58). These investigators used fluorescent selectin-Ig fusion proteins to identify cells expressing selectin ligands by flow cytometry but did not establish the identity of the ligands expressed.

Pertinent to EAE and MS, a CNS-specific ligand for L-selectin has been identified. Using a mouse L-selectin-Ig fusion protein, this ligand was shown to be localized to CNS white matter and, more specifically, to the myelin sheath of CNS axons (74). This ligand was not apparent in PNS myelin (75). The identity of this ligand has yet to be determined.

Of the identified selectin ligands, a role for PSGL-1 in EAE has been specifically addressed (60). Mice were pretreated with LPS or TNF to activate endothelial cells followed by the passive transfer of fluorescently labeled T cells specific for neuroantigen. The migration of cells was followed by invtravital microscopy. The neuroantigen specific T cells were observed to roll and arrest activated endothelium, whereas no such behavior was observed in untreated animals. Furthermore, the percentage of neuroantigen-specific T cells rolling along the vascular endothelium was significantly higher than for unactivated peripheral lymph node cells. The rolling behavior was nearly completely inhibited in the presence of anti-PSGL-1 Ab, but not antibody recognizing L-selectin, indicating a role for PSGL-1 and P/E selectin in the early stages of T cell migration across CNS-EC.

A role for MAdCAM-1 in EAE has also been investigated (76-77). MAdCAM-1 is relevant not only as a ligand for the selectins but also as a ligand for alpha4beta7 integrin (discussed below). Expression of MAdCAM-1 was induced on choroid plexus epithelial cells, but not endothelial cells, in a mouse model of EAE. Expression could be induced on choroid plexus epithelial cells *in vitro* following incubation with proinflammatory cytokines (TNFα, IFNγ, LPS), and MAdCAM-mediated adhesion of lymphocytes using *in vitro* binding assays with frozen tissue was demonstrated (76). It was subsequently shown that anti-MAdCAM-1 antibody, administered in the presence or absence of anti-VCAM-1 and anti-ICAM-1 led to rapid remission of clinical signs when given early in disease progression, although combinatorial therapy was more efficient. Anti-MAdCAM-1 alone or in combination with anti-VCAM-1 and anti-ICAM-1 had no effect on remission when given later in disease (77).

### 4.1.5    Non-adhesion related functions of selectins

Although selectins are predominantly thought of as adhesion molecules, selectin-ligand interaction results in signal transduction in the selectin-expressing cells as evidenced by increases in intracellular calcium (reviewed in 46). Ligation of L-selectin or E-selectin on neutrophils results in aggregation, augmentation of the respiratory burst, oxygen radical production, and upregulation of Mac-1 expression (78-84). Similar activation of monocytes is noted following P-selectin/PSGL-1 interaction, including secretion of chemokines and cytokines involved in inflammation (85-87). Further activation of T cell transmigration results from L-selectin-mediated activation of beta1 and beta2 integrins and increased adhesion to fibronectin (88-89).

## 4.2    Integrins and their Immunoglobulin Superfamily receptors

### 4.2.1    Integrin / Ig Superfamily member interactions

Activated T cells and other leukocytes express the integrins, LFA-1 and VLA-4. These molecules are ligands for the Ig-superfamily molecules, ICAM-1 and –2, and VCAM-1, respectively. Interaction of these receptor-ligand pairs is believed to be critical in T cell adherence to CNS-EC. Resting cultured brain ECs constitutively express low levels of ICAM-1, and minimal VCAM-1 (90). In response to inflammatory cytokines such as IFNγ and TNFα, expression of ICAM-1 and VCAM-1 by cultured rodent and human CNS-EC is greatly enhanced (91-92).

### 4.2.2    Role of ICAM-1 interactions with LFA-1 in EAE

To determine the function of the integrins and their Ig superfamily member receptors in EAE, studies using mAb directed against these molecules were performed. Archelos et al. reported that either intact or F(Ab)$_2$ fractions of a purified antibody to ICAM-1 inhibited actively induced EAE in guinea pigs (93). Inflammatory infiltration of the CNS was greatly reduced. In contrast, the anti-ICAM-1 treatment had minimal effect upon adoptively transferred EAE. Proliferation to antigen was suppressed in T cells derived from anti-ICAM-1-treated animals. Another study of anti-ICAM-1 in a mouse model of adoptively transferred EAE using MBP-sensitized cells had similar results, with minimally decreased severity in anti-ICAM-1-treated mice (94). EAE induction in ICAM-1-deficient mice revealed decreased proliferation and Th1 cytokine production in response to antigen in the mutant mice, but unexpectedly EAE was worse in the ICAM-1 deficient mice (95). These studies suggest that ICAM-1 plays a more important role in the induction phase of EAE than in the adherence and transendothelial migration of leukocytes.

### 4.2.3    Role of α4β1 and αEβ7 integrins in EAE and MS

Several groups have studied the role of α4β1 integrins (VLA-4) in EAE. Yednock et al. found that binding of rat and human mononuclear cells to EAE affected vessels *in vitro* was inhibited by pretreatment of the mononuclear cells with antibody directed toward α4 integrin (96). Moreover, *in vivo* administration of anti-α4 integrin prevented CNS inflammatory cell infiltration and clinical signs of EAE in a dose-dependent manner. Members of this same group later showed that treatment with antibody against α4β1integrin markedly reduced the clinical and pathologic features of EAE in the Hartley guinea pig. Perhaps more significantly, they found that treatment after clinical onset led to rapid recovery (97). Another group of investigators reported that surface expression of α4β1 integrins by Th1 T cells was a critical feature in determining encephalitogenicity. In the latter study of T cell clones and lines directed against MBP peptide Ac1-16, expression of α4β1 integrins correlated with encephalitogenicity, whereas antigen-specific proliferation and production of cytokines IL-2, IFNγ, LT or TNF did not (98).

The β7 subfamily of integrins has two members, α4β7 and αEβ7. Both are expressed solely by leukocytes. The ligand for α4β7 is MAdCAM-1, identified by the antibody MECA89. This molecule is upregulated on CNS-EC with the morphology of high endothelial venular EC during EAE (73). αEβ7 is expressed by intraepithelial lymphocytes and interacts with E-cadherin on gut-associated lymphoid tissue. Expression of E-cadherin can be induced on cultured CNS-EC. Treatment with an antibody directed against β7 integrins leads to remission of established chronic EAE induced with myelin oligodendrocyte glycoprotein (MOG). This remission is even more dramatic when anti-β7 and anti-α4 antibodies are administered together. Mice deficient for β7 integrins display attenuated EAE (99). These findings strongly implicate the interaction of β7 integrins on leukocytes with their ligands expressed by activated CNS-EC, or perhaps other cells, in the pathogenesis of EAE.

The interaction of α4 integrins with their ligands appears critical to multiple sclerosis disease activity. Early phase trials have shown that treatment of MS patients with a humanized monoclonal antibody against α4 integrins (natalizumab) decreases the number of MS lesions on MRI (100). A multi-center Phase III trial of this drug in MS patients is currently underway.

### 4.2.4    PECAM-1 in EAE

At least one member of the Ig superfamily appears to play a key role in the maintenance of the BBB. Platelet/endothelial cell adhesion molecule-1 (PECAM-1, CD31) is a member of the Ig superfamily of transmembrane proteins and is expressed on EC, platelets, PMNs and subsets of lymphocytes. In systemic organs, expression of PECAM-1 by EC and PMNs is important for neutrophil transmigration into tissues.

Ligation of PECAM-1 expressed by leukocytes mediates signal transduction events culminating in the expression of β1 and β2 integrins of increased affinity. This would be expected to increase adhesion of leukocytes, enhancing inflammation. However, inhibiting PECAM-1 with antibodies does not alter the course of EAE in the rat (101). More unexpectedly, PECAM-1 deficient mice display earlier onset of clinical EAE, with earlier CNS infiltration by inflammatory cells and with exaggerated CNS vascular permeability compared to wild-type mice. *In vitro*, T cell migration across artificial BBB made from PECAM-1[neg] EC is enhanced, with PECAM-1[neg] EC displaying increased susceptibility to histamine-induced permeability. Thus, PECAM-1 appears critical to maintaining BBB integrity in the setting of inflammation or histamine challenge (102).

## 4.3    Other adhesion-related molecules

Data exist for the role of adhesion molecules that are not members of the above families in EAE.

### 4.3.1    Leukosialin (CD43)

Leukosialin is a conformationally-extended single-chain leukocyte glycoprotein that binds to ICAM-1 (103). It is expressed by all leukocytes, including T cells, macrophages, neutrophils, and a subpopulation of B cells.

Leukosialin/ICAM-1 interaction leads to intracellular signaling in the leukosialin-expressing cell that synergizes with TCR/CD3 signaling resulting in T cell activation (104). Immunohistochemical analysis of leukosialin[pos] cells over the course of actively induced EAE in mice revealed their infiltration beginning approximately 8 days post immunization, reaching peak levels at days 11-18 and persisting at peak levels through 28 days post immunization. These cells were noted adhering to the lateral ventricular choroid plexus, in periventricular loci in the hippocampus and in perivascular locations near the optic tract. Injection of Ab directed against leukosialin during the onset of EAE resulted in a trend towards more severe clinical EAE, suggesting inhibition of regulatory mechanisms in disease (52).

### 4.3.2    Laminins

Laminins are heterotrimeric glycoproteins that form a major functional component of all basement membranes. At least 12 isoforms exist, and expression can be regulated by proinflammatory cytokines (105). It is known that laminins 8 and 10 are expressed in endothelial basement membranes, and laminins 1 and 2 are expressed by blood vessels in the CNS, but not in other locations (106-108).

Larger blood vessels in the CNS are associated with at least two basement membranes--an endothelial basement membrane and an astroglial basement membrane (109). The astroglial basement membrane and the meningeal epithelial basement membrane together form the "parenchymal basement membrane", which defines the border to the brain parenchyma. The perivascular space, where inflammatory infiltrates accumulate during EAE, is defined by the endothelial and parenchymal basement membranes. Thus, in order to penetrate the brain parenchyma, inflammatory immune cells must penetrate two physiologically distinct basement membranes. That cell migration across the two distinct basement membranes is not mechanistically identical is suggested by EAE experiments in macrophage-depleted mice in which infiltrates are trapped in the perivascular space, failing to enter the brain parenchyma (110). Similar localization of cells to the perivascular space is also noted with EAE experiments in TNF-deficient mice (111).

The role of laminins in directing T cell migration across the BBB and into the brain parenchyma was investigated in actively induced EAE (112). Using immunofluorescent staining of tissue sections, localization of laminins 8 and 10 to the endothelial basement membrane and laminins 1 and 2 to the parenchymal basement membrane was observed. Inflammatory cell migration across the endothelial basement membrane took place in areas where only laminin 8 was expressed, suggesting that laminin 10 is somehow restrictive to cell migration, perhaps through high affinity interaction with T cells. Using neuroantigen-specific T cell lines, activation-dependent integrin-mediated interactions of T cells with laminin 8 and laminin 10 were demonstrated. Penetration of the parenchymal basement membrane was characterized by expression of both laminins 1 and 2 in the absence of smooth muscle actin. However, no physical interaction between activated T cells with either of these lectins was demonstrated.

### 4.3.3    CD44 (Pgp-1)

CD44 is a family of cell surface proteoglycans thought to function as adhesion molecules. Observed roles include binding to extracellular matrix, cell migration, lymphocyte development, tumor cell metastasis, and lymphocyte homing (113-115). CD44 has been shown to interact with extracellular matrix proteins, including hyaluronate, chrondroitin sulfate, collagen, fibronectin, and laminin (116-120). More recently, CD44 has been shown to function as a receptor for L- and E-selectin (121-122).

CD44 has been shown to mediate tethering and rolling under conditions of physiologic flow (115, 123). A role for CD44 in cell migration across the BBB is suggested by experiments in the EAE model (53, 124). In a passive transfer model of disease, pre-coating encephalitogenic T cells with anti-CD44 prior to transfer reduced cell infiltration into the CNS by half when compared to uncoated cells. Furthermore, mice treated with anti-CD44 antibody failed to develop inflammatory lesions in the brain, and EAE was inhibited in a dose-dependent manner by anti-CD44 antibody (53). Similar results were obtained by Brennan et al. (124), who further showed that amelioration of disease was due to antibody-induced shedding of CD44 by encephalitogenic T cells. While trafficking of T cells to the CNS was inhibited, trafficking to the lymph nodes was unaffected in Ab treated mice, suggesting that CD44 is specifically involved in the trafficking of Ag-primed cells to sites of inflammation.

### 4.3.4    MECA-325

MECA-325 defines a marker of high endothelial venules (HEV) in lymphoid tissue of mice and is associated with cell transmigration. The identity of the moiety defined by MECA-325, or its exact function is unknown. Expression of the MECA-325 epitope correlated with disease signs in a relapsing-remitting model of EAE in SJL mice induced by passive transfer of myelin basic protein-sensitized lymphocytes (73). MECA-325 was expressed on CNS venules 24-48 hr prior to initial signs of murine EAE, and at higher levels during acute clinical disease. Normal SJL CNS vascular cells did not express MECA-325 *in situ*. Expression decreased to preclinical levels during remission and rose again during relapses. This expression profile was similar to that of molecules also thought to be involved in EAE pathogenesis (ICAM-1 and MHC II), which fluctuated similarly on and around vascular cells during EAE.

## 4.4     Cytokine regulation of adhesion molecule expression and BBB permeability

### 4.4.1     Regulation by Th2 cells

It is thought that Th1 cytokines (i.e., IFNγ, TNFα) are necessary and responsible for the induction of EAE, while Th2 cytokines (IL-4, IL-10) are protective for disease. Transfer of Th2 neuroantigen specific T cells, either with transfer of encephalitogenic (Th1) neuroantigen specific T cells or with active immunization with neuroantigen, ameliorates EAE (125-126). Organotypic entorhinal-hippocampal slice cultures were used to study the mechanism by which Th2 cells are protective for EAE (127). By culturing tissue slices with Th1 or Th2 cells specific for neuroantigen, it was demonstrated that both cell types could invade the parenchyma and interact with microglia. However, only Th1 cells induced the expression of ICAM-1 by microglia. Furthermore, Th2 cells were able to inhibit and reverse the induction of ICAM-1 by Th1 cells, suggesting that Th2 cells may protect against EAE in part by affecting migration of inflammatory cells across the BBB.

### 4.4.2     Effects of Interferon beta on adhesion molecule expression in EAE

Interferon beta (IFNβ) is of established benefit in relapsing-remitting MS, decreasing relapse rate, MRI activity and accumulation of disability. IFNβ also decreases clinical and pathologic manifestations of EAE (128). The mechanisms by which IFNβ acts are unknown, however in a rat model of EAE, a marked reduction in expression of ICAM-1 and VCAM-1 by CNS-EC was noted, accompanied by decreased mononuclear cell infiltration of the CNS (129). Cultured CNS-EC upregulate ICAM-1 and VCAM-1 upon exposure to proinflammatory cytokines and this upregulation can be inhibited in a dose-dependent manner by co-incubation with IFNβ. IFNβ has a number of other effects that may benefit MS, including the diminution of MMP secretion by T cells (130).

# 5. CONCLUSION

The BBB represents a specialized barrier to inflammatory cells. Migration across the BBB is highly regulated, preventing migration of immune cells during normal circumstances. However, in certain diseases this barrier is breached. The EAE model provides a valuable tool for studying immune cell migration into the CNS and is proving useful in the development of therapies to counteract human CNS inflammation.

# ACKNOWLEDGMENTS

We thank Dr. Jen Stark for critical reading of the manuscript and dedicate this chapter to the memory of our colleague, Dr. John L. Trotter. The electron micrographs were taken while AHC was a post-doctoral fellow in the laboratory of Dr. Cedric S. Raine at Albert Einstein College of Medicine. Supported by the National MS Society USA.

# REFERENCES:

1. Hickey, WF, Hsu, BL, Kimura, J. T-lymphocyte entry into the central nervous system. J Neurosci Res 28:254-260, 1991.
2. Springer, TA. Traffic signals for lymphocyte recirculation and leukocyte emigration: the multistep paradigm. Cell. 76:301-314, 1994.
3. Andjelkovic AV, Spencer DD, Pachter JS. Visualization of chemokine binding sites on human brain microvessels. J Cell Biol 145: 403-412, 1999.
4. Rott, LS, Briskin, MJ, Butcher, EC. Expression of alpha4beta7 and E-selectin ligand by circulating memory B cell: implications for targeted trafficking to mucosal and systemic sites. J. Leuk Biol 68:807-814, 2000.
5. Walcheck, B, Moore, KL, McEver, RP, Kishimoto, TK. Neutrophil-neutrophil interactions under hydrodynamic shear stress involve L-selectin and PSGL-1. A mechanism that amplifies initial leukocyte accumulation of P-selectin in vitro. J Clin Invest 98:1081-1087, 1996.
6. Dwir, O, Shimron, F, Chen, C, Singer MS, Rosen SD, Alon R. GlyCAM-1 supports leukocyte rolling in flow: evidence for a greater dynamic stability of L-selectin rolling of lymphocytes than of neutrophils. Cell Adhes Commun 6:349-370, 1998.
7. Dempsey EW, Wislocki GB. An electron microscopic study of the blood-brain barrier in the rat, employing silver nitrate as a vital stain. J. Biophys Biochem Cytol 1: 245-256, 1955.
8. Lossinsky AS, Badmajew V, Robson JA, Moretz RC, Wisniewski HM. Sites of egress of inflammatory cells and horseradish peroxidase transport across the blood-brain barrier in a

murine model of chronic relapsing experimental allergic encephalomyelitis. Acta Neuropathol 78: 359-371, 1989

9.  Nag S, Ultrastructural localization of lectin receptors on cerebral endothelium Acta Neuropathol 66: 105-110, 1985.

10. Cross AH and Raine CS. Central nervous system endothelial cell-polymorphonuclear cell interactions during autoimmune demyelination. Am J Pathology 139: 1401-1409, 1991.

11. Marchesi VT, Gowens JI. The migration of lymphocytes through the endothelium of venules in lymph nodes: an electronmicroscope study. Proc R Soc Lond, Ser B 159: 283-290, 1964.

12. Claudio L, Kress Y, Norton WT, Brosnan CF. Increased vesicular transport and decreased mitochondrial content in blood-brain barrier endothelial cells during experimental autoimmune encephalomyelitis. Am J Pathol 135: 446-463, 1989.

13. Martiney JA, Litwak M, Berman JW, Arezzo JC, Brosnan CF. Pathophysiologic effect of interleukin-1b in the rabbit retina. Am J Pathol 137: 1411-1423, 1990.

14. Male D, Pryce G, Rahman J. Comparison of the immunological properties of rat cerebral and aortic endothelium. J Neuroimmunol 30: 161-168, 1990.

15. Hughes CCW, Male DK, Lantos PL. Adhesion of lymphocytes to cerebral microvascular cells: effects of interferon-$\gamma$, tumour necrosis factor and interleukin-1. Immunology 64: 677-681, 1988.

16. Simmons RD and Willenborg DO. Direct injection of cytokines into the spinal cord causes autoimmune encephalomyelitis-like inflammation. J Neurol Sci 100: 37-42, 1990.

17. Cotran RS, Pober JS, Gimbrone MA, Springer TA, Wiebke EA, Gaspari AA, Rosenberg SA, Lotze MT. Endothelial activation during interleukin 2 immunotherapy. A possible mechanism of the vascular leak syndrome. J Immunol 139: 1883-1888, 1987.

18. Saija A, Princi P, Lanza M, Scalese M, Aramnejad E, De Sarro A. Systemic cytokine administration can affect blood-brain barrier permeability in the rat. Life Sciences 56: 775-784, 1995.

19. Ellison MD, Krieg RJ, Povlishock JT. Differential central nervous system responses following single and multiple recombinant interleukin-2 infusions. J Neuroimmunol. 28: 249-60, 1990.

20. Pankonin G, Reipert B, Ager A. Interactions between interleukin-2-activated lymphocytes and vascular endothelium: binding to and migration across specialized and non-specialized endothelia. Immunology 77: 51-60, 1992.

21. Trotter JL, Clifford DB, Anderson CB, van der Veen RC, Hicks BC, Banks G: Elevated serum interleukin-2 levels in chronic progressive multiple sclerosis. New Eng J Med 318:1206, 1988.

22. Sharief MK, Hentges R, Ciardi M, Thompson EJ. In vivo relationship of interleukin-2 and soluble IL-2 receptor to blood-brain barrier impairment in patients with active multiple sclerosis. J Neurol 24046-50, 1993.

23. Raine CS, Lee S, Scheinberg LC, Duijvestijn A, Cross AH. Adhesion molecules on endothelial cells in the central nervous system: an emerging area in the Neuroimmunology of multiple sclerosis. Clin Immunol Immunopathol 57: 173-187, 1990.

24. Sobel R, Mitchell ME, Fondren G. Intercellular adhesion molecule-1 (ICAM-1) in cellular immune reactions in the human central nervous system. Am J Pathol 136: 1309-1316, 1990.

25. Washington R, Burton J, Todd RF, Newman W, Dragovic L, Dore-Duffy P. Expression of immunologically relevant endothelial cell activation antigens on isolated central nervous system microvessels from patients with multiple sclerosis. Ann Neurol 36: 89-97, 1994.

26. Harris JO, Frank JA, Patronas N, McFarlin DE, McFarland HF. Serial gadolinium-enhanced magnetic resonance imaging scans in patients with early, relapsing multiple sclerosis: implications for clinical trials and natural history. Ann Neurol 29: 548-555, 1991.
27. Raine CS, Cannella B, Duijvestijn AM, Cross AH. Homing to central nervous system vasculature by antigen-specific lymphocytes. II. Lymphocytes/endothelial cell adhesion during the initial stages of autoimmune demyelination. Lab Invest 63: 476-489, 1990.
28. Durieu-Trautmann O, Chaverot N, Cazaubon S, Strosberg AD, Couraud P-O. Intercellular adhesion molecule 1activation induces tyrosine phosphorylation of the cytoskeleton-associated protein cortactin in brain microvessel endothelial cells. J Biol Chem 269: 12536-12540, 1994.
29. Carpen O, Pallai P, Staunton DE, Springer TA. Association of intercellular adhesion molecule-1 (ICAM-1) with actin-containing cytoskeleton and alpha-actinin. J Cell Biol 118: 1223-34, 1992.
30. Adamson P, Etienne S, Couraud PO, Calder V, Greenwood J. Lymphocyte migration through brain endothelial cell monolayers involves signaling through endothelial ICAM-1 via a rho-dependent J Immunol. 162: 2964-73, 1999.
31. Berman JW, Guida MP, Warren J, Arnat J, Brosnan CR. Localization of monocyte chemoattractant peptide-1 expression in the central nervous system in experimental autoimmune encephalomyelitis and trauma in the rat. J Immunol 156: 3017-3023, 1996.
32. Prat A, Biernacki K, Lavoie JF, Poirier J, Duquette P, Antel JP. Migration of multiple sclerosis lymphocytes through brain endothelium. Arch Neurol 59: 391-7, 2002.
33. Shukaliak JA, Dorovini-Zis K. Expression of the beta-chemokines RANTES and MIP-1beta by human brain microvessel endothelial cells in primary culture. J Neuropathol Exp Neurol 59: 339-352, 2000.
34. Tanaka Y, Adams DH, Hubscher S, Hirano H, Siebenlist U, Shaw S. T cell adhesion induced by proteoglycan-immobilized cytokine MIP-1β. Nature 361: 79-82, 1993.
35. Zhang W, Smith C, Howlett C, Stanimirovic D,. Inflammatory activation of human brain endothelial cells by hypoxic astrocytes in vitro is mediated by IL-1β. J Cerebr Blood Flow Metab 20: 967-78, 2000.
36. Sobel RA. Natale JM. Schneeberger EE. The immunopathology of acute experimental allergic encephalomyelitis. IV. An ultrastructural immunocytochemical study of class II major histocompatibility complex molecule (Ia) expression. J Neuropathol Exp Neurol 46: 239-49, 1987.
37. Omari KI, Dorovini-Zis K. Expression and function of the costimulatory molecules B7-1 and B7-2 in an in vitro model of the human blood-brain barrier. J Neuroimmunol 113: 129-141, 2001.
38. Jemison LM, Williams SK, Lublin FD, Knobler RL, Korngold R. Interferon-gamma-inducible endothelial cell class II major histocompatibility complex expression correlates with strain- and site-specific susceptibility to experimental allergic encephalomyelitis. J Neuroimmunol 47: 15-22, 1993.
39. Cross A.H., Ku G. Astrocytes and central nervous system endothelial cells do not express B7-1 (CD80) or B7-2 (CD86) immunoreactivity during experimental autoimmune encephalomyelitis. J Neuroimmunol 110: 76-82, 2000.
40. Pryce G, Male D, Sedgwick J. Antigen presentation in brain: brain endothelial cells are poor stimulatory of T cell proliferation. Immunology 66: 207-212, 1989.

41. Risau W, Engelhardt B, Wekerle H. Immune function of the blood-brain barrier: Incomplete presentation of protein (auto-) antigen by rat brain microvascular endothelium in vitro. J Cell Biol. 110: 1757-1766, 1990.
42. Girvin, Gordon KB, Welsh CJ, Clipstone NA. Miller SD. Differential abilities of central nervous system resident endothelial cells and astrocytes to serve as inducible antigen-presenting cells. Blood 99: 3692-701, 2002.
43. McCarron RM. IL-1-induced prostacyclin production by cerebral vascular endothelial cells inhibits myelin basic protein-specific lymphocyte proliferation. Cell Immunol 145: 21-29, 1992.
44. Ley K, Bullard DC, Arbones ML, Bosse R, Vestweber D, Tedder TF, Beaubet AL. Sequential contribution of L- and P-selectin to leukocyte rolling in vivo. J Exp Med 181: 669-675, 1995.
45. Cohnheim JF. Lectures on General Pathology. 2nd edition. Vol. I London, The New Sydenham Society. 1889
46. Kansas, GS. Selectins and their ligands: current concepts and controversies. Blood 88: 3259-3287, 1996.
47. Tedder TF, Steeber DA, Chen A, Engel P. The selectins: vascular adhesion molecules. FASEB J 9: 866-873, 1995.
48. Naparstek Y, Ben-Nun A, Holoshitz J, Reshef T, Frenkel A, Rosenberg M, Cohen IR. T lymphocyte lines producing or vaccinating against autoimmune encephalomyelitis (EAE). Functional activation induced peanut agglutinin receptors and accumulation in the brain and thymus of lines cells. Eur J Immunol 13: 418-423, 1983.
49. Simmons RD, Cattle BA, Hugh AR, Willenborg DO. Lymphocytes utilize sialyated surface molecules to accumulate in developing lesions of autoimmune encephalomyelitis. Autoimmunity 14: 17-21, 1992.
50. Rainer TH. L-selectin in health and disease. Resuscitation 52: 127-141, 2002.
51. Engelhardt B, Martin-Simonet MG, Rott LS, Butcher EC, Michie SA. Adhesion molecule phenotype of T lymphocytes in inflamed CNS. J Neuroimmunol 84: 92-104, 1998.
52. Dopp JM, Breneman SM, Olschowka JA. Expression of ICAM-1, VCAM-1, L-selectin, and leukosialin in the mouse central nervous system during the induction and remission stages of experimental allergic encephalomyelitis. J Neuroimmunol 54: 129-144, 1994.
53. Brocke S, Piercy C, Steinman L, Weissman IL, Veromaa T. Antibodies to CD44 and integrin alpha-4, but not L-selectin, prevent central nervous system inflammation and experimental encephalomyelitis by blocking secondary leukocyte recruitment. Proc Natl Acad Sci 96: 6896-6901, 1999.
54. Archelos JJ, Jung S, Rinner W, Lassman H, Miyaska M, Hartung HP. Role of the leukocyte-adhesion molecule L-selectin in experimental autoimmune encephalomyelitis. J Neurol Sci 159: 127-134, 1998.
55. Macey G, McCarthy DA, Vordermeir S, Newland AC, Brown KA. Effects of cell purification methods on CD11b and L-selectin expression as well as the adherence and activation of leukocytes. J Immunol Methods 181: 211-219, 1995.
56. Stibenz D, Buhrer C. Down-regulation of L-selectin surface expression by various leukocyte isolation procedures. Scand J Immunol 39: 59-63, 1994.
57. Grewal IS, Foellmer HG, Grewal KD, Wang H, Lee WP, Tumas D, Janeway CA Jr., Flavell RA. CD62L is required on effector cells for local interactions in the CNS to cause myelin damage in experimental allergic encephalomyelitis. Immunity 14: 291-302, 2001.
58. Engelhardt B, Vestweber D, Hallmann R, Schultz M. E- and P-selectin are not involved in the recruitment of inflammatory cells across the blood-brain barrier in experimental autoimmune encephalomyelitis. Blood 90: 4459-4472, 1997.

59. Carrithers MD, Visintin I, Kang SJ, Janeway, CA Jr. Differential adhesion molecule requirements for immune surveillance and inflammatory recruitment. Brain 123: 1092-1101, 2000.
60. Piccio L, Rossi B, Scarpini E, Laudanna C, Giagulli C, Issekutz AC, Vestweber D, Butcher EC, Constantin G. Molecular mechanisms involved in lymphocyte recruitment in inflamed brain microvessels: critical roles for P-selectin glycoprotein ligand-1 and heterotrimeric Gi-linked receptors. J Immunol 168: 1940-1949, 2002.
61. Tran GT, Hodgkinson SJ, Carter N, Killingsworth M, Spicer ST, Hall BM. Attenuation of experimental allergic encephalomyelitis in complement component 6-deficient rats is associated with reduced complement C9 deposition, P-selectin expression, and cellular infiltrate in spinal cords. J Immunol 168: 4293-4300, 2002.
62. Kerfoot SM, Kubes P. Overlapping roles of P-selectin and alpha4 integrin to recruit leukocytes to the central nervous system in experimental autoimmune encephalomyelitis. J Immunol 169: 1000-1006, 2002.
63. Hindmarsh EJ, Marks RM. Complement activation occurs on subendothelial extracellular matrix in vitro and is initiated by retraction or removal of overlying endothelial cells. J Immunol 160: 6128-6136, 1998.
64. Bernardes-Silva M, Anthony DC, Issekutz AC, Perry VH. Recruitment of neutrophils across the blood-brain barrier: the role of E- and P-selectins. J Cerebral Blood Flow and Metabolism 21: 1115-1124, 2001.
65. Anthony DC, Bolton SJ, Fearn S, Perry VH. Age related effects of interleukin-1 beta on polymorphonuclear neutrophil-dependent increases in blood-brain barrier permeability in rats. Brain 120: 435-444, 1997.
66. Hurwitz AA, Lyman WD, Guida MP, Calderon TM, Berman JW. Tumor necrosis factor alpha induces adhesion molecule expression on fetal astrocytes. J Exp Med 176: 1631-1636, 1992.
67. Rosen SD. Endothelial ligands for L-selectin: From lymphocyte recirculation to allograft rejection. Am J Pathol 155: 1013-1020, 1999.
68. Lasky LA, Singer MS, Dowbenko D, Imai Y, Henzel EJ, Fennie C, Gillett N, Watson SR, Rosen SD. An endothelial ligand for L-selectin in a novel mucin-like molecule. Cell 69: 927-938, 1992.
69. Baumheuter S, Singer MS, Henzel W, Hemmerich S, Renz M. Rosen SD, Lasky LA. Binding of L-selectin to the vascular silaomucin, CD34. Science 262: 436-438, 1993.
70. Berg EL, McEvoy LM, Berlin C, Bargatze RF, Butcher EC L-selectin-mediated lymphocyte rolling on MAdCAM-1. Nature 366: 695-698, 1993.
71. Bargatze RF, Jutila MA, Butcher EC. Distinct roles of L-selectin and integrins alpha4beta7 and LFA-1 in lymphocyte homing to Peyer's patch-HEV in situ: the multistep model confirmed and refined. Immunity 3: 99-108, 1995.
72. Borsig L, Wong R, Hynes RO, Varki NM, Varki A. Synergistic effects of L- and P-selectin in facilitating tumor metastasis can involve non-mucin ligands and implicate leukocytes as enhancers of metastasis. Proc Natl Acad Sci 99: 2193-2198, 2002.
73. Cannella B, Cross AH, Raine CS Relapsing autoimmune demyelination: a role for vascular addressins. J Neuroimmunol 35: 295-300, 1991.
74. Huang K, Kikuta A, Rosen SD. Myelin localization of a central nervous system ligand for L-selectin. J Neuroimmunol 53: 133-141, 1994.
75. Huang K, Geoffroy JS, Singer MS, Rosen SD. A lymphocyte homing receptor (L-selectin) mediates the in vitro attachment of lymphocytes to myelinated tracts of the central nervous system. J Clin Invest 88: 1778-1783, 1991.

76. Steffen BJ, Breier G, Butcher EC, Schulz M, Engelhardt B. ICAM-1, VCAM-1, and MAdCAM-1 are expressed on choroids plexus epithelium but not endothelium and mediate binding of lymphocytes in vitro. Am J Pathol 148: 1819-1838, 1996.
77. Kanwar JR, Kanwar RK, Wang D, Krissansen GW. Prevention of a chronic progressive form of experimental autoimmune encephalomyelitis by an antibody against mucosal addressin cell adhesion molecule-1 given early in the course of disease progression. Immunol Cell Biol 78: 641-645, 2000.
78. Bengtsson T, Grenegard M, Olsson A, Sjogren F, Stendahl O, Zalavary S. Sulfatide-induced L-selectin activation generates intracellular oxygen radicals in human neutrophils: modulation by extracellular adenosine. Biochim Biophys Acta 1313: 119-129, 1996.
79. Crockett-Torabi E, Sulenbarger B, Smith CW, Fantone JC. Activation of human neutrophils through L-selectin and Mac-1 molecules. J Immunol 154: 2291-2302, 1995.
80. Po JL, Mazer B, Jensen GS. The L-selectin antibody FMC46 mediates rapid, transient increase in intracellular calcium in human peripheral blood mononuclear cells and Daudi lymphoma cells. Biochem Biophys Res Commun 217: 1145-1150, 1995.
81. Waddell TK, Fialkow L, Chan CK, Kishimoto TK, Downey GP. Signaling functions of L-selectin. Enhancement of tyrosine phosphorylation and activation of MAP kinase. J Biol Chem 270: 15403-15411, 1995.
82. Laudanna C, Constantin C, Baron P, Scarpini E, Scarlato G, Cabrini G Dechecchi C, Rossi F, Cassatella MA, Berton G. Sulfatides trigger increase of cytosolic free calcium and enhanced expression of tumor necrosis factor-alpha and interleukin-8 mRNA in human neutrophils. Evidence for a role of L-selectin as a signaling molecule. J Biol Chem 269: 4021-4026, 1994.
83. Waddell TK, Fialkow L, Chan CK, Kishimoto TK, Downey GP. Potentiation of the oxidative burst of human neutrophils. A signaling role for L-selectin. J Biol Chem 269: 18485-18491, 1994.
84. Lo SK, Lee S, Ramos RA, Lobb R, Rosa M, Chi-Rosso G, Wright SD. Endothelial-leukocyte adhesion molecule 1 stimulates the adhesive activity of leukocyte integrin CR3 (CD11b/CD18, Mac-1, alpha(m)beta2) on human neutrophils. J Exp Med 173: 1493-1500, 1991.
85. Nagata K, Tsuji T, Todoroki N, Katagiri Y, Tanoue K, Yamazaki H, Hanai N, Irimura T. Activated platelets induce superoxide anion release by monocytes and neutrophils through P-selectin (CD62). J Immunol 151: 3267-3273, 1993.
86. Celi A, Pellegrini G, Lorenzet R, De Blasi A, Ready N, Furie BC, Furie B. P-selectin induces the expression of tissue factor on monocytes. Proc Natl Acad Sci 91: 8767-8771, 1994.
87. Elstad MR, La Pine TR, Cowley FS, McEver RP, McIntyre TM, Prescott SM, Zimmerman GA. P-selectin regulates platelet-activating factor synthesis and phagocytosis by monocytes. J Immunol 155: 2109-2122. 1995.
88. Hwang ST, Singer MS, Giblin PA, Yednock TA, Bacon KB, Simon SI, Rosen SD. GlyCAM-1, a physiological ligand for L-selectin, activates beta2 integrins on naïve peripheral lymphocytes. J Exp Med 184: 1343-1348, 1996.
89. Giblin PA, Hwang ST, Katsumoto TR, Rosen SD. Ligation of L-selectin on T lymphocytes activates beta1 integrins and promotes adhesion to fibronectin. J Immunol 159: 3498-3507, 1997.
90. Prat A, Biernacki K, Wosik K, Antel JP. Glial cell influence on the human blood-brain barrier. GLIA 36: 145-155, 2001.

91. McCarron RM, Wang I, Stanimirovic DB, Spatz M. Differential regulation of adhesion molecule expression by human cerebrovascular and umbilical vein endothelial cells. Endothelium 2: 339-346, 1995.
92. Wong D, Dorovini-Zis K. Expression of vascular cell adhesion molecule-1 by human brain microvessel endothelial cells in primary culture. Microvasc Res 49: 325-339, 1995.
93. Archelos JJ, Maurer M, Jung S, Toyka KV, Hartung HP. Suppression of experimental allergic neuritis by an antibody to the intracellular adhesion molecule ICAM-1. Brain 116: 1043-58, 1993.
94. Cannella B, Cross AH, Raine CS: Anti-adhesion molecule therapy in experimental autoimmune encephalomyelitis. J Neuroimmunol 46: 43-55, 1993.
95. Samoilova EB, Horton JL, Chen Y. Experimental autoimmune encephalomyelitis in intercellular adhesion molecule-1-deficient mice. Cell. Immunol. 190: 83-89, 1998.
96. Yednock TA, Cannon C, Fritz LC, Sanchez-Madrid F, Steinman L, Karin N. Prevention of experimental autoimmune encephalomyelitis by antibodies against $\alpha 4\beta 1$ integrin. Nature 356: 63-66, 1992.
97. Kent SJ, Karlik SJ, Cannon C, Hines DK, Yednock TA, Fritz LC, Horner HC. A monoclonal antibody to $\alpha 4$ integrin suppresses and reverses active experimental allergic encephalomyelitis. J Neuroimmunol 58: 1-10, 1995.
98. Baron JL, Madri JA, Ruddle NH, Hashim G, Janeway CA JR. Surface expression of a4 integrin by CD4 T cells is required for their entry into brain parenchyma. J Exp Med 177: 57-68, 1993.
99. Kanwar JR, Harrison JEB, Wang D, Leung E , Mueller W, Wagner N, Krissansen GW. $\beta 7$ integrins contribute to demyelinating disease of the central nervous system. J Neuroimmunol 103: 146-152, 2000.
100. Tubridy N, Behan PO, Capildeo R, Chaudhuri A, Forbes R, Hawkins CP, Hughes RA, Palace J, Sharrack B, Swingler R, Young C, Moseley IF, MacManus DG, Donoghue A, Miller DH. The effect of anti-alpha 4 integrin antibody on brain lesion activity in MS. Neurology 53: 466-72, 1999.
101. Williams KC, Zhou RW, Ueno K, Hickey WF. PECAM-1 expression in the central nervous system and its role in experimental allergic encephalomyelitis in the rat. J. Neurosci Res 45: 747-757, 1996.
102. Graesser D, Solowiej A, Bruckner M, Osterweil E, Juedes A, Davis S, Ruddle NH, Engelhardt B, Madri JA. Altered vascular permeability and early onset of experimental autoimmune encephalomyelitis in PECAM-1-deficient mice. J Clin Invest 109: 383-392, 2002.
103. Rosenstein, Y, Park J, Hahn W, Rosen F, Bierer BE, Burakoff S. CD43, a molecule defective in Wiskott-Aldrich syndrome, binds ICAM-1. Nature 354: 233-235, 1991.
104. Park J, Rosenstein Y, Remold-O'Donnell E, Bierer B, Rosen F, Burakoff S. Enhancement of T cell activation by the CD43 molecule whose expression is defective in Wiskott-Aldrich syndrome. Nature 350: 706-709, 1991.
105. Frieser M, Nockel H, Pausch F, Roder C, Hahn A, Deutzmann R, Sorokin LM. Cloning of the mouse laminin alpha 4 gene: expression in a subset of endothelium. Eur J Biochem 246: 727-735, 1997.
106. Jucker M, Tian M, Norton DD, Sherman C, Kusiak JW. Laminin alpha2 is a component of brain capillary basement membrane: reduced expression in dystrophic dy mice. Neuroscience 71: 1153-1161, 1996.
107. Powell SK, Kleinman HK. Neuronal laminins and their cellular receptors. Int J Biochem Cell Biol 29: 401-414, 1997.

108. Tian M, Jacobson C, Gee SH, Campbell KP, Carbonetto S, Jucker M. Dystroglycan in the cerebellum is a laminin a2-chain binding protein at the glial-vascular interface and is expressed in purkinje cells. Eur J Neurosci 8: 2739-2747, 1997.
109. Wolburg H, Risau W. Formation of the blood-brain barrier. In Neuroglia. H. Kettenmann and B.R. Ransom, eds. Oxford University Press, Oxford, UK. 763-776, 1995.
110. Tran EH, Hoekstra K, Rooijen NV, Dijkstra CD, Owens, T. Immune evasion of the central nervous system parenchyma and experimental allergic encephalomyelitis, but not leukocyte extravasation from blood, are prevented in macrophage-depleted mice. J Immunol 161: 3767-3775, 1998.
111. Korner H, Riminton DS, Strickland DH, Lemckert FA, Pollard JD, Sedgwick JD. Critical points of tumor necrosis factor action in central nervous system autoimmune inflammation defined by gene targeting. J Exp Med 186: 1585-1590, 1997.
112. Sixt M, Engelhardt B, Pausch F, Hallmann R, Wendler O, Sorokin LM. Endothelial cell laminin isoforms, laminins 8 and 10, play decisive roles in T cell recruitment across the blood-brain barrier in experimental autoimmune encephalomyelitis. J Cell Biol 153: 933-945, 2001.
113. Lesley J, Hyman R, Kincade PW. CD44 and its interaction with extracellular matrix. Adv Immunol 54: 271-335, 1993.
114. Camp RL, Scheynius A, Johansson C, Pure E. CD44 is required for optimal contact allergic responses but is not required for normal leukocyte extravasation. J Exp Med 178: 497-507, 1993.
115. DeGrendele HC, Estess P, Siegelman MH. Requirement for CD44 in activated T cell extravasation into an inflammatory site. Science. 278: 672-5, 1997.
116. Turley E, Moore D. Hyaluronate binding proteins also bind to fibronectin, laminin, and collagen. Biochem Biophys Res Commun 121: 808-814, 1984.
117. Miyake K, Underhill CB, Lesley J, Kincade PW. Hyaluronate can function as a cell adhesion molecule and CD44 participates in hyaluronate recognition. J Exp Med 172: 69-75, 1990.
118. Jalkanen S, Jalkanen M. Lymphocyte CD44 binds the COOH-terminal heparin-binding domain of fibronectin. J Cell Biol 116: 817-25, 1992.
119. Ishii S, Ford R, Thomas P, Nachman A, Steele G Jr., Jessup JM. CD44 participates in the adhesion of human colorectal carcinoma cells to laminin and type IV collagen. Surg Oncol 2: 255-264, 1993.
120. Carter WG, Wayner EA. Characterization of the class III collagen receptor, a phosphorylated, transmembrane glycoprotein expressed in nucleated human cells. J Biol Chem 263: 4193-4201, 1998.
121. Dimitroff CJ, Lee JY, Fuhlbrigge RC, Sackstein R. A distinct glycoform of CD44 is an L-selectin ligand on human hematopoietic cells. Proc Natl Acad Sci 97: 13841-13846, 2000.
122. Dimitroff CJ, Lee JY, Rafii S, Fuhlbrigge RC, Sackstein R. CD44 is a major E-selectin ligand on human hematopoietic progenitor cells. J Cell Biol 153: 1277-1286, 2001.
123. DeGrendele HC, Estess P, Picker LJ, Siegelman MH. CD44 and its ligand hyaluronate mediate rolling under physiologic flow: a novel lymphocyte-endothelial cell primary adhesion pathway. J Exp Med 183: 1119-30, 1996.
124. Brennan FR, O'Neill JK, Allen SJ, Butter CS, Nuki G, Baker D. CD44 is involved in selective leukocyte extravasation during inflammatory central nervous system disease. Immunolgy 98: 427-435, 1999.
125. Shaw MK, Lorens JB, Dhawan A, DalCanto R, Tse HY, Tran AB, Bonpane C, Eswaran SL, Brocke S, Sarvetnick N, Steinman L, Nolan G, Fathman CG. Local delivery of

interleukin 4 by retrovirus-transduced T lymphocytes ameliorates experimental autoimmune encephalomyelitis. J Exp Med 185: 1711-1714, 1997.

126. Kuchroo VK, Das MP, Brown JA, Ranger AM, Zamvil SS, Sobel RA, Weiner HL, Nabavi N, Glimcher LH. B7-1 and B7-2 costimulatory molecules activate differentially the Th1/Th2 developmental pathways: application to autoimmune disease therapy. Cell 80: 707-718, 1995.

127. Gimsa U, Wolf SA, Haas D, Bechmann I, Nitsch R. Th2 cells support intrinsic anti-inflammatory properties of the brain. J Neuroimmunol 119: 73-80, 2001.

128. Yu M, Nishiyama A, Trapp BD, Tuohy VK. Interferon-beta inhibits progression of relapsing-remitting experimental autoimmune encephalomyelitis. J Neuroimmunol 64: 91-100, 1996.

129. Floris S, Ruuls SR, Wierinckx A, van der Pol SM, Dopp E, van der Meide PH, Dijkstra CD, De Vries HE.. Interferon-beta directly influences monocyte infiltration into the central nervous system. J Neuroimmunol 127: 69-79, 2002.

130. Stuve O, Dooley NP, Uhm JH, Antel JP, Williams G, Yong VW. Interferon beta-1b decreases the migration of T lymphocytes in vitro: effects on matrix metalloprotease –9. Ann Neurol 40: 853-863, 1996.

Chapter A10

# GENETICS OF EXPERIMENTAL ALLERGIC ENCEPHALOMYELITIS

**David Baker**
*Department of Neuroinflammation, Institute of Neurology, University College, London, UK*

**Abstract:** The identification of genetic factors that control the pathogenetic pathways of autoimmune neuroinflammation and degeneration may provide targets for new therapeutic strategies in human diseases, such as multiple sclerosis. This is not easily accomplished as significant heterogeneity exists and such genetic dissection may more easily be performed using inbred rodent strains, where the genetic heterogeneity is avoided and variation in the environmental influences is minimized. Chronic relapsing experimental autoimmune encephalomyelitis (EAE) exhibits some histopathological features typical of multiple sclerosis and provides a tool to systemically identify genetic influences controlling susceptibility through genome scanning and selective breeding of congenic animals harbouring disease-related loci. As with MS, susceptibility is controlled by expression of certain major histocompatibility complex allotypes, which in EAE is in response to induction with certain myelin antigens. Both are complex polygenic diseases where disease traits are controlled by a number of polymorphic genes which individually exhibiting modest effect on disease course. A number of as yet largely unknown genes, cluster in certain regions that are sometimes shared between different strain combinations and are also shared by other inflammatory diseases, as they are uncovered the biology understood they may open avenues to treatment.

**Key words:** EAE, MS, immunogenetics, susceptibility, major histocompatibility complex, polygenic disease

**Multiple Sclerosis- a disease with a predisposing genetic susceptibility**
Multiple sclerosis (MS) is the major demyelinating disease of the central nervous system (CNS) affecting about 2-3 million people worldwide, however the distribution of MS is geographically restricted and indicates some influence of both environmental factor(s) and genetic background of the affected individuals. (Compston & Coles 2002). Disease is most prevalent in temperate regions inhabited by descendents of northern

Europeans (Hogancamp *et al.* 1997, Compston & Coles 2002) and indicates some genetic aetiology most notable there is evidence of familial aggregation of MS and increased risks to siblings. This is evidenced by higher risk to monozygotic twins (25-30% concordance), dizygotic twins and non-twin siblings (2-5% concordance) compared to adoptees from non-affected parents (<0.2% concordance) which reflects the recurrence risk in the general population. MS sibling pairs tend to cluster by age of onset, rather than year of onset, and there is no detectable effect of shared environment on MS susceptibility in first-degree non-biological relatives (spouses, adoptees). Furthermore there is racial clustering of MS cases, particularly evidenced relative lack of MS in native African black people resident in high risk white regions and suggestive correlations between certain polymorphic genetic loci. (Ebers *et al.* 1995. Compston & Sawcer 2002). Disease in MS exhibits many different clinical and histopathological phenotypes that may change as disease progresses (Olerup *et al.* 1989, Lublin & Reingold 1996, Kira *et al.*1996, Lucchinetti *et al.* 1996, Thorpe *et al.* 1996. Thompson *et al* 1996, Confavreux *et al.* 2000, Compston & Coles 2002, Bjartmar *et al.,* 2003). The disease may have different aetiologies or represent abroad a spectrum of disease and this heterogeneity makes it difficult to analysed the genetic components of MS. Whilst the precise aetiology is unknown, it is thought that once disease is triggered, that MS represents an autoimmune disease (Compston & Coles 2002). Whilst there are no other non-human animals that spontaneously develop a disease identical to MS, elements of MS can be mimicked in experimental allergic encephalomyelitis (EAE), which is an induced autoimmune disease of the CNS (Brocke *et al.* 1994, Wekerle *et al.* 1994). Just as MS exhibits many clinical phenotypes that include benign disease, which is relatively self-limiting, primary progressive disease from onset, relapsing-remitting MS which show periods of neurological deficit with periods of remission that often leads to secondary progressive disease where disease continues unabated. (Compston & Coles 2002), animals species and strains develop a variety of different clinical phenotypes (Raine et al 1980, Brocke *et al.* 1994, Wekerle *et al.* 1994, Brok *et al.* 2001, t'Hart and Amor 2003). In animals disease susceptibility is clearly under genetic control (Levine & Sowlinski 1973, Raine et al 1980) some strains develop hyperacute and often fatal disease (e.g. Hartley strain guinea pigs), others monophasic (e.g. Lewis rats), relapsing (strain 13 guinea pig) or chronic (e.g. Marmosets) disease following induction with CNS myelin. As similar pathological events appear to occur in EAE and MS, its animal model may be used to tease out some of the genetic checkpoints that control different disease traits.

Most adult-onset autoimmune diseases exhibit a female bias (2:1 in MS. Duquette et al.1992), yet pregnancy is associated with disease quiescence to a level not generally matched by currently available pharmaceutical agents

(Homo-Delarche *et al.*1991, Jansson & Holmdahl 1998). Although females exhibit higher incidence of relapsing remitting disease, males with later onset disease more often develop primary progressive disease which is more neurodegenerative (Compston & Coles 2002, Bjartmar et al 2003), this may suggest that differences in genetic background create physiological differences that can withstand varying degrees of insult. However, once a threshold is exceeded, both primary and secondary progressive disease can develop at similar rates, suggesting that these phenotypes are different poles of the same disease process (Confavreux *et al.* 2000). It is well recognized that sex-hormones can influence a great number of immune parameters (Homo-Delarche *et al.*1991, Jansson & Holmdahl 1998). This is also reflected by sex differences (typically females are more susceptible than males) to susceptibility to the development of CNS autoimmunity in EAE (Baker *et al.* 1995, Bebo *et al.*, 1998,1999). The penetrance of disease expression varies between different experimental paradigms and strain combinations and this influences clinical outcome. (Dalal et al.,1997, Jansson & Holmdahl 1998, Bebo et al.,1998,1999).


**1.1**     **Myelin and the Major Histocompatibility complex in susceptibility to CNS autoimmunity**

In MS the major pathological feature is demyelination and the loss of oligodendrocytes (Lucchinetti *et al.* 1996, Compston & Coles 2002). That EAE develops CNS autoimmunity following sensitization with CNS myelin, suggests that it could be a major target in MS. Myelin contains a number of proteins where proteolipid protein (PLP. ~50% of CNS myelin) and myelin basic protein (MBP. ~30% of myelin protein) are major constituents. Due to the water solubility and ease of purification, early studies concentrated on MBP as a candidate autoantigen along with whole spinal cord homogenates (Fritz *et al.* 1985). The prototypical susceptible EAE strains were SJL (H-2$^s$ haplotype), SWR (H-2$^q$), PL and more recently ABH mice (Levine & Sowinski 1973, Baker *et al.* 1990), whilst C57BL/6 (H-2$^b$) and BALB/c (H-2$^d$) were classically considered resistant (Levine & Sowinski 1973, Raine *et al.* 1980). Using mice congenic for the major histocompatibility complex (MHC), termed H-2 in the mouse, it could be demonstrated that MBP-induced disease was restricted by the MHC class II, notably H-2A, gene products, however there are exceptions such as B10.S (H-2$^s$) and B10.Q (H-2$^s$. Fritz *et al.* **Table** **1**). Furthermore as time has passed novel encephalitogenic antigens have been described that include the CNS-restricted myelin proteins: proteolipid protein (PLP, Tuohy *et al.* 1988) and myelin oligodendrocyte glycoprotein (MOG) in particular (Amor *et al.* 1994, Mendel *et al.* 1995), but includes other myelin antigens such as myelin associated glycoprotein (MAG), myelin oligodendrocyte specific protein (OSP) and non-myelin antigens (Kojima *et al.* 1994, Morris-Downes *et al.* 2002, Furlan *et al.* 2002). Here *H2* allotypes confer a different set of susceptibilities and peptide-epitope restrictions between inbred strains (Tuohy *et al.* 1988, Abdul-Majid *et al.* 2001, Kjellén *et al.* 2001). Notably C57BL/6 mice, often used for the generation of 'gene-knockout mice" are susceptible to EAE induction with MOG and MOG residues 35-55 peptide (Mendel et al 1995, 1999). Thus although the inducing antigens are varied between strains, animals from virtually all MHC haplotypes of mice can be induced to develop CNS autoimmunity (**Table 1**). Therefore suggests that most mouse strains are susceptible and should no longer be considered as resistant but as low responders to certain disease induction protocols. Indeed mapping studies have indicated that genetic elements from "resistant" strains contribute to susceptibility (Sundvall *et al.* 1994, Teuscher *et al.* 1998, Jirholt *et al.* 2000).

**TABLE 1. Susceptibility to myelin antigen induced EAE is genetically controlled**

**Inducing Antigen**

| Strain | H-2A | MBP | PLP | MOG |
|---|---|---|---|---|
| **C57BL/6** | b | - | - | + |
| **BALB/c** | d | - | - | - |
| **SJL** | s | + | + | + |
| **PL/J** | u | + | [+] | [+] |
| **SWR** | q | + | + | + |
| **ABH** | g7 | - | + | + |

Different mouse strains actively induced with myelin basic protein (MBP), Proteolipid protein (PLP) and myelin oligodendrocyte glycoprotein (MOG) proteins, and peptides develop different levels of susceptibility, some are relatively resistant (-), others show limited or delayed susceptibilities ([+]) and others develop full susceptibilities (+) to the development of clinical EAE. (Levine & Sowinski 1973, Raine *et al.* 1980, Fritz *et al.* 1985, Tuohy *et al.* 1998, Pagnany *et al.* 2003

Whilst MHC expression is critical for the generation of autoimmunity, there is variability in the background genes of the susceptible MHC haplotypes and this may mask the importance of this restriction (Levine & Sowinski 1973, Fritz *et al.* 1985). Spontaneous EAE can occur in myelin-specific T cell receptor (TCR), transgenic animals that have a high frequency of activated myelin-specific T cells that are capable of entering the CNS (Governman *et al.* 1993, Brabb *et al.* 1997, Kuchroo *et al.* 2002, Bettelli *et al.* 2003), but EAE does not usually occur spontaneously therefore genetics, biology, environment and chance may determine whether spontaneous disease develops in humans. A key issue is likely to be the elements that activate potentially, autoreactive cells outside the control of normal homeostatic tolerance mechanisms (Governman *et al.* 1993). Many autoimmune diseases in animals use the same induction protocol and it is the nature of the inducing antigen that determines target specificity. Therefore many of the disease processes will be similar between different autoimmune models. It is therefore not surprising that disease loci are mapped to similar regions when different autoimmune diseases are mapped in the same strain combinations, as they are probably involved in the initiation of the conditions that promote (auto)immunity (Dalman *et al.* 1998; Encinas & Kuchroo 2000, Merriman *et al.* 2000, Bergsteinsdottir *et al.* 2000, Becanovi *et al.* 2003).

Although some evidence exists that indicates that polymorphisms in myelin genes such as MBP may be associated with susceptibility in models of MS and a subset of people with MS (Brahic & Bureau 1998, Pihlaja *et al.* 2003), this however, does not appear to be related in the majority of cases (Seboun *et al.* 1999). However, when MBP is naturally catabolised, peptide fragments are created that may not have the necessary chemico-physical properties to form a productive MHC/TCR interaction with many MHC class II antigens. However a major fragment that is generated following MBP cleavage contains the residues MBP 89-101 (Barry *et al.* 1992). This can bind to HLA-DR2 (DRB1*1501–DRB5*0101, DQA1*0102, DQB1*0602) molecules (Li et al. 2000), which are over represented in MS from people of northern Europeans dissent (Barcellos et al., 2003, Hensiek *et al.* 2002) and thus could offer could offer the potential to generate autoimmunity, particularly if the antigen fails to effectively establish thymic tolerance mechanisms (Li *et al.* 2000, Bruno *et al.* 2002). Indeed HLA-DR/DQ transgenic mice of the DR15 haplotype, coupled with transgenic myelin-specific human TCR transgenes spontaneously develop disease (Madden *et al.* 1999). However as immune responses to different myelin antigens can induce the same clinical phenotype it is not surprising that variability of MHC haplotype expression occurs in people with MS, such as DR3/DR4 association observed in Sardinia (Marrosu *et al.* 1997) Furthermore within the human immunological context a significantly greater number of MHC class I and particularly class II molecule variants occur than found in rodents, which allow them to gain protective immunity to pathogens but also offers the, probable evolutionary acceptable, possibility of adult onset autoimmunity. Whilst HLA allotypes form a component of the genetic susceptibility of MS (Haines *et al.* 1996, Sawcer *et al.*1996, Oturai *et al.*1999,) as with EAE disease is clearly polygenic.

**Systematic Genome Scans for Eae Susceptibility loci**

The search for genes controlling human disease genes continues but the only locus that consistently is associated with MS susceptibility are to genes within the MHC, but no non-HLA region has currently been shown unequivocal linkage in disease and putative susceptibility loci vary between regions, when analyzed by systematic genome scanning of individuals (Ebers *et al* 1996, Haines *et al* 1997, Sawcer *et al* 1996, Kuokkanen *et al* 1997, Chataway *et al* 1998) and in pools of affected individuals (Coraddu *et al.* 2001, Broadley et al 2001, Akesson et al. 2002, Ban *et al.* 2002, Sawcer et al. 2002) and even when stratified for DR2 expression (Coraddu *et al.*1998, Larsen *et al.* 2000, Goedde *et al.*2002). Linkage disequilibrium that occurs frequently in human genetics (certain allotypes of spatially related genes tend to be inherited in one supra haplotype, such as occurs in the MHC) makes it easier to locate disease-related genes, however due to linkage disequilibrium few chromosomal crossover events occur during inheritance, which make it difficult to segregate the disease gene. Each inbred strains essentially represents one individual and by selective breeding to segregate disease traits essentially a simple large family can be generated for mapping purposes within a controlled environment. Whilst it is clear that there are often phenotypic differences between susceptible and resistant strains, these may either represent causal factors or be totally unrelated phenotypes to the disease process, which have been acquired through the 20 generations of sibling mating to generate an inbred strain. Conventional tools such as gene knockout mice, transgenes or any other principle for selective targeting, are not particularly useful for the finding of genetically controlled mechanisms regulating disease, since immense numbers of immunopathological pathways or mediators can be hypothesized to be involved. Microsatellittes are stably inherited polynucleotide repeats that can be amplified using polymerase chain reaction using specific flanking primers. These are present throughout the genome and exhibit allelic differences in the length of the polynucleotide repeat and can be used to track the inheritance of specific chromosomal regions (Todd *et al.* 1991). A search for disease-regulating genes can be done by linkage analysis that requires no preformed hypotheses.

## 1.2      Non-MHC autosomal susceptibility loci

Induction of EAE in mice often uses co-injection of *Bordetella pertussis* toxin as a co-adjuvant, and an autosomal dominant *B.pertussis histamine sensitivity (Bphs)* was one of the first non-MHC genes shown to be involved in susceptibility to autoimmune diseases and was mapped to the distal of chromosome 6 in (C3H (resistant) x SJL (susceptible) x C3H mice. (Sudweeks *et al.* 1993). Following the generation a panel of interval specific congenic lines to intergress the susceptible gene on the resistant background in toxin induced vasoactive amine sensitisation controls *Bphs* was recently identified as the histamine $H_1$ receptor gene (*Hrh1*) and differs in 3 amino acids between the resistant and susceptible strains.(Ma *et al.* 2003). Although the actions of *B.pertussis* toxin are often attributed to permeability and vasoactive effects on blood:brain barrier permeability, the significant leucocytosis (10 fold) is more likely to account for this action. The disruption of *Hrh1* leads to immune deviation of T cell responses characterized by the suppression of IFNγ production and an up-regulation in the production of Th2-related cytokines, a paradigm often shown to inhibit the generation of EAE (Ma *et al.* 2003, Goa *et al.* 2003). Although this represents the first case where a non-MHC gene controlling autoimmunity has been identified, many others require identification.

The majority of mapping studies have used strains where the MHC haplotype is shared by both the susceptible and resistant strains of mice/rats to exclude its influence. Through selective crossing between resistant and susceptible strains and genotypic analysis of affected and unaffected progeny it is possible to systematically scan the whole genome to detect linkage to susceptibility to EAE or related traits (**Table 2-4**). So far a number of loci containing disease-modifying genes have been detected and at present there are over 20 EAE-associated loci (*Eae*) in the rat (**Table 2**) and over 27 loci in the mouse (**Table 3**), whose precise identity are unknown/proven. In the rat, EAE quantitative trait loci have been demonstrated in DA (susceptible) x Brown Norway (BN. resistant), Lewis (susceptible) x BN, and E3 x DA intercrosses induced with whole spinal cord homogenate or with MOG as immunogen in DA x ACI and DA x PVG.1AV1 intercrosses (**Table 2**). Whereas in the mouse, studies have been preformed in B10.RIII (susceptible) x RIII, ABH (susceptible) x NOD, ABH x BALB/c or SJL x B10.S intercrosses following immunization with whole spinal cord homogenates, MBP or PLP peptides with or with *B. pertussis* toxin as co-adjuvant (**Table 3**).

These studies have indicated that disease is a complex polygenic trait of dominant, additive (heterozygotes) and recessive alleles whose products may interact in an epistatic fashion, where the detection of susceptibility loci depends on the presence of another interacting locus. Susceptibility to EAE is typically a dominant phenotype, but with various levels of penetrance of genotypic elements and this varies dramatically depending on the parental strains been analysed (Sundvall et al 1994, Baker et al 1995, Encinas et al 1996, Croxford et al 1997, Butterfield et al 1998). Within any strain combination it appears that a limited number of loci (typically <10-15 depending on the type of trait being analysed) are revealing significant segregation and can be compared to expected Mendelian rates of inheritance or to expression with the effected and or resistant populations (**Table 2-4**). At a genetic level there is evidence that different disease phenotypes such as acute monophasic, relapsing remitting and chronic disease can be influenced by segregating loci and may also be dependent on the gender of the animal (Butterfield *et al.* 1998, 1999, Karlsson *et al.* 2003. **Table 3**). However clinical course may not be solely explained by genetic variability. Whilst chronic paralytic disease, which is associated with significant neuronal/axonal loss (Wujek *et al.* 2002), often develops in C57BL/6 mice (Mendel *et al.* 1995, Slavin *et al.* 1998). ABH mice often develop relapsing remitting disease and may be considered as evidence of genetic influence (Baker *et al.* 1990). Due to MHC restrictions, different antigens are often uses to induce disease in different strains. However whilst MOG$^{35\text{-}55}$ induces chronic disease in C57BL/6, NOD and also ABH mice (Amor *et al.* 1994, Slavin *et al.* 1998, Ichikawa et al. 1999), interesting MOG$^{8\text{-}22}$ induces relapsing-remitting disease in ABH mice despite an identical background (Amor *et al.* 1994), unpublished). Likewise although the increased risk of monozygotic twins developing disease clearly indicates a genetic influence, even in twins the disease phenotype may not be concordant. Although there may be other explanations such as antibody isotype responses (Ichikawa et al. 1999), chronic disease in mouse EAE has recently been associated with significant CD8$^{+}$ MOG-specific T cell activation (Sun *et al.* 2001) compared to classical-remitting EAE, which is clearly CD4$^{+}$ T cell mediated (O'Neill et al. 1993). There are a number of loci that control the ratio of CD4:CD8 in the mouse that may be linked to EAE susceptibility, such as a locus in the distal part of chromosome 1 (Karlsson *et al.* 2003). Therefore the way the individual may process and present the antigen may determine clinical course, as will lesion location where strategically important lesions can be related to clinical phenotype (Compston & Coles 2002). However genetic background may also contribute to this process such as through influences on lesion location (Butterfield et al 1999, Karlsson et al 2003. **Table 3**).

Initial genome wide mapping studies in (B10.RIII x RIII)F2 indicated a susceptibility locus (*Eae2-m*) proximal portion of chromosome 15 and a second locus *Eae3-m* on chromosome 3. (Sundvall *et al.* 1995). Although the *Eae2* locus (associated with homozygousity of the allele from the RIII-resistant mouse) on mouse chromosome 15 was homologous with a region of human chromosome 5p identified in mapping in MS (Kuokkanen *et al.* 1996, Oturai *et al.* 1999), following backcross to produce *Eae2-m* congenic mice, influence of this locus could not be demonstrated despite replication of the original observations (Jirholt et al 2000). This is indicative that *Eae2-m* is in epistasis (expression of phenotype is dependent on the presence of other unrelated genetic loci), however in crosses of such congenic animals a locus on chromosome 7 was identified. Similarly susceptibility loci could be mapped to two loci on chromosome 7 (*Eae4, Eae12*) in mouse myelin induced (SJL x B10.S)F2 mice, mapping in PLP peptide induced (SJL x B10.S) x B10.S failed to demonstrate evidence of linkage (Encinas et al 1996, Butterfield *et al.* 1998). The major loci controlling EAE susceptibility in (ABH x NOD) x NOD mice was mapped chromosome 7, where disease inheritance was highly penetrant compared to other mouse mapping studies (Sundvall *et al.* 1994, Baker *et al.* 1995, Encinas *et al.* 1996, Butterfield *et al.* 1998). Further analysis of additional microsatellites and susceptible (ABH x NOD) x NOD females (Table 4), strongly support the involvement of in this case dominant ABH alleles near *Eae4-m* (45-51cM), *Eae26-m* (53cM) and *Eae10-m* (10-16cM) in susceptibility and indicate linkage to an additional locus distal to *Eae10-m* cM, consistent will mapping data in (SJL x B10.S)F2 mice. (Butterfield *et al.* 1998. **Table 3**). The importance of these loci has been definitively demonstrated by the generation of *NOD.ABH.Chr7* mice, where at backcross generation N8, a single copy of ABH chromosome 7 loci conferred virtually full susceptibility to disease on the NOD mouse background and indicates that this chromosome may harbor important loci for the control of EAE **(Table 4)**. It is possible that mouse chromosome 7 harbors a cluster of loci controlling elements of EAE, as has been found for the cluster of genes *Idd3*, *Idd10*, *Idd17* and *Idd18* controlling insulin dependent diabetes (Idd) susceptibility in NOD mice (Podlin *et al.* 1997).

In addition to linkage of *Eae3-m* to susceptibility in (B10.RIII x RIII) F2 mice (Sundvall *et al.* 1994), linkage was also detected on chromosome 3 in (SJL x B10.S) x B10.S backcross and in (SJL x B10.S)F2 both for incidence and histological disease (Encinas *et al.* 1996, 2001, Butterfield *et al.* 1998, 2000) suggesting that this may be important locus. However no evidence of linkage was detected in (ABH x NOD) x NOD backcross mice (Baker *et al.* 1995), which suggested that either this gene product was not important in this strain combination or they were not segregating possibly due to identity of the gene product in both strains. This may be supported by observations in diabetes . In NOD mice is diabetes is strong linked to loci on Chromosome 3 in (B6.H-2A$^{g7}$ x NOD) x NOD mice, however the polymorphisms in *Il2* though to be *Idd3* (Lyons *et al.* 2000) and *Fcrg* gene originally (Prins et al 1993) thought to be *Idd10* (Podolin et al 1997) are identical between ABH and NOD mice as was *Idd1* (H-2Ag7). Furthermore significant linkage was not detected in cyclophosphamide-induced (2 x 200mg/kg day 0 & 14) in diabetic (124/807) (ABH x NOD) x NOD mice (unpublished observations). This suggested that NOD and C57BL mice may differ at these loci and number of congenic mice had been developed for diabetes mapping of *Idd3*, *Idd10*, *Idd17* and *Idd18* of the chromosome region (Lyons et al 2001) and were studied for susceptibility to EAE (Encinas *et al.* 1998). This indicated that the loci for *Eae3* and *Idd3* are in the same 0.15cM interval. Whilst it must be recognized that these may be different but spatially-related loci, this could represent the same gene product involved in autoimmunity. The interleukin 2 gene (*Il2*) in mice has coding microsatellite repeat allelic variants (Lyons *et al.* 2000) and altered glycosylation (Podolin *et al.* 2000) and is a candidate, indeed *Il2* knockout mice demonstrate spontaneous multi-system autoimmunity (Sadlack *et al.* 1995), further studies in knock-out/knock-in mice will be required to definitively address this. Currently the identity of any gene from genome screens of EAE mapping studies has yet to be identified.

Despite large group sizes of many hundred of animals, which is larger probably than any multiplex family used for mapping in humans, modest levels of significance (P<0.05) are often observed (Baker *et al.* 1995, Encinas *et al.* 1996, Croxford *et al.* 1997, Butterfield *et al.* 1998) and few reach P<0.001, which is substantially less than required for significant linkage for mapping the identity of loci coding monogenic traits. Furthermore within any combination of parental strains a number of loci are found, but often differ between strain combination and mode of inheritance (Jirholt *et al.* 2000). Thus based on even simple inbred systems with a relatively homogeneous phenotype, it indicates the problems faced by similar analysis of mixed samples from an outbred human populations with heterogeneous disease profiles, is immense. This will likely to dilute the ability to identify genes and many thousands will need to be studies to have the power to link major loci and importantly identify genes that confer susceptibility to MS (Ebers *et al* 1996, Haines *et al.* 1996, Sawcer *et al* 1996, Kuokkanen *et al* 1997, Chataway *et al* 1998, Coraddu *et al.* 2001, Broadley et al 2001, Akesson et al. 2002, Ban *et al.* 2002, Sawcer et al. 2002). Some favour the candidate gene approach, possibly due to the increasing availability of markers for typing alleles of known genes however this also means that adequate control populations are available and again is often met with lack of confirmation in studies on other populations of MS patients (Reboul *et al.* 2000, Feakes *et al.* 2000). In contrast to monogenic diseases where emphasis is often placed on identifying mutations, within polygenic disease susceptibility genes are probably part of the normal repertoire of the human population. There is often no need for coding mutations as subtle differences in non-coding elements may influence the level of protein expression within a biochemical/immunological pathway and there may be many different genetic pathways to the phenotype. The evolutionary selective pressures on inbred rodents and outbred humans will not be the same, and therefore different genetic pathways probably also exist between the species, furthermore current models which develop over weeks or maximally months (t'Hart & Amor 2003) may not offer access to all the complexities of pathologies of human disease that develops over years and therefore genetic components of disease may be different. However rodents models do clearly exhibit MHC associations (*Eae 1-m, Eae-1-r*) as found

with human disease (chromosome 6p21) and there are rodent loci (Table 2 &
3) in regions complementary to "hot spots" showing suggestive evidence of
linkage to MS such as human chromosomes (Butterfield *et al.* 1998, Olsson
*et al.* 2000) such as, 5 p21-p24 (*Eae2-m*. Ebers *et al.*1996, Kuokkanen *et al.*
1996), chromosome 17q12-q25 (*Eae7-m, Eae-12-r.* Sawcer *et al* 1996,
Kuokkanen *et al.* 1997, Larsen et al. 2000), 18q12-21 (*Eae18-m, Eae18-
r.*Merriman *et al.* 2001), 19q (*Eae12-m*) (Haines *et al.* 1996, Barcellos *et al.*
1997). As genes are identified, rodent studies will undoubtedly play an
important role in understanding the biology of the genetic influence.

## ACKNOWLEDGEMENTS

D.Baker is a Research Fellow of the Multiple Sclerosis Society of Great
Britain and Northern Ireland

# REFERENCES

Abdul-Majid K-B, Jirholt J, Stadelmann C, Stefferl A, Kjellen P,Wallstrom R, Holmdahl R,
Lassmann H, Olsson T, Harris RA. q Screening of several H-2 congenic mouse strains
identified H-2 mice as highly susceptible to MOG-induced EAE with minimal adjuvant
requirement. Journal of Neuroimmunology 2000; 111: 23–33

Akesson E, Oturai A, Berg J, Fredrikson S, Andersen O, Harbo HF, Laaksonen M, Myhr KM,
Nyland HI, Ryder LP, Sandberg-Wollheim M, Sorensen PS, Spurkland A, Svejgaard A,
Holmans P, Compston A, Hillert J, Sawcer S. A genome-wide screen for linkage in Nordic
sib-pairs with multiple sclerosis. Genes Immun. 2002;3:279-85.

Amor, S., Groome, N., Linington, C., Morris, M.M., Dornmair, K., Gardinier, M.V.,
Matthieu, J.M., Baker, D. Identification of epitopes of myelin oligodendrocyte glycoprotein
for the induction of experimental allergic encephalomyelitis in SJL and Biozzi AB/H J.
Immunol. 1994; 153: 4349–4356.

Baker D, O'Neill JK, Gschmeissner SE, Wilcox CE, Butter C, Turk JL. Induction of chronic
relapsing experimental allergic encephalomyelitis in Biozzi mice. J Neuroimmunol.
1990;28:261-70.

Baker D, Rosenwasser OA, O'Neill JK, Turk JL. Genetic analysis of experimental allergic
encephalomyelitis in mice. J Immunol. 1995;155:4046-51.

Ban M, Stewart GJ, Bennetts BH, Heard R, Simmons R, Maranian M, Compston A, Sawcer
SJ. A genome screen for linkage in Australian sibling-pairs with multiple sclerosis. Genes
Immun. 2002; 3:464-9.

Barcellos LF, Thomson G, Carrington M, Schafer J, Begovich AB, Lin P, Xu XH, Min BQ,
Marti D, Klitz W: Chromosome 19 single-locus and multilocus haplotype associations with

multiple sclerosis. Evidence of a new susceptibility locus in Caucasian and Chinese patients. JAMA 1997, 278:1256–1261

Barry R, Bolton C, Glynn P, Groome N. Myelin basic protein peptides in urine. Ann Neurol. 1992; 31:345-8.

Becanovic K, Backdahl L, Wallstrom E, Aboul-Enein F, , Lassmann H, Olsson T, Lorentzen JC. Paradoxical effects of srthritis-regulating chromosome 4 rgions on myelin oligodendrocyte glycoprotein induced. Eur J Immunol. 2003a;33:19076-16.

Becanovic K, Wallstrom E, Kornek B, Glaser A, Broman KW, Dahlman I, Olofsson P, Holmdahl R, Luthman H, Lassmann H, Olsson T. New loci regulating rat myelin oligodendrocyte glycoprotein-induced experimental autoimmune encephalomyelitis. J Immunol. 2003b;170:1062-9.

Bergsteinsdottir K, Yang HT, Pettersson U, Holmdahl R. Evidence for common autoimmune disease genes controlling onset, severity, and chronicity based on experimental models for multiple sclerosis and rheumatoid arthritis. J Immunol. 2000: 164:1564-8.

Bjartmar C, Wujek JR, Trapp BD. Axonal loss in the pathology of MS: consequences for understanding the progressive phase of the disease. J Neurol Sci 2003;206:165-71.

Blankenhorn E P, Butterfield RJ, Rigby R, Cort L, Giambrone D,McDermott R, McEntee K, Solowski N, Meeker ND, Zachary JF, Doerge RW, Teuscher C: Genetic analysis of the influence of pertussis toxin on EAE susceptibility: an environmental agent can override genetic checkpoints. J Immunol 2000, 164:3420–3425

Bebo BF Jr, Zelinka-Vincent E, Adamus G, Amundson D, Vandenbark AA, Offner H: Gonadal hormones influence the immune response to PLP 139–151 and the clinical course of relapsing experimental autoimmune encephalomyelitis. J Neuroimmunol 1998, 84:122–130

Bebo BF Jr, Schuster JC, Vandenbark AA, Offner H: Androgens alter the cytokine profile and reduce encephalitogenicity of myelin-reactive T cells. J Immunol 1999, 162:35–40

Bettelli E, Pagany M, Weiner HL, Linington C, Sobel RA, Kuchroo VK. oligodendrocyte glycoprotein-specific T cell receptor transgenic mice develop spontaneous autoimmune optic neuritis. J Exp Med. 2003;197:1073-81.

Brabb T, Goldrath AW, von Dassow P, Paez A, Liggitt HD, Goverman J Triggers of autoimmune disease in a murine TCR-transgenic model for multiple sclerosis. J Immunol. 1997;159:497-507.

Brahic M, Bureau JF. Genetics of susceptibility to Theiler's virus infection. Bioessays. 1998; 20:627-33.

Broadley S, Sawcer S, D'Alfonso S, Hensiek A, Coraddu F, Gray J, Roxburgh R, Clayton D, Buttinelli C, Quattrone A, Trojano M, Massacesi L, Compston A. A genome screen for multiple sclerosis in Italian families.Genes Immun. 2001; 2:205-10

Brocke S, Gijbels K, Steinman L: Experimental autoimmune encephalomyelitis in the mouse. Autoimmune Disease Models, a Guidebook. Edited by IR Cohen, A Miller. New York, Academic Press, 1994, pp1–14

Brok HP, Bauer J, Jonker M, Blezer E, Amor S, Bontrop RE, Laman JD, 't Hart BA. Non-human primate models of multiple sclerosis.Immunol Rev. 2001;183:173-85.

Butterfield RJ, Sudweeks JD, Blankenhorn EP, Korngold R, Marini JC, Todd JA, Roper RJ, Teuscher C: New genetic loci that control susceptibility and symptoms of experimental allergic encephalomyelitis in inbred mice. J Immunol 1998, 161:1860–1867

Butterfield RJ, Blankenhorn EP, Roper RJ, Zachary JF, Doerge RW, Sudweeks J, Rose J, Teuscher C: Genetic analysis of disease subtypes and sexual dimorphisms in mouse experimental allergic encephalomyelitis (EAE): relapsing/remitting and monophasic remitting/ nonrelapsing EAE are immunogenetically distinct. J Immunol 1999, 162:3096–3102

Butterfield RJ, Blankenhorn EP, Roper RJ, Zachary JF, Doerge RW, Teuscher C. Identification of genetic loci controlling the characteristics and severity of brain and spinal cord lesions in experimental allergic encephalomyelitis. Am. J. Pathol. 2000; 157:637-645.

Bruno R, Sabater L, Sospedra M, Ferrer-Francesch X, Escudero D, Martinez-Caceres E, Pujol-Borrell R. Multiple sclerosis candidate autoantigens except myelin oligodendrocyte glycoprotein are transcribed in human thymus.Eur J Immunol. 2002; 32:2737-47.

Chataway, J., Feakes, R., Coraddu, F., Gray, J., Deans, J., Fraser, M., Robertson, N., Broadley, S., Jones, H., Clayton, D., Goodfellow, P., Sawcer, S., Compston, A., The genetics of multiple sclerosis:
    principles, background and updated results of the United Kingdom systematic genome screen. Brain, 1998; 121, 1869–1887.

Compston A, Coles A. Multiple sclerosis. Lancet 2002; 359: 1221-1231.

Compston A, Sawcer S. Genetic analysis of multiple sclerosis. Curr Neurol Neurosci Rep. 2002;2:259-66.

Confavreux C, Vukusic S, Moreau T, Adeleine P. Relapses and progression of disability in multiple sclerosis. N Engl J Med 2000; 343: 1430-38.

Coraddu F, Sawcer S, Feakes R, Chataway J, Broadley S, Jones HB, Clayton D, Gray J, Smith S, Taylor C, Goodfellow PN, Compston A. HLA typing in the United Kingdom multiple sclerosis genome screen. Neurogenetics. 1998; 2:24-33.

Coraddu F, Sawcer S, D'Alfonso S, Lai M, Hensiek A, Solla E, Broadley S, Mancosu C, Pugliatti M, Marrosu MG, Compston A. A genome screen for multiple sclerosis in Sardinian multiplex families.Eur J Hum Genet. 2001;9:621-6.

Croxford JL, O'Neill JK, Baker D. Polygenic control of experimental allergic encephalomyelitis in Biozzi ABH and BALB/c mice.J Neuroimmunol. 1997; 74:205-11.

Dalal M, Kim S, Voskuhl RR: Testosterone therapy ameliorates experimental autoimmune encephalomyelitis and induces a T helper 2 bias in the autoantigen-specific T lymphocyte response. J Immunol 1997,159:3–6

Dahlman I, Lorentzen JC, de Graaf KL, Stefferl A, Linington C, Luthman H, Olsson T. Quantitative trait loci disposing for both experimental arthritis and encephalomyelitis in the

DA rat; impact on severity of myelin oligodendrocyte glycoprotein-induced experimental autoimmune encephalomyelitis and antibody isotype pattern. Eur J Immunol. 1998; 28:2188-96.

Dahlman I, Jacobsson L, Glaser A, Lorentzen JC, Andersson M, Luthman H, Olsson T. Genome-wide linkage analysis of chronic relapsing experimental autoimmune encephalomyelitis in the rat identifies a major susceptibility locus on chromosome 9. J Immunol. 1999a;162:2581-8

Dahlman I, Wallstrom E, Weissert R, Storch M, Kornek B, Jacobsson L, Linington C, Luthman H, Lassmann H, Olsson T. Linkage analysis of myelin oligodendrocyte glycoprotein-induced experimental autoimmune encephalomyelitis in the rat identifies a locus controlling demyelination on chromosome 18. Hum Mol Genet. 1999b;8:2183-90.

Duquette P, Pleines J, Girard M, Charest L, Senecal-Quevillon M, Masse C: The increased susceptibility of women to multiple sclerosis. Can J Neurol Sci 1992, 19:466–471

Ebers GC, Sadovnick AD, Risch NJ: A genetic basis for familial aggregation in multiple sclerosis. Canadian Collaborative Study Group. Nature 1995, 377:150–151

Ebers GC, Kukay K, Bulman DE, Sadovnick AD, Rice G, Anderson C, Armstrong H, Cousin K, Bell RB, Hader W, Paty DW, Hashimoto S, Oger J, Duquette P, Warren S, Gray T, O'Connor P, Nath A, Auty A, Metz L, Francis G, Paulseth JE, Murray TJ, Pryse-Phillips W, Risch N: A full genome search in multiple sclerosis. Nat Genet 1996, 13:472–476.

Encinas, JA, Lees MB, Sobel RA, Symonowics C, Greer JM, Shovlin CL, Weiner HL, Seidman CE, Seidman JG, Kuchroo VJ. Genetic analysis of susceptibility to experimental autoimmune encephalomyelitis in a cross between SJL/L and B10.S mice. J. Immunol. 1996; 157:2186.

Encinas JA, Wicker LS, Peterson LB, Mukasa A, Teuscher C, Sobel R, Weiner HL, Seidman CE, Seidman JG, Kuchroo VK: QTL influencing autoimmune diabetes and encephalomyelitis map to a 0.15 cM region containing Il2. Nat Genet 1999, 21:158-160.

Encinas JA, Kuchroo VK. Mapping and identification of autoimmunity genes Curr Opin Immunol. 2000;12:691-7.

Encinas, JA, Lees MB, Sobel RA, Symonowics C, Weiner HL, Seidman CE, Seidman JG, Kuchroo VJ. Identification of genetic loci associated with paralysis, inflammation and weight loss in mouse experimental autoimmune encephalomyelitis. Int. Immunol. 2001; 13:257-64

Feakes R, Sawcer S, Smillie B, Chataway J, Broadley S, Compston A. No evidence for the involvement of interleukin 2 or the immunoglobulin heavy chain gene cluster in determining genetic susceptibility to multiple sclerosis. J Neurol Neurosurg Psychiatry. 2000; 68:679.

Fritz, RB, Skeen MJ, Chou CH, Garcia M, Egorov IK. Major histocompatibility complex-linked control of the murine immune response to myelin basic protein. J. Immunol. 1985; 134:2328-2332.

Furlan R, Brambilla E, Sanvito F, Roccatagliata L, Olivieri S, Bergami A, Pluchino S, Uccelli A, Comi G, Martino G. Vaccination with amyloid-beta peptide induces autoimmune encephalomyelitis in C57/BL6 mice.Brain. 2003;126:285-91.

Gao JF, Call SB, Fillmore PD, Watanabe T, Meeker ND, Teuscher C. Analysis of the role of Bphs/Hrh1 in the genetic control of responsiveness to pertussis toxin.Infect Immun. 2003;71:1281-7.

Goedde R, Sawcer S, Boehringer S, Miterski B, Sindern E, Haupts M, Schimrigk S, Compston A, Epplen JT. A genome screen for linkage disequilibrium in HLA-DRB1*15-positive Germans with multiple sclerosis based on 4666 microsatellite markers. Hum Genet. 2002;111:270-7.

Goverman J, Woods A, Larson L, Weiner LP, Hood L, Zaller DM Transgenic mice that express a myelin basic protein-specific T cell receptor develop spontaneous autoimmunity. Cell. 1993;72:551-60.

Haines JL, Ter Minassian M, Bazyk A, Gusella JF, Kim DJ, Terwedow H, Pericak-Vance MA, Rimmler JB, Haynes CS, Roses AD, Lee A, Shaner B, Menold M, Seboun E, Fitoussi RP, Gartioux C, Reyes C, Ribierre F, Gyapay G, Weissenbach J, Hauser SL, Goodkin DE, Lincoln R, Usuku K, Oksenberg JR: A complete genomic screen for multiple sclerosis underscores a role for the major histocompatibility complex. The Multiple Sclerosis Genetics Group. Nat Genet 1996, 13:469–471

Hensiek AE, Sawcer SJ, Feakes R, Deans J, Mander A, Akesson E, Roxburgh R, Coraddu F, Smith S, Compston DA. HLA-DR 15 is associated with female sex and younger age at diagnosis in multiple sclerosis. J Neurol Neurosurg Psychiatry. 2002; 72:184-7.

Hogancamp WE, Rodriguez M, Weinshenker BG: The epidemiology of multiple sclerosis. Mayo Clin Proc 1997, 72:871–878

Homo-Delarche F, Fitzpatrick F, Christeff N, Nunez EA, Bach JF, Dardenne M: Sex steroids, glucocorticoids, stress and autoimmunity. J Steroid Biochem Mol Biol 1991, 40:619–637

Ichikawa M, Koh CS, Inaba Y, Seki C, Inoue A, Itoh M, Ishihara Y, Bernard CC, Komiyama A. IgG subclass switching is associated with the severity of experimental autoimmune encephalomyelitis induced with myelin oligodendrocyte glycoprotein peptide in NOD mice. Cell Immunol. 1999; 191:97-104.

Jansson L, Holmdahl R. Estrogen-mediated immunosuppression in autoimmune diseases. Inflamm. res. 1998, 47:290–301

Jirholt, J., A.-K. Lindqvist, J. Karlsson, Å. Andersson, and R. Holmdahl. Identification of susceptibility genes for experimental autoimmune encephalomyelitis that overcome the effect of protective alleles at the *eae2* locus. Int. Immunol. 2002;14:79-85

Karlsson J, Zhao, Lonskaya I, Neptin M, Holmdahl R, Andersson A. Novel quantitative trait loci controlling development of experimental autoimmune encephalomyelitis and proportion of lymphocyte subpopulations. J. Immunol 2003 170:1019-1026

Kira J, Kanai T, Nishimura Y, Yamasaki K, Matsushita S, Kawano Y, Hasuo K, Tobimatsu S, Kobayashi T: Western *versus* Asian types of multiple sclerosis: immunogenetically and clinically distinct disorders. Ann Neurol 1996, 40:569–574

Kjellén P, Jansson, Vestberg M, A Andersson, MattssonR, Holmdahl. The H2-Ab gene influences the severity of experimental allergic encephalomyelitis induced by proteolipoprotein 103-116 J.Neuroimmunol. 2001;120:25-33

Kojima K, Berger T, Lassmann H, Hinze-Selch D, Zhang Y, Gehrmann J, Reske K, Wekerle H, Linington C. Experimental autoimmune panencephalitis and uveoretinitis transferred to the Lewis rat by T lymphocytes specific for the S100 beta molecule, a calcium binding protein of astroglia. J Exp Med. 1994;180:817-29.

Kuokkanen S, Gschwend M, Rioux JD, Daly MJ, Terwilliger JD, Tienari PJ, Wikstrom J, Palo J, Stein LD, Hudson TJ, Lander ES, Peltonen L: Genomewide scan of multiple sclerosis in Finnish multiplex families. Am J Hum Genet 1997, 61:1379–1387

Kuokkanen S, Sundvall M, Terwilliger JD, Tienari PJ, Wikstrom J, Holmdahl R, Pettersson U, Peltonen L. A putative vulnerability locus to multiple sclerosis maps to 5p14–p12 in a region syntenic to the murine locus Eae2. Nat Genet 1996, 13:477–480

Kuchroo VK, Anderson AC, Waldner H, Munder M, Bettelli E, Nicholson LB. T cell response in experimental autoimmune encephalomyelitis (EAE): role of self and cross-reactive antigens in shaping, tuning, and regulating the autopathogenic T cell repertoire. Annu Rev Immunol. 2002;20:101-23.

Larsen F, Oturai A, Ryder LP, Madsen HO, Hillert J, Fredrikson S, Sandberg-Wollheim M, Laaksonen M, Harbo HF, Sawcer S, Fugger L, Sorensen PS, Svejgaard A. Linkage analysis of a candidate region in Scandinavian sib pairs with multiple sclerosis reveals linkage to chromosome 17q. Genes Immun. 2000;1:456-9.

Levine S, Sowinski R. Experimental allergic encephalomyelitis in inbred and outbred mice. J Immunol. 1973;110:139-43.

Li Y, Li H, Martin R, Mariuzza RA. Structural basis for the binding of an immunodominant peptide from myelin basic protein in different registers by two HLA-DR2 proteins. J Mol Biol. 2000;304:177-88.

Lublin FD, Reingold SC: Defining the clinical course of multiple sclerosis: results of an international survey. National Multiple Sclerosis Society (USA) Advisory Committee on Clinical Trials of New Agents in Multiple Sclerosis. Neurology 1996, 46:907–911

Lucchinetti CF, Bruck W, Rodriguez M, Lassmann H: Distinct patterns of multiple sclerosis pathology indicates heterogeneity on pathogenesis. Brain Pathol 1996, 6:259–274

Lyons PA, Armitage N, Argentina F, Denny P, Hill NJ, Lord CJ, Wilusz MB, Peterson LB, Wicker LS, Todd JA.Congenic mapping of the type 1 diabetes locus, Idd3, to a 780-kb region of mouse chromosome 3: identification of a candidate segment of ancestral DNA by haplotype mapping. Genome Res. 2000;10:446-53.

Lyons PA, Armitage N, Lord CJ, Phillips MS, Todd JA, Peterson LB, Wicker LS. Mapping by genetic interaction: high-resolution congenic mapping of the type 1 diabetes loci Idd10 and Idd18 in the NOD mouse. Diabetes. 2001; 50:2633-7.

Ma RZ, Gao J, Meeker ND, Fillmore PD, Tung KS, Watanabe T, Zachary JF, Offner H, Blankenhorn EP, Teuscher C. Identification of Bphs, an autoimmune disease locus, as histamine receptor H1.Science. 2002;297:620-3.

Madsen LS, Andersson EC, Jansson L, krogsgaard M, Andersen CB, Engberg J, Strominger JL, Svejgaard A, Hjorth JP, Holmdahl R, Wucherpfennig KW, Fugger L. A humanized model for multiple sclerosis using HLA-DR2 and a human T-cell receptor.Nat Genet. 1999;23:343-7.

Marrosu MG, Murru MR, Costa G, Cucca F, Sotgiu S, Rosati G, Muntoni F. Multiple sclerosis in Sardinia is associated and in linkage disequilibrium with HLA-DR3 and -DR4 alleles.Am J Hum Genet. 1997;61:454-7.

Mendel I, Kerlero de Rosbo N, Ben-Nun A. A myelin oligodendrocyte glycoprotein peptide induces typical chronic experimental autoimmune encephalomyelitis in H-2b mice: fine specificity and T cell receptor V beta expression of encephalitogenic cells. Eur. J. Immunol. 1995; 25: 1951–1959.

Mendel I, Gur H, Kerlero de Rosbo N, Ben-Nun A. Experimental autoimmune encephalomyelitis induced in B6.C-H-2bm12 mice by myelin oligodendrocyte glycoprotein: effect of MHC class II mutation on immunodominant epitope selection and fine epitope specificity of encephalitogenic T cells.J Neuroimmunol. 1999;96:9-20.

Merriman TR, Cordell HJ, Eaves IA, Danoy PA, Coraddu F, Barber R, Cucca F, Broadley S, Sawcer S, Compston A, Wordsworth P, Shatford J, Laval S, Jirholt J, Holmdahl R, Theofilopoulos AN, Kono DH, Tuomilehto J, Tuomilehto-Wolf E, Buzzetti R, Marrosu MG, Undlien DE, Ronningen KS, Ionesco-Tirgoviste C, Shield JP, Pociot F, Nerup J, Jacob CO, Polychronakos C, Bain SC, Todd JA. Suggestive evidence for association of human chromosome 18q12-q21 and its orthologue on rat and mouse chromosome 18 with several autoimmune diseases.Diabetes. 2001;50:184-94.

Morris-Downes MM, McCormack K, Baker D, Sivaprasad D, Natkunarajah J, Amor S. Encephalitogenic and immunogenic potential of myelin-associated glycoprotein (MAG), oligodendrocyte-specific glycoprotein (OSP) and 2',3'-cyclic nucleotide 3'-phosphodiesterase (CNPase) in ABH and SJL mice.J Neuroimmunol. 2002;122:20-33.

Olerup O, Hillert J, Fredrikson S, Olsson T, Kam-Hansen S, Moller E, Carlsson B, Wallin J: Primarily chronic progressive and relapsing/remitting multiple sclerosis: two immunogenetically distinct disease entities. Proc Natl Acad Sci USA 1989, 86:7113–7117

Olsson T, Dahlman I, Wallström E, Weissert R, Piehl F. Genetics of rat Neuroinflammation. J. Neuroimmunol 2000; 107:191-200

O'Neill JK, Baker D, Davison AN, Allen SJ, Butter C, Waldmann H, Turk JL.Control of immune-mediated disease of the central nervous system with monoclonal (CD4-specific) antibodies. J Neuroimmunol. 1993; 45:1-14.

Oturai A, Larsen F, Ryder LP, Madsen HO, Hillert J, Fredrikson S, Sandberg-Wollheim M, Laaksonen M, Koch-Henriksen N, Sawcer S, Fugger L, Sorensen PS, Svejgaard A. Linkage and association analysis of susceptibility regions on chromosomes 5 and 6 in 106 Scandinavian sibling pair families with multiple sclerosis. Ann Neurol. 1999; 46:612-6.

Pagany M, Jagodic M, Bourquin C, Olsson T, Linington C. Genetic variation in myelin oligodendrocyte glycoprotein expression and susceptibility to experimental autoimmune encephalomyelitis.              J              Neuroimmunol.              2003;139:1-8.

Pihlaja H, Rantamaki T, Wikstrom J, Sumelahti ML, Laaksonen M, Ilonen J, Ruutiainen J, Pirttila T, Elovaara I, Reunanen M, Kuokkanen S, Peltonen L, Koivisto K, Tienari PJ. Linkage disequilibrium between the MBP tetranucleotide repeat and multiple sclerosis is restricted to a geographically defined subpopulation in Finland.Genes Immun. 2003; 4:138-46.

Podolin PL, Denny P, Lord CJ, Hill NJ, Todd JA, Peterson LB, Wicker LS, Lyons PA. Congenic mapping of the insulin-dependent diabetes (Idd) gene, Idd10, localizes two genes mediating the Idd10 effect and eliminates the candidate Fcgr1. J Immunol. 1997 ;159:1835-43.

Podolin PL, Wilusz MB, Cubbon RM, Pajvani U, Lord CJ, Todd JA, Peterson LB, Wicker LS, Lyons PA. Differential glycosylation of interleukin 2, the molecular basis for the NOD Idd3 type 1 diabetes gene? Cytokine. 2000;12:477-82.

Prins JB, Todd JA, Rodrigues NR, Ghosh S, Hogarth PM, Wicker LS, Gaffney E, Podolin PL, Fischer PA, Sirotina A, et al. Linkage on chromosome 3 of autoimmune diabetes and defective Fc receptor for IgG in NOD mice. Science. 1993;260:695-8.

Raine CS, Barnett LB, Brown A, Behar T, McFarlin DE: Neuropathology of experimental allergic encephalomyelitis in inbred strains of mice. Lab Invest 1980, 43:150–157

Reboul J, Mertens C, Levillayer F, Eichenbaum-Voline S, Vilkoren T, Cournu I, Babron MC, Lyon-Caen O, Clerget-Darpoux F, Edan G, Clanet M, Brahic M, Bureau JF, Fontaine B, Liblau R. Cytokines in genetic susceptibility to multiple sclerosis: a candidate gene approach. French Multiple Sclerosis Genetics Group. *J Neuroimmunol* 2000, 102:107-112.

Roth MP, Viratelle C, Dolbois L, Delverdier M, Borot N, Pelletier L, Druet P, Clanet M, Coppin H. A genome-wide search identifies two susceptibility loci for experimental autoimmune encephalomyelitis on rat chromosomes 4 and 10. J Immunol. 1999;162:1917-22.

Sadlack B, Lohler J, Schorle H, Klebb G, Haber H, Sickel E, Noelle RJ, Horak I. Generalized autoimmune disease in interleukin-2-deficient mice is triggered by an uncontrolled activation and proliferation of CD4+ T cells. Eur J Immunol. 1995;25:3053-9.

Sawcer S, Jones HB, Feakes R, Gray J, Smaldon N, Chataway J, Robertson N, Clayton D, Goodfellow PN, Compston A: A genome screen in multiple sclerosis reveals susceptibility loci on chromosome 6p21 and 17q22. Nat Genet 1996, 13:464–468

Sawcer S, Maranian M, Setakis E, Curwen V, Akesson E, Hensiek A, Coraddu F, Roxburgh R, Sawcer D, Gray J, Deans J, Goodfellow PN, Walker N, Clayton D, Compston A. A whole genome screen for linkage disequilibrium in multiple sclerosis confirms disease associations with regions previously linked to susceptibility. Brain. 2002; 125:1337-47.

Seboun E, Oksenberg JR, Rombos A, Usuku K, Goodkin DE, Lincoln RR, Wong M, Pham-Dinh D, Boesplug-Tanguy O, Carsique R, Fitoussi R, Gartioux C, Reyes C, Ribierre F, Faure S, Fizames C, Gyapay G, Weissenbach J, Dautigny A, Rimmler JB, Garcia ME, Pericak-Vance MA, Haines JL, Hauser SL. Linkage analysis of candidate myelin genes in familial multiple sclerosis. Neurogenetics. 1999;2:155-62.

Slavin A, Ewing C, Liu J, Ichikawa M, Slavin J, Bernard CC. Induction of a multiple sclerosis-like disease in mice with an immunodominant epitope of myelin oligodendrocyte glycoprotein.
Autoimmunity. 1998;28:109-20.

Sudweeks JD, Todd JA, Blankenhorn EP, Wardell BB, Woodward SR, Meeker ND, Estes SS, Teuscher C: Locus controlling *Bordetella* pertussis-induced histamine sensitization (Bphs), an autoimmune disease-susceptibility gene, maps distal to T-cell receptor beta-chain gene on mouse chromosome 6. Proc Natl Acad Sci USA 1993, 90:3700–3704

Sun D, Whitaker JN, Huang Z, Liu D, Coleclough C, Wekerle H, Raine CS. Myelin antigen-specific CD8+ T cells are encephalitogenic and produce severe disease in C57BL/6 mice. J Immunol. 2001; 166:7579-87

Sundvall, M, Jirholt J, Yang HT, Jansson L, Engstrom A, Pettersson U, Holmdahl R. Identification of murine loci associated with susceptibility to chronic experimental autoimmune encephalomyelitis. Nat. Genet. 1995, 10:313-317.

Teuscher C, Butterfield RJ, Ma RZ, Zachary JF, Doerge RW, Blankenhorn EP: Sequence polymorphisms in the chemokines Scya1(TCA-3), Scya2 (monocyte chemoattractant protein (MCP)-1), and Scya12 (MCP-5) are candidates for eae7, a locus controlling susceptibility to monophasic remitting/nonrelapsing experimental allergic encephalomyelitis. J Immunol 1999, 163:2262–2266

t'Hart B, Amor S. The use of animal models to investigate the pathogenesis of neuroinflammatory disorders of the central nervous sysyem. Curr Opin Neurol. 2003;16:375-384

Thompson AJ, Polman CH, Miller DH, McDonald WI, Brochet B, Filippi MMX, De Sa J: Primary progressive multiple sclerosis. Brain 1997, 120:1085–1096

Thorpe JW, Kidd D, Moseley IF, Thompson AJ, MacManus DG, Compston DA, McDonald WI, Miller DH: Spinal MRI in patients with suspected multiple sclerosis and negative brain MRI. Brain 1996, 119:709–714

Todd, JA, Aitman TJ, Cornall RJ, Ghosh S, Hall JRS, Hearne CM, Knight AM, Love JM, McAleer MA, Prins J-B, Rodrigues N, Lathrop M, Pressey A, DeLarato NH, Peterson LB, Wicker LS.
Genetic analysis of autoimmune type 1 diabetes mellitus in mice. Nature. 1991;351:542-547.

Tuohy VK, Sobel RA, Lees MB: Myelin proteolipid protein-induced experimental allergic encephalomyelitis. Variations of disease expression in different strains of mice. J Immunol 1988, 140:1868–1873

Wekerle H, Kojima K, Lannes-Vieira J, Lassmann H, Linington C: Animal models. Ann Neurol 1994, 36:S47–S53

Wujek JR, Bjartmar C, Richer E, Ransohoff RM, Yu M, Tuohy VK, Trapp BD. Axon loss in the spinal cord determines permanent neurological disability in an animal model of multiple sclerosis. J Neuropathol Exp Neurol 2002; 61: 23-32.

**Table 2. Quantitative Trait Loci identified for susceptibility to EAE in the rat.**

| Locus | Chromosome | Markers flanking interval | Traits | Ref |
|---|---|---|---|---|
| *Eae1-r* | 20 | D20Rat41-D20Mgh4 | Severity | (Bergsteinsdottir et al 2000) |
| *Eae2-r* | 4 | D4Mhh1-D4Mgh14 close to IL-6 | Incidence | (Roth *et al.* 1999) |
| *Eae3-r* | 10 | D10Mit 10-D10Mgh10 close IL-4 | Incidence | (Roth *et al.* 1999) |
| *Eae4-r* | 9 | D9Rat29-D9Mit6 | Incidence | (Dahlman *et al.* 1999a) |
| *Eae5-r* | 12 | D12Mit2-D12Rat2 (recessive) | Severity,relapses | (Bergsteinsdottir et al 2000) |
| *Eae6-r* | 1 | D1Mgh9-D1Mit33 (M recessive) | Severity,duration, weight loss | (Bergsteinsdottir et al 2000) |
| *Eae7-r* | 1 | D1Mit7-D1Mgh12 (M recessive) | Severity,duration, relapses,weight loss | (Bergsteinsdottir et al 2000) |
| *Eae8-r* | 19 | D19Mit9-D19Mit5 (recessive) | Acute disease | (Bergsteinsdottir et al 2000) |
| *Eae9-r* | 6 | D6Mgh3-D6Rat10 (additive) | Weight loss,duration, relapse | (Bergsteinsdottir et al 2000) |
| *Eae10-r* | 14 | D14Mit6-D14Wox11 (recessive) | Weight loss,duration, relapse | (Bergsteinsdottir et al 2000) |
| *Eae11-r* | 4 | D4Wox30-D4Mgh14 (F) | Onset | (Bergsteinsdottir et al 2000) |
| *Eae12-r* | 10 | D10Rat7-D10Rat15 | Incidence, | (Dahlman *et al.* 1999b) |
| *Eae13-r* | 12 | D12Rat19-D12Rat23 | Incidence | (Dahlman *et al.* 1999b) |
| *Eae14-r* | 13 | D13Rat23 | Incidence | (Dahlman *et al.* 1999b) |
| *Eae15-r* | 18 | D18Mgh4-D18Mit9 | Incidence | (Dahlman *et al.* 1999b) |
| *Eae16-r* | 8 | D8Rat36-D8Rat11 | Chronicity | (Becanovic et al 2003) |
| *Eae17-r* | 13 | D13Rat85-D8Rat64 | Severity | (Becanovic et al 2003) |
| *Eae18-r* | 18 | D18Mit9 | Demyelination | (Dahlman *et al.* 1999b) |
| *Eaex-r* | 5 | D5Mit10-D5Wox21 | Onset | (Bergsteinsdottir et al 2000) |
| *Eaey-r* | 18 | D18Mgh1-D18Wox13 | Duration | (Bergsteinsdottir et al 2000) |
| *Eaez-r* | 18 | D18Mit7-D18Wox1 | Severity, duration | (Bergsteinsdottir et al 2000) |

Microsatellite quantitative trait locus mapping in a number of rat strains and induction techniques suggest linkage in a number of chromosomal regions controlling the development and severity of clinical signs including the development of acute and relapsing disease, demyelination and duration of disease and time of onset and different histological types of lesions (See Becanovic *et al.* 2003 for relative location (proximal or distal) of chromosomal segments). These *Eae-r* loci controlling EAE in the rat bear no relationship of the corresponding mouse *Eae-m* locus

**Table 3. Quantitative Trait Loci identified for susceptibility to EAE in the mouse**

| Locus | Location (cM) | | Traits | Reference |
|---|---|---|---|---|
| *Eae1* | 17 | 20 (*H2*) | Incidence | (Fritz *et al.* 1985) |
| *Eae2* | 15 | 16 | Incidence | (Sundvall *et al.* 1995) |
| *Eae3* | 3 | 43 | Incidence, Acute | (Sundvall *et al.* 1995, Encinas et al. 1996, Butterfield *et al.* 1998) |
| *Eae4* | 7 | 50 | Incidence, lesion location | (Baker *et al.* 1995, Butterfield *et al.* 1998, 1999) |
| *Eae5* | 17 | 21 | Incidence | (Croxford et al 1997, Butterfield et al 1998) |
| *Eae6a* | 11 | 7 | Severity | (Baker *et al.* 1995, Butterfield et al. 1998, Teuscher *et al.* 1999) |
| *Eae6b* | 11 | 24 | Duration | (Teuscher *et al.* 1999) |
| *Eae7* | 11 | 48 | Severity, relapsing chronic | (Baker *et al.* 1995, Butterfield et al. 1998. Teuscher *et al.* 1999) |
| *Eae8* | 2 | 103 | Incidence, Severity, weight | (Baker *et al.* 1995, Butterfield et al 1998) |
| *Eae9* | 9 | 35 | Incidence, Duration, lesions | (Baker *et al.* 1995, Butterfield et al 1998) |
| *Eae10* | 3 | 72 | Onset | (Butterfield *et al.* 1998) |
| *Eae11* | 16 | 41 | Incidence, B lesions (M) | (Baker *et al.* 1995, Butterfield *et al.* 1999) |
| *Eae12* | 7 | 6 | Incidence, Relapsing Disease(F) | (Baker *et al.* 1995, Butterfield *et al* 1998,1999) |
| *Eae13* | 13 | 37 | Relapsing (Male) | (Butterfield *et al.* 1999) |
| *Eae14* | 8 | 21 | Incidence | (Blankenhorn *et al.* 2000) |
| *Eae15* | 10 | 16 | Lesion location(Male) | (Butterfield *et al.* 2000) |
| *Eae16* | 12 | 12 | Lesion location | (Butterfield *et al.* 2000) |
| *Eae17* | 10 | 44 | Severity, Demyelination F. | (Blakenhorn *et al.* 2000) |
| *Eae18* | 18 | 54 | Incidence, Lesion location M. | (Baker *et al.* 1995, Merriman et al. 2000,Butterfield *et al.* 2000) |
| *Eae19* | 19 | 34 | Demyelination(M) | (Butterfield *et al.* 2000) |
| *Eae20* | 3 | 14 | Demyelination | (Butterfield *et al.* 2000) |
| *Eae21* | 2 | 36 | Lesion location F. | (Butterfield *et al.* 2000) |
| *Eae22* | 11 | 61 | Lesion location (F) | (Butterfield *et al.* 2000) |
| *Eae23* | 11 | 38 | Lesion location (M) | (Butterfield *et al.* 2000) |
| *Eae24* | 8 | 10 | severity | (Encinas *et al.* 2001) |
| *Eae25* | 18 | 54 | severity | (Blakenhorn *et al.* 2000) |
| *Eae26* | 7 | 53 | Susceptibility | (Jirholt *et al.* 2000) |

Microsatellite quantitative trait locus mapping in a number of mouse strains and induction techniques suggest linkage in a number of chromosomal regions controlling the development and severity of clinical signs including the development of acute and relapsing disease, demyelination and duration of disease and time of onset and the presence and location of histological mononuclear cell infiltration and demyelination and disease associated weight loss. Some of the effects were sex related and seen in only males (M) or females (F). These *Eae-m* (Eae loci of mice) loci controlling EAE in the mouse bear no identity relationship of the nomenclature of the rat, although *Eae1* both map to MHC. The chromosomal locations are approximations of the peak location (www.ncbi.org) and the intervals containing the genes are significantly larger.

**TABLE 4.**

**Mouse chromosome 7 harbours dominant susceptibility loci for the development of EAE.**

**A)**

| Locus | Position (cM) | ABH:NOD Heterozygous:Homozygous | $\chi^2 > 3.8 = P < 0.05$ |
|---|---|---|---|
| D7Nds6 (*Ckmm*) | 4 | 90:58 | 6.92 |
| D7Mit20 | 6 | 92:57 | 8.22 |
| D7Mit180 | 6 | 92:57 | 8.22 |
| D7Mit55 | 12 | 96:53 | 12.41* |
| D7Mit227 | 13 | 95:54 | 11.28 |
| D7Mit310 | 15 | 93:57 | 8.64 |
| D7Mit69 | 21 | 95:55 | 10.67 |
| D7Mit211 | 26 | 98:52 | 14.11* |
| D7Mit297 | 26 | 97:51 | 14.30 |
| D7Nds1 | 28 | 97:53 | 12.91 |
| D7Mit92 | 29 | 97:53 | 12.91 |
| D7Mit124 | 37 | 95:55 | 10.67 |
| D7Mit262 | 37 | 94:53 | 11.44 |
| D7Mit37 (*HbbAr*) | 37 | 97:53 | 12.91 |
| D7Mit53 | 38 | 97:53 | 12.91* |
| D7Mit130 | 38 | 97:53 | 12.91 |
| D7Nds14 (*Hbb*) | 39 | 96:54 | 11.76 |
| D7Mit40 | 43 | 95:54 | 11.28 |
| D7Mit101 | 46 | 91:59 | 6.83 |
| D7Mit206 | 46 | 91:59 | 6.83 |
| D7Mit68 | 46 | 89:61 | 5.23 |
| D7Mit44 | 50 | 88:62 | 4.51 |
| D7Mit186 | 50 | 88:62 | 4.51 |
| D7Mit71 | 54 | 88:62 | 4.51 |
| D7Mit12 | 62 | 87:63 | 3.84 |
| D7Mit14 | 65 | 82:68 | |

**B)**

| Strain | No. EAE/ total | Group EAE Score ± SEM | Day of Onset ± SD |
|---|---|---|---|
| ABH x ABH | 7/7 | 4.0 ± 0.0 | 15.1 ± 1.2 |
| NOD x NOD | 1/7 | 0.5 ± 0.5 | 18.0 ± n/a |
| NOD.$ABH^{chr7}$ x NOD.$ABH^{chr7}$ | 10/12 | 3.3 ± 0.4 | 16.1 ± 1.9 |
| NOD.$ABH^{chr7}$ x NOD | 8/8 | 3.6 ± 0.1 | 16.8 ± 1.9 |

Following outcross of male ABH x female NOD mice, male (ABH x NOD) F1 were backcrossed with NOD mice and male progeny were selected. EAE was induced following induction with mouse spinal cord homogenate in Freund's adjuvant. **(A)** animals that developed hindlimb paresis or paralysis (grade 3-4 occurs in 25% of the population) were selected for microsatellite analysis and compared to expected mendelian inheritance distribution of the ABH mouse allele (Baker *et al.* 1995). **(B)** Mice were selected for the expression of ABH chromosome 7 ($ABH^{chr7}$) alleles at D7Nds6, D7Mit211/297 and D7Mit 37/53, D17Mit12 at each generation. Mice expressing the ABH mouse alleles D6Nds3, D9Nds2, D11Nds29, D13Mit9, D15Mit11 and D18Mit8/Mit9 previously implicated in EAE susceptibility in this strain combination (Baker *et al.* 1995), where excluded. Following a further 7 backcross generations selecting for $ABH^{chr7}$ markers mice were intercrossed to produce NOD.$ABH^{chr7}$ homozygous congenic mice. Induction of EAE using spinal cord homogenate (Baker *et al.* 1995) demonstrates that chromosome 7 harbours major susceptibility loci that induce clinical disease (Scale 0-4). and that this or these are autosomal dominant genes (Baker *et al.* 1995).

Chapter A11

# T LYMPHOCYTES IN EAE
*Insights into the development and control of the autoreactive T cell*

Kelli Ryan and Stephen M Anderton
*Univeristy of Edinburgh, Institute of Cell Animal and Population Biology*

**Abstract**: The EAE lesion is initiated by CD4+ T cells. Although this statement still holds, recent work has shown that CD8+ T cells and possibly γδ T cells can also contribute to pathology. The recovery phase of the disease poses more of a challenge with the potential for various regulatory T cell populations to halt the pathology. We know a good deal about the fine-specificity of T cell recognition of epitopes derived from myelin autoantigens. This, coupled with technological refinements in cellular immunology, is allowing us to ask important questions about the generation and control of the autoreactive T cell repertoire. This information also provides the basis for models of how these cells become activated and for advances in antigen-specific therapeutic intervention.

Key words: T cells; autoimmunity; regulation

## 1 EAE AS A T CELL-MEDIATED DISEASE

Since Rivers' early experiments in monkeys[1], experimental autoimmune encephalomyelitis (EAE) has become the most widely used inducible model of autoimmunity, serving as a model for multiple sclerosis (MS)[2]. EAE shares some characteristics with MS involving widespread foci of inflammation and demyelination with perivascular infiltration of mononuclear leukocytes[2]. The disease may show a monophasic acute, chronic, or relapsing-remitting course depending on the species or strain studied and the nature of the autoantigen used. Nevertheless, EAE should be viewed as an imperfect model of a diverse human disease. EAE is driven by activation of autoaggressive myelin-specific T cells and provides several models for the study of various aspects of T cell biology and immune tolerance. The initiation phase of EAE is relatively straightforward, although the precise molecular basis for pathology remains an enigma. However, the many studies that now focus on the regulation of the pathogenic T cells provide a complicated picture of diverse and possibly redundant mechanisms for the control of self–reactivity. The wealth of information on the precise nature of T cell recognition of myelin epitopes now gives a basis for investigation of the breakdown of central tolerance and the generation of the peripheral autoimmune repertoire and for the development of novel therapeutic strategies.

## 1.1   EAE is driven by CD4+ T cells

The development of delayed type hypersensitivity (DTH) reactions to re-challenge with the inducing antigen provided the first evidence that EAE might have an immunological basis[3]. This report also proposed that the humoral immune response was not relevant to the pathology seen. Subsequent experiments showed that transfer of lymph node cells (and not serum) could elicit the disease in non-immunized recipient rats[4]. Neonatal thymectomy was shown to prevent the development of EAE[5], clearly implicating T cells in driving the disease. But which T cell subset – T helper cells or cytotoxic T cells - was responsible? Transfer of disease in mice was prevented by *in vitro* depletion of CD4+ T helper cells (using an antibody to the Lyt1 marker) prior to transfer[6]. Disease could still be transferred, however, if CD8+ cells were depleted (using anti-Lyt2)[6]. Therefore CD4+ T cells appeared to be responsible for disease, a notion supported by the prevention of disease by the *in vivo* administration of antibodies that depleted CD4+ cells prior to active immunization[7]. In contrast, *in vivo* depletion of CD8+ cells again showed no capacity to prevent disease[8]. The ultimate proof came with the demonstration that EAE developed after the transfer of CD4+ T cell clones specific for a CNS autoantigen. More recently, the crucial role of T cells in EAE has been highlighted by prevention of disease with antibodies that block the costimulatory interactions involved in the initiation and maintenance of the T cell response. Thus, disruption of CD28-CD80/86, CD40-CD154, or OX40-OX40Ligand interactions will inhibit disease[9-13]. In contrast, mice lacking ICOS show more severe disease[14].

Although the activation of CD4+ T cells is central to the induction of EAE, the general view is that the effector cell in EAE is of the macrophage lineage with both the CNS resident microglial cells and infiltrating blood monocytes responsible for tissue destruction and demyelination[15,16]. These cells require instruction from CD4+ T cells in the form of cytokines to become activated. Antibodies with specificity for CNS autoantigens are also produced during the course of EAE. Whilst these antibodies are not sufficient to initiate the inflammatory lesion in the CNS, they are believed to play a role in the propagation of disease and may be particularly important to the demyelination process through binding to surface components of the myelin sheath[17]. Clearly CD4+ T cell help is also essential for the production of these autoantibodies.

## 1.2. Identification of myelin autoantigens recognized by T cells

EAE was initially induced using crude preparations of brain or spinal cord tissue. This still remains a very effective antigen for disease induction and its use has advantages in situations when we have no precise information about T cell antigen recognition (when attempting to develop models in previously untested species or strains). Whole myelin is also useful when we wish to provoke a "global" immune response to CNS tissue, for example when we are testing the effects of therapies using individual antigenic epitopes on the entire anti-myelin response[18].

As our understanding of the structure and composition of the myelin sheath has advanced[19-21] our ability to induce EAE with defined CNS autoantigens has developed. Myelin basic protein (MBP) was the first defined CNS autoantigen to be tested[22] and has proved the most reliable antigen for inducing EAE in several species/strains. MBP remains the most widely used antigen, largely because it composes a large fraction (~30%) of the protein content of myelin and is therefore relatively easy to purify (CNS antigens are notoriously difficult to isolate due to the high lipid content of myelin). There are, however, other CNS antigens that are clearly relevant to EAE. Proteolipid protein (PLP)[23] and myelin oligodendrocyte glycoprotein (MOG)[24] are both now commonly used to provoke EAE in appropriate strains of mice. PLP constitutes some 50 % of the protein content of the myelin sheath, whereas MOG makes up less than 1%. Both these antigens are highly hydrophobic and difficult to purify[20,21]. The use of synthetic peptides containing T cell epitopes from PLP[25,26] and MOG[27,28] provide a simpler approach to activating pathogenic T cells. However, immunization with intact antigens does have benefits over the use of individual peptide epitopes to induce EAE. It allows us to study the interaction of one T cell with other T cells or B cells responding to different epitopes within the same antigen. The cloning of the genes encoding MBP[29] and MOG[30] has facilitated the production of large quantities of highly pure antigens to meet this challenge.

## 1.3 From myelin antigens to synthetic peptides

T cells recognize short (9-15 amino acid) peptide fragments generated from the intracellular proteolytic degradation of antigen (antigen processing) that occurs within antigen presenting cells (APC)[31]. These peptides are then displayed on the surface of the APC bound to major histocompatibility complex (MHC) molecules. CD4+ T cells recognize peptides bound to MHC class II molecules that are (under physiological conditions) only expressed by dendritic cells (DC), B cells and cells of the macrophage lineage. If we know the amino acid sequence of a protein, we can produce synthetic peptides containing the amino acid sequence of a region of interest. Such short peptides can bind directly to MHC class II molecules (without the need for antigen processing) for presentation to T cells. This technology has allowed the definition of several T cell epitopes within MBP[32-34], PLP[25,26] and MOG[27,28]. Since EAE is driven by the activation of CD4+ T cells, immunization simply with peptides containing individual T cell epitopes is sufficient to provoke disease.

The first myelin autoantigen in which an encephalitogenic region was defined was MBP[32,35]. These studies used synthetic peptides to induce EAE in guinea pigs. Remarkably these experiments were performed over three decades ago, before the demonstration of MHC restriction and long before our appreciation of the molecular basis of T cell recognition. Moreover, analog peptides with amino acid substitutions were used to demonstrate the requirement for a single tryptophan for disease development. In essence, these workers were using altered peptide ligands (**APL**) two decades before the term was first coined[36] Subsequent studies used peptides to define encephalitogenic regions of MBP for the rhesus monkey[37] and rat[34,38]. After these reports, it took a further decade to identify encephalitogenic epitopes for the mouse[33,39].

The precise peptide sequences from a given autoantigen that induce disease vary between species and strains depending on the requirements for binding to the MHC haplotype expressed. There do appear, however, to be particular regions that contain epitopes recognized in several species. For example the 80-100 region of MBP contains epitopes recognized by SJL mice and Lewis rats[38-40]. This region is of particular interest as it also contains epitopes recognized by T cells from MS patients[41,42]. This immunodominant region may reflect a dominant cross-species processing activity that generates fragments of this region of MBP for binding to MHC and presentation to T cells.

Today there are several epitopes identified in MBP[33,39], PLP[25,26] and MOG[27,28] that can induce EAE in mice. Initially the mouse strains found to be susceptible to EAE were of the $H-2^u$ or $H-2^s$ MHC genetic backgrounds[33,39]. These are not common laboratory strains and so these models remain the preserve of groups that specialize in studying EAE. In recent years, however, this has changed with the identification of peptide 35-55 of MOG as an encephalitogenic epitope in the widely used C567BL/6 (H-2[b]) mouse[28]. This has now become the model of choice for many studies, not least because gene knockout and transgenic technologies are well developed in this strain. There has therefore been a rapid expansion in reports on the influence of gene deficiency in EAE without recourse to laborious backcrossing programs.

## 1.4 Myelin-reactive TCR transgenic mice

The development of transgenic technologies has enabled the generation of mice expressing TCRs specific for myelin antigens. This represents a great advance as it has allowed us to study the activation of naïve autoreactive T cells *in vitro* and *in vivo*. Several different lines of mice have been produced that express TCRs that recognize the acetylated N-terminal epitope of MBP: MBP(Ac1-9)[43-45]. These mice in general do not develop spontaneous encephalomyelitis and disease requires T cell activation following immunization with the Ac1-9 peptide. In some situations maintenance in a dirty environment will provoke disease[44], in other cases giving pertussis toxin alone (without Ac1-9) will suffice[46]. Crossing these TCR transgenic mice with mice that lack recombinase activating gene (RAG) activity, and therefore cannot rearrange endogenous TCRα chains, also leads to an increase in spontaneous encephalomyelitis[43]. This appears to be due to an aberration in immune regulation (see below).

More recently, TCR transgenics specific for PLP(139-151) have been generated that develop severe spontaneous disease[47]. MOG-specific TCR transgenic mice are also now available (V.K. Kuchroo, personal communication). A transgenic line expressing an MBP(85-99)-specific TCR derived from a human T cell clone, together with the human CD4 molecule and the human DR2 class II MHC molecule has been generated[48]. These mice develop encephalomyelitis (with ~4% spontaneous incidence) providing strong evidence that T cells recognizing this region of MBP are important in human MS.

## 1.5   T cells in the etiology of EAE

Clinical signs of EAE usually develop within two weeks of immunization with myelin antigen. T cells primed in the peripheral lymphoid organs need to migrate into the CNS to establish the EAE lesion. Recent advances have allowed us to track disease-relevant cells. The use of fluorescently labeled multimeric peptide-MHC complexes provides a specific reagent for identifying T cells bearing antigen-specific TCRs[49]. Such "tetramers" have been developed to bind T cells that recognize MBP(Ac1-9) in association with the $A^u$ class II molecule[50,51]. The frequency of Ac1-9-specific T cells in non-immunized $H-2^u$ mice was below the level of detection. However, by 10 days after immunization with either whole myelin or the Ac1-9 peptide around 0.5% of CD4+ T cells in the draining lymph nodes were found to bind the tetramer[50]. But how many of these cells go on to enter the CNS? This has not yet been addressed using peptide-MHC tetramers. An ELISPOT system has been used to identify PLP-specific cytokine secreting T cells derived from lymphoid organs and CNS[52]. After immunization with the PLP(139-151) peptide, high frequencies of antigen-specific T cells were found in the draining lymph node, spleen and peritoneal cavity immediately prior to onset of clinical signs. At this time-point no cells were detectable in the CNS. Once EAE had developed, however, the CNS showed the highest frequency of 139-151-specific T cells, although this still only accounted for ~20% of the total number of specific cells in the mouse. At later time-points (by 6-10 weeks after immunization) the total number of 139-151-specific T cells had dropped by around ten-fold compared to the peak numbers seen prior to EAE onset. At these later time-points responses to a second PLP epitope, 178-191, were detectable (due to epitope spreading), mainly in the CNS.

In elegant experiments, MBP-specific rat T cell clones were retrovirally engineered to express green fluorescent protein (GFP) so that they could be followed anatomically after transfer to naïve recipients[53,54]. As a control, GFP-expressing cells specific for chicken ovalbumin (OVA) were transferred. MBP-specific (but not OVA-specific) T cells began to accumulate in the spinal cord around 3 days after transfer. In this lag phase between injection and migration the cells populated the peripheral lymphoid organs. Expression of IL-2 receptor and OX40 was reduced whilst expression of the chemokine receptors involved in migration into the tissues was increased. Once in the CNS the T cells regained an activated phenotype. In these experiments the majority of the myelin reactive T cells entered the CNS. This contrasts with the results found after PLP immunization described above, and may reflect differences between transfer systems and active immunization models.


## 2. CYTOKINES IN EAE

### 2.1 The effector phase is driven by a Th1 response

The effector cell in EAE appears to be an activated macrophage or microglial cell[16]. Macrophage activation is driven by the production of interferon γ (IFN-γ) and tumor necrosis factor α (TNF-α) from T cells[55]. Effector CD4+ cells have been sub-divided based on the cytokines they secrete upon antigenic or mitogenic stimulation[56]. T helper 1 (Th1) cells produce IFN-γ and TNF-α, whereas Th2 cells produce interleukin 4 (IL-4) and IL-10. The general finding that the CD4+ T cells that transfer EAE fall into the Th1 camp therefore comes as no great surprise. The transfer of myelin-specific T cell clones has shown this most convincingly[57]. The Th1 phenotype was found to be required to induce disease. This most probably explains the requirement to mix the autoantigen in complete Freund's adjuvant (CFA) to induce EAE. CFA is a very aggressive adjuvant that primes strongly for Th1 response due to the presence of heat-killed mycobacteria. In general, the promotion of a Th2 response does not lead to EAE. Immunization with autoantigen in incomplete Freund's adjuvant (IFA), lacking mycobacteria, promotes a Th2 response and fails to induce EAE[58]. Differentiation of naïve T cells into Th1 effector cells is driven by DC production of IL-12[59,60]. Microbially derived danger signals such as bacterial DNA stimulate DC activation and IL-12 production[61,62]. Addition of bacterial CpG DNA to MBP in IFA has recently been shown to substitute for mycobacteria in driving Th1 differentiation and therefore EAE[63,64]. The prevention of disease by *in vivo* administration of a neutralizing antibody to IL-12 has demonstrated the key role of this cytokine initiation of EAE induction[65]. Consistent with this, IL-12 deficient mice are totally resistant to EAE[66].

The key cytokines in progression of disease appear to be the Th1-associated cytokines IL-12, IFN-γ, TNF-α and LT-α. Exposure to T cell-derived IFN-γ leads to upregulation of MHC and costimulatory molecules on CNS APC (microglia and astrocytes)[16,67]. IFN-γ also enhances adhesion molecule expression on endothelial cells thereby facilitating the mononuclear infiltration characteristic of EAE[67]. The role of IFN-γ in EAE, however, is not straightforward. Administration of neutralizing antibodies to IFN-γ can, when given at the right time, increase EAE severity[68,69]. Furthermore, several reports have demonstrated in gene-targeted mice that the absence of IFN-γ or the IFN-γ receptor exacerbates rather than prevents disease[70-72]. Interestingly, antibody blockade of IL-12 prevents EAE in IFN-γ deficient mice, suggesting that EAE pathology is not simply driven by an IL-12-IFN-γ axis[66]. IL-12 clearly must drive other pathogenic processes.

TNF-α and LT-α have cytotoxic effects on oligodendrocytes leading to apoptosis[73,74]. TNF-α also promotes the release of nitric oxide from macrophages leading to further damage of oligodendrocytes and demyelination[75]. TNF-α and LT-α may therefore have a more direct role in pathology than IFN-γ. Antibody-mediated neutralization of both of these cytokines has been reported to limit EAE severity[76] and LT-α deficient mice develop mild disease[77]. However, other studies using mice that are deficient in TNF-α or both TNF-α and LT-α have found no effect on susceptibility to EAE[78]. Analysis of TNF-α$^{-/-}$ mice is complicated by a failure in peripheral lymphoid organogenesis. Therefore the influence of T cell-derived cytokines on EAE pathology is far from fully understood and probably involves a complex interplay and a significant level of redundancy in function.

It must be noted that the Th1/Th2 paradigm is not totally secure in relation to pathology in EAE. Activation of Th2 cells after disease onset has been shown to exacerbate disease. Immunomodulatory approaches that have expanded Th2 immunity either deliberately (in primates)[79] or unintentionally (in mice)[80] have been found to worsen disease. Furthermore, using a T cell receptor (TCR) transgenic system, MBP-specific Th2 cells have been shown to induce EAE upon transfer to immunocompromized hosts[81]. Th1 and Th2 cells are believed to have the potential to counter regulate the expansion of each other[56]. Establishment of a Th2 response prior to the induction of EAE can prevent disease development[82,83]. There is little evidence, however, that the activation or transfer of a myelin-reactive Th2 population is able to control an established Th1 response and thereby inhibit EAE[84].

## 2.2   Cytokines controlling EAE remission

Recovery from EAE has been reported to correlate with increased levels of IL-4 and IL-10 mRNA within the CNS[85,86]. Administration of exogenous IL-4 has been found to have some beneficial effect[87] as does the use of myelin-specific T cells transduced with the IL-4 gene to target the CNS[88]. However the use of IL-4 deficient mice has revealed only slight increases in EAE severity[89,90].

In contrast, IL-10 appears to have a dominant role in the regulation of EAE with IL-10 deficient mice suffering a severe non-remitting form of the disease[90,91]. Mice that are transgenic for IL-10 are protected[90,92]. Administration of exogenous IL-10 to treat EAE has proved problematic in rodent models of EAE[93-96]. It appears that the most effective approach requires targeting of the IL-10 to the CNS[94]. This has been achieved either by direct intracranial injection of viral vectors[94] or by transfer of myelin reactive T cells transduced with the IL-10 gene[97]. Both IL-4 and IL-10 are produced by Th2 cells[56]. In fact, the suppressive qualities of IL-10 have been reported to be abrogated by the concurrent administration of IL-4[98].

There is evidence that transforming growth factor β (TGF-β) also plays a regulatory role in EAE since administration TGFβ-blocking antibodies leads to more severe disease[99,100]. Provision of exogenous TGF-β, in contrast, reduces EAE severity[99,100]. These effects may be through inhibition of T cell expansion and/or a direct inhibition of macrophage activity in the EAE lesion that controls the inflammatory response[101-103].

Nevertheless, the consensus is that IL-10 is a more potent regulator of EAE. This has led to the proposal of a cytokine circuitry involving IL-10 and IL-12 as counter-regulators controlling the progression of the disease[66]. Many cell types produce IL-10[104]. The source of the IL-10 that promotes recovery remains an issue under investigation.

The complexities of cytokine networks in EAE is exemplified by a recent report that administration of DC that had been loaded *in vitro* with the MOG(35-55) peptide in the presence of TNF-α could protect against subsequent attempts to induce EAE[105]. These DC led to the expansion of a MOG-specific CD4+ T cell population that produced IL-10. However, the use of IL-10-blocking antibodies led to only a partial restoration of EAE-susceptibility, suggesting the influence of other factors on protection.

# 3. NON-CD4+ T CELLS IN EAE

CD4 T cells have been widely accepted as the critical cell in the pathogenesis of EAE and until recently research attention has been focused almost entirely on these cells. However, in the last few years, evidence has emerged supporting a role for conventional CD8+ αβ T cells as well as T cells expressing the γδ TCR.

## 3.1   CD8+ T cells as effectors in EAE

In the early 1980s it was first reported that both CD4+ and CD8+ T cells are present in CNS lesions of MS patients and mice with EAE[106-109]. However, direct evidence that CD8+ T cells alone can induce EAE has only recently been provided. In 2001, two independent studies demonstrated that CD8+ T cells activated by MHC class I restricted recognition of major myelin antigens can induce EAE when transferred intravenously into normal mice[110,111]. This was accomplished using peptides from either MBP or MOG. Interestingly one of these peptides also contained a previously defined CD4+ T cell epitope[28,110]. Therefore in EAE experiments when we immunize with these peptides we may well be activating CD8+ T cells in addition to CD4+ T cells. Disease produced by MOG-specific CD8+ T cells was more severe than EAE induced by immunization with the MOG peptide in CFA, resulting in massive demyelination and more destructive CNS lesions[110]. EAE resulting from transfer of MBP-specific CD8+ T cells also resulted in severe demyelination, but lesions were confined to the brain[111].

The effector mechanisms of EAE mediated by MBP-specific CD8+ T cells was investigated, specifically the requirements for IFNγ and TNFα. Whilst blocking TNFα appeared to have little effect, administration of an anti-IFNγ antibody *in vivo* significantly reduced severity of disease[111]. This finding contrasts with the CD4+ T cell-mediated models where a genetic deficiency in IFNγ signaling worsens disease[71,72]. Interestingly, in this regard, EAE mediated by CD8+ T cells more closely resembles MS, since administration of both IFNγ and a TNF receptor-Fc fusion protein exacerbates MS[106,112,113]. These CD8+ T cell driven models used transfer of CD8+ T cell lines derived from mice immunized with myelin antigens. They prove that activated CD8+ cells can enter the CNS and initiate pathology without the involvement of CD4+ cells. However, since *in vivo* priming of CD8+ cells usually requires help from CD4+ cells, it remains to be seen whether these pathogenic CD8+ populations can be initially activated *in vivo* in the absence of a CD4+ response.

Whilst these data show CD8+ T cells can have a pathogenic role, other studies have suggested that CD8+ T cells may be a regulatory function promoting recovery from EAE (see below).

## 3.2 γδ T cells in EAE

Only a handful of studies have addressed the role of γδ T cells in EAE or MS, and these reports give conflicting results. Depending on the method of investigation, some studies find that γδ T cells contribute to disease pathogenesis, others report an inhibitory role, while still others find that γδ cells have no impact whatsoever on the course of disease.

Evidence that γδ T cells have no role in EAE comes from investigations using the Lewis rat and δ-chain deficient mice. In Lewis rats, γδ T cells are absent from the CNS and giving anti-TCRγδ fails to exacerbate EAE[114]. Similarly, the progression and severity of EAE in γδ-deficient mice was found to be no different from control mice[115].

In direct contrast to reports assigning no role for γδ T cells, other studies have found γδ cells in MS plaques and in EAE CNS infiltrates[116-119]. By implication, there may be a role for these cells, but their presence in the target tissue could reflect either a pathogenic or regulatory role. Evidence for a regulatory function comes from *in vivo* γδ depletion experiments that showed exacerbation of EAE. This correlated with an increase in antigen-specific proliferation and elevated IFNγ mRNA at certain phases of EAE[120]. Additionally, IL-4-secreting γδ T cells have been found to increase during onset and peak of the disease[121]. This correlation suggests these cells may contribute to a shift away from a destructive Th1 cytokine profile.

Finally, contrasting evidence also exists showing that γδ T cells can help to promote disease. Recent analysis of surface markers on γδ T cells reveals an activated phenotype that correlates with disease, and these cells produce Th1-like cytokines[122]. More significantly, depletion of γδ T cells has been reported to reduce EAE signs considerably and was accompanied by a decrease in mRNA levels for several pro-inflammatory cytokines, including IFNγ[123]. Reduced expression of certain pro-inflammatory chemokines and chemokine receptors was also reported following *in vivo* depletion[124]. In another study disease was found to be significantly reduced in δ-chain knockout mice compared with controls after either immunization with MOG(35-55) or transfer of a MOG(35-55)-specific αβ T cell line[125]. It is important to note that these mice also display a generalized impairment in their ability to mount an immune response. Nevertheless, together these data indicate that γδ T cells may contribute positively to EAE pathogenesis.

# 4. REGULATORY CELLS IN EAE

The pathogenesis of EAE therefore is associated with strong Th1 bias in immune-reactivity to myelin antigens, whilst protection is associated with IL-10 production. Since the molecular targets that drive the disease (myelin autoantigens) are constantly available for immune recognition, what mechanisms prevent the pathogenic Th1 response from developing spontaneously? Furthermore, do these same mechanisms re-balance the immune response and promote recovery from EAE? Over the past decade there has been an exponential rise in the study of lymphocytes with the capacity to down-modulate the responses of other potentially pathogenic lymphocytes. The majority of studies have examined the role of regulatory T cells. These cells are currently attracting intense interest and have been the subject of several recent reviews[126-128].

## 4.1 Spontaneous regulatory T cells.

Experiments using neonatal thymectomy of mice revealed an interesting pathologic outcome. Thymectomy three days after birth (but not at later time points) resulted in the subsequent development of multi-organ immune infiltrations and spontaneous development of high titers of autoantibody[126]. This autoimmunity was proposed to be due to the absence of a suppressive T cell population that exits the thymus at this early time point[129]. Potentially pathogenic autoreactive T cells had already populated the periphery and removal of the regulatory cells by thymectomy prevented their normal control, allowing autoimmunity to develop. Later experiments showed these regulatory T (Treg) cells to reside in the CD4+CD25+ fraction[130]. Transfer of these cells to thymectomized recipients prevented autoimmunity, whilst their depletion from non-thymectomized mice led to a similar multi-organ pathology.

Experiments using defined CD4+ T cell populations in rats and mice revealed that transfer of purified CD45RB/C$^{hi}$ cells into lymphopenic or immunodeficient hosts resulted in organ-specific autoimmunity (diabetes and thyroiditis)[131] or inflammatory colitis[132]. The co-transfer of CD4+ CD45RB/C$^{lo}$ cells could prevent these conditions[132,133]. These suppressive cells had similar qualities to the CD4+CD25+ Treg cells and the two populations have since been shown to largely (but not totally) overlap[128]. Both populations are described as Treg cells. These cells display an anergic phenotype (they do not proliferate in response to antigenic or mitogenic challenge *in vitro*, unless supplied with exogenous IL-2)[134]. Tregs have a potent capacity to inhibit the activation of other T cells *in vitro* as well as *in vivo*. The *in vitro* activity of CD4+CD25+ Tregs has been reported to require TCR ligation and to be dependent on cell contact, but not IL-10 production[126-128,135]. Other *in vivo* studies, however, have concluded that IL-10 and TGF-β are required for protection from autoimmune pathology[136,137].

The autoimmune manifestations seen in the absence of Tregs in most models do not include encephalomyelitis. However, Treg cells do appear to have an important role in the prevention of spontaneous encephalomyelitis in mice expressing transgenic TCRs recognizing the Ac1-9 epitope of MBP. These mice do not normally develop spontaneous disease. They do, however, when crossed onto a recombinase activating gene (RAG) deficient background[43]. In the RAG-sufficient transgenics a limited amount of endogenous TCRα gene rearrangement can occur and CD4+CD25- from these mice prevent the onset of disease when transferred into RAG-/- recipients[138]. The protective cells were found to express both the transgenic MBP-specific TCR and endogenously rearranged receptor[139]. Thus Treg cells with specificity for MBP can prevent the onset of spontaneous CNS autoimmunity. However this system is far from physiological. More experiments are therefore required to assess the role of Tregs in non-transgenic animals. Also, Tregs clearly do not regulate the initiation of autoimmunity in models such as EAE that are induced by active immunization with autoantigen (if they did, we would not see disease!). Whether they play a role in the recovery phase is unlikely but remains to be investigated fully. However, a recent study has reported that prior transfer of CD4+CD25+ from normal mice conferred significant protection against MOG-induced EAE[140]. Why this should be the case when the recipient mice possess their own Treg cells is something of a puzzle.

## 4.2 Tr1 cells

T regulatory 1 (Tr1) cells are another regulatory CD4+ T cell population that have been generated experimentally rather than purified as a naturally occurring population[141]. Tr1 cells produce large amounts of IL-10 (in addition to lower levels of IL-4 and IFN-γ)[141]. They have been generated *in vitro* by chronic stimulation in the presence of IL-10[141], or more recently by stimulation in the presence dexamethasone and vitamin D3[142].

Tr1 cells derived from transgenic mice expressing an ovalbumin (OVA)-specific TCR were able to prevent colitis in immunodeficient mice upon oral administration of OVA[141]. Thus these IL-10 producing T cells can prevent onset of spontaneous disease, but can they influence an induced autoimmune disease? Recent data suggest that they can. The transfer of OVA-specific Tr1 cells was found to suppress the development of EAE when OVA was injected intracranially[142]. This provision of antigen in the CNS was presumably essential for the Tr1 cells to exert their effects on pathogenic immune responses taking place within the CNS or in the lymphoid tissue draining the CNS. This is an exciting development, as the generation of myelin-specific Tr1 cells would obviate the need for the intracranial delivery of exogenous antigen.

A recent report suggests the Tr1 cells specific for *Bordetella pertussis* arise naturally during the course of the immune response to the infection[143]. It might be, therefore, that at least some of the effects of IL-10 in promoting the remission of EAE may be provided through the influence of cells of the Tr1 type arising spontaneously as the immune response to the autoantigen develops. At present, however, no evidence that such a cell type arises spontaneously during EAE has been provided. Whether Tr1 cells can be generated *in vivo* by existing or novel immunotherapeutic approaches will be discussed later.

## 4.3 NKT cells

Natural killer (NK) T cells express surface markers found on both NK cells and αβTCR T cells and have been shown to have regulatory properties in murine models of autoimmunity, particularly diabetes[144]. NKT cells recognize lipid antigens such as α-galactosylceramide (α-GalCer) in association with CD1d[145-147]. This nonpolymorphic MHC class Ib molecule is widely expressed, but is found in high levels on marginal zone B cells and dendritic cells[148,149]. Unlike conventional αβ TCR T cells, NKT cells are able to rapidly produce high levels of IL-4 and IFN-γ upon activation[144]. This IL-4 production has suggested to play a role in driving Th2 responses and thus NKT cells may be important in maintaining balanced immune responses[144]. This would be consistent with a regulatory function in Th1 driven autoimmune responses. A large proportion of NKT cells express invariant TCR α chains (Vα14Jα281 in mice[150], Vα24JαQ in humans[151]). Levels of Vα14Jα281+ T cells have been found to be reduced in non obese diabetic (NOD) mice[152] and transfer of CD1d restricted cells could prevent spontaneous onset of diabetes[152]. Similar results have been found in diabetic BB rats[153].

Do NKT cells have a regulatory role in EAE? The EAE-prone SJL mouse shows quantitative and qualitative defects in NKT cell function[154]. Although a causative link has not been made in this model, it is interesting to note reported reductions in the numbers of Vα24JαQ T cells in MS lesions[155]. Two recent studies have reported that activation of NKT cells via administration of αGalCer, or a synthetic analog that promoted IL-4, but not IFN-γ production could limit the development of EAE[156,157]. These findings reflected reduced Th1 and increased Th2 responses against myelin autoantigens. Protection was not found in mice deficient in either IL-4 or IL-10[156]. NOD mice transgenic for Vα14Jα281 (that possess large numbers of NKT cells) were found to be partially protected from diabetes without exogenous activation of NKT cells. Activation of NKT cells may therefore provide a future therapeutic approach. The contribution of NKT cells to normal regulation of EAE remains in question, however, since CD1d-/- mice did not develop more severe disease than their CD1d+/+ counterparts[156]. Furthermore, mice transgenic for Vα14Jα281 were also found to be fully susceptible to EAE[156].

## 4.4   CD8+ T cells as regulators of EAE

Studies with CD8-/- mice crossed onto an H-2$^u$ background suggest a dual role for CD8+ T cells as both effectors and modulators of EAE since the deficient mice undergo a milder primary course of disease resulting in fewer deaths while later experiencing more spontaneous relapses[158]. The CD8+ T cells appear to act as effectors during the induction of EAE by contributing to the mortality rate, yet help to downmodulate the inflammatory response at the end of the disease course. Lending support to this idea, anti-CD8 depletion *in vivo* rendered mice more susceptible to active re-induction of EAE[159]. Finally, TGF-β producing CD8+ T cells mediate oral tolerance to MBP in rats[160].

Early studies attempting to transfer EAE with T cell lines led to the paradoxical finding that, in certain circumstances, this did not induce disease, but rendered the recipients resistant to subsequent attempts to induce EAE by immunization[161]. This led to the idea of "T-cell vaccination" - variable TCR regions on pathogenic T cells would trigger protective T cells that would destroy these autoaggressive threats[162]. Subsequently, immunization with the appropriate peptides derived from encephalitogenic TCRs was shown to be sufficient to confer protection[163,164].

In H-2$^u$ mice the majority of MBP(Ac1-9)-specific encephalitogenic CD4+ T cells use the Vβ8.2 gene segment[165]. The expansion of this pathogenic T cell population during EAE has been reported to induce a regulatory CD8+ T cell population specific for the Vβ8.2 chain[166]. These cells can recognize activated Vβ8.2-expressing pathogenic cells through their presentation of Vβ8.2 derived peptides in association with MHC class I leading to their removal via apoptosis. Further complexity is added by the activation of Vβ-specific CD4+ regulatory T cells during EAE (these cells express Vβ14)[167]. The CD4+ and CD8+ regulatory cells recognize distinct TCR-derived epitopes, with the CD4+ cells focusing on the TCR framework 3 region which also appears to be a target for regulation in rat EAE[167,168].

A recent study has reported CD8+ T cells that are activated by Vβ8+ Th1, but not Th2 cells and are restricted by the non-classical MHC class I molecule, Qa-1[169]. These cells downregulated the Th1 response and depletion of CD8 cells skewed the cytokine response to the Th1 phenotype. Cell-cell contact between CD4+ and CD8+ cells was required, but this study did not determine whether the CD8+ cells were killing the CD4+ cells directly.

## 4.5   B cells as regulators of EAE

So far we have considered the contribution of T cell populations to the induction and control of EAE. But what of B cells? They have the potential to influence the progression of EAE at three levels:- by the production of pathogenic autoantibodies; by acting as APC to enhance activation of autoaggressive T cells; or by having a regulatory function during recovery from disease. There is evidence in support of all of these. Autoantibodies that bind components of the myelin sheath, particularly MOG, can increase disease severity[17,170].

Early experiments in which mice were depleted of B cells using anti-μ antibodies indicated that B cells played an important role in the development of MBP-induced EAE[171]. This involved an enhancement of T cell priming presumably by the B cells acting as APC (although secreted anti-MBP antibodies may also have contributed). The use of genetically B cell deficient μMT mice allowed for a clearer picture to develop. An early report concluded that B cells were not required for the development of EAE after immunization with the MOG(35-55) peptide[172]. A subsequent report reached this same conclusion, but also that B cells were required for EAE development when intact recombinant human MOG protein was used to immunize[173]. Pathology could be induced in these B cell deficient mice by transfer of serum containing anti-MOG antibodies[174]. However, we have recently shown that μMT mice develop EAE of similar severity after immunization with either the MOG(35-55) peptide or recombinant mouse MOG[175]. The reasons for these contrasting findings remain unclear, but may reflect our use of the true autoantigen (mouse MOG) rather than the xenogeneic human MOG. Both studies used recombinant fragments corresponding to the 121-amino acid extracellular domain of MOG. The mouse and human sequences differ at 11 residues within this region[176]. These differences may be crucial to B cell recognition and hence whether B cells are required as APC.

Janeway's group have reported that μMT mice on the $H-2^u$ background failed to recover from EAE induced with MBP suggesting a role for B cells as regulators of disease[177]. We found the same effect when using $H-2^b$ μMT mice immunized with either MOG or the MOG(35-55) peptide[175]. B cell sufficient mice show a peak of disease in the second and third weeks after immunization and then enter the recovery phase. μMT mice, however, show a non-remitting disease course. Transfer of B cells from recovered mice to naïve recipients confers a level of protection against subsequent EAE induction. These B cells were found to produce IL-10 when stimulated *in vitro* with MOG. Interestingly IL-10-/- mice show the same severe non-remitting disease as μMT mice[90,91]. Did this point to a common deficit – a lack of IL-10 production by B cells?

To address this question, we developed a bone marrow chimera technology that allowed us to generate mice in which IL-10 deficiency was restricted only to B cells[175]. These mice also developed the non-remitting form of EAE. This was associated with an uncontrolled anti-MOG Th1 response compared with fully IL-10 sufficient mice. These findings are consistent with the recent report that IL-10 production by B cells is essential to control a model of chronic bowel disease[178]. Furthermore B cell production of IL-10 seems key to their ability to limit the severity of collagen-induced arthritis (C. Mauri, personal communication). Production of IL-10 by B cells may therefore be an essential factor in counteracting the pathogenic T cell function in autoimmune and inflammatory conditions, i.e. B cells have a regulatory function. The precise mechanisms underlying these effects remain to be dissected.

## 5.    STUDIES ON AUTOREACTIVE T CELL DEVELOPMENT AND ACTIVATION

### 5.1   How do myelin-reactive T cells avoid negative selection?

The cellular and molecular mechanisms of central tolerance – the removal of T cells expressing high affinity TCRs specific for self by negative selection in the thymus – are fairly well understood[179,180]. But how do the T cells that initiate autoimmune damage avoid this central checkpoint? Study of T cell reactivity to several encephalitogenic epitopes is beginning to provide some answers[181].

The simplest explanation would be that the autoreactive T cell does not see the autoantigen during thymic development. However, there is now good evidence that both MBP and PLP are expressed intra-thymically[182,183]. Autoreactivity to PLP seems to be shaped by the form of PLP encountered in the thymus. The PLP(139-151) sequence represents the immunodominant epitope in H-2$^s$ mice[25]. A remarkably high frequency of the resting peripheral T cells of an SJL mouse will respond to this peptide (around 1 in 20,000 T cells)[184]. Why aren't these cells deleted in the thymus?

The gene encoding PLP undergoes alternative splicing producing the DM20 isoform that lacks the 116-150 sequence and therefore the 139-151 epitope[185]. Both full length PLP and DM20 are expressed in the CNS, but DM20 is expressed preferentially in the thymus[184]. The PLP(139-151)-reactive T cells therefore escape negative selection in the thymus because the epitope is absent, or is present at levels that are insufficient to activate the T cell apoptosis pathway. Administration of exogenous PLP139-151 was found to induce negative selection, supporting this conclusion[184]. PLP-deficient C57BL/6 mice mount T cell responses to epitopes contained within DM20 whereas wild type mice do not, indicating that this from of PLP is able to induce tolerance[186]. In this case, the induction of tolerance was localized to the activity of thymic epithelial cells.

In cases where the autoantigen is expressed in the thymus, T cell fate is determined by the strength of the TCR-peptide-MHC interaction with high avidity T cells undergoing negative selection[179,180]. Low avidity interactions fail to trigger T cell death, however. The Ac1-9 peptide of MBP is the immunodominant epitope in H-2$^u$ mice[33]. A remarkable feature of this peptide is that it binds so poorly to the A$^u$ MHC class II molecule that a binding affinity cannot be determined[187,188]. As a consequence the autoreactive T cells that recognize this peptide express very high affinity TCRs[51]. The lysine residue at position 4 of Ac1-9 interacts unfavorably with a hydrophobic pocket within the peptide binding groove of the A$^u$ molecule and accounts for the poor class II binding[188-191]. APL of Ac1-9 in which the lysine residue at position 4 is changed, for example to a tyrosine (Ac1-9[4Tyr]), show greatly enhanced binding to A$^{u}$[45,51,192] and hence induce ꞌT cell responses *in vitro* at concentrations up to a million times lower than are required when using the wild type Ac1-9.

The wild type version and APL of the Ac1-9 peptide were tested for the ability to induce thymic negative selection in a TCR transgenic mouse (Tg4) with specificity for Ac1-9. Systemic administration of high doses of wild type Ac1-9 failed to induce death of Ac1-9-specific thymocytes. The Ac1-9[4Tyr] APL, however, induced substantial negative selection[45].

These findings are consistent with avidity-based models of thymic selection[193,194]. When the developing Ac1-9-specific thymocytes receive a strong signal through Ac1-9[4Tyr] they die. They avoid this fate, however, when exposed to weak stimulation from the wild type autoantigen. Thus low avidity interactions allow escape from negative selection. A similar mechanism has been proposed to account for the ability of autoaggressive T cells to escape negative selection in NOD mice[195]. Here, the $A^{g7}$ class II molecule is believed to make generally unstable interactions with many peptides.

To stimulate T cell activation the number of peptide-MHC complexes displayed by the APC must reach a threshold level[181]. This in turn is influenced by the ability to release the correct peptide fragment during antigen processing. Analysis of T cell reactivity to the immunodominant MBP(80-100) region in SJL mice has revealed that MBP is probably cleaved after amino acid 94Asn during antigen processing[196]. This is driven by the enzyme asparagine endopeptidase (AEP), which cuts after this Asn residue[197,198]. The MBP(85-99) epitope that is recognized by T cells from MS patients is therefore destroyed by AEP and inhibitors of AEP activity greatly increase the response of these T cells in assays that require antigen processing of native MBP[198]. Thus the action of the AEP enzyme limits the availability of MBP peptide-MHC complexes. Importantly AEP activity was shown in purified thymic DC that induce negative selection[198]. Therefore the ability of MBP(89-99)-specific T cells to avoid negative selection may be linked directly to the action of a single enzyme during antigen processing.

## 5.2   Promiscuous T cell reactivity: implications for autoimmunity

The old view that each T cell bears a TCR that confers unique specificity for a single peptide-MHC complex has become untenable. Mathematical modeling studies indicate that if this were the case, the shear number of potential pathogen-derived epitopes would leave the immune system, with its limited number of T cells, highly ineffective[199,200]. The answer is simple, TCRs must be able to cross-react with many related but different peptide-MHC complexes. This has been shown clearly in many studies over the past decade since the advent of APL[201,202]. An unavoidable result of this is that a T cell that can mount a desirable response to a pathogen might also be able to respond to an autoantigen.

## 5.3 Molecular mimicry

TCR cross-reactivity gives the T cell the potential to respond to structurally similar antigenic peptides derived from an infectious agent and self-tissue. Once fully activated in response to infection these cells would have the potential to mount an autoaggressive attack in the organ(s) expressing the cross-reactive self-antigen. This idea of "molecular mimicry"[203] has, over the past decade, become the dominant theory for how autoimmunity arises. Although elegant experimental systems have been devised to provide supportive data[204,205], this remains controversial with limited unequivocal evidence[206,207]. Several studies in EAE have provided data pertinent to this issue.

Database searches have revealed sequence similarities between MBP and various viral peptides. Indeed, human MBP specific T cell clones were able to respond *in vitro* when cultured with peptides corresponding to some of these viral sequences[208]. However, the response pattern of the clones was variable with no viral peptide able to stimulate all clones. Since T cell cross-reactivity has been demonstrated using peptides with minimal similarities, the database approach may not pick out all the relevant microbial peptides[209].

Nevertheless this approach has also been used in EAE. Peptides derived from murine hepatitis virus (MHV) and haemophilus influenzae type B (HAE) could stimulate PLP(139-151)-specific T cells[210]. Whilst immunization with these viral peptides could not provoke EAE directly, it did enhance disease upon subsequent immunization with the PLP peptide. Other studies have reported the development of EAE after immunization with microbial peptides that shared sequence homology with MBP[211-213]. In other studies protection against EAE has been afforded by immunization with viral mimic peptides[214], or after infection with vaccinia virus containing MBP sequences[215]. Studies using immunization are not ideal strategies for analysis of molecular mimicry as the theory depends on infection. Giving a large dose of peptide antigen in CFA is very different from an infection. Also the source of the homologous peptide is not a natural mouse pathogen. Moreover, the use of virus expressing the normal self-sequence does not show molecular mimicry, but molecular identity.

To overcome these criticisms, the HAE sequence that mimics PLP(139-151) was introduced into a non-encephalitogenic mutant strain of Theiller's murine encephalomyelitis virus (TMEV), a natural mouse virus that shows tropism for the CNS. Infection with this virus induced PLP(139-151)-reactive T cell immunity and severe encephalomyelitis[216]. The experiments with TMEV used direct intracranial injection. The CNS inflammatory response associated with this may well play an important role. Since the molecular mimicry theory does not necessarily require the infection to be in the target organ, it will be interesting to see if the use of murine viruses that are not tropic for the CNS can be used. This has been partially addressed by peripheral infection with mutant TMEV expressing PLP(139-151), which again led to EAE[217].

If molecular mimicry is the dominant factor driving autoimmune responses we are left with some important questions. We all have a large population of self-reactive T cells. We all have infections. Why don't we all develop autoimmune disease? Clearly genetic factors must play a role, but is there any immunological reason?

## 5.4   Do all autoreactive T cells matter? The threshold for harm.

It is probable that the majority of, if not all, T cells show some degree of self-reactivity[199]. Clearly thymic negative selection cannot be allowed to delete all of these T cells and the threshold for this process seems to be set such that only the T cells with high affinity self-reactive TCRs are removed[179,218]. Therefore we have a peripheral T cell repertoire full of self-reactive cells. How are they controlled? A body of evidence shows that T cells can adjust their threshold for activation so that they become relatively insensitive after sustained exposure to antigen. This can be achieved in two ways:- by the biochemical "tuning" of individual T cells through alterations in their TCR proximal signaling machinery[219-221]; or, within a heterogeneous T cell population, by the activation-induced cell death of those T cells expressing TCRs with high affinity for antigen[181].

Evidence for both these processes has been obtained from the study of responses of encephalitogenic T cells to "superagonist" APL of MBP or PLP peptides. The PLP(139-151) system suggests biochemical tuning *in vitro*[222] whereas the MBP(Ac1-9) model shows that Ac1-9-reactive T cells that bear high affinity TCRs are deleted after immunization with the strong MHC binding Ac1-9[4Tyr] APL[51]. Thus mice do not develop EAE after immunization with superagonist APL. They do however mount T cell responses to the APL, but these T cells use lower affinity TCRs and therefore respond very poorly to wild type Ac1-9. Residual autoreactivity is maintained in this situation, but the T cells cannot initiate EAE because the very high levels of Ac1-9-A$^u$ complexes that are required for their activation cannot be reached in the CNS. So can we determine a threshold of sensitivity for antigen that confers the ability to damage self - that is a "threshold for harm"[196]?

The T cells that induce EAE in the MBP system seem to respond to wild type Ac1-9 *in vitro* at or below a concentration of 1μM[51]. T cells that require concentrations above this seem to be unable to induce EAE. Interestingly, studies using semi-TCR transgenic systems show that T cells can survive happily in mice expressing their cognate antigen without causing pathology as long as their threshold for activation is above 1µM[223]. Furthermore transfer of resting TCR transgenic T cells into mice expressing transgenic antigen lead to the progressive desensitization of the T cells again leaving them unable to respond to antigen concentrations below 1µM[224]. We can therefore propose a threshold for harm of 1µM. CD4+ T cells generally respond to their cognate antigen at doses of ~0.001µM.

Therefore mechanisms appear to be in place to allow autoreactive T cells to populate the peripheral pool, but to have at least 100 lower sensitivity than "normal" To give this some meaning, *in vitro* studies suggest that 1µM equates to more than one thousand peptide-MHC complexes on the surface of the APC[225-227].

Loss of self-reactivity in response to APL has been shown using PLP-reactive T cells[222,228]. For the 1B6 PLP-reactive TCR transgenic mouse, the wild-type 139-151 peptide behaves as a weak agonist[222]. The Q144 APL (with a W→Q substitution at residue 144) is the full agonist and the L144 APL behaves as a superagonist. Stimulation of 1B6 T cells *in vitro* with the L144 superagonist leads to their desensitization such that they are still able to respond to (higher than normal concentrations of) L144 but can no longer respond to the wild type PLP peptide. This "tuning" of sensitivity was associated with reductions in TCR-proximal phoshorylation and in calcium influx upon TCR stimulation[222]. Therefore after exposure to superagonist these cells that express a self-reactive TCR, are functionally not self-reactive.

Cross-reactive T cells that have been filtered through the selective pressures in the thymus and periphery will see self-antigen with low avidity (i.e. the self peptide is a weak agonist, requiring high doses to induce activation). To mount an effective anti-non-self response, however the cell must have medium-high avidity for the foreign antigen(s). This differential avidity keeps the T cell repertoire as diverse as possible while reserving the ultimate sanction of negative selection (either in the thymus or periphery) as a last resort for T cells expressing high affinity autoreactive TCRs.

## 5.5 Other factors relevant to autoimmune T cell activation

The "tuning" of T cell activation thresholds would serve to limit the risk of autoimmunity as a result of molecular mimicry as the self antigen would need to be displayed with MHC at unusually high levels to become a target that will interest the T cell. There may be situations when this is achieved probably involving a combination of factors[181]. For example, there is now evidence that the cross-reactive potential of PLP-specific and MBP-specific

T cells is enhanced by increased expression of costimulatory molecules by APC[46,229]. This has been achieved either by transfection of "artificial" APC with high levels of CD80/CD86[229], or by *in vitro* activation of splenic APC[46]. In both cases autoreactive T cells showed a broader response pattern to APL than was observed using "resting" APC with low levels of CD80/CD86 expression.

Alterations in antigen processing upon activation of APC might increase the peptide load available for binding to MHC[230]. This might be the result of enhanced activity of a particular processing enzyme, or the inhibition of enzymes that destroy epitopes. As mentioned above, the AEP enzyme cleaves MBP and destroys the 85-99 epitope and inhibition of AEP improves the ability of 85-99-specific T cells to respond to MBP[198]. Therefore suppression of AEP activity *in vivo* could allow generation of sufficient peptide-MHC complexes to activate the T cell. Interestingly, a parasite-derived inhibitor has recently been described[231]. A possible role for infection could therefore involve release of factors that modulate the activity of enzymes crucial to antigen processing.In addition, there may be a requirement for the natural mechanisms that dampen the T cell's inherent avidity for self to be (perhaps temporarily) subverted.

## 6. T CELL-DIRECTED PEPTIDE THERAPY OF EAE

Because we know so much about the epitopes recognized by encephalitogenic CD4+ T cells EAE has proved a popular testing ground for peptide-based antigen-specific immunotherapy. This has been approached in two ways: by using peptides corresponding to the wild type sequences, or by using APL based on these sequences[232].

## 6.1 APL-based approaches

Experiments in the PLP(139-151)-induced model in SJL mice have proved very successful at inhibiting EAE with APL. These studies used APL that were initially identified as TCR antagonists *in vitro* and proved effective blockers of EAE when given at the same time as the wild type PLP peptide[233,234]. It was subsequently found that the most effective APL, rather than simply blocking T cell priming, was inducing PLP-reactive Th2 cells[235,236]. Furthermore, immunization with this APL was able to inhibit EAE induced with different epitopes from PLP, MOG, or MBP via "bystander suppression"[236]. Along similar lines, APL based on the MBP(87-99) sequence have been shown to inhibit EAE in mice and rats[237,238]. This inhibition of disease was reported to be associated with a reduction in MBP-specific TNF-$\alpha$ production and increased production of IL-4.

An alternative approach is to generate superagonist APL that induce apoptosis the autoaggressive T cells. As described earlier, APL of MBP(Ac1-9) behave as superagonists *in vitro*, but do not induce EAE *in vivo*. Instead Ac1-9-specific T cells are deleted after immunization with these APL[51]. Similar induction of apoptosis has been reported using superagonist APL of MBP(87-99)[83]. These approaches probably reflect a similar effect to that seen when high doses of wild-type antigen are given. This leads to activation-induced T cell death and prevention of EAE[239,240].

A note of caution should be added over the use of APL. We identified TCR antagonist APL based on the MBP(Ac1-9) sequence that were effective at blocking *in vitro* responses of Tg4 transgenic cells. However, these TCR antagonist APL were fully capable of inducing EAE when used on their own to immunize non-transgenic H-2$^u$ mice[241]. This was found to be due to the previously unappreciated complexities of the Ac1-9-reactive T cell repertoire that was able to induce EAE. These cells showed subtle differences in antigen fine-specificity so that not all of them responded to APL in the same way. Thus the APL that were TCR antagonists for Tg4 cells behaved as full agonists capable of activating other Ac1-9-reactive T cells *in vivo* leading to EAE.

Subsequent to these findings in the mouse, reports on phase II clinical trials in MS patients using APL based on the MBP(83-99) sequence suggested similar effects[242,243]. At least in one study, administration of the APL was found to boost for Th1 responses that cross-reacted with the wild-type MBP sequences and two of eight patients showed clinical exacerbations[242]. When attempting to predict the results of an APL-based therapy we must therefore consider the variability of autoreactive T cell repertoire in addition to potential adverse effects arising from hyper-reactivity to the APL itself.

## 6.2  Therapy using native peptide sequences

The use of peptides based on natural sequences recognized by the autoreactive T cells should give more predictable results. These peptides should be able to ligate all the potential autoreactive T cells and, if given in the appropriate way, inactivate them. We have known for sometime that the form in which an antigen is administered has a major bearing on the subsequent immune response[244]. Antigen that is aggregated, or mixed with adjuvant will provoke a response. On the other hand, antigen in monomeric form, in the absence of adjuvant, will lead to a specific state of antigen-unresponsiveness, or tolerance. Thus the systemic administration of peptide antigens via various routes have been shown to inactivate myelin-reactive T cells and inhibit EAE[18,196,245-247].

This ability to switch off autoreactive T cells using soluble peptides has been reported when using peptides derived from MBP[18,245], PLP[18,247,248] and MOG[249] (and S.M.A., unpublished observations). Since there are many potential targets for autoimmune T cell attack in MS, how would we be able to treat the disease with a single peptide? Several studies have tried to address this in EAE by administering a single peptide to induce tolerance and then immunizing with complex mixtures of antigens (usually whole myelin) and looking for bystander suppression[232]. T cell reactivity to several epitopes is assessed and bystander suppression appears as a general down-regulation of responses to epitopes other than the one used for tolerance induction. An alternative is to assess bystander suppression in terms of disease. Animals are tolerized to one epitope and immunized with another. If disease is inhibited, bystander suppression is concluded to be at work. Several studies have successfully induced bystander suppression in different EAE models[18,236,247,248].

The induction of bystander suppression may therefore provide attractive future therapies in human autoimmune disorders. But how does the suppression work? Perhaps the most beneficial scenario would be for the tolerance process to boost the natural disease suppressive effects of Treg or Tr1 cells. Recent experiments analyzing T cell tolerance to OVA peptide suggest that this may be possible. A limited number of OVA-specific TCR transgenic cells were adoptively transferred to non-transgenic recipients prior to either intravenous or oral administration of soluble OVA peptide[250]. Subsequent analysis revealed that around 20% of the OVA-specific T cells had converted to the CD4+CD25+ phenotype (such cells were at a much lower frequency in mice that had been immunized with the OVA peptide in CFA). These cells had the typical anergic phenotype of Treg cells and were able to suppress the activation of naïve OVA-specific T cells *in vitro*. This suppression was independent of IL-10 and presumably involved a cell-contact dependent mechanism.

Similar induction of CD4+CD25+ Treg cells has not, so far, been reported from studies of peptide-induced tolerance in EAE models. Instead, recent data point to the generation of IL-10 producing T cells that would more likely correspond to a regulatory population of the Tr1 type[251]. In our early experiments describing bystander suppression in EAE we were unable to detect any qualitative shift in antigen-specific cytokine production in peptide-treated mice[18]. Thus production of all the cytokines that could be detected in control mice was reduced in the peptide treated mice and IL-10 could not be detected in either group.

Subsequent experiments using the MBP(Ac1-9)-specific Tg4 transgenic mice did reveal production of IL-10, however[251]. Although these mice possess very high numbers of Ac1-9-reactive T cells, they can be rendered resistant to EAE-induction by repeated (five or more) doses of the Ac1-9[4Tyr] APL in soluble form. Analysis of Ac1-9-specific cytokine production revealed that control mice made IL-2, IL-4, IFN$\gamma$ and low levels of IL-10. In peptide-treated mice, however, production of IL-2, IL-4 and IFN$\gamma$ were greatly reduced, whereas IL-10 production was greatly increased. This implicated IL-10 in the tolerance process a view supported by experiments that showed administration of neutralizing anti-IL-10 antibodies could block tolerance induction and restore susceptibility of the Tg4 mice to EAE.

Analysis of these IL-10 producing Tg4 cells has revealed a capacity to block activation of naïve Tg4 T cells both *in vitro* and *in vivo*. This inhibition was reversed *in vivo* by administration of anti-IL-10. Importantly, these peptide induced regulatory cells did not reside in the CD25+ Treg population. They could be generated readily from RAG-/- mice and depletion of CD25+ cell did not abrogate their suppressive activity (D.C. Wraith, personal communication). Other recent experiments using the MOG(35-55) model in C57BL/6 mice have revealed a crucial role for IL-10, since IL-10 deficient mice are not protected from EAE by peptide treatment[252].

These data clearly implicate IL-10 as a crucial regulatory cytokine controlling protection from EAE following peptide-induced tolerance induction. This fits with the data showing that IL-10 is probably the most important cytokine in the natural resolution of EAE[90-92]. However, activation of T cells that produce large amounts of TGF$\beta$ also provides effective resistance to EAE in studies of oral tolerance[253]. Prior expansion of a myelin-reactive Th2 populations can also limit development of the disease[58,235]. Therefore there is more than one way of eliciting an antigen-specific protective immune response.

Although peptide-induced regulatory T cell populations hold promise as future therapeutic approaches, there is still lot we do not understand. For example, the ability to provoke bystander suppression is dependent on which myelin epitope is used[18]. Furthermore, we are only beginning to appreciate the complexities that arise from the potential ways in which an exogenous peptide can interact with MHC compared with the peptide fragments that are generated through antigen processing of the intact autoantigen[196]. There are therefore considerable challenges remaining to the development of effective and predictable peptide-based strategies for treatment of human autoimmune disorders.

## 7.  CONCLUDING REMARKS

The various models of EAE that are available provide many opportunities for the study of autoreactive T cell development, activation and manipulation. The key to this is our understanding of the epitopes recognized in several myelin autoantigens. The use of transgenic and knockout technologies allows us to dissect the role of individual molecules in the etiology of the disease. EAE is providing insights into the molecular mechanisms that allow autoreactive T cells to avoid negative selection and processes that might control their sensitivity once they reach the periphery. Progress has also been made in developing antigen-specific therapies. The complexities of the immune processes that turn autoreactive T cells on or off are starting to be unraveled. We still have a long way to go, however, and the study of EAE will be central to efforts to meet these future challenges.

## REFERENCES


1.      Rivers, T. M., Sprunt, D. H. & Berry, G. P. Observations on attempts to produce acute disseminated encephalomyelitis in monkeys. *J. Exp. Med.* **58**, 39-53 (1933).

2.      Martin, R. & McFarland, H. F. Immunological Aspects of Experimental Allergic Encephalomyelitis and Multiple-Sclerosis. *Crit. Rev. Clin. Lab. Sci.* **32**, 121-182 (1995).

3.      Kabat, E. A., Wolf, A. & Bezer, A. E. The rapid production of acute disseminated encephalomyelitis in rhesus monkeys by injection of heterozygous and homologous brain tisuue in adjuvants. *J. Exp. Med.* **85**, 117-130 (1947).

4.      Paterson, P. Y. Transfer of allergic encephalomyelitis in rats by means of lymph node cells. *J. Exp. Med.* **111**, 119-135 (1960).

5.      Arnason, B. G., Jankovic, B. D., Waksman, B. H. & Wennerston, C. Role of the thymus in immune reactions in rats. II. Suppressive effect of thymectomy at birth

on reactions of delayed (cellular) hypersensitivity and the circulating small lymphocyte. *J. Exp. Med.* **116**, 177-186 (1962).

6.  Pettinelli, C. B. & McFarlin, D. E. Adoptive transfer of experimental allergic encephalomyelitis in SJL/J mice after in vitro activation of lymph node cells by myelin basic protein: requirement for Lyt 1+ 2- T lymphocytes. *J. Immunol.* **127**, 1420-3 (1981).

7.  Brostoff, S. W. & Mason, D. W. Experimental allergic encephalomyelitis: successful treatment in vivo with a monoclonal antibody that recognizes T helper cells. *J. Immunol.* **133**, 1938-42 (1984).

8.  Zamvil, S. et al. T-cell clones specific for myelin basic protein induce chronic relapsing paralysis and demyelination. *Nature* **317**, 355-8 (1985).

9.  Karandikar, N. J., Vanderlugt, C. L., Bluestone, J. A. & Miller, S. D. Targeting the B7/CD28:CTLA-4 costimulatory system in CNS autoimmune disease. *J. Neuroimmunol.* **89**, 10-8 (1998).

10. Grewal, I. S. et al. Requirement for CD40 ligand in costimulation induction, T cell activation, and experimental allergic encephalomyelitis. *Science* **273**, 1864-7 (1996).

11. Howard, L. M. et al. Mechanisms of immunotherapeutic intervention by anti-CD40L (CD154) antibody in an animal model of multiple sclerosis. *J. Clin. Invest.* **103**, 281-90 (1999).

12. Ndhlovu, L. C., Ishii, N., Murata, K., Sato, T. & Sugamura, K. Critical involvement of OX40 ligand signals in the T cell priming events during experimental autoimmune encephalomyelitis. *J. Immunol.* **167**, 2991-9 (2001).

13. Weinberg, A. D., Wegmann, K. W., Funatake, C. & Whitham, R. H. Blocking OX-40/OX-40 ligand interaction in vitro and in vivo leads to decreased T cell function and amelioration of experimental allergic encephalomyelitis. *J. Immunol.* **162**, 1818-26 (1999).

14. Dong, C. et al. ICOS co-stimulatory receptor is essential for T-cell activation and function. *Nature* **409**, 97-101 (2001).

15. Dijkstra, C. D., De Groot, C. J. & Huitinga, I. The role of macrophages in demyelination. *J. Neuroimmunol.* **40**, 183-8 (1992).

16. Bauer, J. et al. The role of macrophages, perivascular cells, and microglial cells in the pathogenesis of experimental autoimmune encephalomyelitis. *Glia* **15**, 437-46 (1995).

17. Stefferl, A., Brehm, U. & Linington, C. The myelin oligodendrocyte glycoprotein (MOG): a model for antibody-mediated demyelination in experimental autoimmune encephalomyelitis and multiple sclerosis. *J. Neur. Trans.*, 123-33 (2000).

18. Anderton, S. M. & Wraith, D. C. Hierarchy in the ability of T cell epitopes to induce peripheral tolerance to antigens from myelin. *Eur. J. Immunol.* **28**, 1251-61 (1998).

19. Deber, C. M. & Reynolds, S. J. Central nervous system myelin: structure, function, and pathology. *Cell. Biochem.* **24**, 113-34 (1991).

20.    Greer, J. M. & Lees, M. B. Myelin proteolipid protein--the first 50 years. *Int. J. Biochem.* **34**, 211-5 (2002).

21.    della Gaspera, B., Pham-Dinh, D., Roussel, G., Nussbaum, J. L. & Dautigny, A. Membrane topology of the myelin/oligodendrocyte glycoprotein. *Eur. J. Biochem.* **258**, 478-84 (1998).

22.    Kies, M. W., Murphy, J. B. & Alvord, E. C. Fractionation of guinea pig proteins with encephalitogenic activity. *Fed. Proc.* **19**, 207 (1960).

23.    Williams, R. M., Lees, M. B., Cambi, F. & Macklin, W. B. Chronic experimental allergic encephalomyelitis induced in rabbits with bovine white matter proteolipid apoprotein. *J. Neuropath. Exp. Neurol.* **41**, 508-21 (1982).

24.    Adelmann, M. et al. The N-terminal domain of the myelin oligodendrocyte glycoprotein (MOG) induces acute demyelinating experimental autoimmune encephalomyelitis in the Lewis rat. *J. Neuroimmunol.* **63**, 17-27 (1995).

25.    Tuohy, V. K., Lu, Z., Sobel, R. A., Laursen, R. A. & Lees, M. B. Identification of an encephalitogenic determinant of myelin proteolipid protein for SJL mice. *J. Immunol.* **142**, 1523-7 (1989).

26.    Amor, S., Baker, D., Groome, N. & Turk, J. L. Identification of a major encephalitogenic epitope of proteolipid protein (residues 56-70) for the induction of experimental allergic encephalomyelitis in Biozzi AB/H and nonobese diabetic mice. *J. Immunol.* **150**, 5666-72 (1993).

27.    Amor, S. et al. Identification of epitopes of myelin oligodendrocyte glycoprotein for the induction of experimental allergic encephalomyelitis in SJL and Biozzi AB/H mice. *J. Immunol.* **153**, 4349-56 (1994).

28.    Mendel, I., Derosbo, N. K. & Bennun, A. A Myelin Oligodendrocyte Glycoprotein Peptide Induces Typical Chronic Experimental Autoimmune Encephalomyelitis in H-2(B) Mice - Fine Specificity and T-Cell Receptor V-Beta Expression of Encephalitogenic T-Cells. *Eur. J. Immunol.* **25**, 1951-1959 (1995).

29.    Bates, I. R. et al. Characterization of a recombinant murine 18.5-kDa myelin basic protein. *Prot. Exp. Purif.* **20**, 285-99 (2000).

30.    Pham-Dinh, D. et al. Characterization and expression of the cDNA coding for the human myelin/oligodendrocyte glycoprotein. *J. Neurochem.* **63**, 2353-6 (1994).

31.    Watts, C. Capture and processing of exogenous antigens for presentation on MHC molecules. *Annu. Rev. Immunol.* **15**, 821-50 (1997).

32.    Eylar, E. H., Caccam, J., Jackson, J. J., Westall, F. C. & Robinson, A. B. Experimental allergic encephalomyelitis: synthesis of disease-inducing site of the basic protein. *Science* **168**, 1220-3 (1970).

33.    Zamvil, S. S. et al. T-cell epitope of the autoantigen myelin basic protein that induces encephalomyelitis. *Nature* **324**, 258-60 (1986).

34.    Chou, C. H., Chou, F. C., Kowalski, T. J., Shapira, R. & Kibler, R. F. The major site of guinea-pig myelin basic protein encephalitogenic in Lewis rats. *J. Neurochem.* **28**, 115-9 (1977).

35.   Westall, F. C., Robinson, A. B., Caccam, J., Jackson, J. & Ylar, E. H. Essential chemical requirements for induction of allergic encephalomyelitis. *Nature* **229**, 22-4 (1971).

36.   Evavold, B. D. & Allen, P. M. Separation of Il-4 Production from Th-Cell Proliferation by an Altered T-Cell Receptor Ligand. *Science* **252**, 1308-1310 (1991).

37.   Karkhanis, Y. D., Carlo, D. J., Brostoff, S. W. & Eylar, E. H. Allergic encephalomyelitis. Isolation of an encephalitogenic peptide active in the monkey. *J. Biol. Chem.* **250**, 1718-22 (1975).

38.   Offner, H. et al. T cell determinants of myelin basic protein include a unique encephalitogenic I-E-restricted epitope for Lewis rats. *J. Exp. Med.* **170**, 355-67 (1989).

39.   Sakai, K. et al. Characterization of a major encephalitogenic T cell epitope in SJL/J mice with synthetic oligopeptides of myelin basic protein. *J. Neuroimmunol.* **19**, 21-32 (1988).

40.   Chou, Y. K., Vandenbark, A. A., Jones, R. E., Hashim, G. & Offner, H. Selection of encephalitogenic rat T-lymphocyte clones recognizing an immunodominant epitope on myelin basic protein. *J. Neurosci. Res.* **22**, 181-7 (1989).

41.   Martin, R. et al. Fine specificity and HLA restriction of myelin basic protein-specific cytotoxic T cell lines from multiple sclerosis patients and healthy individuals. *J. Immunol.* **145**, 540-8 (1990).

42.   Ota, K. et al. T-cell recognition of an immunodominant myelin basic protein epitope in multiple sclerosis. *Nature* **346**, 183-7 (1990).

43.   Lafaille, J. J., Nagashima, K., Katsuki, M. & Tonegawa, S. High incidence of spontaneous autoimmune encephalomyelitis in immunodeficient anti-myelin basic protein T cell receptor transgenic mice. *Cell* **78**, 399-408 (1994).

44.   Goverman, J. et al. Transgenic mice that express a myelin basic protein-specific T cell receptor develop spontaneous autoimmunity. *Cell* **72**, 551-60 (1993).

45.   Liu, G. Y. et al. Low avidity recognition of self-antigen by T cells permits escape from central tolerance. *Immunity* **3**, 407-15 (1995).

46.   Kissler, S., Anderton, S. M. & Wraith, D. C. Antigen-presenting cell activation: a link between infection and autoimmunity? *J. Autoimmun.* **16**, 303-8 (2001).

47.   Waldner, H., Whitters, M. J., Sobel, R. A., Collins, M. & Kuchroo, V. K. Fulminant spontaneous autoimmunity of the central nervous system in mice transgenic for the myelin proteolipid protein-specific T cell receptor. *Proc. Natl. Acad. Sci. USA* **97**, 3412-7 (2000).

48.   Madsen, L. S. et al. A humanized model for multiple sclerosis using HLA-DR2 and a human T-cell receptor. *Nature Gen.* **23**, 343-7 (1999).

49.   Klenerman, P., Cerundolo, V. & Dunbar, P. R. Tracking T cells with tetramers: new tales from new tools. *Nat. Rev. Immunol.* **2**, 263-72 (2002).

50.     Radu, C. G., Anderton, S. M., Firan, M., Wraith, D. C. & Ward, E. S. Detection of autoreactive T cells in H-2u mice using peptide-MHC multimers. *Int. Immunol.* **12**, 1553-60 (2000).

51.     Anderton, S. M., Radu, C. G., Lowrey, P. A., Ward, E. S. & Wraith, D. C. Negative selection during the peripheral immune response to antigen. *J. Exp. Med* **193**, 1-11 (2001).

52.     Targoni, O. S. et al. Frequencies of neuroantigen-specific T cells in the central nervous system versus the immune periphery during the course of experimental allergic encephalomyelitis. *J. Immunol.* **166**, 4757-64 (2001).

53.     Flugel, A., Willem, M., Berkowicz, T. & Wekerle, H. Gene transfer into CD4+ T lymphocytes: green fluorescent protein-engineered, encephalitogenic T cells illuminate brain autoimmune responses. *Nature Med.* **5**, 843-7 (1999).

54.     Flugel, A. et al. Migratory activity and functional changes of green fluorescent effector cells before and during experimental autoimmune encephalomyelitis. *Immunity* **14**, 547-60 (2001).

55.     Elenkov, I. J. & Chrousos, G. P. Stress hormones, proinflammatory and antiinflammatory cytokines, and autoimmunity. *Ann. NY. Acad. Sci.* **966**, 290-303 (2002).

56.     Mosmann, T. R. & Coffman, R. L. TH1 and TH2 cells: different patterns of lymphokine secretion lead to different functional properties. *Annu. Rev. Immunol.* **7**, 145-73 (1989).

57.     Kuchroo, V. K. et al. Cytokines and adhesion molecules contribute to the ability of myelin proteolipid protein-specific T cell clones to mediate experimental allergic encephalomyelitis. *J. Immunol.* **151**, 4371-82 (1993).

58.     Heeger, P. S. et al. Revisiting tolerance induced by autoantigen in incomplete Freund's adjuvant. *J. Immunol.* **164**, 5771-81 (2000).

59.     Macatonia, S. E. et al. Dendritic cells produce IL-12 and direct the development of Th1 cells from naive CD4+ T cells. *J. Immunol.* **154**, 5071-9 (1995).

60.     Seder, R. A., Gazzinelli, R., Sher, A. & Paul, W. E. Interleukin 12 acts directly on CD4+ T cells to enhance priming for interferon gamma production and diminishes interleukin 4 inhibition of such priming. *Proc. Natl. Acad. Sci. USA* **90**, 10188-92 (1993).

61.     Halpern, M. D., Kurlander, R. J. & Pisetsky, D. S. Bacterial DNA induces murine interferon-gamma production by stimulation of interleukin-12 and tumor necrosis factor-alpha. *Cell. Immunol.* **167**, 72-8 (1996).

62.     Klinman, D. M., Yi, A. K., Beaucage, S. L., Conover, J. & Krieg, A. M. CpG motifs present in bacteria DNA rapidly induce lymphocytes to secrete interleukin 6, interleukin 12, and interferon gamma. *Proc. Natl. Acad. Sci. USA* **93**, 2879-83 (1996).

63.     Segal, B. M., Chang, J. T. & Shevach, E. M. CpG oligonucleotides are potent adjuvants for the activation of autoreactive encephalitogenic T cells in vivo. *J. Immunol.* **164**, 5683-8 (2000).

64.   Ichikawa, H. T., Williams, L. P. & Segal, B. M. Activation of APCs through CD40 or Toll-like receptor 9 overcomes tolerance and precipitates autoimmune disease. *J. Immunol.* **169**, 2781-7 (2002).

65.   Leonard, J. P., Waldburger, K. E. & Goldman, S. J. Prevention of experimental autoimmune encephalomyelitis by antibodies against interleukin 12. *J. Exp. Med* **181**, 381-6 (1995).

66.   Segal, B. M., Dwyer, B. K. & Shevach, E. M. An interleukin (IL)-10/IL-12 immunoregulatory circuit controls susceptibility to autoimmune disease. *J. Exp. Med* **187**, 537-46 (1998).

67.   Dore-Duffy, P., Balabanov, R., Rafols, J. & Swanborg, R. H. Recovery phase of acute experimental autoimmune encephalomyelitis in rats corresponds to development of endothelial cell unresponsiveness to interferon gamma activation. *J. Neurosci. Res.* **44**, 223-34 (1996).

68.   Duong, T. T., Finkelman, F. D., Singh, B. & Strejan, G. H. Effect of anti-interferon-gamma monoclonal antibody treatment on the development of experimental allergic encephalomyelitis in resistant mouse strains. *J. Neuroimmunol.* **53**, 101-7 (1994).

69.   Lublin, F. D. et al. Monoclonal anti-gamma interferon antibodies enhance experimental allergic encephalomyelitis. *Autoimmunity* **16**, 267-74 (1993).

70.   Ferber, I. A. et al. Mice with a disrupted IFN-gamma gene are susceptible to the induction of experimental autoimmune encephalomyelitis (EAE). *J. Immunol.* **156**, 5-7 (1996).

71.   Krakowski, M. & Owens, T. Interferon-gamma confers resistance to experimental allergic encephalomyelitis. *Eur. J. Immunol.* **26**, 1641-6 (1996).

72.   Willenborg, D. O., Fordham, S., Bernard, C. C., Cowden, W. B. & Ramshaw, I. A. IFN-gamma plays a critical down-regulatory role in the induction and effector phase of myelin oligodendrocyte glycoprotein-induced autoimmune encephalomyelitis. *J. Immunol.* **157**, 3223-7 (1996).

73.   Selmaj, K. & Raine, C. S. Tumor necrosis factor mediates myelin damage in organotypic cultures of nervous tissue. *Ann. NY. Acad. Sci.* **540**, 568-70 (1988).

74.   Selmaj, K., Raine, C. S., Farooq, M., Norton, W. T. & Brosnan, C. F. Cytokine cytotoxicity against oligodendrocytes. Apoptosis induced by lymphotoxin. *J. Immunol.* **147**, 1522-9 (1991).

75.   Merrill, J. E., Ignarro, L. J., Sherman, M. P., Melinek, J. & Lane, T. E. Microglial cell cytotoxicity of oligodendrocytes is mediated through nitric oxide. *J. Immunol.* **151**, 2132-41 (1993).

76.   Ruddle, N. H. et al. An antibody to lymphotoxin and tumor necrosis factor prevents transfer of experimental allergic encephalomyelitis. *J. Exp. Med* **172**, 1193-200 (1990).

77.   Suen, W. E., Bergman, C. M., Hjelmstrom, P. & Ruddle, N. H. A critical role for lymphotoxin in experimental allergic encephalomyelitis. *J. Exp. Med* **186**, 1233-40 (1997).

78.  Frei, K. et al. Tumor necrosis factor alpha and lymphotoxin alpha are not required for induction of acute experimental autoimmune encephalomyelitis. *J. Exp. Med* **185**, 2177-82 (1997).

79.  Genain, C. P. et al. Late complications of immune deviation therapy in a nonhuman primate. *Science* **274**, 2054-7 (1996).

80.  Pedotti, R. et al. An unexpected version of horror autotoxicus: anaphylactic shock to a self-peptide. *Nature Immunol.* **2**, 216-22 (2001).

81.  Lafaille, J. J. et al. Myelin basic protein-specific T helper 2 (Th2) cells cause experimental autoimmune encephalomyelitis in immunodeficient hosts rather than protect them from the disease. *J. Exp. Med* **186**, 307-12 (1997).

82.  Heeger, P. S. et al. Revisiting tolerance induced by autoantigen in incomplete Freund's adjuvant. *J. Immunol.* **164**, 5771-5781 (2000).

83.  Gaur, A. et al. Amelioration of relapsing experimental autoimmune encephalomyelitis with altered myelin basic protein peptides involves different cellular mechanisms. *J. Neuroimmunol.* **74**, 149-58 (1997).

84.  Di Rosa, F. et al. Lack of Th2 cytokine increase during spontaneous remission of experimental allergic encephalomyelitis. *Eur. J. Immunol.* **28**, 3893-903 (1998).

85.  Kennedy, M. K., Torrance, D. S., Picha, K. S. & Mohler, K. M. Analysis of cytokine mRNA expression in the central nervous system of mice with experimental autoimmune encephalomyelitis reveals that IL-10 mRNA expression correlates with recovery. *J. Immunol.* **149**, 2496-505 (1992).

86.  Khoury, S. J., Hancock, W. W. & Weiner, H. L. Oral tolerance to myelin basic protein and natural recovery from experimental autoimmune encephalomyelitis are associated with downregulation of inflammatory cytokines and differential upregulation of transforming growth factor beta, interleukin 4, and prostaglandin E expression in the brain. *J. Exp. Med* **176**, 1355-64 (1992).

87.  Racke, M. K. et al. Cytokine-induced immune deviation as a therapy for inflammatory autoimmune disease. *J. Exp. Med* **180**, 1961-6 (1994).

88.  Shaw, M. K. et al. Local delivery of interleukin 4 by retrovirus-transduced T lymphocytes ameliorates experimental autoimmune encephalomyelitis. *J. Exp. Med* **185**, 1711-4 (1997).

89.  Liblau, R., Steinman, L. & Brocke, S. Experimental autoimmune encephalomyelitis in IL-4-deficient mice. *Int. Immunol.* **9**, 799-803 (1997).

90.  Bettelli, E. et al. IL-10 is critical in the regulation of autoimmune encephalomyelitis as demonstrated by studies of IL-10- and IL-4-deficient and transgenic mice. *J. Immunol.* **161**, 3299-306 (1998).

91.  Samoilova, E. B., Horton, J. L. & Chen, Y. Acceleration of experimental autoimmune encephalomyelitis in interleukin-10-deficient mice: roles of interleukin-10 in disease progression and recovery. *Cell. Immunol.* **188**, 118-24 (1998).

92. Cua, D. J., Groux, H., Hinton, D. R., Stohlman, S. A. & Coffman, R. L. Transgenic interleukin 10 prevents induction of experimental autoimmune encephalomyelitis. *J. Exp. Med* **189**, 1005-10 (1999).

93. Xiao, B. G., Bai, X. F., Zhang, G. X. & Link, H. Suppression of acute and protracted-relapsing experimental allergic encephalomyelitis by nasal administration of low-dose IL-10 in rats. *J. Neuroimmunol.* **84**, 230-7 (1998).

94. Croxford, J. L., Feldmann, M., Chernajovsky, Y. & Baker, D. Different therapeutic outcomes in experimental allergic encephalomyelitis dependent upon the mode of delivery of IL-10: a comparison of the effects of protein, adenoviral or retroviral IL-10 delivery into the central nervous system. *J. Immunol.* **166**, 4124-30 (2001).

95. Cannella, B., Gao, Y. L., Brosnan, C. & Raine, C. S. IL-10 fails to abrogate experimental autoimmune encephalomyelitis. *J. Neurosci. Res.* **45**, 735-46 (1996).

96. Rott, O., Fleischer, B. & Cash, E. Interleukin-10 prevents experimental allergic encephalomyelitis in rats. *Eur. J. Immunol.* **24**, 1434-40 (1994).

97. Mathisen, P. M., Yu, M., Johnson, J. M., Drazba, J. A. & Tuohy, V. K. Treatment of experimental autoimmune encephalomyelitis with genetically modified memory T cells. *J. Exp. Med* **186**, 159-64 (1997).

98. Nagelkerken, L., Blauw, B. & Tielemans, M. IL-4 abrogates the inhibitory effect of IL-10 on the development of experimental allergic encephalomyelitis in SJL mice. *Int. Immunol.* **9**, 1243-51 (1997).

99. Racke, M. K. et al. Prevention and treatment of chronic relapsing experimental allergic encephalomyelitis by transforming growth factor-beta 1. *J. Immunol.* **146**, 3012-7 (1991).

100. Kuruvilla, A. P. et al. Protective effect of transforming growth factor beta 1 on experimental autoimmune diseases in mice. *Proc. Natl. Acad. Sci. USA* **88**, 2918-21 (1991).

101. Kehrl, J. H. et al. Production of transforming growth factor beta by human T lymphocytes and its potential role in the regulation of T cell growth. *J. Exp. Med* **163**, 1037-50 (1986).

102. Espevik, T. et al. Inhibition of cytokine production by cyclosporin A and transforming growth factor beta. *J. Exp. Med* **166**, 571-6 (1987).

103. Tsunawaki, S., Sporn, M., Ding, A. & Nathan, C. Deactivation of macrophages by transforming growth factor-beta. *Nature* **334**, 260-2 (1988).

104. Moore, K. W., de Waal Malefyt, R., Coffman, R. L. & O'Garra, A. Interleukin-10 and the interleukin-10 receptor. *Annu. Rev. Immunol.* **19**, 683-765 (2001).

105. Menges, M. et al. Repetitive injections of dendritic cells matured with tumor necrosis factor alpha induce antigen-specific protection of mice from autoimmunity. *J. Exp. Med* **195**, 15-21 (2002).

106. Steinman, L. Myelin-specific CD8 T cells in the pathogenesis of experimental allergic encephalitis and multiple sclerosis. *J. Exp. Med* **194**, F27-30 (2001).

107. Traugott, U., Reinherz, E. L. & Raine, C. S. Multiple sclerosis: distribution of T cell subsets within active chronic lesions. *Science* **219**, 308-10 (1983).

108.  Hauser, S. L. et al. Immunohistochemical analysis of the cellular infiltrate in multiple sclerosis lesions. *Ann. Neurol.* **19**, 578-87 (1986).

109.  Sriram, S., Solomon, D., Rouse, R. V. & Steinman, L. Identification of T cell subsets and B lymphocytes in mouse brain experimental allergic encephalitis lesions. *J. Immunol.* **129**, 1649-51 (1982).

110.  Sun, D. et al. Myelin antigen-specific CD8+ T cells are encephalitogenic and produce severe disease in C57BL/6 mice. *J. Immunol.* **166**, 7579-87 (2001).

111.  Huseby, E. S. et al. A pathogenic role for myelin-specific CD8(+) T cells in a model for multiple sclerosis. *J. Exp. Med* **194**, 669-76 (2001).

112.  The Lenercept Multiple Sclerosis Study Group and The University of British Columbia MS/MRI Analysis Group. TNF neutralization in MS: results of a randomized, placebo-controlled multicenter study. *Neurol.* **53**, 457-65 (1999).

113.  Panitch, H. S., Hirsch, R. L., Schindler, J. & Johnson, K. P. Treatment of multiple sclerosis with gamma interferon: exacerbations associated with activation of the immune system. *Neurol.* **37**, 1097-102 (1987).

114.  Matsumoto, Y. et al. Role of natural killer cells and TCR gamma delta T cells in acute autoimmune encephalomyelitis. *Eur. J. Immunol.* **28**, 1681-8 (1998).

115.  Clark, R. B. & Lingenheld, E. G. Adoptively transferred EAE in gamma delta T cell-knockout mice. *J. Autoimmun.* **11**, 105-10 (1998).

116.  Olive, C. Gamma delta T cell receptor variable region usage during the development of experimental allergic encephalomyelitis. *J. Neuroimmunol.* **62**, 1-7 (1995).

117.  Selmaj, K., Brosnan, C. F. & Raine, C. S. Colocalization of lymphocytes bearing gamma delta T-cell receptor and heat shock protein hsp65+ oligodendrocytes in multiple sclerosis. *Proc. Natl. Acad. Sci. USA* **88**, 6452-6 (1991).

118.  Wucherpfennig, K. W. et al. Gamma delta T-cell receptor repertoire in acute multiple sclerosis lesions. *Proc. Natl. Acad. Sci. USA* **89**, 4588-92 (1992).

119.  Rajan, A. J., Gao, Y. L., Raine, C. S. & Brosnan, C. F. A pathogenic role for gamma delta T cells in relapsing-remitting experimental allergic encephalomyelitis in the SJL mouse. *J. Immunol.* **157**, 941-9 (1996).

120.  Kobayashi, Y. et al. Aggravation of murine experimental allergic encephalomyelitis by administration of T-cell receptor gammadelta-specific antibody. *J. Neuroimmunol.* **73**, 169-74 (1997).

121.  Jensen, M. A., Dayal, A. & Arnason, B. G. Cytokine secretion by deltagamma and alphabeta T cells in monophasic experimental autoimmune encephalomyelitis. *J. Autoimmun.* **12**, 73-80 (1999).

122.  Gao, Y. L., Rajan, A. J., Raine, C. S. & Brosnan, C. F. gammadelta T cells express activation markers in the central nervous system of mice with chronic-relapsing experimental autoimmune encephalomyelitis. *J. Autoimmun.* **17**, 261-71 (2001).

123.  Rajan, A. J., Klein, J. D. & Brosnan, C. F. The effect of gammadelta T cell depletion on cytokine gene expression in experimental allergic encephalomyelitis. *J. Immunol.* **160**, 5955-62 (1998).

124. Rajan, A. J., Asensio, V. C., Campbell, I. L. & Brosnan, C. F. Experimental autoimmune encephalomyelitis on the SJL mouse: effect of gamma delta T cell depletion on chemokine and chemokine receptor expression in the central nervous system. *J. Immunol.* **164**, 2120-30 (2000).

125. Spahn, T. W., Issazadah, S., Salvin, A. J. & Weiner, H. L. Decreased severity of myelin oligodendrocyte glycoprotein peptide 33 - 35-induced experimental autoimmune encephalomyelitis in mice with a disrupted TCR delta chain gene. *Eur. J. Immunol.* **29**, 4060-71 (1999).

126. Sakaguchi, S. et al. Immunologic tolerance maintained by CD25+ CD4+ regulatory T cells: their common role in controlling autoimmunity, tumor immunity, and transplantation tolerance. *Immunol. Rev.* **182**, 18-32 (2001).

127. Shevach, E. M. CD4+ CD25+ suppressor T cells: more questions than answers. *Nat. Rev. Immunol.* **2**, 389-400 (2002).

128. Maloy, K. J. & Powrie, F. Regulatory T cells in the control of immune pathology. *Nature Immunol.* **2**, 816-822 (2001).

129. Asano, M., Toda, M., Sakaguchi, N. & Sakaguchi, S. Autoimmune disease as a consequence of developmental abnormality of a T cell subpopulation. *J. Exp. Med* **184**, 387-96 (1996).

130. Sakaguchi, S., Sakaguchi, N., Asano, M., Itoh, M. & Toda, M. Immunologic self-tolerance maintained by activated T cells expressing IL-2 receptor alpha-chains (CD25). Breakdown of a single mechanism of self-tolerance causes various autoimmune diseases. *J. Immunol.* **155**, 1151-64 (1995).

131. Fowell, D. & Mason, D. Evidence that the T cell repertoire of normal rats contains cells with the potential to cause diabetes. Characterization of the CD4+ T cell subset that inhibits this autoimmune potential. *J. Exp. Med* **177**, 627-36 (1993).

132. Powrie, F. et al. Inhibition of Th1 responses prevents inflammatory bowel disease in scid mice reconstituted with CD45RBhi CD4+ T cells. *Immunity* **1**, 553-62 (1994).

133. Powrie, F. & Mason, D. OX-22high CD4+ T cells induce wasting disease with multiple organ pathology: prevention by the OX-22low subset. *J. Exp. Med* **172**, 1701-8 (1990).

134. Papiernik, M., de Moraes, M. L., Pontoux, C., Vasseur, F. & Penit, C. Regulatory CD4 T cells: expression of IL-2R alpha chain, resistance to clonal deletion and IL-2 dependency. *Int. Immunol.* **10**, 371-8 (1998).

135. Maloy, K. J. & Powrie, F. Regulatory T cells in the control of immune pathology. *Nature Immunol.* **2**, 816-22 (2001).

136. Asseman, C., Mauze, S., Leach, M. W., Coffman, R. L. & Powrie, F. An essential role for interleukin 10 in the function of regulatory T cells that inhibit intestinal inflammation. *J. Exp. Med* **190**, 995-1004 (1999).

137. Powrie, F., Carlino, J., Leach, M. W., Mauze, S. & Coffman, R. L. A critical role for transforming growth factor-beta but not interleukin 4 in the suppression of T

helper type 1-mediated colitis by CD45RB(low) CD4+ T cells. *J. Exp. Med* **183**, 2669-74 (1996).

138.  Olivares-Villagomez, D., Wang, Y. & Lafaille, J. J. Regulatory CD4(+) T cells expressing endogenous T cell receptor chains protect myelin basic protein-specific transgenic mice from spontaneous autoimmune encephalomyelitis. *J. Exp. Med* **188**, 1883-94 (1998).

139.  Olivares-Villagomez, D., Wensky, A. K., Wang, Y. & Lafaille, J. J. Repertoire requirements of CD4+ T cells that prevent spontaneous autoimmune encephalomyelitis. *J. Immunol.* **164**, 5499-507 (2000).

140.  Kohm, A. P., Carpentier, P. A., Anger, H. A. & Miller, S. D. Cutting edge: CD4+CD25+ Regulatory T cells suppress antigen-specific autoreactive immune responses and central nervous system inflammation during active experimental autoimmune encephalomyelitis. *J. Immunol.* **169**, 4712-4716 (2002).

141.  Groux, H. et al. A CD4+ T-cell subset inhibits antigen-specific T-cell responses and prevents colitis. *Nature* **389**, 737-42 (1997).

142.  Barrat, F. J. et al. In vitro generation of interleukin 10-producing regulatory CD4(+) T cells is induced by immunosuppressive drugs and inhibited by T helper type 1 (Th1)- and Th2-inducing cytokines. *J. Exp. Med* **195**, 603-16 (2002).

143.  McGuirk, P., McCann, C. & Mills, K. H. Pathogen-specific T regulatory 1 cells induced in the respiratory tract by a bacterial molecule that stimulates interleukin 10 production by dendritic cells: a novel strategy for evasion of protective T helper type 1 responses by Bordetella pertussis. *J. Exp. Med* **195**, 221-31 (2002).

144.  Bendelac, A., Rivera, M. N., Park, S. H. & Roark, J. H. Mouse CD1-specific NK1 T cells: development, specificity, and function. *Annu. Rev. Immunol.* **15**, 535-62 (1997).

145.  Kawano, T. et al. CD1d-restricted and TCR-mediated activation of valpha14 NKT cells by glycosylceramides. *Science* **278**, 1626-9 (1997).

146.  Brossay, L. et al. CD1d-mediated recognition of an alpha-galactosylceramide by natural killer T cells is highly conserved through mammalian evolution. *J. Exp. Med* **188**, 1521-8 (1998).

147.  Naidenko, O. V. et al. Binding and antigen presentation of ceramide-containing glycolipids by soluble mouse and human CD1d molecules. *J. Exp. Med* **190**, 1069-80 (1999).

148.  Roark, J. H. et al. CD1.1 expression by mouse antigen-presenting cells and marginal zone B cells. *J. Immunol.* **160**, 3121-7 (1998).

149.  Pulendran, B. et al. Developmental pathways of dendritic cells in vivo: distinct function, phenotype, and localization of dendritic cell subsets in FLT3 ligand-treated mice. *J. Immunol.* **159**, 2222-31 (1997).

150.  Koseki, H., Imai, K., Ichikawa, T., Hayata, I. & Taniguchi, M. Predominant use of a particular alpha-chain in suppressor T cell hybridomas specific for keyhole limpet hemocyanin. *Int. Immunol.* **1**, 557-64 (1989).

151.    Porcelli, S., Yockey, C. E., Brenner, M. B. & Balk, S. P. Analysis of T cell antigen receptor (TCR) expression by human peripheral blood CD4-8- alpha/beta T cells demonstrates preferential use of several V beta genes and an invariant TCR alpha chain. *J. Exp. Med* **178**, 1-16 (1993).

152.    Baxter, A. G., Kinder, S. J., Hammond, K. J., Scollay, R. & Godfrey, D. I. Association between alphabetaTCR+CD4-CD8- T-cell deficiency and IDDM in NOD/Lt mice. *Diabetes* **46**, 572-82 (1997).

153.    Iwakoshi, N. N., Greiner, D. L., Rossini, A. A. & Mordes, J. P. Diabetes prone BB rats are severely deficient in natural killer T cells. *Autoimmunity* **31**, 1-14 (1999).

154.    Yoshimoto, T., Bendelac, A., Hu-Li, J. & Paul, W. E. Defective IgE production by SJL mice is linked to the absence of CD4+, NK1.1+ T cells that promptly produce interleukin 4. *Proc. Natl. Acad. Sci. USA* **92**, 11931-4 (1995).

155.    Illes, Z. et al. Differential expression of NK T cell V alpha 24J alpha Q invariant TCR chain in the lesions of multiple sclerosis and chronic inflammatory demyelinating polyneuropathy. *J. Immunol.* **164**, 4375-81 (2000).

156.    Singh, A. K. et al. Natural killer T cell activation protects mice against experimental autoimmune encephalomyelitis. *J. Exp. Med* **194**, 1801-11 (2001).

157.    Jahng, A. W. et al. Activation of natural killer T cells potentiates or prevents experimental autoimmune encephalomyelitis. *J. Exp. Med* **194**, 1789-99 (2001).

158.    Koh, D. R. et al. Less mortality but more relapses in experimental allergic encephalomyelitis in CD8-/- mice. *Science* **256**, 1210-3 (1992).

159.    Jiang, H., Zhang, S. I. & Pernis, B. Role of CD8+ T cells in murine experimental allergic encephalomyelitis. *Science* **256**, 1213-5 (1992).

160.    Miller, A., Lider, O., Roberts, A. B., Sporn, M. B. & Weiner, H. L. Suppressor T cells generated by oral tolerization to myelin basic protein suppress both in vitro and in vivo immune responses by the release of transforming growth factor beta after antigen-specific triggering. *Proc. Natl. Acad. Sci. USA* **89**, 421-5 (1992).

161.    Ben-Nun, A., Wekerle, H. & Cohen, I. R. Vaccination against autoimmune encephalomyelitis with T-lymphocyte line cells reactive against myelin basic protein. *Nature* **292**, 60-1 (1981).

162.    Zhang, J. & Raus, J. T-cell vaccination in autoimmune diseases. From laboratory to clinic. *Hum. Immunol.* **38**, 87-96 (1993).

163.    Vandenbark, A. A., Hashim, G. & Offner, H. Immunization with a synthetic T-cell receptor V-region peptide protects against experimental autoimmune encephalomyelitis. *Nature* **341**, 541-4 (1989).

164.    Howell, M. D. et al. Vaccination against experimental allergic encephalomyelitis with T cell receptor peptides. *Science* **246**, 668-70 (1989).

165.    Acha-Orbea, H. et al. Limited heterogeneity of T cell receptors from lymphocytes mediating autoimmune encephalomyelitis allows specific immune intervention. *Cell* **54**, 263-73 (1988).

166.    Kumar, V. & Sercarz, E. An integrative model of regulation centered on recognition of TCR peptide/MHC complexes. *Immunol. Rev.* **182**, 113-21 (2001).

167.  Kumar, V., Tabibiazar, R., Geysen, H. M. & Sercarz, E. Immunodominant framework region 3 peptide from TCR V beta 8.2 chain controls murine experimental autoimmune encephalomyelitis. *J. Immunol.* **154**, 1941-50 (1995).

168.  Vainiene, M. et al. Neonatal injection of Lewis rats with recombinant V beta 8.2 induces T cell but not B cell tolerance and increased severity of experimental autoimmune encephalomyelitis. *J. Neurosci. Res.* **45**, 475-86 (1996).

169.  Jiang, H., Braunstein, N. S., Yu, B., Winchester, R. & Chess, L. CD8+ T cells control the TH phenotype of MBP-reactive CD4+ T cells in EAE mice. *Proc. Natl. Acad. Sci. USA* **98**, 6301-6 (2001).

170.  Litzenburger, T. et al. B lymphocytes producing demyelinating autoantibodies: development and function in gene-targeted transgenic mice. *J. Exp. Med* **188**, 169-80 (1998).

171.  Myers, K. J., Sprent, J., Dougherty, J. P. & Ron, Y. Synergy between encephalitogenic T cells and myelin basic protein-specific antibodies in the induction of experimental autoimmune encephalomyelitis. *J. Neuroimmunol.* **41**, 1-8 (1992).

172.  Hjelmstrom, P., Juedes, A. E., Fjell, J. & Ruddle, N. H. B-cell-deficient mice develop experimental allergic encephalomyelitis with demyelination after myelin oligodendrocyte glycoprotein sensitization. *J. Immunol.* **161**, 4480-3 (1998).

173.  Lyons, J. A., San, M., Happ, M. P. & Cross, A. H. B cells are critical to induction of experimental allergic encephalomyelitis by protein but not by a short encephalitogenic peptide. *Eur. J. Immunol.* **29**, 3432-9 (1999).

174.  Lyons, J. A., Ramsbottom, M. J. & Cross, A. H. Critical role of antigen-specific antibody in experimental autoimmune encephalomyelitis induced by recombinant myelin oligodendrocyte glycoprotein. *Eur. J. Immunol.* **32**, 1905-13 (2002).

175.  Fillatreau, S., Sweenie, C. H., McGeachy, M. J., Gray, D. & Anderton, S. M. B cells regulate autoimmunity by provision of IL-10. *Nature Immunol.* **3**, 944-50 (2002).

176.  Pham-Dinh, D. et al. Myelin/oligodendrocyte glycoprotein is a member of a subset of the immunoglobulin superfamily encoded within the major histocompatibility complex. *Proc. Natl. Acad. Sci. USA* **90**, 7990-4 (1993).

177.  Wolf, S. D., Dittel, B. N., Hardardottir, F. & Janeway, C. A., Jr. Experimental autoimmune encephalomyelitis induction in genetically B cell-deficient mice. *J. Exp. Med* **184**, 2271-8 (1996).

178.  Mizoguchi, A., Mizoguchi, E., Takedatsu, H., Blumberg, R. S. & Bhan, A. K. Chronic intestinal inflammatory condition generates IL-10-producing regulatory B cell subset characterized by CD1d upregulation. *Immunity* **16**, 219-30 (2002).

179.  Sebzda, E. et al. Selection of the T cell repertoire. *Annu. Rev. Immunol.* **17**, 829-74 (1999).

180.  Goldrath, A. W. & Bevan, M. J. Selecting and maintaining a diverse T-cell repertoire. *Nature* **402**, 255-62 (1999).

181.    Anderton, S. M., Wraith, D.C. Selection and Fine Tuning of the Autoreactive T cell repertoire. *Nat. Rev. Immunol.* **2**, 487-498 (2002).

182.    Heath, V. L., Moore, N. C., Parnell, S. M. & Mason, D. W. Intrathymic expression of genes involved in organ specific autoimmune disease. *J. Autoimmun.* **11**, 309-18 (1998).

183.    Fritz, R. B. & Zhao, M. L. Thymic expression of myelin basic protein (MBP). Activation of MBP-specific T cells by thymic cells in the absence of exogenous MBP. *J. Immunol.* **157**, 5249-53 (1996).

184.    Anderson, A. C. et al. High frequency of autoreactive myelin proteolipid protein-specific T cells in the periphery of naive mice: mechanisms of selection of the self-reactive repertoire. *J. Exp. Med* **191**, 761-70 (2000).

185.    Nave, K. A., Lai, C., Bloom, F. E. & Milner, R. J. Splice site selection in the proteolipid protein (PLP) gene transcript and primary structure of the DM-20 protein of central nervous system myelin. *Proc. Natl. Acad. Sci. USA* **84**, 5665-9 (1987).

186.    Klein, L., Klugmann, M., Nave, K. A., Tuohy, V. K. & Kyewski, B. Shaping the autoreactive T-cell repertoire by a splice variant of self protein expressed in thymic epithelial cells. *Nature Med.* **6**, 56-61 (2000).

187.    Fugger, L., Liang, J., Gautam, A., Rothbard, J. B. & McDevitt, H. O. Quantitative analysis of peptides from myelin basic protein binding to the MHC class II protein, I-Au, which confers susceptibility to experimental allergic encephalomyelitis. *Mol. Med.* **2**, 181-8 (1996).

188.    Fairchild, P. J., Wildgoose, R., Atherton, E., Webb, S. & Wraith, D. C. An autoantigenic T cell epitope forms unstable complexes with class II MHC: a novel route for escape from tolerance induction. *Int. Immunol.* **5**, 1151-8 (1993).

189.    Fairchild, P. J., Thorpe, C. J., Travers, P. J. & Wraith, D. C. Modulation of the immune response with T-cell epitopes: the ultimate goal for specific immunotherapy of autoimmune disease. *Immunol.* **81**, 487-96 (1994).

190.    Pearson, C. I., Gautam, A. M., Rulifson, I. C., Liblau, R. S. & McDevitt, H. O. A small number of residues in the class II molecule I-Au confer the ability to bind the myelin basic protein peptide Ac1-11. *Proc. Natl. Acad. Sci. USA* **96**, 197-202 (1999).

191.    Lee, C. et al. Evidence that the autoimmune antigen myelin basic protein (MBP) Ac1-9 binds towards one end of the major histocompatibility complex (MHC) cleft. *J. Exp. Med* **187**, 1505-16 (1998).

192.    Wraith, D. C., Smilek, D. E., Mitchell, D. J., Steinman, L. & McDevitt, H. O. Antigen recognition in autoimmune encephalomyelitis and the potential for peptide-mediated immunotherapy. *Cell* **59**, 247-55 (1989).

193.    Ashtonrickardt, P. G. & Tonegawa, S. A Differential-Avidity Model for T-Cell Selection. *Immunol. Today* **15**, 362-366 (1994).

194.    Sebzda, E. et al. Positive and negative thymocyte selection induced by different concentrations of a single peptide. *Science* **263**, 1615-8 (1994).

195.   Ridgway, W. M., Ito, H., Fasso, M., Yu, C. & Fathman, C. G. Analysis of the role of variation of major histocompatibility complex class II expression on nonobese diabetic (NOD) peripheral T cell response. *J. Exp. Med* **188**, 2267-75 (1998).

196.   Anderton, S. M., Viner, N. J., Matharu, P., Lowrey, P. A. & Wraith, D. C. Influence of a dominant cryptic epitope on autoimmune T cell tolerance. *Nature Immunol.* **3**, 175-181 (2002).

197.   Manoury, B. et al. An asparaginyl endopeptidase processes a microbial antigen for class II MHC presentation. *Nature* **396**, 695-9 (1998).

198.   Manoury, B. et al. Destructive processing by asparagine endopeptidase limits presentation of a dominant T cell epitope in MBP. *Nature Immunol.* **3**, 169-174 (2002).

199.   Mason, D. A very high level of crossreactivity is an essential feature of the T-cell receptor. *Immunol. Today* **19**, 395-404 (1998).

200.   Borghans, J. A., Noest, A. J. & De Boer, R. J. How specific should immunological memory be? *J. Immunol.* **163**, 569-75 (1999).

201.   Kersh, G. J. & Allen, P. M. Structural basis for T cell recognition of altered peptide ligands: A single T cell receptor can productively recognize a large continuum of related ligands. *J. Exp. Med* **184**, 1259-1268 (1996).

202.   Kersh, G. J. & Allen, P. M. Essential flexibility in the T-cell recognition of antigen. *Nature* **380**, 495-498 (1996).

203.   Fujinami, R. S. & Oldstone, M. B. Amino acid homology between the encephalitogenic site of myelin basic protein and virus: mechanism for autoimmunity. *Science* **230**, 1043-5 (1985).

204.   Oldstone, M. B., Nerenberg, M., Southern, P., Price, J. & Lewicki, H. Virus infection triggers insulin-dependent diabetes mellitus in a transgenic model: role of anti-self (virus) immune response. *Cell* **65**, 319-31 (1991).

205.   Ohashi, P. S. et al. Ablation of "tolerance" and induction of diabetes by virus infection in viral antigen transgenic mice. *Cell* **65**, 305-17 (1991).

206.   Benoist, C. & Mathis, D. Autoimmunity provoked by infection: how good is the case for T cell epitope mimicry? *Nature Immunol.* **2**, 797-801 (2001).

207.   Panoutsakopoulou, V. & Cantor, H. On the relationship between viral infection and autoimmunity. *J. Autoimmun.* **16**, 341-5 (2001).

208.   Wucherpfennig, K. W. & Strominger, J. L. Molecular mimicry in T cell-mediated autoimmunity: viral peptides activate human T cell clones specific for myelin basic protein. *Cell* **80**, 695-705 (1995).

209.   Gautam, A. M. et al. Minimum structural requirements for peptide presentation by major histocompatibility complex class II molecules: implications in induction of autoimmunity. *Proc. Natl. Acad. Sci. USA* **91**, 767-71 (1994).

210.   Carrizosa, A. M. et al. Expansion by self antigen is necessary for the induction of experimental autoimmune encephalomyelitis by T cells primed with a cross-reactive environmental antigen. *J. Immunol.* **161**, 3307-14 (1998).

211.    Ufret-Vincenty, R. L. et al. In vivo survival of viral antigen-specific T cells that induce experimental autoimmune encephalomyelitis. *J. Exp. Med* **188**, 1725-38 (1998).

212.    Gautam, A. M., Liblau, R., Chelvanayagam, G., Steinman, L. & Boston, T. A viral peptide with limited homology to a self peptide can induce clinical signs of experimental autoimmune encephalomyelitis. *J. Immunol.* **161**, 60-4 (1998).

213.    Grogan, J. L. et al. Cross-reactivity of myelin basic protein-specific T cells with multiple microbial peptides: experimental autoimmune encephalomyelitis induction in TCR transgenic mice. *J. Immunol.* **163**, 3764-70 (1999).

214.    Ruiz, P. J. et al. Microbial epitopes act as altered peptide ligands to prevent experimental autoimmune encephalomyelitis. *J. Exp. Med* **189**, 1275-84 (1999).

215.    Barnett, L. A., Whitton, J. L., Wang, L. Y. & Fujinami, R. S. Virus encoding an encephalitogenic peptide protects mice from experimental allergic encephalomyelitis. *J. Neuroimmunol.* **64**, 163-73 (1996).

216.    Olson, J. K., Croxford, J. L., Calenoff, M. A., Dal Canto, M. C. & Miller, S. D. A virus-induced molecular mimicry model of multiple sclerosis. *J. Clin. Invest.* **108**, 311-8 (2001).

217.    Olson, J. K., Eagar, T. N. & Miller, S. D. Functional activation of myelin-specific T cells by virus-induced molecular mimicry. *J. Immunol.* **169**, 2719-26 (2002).

218.    Savage, P. A. & Davis, M. M. A kinetic window constricts the T cell receptor repertoire in the thymus. *Immunity* **14**, 243-52 (2001).

219.    Lucas, B., Stefanova, I., Yasutomo, K., Dautigny, N. & Germain, R. N. Divergent changes in the sensitivity of maturing T cells to structurally related ligands underlies formation of a useful T cell repertoire. *Immunity* **10**, 367-76 (1999).

220.    Grossman, Z. & Paul, W. E. Autoreactivity, dynamic tuning and selectivity. *Curr. Opin. Immunol.* **13**, 687-698 (2001).

221.    Germain, R. N. & Stefanova, I. The dynamics of T cell receptor signaling: Complex orchestration and the key roles of tempo and cooperation. *Annu. Rev. Immunol.* **17**, 467-522 (1999).

222.    Munder, M. et al. Reduced self-reactivity of an autoreactive T cell after activation with cross-reactive non-self-ligand. *J. Exp. Med.* **196**, 1151-1162 (2002).

223.    Bouneaud, C., Kourilsky, P. & Bousso, P. Impact of negative selection on the T cell repertoire reactive to a self-peptide: a large fraction of T cell clones escapes clonal deletion. *Immunity* **13**, 829-40 (2000).

224.    Tanchot, C., Barber, D. L., Chiodetti, L. & Schwartz, R. H. Adaptive tolerance of CD4(+) T cells in vivo: Multiple thresholds in response to a constant level of antigen presentation. *J. Immunol.* **167**, 2030-2039 (2001).

225.    Valitutti, S., Muller, S., Cella, M., Padovan, E. & Lanzavecchia, A. Serial triggering of many T-cell receptors by a few peptide-MHC complexes. *Nature* **375**, 148-51 (1995).

226.    Harding, C. V. & Unanue, E. R. Quantitation of antigen-presenting cell MHC class II/peptide complexes necessary for T-cell stimulation. *Nature* **346**, 574-6 (1990).

227.     Kimachi, K., Croft, M. & Grey, H. M. The minimal number of antigen-major histocompatibility complex class II complexes required for activation of naive and primed T cells. *Eur. J. Immunol.* **27**, 3310-7 (1997).

228.     Nicholson, L. B., Anderson, A. C. & Kuchroo, V. K. Tuning T cell activation threshold and effector function with cross-reactive peptide ligands. *Int. Immunol.* **12**, 205-13 (2000).

229.     Murtaza, A., Kuchroo, V. K. & Freeman, G. J. Changes in the strength of co-stimulation through the B7/CD28 pathway alter functional T cell responses to altered peptide ligands. *Int. Immunol.* **11**, 407-16 (1999).

230.     Drakesmith, H., Chain, B. & Beverley, P. How can dendritic cells cause autoimmune disease? *Immunol. Today* **21**, 214-7 (2000).

231.     Manoury, B., Gregory, W. F., Maizels, R. M. & Watts, C. Bm-CPI-2, a cystatin homolog secreted by the filarial parasite Brugia malayi, inhibits class II MHC-restricted antigen processing. *Current Biology* **11**, 447-51 (2001).

232.     Anderton, S. M. Peptide-based immunotherapy of autoimmunity: a path of puzzles, paradoxes and possibilities. *Immunol.* **104**, 367-376 (2001).

233.     Franco, A. et al. T cell receptor antagonist peptides are highly effective inhibitors of experimental allergic encephalomyelitis. *Eur. J. Immunol.* **24**, 940-6 (1994).

234.     Kuchroo, V. K. et al. A single TCR antagonist peptide inhibits experimental allergic encephalomyelitis mediated by a diverse T cell repertoire. *J. Immunol.* **153**, 3326-36 (1994).

235.     Nicholson, L. B., Greer, J. M., Sobel, R. A., Lees, M. B. & Kuchroo, V. K. An Altered Peptide Ligand Mediates Immune Deviation and Prevents Autoimmune Encephalomyelitis. *Immunity* **3**, 397-405 (1995).

236.     Nicholson, L. B., Murtaza, A., Hafler, B. P., Sette, A. & Kuchroo, V. K. A T cell receptor antagonist peptide induces T cells that mediate bystander suppression and prevent autoimmune encephalomyelitis induced with multiple myelin antigens. *Proc. Natl. Acad. Sci. USA* **94**, 9279-84 (1997).

237.     Brocke, S. et al. Treatment of experimental encephalomyelitis with a peptide analogue of myelin basic protein. *Nature* **379**, 343-6 (1996).

238.     Karin, N., Mitchell, D. J., Brocke, S., Ling, N. & Steinman, L. Reversal of experimental autoimmune encephalomyelitis by a soluble peptide variant of a myelin basic protein epitope: T cell receptor antagonism and reduction of interferon gamma and tumor necrosis factor alpha production. *J. Exp. Med* **180**, 2227-37 (1994).

239.     Critchfield, J. M. et al. T cell deletion in high antigen dose therapy of autoimmune encephalomyelitis. *Science* **263**, 1139-43 (1994).

240.     Lenardo, M. et al. Mature T lymphocyte apoptosis--immune regulation in a dynamic and unpredictable antigenic environment. *Annu. Rev. Immunol.* **17**, 221-53 (1999).

241.   Anderton, S. M. et al. Fine specificity of the myelin-reactive T cell repertoire: implications for TCR antagonism in autoimmunity. *J. Immunol.* **161**, 3357-64 (1998).

242.   Bielekova, B. et al. Encephalitogenic potential of the myelin basic protein peptide (amino acids 83-99) in multiple sclerosis: results of a phase II clinical trial with an altered peptide ligand. *Nature Med.* **6**, 1167-75 (2000).

243.   Kappos, L. et al. Induction of a non-encephalitogenic type 2 T helper-cell autoimmune response in multiple sclerosis after administration of an altered peptide ligand in a placebo-controlled, randomized phase II trial. The Altered Peptide Ligand in Relapsing MS Study Group. *Nature Med.* **6**, 1176-82 (2000).

244.   Weigle, W. O. Immunological unresponsiveness. *Adv. Immunol.* **16**, 61-122 (1973).

245.   Metzler, B. & Wraith, D. C. Inhibition of experimental autoimmune encephalomyelitis by inhalation but not oral administration of the encephalitogenic peptide: influence of MHC binding affinity. *Int. Immunol.* **5**, 1159-65 (1993).

246.   Liu, G. Y. & Wraith, D. C. Affinity for class II MHC determines the extent to which soluble peptides tolerize autoreactive T cells in naive and primed adult mice--implications for autoimmunity. *Int. Immunol.* **7**, 1255-63 (1995).

247.   Leadbetter, E. A. et al. Experimental autoimmune encephalomyelitis induced with a combination of myelin basic protein and myelin oligodendrocyte glycoprotein is ameliorated by administration of a single myelin basic protein peptide. *J. Immunol.* **161**, 504-12 (1998).

248.   al-Sabbagh, A., Miller, A., Santos, L. M. & Weiner, H. L. Antigen-driven tissue-specific suppression following oral tolerance: orally administered myelin basic protein suppresses proteolipid protein-induced experimental autoimmune encephalomyelitis in the SJL mouse. *Eur. J. Immunol.* **24**, 2104-9 (1994).

249.   Gonnella, P. A., Waldner, H. P. & Weiner, H. L. B cell-deficient (mu MT) mice have alterations in the cytokine microenvironment of the gut-associated lymphoid tissue (GALT) and a defect in the low dose mechanism of oral tolerance. *J. Immunol.* **166**, 4456-64 (2001).

250.   Thorstenson, K. M. & Khoruts, A. Generation of anergic and potentially immunoregulatory CD25+CD4 T cells in vivo after induction of peripheral tolerance with intravenous or oral antigen. *J. Immunol.* **167**, 188-95 (2001).

251.   Burkhart, C., Liu, G. Y., Anderton, S. M., Metzler, B. & Wraith, D. C. Peptide-induced T cell regulation of experimental autoimmune encephalomyelitis: a role for IL-10. *Int. Immunol.* **11**, 1625-34 (1999).

252.   Massey, E. J. et al. Intranasal peptide-induced peripheral tolerance: the role of IL-10 in regulatory T cell function within the context of experimental autoimmune encephalomyelitis. *Vet. Immunol. Immunopathol.* **87**, 357-72 (2002).

253.   Chen, Y. H., Kuchroo, V. K., Inobe, J., Hafler, D. A. & Weiner, H. L. Regulatory T-Cell Clones Induced by Oral Tolerance - Suppression of Autoimmune Encephalomyelitis. *Science* **265**, 1237-1240 (1994).

Chapter A12

# THE ROLE OF COMPLEMENT IN EAE

Johan van Beek and B. Paul Morgan
*Department of Medical Biochemistry and Immunology, University of Wales College of Medicine, Cardiff CF14 4XX, United Kingdom*

Abstract:     The complement system is an essential effector of the humoral and cellular immunity involved in cytolysis and immune/inflammatory responses. Complement participates in host defence against pathogens by triggering the formation of the membrane attack complex. Complement opsonins (C1q, C3b, iC3b) interact with surface complement receptors to promote phagocytosis while complement anaphylatoxins C3a and C5a initiate local pro-inflammatory responses that, ultimately, contribute to the protection and healing of the host. However, activation of complement to an inappropriate extent has been proposed to promote tissue injury. There is now compelling evidence that complement is implicated in the pathogenesis of several neurological disorders including the human demyelinating disease multiple sclerosis and experimental allergic encephalomyelitis (EAE), an animal model that mimics the demyelination seen in multiple sclerosis. Deposition of complement proteins correlates with areas of demyelination and axonal loss observed in EAE and complement inhibition ameliorates disease. However, the precise mechanisms underlying complement-mediated damage are still largely unknown. The recent use of transgenic animals is beginning to shed light on the relative contributions of the different complement activation pathways in the pathogenesis of experimental demyelination. These studies will provide the basis for the development of novel drugs aiming to inhibit complement in the the central nervous system.

Key words:     Complement; experimental allergic encephalomyelitis; inflammation; anaphylatoxins; opsonins; membrane attack complex; demyelination; axonal damage; therapy

# 1.      THE COMPLEMENT SYSTEM

## 1.1     A crucial effector arm of the innate immune response

Functions of the complement system include the recognition and killing of invading pathogens (e.g. bacteria, virus infected cells, parasites) while preserving normal 'self' cells. The complement system also contributes through its activation products to the release of inflammatory mediators, promoting tissue injury at sites of inflammation (1). Having preceded the emergence of adaptive immunity, the complement system has maintained a degree of phylogenic conservation among both invertebrates and mammals, underlying the critical role of complement in tissue homeostasis (2).

The complement system consists of some 30 fluid-phase and cell-membrane associated proteins. Complement can be activated by three distinct routes (Figure 1). The classical pathway (involving C1q, C1r, C1s, C4, C2 and C3 components) is activated primarily by the interaction of C1q with immune complexes (antibody-antigen) but activation can also be achieved after interaction of C1q with non-immune molecules such as DNA, RNA, C-reactive protein, serum amyloid P, bacterial lipopolysaccharides, and some fungal and virus membranes. The initiation of the alernative pathway (involving C3, factor B, factor D, and properdin) does not require the presence of immune complexes and leads to the deposition of C3 fragments on target cells. Mannose-binding lectin (MBL), a lectin homologous to C1q, can recognize carbohydrates such as mannose and N-acetylglucosamine on pathogens and initiate the complement pathway independently of both the classical and alternative activation pathways. MBL is associated with two serine-proteases, mannose-binding lectin-associated serine proteases (MASP)-1 and 2, that, like the C1 complex in the classical pathway, cleaves and C4 and C2 components, leading to the formation of the classical C3 convertase.

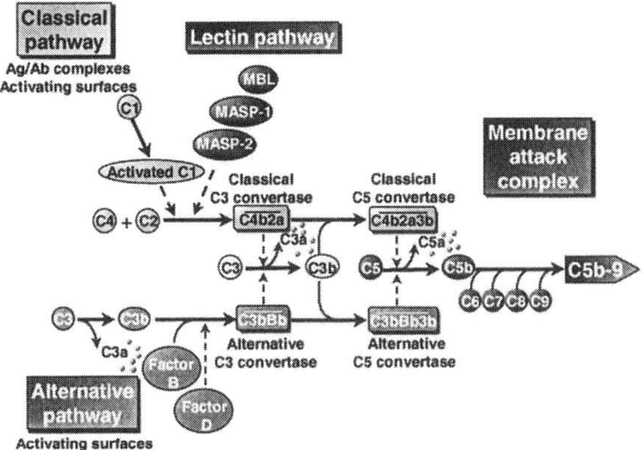


*Figure 1.* Schematic representation of the complement cascade. The complement system can be activated through the classical, the alternative or the lectin pathway, leading to a common terminal pathway: the cytolytic membrane attack complex. Proteolytic cleavage of components C3 and C5 leads to the release of pro-inflammatory anaphylatoxins C3a and C5a respectively. MBL: mannose-binding lectin; MASP: MBL-associated serine protease.

## 1.2 Complement receptors involved in phagocytosis

The target pathogen coated with complement opsonins (C1q and C3 activation by-products: C3b, iC3b) is specifically recognised and phagocytosed by macrophages expressing complement receptors (C1qRp (CD93), complement receptor (CR) type 1, 3 and 4). In addition to its role in pathogen clearance, C1q has been shown to play an important role in the clearance of apoptotic cells. C1q binds directly to surface blebs of apoptotic keratinocytes and T cells and there can initiate complement opsonisation (3-5). Moreover, transgenic mice lacking C1q display an impairment in the clearance of apoptotic cells in the kidney, leading to severe glomerulonephritis with immune deposits (6). These obsevations support the hypothesis that C1q serves as an opsonin in the recognition and clearance of apoptotic cells. The transmembrane glycoprotein termed "C1q receptor that enhances phagocytosis" (C1qRp), recently shown to be identical to CD93 (7), has emerged as a defense collagen receptor. Monocytes that have adhered on C1q-coated surfaces display a 4 to 10 fold enhancement of phagocytosis of targets coated with IgG or complement (8). However, the potential involvement of C1q as a ligand for C1qRp/CD93 remains controversial (7) and the cellular and molecular properties of C1qRp/CD93 are yet to be elucidated.

C3, when activated on a cell surface, binds covalently (opsonisation) as C3b and is subsequently cleaved into a very stable fragment iC3b. CR3 (CD11b/CD18) and CR4 (CD11c/CD18) are involved in the phagocytosis process of targets opsonised with C3b and iC3b fragments (9). CR1 (CD35) is also a multifunctional receptor binding to C4b, C3b, iC3b and C1q, and as such, has been implicated in phagocytic and complement regulatory activities (10).

## 1.3     The lytic terminal complex and complement regulators

Ultimately, activation of the complement system leads to the formation, through the complement terminal pathway (involving C5, C6, C7, C8 and C9 components), of the membrane attack complex (MAC) which forms a pore in the phosholipid bilayer to lyse the target cell. However, activation of the complement system at inappropriate sites and/or to an inappropriate extent can lead to host tissue damage.

To protect 'self' cells against by-stander lysis, host cells express a battery of regulatory proteins (complement inhibitors) which inhibit either assembly of the C3-cleaving enzymes or the formation of the MAC (11; Figure 2). C1 inhibitor (C1-INH), C4 binding protein (C4bp), factor H, factor I, S protein and clusterin are soluble inhibitors secreted and released in the fluid phase. Membrane-associated complement inhibitors include membrane cofactor protein (MCP, CD46), decay-accelerating factor (DAF, CD55), and CD59. Of note, an inhibitor of complement activation termed "complement receptor related protein y" (Crry) is expressed on rodent, but not human, cell membranes. Crry is broadly distributed and is a functional and structural analog of human DAF and MCP.

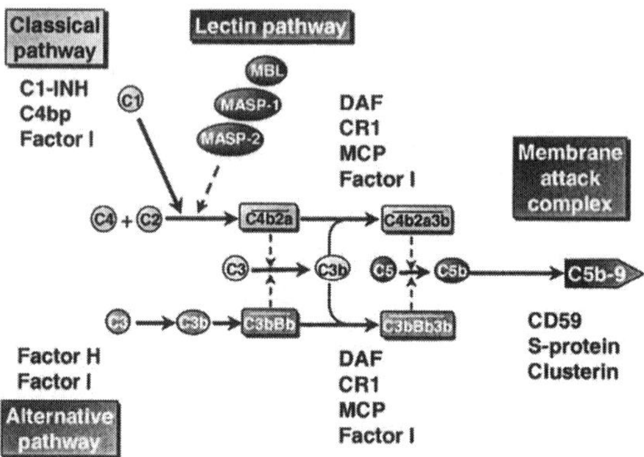


Figure 2. The different pathways of the complement cascade are tightly regulated by membrane-associated (DAF, MCP, CD59, CR1) or soluble (C1-INH, C4bp, Factor I, Factor H, S-protein, clusterin) inhibitors. C1-INH: C1 inhibitor; C4bp: C4 binding protein; DAF: decay-accelerating factor; MCP: membrane cofactor protein; CR1: complement receptor 1.

## 1.4    Sublytic effects of MAC

Assembly of the MAC is, by definition, involved in cytolytic activities. However, when the number of C5b-9 molecules is limited, cells are able to escape cell death by rapidly eliminating membrane-inserted C5b-9 by endocytosis and membrane shedding (12). When it does not cause death, the MAC is involved in cell stimulation (12, 13). Sublytic levels of C5b-9 have been shown to increase $Ca^{2+}$ influx, to activate phospholipases and protein kinases, and to induce the generation of arachidonic acid-derived inflammatory mediators (14-19). The MAC at sublytic level has also been shown to stimulate endothelial cells to express complement regulatory proteins and to protect against secondary damage (20). Taken together, these data indicate that the MAC at sublytic level could act as a stress signal to activate cells to express increased level of complement inhibitors and protect against subsequent damage.

## 1.5    Anaphylatoxin receptors

The most potent pro-inflammatory molecules generated upon complement activation are the anaphylatoxins C3a and C5a. C3a and C5a are small polypeptides (~9-10 kDa) that are highly pleiotropic in function and exert their effect at picomolar to nanomolar concentrations (21). C3a and C5a derive from the enzymatic cleavage of C3 and C5 respectively and are involved in the stimulation and chemotaxis of myeloid cells. The functional responses induced by C3a and C5a are mediated through their binding to specific receptors, C3a and C5a receptors (C3aR and C5aR) respectively (21). These receptors are members of the rhodopsin family of seven transmembrane G-protein coupled receptors (22-26).

C5a is an important chemoattractant molecule and stimulates cells to express increased levels of cytokines, chemokines, adhesion molecules and complement components (21). In contrast to the broad pro-inflammatory effects of C5a, the effects of C3a appear to be more selective and rather anti-inflammatory. C3a is chemoattractant but only for mast cells and eosinophils and not for either macrophages or microglia. Recent data indicate that C3a can decrease the production of pro-inflammatory cytokines by LPS-stimulated macrophages and, on the other hand, can induce the production of immunosuppressive cytokines such as IL-10 (21). Furthermore, mice with targeted disruption of C3aR exhibited an enhanced lethality when subjected to endotoxin shock, underlying the potential anti-inflammatory properties of anaphylatoxin C3a (27).

# 2.        COMPLEMENT AND BRAIN INFLAMMATION

## 2.1        Complement expression by brain cells

In mammals, the liver is the major source of most complement proteins with the exception of C1q, C7 and FD. The brain is separated from the plasma by the blood-brain barrier (BBB), formed by endothelial cells of microvessels, the smooth muscle cells  (also called pericytes), and astrocytes. The BBB acts as a molecular sieve and isolates the brain parenchyma from plasma poroteins, including complement components, as well as circulating immunocompetent cells such as lymphocytes, macrophages and natural killer cells. However, transudation of plasma proteins through a damaged BBB can contribute to the deposition of potentially cytotoxic and cytolytic complement components in the brain parenchyma.

In addition, we and others have proposed that brain cells can produce complement proteins to recognize and kill pathogens locally while preserving normal cells in the central nervous system (CNS) (28). Levi-Strauss and Mallat were the first to demonstrate that brain cells can synthesise complement components (29). They showed that cultured rodent astrocyte cell lines and primary cultures of mouse astrocytes produced C3 and FB and that the expression of complement was increased upon stimulation with lipopolysaccharide (LPS). In the last decade, this observation has been extended to include astrocytes, microglia, neurons and oligodendrocytes (30, 31). Cell lines and primary cultures of human origin were used to show that glial and neuronal cells were able to produce most complement proteins, particularly after stimulation with inflammatory cytokines (32-35). In addition, there is now considerable evidence that local expression of complement by resident cells can be dramatically increased in CNS disorders including Alzheimer's disease, brain infection or stroke (36-39). From these studies, it has been proposed that uncontrolled complement activation within the brain parenchyma may contribute to the propagation of the intracerebral inflammatory response and participate in the development of tissue damage.

## 2.2        Anaphylatoxin receptors in the brain

The expression of C3aR and C5aR was thought to be restricted largely to cells of the myeloid lineage such as macrophages, eosinophils, basophils and mast cells (40-42). However, in recent years, studies have reported widespread expression of these receptors throughout many tissues and cell types outside the immune system, supporting the view that anaphylatoxins might even exert effects unrelated to inflammation (43-45). If the presence of C3aR and C5aR on microglia was expected given the monocytic origin of these cells, their expression on astrocytes and neurons was not anticipated. The functional significance of the presence of anaphylatoxin receptors on neurons remains ill-defined. Upregulated expression of anaphylatoxin receptors by brain cells has been described in a large variety of CNS disorders, including Alzheimer's disease, brain infection, cerebral trauma and stroke (46-51).

Human astrocyte cell lines stimulated with C5a produce increased amounts of IL-6 while the levels of IL-1, TNF-$\alpha$ and TGF-$\beta$ remain unchanged (52). Unexpectedly, C5a has been shown to induce apoptosis of the human neuroblastoma cell line TGW (53, 54). Taken together, these data suggest that C5a release during CNS inflammation could contribute to the exacerbation of tissue damage. Interestingly, Heese and colleagues have shown that a human microglial cell line exposed to C3a expressed de novo nerve growth factor (NGF), a molecule involved in neuronal growth and survival (55). These data would support the view that C3a, in contrast to C5a, acts as an anti-inflammatory signal promoting the resolution of the inflammatory process. Unlike C5a, C3a failed to display any effect on apoptosis-mediated neuronal cell death; however, C3a was shown to protect neurons against N-methyl-D-aspartate toxicity, extending its role beyond immune functions (56).


# 3.     COMPLEMENT IN EAE


## 3.1     Inhibition of complement ameliorates disease

Early work using in vitro models of demyelination demonstrated that the demyelinating component found in sera from animals with experimental allergic encephalomyelitis (EAE), the animal model of multiple sclerosis (MS), was heat labile, a well known characteristic of the complement system (57). The concept that complement activation may contribute to the pathogenesis of demyelinating diseases then developed from key observations in the early 1970s using EAE models. Treatment with cobra venom factor (CVF), which functions as an unregulated C3 convertase, thus activating and depleting C3 and C5 via the alternative pathway, delayed the onset of EAE and reduced demyelination in rodents without affecting cellular infiltration (58-60). Soluble CR1 (sCR1), a potent inhibitor of early complement activation, was found to inhibit pathology in Lewis rats induced to develop a modified antibody-mediated demyelinating EAE (61). Histologically, there was less inflammation and almost no demyelination or complement deposition in treated rats which were also protected from the paralytic symptoms (Figure 3).

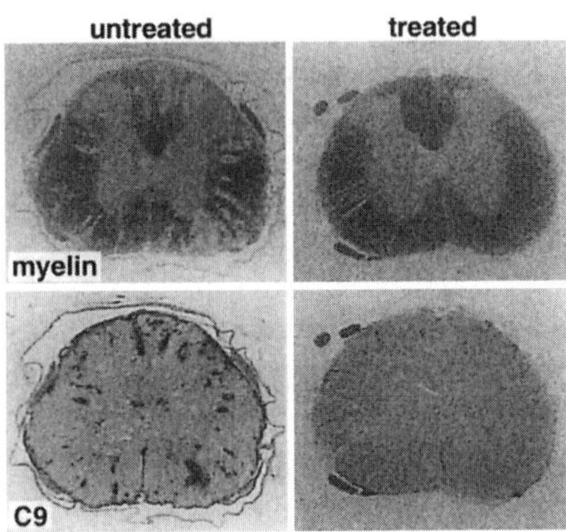


*Figure 3.* Protective effects of complement inhibition by soluble CR1 (sCR1) in antibody-mediated demylination EAE rats. Rats treated with sCR1 display reduced levels of demyelination and C9 deposition, a maker for the membrane attack complex.

Barnum and colleagues have engineered mice expressing a soluble form of the rodent complement regulator Crry under the control of the glial fibrillary acidic protein (GFAP) promotor to express specifically in the CNS. These mice were protected from MOG peptide-induced EAE, indicating that complement activation plays a critical role in the development of the pathology in this model (62). Studies carried out in C3 gene-targeted mice (C3-/-) have been contradictory. One study reported a protective effect of C3 deficiency. C3-/- mice induced to develop EAE by MOG peptide immunisation displayed reduced levels of disease, inflammation and demyelination, implicating complement in the pathology (63). In contrast, Calida and colleagues found no effect of C3 deficiency on the course of MOG-induced EAE in C3-/- mice (64). The reason for these contradictory results remains unclear but may relate to the genetic backgrounds of the mice studied.

## 3.2     Complement components are found in EAE lesions

Due to the lack of specific antibodies against rodent complement activation products, the distribution of complement proteins has been studied much less extensively in EAE than in MS. Using an acute antibody-mediated EAE model in the rat, Linington and colleagues demonstrated the presence of C9, an indicator of MAC deposition, within demyelinating lesions (65). More recently, using EAE in the common marmoset, a model that resembles a chronic MS pathology (66, 67), and EAE in the rhesus monkey, a model that resembles acute disseminated encephalomyelitis (68, 69), we have demonstrated that complement components including C1q and C3b neoepitope are present in foci of inflammatory infiltrate (70; figure 4).

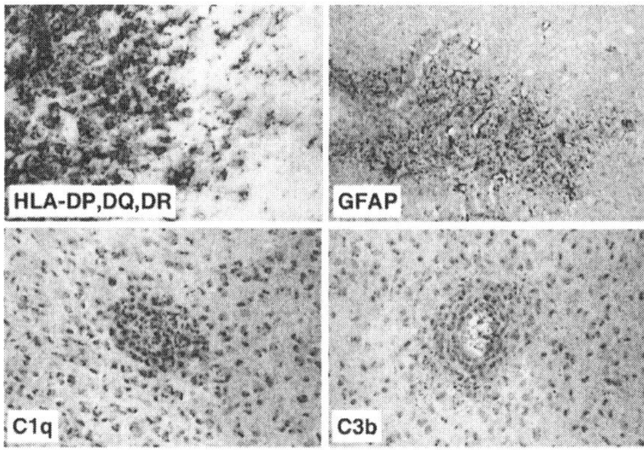


*Figure 4.* Glial reponses and complement deposition in marmoset EAE. A perivascular demyelinating lesion is diffusely infiltrated by macrophages stained for HLA-DP,DQ,DR. Note the presence of ramified HLA-DP,DQ,DR-positive microglial cells at the edge of the lesion. Reactive astrocytes (antibody to glial fibrillary acidic protein, GFAP) are found in areas surrounding the demyelinating plaque. Complement components C1q and C3b are present in pericocal inflammatory infiltrates.

In line with these findings in EAE models, early observations reported the presence of several activation products of the cytotoxic/cytolytic complement system in the cerebrospinal fluid of multiple sclerosis (71-73). More recently, a correlation of neurological disability and CSF concentration of MAC has been reported (74). Complement activation products have also been localized within the CNS parenchyma of MS patients, providing a strong body of evidence that complement-mediated cell death is prominent in areas of active demyelination in MS (75-77). In these studies, complement activation products such as C3b and C3d, as well as MAC, were found on myelin sheaths and altered myelin within the lesion. Finding deposition of opsonin C3b and C3d in close proximity to disrupted myelin strongly suggests that recognition and uptake, or attempted uptake, of myelin by microglia and/or macrophages is mediated by complement.

## 3.3     Expression of anaphylatoxin receptors is upregulated in EAE

C3aR and C5aR distribution pattern in EAE has only been documented recently. Expression of the gene encoding C5aR was found to be upregulated in the white matter of EAE rats (78). Cells expressing C5aR were identified as reactive astrocytes, microglia and infiltrating mononuclear cells. In addition, in the spinal cord of affected animals, a subset of motor neurons adjacent to inflammatory infiltrates were found to express C5a mRNA. Interestingly, mice in which the expression of IL-3 is driven by the GFAP promotor display clinical and histopathological features that are relevant for MS (79). In these mice, numerous foci of infitrating cells as well as demyelination were observed, along with a dramatic upregulated expression of C5aR mRNA by glial cells and neurons. In contrast, expression of C3aR on neurons was found to be unchanged during the course of EAE (80). As documented for C5aR, increased expression of C3aR was seen on infiltating mononuclear cells, microglia and astrocytes in both EAE and multiple sclerosis (50, 80).

It is still unclear whether anaphylatoxins drive inflammation in models of EAE. Recent data indicate that the C5aR is not an essential mediator in the induction and progression of EAE. In a study carried out by Reiman and colleagues, C5aR-deficient mice (C5aR-/-) and wild-type mice were induced to develop EAE using a myelin oligodendrocyte glycoprotein (MOG) peptide as the immunogen (81). They found that C5aR-/- mice were fully susceptible to MOG-induced EAE with no difference in disease onset or severity when compared to control mice.

## 3.4     Mechanisms involved in complement-mediated demyelination in EAE and MS

### 3.4.1     Is complement activated on myelin in an antibody-dependent manner?

Taken together, data obtained from animal models of MS and the human disease unequivocally implicate complement in the pathology of experimental and clinical demyelination. However, the mechanisms responsible for activating complement and myelinotoxic products of complement on myelin are still uncertain.

Complement activation is classically initiated via C1q interacting with bound immunoglobulin to produce opsonins, anaphylatoxins and MAC. Antibodies against a number of myelin antigens including MOG have been detected in the blood, cerebrospinal fluid, and affected tissue of MS patients (82, 83), and have been shown to be directly involved in the demyelination process in some models of EAE (83, 84). Interestingly, the demyelinating potential of a given anti-MOG monoclonal antibody in a model of antibody-mediated EAE has been correlated with its ability to fix complement (85). Furthermore, disrupted myelin within active lesions in acute MS is sometimes found to be immunoreactive for IgG and MAC, along with C3d (76, 77, 86). However, the significance of such observations remains uncertain since, in contrast to what observed in EAE, sera from MS patients were found not to be more opsonic for isolated myelin than sera from normal controls (87).

### 3.4.2    Is complement activated on myelin in an antibody-independent manner?

Physically disrupted myelin itself has been shown to activate spontaneously the complement system via the classical pathway (88, 89). A putative candidate for direct complement binding is MOG, a molecule expressed on the outermost lamellae of myelin and which contains a conserved motif similar to the C1q-binding site of IgG (90). Accordingly, it has been reported that the extracellular domain of MOG can fix C1q in a calcium- and dose-dependent manner (91). Another candidate is 2',3'-cyclic nucleotide 3' phosphodiesterase (CNPase), which has been shown to bind C3 (92). The first observation that oligodendrocytes are themselves extremely susceptible to complement lysis was made by Scolding and colleagues in 1989. They demonstrated that antibody-independent activation occured in vitro on rat oligodendrocyte cell membranes, while O-2A progenitors and astrocytes type I and II remained unaffected (93, 94). In this particular scenario, complement activation was taking place through the classical pathway although the activating molecules were not identified. Further studies have demonstrated that rat oligodendrocytes were lacking CD59, the major inhibitor of lysis, suggesting that their susceptibility to complement-mediated lysis was due in part to their poor ability to control complement activation (85). However, the situation in man appears to be different. Human oligodendrocytes and human oligodendroglioma cell lines were found to express high level of complement regulators, including CD59, and failed to activate spontaneously the complement system (95-97).

### 3.4.3    An essential role for the MAC in the demyelination process?

Activation of the complement cascade leads to a range of biological outcomes driven either by the generation of anaphylatoxins C3a and C5a, deposition of complement fragments on cell surfaces, or formation of the cytolytic MAC. The relative contributions of the different pathways in the pathogenesis of experimental and clinical demyelination is still not fully understood.

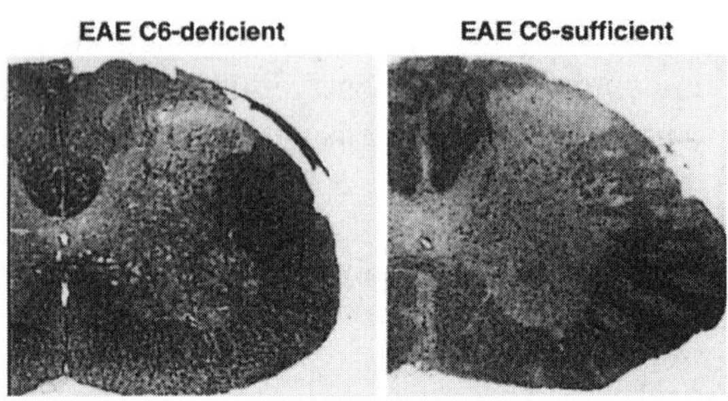


*Figure 5.* Histological examination of cervical cord sections from C6-deficient and C6-sufficient rats with antibody-mediated demyelination EAE using luxol fast blue/cresyl violet staining. Significant demyelination can be seen in the C6-sufficient rats with EAE, whereas no myelin loss can be seen in the C6-deficient rats.

Although it has been proposed that early components of the complement cascade mediate demyelination via the opsonization of myelin by C3b and C4b and subsequent interaction of these opsonins with complement receptors on microglia/macrophages (98), recent data suggest that the MAC is an essential effector of the demyelination process. Indeed, demyelination was dramatically reduced in C6-deficient rats induced to develop EAE by injection of antibodies to MOG (99; Figure 5). These rats have a normal complement system except that they are unable to form the MAC, suggesting that demyelination in this animal model requires activation of the entire complement cascade, including MAC deposition. This observation also demonstrates that in the absence of MAC deposition, opsonization of myelin and generation of anaphylatoxins C3a and C5a is not sufficient to induce demyelination. Interestingly, sCR1 markedly reduced both demyelination and inflammation in antibody-mediated EAE (61), whereas C6-deficient and C6-sufficient rats had similar degrees of inflammatory infiltrate (99). Because CR1 acts early in the complement cascade, this would suggest that complement anaphylatoxins C3a and C5a drive the inflammatory process in this model but are not responsible for myelin loss. Recent data obtained by inducing EAE in CD59-deficient mice further support the suggestion of a critical role of the MAC in demyelination. CD59 deficiency was shown to markedly exacerbate disease, demyelination and CNS inflammation (100).

### 3.4.4    A role for complement in axonal damage?

Axonal damage and neuronal injury in multiple sclerosis, first described by Charcot in 1868, have been emphasized recently as hallmarks additional to myelin loss, as evidenced by magnetic resonance imaging (101) and histopathological studies (102, 103). Axonal damage as a consequence of demyelination is also a common observation in rodent (99, 104, 105; Figure 6) and nonhuman primate (106) EAE. Although the exact mechanisms leading to axonal loss are still poorly understood, immunostaining for C9 as an indicator of MAC deposition was found to correlate with areas of axonal damage (99), suggesting a possible role for the cytolytic MAC in this pathological process. However, whether axonal loss is directly mediated by the MAC remains to be ascertained. It is possible that axons, exposed after demyelination, activate complement in an antibody-independent manner. Indeed, human neurons in vitro have the propensity to activate spontaneously the complement system and, consequently, are suceptible to complement-mediated lysis (107). The poor capacity of neurons to control complement activation has been attibuted to the fact that these cells express low levels of complement regulators (107, 108).

Figure 7 summarizes the putative roles of complement activation in inflammation mechanisms, myelin loss and axonal damage.

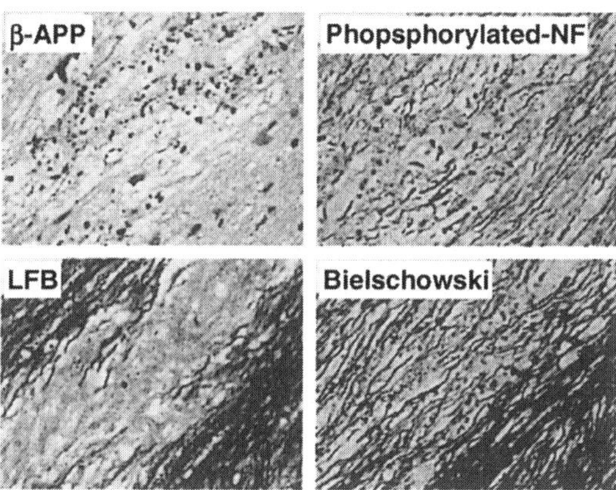


*Figure 6.* Axonal pathology in antibody-mediated demyelinating EAE rats. Serial transvere sections of spinal cord from EAE rats centered around an inflamed vessel stained for β-APP (β-amyloid precursor protein), phosphorylated-NF (neurofilament), LFB (luxol fast blue), and Bielschowski's silver stain showing axonal fragmentation in the demyelinated area.

### 3.4.5    Intrinsic neuroprotective mechanisms against complement activation

Although the neurotoxic effects of complement activation are well established in demyelinating diseases, the protective roles of complement regulatory proteins on neurons remain ill-characterized. We found that the complement classical pathway is activated on neurons located in close proximity to inflammatory infiltrates during the course of EAE in the marmoset (70). However, these neurons were found to be relatively spared, suggesting the existence of intrinsic immunoregulatory mechanisms involved in neuronal survival. In response to the insult, these neurons were found to express de novo DAF, a critical regulator of the complement system (70). This observation underlines the ability of neurons to mobilize defense mechanisms to dampen the potentially harmful effects of localised complement activation. In this vein, Stadelmann et al. recently reported that brain-derived neurotrophic factor and its receptor gp145trkB are expressed by neurons within or close to MS lesions, suggesting a role for neurotrophins in neuroprotective mechanisms in demyelinating diseases (109).

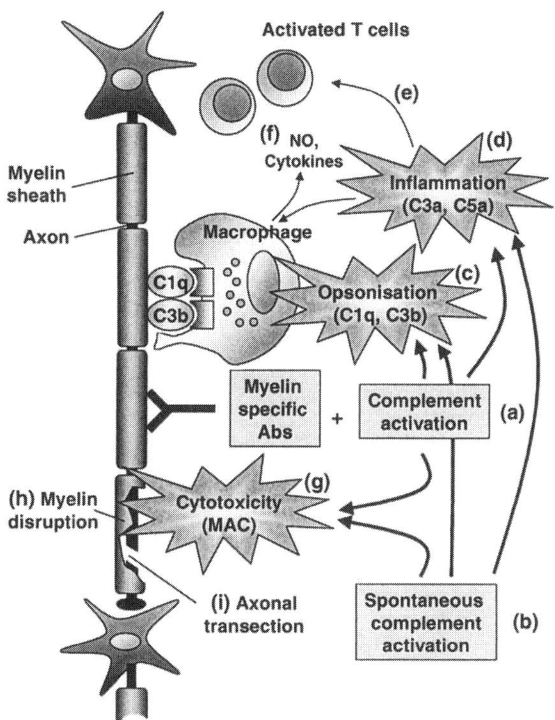


*Figure 7.* Proposed roles of complement activation in mechanisms leading to demyelination and axonal loss. Complement can be activated on myelin in an antibody-dependent manner (a) or spontaneously (b). Opsonins will decorate myelin sheaths (c), leading to macrophage binding through specific complement receptors. Complement activation can also induce the release of the pro-inflammatory anaphylatoxins (d), which in turn will stimulate inflammatory cells to produce toxic signals (f). Ultimately, the complement cascade culminates in the formation of the lytic membrane attack complex (MAC; (g)), leading to myelin disruption (h) as well as axonal damage (i).

### 3.4.6     Effects of sublytic assembled MAC on oligodendrocytes

Data indicate that the number of membrane-inserted MAC complexes determines whether complement activation induces or protects from cell death. As an example of their protective role, postmitotic oligodendrocytes exposed to sublytic MAC have been shown to enter cell cycle (110, 111). A novel gene named "response gene to complement" (RGC)-32, induced in rat oligodendrocytes and involved in cell cycle progression, has recently been cloned by differential display and implicated in this phenomenom (112). Moreover, exposure to sublytic MAC has been demonstrated to protect oligodendrocytes from apoptosis induced by serum deprivation (110) or exposure to tumor necrosis factor-$\alpha$ (113). The question of whether sublytic assembly of the MAC could also modulate the outcome of damage during the course of EAE has been adressed recently. C5-sufficient mice with EAE were found to develop less pronounced levels of inflammation when compared to C5-deficient mice, unable to form the MAC (114). In addition, active remyelination with restoration of tissue integrity was observed in chronic lesions only in C5-sufficient mice (114). Taken together, these data suggest a possible role for the sublytic assembly of MAC in the modulation of tissue inflammation and the initiation of tissue repair in demyelinating diseases.

## 4.  THERAPEUTIC IMPLICATIONS OF COMPLEMENT INVOLVEMENT IN DEMYELINATION

The case for an important role of complement in experimental and clinical demyelination made above has implications for therapeutic strategies to inhibit myelin loss. If drugs can be developed that significantly inhibit complement activation in the CNS then a substantial benefit might be anticipated in demyelinating disease. Complement therapeutics have undergone a mini-revolution over the last decade with the development of agents, based upon the naturally occurring regulators of complement, that efficiently inhibit complement when administered to man or animal (115). As noted earlier, one of these agents, sCR1, has been shown to markedly reduce myelin loss in rodent EAE models (61). None of the current crop of complement therapeutics is suited to treatment of a chronic disease. All are large molecules, active only when administered systemically; all have relatively short half lives and all cause global inhibition of complement with concomitant loss of the protective physiological roles of complement. These shortcomings effectively restrict use to acute disorders. Current research aims to make better complement therapeutics by creating small molecule inhibitors, by targeting complement inhibition to the relevant site in the body or by using a gene therapy approach to the inhibition of complement (116). Again, EAE provides the test-bed for these new approaches. Barnum and colleagues took the ingenious approach of using transgenesis to achieve CNS-specific expression of the powerful rodent regulator Crry. Soluble Crry generated in the CNS rendered the transgenic mouse resistant to disease and demyelination in the EAE model (62). Although transgenesis is not directly applicable to therapy in human disease, the profound effect of local inhibition of complement in the CNS in this model provides a powerful rationale for pursuing more relevant methods for targeting complement regulators to the CNS. The recognition that complement is a major drive to pathology in EAE models has thus opened a new direction to be explored in the search for effective treatments for human demyelinating diseases.

# REFERENCES


1. Frank, M. M., and L. F. Fries. 1991. The role of complement in inflammation and phagocytosis. *Immunol Today 12:322.*
2. Sunyer, J. O., I. K. Zarkadis, and J. D. Lambris. 1998. Complement diversity: a mechanism for generating immune diversity? *Immunol Today 19:519.*
3. Korb, L. C., and J. M. Ahearn. 1997. C1q binds directly and specifically to surface blebs of apoptotic human keratinocytes: complement deficiency and systemic lupus erythematosus revisited. *J Immunol 158:4525.*
4. Mevorach, D., J. O. Mascarenhas, D. Gershov, and K. B. Elkon. 1998. Complement-dependent clearance of apoptotic cells by human macrophages. *J Exp Med 188:2313.*
5. Taylor, P. R., A. Carugati, V. A. Fadok, H. T. Cook, M. Andrews, M. C. Carroll, J. S. Savill, P. M. Henson, M. Botto, and M. J. Walport. 2000. A hierarchical role for classical pathway complement proteins in the clearance of apoptotic cells in vivo. *J Exp Med 192:359.*
6. Botto, M., C. Dell'Agnola, A. E. Bygrave, E. M. Thompson, H. T. Cook, F. Petry, M. Loos, P. P. Pandolfi, and M. J. Walport. 1998. Homozygous C1q deficiency causes glomerulonephritis associated with multiple apoptotic bodies. *Nat Genet 19:56.*
7. McGreal, E. P., N. Ikewaki, H. Akatsu, B. P. Morgan, and P. Gasque. 2002. Human C1qRp is identical with CD93 and the mNI-11 antigen but does not bind C1q. *J Immunol 168:5222.*
8. Guan, E., S. L. Robinson, E. B. Goodman, and A. J. Tenner. 1994. Cell-surface protein identified on phagocytic cells modulates the C1q-mediated enhancement of phagocytosis. *J Immunol 152:4005.*
9. Ehlers, M. R. 2000. CR3: a general purpose adhesion-recognition receptor essential for innate immunity. *Microbes Infect 2:289.*
10. Krych-Goldberg, M., and J. P. Atkinson. 2001. Structure-function relationships of complement receptor type 1. *Immunol Rev 180:112.*
11. Morgan, B. P., and C. L. Harris. 1999. *Complement Regulatory Proteins.* Academic Press, San Diego.
12. Morgan, B. P. 1989. Complement membrane attack on nucleated cells: resistance, recovery and non-lethal effects. *Biochem J 264:1.*
13. Rus, H. G., F. I. Niculescu, and M. L. Shin. 2001. Role of the C5b-9 complement complex in cell cycle and apoptosis. *Immunol Rev 180:49.*
14. Wiedmer, T., B. Ando, and P. J. Sims. 1987. Complement C5b-9-stimulated platelet secretion is associated with a Ca2+-initiated activation of cellular protein kinases. *J Biol Chem 262:13674.*
15. Niculescu, F., H. Rus, S. Shin, T. Lang, and M. L. Shin. 1993. Generation of diacylglycerol and ceramide during homologous complement activation. *J Immunol 150:214.*

16. Imagawa, D. K., N. E. Osifchin, W. A. Paznekas, M. L. Shin, and M. M. Mayer. 1983. Consequences of cell membrane attack by complement: release of arachidonate and formation of inflammatory derivatives. *Proc Natl Acad Sci U S A 80:6647.*

17. Shirazi, Y., F. A. McMorris, and M. L. Shin. 1989. Arachidonic acid mobilization and phosphoinositide turnover by the terminal complement complex, C5b-9, in rat oligodendrocyte x C6 glioma cell hybrids. *J Immunol 142:4385.*

18. Shirazi, Y., D. K. Imagawa, and M. L. Shin. 1987. Release of leukotriene B4 from sublethally injured oligodendrocytes by terminal complement complexes. *J Neurochem 48:271.*

19. Carney, D. F., T. J. Lang, and M. L. Shin. 1990. Multiple signal messengers generated by terminal complement complexes and their role in terminal complement complex elimination. *J Immunol 145:623.*

20. Mason, J. C., H. Yarwood, K. Sugars, B. P. Morgan, K. A. Davies, and D. O. Haskard. 1999. Induction of decay-accelerating factor by cytokines or the membrane-attack complex protects vascular endothelial cells against complement deposition. *Blood 94:1673.*

21. Ember, J. A., M. A. Jagels, and T. E. Hugli. 1998. Characterization of complement anaphylatoxins and biological responses. In *The Human Complement System in Health and Disease.* J. E. Volanakis, and M. M. Frank, eds. Marcel Dekker, New York, NY, p. 241.

22. Ames, R. S., Y. Li, H. M. Sarau, P. Nuthulaganti, J. J. Foley, C. Ellis, Z. Zeng, K. Su, A. J. Jurewicz, R. P. Hertzberg, D. J. Bergsma, and C. Kumar. 1996. Molecular cloning and characterization of the human anaphylatoxin C3a receptor. *J Biol Chem 271:20231.*

23. Crass, T., U. Raffetseder, U. Martin, M. Grove, A. Klos, J. Kohl, and W. Bautsch. 1996. Expression cloning of the human C3a anaphylatoxin receptor (C3aR) from differentiated U-937 cells. *Eur J Immunol 26:1944.*

24. Gerard, N. P., and C. Gerard. 1991. The chemotactic receptor for human C5a anaphylatoxin. *Nature 349:614.*

25. Boulay, F., L. Mery, M. Tardif, L. Brouchon, and P. Vignais. 1991. Expression cloning of a receptor for C5a anaphylatoxin on differentiated HL-60 cells. *Biochemistry 30:2993.*

26. Roglic, A., E. R. Prossnitz, S. L. Cavanagh, Z. Pan, A. Zou, and R. D. Ye. 1996. cDNA cloning of a novel G protein-coupled receptor with a large extracellular loop structure. *Biochim Biophys Acta 1305:39.*

27. Kildsgaard, J., T. J. Hollmann, K. W. Matthews, K. Bian, F. Murad, and R. A. Wetsel. 2000. Cutting edge: targeted disruption of the C3a receptor gene demonstrates a novel protective anti-inflammatory role for C3a in endotoxin-shock. *J Immunol 165:5406.*

28. Morgan, B. P., and P. Gasque. 1996. Expression of complement in the brain: role in health and disease. *Immunol Today 17:461.*

29. Levi-Strauss, M., and M. Mallat. 1987. Primary cultures of murine astrocytes produce C3 and factor B, two components of the alternative pathway of complement activation. *J Immunol 139:2361.*

30. Gasque, P., Y. D. Dean, E. P. McGreal, J. van Beek, and B. P. Morgan. 2000. Complement components of the innate immune system in health and disease in the CNS. *Immunopharmacology 49:171.*

31. Barnum, S. R. 1995. Complement biosynthesis in the central nervous system. *Crit Rev Oral Biol Med 6:132.*

32. Gasque, P., M. Fontaine, and B. P. Morgan. 1995. Complement expression in human brain. Biosynthesis of terminal pathway components and regulators in human glial cells and cell lines. *J Immunol 154:4726.*

33. Gasque, P., A. Ischenko, J. Legoedec, C. Mauger, M. T. Schouft, and M. Fontaine. 1993. Expression of the complement classical pathway by human glioma in culture. A model for complement expression by nerve cells. *J Biol Chem 268:25068.*
34. Gasque, P., N. Julen, A. M. Ischenko, C. Picot, C. Mauger, C. Chauzy, J. Ripoche, and M. Fontaine. 1992. Expression of complement components of the alternative pathway by glioma cell lines. *J Immunol 149:1381.*
35. Thomas, A., P. Gasque, D. Vaudry, B. Gonzalez, and M. Fontaine. 2000. Expression of a complete and functional complement system by human neuronal cells in vitro. *Int Immunol 12:1015.*
36. van Beek, J., P. Chan, M. Bernaudin, E. Petit, E. T. MacKenzie, and M. Fontaine. 2000. Glial responses, clusterin, and complement in permanent focal cerebral ischemia in the mouse. *Glia 31:39.*
37. Shen, Y., R. Li, E. G. McGeer, and P. L. McGeer. 1997. Neuronal expression of mRNAs for complement proteins of the classical pathway in Alzheimer brain. *Brain Res 769:391.*
38. Fischer, B., H. Schmoll, P. Riederer, J. Bauer, D. Platt, and A. Popa-Wagner. 1995. Complement C1q and C3 mRNA expression in the frontal cortex of Alzheimer's patients. *J Mol Med 73:465.*
39. Dietzschold, B., W. Schwaeble, M. K. Schafer, D. C. Hooper, Y. M. Zehng, F. Petry, H. Sheng, T. Fink, M. Loos, H. Koprowski, and et al. 1995. Expression of C1q, a subcomponent of the rat complement system, is dramatically enhanced in brains of rats with either Borna disease or experimental allergic encephalomyelitis. *J Neurol Sci 130:11.*
40. Zwirner, J., O. Gotze, G. Begemann, A. Kapp, K. Kirchhoff, and T. Werfel. 1999. Evaluation of C3a receptor expression on human leucocytes by the use of novel monoclonal antibodies. *Immunology 97:166.*
41. Martin, U., D. Bock, L. Arseniev, M. A. Tornetta, R. S. Ames, W. Bautsch, J. Kohl, A. Ganser, and A. Klos. 1997. The human C3a receptor is expressed on neutrophils and monocytes, but not on B or T lymphocytes. *J Exp Med 186:199.*
42. Daffern, P. J., P. H. Pfeifer, J. A. Ember, and T. E. Hugli. 1995. C3a is a chemotaxin for human eosinophils but not for neutrophils. I. C3a stimulation of neutrophils is secondary to eosinophil activation. *J Exp Med 181:2119.*
43. Ember, J. A., and T. E. Hugli. 1997. Complement factors and their receptors. *Immunopharmacology 38:3.*
44. Zwirner, J., A. Fayyazi, and O. Gotze. 1999. Expression of the anaphylatoxin C5a receptor in non-myeloid cells. *Mol Immunol 36:877.*
45. Wetsel, R. A. 1995. Structure, function and cellular expression of complement anaphylatoxin receptors. *Curr Opin Immunol 7:48.*
46. van Beek, J., M. Bernaudin, E. Petit, P. Gasque, A. Nouvelot, E. T. MacKenzie, and M. Fontaine. 2000. Expression of receptors for complement anaphylatoxins C3a and C5a following permanent focal cerebral ischemia in the mouse. *Exp Neurol 161:373.*
47. Barnum, S. R., R. S. Ames, P. R. Maycox, S. J. Hadingham, J. Meakin, D. Harrison, and A. A. Parsons. 2002. Expression of the complement C3a and C5a receptors after permanent focal ischemia: An alternative interpretation. *Glia 38:169.*
48. Stahel, P. F., K. Kariya, E. Shohami, S. R. Barnum, H. Eugster, O. Trentz, T. Kossmann, and M. C. Morganti-Kossmann. 2000. Intracerebral complement C5a receptor (CD88) expression is regulated by TNF and lymphotoxin-alpha following closed head injury in mice. *J Neuroimmunol 109:164.*
49. Stahel, P. F., K. Frei, H. P. Eugster, A. Fontana, K. M. Hummel, R. A. Wetsel, R. S. Ames, and S. R. Barnum. 1997. TNF-alpha-mediated expression of the receptor for anaphylatoxin C5a on neurons in experimental Listeria meningoencephalitis. *J Immunol 159:861.*

50. Gasque, P., S. K. Singhrao, J. W. Neal, P. Wang, S. Sayah, M. Fontaine, and B. P. Morgan. 1998. The receptor for complement anaphylatoxin C3a is expressed by myeloid cells and nonmyeloid cells in inflamed human central nervous system: analysis in multiple sclerosis and bacterial meningitis. *J Immunol 160:3543*.
51. Gasque, P., S. K. Singhrao, J. W. Neal, O. Gotze, and B. P. Morgan. 1997. Expression of the receptor for complement C5a (CD88) is up-regulated on reactive astrocytes, microglia, and endothelial cells in the inflamed human central nervous system. *Am J Pathol 150:31*.
52. Sayah, S., A. M. Ischenko, A. Zhakhov, A. S. Bonnard, and M. Fontaine. 1999. Expression of cytokines by human astrocytomas following stimulation by C3a and C5a anaphylatoxins: specific increase in interleukin-6 mRNA expression. *J Neurochem 72:2426*.
53. Farkas, I., L. Baranyi, Z. S. Liposits, T. Yamamoto, and H. Okada. 1998. Complement C5a anaphylatoxin fragment causes apoptosis in TGW neuroblastoma cells. *Neuroscience 86:903*.
54. Farkas, I., L. Baranyi, M. Takahashi, A. Fukuda, Z. Liposits, T. Yamamoto, and H. Okada. 1998. A neuronal C5a receptor and an associated apoptotic signal transduction pathway. *J Physiol 507:679*.
55. Heese, K., C. Hock, and U. Otten. 1998. Inflammatory signals induce neurotrophin expression in human microglial cells. *J Neurochem 70:699*.
56. van Beek, J., O. Nicole, C. Ali, A. Ischenko, E. T. MacKenzie, A. Buisson, and M. Fontaine. 2001. Complement anaphylatoxin C3a is selectively protective against NMDA-induced neuronal cell death. *Neuroreport 12:289*.
57. Appel, S. H., and M. B. Bornstein. 1964. The application of tissue culture to the study of experimental allergic encephalomyelitis. II. Serum factors reponsible for demyelination. *J Exp Med 119:119*.
58. Levine, S., C. G. Cochrane, C. B. Carpenter, and P. O. Behan. 1971. Allergic encephalomyelitis: effect of complement depletion with cobra venom. *Proc Soc Exp Biol Med 138:285*.
59. Morariu, M. A., and A. P. Dalmasso. 1978. Experimental allergic encephalomyelitis in cobra venom factor-treated and C4-deficient guinea pigs. *Ann Neurol 4:427*.
60. Linington, C., B. P. Morgan, N. J. Scolding, P. Wilkins, S. Piddlesden, and D. A. Compston. 1989. The role of complement in the pathogenesis of experimental allergic encephalomyelitis. *Brain 112:895*.
61. Piddlesden, S. J., M. K. Storch, M. Hibbs, A. M. Freeman, H. Lassmann, and B. P. Morgan. 1994. Soluble recombinant complement receptor 1 inhibits inflammation and demyelination in antibody-mediated demyelinating experimental allergic encephalomyelitis. *J Immunol 152:5477*.
62. Davoust, N., S. Nataf, R. Reiman, M. V. Holers, I. L. Campbell, and S. R. Barnum. 1999. Central nervous system-targeted expression of the complement inhibitor sCrry prevents experimental allergic encephalomyelitis. *J Immunol 163:6551*.
63. Nataf, S., S. L. Carroll, R. A. Wetsel, A. J. Szalai, and S. R. Barnum. 2000. Attenuation of experimental autoimmune demyelination in complement-deficient mice. *J Immunol 165:5867*.
64. Calida, D. M., C. Constantinescu, E. Purev, G. X. Zhang, E. S. Ventura, E. Lavi, and A. Rostami. 2001. Cutting edge: C3, a key component of complement activation, is not required for the development of myelin oligodendrocyte glycoprotein peptide-induced experimental autoimmune encephalomyelitis in mice. *J Immunol 166:723*.
65. Linington, C., H. Lassmann, B. P. Morgan, and D. A. Compston. 1989. Immunohistochemical localisation of terminal complement component C9 in experimental allergic encephalomyelitis. *Acta Neuropathol (Berl) 79:78*.

66. 't Hart, B. A., J. Bauer, H. J. Muller, B. Melchers, K. Nicolay, H. Brok, R. E. Bontrop, H. Lassmann, and L. Massacesi. 1998. Histopathological characterization of magnetic resonance imaging-detectable brain white matter lesions in a primate model of multiple sclerosis: a correlative study in the experimental autoimmune encephalomyelitis model in common marmosets (Callithrix jacchus). *Am J Pathol 153:649.*

67. 't Hart, B. A., M. van Meurs, H. P. Brok, L. Massacesi, J. Bauer, L. Boon, R. E. Bontrop, and J. D. Laman. 2000. A new primate model for multiple sclerosis in the common marmoset. *Immunol Today 21:290.*

68. Brok, H. P., J. Bauer, M. Jonker, E. Blezer, S. Amor, R. E. Bontrop, J. D. Laman, and B. A. t Hart. 2001. Non-human primate models of multiple sclerosis. *Immunol Rev 183:173.*

69. Kerlero de Rosbo, N., H. P. Brok, J. Bauer, J. F. Kaye, B. A. t Hart, and A. Ben-Nun. 2000. Rhesus monkeys are highly susceptible to experimental autoimmune encephalomyelitis induced by myelin oligodendrocyte glycoprotein: characterisation of immunodominant T- and B-cell epitopes. *J Neuroimmunol 110:83.*

70. van Beek, J., M. van Meurs, B. A. 't Hart, H. Brok, J. W. Neal, B. P. Morgan, J. D. Laman, and P. Gasque. 2002. Decay-accelerating factor (CD55) is abundantly expressed by reactive neurons in response to chronic but not acute inflammatory insults associated with strong complement activation. *Int Immunol 2:1364.*

71. Link, H. 1972. Complement factors in multiple sclerosis. *Acta Neurol Scand 48:521.*

72. Trouillas, P., J. Moindrot, H. Betuel, C. Quincy, G. Aimard, and M. Devic. 1976. Multiple sclerosis with reduced and with normal levels of complement in the blood. Clinical and genetic correlation. *Rev Neurol (Paris) 132:684.*

73. Morgan, B. P., A. K. Campbell, and D. A. Compston. 1984. Terminal component of complement (C9) in cerebrospinal fluid of patients with multiple sclerosis. *Lancet 2:251.*

74. Sellebjerg, F., I. Jaliashvili, M. Christiansen, and P. Garred. 1998. Intrathecal activation of the complement system and disability in multiple sclerosis. *J Neurol Sci 157:168.*

75. Compston, D. A., B. P. Morgan, A. K. Campbell, P. Wilkins, G. Cole, N. D. Thomas, and B. Jasani. 1989. Immunocytochemical localization of the terminal complement complex in multiple sclerosis. *Neuropathol Appl Neurobiol 15:307.*

76. Storch, M. K., S. Piddlesden, M. Haltia, M. Iivanainen, P. Morgan, and H. Lassmann. 1998. Multiple sclerosis: in situ evidence for antibody- and complement-mediated demyelination. *Ann Neurol 43:465.*

77. Lucchinetti, C., W. Bruck, J. Parisi, B. Scheithauer, M. Rodriguez, and H. Lassmann. 2000. Heterogeneity of multiple sclerosis lesions: implications for the pathogenesis of demyelination. *Ann Neurol 47:707.*

78. Nataf, S., N. Davoust, and S. R. Barnum. 1998. Kinetics of anaphylatoxin C5a receptor expression during experimental allergic encephalomyelitis. *J Neuroimmunol 91:147.*

79. Paradisis, P. M., I. L. Campbell, and S. R. Barnum. 1998. Elevated complement C5a receptor expression on neurons and glia in astrocyte-targeted interleukin-3 transgenic mice. *Glia 24:338.*

80. Davoust, N., J. Jones, P. F. Stahel, R. S. Ames, and S. R. Barnum. 1999. Receptor for the C3a anaphylatoxin is expressed by neurons and glial cells. *Glia 26:201.*

81. Reiman, R., C. Gerard, I. L. Campbell, and S. R. Barnum. 2002. Disruption of the C5a receptor gene fails to protect against experimental allergic encephalomyelitis. *Eur J Immunol 32:1157.*

82. Xiao, B. G., C. Linington, and H. Link. 1991. Antibodies to myelin-oligodendrocyte glycoprotein in cerebrospinal fluid from patients with multiple sclerosis and controls. *J Neuroimmunol 31:91.*

83. Raine, C. S., B. Cannella, S. L. Hauser, and C. P. Genain. 1999. Demyelination in primate autoimmune encephalomyelitis and acute multiple sclerosis lesions: a case for antigen-specific antibody mediation. *Ann Neurol 46:144.*

84. Linington, C., M. Bradl, H. Lassmann, C. Brunner, and K. Vass. 1988. Augmentation of demyelination in rat acute allergic encephalomyelitis by circulating mouse monoclonal antibodies directed against a myelin/oligodendrocyte glycoprotein. *Am J Pathol 130:443.*

85. Piddlesden, S. J., H. Lassmann, F. Zimprich, B. P. Morgan, and C. Linington. 1993. The demyelinating potential of antibodies to myelin oligodendrocyte glycoprotein is related to their ability to fix complement. *Am J Pathol 143:555.*

86. Gay, F. W., T. J. Drye, G. W. Dick, and M. M. Esiri. 1997. The application of multifactorial cluster analysis in the staging of plaques in early multiple sclerosis. Identification and characterization of the primary demyelinating lesion. *Brain 120:1461.*

87. Goldenberg, P. Z., R. A. Troiano, E. E. Kwon, and J. W. Prineas. 1990. Sera from MS patients and normal controls opsonize myelin. *Neurosci Lett 109:353.*

88. Vanguri, P., C. L. Koski, B. Silverman, and M. L. Shin. 1982. Complement activation by isolated myelin: activation of the classical pathway in the absence of myelin-specific antibodies. *Proc Natl Acad Sci U S A 79:3290.*

89. Silverman, B. A., D. F. Carney, C. A. Johnston, P. Vanguri, and M. L. Shin. 1984. Isolation of membrane attack complex of complement from myelin membranes treated with serum complement. *J Neurochem 42:1024.*

90. Hilton, A. A., A. J. Slavin, D. J. Hilton, and C. C. Bernard. 1995. Characterization of cDNA and genomic clones encoding human myelin oligodendrocyte glycoprotein. *J Neurochem 65:309.*

91. Johns, T. G., and C. C. Bernard. 1997. Binding of complement component Clq to myelin oligodendrocyte glycoprotein: a novel mechanism for regulating CNS inflammation. *Mol Immunol 34:33.*

92. Walsh, M. J., and J. M. Murray. 1998. Dual implication of 2',3'-cyclic nucleotide 3' phosphodiesterase as major autoantigen and C3 complement-binding protein in the pathogenesis of multiple sclerosis. *J Clin Invest 101:1923.*

93. Scolding, N., C. Linington, and A. Compston. 1989. Immune mechanisms in the pathogenesis of demyelinating diseases. *Autoimmunity 4:131.*

94. Scolding, N. J., B. P. Morgan, W. A. Houston, C. Linington, A. K. Campbell, and D. A. Compston. 1989. Vesicular removal by oligodendrocytes of membrane attack complexes formed by activated complement. *Nature 339:620.*

95. Zajicek, J., M. Wing, J. Skepper, and A. Compston. 1995. Human oligodendrocytes are not sensitive to complement. A study of CD59 expression in the human central nervous system. *Lab Invest 73:128.*

96. Gasque, P., and B. P. Morgan. 1996. Complement regulatory protein expression by a human oligodendrocyte cell line: cytokine regulation and comparison with astrocytes. *Immunology 89:338.*

97. Scolding, N. J., B. P. Morgan, and D. A. Compston. 1998. The expression of complement regulatory proteins by adult human oligodendrocytes. *J Neuroimmunol 84:69.*

98. Smith, M. E. 1999. Phagocytosis of myelin in demyelinative disease: a review. *Neurochem Res 24:261.*

99. Mead, R. J., S. K. Singhrao, J. W. Neal, H. Lassmann, and B. P. Morgan. 2002. The membrane attack complex of complement causes severe demyelination associated with acute axonal injury. *J Immunol 168:458.*

100. Mead, R. J., J. W. Neal, and B. P. Morgan. 2002. CD59-deficiency exacerbates clinical disease and central nervous system inflammation in murine experimental allergic encephalomyelitis. *Int Immunol 2:1248.*

101. Kidd, D., F. Barkhof, R. McConnell, P. R. Algra, I. V. Allen, and T. Revesz. 1999. Cortical lesions in multiple sclerosis. *Brain 122:17.*

102. Bjartmar, C., and B. D. Trapp. 2001. Axonal and neuronal degeneration in multiple sclerosis: mechanisms and functional consequences. *Curr Opin Neurol 14:271.*

103. Ferguson, B., M. K. Matyszak, M. M. Esiri, and V. H. Perry. 1997. Axonal damage in acute multiple sclerosis lesions. *Brain 120:393.*

104. White, S. R., G. K. Samathanam, R. M. Bowker, and M. W. Wessendorf. 1990. Damage to bulbospinal serotonin-, tyrosine hydroxylase-, and TRH-containing axons occurs early in the development of experimental allergic encephalomyelitis in rats. *J Neurosci Res 27:89.*

105. Kornek, B., M. K. Storch, R. Weissert, E. Wallstroem, A. Stefferl, T. Olsson, C. Linington, M. Schmidbauer, and H. Lassmann. 2000. Multiple sclerosis and chronic autoimmune encephalomyelitis: a comparative quantitative study of axonal injury in active, inactive, and remyelinated lesions. *Am J Pathol 157:267.*

106. Mancardi, G., B. Hart, L. Roccatagliata, H. Brok, D. Giunti, R. Bontrop, L. Massacesi, E. Capello, and A. Uccelli. 2001. Demyelination and axonal damage in a non-human primate model of multiple sclerosis. *J Neurol Sci 184:41.*

107. Singhrao, S. K., J. W. Neal, N. K. Rushmere, B. P. Morgan, and P. Gasque. 2000. Spontaneous classical pathway activation and deficiency of membrane regulators render human neurons susceptible to complement lysis. *Am J Pathol 157:905.*

108. Singhrao, S. K., J. W. Neal, N. K. Rushmere, B. P. Morgan, and P. Gasque. 1999. Differential expression of individual complement regulators in the brain and choroid plexus. *Lab Invest 79:1247.*

109. Stadelmann, C., M. Kerschensteiner, T. Misgeld, W. Bruck, R. Hohlfeld, and H. Lassmann. 2002. BDNF and gp145trkB in multiple sclerosis brain lesions: neuroprotective interactions between immune and neuronal cells? *Brain 125:75.*

110. Rus, H. G., F. Niculescu, and M. L. Shin. 1996. Sublytic complement attack induces cell cycle in oligodendrocytes. *J Immunol 156:4892.*

111. Rus, H., F. Niculescu, T. Badea, and M. L. Shin. 1997. Terminal complement complexes induce cell cycle entry in oligodendrocytes through mitogen activated protein kinase pathway. *Immunopharmacology 38:177.*

112. Badea, T. C., F. I. Niculescu, L. Soane, M. L. Shin, and H. Rus. 1998. Molecular cloning and characterization of RGC-32, a novel gene induced by complement activation in oligodendrocytes. *J Biol Chem 273:26977.*

113. Soane, L., H. Rus, F. Niculescu, and M. L. Shin. 1999. Inhibition of oligodendrocyte apoptosis by sublytic C5b-9 is associated with enhanced synthesis of bcl-2 and mediated by inhibition of caspase-3 activation. *J Immunol 163:6132.*

114. Rus, H., S. Weerth, L. Soane, F. Niculescu, V. Rus, C. S. Raine, and M. L. Shin. 2002. The effect of terminal complement complex on apoptosis gene expression in experimental autoimmune encephalomyelitis. *Int Immunol 2:1248.*

115. Morgan, B. P. 1996. Intervention in the complement system: a therapeutic strategy in inflammation. *Biochem Soc Trans 24:224.*

116. McGrath, Y., G. W. Wilkinson, O. B. Spiller, and B. P. Morgan. 1999. Development of adenovirus vectors encoding rat complement regulators for use in therapy in rodent models of inflammatory diseases. *J Immunol 163:6834.*

Chapter A13

# B CELLS AND ANTIBODIES IN EXPERIMENTAL AUTOIMMUNE ENCEPHALOMYELITIS


Xia Lin and Cris S Constantinescu
*Division of Clinical Neurology, University of Nottingham, UK*



**Abstract:** EAE is a T cell mediated disease, but increasing evidence shows an important role for B cells and antibodies in its pathogenesis. This also reflects a role for humoral factors in the pathogenesis of multiple sclerosis. The main functions of B cells and antibodies in EAE are discussed.

**Key words:** EAE, MS, B cells, antibody, Immunoglobulin, antigen presentation, demyelination


B-lymphocytes are the major types of immune cells that mediate antigen-specific humoral immune responses. In the past two decades, the research based on experimental autoimmune encephalomyelitis (EAE) and multiple sclerosis MS has solidified the concept that T-helper cell-driven immune processes are critical for the pathogenesis of MS and MS has been regarded as a T-cell-mediated disease. However, an early support for a role of humoral immunity in the pathogenesis of MS followed Kabat's report on the elevated levels of immunoglobulins in the cerebrospinal fluid (CSF) of patients with this disease[1]. Oligoclonal bands (OCB) (oligoclonal IgG) is present in the CSF of the majority of MS patients and are produced by CNS-resident B cells. Underscoring the utility of the EAE model, both early[2,3] and very recent studies[4] have also been shown in several rodent EAE models and are also present in primate models[5]. In recent years, evidence derived from studies in humans, and in animal models such as knock-out or transgenic mice and autoimmune and virus-induced variants of encephalomyelitis and EAE, has led to a renewed interest in a role for B cells and the humoral immune system in the pathogenesis of MS. Studies show that antibody-dependent effector mechanisms are also involved in the immunopathogenesis of demyelination[6,7]. It emerges that mutually interacting cellular and humoral immune components may contribute to immune-mediated demyelination in human disorders. In this chapter, a general role of B cells and auto-reactive antibodies in EAE pathogenesis is reviewed.

**B cells act as antigen-presenting cells for CD4+ T-cells**
In most immune reactions, B and T cells closely co-operate. The important contribution of B cell antigen presentation to T cell activation has been demonstrated in studies in which the absence or depletion of B cells decreases antigen-induced T cell activation[8-10]. It has also been shown in experiments in which B cell presentation of antigen to primed or already activated T cells further stimulates the T cells[11]. In brief, naïve B cells develop into memory B cells with the help of CD4[+] T-helper lymphocytes and follicular dendritic cells after an antigen challenge in the secondary lymphoid organs. Memory B cells then can take up antigen via their specific surface receptors (BCR) consisting of the membrane-bound immunoglobulin (mIg) molecule that mediates the binding to antigen, and the mIg associated Ig-α/Ig-β heterodimer [12]that functions as signaling subunits of the BCR. This antigen-induced signaling is modulated by the activity of B-cell co-receptor molecules such as CD19[13]or CD22[14]. The captured antigen is then fragmented by endosomal proteases and presented with MHC-class II molecules to T-cell receptor on corresponding helper T cells. Contact between the antigen-presenting B cell and a specifically primed T cell leads to T-cell activation via engagement of the T-cell receptor and a number of accessory molecules including T-cell costimulatory ligands, such as B7.1 and B7.2 on the antigen-presenting B-cell, and CD28 on the T cell[15,16]. Activation of T cells by B cells is accompanied by secretion of lymphokines and transient expression of new cell surface molecules, particularly the ligand for CD40[17-19]. In this step, B cells are induced to proliferate after contact with the activated T-cell surface[20]. Interaction between CD40 and its ligand on the T-cell surface is essential for this cell-contact-mediated event[18]. Finally, exposure of cell-contact-activated B cells to T-cell-derived lymphokines leads to their differentiation into antibody-secreting cells[20]. This reciprocal activation of T and B cells represents a key immunological event that results in cytokine release, immunoglobulin isotype class swith, somatic hypermutations in the complementarity determining regions, and affinity maturation[21].

**Antibodies**
The key feature of the humoral immune response is the production of antibodies. Antibodies are immunoglobulins of several classes which play specific roles in the immune response. An important immunoglobulin class

for the pathogenesis of EAE and MS is IgG, but IgM and IgA have also been implicated. IgE is involved in reactions to allergens.

Immunoglobulins consist of heavy and light chains containing the variable (V), diversity (D) and joining (J) regions. The V segments on the heavy and light chains ($V_H$ and $V_L$, respectively) contain 3 regions that bind the antigen called the complementarity-determining regions (CDR). Each B cells has a specific pair of heavy and light V region, ensuring that each B cell have a single antibody specificity.

## Cytokines

B cells can be a source of cytokines, most notably IL-6, IL-10 and IL-12. These cytokines and others potentially produced by B cells play an important role in EAE pathogenesis (see chapter on cytokines).

## Self tolerance and mechanism of B-cell autoimmunity

The ability to distinguish between self and foreign antigens is a central feature of immune recognition, allowing immunity to be acquired against foreign organisms while avoiding destructive auto-immunity. Many of the negative selection steps in the generation of B-cell repertoire occur early in the ontogeny of these cells in the bone marrow[22]. However, several additional mechanisms of self-tolerance in the periphery have evolved to prevent B cells from reacting against self-antigen and from producing antibodies that contribute to tissue damage in immune-mediated disorders[23]. These are largely attributable to the degree of Ig receptor engagement and the availability of T cell help[15,21,24,25]. Under certain circumstances, some B cells may escape established tolerance and start to produce autoantibodies that contribute to tissue damage in immune-mediated disorders. The mechanisms that operate in generating B-cell autoimmunity in humans are still obscure. From studies on transgenic and knock-out mice and in vitro investigations, dysregulation at different molecular levels of B-cell function appears to result in B-cell autoimmunity.

## Roles of B cells in EAE pathogenesis

There is increasing evidence that antibodies are involved in the development of the demyelinating plaques[26,27]. It is generally accepted that B cells and antobodies, although incapable of initiating EAE, are likely important contributors to the disease process. Possible roles for B cells, plasma cells and antibodies in EAE include: 1). Antigen uptake, processing and presentation that involves B cells acting as APC; 2). Antibody and complement-mediated demyelination 3). Co-stimulation of T-cells; 4). Influence on the cytokine production balance; 5). Potential beneficial immunoregulatory effects of autoantibodies.

***B lymphocytes as antigen-presenting cells and as source of antibodies in EAE***

Studies both in vitro[28,29] and in vivo[8] show that previously activated T cells can be stimulated to proliferate and differentiate using B cells as APC. An in vivo study has demonstrated that the uptake of antigen through Ig receptor results in B cells having an activated phenotype, i.e. up-regulation of surface B7-2 and MHC class II, and that they are competent APC for naïve CD4[+] T cells. Subsequently, the presentation of a foreign soluble protein by APC in the absence of appropriate antigen-presenting B cells was limited, suggesting a major contribution of B cells for T cell priming[8,10]. The role of B cells as APC in EAE is suggested in B cell-depleted mice. Studies have shown that treatment of animals with anti-rat IgM antibody to deplete B cells induces resistance to EAE, suggesting B cells and/or their products were necessary for EAE in Lewis rats[30,31] and in mice[32]. Further studies show that B cells are critical to the induction of EAE, B cell deficient mice being resistant to EAE induced by immunisation with the recombinant form of MOG[33,34]. Transfer of B cells restores the ability of recombinant MOG to induce clinical EAE in B cell deficient mice[34]. A role for B cells in enhancing the priming of autoreactive T cells has also been proposed in EAE induced with myelin basic protein[32]. However, some studies suggest that the importance of B cells in the pathogenesis of MOG-induced EAE is highly dependent on the nature of the MOG, EAE could be induced in the absence of B cells after immunisation with short peptide, but not with the whole protein[33-35]. Similarly, Dittel and colleagues failed to find a requirement for B cells in MBP-peptide induced EAE[36]. Thus, the role of B cells in MOG-induced EAE depends on the antigen used for sensitisation. B cells are critical for human and rat MOG protein-induced EAE but are not critical for EAE induced by a short encephalitogenic MOG peptide, where other mechanisms may be operative[33,37].


***Autoantibody-mediated demyelination in EAE***

The identification of the myelin oligodendrocyte glycoprotein (MOG) as a major target for auto-antibody-mediated demyelination in EAE[38] received interest in the role of antibody in the pathogenesis of MS. It is now apparent that antibody-dependent effector mechanisms are involved in the immunopathgenesis of demyelination[6,7,39]. MOG is a minor component of CNS myelin that is exclusively expressed at the outer surface of the myelin sheath and oligodendrocyte plasma membrane[40]. Sera from animals with chronic EAE injected into the CSF of normal animal induces demyelination[41]. In some animal models that normally show little or no demyelination, such as guinea pigs immunized with MBP without lipids demyelination can be observed following immunisation with both MBP and GALC[42]. Administration of antibody to MOG to rats with EAE converts the disease to a more severe clinical form and produces more extensive demyelination[43]. Further, study of pathology of MS that examined a large number of samples obtained by biopsy or at autopsy showed that the inflammation was associated with antibody binding to myelin, suggesting an antibody-mediated process producing damage to myelin[44]. These findings are further extended by other reports showing MOG-specific antibodies present in actively demyelinating lesions in MOG-induced marmoset EAE as well as in MS, supporting the role of antibody in myelin damage[7,45]. In humans, elevated levels of anti-MOG antibodies and increased frequencies of anti-MOG antibody IgG-secreting B cells have been demonstrated in the CSF of MS patients[46,47] and in the MS lesions[7,45]. Extensive evidence of the role of autoantibodies in the pathological process of EAE and MS comes from in vitro studies showing demyelination and inhibition of myelination with either sera or immunoglobulin fractions from patients with MS or from animals with EAE in organotypic cultures[48].

### B cells express B7-1 and B7-2 and co-stimulate T-cells

The activation of naïve T cells requires not only the recognition of a peptide-MHC complex, but also the interaction of co-stimulatory molecules expressed by antigen presenting B cells. In humans, the number of circulating B cells that express B7.1 are significantly increased during MS exacerbations and reduced with interferon-$\beta$ treatment[49]. Recent reports have also shown that EAE is markedly reduced in B7.1 and B7.2 deficient mice following adoptive transfer of wild type MOG short peptide activated T cells, suggesting that B7 co-stimulatory molecules, possibly including those expressed by B cells, have a critical role in the effector phase of EAE[50]. Taken together, these studies indicate that B7 ligands on B cells are part of the costimulatory interactions that play a critical role in the initiation of EAE.

### Influence of B cells on cytokine balance

Cytokines play a key role in the development and remission of EAE. The inflammatory lesion in the CNS requires a type 1 autoreactive response, producing the pro-inflammatory cytokines interferon-$\gamma$ and tumour necrosis factor-$\alpha$.[51]Recovery is associated with production of the type 2 cytokines interleukin 4 (IL-4) and IL-10[52,53]. B cells can regulate immune responses *in vitro* through the production of IL-10[54]. Studies have shown that B cells from normal B6 mice recovered from EAE produced IL-10 when stimulated with the auto-antigen and anti-CD40, while mice with B cells that could not be activated through CD40 or could not synthesise IL-10 suffered severe non-remitting EAE[55]. In addition, transfer of B cells from mice that had recovered from EAE could rescue this defect[55]. These results are consistent with the previous reports showing that mice do not recover from EAE if they lack B cells[56] or the ability to produce IL-10[57,58]. The effects of B cells in resolving EAE may be attributed to the direct B cell and T cell interaction with B cell antigen presentation favouring Th2 rather than Th1 differentiation[59]. This concept is supported by the previous finding in which targeting auto-antigen to B cells before active induction of EAE leads to protection against disease that correlates with enhanced Th2 responses or with induction of antigen-specific anergy[60,61].

### *Beneficial role of B cells and immunoglobulins in demyelination*
There is evidence that humoral mechanisms may mediate myelin repair. It has shown that administration of either whole antiserum[62] or purified IgG[63,64] directed against unidentified antigens enhances CNS remyelination in the Theiler's virus and EAE models. This property is postulated to be one of the potential explanations for the beneficial effect of intravenous immune globulin in MS.

### *Antibody-Dependent Cell Mediated Cytotoxicity*
Antibody-Dependent Cell Mediated Cytotoxicity is a process whereby specific antibodies contibute to the lysis of cells expressing the cognate antigen. This phenomenon is known to occur in EAE[65] and may explain demyelination and oligodendrocyte loss in autoimmune demyelinating disease.

### *Opsonization*
Myelin phagocytosis occurs during autoimmune demyelination and can be reproduced in *in vitro* models[66]. In MBP-induced, IL-12 enhanced Lewis rat relapsing EAE, MBP-specific IgG1 and IgG2b rise over time and myelin-positive microglia appears increasingly in the CNS[67]. Phagocytosis is enhanced by opsonization with antibodies or complement. Anti-myelin antibodies enhance opsonization and increase the efficiency of myelin phagocytosis, contributing to demyelination.

### *Complement*
A major role of antibodies is to fix complement, which contributes to many beneficial and pathological immune responses. Complement is discussed in a separate chapter.

## Conclusion

In conclusion, B cells and antibodies are major components of the immune system that contribute significantly both to the pathogenesis and to the modulation of EAE.

## ACKNOWLEDGEMENTS

Dr. Constantinescu's research has been supported in part by the NIH, University of Nottingham, The Stanhope Trust and the Multiple Sclerosis Society of the United Kingdom and Northern Ireland.

## REFERENCES

1. Kabat, E.A., M. Glusman, and V. Knaub, *Quantitative estimation of albumin and gamma globulin in normal and pathologic cerebrospinal fluid by immunochemical methods.* Am J Med, 1948. **4**: 653-62.

2. Mehta, P.D., H. Lassmann, and H.M. Wisniewski, *Immunologic studies of chronic relapsing EAE in guinea pigs: similarities to multiple sclerosis.* J Immunol, 1981. **127**: 334-8.

3. Rostami, A., R.P. Lisak, N. Blanchard, et al., *Oligoclonal IgG in the cerebrospinal fluid of guinea pigs with acute experimental allergic encephalomyelitis.* J Neurol Sci, 1982. **53**: 433-41.

4. Rostrom, B., A. Grubb, and R. Holmdahl, *Oligoclonal IgG bands synthesized in the central nervous system are present in rats with experimental autoimmune encephalomyelitis.* Acta Neurol Scand, 2004. **109**: 106-12.

5. Gallo, P., D. Cupic, F. Bracco, et al., *Experimental allergic encephalomyelitis in the monkey: humoral immunity and blood-brain barrier function.* Ital J Neurol Sci, 1989. **10**: 561-5.

6. Storch, M.K., S. Piddlesden, M. Haltia, et al., *Multiple sclerosis: in situ evidence for antibody- and complement-mediated demyelination.* Ann Neurol, 1998. **43**: 465-71.

7. Raine, C.S., B. Cannella, S.L. Hauser, et al., *Demyelination in primate autoimmune encephalomyelitis and acute multiple sclerosis lesions: a case for antigen-specific antibody mediation.* Ann Neurol, 1999. **46**: 144-60.

8.      Ron, Y. and J. Sprent, *T cell priming in vivo: a major role for B cells in presenting antigen to T cells in lymph nodes.* J Immunol, 1987. **138**: 2848-56.

9.      Morris, S.C., A. Lees, and F.D. Finkelman, *In vivo activation of naive T cells by antigen-presenting B cells.* J Immunol, 1994. **152**: 3777-85.

10.     Constant, S.L., *B lymphocytes as antigen-presenting cells for CD4+ T cell priming in vivo.* J Immunol, 1999. **162**: 5695-703.

11.     Ashwell, J.D., *Are B lymphocytes the principal antigen-presenting cells in vivo?* J Immunol, 1988. **140**: 3697-700.

12.     Gold, M.R. and A.L. DeFranco, *Biochemistry of B lymphocyte activation.* Adv Immunol, 1994. **55**: 221-95.

13.     Inaoki, M., S. Sato, B.C. Weintraub, et al., *CD19-regulated signaling thresholds control peripheral tolerance and autoantibody production in B lymphocytes.* J Exp Med, 1997. **186**: 1923-31.

14.     O'Keefe, T.L., G.T. Williams, F.D. Batista, et al., *Deficiency in CD22, a B cell-specific inhibitory receptor, is sufficient to predispose to development of high affinity autoantibodies.* J Exp Med, 1999. **189**: 1307-13.

15.     Cooke, M.P., A.W. Heath, K.M. Shokat, et al., *Immunoglobulin signal transduction guides the specificity of B cell-T cell interactions and is blocked in tolerant self-reactive B cells.* J Exp Med, 1994. **179**: 425-38.

16.     Ho, W.Y., M.P. Cooke, C.C. Goodnow, et al., *Resting and anergic B cells are defective in CD28-dependent costimulation of naive CD4+ T cells.* J Exp Med, 1994. **179**: 1539-49.

17.     Armitage, R.J., W.C. Fanslow, L. Strockbine, et al., *Molecular and biological characterization of a murine ligand for CD40.* Nature, 1992. **357**: 80-2.

18.     Noelle, R.J., M. Roy, D.M. Shepherd, et al., *A 39-kDa protein on activated helper T cells binds CD40 and transduces the signal for cognate activation of B cells.* Proc Natl Acad Sci U S A, 1992. **89**: 6550-4.

19.     Castle, B.E., K. Kishimoto, C. Stearns, et al., *Regulation of expression of the ligand for CD40 on T helper lymphocytes.* J Immunol, 1993. **151**: 1777-88.

20.     Hodgkin, P.D., L.C. Yamashita, R.L. Coffman, et al., *Separation of events mediating B cell proliferation and Ig production by using T cell membranes and lymphokines.* J Immunol, 1990. **145**: 2025-34.

21.     Fulcher, D.A. and A. Basten, *B-cell activation versus tolerance--the central role of immunoglobulin receptor engagement and T-cell help.* Int Rev Immunol, 1997. **15**: 33-52.

22.     Pulendran, B., R. van Driel, and G.J. Nossal, *Immunological tolerance in germinal centres.* Immunol Today, 1997. **18**: 27-32.

23.     Goodnow, C.C., J.G. Cyster, S.B. Hartley, et al., *Self-tolerance checkpoints in B lymphocyte development.* Adv Immunol, 1995. **59**: 279-368.

24.     Eris, J.M., A. Basten, R. Brink, et al., *Anergic self-reactive B cells present self antigen and respond normally to CD40-dependent T-cell signals but are defective in antigen-receptor-mediated functions.* Proc Natl Acad Sci U S A, 1994. **91**: 4392-6.

25.     Fulcher, D.A., A.B. Lyons, S.L. Korn, et al., *The fate of self-reactive B cells depends primarily on the degree of antigen receptor engagement and availability of T cell help.* J Exp Med, 1996. **183**: 2313-28.

26.     Storch, M. and H. Lassmann, *Pathology and pathogenesis of demyelinating diseases.* Curr Opin Neurol, 1997. **10**: 186-92.

27.     Lucchinetti, C., W. Bruck, J. Parisi, et al., *A quantitative analysis of oligodendrocytes in multiple sclerosis lesions. A study of 113 cases.* Brain, 1999. **122 ( Pt 12)**: 2279-95.

28.     Inaba, K. and R.M. Steinman, *Resting and sensitized T lymphocytes exhibit distinct stimulatory (antigen-presenting cell) requirements for growth and lymphokine release.* J Exp Med, 1984. **160**: 1717-35.

29.     Croft, M. and S.L. Swain, *B cell response to fresh and effector T helper cells. Role of cognate T-B interaction and the cytokines IL-2, IL-4, and IL-6.* J Immunol, 1991. **146**: 4055-64.

30.     Gausas, J., P.Y. Paterson, E.D. Day, et al., *Intact B-cell activity is essential for complete expression of experimental allergic encephalomyelitis in Lewis rats.* Cell Immunol, 1982. **72**: 360-6.

31.     Willenborg, D.O., P. Sjollema, and G. Danta, *Immunoglobulin deficient rats as donors and recipients of effector cells of allergic encephalomyelitis.* J Neuroimmunol, 1986. **11**: 93-103.

32.     Myers, K.J., J. Sprent, J.P. Dougherty, et al., *Synergy between encephalitogenic T cells and myelin basic protein-specific antibodies in the induction of experimental autoimmune encephalomyelitis.* J Neuroimmunol, 1992. **41**: 1-8.

33.     Lyons, J.A., M. San, M.P. Happ, et al., *B cells are critical to induction of experimental allergic encephalomyelitis by protein but not by a short encephalitogenic peptide.* Eur J Immunol, 1999. **29**: 3432-9.

34.     Lyons, J.A., M.J. Ramsbottom, and A.H. Cross, *Critical role of antigen-specific antibody in experimental autoimmune encephalomyelitis induced by recombinant myelin oligodendrocyte glycoprotein.* Eur J Immunol, 2002. **32**: 1905-13.

35.     Hjelmstrom, P., A.E. Juedes, J. Fjell, et al., *B-cell-deficient mice develop experimental allergic encephalomyelitis with demyelination after myelin oligodendrocyte glycoprotein sensitization.* J Immunol, 1998. **161**: 4480-3.

36.     Dittel, B.N., T.H. Urbania, and C.A. Janeway, Jr., *Relapsing and remitting experimental autoimmune encephalomyelitis in B cell deficient mice.* J Autoimmun, 2000. **14**: 311-8.

37.     Oliver, A.R., G.M. Lyon, and N.H. Ruddle, *Rat and human myelin oligodendrocyte glycoproteins induce experimental autoimmune encephalomyelitis by different mechanisms in C57BL/6 mice.* J Immunol, 2003. **171**: 462-8.

38.     Lebar, R., C. Lubetzki, C. Vincent, et al., *The M2 autoantigen of central nervous system myelin, a glycoprotein present in oligodendrocyte membrane.* Clin Exp Immunol, 1986. **66**: 423-34.

39.     Iglesias, A., J. Bauer, T. Litzenburger, et al., *T- and B-cell responses to myelin oligodendrocyte glycoprotein in experimental autoimmune encephalomyelitis and multiple sclerosis.* Glia, 2001. **36**: 220-34.

40.     Brunner, C., H. Lassmann, T.V. Waehneldt, et al., *Differential ultrastructural localization of myelin basic protein, myelin/oligodendroglial glycoprotein, and 2',3'-cyclic nucleotide 3'-phosphodiesterase in the CNS of adult rats.* J Neurochem, 1989. **52**: 296-304.

41.     Lassmann, H., K. Kitz, and H.M. Wisniewski, *In vivo effect of sera from animals with chronic relapsing experimental allergic encephalomyelitis on central and peripheral myelin.* Acta Neuropathol (Berl), 1981. **55**: 297-306.

42.     Moore, G.R., U. Traugott, M. Farooq, et al., *Experimental autoimmune encephalomyelitis. Augmentation of demyelination by different myelin lipids.* Lab Invest, 1984. **51**: 416-24.

43.     Linington, C., M. Bradl, H. Lassmann, et al., *Augmentation of demyelination in rat acute allergic encephalomyelitis by circulating mouse monoclonal antibodies directed against a myelin/oligodendrocyte glycoprotein.* Am J Pathol, 1988. **130**: 443-54.

44.  Lucchinetti, C.F., W. Bruck, M. Rodriguez, et al., *Distinct patterns of multiple sclerosis pathology indicates heterogeneity on pathogenesis.* Brain Pathol, 1996. **6**: 259-74.

45.  Genain, C.P., B. Cannella, S.L. Hauser, et al., *Identification of autoantibodies associated with myelin damage in multiple sclerosis.* Nat Med, 1999. **5**: 170-5.

46.  Sun, J., H. Link, T. Olsson, et al., *T and B cell responses to myelin-oligodendrocyte glycoprotein in multiple sclerosis.* J Immunol, 1991. **146**: 1490-5.

47.  Xiao, B.G., C. Linington, and H. Link, *Antibodies to myelin-oligodendrocyte glycoprotein in cerebrospinal fluid from patients with multiple sclerosis and controls.* J Neuroimmunol, 1991. **31**: 91-6.

48.  Seil, F.J., *Effects of humoral factors on myelin in organotypic cultures*, in *Multiple Sclerosis: Current Status of Research and Treatment*, R.M. Herndon and F.J. Seil, Editors. 1994, Demos Publications: New York. p. 33-50.

49.  Genc, K., D.L. Dona, and A.T. Reder, *Increased CD80(+) B cells in active multiple sclerosis and reversal by interferon beta-1b therapy.* J Clin Invest, 1997. **99**: 2664-71.

50.  Chang, T.T., C. Jabs, R.A. Sobel, et al., *Studies in B7-deficient mice reveal a critical role for B7 costimulation in both induction and effector phases of experimental autoimmune encephalomyelitis.* J Exp Med, 1999. **190**: 733-40.

51.  Kuchroo, V.K., C.A. Martin, J.M. Greer, et al., *Cytokines and adhesion molecules contribute to the ability of myelin proteolipid protein-specific T cell clones to mediate experimental allergic encephalomyelitis.* J Immunol, 1993. **151**: 4371-82.

52.  Kennedy, M.K., D.S. Torrance, K.S. Picha, et al., *Analysis of cytokine mRNA expression in the central nervous system of mice with experimental autoimmune encephalomyelitis reveals that IL-10 mRNA expression correlates with recovery.* J Immunol, 1992. **149**: 2496-505.

53.  Khoury, S.J., W.W. Hancock, and H.L. Weiner, *Oral tolerance to myelin basic protein and natural recovery from experimental autoimmune encephalomyelitis are associated with downregulation of inflammatory cytokines and differential upregulation of transforming growth factor beta, interleukin 4, and prostaglandin E expression in the brain.* J Exp Med, 1992. **176**: 1355-64.

54.    Skok, J., J. Poudrier, and D. Gray, *Dendritic cell-derived IL-12 promotes B cell induction of Th2 differentiation: a feedback regulation of Th1 development.* J Immunol, 1999. **163**: 4284-91.

55.    Fillatreau, S., C.H. Sweenie, M.J. McGeachy, et al., *B cells regulate autoimmunity by provision of IL-10.* Nat Immunol, 2002. **3**: 944-50.

56.    Wolf, S.D., B.N. Dittel, F. Hardardottir, et al., *Experimental autoimmune encephalomyelitis induction in genetically B cell-deficient mice.* J Exp Med, 1996. **184**: 2271-8.

57.    Bettelli, E., M.P. Das, E.D. Howard, et al., *IL-10 is critical in the regulation of autoimmune encephalomyelitis as demonstrated by studies of IL-10- and IL-4-deficient and transgenic mice.* J Immunol, 1998. **161**: 3299-306.

58.    Samoilova, E.B., J.L. Horton, and Y. Chen, *Acceleration of experimental autoimmune encephalomyelitis in interleukin-10-deficient mice: roles of interleukin-10 in disease progression and recovery.* Cell Immunol, 1998. **188**: 118-24.

59.    Macaulay, A.E., R.H. DeKruyff, C.C. Goodnow, et al., *Antigen-specific B cells preferentially induce CD4+ T cells to produce IL-4.* J Immunol, 1997. **158**: 4171-9.

60.    Day, M.J., A.G. Tse, M. Puklavec, et al., *Targeting autoantigen to B cells prevents the induction of a cell-mediated autoimmune disease in rats.* J Exp Med, 1992. **175**: 655-9.

61.    Saoudi, A., S. Simmonds, I. Huitinga, et al., *Prevention of experimental allergic encephalomyelitis in rats by targeting autoantigen to B cells: evidence that the protective mechanism depends on changes in the cytokine response and migratory properties of the autoantigen-specific T cells.* J Exp Med, 1995. **182**: 335-44.

62.    Rodriguez, M., V.A. Lennon, E.N. Benveniste, et al., *Remyelination by oligodendrocytes stimulated by antiserum to spinal cord.* J Neuropathol Exp Neurol, 1987. **46**: 84-95.

63.    Rodriguez, M. and V.A. Lennon, *Immunoglobulins promote remyelination in the central nervous system.* Ann Neurol, 1990. **27**: 12-7.

64.    Rodriguez, M., *Immunoglobulins stimulate central nervous system remyelination: electron microscopic and morphometric analysis of proliferating cells.* Lab Invest, 1991. **64**: 358-70.

65.    Varitek, V.A., Jr. and E.D. Day, *Studies of rat antibodies specific for myelin basic protein (MBP). Antibody-dependent cell-mediated lysis of MBP-sensitized targets in vitro.* J Neuroimmunol, 1981. **1**: 27-39.

66. Smith, M.E., *Phagocytic properties of microglia in vitro: implications for a role in multiple sclerosis and EAE.* Microsc Res Tech, 2001. **54**: 81-94.

67. Ahmed, Z., D. Gveric, G. Pryce, et al., *Myelin/axonal pathology in interleukin-12 induced serial relapses of experimental allergic encephalomyelitis in the Lewis rat.* Am J Pathol, 2001. **158**: 2127-38.

Chapter A14

# CYTOKINES IN EXPERIMENTAL AUTOIMMUNE ENCEPHALOMYELITIS

Cris S Constantinescu and David Baker
*Division of Clinical Neurology, University of Nottingham, UK and Department of Neuroinflammation, Institute of Neurology, University College, London, UK*

**Abstract:** The cytokine network governing the processes of inflammation, demyelination, relapses and recovery in EAE is very complex. Here we review the major proinflammatory and anti-inflammatory cytokines and their role in EAE, as shown by expression in nervous tissue, production by cells of the immune and nervous system, experiments with exogenous administration, overexpression, antibody neutralization or knockout mice.

**Key words**: *EAE, cytokines, proinflammatory, anti-inflammatory, Th1/Th2, interleukins, tumor necrosis factor*

Cytokines are soluble proteins that mediate interactions between cells of the immune system and thus regulate immune functions. The term cytokine includes factors produced by lymphocytes (previously named lymphokines), monocytes (monokines) and other cells. The term interleukin (IL) originally denoted cytokines mediating interaction between lymphocytes or monocytes and other lymphocytes but is currently more inclusive and in the current nomenclature most cytokines are interleukins.

## Cellular sources of cytokines

The primary cellular sources of cytokines are cells of the immune system, including T and B cells, NK cells, monocytes and macrophages, neutrophils, eosinophils and basophils and mast cells. In addition, other cells in the body produce cytokines, including cells of the central nervous system such as neurons, astrocytes, microglia and oligodendrocytes.

In terms of their effect on the immune responses, cytokines have either predominantly proinflammatory or anti-inflammatory effects. However, there is a great deal of overlap. In experimental autoimmune encephalomyelitis (EAE), the ability of an animal to respond to immunization by producing proinflammatory or anti-inflammatory cytokines correlates with the ability to resist or be susceptible to the development of disease. Numerous cytokine manipulations can influence the ability to develop EAE, its severity and its resolution.

Another classification takes into account the main function of cytokines in the immune system. The haematopoietic cytokines include IL-3, IL-7, IL-9, IL-11 and GM-CSF (granulocyte-macrophage colony stimulating factor). Cytokines involved in innate immunity include type 1 interferons (IFN-1), IL-12 and TNF. Cytokines mediating adaptive (specific immunity) include IL-2, IL-4 and IL-5.

There is extensive redundancy in the cytokine network, and the immune response in EAE is extremely complex. With the continuing discovery of new cytokines, this complexity becomes ever more obvious. The majority of cytokines described to date have been investigated in the context of EAE (and many in the context of MS as well).

The Table lists the main cytokines with main cellular sources and effects

Table I. Cytokines

| Cytokine | Main cellular sources | Main cellular targets | Effects | Pro/anti-inflammatory net effect |
|---|---|---|---|---|
| IL-1 | Mono, Mac, astr, mgl | Many cells | Fever, inflammation, tissue injury | pro |
| IL-2 | Mainly T | T, NK | Growth factor, antiapoptotic for T, B | pro |
| IL-3 | T | Immature progenitors | Growth and differentiation | |
| IL-4 | T, baso, ?eos, B | T, B, mast, eos | Th2, IgE, mast, inhibits Th1 and IgG2a switch | anti |

| | | | | |
|---|---|---|---|---|
| IL-5 | T | Eos, B, ?T | Eos activation, growth factor for B | anti |
| IL-6 | Mono, fibr, EC, neutr | B cells, hepatocytes | growth factor for B, acute phase protein synthesis | pro |
| IL-7 | Fibr, bone marrow stromal cells | Immature progenitors | Growth and differentiation | |
| IL-8 | EC, Mono | Neutr | Recruitment of neutr | pro |
| IL-9 | | T | Growth of T | pro |
| IL-10 | Mono, Mac, T, B, astr, mgl | Mac | Inhibits mac cytokines, MHC, B7 | anti |
| IL-11 | Bone marrow stromal cells | Platelets, megakaryo | Megakaryopoiesis | |
| IL-12 | p35 many cells p40 DC, mac, APC | T, NK, B | IFN-gamma production, Th1 development, cytotoxicity | pro |
| IL-13 | Mac, EC | EC, Mac | Induction of adhesion molec, Th2, inhibits mac | anti |
| IL-15 | Mono, Mac | T, NK | NK proliferation, cytotox, Th1 | pro |
| IL-16 | T | Eos, T? | Eos recruitment, T | pro? |
| IL-17 | T | Many | Cytokine, adhesion | pro |
| IL-18 | Mac | T, NK | Th1, NK activation | pro |
| IL-23 | Mac, DC | T memory | Th1 | pro |
| TNF-$\alpha$ | Mono, Mac, mast, NK, T, astr, mgl, EC | Many | Activates Mac, NK, IgG2a switch, neutr, adhesion, apoptosis | pro |
| LT | T | Neutr, EC | Activates EC, neutr | pro |
| TGF-$\beta$ | T, Mono, Mac, astr, mgl | T, other cells | Inhibits T activation; inhibits inflammation; promotes fibrosis | anti |
| IL-27 | | | | |

APC=antigen presenting cells; astr=astrocytes; baso=basophils; DC=dendritic cells; EC=endothelial cells; eos=eosinophils; fibr=fibroblasts; mac=macrophages; mgl=microglia; mono=monocytes; neutr=neutrophils.

**Cytokine receptors**
Cytokine receptors are transmembrane proteins capable of signal transduction. There are several families of cytokine receptors. These include:

1. Immunoglobulin superfamily members, for example type I and II IL-1 receptor.
2. Type I family cytokine receptor, characterized by a Trp-Ser-X-Trp-Ser linked to two Cys residues, present in many cytokine receptors including IL-2, IL-4, IL-5, IL-6, IL-7, IL-9, IL-11, IL-13, IL-15 and GM-CSF.
3. Type II family cytokine receptor, for example the receptors for IFNs.
4. Type III family cytokine receptor, for example TNF receptor I and II, members of the TNF receptor family.
5. Multichain receptor complexes, for example IL-2, IL-12 and IL-15. The receptor consists of several receptor subunits (some shared with other cytokines), each of which serves different functions.
6. Transmembrane $\alpha$-helix recptors, for example many chemokine receptors.

Intracellular signaling through the receptors of many cytokines is mediated through the Jak (Janus kinase)/ STAT (signal transducer and activator of transcription) pathways. There are four known Jak and 6 STAT molecules. Different cytokine receptors are associated with different Jak/STAT proteins. Ligand binding results in Jak activation and phosphorylation of Tyr residues. This in turn allows binding of STAT protein. STAT becomes phosphorylated, dissociates from the receptor and dimerizes, and activates gene transcription.

**T cell subsets**
Two major types of T cells are distinguished based on their cytokine production and their ability to stimulate cell mediated v. humoral immune responses, respectively [1]. Th1 cells produce IL-2, TNFß and IFN-γ and Th2 cells produce IL-4, IL-5, IL-10 and IL-13. In EAE, generally Th1 and other proinflammatory cytokines are associated with disease activity, while Th2 and anti-inflammatory cytokines with recovery, resistance and suppression of EAE. Th1 and Th2 responses are largely reciprocally inhibitory.

The role of cytokines in EAE can be studied in several approaches. The expression of these cytokines at the mRNA and protein levels in the CNS, cerebrospinal fluid (CSF), lymphoid organs and blood during the course of

EAE can be investigated. The effects of exogenous administration of the cytokine on the course of EAE, or in the adoptive transfer models, on the encephalitogenicity of T cells can be studied. Also, the effects of neutralization of the cytokine by antibodies or soluble receptors can be assessed. EAE can be induced in gene-deficient mice or mice transgenic for a cytokine. In addition, pharmacological or biological treatments that specifically inhibit a cytokine or a set of cytokines can be utilized.

## Proinflammatory cytokines in EAE

**IL-1** is a well-characterized proinflammatory cytokine with multiple effects on EAE. It is largely produced by activated monocytes/macrophages. There are two forms IL-1$\alpha$ and IL-1$\beta$, with similar actions and using the same receptors. IL-1$\beta$ requires processing for activation by IL-1 converting enzyme (ICE or caspase-1). IL-1 is responsible for sickness behaviour (fever, withdrawal, shaking, lassitude, weight loss) associated typically with systemic infections but also with EAE [2]. IL-1 is present in brain [3, 4] and CSF [5] during the induction phase of EAE. Administration of IL-1 enhances disease severity [6] as does exposure of encephalitogenic T cells to IL-1 prior to transfer [7], a process that also inhibits T cell anergy [8]. IL-1 receptor antagonist in EAE successfully prevented or suppressed disease [6, 9, 10], an effect that, however, was not reproduced in an MS trial. Some immunomodulatory drugs such as linomide, which suppress EAE, suppress IL-1 production [11, 12]. Recently it was shown that IL-1 receptor gene deficient mice are resistant to EAE [13].

**IL-2** is a pleiotropic cytokine is a key factor for expansion and survival of T cells. Th1 cells produce IL-2 but IL-2 may be produced in early, proliferative stages by T cells that have not differentiated to Th1 or Th2. In Th1 cells, IL-2 production decreases as IFN-$\gamma$ production increases. Anergic T cells show repressed IL-2 gene expression [14] and costimulation acts in part by stabilizing IL-2 mRNA [15]. IL-2 is expressed together with other proinflammatory cytokines in the CNS during EAE. IL-2 stimulation of encephalitogenic T cells aggravates EAE [16] [17]. Antibody neutralization of IL-2 has had different results in the adoptive transfer and active model. In the former, it effectively blocked disease, while in the latter it had no significant effect. The same divergent results were noted when an anti-IL-2 receptor antibody was utilized [18]. Other therapies targeting the IL-2 receptor IL-2 bearing cells have used immunotoxin such as the chimeric fusion protein between IL-2 and pseudomonas endotoxin, which successfully suppressed EAE [19]. We also have used an IL-2-diphtheria toxin fusion toxin and significantly inhibited EAE in Lewis rats (Rostami et al, unpublished observations). By increasing susceptibility of IL-2 stimulated T cells to apoptosis induced by high doses of antigen, IL-2 plays a role in this form of peripheral tolerance.  On the other hand when supplied exogenously along with high doses of antigen, it exacerbates EAE.  Recent small clinical trials of IL-2 receptor antibodies in MS have been promising [20].

**IL-3** is a cytokine, which functions as a haematopoietic growth factor for cells of many lineages. It is produced by T cells and mast cells. The role of IL-3 in EAE has not been extensively investigated. IL-3 mRNA is identified in the central nervous system (CNS) of mice along with message for other proinflammatory cytokines at peak of disease and in relapse [21]. Encephalitogenic T cells produce higher amounts of IL-3 along with Th1 cytokines [22]. Moreover, IL-3 stimulation of donor T cells before adoptive transfer enhances their encephalitogenicity [23]. Experimental T-cell receptor-directed immunomodulatory interventions decrease IL-3 production of by encephalitogenic T cells, suggesting a role for IL-3 in disease pathogenesis [24]. Recently IL-3 was shown to stimulate an indirect recall antigen-specific IL-4 response in the spleen and in the blood, but not in lymph nodes of mice, via stimulation of mast cells to produce IL-4 [25].

**IL-6** is a pleiotropic cytokine.  It is produced by mononuclear phagocytes, T-cells, endothelial cells, astrocytes, fibroblasts, and other cells. It has been shown that astrocytes can also produce IL-6. Inducing stimuli include bacterial lipopolysaccharide (LPS) and IL-1. Although it is a growth factor for B cells, in the context of EAE, the main function of IL-6 is that of a proinflammatory cytokine.  IL-6 is expressed in the CNS of mice or rats with EAE during the induction phase [4, 26-29].  Treatment of EAE with antibodies against IL-6 yielded conflicting results. Gijbels at al suppressed EAE using an anti-IL-6 antibody [30]. Interestingly this was not associated with IL-6 neutralization, as the IL-6 levels were actually increased. Willenborg et al, on the other hand, found no effect of anti-IL-6 antibody treatment on EAE [31].  Several studies have shown that IL-6 knockout mice are resistant to EAE [32-34]. This was shown to be due to a deficit in the production of Th1 and Th2 cytokines and in the expression of adhesion molecules. Administration of exogenous IL-6 to the knockout mice in the early stage restored typical EAE.  Likewise, administration of the superantigen, staphylococcal enterotoxin, led to development of typical EAE in IL-6-deficient mice, via TNFR1 [35].

**IL-7** was originally described as a growth factor for B cells, but more recently, it has been shown that IL-7 plays an important role in the development and function of both T and B cells.  Little is known about IL-7 in EAE; however, the studies indicate a proinflammatory function in EAE. IL-7 is expressed in both spleen and lymph nodes of mice with chronic relapsing EAE [36]. Stimulation of proteolipid protein (PLP)-reactive T cells with IL-7 enhanced proliferation and IL-2 production. Moreover, when transferred to naïve animals, these cells showed significantly enhanced encephalitogenicity [37].

**IL-9** has been shown to be co-expressed with many other proinflammatory cytokines in EAE during the acute onset phase [21].

**IL-12** was originally discovered as a cytotoxic lymphocyte maturation factor and as a natural killer (NK) cell stimulatory factor. However, its major role is in the development of Th1 responses. IL-12 bridges innate and adaptive immunity and has been involved in numerous normal and pathological immune responses. IL-12 is a disulfide-linked heterodimer with two subunits, p35 and p40 encoded by genes on different chromosomes. The p35 subunit is constitutively expressed at low levels in a variety of cells, while p40 is inducible in a restricted class of cells with antigen presenting function. IL-12 p40 is produced in excess, the ratio being up to 1000. Both subunits are necessary for IL-12 biological activity and IL-12p40 alone may be antagonistic. IL-12 is critical in the differentiation of T helper cells towards the Th1 phenotype. The maintenance of the Th1 responses also requires the synergy of IL-12 and costimulatory interactions between T cells and antigen presenting cells. IL-12 binds its receptor (IL-12Rb1 and IL-12Rb2, the latter being associated with the full-blown Th1 response) and triggers phosphorylation of the STAT-4 signal transduction molecule. STAT-4-deficient mice fail to respond to IL-12 and thus to develop efficient Th1 responses. The major effect of IL-12 on T cells is the stimulation of IFN-gamma production [38]. A solid body of evidence points to EAE and MS being diseases largely mediated by Th1 responses [4, 39-45]. Therefore, since the demonstration of the major role of IL-12 in Th1 responses, its involvement in EAE and MS has been investigated.

Expression of IL-12 in the acute lesions of MS was identified by Windhagen et al [46]. In monophasic EAE, IL-12p40 mRNA was found by *in situ* hybridisation and by RT-PCR in the CNS of Lewis rats only during the initial inflammation period [42, 47]. Jander and Stoll found that, IL-12p35 mRNA did not change throughout the course of EAE. IL-12 mRNA expression was found in the CNS and lymphoid organs in mouse EAE as well [48]. We also detected IL-12p40 mRNA in the CNS of PL/J x SJL/J F1 mice during the initial inflammatory phase and during the relapses. The identity of producing cells is unclear. Microglia, the CNS macrophages, are likely to represent a source of IL-12. We have also shown that astrocytes express IL-12p40 *in vitro* after stimulation with LPS or LPS plus IFN-gamma [49]. Other groups have also detected IL-12p40 mRNA by RT-PCR, but failed to detect IL-12p35, and postulated that the modulatory actions of astrocytes on immune responses in the CNS may be due to astrocytes production of p40 without p35 [50]. However, in our experiments, we detected mRNA for IL-12p35 in astrocytes from different strains of mice, with our without stimulation with LPS (Constantinescu CS et al, manuscript in preparation).

Neutralization of IL-12 by antibodies has also been investigated in EAE. Anti-IL-12 antibodies (directed against the p40 subunit) prevented or suppressed disease in several models [51-53]. Moreover, the relapses (both spontaneous and superantigen-induced) and progression of disease were also suppressed by anti-IL-12 antibody treatment [52-55]. Agents that are known to suppress IL-12 have also been investigated in EAE. Pentoxifylline and related agents suppress EAE and are known to suppress a number of proinflammatory/Th1 cytokines including IL-12 [56, 57]. Captopril, which suppresses EAE [58] also suppresses proinflammatory cytokines including IL-12 [59]. Curcumin inhibits EAE by interfering with IL-12 signaling [60]. Atorvastatin, a cholesterol-lowering agent, has multiple effects on the immune system, including the ability to inhibit Th1 responses, via reduction in IL-12 production. Atorvastatin successfully blocked EAE [61].

IL-12 signaling is critical for EAE development. This has been demonstrated in STAT-4 deficient mice, which fail to develop robust Th1 responses and are resistant to EAE [62].

Another line of evidence for the role of IL-12 in EAE comes from experiments with exogenous IL-12 administration. Stimulation with IL-12 of antigen-reactive T cells prior to transfer rendered them more encephalitogenic [51, 63]. Administration of IL-12 induces relapses both in Lewis rats [64] and in the mouse model [52]. The mechanisms of induced relapses is unclear, but an upregulation of NO and proteases has been noted [65, 66]. Rats may have multiple relapses following exogenous IL-12, and eventually this leads not only to demyelination but also to axonal loss [65]. More than altering the severity and clinical course of EAE, exogenous IL-12 may also modify susceptibility to EAE. Resistant mice are rendered susceptible to adoptive transfer EAE by treatment of T cells with IL-12 prior to transfer [63]. Moreover, CpG-rich bacterial DNA overcome EAE resistance via IL-12 induction [67]. Experiments with IL-12p40 knockout mice showed that IL-12p40 is essential for EAE development [68]. The role of IL-12p35 is discussed below in the context of IL-23. An important question is: what are the mechanisms of induction of IL-12 in EAE? In addition to induction by bacterial products, induction by the interaction of T cells and antigen presenting cells is obviously important. In this context, the interaction of CD40 and CD40 ligand is probably a critical mechanism. Blockade of this interaction has been shown to suppress EAE and CD40Ligand knockout mice are resistant to EAE [69-71]. We also demonstrated that induction of IL-12 via this interaction plays a significant role in EAE, because exogenous IL-12 reversed the suppression of EAE achieved by the blockade of the CD40-CD40 ligand [72]. The effects of IL-12 in EAE are also unclear. IL-12 induces IFN-gamma. However, IFN-gamma-deficient mice remain susceptible to EAE [73]. The role of IFN in EAE is discussed in a separate chapter. In IFN-gamma knockout mice, IL-12 neutralization prevented the induction of EAE, suggesting that the role of IL-12 in EAE may involve IFN-gamma-independent actions [68]. One candidate is NO, which is upregulated by IL-12. Other effects of IL-12 may involve the breakdown of tolerance. Exogenous IL-12 reverses both oral [74] and intravenous tolerance [75]. This effect may be more prominent in combination with IL-18 [76]. Also, neutralization of IL-12 facilitates oral tolerance [77].

**IL-17** is a recently discovered cytokine with prominent proinflammatory function. IL-17 appears to play a significant role in a variety of inflammatory conditions [78-80]. Its role in EAE has not been amply investigated. It is expressed in EAE in IL-12Rb2 knockout mice during the inflammatory phase of the disease [81].

**IL-18** has been identified as a major IFN-gamma inducing factor [82]. It has some similarity to IL-1 and needs processing by caspase 1 (ICE) for biological activity. IL-18 synergizes with IL-12 for many of its actions. The signalling pathways may be different. Both IL-18 and IL-12 may require each other for optimal activity as it has been shown that IL-18 Th1 induction is IL-12-dependent and *vice versa* [83-86]. IL-18 is expressed in the CNS as shown by RT-PCR in parallel to the disease course [47] and in proportion to the inflammatory infiltrate. Neutralizing antibodies against IL-18 suppressed EAE [87].

**IL-23.** Recently IL-23 was discovered as a cytokine involved in Th1 responses, which stimulates T cell for production of IFN-gamma with preferential effect on memory T cells. IL-23 shares the p40 subunit with IL-12 and has a unique p19 subunit. Producer cells are primarily mononuclear phagocytic cells and dendritic cells. IL-23 shares the IL-12Rb1 and binds another specific receptor. It activates STAT-4, like IL-12 [88]. The discovery of IL-23 led to re-evaluation of the previous work done with IL-12. The neutralizing experiments had used antibodies against IL-12p40 subunit and knockout experiments used IL-12p40 crossed with IL-12p35 knockouts. To determine whether neutralization of IL-12p35 affects EAE, we utilized an antibody against IL-12p35, which had previously been shown to have neutralizing activity and thus inhibit IL-12-induced IFN-gamma *in vitro* and *in vivo*. While ABH mice immunized for EAE treated with anti-IL-12p40 antibodies developed no disease, mice treated with either anti-IL-

12p35 antibodies or control rat IgG developed EAE with the maximal score of 4 (Baker, Constantinescu, Wysocka, unpublished observations) (Figure 1).

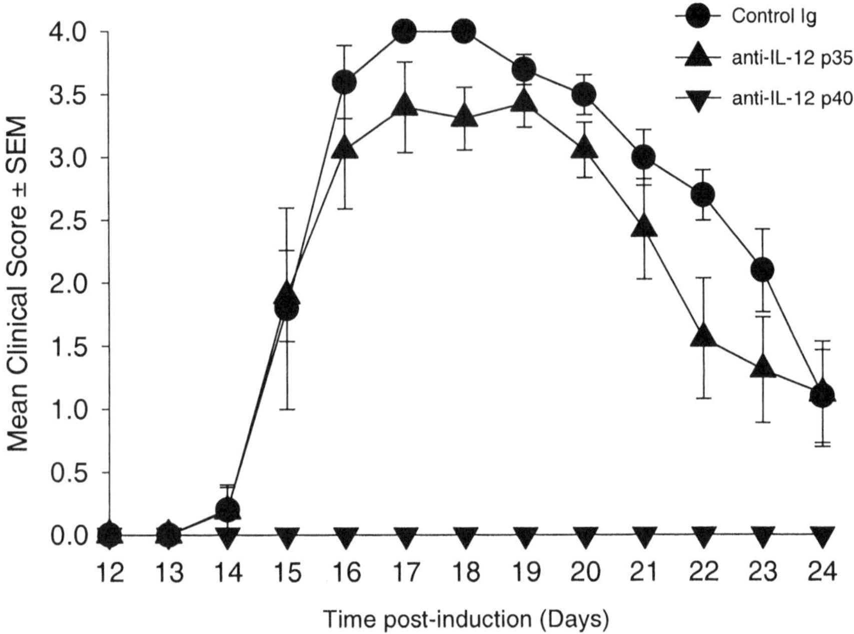


**Figure 1**. Clinical course of EAE in ABH mice treated with anti-IL-12 p35 or p40 antibodies, showing no suppression of EAE with anti-IL-12 p35 antibodies.

This indicates that in the effects of IL-12 in EAE the p40 subunit is essential. Experiments on knockout mice [89, 90] showed that IL-12p40 knockouts are resistant to MOG-induced EAE while IL-12p35 knockout mice are susceptible. Finally, experiments using the recently generated IL-23p19 knockout mice have confirmed that IL-23p19 is required for EAE development [91]. In these experiments, exogenous IL-12 in IL-12p40 knockout mice did not overcome the resistance to EAE, similar to our own experiments (Gran, Rostami, Constantinescu, unpublished observations). Further work will investigate the relative role of IL-12 and IL-23 in EAE. Experiments with cytokine knockouts are difficult to interpret because of the great redundancy in the cytokine network and the likely development of compensatory mechanisms in the presence of a deleted gene. The evidence from exogenous administration of IL-12 and that from its overexpression in CNS leading to inflammation suggests that IL-12 is, nevertheless important. Immune interventions targeting IL-12p40 are likely to be beneficial in EAE, regardless on the relative proportion of IL-12 and IL-23 they affect.

**IL-27.** The IL-12 family of cytokines is expanding. IL-27 was discovered as another proinflammatory cytokine with an IL-12p35-like subunit [92]. Its role in EAE has not yet been investigated.

**TNF.** This is a highly pleiotropic cytokine to which a multitude of effects of relevance to the pathology of EAE ad MS have been attributed. The tumor necrosis factor family of cytokines comprises TNF-$\alpha$, lymphotoxin (LT)- $\alpha$ (LT) or TNF-$\beta$ and LT-$\beta$. They are encoded by neighbouring genes forming part of the major histocompatibility complex. There are no major differences in the biological activities of TNF-$\alpha$ and lymphotoxin but TNF-$\alpha$ is predominantly produced by monocytes/macrophages (but also other cells including astrocytes and microglia) whereas TNF-$\beta$ ( LT-$\alpha$) is produced by T cells. LT-$\beta$ is a membrane protein that can bind LT to form a receptor related to the two TNF receptors (TNF-RI and TNF-RII). TNF (here used collectively for both $\alpha$ and $\beta$) is expressed in the CNS in MS [93] and can induce apoptosis in oligodendrocytes with potential resultant demyelination [93]. In EAE, TNF expression parallels the course of the disease [42, 93, 94]. Encephalitogenic T cells produce more TNF than nonencephalitogenic counterparts [43, 44]. Exogenous TNF administration can exacerbate the course of EAE [43, 95, 96], as does its overexpression in the CNS [97]. Moreover, overexpression of TNF-$\alpha$ in the CNS can lead to a spontaneous inflammatory disease with lymphocytic infiltration, gliosis and demyelination [98]. This overexpression can also lead to a primary oligodendrogliopathy with demyelination [99]. Neutralization of TNF by antibodies suppressed adoptive transfer EAE [100]. However, anti-TNF antibodies did not inhibit actively induced EAE in another experiment [101]. Soluble TNF receptor I successfully treated EAE. Moreover, this approach led to significant suppression of disease in mice with established EAE [100, 102]. In addition, a variety of pharmacological agents known to suppress production of TNF have been used successfully to treat EAE, although such agents may also suppress other cytokines and immunologically active molecules [103, 104]. The results from experiments with TNF knockout mice are interesting. In TNF-$\alpha$ and LT double knockouts on 129 x C57BL/6 backgrounds, EAE susceptibility and severity was not changed [105]. However, MOG induced EAE on the 129 x C57BL/6 background was delayed and milder, suggesting a role for TNF in the early inflammatory events [106]. The mice had less demyelination but more inflammatory infiltrates in the CNS than the wild-type littermates. On the other hand, single TNF-$\alpha$ knockouts on 129 or C57BL/6 background develop more severe EAE with enhanced inflammation and demyelination [107]. Investigating the role of the TNF-RI and TNF-RII receptors in the pathogenesis of EAE using knockout mice has revealed that the susceptibility to EAE depends on TNF-RI (p55). In contrast, TNF-RII deficient mice exhibited more severe disease [106]. In summary, there is variation in the results with TNF knockout mice, which, in part, may be explained by the genetic background and the immunizing antigen used in the experiments. On the other hand, it is likely that the role of the TNF family of cytokine, a pleiotropic set of immune molecules, in EAE, is very complex and not restricted to enhancement of inflammation. For example, TNF is involved in apoptosis, in particular via TNF-RI. Genetic deletion of TNF or TNF-RI enhances survival of lymphocytes and may contribute to the inflammatory damage [108]. Moreover, TNF has multiple direct effects on CNS cells and its genetic absence may have numerous non-immunological consequences, which may influence the course of EAE. For example, TNF may be neuroprotective [109, 110] and it may contribute to synaptic strength [111].

**Osteopontin.** This proinflammatory molecule has been shown to induce Th1 responses. Osteopontin is expressed in the CNS in several disease states and it was identified in MS lesions by DNA microarray analysis. Because of its marker proinflammatory role, the finding was validated in EAE by assessing susceptibility in osteopontin knockout mice. It was found that osteopontin is required for induction of EAE [112]. This was confirmed in further experiments [113].

**Anti-inflammatory cytokines in EAE.**
While Th1 cytokines have been associated with EAE induction and severity, Th2 cytokines are associated with remission and suppression of EAE.

**IL-4** is the critical cytokine for the development of Th2 responses [114]. IL-4 expression in the CNS has been associated with the remission stage of EAE [40] [94] [115]. Most Th2 cells producing IL-4 are nonencephalitogenic [44, 95, 116-118]. However, all such Th2 cells may not be efficient in their ability to suppress EAE [119]. Exogenous IL-4 administration or delivery by transfected T cells significantly ameliorates EAE [31, 117, 120, 121]. A simultaneous, competing Th2 response not directed against neuroantigens, characterized by high IL-4 (and other Th2 cytokines) production also ameliorates EAE [122, 123]. In addition, experimental immunomodulatory treatments of EAE including the induction of peripheral tolerance or agents such as retinoids are mediated via IL-4. [115, 124, 125]. Similarly, altered peptide ligands and glatiramer acetate induce IL-4 [126-128]. There is also evidence supporting a role for IL-4 in governing the resistance to EAE in mice. Studies using IL-4 knockouts are discussed below. We have shown that MBP-stimulated lymphocytes from relatively genetically resistant strains of mice immunized with mouse spinal cord homogenate produce IL-4, while those of resistant mice produce IFN-γ. Moreover, *in vivo* neutralization of IL-4 reversed EAE resistance [129]. Experiments with IL-4 knockouts yielded conflicting results. Some investigators showed IL-4 knockouts to have more severe disease [130, 131], and mice deficient in STAT-6, the signal transduction molecule through which IL-4 acts, also had more severe disease [62] while others showed no differences from wild-type mice [68, 132, 133]. These differences may be related to different experimental systems or to compensatory mechanisms with utilization of default pathways triggered by the inactivation of cytokine genes [134]. IL-4 knockout mice do not present a deficit in entering the recovery stage of EAE. The results indicate that, while IL-4 is an important immunoregulatory cytokine in MS, it is not the only factor involved in the downregulation of the immune response in EAE.

**IL-5** is a Th2 cytokine produced largely by T cells. Its main role is in eosinophil activation, and it acts as a B cell growth factor as well in mice, but not in humans. Because of the sensitivity of IL-5 assays and the more restricted cellular sources (more literally a Th2 cytokine), IL-5 has been used as a marker for Th2 responses [135]. It has been used in EAE in this context, for the determination of Th1 versus Th2 cytokine expression pattern. In contrast to IL-4, IL-13 and TGF-b (see below), addition of IL-5 to PLP-reactive T cells did not significantly modify their ability to induce transfer EAE [136]. IL-5 appears to be expressed along with other anti-inflammatory cytokines in the CNS in the resolution phase of EAE [40]. Some immunomodulatory treatments beneficial for EAE that suppress Th1 and induce Th2 responses (immune-deviation) either do not affect or enhance IL-5 with other Th2 cytokines [61, 104, 127]. In contrast, other strategies, including those inducing peripheral tolerance by anergy or deletion may downregulate IL-5 along with other Th2 cytokines [137]. Another potential immunomodulatory role of IL-5 in EAE may be the stimulation of CD5+ B cells, which have been attributed a regulatory function in EAE [138].

**IL-10** is a thoroughly anti-inflammatory cytokine. It suppresses many proinflammatory cytokines, antigen presentation and NO production [139]. It is produced by T cells with the Th2 phenotype and by macrophages, but other cells including astrocytes and microglia can produce it. IL-10 is also induced by inflammatory stimuli, for example by LPS in endotoxic shock, and by IL-12 [140]. In most studies, IL-10 expression in EAE CNS corresponded to the onset of the recovery phase [4, 42]. Low expression of IL-10 in EAE is associated with a protracted course [29]. IL-10 may participate in a regulatory loop that governs susceptibility versus resistance to EAE [68]. *In vivo* delivery of IL-10 via an IL-10-expressing Vaccinia vector [31] and adenovirus inhibited EAE. Also, exogenous *in vivo* IL-10 ameliorated EAE in Lewis rats [141]. Nasal administration of IL-10 also was successful in treating EAE in Lewis rats [142]. Immunomodulation of EAE with altered peptide ligands induces Th2 or Th0 nonencephalitogenic and protective cells producing IL-10 [143, 144]. Other therapeutic approaches including peipheral tolerance, IFN-β or –τ and glatiramer acetate (copolymer-1) treatment induce IL-10 [128, 145-148]. IL-10 is also protective against TNF induced relapses of EAE [96]. On the other hand, IL-10 given *in vivo* after transfer of encephalitogenic T cells failed to protect SJL/L mice against EAE and led to slightly more severe disease [149]. Similarly, antibody neutralization of IL-10 in this model immediately after cell transfer enhanced disease, while treatment of donor cells with IL-10 prior to transfer had no effect. These results need to be balanced against those using transgenic and knockout mice. Bettelli et al. showed that transgenic mice overexpressing IL-10 were resistant to EAE while IL-10 knockout mice develop more severe disease [133, 150]. Transfer of transgenic memory T cells overexpressing IL-10 under the control of the activation-inducible promoter of IL-2 both prevented and reversed EAE [151]. In another transgenic model expressing human IL-10 under the control of MHC class II, EAE was successfully prevented despite the development of neuroantigen-reactive T cells, indicating that the inhibitory effect of IL-10 is not on the generation but on the effector mechanisms of Th1 responses [152].

**IL-13** is classed under the Th2 cytokines and has many similarities with IL-4 including activation of the STAT-6 signaling pathway. IL-13 is expressed in the CNS in the recovery phase of EAE [115]. It is also expressed in lymph nodes of rats protected from EAE by targeting the autoantigen to B cells [153]. *In vivo* delivery of human IL-13 via an IL-13 expressing vector significantly suppressed MBP-induced EAE. The suppression was not due to a major inhibition of development and responses of T cell reactivity to autoantigen, nor due to a blockade of autoantibodies against MBP. The mechanism involved was an inhibition of macrophage activation [154].

**The IL-10 family of cytokines.** A whole family of IL-10-related cytokines has been recently discovered, including IL-19, IL-20, IL-22 (IL-10-related T-cell-derived inducible factor or IL-TIF), IL-24 (also known as melanoma differentiation-associated antigen 7 or MDA-7) and IL-26 (AK155) [155]. Their function is not well known and their involvement in EAE has not yet been investigated. Not all of these cytokines may be immunosuppressive/immunomodulatory like IL-10. IL-19 and IL-20, for example, have been implicated in the pathogenic mechanisms of psoriasis, a Th1-mediated autoimmune disease, while IL-22/IL-TIF mediates acute-phase response signals in hepatocytes.

**IL-25.** Although related to IL-17, IL-25 is an inducer of Th2 responses [156]. Its role in EAE is yet to be determined.

**TGF-β.** This pleiotropic cytokine has multiple, largely anti-inflammatory, effects on the immune system. TGF-β is produced by T cells, macrophages and many other cells including astrocytes of the brain. TGF-β inhibits T cell activation, down-regulates expression of MHC class II [157, 158] and proinflammatory cytokines and their receptors, and induces T cell apoptosis [159, 160]. TGF-β is expressed in brain during both the acute and chronic phases of EAE [161]. Treatment of neuroantigen-reactive T cells with TGF-β prior to adoptive transfer prevents passive EAE [162, 163]. TGF-β administered systemically suppresses acute and chronic EAE [164-166]. The improvement occurs even when treatment is initiated in established disease. TGF-β also suppresses relapses of EAE [96]. Antibody neutralization of TGF-β enhances the severity of EAE both clinically and histopathologically [161, 167]. TGF-β has been implicated as an important cytokine of regulatory T cells, likely to play an important role in immunoregulation of EAE [168]. A variant of such cells, CD4+ T cells involved in the recovery after the acute phase of EAE in rats, the so-called post-recovery suppressor cells, secrete TGF-β, which down-regulates cytokine production by EAE effector CD4+ T cells [169]. In addition, the hallmark of Th3 cell clones generated by peripheral tolerance is the secretion of TGF-β [170]. The above works strongly support the concept of a prominent immunomodulatory and anti-inflammatory role for TGF-β in EAE. The exact action(s) through which it exerts this role or the sites of action are not known. It is likely that TGF-β operates at multiple levels and possible via multiple mechanisms.

**Interferons.** These subgroup of cytokines with multiple, important effects on EAE are discussed in another chapter.

**Paradoxical effects of cytokines.**
The EAE literature contains numerous paradoxes, in particular with respect to effects and mechanisms of actions of cytokines. This underscores the complexity and the redundancy of the cytokine network.

**Anti-inflammatory effects of proinflammatory cytokines.**
Several proinflammatory/Th1 cytokines have been found to have paradoxical and unexpected anti-inflammatory effects. IFN-γ is discussed in the chapter on interferons in EAE.

**IL-1.** Administration of IL-1 prior to induction of EAE down-modulates the disease. The mechanism probably involves increasing resting levels of endogenous corticosteroids [171]. **IL-2.** The high-affinity receptor for this cytokine is a putative marker for regulatory T cells. Immunological manipulations in EAE targeting the IL-2R may theoretically impair the function of regulatory T cells and thus enhance disease. In addition, the survival of regulatory T cells in the periphery is dependent upon IL-2, as IL-2 knockouts are deficient in regulatory T cells and are prone to various autoimmune/inflammatory phenomena [172-174]. **TNF.** The experiments with TNF knockout mice are discussed above. Maturation of dendritic cells with TNF led to their development into a tolerogenic population capable of inducing regulatory CD4+IL-10+ T cells [175]. TNF administration rescued TNF knockout mice from severe EAE [107]. These results may be explained in part by the proapoptotic activity of TNF. **IL-6.** Although antibody neutralization of IL-6 suppresses EAE, it does so by enhancing IL-6 levels [30]. **IL-12.** In IL-12p35-deficient mice, there is a significant up-regulation of many proinflammatory cytokines and NO [90]. The mechanisms have not yet been clarified. On the other hand, in vivo administration of IL-12 has enhanced the severity of disease, unlike in experimental autoimmune uveitis (EAU) where IL-12 can have a paradoxically protective role involving IFN-γ, NO and apoptosis [176].

**Proinflammatory effects of anti-inflammatory cytokines.**
Conversely, some paradoxical effects of Th2/anti-inflammatory cytokines in EAE are also known. **IL-4.** While in many studies neuroantigen-specific Th2 cells are nonencephalitogenic, there are exceptions: IL-4-producing Th2 cells transferred to naïve animals induced EAE-like CNS inflammation [177]. **IL-10.** *In vivo* treatment of SJL/J mice with IL-10 enhanced EAE severity [149]. **TGF-β.** Treatment of encephalitogenic T cells with TGF-β increased the severity of transfer EAE [178]. Moreover, in mice overexpressing in the TGF-β CNS, EAE was more severe [179]. These results may be tentatively explained by the multiple direct effects of TGF-β in the CNS including enhanced gliosis and amyloidogenesis and also by the

possibility that the beneficial effect of in EAE involves the peripheral immune cells before CNS invasion.

In conclusion, many proinflammatory cytokines are involved in the pathogenesis of EAE. Modulation of cytokine production during EAE may provide therapeutic intervention.


# ACKNOWLEDGEMENTS

Dr. Constantinescu's research has been supported in part by the NIH, University of Nottingham, The Stanhope Trust and the Multiple Sclerosis Society of the United Kingdom and Northern Ireland.  Dr. Baker is a Research Fellow of the Multiple Sclerosis Society of Great Britain and Northern Ireland


# REFERENCES

1.   Mosmann, T.R. and R.L. Coffman, *TH1 and TH2 cells: different patterns of lymphokine secretion lead to different functional properties.* Annu Rev Immunol, 1989. **7**: p. 145-73.

2.   Pollak, Y., H. Ovadia, E. Orion, J. Weidenfeld, and R. Yirmiya, *The EAE-associated behavioral syndrome. I. Temporal correlation with inflammatory mediators.* J Neuroimmunol, 2003. **137**(1-2): p. 94-9.

3.   Bauer, J., F. Berkenbosch, A.M. Van Dam, and C.D. Dijkstra, *Demonstration of interleukin-1 beta in Lewis rat brain during experimental allergic encephalomyelitis by immunocytochemistry at the light and ultrastructural level.* J Neuroimmunol, 1993. **48**(1): p. 13-21.

4.   Kennedy, M.K., D.S. Torrance, K.S. Picha, and K.M. Mohler, *Analysis of cytokine mRNA expression in the central nervous system of mice with experimental autoimmune encephalomyelitis reveals that IL-10 mRNA expression correlates with recovery.* J Immunol, 1992. **149**(7): p. 2496-505.

5.   Symons, J.A., R.V. Bundick, A.J. Suckling, and M.G. Rumsby, *Cerebrospinal fluid interleukin 1 like activity during chronic relapsing experimental allergic encephalomyelitis.* Clin Exp Immunol, 1987. **68**(3): p. 648-54.

6.   Jacobs, C.A., P.E. Baker, E.R. Roux, K.S. Picha, B. Toivola, S. Waugh, and M.K. Kennedy, *Experimental autoimmune encephalomyelitis is exacerbated by IL-1 alpha and suppressed by soluble IL-1 receptor.* J Immunol, 1991. **146**(9): p. 2983-9.

7.   Mannie, M.D., C.A. Dinarello, and P.Y. Paterson, *Interleukin 1 and myelin basic protein synergistically augment adoptive transfer activity of lymphocytes mediating experimental autoimmune encephalomyelitis in Lewis rats.* J Immunol, 1987. **138**(12): p. 4229-35.

8.      Bourdoulous, S., E. Beraud, C. Le Page, A. Zamora, A. Ferry, D. Bernard, A.D. Strosberg, and P.O. Couraud, *Anergy induction in encephalitogenic T cells by brain microvessel endothelial cells is inhibited by interleukin-1.* Eur J Immunol, 1995. **25**(5): p. 1176-83.

9.      Martin, D. and S.L. Near, *Protective effect of the interleukin-1 receptor antagonist (IL-1ra) on experimental allergic encephalomyelitis in rats.* J Neuroimmunol, 1995. **61**(2): p. 241-5.

10.     Badovinac, V., M. Mostarica-Stojkovic, C.A. Dinarello, and S. Stosic-Grujicic, *Interleukin-1 receptor antagonist suppresses experimental autoimmune encephalomyelitis (EAE) in rats by influencing the activation and proliferation of encephalitogenic cells.* J Neuroimmunol, 1998. **85**(1): p. 87-95.

11.     Xiao, B.G., X.F. Bail, G.X. Zhang, G. Hedlund, and H. Link, *Linomide-mediated protection of oligodendrocytes is associated with inhibition of nitric oxide production and IL-1beta expression in Lewis rat glial cells.* Neurosci Lett, 1998. **249**(1): p. 17-20.

12.     Weiss, L., V. Barak, M. Zeira, A. Abdul-Hai, I. Raibstein, S. Reich, E. Hirschfeld, D. Gross, and S. Slavin, *Cytokine production in Linomide-treated nod mice and the potential role of a Th (1)/Th(2) shift on autoimmune and anti-inflammatory processes.* Cytokine, 2002. **19**(2): p. 85-93.

13.     Schiffenbauer, J., W.J. Streit, E. Butfiloski, M. LaBow, C. Edwards, 3rd, and L.L. Moldawer, *The induction of EAE is only partially dependent on TNF receptor signaling but requires the IL-1 type I receptor.* Clin Immunol, 2000. **95**(2): p. 117-23.

14.     Schwartz, R.H., *T cell anergy.* Annu Rev Immunol, 2003. **21**: p. 305-34.

15.     Gonsky, R., R.L. Deem, D.H. Lee, A. Chen, and S.R. Targan, *CD28 costimulation augments IL-2 secretion of activated lamina propria T cells by increasing mRNA stability without enhancing IL-2 gene transactivation.* J Immunol, 1999. **162**(11): p. 6621-9.

16.     Schluesener, H.J. and H. Lassmann, *Recombinant interleukin 2 (IL-2) promotes T cell line-mediated neuroautoimmune disease.* J Neuroimmunol, 1986. **11**(1): p. 87-91.

17.     Racke, M.K., J.M. Critchfield, L. Quigley, B. Cannella, C.S. Raine, H.F. McFarland, and M.J. Lenardo, *Intravenous antigen administration as a therapy for autoimmune demyelinating disease.* Ann Neurol, 1996. **39**(1): p. 46-56.

18.     Engelhardt, B., T. Diamantstein, and H. Wekerle, *Immunotherapy of experimental autoimmune encephalomyelitis (EAE): differential effect of anti-IL-2 receptor antibody therapy on actively induced and T-line mediated EAE of the Lewis rat.* J Autoimmun, 1989. **2**(1): p. 61-73.

19.     Rose, J.W., H. Lorberboum-Galski, D. Fitzgerald, R. McCarron, K.E. Hill, J.J. Townsend, and I. Pastan, *Chimeric cytotoxin IL2-PE40 inhibits relapsing experimental allergic encephalomyelitis.* J Neuroimmunol, 1991. **32**(3): p. 209-17.

20.     Bielekova, B., S. Reichert-Scrivner, H.F. McFarland, and R. Martin, *Mechanism of action of multiple sclerosis treatment by Zenapax, a humanized monoclonal antibody against the interleukin-2 receptor alpha chain.* Neurology, 2003. **60(suppl1)**: p. A434-5.

21.     Lyons, J.A., M.L. Zhao, and R.B. Fritz, *Pathogenesis of acute passive murine encephalomyelitis II. Th1 phenotype of the inducing population is not sufficient to cause disease.* J Neuroimmunol, 1999. **93**(1-2): p. 26-36.

22.     Weinberg, A.D., J.J. Wallin, R.E. Jones, T.J. Sullivan, D.N. Bourdette, A.A. Vandenbark, and H. Offner, *Target organ-specific up-regulation of the MRC OX-40 marker and selective production of Th1 lymphokine mRNA by encephalitogenic T helper cells isolated from the spinal cord of rats with experimental autoimmune encephalomyelitis.* J Immunol, 1994. **152**(9): p. 4712-21.

23.  Zhao, M.L., J.Q. Xia, and R.B. Fritz, *Interleukin-3 and encephalitogenic activity of SJL/J myelin basic protein-specific T cell lines.* J Neuroimmunol, 1993. **43**(1-2): p. 69-78.

24.  Offner, H., M. Vainiene, B. Celnik, A.D. Weinberg, A. Buenafe, and A.A. Vandenbark, *Coculture of TCR peptide-specific T cells with basic protein-specific T cells inhibits proliferation, IL-3 mRNA, and transfer of experimental autoimmune encephalomyelitis.* J Immunol, 1994. **153**(11): p. 4988-96.

25.  Karulin, A.Y., M.D. Hesse, H.C. Yip, and P.V. Lehmann, *Indirect IL-4 pathway in type 1 immunity.* J Immunol, 2002. **168**(2): p. 545-53.

26.  Gijbels, K., J. Van Damme, P. Proost, W. Put, H. Carton, and A. Billiau, *Interleukin 6 production in the central nervous system during experimental autoimmune encephalomyelitis.* Eur J Immunol, 1990. **20**(1): p. 233-5.

27.  Okuda, Y., S. Sakoda, and T. Yanagihara, *The pattern of cytokine gene expression in lymphoid organs and peripheral blood mononuclear cells of mice with experimental allergic encephalomyelitis.* J Neuroimmunol, 1998. **87**(1-2): p. 147-55.

28.  Okuda, Y., S. Sakoda, C.C. Bernard, and T. Yanagihara, *The development of autoimmune encephalomyelitis provoked by myelin oligodendrocyte glycoprotein is associated with an upregulation of both proinflammatory and immunoregulatory cytokines in the central nervous system.* J Interferon Cytokine Res, 1998. **18**(6): p. 415-21.

29.  Diab, A., J. Zhu, B.G. Xiao, M. Mustafa, and H. Link, *High IL-6 and low IL-10 in the central nervous system are associated with protracted relapsing EAE in DA rats.* J Neuropathol Exp Neurol, 1997. **56**(6): p. 641-50.

30.  Gijbels, K., S. Brocke, J.S. Abrams, and L. Steinman, *Administration of neutralizing antibodies to interleukin-6 (IL-6) reduces experimental autoimmune encephalomyelitis and is associated with elevated levels of IL-6 bioactivity in central nervous system and circulation.* Mol Med, 1995. **1**(7): p. 795-805.

31.  Willenborg, D.O., S.A. Fordham, W.B. Cowden, and I.A. Ramshaw, *Cytokines and murine autoimmune encephalomyelitis: inhibition or enhancement of disease with antibodies to select cytokines, or by delivery of exogenous cytokines using a recombinant vaccinia virus system.* Scand J Immunol, 1995. **41**(1): p. 31-41.

32.  Samoilova, E.B., J.L. Horton, B. Hilliard, T.S. Liu, and Y. Chen, *IL-6-deficient mice are resistant to experimnetal autoimmune encephalomyelitis: roles of IL-6 in the actiavtion and differentiation of autoreactive T cells.* J Immunol, 1998. **161**: p. 6480-6.

33.  Eugster, H.P., K. Frei, M. Kopf, H. Lassmann, and A. Fontana, *IL-6-deficient mice resist myelin oligodendrocyte glycoprotein-induced autoimmune encephalomyelitis.* Eur J Immunol, 1998. **28**: p. 2178-87.

34.  Mendel, I., A. Katz, N. Kozak, A. Ben-Nun, and M. Revel, *Interleukin-6 functions in autoimmune encephalomyelitis: a study in gene-targeted mice.* Eur J Immunol, 1998. **28**(5): p. 1727-37.

35.  Eugster, H.P., K. Frei, F. Winkler, U. Koedel, W. Pfister, H. Lassmann, and A. Fontana, *Superantigen overcomes resistance of IL-6-deficient mice towards MOG-induced EAE by a TNFR1 controlled pathway.* Eur J Immunol, 2001. **31**: p. 2302-12.

36.  Yang, J., P.J. Lindsberg, V. Hukkanen, R. Seljelid, C.G. Gahmberg, and S. Meri, *Differential expression of cytokines (IL-2, IFN-gamma, IL-10) and adhesion molecules (VCAM-1, LFA-1, CD44) between spleen and lymph nodes associates with remission in chronic relapsing experimental autoimmune encephalomyelitis.* Scand J Immunol, 2002. **56**(3): p. 286-93.

37.    Bebo, B.F., Jr., J.C. Schuster, K. Adlard, A.A. Vandenbark, and H. Offner, *Interleukin 7 is a potent co-stimulator of myelin specific T cells that enhances the adoptive transfer of experimental autoimmune encephalomyelitis.* Cytokine, 2000. **12**(4): p. 324-31.

38.    Trinchieri, G., *Interleukin-12 and the regulation of innate resistance and adaptive immunity.* Nat Rev Immunol, 2003. **3**(2): p. 133-46.

39.    Baker, D., J.K. O'Neill, and J.L. Turk, *Cytokines in the central nervous system of mice during chronic relapsing experimental allergic encephalomyelitis.* Cell Immunol, 1991. **134**(2): p. 505-10.

40.    Merrill, J.E., D.H. Kono, J. Clayton, D.G. Ando, D.R. Hinton, and F.M. Hofman, *Inflammatory leukocytes and cytokines in the peptide-induced disease of experimental allergic encephalomyelitis in SJL and B10.PL mice.* Proc Natl Acad Sci U S A, 1992. **89**(2): p. 574-8.

41.    Renno, T., J.Y. Lin, C. Piccirillo, J. Antel, and T. Owens, *Cytokine production by cells in cerebrospinal fluid during experimental allergic encephalomyelitis in SJL/J mice.* J Neuroimmunol, 1994. **49**(1-2): p. 1-7.

42.    Issazadeh, S., A. Ljungdahl, B. Hojeberg, M. Mustafa, and T. Olsson, *Cytokine production in the central nervous system of Lewis rats with experimental autoimmune encephalomyelitis: dynamics of mRNA expression for interleukin-10, interleukin-12, cytolysin, tumor necrosis factor alpha and tumor necrosis factor beta.* J Neuroimmunol, 1995. **61**(2): p. 205-12.

43.    Powell, M.B., D. Mitchell, J. Lederman, J. Buckmeier, S.S. Zamvil, M. Graham, N.H. Ruddle, and L. Steinman, *Lymphotoxin and tumor necrosis factor-alpha production by myelin basic protein-specific T cell clones correlates with encephalitogenicity.* Int Immunol, 1990. **2**(6): p. 539-44.

44.    Ando, D.G., J. Clayton, D. Kono, J.L. Urban, and E.E. Sercarz, *Encephalitogenic T cells in the B10.PL model of experimental allergic encephalomyelitis (EAE) are of the Th-1 lymphokine subtype.* Cell Immunol, 1989. **124**(1): p. 132-43.

45.    Voskuhl, R.R., R. Martin, C. Bergman, M. Dalal, N.H. Ruddle, and H.F. McFarland, *T helper 1 (Th1) functional phenotype of human myelin basic protein-specific T lymphocytes.* Autoimmunity, 1993. **15**(2): p. 137-43.

46.    Windhagen, A., J. Newcombe, F. Dangond, C. Strand, M.N. Woodroofe, M.L. Cuzner, and D.A. Hafler, *Expression of costimulatory molecules B7-1 (CD80), B7-2 (CD86), and interleukin 12 cytokine in multiple sclerosis lesions.* J Exp Med, 1995. **182**(6): p. 1985-96.

47.    Jander, S. and G. Stoll, *Differential induction of interleukin-12, interleukin-18, and interleukin-1beta converting enzyme mRNA in experimental autoimmune encephalomyelitis of the Lewis rat.* J Neuroimmunol, 1998. **91**(1-2): p. 93-9.

48.    Bright, J.J., B.F. Musuro, C. Du, and S. Sriram, *Expression of IL-12 in CNS and lymphoid organs of mice with experimental allergic encephalitis.* J Neuroimmunol, 1998. **82**(1): p. 22-30.

49.    Constantinescu, C.S., K. Frei, M. Wysocka, G. Trinchieri, U. Malipiero, A. Rostami, and A. Fontana, *Astrocytes and microglia produce interleukin-12 p40.* Ann N Y Acad Sci, 1996. **795**: p. 328-33.

50.    Aloisi, F., G. Penna, J. Cerase, B. Menendez Iglesias, and L. Adorini, *IL-12 production by central nervous system microglia is inhibited by astrocytes.* J Immunol, 1997. **159**(4): p. 1604-12.

51.    Leonard, J.P., K.E. Waldburger, and S.J. Goldman, *Prevention of experimental autoimmune encephalomyelitis by antibodies against interleukin 12.* J Exp Med, 1995. **181**(1): p. 381-6.

52.    Constantinescu, C.S., M. Wysocka, B. Hilliard, E.S. Ventura, E. Lavi, G. Trinchieri, and A. Rostami, *Antibodies against IL-12 prevent superantigen-induced and*

*spontaneous relapses of experimental autoimmune encephaloomyelitis.* J Immunol, 1998. **161**: p. 5097-5104.

53.  Ichikawa, M., C.S. Koh, A. Inoue, J. Tsuyusaki, M. Yamazaki, Y. Inaba, Y. Sekiguchi, M. Itoh, H. Yagita, and A. Komiyama, *Anti-IL-12 antibody prevents the development and progression of multiple sclerosis-like relapsing--remitting demyelinating disease in NOD mice induced with myelin oligodendrocyte glycoprotein peptide.* J Neuroimmunol, 2000. **102**(1): p. 56-66.

54.  Heremans, H., C. Dillen, M. Groenen, P. Matthys, and A. Billiau, *Role of endogenous interleukin-12 (IL-12) in induced and spontaneous relapses of experimental autoimmune encephalomyelitis in mice.* Eur Cytokine Netw, 1999. **10**(2): p. 171-80.

55.  Brok, H.P., M. van Meurs, E. Blezer, A. Schantz, D. Peritt, G. Treacy, J.D. Laman, J. Bauer, and B.A. t Hart, *Prevention of experimental autoimmune encephalomyelitis in common marmosets using an anti-IL-12p40 monoclonal antibody.* J Immunol, 2002. **169**(11): p. 6554-63.

56.  Moller, D.R., M. Wysocka, B.M. Greenlee, X. Ma, L. Wahl, G. Trinchieri, and C.L. Karp, *Inhibition of human interleukin-12 production by pentoxifylline.* Immunology, 1997. **91**(2): p. 197-203.

57.  Du, C., J.C. Cooper, S.J. Klaus, and S. Sriram, *Amelioration of CR-EAE with lisofylline: effects on mRNA levels of IL-12 and IFN-gamma in the CNS.* J Neuroimmunol, 2000. **110**(1-2): p. 13-9.

58.  Constantinescu, C.S., E. Ventura, B. Hilliard, and A. Rostami, *Effects of the angiotensin converting enzyme inhibitor captopril on experimental autoimmune encephalomyelitis.* Immunopharmacol Immunotoxicol, 1995. **17**(3): p. 471-91.

59.  Constantinescu, C.S., D.B. Goodman, and E.S. Ventura, *Captopril and lisinopril suppress production of interleukin-12 by human peripheral blood mononuclear cells.* Immunol Lett, 1998. **62**(1): p. 25-31.

60.  Natarajan, C. and J.J. Bright, *Curcumin inhibits experimental allergic encephalomyelitis by blocking IL-12 signaling through Janus kinase-STAT pathway in T lymphocytes.* J Immunol, 2002. **168**(12): p. 6506-13.

61.  Youssef, S., O. Stuve, J.C. Patarroyo, P.J. Ruiz, J.L. Radosevich, E.M. Hur, M. Bravo, D.J. Mitchell, R.A. Sobel, L. Steinman, and S.S. Zamvil, *The HMG-CoA reductase inhibitor, atorvastatin, promotes a Th2 bias and reverses paralysis in central nervous system autoimmune disease.* Nature, 2002. **420**(6911): p. 78-84.

62.  Chitnis, T., N. Najafian, C. Benou, A.D. Salama, M.J. Grusby, M.H. Sayegh, and S.J. Khoury, *Effect of targeted disruption of STAT4 and STAT6 on the induction of experimental autoimmune encephalomyelitis.* J Clin Invest, 2001. **108**(5): p. 739-47.

63.  Segal, B.M. and E.M. Shevach, *IL-12 unmasks latent autoimmune disease in resistant mice.* J Exp Med, 1996. **184**(2): p. 771-5.

64.  Smith, T., A.K. Hewson, C.I. Kingsley, J.P. Leonard, and M.L. Cuzner, *Interleukin-12 induces relapse in experimental allergic encephalomyelitis in the Lewis rats.* Am J Pathol, 1997. **150**: p. 1909-17.

65.  Ahmed, Z., A.I. Doward, G. Pryce, D.L. Taylor, J.M. Pocock, J.P. Leonard, D. Baker, and M.L. Cuzner, *A role for caspase-1 and -3 in the pathology of experimental allergic encephalomyelitis : inflammation versus degeneration.* Am J Pathol, 2002. **161**(5): p. 1577-86.

66.  Ahmed, Z., D. Gveric, G. Pryce, D. Baker, J.P. Leonard, M.L. Cuzner, and L.T. Diemel, *Myelin/axonal pathology in interleukin-12 induced serial relapses of experimental allergic encephalomyelitis in the Lewis rat.* Am J Pathol, 2001. **158**(6): p. 2127-38.

67.     Segal, D.M., J.T. Chang, and E.M. Shevach, *CpG oligonucleotides are potent adjuvants for the activation of autoreactive encephalitogenic T cells in vivo.* J Immunol, 2000. **164**: p. 5683-8.

68.     Segal, B.M., B.K. Dwyer, and E.M. Shevach, *An interleukin (IL)-10/IL-12 immunoregulatory circuit controls susceptibility to autoimmune disease.* J Exp Med, 1998. **187**(4): p. 537-46.

69.     Gerritse, K., J.D. Laman, R.J. Noelle, A. Aruffo, J.A. Ledbetter, W.J. Boersma, and E. Claassen, *CD40-CD40 ligand interactions in experimental allergic encephalomyelitis and multiple sclerosis.* Proc Natl Acad Sci U S A, 1996. **93**(6): p. 2499-504.

70.     Samoilova, E.B., J.L. Horton, H. Zhang, and Y. Chen, *CD40L blockade prevents autoimmune encephalomyelitis and hampers TH1 but not TH2 pathway of T cell differentiation.* J Mol Med, 1997. **75**(8): p. 603-8.

71.     Grewal, I.S., H.G. Foellmer, K.D. Grewal, J. Xu, F. Hardardottir, J.L. Baron, C.A. Janeway, Jr., and R.A. Flavell, *Requirement for CD40 ligand in costimulation induction, T cell activation, and experimental allergic encephalomyelitis.* Science, 1996. **273**(5283): p. 1864-7.

72.     Constantinescu, C.S., B. Hilliard, M. Wysocka, E.S. Ventura, M.K. Bhopale, G. Trinchieri, and A.M. Rostami, *IL-12 reverses the suppressive effect of the CD40 ligand blockade on experimental autoimmune encephalomyelitis (EAE).* J Neurol Sci, 1999. **171**(1): p. 60-4.

73.     Ferber, I.A., S. Brocke, C. Taylor-Edwards, W. Ridgway, C. Dinisco, L. Steinman, D. Dalton, and C.G. Fathman, *Mice with a disrupted IFN-gamma gene are susceptible to the induction of experimental autoimmune encephalomyelitis (EAE).* J Immunol, 1996. **156**(1): p. 5-7.

74.     Scheper, R.J., M.E. von Blomberg, J. de Groot, D.A. Wolvers, G. Kraal, and A.M. Claessen, *Reversal of orally induced T-cell tolerance by subcutaneous administration of interleukin-12 at the site of attempted sensitization.* Ann N Y Acad Sci, 1996. **795**: p. 403-9.

75.     Zhang, G.X., H. Xu, M. Kishi, D. Calida, and A. Rostami, *The role of IL-12 in the induction of intravenous tolerance in experimental autoimmune encephalomyelitis.* J Immunol, 2002. **168**(5): p. 2501-7.

76.     Eaton, A.D., D. Xu, and P. Garside, *Administration of exogenous interleukin-18 and interleukin-12 prevents the induction of oral tolerance.* Immunology, 2003. **108**(2): p. 196-203.

77.     Marth, T., W. Strober, and B.L. Kelsall, *High dose oral tolerance in ovalbumin TCR-transgenic mice: systemic neutralization of IL-12 augments TGF-beta secretion and T cell apoptosis.* J Immunol, 1996. **157**(6): p. 2348-57.

78.     Chevrel, G., G. Page, C. Granet, N. Streichenberger, A. Varennes, and P. Miossec, *Interleukin-17 increases the effects of IL-1beta on muscle cells: arguments for the role of T cells in the pathogenesis of myositis.* J Neuroimmunol, 2003. **137**(1-2): p. 125-33.

79.     Nielsen, O.H., I. Kirman, N. Rudiger, J. Hendel, and B. Vainer, *Upregulation of interleukin-12 and -17 in active inflammatory bowel disease.* Scand J Gastroenterol, 2003. **38**(2): p. 180-5.

80.     Kotake, S., N. Udagawa, N. Takahashi, K. Matsuzaki, K. Itoh, S. Ishiyama, S. Saito, K. Inoue, N. Kamatani, M.T. Gillespie, T.J. Martin, and T. Suda, *IL-17 in synovial fluids from patients with rheumatoid arthritis is a potent stimulator of osteoclastogenesis.* J Clin Invest, 1999. **103**(9): p. 1345-52.

81.     Zhang, G.X., B. Gran, S. Yu, J. Li, I. Siglienti, X. Chen, M. Kamoun, and A. Rostami, *Induction of experimental autoimmune encephalomyelitis in IL-12 receptor-beta 2-deficient mice: IL-12 responsiveness is not required in the*

pathogenesis of inflammatory demyelination in the central nervous system. J Immunol, 2003. **170**(4): p. 2153-60.

82.    Okamura, H., H. Tsutsi, T. Komatsu, M. Yutsudo, A. Hakura, T. Tanimoto, K. Torigoe, T. Okura, Y. Nukada, K. Hattori, and et al., *Cloning of a new cytokine that induces IFN-gamma production by T cells.* Nature, 1995. **378**(6552): p. 88-91.

83.    Yoshimoto, T., K. Takeda, T. Tanaka, K. Ohkusu, S. Kashiwamura, H. Okamura, S. Akira, and K. Nakanishi, *IL-12 up-regulates IL-18 receptor expression on T cells, Th1 cells, and B cells: synergism with IL-18 for IFN-gamma production.* J Immunol, 1998. **161**(7): p. 3400-7.

84.    Xu, D., W.L. Chan, B.P. Leung, D. Hunter, K. Schulz, R.W. Carter, I.B. McInnes, J.H. Robinson, and F.Y. Liew, *Selective expression and functions of interleukin 18 receptor on T helper (Th) type 1 but not Th2 cells.* J Exp Med, 1998. **188**(8): p. 1485-92.

85.    Chang, J.T., B.M. Segal, K. Nakanishi, H. Okamura, and E.M. Shevach, *The costimulatory effect of IL-18 on the induction of antigen-specific IFN-gamma production by resting T cells is IL-12 dependent and is mediated by up-regulation of the IL-12 receptor beta2 subunit.* Eur J Immunol, 2000. **30**(4): p. 1113-9.

86.    Fantuzzi, G., D.A. Reed, and C.A. Dinarello, *IL-12-induced IFN-gamma is dependent on caspase-1 processing of the IL-18 precursor.* J Clin Invest, 1999. **104**(6): p. 761-7.

87.    Wildbaum, G., S. Youssef, N. Grabie, and N. Karin, *Neutralizing antibodies to IFN-gamma-inducing factor prevent experimental autoimmune encephalomyelitis.* J Immunol, 1998. **161**(11): p. 6368-74.

88.    Oppmann, B., R. Lesley, B. Blom, J.C. Timans, Y. Xu, B. Hunte, F. Vega, N. Yu, J. Wang, K. Singh, F. Zonin, E. Vaisberg, T. Churakova, M. Liu, D. Gorman, J. Wagner, S. Zurawski, Y. Liu, J.S. Abrams, K.W. Moore, D. Rennick, R. de Waal-Malefyt, C. Hannum, J.F. Bazan, and R.A. Kastelein, *Novel p19 protein engages IL-12p40 to form a cytokine, IL-23, with biological activities similar as well as distinct from IL-12.* Immunity, 2000. **13**(5): p. 715-25.

89.    Becher, B., B.G. Durell, and R.J. Noelle, *Experimental autoimmune encephalitis and inflammation in the absence of interleukin-12.* J Clin Invest, 2002. **110**(4): p. 493-7.

90.    Gran, B., G.X. Zhang, S. Yu, J. Li, X.H. Chen, E.S. Ventura, M. Kamoun, and A. Rostami, *IL-12p35-deficient mice are susceptible to experimental autoimmune encephalomyelitis: evidence for redundancy in the IL-12 system in the induction of central nervous system autoimmune demyelination.* J Immunol, 2002. **169**(12): p. 7104-10.

91.    Cua, D.J., J. Sherlock, Y. Chen, C.A. Murphy, B. Joyce, B. Seymour, L. Lucian, W. To, S. Kwan, T. Churakova, S. Zurawski, M. Wiekowski, S.A. Lira, D. Gorman, R.A. Kastelein, and J.D. Sedgwick, *Interleukin-23 rather than interleukin-12 is the critical cytokine for autoimmune inflammation of the brain.* Nature, 2003. **421**(6924): p. 744-8.

92.    Pflanz, S., J.C. Timans, J. Cheung, R. Rosales, H. Kanzler, J. Gilbert, L. Hibbert, T. Churakova, M. Travis, E. Vaisberg, W.M. Blumenschein, J.D. Mattson, J.L. Wagner, W. To, S. Zurawski, T.K. McClanahan, D.M. Gorman, J.F. Bazan, R. de Waal Malefyt, D. Rennick, and R.A. Kastelein, *IL-27, a heterodimeric cytokine composed of EBI3 and p28 protein, induces proliferation of naive CD4(+) T cells.* Immunity, 2002. **16**(6): p. 779-90.

93.    Selmaj, K., C.S. Raine, B. Cannella, and C.F. Brosnan, *Identification of lymphotoxin and tumor necrosis factor in multiple sclerosis lesions.* J Clin Invest, 1991. **87**(3): p. 949-54.

94.     Begolka, W.S., C.L. Vanderlugt, S.M. Rahbe, and S.D. Miller, *Differential expression of inflammatory cytokines parallels progression of central nervous system pathology in two clinically distinct models of multiple sclerosis.* J Immunol, 1998. **161**(8): p. 4437-46.

95.     Karpus, W.J., K.E. Gould, and R.H. Swanborg, *CD4+ suppressor cells of autoimmune encephalomyelitis respond to T cell receptor-associated determinants on effector cells by interleukin-4 secretion.* Eur J Immunol, 1992. **22**(7): p. 1757-63.

96.     Crisi, G.M., L. Santambrogio, G.M. Hochwald, S.R. Smith, J.A. Carlino, and G.J. Thorbecke, *Staphylococcus eneterotoxin B and tumor-necrosis factor-$\alpha$ induced relapses of experimental allergic encephalomyelitis: protection by transforming growth factor-b and interleukin-10.* Eur J Immunol, 1995. **25**: p. 3035-40.

97.     Taupin, V., T. Renno, L. Bourbonniere, A.C. Peterson, M. Rodriguez, and T. Owens, *Increased severity of experimental autoimmune encephalomyelitis, chronic macrophage/microglial reactivity, and demyelination in transgenic mice producing tumor necrosis factor-alpha in the central nervous system.* Eur J Immunol, 1997. **27**(4): p. 905-13.

98.     Probert, L., K. Akassoglou, M. Pasparakis, G. Kontogeorgos, and G. Kollias, *Spontaneous inflammatory demyelinating disease in transgenic mice showing central nervous system-specific expression of tumor necrosis factor alpha.* Proc Natl Acad Sci U S A, 1995. **92**(24): p. 11294-8.

99.     Akassoglou, K., J. Bauer, G. Kassiotis, M. Pasparakis, H. Lassmann, G. Kollias, and L. Probert, *Oligodendrocyte apoptosis and primary demyelination induced by local TNF/p55TNF receptor signaling in the central nervous system of transgenic mice: models for multiple sclerosis with primary oligodendrogliopathy.* Am J Pathol, 1998. **153**(3): p. 801-13.

100.    Ruddle, N.H., C.M. Bergman, K.M. McGrath, E.G. Lingenheld, M.L. Grunnet, S.J. Padula, and R.B. Clark, *An antibody to lymphotoxin and tumor necrosis factor prevents transfer of experimental allergic encephalomyelitis.* J Exp Med, 1990. **172**(4): p. 1193-200.

101.    Teuscher, C., W.F. Hickey, and R. Korngold, *An analysis of the role of tumor necrosis factor in the phenotypic expression of actively induced experimental allergic orchitis and experimental allergic encephalomyelitis.* Clin Immunol Immunopathol, 1990. **54**(3): p. 442-53.

102.    Baker, D., D. Butler, B.J. Scallon, J.K. O'Neill, J.L. Turk, and M. Feldmann, *Control of established experimental allergic encephalomyelitis by inhibition of tumor necrosis factor (TNF) activity within the central nervous system using monoclonal antibodies and TNF receptor-immunoglobulin fusion proteins.* Eur J Immunol, 1994. **24**(9): p. 2040-8.

103.    Sommer, N., P.A. Loschmann, G.H. Northoff, M. Weller, A. Steinbrecher, J.P. Steinbach, R. Lichtenfels, R. Meyermann, A. Riethmuller, A. Fontana, and et al., *The antidepressant rolipram suppresses cytokine production and prevents autoimmune encephalomyelitis.* Nat Med, 1995. **1**(3): p. 244-8.

104.    Dinter, H., J. Tse, M. Halks-Miller, D. Asarnow, J. Onuffer, D. Faulds, B. Mitrovic, G. Kirsch, H. Laurent, P. Esperling, D. Seidelmann, E. Ottow, H. Schneider, V.K. Tuohy, H. Wachtel, and H.D. Perez, *The type IV phosphodiesterase specific inhibitor mesopram inhibits experimental autoimmune encephalomyelitis in rodents.* J Neuroimmunol, 2000. **108**(1-2): p. 136-46.

105.    Frei, K., H.P. Eugster, M. Bopst, C.S. Constantinescu, E. Lavi, and A. Fontana, *Tumor necrosis factor alpha and lymphotoxin alpha are not required for induction of acute experimental autoimmune encephalomyelitis.* J Exp Med, 1997. **185**(12): p. 2177-82.

106.     Eugster, H.P., K. Frei, R. Bachmann, H. Bluethmann, H. Lassmann, and A. Fontana, *Severity of symptoms and demyelination in MOG-induced EAE depends on TNFR1.* Eur J Immunol, 1999. **29**(2): p. 626-32.

107.     Liu, J., M.W. Marino, G. Wong, D. Grail, A. Dunn, J. Bettadapura, A.J. Slavin, L. Old, and C.C. Bernard, *TNF is a potent anti-inflammatory cytokine in autoimmune-mediated demyelination.* Nat Med, 1998. **4**(1): p. 78-83.

108.     Speiser, D.E., E. Sebzda, T. Ohteki, M.F. Bachmann, K. Pfeffer, T.W. Mak, and P.S. Ohashi, *Tumor necrosis factor receptor p55 mediates deletion of peripheral cytotoxic T lymphocytes in vivo.* Eur J Immunol, 1996. **26**(12): p. 3055-60.

109.     Shinpo, K., S. Kikuchi, F. Moriwaka, and K. Tashiro, *Protective effects of the TNF-ceramide pathway against glutamate neurotoxicity on cultured mesencephalic neurons.* Brain Res, 1999. **819**(1-2): p. 170-3.

110.     Cheng, B., S. Christakos, and M.P. Mattson, *Tumor necrosis factors protect neurons against metabolic-excitotoxic insults and promote maintenance of calcium homeostasis.* Neuron, 1994. **12**(1): p. 139-53.

111.     Beattie, E.C., D. Stellwagen, W. Morishita, J.C. Bresnahan, B.K. Ha, M. Von Zastrow, M.S. Beattie, and R.C. Malenka, *Control of synaptic strength by glial TNFalpha.* Science, 2002. **295**(5563): p. 2282-5.

112.     Chabas, D., S.E. Baranzini, D. Mitchell, C.C. Bernard, S.R. Rittling, D.T. Denhardt, R.A. Sobel, C. Lock, M. Karpuj, R. Pedotti, R. Heller, J.R. Oksenberg, and L. Steinman, *The influence of the proinflammatory cytokine, osteopontin, on autoimmune demyelinating disease.* Science, 2001. **294**(5547): p. 1731-5.

113.     Jansson, M., V. Panoutsakopoulou, J. Baker, L. Klein, and H. Cantor, *Cutting edge: Attenuated experimental autoimmune encephalomyelitis in eta-1/osteopontin-deficient mice.* J Immunol, 2002. **168**(5): p. 2096-9.

114.     O'Garra, A., S.E. Macatonia, C.S. Hsieh, and K.M. Murphy, *Regulatory role of IL4 and other cytokines in T helper cell development in an alpha beta TCR transgenic mouse system.* Res Immunol, 1993. **144**(8): p. 620-5.

115.     Khoury, S.J., W.W. Hancock, and H.L. Weiner, *Oral tolerance to myelin basic protein and natural recovery from experimental autoimmune encephalomyelitis are associated with downregulation of inflammatory cytokines and differential upregulation of transforming growth factor beta, interleukin 4, and prostaglandin E expression in the brain.* J Exp Med, 1992. **176**(5): p. 1355-64.

116.     Baron, J.L., J.A. Madri, N.H. Ruddle, G. Hashim, and C.A. Janeway, Jr., *Surface expression of alpha 4 integrin by CD4 T cells is required for their entry into brain parenchyma.* J Exp Med, 1993. **177**(1): p. 57-68.

117.     Racke, M.K., A. Bonomo, D.E. Scott, B. Cannella, A. Levine, C.S. Raine, E.M. Shevach, and M. Rocken, *Cytokine-induced immune deviation as a therapy for inflammatory autoimmune disease.* J Exp Med, 1994. **180**(5): p. 1961-6.

118.     van der Veen, R.C. and S.A. Stohlman, *Encephalitogenic Th1 cells are inhibited by Th2 cells with related peptide specificity: relative roles of interleukin (IL)-4 and IL-10.* J Neuroimmunol, 1993. **48**(2): p. 213-20.

119.     Khoruts, A., S.D. Miller, and M.K. Jenkins, *Neuroantigen-specific Th2 cells are inefficient suppressors of experimental autoimmune encephalomyelitis induced by effector Th1 cells.* J Immunol, 1995. **155**(10): p. 5011-7.

120.     Inobe, J.I., Y. Chen, and H.L. Weiner, *In vivo administration of IL-4 induces TGF-beta-producing cells and protects animals from experimental autoimmune encephalomyelitis.* Ann N Y Acad Sci, 1996. **778**: p. 390-2.

121.     Shaw, M.K., J.B. Lorens, A. Dhawan, R. DalCanto, H.Y. Tse, A.B. Tran, C. Bonpane, S.L. Eswaran, S. Brocke, N. Sarvetnick, L. Steinman, G.P. Nolan, and C.G. Fathman, *Local delivery of interleukin 4 by retrovirus-transduced T*

*lymphocytes ameliorates experimental autoimmune encephalomyelitis.* J Exp Med, 1997. **185**(9): p. 1711-4.

122.   Falcone, M. and B.R. Bloom, *A T helper cell 2 (Th2) immune response against non-self antigens modifies the cytokine profile of autoimmune T cells and protects against experimental allergic encephalomyelitis.* J Exp Med, 1997. **185**(5): p. 901-7.

123.   Sewell, D., Z. Qing, E. Reinke, D. Elliot, J. Weinstock, M. Sandor, and Z. Fabry, *Immunomodulation of experimental autoimmune encephalomyelitis by helminth ova immunization.* Int Immunol, 2003. **15**(1): p. 59-69.

124.   Ratts, R.B., L.R. Arredondo, P. Bittner, P.J. Perrin, A.E. Lovett-Racke, and M.K. Racke, *The role of CTLA-4 in tolerance induction and T cell differentiation in experimental autoimmune encephalomyelitis: i.p. antigen administration.* Int Immunol, 1999. **11**(12): p. 1881-8.

125.   Racke, M.K., D. Burnett, S.H. Pak, P.S. Albert, B. Cannella, C.S. Raine, D.E. McFarlin, and D.E. Scott, *Retinoid treatment of experimental allergic encephalomyelitis. IL-4 production correlates with improved disease course.* J Immunol, 1995. **154**: p. 450-8.

126.   Brocke, S., K. Gijbels, M. Allegretta, I. Ferber, C. Piercy, T. Blankenstein, R. Martin, U. Utz, N. Karin, D. Mitchell, T. Veromaa, A. Waisman, A. Gaur, P. Conlon, N. Ling, P.J. Fairchild, D.C. Wraith, A. O'Garra, C.G. Fathman, and L. Steinman, *Treatment of experimental encephalomyelitis with a peptide analogue of myelin basic protein.* Nature, 1996. **379**(6563): p. 343-6.

127.   Aharoni, R., D. Teitelbaum, O. Leitner, A. Meshorer, M. Sela, and R. Arnon, *Specific Th2 cells accumulate in the central nervous system of mice protected against experimental autoimmune encephalomyelitis by copolymer 1.* Proc Natl Acad Sci U S A, 2000. **97**(21): p. 11472-7.

128.   Aharoni, R., D. Teitelbaum, M. Sela, and R. Arnon, *Copolymer 1 induces T cells of the T helper type 2 that crossreact with myelin basic protein and suppress experimental autoimmune encephalomyelitis.* Proc Natl Acad Sci U S A, 1997. **94**(20): p. 10821-6.

129.   Constantinescu, C.S., B. Hilliard, E. Ventura, M. Wysocka, L. Showe, E. Lavi, T. Fujioka, P. Scott, G. Trinchieri, and A. Rostami, *Modulation of susceptibility and resistance to an autoimmune model of multiple sclerosis in prototypically susceptible and resistant strains by neutralization of interleukin-12 and interleukin-4, respectively.* Clin Immunol, 2001. **98**(1): p. 23-30.

130.   Falcone, M., A.J. Rajan, B.R. Bloom, and C.F. Brosnan, *A critical role for IL-4 in regulating disease severity in experimental allergic encephalomyelitis as demonstrated in IL-4-deficient C57BL/6 mice and BALB/c mice.* J Immunol, 1998. **160**(10): p. 4822-30.

131.   Zhao, M.L. and R.B. Fritz, *Acute and relapsing experimental autoimmune encephalomyelitis in IL-4- and alpha/beta T cell-deficient C57BL/6 mice.* J Neuroimmunol, 1998. **87**(1-2): p. 171-8.

132.   Liblau, R., L. Steinman, and S. Brocke, *Experimental autoimmune encephalomyelitis in IL-4-deficient mice.* Int Immunol, 1997. **9**(5): p. 799-803.

133.   Bettelli, E., M.P. Das, E.D. Howard, H.L. Weiner, R.A. Sobel, and V.K. Kuchroo, *IL-10 is critical in the regulation of autoimmune encephalomyelitis as demonstrated by studies of IL-10- and IL-4-deficient and transgenic mice.* J Immunol, 1998. **161**(7): p. 3299-306.

134.   Steinman, L., *Some misconceptions about understanding autoimmunity through experiments with knockouts.* J Exp Med, 1997. **185**(12): p. 2039-41.

135.   Forsthuber, T., H.C. Yip, and P.V. Lehmann, *Induction of TH1 and TH2 immunity in neonatal mice.* Science, 1996. **271**(5256): p. 1728-30.

136. Young, D.A., L.D. Lowe, S.S. Booth, M.J. Whitters, L. Nicholson, V.K. Kuchroo, and M. Collins, *IL-4, IL-10, IL-13, and TGF-beta from an altered peptide ligand-specific Th2 cell clone down-regulate adoptive transfer of experimental autoimmune encephalomyelitis.* J Immunol, 2000. **164**(7): p. 3563-72.

137. Benson, J.M., S.S. Stuckman, K.L. Cox, R.M. Wardrop, I.E. Gienapp, A.H. Cross, J.L. Trotter, and C.C. Whitacre, *Oral administration of myelin basic protein is superior to myelin in suppressing established relapsing experimental autoimmune encephalomyelitis.* J Immunol, 1999. **162**(10): p. 6247-54.

138. Abramsky, O., D.M. Karussis, R. Mizrachi-Koll, and D. Lehmann, *Inoculation of BCL1 lymphoma cells into CSJL/J F1 mice inhibits acute experimental autoimmune encephalomyelitis.* Isr J Med Sci, 1994. **30**(1): p. 1-6.

139. Moore, K.W., R. de Waal Malefyt, R.L. Coffman, and A. O'Garra, *Interleukin-10 and the interleukin-10 receptor.* Annu Rev Immunol, 2001. **19**: p. 683-765.

140. Gerosa, F., C. Paganin, D. Peritt, F. Paiola, M.T. Scupoli, M. Aste-Amezaga, I. Frank, and G. Trinchieri, *Interleukin-12 primes human CD4 and CD8 T cell clones for high production of both interferon-gamma and interleukin-10.* J Exp Med, 1996. **183**(6): p. 2559-69.

141. Rott, O., B. Fleischer, and E. Cash, *Interleukin-10 prevents experimental allergic encephalomyelitis in rats.* Eur J Immunol, 1994. **24**(6): p. 1434-40.

142. Xiao, B.G., X.F. Bai, G.X. Zhang, and H. Link, *Suppression of acute and protracted-relapsing experimental allergic encephalomyelitis by nasal administration of low-dose IL-10 in rats.* J Neuroimmunol, 1998. **84**(2): p. 230-7.

143. Nicholson, L.B., J.M. Greer, R.A. Sobel, M.B. Lees, and V.K. Kuchroo, *An altered peptide ligand mediates immune deviation and prevents autoimmune encephalomyelitis.* Immunity, 1995. **3**(4): p. 397-405.

144. Burkhart, C., G.Y. Liu, S.M. Anderton, B. Metzler, and D.C. Wraith, *Peptide-induced T cell regulation of experimental autoimmune encephalomyelitis: a role for IL-10.* Int Immunol, 1999. **11**(10): p. 1625-34.

145. Yasuda, C.L., A. Al-Sabbagh, E.C. Oliveira, B.M. Diaz-Bardales, A.A. Garcia, and L.M. Santos, *Interferon beta modulates experimental autoimmune encephalomyelitis by altering the pattern of cytokine secretion.* Immunol Invest, 1999. **28**(2-3): p. 115-26.

146. Massey, E.J., A. Sundstedt, M.J. Day, G. Corfield, S. Anderton, and D.C. Wraith, *Intranasal peptide-induced peripheral tolerance: the role of IL-10 in regulatory T cell function within the context of experimental autoimmune encephalomyelitis.* Vet Immunol Immunopathol, 2002. **87**(3-4): p. 357-72.

147. Soos, J.M., O. Stuve, S. Youssef, M. Bravo, H.M. Johnson, H.L. Weiner, and S.S. Zamvil, *Cutting edge: oral type I IFN-tau promotes a Th2 bias and enhances suppression of autoimmune encephalomyelitis by oral glatiramer acetate.* J Immunol, 2002. **169**(5): p. 2231-5.

148. Faria, A.M., R. Maron, S.M. Ficker, A.J. Slavin, T. Spahn, and H.L. Weiner, *Oral tolerance induced by continuous feeding: enhanced up-regulation of transforming growth factor-beta/interleukin-10 and suppression of experimental autoimmune encephalomyelitis.* J Autoimmun, 2003. **20**(2): p. 135-45.

149. Cannella, B., Y.L. Gao, C. Brosnan, and C.S. Raine, *IL-10 fails to abrogate experimental autoimmune encephalomyelitis.* J Neurosci Res, 1996. **45**(6): p. 735-46.

150. Samoilova, E.B., J.L. Horton, and Y. Chen, *Acceleration of experimental autoimmune encephalomyelitis in interleukin-10-deficient mice: roles of interleukin-10 in disease progression and recovery.* Cell Immunol, 1998. **188**(2): p. 118-24.

151.    Mathisen, P.M., M. Yu, J.M. Johnson, J.A. Drazba, and V.K. Tuohy, *Treatment of experimental autoimmune encephalomyelitis with genetically modified memory T cells.* J Exp Med, 1997. **186**(1): p. 159-64.

152.    Cua, D.J., H. Groux, D.R. Hinton, S.A. Stohlman, and R.L. Coffman, *Transgenic interleukin 10 prevents induction of experimental autoimmune encephalomyelitis.* J Exp Med, 1999. **189**(6): p. 1005-10.

153.    Saoudi, A., S. Simmonds, I. Huitinga, and D. Mason, *Prevention of experimental allergic encephalomyelitis in rats by targeting autoantigen to B cells: evidence that the protective mechanism depends on changes in the cytokine response and migratory properties of the autoantigen-specific T cells.* J Exp Med, 1995. **182**(2): p. 335-44.

154.    Cash, E., A. Minty, P. Ferrara, D. Caput, D. Fradelizi, and O. Rott, *Macrophage-inactivating IL-13 suppresses experimental autoimmune encephalomyelitis in rats.* J Immunol, 1994. **153**(9): p. 4258-67.

155.    Fickenscher, H., S. Hor, H. Kupers, A. Knappe, S. Wittmann, and H. Sticht, *The interleukin-10 family of cytokines.* Trends Immunol, 2002. **23**(2): p. 89-96.

156.    Fort, M.M., J. Cheung, D. Yen, J. Li, S.M. Zurawski, S. Lo, S. Menon, T. Clifford, B. Hunte, R. Lesley, T. Muchamuel, S.D. Hurst, G. Zurawski, M.W. Leach, D.M. Gorman, and D.M. Rennick, *IL-25 induces IL-4, IL-5, and IL-13 and Th2-associated pathologies in vivo.* Immunity, 2001. **15**(6): p. 985-95.

157.    Panek, R.B. and E.N. Benveniste, *Class II MHC gene expression in microglia. Regulation by the cytokines IFN-gamma, TNF-alpha, and TGF-beta.* J Immunol, 1995. **154**(6): p. 2846-54.

158.    Panek, R.B., Y.J. Lee, and E.N. Benveniste, *TGF-beta suppression of IFN-gamma-induced class II MHC gene expression does not involve inhibition of phosphorylation of JAK1, JAK2, or signal transducers and activators of transcription, or modification of IFN-gamma enhanced factor X expression.* J Immunol, 1995. **154**(2): p. 610-9.

159.    Santambrogio, L., G.M. Hochwald, C.H. Leu, and G.J. Thorbecke, *Antagonistic effects of endogenous and exogenous TGF-beta and TNF on auto-immune diseases in mice.* Immunopharmacol Immunotoxicol, 1993. **15**(4): p. 461-78.

160.    Santambrogio, L., G.M. Hochwald, B. Saxena, C.H. Leu, J.E. Martz, J.A. Carlino, N.H. Ruddle, M.A. Palladino, L.I. Gold, and G.J. Thorbecke, *Studies on the mechanisms by which transforming growth factor-beta (TGF-beta) protects against allergic encephalomyelitis. Antagonism between TGF-beta and tumor necrosis factor.* J Immunol, 1993. **151**(2): p. 1116-27.

161.    Racke, M.K., B. Cannella, P. Albert, M. Sporn, C.S. Raine, and D.E. McFarlin, *Evidence of endogenous regulatory function of transforming growth factor-beta 1 in experimental allergic encephalomyelitis.* Int Immunol, 1992. **4**(5): p. 615-20.

162.    Racke, M.K., S. Dhib-Jalbut, B. Cannella, P.S. Albert, C.S. Raine, and D.E. McFarlin, *Prevention and treatment of chronic relapsing experimental allergic encephalomyelitis by transforming growth factor-beta 1.* J Immunol, 1991. **146**(9): p. 3012-7.

163.    Stevens, D.B., K.E. Gould, and R.H. Swanborg, *Transforming growth factor-beta 1 inhibits tumor necrosis factor-alpha/lymphotoxin production and adoptive transfer of disease by effector cells of autoimmune encephalomyelitis.* J Neuroimmunol, 1994. **51**(1): p. 77-83.

164.    Kuruvilla, A.P., R. Shah, G.M. Hochwald, H.D. Liggitt, M.A. Palladino, and G.J. Thorbecke, *Protective effect of transforming growth factor beta 1 on experimental autoimmune diseases in mice.* Proc Natl Acad Sci U S A, 1991. **88**(7): p. 2918-21.

165.    Johns, L.D., K.C. Flanders, G.E. Ranges, and S. Sriram, *Successful treatment of experimental allergic encephalomyelitis with transforming growth factor-beta 1.* J Immunol, 1991. **147**(6): p. 1792-6.

166.    Racke, M.K., S. Sriram, J. Carlino, B. Cannella, C.S. Raine, and D.E. McFarlin, *Long-term treatment of chronic relapsing experimental allergic encephalomyelitis by transforming growth factor-beta 2.* J Neuroimmunol, 1993. **46**(1-2): p. 175-83.

167.    Johns, L.D. and S. Sriram, *Experimental allergic encephalomyelitis: neutralizing antibody to TGF beta 1 enhances the clinical severity of the disease.* J Neuroimmunol, 1993. **47**(1): p. 1-7.

168.    Chen, W. and S.M. Wahl, *TGF-beta: the missing link in CD4(+)CD25(+) regulatory T cell-mediated immunosuppression.* Cytokine Growth Factor Rev, 2003. **14**(2): p. 85-9.

169.    Karpus, W.J. and R.H. Swanborg, *CD4+ suppressor cells inhibit the function of effector cells of experimental autoimmune encephalomyelitis through a mechanism involving transforming growth factor-beta.* J Immunol, 1991. **146**(4): p. 1163-8.

170.    Chen, Y., V.K. Kuchroo, J. Inobe, D.A. Hafler, and H.L. Weiner, *Regulatory T cell clones induced by oral tolerance: suppression of autoimmune encephalomyelitis.* Science, 1994. **265**(5176): p. 1237-40.

171.    Huitinga, I., E.D. Schmidt, M.J. van der Cammen, R. Binnekade, and F.J. Tilders, *Priming with interleukin-1beta suppresses experimental allergic encephalomyelitis in the Lewis rat.* J Neuroendocrinol, 2000. **12**(12): p. 1186-93.

172.    Papiernik, M., M.L. de Moraes, C. Pontoux, F. Vasseur, and C. Penit, *Regulatory CD4 T cells: expression of IL-2R alpha chain, resistance to clonal deletion and IL-2 dependency.* Int Immunol, 1998. **10**(4): p. 371-8.

173.    Bassiri, H. and S.R. Carding, *A requirement for IL-2/IL-2 receptor signaling in intrathymic negative selection.* J Immunol, 2001. **166**(10): p. 5945-54.

174.    Contractor, N.V., H. Bassiri, T. Reya, A.Y. Park, D.C. Baumgart, M.A. Wasik, S.G. Emerson, and S.R. Carding, *Lymphoid hyperplasia, autoimmunity, and compromised intestinal intraepithelial lymphocyte development in colitis-free gnotobiotic IL-2-deficient mice.* J Immunol, 1998. **160**(1): p. 385-94.

175.    Menges, M., S. Rossner, C. Voigtlander, H. Schindler, N.A. Kukutsch, C. Bogdan, K. Erb, G. Schuler, and M.B. Lutz, *Repetitive injections of dendritic cells matured with tumor necrosis factor alpha induce antigen-specific protection of mice from autoimmunity.* J Exp Med, 2002. **195**(1): p. 15-21.

176.    Tarrant, T.K., P.B. Silver, J.L. Wahlsten, L.V. Rizzo, C.C. Chan, B. Wiggert, and R.R. Caspi, *Interleukin 12 protects from a T helper type 1-mediated autoimmune disease, experimental autoimmune uveitis, through a mechanism involving interferon gamma, nitric oxide, and apoptosis.* J Exp Med, 1999. **189**(2): p. 219-30.

177.    Lafaille, J.L., F. Van de Keere, A.L. Hsu, and a.l. et, *Myelin basic protein-specific T helper 2 cells cause experimental autoimmune encephalomyelitis in immunodeficient hosts rather than protect them from the disease.* J Exp Med, 1997. **186**: p. 307-12.

178.    Weinberg, A.D., R. Whitham, S.L. Swain, W.J. Morrison, G. Wyrick, C. Hoy, A.A. Vandenbark, and H. Offner, *Transforming growth factor-beta enhances the in vivo effector function and memory phenotype of antigen-specific T helper cells in experimental autoimmune encephalomyelitis.* J Immunol, 1992. **148**(7): p. 2109-17.

179.    Wyss-Coray, T., P. Borrow, M.J. Brooker, and L. Mucke, *Astroglial overproduction of TGF-beta 1 enhances inflammatory central nervous system disease in transgenic mice.* J Neuroimmunol, 1997. **77**(1): p. 45-50.

Chapter A15

# THE ROLE OF INTERFERONS IN EXPERIMENTAL AUTOIMMUNE ENCEPHALOMYELITIS

Hubertine Heremans & Alfons Billiau
*Rega Institute for Medical Research, University of Leuven, Belgium*

**Abstract**: Interferons, despite their common name, comprise a group of cytokines with quite different molecular structures, cellular receptors, biological effects, functions and applications. We here review studies directed at defining the role played by these molecules when they are produced endogenously or administered exogenously in the course of experimental autoimmune encephalomyelitis (EAE), a model disease considered relevant for the pathogenesis of multiple sclerosis (MS).

Studies on the role of Type II interferon, i.e. interferon-γ (IFN-γ), are almost unanimously indicative of a beneficial role, though this is in contrast to clinical observations pertaining to the role of IFN-γ in MS. Possible explanations for this discrepancy are considered in this review.

Interferon-β (IFN-β), one of the Type I interferons, also exerts a beneficial effect in EAE. In this case there is good correspondence with clinical observations, and studies in the animal model have contributed to providing possible explanations for these beneficial effects.

Our review also covers a limited number of studies on the apparently beneficial effect of treatment with another Type I interferon, namely IFN-τ, originally discovered as a pregnancy recognition hormone in sheep.

**Keywords**: interferon, IFN-β (interferon-beta), IFN-γ (interferon-gamma), IFN-τ (inteferon-tau), multiple sclerosis, autoimmune disease, autoimmunity, autoimmune encephalomyelitis.

# 1.    INTRODUCTION

Interferons (IFNs) are major contributors to the first line of antiviral defence. However, soon after their discovery in 1957 as antiviral factors, it became clear that they exert many other important effects on cells and organs. In fact interferons belong to the network of cytokines involved in the control of cellular functions, replication, and the immune response.

Over the years it became clear that interferons represent a large group of different proteins and glycoproteins some of which are products of a multigene family, whereas others are the products of separate genes (reviewed by De Maeyer & De Maeyer-Guignard [1]). The current nomenclature for interferons is based on sequence analysis of the IFN genes. Two major interferon superfamilies can be distinguished. Type I IFNs are all derived from the same ancestral gene represented as a cluster on the same chromosome and have sufficient structural homology to act via a common cell surface receptor. They comprise IFN-$\alpha$, IFN-$\beta$, IFN-$\omega$ and IFN-$\tau$. Details on the different species of Type I interferons and their cognate cell surface complex have been reviewed [1-3]. Type II interferon or interferon-$\gamma$ (IFN-$\gamma$), originally discovered by Wheelock in 1965 [4] as an antiviral factor, is a lymphokine encoded by a single gene that displays no molecular homology with type I IFN's and uses a receptor on the cell membrane distinct from the common receptor for IFN-$\alpha$ and IFN-$\beta$, while sharing some important biological activities with the "classical" interferons. IFN-$\gamma$ exerts multiple regulatory functions on cells of the immune system distinct from these of IFN-$\alpha$ and IFN-$\beta$. To date considerable progress has been made in the knowledge of the IFN receptors, molecular mechanisms of induction and of signal transduction in IFN-activated cells. For references the reader is referred to extensive reviews [2, 5-7].

Another aspect that became clear over the years is that interferons are involved in the pathogenesis of many diseases (reviewed in references [6, 8, 9]). Medical interest in interferons stems from the fact that they represent an important mechanism of nonspecific resistance as well as the specific immune response against a variety of micro-organisms. A deficient function of the sysem is generally believed to be instrumental in increased susceptibility to infection and cancer. Conversely, inappropriately high-level production of IFN-$\gamma$ is considered to potentially act as a disease-promoting factor in inflammatory and autoimmune diseases such as MS, Crohn's disease, rheumatoid arthritis, myasthenia gravis and others.

Studies on the role of IFN in experimental autoimmune encephalomyelitis (EAE), which are the subject of this review, have made a major contribution to our current knowledge on the latter subject. EAE is a widely used model for multiple sclerosis (MS) which can be induced in several genetically susceptible rodents and primates by active immunisation with myelin components together with one or several immune adjuvants or by adoptive transfer of encephalitogenic T lymphocytes, T cell lines or T cell clones sensitized to myelin basic protein (MBP) or myelin proteolipid protein (reviewed by Zamvill and Steinman [10] and Martin et al. [11]). The cytokine network plays a central role in initiation and regulation of the disease process. For some cytokines, there is unanimity in literature that they act either as disease promoters (e.g. TNF-$\alpha$), for others that they downregulate the disease process. IFN-$\gamma$ is rather exceptional in that both types of effects have been noted.

## 2.        IFN-$\gamma$ IN EAE

## 2.1        Effects of IFN-$\gamma$ relevant to the pathogenesis of EAE

IFN-$\gamma$ is a cytokine with several immunoregulatory activities (reviewed in ref. [6, 7]) that could in different ways contribute to the pathogenesis of EAE. IFN-$\gamma$ is known as the most potent inducer of major histocompatibility complex (MHC) antigen expression on a variety of cell types in vitro, including cerebral vascular endothelial cells, astrocytes, and microglia, which in EAE have been shown to play an important role as presenters of central nervous system (CNS) antigen to encephalitogenic T cells (reviewed by [12, 13]). Class I and Class II MHC antigens are barely detectable in the normal CNS. By contrast, in animals with EAE, a marked elevation of these antigen expression mainly on microvascular endothelial cells, but also on infiltrating macrophages and microglia is observed, possibly induced by IFN-$\gamma$ that is produced by the invading T cells. Following intrathecal injection of IFN-$\gamma$, MHC class II antigen expression was found to be increased on perivascular cells and various neuronal cells that do not normally express these proteins at detectable levels [14]. Although class II antigen expression on astrocytes has been conclusively demonstrated in vitro, in vivo studies have yielded conflicting results (see [12] for references). With some exceptions, most studies failed to detect class II antigen-positive astrocytes in the CNS of rats and mice during the disease process. However, class II antigen expression on astrocytes was reported in active MS lesions suggesting that the ability of astrocytes to function as antigen-presenting cells in vivo differs between rodents and humans.

Another activity of IFN-$\gamma$ consists in influencing production as well as activity of other cytokines (TNF-$\alpha$, IL-1) (see [5]) and other mediators of inflammation that play an important role in the pathogenesis of EAE (reviewed by Hartung et al. [15]). This is of particular importance for mediators released by invading MPC, by endothelial cells and by resident astrocytes and microglia. For instance, production of proteases by these cells is thought to be responsible for destruction of myelin (reviewed by Opdenakker & Van Damme [16]. In addition, IFN-γ synergizes with TNF-α that is recognized to exert a disease-promoting effect in EAE, probably by increasing the number of receptors for TNF-α on a variety of different cells [5].

Adhesion to the endothelial cell wall and breakdown of the blood brain barrier are critical steps in regulating the recruitment of activated encephalitogenic T cells and the subsequent influx of additional inflammatory cells (mononuclear and polynuclear phagocytes) from the circulation into the CNS (reviewed in ref. [11, 17]). In vitro IFN-γ causes increased lymphocyte adhesion to microvascular endothelial cells derived from the CNS [18, 19] and may even promote migration through endothelial cell layers in vitro [20]. After intracerebral injection or microinjection of IFN-γ into the lumbosacral spinal cords of normal rats, an influx of lymphocytes and other inflammatory cells into the CNS was noted similar in pattern to that observed in early EAE [14, 21]. A compromised blood-brain barrier may underlie this action of IFN-γ. Morphological and permeability changes on primary cultures of human brain microvessel endothelial cells have been reported [22]. The process of transmigration is governed by adhesion molecules that are reciprocally expressed on endothelial cells and leukocytes (reviewed by Hickey [23]). There have been several reports of fluctuating patterns of elevated levels of the intercellular adhesion molecule-1 (ICAM-1) and vascular cell-adhesion molecule-1 (VCAM-1) on CNS endothelium during EAE and intergrins on leukocytes, corresponding to periods of clinical disease [24-26, 28] (and reviewed by Hartung et al. [17]. IFN-γ is among the cytokines that augment expression of the intercellular adhesion molecule-1 (ICAM-1) on cultured brain microvascular endothelial cells in vitro resulting in increased adhesion of MBP-specific encephalitogenic T cells to these endothelial cells [27, 28] expressing the integrin LFA-1 [29]. The involvement of IFN-γ in the expression of VCAM-1 in vivo has been demonstrated by the observation that anti-IFN-γ antibody treatment inhibits the upregulation of VCAM-1 in a passive-transfer EAE model [30].

As soon as leukocytes have transgressed the endothelial barrier, they encounter chemokines which control their further movement to the target and cause them to release proteases leading to breakdown of myelin. Chemokines, including RANTES, macrophage inflammatory protein (MCP-1) α, macrophage chemoattractant protein-1 (MCP-1), IFN-γ-inducible protein-10 (IP-10) and chemokine receptors are upregulated in EAE, and their expression has been shown to correlate with disease severity and distribution of inflammatory infiltrates in the CNS (reviewed in ref. [31-33]). Chemokine secretion by endothelial cells and astrocytes is under the regulatory control by cytokines, particularly IFN-γ and TNF-α. The combination of IFN-γ and TNF-α was highly effective in inducing RANTES production *in vitro*: pretreatment with IFN-γ sensitized human glial cells to the induction of RANTES by TNF-α ([34, 35]. IP-10 which is highly expressed in the CNS of mice with an intact IFN-γ gene and EAE, was strikingly absent in IFN-γ knock-out mice. In vitro experiments confirmed that IFN-γ selectively stimulates astrocytes for IP-10 expression [36]. IFN-γ also upregulates MCP-1 gene transcription in vivo [37].

Several lines of evidence suggest that IFN-γ plays a deleterious role in immune-mediated demyelination (reviewed by Popko [38]). IFN-γ may have an indirect effect on the myelination process through the activation of invading MPC and microglial cells e.g. by release of TNF-α and NO [39, 40]. IFN-γ may also exert a direct deleterious effect on oligdendrocytes. Thus, application of IFN-γ to oligodendrocyte cell lines was found to cause molecular and morphological changes, and application to primary oligodendrocyte cultures resulted in apoptotic cell death [41, 42]. Furthermore, transgenic mice expressing IFN-γ in the CNS displayed a reduced myelination, reactive gliosis, macrophage/microglial stimulation, and an increase in MHC class I and class II expression in the CNS [43, 44] as well as increased susceptibility to EAE in mice that are normally resistant to EAE [45]

Finally, IFN-γ facilitates proliferation of the Th1 subset of murine CD4+ cells (see [46]), whereas it has a profound antiproliferative effect on the Th2 subset thereby reducing a key cellular source for IL-4 and IL-10. It has also been shown that IFN-γ blocks production of IL-10 by monocytes [47] and diminishes the ability to produce IL-4 in an immune response (see [46]), thereby reducing or interfering with the anti-inflammatory properties of these cytokines.

Taken together, IFN-γ exerts many effects on MPC, endothelial and resident cells of the CNS that are suggestive for a pro-inflammatory and thus predominantly disease-promoting role in EAE. Many of the pathological effects in the CNS that are observed in EAE and MS, including decreased myelination, increased gliosis, disruption of the blood-brain barrier function, increased MHC expression, stimulation of macrophages/microglial activation (reviewed by Popko [38]) are consistent with the known effects of IFN-γ.

## 2.2      IFN-γ production in EAE

Proliferation and activity of CNS antigen-specific CD4[+] lymphocytes of the Th1 phenotype are assumed to be the driving force in the development of tissue damage in EAE. Most significantly, EAE can be transferred to naïve animals using CD4[+] antigen specific Th1 cells isolated from animals suffering from EAE and anti-CD4[+] antibodies can inhibit induction of disease [10, 48]. IFN-γ is one of the cytokines produced by these cells after exposure to specific antigens or other stimuli, although in mice the suppressor/cytotoxic CD8+ subset of T cells and NK cells are also able to produce IFN-γ (reviewed by Young and Hardy [7]). In vitro studies have demonstrated that the effector T cells from rats previously challenged with MBP in complete Freunds adjuvant produce IFN-γ when cultured in the presence of the protein. As a contrast, disease suppressing splenic T cells found during the recovery phase of the disease failed to produce IFN-γ and inhibited IFN-γ production when added to challenged effector T cells [49].

Normally these IFN-γ producing cells do not transgress the blood-brain barrier at considerable levels and as such IFN-γ is generally undetectable within the CNS. Nevertheless, in response to infections of the CNS as well as in many immune-demyelinating disorders, CD4+ T cell-traffic across the blood-brain barrier does increase considerably (reviewed in ref. [23, 50]). Before performing their effector functions in the target organ, T cells must recognize Ag presented by antigen-presenting cells in the CNS. It has been shown that both infiltrating macrophages and microglia cells can efficiently present endogenous and exogenous Ags to T cells during the course of MOG-induced EAE, resulting in IFN-γ production [51]. Thus, the CNS-infiltrating CD4+ T-cells are the major source of local IFN-γ, thereby exposing neuronal and glial cells to the pleiotropic effects of IFN-γ. However, IFN-γ may have neuronal origins as well [52]. It should be noted that IFN-γ production by Th1 and NK cells is regulated by IL-12 and IL-18 in a synergistic manner and that the generation of CD4+ Th1 cells depends on the coordinate action of IL-12 and IL-18 with IFN-γ (reviewed in ref. [53-55]). An immunoregulatory role of IL-12 and IL-18 in EAE was demonstrated. IL-12 was found to enhance the in vitro activation of encephalitogenic T cells, and administration of IL-12 following adoptive transfer of PLP-stimulated lymph node cells accelerated, whereas blocking of endogenous IL-12 completely prevented disease progression [56-58]. IL-12 rendered otherwise resistant mice susceptible to EAE [59]. In vitro, neutralizing antibodies to IL-18 reduced T cell production of IFNγ in response to MBP epitopes. Moreover, neutralizing ant-IL-18 antibodies partially protected rats from EAE while shifting the Th1/Th2 balance in favor of Th2 selection, accompanied by a marked reduction in the production of IFN-γ and TNF-α ([60]).

With sensitive methods, individual IFN-γ-secreting cells could be detected in peripheral lymphoid organs and in the CNS of animals with acute and passively induced EAE [61-68]. The presence of IFN-γ within the CNS paralleled disease severity and occurred predominantly on inflammatory cells, consistent with previous observations correlating disease progression with infiltrations of effector CD4$^+$ T cells, a major source of this cytokine. The highest levels of IFN-γ were observed at the peak of acute disease activity and during relapses [51, 69-71, 71]. IFN-γ levels were very low or absent in recovered and refractory animals. Thus, local IFN-γ production seems to correlate with the disease induction phase. Accordingly, one would tend to postulate that endogenous IFN-γ, produced during EAE, promotes disease, a prediction thas is supported by the various pro-inflammatory effects of IFN-γ (*vide supra*) and the observation that exogenous IFN-γ augments demyelination in rats given anti-myelin/oligodendrocyte glycoprotein antibodies [72]. Moreover, in a single study on patients with MS, systemic treatment with IFN-γ was associated with exacerbated disease symptoms [73-75]. In reality, however, direct and indirect evaluation of the in vivo effects of endogenous IFN-γ in EAE has yielded contradictory results.

## 2.3    In vivo evidence for the involvement of IFN-γ in EAE

Most of our current knowledge on the role of IFN-γ in EAE stems from experiments employing monoclonal antibodies to neutralize IFN-γ in vivo, and from experiments involving the use of IFN-γ or IFN-γR knock-out mice. In mice, treatment with neutralizing anti-IFN-γ antibodies led to a significant exacerbation of EAE induced by either active immunization or passive transfer and increased mortality even in strains which are relatively resistant to the disease [76-80] (reviewed by Billiau [81]). The disease-enhancing effect was observed only when the antibodies were given early i.e. shortly after immunization or at the time of cell transfer, but not at later times, indicating that the effect of anti-IFN-γ antobody is exerted early in the induction phase of the disease at a stage which follows the appearance of activated CNS-reactive T cells. Also, relapses whether actively induced or occurring spontaneously, are facilitated by administration anti-IFN-γ antibody in the disease-free interval [57]. Studies using both IFN-γ ligand and IFN-γ receptor knock-out mice have been unanimous in reporting that such mice are more sensitive to induction of EAE than their wild-type littermates [82-85]. These data indicate that, although local production of

IFN-γ in the CNS is maximal during the clinically active phases of the disease, the overall role of IFN-γ is to inhibit (limit) EAE contrary to what was expected from a pro-inflammatory and immunostimulatory cytokine. The enhancing effect of anti-IFN-γ is consistent with the observation that, in mice, exogenously administered IFN-γ ameliorated the actively induced EAE reaction and alleviated or delayed disease attacks [57, 76], again suggesting that systemic IFN-γ is protective. This protective role was not observed when EAE was passively transferred [80].

None of these experiments allow dissociating the effects of IFN-γ that is present in the periphery and that produced in the CNS (which may affect EAE pathogenesis in opposite ways). Approaches to deliver the cytokine to the CNS have yielded contradictory results. Local, intraventricular injection of IFN-γ in a rat model for EAE prior to the onset of disease signs failed to affect primary induction of EAE but facilitated relapse induction, suggesting that local IFN-γ promotes disease development [69]. Contradictory to these observations was a result reported by Voorthuis et al. [79], showing no change in disease course following systemic administration, whereas ventricular administration resulted in a complete suppression of symptoms. Intrathecal delivery of an HSV-1-derived vector engineered with the mouse IFN-γ gene protected C57BL/6 mice from chronic progressive EAE induced with myelin oligodendrocyte glycoprotein (MOG) 35-55 [86]. Mice treated before onset of EAE with the IFN-γ-containing vector showed a shorter duration of the disease with full recovery and a significant reduction of neuropathological signs of the disease such as demyelination and axonal loss. Mice treated with the IFN-γ-containing vector within 1 week after EAE onset partially recovered from the disease. In mice, overexpression of IFN-γ in oligodendrocytes, incidence, severity and histopathology of EAE were found similar to those in wild-type controls, but the transgenic mice did develop chronic neurologic deficits, whereas the disease completely resolved in the controls. [43, 45].

## 2.4    Mechanisms involved

The overall disease-preventing effect of endogenous IFN-γ in murine EAE is suggestive of an immunosuppressive or anti-inflammatory pathway, which supersedes the 'classical' pro-inflammatory, and immunostimulatory pathways. By what mechanism IFN-γ exerts this action in EAE remains unclear. The discrepancy in results coming from the in vivo studies has been rationalized by invoking a differential action of IFN-γ in the periphery and in the CNS [76, 83]. At the local level IFN-γ may promote inflammation, while at the systemic level it may act in an anti-inflammatory fashion. It was proposed that circulating IFN-γ restrains Ag-triggered T lymphocyte proliferation and may promote apoptosis [83]. Therefore, it is plausible that elimination of IFN-γ from the circulation would have the effect of upregulating autoaggressive T cells and worsening the EAE disease process. In contrast, intrathecal IFN-γ is considered likely to activate resident microglia and infiltrating MPCs cumulating in increased tissue injury [83]. This concept is strongly supported by the studies in transgenic mice that expressed IFN-γ in the CNS under control of an oligodendrocyte-specific promoter. [43, 44].

Several studies have documented a role for IFN-γ in the generation and action of MPC-like cells that can suppress proliferative responses of T cells to mitogens (reviewed by Billiau [81]. For instance, it has been shown that T cells from IFN-γ knockout mice display a markedly enhanced in vitro proliferative response to Con A and alloantigens, suggesting a critical role of IFN-γ in suppressing T cell proliferation [87]. Anti-IFN-γ antibody enhanced in vitro proliferation of cloned T cells and of T cell lines specific to myelin proteins [77, 88]. In addition, lymph node and spleen cell cultures derived from either IFN-γ knockout or IFN-γR knockout mice showed enhanced proliferation to Ag restimulation in vitro [84, 85, 89]. More recently, in vivo evidence was presented that IFN-γ downregulates EAE by inhibiting proliferation of activated CD4+ T cells [90]. A possible mechanism of IFN-γ-dependent suppression of T cell proliferation through mononuclear phagocytes is generation of $H_2O_2$ and prostaglandins, since both catalase and indomethacin can alleviate suppression [91]. IFN-γ also enhances the release of the active form of TGF-β by human mononuclear phagocytes [92]. On the other hand, TGF-β downregulates IFN-γ production by T effector cells [93]. It is therefore tempting to speculate that a feedback regulatory loop exists between IFN-γ and TGF-β, and that neutralizing IFN-γ may prevent TGF-β release.

Another pathway used by suppressor macrophages that can be activated by IFN-γ, is the generation of nitric oxide (NO). NO is considered as one of the major effector pathways in inflammatory demyelination (reviewed by Smith [94]. However, evidence for its role in the pathogenesis of EAE is fraught with internal contradiction. This is well illustrated by studies using selective inhibitors of inducible nitric oxide synthetase (iNOS) (reviewed by Willenborg [95]). These studies showed that iNOS-derived NO can in fact be both deleterious as well as protective in EAE. In general, for both rats and mice, NO-inhibitory treatments targeting passive EAE or the effector phase of actively induced EAE resulted in amelioration of disease, whereas treatments targeting the afferent arm of the response aggravated disease. However, the increased EAE severity in iNOS-knockout mice forms a clear indication of a predominantly protective role for NO [96, 97]. As a possible mechanism it is suggested that IFN-γ-driven NO production by peripheral MPCs as well as by CNS-resident cells, such as microglia and astrocytes, might downregulate the development of EAE by limiting proliferation of autoreactive cells [51, 95]. MPC-derived NO is recognized as a strong inhibitor of T cell proliferation [98-100]. Inhibition of T cell proliferation in the CNS by NO might account for less severe EAE in wild-type than in IFN-γ knock-out and IFN-γR knockout mice. Consistent with this, peritoneal exudate cells from IFN-γR knock-out mice induced enhanced Ag-specific T cell proliferation compared with wild type mice, apparently because of their reduced production of NO [85]. In addition, T cells from IFN-γ knockout mice demonstrated enhanced in vivo proliferation as measured by BrdU incorporation [90]. APCs in the CNS may also limit the expansion of autoreactive T cells late in disease by downregulation of their MHC class II and costimulatory molecules through NO. [51]. IFN-γ-induced MHC-II expression on macrophages is inhibited by NO in vitro [101].

Another possible down-regulatory mechanism of IFN-γ in EAE is by induction or promotion of apoptosis in T lymphocytes. Many studies have demonstrated apoptosis of infiltrating cells in the CNS of animals with different models of EAE [102-105]. Blockade of IFN-γ inhibits cell death induced in effector cells by TCR linkage in the absence of accessory cells [106]. Furthermore, both in normal and in cultured malignant lympocytes, IFN-γ has been shown to exert contrasting effects, i.e. apoptosis or proliferation, depending on the level of expression of IFN-γ receptors: high level expression is associated with an apoptotic response, low level expression with a proliferative one [107]. Chu et al (2000) [90] noted that during EAE IFN-γ knock-out mice accumulate larger numbers of activated CD4+ T cells in their CNS than wild-type mice. The authors related this to increased proliferation and reduced apoptosis of these cells when exposed to antigen *ex vivo*. A similar situation was found to occur in mycobacterial infection in IFN-γ knock-out and wild type mice, and in this case it appeared that less proliferation and more apoptosis in the IFN-γ-competent mice was not a direct effect of IFN-γ, but was rather mediated by IFN-γ-activated MPCs [108]. On the other hand, by using an HSV-1-derived vector engineered with the mouse IFN-γ gene it has been shown that intrathecal delivery of IFN-γ can protect mice from EAE progression by inducing a fast clearance of encephalitogenic T cells infiltrating the CNS parenchyma via an apoptotic pathway [86].

IFN-γ produced locally in the CNS during EAE may also constitute a feedback control for the efficient clearance of the inflammatory infiltrate in the CNS. A fast and efficient phagocytosis of cells undergoing apoptosis by tissue-specific macrophages may limit direct tissue injury by preventing leakage of harmful (noxious) contents from the dying cells (reviewed in ref.

[109, 110]). Phagocytosis of apoptotic lymphocytes by macrophages, microglia, oligodendrocytes, and astrocytes has been described in histological sections of Lewis rat with EAE [111]. Different pro-inflammatory cytokines, including IFN-γ, have been shown to regulate the uptake of apoptotic cells by human macrophages *in vitro* [112, 113]. Using an *in vitro* phagocytosis assay, Chan et al. [113] demonstrated that pretreatment of microglia cells with IFN-γ led to an enhanced phagocytosis of apoptotic encephalitogenic T cells and to an increase of the proportion of microglial cells ingesting apoptotic cells.

Finally, the effect of IFN-γ on secretion of chemokines implicated in trafficking of leukocytes to the CNS may be involved. It was found that IFN-γ downregulates the induction of IL-8 by IL-1 and IL-2 in monocytes in vitro and may thereby act as a repressor of neutrophil chemotaxis in EAE [114]. In mice with intact IFN-γ response there is a strong up-regulation of the typical T cell and macrophage chemoattractants RANTES, IP-10, MCP-1α and MCP-1, in the CNS during EAE (reviewed in ref. [31, 71, 115]). In mice lacking the IFN-γ gene or the receptor for IFN-γ, MIP-2 and T cell activation gene–3 (TAC-3), both neutrophil-attracting chemokines, were strongly up-regulated, while RANTES and MCP-1 remained undetectable [89]. These mice showed an acute, lethal disease progression compared to limited disease in their wild-type littermates. McColl et al [116] reported that neutrophil depletion inhibits MBP-induced EAE in SJL/J mice and in MOG-induced EAE in IFN-γR knockout mice, suggesting a crucial role for PMN in the pathogensis of this inflammatory disease. Thus, the absence of down-regulation by IFN-γ may lead to a chemokine imbalance and lethal neutrophil invasion.

Taken together, despite many studies the protective effect of IFN-γ on EAE remains enigmatic. Several immunosuppressive effects of IFN-γ have been considered as possible explanations, but there is no clear cut evidence in favor of one of them no more than there is a claer-cut explanation for the contrasting effect of IFN-γ treatment in MS patients [73, 74]. The discrepancy may be due to differences in dosage, treatment schedules or timing. It may be connected to the fact that these patients had already undergone some overt clinical episodes and were in remission at the time of IFN-γ treatment. Further experimentation is required to determine the molecular mechanisms responsible for the seemingly contradictory roles of IFN-γ in EAE and MS.

## 3.    INTERFERON-α AND -β IN EAE

Many different cells can produce type I interferons (IFN-α or -β, or both) during viral infection. IFN-α is a product mainly of monocytes and a subset of immature dendritic cells (reviewed by Liu et al. [117] and Asselin-Paturel et al. [118]. IFN-β can be produced by virtually any cell infected with virus or given another appropiate stimulus. In addition, several growth factors and cytokines have been shown to induce the synthesis of IFN-α and IFN-β (reviewed by De Maeyer & De Maeyer-Guignard [1, 119]). For example IL-2 can induce the production of murine IFN-α,β in mouse bone marrow cells. IFN-γ can also act as an inducer of IFN-α or IFN-β. IFN-α and β are comparable in their actions and interact with the same type I IFN receptor to transduce signals across cell membranes. IFN-α/β are implicated in both normal and neoplastic cell growth regulation and in modulating both innate and adaptive immune responses (reviewed by Brierley & Fish [120]).

## 3.1    In vivo evidence

Over the past years clinical trials have shown IFN-β to be beneficial in the treatment of relapsing-remitting MS (reviewed by Chofflon [121]). It ameliorates disease activity, reduces relapse frequency, decreases the number of inflammatory CNS lesions, but fails to achieve a satisfactory control of the disease evolution. Clinical trials with IFN-α have given varying results probably because differences in study design (study populations, type, dose and route of IFN-α, length of treatment) and seems to be less effective than IFN-β treatment (e.g. [122-124]).

In animals, many studies have shown that IFN-β can exert a down-regulating effect on the development of EAE. More than 15 years ago both natural and purified IFN-α,β were reported to partially inhibit both the active form of EAE and the adoptively transferred disease in rats when administered during the induction phase of the disease [125-127]. A similar effect was reported to occur in EAE induced in SJL/J mice [76, 128]. In these studies inhibition was obtained by systemic treatment or by adding the interferon to the T cell cultures before transfer. Even a decreased relapse rate was observed [128]. In addition, infection of mice with lactic dehydrogenase virus known to act as an inducer of IFN-α,β had a similar inhibitory effect [76, 129]. Human IFN-β on the contrary had no effect when similarly tested in the rat model, probably due to the known species specificity of interferons [127]. More recently, low doses of orally administered natural IFN-α,β has been shown to suppress clinical relapses and diminished inflammation in a murine model for chronic relapsing EAE [130, 131]. It should be noted that whether or not oral IFN administration can exert systemic effects is a controversial issue and that the underlying mechanisms, if any, are unknown. IFN-β, given orally, also enhanced the suppressive effects of oral tolerance to MBP and PLP in both rats and mice with EAE [132]. In contrast, studies with intracerebrally injected IFN were inconclusive. Highly purified rat IFN-α,β was not effective in preventing or ameliorating the disease process in rats [133], while natural fibroblast-derived interferon exerted a beneficial effect [127]. This discrepancy may possibly have been due to the fact that the natural rat IFN used was of low purity and may therefore have contained several other effector molecules. Studies using recombinant IFN-β indicated that the timing and length of treatment determines wether IFN-β prevents or aggravates EAE in rats [134, 135]. Recombinant IFN-β effectively prevented acute EAE in rats provided that treatment was continued over a long period of time. Contrastingly, discontinuation of treatment during the induction phase or during the recovery phase led to the rapid onset of EAE with strongly enhanced severity and prolonged duration [135]. Finally, gene therapy by direct injection of IFN-β plasmid DNA–liposome complex into the CNS of mice with established EAE reduced the severity and onset of clinical disease [136].

Studies on the immunoregulatory activity of IFN-α in EAE are limited in number. Small doses of murine natural IFN-α or a cross-species-reactive human recombinant IFN-α, administered orally 3 times a week was found to suppress acute EAE in rats and clinical relapses in mice with CREAE and also to suppress adoptively transferred EAE [130, 137, 138]. Oral administration of IFN-α is superior to equivalent or higher doses of parenteral administration in suppression of clinical relapse in EAE [137, 138].

## 3.2    Mechanisms involved

The protective effect of IFN-β in EAE may stem from the immunomodulatory and anti-inflammatory properties of this type of interferon (reviewed by Yong et al ([139]). Among these properties is its ability to counteract the production and effects of IFN-γ. IFN-β inhibits IFN-γ secretion by activated lymphocytes. Suppression of EAE relapses by IFN-α,β and recombinant human IFN-α correlated with a decreased Con A stimulated IFN-γ secretion in spleen and lymph node cells [130, 131, 137, 138]. Similarly, Luca et al. [140] showed decreased production of IFN-γ by antigen-specific stimulated spleen and lymph node cells derived from mice with EAE after in vitro treatment with IFN-β. Likewise, IFN-β was shown to limit IFN-γ production by activated lymphocytes from both MS patients and controls [141]. IFN-α/β also antagonizes the effects of IFN-γ on mononuclear phagocytes. In this respect, IFN-β inhibits IFN-γ-induced class II MHC antigen expression on the surface of antigen presenting cells in vitro, such as macrophages/monocytes, astrocytes and endothelial cells [142-145]. However, IFN-β failed to downregulate basal or IFN-γ-induced MHC class II expression on primary human astrocytes and microglial cells except at high concentrations. Thus, IFN-β lessens the capacity for antigen presentation to T cells and so might be expected to attenuate T-cell response.

Another anti-inflammatory effect of IFN-β is its ability to interfere with the process of recruitment and entry of inflammatory cells into the CNS, events which are regulated by adhesion molecules, cytokines, chemokines and metalloproteinases and are compromised in MS and EAE (reviewed by Brown [146]). Diminishment of clinical signs of EAE in IFN-β-treated mice was found to be associated with a major reduction in the formation of cellular infiltrates in the CNS [128, 134, 147]. IFN-β may directly act on the endothelium by reducing the expression of cytokine-induced (IFN-γ, TNF-α) adhesion molecules, such as ICAM-1 and VCAM-1, as observed in vitro on cerebral and extracerebral vascular endothelial cells and in vivo in rats with acute EAE [144, 145, 147-149]). On the other hand, IFN-β may also influence the interaction of activated leukocytes with brain endothelium [149]. It has been reported that pretreatment of activated leukocytes with IFN-β inhibited their transendothelial migration by down regulation of the expression of the integrins VLA-4 and LFA-1 [144]. In MS-patients treated with IFN-β, VLA-4 expression on selected T cell subsets from blood is also downregulated possibly due to the increased levels of soluble VCAM-1 observed in these patients [150, 151]. Furthermore, IFN-β may antagonize migration of leukocytes through brain vascular endothelium by down-regulating the production and activity of matrix metalloproteinases (MMPs), enzymes implicated in breakdown of the blood-brain barrier, activation and degradation of disease-modifying cytokines and myelin degradation in MS and EAE (review by Yong et al [152, 153]. Indeed, in vitro data have demonstrated that exposing activated T cells to IFN-β resulted in a significant decrease in MMP-9 expression associated with inhibited migratory capacity of T cells [149, 154, 155]. In addition, peripheral blood leukocytes from MS patients treated with IFN-β showed decreased levels of distinct MMP mRNA expression [156-158]. Chemokines that are specifically expressed in EAE tissue at the time of lesion formation (reviewed by Ransohoff [32]) may also be susceptible to IFN-β action and may hence account for reduced infiltration of the CNS by specific subsets of leukocytes. Recently, it was demonstrated that in vitro exposure of stimulated human peripheral blood monocytes to IFN-β differentially affects chemokine production. IFN-β decreased the production by monocytes/macrophages of IL-8, IP-10, MCP-1, MCP-3 and the monokine induced by IFN-γ (Mig), but not of MIP-1α [159]. In addition, IFN-β has been found to selectively inhibit expression of mRNA for RANTES and MIP-1α and for their receptor, CCR5, on activated peripheral T cells obtained from MS patients [160]. Significantly reduced expression and production of chemokines and the chemokine receptors CCR5 and CXCR3 were found in sera and peripheral blood mononuclear cells from MS patients under IFN-β therapy [159-162]. Data on the regulation of chemokine expression by IFN-α or -β in relevant cell types during EAE are not available. Thus, an important part of the action of IFN-β may be its effect on the complex interaction between adhesion molecules, cytokines, chemokines and MMPs implicated in the blood-brain barrier function.

Autoreactive T cell proliferation in response to antigen presentation is a prominent feature of the induction phase of immune responses. One mechanism of protection by IFN-β may reside in its antiproliferative activity [131, 141], by which it may inhibit the expansion of autoaggressive T cells. Mice induced for EAE and fed murine IFN-α or cross-reactive recombinant human IFN-α showed decreased mitogen- and antigen-specific proliferative response of spleen cells ex vivo [130, 131, 137]. Van der Meide et al [135] clearly showed that administration of rIFN-β to EAE rats in the early phase of the disease resulted in a strong (50 to 80%) reduction of encephalitogenic T cells in both spleen and inguinal lymph nodes. Discontinuation of treatment led to a rapid abrogation of the antiproliferative activity of rIFN-β followed by excessive expansion of encephalitogenic T cells, more severe perivascular inflammation in spinal cord tissue relative to control rats. Type I interferons may also serve as growth regulators of astrocytes at sites of reactive gliosis. An antagonistic effect of IFN-α and IFN-β on IFN-γ induced proliferation of human astrocytes in vitro has been reported [163].

Some studies have addressed the question of whether IFN-β could also act by inducing apoptosis. In fact, anti-apoptotic effects of both IFN-α and IFN-β on activated CD4[+] and CD8[+] T cells have been descibed (reviewed by Akbar et al. [164]), though these findings were not confirmed in more recent studies. Although IFN-β enhanced the expression of the death receptor CD94 (Fas/APO-1) on human antigen-specific T cell lines from MS patients, no clear evidence was found for direct induction of apoptosis as assessed by caspase activity [165]. In adoptively transferred EAE, T cell infiltration was reduced by IFN-β without apparent induction of apoptosis [166].

IFN-β might also redirect cytokine production by immune cells during EAE from a Th1 towards a Th2 profile and this seems to occur in the circulation as well as within the CNS (reviewed by Yong et al. [139]). Thus, IFN-β may lessen tissue damage by inhibiting release of TNF-α and LT by blood mononuclear cells, microglia and infiltrating macrophages [137, 166-

168], molecules reported to be toxic for oligodendrocytes [169, 170] and to be implicated in demyelination [171]. Adoptive transfer of T cells and CD8[+] T cells from non-immunized naïve donor mice fed with IFN-α suppressed actively induced EAE in recipients while decreasing TNF-α secretion in these recipients [172].

Alternatively, IFN-β may induce or promote secretion of immunosuppressive cytokines, such as IL-10 or TGF-β. A decreased production of IFN-γ and an increased production of IL-10 and/or TGF-β by T cells from EAE-induced mice treated with IFN-β or IFN-α have been reported [140, 173-175]. Furthermore, PBMC from MS patients treated with IFN-β showed enhanced gene expression of TGF-β and increased secretion of IL-10 [167, 176-178]. Both cytokines can upregulate TIMPs and decrease MMP biosynthesis (reviewed by Yong et al. [152]). IFN-β may also act by modulating the IL-10/IL-12 cytokine circuit in vivo, which has been suggested to control susceptibility to autoimmune disease [179]. McRae et al [180, 181] indicated that IFN-β abrogates Th1 cell differentiation in vitro while inhibiting T cell–dependent CD40-induced production of IL-12 from dendritic cells. In addition, in IFN-β-treated mice the cytokine profile in response to the priming immunogen was significantly skewed toward an increased production of IL-10 and concurrent decreased production of IL-12 [182]. This was accompanied by an aborted development of epitope spreading and disease progression.

Finally, IFN-β may keep in check ongoing immune response by restoring T suppressor cell function. It has been shown that human IFN-β augments suppressor cell activity in MS at least in vitro [183, 184]. A similar positive effect on suppression was reported with IFN-α.

Thus, in EAE and MS IFN-β may curtail immune activation by decreasing T-cell proliferation, by blocking IFN-γ production and by counteracting the peripheral actions of circulating IFN-γ directly, by downregulating the entry of encephalitogenic T cells and MPCs into the CNS and by downregulating the levels of pro-anti-inflammatory and upregulating anti-inflammatory molecules. All these biological activities of IFN-β are likely to act in concert to produce a therapeutic outcome.


## 4.       IFN-TAU IN EAE

Interferon tau (IFN-τ), also called ovine trophoblast protein-1, is a member of the type I IFN family that was originally identified as a pregnancy recognition hormone, constitutively produced in high levels by trophoblast cells in ruminants during a very short period in early pregnancy (reviewed in ref. by [185, 186]). Its major function is to signal receptors in the endometrium to maintain an appropriate milieu for the embryo. IFN-τ is not induced by viruses or double-stranded RNA, and its production is regulated differently from other Type I interferons. Like IFN-α and IFN-β it possesses antiviral, antiproliferative, and immunomodulatory activities (reviewed by Martal et al. [185]). In contrast to other type I IFNs, IFN-τ does cross species barriers and lacks the toxicity normally associated with high concentrations of these IFNs in vitro as well as in vivo [187]. It has been suggested that differential recognition by IFN-τ of the receptor complex at its N-terminus as well as interaction with different subunits of the type I IFN receptor accounts for its high biologic potency in the absence of associated toxicity [188]. Thus, IFN-τ seems promising as a therapeutic agent in cancer, viral infections and autoimmune diseases as it may be devoid of the adverse side effects associated with high doses of IFN-α and IFN-β.

IFN-τ was explored for its ability to prevent the development of EAE. Ovine IFN-τ was shown to provide protection as effectively as murine IFN-β against development of the acute form of EAE in NZW mice and chronic relapsing form of EAE in SJL/J mice, but without associated toxicity such as weight loss and leukopenia seen with IFN-α and IFN-β [186, 187]. The onset of EAE can be prevented by treating mice with IFN-τ before, during and shortly after immunization with MBP. However, when IFN-τ treatment was discontinued onset of disease was observed within 6-12 days [189]. In addition to protecting against both acute and chronic forms of EAE, parenteral or oral administration of IFN-τ was found to reverse ongoing relapsing EAE in SJL/J mice [190]. A histological evaluation of the CNS of these mice showed a subsiding lymphocyte infiltration and microglial activation. Finally, IFN-τ was also able to prevent superantigen-induced exacerbations of EAE [186, 187].

Evidence for mechanisms by which IFN-τ may prevent EAE include reduced proliferation of T cells specific for the autoantigen MBP [187]. Furthermore, anti-MBP antibody levels and proliferation of B cells in response to MBP were found to be reduced by IFN-τ either in vitro or in vivo in both chronic and acute forms of EAE [190]. Thus, IFN-τ seems to be able to suppress both the humoral and cellular responses in EAE. It has also been reported that amelioration of EAE by INF-τ is associated with induction of suppressor cells and suppressor factors [191]. Suppressor cells of the CD4 phenotype were found to be induced in vitro and in vivo after parenteral or oral administration of IFN-τ. Such IFN-τ-induced suppressor cells were found able to inhibit the MBP-induced proliferation of MBP-sensitized T cells, and injection of these suppressor cells into MBP-immunized NZW mice blocked induction of EAE [191]. In addition, the

CD4$^+$ suppressor cells produced both IL-10 and TGF-$\beta$, cytokines previously found to synergistically inhibit activation of MBP-specific T cells derived from EAE mice [191]. It was suggested that the amelioration of superantigen-induced reactivation of EAE seen after treatment with IFN-$\tau$ may also be related to induction of suppressor cells [187]. Like other type I interferons, IFN-$\tau$ can alter cytokine profiles as evidenced by its ability to reduce production of TNF-$\alpha$, that has been suggested to be a mediator of EAE [170, 192]. IFN-$\tau$ also induces enhanced production of IL-10. Prolonged IFN-$\tau$ treatment of mice with EAE was shown to result in production of IL-10 in the circulation [189]. In addition, prevention of EAE by IFN-$\tau$ correlated with the induction of IL-10 and a concomitant reduction of the Th1 cytokines IFN-$\gamma$, TNF-$\alpha$ and IL-2 by MBP-stimulated spleen and lymph node cells [175]. Another proposed action by which IFN-$\tau$ reduces EAE is by interfering with MHC class II expression: IFN-$\tau$ was shown to downregulate IFN-$\gamma$-induced expression on murine cerebrovascular endothelial cells [193].

The lack of toxicity of IFN-$\tau$ and its ability to ameliorate EAE may have important implications for its potential use as an immunotherapy for inflammatory and autoimmune diseases such as MS.

## ACKNOWLEDGMENTS
Work in the authors' laboratory is supported by grants from the Fund for Medical Scientific Research of Flanders (FWO), from the Regional Government of Flanders (GOA-program) and from the Belgian Federal Government (IUAP-program).

## REFERENCE LIST


1.  De Maeyer,E. and J.De Maeyer-Guignard. 1998. Type I interferons. *Int.Rev.Immunol.* 17:53-73.
2.  Stark,G.R., I.M.Kerr, B.R.G.Williams, R.H.Silverman, and R.D.Schreiber. 1998. How cells respond to interferons. *Annu.Rev.Biochem.* 67:227-264.
3.  Domanski,P. and O.R.Colamonici. 1996. The type I interferon receptor. The long and short of it. *Cytokine Growth Factor Rev.* 7:143-151.
4.  Wheelock,E.F. 1965. Interferon-like virus inhibitor induced in human leukocytes by phytohemagglutinin. *Science* 141:30-311.
5.  Farrar,M.A. and R.D.Schreiber. 1993. The molecular cell biology of interferon-$\gamma$ and its receptor. *Annu.Rev.Immunol.* 11:571-611.
6.  Billiau,A. 1996. Interferon-$\gamma$: biology and role in pathogenesis. *Adv.Immunol.* 62:61-130.
7.  Young,H.A. and K.J.Hardy. 1995. Role of interferon-$\gamma$ in immune cell regulation. *J.Leukocyte Biol.* 58:373-381.
8.  Gresser,I. and F.Belardelli. 2002. Endogenous type 1 interferons as a defence against tumors. *Cytokine Growth Factor Rev.* 13:111-118.

9.      Bogdan,C. 2000. The function of type I interferons in antimicrobial immunity. *Curr.Opin.Immunol.* 12:419-424.

10.     Zamvil,S.S. and L.Steinman. 1990. The T lymphocyte in experimental allergic encephalomyelitis. *Annu.Rev.Immunol.* 8:579-621.

11.     Martin,R., H.F.McFarland, and D.E.McFarlin. 1992. Immunological aspects of demyelinating diseases. *Annu.Rev.Immunol.* 10:153-187.

12.     Benveniste,E.N. 1992. Inflammatory cytokines within the central nervous system: sources, function, and mechanism of action. *Am.J.Physiol.* 263:C1-C16.

13.     Benveniste,E.N. and D.J.Benos. 1995. TNF-α- and IFN-γ-mediated signal transduction pathways: effects on glial gene expression and function. *FASEB J* 9:1577-1584.

14.     Sethna,M.P. and L.A.Lampson. 1991. Immune modulation within the brain:recruitment of inflammatory cells and increased major histocompatibility antigen expression following intracerebral injection of interferon-γ. *J.Neuroimmunol.* 34:121-132.

15.     Hartung,H.P., S.Jung, G.Stoll, J.Zielasek, B.Schmidt, J.Archelos, and K.Toyka. 1992. Inflammatory mediators in demyelinating disorders of the CNS and PNS. *J.Neuroimmunol.* 40:197-210.

16.     Opdenakker,G. and J.Van Damme. 1994. Cytokine-regulated proteases in autoimmune diseases. *Immunol.Today* 15:103-107.

17.     Hartung,H.P., J.J.Archelos, J.Zielasek, R.Gold, M.Koltzenburg, K.H.Reiners, and K.V.Toyka. 1995. Circulating adhesion molecules and inflammatory mediators in demyelination: a review. *Neurology* 45:S22-S32.

18.     Hughes,C.C.W., D.K.Male, and P.L.Lantos. 1988. Adhesion of lymphocytes to cerebral microvascular cells: effects of interferon- γ, tumor necrosis factor and interleukin-1. *Immunology* 64:677-681.

19.     Issekutz,T.B. 1990. Effects of six different cytokines on lymphocyte adherence to microvascular endothelium and in vivo lymphocyte migration in the rat. *J.Immunol* 144:2140-2146.

20.     Oppenheimer-Marks,N. and M.Ziff. 1988. Migration of lymphocytes through endothelial cell monolayers: augmentation by interferon-γ. *Cell.Immunol.* 114:307-323.

21.     Simmons,R.D. and D.O.Willenborg. 1990. Direct injection of cytokines into the spinal cord causes autoimmune encephalomyelitis-like inflammation. *J.Neurol.Sci.* 100:37-42.

22.     Huynh,H.K. and K.Dorovini-Zis. 1993. Effects of interferon-γ on primary cultures of human brain microvessel endothelial cells. *Am.J.Pathol.* 142:1265-1278.

23.     Hickey,W.F. 1999. Leukocyte traffic in the central nervous system: the participants and their roles. *Seminars Immunol.* 11:125-137.

24.     Cannella,B., A.H.Cross, and C.S.Raine. 1990. Upregulation and coexpression of adhesion molecules correlate with relapsing autoimmune demyelination in the central nervous system. *J.Exp.Med.* 172:1521-1523.

25.     Lindsey,J.W. and L.Steinman. 1993. Competitive PCR quantification of CD4, CD8, ICAM-1, VCAM-1, and MHC class II mRNA in the central nervous system during development and resolution of experimental allergic encephalomyelitis. *J.Neuroimmunol.* 48:227-234.

26.     Steffen,B.J., E.C.Butcher, and B.Engelhardt. 1994. Evidence for involvement of ICAM-1 and VCAM-1 in lymphocyte interaction with endothelium in experimental

autoimmune encephalomyelitis in the central nervous system in the SJL/J mouse. *Am.J.Pathol.* 145:189-201.

27. McCarron,R.M., L.Wang, M.K.Racke, D.E.McFarlin, and M.Spatz. 1993. Cytokine-regulated adhesion between encephalitogenic T lymphocytes and cerebrovascular endothelial cells. *J.Neuroimmunol.* 43:23-30.

28. Fabry,Z., M.M.Waldschmidt, D.Hendrickson, J.Keiner, L.Love-Homan, F.Takei, and M.N.Hart. 1992. Adhesion molecules on murine brain microvascular endothelial cells: expression and regulation of ICAM-1 and Lgp 55. *J.Neuroimmunol.* 36:1-11.

29. Yu,C.L., D.O.Haskard, D.Cavender, A.Johnson, and M.Ziff. 1985. Human γ interferon increases the binding of T lymphocytes to endothelial cells. *Clin.Exp.Immunol.* 62:554-560.

30. Barten,D.M. and N.H.Ruddle. 1994. Vascular adhesion molecule-1 modulation by tumor necrosis factor in experimental allergic encephalomyelitis. *J.Neuroimmunol.* 51:123-133.

31. Kennedy,K.J. and W.J.Karpus. 1999. Role of chemokines in the regulation of Th1/Th2 and autoimmune encephalomyelitis. *J.Clin.Immunol* 19:273-279.

32. Ransohoff,R.M. 1999. Mechanisms of inflammation in MS tissue: adhesion molecules and chemokines. *J.Neuroimmunol.* 98:57-68.

33. Glabinski,A.R., S.O'Bryant, K.Selmaj, and R.M.Ransohoff. 2000. CXC chemokine receptors expression during chronic relapsing experimental autoimmune encephalomyelitis. *Ann.NY Acad.Sci* 917:135-144.

34. Marfaing-Koka,A., O.Devergne, G.Gorgone, A.Portier, T.J.Schall, P.Galanaud, and D.Emilie. 1995. Regulation of the production of the RANTES chemokine by endothelial cells: Synergistic induction by IFN-γ plus TNF-α and inhibition by IL-4 and IL-13. *J.Immunol.* 154:1870-1878.

35. Janabi,N., I.Hau, and M.Tardieu. 1999. Negative feedback between prostaglandin and α- and β-chemokine synthesis in human microglial cells and astrocytes. *J.Immunol* 162:1701-1706.

36. Glabinski,A.R., M.Krakowski, Y.Han, T.Owens, and R.M.Ransohoff. 1999. Chemokine expression in GKO mice (lacking interferon-γ) with experimental autoimmune encephalomyelitis. *J.Neurovirology* 5:95-101.

37. Valente,A.J., J.F.Xie, M.A.Abramova, U.O.Wenzel, H.E.Abboud, and D.T.Graves. 1998. A complex element regulates IFN-γ-stimulated monocyte chemoattractant protein-1 gene transcription. *J.Immunol.* 161:3719-3728.

38. Popko,B., J.G.Corbin, K.D.Baerwald, J.Dupree, and A.M.Garcia. 1997. The effects of interferon-γ on the central nervous system. *Mol.Neurobiol.* 14:19-35.

39. Merrill,J.E., L.J.Ignarro, M.P.Sherman, J.Melinek, and T.E.Lane. 1993. Microglial cell cytotoxicity of oligodendrocytes is mediated through nitric oxide. *J.Immunol.* 151:2132-2141.

40. Sherman,M.P., J.M.Griscavage, and L.J.Ignarro. 1992. Nitric oxide-mediated neuronal injury in multiple sclerosis. *Med.Hypotheses* 39:143-146.

41. Li,Y., J.Atashi, C.Hayes, E.Reap, S.Hunt, III, and B.Popko. 1995. Morphological and molecular response of the MOCH-1 oligodendrocyte cell line to serum and interferon-γ: Possible implications for demyelinating disorders. *J.Neurosci.Res.* 40:189-198.

42. Vartanian,T., Y.Li, M.Zhao, and K.Stefansson. 1995. Interferon-γ-induced oligodendrocyte cell death: implications for the pathogenesis of multiple sclerosis. *Mol.Med.* 1:732-743.

43.  Horwitz,M.S., C.F.Evans, D.B.McGavern, M.Rodriguez, and M.B.A.Oldstone. 1997. Primary demyelination in transgenic mice expressing interferon-γ. *Nature Med.* 3:1037-1041.

44.  Corbin,J.G., D.Kelly, E.M.Rath, K.D.Baerwald, K.Suzuki, and B.Popko. 1996. Targeted CNS expression of interferon-γ in transgenic mice leads to hypomyelination, reactive gliosis, and abnormal cerebellar development. *Mol.Cell.Neurosci.* 7:354-370.

45.  Renno,T., V.Taupin, L.Bourbonniere, G.Verge, E.Tran, R.De Simone, M.Krakowski, M.Rodriguez, A.Peterson, and T.Owens. 1998. Interferon-γ in progression to chronic demyelination and neurological deficit following acute EAE. *Mol.Cell.Neurosci.* 12:376-389.

46.  Seder,R.A. and W.E.Paul. 1994. Acquisition of lympokine-producing phenotypes by CD4+ T cells. *Annu.Rev.Immunol.* 12:635-673.

47.  Chomarat,P., M.-C.Rissoan, J.Banchereau, and P.Miossec. 1993. Interferon γ inhibits interleukin 10 production by monocytes. *J.Exp.Med.* 177:523-527.

48.  Ando,D.G., J.Clayton, D.Kono, J.L.Urban, and E.E.Sercarz. 1989. Encephalitogenic T cells in the B10.PL model of experimental allergic encephalomyelitis (EAE) are of the Th-1 lymphokine subtype. *Cell.Immunol.* 132-143.

49.  McDonald,A.H. and R.H.Swanborg. 1988. Antigen-specific inhibition of immune interferon production by suppressor cells of autoimmune encephalomyelitis. *J.Immunol.* 140:1132-1138.

50.  Raine,C.S. 1994. The Dale E. McFarlin memorial lecture: The immunology of multiple sclerosis. *Ann.Neurol.* 36:S61-S72.

51.  Juedes,A.E. and N.H.Ruddle. 2001. Resident and infiltrating central nervous system APCs regulate the emergence and resolution of experimental autoimmune encephalomyelitis. *J.Immunol.* 166:5168-5175.

52.  Neumann,H., H.Schmidt, E.Wilharm, L.Behrens, and H.Wekerle. 1997. Interferon γ gene expression in sensory neurons: evidence for autocrine gene regulation. *J.Exp.Med.* 186:2023-2031.

53.  Adorini,L. 1999. Interleukin-12, a key cytokine in Th1-mediated autoimmune diseases. *Cell.Mol.Life Sci.* 55:1610-1625.

54.  Dinarello,C.A. 1999. Interleukin-18. *Methods* 19:121-132.

55.  Nakanishi,K., T.Yoshimoto, H.Tsutsui, and H.Okamura. 2001. Interleukin-18 regulates both Th1 and Th2 responses. *Annu.Rev.Immunol.* 19:423-474.

56.  Leonard,J.P., K.E.Waldburger, and S.J.Goldman. 1995. Prevention of experimental autoimmune encephalomyelitis by antibodies against interleukin 12. *J.Exp.Med.* 181:381-386.

57.  Heremans,H., C.Dillen, M.Groenen, E.Martens, and A.Billiau. 1996. Chronic relapsing experimental autoimmune encephalomyelitis (CREAE) in mice: enhancement by monoclonal antibodies against IFN-γ. *Eur.J.Immunol.* 26:2393-2398.

58.  Heremans,H., C.Dillen, and A.Billiau. 1996. Role of IFN-γ and IL-12 in a model for chronic relapsing EAE in Biozzi mice. *Eur.Cytokine Network* 7:458. (Abstr.)

59.  Segal,B.M. and E.M.Shevach. 1996. IL-12 unmasks latent autoimmune disease in resistant mice. *J.Exp.Med.* 184:771-775.

60.  Wildbaum,G., S.Youssef, N.Grabie, and N.Karin. 1998. Neutralizing antibodies to IFN-γ–inducing factor prevent experimental autoimmune encephalomyelitis. *J.Immunol.* 161:6368-6374.

61.  Kennedy,M.K., D.S.Torrance, K.S.Picha, and K.M.Mohler. 1992. Analysis of cytokine mRNA expression in the central nervous system of mice with experimental

autoimmune encephalomyelitis reveals that IL-10 mRNA expression correlates with recovery. *J.Immunol.* 149:2496-2505.

62.   Merrill,J.E., D.H.Kono, J.Clayton, D.G.Ando, D.R.Hinton, and F.M.Hofman. 1992. Inflammatory leukocytes and cytokines in the peptide-induced disease of experimental allergic encephalomyelitis in SJL/J and B10.PL mice. *Proc.Natl.Acad.Sci.USA* 89:574-578.

63.   Mustafa,M., P.Diener, B.Höjeberg, P.Van der Meide, and T.Olsson. 1991. T cell immunity and interferon-γ secretion during experimental allergic encephalomyelitis in Lewis rats. *J.Neuroimmunol.* 31:165-177.

64.   Baker,D., J.K.O'Neill, and J.L.Turk. 1991. Cytokines in the central nervous system of mice during chronic relapsing experimental allergic encephalomyelitis. *Cell.Immunol.* 134:505-510.

65.   Renno,T., J.-Y.Lin, C.Piccirillo, J.Antel, and T.Owens. 1994. Cytokine production by cells in cerebrospinal fluid during Experimental Allergic Encephalomyelitis in SJL/J mice. *J.Neuroimmunol.* 49:1-7.

66.   Stoll,G., S.Müller, B.Schmidt, P.Van der Meide, S.Jung, K.V.Toyka, and H.-P.Hartung. 1993. Localization of interferon-γ and Ia-antigen in T cell line-mediated experimental autoimmune encephalomyelitis. *Am.J.Pathol.* 142:1866-1875.

67.   Young,H.A. and K.J.Hardy. 1990. Interferon-γ: producer cells, activation stimuli, and molecular genetic regulation. *Pharmac.Ther.* 45:137-151.

68.   Khoury,S.J., W.W.Hancock, and H.L.Weiner. 1992. Oral tolerance to myelin basic protein and natural recovery from experimental autoimmune encephalomyelitis are associated with downregulation of inflammatory cytokines and differential upregulation of transforming growth factor β, interleukin 4, and prostaglandin E expression in the brain. *J.Exp.Med.* 176:1355-1364.

69.   Tanuma,N., M.L.Shin, T.Shin, T.Koga, K.Kogure, and Y.Matsumoto. 1999. Differential role of TNF-α and IFN-γ in the brain of rats with chronic relapsing autoimmune encephalomyelitis. *J.Neuroimmunol.* 96:73-79.

70.   Begolka,W.S., C.L.Vanderlugt, S.M.Rahbe, and S.D.Miller. 1998. Differential expression of inflammatory cytokines parallels progression of central nervous system pathology in two clinically distinct models of multiple sclerosis. *J.Immunol.* 161:4437-4446.

71.   Juedes,A.E., P.Hjelström, C.M.Bergman, A.L.Neild, and N.H.Ruddle. 2000. Kinetics and cellular origin of cytokines in the central nervous system: insight into mechanisms of myelin oligodendrocyte glycoprotein-induced experimental autoimmune encephalomyelitis. *J.Immunol.* 164:419-426.

72.   Vass,K., K.Heininger, B.Schäfer, C.Linington, and H.Lassman. 1992. Interferon-γ potentiates antibody-mediated demyelination in vivo. *Ann.Neurol.* 32:198-206.

73.   Panitch,H.S., R.L.Hirsch, J.Schindler, and K.P.Johnson. 1987. Treatment of multiple sclerosis with γ interferon: exacerbations associated with activation of the immune system. *Neurology* 37:1097-1102.

74.   Panitch,H.S., A.S.Haley, R.L.Hirsch, and K.P.Johnson. 1987. Exacerbations of multiple sclerosis in patients treated with γ interferon. *Lancet* i:893-895.

75.   Panitch,H.S. 1992. Interferons in multiple sclerosis. *Drugs* 44:946-962.

76.   Billiau,A., H.Heremans, F.Vandekerckhove, R.Dijkmans, H.Sobis, E.Meulepas, and H.Carton. 1988. Enhancement of experimental allergic encephalomyelitis in mice by antibodies against IFN-γ. *J.Immunol.* 140:1506-1510.

77.   Duong,T.T., J.St.Louis, J.J.Gilbert, F.D.Finkelman, and G.H.Strejan. 1992. Effect of anti-interferon-γ and anti-interleukin-2 monoclonal antibody treatment on the

development of actively and passively induced experimental allergic encephalomyelitis in the SJL/J mouse. *J.Neuroimmunol.* 36:105-115.

78. Duong,T.T., F.D.Finkelman, B.Singh, and G.H.Strejan. 1994. Effect of anti-interferon-$\gamma$ monoclonal antibody treatment on the development of experimental allergic encephalomyelitis in resistant mouse strains. *J.Neuroimmunol.* 53:101-107.

79. Voorthuis,J.A.C., B.M.J.Uitdehaag, C.J.A.De Groot, P.H.Goede, P.H.Van der Meide, and C.D.Dijkstra. 1990. Suppression of experimental allergic encephalomyelitis by intraventricular administration of interferon-$\gamma$ in rats. *Clin.Exp.Immunol.* 81:183-188.

80. Lublin,F.D., R.L.Knobler, B.Kalman, M.Goldhaber, J.Marini, M.Perrault, C.D'Imperio, J.Joseph, S.S.Alkan, and R.Korngold. 1993. Monoclonal anti-$\gamma$ interferon antibodies enhance experimental allergic encephalomyelitis. *Autoimmunity* 16:264-374.

81. Billiau,A. 1996. Interferon-$\gamma$ in autoimmunity. *Cytokine Growth Factor Rev.* 7:25-34.

82. Ferber,I.A., S.Brocke, C.Taylor-Edwards, W.Ridgway, C.Dinisco, L.Steinman, D.Dalton, and C.G.Fathman. 1996. Mice with a disrupted IFN-$\gamma$ gene are susceptible to the induction of experimental autoimmune encephalomyelitis (EAE). *J.Immunol.* 156:5-7.

83. Krakowski,M. and T.Owens. 1996. Interferon-$\gamma$ confers resistance to experimental allergic encephalomyelitis. *Eur.J.Immunol.* 26:1641-1646.

84. Willenborg,D.O., S.Fordham, C.C.A.Bernard, W.B.Cowden, and I.A.Ramshaw. 1996. IFN-$\gamma$ plays a critical down-regulatory role in the induction and effector phase of myelin oligodendrocyte glycoprotein-induced autoimmune encephalomyelitis. *J.Immunol.* 157:3223-3227.

85. Willenborg,D.O., S.A.Fordham, M.A.Staykova, I.A.Ramshaw, and W.B.Cowden. 1999. IFN-$\gamma$ is critical to the control of murine autoimmune encephalomyelitis and regulates both in the periphery and in the target tissue: a possible role for nitric oxide. *J.Immunol.* 163:5278-5286.

86. Furlan,R., E.Brambilla, F.Ruffini, P.L.Poliani, A.Bergami, P.C.Marconi, D.M.Franciotta, G.Penna, G.Comi, L.Adorini, and G.Martino. 2001. Intrathecal delivery of IFN-$\gamma$ protects C57BL/6 mice from chronic-progressive experimental autoimmune encephalomyelitis by increasing apoptosis of central nervous system-infiltrating lympocytes. *J.Immunol.* 167:1821-1829.

87. Dalton,D.K., S.Pitts-Meek, S.Keshav, I.S.Figari, A.Bradley, and T.A.Stewart. 1993. Multiple defects of immune cell function in mice with disrupted interferon-$\gamma$ genes. *Science* 259:1739-1742.

88. van der Veen,R.C., T.A.Dietlin, J.D.Gray, and W.Gilmore. 2000. Macrophage-derived nitric oxide inhibits the proliferation of activated T helper cells and is induced during antigenic stimulation of resting T cells. *Cell.Immunol.* 199:43-49.

89. Tran,E.H., E.N.Prince, and T.Owens. 2000. IFN-$\gamma$ shapes immune invasion of the central nervous system via regulation of chemokines. *J.Immunol.* 164:2759-2768.

90. Chu,C.-Q., S.Wittmer, and D.K.Dalton. 2000. Failure to suppress the expansion of the activated CD4 T cell population in interferon $\gamma$-deficient mice leads to exacerbation of experimental autoimmune encephalomyelitis. *J.Exp.Med.* 192:123-128.

91. Metzger,Z., J.T.Hoffeld, and J.J.Oppenheim. 1980. Macrophage-mediated suppression. I. evidence for participation of both hydrogen peroxide and prostaglandin in suppression of murine lymphocyte proliferation. *J.Immunol.* 124:983-988.

92. Twardzik,D.R., J.A.Mikovits, J.E.Ranchalis, A.F.Purchio, L.Ellingswaorth, and F.W.Ruscetti. 1990. $\gamma$-Interferon-induced activation of latent transforming growth-factor-$\beta$ by human monocytes. *Ann.NY Acad.Sci.* 593:276-284.

93.   Karpus,W.J. and R.H.Swanborg. 1991. CD4+ suppressor cells inhibit the function of effector cells of experimental autoimmune encephalomyelitis through a mechanism involving transforming growth factor-β. *J.Immunol* 146:1163-1168.
94.   Smith,K.J., R.Kapoor, and P.A.Felts. 1999. Demyelination: the role of reactive oxygen and nitrogen species. *Brain Pathol.* 9:69-92.
95.   Willenborg,D.O., M.A.Staykova, and W.B.Cowden. 1999. Our shifting understanding of the role of nitric oxide in autoimmune encephalomyelitis: a review. *J.Neuroimmunol.* 100:21-35.
96.   Fenyk-Melody,J.E., A.E.Garrison, S.R.Brunnert, J.R.Weidner, F.Shen, B.A.Shelton, and J.S.Mudgett. 1998. Experimental autoimmune encephalomyelitis is exacerbated in mice lacking the *NOS2* gene. *J.Immunol.* 160:2940-2946.
97.   Sahrbacher,U.C., F.Lechner, H.P.Eugster, K.Frei, H.Lassmann, and A.Fontana. 1998. Mice with an inactivation of the inducible nitric oxide synthase gene are susceptible to experimental autoimmune encephalomyelitis. *Eur.J.Immunol.* 28:1332-1338.
98.   Albina,J.E., J.A.Abate, and W.L.J.Henry. 1991. Nitric oxide production is required for murine resident peritoneal macrophages to suppress mitogen-stimulated T cell proliferation. Role of IFN-γ in the induction of nitric oxide- synthesizing pathway. *J.Immunol.* 147:144-148.
99.   Mills,C.D. 1991. Molecular basis of suppressor macrophages. Arginine metabolism via the nitric oxide synthetase pathway. *J.Immunol.* 146:2719-2723.
100.  Krenger,W., G.Falzarano, J.Delmonte, Jr., K.M.Snyder, J.C.H.Byon, and J.L.M.Ferrara. 1996. Interferon-γ suppresses T-cell proliferation to mitogen via the nitric oxide pathway during experimental acute graft-versus- host disease. *Blood* 88:1113-1121.
101.  Sicher,S.C., M.A.Vazquez, and C.Y.Lu. 1994. Inhibition of macrophage Ia expression by nitric oxide. *J.Immunol.* 153:1293-130.
102.  Schmied,M., H.Breitschopf, R.Gold, H.Zischer, G.Rothe, and H.Wekerle. 1993. Apoptosis of T lymphocytes in experimental autoimune encephalomyelitis. Evidence for programmed cell death as a mechanism to control inflammation in the brain. *Am.J.Pathol.* 143:446-452.
103.  Bonetti,B., J.Pohl, Y.L.Gao, and C.S.Raine. 1997. Cell death during autoimmune demyelination: effector but not target cells are eliminated by apoptosis. *J.Immunol.* 159:5733-5741.
104.  Pender,M.P., K.B.Nguyen, P.A.McCombe, and J.F.Kerr. 1991. Apoptosis in the nervous system in experimental allergic encephalomyelitis. *J.Neurol.Sci.* 104:81-87.
105.  Bauer,J., M.Bradl, W.F.Hickey, S.Forss-Petter, H.Breitschopf, C.Linington, H.Wekerle, and H.Lassmann. 1998. T-cell apoptosis in inflammatory brain lesions. Destruction of T cells does not depend on antigen recognition. *Am.J.Pathol.* 153:715-724.
106.  Liu,Y. and C.A.Jr.Janeway. 1990. Interferon γ plays a critical role in induced cell death of effector T cell: a possible third mechanism of self-tolerance. *J.Exp.Med.* 172:1735-1739.
107.  Novelli,F., F.Di Pierro, P.F.Di Celle, S.Bertini, P.Affaticati, G.Garotta, and G.Forni. 1994. Environmental signals influencing expression of the IFN-γ receptor on human T cells control whether IFN-g promotes proliferation or apoptosis. *J.Immunol.* 152:496-504.
108.  Dalton,D.K., L.Haynes, C.-Q.Chu, S.L.Swain, and S.Wittmer. 2000. Interferon γ eliminates responding CD4 T cells during mycobacterial infection by inducing apoptosis of activated CD4 T cells. *J.Exp.Med.* 192:117-122.

109.  Ren,Y. and J.Savill. 1998. Apoptosis: the importance of being eaten. *Cell Death Differentiation* 5:563-568.

110.  Platt,N., R.P.da Silva, and S.Gordon. 1998. Recognizing death: the phagocytosis of apoptitic cells. *Trends Cell Biol.* 8:368-372.

111.  Nguyen,K.B. and M.P.Pender. 1998. Phagocytosis of apoptotic lymphocytes by oligodendrocytes in experimental autoimmune encephalomyelitis. *Acta Neuropathol.* 95:40-46.

112.  Ren,Y. and J.Savill. 1995. Proinflammatory cytokines potentiate thrombospondin-mediated phagocytosis of neutrophils undergoing apoptosis. *J.Immunol.* 154:2366-2374.

113.  Chan,A., T.Magnus, and R.Gold. 2001. Phagocytosis of apoptotic inflammatory cells by microglia and modulation by different cytokines: mechanisms for removal of apoptotic cells in the inflamed nervous system. *Glia* 33:87-95.

114.  Gusella,G.L., T.Musso, M.C.Bosco, I.Espinoza-Delgado, K.Matsushima, and L.Varesio. 1993. IL-2 up-regulates but IFN-γ suppresses IL-8 expression in human monocytes. *J.Immunol.* 151:2725-2732.

115.  Asensio,V.C., H.Lassmann, A.Pagenstecher, S.C.Steffensen, S.J.Henriksen, and I.Campbell. 1999. C10 is a novel chemokine expressed in experimental inflammatory demyelinating disorders that promotes recruitment of macrophages to the central nervous system. *Am.J.Pathol.* 154:1181-1191.

116.  McColl,S.R., M.A.Staykova, A.Wozniak, S.Fordham, J.Bruce, and D.O.Willenborg. 1998. Treatment with anti-granulocyte antibodies inhibits the effector phase of experimental autoimmune encephalomyelitis. *J.Immunol.* 161:6421-6426.

117.  Liu,Y.J., H.Kanzler, V.Soumelis, and M.Gilliet. 2001. Dendritic cell lineage, plasticity and cross-regulation. *Nature Immunol.* 2:585-589.

118.  Asselin-Paturel,C., A.Boonstra, M.Dalod, I.Durand, N.Yessaad, C.Dezutter-Dambuyant, A.Vicari, A.O'Garra, C.Biron, and G.Trinchieri. 2001. Mouse type I interferon producing cells are immature antigen-presenting cells with plasmacytoid morphology. *Nature Immunol.* 2:1144-1150.

119.  De Maeyer,E. and J.De Maeyer-Guignard. 1994. Interferons. *In* The Cytokine Handbook. A.Thompson, editor. Academic Press, 265-288.

120.  Brierley,M.M. and E.N.Fish. 2002. IFN-α/β receptor interactions to biologic outcomes: understanding the circuitry. *J.Interferon Cytokine Res.* 22:835-845.

121.  Chofflon,M. 2000. Recombinant human interferon β in relapsing-remitting multiple sclerosis: a review of the major clinical trials. *Eur.J.Neurol.* 7:369-380.

122.  Durelli,L., M.R.Bongioanni, R.Cavallo, B.Ferrero, R.Ferri, E.Verdun, G.B.Bradac, A.Riva, M.Geuna, L.Bergamini, and et al. 1995. Interferon α treatment of relapsing-remitting multiple sclerosis: long-term study of the correlations between clinical and magnetic resonance imaging results and effects on the immune function. *Multiple Sclerosis* 1:S32-S37.

123.  Myhr,K.M., T.Riise, F.E.Green Lilleas, T.G.Beiske, E.G.Celius, A.Edland, D.Jensen, J.P.Larsen, R.Nilsen, M.W.Nortvedt, A.I.Smievoll, C.Vedeler, and H.I.Nyland. 1999. Interferon-α2a reduces MRI disease activity in relapsing-remitting multiple sclerosis. *Neurology* 52:1049-1056.

124.  Brod,S.A., J.W.Lindsey, F.S.Vriesendorp, C.Ahn, E.Henninger, P.A.Narayana, and J.S.Wolinsky. 2001. Ingested IFN-α. *Neurology* 57:845-852.

125.  Abreu,S.L., J.Tondreau, S.Levine, and R.Sowinski. 1983. Inhibition of passive localized experimental allergic encephalomyelitis by interferon. *Int.Arch.Allergy Appl.Immunol.* 72:30-33.

126.    Abreu,S.L. 1982. Suppression of experimental allergic encephalomyelitis by interferon. *Immunol.Commun.* 11:1-7.

127.    Hertz,F. and R.Degheni. 1985. Effect of rat and β-human interferons on hyperacute experimental allergic encephalomyelitis in rats. *Agents and Actions* 16:397-403.

128.    Yu,M., A.Nishiyama, B.D.Trapp, and V.K.Tuohy. 1996. Interferon-β inhibits progression of relapsing-remitting experimental autoimmune encephalomyelitis. *J.Neuroimmunol.* 64:91-100.

129.    Inada,T. and C.A.Mims. 1986. Infection of mice with lactic dehydrogenase virus prevents development of experimental allergic encephalomyelitis. *J.Neuroimmunol.* 11:53-56.

130.    Brod,S.A., M.Khan, R.H.Kerman, and M.Pappolla. 1995. Oral administration of human or murine interferon α suppresses relapses and modifies adoptive transfer in experimental autoimmune encephalomyelitis. *J.Neuroimmunol.* 58:61-69.

131.    Brod,S.A. and D.K.Burns. 1994. Suppression of relapsing experimental autoimmune encephalomyelitis in the SJL/J mouse by oral administration of type I interferons. *Neurology* 44:1144-1148.

132.    Nelson,P.A., Y.Akselband, S.M.Dearborn, A.Al-Sabbagh, Z.J.Tian, P.A.Gonnella, S.S.Zamvil, Y.Chen, and H.L.Weiner. 1996. Effect of oral β interferon on subsequent immune responsiveness. *Ann.NY Acad.Sci* 778:145-155.

133.    Abreu,S.L., I.Thampoe, and P.Kaplan. 1986. Interferon in experimental autoimmune encephalomyelitis: intraventricular administration. *J.Interferon Res.* 6:627-632.

134.    Ruuls,S.R., M.C.D.C.de Labie, K.S.Weber, C.A.D.Botman, R.J.Groenestein, C.D.Dijkstra, T.Olsson, and P.H.Van der Meide. 1996. The length of treatment determines whether IFN-β prevents or aggravates experimental autoimmune encephalomyelitis in Lewis rats. *J.Immunol.* 157:5721-5731.

135.    Van der Meide,P.H., M.C.de Labie, S.R.Ruuls, R.J.Groenestein, C.A.Botman, T.Olsson, and C.D.Dijkstra. 1998. Discontinuation of treatment with IFN-β leads to exacerbation of experimental autoimmune encephalomyelitis in Lewis rats. Rapid reversal of the antiproliferative activity of IFN-β and excessive expansion of autoreactive T cells as disease promoting mechanisms. *J.Neuroimmunol.* 84:14-23.

136.    Croxford,J.L., K.Triantaphyllopoulos, O.L.Podhajcer, M.Feldmann, D.Baker, and Y.Chernajovsky. 1998. Cytokine gene therapy in experimental allergic encephalomyelitis by injection of plasmid DNA-cationic liposome complex into the central nervous system. *J.Immunol.* 160:5181-5187.

137.    Brod,S.A. and M.Khan. 1996. Oral administration of IFN-α is superior to subcutaneous administration of IFN-α in the suppression of chronic relapsing experimental autoimmune encephalomyelitis. *J.Autommunity* 9:11-20.

138.    Brod,S.A., M.Scott, D.K.Burns, and J.T.Phillips. 1995. Modification of acute experimental autoimmune encephalomyelitis in the Lewis rat by oral administration of type 1 Interferons. *J.Interferon Res.* 15:115-122.

139.    Yong,V.W., S.Chabot, O.Stuve, and G.Williams. 1998. Interferon β in the treatment of multiple sclerosis: mechanisms of action. *Neurology* 51:682-689.

140.    Luca,M.E., L.Visser, C.J.Lucas, and L.Nagelkerken. 1999. IFN-β modulates specific T cell responses in vitro but does not affect experimental autoimmune encephalomyelitis in the SJL mouse. *J.Neuroimmunol.* 100:190-196.

141.    Noronha,A., A.Toscas, and M.A.Jensen. 1993. Interferon β decreases T cell activation and interferon γ production in multiple sclerosis. *J.Neuroimmunol.* 46:145-153.

142.    Ling,P.D., M.K.Warren, and S.N.Vogel. 1985. Antagonistic effect of interferon-α/β on the interferon-γ- induced expression of Ia antigen in murine macrophages. *J.Immunol.* 135:1857-1863.

143. Ransohoff,R.M., C.Devajyothi, M.L.Estes, G.Babcock, R.A.Rudick, E.M.Frohman, and B.P.Barna. 1991. Interferon-β specifically inhibits interferon-γ-induced class II major histocompatibility complex gene transcription in a human astrocytoma cell line. *J.Neuroimmunol.* 33:103-112.

144. Soilu-Hänninen,M., A.Salmi, and R.Salonen. 1995. Interferon-β downregulates expression of VLA-4 antigen and antagonizes interferon-γ-induced expression of HLA-DQ on human peripheral blood monocytes. *J.Neuroimmunol.* 60:99-106.

145. Miller,A., N.Lanir, S.Shapiro, M.Revel, S.Honigman, A.Kinarty, and N.Lahat. 1996. Immunoregulatory effects of interferon-β and interacting cytokines on human vascular endothelial cells. Implications for multiple sclerosis and other autoimmune diseases. *J.Neuroimmunol.* 64:151-161.

146. Brown,K.A. 2001. Factors modifying the migration of lymphocytes across the blood-brain barrier. *Int.Immunopharmacol.* 1:2043-2062.

147. Floris,S., S.R.Ruuls, A.Wierinckx, S.M.A.van der Pol, E.Döpp, P.H.Van der Meide, C.D.Dijkstra, and H.E.De Vries. 2002. Interferon-β directly influences monocyte infiltration into the central nervous system. *J.Neuroimmunol.* 127:69-79.

148. Defazio,G., M.Gelati, E.Corsini, B.Nico, A.Dufour, G.Massa, and A.Salmaggi. 2002. In vitro modulation of adhesion molecules, adhesion phenomena, and fluid phase endocytosis on human umbilical vein endothelial cells and brain-derived microvascular endothelium by IFN-β1a. *J.Interferon Cytokine Res.* 21:267-272.

149. Lou,J., Y.Gasche, L.Zheng, C.Giroud, P.Morel, J.Clements, A.Ythier, and G.E.Grau. 1999. Interferon-β inhibits activated leukocyte migration through human brain microvascular endothelial cell monolayer. *Lab.Invest.* 79:1015-1025.

150. Calabresi,P.A., C.M.Pelfrey, L.R.Tranquill, H.Maloni, and H.F.McFarland. 1997. VLA-4 expression on peripheral blood lymphocytes is downregulated after treatment of multiple sclerosis with interferon β. *Neurology* 49:1111-1116.

151. Muraro,P.A., T.Leist, B.Bielekova, and H.F.McFarland. 2000. VLA-4/CD49d downregulated on primed T lymphocytes during interferon-β therapy in multiple sclerosis. *J.Neuroimmunol.* 111:186-194.

152. Yong,V.W., C.A.Krekoski, P.A.Forsyth, R.Bell, and D.R.Edwards. 1998. Matrix metalloproteinases and diseases of the CNS. *Trends Neurosci.* 21:75-80.

153. Leppert,D., R.L.P.Lindberg, L.Kappos, and S.L.Leib. 2001. Matrix metalloproteinases: multifunctional effectors of inflammation in multiple sclerosis and bacterial meningitis. *Brain Res.Rev.* 36:249-257.

154. Stuve,O., N.P.Dooley, J.H.Uhm, J.P.Antel, G.S.Francis, G.Williams, and V.W.Yong. 1996. Interferon β-1b decreases the migration of T lymphocytes in vitro: effects on matrix metalloproteinase-9. *Ann.Neurol.* 40:853-863.

155. Leppert,D., E.Waubant, M.R.Burk, J.R.Oksenberg, and S.L.Hauser. 1996. Interferon β-1b inhibits gelatinase secretion and in vitro migration of human T cells: a possible mechanism for treatment efficacy in multiple sclerosis. *Ann.Neurol.* 40:846-852.

156. Ozenci,V., M.Kouwenhoven, N.Teleshova, M.Pashenkov, S.Fredrikson, and H.Link. 2000. Multiple sclerosis: pro- and anti-inflammatory cytokines and metalloproteinases are affected differentially by treatment with IFN-β. *J.Neuroimmunol.* 108:236-243.

157. Galboiz,Y., S.Shapiro, N.Lahat, H.Rawashdeh, and A.Miller. 2001. Matrix metalloproteinases and their tissue inhibitors as markers of disease subtype and response to interferon-β therapy in relapsing and secondary-progressive multiple sclerosis patients. *Ann.Neurol.* 50:443-451.

158.  Uhm,J.H., N.P.Dooley, O.Stuve, G.S.Francis, P.Duquette, J.P.Antel, and V.W.Yong. 1999. Migratory behavior of lymphocytes isolated from multiple sclerosis patients: effects of interferon β-1b therapy. *Ann.Neurol.* 46:319-324.

159.  Comabella,M., J.Imitola, H.L.Weiner, and S.J.Khoury. 2002. Interferon-β treatment alters peripheral blood monocytes chemokine production in MS patients. *J.Neuroimmunol.* 127:205-212.

160.  Zang,Y.C.Q., J.B.Halder, A.K.Samanta, J.Hong, V.M.Rivera, and J.Z.Zhang. 2001. Regulation of chemokine receptor CCR5 and production of RANTES and MIP-1α by interferon-β. *J.Neuroimmunol.* 112:174-180.

161.  Iarlori,C., M.Reale, A.Lugaresi, G.De Luca, L.Bonanni, A.Di-Iorio, C.Feliciani, P.Conti, and D.Gambi. 2000. RANTES production and expression is reduced in relapsing-remitting multiple sclerosis patients treated with interferon-β1b. *J.Neuroimmunol.* 107:100-107.

162.  Teleshova,N., M.Pashenkov, Y.M.Huang, M.Soderstrom, P.Kivisakk, V.Kostulas, M.Haglund, and H.Link. 2002. Multiple sclerosis and optic neuritis: CCR5 and CXCR3 expressing T cells are augmented in blood and cerebrospinal fluid. *J.Neurology* 249:723-729.

163.  Satoh,J., D.W.Paty, and S.U.Kim. 1996. Counteracting effect of IFN-β on IFN-γ-induced proliferation of human astrocytes in culture. *Multiple Sclerosis* 1:279-287.

164.  Akbar,A.N., J.M.Lord, and M.Salmon. 2000. IFN-α and IFN-β: a link between immune memory and chronic inflammation. *Immunol.Today* 21:337-342.

165.  Zipp,F., M.Beyer, H.Gelderblom, D.Wernet, R.Zschenderlein, and M.Weller. 2000. No induction of apoptosis by IFN-β in human antigen-specific T cells. *Neurology* 54:485-487.

166.  Schmidt,J., S.Stürzebecher, K.V.Toyka, and R.Gold. 2001. Interferon-β treatment of experimental autoimmune encephalomyelitis leads to rapid nonapototic termination of T cell infiltration. *J.Neurosci.Res.* 65:59-67.

167.  Ozenci,V., M.Kouwenhoven, Y.M.Huang, P.Kivisakk, and H.Link. 2000. Multiple sclerosis is associated with an imbalance between tumour necrosis factor-α (TNF-α)- and IL-10-secreting blood cells that is corrected by interferon-β (IFN-β) treatment. *Clin.Exp.Immunol* 120:147-153.

168.  Abu-khabar,K.S., J.A.Armstrong, and M.Ho. 1992. Type I interferons (IFN-α and -β) suppress cytotoxin (tumor necrosis factor-α and lymphotoxin) production by mitogen-stimulated human peripheral blood mononuclear cells. *J.Leukocyte Biol.* 52:165-172.

169.  Selmaj,K., C.S.Raine, M.Farooq, W.T.Norton, and C.F.Brosnan. 1991. Cytokine cytotoxicity against oligodendrocytes. Apoptosis induced by Lymphotoxin. *J.Immunol.* 147:1522-1529.

170.  Selmaj,K. and C.S.Raine. 1988. Tumor necrosis factor mediates myelin and oligodendrocyte damage in vitro. *Ann.Neurol.* 23:339-346.

171.  Ruddle,N.H., C.M.Bergman, K.M.McGrath, E.G.Lingenheld, M.L.Grunnet, S.J.Padula, and R.B.Clark. 1990. An antibody to lymphotoxin and tumor necrosis factor prevents transfer of experimental allergic encephalomyelitis. *J.Exp.Med* 172:1193-1200.

172.  Brod,S.A., M.Khan, L.D.Nelson, B.Decuir, M.Malone, and E.Henninger. 2000. Adoptive transfer from interferon-α-fed mice is associated with inhibition of active experimental autoimmune encephalomyelitis by decreasing recipient tumor necrosis factor-α secretion. *J.Immunotherapy* 23:235-245.

173.  Rudick,R.A., R.M.Ransohoff, R.Peppler, S.Vanderbrug-Medendorp, P.Lehmann, and J.Alam. 1996. Interferon β induces interleukin-10 expression: relevance to multiple sclerosis. *Ann.Neurol.* 40:618-627.

174. Yasuda,C.L., A.Al-Sabbagh, E.C.Oliveira, B.M.Diaz-Bardales, A.A.Garcia, and L.M.Santos. 1999. Interferon β modulates experimental autoimmune encephalomyelitis by altering the pattern of cytokine secretion. *Immunol.Invest.* 28:126.

175. Soos,J.M., O.Stuve, S.Youssef, M.Bravo, H.M.Johnson, H.L.Weiner, and S.S.Zamvil. 2002. Oral type I IFN-τ promotes Th2 bias and enhances suppression of autoimmune encephalomyelitis by oral glatiramer acetate. *J.Immunol.* 169:2235.

176. Rudick,R.A., R.M.Ransohoff, J.C.Lee, R.Peppler, M.Yu, P.M.Mathisen, and V.K.Tuohy. 1998. In vivo effects of interferon β-1a on immunosuppressive cytokines in multiple sclerosis. *Neurology* 50:1294-1300.

177. Ossege,L.M., E.Sindern, T.Patzold, and J.P.Malin. 1998. Immunomodulatory effects of interferon-β-1b in vivo: induction of the expression of transforming growth factor-β1 and its receptor type II. *J.Neuroimmunol.* 91:73-81.

178. Liu,Z., C.M.Pelfrey, A.Cotleur, J.C.Lee, and R.A.Rudick. 2001. Immunomodulatory effects of interferon β-1a in multiple sclerosis. *J.Neuroimmunol.* 112:153-162.

179. Segal,B.M., B.K.Dwyer, and E.M.Shevach. 1998. An interleukin (IL)-10/IL-12 immunoregulatory circuit controls susceptibility to autoimmune disease. *J.Exp.Med.* 187:537-546.

180. McRae,B.L., R.T.Semnani, M.P.Hayes, and G.A.van Seventer. 1998. Type I IFNs inhibit human dendritic cell IL-12 production and Th1 cell development. *J.Immunol* 160:4298-4304.

181. McRae,B.L., B.A.Beilfuss, and G.A.van Seventer. 2000. IFN-β differentially regulates CD40-induced cytokine secretion by human dendritic cells. *J.Immunol.* 164:23-28.

182. Tuohy,V.K., M.Yu, L.Yin, P.M.Mathisen, J.M.Johnson, and J.A.Kawczak. 2000. Modulation of the IL-10/IL-12 cytokine circuit by interferon-β inhibits the development of epitope spreading and disease progression in murine autoimmune encephalomyelitis. *J.Neuroimmunol.* 111:55-63.

183. Noronha,A., A.Toscas, and M.A.Jensen. 1992. Contrasting effects of α, β and γ interferons on nonspecific suppressor function in multiple sclerosis. *Ann.Neurol.* 31:103-106.

184. Noronha,A., A.Toscas, and M.A.Jensen. 1990. Interferon β augments suppressor cell function in multiple sclerosis. *Ann.Neurol.* 27:207-210.

185. Martal,J.L., N.M.Chene, L.P.Huynh, R.M.L'Haridon, P.B.Reinaud, M.W.Guillomot, M.A.Charlier, and S.Y.Charpigny. 1998. IFN-τ: a novel subtype I IFN1. Structural characteristics, non-ubiquitous expression, structure-function relationships, a pregnancy hormonal embryonic signal and cross-species therapeutic potentialities. *Biochimie* 80:755-777.

186. Soos,J.M. and H.M.Johnson. 1999. Interferon-τ. Prospects for clinical use in autoimmune disease. *Biodrugs* 11:125-135.

187. Soos,J.M., P.S.Subramaniam, A.C.Hobeika, J.Schiffenbauer, and H.M.Johnson. 1995. The IFN pregnancy recognition hormone IFN-τ blocks both development and superantigen reactivation of experimental allergic encephalomyelitis without associated toxicity. *J.Immunol.* 155:2747-2753.

188. Subramaniam,P.S., S.A.Khan, C.H.Pontzer, and H.M.Johnson. 1995. Differential recognition of the type 1 interferon receptor by interferons τ and α is responsible for their disparate cytotoxicities. *Proc.Natl.Acad.Sci.USA* 92:12270-12274.

189. Soos,J.M., M.G.Mujtaba, P.S.Subramaniam, W.J.Streit, and H.M.Johnson. 1997. Oral feeding of interferon τ can prevent the acute and chronic relapsing forms of experimental allergic encephalomyelitis. *J.Neuroimmunol.* 75:43-50.

190.  Mujtaba,M.G., W.J.Streit, and H.M.Johnson. 1998. IFN-τ suppresses both the autoreactive humoral and cellular immune responses and induces stable remission in mice with chronic experimental allergic encephalomyelitis. *Cell.Immunol.* 186:94-102.

191.  Mujtaba,M.G., J.M.Soos, and H.M.Johnson. 1997. CD4 T suppressor cells mediate interferon tau protection against experimental allergic encephalomyelitis. *J.Neuroimmunol.* 75:35-42.

192.  Powell,H.S., D.Mitchell, J.Lederman, J.Bucklmeier, S.S.Zamvil, M.Graham, N.H.Ruddle, and L.Steinman. 1990. Lympotoxin and tumor necrosis factor α production by myelin basic protein specific T cells clones correlates with encephalogenicity. *Int.Immunol.* 2:539.

193.  Tennakoon,D.K., R.Smith, M.D.Stewart, T.E.Spencer, M.Nayak, and C.J.R.Welsh. 2001. Ovine IFN-τ modulates the xpression of MHC antigens on murine cerebrovascular endothelial cells and inhibits replication of Theiler's virus. *J.Interferon Cytokine Res.* 21:785-792.

Chapter A16

# THE ROLE OF GROWTH FACTORS IN EXPERIMENTAL AUTOIMMUNE ENCEPHALOMYELITIS

Amy E. Lovett-Racke and Michael K. Racke*
*Department of Neurology and *The Center for Immunology, UT Southwestern Medical Center, Dallas, Texas, USA*

**Abstract:** This chapter reviews the research done on the role of growth factor in experimental autoimmune encephalomyelitis and demonstrates the increasingly recognized therapeutic potential of growth factors in inflammatory demyelinating disease.

**Key words:** EAE, growth factors, demyelination, Neurotrophins, insulin-like growth factor-1, glial growth factors, fibroblast growth factors, ciliary neurotrophic factor.

## 1. INTRODUCTION

The development of tissues, maintenance of homeostasis, and responses to disease and injury are all dependent on effective communication between cells. Cells produce intercellular signaling molecules that provide this function. Peptide hormones, cytokines and growth factors are the three major categories of signaling molecules. This review will focus on the role of growth factors in the inflammatory demyelinating disease, experimental autoimmune encephalomyelitis (EAE). In particular, we will evaluate growth factors that were originally characterized in cells or tissues outside of the immune system. Neurotrophins (NT), insulin-like growth factor-1 (IGF-1), glial growth factors (GGF), fibroblast growth factors (FGF), and ciliary neurotrophic factor (CNTF) are among many growth factors that have recently been studied in EAE. These molecules have the potential to protect the CNS from damage caused by inflammation, as well as the potential to repair the CNS after damage has occurred. Interestingly, many of these growth factors have been shown to effect the function of lymphocytes and/or be produced by lymphocytes which makes understanding their role in neuroimmune interactions valuable in understanding inflammatory demyelinating diseases, such as EAE and multiple sclerosis (MS).

## 2.        NEUROTROPHINS

Neurotrophins are a family of neurotrophic factors that promote the survival, differentiation and maintenance of neurons. There are currently seven described neurotrophins, nerve growth factor (NGF), brain-derived neurotrophin factor (BDNF), neurotrophin (NT)-3, NT-4, NT-5, NT-6, and NT-7. The biologically active form of these molecules are non-covalent homodimers which interact with two classes of cell surface tyrosine kinase receptors. The high affinity receptors include trkA, trkB and trkC. NGF binds trkA, BDNF and NT-4/5 bind trkB, and NT-3 binds trkC. The low affinity tyrosine kinase receptor, p75, binds all of the neurotrophins with similar affinity, yet with different rate constants. The distribution of these receptors on different populations of neurons in both the central and peripheral nervous system determines the responsiveness of neurons to the various neurotrophins.

Although neurotrophins were originally described because of their effects on neurons, it has become increasely clear that their biological effects extend well beyond the nervous system. Recent studies have clearly shown a bi-directional link between the nervous system and the immune system . In particular, there is significant data illustrating the effects of NGF on both the innate and adaptive arms of the immune system. NGF has been shown to affect mast cells (Aloe 1977, Horigome 1993, Mazurek 1986), basophils (Bischoff 1992, Burgi 1996, Miura 2001), neutrophils (Boyle 1985), granulocytes (Matsuda 1988), and monocytes/macrophages (Matsuda 1988, Chevalier 1994, Ehrhard 1993, Flugel 2001, Kobayashi 2001) . In addition, NGF has been shown to affect the proliferation, differentiation and trk receptor expression by T cells and B cells (Otten 1989, Manning 1985, Thorpe 1987, Kimata 1991, Brodie 1992, Bodie 1994, Ehrhard 1993, Ehrhard 1994, Garacia 2001, Torcia 2001, Bracci-Laudiero 2002). These studies led to interest in the role of neurotrophins in EAE and MS.

Initial studies attempting to elucidate the role of NGF in EAE examined the expression of NGF and its receptors in the CNS during the course of EAE. In the Lewis rat model of EAE, the low affinity NGF receptor (p75) was found to be dramatically upregulated on the walls of blood vessels and the surrounding neurons in perivascular regions during the acute phase of the disease (Calza 1997, 1998). Nataf's (1998) longitudinal study of p75 expression in the CNS of EAE-affected Lewis rats showed that p75 was first upregulated on the Purkinje cells and later on endothelial and perivascular cells, as well as myelinating cells, suggesting that p75/NGF may play a role in lymphocyte-endothelial cell interaction during EAE. Calza (1997) found that NGF content increased in some areas of the brain, but was decreased in the spinal cord during the peak of inflammation. Further analysis demonstrated that astrocytes and oligodendrocytes in the white matter of EAE-affected brains had an increased expression of NGF (Micera 1998). To determine if NGF was providing a protective effect in EAE, rats were immunized with NGF/CFA twice prior to EAE induction (Micera 2000). This resulted in the production of anti-NGF antibodies capable of binding endogenous NGF. The NGF-immunized rats had more severe disease, more inflammation, and had a reduction of NGF in the CNS, demonstrating the negative effect of NGF deprivation on EAE.

The obvious next step was to determine if administration of NGF might be beneficial in the treatment of EAE. Using a recombinant retrovirus encoding NGF, encephalitogenic T cells were generated that produced high levels of NGF and transferred into Lewis rats (Flugel 2000). Even though these cells maintained a Th1 phenotype, they were unable to induce disease and these rats had very little CNS inflammation. In addition, NGF was shown to upregulate p75 NGF receptor on monocytes and inhibited monocyte migration through TNF$\alpha$-activated endothelial cells derived from brain vasculature in vitro.

The therapeutic effects of NGF were investigated in a primate model of EAE using the common marmoset (Genain 1997, Villoslada 2000). NGF was administered via continuous intracerebroventricular infusion beginning 7 days after immunization with recombinant MOG/CFA. NGF-treated marmosets had a lower incidence of disease, a delay in disease onset, and a markedly milder clinical course. Histopathological examination of the CNS showed minimal demyelination and inflammation. Both glial and inflammatory cells within the CNS were found to express the p75 and trkA NGF receptors. NGF-treated marmosets had similar lymphoproliferative responses to MOG and similar anti-MOG antibody titers, suggesting that NGF did not interfere with peripheral T and B cell responses. However, cytokine expression in perivascular inflammatory cuffs in NGF-treated primates showed minimal IFN$\gamma$ and a significant increase in IL-10, suggesting that the phenotype of the infiltrating cells had shifted from Th1 to Th2.

The systemic administration of NGF was evaluated in the B10.PL mouse model of chronic-relapsing EAE (Arredondo 2001). Intraperitoneal injection of NGF during the first two weeks post immunization or adoptive transfer resulted in a delay in disease onset and decreased severity of disease. In vitro stimulation of encephalitogenic T cells with antigen in the presence of NGF did not alter the proliferation, cytokine production or the ability of these cells to transfer disease, suggesting that NGF alone was not capable of altering the phenotype of encephalitogenic T cells. However, when MBP-specific T cells were cultured in the presence of IL-4, NGF gene expression was increased similar to what had previously seen in cultured astrocytes (Brodie 1998). To determine if this was relavant in vivo, NGF levels were measured in splenocytes from mice who were given i.p. administration of antigen which is an immunization protocol that results in Th2 responses. Both NGF and IL-4 were produced by the splenocytes of these mice with a concommitant decrease in IFN$\gamma$, demonstrating that NGF production occurs in response to Th2-associated stimuli in vivo. Together, these studies provide several potential mechanisms by which NGF may ameliorate EAE, including inhibiting migration of inflammatory cells across the blood-brain barrier and altering the phenotype of potentially encephalitogenic T cells from Th1 to Th2.


## 3.      CILIARY NEUROTROPHIC FACTOR


Ciliary neurotrophic factor (CNTF) was originally identified as a survival factor for neurons (Alder 1979). Subsequently, it has been shown to promote the differentiation, maturation and survival of oligodendrocytes (Barres 1993 and 1996). This observation led to the investigation of the role of CNTF in EAE. MOG35-55 induced EAE was more severe in CNTF-deficient mice and recovery was poor, but the incidence of disease was similar (Linker 2002). Histopathological analysis of the CNS following EAE induction in CNTF-deficient mice showed a 60% decrease in the number of proliferating oligodendrocyte precursor cells and apoptosis of oligodendrocytes was increased by 50% compared to wild-type mice. CNTF-deficient mice also had more vacuolar degeneration of myelin which is consistent with the loss of oligodendrocytes. Since TNF$\alpha$ had been previously shown to mediate oligodendrocyte apoptosis and loss of myelin, the effect of in vivo administration of anti-TNF$\alpha$ was evaluated in the CNTF-deficient mice. Anti-TNF$\alpha$ treatment did not alter the onset of the disease, but there was significantly less myelin vacuolation and oligodendrocyte apoptosis. Interestingly, it was recently shown that a null mutation in the CNTF gene is associated with early onset MS with predominantly motor symptoms (Giess 2002). Consistent with the observation in the CNTF-deficient mice, there was no increase in the incidence of MS in CNTF-deficient individuals.

# 4.    GLIAL GROWTH FACTOR

Since oligodendrocytes produce myelin in the CNS, growth factors that stimulate oligodendrocytes are obvious candidates as potential therapies for demyelinating diseases. Glial growth factor (GGF2) has been shown to promote the proliferation and survival of oligodendrocytes (Marchionni 1993, Rutkowski 1995, Canoll 1996, Milner 1997). GGF2 is a member of the neuregulin family of proteins within the epidermal growth factor superfamily (Gassmann 1997). The effect of human recombinant GGF2 on chronic relapsing EAE in mice was investigated (Cannella 1998, Marchionni 1999). Administration of GGF2 to mice during the first ten days after adoptive transfer of encephalitogenic T cells resulted in a delay in disease onset and reduced disease severity. If treatment with GGF2 was begun during the chronic phase of the disease, there was a decrease in the mean clinical score compared to the vehicle-treated mice. In addition, the relapse rate
in these mice was reduced and this was seen even after treatment was terminated. As anticipated, GGF2-treated mice showed more remyelination than control mice and had an increase in gene expression of MBP exon 2 which is only contained in an isoform of MBP produced during myelination. Analysis of cytokine expression in the CNS of GGF2-treated mice showed an increase in IL-4 in mice treated early and an increase in IL-10 in mice treated during the chronic phase of the disease, suggesting the GGF2 may enhance the expression of Th2 (regulatory) cytokines in the CNS.

# 5.    INSULIN-LIKE GROWTH FACTOR-1

Insulin-like growth factor-1 (IGF-1) is a single-chain polypeptide with structural homology to pro-insulin. It is abundant in the circulation and is produced by many tissues, and it regulates the proliferation and differentiation of many cell types. As reviewed in Folli (1996), IGF-1 has been shown to play a role in CNS development, injury repair, neurodegeneration, as well as a neuromodulator of some higher brain functions. However, it was the ability of IGF-1 to promote the survival of oligodendrocytes and the formation of myelin sheaths in vitro that prompted the investigation of IGF-1 in demyelinating disease (Mozell 1991, Barres 1992, 1993). The potential of IGF-1 to enhance myelin production was confirmed in transgenic mice that overexpressed IGF-1. These mice produced significantly more CNS myelin than wild-type mice (Carson 1993). In situ hybridization analysis of IGF-1 and associated receptors in EAE-affected rats showed that reactive astrocytes produced high levels of IGF1 and oligodendrocytes expressed IGF-1 receptors in areas of lesions (Liu 1994). This observation was also seen in cuprizone-induced demyelination and cryogenic spinal cord injury (Komoly 1992, Yao 1995).

The therapeutic potential of IGF-1 in EAE was first examined in the Lewis rat model. In the initial studies, IGF-1 was administered intravenously at the first signs of tail and hind limb weakness. This resulted in lower maximum clinical deficit scores and a faster recovery than the control animals. IGF-1 treatment appeared to reduce the permeability of the blood-spinal cord barrier as demonstrated by Evans-blue albumin transport. Further analysis illustrated that there were fewer and smaller inflammatory lesions in the spinal cord. Since IGF-1 had previously been shown to enhance myelin formation and promote oligodendrocyte survival in vitro, it seemed possible that IGF-1's therapeutic effects may be due to remyelination and oligodendrocyte proliferation. Following EAE induction in Lewis rats, the gene expression of several myelin proteins was up-regulated, including MBP, PLP and CNPase (Yao 1995). In addition, there appeared to be proliferating oligodendrocyte-like cells in the area of lesions. In the adoptive transfer model of Lewis rat EAE, which has CNS inflammation in the absence of demyelination, IGF-1 treatment decreased disease severity, lesion numbers, lesion areas, and the number of infiltrating cells (Liu 1997). This observation suggested that IGF-1-mediated remyelination was not responsible for the therapeutic effect of IGF-1 in the Lewis rat.

To determine how effective IGF-1 may be on relapsing-remitting demyelinating disease which is more reminiscent of multiple sclerosis, two murine models of EAE were evaluated. Li (1998) adoptively transferred MBP-specific T cells into SJL/J mice and administered IGF-1 on days 7-16 post-transfer. Again, there was a reduction in disease severity, number of lesions and size of lesions. Three-dimensional MR microscopy of placebo-treated mice showed abnormal signal throughout the brain that was not present in the IGF-1-treated mice or the normal control mice, which may reflect changes in the permeability of cell membranes and/or the blood-brain barrier (Xu 1998).

Since IGF-1 can bind to insulin receptors, as well as IGF-1 receptors, the effect of IGF-1 on EAE in (PLXSJL)F1 mice was analyzed using both free IGF-1 and IGF-1 bound to IGF-1 binding protein-3 (IGF-1/IGFBP3) to determine if the dose of IGF-1 could be optimized without the development of hypoglycemia due to IGF-1 occupation of insulin receptors (Lovett-Racke 1998). Pharmacodynamic studies showed that free IGF-1 could be given at 10 mg/kg/day and IGF-1/IGFPB3 complex given at 100 mg/kg/day, which is equivalent to 20 mg/kg/day free IGF-1, without clinical signs of hypoglycemia. In the initial experiments, IGF-1/IGFBP3 complex was given twice daily during the first three weeks following transfer of encephalitogenic T cells. The disease onset was delayed in a dose-dependent manner, but disease severity was dramatically enhanced in the IGF-1/IGFBP3-treated mice. The histopathology of the CNS of these animals paralleled the clinical picture with minimal inflammation prior to the onset of clinical symptoms followed by increased inflammation, demyelination and axonal degeneration compared to placebo-treated animals. To determine if this increased disease severity was due to IGF-1 being bound to IGFBP3, treatment with free IGF-1 and the complex were compared. Comparison of equimolar amounts of IGF-1 or IGF-1/IGFBP3 administered during the first week post-transfer of encephalitogenic cells showed that free IGF-1 was only mildly beneficial and the complex provided no clinical benefit. When equimolar amount of IGF-1 or IGF-1/IGFBP3 were given at the onset of disease (days 12-20), IGF-1 had a similar disease course as controls and the complex-treated mice had more severe disease. Analysis of adhesion molecule expression within the CNS showed a decrease in ICAM-1 expression in IGF-1 and complex treated mice which may have played a role in the inhibition of inflammation early in the disease. In vitro culture of encephalitogenic T cells with IGF-1/IGFBP3 resulted in increased proliferation of these cells which may explain why disease was exacerbated in mice treated with this complex. In a second study that examined IGF-1 treatment in SJL/J mice, there was only a transient clinical amelioration of disease and a low level of remyelination after IGF-1 treatment during the acute phase (Cannella 2000). During the chronic phase of the disease, there was no evidence of enhanced remyelination or alterations in oligodendrocyte progenitor populations.
The previous studies in rodents showed that timing, dosage, and species may all be critical factors in using IGF-1 as a therapy in inflammatory demyelinating disease. An open-label, crossover study was performed in 7 MS patients to determine the safety and efficacy of recombinant human IGF-

1 (Frank 2002). After 6 months of therapy, there was no significant difference in MRI or clinical measures of disease activity in these patients. The drug was well tolerated, but accumulating data suggest that IGF-1 may not be as beneficial as anticipated in the treatment of inflammatory demyelinating disease.

## 6. FIBROBLAST GROWTH FACTOR

Like IGF-1, basic fibroblast growth factor (FGF-2) has been shown to stimulate the differentiation of precursor cells to mature oligodendrocytes in vitro (McKinnon 1990). FGF-2 is ubiquitously expressed in the periphery and is expressed by several types of neurons and astrocytes (Grothe 1991, Gomez-Pinilla 1992, Koshinaga 1993, Eckenstein 1994, Follesa 1994, Liu 1994, Reilly 1996). FGF-2 promotes neuronal survival and neurite outgrowth (Walicke 1986, Unsicker 1986). FGF-2 expression in the CNS of EAE-affected Lewis rats was analyzed as a means to assess the potential role of this molecule in inflammatory demyelinating disease (Liu 1998). FGF-2 gene expression was dramatically increased in the spinal cords of EAE-affected rats compared to normal rats. FGF-2 expression was limited to neurons in normal CNS tissue, but was prominent in microglia in the perivascular regions of EAE-affected tissues. Similarly, the FGF-2 receptor (FGFR1) was induced in activated microglia. The expression of FGF-2 on microglia had previously only been seen in vitro (Shimojo 1991, Presta 1995). Liu (1998) gave several potential roles for FGF-2 in EAE. First, it may regulate the activity of microglia in an autocrine manner. Second, it may induce the expression of IGF-1 which may enhance remyelination. Third, it may promote the proliferation of oligodendrocytes in the areas of demyelination. The effect of FGF-2 as a therapy has not yet been published.

## 7. NEUROPROTECTION BY ENCEPHALITOGENIC T CELLS

Both MS patients and healthy individuals have myelin-reactive T cells in their peripheral blood, suggesting that the presence of these autoreactive T cells is not sufficient to confer susceptibility to MS (Burns 1983, Martin 1990, Zhang 1992, Joshi 1993). It has been shown that the majority of MBP-reactive T cells in MS patients are less dependent on CD28-mediated costimulation which is more characteristic of memory cells (Lovett-Racke 1998, Scholz 1998). However, several studies have shown that insults to the CNS such as peripheral nerve injury (Olsson 1992 and 1993), cerebrovascular disease (Wang 1992), and viral infection (Miller 1997) can lead to the expansion of myelin-reactive T cells in vivo. These findings have shifted our focus on the role of these cells from the belief that autoreactive T cells should be eliminated during thymic selection to understanding how these cells may become encephalitogenic in only a relatively small percentage of the population. There is increasing evidence to indicate that T cells may actually provide a neuroprotective effect in the event of nerve injury, thus providing a possible explanation why it may be beneficial to maintain a population of myelin-reactive T cells.

To evaluate whether lymphocytes effect the severity of motoneuron loss during nerve injury, Serpe (1999) performed facial nerve transections in severe combined immunodeficient *(scid)* mice and found that these mice had more motoneuron loss than the wild-type controls. Moalem (1999) showed that the protection against neuronal loss following mechanical nerve injury was due specifically to the presence of myelin-reactive T cells. This was demonstrated by showing that the transfer of MBP-specific, but not ovalbumin-specific, T cells protected retinal ganglion cells after optic nerve crush. These studies, as well as the observation that lymphocytes express many growth factors known to affect cells of the CNS, led to the hypothesis that lymphocytes may express molecules that protect neurons from damage. Since it was known that treatment with neurotrophins enhanced neuronal survival after injury, the role of neurotrophins produced by infiltrating lymphocytes following nerve injury was investigated. In rats that were immunized with MBP one day prior to ventral root avulsion, there was a 50% higher survival rate of motoneurons in immunized animals versus controls (Hammarberg 2000). Both BDNF, NT-3 and glial cell line-derived neurotrophic factor were expressed by T and NK cells in the spinal cord. The expression of these neurotrophins was not limited to the encephalitogenic T cells, but was actually higher in the non-specifically recruited T cells. In the optic nerve crush model of nerve injury, it was shown that after local administration of a tyrosine kinase inhibitor known to prevent autophosphorylation of NT receptors, the neuroprotective effect of anti-MBP T cells was diminished (Moalem 2000). These studies suggests that production of neutrophins by CNS infiltrating T cells may limit neuronal damage during injury.

## 8.    CONCLUSION

Theoretically, several of the growth factors previously discussed seem like ideal candidates as therapeutic agents in the treatment of inflammatory demyelinating disease. IGF-1, GGF2, FGF are all capable of influencing the function of oligodendrocytes, which are the myelinating cells of the CNS. Since the myelin sheath appears to be the primary target of inflammatory cells in EAE and MS, these molecules possess the potential to enhance remyelination and oligodendrocyte proliferation. NGF and CNTF are potent neurotrophic factors that promote the survival of neurons. Since it appears that axonal loss following demyelination may be responsible for the permanent deficits seen during the progressive phase of MS, these agents seem like obvious candidates to limit axonal damage. However, these initial studies summarized here using these growth factors as treatments for EAE have demonstrated how vast the effects of these molecules are outside of the nervous system. In particular, these growth factors can have profound effects on immune cells which may alter the encephalitogenicity of lymphocytes in a positive or negative manner. Consequently, it has become increasingly clear that understanding the role that these growth factors play in both the nervous system and immune system are critical in determining how effective they may be as therapies for inflammatory demyelinating disease.

**ACKNOWLEDGMENTS**
This work was supported by NIH grant RO1-NS-37513 (MKR). This work was also supported by grants from the National Multiple Sclerosis Society to MKR. AEL-R and MKR were also supported by the Yellow Rose Foundation.

## 9.    REFERENCE LIST

Adler R, Landa KB, Manthorpe M, Varon S (1979) Cholinergic Neuronotrophic factors: intraocular distribution of trophic activity for ciliary neurons. Science 204:1434-1436.

Aloe L and Levi-Montalcini R (1977) Mast cell increase in tissues of neonatal rats injected with the nerve growth factor. Brain Res 133:358-366.

Arredondo LR, Deng C, Ratts RB, Lovett-Racke AE, Holtzman DM, Racke MK. (2001) Role of nerve growth factor in experimental autoimmune encephalomyelitis. Eur J Immunol 31:625-633.

Barres BA, Hart IK, Coles HSR, Burne JF, Voyvodic JT, Richardson WD, Raff MC. (1992) Cell death and control of cell survival in the oligodendrocyte lineage. Cell 70:31-46.

Barres BA, Schmid R, Sendtner M, Raff MC. (1993) Multiple extracellular signals are required for long-term oligodendrocyte survival. Development 118:283-295.

Barres BA, Burne JF, Holtmann B, Thoenen H, Sendtner M, Raff MC. (1996) Ciliary neurotrophic factor enhances the rate of oligodendrocyte generation. Mol Cell Neurosci 8:146-156.

Bischoff SC and Dahinden CA (1992) Effect of nerve growth factor on the release of inflammatory mediators by mature human basophils. Blood 79:2662-2669.

Boyle MDP, Lawman MJP, Gee AP, Young M (1985) Nerve growth factor: a chemotactic factor for polymorphonuclear leukocytes in vivo. J Immunol 134:564-568.

Bracci-Laudiero L, Aloe L, Buanne P, Finn A, Stenfors C, Vigneti E, Theodorssson E, Lundeberg T. (2002) NGF modulates CGRP synthesis in human B-lymphocytes: a possible anti-inflammatory action of NGF? J Neuroimmunol 123:58-65.

Brodie C and Gelfand EW (1992) Functional nerve growth factor receptors on human B lymphocytes – interaction with IL-2. J Immunol 148:3492-3497.

Brodie C, Gelfand E. (1994) Regulation of immunoglobulin production by nerve growth factor: comparison with anti-CD40. J Neuroimmunol 52:87-96.

Brodie C, Goldreich N, Haiman T, Kazimirsky G. (1998) Functional IL-4 receptors on mouse astrocytes: IL-4 inhibits astrocyte activation and induces NGF secretion. J Neuroimmunol 81:20-30.

Burgi B, Otten UH, Oschensberger B, Rihs S, Heese K, Ehrhard PB, Ibanez CF, Dahinden CA (1996) Basophil priming by neurotrophic factors. Activation through the trk receptor. J Immunol 157:5582-5588.

Burns J, Rosenzweig A, Zweiman B, Lisak RP. (1983) Isolation of myelin basic protein-reactive T-cell lines from normal human blood. Cell Immunol 81:435-440.

Calza L, Giardino L, Pozza M, Micera A, Aloe L. (1997) Time-course changes of nerve growth factor, corticotropin-releasing hormone, and nitric oxide isoforms and their possible role in the development of inflammatory response in experimental allergic encephalomyelitis. Proc Natl Acad Sci 94:3368-3373.

Calza L, Giardino L, Pozza M, Bettelli C, Micera A, Aloe L. (1998) Proliferation and phenotype regulation in the subventricular zone during experimental allergic encephalomyelitis: in vivo evidence of a role for nerve growth factor. Proc Natl Acad Sci 95:3209-3214.

Cannella B, Hoban CJ, Gao Y-L, Garcia-Arenas R, Lawson D, Marchionni M, Gwynne D, Raine CS. (1998) The neuregulin, glial growth factor 2, diminishes autoimmune demyelination and enhances remyelination in a chronic relapsing model for multiple sclerosis. Proc Natl Acad Sci 95:10100-10105.

Cannella B, Pitt D, Capello E, Raine CS. (2000) Insulin-like growth factor-1 fails to enhance central nervous system myelin repair during autoimmune demyelination. Am J Pathol 157:933-943.

Canoll PD, Musacchio JM, Hardy R, Reynolds R, Marchionni MA, Salzer JL. (1996) GGF/neuregulin is a neuronal signal that promotes the proliferation oand survival and inhibits the differentiation of oligodendrocyte progenitors. Neuron 17:229-243.

Carson MJ, Behringer RR, Brinster RL, McMorris FA (1993) Insulin-like growth factor 1 increases brain growth and central nervous system myelination in transgenic mice. Neuron 10:729-740.

Chevalier S, Praloran V, Smith C, McGogan D, Ip N, Yancopoulos GD, Brachet P, Pouplard A, Gascan H. (1994) Expression and functionality of the trkA protooncogene product/NGF receptor in undifferentiated hematopoietic cells. Blood 83:1479-1485.

Eckenstein FP, (1994) Fibroblast growth factors in the nervous system. J Neurobiol 25:1467-1480.

Ehrhard PB, Ganter U, Bauer J, Otten U. (1993) Expression of functional trk protooncogene in human monocytes. Proc Natl Acad Sci 90:5423-5427.

Ehrhard PB, Erb P, Graumann U, Otten U. (1993) Expression of nerve growth factor and nerve growth factor receptor tyrosine kinase trk in activated CD4-positive T-cell clones. Proc Natl Acad Sci 90:10984-10988.

Ehrhard PB, Erb P, Graumann U, Schmutz B, Otten U. (1994) Expression of functional trk tyrosine kinase receptors after T cell activation. J Immunol 152:2705-2709.

Flugel A, Matsumuro K, Neumann H, Klinkert WEF, Birnbacher R, Lassmann H, Otten U, Wekerle H.  (2000)  Anti-inflammatory activity of nerve growth factor in experimental autoimmune encephalomyelitis: inhibition of monocyte transendothelial migration.  Eur J Immunol 31:11-22.

Flugel A, Matsumuro K, Neumann H, Klinkert WE, Birnbacher R, Lassmann H, Otten U, Wekerle H.  (2001)  Anti-inflammatory activity of nerve growth factor in experimental autoimmune encephalomyelitis: inhibition of monocyte transendothelial migration.  Eur J Immunol 31:11-22.

Follesa F, Wrathall J, Mocchetti M.  (1994)  Regional and temporal pattern of expression of nerve growth factor and basic fibroblast growth factor mRNA in rat brain following electroconvulsive shock.  Brain Res Mol Brain Res 22:1-8.

Folli F, Ghidella S, Bonfanti L, Kahn CR, Merighi A  (1996)  The early intracellular signaling pathway for the insulin/insulin-like growth factor receptor family in the mammalian central nervous system.  Mol Neurobiol 13:155-183.

Frank JA, Richert N, Lewis B, Bash C, Howard T, Civil R, Stone R, Eaton J, McFarland H, Leist T.  (2002)  A pilot study of recombinant insulin-like growth factor-1 in seven multiple sclerosis patients.  Mult Scler 8:24-29.

Garaci E, Cozzolino F.  (2001)  Nerve growth factor inhibits apoptosis in memory B lymphocytes via inactivation of p38 MAPK, prevention of Bcl-2 phosphorylation, and cytochrome c release.  J Biol Chem 276:39027-39036.

Gassmann M, Lemke G.  (1997)  Neuregulins and neuregulin receptors in neural development.  Curr Opin Neurobiol 7:87-92.

Geiss R, Maurer M, Linker R, Gold R, Warmuth-Metz M, Toyka KV, Sendtner M, Rieckmann P  (2002)  A null mutation in the CNTF gene is associated with early onset of multiple sclerosis  Arch Neurol 59:407-409.

Genain CP, Hauser SL  (1997)  Creation of a model for multiple sclerosis in *Callithrix jacchus* marmosets.  J Mol Med 75:187-197.

Gomez-Pinilla F, Lee JW, Cotman CW  (1992)  Basic FGF in adult rat brain: Cellular distribution and response to entorhinal lesion and fimbria-fornix transection.  J Neurosci 12:345-355.

Grothe C, Zachmann K, Unsicker K. (1991) Basic FGF-like immunoreactivity in the developing and adult rat brainstem. Comp Neurol 305:328-336.

Hammarberg H, Lidman O, Lundberg C, Eltayeb SY, Gielen AW, Muhallab S, Svenningsson A, Linda H, van der Meide PH, Cullheim S, Olsson T, Piehl F. (2000) Neuroprotection by encephalomyelitis: rescue of mechanically injured neurons and neurotrophin production by CNS-infiltrating T and natural killer cells. J Neurosci 20:5283-5291.

Horigome K, Pryor JC, Bullock ED, Johnson EM. (1993) Mediator release from mast cells by nerve growth factor. J Biol Chem 268:14881-14887.

Joshi N, Usuku K, Hauser SL (1993) The T cell response to myelin basic protein in familial multiple sclerosis: diversity of fine specificity, restricting elements, and T cell receptor usage. Ann Neurol 34:385-393.

Kimata H, Yoshida A, Ishioka C, Kusunoki T, Hosoi S, Mikawa H. (1991) Nerve growth factor specifically induces human IgG4 production. Eur J Immunol 21:137-141.

Kobayashi H, Mizisin AP. (2001) Nerve growth factor and neurotrophin-3 promote chemotaxis of mouse macrophages in vitro. Neurosci Lett 305:157-160.

Komoly S, Hudson LD, Webster HD, Bondy CA (1992) Insulin-like growth factor 1 gene expression is induced in astrocytes during experimental demyelination. Proc Natl Acad Med 89:1894-1898.

Koshinaga M, Sanon HR, Whitemore SR. (1993) Altered acidic and basic fibroblast growth factor expression following spinal cord injury. Exp Neurol 120:32-48.

Li W, Quigley L, Yao DL, Hudson LD, Brenner M, Zhang BJ, Brocke S, McFarland HF, Webster HD (1998) Chronic relapsing experimental autoimmune encephalomyelitis: effects of insulin-like growth factor-1 treatment on clinical deficits, lesion severity, glial responses, and blood brain barrier defects. J Neuropathol Exp Neurol 57:426-438.

Linker RA, Maurer M, Gaupp S, Martini R, Holtmann B, Giess R, Rieckmann P, Lassman H, Toyka KV, Sendtner M, Gold R. (2002) CNTF is a major protective factor in demyelinating CNS disease: a neurotrophic cytokine as modultor in neuroinflammation. Nat Med 8:620-624.

Liu X, Yao D-L, Bondy CA, Brenner M, Hudson LD, Zhou J, Webster HD. (1994) Astrocytes express insulin-like growth factor-1 (IGF-1) and its binding protein, IGFBP-2, during demyelination induced by experimental autoimmune encephalomyelitis.  Mol Cell Neurosci 5:418-443.

Liu X, Yao D-L, Webster HD  (1995)  Insulin-like growth factor 1 treatment reduces clinical deficits and lesion severity in acute demyelinating experimental autoimmune encephalomyelitis.  Mult Scler 1:2-9.

Liu X, Linnington C, Webster HD, Lassmann S, Yao DL, Hudson LD, Wekerle H, Kreutzberg GW.  (1997)  Insulin-like growth factor-1 treatment reduces immune cell responses in acute non-demyelinative experimental autoimmune encephalomyelitis.  J Neurosci Res 47:531-538.

Liu X, Mashour GA, Webster HD, Kurtz A.  (1998)  Basic FGF and FGF receptor 1 are expressed in microglia during experimental autoimmune encephalomyelitis: temporally distinct expression of midkine and pleiotrophin.  Glia 24:390-397.

Lovett-Racke AE, Bittner P, Cross AH, Carlino JA, Racke MK.  (1998)  Regulation of experimental autoimmune encephalomyelitis with insulin-like trowth factor (IGF-1) and IGF-1/IGF-binding protein-3 complex (IGF-1/IGFBP3).  J Clin Invest 101:1797-1804.

Lovett-Racke AE, Trotter JL, Lauber J, Perrin PJ, June CH, Racke MK.  (1998) Decreased dependence of myelin basic protein-reactive T cells on CD28-mediated costimulation in multiple sclerosis patients: a marker of activated/memory T cells.  J Clin Invest 101:725-730.

Manning PT, Russell JH, Simmons B, Johnson EM  (1985)  Protection from guanethidine-induced neuronal destruction by nerve growth factor: effects of NGF on immune function.  Brain Res 340:61-69.

Marchionni MA, Goodearl AD, Chen MS, Bermingham-McDonogh O, Kirk C, Hendricks M, Danehy D, Misumi D, Sudhalfer J, Kobayashi D, Wroblewski D, Lynch C, Baldassare M, Hiles I, Davis JB, Hsuan JJ, Totty NF, Otsu M, McBurney RN, Waterfield MD, Stroobant P, Gwynne D.  (1993)  Glial growth factors are alternatively spliced erbB2 ligands expressed in the nervous system.  Nature 362:312-318.

Marchionni MA, Cannella B, Hoban C, Gao YL, Garcia-Arenas R, Lawson D, happel E, Noel F, Tofilon P, Gwynne D, Raine CS. (1999) Neuregulin in neuron/glial interaction in the central nervous system.  GGF2 diminishes autoimmune demyelination, promotes oligodendrocyte progenitor expansion, and enhances remyelination.  Adv Exp Med Biol 468:283-295.

Martin R, Jaraquemada D, Flerlage M, Richert J, Whitaker J, Long EO, McFarlin DE, McFarland HF. (1990) Fine specificity and HLA restriction of myelin basic protein-specific cytotoxic T-cell lines from multiple sclerosis patients and healthy individuals. J Immunol 145:540-548.

Matsuda H, Coughlin MD, Bienenstock J, Denburg J (1988) Nerve growth factor promotes human hemopoietic colony growth and differentiation. Proc Natl Acad Sci 85:6508-6512.

Mazurek N, Weskamp G, Erne E, Otten U (1986) Nerve growth factor induces mast cell degranulation without changing intrcellular calcium levels. FEBS Lett 198:315-320.
McKinnon RD, Matsui T, Dubois-Dalcq M, Aaronson SA. (1990) FGF modulates the PDGF-driven pathway of oligodendrocyte development. Neuron 5:603-614.

McKinnon RC, Matsui T, Dubois-Dalcq M, Aaronson SA (1990) FGF modulates the PDGF-driven pathway of oligodendrocyte development. Neuron 5:603-614.

Micera A, Vigneti E, Aloe L. (1998) Changes of NGF presence in nonneuronal cells in response to experimental allergic encephalomyelitis in Lewis rats. Exp Neurol 154:41-46.

Micera A, Properzi F, Triaca V, Aloe L. (2000) Nerve growth factor antibody exacerbates neuropathological signs of experimental allergic encephalomyelitis in adult Lewis rats. J Neuroimmunol 104:116-123.

Miura K, Saini SS, Gauvreau G, MacGlashan DW (2001) Differences in functional consequences and signal transduction induced by IL-3, IL-5, and nerve growth factor in human basophils. J Immunol 167:2282-2291.

Miller SD, Vanderlugt CL, Begolka WS, Pao W, Yauch RL, Neville KL, Katz-Levy Y, Carrizosa A, Kim BS. (1997) Persistent infection with Theiler's virus leads to CNS autoimmunity via epitope spreading. Nat Med 3:1133-1136.

Milner R, Anderson HJ, Rippon RF, McKay JS, Franklin RJ, Marchionni MA, Reynolds R, and ffrench-Constant C. (1997) Contrasting effects of mitogenic growth factors on oligodendrocyte precursor cell migration. Glia 19:85-90.

Moalem G, Leibowitz-Amit R, Yoles E, Mor F, Cohen IR, Schartz M. (1999) Autoimmune T cells protect neurons from secondary degeneration after central nervous system axotomy. Nat Med 5:49-55.

Moalem G, Yoles E, Leibowitz-Amit R, Muller-Gilor S, Mor F, Cohen IR, Schwartz M (2000) Autoimmune T cells retard the loss of function in injured rat optic nerves. J Neuroimmunol 106:189-197.

Mozell RL, McMorris FA (1991) Insulin-like growth factor 1 stimulates oligodendrocyte development and myelination in rat brain aggregate cultures. J Neurosci Res 30:382-390.

Muhallab S, Lundberg C, Gielen AW, Lidman O, Svenningsson A, Piehl F, Olsson T (2002) Differential expression of neurotrophic factors and inflammatory cytokines by myelin basic protein-specific and other recruited T cells infiltrating the central nervous system during experimental autoimmune encephalomyelitis. Scand J Immunol 55:264-273.

Nataf S, Naveilhan P, Sindji L, Darcy F, Brachet P, Montero-Menei CN. (1998) Low affinity NGF receptor expression in the central nervous system during experimental allergic encephalomyelitis. J Neurosci Res 52:83-92.

Presta M, Urbinati C, Dell-era P, et al (1995) Expression of basic fibroblast growth factor and its receptors in human fetal microglia. Int J Dev Neurosci 13:29-39.

Olsson T, Diener P, Ljungdahl A, Hojeberg B, van der Meide P, Kristensson K (1992) Facial nerve transection causes expansion of myelin autoreactive T cells in regional lymph nodes and T cell homing to the facial nucleus. Autoimmunity 13:117-126.

Olsson T, Sun J, Solders G, Ekre H-P, Link H. (1993) Autoreactive T- and B-cell responses to myelin antigens after diagnostic sural nerve biopsy. J Neurol Sci 117:130-139.

Otten U, Ehrhard P, Peck R. (1989) Nerve growth factor induces growth and differentiation of human B lymphocytes. Proc Natl Acad Sci 86:10059-10063.

Reilly JF, Kumari VG. (1996) Alterations in fibroblast growth factor receptor expression following brain injury. Exp Neurol 140:139-150.

Rutkowski JL, Kirk CJ, Lerner MA, Tennekoon GI. (1995) Purification and expansion of human Schwann cells in vitro. Nat Med 1:80-83.

Scholz C, Patton KT, Anderson DE, Freeman GF, Hafler DA (1998) Expansion of autoreactive T cells in multiple sclerosis is independent of exogenous B7 costimulation. J Immunol 160:1532-1538.

Serpe CJ, Kohm AP, Huppenbauer CB, Sanders VJ, Jones KJ (1999) Exacerabation of facial motorneuron loss after facial nerve transection in severe combined immunodeficient (SCID) mice. J Neurosci RC7:1-5.

Shimojo M, Nakajima K, Takei N, Hamanoue M, Koshaka S. (1991) Production of fibroblast growth factor in cultured rat brain microglia. Neurosci Lett 123:229-231.

Thorpe LW, Perez-Polo JR (1987) The influence of nerve growth factor on the in vitro proliferative response of rat splee lymphocytes. J Neurosci Res 18:134-139.

Torcia M, De Chiara G, Nencioni L, Ammendola S, Labardi D, Lucibello M, Rosini P, Marlier LN, Bonini P, Dello Sbarba P, Palamara AT, Zambrano N, Russo T, Garaci E, Cozzolino F. (2001) Nerve growth factor inhibits apoptosis in memory B lymphocytes via inactivation of p38 MAPK, prevention of Bcl-2 phosphorylation, and cytochrome c release. J Biol Chem 276:39027-39036.

Unsicker K, Reichert H, Schmidt R, Pettmann B, Labourdette G, Sensenbrenner M. (1986) Astroglial and fibroblast growth factors have neurotrophic functions for cultured perpheral and central nervous system neurons. Proc Natl Acad Sci 84:5459-5463.

Villoslada P, Hauser SL, Bartke I, Unger j, Heald N, Rosenberg D, Cheung SW, Mobley WC, Fisher S, Genain CP. (2000) Human nerve growth factor protects common marmosets against autoimmune encephalomyelitis by switching the balance of T helper cell type 1 and 2 cytokine within the central nervous system. J Exp Med 191:1799-1806.

Walicke PA, Cowan WM, Baird NU, Guillemin R. (1986) Fibroblast growth factor promotes survival of dissociated hippocampal neurons and enhances neurite extension. Proc Natl Acad Sci 83:3012-3016.

Wang W, Olsson T, Kostulas V, Hojeberg B, Ekre H, Link H. (1992) Myelin antigen reactive T cells in cerebrovascular diseases. Clin Exp Immunol 88:157-162.

Xu S, Jordan EK, Li W, Yang Y, Chesnick SA, Webster HD, Brocke S, Quigley L, McFarland HF, Frank JA. (1998) In vivo three-dimensional MR microscopy of mice with chronic relapsing experimental autoimmune encephalomyelitis after treatment with insulin-like growth factor-1. Am J Neuroradiol 19:653-658.

Yao D-L, West NR, Bondy CA, Brenner M, Hudson LD, Zhou J, Collins GH, Webster HD. (1995) Cryogenic spinal cord injury induces astrocytic gene expression of insulin-like growth factor I and insulin-like growth factor binding protein 2 during myelin regeneration. J Neurosci Res 40:647-659.

Yao D-L, Liu X, Hudson LD, Webster HD. (1995) Insulin-like growth factor 1 treatment reduces demyelination and up-regulates gene expression of myelin-related proteins in experimental autoimmune encephalomyelitis. Proc Natl Acad Sci 92:6190-6194.

Zhang JW, Medaer R, Chin Y, van den Berg-Loonen E, Raus JCM (1992) Myelin basic protein-specific T lymphocytes in multiple sclerosis and controls: precursor frequency, fine specificity, and cytotoxicity. Ann Neurol 32:330-339.

Chapter  A17

# THE CHEMOKINE SYSTEM IN EXPERIMENTAL AUTOIMMUNE ENCEPHALOMYELITIS


Andrzej R Glabinski[1] and Richard M. Ransohoff[2]
*1) Department of Neurology, Medical University, Lodz, Poland, 2) Department of Neurosciences, Cleveland Clinic Foundation, Cleveland, OH, USA*



Abstract:    Chemokines chemoattract selected populations of inflammatory cells towards sites of inflammation in a gradient-dependent fashion, and also activate both recruited and resident inflammatory cells. Chemokines act on target cells through G-protein-coupled seven-transmembrane-domain receptors. High expression of several chemokines was found in the CNS during EAE. Cells expressing these chemokines were predominantly astrocytes and macrophages/microglia. In addition to chemokines, expression of several chemokine receptors was reported in EAE. Amelioration of EAE by anti-chemokine antibodies and studies in knock-out mice confirm the important roles of some chemokines in EAE pathogenesis. In the last several years many reports have been published addressing chemokine expression in multiple sclerosis. These results resemble results obtained earlier in EAE. Taken together, these data suggest that chemokine system may be a promising target for future treatment methods of multiple sclerosis.



Key words:    Chemokines, chemokine receptors, experimental autoimmune encephalomyelitis, neuroinflammation, inflammatory cell migration


Experimental autoimmune encephalomyelitis (EAE) is a model autoimmune disease of the central nervous system (CNS) induced in susceptible strains of experimental animals with myelin, myelin proteins or peptides. The classical autoantigens used for EAE induction are myelin basic protein (MBP), proteolipid protein (PLP) and myelin oligodendrocyte glycoprotein (MOG). This model may be induced in appropriate strains of mice, rats, guinea pigs and nonhuman primates including rhesus and marmoset species. The advances made by studying this experimental disease resulted from the observation that the pathology of EAE resembles in some respects the pathology of the major human demyelinating disease – multiple sclerosis (MS). Typical histopathological hallmarks of the initial phase of both diseases are perivascular mononuclear inflammatory foci disseminated in the CNS. The mechanism of migration of inflammatory cells from blood to the CNS has been studied for many years and it remains incompletely understood. Our knowledge about this process has significantly grown since the observation that chemoattractant cytokines – chemokines may play an important role in inflammatory cells homing to the CNS.

## CHEMOKINES AND THEIR RECEPTORS

Chemokines and their receptors are the principal determinants of leukocyte-type and organ specificity within inflammatory foci. Chemokines chemoattract selected populations of inflammatory cells towards sites of inflammation in a gradient-dependent fashion, and also activate both recruited and resident inflammatory cells. The chemokine family can be divided into four separate subfamilies (CXC, CC, C, CXXXC) according to the position of the first two cysteines near the N-terminus. Beyond these differences in structure, there are also functional differences between chemokine subfamilies. The largest subfamilies are CXC and CC chemokines. In CXC chemokines the first two cysteines are separated by one additional amino acid. The CXC subfamily can be further divided into two groups, one with ELR (glutamate-leucine-arginine) motif preceding the first cysteine and another without it. CXC chemokines with ELR are primarily chemoattractant for neutrophils, while several other non-ELR chemokines attract mainly activated lymphocytes and monocytes. SDF-1/CXCL12 is an outlier among chemokines in that it acts towards many non marrow-derived cells in addition to most leukocytes. CC chemokines have the first two cysteines adjacent to one another. They attract principally mononuclear leukocytes. The C subfamily consists of lymphotactin which is a potent T cell chemoattractant. The single member of CXXXC subfamily is fractalkine, a chemokine with three amino acids intervening between the first two cysteines. Fractalkine is the prototype tethered chemokine with a long stalk attached to the cell membrane; CXCL16 also possesses this unusual structure. After cleavage fractalkine may chemoattract mononuclear inflammatory cells [1, 2]. In the last few years several new chemokines were described by independent research teams. They have been known by different names. This complicated chemokine nomenclature. To solve this problem, a consensus meeting at the Keystone Chemokine Conference (Keystone, CO, January 18-23, 1999) proposed a new nomenclature for the chemokines (Table 1).

*Table 1.* New nomenclature for mouse chemokines (Keystone Chemokine Conference, Keystone, 01.18-23,1999).

| Systematic name | Mouse ligand | Receptor(s) | Systematic name | Mouse ligand | Receptor(s) |
|---|---|---|---|---|---|
| **CC CHEMOKINE FAMILY** | | | **CXC CHEMOKINE FAMILY** | | |
| CCL1 | TCA-3, P500 | CCR8 | CXCL1 | GRO(KC) | CXCR2, CXCR1 |
| CCL2 | MCP-1, JE | CCR2 | CXCL2 | GRO(KC) | CXCR2 |
| CCL3 | MIP-1α | CCR1, CCR5 | CXCL3 | GRO(KC) | CXCR2 |
| CCL4 | MIP-1β | CCR5 | CXCL4 | PF4var1, PF4alt | Unknown |
| CCL5 | RANTES | CCR1, CCR3, CCR5 | CXCL5 | LIX | CXCR2 |
| CCL6 | C10, MRP-1 | Unknown | CXCL6 | CKα–3 | CXCR1, CXCR2 |
| CCL7 | MARC, FIC, NC28 | CCR1, CCR2, CCR3 | CXCL7 | Unknown | Unknown |
| CCL8 | MCP-2 | CCR2, CCR3 | CXCL8 | Unknown | Unknown |
| CCL9, CCL10 | MRP-2, CCF18, MIP-1γ | Unknown | CXCL9 | Mig | CXCR3 |
| CCL11 | Eotaxin | CCR3 | CXCL10 | crg-2, mob-1 | CXCR3 |
| CCL12 | MCP-5 | CCR2 | CXCL11 | Unknown | Unknown |
| CCL13 | Unknown | Unknown | CXCL12 | SDF-1α/β | CXCR4 |
| CCL14 | Unknown | Unknown | CXCL13 | BLR1L, Angie | CXCR5 |
| CCL15 | Unknown | Unknown | CXCL14 | BRAK | Unknown |
| CCL16 | LCC-1 | CCR1 | CXCL15 | Lungkine | Unknown |
| CCL17 | TARC | CCR4 | | | |
| CCL18 | Unknown | Unknown | **C CHEMOKINE FAMILY** | | |
| CCL19 | MIP-3β | CCR7 | | | |
| CCL20 | MIP-3α | CCR6 | XCL1 | Lymphotactin | XCR1 |
| CCL21 | 6Ckine | CCR7 | | | |
| CCL22 | ABCD-1 | CCR4 | | | |
| CCL23 | Unknown | Unknown | **CX3C CHEMOKINE FAMILY** | | |
| CCL24 | Unknown | Unknown | | | |
| CCL25 | TECK | CCR9 | CX3CL1 | Fractalkine | CX3CR1 |
| CCL26 | Unknown | Unknown | | | |
| CCL27 | CTACK, ALP | Unknown | | | |

Chemokines act on target cells through G-protein-coupled seven-transmembrane-domain receptors. They are designated as CXCR, CCR, XCR and CXXXCR. The largest family of chemokine receptors are CCR receptors consisting at present of ten well-characterized members. CXCR family of receptors consists of five receptors described to date (CXCR1-CXCR5). There are only two XCR receptors and one CXXXCR receptor. Classically chemokine receptors are divided into four different subgroups: shared, specific, promiscuous and viral [3]. Most of the chemokine receptors belong to the first group because they bind more than one chemokine ligand. A typical example of this kind of chemokine receptor is CXCR2 that binds as many as seven CXC chemokines with ELR motif. The example of specific receptor is CXCR4 that binds SDF-1/CXCL12. Duffy blood group antigen (DARC) is a promiscuous chemokine-binding molecule (distinct from a receptor), which can bind several chemokines belonging to CXC and CC subfamilies but does not signal. Many viruses including *Cytomegalovirus CMV US28* and *Herpes virus saimiri HSV ECRF3* encode chemokine receptors. The role of virally encoded chemokine receptors is unknown [4].


## CHEMOKINES IN EAE


Initial studies describing chemokine expression in EAE were published less than ten years ago [5, 6]. They showed that chemokines MCP-1/CCL2 and IP-10/CXCL10 were transiently upregulated in mouse and rat models. In the mouse model astrocytes were the cellular source of chemokine expression confirmed by colocalization studies using *in situ* hybridization and immunohistochemistry for GFAP [7, 8]. Later other chemokines including RANTES/CCL5, MIP-1α/CCL3, MIP-1β/CCL4, TCA-3/CCL1, IP-10/CXCL10, MCP-1/CCL2, KC/CXCL1-3 and MCP-3/CCL7 were also detected in the spinal cord during passive transfer mouse EAE [9]. Encephalitogenic T cells expressed RANTES/CCL5, MIP-1α/CCL3, MIP-1β/CCL4 and TCA-3/CCL1 in that study [9]. In murine EAE expression of chemokines RANTES/CCL5, MIP-1α/CCL3 and GRO-α/CXCL1 correlated with the intensity of CNS inflammation [10]. Analyzing the kinetics of chemokine gene expression at the beginning of EAE we suggested that chemokines amplify but not initiate invasion of CNS by inflammatory cells from the blood [11]. In chronic relapsing EAE we observed increased expression of MCP-1/CCL2, IP-10/CXCL10, MIP-1α/CCL3, RANTES/CCL5 and GRO-α/CXCL1 at the beginning of spontaneous disease relapse. Chemokines MCP-1/CCL2, IP-10/CXCL10 and GRO-α/CXCL1 were expressed by astrocytes in the vicinity of inflammatory foci, whereas MIP-1α/CCL3 and RANTES/CCL5 were expressed by inflammatory cells [12]. In variant of EAE in BALBc mice characterized by pronounced neutrophil accumulation upregulation of MIP-1α/CCL3, MIP-2/CXCL2-3 and MCP-1/CCL2 was reported. The main cellular sources of MIP-1α/CCL3 and MIP-2/CXCL2-3 in this model were activated astrocytes. Neutrophils produced MIP-1α/CCL3 and MCP-1/CCL2 [13]. In an EAE model enhanced by focal brain injury (cryolesion) significant upregulation of chemokines MCP-1/CCL2 and MCP-5/CCL12 was observed at the beginning of the disease. Interestingly, expression of RANTES/CCL5, GRO-α/CXCL1 and MIP-1α/CCL3 was not elevated in these animals. These results suggest that MCP-1/CCL2 and MCP-5/CCL12 are selectively involved in enhancement of autoimmune inflammation by focal cortical injury [14].

Recently expression of relatively new chemokines was analyzed in EAE. High expression of C10/CCL6 was found in the CNS during MBP-induced murine EAE. Cells expressing this chemokine were predominantly macrophage/microglia and foamy macrophages present in perivascular cuffs and within the demyelinating lesions [15]. Expression of CXC chemokine I-TAC/CXCL11 was very low early after induction of EAE, but was upregulated at day 14 post-immunization. By day 20, expression of I-TAC/CXCL11 in EAE returned to low baseline levels [16].

## CHEMOKINE RECEPTORS IN EAE

In addition to chemokines, expression of several chemokine receptors was reported in EAE. In the first study in acute rat EAE upregulation of CXCR4, CCR2, CCR5, CCR6 and CX3CR1 was observed in the lumbar spinal cords [17]. In chronic-relapsing EAE expression of CXCR2 and CXCR4 was significantly increased in the spinal cord during relapses. It was suggested that CXCR2 is expressed by migrating inflammatory cells, whereas CXCR4 is expressed mainly by CNS parenchymal cells [18]. We also observed upregulation of CC chemokine receptors CCR1, CCR2 and CCR5 during the initial attack and relapse of the same chronic EAE model [19]. CCR5 was expressed by inflammatory cells invading the spinal cord. CC chemokine receptors associated with Th2 T-cells (CCR3 and CCR4) were not detected during the active stages of this model [19].

It was reported recently that CCR6 and its ligand MIP-3α/CCL20 are upregulated in the CNS during EAE. MIP-3α/CCL20 is a potent chemoattractant for dendritic cells, which infiltrate the spinal cord in this model [20]. Moreover, in the rat acute EAE model CXCR3 and all its ligands IP-10/CXCL10, Mig/CXCL9 and I-TAC/CXCL11 were detected in the spinal cord. In the draining lymph nodes only expression of CXCR3, IP-

10/CXCL10 and Mig/CXCL9 was detected [21]. A CC chemokine receptor CCR7 associated with lymph-node trafficking, as well as CXCR3, were found by immunostaining on encephalitogenic T cells isolated from mice with EAE [22]. Immunohistochemistry localized ligands for these receptors, ELC/CCL19 and SLC/CCL21 within inflammatory cuffs in CNS parenchyma. This is the first observation suggesting that these chemokines are also involved in lymphocyte migration to the CNS during EAE, in addition to their documented role in regulation of lymphocyte homing to lymphoid tissue [22].

In acute EAE encephalitogenic CD4+ T cells isolated from the CNS expressed mRNA for chemokine receptors CCR1 – CCR5 [23]. Encephalitogenic T cells isolated from spleens of the same animals did not express CCR1, but expressed other analyzed CC chemokine receptors. Neutralization of the CCR1 ligand, MIP-1α/CCL3 resulted in lower accumulation of encephalitogenic T cells in the CNS [23].

Depletion of gamma-delta T cells reduced expression of CCR1, CCR2, CCR3 and CCR5 at the beginning of EAE. Later levels of CCR2, CCR3 and CCR5 were higher than in control EAE mice. Similarly expression of chemokines RANTES/CCL5, MIP-1α/CCL3, MIP-1β/CCL4, MIP-2/CXCL2-3, MCP-1/CCL2, eotaxin in gamma-delta depleted mice was reduced at the onset of EAE [24]. Levels of CXCR2, CXCR3, CCR1, CCR5 and CCR8 in spinal cords were significantly reduced by protection from EAE with altered peptide ligand (APL). In APL-treated mice CCR3 and CCR4 epitopes were found mostly on lymphocytes supporting the hypothesis that Th1 cells are down-regulated by APL treatment [25].


## CHEMOKINES IN GENETICALLY MODIFIED ANIMALS WITH EAE


In interferon gamma knockout (GKO) mice with EAE expression of IP-10/CXCL10 was selectively diminished in the spinal cord. This observation suggests that expression of IP-10/CXCL10 in EAE is induced by IFN-γ [26]. Upregulation of neutrophil-attracting chemokines MIP-2/CXCL2-3 and TCA-3/CCL1 was observed in IFN-γ and IFN-γ receptor-deficient mice. In those models expression of MCP-1/CCL2 and RANTES/CCL5 was undetectable during EAE [27]. Moreover, in the CNS of GKO mice PCR analysis showed very low levels of I-TAC/CXCL11 throughout the time course of the disease, although these mice developed severe EAE [16].

It has been recently demonstrated that chemokine and chemokine receptor expression during EAE follows a distinct pattern in cytokine deficient mice [28]. In TNF-α knockout mice expression of MIP-1α/CCL3 and MIP-2/CXCL2-3 was almost completely absent, while expression of RANTES/CCL5 and IP-10/CXCL10 revealed strong reduction, and the decrease of MCP-1/CCL2 was moderate (50%). IFN-γ deficient mice displayed pronounced reduction of RANTES/CCL5 and IP-10/CXCL10 and moderate decrease in MIP-1α/CCL3, MIP-2/CXCL2-3 and MCP-1/CCL2 [28]. Other analyzed knockout mice (IL-4 and IL-10 KO) showed less remarkable diminution of chemokine expression during EAE. In IL-4 deficient mice only a 50-60% decrease of RANTES/CCL5 and MIP-1α/CCL3 was found, whereas in IL-10 KO mice 50-60% reduction was detected for RANTES/CCL5, IP-10/CXCL10, MIP-1α/CCL3 and MIP-2/CXCL2-3 [28]. In addition to chemokine expression, expression of chemokine receptors was also analyzed in those experiments. Expression of CCR1, CCR2 and CCR5 receptors was diminished in TNF-α and IL-10 knockout mice. Lack of IL-4 or IFN-γ weakly affected expression of these three chemokine receptors [28].

Chemokine and chemokine receptor expression was also analyzed in the CNS of TCR BV8S2 transgenic mice expressing BV8S2 specific for the myelin basic protein (MBP)-NAc1-11 peptide [29]. In the spinal cord of these mice increased expression of chemokines RANTES/CCL5, IP-10/CXCL10 and MCP-1/CCL2 as well as chemokine receptors CCR1, CCR2 and CCR5 was found. This expression correlated with pronounced cellular infiltration of the target organ. Transgenic mice protected from EAE by vaccination with heterologous rat BV8S2 protein showed amelioration of inflammatory infiltrates in the spinal cord. This amelioration correlated with a profoundly reduced expression of chemokines and their cognate receptors in the spinal cords of treated animals [29].

# EAE IN ANIMALS WITH GENETICALLY MODIFIED CHEMOKINE SYSTEM

Recently two publications showed that CCR2 plays an important role in the pathogenesis of EAE [30, 31]. Mice without CCR2 did not develop clinical signs of EAE and did not accumulate mononuclear inflammatory cells in the CNS. CCR2 knockout mice failed to upregulate RANTES/CCL5, MCP-1/CCL2, IP-10/CXCL10, CCR1, CCR2 and CCR5 during EAE. In addition, their T cells displayed decreased proliferation in response to autoantigen and lower production of IFN-γ than T cells from control wild-type animals [30]. MOG-specific T cells from CCR2-deficient mice were able to induce passive-transfer EAE but autoimmune T cells from wild-type animals transferred to CCR-2 knockout recipients could not induce EAE

[31]. Complementing these results, CCR2 knockout mice showed impaired recruitment of monocytes and degradation of myelin at the site of a spinal cord contusion injury [32].

C57BL/6 mice lacking CCR2 ligand - MCP-1/CCL2, are also resistant to EAE induction and show impaired recruitment of macrophages to the CNS [33]. Similarly like in CCR2 knockout mice, T cells from MCP-1/CCL2 deficient animals were able to induce passive-transfer EAE in naive recipients, whereas autoimmune T cells from wild-type mice transferred to mice lacking MCP-1/CCL2 did not trigger clinical EAE [33]. Disruption of MCP-1/CCL2 gene in SJL mice produced milder EAE. Expression of MCP-2/CCL8, MCP-3/CCL7 and MCP-5/CCL12 was not upregulated in MCP-1/CCL2 knockout mice with EAE but expression of IFN-γ in CNS and draining lymph nodes was reduced in those animals [33].

CCR1 knockout mice developed less severe EAE [34]. Disease incidence and clinical score were significantly lower than in wild-type controls. Moreover, only expression of IP-10/CXCL10 was elevated in CCR1-deficient mice, whereas in controls expression of IP-10/CXCL10, RANTES/CCL5 and MCP-1/CCL2 was elevated. Lymphocyte proliferation, cytokine production upon stimulation with MOG and cutaneous hypersensitivity were similar in both groups showing that CCR1 knockout mice were not immunosuppressed [34].

Mice with lack of MIP-1α/CCL3 and its receptor CCR5 are fully susceptible to development of acute EAE [35]. Clinical and histopathological signs of EAE in these knockout mice were indistinguishable from those observed in wild-type control animals with EAE. IP-10/CXCL10, RANTES/CCL5, MCP-1/CCL2, MIP-1β/CCL4, MIP-2/CXCL2-3, Lymphotactin/XCL1 and TCA-3/CCL1 gene expression during EAE was similar in MIP-1α/CCL3 deficient mice and in controls. Additionally, comparable Th1 responses were present in the spinal cord of MIP-1α/CCL3 knock-out mice and wild-type animals [35].

## TREATMENT OF EAE WITH ANTICHEMOKINE STRATEGY

Initial attempts to treat EAE with anti-chemokine strategies came from Karpus and colleagues. They showed that anti-MIP-1α/CCL3 antibodies prevent development of an initial attack of EAE after adoptive transfer of activated antigen-specific T cells. Moreover, anti-MIP-1α/CCL3 antibodies may also ameliorate ongoing disease [36]. This therapy does not influence the cytokine production by encephalitogenic T cells suggesting that they rather influence inflammatory cell recruitment to the CNS [36]. Later, the same group showed that anti-MCP-1/CCL2 antibodies, which were inert towards initial attacks, could reduce relapses of chronic-relapsing disease, which were unaffected by anti-MIP-1α/CCL3 [37]. In another study anti-MIP-1β/CCL4 and anti-MCP-1/CCL2 antibodies influenced the onset of EAE and disease severity [38]. Expression of MIP-1β/CCL4 was not detected in the spinal cord of antibody-treated animals, but expression of RANTES/CCL5, MCP-1/CCL2 and MIP-1α/CCL3 was not decreased in these animals. Anti-RANTES/CCL5 and anti-MIP-1α/CCL3 treatment had no significant effect on amelioration of EAE [38].

There are unclear observations regarding treatment of EAE with anti-IP-10/CXCL10 antibodies. In a murine model induced with 139-151 PLP peptide anti-IP-10/CXCL10 polyclonal antibody decreased clinical and histological EAE [39]. In that study anti-IP-10/CXCL10 treatment was shown to decrease accumulation of mononuclear cells in the CNS. Specifically accumulation of encephalitogenic T cells in the spinal cord was diminished. Anti-IP-10/CXCL10 did not influence the peripheral activation of encephalitogenic T cells [39]. Opposite results were reported in acute EAE induced by immunizing rats with guinea pigs brain homogenate [21]. In that model, treatment with neutralizing antibody against IP-10/CXCL10 resulted in exacerbation of EAE with increased migration of T cells to the CNS. In the group treated with anti-IP-10/CXCL10, draining popliteal lymph nodes were smaller, with lower cell number than in the control group. In the spinal cords of these mice, larger numbers of CD4+ infiltrating cells were found [21].

A newer strategy to block chemokine-chemokine receptor interactions was the usage of chemokines with amino-terminal modifications converting agonists to specific antagonists. An example of this type of compound is Met-RANTES/CCL5. However, recent study showed that Met-RANTES/CCL5 is ineffective in prevention and treatment of acute EAE. In a chronic–relapsing model this treatment showed modest amelioration of neurological disability. Histological analysis of treated mice did not show reduction of inflammatory cell accumulation in the CNS [40].

Several other therapeutic strategies of EAE were shown to affect the chemokine system. Oral administration of myelin basic protein (MBP) conjugated to cholera toxin B subunit before the appearance of the first signs of EAE prevented the development of the disease in Lewis rats. This treatment was associated with a significant reduction of inflammation in the CNS. *In situ* hybridization analysis showed evident reduction of MCP-1/CCL2 and RANTES/CCL5 expression in spinal cords of treated animals [41]. Treatment of ongoing EAE with granulocyte-CSF (G-CSF) resulted in significantly reduced expression of chemokines RANTES/CCL5, MIP-1α/CCL3, MIP-1β/CCL4, MCP-1/CCL2 and MIP-2/CXCL2-3 in spinal cords [42]. G-CSF-treated mice displayed reduced recruitment of T cells to the CNS and limited demyelination. G-CSF induced a cytokine imbalance in periphery characterized by deviation towards a type 2 immune response, with reduced IFN-γ and increased IL-4 levels. Gene therapy of EAE with an HSV vector containing IL-4 gene reduced severity of this disease. Amelioration of EAE was associated with decreased expression of MCP-1/CCL2 and RANTES/CCL5 in the CNS of treated mice [43].

## CHEMOKINES AND CHEMOKINE RECEPTORS IN MULTIPLE SCLEROSIS

In the last several years many reports have been published addressing chemokine expression in MS. These results resemble results obtained earlier in EAE. In one of the first studies, expression of RANTES/CCL5 was described in perivascular inflammatory cells [44]. Expression of MCP-1/CCL2 by astrocytes and inflammatory cells was demonstrated in another study. In that study, MIP-1α/CCL3 and MIP-1β/CCL4 were expressed in inflammatory cells in active MS plaques [45]. MCP-1/CCL2, MCP-2/CCL8 and MCP-3/CCL7 were detected in reactive astrocytes and inflammatory cells in acute and chronic MS lesions [46]. Recently hematogenous monocytes expressing CCR1 and CCR5 were found in perivascular cuffs in active MS lesions. In later stages macrophages CCR1-/CCR5+ predominated [47].

In cerebrospinal fluid from MS patients with relapse we found increased levels of IP-10/CXCL10, Mig/CXCL9 and RANTES/CCL5 [48]. IP-10/CXCL10 was detected in all CSF samples from active MS. Cells bearing receptors for IP-10/CXCL10 were localized in MS plaques. CXCR3, the receptor for IP-10/CXCL10, was detected on lymphocytes in perivascular inflammatory cuffs as well as on approximately 90% of T cells from the cerebrospinal fluid of MS patients. CCR5, receptor for RANTES/CCL5 was present on lymphocytes, macrophages and microglial cells in active MS lesions. MCP-1/CCL2, chemokine associated with Th2 responses was significantly reduced in CSF of patients with MS attacks [48]. Balashov and coworkers reported similar findings and additionally demonstrated increased production of IFN-γ by circulating CCR5+ T cells from MS patients [49]. Another group reported decreased levels of MCP-1/CCL2 in CSF in acute MS, whereas IP-10/CXCL10 levels were significantly increased in this group. Therapy of MS with 6-methylprednisolone or IFN-β did not modify the levels of these chemokines [50].

In very small pilot studies, expression of CCR2 and CCR5 on CD4+ cells was significantly higher in MS patients at relapse than in controls and expression of CCR3 and CCR4 (Th2-type receptors) on CD4+ cells was significantly lower in MS patients than in control subjects [51]. Similarly, in the population of CD8+ cells from MS patients at relapse, expression of

CCR5 was higher than in controls, and expression of CCR4 was significantly lower than in control subjects [51]. Similar preliminary results in small groups of patients suggested that treatment of MS with IFN-β reduced RANTES/CCL5 levels in sera and its production by blood mononuclear cells from MS patients [52]. There are formidable technical challenges to performing such studies in human material and these results await confirmation in larger, prospective studies.

## CONCLUSIONS

The presented data obtained to date show relationships between expression of some chemokines and disease activity in EAE. Amelioration of EAE by anti-chemokine antibodies and studies in knockout mice confirm the important roles of some chemokines in EAE pathogenesis. Involvement of chemokines in development of CNS autoimmune inflammation is further supported by recent results from MS studies. Taken together, these data suggest that the chemokine system may be a promising target for future treatment methods of MS.

## Acknowledgements:

Results from authors laboratories were financially supported by the following grants, RMR/AG: RO3 TW00784 and RO3 TW06012, and RO1NS32151 (to RMR), ARG: KBN 4P05A 05913 and KBN 6P05A 07021.

## REFERENCES

1. Rollins BJ: Chemokines. *Blood.* (1997) **90**: 909-928.
2. Luster F: Chemokines - chemotactic cytokines that mediate inflammation. *New Engl J Med.* (1998) **338**: 436-445.
3. Premack BA and Schall TJ: Chemokine receptors: gateways to inflammation and infection. *Nat Med.* (1996) **2**: 1174-8.
4. Murphy P: Chemokine receptors: structure, function and role in microbial pathogenesis. *Cytokine Growth Factor Rev.* (1996) **7**: 47-65.
5. Ransohoff RM, Hamilton TA, Tani M, Stoler MH, Shick HE, Major JA, Estes ML, Thomas DM, and Tuohy VK: Astrocyte expression of mRNA encoding cytokines IP-10 and JE/MCP-1 in experimental autoimmune encephalomyelitis. *FASEB J .* (1993) **7**: 592-602.
6. Hulkower K, Brosnan CF, Aquino DA, Cammer W, Kulshrestha S, Guida MP, Rapoport DA, and Berman JW: Expression of CSF-1, c-fms, and MCP-1 in the central nervous system of rats with experimental allergic encephalomyelitis. *J. Immunol.* (1993) **150**: 2525-2533.

7.      Glabinski A, Tani M, Aras S, Stoler M, Tuohy V, and Ransohoff R: Regulation and function of central nervous system chemokines. *Internat J Dev Neurosci.* (1995) **13**: 153–165.

8.      Tani M, Glabinski AR, Tuohy VK, Stoler MH, Estes ML, and Ransohoff RM: In situ hybridization analysis of glial fibrillary acidic protein mRNA reveals evidence of biphasic astrocyte activation during acute experimental autoimmune encephalomyelitis. *Am J Pathol.* (1996) **148**: 889-96.

9.      Godiska R, Chantry D, Dietsch G, and Gray P: Chemokine expression in murine experimental autoimmune encephalomyelitis. *J Neuroimmunol.* (1995) **58**: 167-176.

10.     Glabinski A, Tuohy V, and Ransohoff R: Expression of chemokines RANTES, MIP-1 alpha and GRO-alpha correlates with inflammation in acute experimental autoimmune encephalomyelitis. *Neuroimmunomodulation.* (1998) **5**: 166-171.

11.     Glabinski A, Tani M, Tuohy VK, Tuthill R, and Ransohoff RM: Central nervous system chemokine gene expression follows leukocyte entry in acute murine experimental autoimmune encephalomyelitis. *Brain Behav Immun.* (1995) **9**: 315-330.

12.     Glabinski A, Tani M, Strieter R, Tuohy V, and Ransohoff R: Synchronous synthesis of α- and β-chemokines by cells of diverse lineage in the central nervous system of mice with relapses of experimental autoimmune encephalomyelitis. *Am J Pathol.* (1997) **150**: 617-630.

13.     Nygardas PT, Maatta JA, and Hinkkanen AE: Chemokine expression by central nervous system resident cells and infiltrating neutrophils during experimental autoimmune encephalomyelitis in the BALB/c mouse. *Eur J Immunol.* (2000) **30**: 1911-8.

14.     Sun D, Tani M, Newman TA, Krivacic K, Phillips M, Chernosky A, Gill P, Wei T, Griswold KJ, Ransohoff RM, and Weller RO: Role of chemokines, neuronal projections, and the blood-brain barrier in the enhancement of cerebral EAE following focal brain damage. *J Neuropathol Exp Neurol.* (2000) **59**: 1031-43.

15.     Asensio VC, Lassmann S, Pagenstecher A, Steffensen SC, Henriksen SJ, and Campbell IL: C10 is a novel chemokine expressed in experimental inflammatory demyelinating disorders that promotes recruitment of macrophages to the central nervous system. *Am J Pathol.* (1999) **154**: 1181-91.

16.     Hamilton N, Banyer J, Hapel A, Mahalingam S, Ramsay A, Ramshaw I, and Thomson S: IFN-gamma regulates murine interferon-inducible T cell alpha chemokine (I-TAC) expression in dendritic cell lines and during experimental autoimmune encephalomyelitis. *Scand J Immunol.* (2002) **55**: 171-177.

17.     Jiang Y, Salafranca M, Adhikari S, Xia Y, Feng L, Sonntag M, deFiebre C, Pennell N, Streit W, and Harrison J: Chemokine receptor expression in cultured glia and rat experimental allergic encephalomyelitis. *J Neuroimmunol.* (1998) **86**: 1-12.

18.     Glabinski AR, O'Bryant S, Selmaj K, and Ransohoff RM: CXC chemokine receptor expression during chronic relapsing experimental autoimmune encephalomyelitis. *Ann New York Acad Sci.* (2000) **917**: 135-144.

19.     Glabinski A, Bielecki B, O'Bryant S, Selmaj K, and Ransohoff R: Experimental autoimmune encephalomyelitis: CC chemokine receptor expression by trafficking cells. *J Autoimmunity.* (2002) **in press**:

20.     Serafini B, Columba-Cabezas S, Di Rosa F, and Aloisi F: Intracerebral recruitment and maturation of dendritic cells in the onset and progression of experimental autoimmune encephalomyelitis. *Am J Pathol.* (2000) **157**: 1991-2002.

21.     Narumi S, Kaburaki T, Yoneyama H, Iwamura H, Kobayashi Y, and Matsushima K: Neutralization of IFN-inducible protein 10/CXCL10 exacerbates experimental autoimmune encephalomyelitis. *Eur J Immunol.* (2002) **32**: 1784-1791.

22.	Alt C, Laschinger M, and Engelhardt B: Functional expression of the lymphoid chemokines CCL19 (ELC) and CCL21 (SLC) at the blood-brain barrier suggests their involvement in G-protein-dependent lymphocyte recruitment into the central nervous system during experimental autoimmune encephalomyelitis. *Eur J Immunol.* (2002) **32**: 2133-2144.

23.	Fife B, Paniagua M, Lukacs N, Kunkel S, and Karpus W: Selective CC chemokine receptor expression by central nervous system-infiltrating encephalitogenic T cells during experimental autoimmune encephalomyelitis. *J Neurosci Res.* (2001) **66**: 705-714.

24.	Rajan AJ, Asensio VC, Campbell IL, and Brosnan CF: Experimental autoimmune encephalomyelitis on the SJL mouse: effect of gamma delta T cell depletion on chemokine and chemokine receptor expression in the central nervous system. *J Immunol.* (2000) **164**: 2120-30.

25.	Fischer FR, Santambrogio L, Luo Y, Berman MA, Hancock WW, and Dorf ME: Modulation of experimental autoimmune encephalomyelitis: effect of altered peptide ligand on chemokine and chemokine receptor expression. *J Neuroimmunol.* (2000) **110**: 195-208.

26.	Glabinski A, Krakowski M, Han Y, Owens T, and Ransohoff R: Chemokine expression in GKO mice (lacking interferon-gamma) with experimental autoimmune encephalomyelitis. *J Neurovirol.* (1999) **5**: 95-101.

27.	Tran EH, Prince EN, and Owens T: IFN-gamma shapes immune invasion of the central nervous system via regulation of chemokines. *J Immunol.* (2000) **164**: 2759-68.

28.	Matejuk A, Dwyer J, Ito A, Bruender Z, Vandenbark A, and Offner H: Effects of cytokine deficiency on chemokine expression in CNS of mice with EAE. *J Neurosci Res.* (2002) **67**: 680-688.

29.	Matejuk A, Vandenbark A, Burrows G, Bebo B, and Offner H: Reduced chemokine and chemokine receptor expression in spinal cords of TCR BV8S2 transgenic mice protected against experimental autoimmune encephalomyelitis with BV8S2 protein. *J Immunol.* (2000) **164**: 3924-3931.

30.	Izikson L, Klein RS, Charo IF, Weiner HL, and Luster AD: Resistance to experimental autoimmune encephalomyelitis in mice lacking the CC chemokine receptor (CCR)2. *J Exp Med.* (2000) **192**: 1075-80.

31.	Fife BT, Huffnagle GB, Kuziel WA, and Karpus WJ: CC chemokine receptor 2 is critical for induction of experimental autoimmune encephalomyelitis. *J Exp Med.* (2000) **192**: 899-906.

32.	Ma M, Wei T, Boring L, Charo I, Ransohoff R, and Jakeman L: Monocyte recruitment and myelin removal are delayed following spinal cord injury in mice with CCR2 chemokine receptor deletion. *J Neurosci Res.* (2002) **68**: 691-702.

33.	Huang D, Wang J, Kivisakk P, Rollins BJ, and Ransohoff RM: Absence of Monocyte Chemoattractant Protein 1 in mice leads to decreased local macrophage recruitment and antigen-specific T helper cell type 1 immune response in experimental autoimmune encephalomyelitis. *J Exp Med.* (2001) **193**: 713-725.

34.	Rottman J, Slavin A, Silva R, Weiner H, Gerard C, and Hancock W: Leukocyte recruitment during onset of experimental autoimmune encephalomyelitis is CCR1 dependent. *Eur J Immunol.* (2000) **30**: 2372-2377.

35.	Tran EH, Kuziel WA, and Owens T: Induction of experimental autoimmune encephalomyelitis in C57BL/6 mice deficient in either the chemokine macrophage inflammatory protein-1alpha or its CCR5 receptor. *Eur J Immunol.* (2000) **30**: 1410-5.

36.	Karpus WJ, Lukacs NW, McRae BL, Strieter RM, Kunkel SL, and Miller SD: An important role for the chemokine macrophage inflammatory protein–1α in the pathogenesis of the T cell–mediated autoimmune disease, experimental auotimmune encephalomyelitis. *J Immunol.* (1995) **155**: 5003–5010.

37. Kennedy K, Strieter R, Kunkel S, Lukacs N, and Karpus W: Acute and relapsing experimental autoimmune encephalomyelitis are regulated by differential expression of the CC chemokines macrophage inflammatory protein-1 and monocyte chemotactic protein 1. *J Neuroimmunol.* (1998) **92**: 98-108.

38. Adamus G, Manczak M, and Machnicki M: Expression of CC chemokines and their receptors in the eye in autoimmune anterior uveitis associated with EAE. *Invest Ophthalmol Vis Sci.* (2001) **42**: 2894-2903.

39. Fife B, Kennedy K, Paniagua M, Lukacs N, Kunkel S, Luster A, and Karpus W: CXCL10 (IFN-gamma-inducible protein-10) control of encephalitogenic CD4+ T cell accumulation in the central nervous system during experimental autoimmune encephalomyelitis. *J Immunol.* (2001) **166**: 7617-7624.

40. Matsui M, Weaver J, Proudfoot A, Wujek J, Wei T, Richer E, Trapp B, Rao A, and Ransohoff R: Treatment of experimental autoimmune encephalomyelitis with the chemokine receptor antagonist Met-RANTES. *J Neuroimmunol.* (2002) **128**: 16-22.

41. Sun J, Xiao B, Lindblad M, Li B, Link H, Czerkinsky C, and Holmgren J: Oral administration of cholera toxin B subunit conjugated to myelin basic protein protects against experimental autoimmune encephalomyelitis by inducing transforming growth factor-beta-secreting cells and suppressing chemokine expression. *Int Immunol.* (2000) **12**: 1449-57.

42. Zavala F, Abad S, Ezine S, Taupin V, Masson A, and Bach J: G-CSF therapy of ongoing experimental allergic encephalomyelitis via chemokine- and cytokine-based immune deviation. *J Immunol.* (2002) **168**: 2011-2019.

43. Furlan R, Poliani P, Marconi P, Bergami A, Ruffini F, Adorini L, Glorioso J, Comi G, and Martino G: Central nervous system gene therapy with interleukin-4 inhibits progression of ongoing relapsing-remitting autoimmune encephalomyelitis in Biozzi AB/H mice. *Gene Ther.* (2001) **8**: 13-9.

44. Hvas J, McLean C, Justesen J, Kannourakis G, Steinman L, Oksenberg JR, and Bernard CC: Perivascular T cells express the pro-inflammatory chemokine RANTES mRNA in multiple sclerosis lesions. *Scand J Immunol.* (1997) **46**: 195-203.

45. Simpson J, Newcombe J, Cuzner M, and Woodroofe M: Expression of monocyte chemoattractant protein-1 and other beta-chemokines by resident glia and inflammatory cells in multiple sclerosis lesions. *J Neuroimmunol.* (1998) **84**: 238-249.

46. McManus C, Berman J, Brett F, Staunton H, Farrell M, and Brosnan C: MCP-1, MCP-2 and MCP-3 expression in multiple sclerosis lesions: an immunohistochemical and in situ hybridization study. *J Neuroimmunol.* (1998) **86**: 20-29.

47. Trebst C, Sorensen T, Kivisakk P, Cathcart M, Hesselgesser J, Horuk R, Sellebjerg F, Lassmann H, and Ransohoff R: CCR1+/CCR5+ mononuclear phagocytes accumulate in the central nervous system of patients with multiple sclerosis. *Am J Pathol.* (2001) **159**: 1701-1710.

48. Sorensen T, Tani M, Jensen J, Pierce V, Lucchinetti C, Folcik V, Qin S, Rottman J, Sellebjerg F, Strieter R, Frederiksen J, and Ransohoff R: Expression of specific chemokines and chemokine receptors in the central nervous system of multiple sclerosis patients. *J Clin Invest.* (1999) **103**: 807-815.

49. Balashov KE, Rottman JB, Weiner HL, and Hancock WW: CCR5(+) and CXCR3(+) T cells are increased in multiple sclerosis and their ligands MIP-1alpha and IP-10 are expressed in demyelinating brain lesions. *Proc Natl Acad Sci U S A.* (1999) **96**: 6873-8.

50.     Franciotta D, Martino G, Zardini E, Furlan R, Bergamaschi R, Andreoni L, and Cosi V: Serum and CSF levels of MCP-1 and IP-10 in multiple sclerosis patients with acute and stable disease and undergoing immunomodulatory therapies. *J Neuroimmunol.* (2001) **115**: 192-198.

51.     Misu T, Onodera H, Fujihara K, Matsushima K, Yoshie O, Okita N, Takase S, and Itoyama Y: Chemokine receptor expression on T cells in blood and cerebrospinal fluid at relapse and remission of multiple sclerosis: imbalance of Th1/Th2-associated chemokine signaling. *J Neuroimmunol.* (2001) **114**: 207-212.

52.     Iarlori C, Reale M, Lugaresi A, De Luca G, Bonanni L, Di Iorio A, Feliciani C, Conti P, and Gambi D: RANTES production and expression is reduced in relapsing-remitting multiple sclerosis patients treated with interferon-beta-1b. *J Neuroimmunol.* (2000) **107**: 100-7.

Chapter A18

# FREE RADICALS AND EXPERIMENTAL AUTOIMMUNE ENCEPHALOMYELITIS

Sean Murphy

*Institute of Cell Signalling, Medical School, QMC, University of Nottingham, UK*

**Abstract:**     This chapter reviews the major reactive oxygen and nitrogen species, their generation and their relevance for EAE. The relative contribution of NO and ONOO- species is discussed and the current literature EAE reviewed. The dual nature of free radicals, pathogenic and protective, and the correspondingly conflicting results of therapeutic interventions aimed at diminishing free radicals (in particular reactive nitrogen species), are also discussed in this chapter.

**Key words:**     free radicals, EAE, MS, reactive oxygen species, reactive nitrogen species, peroxynitrite, nitric oxide

## Introduction

Inflammation is associated with demyelinating disease in humans. The generation of reactive oxygen (ROS) and nitrogen species (RNS) is increased dramatically in conditions involving inflammation, overwhelming intrinsic anti-oxidant defences, and causing damage to the lipid, protein and nucleic acid constituents of cells and mitochondria. Oligodendrocytes, with little in the way of antioxidant defence and also a high iron content, are particularly sensitive to oxidative and nitrative stress. In addition, reactive species can damage the myelin sheath directly, promoting the attention of macrophages.

Consequently, there has been intense interest in the contribution of free radicals to myelin disorders, summarized recently in a collection of useful reviews (1-4). The tone of their collective conclusions is that, while the neurological deficits associated with demyelination can be ameliorated by reducing ROS, the outcomes of therapies aimed at diminishing RNS are more variable, and often exacerbate disease. The purpose of this chapter is to try and resolve roles for RNS in EAE.

**Nitric oxide generation**

There are three major isoforms of NO synthase (NOS). While these are products of distinct genes, the cDNAs share 50-60% homology at the nucleotide and amino acid levels (5). Two of these gene products, nitric oxide synthase (NOS)-1 and NOS-3, are constitutively expressed enzymes that are calcium-calmodulin dependent, producing small amounts of NO in response to transient elevations in intracellular calcium level. Both NOS-1 and NOS-3 are catalytically functional only as dimers (head-to-head), exhibit alternative splice products, and display protein- and lipid-interacting domains in their N-terminal regions. To date, the understanding of transcriptional regulation of these isoforms is rudimentary. For NOS-1, first characterized in the nervous system, activators of the gene include mechanical injury, hypoxia, and sex steroids. For NOS-3, responsible for the production of endothelium-derived relaxing factor, transcriptional activation is associated with shear stress, hypoxia, sex steroids and growth factors (6).

The NOS-2 isoform is also active only as a dimer, but functions independently of a rise in intracellular calcium and is capable of producing a large and continuous flux of NO. This isoform is not normally expressed but can be transcriptionally induced in response to proinflammatory cytokines, bacterial endotoxins, hypoxia, and viral coat proteins. Upon activation of the NOS-2 gene, the transcript is very unstable with a $t_{1/2}$ of 2-3 hours. Clearly there are proteins co-induced which are responsible for this RNA instability, as $t_{1/2}$ is prolonged by inhibitors of transcription and also protein synthesis. The relatively large amount of NO generated by NOS-2 over a sustained period has been implicated in various pathologies (7), and also as part of the host response to infectious organisms (8).

The NOS isoforms have similar catalytic domains: a C-terminal reductase (homologous with cytochrome P-450 reductase) exhibiting binding sites for flavins and NADPH, and a N-terminal oxygenase domain which contains bound heme and the site for H4 biopterin. The reductase domain transfers electrons from NADPH to heme. Calmodulin binds just at the N-terminal side of the reductase. The regulation of dimer assembly is of interest, as this leads to functional NOS. The degree to which $H_4$ biopterin is required depends on the isoform and the cell type in which the enzyme is expressed. Whether arginine and/or heme limits intracellular NOS assembly is still unclear. Certainly, in activated cells, NOS-2 exists as a mixture of dimer and monomer, implying either that dimers are unstable or that assembly is limited. Cell culture studies point to the former. The accumulating NO limits dimerization, perhaps by preventing heme insertion into the protein.

In active MS lesions multiple cell types express NOS-2, and there is abundant evidence for the local generation of NO oxidation products, such as peroxynitrite (ONOO-) (9). Elevated levels of CSF nitrate ($NO_3$-) correlate with clinical relapse. Calabrese et al. (10) report NOS-2 and NOS activity in the CSF from patients with MS, and detect nitrotyrosine (NT) immunostaining of CSF proteins. There is also evidence for nitrosylation (11). In MS there is greater concordance among monozygotic (30%) that dizygotic (2.5%) twins, and involvement of the NOS genes has been investigated. The NOS-2 gene is situated on the long arm of chromosome 17, which is linked to MS in British and Finnish populations. The NOS-3 gene is on 7q36, a locus for induced arthritis. The NOS-1 gene is on the long arm of chromosome 12, which is linked to MS. Modin et al. (12) investigated five markers within the three NOS genes with regard to MS susceptibility and disease course in 156 affected sib-pairs, and in 96 benign and 96 severe definite MS patients in the Nordic population. However, genetic variation of these genes neither contributes directly to susceptibility to MS, nor does it influence disease severity.

**Free radical chemistry**

Reactive nitrogen species are oxidation states and adducts of the products of NOS that arise in physiological environments, and include NO-, $.NO_2$, $NO_2$-, $N_2O_3$, $N_2O_4$, S-nitrothiols, ONOO-, and dinitrosyl-iron complexes. Reactive oxygen species are intermediate reduction products of $O_2$ en route to water, namely, $O_2$-, $H_2O_2$ and OH (Figure 1). What follows is a synopsis, and more complete coverage can be found (13, 14).

There are three major classes of pro-oxidant enzymes: NOS, cyclooxygenase (COX) and NADPH oxidase, and myeloperoxidase (MPO). It is estimated that between 2-5% of electron flow in brain mitochondria produces $O_2$-. This is scavenged by superoxide dismutases (SOD), glutathione peroxidase and catalase. Cellular antioxidants include glutathione, ascorbate, and α-tocopherol.

All eukaryotic cells, as well as simpler organisms, are capable of synthesizing NO (with L-citrulline as the co-product) from the five electron oxidation of L-arginine. The reaction is catalyzed by NOS, which employs components familiar from other enzymes but in novel and somewhat unconventional ways. As a radical, NO reacts rapidly with other species that contain unpaired electrons. In a reaction with $O_2$, NO can form an

Figure 1 Production of reactive oxygen and nitrogen species. Modified from (8).

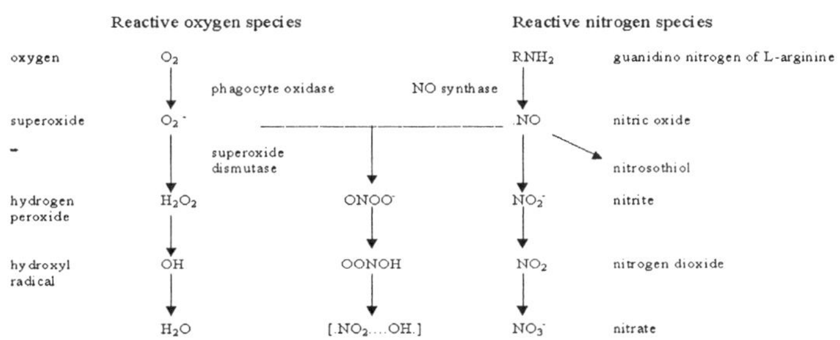

intermediate species ($N_2O_3$?) that efficiently nitrosates thiols and amines. Out-competing SOD, NO reacts with $O_2^-$ to produce $ONOO^-$ which, if protonated, decomposes within seconds to a hydroxyl radical. This $ONOO^-$ reacts rapidly with $CO_2$ (acting as a catalyst) in a complex manner that produces several short-lived reaction intermediates, such as $NO_2$ and $CO_3^-$. These intermediates are probably responsible for many of the reported toxic effects of NO. In the absence of arginine and $BH_4$, the NOS enzymes can catalyze the production of $O_2$- and, consequently, $H_2O_2$ when the $O_2$- is dismuted by SOD. Therefore, the NOS enzymes can produce ONOO-, and the decomposition of $H_2O_2$ or ONOO- can lead to the production of HO (15).

**Molecular targets for RNS**

The chemical alteration of DNA underlies a variety of pathological states. Nitric oxide can potentially damage DNA either through RNS ($ONOO^-$, $N_2O_3$), the inhibition of DNA repair processes, or by increased production of genotoxic (alkylating) agents. A species such as $N_2O_3$ causes mutations in cells, chemically altering DNA. Deamination of cytosine, adenine, and guanine results in conversion to uracil, hypoxanthine and xanthine, respectively. As well as modification, NO can form or modulate the activity of carcinogens. Nitrosamines are metabolized to alkylating species which lesion DNA.

The ability to nitrosate and to nitrate key amino acids in proteins (transcriptional regulators, enzymes, receptors) explains many of the actions of RNS, in terms of gene regulation and alterations in cell signalling pathways. Transcription factors, the regulatory proteins that bind to gene promoters and recruit RNA polymerase, display sequences specific for DNA binding and also a transactivation domain. A number of transcriptional activators have been shown to be regulated by NO, either post-translationally through direct modification of the protein, or indirect regulation via alteration in the rate of their own transcription. In so doing, NO can turn off constitutively expressed genes, activate transcriptionally regulated genes, or prevent their activation. The best examples to date are the transcription factors NF-kB and AP-1. Their activation can be blocked by NO and the proteins can also be directly nitrosated, with a reduction in promoter association. These two transcription factors are involved in the regulation of expression of a very large number of genes. Serine substitution of a cysteine residue at the N-terminal region of the p50 NF-kB subunit reduces DNA binding. Specific binding of NF-kB is also inhibited reversibly by NO, and these effects appear to be mediated by nitrosation of the same cysteine residue. There is also good evidence that NO stabilizes the NF-kB inhibitory protein IkB$\alpha$ and activates its transcription, thus inhibiting NF-kB translocation from the cytoplasm to the nucleus. AP-1 is a dimeric complex of Jun-Jun or Jun-Fos proteins. The key to dimer formation resides in two cysteine residues in the leucine zipper and basic regions of Fos and Jun, and nitrosation of cys 252 in the DNA binding domain decreases AP-1 binding to DNA.

Critical signalling proteins can be influenced by RNS, functioning at the transcriptional and/or post-transcriptional level either as an activator (poly ADP-ribose synthetase, p21ras, sGC) or inhibitor (adenylyl cyclase Type I, protein kinase C, cytochrome P450, nitric oxide synthases, lipoxygenase). The heme protein cyclooxygenase (COX) is inhibited by NO at high concentrations but activated by low concentrations. The mitochondrial respiratory chain is also susceptible. While NO inhibits cytochrome c oxidase (complex IV) in a reaction that is reversible and competitive with oxygen, ONOO⁻ irreversibly inhibits respiratory complex I-III as glutathione levels decrease. Nitric oxide can have diverse effects on cell fate, initiating or protecting against apoptosis depending on the cell type, NO concentration and redox environment. The family of protein-cleaving enzymes known as caspases are targets for NO, which inhibits their activity in a reversible manner.

Exposing cells to NO, coupled with the use of NOS inhibitors, has revealed complex effects on the production of cytokines. Some members of the caspase family participate in the maturation of cytokines. Precursors of IL-1$\beta$ and IL-18 are cleaved by caspase-1, and NO reversibly blocks this process, thereby acting as an anti-inflammatory agent. However, NO can be proinflammatory, increasing release of other cytokines such as TNF$\alpha$ and IL-6. Chemokines, small chemoattractant cytokines, can be quite specific in determining the immigration patterns of lymphocytes and macrophages to certain inflammatory regions. There appears to be an interesting cross-talk between NO and chemokines in that the expression of IL-8 and MCP-1 are inhibited by NO, whereas MCP-1 can prevent the expression of NOS-2.

## RNS and EAE

It is 10 years since RNS and ROS were first described as products of the inflammatory cells seen in rats with hyperacute EAE (16). Associated with increased pro-inflammatory gene expression in EAE, there is expression of NOS-2 mRNA, protein, and evidence for the generation of RNS. There is also evidence for the formation of NT and nitroso-S-cysteine (11), indicating interaction of RNS with proteins. While tyrosine nitration has been assumed to result from the generation of ONOO-, this view has recently been challenged, and .NO$_2$ is now suggested to be the culprit (17).

In 1999, Willenborg et al. (4) reviewed the 45 papers that had appeared on the role of NO in CNS autoimmune inflammatory disease. Predictably, the first group of publications all concluded that NO contributed to EAE pathology. This was based on observation that anti-inflammatory steroids block cytokine-mediated NO production and reduce EAE symptoms, NOS inhibitors reduce passively transferred EAE, encephalitogenic T cells cause macrophages to produce NO, and NOS-2 expression correlates with disease severity. However, a study of the therapeutic effects of five different NOS inhibitors in both adoptive transfer and actively induced EAE revealed no protection, and some exacerbation of disease (18). Okuda et al. (19) noted that the inflammatory cells dying in spinal cord lesions of EAE mice were juxtaposed to NOS-2 expressing cells, suggesting that NO was eliminating them. The phosphodiesterase inhibitor, Rolipram, elevated NOS-2 expression in monocytes and ameliorated disease (20). With one of the first NOS-2 selective inhibitors (L-NIL), Gold et al found protection in adoptive EAE and exacerbation of actively induced disease (21). However, the emergence of NOS-2 knockout mice, and the observation that NO is important in recovery and refractoriness to active reinduction, again pointed to the potential protective role of NO in actively induced EAE. Thus, while the majority verdict was to maintain that NO was contributing to disease, by 1999 fully one third of publications were leaning towards NO having a protective role.

It is well known that NO has a number of effects on immune responses, such as the inhibition of macrophage Ia expression, which would have the effect of preventing T cell expansion. This inhibition of T cell proliferation appears to be a specific impairment of Th1, while sparing Th2 cells. Since EAE is a function of Th1 cells, the increase in NO may selectively limit proliferation of the encephalitogenic effector population. In addition, the expression of VCAM and ICAM-1 are downregulated by NO, significantly altering lymphocyte migration. In target tissue, NO can induce both apoptosis and necrosis in T effector cells and protect oligodendrocytes against destruction by lipid peroxidation (1).

So, what are the roles of RNS in EAE? More recent publications have clarified their effects.

## ONOO- vs. NO in EAE

The group at Thomas Jefferson University in Philadelphia have long claimed a destructive role for ONOO- in MS and EAE, based on the observation that a natural antioxidant scavenger, uric acid (UA), ameliorates disease. They postulate that the normal levels of UA in humans confers enhanced protection against free radical-mediated cell injury, and that low UA would predispose towards the development of CNS diseases involving ONOO-, which include not only MS but also amyotrophic lateral sclerosis (ALS), Parkinson's and Alzheimer's diseases.

To understand how ONOO- is involved, they examined the effects of UA on blood-brain barrier permeability, and on the accumulation and activity of inflammatory cells in the spinal cord (22). Induction of EAE in female mice was by direct immunization with MBP. They reasoned that, if UA acts solely by inactivating ONOO-, then raising UA levels should not directly affect the activation of inflammatory cells (eg. NOS-2 expression) but would reduce the subsequent inflammation (eg. NT accumulation). In mice that received UA before the onset of clinical signs of EAE there was reduced invasion of inflammatory cells into the CNS and disease prevention. In mice with active EAE, and therefore a compromised blood-brain barrier (BBB), the administration of UA blocked NT, reduced apoptosis in the lesions, and promoted recovery. The UA treatment also reversed the increase in BBB permeability observed in EAE. This action could explain why UA administration prevents the onset of EAE. While UA does not affect the specific T cell proliferative response, the levels of IgG antibodies, or appearance of cells expressing NOS-2 in lymph nodes and peripheral blood of MBP-immunized mice (23), it does reduce the levels of NOS-2 expression in spinal cord (24). This group has also looked at novel ONOO-scavengers, mercaptoethylguanidines, that readily cross the BBB (25). These compounds delayed the onset, and reduced the incidence of MBP-induced EAE. Surprisingly, the scavengers did not alter the severity of disease or overall mortality, but this could be due to the lower efficiency with which these compounds inactivate ONOO-, as compared with UA. Alternatively, it could be because these scavengers also inhibit NOS-2 activity, removing the protective influence of NO. Interestingly, the radical scavenger ascorbic acid (AA) can also block nitration, but does not protect against EAE (26).

There have been concerns raised about the interpretation of the UA effects in EAE, and its role as a ONOO- scavenger. Uric acid causes an inflammatory response in the peritoneum, characterized by neutrophil accumulation, and it is suggested that this could reduce the number of cells available for extravasation. In addition, UA was found to be effective even in EAE induced in IFNγ receptor knockout mice, where very little NO (and therefore ONOO-) is generated (27). However, catalysts specific for the decomposition of ONOO- have recently been developed, and these promote the isomerization of ONOO- to $NO_3$ without directly affecting NO levels. Mice receiving catalyst displayed less severe clinical disease, and less inflammation and demyelination, but encephalitogenic T cells could still be recovered from catalyst-treated mice (28).

**Evidence for beneficial NO and/or against NOS-2 derived NO being harmful**

1.  The oral administration of L-arginine (the NOS substrate) was found to delay disease onset and to reduce the severity of neurological and histological signs of EAE, and yet there was a significant elevation in CNS nitrite, and decreases in $O_2$- and $H_2O_2$ (29).

2.  Female PVG rats are resistant to actively-induced EAE, their spleens have more monocytes and fewer RBC than males, and produce more RNS. If splenectomised, these animals are no longer resistant. Immunized intact females do generate EAE effector cells, and are also susceptible to passively transferred disease. Therefore, in actively-induced EAE in these animals, there must be a block at the effector cell expansion phase, or of transmigration into target tissue, and this could be related to the higher RNS levels (30).

3.  Mice deficient in glucocorticoid receptors (GR) are resistant to actively induced EAE, because of a reduction in spleen and lymph node cell proliferation. These GR-deficient mice have much higher $NO_2$ levels, due to unsuppressed NOS expression. When treated with NOS inhibitors, the level of $NO_2$ decreases, there is increased effector cell proliferation, and mice are now susceptible to EAE. Thus, NO appears to be the immunosuppressant (31).

4.  The administration of a low dose of IL-4 prior to immunization improves clinical scores of EAE. The dendritic cell population proliferates, secreting high levels of IFNγ, NO and IL-10, and the NO promotes apoptosis of autoreactive T cells (32).

5.  Xu et al. (33) used 3-morpholinosydnonimine (SIN-1) to enhance NO levels during the development of EAE. This compound can release ONOO-, but they found no changes in $O_2$- or ONOO- formation. Given within the first week following immunization, SIN-1 enhanced IFNγ levels, reduced expression of MHC class II, and augmented apoptosis of blood mononuclear cells. This resulted in less cellular infiltration into the CNS and, therefore, reduced clinical signs of disease.

6.  High dose of the specific antigen can induce apoptotic T cell death and so suppress immune responses. Weishaupt et al. (34) looked at levels of TNFα, IFNγ, and NOS-2, and found them highly elevated in this response. Apoptosis of T cells was evident, but a potentially harmful side effect was the observation that oligodendrocytes were also undergoing apoptosis. Interestingly, there still remained a clinical benefit, and this was lost when TNFα activity was neutralized.

7.  The injection of mice with staphylococcal enterotoxin B prior to immunization for EAE provides protection against disease, and yet more NOS-2 is evident in the brains of treated animals (35).

8.  Mice immunized with complete Freund's adjuvant (CFA) containing mycobacterial extract (which can transcriptionally activate NOS-2) are protected against EAE, but NOS-2 knockout mice are not protected. In the wildtype mice there was augmentation of the IgG1 response, a decrease in IL-6 production by T cells, and a marked decrease in monocyte infiltration into the CNS (36).

**Multiple roles for NO in EAE ?**

The addition of pertussis toxin (PT) when EAE is triggered worsens the CNS effects. Both NOS-2 and TNFα are increased, over and above the levels seen in animals not receiving PT (37). Pozza et al. (38) showed that a NOS inhibitor, given prior to immunization, caused complete recovery from the acute phase of EAE, and a delayed and milder relapse. The benzoquinoid ansamycins (anti-fungals) induce heat shock proteins, and reduce NOS-2 expression and $NO_2$ accumulation. Given after immunization, these agents reduce EAE disease onset by >50% (39).

Lewis rats, immunized with MBP-CFA, recover from EAE. They do not relapse, and develop a resistance to further induction. The serum levels of RNS are increased during recovery and, if challenged again with MBP-CFA, increase further. If a NOS inhibitor is given (even in the absence of MBP-CFA), then 100% enter a second episode of disease. This is consistent with NO inhibiting T cell proliferation, adhesion and migration into the CNS. However, this second episode is chronic, with periods of remission and relapse (40). Why is NO unable to prevent relapse? The authors conclude that there is primary disease, limited in time by the production of NO, and then relapsing disease, perhaps as a result of excessive production of RNS.

**Conclusions**

In their review Willenborg et al. (4) concluded that, for both rats and mice, treatments that target passive EAE, or the effector phase of actively induced EAE, result in amelioration of disease. However, treatments that target the afferent arm of the response lead to the aggravation of pathology. Simply put, an increase in RNS levels in the target tissue is bad, but an increase during immunization is good. Much of the recent data support this notion, and provide a strong hint that it is ONOO-, and not .NO, that causes the tissue damage. If so, then EAE joins a growing list of pathologies in which RNS have causal and protective functions. The trick for therapeutics will be to devise means to eliminate one but preserve the other.

## ACKNOWLEDGEMENTS

Work in the author's laboratory on NO and CNS pathophysiology is supported by the NIH-NINDS (USA), the Wellcome Trust, and the Biotechnology and Biological Sciences Research Council (UK)

## REFERENCES

1. Kolb H, Kolb-Bachofen V (1998) NO in autoimmune disease: cytotoxic or regulatory mediator? Immunol Today 19, 556-561
2. Merrill JE, Scolding NJ (1999) Mechanism of damage to myelin and oligodendrocytes and their relevance to disease. Neuropathol App Neurobiol 2, 435-458
3. Smith KJ, Kapoor R, Felts PA (1999) Demyelination:The role of reactive oxygen and nitrogen species. Brain Pathol 9, 69-92
4. Willenborg DO, Staykova MA, Cowden WB (1999) Our shifting understanding of the role of nitric oxide in autoimmune encephalomyelitis. J Neuroimmunol 100: 21-35
5. Alderton WK, Cooper CE, Knowles RG (2001) NO synthase: structure, function and inhibition.  Biochem J 357, 593-615
6. Forstermann U, Boissel J-P, Kleinert H (1998) Expressional control of the constitutive isoforms of nitric oxide synthase. FASEB J. 12, 773-790.
7. Murphy S (2000) Production of nitric oxide by glial cells. Glia 29, 1-14
8. Nathan C, Shiloh MU (2000) Reactive oxygen and nitrogen intermediates in the relationship between mammalian hosts and microbial pathogens. Proc Natl Acad Sci USA 97, 8841-8848
9. Liu J, Zhao M-L, Brosnan CF, Lee SC (2001) Expression of inducible NO synthase and nitrotyrosine in multiple sclerosis lesions. Am J Pathol 158, 2057-2066
10. Calabrese V, Scapagnini G, Ravagna A, Bella R, Foresti R, Bates TE, Giuffrida Stella A-M, Pennisi G (2002) NO synthase is present in the CSF of patients with active MS and is associated with increases in CSF protein nitrotyrosine and S-nitrosothiols with changes in glutathione levels. J Neurosci Res 70, 580-587
11. Boullerne AI, Rodriguez JJ, Touil T, Brochet B, Schmidt S, Abrous ND, Le Moal M, Pua JR, Jensen MA, Mayo W, Arnason BG, Petry KG  (2002) Anti-S-nitrosocysteine antibodies are a predictive marker for demyelination in experimental autoimmune encephalomyelitis: implications for multiple sclerosis. J Neurosci  22: 123-132
12. Modin H, Dai Y, Masterman T, Svejgaard A, Sorensen PS, Oturai A, Ryder LP, Spurkland A, Vartdal F, Laaksonen M, Sandberg-Wollheom M, Myhr KM, Nyland H, Hillert J ( 2001) No linkage or association of the NO synthase genes to multiple sclerosis. J Neuroimmunol 119, 95-100
13. Chan PH (2001) Reactive oxygen radicals in signaling and damage in the ischemic brain. J. Cerebral Blood Flow Metab. 21, 2-14
14. Davis KL, Martin E, Turko IV, Murad F (2001) Novel effects of nitric oxide Annu. Rev. Pharmacol. Toxicol. 41: 203-236.
15. Stuehr D, Pou S, Rosen GM (2001) Oxygen reduction by NO synthase. J Biol Chem, 276, 14533-14536
16. MacMicking JD, Willenborg DO, Weidemann MJ, Rockett KA, Cowden WB (1992) Elevated secretion of reactive nitrogen and oxygen intermediates by inflammatory leukocytes in hyperacute EAE. J Exp Med 176, 303-307

17. Pfeiffer A, Lass A, Schmidt K, Mayer B (2001) Protein tyrosine nitration in cytokine-activated murine macrophages – involvement of a peroxidase/nitrite pathway rather than peroxynitrite. J Biol Chem 276, 34051-34058

18. Zielasek J, Jung S, Gols R, Liew FY, Toyka KV, Hartung HP (1995) Administration of nitric oxide inhibitors in experimental autoimmune neuritis and EAE. J Neuroimmunol 58, 81-88

19. Okuda Y, Sakoda S, Fujimura H, Yanagihara T (1997) Nitric oxide via an inducible isoform of NOS is a possible factor to eliminate inflammatory cells from the CNS of mice with EAE. J Neuroimmunol 73, 107-116

20. Martinez I, Puerta C, Redondo C, Garcia_Merino A (1999) Type IV phosphodiesterase inhibition in experimental allergic encephalomyelitis of Lewis rats: sequential gene expression analysis of cytokines, adhesion molecules and the inducible nitric oxide synthase. J Neurol Sci  164, 13-23

21. Gold DP, Schroder K, Powell HC, Kelly CJ (1997) Nitric oxide and the immunomodulation of EAE. Europ J Immunol 27, 2863-2869

22. Hooper DC, Scott GS, Zborek A, Mikheeva T, Kean RB, Koprowski H, Spitsin SV (2000) Uric acid, a peroxynitrite scavenger, inhibits CNS inflammation, blood-CNS barrier permeability changes, and tissue damage in a mouse model of multiple sclerosis. FASEB J 14, 691-698

23. Kean RB, Spitsin SV, Mikheeva T, Scott GS, Hooper DC (2000) The peroxynitrite scavenger uric acid prevents inflammatory cell invasion into the CNS in EAE through maintenance of blood-CNS barrier integrity. J Immunol 165, 6511-6518

24. Spitsin SV, Scott GS, Kean RB, Mikheeva T, Hooper DC (2000) Protection of myelin basic protein immunized mice from free-radical mediated inflammatory cell invasion of the central nervous system by the natural peroxynitrite scavenger uric acid. Neurosci Lett 292, 137-41

25. Scott GS, Kean RB, Southan GJ, Szabo C, Hooper DC (2001) Effect of mercaptoethylguanidine scavengers of peroxynitrite on the development of experimental allergic encephalomyelitis in PLSJL mice. Neurosci Lett 311, 125-128

26. Spitsin SV, Scott GS, Mikheeva T, Zborek A, Kean RB, Brimer CM, Koprowsi H, Hooper DC (2002) Comparison of uric acid and ascorbic acid in protection against EAE. Free Rad Biol Med 33, 1363-1371

27. Hooper DC, Spitsin SV, Kean RB, Champion JM, Dickson GM, Chaudhry I, Koprowsi H, (1998) Uric acid, a natural scavenger of peroxynitrite, in EAE and MS. Proc Natl Acad Sci USA 95, 675-680

28. Cross AH, San M, Stern MK, Keeling RM, Salvemini D, Misko TP (2000) A catalyst of peroxynitrite decomposition inhibits murine experimental autoimmune encephalomyelitis. J Neuroimmunol 107, 21-28

29. Scott GS, Bolton C (2000) L-arginine modifies free radical production and the development of experimental allergic encephalomyelitis. Inflamm Res 49, 720-726

30. Staykova MA, Cowden W, Willenborg DO (2002) Macrophages and nitric oxide as the possible cellular and molecular basis for strain and gender  differences in susceptibility to autoimmune central nervous system inflammation. Immunol Cell Biol 80, 188-197

31. Marchetti B, Morale MC, Brouwer J, Tirolo C, Testa N, Caniglia S, Barden N, Amor S, Smith PA, Dijkstra CD (2002) Exposure to a dysfunctional glucocorticoid receptor from early embryonic life programs the resistance to experimental autoimmune encephalomyelitis via nitric oxide-induced immunosuppression. J Immunol 16, 5848-5859

32. Xu L, Huang Y, Yang J, Van Der Meide PH, Levi M, Wahren B, Link H, Xiao B (1999) Dendritic cell-derived nitric oxide is involved in IL-4-induced suppression of experimental allergic encephalomyelitis (EAE) in Lewis rats. Clin Exp Immunol 118, 115-121

33. Xu LY, Yang JS, Link H, Xiao BG (2001) SIN-1, a nitric oxide donor, ameliorates experimental allergic encephalomyelitis in Lewis rats in the incipient phase: the importance of the time window. J Immunol  166: 5810-5816
34. Weishaupt A, Jander S, Bruck W, Kuhlmann T, Stienekemeier M, Hartung T, Toyka KV, Stoll G, Gold R (2000) Molecular mechanisms of high-dose antigen therapy in experimental autoimmune encephalomyelitis: rapid induction of Th1-type cytokines and inducible nitric oxide synthase. J Immunol 165, 7157-7163
35. Kuschnaroff LM, Overbergh L, Sefriouni H, Sobis H, Vandeputte M, Waer M (1999) Effect of staphylococcal enterotoxin B injection on the development of experimental autoimmune encephalomyelitis: influence of cytokine and inducible nitric oxide synthase production. J Neuroimmunol 99, 157-168
36. Kahn DA, Archer DC, Gold DP, Kelly CJ (2001) Adjuvant immunotherapy is dependent on inducible nitric oxide synthase. J Exp Med 193, 1261-8
37. Ahn M, Kang J, Lee Y, Riu K, Kim Y, Jee Y, Matsumoto Y, Shin T  (2001) Pertussis toxin-induced hyperacute autoimmune encephalomyelitis in Lewis rats is correlated  with increased expression of inducible nitric oxide synthase and tumor necrosis factor alpha. Neurosci Lett 308, 41-44
38. Pozza M, Bettelli C, Aloe L, Giardino L, Calza L (2000) Further evidence for a role of nitric oxide in experimental allergic encephalomyelitis: aminoguanidine treatment modifies its clinical evolution. Brain Res  855, 39-46
39. Murphy P, Sharp A, Shin J, Gavrilyuk V, Dello Russo C, Weinberg G, Sharp FR, Lu A, Heneka MT, Feinstein DL (2002) Suppressive effects of ansamycins on inducible nitric oxide synthase expression and the development of experimental autoimmune encephalomyelitis. J Neurosci Res  67, 461-470
40. O_Brien NC, Charlton B, Cowden WB, Willenborg DO (2001) Inhibition of nitric oxide synthase initiates relapsing remitting experimental autoimmune encephalomyelitis in rats, yet nitric oxide appears to be essential for clinical expression of disease. J Immunol 167, 5904-5912

Chapter A19

# PROTEASES AND PEPTIDASES IN EAE

M Nicola Woodroofe and Rowena AD Bunning
*Biomedical Research Centre, School of Science and Mathematics, Sheffield Hallam
University, Howard Street, Sheffield S1 1WB, UK*

Abstract:    The role of proteases and peptidases in the pathogenesis of EAE has been
extensively studied. In particular research has focussed on the matrix
metalloproteinase (MMP) family of enzymes as well the plasminogen
activator/plasmin system. These enzymes have been proposed to have a
number of functional roles in disease pathogenesis including: breakdown of
the blood brain barrier and the extracellular matrix, degradation of the myelin
sheath with the generation of further encephalitogenic peptides and
amplification of the inflammatory process in the CNS, via activation of pro-
inflammatory cytokines. The administration of proteolytic enzyme inhibitors
to animals with EAE resulting in clinical improvement, has provided further
support for the MMP/plasminogen activator/plasmin system in the disease
process. The involvement of other enzymes including, myelencephalon
specific protease, calpain and dipeptidyl petidase IV in EAE pathogenesis is
also discussed.

Key words:    metalloproteinases, serine proteases, demyelination, blood brain barrier,
inflammation

## 1.    INTRODUCTION

Proteinases and peptidases involved in neuroinflammation can be divided
into those with intracellular or extracellular activity. Intracellular proteases,
include calpains and caspases and the extracellular proteases, which have
been extensively investigated in EAE pathogenesis, consist of two major
groups: the serine proteases, plasminogen activators (PAs), including
urokinase-type PA (uPA) (EC 3.4.21.73) and tissue type PA (tPA) (EC
3.4.21.68) and plasmin and the matrix metalloproteinases (MMPs)[1,2].

The MMPs are a family of $Zn^{2+}$ dependent endopeptidases, which are capable of degradation of the extracellular matrix (ECM) and have been implicated in the degradation of myelin. This family of at least 25 proteins, includes collagenases, gelatinases, stromelysins and membrane-type MMPs (MT-MMPs), based on their substrate specificity and domain structures, however there is considerable overlap in substrate specificities between sub groups. MMPs are capable of degradation of all protein components of the ECM including collagen, elastin, fibronectin and laminin[3,4,5]. Members of the related ADAMs (A Disintegrin And Metalloproteinase) family of proteins are important sheddases for the release of membrane bound proteins e.g. TNF and its receptor are cleaved by ADAM 10 and 17 (TACE, tumour necrosis factor converting enzyme)[5]. These enzymes are membrane bound and possess an integrin binding, disintegrin domain, which may be important in cell-cell interactions[6]. Another group of ADAMs family proteins, recently identified, are the ADAMTSs, which are secreted proteins possessing one or more thrombospondin motif and no transmembrane or cytoplasmic domains. ADAMs and ADAMTSs have to date not been investigated in EAE.

The activity of these two main families of proteases, PAs and MMPs, is tightly controlled, to prevent any unwanted tissue damage, in a number of ways: the enzymes are synthesised as pro-forms which require proteolytic cleavage to produce the active enzyme; the presence of endogenous inhibitors, PAIs and TIMPs (Tissue Inhibitors of Metalloproteinases) which bind to the enzymes in a 1:1 stoichiometry; regulation of gene transcription, and *de novo* synthesis by secretion. The transcriptional regulation of these enzymes and their inhibitors by chemokines[7] and cytokines via AP1 transcription factors is also critical in chronic inflammation[6]. Thus the resultant activity of these proteolytic enzymes is dependent on their levels within a tissue, the concentration of the endogenous inhibitors at the same site as well as the presence of pro- or anti-inflammatory cytokines.

PAs catalyse the conversion of plasminogen to plasmin (EC 3.4.21.7)[8]. Plasminogen is present in most body fluids and plasmin generated from it has a wide range of substrates including not only fibrin but components of the  basal lamina, fibronectin and laminin, tenascin-C, components of the ECM of the brain, tenascin-C, proteoglycans and myelin[9,10,11,12,13,14,15,16]. Thus plasmin may play a role in ECM breakdown as occurs in damage to the blood brain barrier (BBB) and during cell migration in the central nervous system (CNS) parenchyma as well as in myelin breakdown. In addition, plasmin is capable of activating the proforms of several MMPs[17,18,19]. Thus plasmin provides a link between the two systems whereby plasmin is a limiting factor on MMP activation. It can also activate latent transforming growth factor $\beta$ (TGF$\beta$), reducing inflammation[20].

These proteolytic enzyme systems have been implicated in a number of pathological events during the pathogenesis of EAE including[1,2]:

o   Disruption of the BBB via proteolysis of the basement membrane, permitting inflammatory cell migration across the blood brain barrier
o   Myelin degradation and generation of further myelin basic protein (MBP) fragments to amplify the immune response
o   Disruption of cell-cell or cell-ECM interactions within the CNS parenchyma
o   Amplification of inflammation via cytokine processing and activation
o   Further cleavage of pro-MMPs into the active enzyme

Paradoxically they have also been reported to play a beneficial role in recovery following an inflammatory insult to the CNS. For example, degradation of the ECM may be required to allow remodelling following CNS injury, migration of phagocytic cells, i.e. microglia and macrophages, within the CNS. To remove cellular debris will require the secretion of MMPs by these cells. Similarly neural cell precursors may require MMPs to aid their migration into the damaged area, to replenish cells damaged as a result of the inflammation and allow axonal elongation, finally they may release growth factors anchored to the ECM, which could then promote repair within the lesion[4].

Here we review the research to date on the role of these proteases and others in the pathogenesis of EAE, as well as summarising evidence which suggests that inhibition of these enzymes is a potential therapeutic strategy in the treatment of CNS inflammation.

**<u>Matrix metalloproteinases</u>**

The structure of the MMP family of enzymes is similar, each having five protein domains, a classical signal peptide is recognised by the ER docking enzyme and leads to secretion, the pro-peptide domain with a single cysteine sulphydryl group forms the amino terminal part of the secreted protein, the presence of this sulphydryl group determines the latent state of the pro-enzyme. The active enzyme domain, and the zinc-binding domain form the catalytic site. The hemopexin domain is located at the carboxy terminus. In addition, the MT MMPs possess a short chain of hydrophobic amino acids, which anchor the enzymes in this group to the cell membrane. MMP2 (EC 3.4.24.24) and 9 (EC 3.4.24.35) have an extra fibronectin-like domain and MMP9 also has an additional type V collagen domain. Activation of the inactive zymogen to the active enzyme occurs via a cysteine switch mechanism, which involves removal of the cysteine, sulphydryl group and replacement with water[21]. This activation occurs downstream of the proteolytic cleavage of plasminogen by uPA and tPA to plasmin, although MT-MMPs can cleave pro MMP2 to the active form and pro MMP9 has been reported to be cleaved by MMP2[1]. Cytokines modulate the expression and regulation of MMPs, via expression of proto-oncogenes of the c-*fos* and c-*jun* family which in turn contributes to formation of dimeric forms of AP1 transcription factors that bind to specific promoter sequences which are present in the promoter region of most MMP genes[5]. MMPs in turn can act as sheddases for membrane bound cytokines, their receptors and adhesion molecules. Although, ADAM10 and 17 (TACE) are more effective sheddases for TNFα, a key cytokine in the CNS immune response in EAE[22].

### MMPs and EAE

Studies on the expression of MMPs in EAE have included a wide range of different models of the disease, including rat and mouse models induced by   active immunisation with myelin antigens, usually MBP, myelin oligodendrocyte glycoprotein (MOG), proteolipid protein (PLP) or spinal cord homogenate as well as passive transfer models using CNS antigen-specific T cells. More recently, genetically modified animals have also provided further information on the role of MMPs in CNS inflammation. For example, mice over expressing cytokines, in particular TNF, in the CNS and mice under expressing  MMP9, have been used to elucidate the role of these enzymes in CNS inflammation[23,24].

Reports on the expression of MMPs during the course of EAE are generally in agreement with a dominant role being assigned to MMPs 7 (EC 3.4.24.23) and 9. These enzymes are expressed by both inflammatory cells in perivascular locations as well as by resident glia. However other MMPs including MMPs 1, 3, 10, 12 (EC 3.4.24.7, EC 3.4.24.17, 3.4.24.22 and 3.4.24.65 respectively) and MMP 14   have also been demonstrated in the CNS in EAE[25]. Elevated levels of MMP 9 were found in the cerebrospinal fluid (CSF) of mice with EAE and correlated with clinical disease score[26]. In acute EAE in the Lewis rat, development of clinical disease was associated with a 3 fold increase in MMP activity in the CSF, measured by a coumarin-labelled peptide substrate assay. Using quantitative PCR, mRNA for at least seven MMPs were expressed in normal and EAE spinal cord. However MMP 7, localised to macrophage cells,  showed a dramatic increase of over 500 fold in EAE animals compared to normal control animals and the levels remained high during the recovery period. MMP 7 has been shown to degrade MBP and this may be one of the actions of this enzyme in EAE. MMP 9 showed only a moderate, five fold increase, which peaked at disease onset but declined thereafter[27].  In an adoptive transfer model of EAE in Lewis rats, MMPs 7 and 9 increased, 100 fold and ten fold respectively, as assessed by a semi-quantitative competitive reverse transcriptase PCR assay, which was in line with the increasing clinical severity.  Also using gelatin zymography, the authors demonstrated that the increase in mRNA was associated with an increase in MMP 9 activity. In contrast to the active disease model, MMP 7 levels declined in the recovery period. Expression was localised by immunohistochemistry to infiltrating mononuclear cells and to the perivascular space as well as within the parenchyma. The localisation of the protein was predominantly extracellular although the most likely source of the enzyme was reported to be from the infiltrating T cells and macrophages (Figure 1).

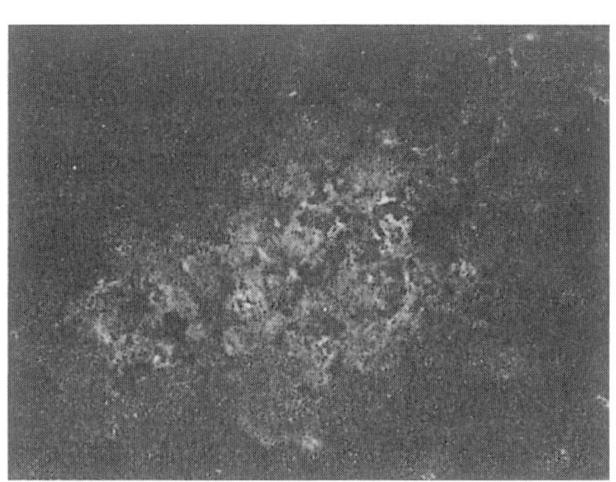

Figure 1. Expression of MMPs 2, 7 and 9 in perivascular inflammatory cuffs within the spinal cords of rats with experimental autoimmune encephalomyelitis.

MMP-2

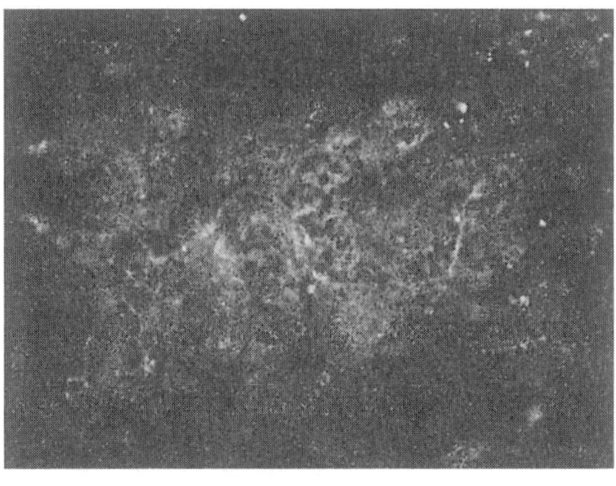

MMP-7

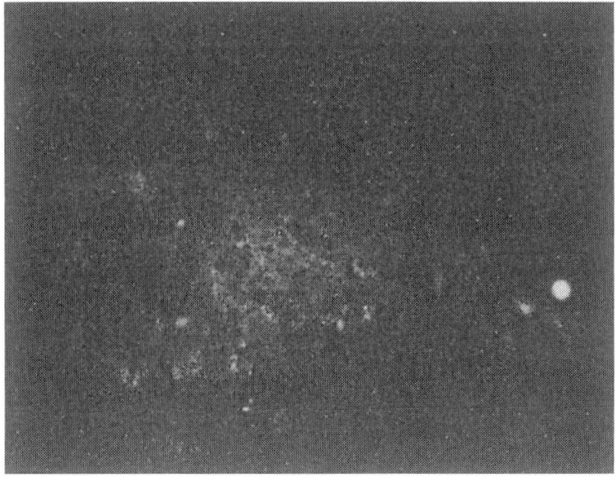

MMP-9

ADAM 17 expression in normal control rat brain parenchyma

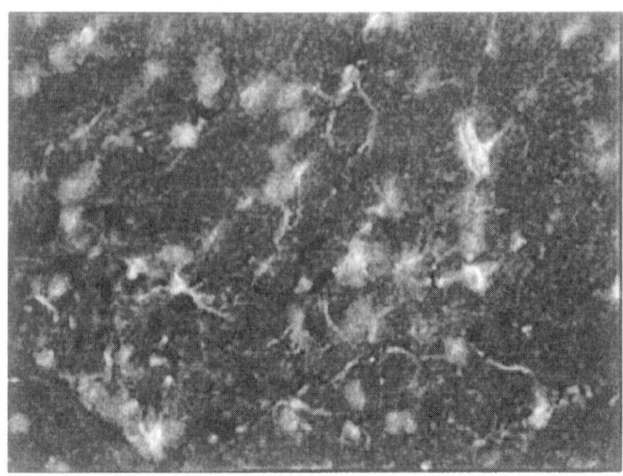

ADAM 17 (TACE) expression by perivascular astrocyctes in spinal cord of EAE rats

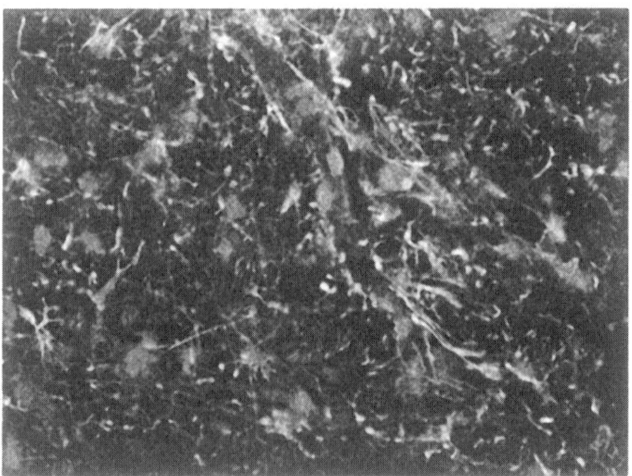

ADAM 17 expression by endothelial cells and astrocytes within spinal cord of EAE rats.

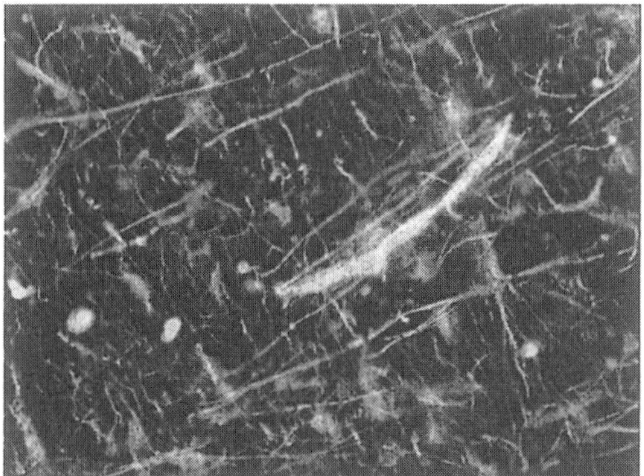

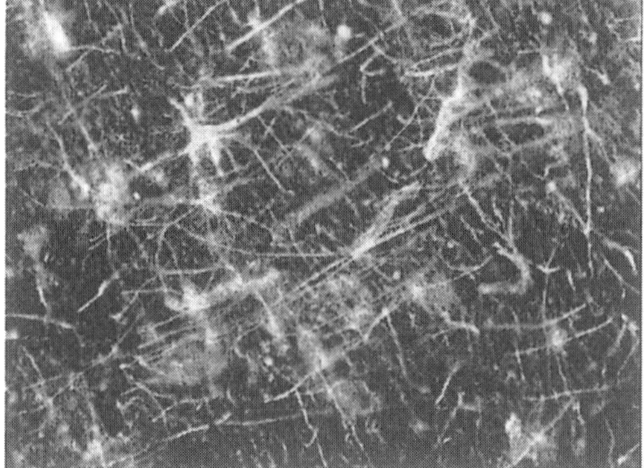

Frozen sections (10μm) were stained for the various MMPs and ADAM17 using monoclonal primary antibodies with an FITC labelled secondary antibody(green). Sections were counter stained with propidium iodide (red) to demonstrate cell nuclei (magnification x 400). The photographs were kindly provided by Mr Jonnie Plumb, Biomedical Research Centre, Sheffield Hallam University

MMPs 2, 3, 11 and 13 were unchanged during the course of the disease[28]. MMP9 activity is essential for degradation of type IV collagen within the basal lamina that surrounds blood vessels, thus local expression of this enzyme would permit an influx of inflammatory cells into the CNS across a damaged BBB. Sustained production of MMPs throughout the course of chronic EAE may be a key factor in the neuronal loss that occurs in the later stages[5].

The expression of nine different MMPs and three TIMPs was examined in normal SWR mice and mice with EAE. Upregulation of expression of three MMPs, MMP 3, MMP 9 and MMP 12, were the major changes reported[23]. In EAE animals an inverse relationship was seen between MMP expression and TIMP1 expression, the former was localised to leukocytes and possibly microglia, whereas TIMP-1 was restricted to the astrocytes surrounding the inflammatory lesions. The balance between the level of MMPs and TIMPs will determine the outcome of the inflammatory lesion. TIMPs 2 and 3 levels remained at constitutive levels in the choroid plexus and neurons.

The role of MMPs in induction of EAE in BALB/c mice with spinal cord homogenate, a strain more resistant to induction of EAE, has been assessed[29]. This is a monophasic disease with peak clinical scores at days 10-14. Although mRNA for MMP 9 was undetectable in the CNS in this model, enzyme activity was detected by in situ zymography in the meninges, around blood vessels and in the parenchyma in areas of inflammatory cell accumulation, particularly in the lumbar and cervical spinal cord, which was confirmed by gelatin zymography. MMP 2 was constitutively expressed in both control and EAE mice

Experiments undertaken on the role of MMP 2 in T cell extravasation in EAE demonstrated a role for alpha 4 integrin in the induction of MMP2[30]. This work agreed with the earlier study of Pagenstcher98 highlighting a role for inflammmatory cell production of MMPs, MT1-MMP, MMP 9 and MMP 12 and astrocytic cell production of the endogenous inhibitor TIMP-1. In a more recent study EAE was induced in BALB/c mice using a sonicated preparation of spinal cord homogenate with complete Freund's adjuvant. Both MMP 8 and 9 activity was demonstrated to correlate with disease severity[31]. MMP 8 has not been previously reported in sensitive strains and its expression was co-localised to neutrophil infiltration into the CNS, which is a unique feature of this model, particularly associated with cortex, cerebellum and submeningeal space of the spinal cord. MMP 9 activity was localised to perivascular infiltrates, in agreement with studies on sensitive strains. The source of MMP 9 appeared to be a result of release from stores within leukocytes, since mRNA for the protein was not detected. In contrast, increased TIMP-1 protein and mRNA were expressed by astrocytes in the surrounding tissue in the mice injected with sonicated antigen, possibly serving to prevent excessive tissue destruction by the released MMPs, as reported previously in SWR mice[23].

We are currently undertaking a study of the distribution of ADAM 17 in the spinal cord of Lewis rats with EAE. Preliminary results demonstrate expression of ADAM 17 in astrocytes and endothelial cells in both normal spinal cord and in animals with EAE, at a single time point (Figure 1). Studies are ongoing to examine the expression of the enzyme throughout the disease course as well as the relationship between the expression of TIMP-3 and ADAM 17, the ration of which will ultimately determine the level of ADAM 17 activity.

## MMP expression in transgenic mice with EAE

In a comparative study on EAE in SWR mice and mice transgenic for one of the cytokines, IL-3, IL-6 or TNFα linked to a GFAP promoter, there was increased expression of MMP 9, MMP 12 and MMP 3 mRNA with increasing disease score. MMP 12 was expressed by foamy macrophages and lymphocytes in the CNS in mice with EAE. There were striking similarities between the mice over expressing TNF in the CNS and mice with EAE except that MMP 2 was not expressed in EAE animals. The changes in level of mRNA were accompanied by increased enzyme activity and immune reactivity for the proteins and expression was localised to lesional areas and in particular infiltrating leukocytes and microglia[23]. In mice in which the gene for MMP 2 had been inactivated, EAE could still be induced, however in the MMP 9 knock out mouse there was resistance to induction of EAE, compared with age-matched controls. When EAE was actively induced in C57Bl/6 mice less than 4 weeks old, in which the MMP 9 gene had been deleted, the animals were less susceptible to EAE induced by spinal cord homogenate, and the mean disease score was significantly lower than the wild type controls, although this was not the case in adult KO mice[24].

In an analogous experimental approach  MBP specific T cells were genetically manipulated in vitro to over-express MMP 2 and injected into PL/J x SJL/J F1 mice, to determine the effect on induction of EAE. An enhanced T cell migration *in vivo* with this T cell line was reported and a hydroxamate inhibitor of  MMPs (GM6001 see below), administered daily i.p. from the day of induction of EAE, delayed T cell entry into the CNS parenchyma. However if administered 48 hours post adoptive transfer of T cells, the treatment had no effect on the disease course. Thus, the data obtained using adoptive transfer of T cells over-expressing MMP 2 provides evidence for a role for MMP 2 in the entry of T cells into the CNS. The results of the inhibitor studies in this model suggest that although autoreactive T cell expression of MMP 2 is important for T cell entry into the CNS, expression of MMPs or other enzymes by cells recruited to the inflammatory lesion are important in invasion of the cells into the parenchyma[30].

## Inhibition of MMP activity in EAE

There are a number of key reports on the use of synthetic inhibitors of MMPs to ameliorate the signs of EAE, indeed as early as 1982 protease inhibitors have been reported to ameliorate or treat EAE in animals models[32].

GM6001, a hydroxamate inhibitor, suppressed the development of EAE in SJL/J mice and reversed clinical symptoms when administered after disease onset in a dose dependent manner[33]. A reduction in MMP9 activity in the CSF of treated animals was noted. The beneficial effects of the drug were a result of decreased inflammation as a result of a restored BBB in treated animals, rather than any inhibitory effects on demyelination as this was not a feature of this EAE model. Since GM6001 is a broad spectrum inhibitor of MMPs, enzymes other than MMP 9 could be inhibited and these may also play a role in induction and progression of EAE. In a separate study, another hydroxamate inhibitor, Ro31-9790, reduced the severity of clinical signs in both active and passive transfer models of EAE in the Lewis rat, when administered on the day of disease induction or 3 days post induction. However in animals with more severe clinical signs the inhibitor was less effective in controlling the disease[34]. Similarly BB1101, a broad spectrum MMP inhibitor, reduced disease severity in Lewis rats with active EAE[27] and prevented onset of EAE and reversed acute disease in SJL/J mice[35].

Chronic relapsing EAE in SJL/J mice, induced by adoptive transfer, was attenuated following BB1101 administration and this agreed with histological evidence of reduced demyelination and glial scarring in treated animals. There was also a shift in cytokine profiles in these animals from pro-inflammatory, TNF, to anti inflammatory, IL-4, relative to untreated controls. The transcription of FasL was also down regulated by MMP inhibition suggesting a role in prevention of Fas induced cell death which has been proposed as a mechanism for oligodendrocyte death[35]. Fas ligand can be cleaved by MMP 7 resulting in soluble FasL which is a potent inducer of cell death by apoptosis[36]. Thus decreased MMP7 activity would lead to decreased apoptosis mediated via Fas-FasL. However recovery from the acute stage of EAE has been associated with the apoptotic deletion of autoreactive T cells[37]. These apparently contradictory functions of apoptosis in EAE need to be more fully understood before the effects of MMP 7 on the Fas system can be fully elucidated.

D-Penicillamine, an anti-rheumatic drug, was partially effective in treating EAE in SJL mice and in Biozzi AB/H mice. In chronic relapsing EAE in the Biozzi mouse model, treatment attenuated exacerbations even when the drug was administered after disease induction[38]. However, other investigators have questioned whether D-penicillamine is acting as a proteinase inhibitor in this model[35].

The use of Minocycline, a semi-synthetic analogue of tetracycline, has been assessed for its effectiveness in treatment of $MOG_{35-55}$ petide-induced EAE in C57BL/6 mice. Two forms of the disease were induced, mild (peptide alone) and severe (peptide plus pertussis toxin i.p.). The drug was administered daily ip from the day of MOG immunisation. Minocycline was found to delay the onset of the severe form of the disease and increased the time by 6-8 days for animals to become moribund, however treatment did not prevent the disease. In the milder form of the disease, the drug was effective in reducing the severity throughout the disease course. The beneficial effects of Minocycline were mediated by inhibition of MMP 9, although MMP 2 was also inhibited, the significance of this finding was unclear[39].

An antioxidant, alpha lipoic acid, has been used as a treatment for EAE and it was found to suppress EAE both clinically and histologically as well as being effective in reducing disease severity, following its administration after disease induction. Since the entry of T cells into the CNS was prevented by this treatment it was proposed that the effect was through the inhibitory actions of alpha lipoic acid on MMP9 which was inhibited in a dose dependent manner[40].

Due to the multiple effects of the MMP inhibitors tested to date it is unlikely that a single molecular mechanism will explain their actions in ameliorating or preventing EAE[35]. The beneficial effects of these broad spectrum MMP inhibitors therefore may be a result of preventing transmigration of cells at the BBB, inhibiting demyelination[1] and also reducing the level of active TNFα within the CNS, through inhibition of ADAM17, thus preventing exacerbation of the inflammatory process. It has been shown that TNFα exacerbates EAE and preventing its actions with specific antibodies or soluble receptors is beneficial in treatment of mice with EAE[22].

### The PA/Plasmin System

The PA/plasmin system has received much less attention than the MMPs in EAE, despite the interactions between the two systems and the potentially important role of plasmin, generated by plasminogen activation inEAE. Both tPA and uPA are produced as single chain proforms requiring activation to two chain forms by proteolytic cleavage by enzymes such as plasmin or cathepsin B, or in the case of tPA by binding to fibrin[41,42]. These enzymes differ in their extracellular location which influences their function. tPA binds to extracellular components whereas uPA is predominantly cell associated, bound to specific uPA receptors (uPARs)[42,43]. uPA is therefore thought to be involved in processes requiring pericellular proteolysis such as tissue remodelling and cell migration/invasion and tPA in fibrinolysis.

uPARs concentrate uPA on cell surfaces and enable more efficient plasminogen activation when plasminogen is also bound to the cell surface[44,45]. uPAR binding of pro-uPA also enhances its activation[44]. Plasmin generated at cell surfaces in this way is not susceptible to inhibition by its major inhibitor α2-antiplasmin[45]. The receptor binding of uPA may also facilitate other direct catalytic activities such as its ability to degrade fibronectin, activation of hepatocyte growth factor, an activity shared with tPA, and activation of procathepsin B[46,47,48]. Furthermore, uPARs can bind to vitronectin in the ECM promoting cell adhesion and migration[49] and signalling can occur via uPA receptors by association of receptor molecules with adaptor molecules such as integrins[50,51].

The activity of PAs and plasmin is controlled not only by their production as proenzymes requiring activation but also by the presence of endogenous inhibitors. The major inhibitor of plasmin is α2-antiplasmin and there are three specific inhibitors of PAs, PA inhibitors 1, 2 and 3 (PAI-1, 2 and 3) which bind only the active two chain forms of the enzymes[52].Two less specific inhibitors are the protease nexin and neuroserpin, which inhibit plasmin and thrombin and trypsin respectively, in addition to the Pas[53,54]. PAI-1, in addition to binding to PAs, also binds to vitronectin and can compete with uPARs for vitronectin binding, hence modulating cell adhesion and movement[49,50].

### The PA/plasmin system in EAE

An early insight into the role of the PA/plasmin system in EAE was obtained from investigations on the effects of a range of proteinase inhibitors on the development of disease. It was thought that a major source of these enzymes was activated macrophages within the inflammatory lesion[32,55,56]. These studies used an active model of EAE in Lewis rat. Inhibitor treatment was started on day 6 or 7 after induction and continued until around day 16 when the experiments were terminated. In these studies, inhibitors of PA and plasmin, aminomethylcyclohexane carboxylic acid (AMCA), ε-aminocaproic acid (EACA) and ρ-nitrophenylguanidinobenzoate (NPGB) gave protection against EAE in terms of reduction of paralysis and weight loss. Histologically only a slight reduction in infiltration by inflammatory cells was seen in rats treated with AMCA and EACA but the extent of infiltration and the degree of demyelination surrounding these cells was reduced. Interestingly trasylol, an inhibitor of PA and plasmin, exacerbated EAE in both studies. The effects of AMCA have also been investigated in a a passive transfer model of EAE in the Lewis rat. The inhibitor was started on day 2 after cell transfer and continued until the end of the experiment on day 9 or 10. Again clinical signs of EAE were reduced in the treated rats, however cellular infiltrates were still observed. These workers also studied the effect of AMCA on the permeability of the BBB using a double isotope method to study the distribution of $^{125}$I and $^{135}$I-labelled human serum albumin. They found a reduction in BBB permeability in AMCA treated rats compared with saline treated controls which appeared to be coincident with clincal improvement. They suggested that the BBB barrier permeability was the result of reduced fibrinolysis. Breakdown of fibrin, deposited on endothelial cells of CNS blood vessels as a result of inflammatory processes would, in the absence of AMCA, produce vasoactive peptides which could increase permeability of the BBB.

Effects of D-penicillamine, an inhibitor of MMP9 and tPA and uPA have been investigated in mouse models of acute and chronic relapsing EAE. Penicillamine inhibited disease development and attenuated disease recurrence[38]. It was suggested that penicillamine modulated MMP9 activity directly and indirectly via PA/plasmin mediated activation.

These observations provide circumstantial evidence for the involvement of the PA/plasmin system in the pathogenesis of EAE, however the inhibitors used may not be absolutely specific for PAs and plasmin and could act via other serine proteinases. More recent studies have investigated the induction of components of the PA/plasmin system in EAE. Koh and coworkers studied PA activity in normal rats and the Lewis rat cell transfer model of EAE[57]. Activity was determined using a fibrin slide technique on sections of spinal cord. Fibrin degradation which occurred only in the presence of plasminogen was taken as a measure of PA activity. Normal rats showed PA activity associated with vascular endothelial cells, particularly meningeal endothelial cells. In EAE, PA activity decreased in the early phase of disease, compared to normal rats and remained low during clinical disease and increased with remission. Little or no activity was associated with vessels containing inflammatory cells, however this was suggested to be an artefact resulting from cell binding of PA which limited its diffusion into the fibrin overlay. This drop in PA activity may be the result of:- (1) exhaustion of endothelial cell PA content in the presence of fibrin deposition in blood vessels at lesion sites; (2) release of PAI-2 by infiltrating monocyte-macrophages and (3) stimulation of PAI-1 production by endothelial cells in response to inflammatory cytokines, IL-1 and TNF. It was suggested that rather than promoting demyelination, infiltrating monocyte-macrophages may promote remission by down-regulating fibrinolysis and release of fibrin degradation products, some of which are chemotactic and may contribute to disruption of the BBB.

More recent studies have however shown induction of elements of the PA/pasmin system in EAE. Teesalu et al investigated the expression of tPA, uPA, PAI-1 and uPAR in spinal cord and brain from a BALB/c mouse model of EAE, induced with an emulsion of mouse spinal cord in complete Freund's adjuvant followed by an injection of pertussis toxin 24h later[29]. In situ hybridisation and zymography were used to detect gene expression of these components and PA activity respectively. In situ hybridisation in normal animals indicated little or no transcripts for tPA, uPA, PAI-1 or uPAR. Spinal cords from EAE mice showed tPA transcripts on the edge of the white matter in activated astrocytes. The presence of tPA protein in activated astrocytes was confirmed by immunohistochemistry. PAI-1 transcripts had a similar distribution to tPA and both were also increased in the brainstem and pons. uPA transcripts were undetectable, but uPAR transcripts[3] were present in immune cells in inflammatory lesions. This contrasts with  PA activity, determined by in situ zymography, which was detected in spinal cords, this was weak in control animals but increased in EAE and was present in areas of the dorsal horn, central spinal cord and along the meninges. A slight reduction of activity was noted in the presence of amiloride, an inhibitor of uPA, indicating the presence of both tPA and uPA, though tPA appears the predominant enzyme in this model.

Ahmed and coworkers, using a relapsing remitting model of EAE in Lewis rat, with relapses induced with IL-12, saw an increase in tPA, determined by immunostaining, during the second and third relapse[58]. It was localised to macrophages and activated microglia in inflammatory cuffs and microglia and axons in grey and white matter. Whereas in first relapse animals, tPA was confined to inflammatory cells in cuffs and microglia in the grey matter but not in axons. PAI-levels, also determined by immunostaining, appeared to decrease as tPA increased.

There are a number of speculations about the role of the PA/plasmin system in these latter two models[29,58]. The increase in tPA may be both detrimental and beneficial. tPA/plasmin mediated ECM degradation may lead to damage to the BBB and neuronal damage[36]. Neuronal damage and demyelination is more evident in relapsing remitting models of EAE. tPA is thought to promote neuronal damage by mediating the degradation of laminin, the loss of substratum causing cell death via apoptosis[12,59] and by the activation of microglia via a non-proteolytic mechanism[60]. Activation of protease-activated receptors (PARs) by serine proteinases such as plasmin, leading to astrocyte proliferation and microglial activation may also contribute to tissue damage[61,62,63].

tPA may have beneficial effects via tPA mediated degradation of fibrin deposits in areas of perivascular inflammation. The clearance of fibrin may limit the extent of immune cell infiltration by removing a substrate for immune cell adhesion and invasion[29]. tPA mediated fibrinolysis has also been shown to reduce axonal degeneration and demyelination after damage[64]. Activation of TGF-β by the PA/plasmin system may also be beneficial, reducing the production of NO by microglia which is toxic to microglia and oligodendrocytes[20]. uPA, though less evident in these models of EAE, may contribute to plasmin generation and its consequences. uPARs could promote adhesion and movement of inflammatory cells on perivascular deposits of vitronectin. PAI-1, in addition to modulating PA activity, modulates uPAR mediated cell adhesion to vitronectin.

The evidence available strongly suggests a role for the PA/plasmin system in EAE, though this would appear to differ in different models and is complicated by the range of effects mediated by this system which may exacerbate or alleviate EAE.

## <u>Myelencephalon-specific protease</u>

In addition to the PA system proteinases, another trypsin-like serine proteinase, rat myelencephalon-specific protease (MSP), (human homologue, kallikrein 6), has been implicated in demyelination in EAE and MS[65]. MSP is more highly expressed in the CNS compared to other non-neuronal tissues where it is present in neurons and oligodendroglia[66]. In a marmoset model of EAE, MSP was found at lesion sites in infiltrating mononuclear cells where it may be secreted from these cells and activated by proteolysis[65,67].

MSP has a broad substrate specificity including MBP, MOG and ECM components laminin, fibronectin, and collagen[65,67]. In EAE, inflammatory cell MSP may therefore contribute to demyelination and immunogenic peptide production and may facilitate their infiltration into the CNS by degradation of the BBB basal lamina. Addition of MSP to oligodendrocytes in vitro has also been shown to interfere with process formation, reducing formation in immature cells and causing loss of processes in mature cells. Such effects may be detrimental to repair processes[65]. It has been suggested that the predominant CNS distribution of MSP, makes it a good potential therapeutic target for MS, however it requires further characterisation in terms of its regulation.

# <u>CALCIUM-ACTIVATED NEUTRAL PROTEINASE</u>

Calcium-activated neutral proteinase (calpain) is another enzyme thought to have a role in the pathogenesis of EAE[68]. Calpain is a cytosolic cysteine endopeptidase, active at neutral pH and with activities against a range of proteins including neurofilament protein, myelin and its own endogenous inhibitor calpastatin[69]. There are two main subclasses of calpain, ubiquitous and tissue specific[70] and it is the ubiquitous subclass which has been implicated in EAE[68,71]. Ubiquitous calpain has microcalpain and millicalpain isoforms (EC 3.4.22.17) so called because they are activated by μM or mM concentrations of calcium respectively.

Calpain level and activity and calpastatin levels were studied in the spinal cords of Lewis rats, 10 – 12 days after induction of EAE by immunisation with guinea pig MBP. Calpain, determined as millicalpain, and calpastatin levels were increased in EAE as was the activity of calpain, determined indirectly by the presence of calpain-specific breakdown products of neurofilament protein and myelin associated glycoprotein. There was no significant increase in mRNA for calpain or calpastatin, suggesting that their regulation was at the post-translational level[68]. In view of its substrate specificity, calpain is thought to be involved in demyelination and axonal degeneration via breakdown of neurofilaments[72,73]. Increased expression in activated glial and inflammatory cells in EAE support this[71]. Its degradation of MBP is only partial but may lead to the production of immunogenic peptides, exacerbating EAE.

In order to degrade myelin or neurofilaments calpain requires activation by calcium. Activation of either of the calpain isoforms intracellularly would require a calcium influx as intracellular calcium is ~ 100 nM. In EAE this could result from complement fixation at the cell surface or perforin damage[74,75] and would enable intracellular degradation of myelin in the myelin sheath. Alternatively calpain may be secreted and function extracellularly where calcium levels are ~2 mM. Calpain can be secreted from activated macrophages and T cells[76,77,78].

Calpain is also thought to be involved in T cell migration from the spleen to the brain in EAE, promoting disease progression. Calpain activity and levels are increased in spleen T cells and macrophages in developing EAE[79]. However this may not be specific to EAE, but may be a more generalised response to immunisation as it was also seen in rats given complete Freund's adjuvant containing *Mycobacterium tuberculosis* but no MBP. Calpain may facilitate T cell migration by disruption of cytoskeleton/integrin links enabling clustering of LFA-1 and enhanced binding to adhesion molecules[80,81]. This may be important in enabling the passage of T cells across the BBB.

### Caspases

Caspases are involved in apoptosis and cytokine activation[82]. Their role in apoptosis is discussed in detail in the chapter by Gold and Chan. Caspase-1 (EC 3.4.22.36) is involved in the activation of IL-1β and was formerly known as IL-1β converting enzyme (ICE). It is therefore a key early component in inflammatory processes. Caspase-1 has been shown to be induced in a mouse model of EAE, induced by immunisation with incomplete Freund's adjuvant supplemented with *Mycobacterium tuberculosis* and MOG peptide 35-55 or spinal cord homogenate[83]. Transcription levels of caspase-1 correlated with the severity of disease and the transcription rates of pro-inflammatory cytokines such as IL-1β and TNFα. The caspase inhibitor, Z-Val-Ala-DL-Asp-fluoromethylketone reduced the incidence of EAE in mice, when administered from the time of immunisation However when administered 7 days post immunisation, it had little therapeutic effect. Caspase-1 null mice showed a reduced severity and susceptibility to induction of EAE. These results indicated the importance of caspase-1 in the early stages of development of EAE[83].

## **Mast cell proteinases**

Early work showed that activated mast cells produced proteinases capable of degrading rat myelin to produce an encephalitogenic peptide. Immunisation of Lewis rats with this peptide or adoptive transfer of T cell lines stimulated with the peptide produced EAE[84]. Mast cells have been localised in MS brain, in normal rat brain and in brain from a Lewis rat model of EAE. In the CNS of EAE rats there was an apparent increase in mucosal-type mast cells fairly late on in the disease process, 13 days post immunisation. This was determined by the mRNA expression and immunoreactivity of rat mast cell protease II (RMCPII) (EC ), a marker for this type of mast cell. The RMCPII content of these cells was also increased. These mast cells were localised adjacent to inflamed blood vessels[85]. Mast cells and their proteinases may therefore contribute to EAE via demyelination and the generation of immunogenic peptides.

## **Peptidases**

There have been few studies on the role of peptidases in EAE. Two peptidases implicated in the disease process are angiotensin converting enzyme I (ACE) (EC 3.4.15.1) and dipeptidyl peptidase IV (DPIV)(CD 26) (EC 3.4.14.5). ACE is a membrane bound peptidase which mediates inflammation, participates in T cell activation via its generation of angiotensin II and has effects on the permeability of the BBB[86,87,88,89]. Evidence for the involvement of ACE in EAE comes from a study of the effects of an ACE inhibitor, captopril, in a Lewis rat model of EAE. It was shown that when captopril was administered from the day of immunisation, it reduced both disease severity and the  in vitro responsiveness of lymphocytes from captopril treated rats to MBP[90].

DPIV is also a membrane bound peptidase which is present on T cells and is involved in T cell activation and function[91]. It has been investigated in a mouse model of EAE induced by adoptive transfer of lymph node cells from mice immunised with either guinea pig  MBP or peptide 139-151 from PLP[92]. The effects of DPIV inhibitor Lys[Z(NO$_2$)]-pyrrolidide (I40) were studied. When administered from the day of cell transfer a reduced incidence and severity of  disease was observed. It was also effective in reducing disease severity if administered following development of clinical signs which agreed with histological evidence of reduced inflammatory lesions and myelin loss. In vitro studies showed a reduction in proliferation and pro-inflammatory cytokine secretion by MBP primed lymph node cells and an increase in TGFβ1 secretion. Thus it was suggested that the effects were mediated by TGFβ1 production, which in turn suppressed disease[93].

Thus DPIV appears to have an important role in the activation and function of T cells in EAE and could constitute an important new therapeutic target.


## SUMMARY

The evidence reviewed here provides strong support for an effector function for proteases and peptidases in the pathogenesis of EAE in a range of rodent species using both active and passive disease transfer models as well as relapsing remitting forms. Inhibition of these enzymes has resulted in amelioration or prevention of the disease. Effective inhibition in some cases required administration before disease onset, whereas others were effective when given after disease onset. The PA/plasmin/MMP enzymes have been most specifically identified as the major contributors to the pathological process including increasing the permeability of the BBB, degradation of the myelin sheath resulting in axonal damage, generation of active pro-inflammatory cytokines by cleavage of the pro forms, further propagating inflammation within the CNS and finally generation of further encephalogenic peptides which may promote the autoimmune. However other enzymes such as MSP may provide more specific potential targets for new drug development.


**REFERENCE LIST:**

1.  Cuzner ML, Opdenakker G (1999). Plasminogen activators and matrix metalloproteases, mediators of extracellular proteolysis in inflammatory demyelination of the central nervous system. J of Neuroimmunol, 94:1-14

2.  Rosenberg GA (2002). Matrix metalloproteinases in neuroinflammation. GLIA, 39: 279-291

3.  Chandler S, Miller KM, Clements JM, Lury J, Corkill D, Anthony DC, Adams SE, Gearing AJ (1997). Matrix metalloproteinases, tumor necrosis factor and multiple sclerosis: an overview. J of Neuro, 72: 155-161

4.  Yong V W, Krekoski CA, Forsythe PA, Bell R and Edwards DR (1998). Matrix metalloproteinases and diseases of the CNS. Trends Neurosci 21, 75-80.

5.  Leppert D, Lindberg RL, Kappos L, Leib SL (2001). Matrix metalloproteinases: multifunctional effectors of inflammation in multiple sclerosis and bacterial meningitis. Brain Res Revs, 36: 249-257

6.  Yong VW, Power C, Forsyth P and Edwards DR. (2001). Metalloproteinases in biology and pathology of the nervous system. Nature Reviews, Neurosci 2,:502-511

7.  Cross AK, Woodroofe MN (1999). Chemokine modulation of matrix metalloproteinase and TIMP production in adult rat brain miroglia and a human microglial cell line in vitro. GLIA, 28: 183-189

8.  Andreasen PA, Kjoller L, Christensen L, Duffy M (1997). The urokinase-type plasminogen activator system in cancer metastasis: A review. Int J of Cancer

72(1): 1-22

9. Cammer W, Bloom BR, Norton WT and Gordon S (1978). Degradation of basic protein in myelin by neutral proteases secreted by stimulated macrophages: A possible mechanism of inflammatory demyelination. Proc Natl Acad Sci USA, 75, : 1554-1558

10. Lijnen HR (2001). Elements of the fibrinolytic system. Ann of NY Ac of Sci, 936:226-236

11. Liotta LA, Goldfarb RH, Brundage R, Siegal GP, Terranova V, Garbisa S (1981). Effect of plasminogen-activator (urokinase), plasmin and thrombin on glycoprotein and collagenous components of basement-membrane. Cancer Research, 41: 4629-4636

12. Chen Z, Strickland S (1997). Neuronal death in the hippocampus is promoted by plasmin-catalyzed degradation of laminin. Cell, 91: 917-925.

13. Gunderson D, Trân-Thang C, Sordat B, Mourali F and Rüegg C (1997). Modulation by Tlymphocyte-derived urokinase-type Plasminogen activator and effect on T Lymphocyte adhesion, activation and cell clustering. J of Immunol, 158: 1051-1060.

14. Yamaguchi Y (2000). Lecticans: organiser of brain extracellular matrix. Cell and Mol Life Sci 57: 276-289

15. Sobel RA (2001). The extracellular matrix in multiple sclerosis: an update. Braz J Med Biol Res, 34: 603-609

16. Mochan E, Keler T (1984). Plasmin degradation of cartilage proteoglycan. Biochim Biphys Acta 800: 312-315

17. Ramos-DeSimone N, Hahn-Dantona E, Sipley J, Nagase H, French DL, Quigley JP (1999). Activation of matrix metalloproteinase-9 (MMP-9) via a converging plasmin/stromelysin-1 cascade enhances tumor cell invasion. J of Biol Chem, 274(19): 13066-13076

18. Werb Z, Mainardi CL, Vater CA, Harris ED (1977). Endogenous activation of latent collagenase by rheumatoid synovial cells. Evidence for a role of plasminogen activator. New Engl J Med, 296: 1017-1023.

19. Baramova EN, Najou K, Remacle A, Lhoir C, Krell HW, Weidel UH, Noel A, Foidart JM (1997). FEBS Letters, 405: 157-162

20. Vincent VA, Lowik CW, Verheijen JH, de Bart AC, Tilders FJ, Van Dam AM (1998). Role of astrocyte derived tissue - type plasminogen activator in the regulation of enxotoxin-stimulated nitric oxide production by microglial cells. Glia, 22: 130-137

21. Hartung HP, Kieseier B (2000). The role of matrix metalloproteinases in autoimmune damage to the central and peripheral nervous system. J Neuroimmunol, 107: 140-147

22. Baker D, Butler D, Scallon BJ, O'Neill JK, Turk JL, Feldmann M (1994). Control of established experimental allergic encephalomyelitis by inhibition of tumor necrosis factor (TNF) activity within the central nervous system using moclonal antibodies and TNF receptor-immunoglobulin fusion proteins. Euro J of Imm, 24: 2040-2048

23. Pagenstecher A, Stalder AK, Kincaid CL, Shapiro SD and Campbell IL (1998). Differential expression of matrix metalloproteinase and tissue inhibitor of matrix metalloproteinase genes in the mouse central nervous system in normal and inflammatory states. Am Journal of Pathology, 152, 52: 729-741

24. Dubois B, Masure S, Hurtenbach U, Paemen L, Heremans H, Van den Oord J, Sciot R, Meinhardt T, Hämerling G, Opdenakker G and Arnold B (1999). Resistance of young gelatinase B-deficient mice to experimental autoimmune encephalomyelitis and necrotizing tail lesions. J Clin Investig, 104: 1507-1515

25. Opdenakker G, Van den Steen PE, Van Damme J (2001). Gelatinase B: a tuner

and amplifier of immune functions, Trends in Immuno, 22: 571-579

26.  Gijbels K, Proost P, Masure S, Carton H, Billian H and Opdenakker G (1993). Geltanise ß is present in the cerebrospinal fluid during EAE and cleaves MBP. J Neurosci Res 36, 432-440

27.  Clements JM, Cossins JA, Wells GM, Corkill DJ, Helfrich K, Wood LM, Pigott R, Stabler G, Ward GA, Gearing et al (1997). Matrix metalloproteinase expression during experimental autoimmune encephalomyelitis and effects of a combined matrix metalloproteinase and tumour necrosis factor-alpha inhibitor. J of Neuroimm, 74: 85-94

28.  Kieseier BC, Storch MK, Hartung HP (2000). Toxic effector molecules in the pathogenesis of immune-mediated disorders of the central nervous system. Journal of Neural Transmission, 59; 69-80

29.  Teesalu T, Einkkanen AE and Vaheri A (2001). Coordinated induction of extracellular proteolysis systems during experimental autoimmune encephalomyelitis in mice. Am J Pathol 159: 2227-2237

30.  Graesser D, Mahooti S, Madri J (2000). Distinct roles for matrix metalloproteinase-2 and α4 integrin in autoimmune T cell extravasation and residency in brain parenchyma during experimental autoimmune encephalomyelitis. J Neuroimmunol 109: 121-131

31.  Nygardas PT and Hinkkanen AE (2002). Up-regulation of MMP-8 and MMP-9 in the BALB/c mouse spinal cord correlates with the severity of experimental autoimmune encephalomyelitis. Clin Exp Immunol, 128: 245-254

32.  Smith ME, Amaducci LA (1982). Observations on the effects of inhibitors on the suppression of experimental allergic encephalomyelitis. Neurochem Res, 7: 541-554

33.  Smith ME, Van der Maesen K and Somera FP (1998). Macrophage and Microglial responses to cytokines in vitro: phagocytic activity, proteolytic enzyme release and free radical production. J Neurosc Res, 54: 68-78

34.  Hewson AK, Smith T, Leonard JP, Cuzner ML (1995). Suppression of experimental allergic encephalomyelitis in the Lewis rat by the matrix metalloproteinase inhibitor Ro31-9790. Official Journal of the European Histamine Res Soc, Vol 44, : 345-349

35.  Liedtke W, Cannella B, Mazzaccaro RJ, Clements JM, Miller KM, Wucherpfennig KW, Gearing AJ, Raine CS (1998). Effective treatment of models of multiple sclerosis by matrix metalloproteinase inhibitors. Annals of Neurology,1: 35-46.

36.  Lo EG, Wang X, Cuzner ML (2002). Extracellular proteolysis in brain injury and inflammation: Role for plasminogen activators and matrix metalloproteinases. J of Neuroscience Res, 69:1-9

37.  Pender MP, Nguyen KB, McCombe PA, Kerr JF (1991). Apoptosis in the nervous system in experimental autoimmune encephalomyelitis. Neurosci Ress, 104: 81-87

38.  Norga K, Paemen L Masure S, Dillen C, Heremans H, Billiau A, Carton H, Cuzner L, Olsson T, Van Damme J (1995). Prevention of acute autoimmune encephalomyelitis and abrogation of relapses in murine models of multiple sclerosis by the protease inhibitor D-pencillamine. Official Journal of the European Histamine Res Soc., 44: 529-534

39.  Brundula V, Rewcastle NB, Metz LM, Bernard CC and Yong VW (2002). Targeting leukocyte MMPs and transmigration Minocycline as a potential therapy for multiple sclerosis. Brain, 125: 1297-1308

40.  Marracci GH, Jones RE, McKeon GP, Boudette DN (2002). Alpha lipoic acid inhibits T cell migration into the spinal cord and suppresses and treats experimental autoimmune encephalomyelitis. Journal of Neuroimmuno 131;

104-114

41.  Hart DA, Rehemtulla A (1988). Plasminogen activators and their inhibitors: regulators of extracellular proteolysis and cell function. Comp Biochem Physiol 90B: 691-708

42.  Oates JA, Wood JJ (1988). Tissue plasminogen activator. New Eng J Med, 319: 925-931

43.  Blasi F (1993), Urokinase and urokinase receptor: a paracrine/autocrine system regulating cell migration and invasiveness. Bio Essays 15: 105-111.

44.  Ellis V, Behrendt N, Dano K (1991). Plasminogen activation by receptor-bound urokinase, a kinetic study with both cell associated and isolated receptor. J Biol Chem 266: 12752-12758

45.  Plow EF, Freaney DE, Plescia J, Miles LA (1986). The plasminogen system and cell surfaces: evidence for plasminogen and urokinase receptors on the same cell type. J Cell Biol 103: 2411-2420.

46   Gold LI, Schwimmer R, Quigley JP (1989). Human fibronectin as a substrate for human urokinase. Biochem J, 262: 529-534

47.  Dalet-Fumeron V, Guinec N, Pagano M (1993). In vitro activation of pro-cathepsin B by 3 serine proteinases, leukocyte elastase, cathepsin G and the urokinase-type plasminogen activator.  FEBS Lett 332: 251-254

48.  Mars WM, Zarnegar R, Michalopoulos GK (1993). Activation of hepatocyte growth factor by the plasminogen activators uPA and tPA. Amer J Pathol 143,: 949-958.

49.  Blasi F (1997). UPA, UPAR, PAI-1: key intersection of proteolytic, adhesive and chemotactic highways? Immunol Today, 18: 415-417

50.  Chapman A (1997). Plasminogen activators, integrins and the coordinated regulations of cell adhesion and migration. Curr Opin Cell Biol 9: 714-724

51.  Dummler I, Weis A, Mayboroda OA, Maasch C, Jerke U, Hallert H and Gulba C (1998).  The Jak/Stat pathway and urokinase receptor signalling in human aortic vascular smooth muscle cells. J Biol Chem, Vol 273, I

52.  Romanaic AM and Madri J (1994). Extracellular matrix-degrading proteinases in the nervous system. Brain Path 4: 145-156

53.  Sprengers ED and Kluft C (1987). Plasminogen activator inhibitors. Blood 69: 381-387.

54.  Silverman GA, Bird PI, Carrell RW, Church FC, Coughlin PB, Gettins PGW, Irving JA, Lomas DA, Luke CJ,Moyer RW, Pemberton PA, Remold-O'Donnell E, Salveson GS Travis J and Whisstock JC (2001). The Serpins are an expanding superfamily of structurally similar but functionally diverse proteins. J Biol Chem, 276,33293-33296.

55.  Brosnan CF, Cammer W, Norton WT and Bloom BR (1980). Proteinase inhibitors suppress the development of experimental allergic encephalomyelitis. Nature, 285: 235-237

56.  Koh CS, Paterson PY (1987).  Suppression of clinical signs of cell-transferred experimental allergic encephalomyelitis and altered cerebrovascular permeability in Lewis rats treated with a plasminogen activator inhibitor. Cell Imm,107: 52-63

57.  Koh CS, Kwaan HC, Paterson PY (1990).  Neurovascular fibrinolytic activity in normal Lewis rats and rats with cell-transferred experimental allergic encephalomyelitis. J Neuroimmuno, 28, :189-200

58.  Ahmed Z, Gveric D, Pryce G, Baker D, Leonard JP and Cuzner ML (2001). Myelin/Axonal Pathology in Interleukin-12 Induced serial relapses of experimental Allergic Encephalomyelitis in the Lewis rat. Am J of Pathology, Vol 158,:2127-2138

59.  Flavin MP, Zhao G and Ho LT (2000). Microglial tissue plasminogen activator

(TPA) triggers neuronal apoptosis In vitro. GLIA 29: 347-354.

60. Siao C-J, Tsirka SE (2002). Tissue plasminogen activator mediates microglial activation via its finger domain through annexin II. J of Neurosci, 22: 3352-3358

61. Gingrich MB and Traynelis (2000). Serine proteases and brain damage - is there a link? TINS 23 (9): 39 9-407.

62. Wang H, Joachim U, Reiser G (2002). Four subtypes of protease-activated receptors, co-expressed in rat astrocytes, evoke different physiological signaling. Glia 37: 53-63

63. Suo Z, Wu M, Ameenuddin S, Anderson HE, Zoloty JE, Citron BA, Andrade-Gordon P, Festoff BW (2002). Participation of protease-activated receptor-1 in thrombin-induced microglial activation. J of Neurochem, 80: 655-666.

64. Akassoglou K, Kombrinck KW, Degen JL, Strickland S (2000). Tissue plasminogen activator-mediated fibrinolysis protects against axonal degeneration and demyelination after sciatic nerve injury. J of Cell Biol, 149: 1157-1166

65. Scarisbrick IA, Blaber SI, Lucchinetti CF, Genain CP, Blaber M and Rodriguez M (2002). Activity of a newly identified serine protease in CNS demyelination. Brain, 125: 1283-1296

66. Scarisbrick I, Isackson PJ, Ciric B, Windebank AJ, Rodriguez M (2001). MSP; a trypsin-like serine protease. J Comp Neurol 431: 347-361

67. Blaber, SI, Scarisbrick IA, Bernett MJ, Dhanarajan P, Seavy M, Yonghao J, Schwartz MA, Moses R, Blaber M (2002). Enzymatic properties of rat myelencephalon-specific protease. Biochemistry, 41: 1165-1173.

68. Shields DC, Banik NL (1998). Upregulation of calpain activity and expression in experimental allergic encephalomy elitis: a putative role for calpain in demyelination. Brain Res, 794: 68-74

69. Suzuki K, Imajob S, Emori Y, Kawasaki H, Minami Y, Ohno S (1987). Calcium-activated neutral protease and its endogenous inhibitor. FEBS Letts 220: 271-277

70. Suzuki K, Sorimachi H, Yoshizawa T, Kinbara K, Ishiura S (1995). Calpain: novel family members activation and physiological function. J Biol Chem 376: 523-529

71. Shields DC, Tyor WR, Deibler GE, Hogan EK and Banik NL (1998). Increased calpain expression in activated glial and inflammatory cells in experimental allergic encephalomyelitis. Neurobiology, 95: 5768-5772.

72. Shields DC, Banilk NL (1999(a)). A Pathophysiological role of calpain in experimental demyelination. J of Neuroscience Res, 55: 533-541

73. Schaecher KE, Shields DC, Banik NL (2001). Mechanism of myelin breakdown in experimental demyelination: A putative role for calpain. Neurochem Res 26:731-737

74. Schaecher KE, Shields DC, Banik NL (2001). Mechanism of myelin breakdown in experimental demyelination: A putative role for calpain. Neurochem Res 26:731-737

75. Zajicek T, Wing M, Scolding W, Compston D (1992). Interactions between oligodenrocytes and microglia: A major role for complement and tumour necrosis factor in oligodendrocyte adherence and killing. Brain 115: 1611-1631

76. Saido T, Suzuki H, Yamazalci H, Tanoue K, Suzuki K (1993). J Biol Chem 268: 7422-7426

77. Deshpande R, Goust T, Hogan G, Banik NL (1995). Calpain secreted by activated human lymphoid cells degrades myelin.J Neurosci Res 42: 259-265

78. Smith ME, Van der Maesen K and Somera FP (1998). Macrophage and Microglial responses to cytokines in vitro: phagocytic activity, proteolytic

enzyme release and free radical production. J Neurosc Res, 54: 68-78

79.  Shields DC, Schaecher KE, Goust J-M, Banik NL (1999(b)). Calpain activity and expression are increased in splenic inflammatory cells associated with experimental allergic encephalomyelitis. J Neuroimmunol, 99:1-12

80.  Potter D, Tirnauer T, Janssen R, Croall D, Hughes C, Fiacco K, Mier T, Maki M, Herman I (1988). Calpain regulates actin remodelling ? cell spreading. T Cell Biol 141: 647-662

81.  Stewart MP, McDowall A, Hogg N (1998). LFA-1-Mediated adhesion is regulated by cytoskeletal restraint and by a $Ca^{2+}$-dependent protease, calpain. J Cell Biol 140: 699-707

82.  Salvesen GS and Dixit VM (1997). Caspases: Intracellular Signalling by Proteolysis. Cell 91: 443-446.

83.  Furlan R, Martino G, Galbiati F, Poliani PL, Smiroldo S, Bergami A, Desina G, Comi G, Flavell R, Michael SS and Ardorini L (1999).  Caspase-1 regulates the inflammatory process leading to autoimmune demyelination[1]. J Immunol, 163: 2403-2409.

84.  Dietsch GN, Hinrichs DJ (1991). Mast cell proteases liberate stable encephalitogenic fragments from intact myelin. Cell Immm 135: 541-548

85.  Rouleau A, Dimitriadou V, Tuong MDT, Newlands GFJ, Miller HRP, Schwartz JC, Garbarg M (1997).  Mast cell specific proteases in rat brain: change in rats with experimental allergic encephalomyelitis. J Neural Transm 104: 399-417

86.  Foris G, Dezsö B, Medgyesi GA, Füst G (1983). effect of angiotensin II on macrophage junctions.  Immunology 48: 529-535

87.  Weinstock JV, Ehrinpreis MN, Boros DL, Gee JB (1981). Effect of S$\theta$14225, an inhibitor of angiotensin I converting enzyme, on the granulomatons response to *Schisto sand man sani eggs in mice.* J Clin Invest 67: 931-936.


88.  Eisenlohr LC, Bacik I, Bennink JR, Bernstein K, Yewdell JW (1992). Expression of a membrane protease enhances prsentation of endogenous anFigens to MHC class I - restricted T lymphocytes. Cell 71: 963-972.


89.  Brownlees T, Williams CH (1993). Peptidases, peptides and the mammalian blood brain barrier. J Neurochem 60: 793-803

90.  Constantinescu CS, Ventura E, Hilliard B, Rostami A (1995). Effects of the angiotensin converting enzyme inhibitor captopril on experimental autoimmune encephalomyelitis. Immunopharma & Immunotoxi, (3): 471-491


91.  Masimoto C, Schlossman SF (1998). The structure and function of CD26 in the T cell immune response.  Immunol Rev 161: 55

92.  Steinbrecher A, Reinhold D, Quigley D, Gado A, Tresser N, Izikson L, Born I, Faust J, Neubert K, Martin R, Ansorge S and Brocke S (2001). Targeting dipeptidyl peptidise IV (CD26) suppresses autoimmune encephalomyelitis and up-regulates TGF-ß1 secretion In vivo. J of Immunol, 166: 2041-2048


93.  Racke MK, Dhib-Jalbut S, Cannella B, Albert PS, Raine CS, McFarlin DE (1991). Prevention and treatment of chronic relapsing experimental allergic encephalomyelitis by transforming growth factor-beta 1. J of Immuno, 146: 3012-3017

Chapter A20

# THE BLOOD-BRAIN BARRIER IN EAE

Britta Engelhardt[1] and Hartwig Wolburg[2]
*[1]Theodor-Kocher Institute, Bern, Switzerland and [2]Institute for Pathology, University of Tübingen, Germany*

**Abstract:** The normal function of the blood brain barrier (BBB) is the maintenance of the central nervous system homeostasis. The normal structure, development, differentiation and maintenance (and in particular the role of tight junctions) of the BBB are discussed in this chapter, followed by a description of the pathological changes in the BBB during EAE: leakiness, molecular changes in the tight junctions, the crossing of encephalitogenic T cells and the immunological factors responsible for the BBB breakdown during EAE.

**Key words:** EAE, cerebrospinal fluid, blood brain barrier, tight junctions, endothelium

## 1. INTRODUCTION

Homeostasis of the central nervous system (CNS) microenvironment is essential for its normal function and is maintained by the blood-brain barrier (BBB). The BBB is formed by highly specialized capillary endothelial cells, which inhibit transcellular passage of molecules across the barrier by an extremely low pinocytotic activity. The lack of fenestrae and an elaborate network of complex tight junctions (TJ) between the endothelial cells restrict the paracellular diffusion of hydrophilic molecules. On the other hand, in order to meet the high metabolic needs of the CNS tissue, specific transport systems selectively expressed in the capillary brain endothelial cell membranes mediate the directed transport of nutrients into the CNS or of toxic metabolites out of the CNS.

Because of the presence of the BBB, the lack of lymphatic vessels and the absence of classical MHC-positive antigen presenting cells the CNS is considered an immunologically privileged site. In fact, under physiological conditions lymphocyte entry into the healthy CNS across the BBB is kept at a low level. During inflammatory diseases of the CNS such as in multiple sclerosis (MS) or its prototype animal model experimental autoimmune encephalomyelitis (EAE), however, circulating immunocompetent cells readily get access to the CNS. In EAE, CD4$^+$ autoaggressive T cells are activated outside the CNS, migrate across the healthy BBB into the CNS and start the molecular events leading to inflammation, loss of barrier properties and subsequently edema formation and finally demyelination.

During inflammatory demyelinating diseases the BBB might be considered as performing both, a passive and an active role. Loss of barrier function during EAE might be considered as a passive involvement of the BBB leading to edema formation and exacerbation of the disease. In contrast, the BBB is actively involved in such disease processes as the altered phenotype of BBB endothelium during EAE that is characterized by enhanced expression of traffic signals like adhesion molecules and chemokines leads to enhanced leukocyte traffic into the CNS and consecutively chronic inflammation. In this chapter we will describe the molecular mechanisms involved in altered BBB function during EAE and consider their impact on the initiation and progression of the disease process.

## 2. THE PHYSIOLOGICAL BLOOD-BRAIN BARRIER

### 2. 1 History

The discovery of a vascular barrier between blood circulation and the CNS dates back more than 100 years, when in the 1880s Paul Ehrlich and Edwin E. Goldman discovered that upon the injection of protein-bound dyes into the vascular system they were rapidly taken up by all organs with the exception of the brain and spinal cord (Ehrlich, 1904), whereas the very same dyes when injected into the cerebrospinal fluid readily stained neural tissue but were excluded from all the other organs (Goldmann, 1913). Obviously these dyes could not penetrate the wall of CNS blood vessel. The concept of a vascular blood-brain barrier, which also functions as a brain-blood barrier was born. It took an additional 70 years until Reese, Karnovsky and Brightman (Reese and Karnovsky, 1967; Brightman, 1968) localized the barrier ultrastructurally to the capillary endothelial cells within the brain by electron microscopy studies. Following injection into the vasculature or into the CNS, the electron-dense tracer horseradish peroxidase distributed into the intercellular clefts of brain endothelial cells up to the tight junctions (TJs) between the endothelial cells. Thus, the interendothelial TJs were recognized as the morphological correlate of the blood-brain barrier (BBB; Figure 1).

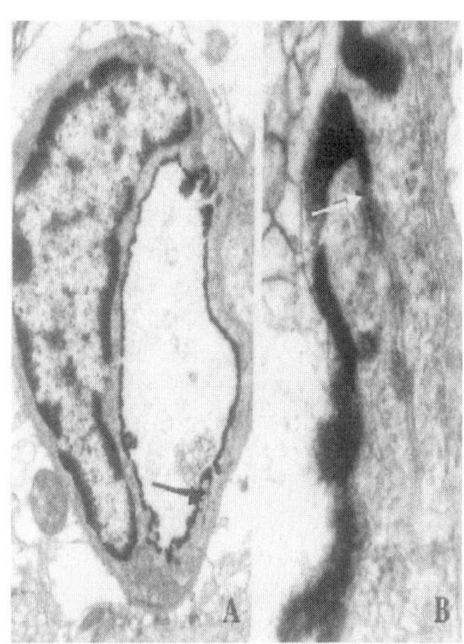


**Figure 1:**
*Ultrastructural localization of the BBB*
Diffusion of the electondense tracer lanthanum localizes the BBB to the endothelial TJs
A: Rat brain capillary with lanthanum deposited on the luminal surface of the endothelium. The arrow points out the TJ where diffusion of lanthanum stops. B: higher magnification of the TJs pointed out in A.

## 2.2 Development, differentiation and maintenance of the BBB

During embryogenesis the development of the brain vascular system begins when angioblasts, which differentiate from the mesoderm, enter the head region and form the perineural vascular plexus that covers the entire surface of the neural tube. From the perineural plexus vascular sprouts invade the proliferating neuroectoderm. This mechanism, whereby new vessels are formed from pre-existing vessels, is called angiogenesis (Risau, 1997). These observations suggested that the embryonic brain produces angiogenic factors that bind to specific receptors expressed on endothelial cells thus leading to migration and proliferation of these cells. To date, the probably most critical of these factors have been characterized largely by analyzing genetically modified mice. By using these mouse models, the essential roles of numerous gene products for blood vessel development and morphogenesis in general including the brain have thus been established. The most prominent examples of genes, the function of which is essential during development already, include VEGF-A and its endothelial tyrosine kinase receptors VEGF-R1 (Flt-1) and -R2 (Flk-1), the receptor tyrosine kinases Tie-1 and Tie-2 as well as the Tie-2 ligands Angiopoietin-1 and -2, components of the TGFβ signalling pathway (such as TGFβ1, ALK-1, endoglin, SMAD5 and SMAD6) and more recently members of the EphB-receptor / ephrinB-ligand family such as EphB4 and ephrinB2 and also select members of the Notch receptor / jagged and delta-like ligand families (for review see Risau 1997; Gale and Yancopoulos 1999; Carmeliet and Collen, 2000; Adams 2002). By virtue of gene targeting technology it has been demonstrated that the VEGF-A ligand and its receptors VEGF-R1 and -R2 are essential for endothelial cell lineage specification, differentiation and blood vessel formation. Based on the spatial and temporal relationship of VEGF and VEGF-R expression during brain angiogenesis, it seems that capillary sprouts originating from the perineural vascular plexus expressing VEGF-R1 and-R2 radially invade into the developing neuroectoderm along a concentration gradient of VEGF, which is highly expressed by cells in the ventricular layer of the developing neuroectoderm (summarized in Engelhardt and Risau, 1995). There is a precise regulation of the timing and pattern formation in brain angiogenesis in mammals with brain angiogenesis being maximal in early post-natal days and downregulated in the adult (Bär, 1980). Obviously, additional signals are required to produce the complex architecture of a mature vascular tree within the CNS. In recent years, two endothelial cell specific receptor tyrosine kinases, Tie-1 and Tie-2 have been identified, the activities of which seem to govern processes that allow for endothelial cell sprouting, remodelling of primitive vascular networks and blood vessel maturation and integrity (Dumont et al., 1994). Tie-2 and at least two of its angiopoietin ligands (Ang-1 and Ang-2; Davis and Yancopoulos, 1999) are involved in the regulation of angiogenic sprouting, vascular remodelling and recruitment of perivascular cells during embryonic development. A common feature of Tie-2 and Ang-1 deficient embryos is a defect in the association of endothelial cells with the underlying extracellular matrix and with perivascular cells. Based on these findings, it has been suggested that Tie-2 signalling could be required for upregulation of factors in endothelial cells that are chemotactic for pericytes and smooth muscle cells leading to the migration of these cells towards the endothelial cell wall and subsequent maturation of the vessels by an increased production of extracellular matrix components elicited by the action of activated TGF-β and other proteins (Folkman and D'Amore, 1996). Amongst these, platelet-derived growth factor (PDGF)-B, a high affinity ligand for the receptor tyrosine kinase PDGF-Rβ present on perivascular mesenchymal cells is produced by endothelial cells during development. PDGF-B has been shown to be involved in vascularization of the brain as disruption of the PDGF-B gene leads to pericyte loss and lethal micoraneurysm formation during late embryogenesis (Lindahl et al., 1997).

Even though many of the molecular components involved in brain angiogenesis have now been identified, their exact mechanisms of action are still not fully understood. Furthermore, as these mechanisms also apply outside the CNS it seems unlikely that they are involved in BBB differentiation. However, the biggest remaining obstacle to the understanding of the roles of those proteins for the development of specific vascular beds is the fact that mutations in their genes invariably lead to lethal phenotpyes during early embryogenesis before BBB differentiation starts. Therefore, meaningful *in vivo* analysis of their possible involvement in BBB differentiation or their contribution during pathological alterations of the BBB has to await the development of more sophisticated genetic models such as conditional and inducible mouse mutants.

In any case, additional molecules specifically expressed in brain endothelial cells have to be involved in the differentiation of the BBB. To date, specific expression of the non-receptor tyrosine kinase *lyn* (Achen et al., 1995) and of gene products encoding P-glycoprotein has been demonstrated early during brain angiogenesis (Qin and Sato, 1995). Whereas the relevance of *lyn* expression during brain angiogenesis remains to be determined, expression of P-glycoprotein is required for the differentiation of the BBB (Schinkel et al., 1994) and seems to ensure the rapid removal of toxic metabolites from the neuroectoderm before the BBB has fully differentiated. In the developing chicken CNS, it has been shown that angiogenic vessels invading the neuroectoderm express N-cadherin between endothelial cells and pericytes. With the onset of barrier differentiation, N-cadherin labelling decreased suggesting that transient N-cadherin expression in endothelial and perivascular cells may represent an initial signal which may be involved in the commitment of early blood vessels to express BBB properties (Gerhardt et al., 1999). Thus establishing of the barrier is accompanied by further changes in the phenotype of the brain endothelial cells such as upregulation of the HT/7-antigen/basigin (Seulberger et al., 1992) or downregulation of the MECA-32 antigen (Hallmann et al., 1995), the mouse homologue of the rat VP-1 (Stan et al., 1999) besides the expression of specific transporters and metabolic pathways can be observed (Pardridge, 1988; Pardridge, 1991). At present, the development of the barrier function of brain capillaries is basically still investigated as done by Paul Ehrlich over 100 years ago by measuring the decreasing permeability for vascular tracers. Although different approaches gave somewhat different results (Wakai and Hirokawa, 1978; Risau, 1991; Stewart, 2000; Dziegielewska et al., 2001) it has become clear that BBB tightness is not just "switched on" at a specific time point during brain angiogenesis but rather that the tightening of the barrier occurs as a gradual process, which is independent from vascular proliferation and begins late during embryogenesis when angiogenesis is not complete (Risau et al., 1986). The molecular mechanisms involved in this process are not understood until today. Based on the elegant transplantation studies of Stewart and Wiley, who were first to show that vessels derived from the coelomic cavity when growing into an ectopic brain transplant gain BBB characteristics (Stewart and Wiley, 1981), it is known that the development of BBB characteristics in endothelial cells is not pre-determined but rather induced by the neuroectoderm.

The question remains, which cell or combination of cell types within the neuroectoderm induces BBB characteristics in endothelium. Astrocytic endfeet are in close proximity to the endothelial cells' plasma membrane and are separated only by the basal lamina. As the astroglial perivascular sheet is a unique feature of CNS capillaries and forms at around the same time as the permeability barrier develops (Phelps, 1972), astrocytes and their precursors have been implicated in the induction of the BBB (Goldstein, 1988). First hints have been provided by Janzer and Raff (Janzer and Raff, 1987), who were the fist to demonstrate that purified astrocytes induce BBB-like permeability changes in invading endothelial cells *in vivo*. However, as some of the BBB characteristics in brain endothelial cells appear very early and prior to astrocyte differentiation, other cell types might be involved in BBB differentiation in addition to astrocytes. The fully differentiated BBB consists of a complex cellular system of highly specialized endothelial cells, pericytes embedded in the basal membrane, perivascular cells and astrocytic endfeet (Figure 2).

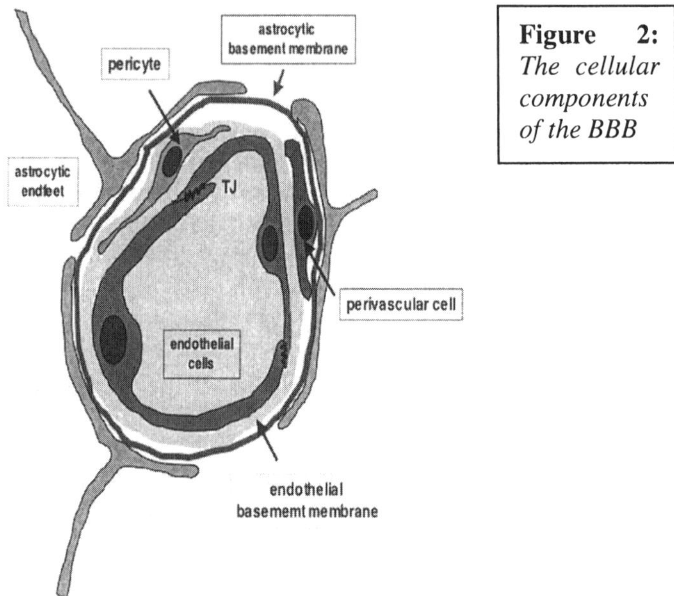


**Figure 2:** *The cellular components of the BBB*

While the endothelial cells form the barrier proper, the interaction with adjacent cells seems to be a prerequisite for barrier function. Pericytes are found in close association with endothelial cells even at very early stages of development and seem to be more prevalent on neural capillaries than on other capillaries (Simionescu et al., 1988). The function of pericytes *in vivo* has been unclear for a long time (Sims, 1986), but recently it has become evident that they are required for vessel maturation (Lindahl et al., 1997).

The role of the extracellular matrix in the induction and maintenance of the BBB has been insufficiently investigated so far. This may partly be due to the enormous molecular heterogeneity including different laminins, collagens, fibronectin etc. (see below). In particular agrin, a heparan sulfate proteoglycan, was found in the basement membranes of vessels with special barrier properties like those in the brain, thymus and testis, and it has been proposed to be involved in the development of the BBB (Barber and Lieth, 1997). Agrin was first isolated from the electric organ of *Torpedo californica* and identified to be essential for clustering acetylcholine receptors in the postsynaptic membrane of the motor endplate (Nitkin et al., 1987).

Astrocytes have been shown to express agrin *in vivo* and *in vitro*, which is why agrin could serve as a basal membrane factor inducing barrier functions in brain endothelium *in vivo*. Since agrin is linked to the cytoskeleton via dystroglycan and the dystrophin-glycoprotein complex (Blake and Kröger, 2000), it could participate in establishing cellular polarity, which is a characteristic feature of perivascular astrocytes. In any case, it is noteworthy that the extracellular matrix of leaky blood vessels in malignant human brain tumors was found to be devoid of agrin (Rascher et al., 2002). In multiple sclerosis or in EAE, BBB capillaries also become generally permeable leading to edema formation and exacerbation of the disease. These observations suggest that a continuing regulation of BBB maintenance is provided by the tissue microenvironment. Disturbances in the molecular interplay between the CNS microenvironment and the BBB might therefore lead to BBB dysfunction.

Taking these observations together, based on the observations that early during embryogenesis endothelial-neuroectodermal interactions induce a "commitment" in endothelial cells to form a BBB lineage, we have previously proposed two phases of endothelial –neuroectodermal interactions leading to BBB differentiation (Engelhardt and Risau, 1995). This is documented by the early induction of specific genes in those endothelial cells such as *lyn* and P-glycoportein (see above). In a subsequent phase, secondary interactions of "committed" brain endothelial cells with the developing neuroectoderm elements induce further endothelial differentiation, which leads to the fully functional BBB. Whereas the "commitment" of brain endothelial cells to form a BBB remains stable, the second phase might be reversible. Maintenance of a functional BBB thus strictly depends on the permanent interaction of the BBB endothelium with the surrounding cellular and extracellular elements. Molecular changes in the CNS within close proximity of the vascular barrier will thus lead to changes in BBB phenotype and function. Thus immigration of encephalitogenic cells across the BBB during EAE would induce an alteration in the "molecular communication" of the perivascular cells with BBB endothelium and ultimately lead to altered BBB function (see below).

## 2. 3 In vitro models of the BBB

Further evidence for the necessitiy of continuous signals provided by the CNS microenvironment for brain endothelial cells to maintain barrier properties is provided by *in vitro* observations. When brain endothelial cells are isolated and cultured *in vitro* they rapidly lose many of their BBB characteristics including formation of proper TJs and a permeability barrier, indicating that integrity of the BBB strictly depends on signals provided by the CNS microenvironment (Figure 3).

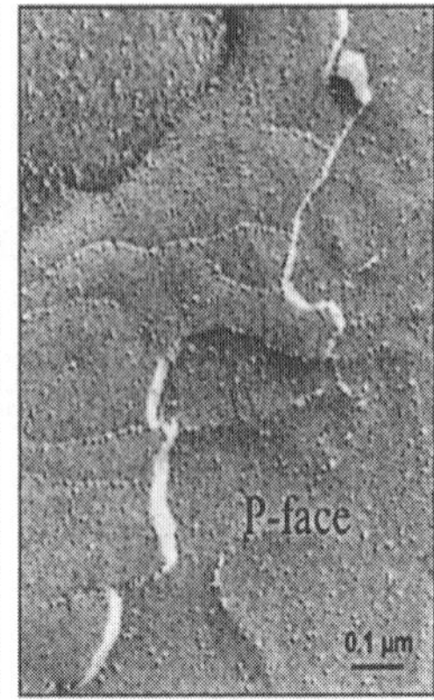



**Figure 3:** *Freeze fracture analysis of BBB TJs* Association of the P-face association of TJs particles as present *in vivo* is rapidly lost from BBB endothelium cultured *in vitro*

This is in contrast to epithelial cells, which when cultured *in vitro* still form proper TJs emphasizing that the formation of epithelial TJs is less dependent on extrinsic factors than the formation of endothelial TJs (Kniesel and Wolburg, 2000). This might be due to the fact that the architecture of epithelial and endothelial TJs differ with respect to their spatial and molecular organisation. In endothelial cells components of adherens junctions and TJs can be intermingled along the intercellular cleft, whereas they remain spatially separated in epithelial cells (Schulze and Firth, 1993). Thus, *in vitro* BBB models employing epithelial cells have to be considered as not suitable for the study of BBB characteristics. Also, based on the observations described above, namely that brain endothelial cells become "committed" to form a BBB during emryogenesis, *in vitro* BBB models employing non-cerebral endothelial cells must be considered less appropriate, as these endothelial cells lack expression of genes specifically induced in brain endothelial cells during development.

Several *in vitro* BBB models have been established which utilize the co-culture of cerebral porcine or bovine brain endothelial cells with primary glial cells mimicking the *in vivo* BBB microenvironment (Meresse et al., 1989; Rubin et al., 1991; Wolburg et al., 1994). In most of those models the permeability barrier and high electrical resistance of the cerebral endothelial cells have been extensively characterized. *In vitro* establishment of other biological aspects of BBB endothelium, such as expression of surface markers (i.e. HT7/basigin) (Prat et al., 2001) or transcytosis of LDL or transferrin (Cecchelli et al., 1999) across the cerebral monolayer has, however, been investigated less frequently. In agreement with the literature (reviewed in (Wolburg and Lippoldt, 2002), when employing the *in vitro* BBB model by Cecchelli (Cecchelli et al., 1999) we observed that bovine brain capillary endothelial (BBCE) cells grown in mono-culture *in vitro* did not build a permeability barrier to tracers, whereas co-culture of BBCE cells with primary glial cells for 10 days induced a tight permeability barrier (Hamm et al. manuscript submitted). In that study, we could demonstrate that removal of glial cells from the co-culture resulted in a modulation of the permeability barrier. Therefore, in this *in vitro* system the permeability barrier of the BBCE cells is directly dependent on humoral factors present in the co-culture only. Modulation of the BBB *in vitro* was due to the opening of the endothelial TJs as demonstrated by the increased permeability to $^3$H-inulin and $^{14}$C-sucrose and most significantly by their opening to horseradish peroxidase at the ultrastructural level. An increase in the number of pinocytotic vesicles leading to increased transendothelial transport of tracers did not occur.

Our observations emphasize that leakiness of the BBB most probably occurs via modulations at the TJ level. To date, no *in vitro* system has yet been established that allowed the induction of TJ complexes between the cultured endothelial cells in such a way that they mimicked barrier properties leading to high electrical resistance measurements as *in vivo*.

In contrast to their compromized usefulness for investigations of the molecular mechanisms involved in barrier formation or barrier breakdown (i.e. the passive involvement of the BBB in EAE pathogeneis), *in vitro* cultured brain endothelial cells or even brain endothelial cell lines have proven to be valuable tools to study traffic signals involved in leukocyte/endothelial interactions *in vitro* (active contribution of the BBB in EAE pathogenesis), which have proven to be valid *in vivo* as well (Greenwood et al., 1995b; Laschinger and Engelhardt, 2000; see below ).

## 2. 4 Tight junctions (TJs) of the BBB

### 2.4.1 Morphology

Although BBB TJs are unique as they are very sensitive to ambient factors, within ultrathin section electron micrographs their morphology resembles those of epithelial TJs as they appear as a chain of fusion points - "kisses" - of the outer plasma membrane leaflet of adjacent cells (Figure 4). Additionally, TJs between the endothelial cells of the brain microvessel have been analysed by qualitative freeze-fracture electron microscopic studies. TJs function as a seal only if they are continuous and branched. The complex network of TJs between brain endothelial cells is therefore primarily responsible for the paracellular impermeablity and, unlike simple TJs present between endothelial cells elsewhere in the organism, provides a high electrical resistance.

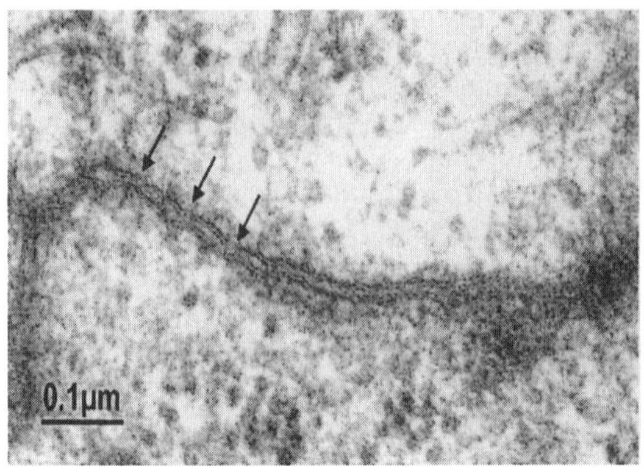

**Figure 4:** *TJ "kisses" between endothelial cells of the BBB*
Ultrathin section of a BBB TJ. Arrows point to the characteristic kissing points of the outer leaflets of the plasma membranes of adjacent cerebral endothelial cells.

In freeze-fracture electron microscopy studies the BBB TJs have been demonstrated to be more complex than TJs in other endothelial cells in the body appearing as a continuous network of parallel and highly interconnected strands that circumscribe the apex of lateral membranes of adjacent cells (Figure 3; Nagy et al., 1984; Nico et al., 1992). However, it turned out that the complexity of strands is only one feature of the tight junctional network and that the association of the TJ particles with the protoplasmic (P-face) or external leaflet of the membrane (E-face) is another criterion to correlate morphology and physiology of TJs. In the endothelial cells that form the BBB, the association of TJ particles with the P-face is higher than in all other endothelial cells of the entire vasculature of the body. It was shown that the P-face association correlates with the barrier function of the BBB endothelium (Figure 3; (Wolburg et al., 1994). This observation is also consistent with the low P-face association in TJs of peripheral, non-barrier endothelial cells, where the E-face associated tight junctional particles clearly predominate (Mühleisen et al., 1989; Simionescu et al., 1988). It has been hypothesized that there is an important functional role for the P-face association, and thus presumably the cytoplasmic anchoring of the TJ particles, particularly for BBB function (Wolburg et al., 1994). This is further supported by the finding that a molecular component, which was identified to associate with the P-face, claudin-3, is predominantly incorporated into BBB endothelial cell tight junctions (see below).

### 2.4.2 Molecular composition of BBB TJs

In recent years, several proteins have been identified that are asociated with epithelial and endothelial TJs (Figure 5). This group of TJ associated proteins include cytoplasmic peripheral membrane proteins of the MAGUK family, such as ZO-1, ZO-2 and ZO-3 (reviewed in Tsukita et al, 1999 and Wolburg and Lippoldt, 2002). These proteins link integral membrane proteins of the TJs to the cytoskeleton. Occludin was the first integral membrane protein found to be exclusively localized within TJs (Furuse et al., 1993). However, mice carrying a null mutation in the occludin gene are viable and develop morphologically normal TJs in most tissues including the brain (Saitou et al., 2000) proving that occludin is not essential for proper TJ formation. In contrast to occludin, the claudins, which exhibit no sequence-homology to occludin, comprise a novel gene-family of integral membrane TJ proteins with more than 20 members to date and have been shown to be sufficient for the formation of TJ strands (Morita et al., 1999a). Transfection of claudin-1 or claudin-3 into fibroblasts lacking TJs, induced P-face associated TJs (Furuse et al., 1999), whereas transfection of fibroblasts with claudin-2 or claudin-5 induced E-face associated TJs in the absence of occludin (Morita et al., 1999c) demonstrating that different claudins induce structurally different TJs. Claudins are not randomly distributed throughout the tissues. Using immunohistochemistry we could demonstrate that besides claudin-5, claudin-3 is localized in endothelial TJs in the CNS of mice and man (Wolburg et al., 2003). Additionally, we and others have shown that occludin and ZO-1 are localized within BBB TJs (Bolton et al., 1998; Wolburg et al., 2003). Furthermore, the Ig supergene family members junctional adhesion molecule (JAM) (Martin-Padura et al., 1998) and the recently discovered endothelial cell-selective adhesion molecule (ESAM) are localized in TJs (Nasdala et al., 2002) of the BBB. Via interaction with ZO-1 these integral membrane proteins interact with the actin cytoskeleton thus providing the molecular link for TJ association with the cytoskeleton. In addition, adherens junctions, which are strongly associated with the cytoskeleton, e.g. via cadherin/catenin interactions seem to be intermingled with TJs at the BBB (Schulze and Firth, 1993).

### 2.4.3 Molecular composition of adherens junctions and cell-to-cell contacts in BBB endothelium

In cerebral endothelial cells non-occluding adherens junctions are found in close proximity to TJs (Schulze and Firth, 1993), Figure 5). In adherens junctions the endothelial specific integral membrane protein VE-cadherin (Corada et al., 1999), is linked to the cytoskeleton via catenins, which belong to the family of armadillo proteins (Vleminckx and Kemler, 1999). In endothelial cells expression and localization of $\beta$-catenin, $\gamma$-catenin and p120[cas] have been described to be crucial for the functional state of adherens junctions (Dejana, 1996).

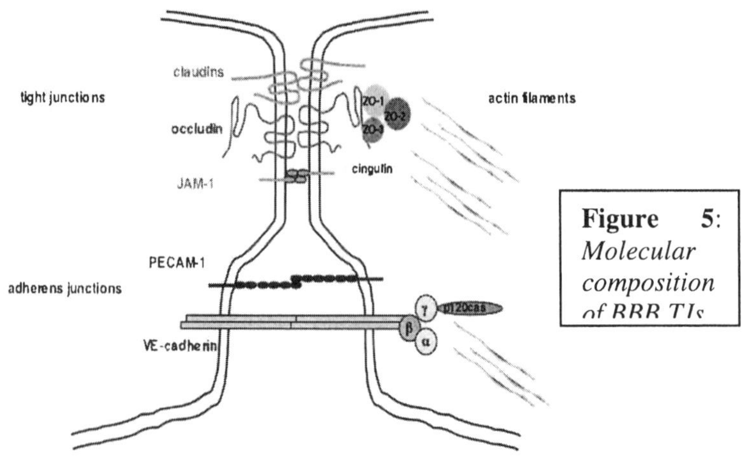



**Figure 5**: *Molecular composition of BBB TJs*

This is supported by the observation that application of monoclonal antibodies directed against VE-cadherin increase vascular permeability *in vivo* (Corada et al., 2001). Whether VE-cadherin is also required to maintain integrity of the BBB remains to be investigated.

Finally, PECAM-1 has been demonstrated to be localized in endothelial cell contacts outside of either TJs or adherens junctions. In mice deficient for PECAM-1 no defect in BBB integrity has been reported (Duncan et al., 1999).

## 3. THE BLOOD-BRAIN BARRIER DURING EAE

### 3.1 BBB leakiness

Breakdown of the BBB was shown to precede symptoms and other MRI signs of new lesion development in MS (Kermode et al., 1990). These findings are modelled in EAE, where the impaired integrity of the BBB has been demonstrated to precede clinical signs of the disease (Claudio et al., 1989). Using several tracers to investigate BBB integrity during EAE, increased transport of sodium, chloride, sucrose and inulin could be demonstrated in different areas of the diseased brain (Juhler, 1988). Studies on the time course of BBB breakdown in rodent models during progress of EAE support a caudal-to-rostral gradient preceeding the progression of the gradient of severity of the inflammatory disease (Butter et al., 1991). However, more recently it was demonstrated that during EAE the cerebellar BBB is dramatically and briefly compromised even before breakdown of the BBB in the spinal cord prior to disease onset (Tonra et al., 2001). Disruption of the BBB during CNS inflammation can lead to the immediate problem of vasogenic edema and the clinical problems related to this condition. Active lesions are characterized by the perivascular accumulation of serum components such as albumin or fibrin and inflammatory cells (Figure 6).

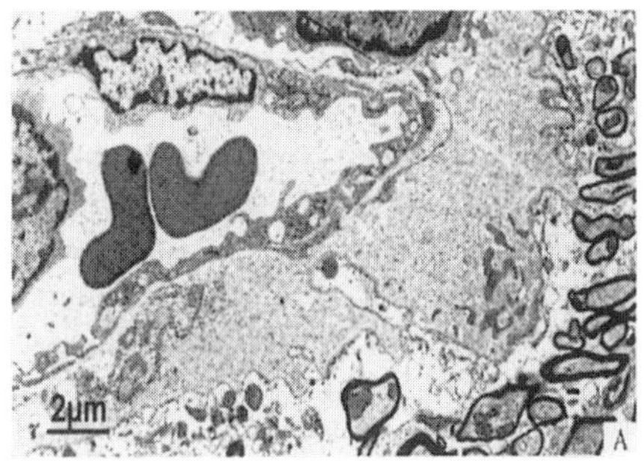
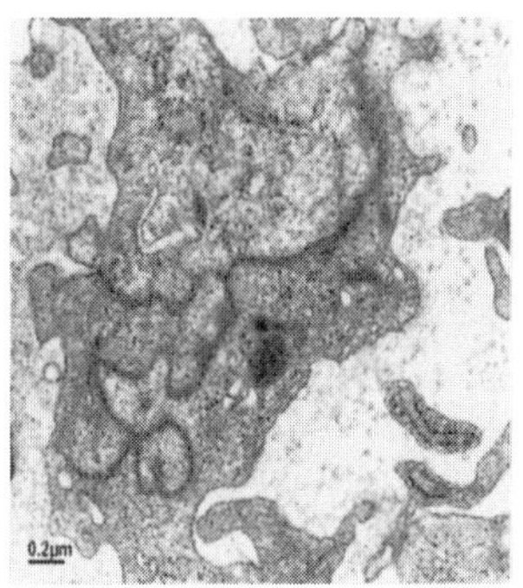

**Figure 6:** *CNS microvessels during EAE*
A) (left) Leakiness of the BBB during EAE . Upper left: CNS micovessel with erythrocytes, center: perivascular space with massive protein deposits, lower right: myelinated nerve fibers. Note the protrusion of the glia limitans from the left to the right.
B) (right) Characteristic "tortured" TJ between cerebral endothelial cells in an inflamed vessel during EAE is shown

Whereas the accumulation of serum components in the perivascular space of CNS microvessels demonstrates that passive diffusion of these proteins can occur across the BBB, at the same time the absence of neutrophils and red blood cells within the CNS suggests that even the impaired BBB still plays an active role in the recruitment of inflammatory cells into the CNS. In EAE, leakiness of the BBB occurs mainly at the level of postcapillary venules, which are surrounded by inflammatory cuffs. CNS postcapillary venules are thin walled and mainly comprised of endothelial cells and function as exchange microvessels, much like capillaries (Fenstermacher et al., 2001). Several observations suggest that these small venules are the microvessels that are most prone to disruption and BBB breakdown. Eventually, leakiness of the BBB is accompanied by the loss of the differentiated phenotype of BBB endothelial cells as supported by the re-expression of the MECA-32 antigen on brain endothelial cell in inflamed vessels surrounded by inflammatory cuffs (Engelhardt et al., 1994).

Leakiness of the BBB could be a result of either TJ opening or enhanced pinocytotic activity of the endothelial cells. Whereas the latter could not be observed by us, in the CNS of mice afflicted with EAE, vessel profiles surrounded by inflammatoty cuffs were characterized by an increase of the luminal surface of postcapillary endothelium and by a complicated and branched course of the tight junctional domains (Figure 6B). This suggests that BBB leakiness mainly occurs via an opening of the BBB TJs. This observation on the ultrastructural level seems to be confirmed by immunocytochemical studies demonstrating that inflamed CNS vessels exhibit a seemingly disrupted immunostaining for TJ molecules, while staining in the healthy BBB is characterized by straight and continuous patterns (Wolburg et al., 2003).

## 3. 2 Molecular alterations of BBB TJs during EAE

Using immunohistochemistry, we recently demonstrated that during the clinical disease of EAE the TJ molecule claudin-3 is selectively lost from vessels surrounded by inflammatory cuffs (Wolburg et al., 2003). In contrast to claudin-3, the presence of other TJ molecules like claudin-5, occludin or ZO-1 in cerebral vessels was not affected during EAE. These observations differ from a previous report, where loss of ZO-1 and occludin from cerebral vascular endothelium was observed during CNS inflammation caused by injection of LPS in juvenile rats (Bolton et al., 1998). However, in this study the cellular infiltrate was dominated by neutrophils, which is in contrast to EAE, where the cellular infiltrate is mostly composed of mononuclear cells. Interestingly, only juvenile rats younger than 3 weeks developed LPS induced neutrophil mediated inflammation in the CNS, whereas injection of LPS in older rats did not lead to any CNS inflammation or alterations at the BBB, indicating that maturation of the BBB was not complete at that age. Thus, loss of occludin and ZO-1 from cerebral vessels upon neutrophil recruitment might be specific to the immature BBB. Alternatively, recruitment of different leukocyte populations might differently affect the TJs due to usage of distinct routes of transendothelial migration, i.e. transcellular versus paracellular (Faustmann and Dermietzel, 1985, Greenwood et al., 1994). Interestingly, selective loss of claudin-3 can also be observed in leaky vessels in glioblastoma multiforme in the absence of any inflammation, suggesting that *in vivo* the loss of claudin-3 from BBB TJs correlates with BBB breakdown at these sites and that this is not necessarily dependent on the presence of inflammatory cells (Wolburg et al., 2003). Claudin-3 might therefore be a key component determining the permeability of BBB endothelial TJs *in vivo*. Future studies will have to demonstrate the signals determining the subcellular distribution of claudin-3. Defining these signals might allow to develop new therapeutic strategies to reduce BBB permeability and thus severity of brain edema in CNS diseases.

An alteration of the TJ structure during EAE is further supported by our findings that during EAE the TJ molecules JAM-1 becomes accessible at the luminal surface of BBB endothelium whereas in the healthy CNS JAM-1 is confined to the BBB TJs and cannot be accessed for immunostaining by antibodies injected into the blood-stream (Dehouck and Engelhardt, unpublished observations). In fact, *in vitro* studies demonstrated that stimulation of endothelial cells by pro-inflammatory cytokines will lead to a release of JAM-1 from endothelial TJs and its availability on the endothelial luminal surface, where it can then mediate the binding of highly activated T cells via the integrin LFA-1 (Ostermann et al., 2002). These observations suggest, that during inflammation, TJ proteins such as JAM-1 might be involved in guiding inflammatory cells across the BBB TJs.

## 3.3 Contribution of endothelial PECAM-1

Platelet/endothelial cell adhesion molecule-1 (PECAM-1, CD31) is a member of the Ig superfamily of transmembrane proteins and is concentrated at cell contact sites in all continuous endothelial cell linings *in vivo* including the endothelial cells of the BBB and is proximal to adherens junctions (Figure 5).

The lack of endothelial PECAM-1 expression in knock-out mice results in an earlier onset of clinical symptoms during EAE, associated with an increased inflammatory cell migration across the BBB (Graesser et al., 2002). Lack of endothelial PECAM-1 was additionally shown to lead to an impaired ability to restore vascular integrity after alteration of the endothelium in PECAM-1 deficient mice or of cultured PECAM-1$^{-/-}$ endothelial cells (Graesser et al., 2002). The resulting increased and prolonged vascular permeability might lead to enhanced transendothelial migration of inflammatory cells and thus early onset of CNS inflammation during EAE. We hypothesized a role for PECAM-1 in the restoration of inter-endothelial junctional integrity following disrupting stimuli, while normal, homeostatic endothelium is not affected. The prolonged vascular permeability in the PECAM-1-deficient mice could be accounted for if PECAM-1 plays a role in re-establishing intact junctions following inflammatory stimuli and transendothelial migration of leukocytes. In turn, disruption of endothelial cell-cell junctions has been shown to facilitate transmigration of leukocytes, which may contribute to a cycle of increased migration and decreased junctional integrity in the PECAM-1-deficient mice. However, the endothelial permeability barrier was eventually restored, showing that the requirement for PECAM-1 at inter-endothelial junctions is not absolute. One might speculate on an involvement of β-catenin, which normally associates with the transmembrane junctional protein VE-cadherin, linking it to cytoskeletal proteins, but can also associate with PECAM-1 (Ilan et al., 2000). When, through certain stimuli, such as inflammation in EAE, junctional disassembly is induced, β–catenin becomes tyrosine-phosphorylated, dissociates from VE-cadherin but associates with PECAM-1 (Ilan et al. 1999). The presence of phosphorylated β-catenin and protein tyrosine phosphatases such as SHP-2 on the same scaffolding molecule may induce the dephosphorylation of β–catenin, allowing it to re-associate with VE-cadherin, thus re-assembling intact adherens junctional complexes (Ilan et al. 1999). Perhaps in the PECAM-1-deficient mice, PECAM-1 cannot act as a reservoir for phosphorylated β-catenin, resulting in an increased cytoplasmic pool of the phosphorylated protein, which cannot readily re-associate with VE-cadherin in the junctional complex. Ultimately, β-catenin may become dephosphorylated by other mechanisms and resume its association with VE-cadherin and other junctional proteins, thus eventually restoring integrity of the endothelial cell layer. Further elucidation of the underlying mechanism(s) by which PECAM-1 modulates vascular integrity will aid in our understanding of the roles PECAM-1 plays during inflammatory responses. Although this does not discount the possibility that signals mediated via PECAM-1 may contribute to the generation or propagation of the immune response during EAE, the impairment of dynamic vascular junctional integrity is likely to be one of the driving forces in the early EAE phenotype of the PECAM-1-deficient mice.

## 3.4 Inflammatory mediators/cytokines involved in BBB leakiness during EAE

Inflammatory cytokines play an active part in the mediation of the immune response during EAE. Proinflammatory cytokines such as tumor necrosis factor (TNF)-α, interferon (IFN)-γ, interleukin (IL-1) and IL-6 have been detected in the cerebrospinal fluid of MS patients and in CNS tissue of mice suffering from EAE (Gijbels et al., 1990). These cytokines have been shown to be involved in the immunopathogenesis of EAE and based on *in vitro* studies demonstrating that TNF-α, IL-1 and IL-6 induce increased vascular permeability of tracers across monolayers of endothelial cells (de Vries et al., 1996), it has been suggested that they also impair BBB function *in vivo*. In fact, administration of TNF-α to an *in vitro* model of the BBB resulted in increased paracellular permeability to small tracers and was accompanied by the reorganization of actin filaments within the brain endothelial cells (Deli et al., 1995), supporting the possibility that locally increased levels of TNF-α in the CNS during EAE might directly lead to increased permeability of the BBB. Intravital microscopy studies demonstated that *in vivo* TNF-α increases the permeability of pial microvessels to tracers via a mechanism involving the activation of soluble guanylate cyclase and protein tyrosine kinases (Mayhanm, 2002). Additionally, *in vivo* application of IL-1 was demonstrated to impair barrier properties of the BBB (Blamire et al., 2000).

In addition to locally produced cytokines affecting BBB function, impairment of BBB in EAE was shown to be accompanied by an increased transport of TNF-α from the blood into the CNS by a saturable system (Gutierrez et al., 1993; Pan et al., 1996). Increased transport of TNF-α was observed throughout the CNS, i.e. brain and cervical, thoracic, and lumbar spinal cord with a distinct time course and reversibility (Pan and Kastin, 2001).

Exposure of cerebral endothelial cells to pro-inflammatory cytokines can lead to the release/expression of various products of the arachidonic acid cascade with both vasoactive and pro-inflammatory properties, including prostaglandins, leukotrienes, and platelet-activating factor (PAF) by the BBB endothelial cells themselves leading to changes in local cerebral blood flow and blood rheology, and finally increases in BBB permeability (Stanimirovic and Satoh, 2000). Whereas involvement of these mechanisms and of classical vasoactive mediators such as nitric oxide or endothelin in BBB breakdown is generally accepted, their involvement during MS is not well studied. Recently, VEGF was suggested to be involved in BBB breakdown during EAE/MS (Proescholdt et al., 2002). As described above, VEGF is a major inducer of angiogenesis, but originally, it was described as vascular permeability factor (Senger et al., 1990). In both acute and chronic MS plaques as well as during EAE, VEGF expression was found to be upregulated in astrocytes in association with inflammatory cuffs (Proescholdt et al., 2002), suggesting that VEGF might exacerbate the inflammatory response in autoimmune diseases of the CNS by inducing focal BBB breakdown. To date, involvement of VEGF in the development of vascular leakage in the CNS has rather been associated with stroke, head injury and high-altitude illness, disorders that are associated with tissue hypoxia. As VEGF gene expression was shown to be upregulated by hypoxia, increased VEGF expression may link hypoxia and vascular leakage in the CNS *in vivo*. Thus it remains to be shown whether hypoxic conditions might occur in inflammatory cuffs during EAE or whether VEGF might be induced by other means.

Last but not least, pro-inflammatroy cytokines lead to the activation of BBB endothelium, which is characterized by the upregulated expression of traffic signals for leukocytes such as adhesion molecules and chemokines that govern inflammatory cell recruitment across the BBB during EAE (see below).


# 4. RECRUITMENT OF ENCEPHALITOGENIC T CELLS ACROSS THE BBB

## 4.1 Antigen presentation by BBB endothelium

The question whether BBB endothelium would present antigen has been addressed extensively, when it was still assumed that the BBB is impermeable for lymphocytes. It was thought that "accidental" presentation of myelin-antigens on the luminal surface of the BBB endothelial cells would lead to local activation and thus transmigration of activated T cells across the BBB. In order to present antigen to encephalitogenic $CD4^+$ T cells, endothelial cells must present MHC class II antigens. Although it was repeatedly claimed that during EAE there is induction of MHC class II products on BBB endothelium, demonstration of MHC class II on BBB endothelium at the electron microscopic level *in vivo* was achieved only in the guinea pig model of EAE (Sobel et al., 1987; Wilcox et al., 1989). In contrast, the inducibility of MHC class II products on brain endothelial cells from different species *in vitro* could be demonstrated by several groups by using pure cultures of BBB endothelial cells. Furthermore expression of the costimulatory molecules B7-1 and B7-2 was demonstrated on human brain endothelial cells in an *in vitro* BBB model (Omari and Dorovini-Zis, 2001). Inflammatory mediators such as TNF-$\alpha$ or $\gamma$-IFN were shown to induce MHC class II expression on those cultured endothelial cells *in vitro* (Male and Pryce, 1988; Risau et al., 1990). Direct investigation of the antigen-presenting potential of MHC class II-positive brain endothelial cells *in vitro*, however, demonstrated that T cells cocultured with MHC class II positive

mediated by CD4[+], MBP-specific T cells in those cultures (Risau et al., 1990; Sedgwick et al., 1990; Wang et al., 1995), which was, however, dependent on the presence of the specific antigen and the appropriate MHC class II molecule. This apparent paradox might be attributed to the production of prostaglandins by brain endothelial cells inhibiting T cell proliferation (McCarron, 1992). More recently it was shown that CD4[+] T cell proliferation and cytokine production is inhibited by a factor derived from human brain endothelial cells. Addition of exogenous Il-2 prevented T cell unresponsiveness reinforcing the notion that brain endothelial cells can induce T cell anergy (Prat et al., 2000).

Taken together, based on the observations that T cell blasts can migrate through the BBB endothelium independent of their antigen specificity and MHC-restriction as shown using ovalbumin-specific T cells and bone-marrow chimeras (Hickey et al., 1991) and that traffic signals guiding T cells across the BBB were identified (see below) it is highly unlikely that antigen-presentation is required for T cell migration across the BBB. It might well be though that later during ongoing EAE, when high levels of cytokines and myelin degradation products are present within the inflammatory cuffs, antigen-presentation by brain endothelial cells might contribute to the destruction of the BBB, which is typical for acute EAE and MS (Tsukada et al., 1993).

## 4. 2 Traffic signals provided by BBB endothelium

In general, lymphocyte recruitment across the vascular wall is regulated by the sequential interaction of different adhesion or signaling molecules on lymphocytes and endothelial cells lining the vessel wall (Butcher et al., 1999). An initial transient contact of the circulating leukocyte with the vascular endothelium, generally mediated by adhesion molecules of the selectin-family and their respective carbohydrate ligands, slows down the leukocyte in the bloodstream. Subsequently, the leukocyte rolls along the vascular wall with greatly reduced velocity. The rolling leukocyte can receive endothelial signals resulting in its firm adhesion to the endothelial surface. These signals are transduced by chemokines via G-protein coupled receptors on the leukocyte surface. Binding of a chemokine to its receptor results in a pertussis toxin sensitive activation of integrins on the leukocyte surface. Only activated integrins mediate the firm adhesion of the leukocytes to the vascular endothelium by binding to their endothelial ligands, which belong to the immunoglobulin (Ig)-superfamily. This ultimately leads to the extravasation of the leukocyte. Successful recruitment of circulating leukocytes into the tissue depends on the productive leukocyte/endothelial interaction during each of these sequential steps.

We and others have investigated the expression of adhesion molecules on CNS endothelium during EAE in the SJL/N mouse by means of *in situ* hybridization and immunohistology and found induction of ICAM-1 and VCAM-1 but not E- and P-selectin on CNS endothelium (Engelhardt et al., 1997; Cannella et al., 1991; Steffen et al., 1994; Wilcox et al., 1990). ICAM-1 and VCAM-1 were shown to mediate adhesion of lymphocytes to inflamed cerebral vessels on frozen brain sections *in vitro* (Steffen et al., 1994; Yednock et al., 1992). *In vivo* monoclonal antibody inhibition studies confirmed the involvement of VCAM-1 and its ligand $\alpha$4-integrin but not E- and P-selectin in the pathogenesis of EAE as antibodies directed against VCAM-1 and its ligand $\alpha$4-integrin but not against E- and P-selectin successfully blocked the development of clinical EAE in the SJL/N mouse (Engelhardt et al., 1997 and 1998). Although we have discounted a role for endothelial selectins in EAE, recent evidence was provided demonstrating that during ongoing EAE P-selectin is involved in leukocyte recruitment at least into the meningeal vessels (Kerfoot and Kubes, 2002). Even though it was previously reported that BBB endothelial cells lack storage of P-selectin in Weibel-Palade bodies (Barkalow et al., 1996), P-selectin was recently implicated in the initial recruitment of activated encephalitogenic T cells into the healthy CNS (Carrithers et al., 2000). These apparently contradictory results can only be investigated by direct observation of leukocyte-endothelial interaction *in vivo*.

In order to gain direct evidence for the molecular mechanisms of leukocyte interaction with the vascular wall within the CNS, observation of the CNS microcirculation by intravital fluorescence microscopy is necessary. This approach is hampered, however, by the protected localization of the brain and spinal cord within the skull and the spinal column, respectively. In order to gain intravital microscopic access to the CNS microcirculation, acute and chronic cranial window preparations have been developed for rodents, which allow observation of the pial and cortical, i.e. CNS grey matter, microcirculation respectively (Uhl et al., 1999; Vajkoczy et al., 2000). In EAE, however, CNS white matter is the battle field of the inflammatory attack with preference of the spinal cord. Therefore, we recently developed a novel spinal cord window preparation, which enabled us to directly visualize CNS white matter microcirculation by intravital fluorescence videomicroscopy (Vajkoczy et al., 2001). We gained direct evidence that T lymphoblast interaction with the BBB is unique due to the lack of rolling. Constitutively expressed VCAM-1 mediates the G-protein independent prompt arrest (capture) of circulating encephalitogenic T cell blasts via $\alpha$4-integrin to the endothelium of the healthy BBB. Transient capture is followed by G-protein dependent $\alpha$4-integrin /VCAM-1 mediated adhesion strengthening and subsequent LFA-1 mediated migration of lymphocytes across the BBB into the spinal cord white matter (Vajkoczy et al., 2001, Laschinger et al., 2002). Requirement for $\alpha$4-integrin in leukocyte interaction with the inflamed BBB during ongoing EAE is maintained as demonstrated by intravital microsocpy of CNS microvessels in the brain (Kerfoot and Kubes, 2002). Furthermore, as recently demonstrated the requirement for $\alpha$4-integrin in inflammatory cell recruitment across the BBB seems to translate to the mechanisms invoved in inflammatory cell recruitment across the BBB in MS, where therapy using a humanized anti-$\alpha$4-integrin antibody has proven to be beneficial (von Andrian and Engelhardt, 2003).

A crucial role for chemokines in successful recruitment of encephalitogenic T cells across the BBB is suggested by the requirement for signalling via G-protein coupled, PTX-sensitive receptors on encephalitogenic T cells to firmly arrest on BBB endothelium *in vivo*. We found expression of the lymphoid chemokine "EBV-induced molecule 1 ligand chemokine" (ELC)/CCL19 at the level of the BBB endothelium on a subpopulation of CNS venules and induced expression of the "secondary lymphoid chemokine" (SLC)/CCL21 in inflamed CNS venules during EAE (Alt et al., 2002). Expression of their common receptor CCR7 was detected in a subpopulation of cells present in the perivascular inflammatory cuffs. Encephalitogenic T cells show surface expression of CCR7 and the alternative receptor for CCL21, CXCR3. They chemotax specifically towards both, CCL19 or CCL21 in a concentration dependent and pertussis toxin-sensitive manner, comparable to naive lymphocytes *in vitro*. Additionally, in frozen section assays functional ablation of CCR7 and CXCR3 or blocking CCL19 and CCL21 reduced binding of encephalitogenic T lymphocytes to inflamed venules in the brain. Absence of rolling, the involvement of the lymphoid chemokines CCL19 and CCL21 and the predominant involvement of α4-integrin and VCAM-1 make this T lymphoblast interaction with the BBB unique and suggest that these specific mechanisms operate to limit leukocyte travel to this immunoprivileged site (Figure 7).

Interestingly, although we observed T cell interaction with the vascular wall in CNS capillaries and post-capillary venules, extravasation of encephalitogenic T cells was only observed at the level of post-capillary venules. This is consistent with the observations of the localization of inflammatory cuffs, which characteristically form around the post-capillary venules but not around capillaries, indicating that extravasation of autoaggressive T cells and inflammatory cells preferentially occurs at this level. In contrast to peripheral organs, where extravasation of leukocytes takes several minutes, we observed that extravasation of encephalitogenic T cells across the BBB takes several hours (Laschinger et al., 2002; Vajkoczy et al., 2001), supporting the notion that transendothelial migration across the BBB follows unique mechanisms. In this context, it should be noted that initial T cell recruitment across the healthy BBB is not accompanied by BBB leakiness as observed by intravital microscopy (Vajkoczy et al., 2001).

In any case the passageway of leukocytes in transendothelial migration across the BBB is still a matter of debate. The general belief is that outside the CNS leukocytes migrate across the interendothelial junctions. This has been supported by observations that antibodies directed against VE-cadherin accelerate neutrophil recruitment into the inflamed peritoneal cavity of mice (Gotsch et al., 1997). Involvement of the TJ molecule JAM-1 was described in transendothelial migration of monocytes at least *in vitro* (Martin-Padura et al., 1998) and supported by the recent findings that JAM-1 can function as a ligand for LFA-1 on highly activated T cells (Ostermann et al., 2002). Involvement of JAM-1 in leukocyte recruitment into the CNS *in vivo*, however, is still controversial (Del Maschio et al., 1999; Lechner et al.,

2000). It is possible that differences in the usage of passageways across endothelial cells are chosen by different leukocyte subpopulations as suggested by the observation that transendothelial migration of neutrophils and monocytes but not of lymphocytes can be blocked by pre-treatment of endothelial cells with antibodies directed against PECAM-1(Muller et al., 1993; Nakada et al., 2000). The highly specialized BBB might even lead to BBB specifc passageways for leukocyte subpopulations during immunosurveillance and inflammatory disease. It has been suggested previously that leukocyte recruitment across the BBB occurs rather parajunctional than through the TJs (Greenwood et al., 1995a), which is supported by our own observations (Wolburg and Engelhardt, unpublished). Therefore understanding the entire sequence of traffic signals involved in the recruitment of inflammatory cells across the BBB will be necessary to understand CNS inflammation and whether these events will eventually lead to BBB breakdown during EAE.

## 4. 3 The BBB basement membrane

After their passage across the endothelial cell monolayer forming the BBB inflammatory cells still face the endothelial cell basement membrane and the subjacent glia limitans consisting of astrocytic endfeet and associated basement membrane before entering the CNS parenchyma (Figures 7 and 8).

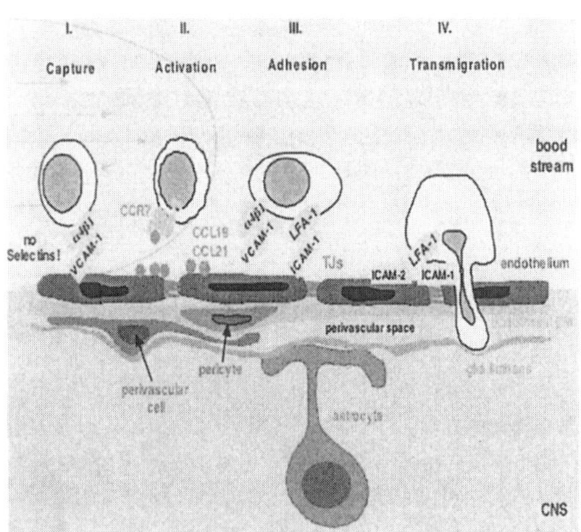


**Figure 7:** *Multi-step cascade of encephalitogenic T cell recruitment across the BBB during EAE*

During EAE the endothelial cell and astroglial basement membranes clearly define the inner and outer limits of the perivascular space around post-capillary venules, where inflammatory cells accumulate during acute EAE before infiltrating the brain parenchyma (Figure 8). However, the molecular mechanisms involved in T cell penetration of the endothelial cell basement membrane and the process of invasion across the glia limitans have not been investigated in great detail until recently.

basement membrane and the process of invasion across the glia limitans have not been investigated in great detail until recently.

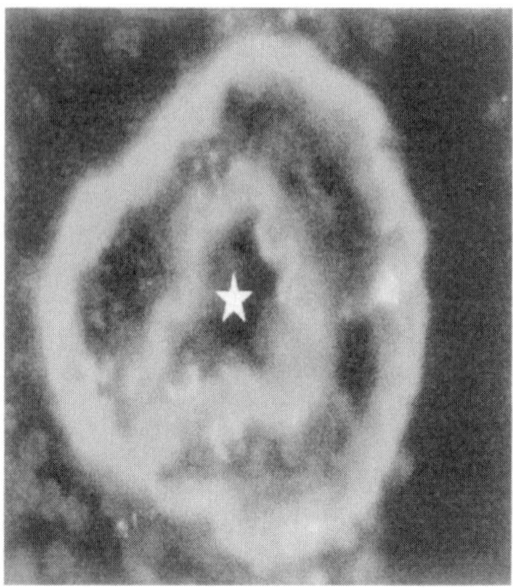

**Figure 8:** *Basement membranes at the inflamed BBB* Immunofluorescence staining for laminin of a postcapillary venule surrounded by inflammatory cells in the brain of a mouse afflicted with EAE. Laminin staining marks the borders of the perivascular space with the inner endothelial basement membrane and the outer astrocytic basement membrane. I = vessel lumen.

There is evidence suggesting that these two steps are distinct and independent of one another: In EAE induced in macrophage depleted mice (Tran et al., 1998) or in TNF-α-deficient mice (Korner et al., 1997) the inflammatory infiltrate becomes entrapped in the perivascular space and parenchymal infiltration is prevented, indicating that progression through the astrocytic basement membrane is functionally distinct from endothelial cell basement membrane transmigration.

The major proteins of all basement membranes belong to the laminin family of glycoproteins. Laminins are heterotrimers composed of an α, β and γ chain. Only two isoforms, namely laminin 8 (composed of laminin $\alpha4$, $\beta1$ and $\gamma1$) and laminin 10 (composed of laminin $\alpha5$, $\beta1$ and $\gamma1$) are found in endothelial basement membranes of most tissues including the CNS (Sorokin et al., 1994). Blood vessels in the CNS have also been reported to express laminins 1 and 2 (Jucker et al., 1996; Virtanen et al., 2000), which are not detected in blood vessel basement membranes elsewhere. Recently, laminin distribution in blood vessels in the CNS could be clarified by investigating the laminin isoforms of BBB basal membranes during EAE (Sixt et al., 2001). The localization of inflammatory cuffs surrounding post-capillary venules during EAE allowed a clear distinction of endothelial cell and astroglial basement membranes, demonstrating that the endothelial cell basement membrane contains laminins 8 and 10 whereas the astroglial basement membrane contains laminins 1 and 2. In the absence of inflammation the endothelial and astroglial basement membranes lie in such close proximity that at the light microscopic level they may be mistaken for a single structure leading to the mis-conception that laminin 1 and 2 are present in endothelial cell basement membranes.

membrane. Perivascular cuffs were only observed around endothelial cell basement membranes containing laminin 8, while in the presence of laminin 10 no infiltration was detectable. *In vitro*, encephalitogenic T-cells could bind to both laminins 8 and 10 present in the endothelial basement membrane via $\alpha6\beta1$-integrin, however, they were unable to bind the laminins 1 and 2 present in the astroglial basment membranes, suggesting that encephalitogenic T cells might not be able to penetrate the astrocytic basement membrane (Sixt et al., 2001). This is supported by previous observations that encephalitogenic T cell blasts injected to induce EAE in naive recipients remain in the perivasular space of CNS venules and do not immigrate into the CNS parenchyma (Cross et al., 1990).

These data suggest that the astrocytic basement membrane breaks down in more advanced stages of the disease process during EAE and is likely to be dependent on the proteolytic activity of matrix metalloproteinases (MMPs). MMPs have been demonstrated to be involved in T-cell entry into and residency in the parenchyma of the CNS, as well as in demyelination (Graesser et al., 1998; Graesser et al., 2000). While protease inhibitors have been shown to reduce the severity or delay the onset of EAE, to date no protease inhibitor has been shown to completely ablate T-cell migration into the perivascular space or the brain parenchyma (Graesser et al., 2000). Until now it was not clear that emigrating T-cells face more than just the subendothelial cell basement membrane. The results of the MMP inhibitor experiments were, therefore, interpreted as reduced transmigration across the endothelial cell basement membrane. However, they can equally well be explained by inhibition of transmigration across the parenchymal basement membrane. It now remains to be determined whether the main targets of MMP activity are components of the endothelial cell or the parenchymal basement membrane.

Investigations of the distribution of the major laminin receptors, $\alpha6\beta1$-integrin and $\alpha$–dystrophin, demonstrated that $\alpha6\beta1$-integrin occurs predominantly on the endothelial cells in the brain mediating interactions with the endothelial cell laminins 8 and 10, while $\alpha$-dystroglycan is expressed on the astrocyte endfeet and probably mediates binding to the parenchymal laminins 1 and 2 (Sixt et al., 2001). It has been shown that during the course of MS, $\alpha6\beta1$-integrin is downregulated on the endothelial cells at sites of infiltration (Sobel et al., 1998). *In vitro*, laminin $\alpha4$ expression was shown to be strongly upregulated in endothelial cells by proinflammatory cytokines such as interleukin-1 (IL-1) (Frieser, 1997), suggesting a role in inflammatory events. Thus, selective upregulation of laminin $\alpha4$ expression by cytokines such as IL-1 or TNF-$\alpha$ which have been shown to play a role in EAE, in combination with a high turnover of laminin $\alpha4$ in endothelial cell basement membrane in the brain may lead to a loosening of the endothelial cell-basement membrane interaction at sites of T-cell infiltration in EAE facilitating the further development of perivascular cuffs during EAE.

## 5. THE BLOOD-CEREBROSPINAL FLUID BARRIER IN EAE

A role of the choroid plexus in the pathogenesis of EAE or MS , i.e. as an alternative entry site for circulating lymphocytes directly into the CSF, has not been seriously considered before. In the choroid plexus, the epithelium forms the blood-cerebrospinal fluid (CSF) barrier and is responsible for the secretion of the CSF from the blood. The morphological correlate of the blood-CSF barrier are the TJs of choroid plexus epithelium. Interestingly, in EAE the adhesion molecules ICAM-1 and VCAM-1 are upregulated on the choroid plexus epithelium in parallel to their upregulation on BBB endothelium (Steffen et al., 1996). Ultrastructural studies revealed polar localization of ICAM-1, VCAM-1 and MAdCAM-1 on the apical surface of choroid plexus epithelial cells and their complete absence on the fenestrated endothelial cells within the choroid plexus parenchyma (Wolburg et al., 1999). Furthermore, ICAM-1, VCAM-1 and MAdCAM-1 expressed in choroid plexus epithelium mediate binding of lymphocytes via their known ligands (Steffen et al., 1996).

Massive ultrastructural changes can be observed during EAE within the choroid plexus (Engelhardt et al., 2001), however, at the ultrastructural level the TJs seem to remain intact (Wolburg et al., 2001). Choroid plexus epithelial TJs are morphologically unique resembling the TJs of the central myelin sheaths, which are characterized by their parallel and poorly anastomosing strands (Dermietzel et al., 1980). These morphologically unique TJs of CNS myelin sheaths were recently found to be exclusively constituted by oligodendrocyte specific protein/claudin-11 (Morita et al., 1999b). The central role of claudin-11 for the formation of parallel-array TJs was confirmed in the OSP/claudin-11 null mouse (Gow et al., 2000), in which myelin sheaths of oligodendrocytes were shown to be devoid of any TJs. We found that the TJs in the choroid plexus are unique with respect to the presence of claudins as claudin-11 is found to be accompanied by claudin-1 and claudin-2 (Wolburg et al., 2001). A possible role for OSP/claudin-11 in the development of EAE was demonstrated by the induction of the disease in susceptible mice after immunization with OSP/claudin-11 (Stevens et al., 1999) and the demonstration of antibodies directed against OSP/claudin-11 in the CSF of patients suffering from multiple sclerosis (Bronstein et al., 2000). Whereas presence of claudin-11 in the CSF of MS patients is taken as a measure for myelin degradation until the present day, it might well reflect a modulation of the blood-CSF barrier localized at the choroid plexus epithelium.

In addition, as choroid plexus epithelial cells have been found to express MHC class I and II molecules on their surface (for review see (Engelhardt et al., 2001) it will be interesting to see the contribution of this barrier to autoimmune inflammation within the CNS.


## 6. OUTLOOK

The endothelial cells of the BBB and probably the epithelial cells of the blood-CSF barrier are involved in the pathogenesis of inflammatory demyelinatingdiseases such as MS or its animal model EAE. Contribution of the BBB can be considered twofold, an active role by guiding inflammatory cells across the BBB and a passive role characterized by breakdown of the BBB.

It has been more than 100 years since the discovery of the BBB. During the last 10 years great progress has been made in understanding the active contribution of the BBB by characterization of the traffic signals involved in leukocyte recruitment into the CNS. In contrast, the molecular mechanisms leading to BBB breakdown are less well understood.

For this purpose it will be mandatory to characterize the mechansims involved in BBB differentiation during development and those involved in maintaining BBB characeristics in endothelial cells in the adult CNS.

## 7. ACKNOWLEDGEMENTS

We owe special thanks to Karen Wolburg-Buchholz for providing and preparing the figures on the ultrastructure of the BBB. We also thank Urban Deutsch for his devoted discussion of this chapter.

## 8. REFERENCES

Achen, M. G., Clauss, M., Schnürch, H., and Risau, W. (1995). The non-receptor tyrosine kinase Lyn is localized in the developing blood-brain barrier. Diff. *59*,15-24.

Adams, R. H. (2002). Vascular patterning by Eph receptor tyrosine kinases and ephrins. Semin Cell Dev Biol *13*, 55-60.

Alt, C., Laschinger, M., and Engelhardt, B. (2002). Functional expression of the lymphoid chemokines CCL19 (ELC) and CCL 21 (SLC) at the blood-brain barrier suggests their possible involvement in lymphocyte recruitment into the central nervous system during experimental autoimmune encephalomyelitis. Eur J Immunol *32*, 2133-2144.

Bär, T. (1980). The vascular system of the cerebral cortex. Adv AnatEmbryolCell Biol *59*, 1-62.
Barber, A. J., and Lieth, E. (1997). Agrin accumulates in the brain microvascular basal lamina during development of the blood-brain barrier. Dev Dyn *208*, 62-74.

Barkalow, F., Goodman, M., Gerritsen, M., and TN, M. (1996). Brain endothelium lack one of two pathways of P-selectin - mediated neutrophil adhesion. Blood *88*, 4585-4593.

Blake, D. J., and Kröger, S. (2000). The neurobiology of Duchenne muscular dystrophy: learning lessons from muscle? Trends Neurosci *23*, 92-99.

Blamire, A. M., Anthony, D. C., Rajagopalan, B., Sibson, N. R., Perry, V. H., and Styles, P. (2000). Interleukin-1beta -induced changes in blood-brain barrier permeability, apparent diffusion coefficient, and cerebral blood volume in the rat brain: a magnetic resonance study. J Neurosci *20*, 8153-8159.

Bolton, S. J., Anthony, D. C., and Perry, V. H. (1998). Loss of tight junction proteins occludin and zonula occludens-1 from cerebral vascular endothelium during neutrophil-induced blood-brain barrier breakdown in vivo. Neurosci *86*, 1245-1257.

Brightman, M. W. (1968). The intracerebral movement of proteins injected into blood and cerebrospinal fludid of mice. Prog Brain Res *29*, 19-31.

Bronstein, J. M., Tiwari-Woodruff, S., Buznikov, A. G., and Stevens, D. B. (2000). Involvement of OSP/Claudin-11 in oligodendrocyte membrane interactions: Role in biology and disease [Review]. J Neurosci Res *59*, 706-711.

Butcher, E. C., Williams, M., Youngman, K., Rott, L., and Briskin, M. (1999). Lymphocyte trafficking and regional immunity. Adv Immunol *72*, 209-253.

Butter, C., Baker, D., O'Neill, J. K., and Turk, J. L. (1991). Mononuclear cell trafficking and plasma protein extravasation into the CNS during chronic relapsing experimental allergic encephalomyelitis in Biozzi AB/H mice. J Neurol Sci *104*, 9-12.

Cannella, B., Cross, A. H., and Raine, C. S. (1991). Adhesion-related molecules in the central nervous system. Upregulation correlates with inflammatory cell influx during relapsing experimental autoimmune encephalomyelitis. Lab Invest *65*, 23-31.

Carmeliet, P., and Collen, D. (2000). Molecular basis of angiogenesis. Role of VEGF and VE-cadherin. Ann N Y Acad Sci *902*, 49-62.

Carrithers, M. D., Visintin, I., Kang, S. J., and Janeway, C. A., Jr. (2000). Differential adhesion molecule requirements for immune surveillance and inflammatory recruitment. Brain *123*, 1092-1101.

Cecchelli, R., Dehouck, B., Descamps, L., Fenart, L., Buee-Scherrer, V., Duhem, C., Lundquist, S., Rentfel, M., Torpier, G., and Dehouck, M. P. (1999). In vitro model for evaluating drug transport across the blood-brain barrier. Adv Drug Deliv Rev *36*, 165-178.

Claudio, L., Kress, Y., Norton, W. T., and Brosnan, C. F. (1989). Increased vesicular transport and decreased mitochondrial content in blood-brain barrier endothelial cells during experimental autoimmune encephalomyelitis. Am J Pathol *135*, 1157-1168.

Corada, M., Liao, F., Lindgren, M., Lampugnani, M. G., Breviario, F., Frank, R., Muller, W. A., Hicklin, D. J., Bohlen, P., and Dejana, E. (2001). Monoclonal antibodies directed to different regions of vascular endothelial cadherin extracellular domain affect adhesion and clustering of the protein and modulate endothelial permeability. Blood *97*, 1679-1684.

Corada, M., Mariotti, M., Thurston, G., Smith, K., Kunkel, R., Brockhaus, M., Lampugnani, M. G., Martin-Padura, I., Stoppacciaro, A., Ruco, L., *et al.* (1999). Vascular endothelial-cadherin is an important determinant of microvascular integrity in vivo. Proc Natl Acad Sci (USA) *96*, 9815-9820.

Cross, A. H., Cannella, B., Brosnan, C. F., and Raine, C. S. (1990). Homing to central nervous system vasculature by antigen-specific lymphocytes. I. Localization of 14C-labeled cells during acute, chronic, and relapsing experimental allergic encephalomyelitis. Lab Invest *63*, 162-170.

Davis, S., and Yancopoulos, G. D. (1999). The angiopoietins: Yin and Yang in angiogenesis. Curr Top Microbiol Immunol *237*, 173-185.

de Vries, H. E., Blom-Roosemalen, M. C., van Oosten, M., de Boer, A. G., van Berkel, T. J., Breimer, D. D., and Kuiper, J. (1996). The influence of cytokines on the integrity of the blood-brain barrier in vitro. J Neuroimmunol *64*, 37-43.

Dejana, E. (1996). Endothelial adherens junctions: implications in the control of vascular permeability and angiogenesis. J Clin Invest *98*, 1949-1953.

Del Maschio, A., De Luigi, A., Martin-Padura, I., Brockhaus, M., Bartfai, T., Fruscella, P., Adorini, L., Martino, G., Furlan, R., De Simoni, M. G., and Dejana, E. (1999). Leukocyte recruitment in the cerebrospinal fluid of mice with experimental meningitis is inhibited by an antibody to junctional adhesion molecule (JAM). J Exp Med *190*, 1351-1356.

Deli, M. A., Descamps, L., Dehouck, M. P., Cecchelli, R., Joo, F., Abraham, C. S., and Torpier, G. (1995). Exposure of tumor necrosis factor-alpha to luminal membrane of bovine brain capillary endothelial cells cocultured with astrocytes induces a delayed increase of permeability and cytoplasmic stress fiber formation of actin. J Neurosci Res *41*, 717-726.

Dermietzel, R., Leibstein, A. G., and Schünke, D. (1980). Interlamellar tight junctions of central myelin. II. A freeze-fracture and cytochemical study on their arrangement and composition,. Cell Tissue Res *213*, 95-108.

Dumont, D. J., Gradwohl, G., Fong, G. H., Puri, M. C., Gertsenstein, M., Auerbach, A., and Breitman, M. L. (1994). Dominant-negative and targeted null mutations in the endothelial receptor tyrosine kinase, tek, reveal a critical role in vasculogenesis of the embryo. Genes Dev *8*, 1897-1909.

Duncan, G. S., Andrew, D. P., Takimoto, H., Kaufman, S. A., Yoshida, H., Spellberg, J., Luis de la Pompa, J., Elia, A., Wakeham, A., Karan-Tamir, B., *et al.* (1999). Genetic evidence for functional redundancy of Platelet/Endothelial cell adhesion molecule-1 (PECAM-1): CD31-deficient mice reveal PECAM-1-dependent and PECAM-1-independent functions. J Immunol *162*, 3022-3030.

Dziegielewska, K. M., Ek, J., Habgood, M. D., and Saunders, N. R. (2001). Development of the choroid plexus. Microsc Res Tech *52*, 5-20.

Ehrlich, P. (1904). Über die Beziehung chemischer Constitution, Vertheilung, und pharmakologischer Wirkung. (Berlin).

Engelhardt, B., Conley, F. K., and Butcher, E. C. (1994). Cell adhesion molecules on vessels during inflammation in the mouse central nervous system. J Neuroimmunol *51*, 199-208.

Engelhardt, B., Laschinger, M., Schulz, M., Samulowitz, U., Vestweber, D., and Hoch, G. (1998). The development of experimental autoimmune encephalomyelitis in the mouse requires alpha4-integrin but not alpha4beta7-integrin. J Clin Invest *102*, 2096-2105.

Engelhardt, B., and Risau, W. (1995). The development of the blood-brain barrier. In New concepts of a blood-brain barrier, J. Greenwood, D. Begley, and M. Segal, eds. (London, Plenum Press).

Engelhardt, B., Vestweber, D., Hallmann, R., and Schulz, M. (1997). E- and P-selectin are not involved in the recruitment of inflammatory cells across the blood-brain barrier in experimental autoimmune encephalomyelitis. Blood *90*, 4459-4472.

Engelhardt, B., Wolburg-Buchholz, K., and Wolburg, H. (2001). Involvement of the choroid plexus in central nervous system inflammation. Microsc Res Tech *52*, 112-129.

Faustmann, P. M., and Dermietzel, R. (1985). Extravasation of polymorphonuclear leukocytes from the cerebral microvasculature. Cell Tiss Res *242*, 399-407.

Fenstermacher, J. D., Nagaraja, T., and Davies, K. R. (2001). Overview of the structure and function of the blood-brain barrier in vivo. In Blood-Brain Barrier: Drug delivery and brain pathology, D. Kobiler, S. Lustig, and S. Shapira, eds. (New York, Kluwer Academic Plenum Publishers), pp. 1-7.

Folkman, J., and D'Amore, P. A. (1996). Blood vessel formation: what is its molecular basis? Cell *87*, 1153-1155.

Furuse, M., Hirase, T., Itoh, M., Nagafuchi, A., Yonemura, S., and Tsukita, S. (1993). Occludin - a novel integral membrane-protein localizing at tight junctions. J Cell Biol *123*, 1777-1788.

Furuse, M., Sasaki, H., and Tsukita, S. (1999). Manner of interaction of heterogeneous claudin species within and between tight junction strands. J Cell Biol *147*, 891-903.

Gale, N. W., and Yancopoulos, G. D. (1999). Growth factors acting via endothelial cell-specific receptor tyrosine kinases: VEGFs, angiopoietins, and ephrins in vascular development. Genes Dev *13*, 1055-1066.

Gerhardt, H., Liebner, S., Redies, C., and Wolburg, H. (1999). N-cadherin expression in endothelial cells during early angiogenesis in the eye and brain of the chicken: relation to blood-retina and blood-brain barrier development. Eur J Neurosci *11*, 1191-1201.

Gijbels, K., Van Damme, J., Proost, P., Put, W., Carton, H., and Billiau, A. (1990). Interleukin 6 production in the central nervous system during experimental autoimmune encephalomyelitis. Eur J Immunol *20*, 233-235.

Goldmann, E. E. (1913). Vitalfärbung am Zentralnervensystem. Abh Preuss Wissensch Phys-Math *1*, 1-60.

Goldstein, G. W. (1988). Endothelial cell-astrocyte interactions. A cellular model of the blood-brain barrier. Ann N Y Acad Sci *529*, 31-39.

Gotsch, U., Borges, E., Bosse, R., Böggemeyer, E., Simon, M., Mossmann, H., and Vestweber, D. (1997). VE-cadherin antibody accelerates neutrophil recruitment in vivo. J Cell Sci *110*, 583-588.

Gow, A., Southwood, C. M., Li, J. S., Pariali, M., Riordan, G. P., Danias, J., Bronstein, J. M., Brodie, S. E., Kachar, B., and Lazzarini, R. A. (2000). CNS myelin and Sertoli cell tight junction stands are absent in Osp/claudin 11-null mice. J Neurochem *74*, S35.

Graesser, D., Mahooti, S., Haas, T., Davis, S., Clark, R. B., and Madri, J. A. (1998). The interrelationship of alpha4 integrin and matrix metalloproteinase-2 in the pathogenesis of experimental autoimmune encephalomyelitis. Lab Invest *78*, 1445-1458.

Graesser, D., Mahooti, S., and Madri, J. A. (2000). Distinct roles for matrix metalloproteinase-2 and alpha4 integrin in autoimmune T cell extravasation and residency in brain parenchyma during experimental autoimmune encephalomyelitis. J Neuroimmunol *109*, 121-131.

Graesser, D., Solowiej, A., Bruckner, M., Osterweil, E., Juedes, A., Davis, S., Ruddle, N., Engelhardt, B., and Madri, J. M. (2002). Changes in vascular permeability and early onset of experimental autoimmune encephalomyelitis in PECAM-1 (CD31) deficient mice. J Clin Invest *109*, 383-392.

Greenwood, J., Bamforth, S. D., Wang, Y., and Devine, L. (1995a). The blood-retinal barrier in immune-mediated diseases of the retina. In New Concepts of a Blood-Brain Barrier, J. Greenwood, ed. (New York, Plenum Press), pp. 315-326.

Greenwood, J., Howes, R., and Lightman, S. (1994). The blood-retinal barrier in experimental autoimmune uveoretinitis - leukocyte interactions and functional damage. Lab Invest *70*, 39-52.

Greenwood, J., Wang, Y., and Calder, V. L. (1995b). Lymphocyte adhesion and transendothelial migration in the central nervous system: the role of LFA-1, ICAM-1, VLA-4 and VCAM-1. Immunol *86*, 408-415.

Gutierrez, E. G., Banks, W. A., and Kastin, A. J. (1993). Murine tumor necrosis factor alpha is transported from blood to brain in the mouse. J Neuroimmunol *47*, 169-176.

Hallmann, R., Mayer, D. N., Berg, E. L., Broermann, R., and Butcher, E. C. (1995). Novel mouse endothelial cell surface marker is suppressed during differentation of the blood-brain barrier. Dev Dyn *202*, 325-332.

Hickey, W. F., Hsu, B. L., and Kimura, H. (1991). T-lymphocyte entry into the central nervous system. J Neurosci Res *28*, 254-260.

Ilan, N., Cheung, L., Pinter, E., and Madri, J. A. (2000). Platelet-endothelial cell adhesion molecule-1 (CD31), a scaffolding molecule for selected catenin family members whose binding is mediated by different tyrosine and serine/threonine phosphorylation. J Biol Chem *275*, 21435-21443.

Ilan, N., Mahooti, S., Rimm, D. L., and Madri, J. A. (1999) PECAM-1 (CD31) functions as a reservoir for and a modulator of tyrosine-phosphorylated beta-catenin. J Cell Sci *112*, 3005-3014.

Janzer, R. C., and Raff, M. C. (1987). Astrocytes induce blood-brain barrier properties in endothelial cells. Nature *325*, 253-257.

Jucker, M., Tian, M., Norton, D. D., Sherman, C., and Kusiak, J. W. (1996). Laminin alpha 2 is a component of brain capillary basement membrane: reduced expression in dystrophic dy mice. Neurosci *71*, 1153-1161.

Juhler, M. (1988). Pathophysiological aspects of acute experimental allergic encephalomyelitis. Acta Neurol Scand Suppl *119*, 1-21.

Kerfoot, S., and Kubes, P. (2002). Overlapping roles of P-selectin and alpha 4 integrin to recruit leukocytes to the central nervous system in experimental autoimmune encephalomyelitis. J Immunol *169*, 1000-1006.

Kermode, A. G., Thompson, A. J., Tofts, P., MacManus, D. G., Kendall, B. E., Kingsley, D. P. E.,, Moseley, I. F., Rudge, P., and McDonald, W. I. (1990). Breakdown of the blood-brain barrier precedes symptoms and other MRI signs of new lesions in multiple sclerosis. Brain *113*, 1477-1489.

Kniesel, U., and Wolburg, H. (2000). Tight junctions of the blood-brain barrier. Cell Mol Neurobiol *20*, 57-76.

Korner, H., Riminton, D. S., Strickland, D. H., Lemckert, F. A., Pollard, J. D., and Sedgwick, J. D. (1997). Critical points of tumor necrosis factor action in central nervous system autoimmune inflammation defined by gene targeting. J Exp Med *186*, 1585-1590.

Laschinger, M., and Engelhardt, B. (2000). Interaction of alpha4-integrin with VCAM-1 is involved in adhesion of encephalitogenic T cell blasts to brain endothelium but not in their transendothelial migration in vitro. J Neuroimmunol *102*, 32-43.

Laschinger, M., Vajkoczy, P., and Engelhardt, B. (2002). LFA-1 is not involved in G-protein dependent adhesion of encephalitogenic T cell blasts to CNS microvessels *in vivo*. Eur. J. Immmunol. 32:3598-3606.

Lechner, F., Sahrbacher, U., Suter, T., Frei, K., Brockhaus, M., Koedel, U., and Fontana, A. (2000). Antibodies to the junctional adhesion molecule cause disruption of endothelial cells and do not prevent leukocyte influx into the meninges after viral or bacterial infection. J Infect Dis *182*, 978-982.

Lindahl, P., Johansson, B. R., Leveen, P., and Betsholtz, C. (1997). Pericyte loss and microaneurysm formation in PDGF-B-deficient mice. Science *277*, 242-245.

Male, D., and Pryce, G. (1988). Kinetics of MHC gene expression and mRNA synthesis in brain endothelium. Immunol *63*, 37-42.

Martin-Padura, I., Lostaglio, S., Schneemann, M., Williams, L., Romano, M., Fruscella, P., Panzeri, C., Stoppacciaro, A., Ruco, L., Villa, A., *et al.* (1998). Junctional adhesion molecule, a novel member of the immunoglobulin superfamily that distributes at intercellular junctions and modulates monocyte transmigration. J Cell Biol *142*, 117-127.

Mayhanm, W. G. (2002). Cellular mechanisms by which tumor necrosis factor-alpha produces disruption of the blood-brain barrier. Brain Res *927*, 144-152.

McCarron, R. M. (1992). IL-1-induced prostacyclin production by cerebral vascular endothelial cells inhibits myelin basic protein-specific lymphocyte proliferation. Cell Immunol *145*, 21-29.

Meresse, S., Dehouck, M. P., Delorme, P., Bensaid, M., Tauber, J. P., Delbart, C., Fruchart, J. C., and Cecchelli, R. (1989). Bovine brain endothelial cells express tight junctions and monoamine oxidase activity in long-term culture. J Neurochem *53*, 1363-1371.

Morita, K., Furuse, M., Fujimoto, K., and Tsukita, S. (1999a). Claudin multigene family encoding four-transmembrane domain protein components of tight junction strands. Proc Natl Acad Sci (USA) *96*, 511-516.

Morita, K., Sasaki, H., Fujimoto, K., Furuse, M., and Tsukita, S. (1999b). Claudin-11/OSP-based tight junctions of myelin sheaths in brain and Sertoli cells in testis. J Cell Biol *145*, 579-588.

Morita, K., Sasaki, H., Furuse, M., and Tsukita, S. (1999c). Endothelial claudin: claudin-5/TMVCF constitutes tight junction strands in endothelial cells. J Cell Biol *147*, 185-194.

Mühleisen, H., Wolburg, H., and Betz, E. (1989). Freeze-fracture analysis of endothelial cell membranes in rabbit carotid arteries subjected to short-term atherogenic stimuli. Virch Arch Cell Pathol *56*, 413-417.

Muller, W. A., Weigl, S. A., Deng, X., and Phillips, D. M. (1993). PECAM-1 is required for transendothelial migration of leukocytes. J Exp Med *178*, 449-460.

Nagy, Z., Peters, H., and Hüttner, I. (1984). Fracture faces of cell junctions in cerebral endothelium during normal and hyperosmotic conditions. Lab Invest *50*, 313-322.

Nakada, M. T., Amin, K., Christofidou-Solomidou, M., O'Brien, C. D., Sun, J., Gurubhagavatula, I., Heavner, G. A., Taylor, A. H., Paddock, C., Sun, Q. H., *et al.* (2000). Antibodies against the first Ig-like domain of human platelet endothelial cell adhesion molecule-1 (PECAM-1) that inhibit PECAM-1-dependent homophilic adhesion block in vivo neutrophil recruitment. J Immunol *164*, 452-462.

Nasdala, I., Wolburg-Buchholz, K., Wolburg, H., Kuhn, A., Ebnet, K., Brachtendorf, G., Samulowitz, U., Kuster, B., Engelhardt, B., Vestweber, D., and Butz, S. (2002). A transmembrane tight junction protein selectively expressed on endothelial cells and platelets. J Biol Chem *277*, 16294-16303.

Nico, B., Cantino, D., Bertossi, M., Ribatti, D., Sassoe, M., and Roncali, L. (1992). Tight endothelial junctions in the developing microvasculature - a thin-section and freeze-fracture study in the chick-embryo optic tectum. J Submicr Cytol *24*, 85-95.

Nitkin, R. M., Smith, M. A., Magil, C., Fallon, J. R., Yao, Y. M., Wallace, B. G., and McMahan, U. J. (1987). Identification of agrin, a synaptic organizing protein from Torpedo electric organ. J Cell Biol *105*, 2471-2478.

Omari, K. I., and Dorovini-Zis, K. (2001). Expression and function of the costimulatory molecules B7-1 (CD80) and B7-2 (CD86) in an in vitro model of the human blood--brain barrier. J Neuroimmunol *113*, 129-141.

Ostermann, G., Weber, K. S. C., Zernecke, A., Schroder, A., and Weber, C. (2002). JAM-1 is a ligand of the beta(2) integrin LFA-1 involved in transendothelial migration of leukocytes. Nature Immunology *3*, 151-158.

Pan, W., Banks, W. A., Kennedy, M. K., Gutierrez, E. G., and Kastin, A. J. (1996). Differential permeability of the BBB in acute EAE: enhanced transport of TNF-alpha. Am J Physiol *271*, E636-642.

Pan, W., and Kastin, A. J. (2001). Upregulation of the transport system for TNFalpha at the blood-brain barrier. Arch Physiol Biochem *109*, 350-353.

Pardridge, W. M. (1988). Recent advances in blood-brain barrier transport. Annu Rev Pharmacol Toxicol *28*, 25-39.

Pardridge, W. M. (1991). Advances in cell biology of blood-brain barrier transport. Semin Cell Biol *2*, 419-426.

Phelps, C. H. (1972). The development of glio-vascular relationships in the rat spinal cord. ZZellforsch *128*, 555-563.

Prat, A., Biernacki, K., Becher, B., and Antel, J. P. (2000). B7 expression and antigen presentation by human brain endothelial cells: requirement for proinflammatory cytokines. J Neuropathol Exp Neurol *59*, 129-136.

Prat, A., Biernacki, K., Wosik, K., and Antel, J. P. (2001). Glial cell influence on the human blood-brain barrier. Glia *36*, 145-155.

Proescholdt, M. A., Jacobson, S., Tresser, N., Oldfield, E. H., and Merrill, M. J. (2002). Vascular endothelial growth factor is expressed in multiple sclerosis plaques and can induce inflammatory lesions in experimental allergic encephalomyelitis rats. J Neuropathol Exp Neurol *61*, 914-925.

Qin, Y., and Sato, T. N. (1995). Mouse multidrug resistance 1a/3 gene is the earliest known endothelial cell differentiation marker during blood-brain barrier development. Dev Dyn *202*, 172-180.

Rascher, G., Fischmann, A., Kröger, S., Duffner, F., Grote, E.-H., and Wolburg, H. (2002). Extracellular matrix and the blood-brain barrier in glioblastoma multiforme: spatial segregation of tenascin and agrin. Acta Neuropathol *104*, 85-91.

Reese, T. S., and Karnovsky, M. J. (1967). Fine structural localization of a blood-brain barrier to exogenous peroxidase. J Cell Biol *34*, 207-217.

Risau, W. (1991). Induction of blood-brain barrier endothelial cell differentiation. Ann N Y Acad Sci *633*, 405-419.

Risau, W. (1997). Mechanisms of angiogenesis. Nature *386*, 671-674.

Risau, W., Engelhardt, B., and Wekerle, H. (1990). Immune function of the blood-brain barrier: Incomplete presentation of protein (auto-)antigens by rat brain microvascular endothelium in vitro. J Cell Biol *110*, 1757-1766.

Risau, W., Hallmann, R., and Albrech, U. (1986). Differentiation-dependent expression of protein in brain endothelium during development of the blood-brain barrier. Dev Biol *117*, 537-545.

Rubin, L. L., Hall, D. E., Porter, S., Barbu, K., Cannon, C., Horner, H. C., Janatpour, M., Liaw, C. W., Manning, K., Morales, J., *et al.* (1991). A cell-culture model of the blood-brain-barrier. J Cell Biol *115*, 1725-1735.

Saitou, M., Furuse, M., Sasaki, H., Schulzke, J.-D., Fromm, M., Takano, H., Noda, T., and Tsukita, S. (2000). Complex phenotype of mice lacking occludin, a component of tight junction strands. Mol Biol Cell *22*, 4131-4142.

Schinkel, A. H., Smit, J. J. M., van Tellingen, O., Beijnen, J. H., Wagenaar, E., van Deemter, L., Mol, C. A. A. M., van der Walk, M. A., Robanus-Maandag, E. C., te Riele, H. P. J., *et al.* (1994). Disruption of the mouse mdr1a P-glycoprotein gene leads to a deficiency in the blood-brain barrier and to increased sensitivity to drugs. Cell *77*, 491-502.

Schulze, C., and Firth, J. A. (1993). Immunohistochemical localization of adherens junction components in blood-brain-barrier microvessels of the rat. JCell Sci *104*, 773-782.

Sedgwick, J. D., Hughes, C. C., Male, D. K., MacPhee, I. A., and terMeulen, V. (1990). Antigen-specific damage to brain vascular endothelial cells mediated by encephalitogenic and nonencephalitogenic CD4+ T cell lines in vitro. J Immunol *145*, 2474-2481.

Senger, D. R., Connolly, D. T., Van de Water, L., Feder, J., and Dvorak, H. F. (1990). Purification and NH2-terminal amino acid sequence of guinea pig tumor-secreted vascular permeability factor. Cancer Res *50*, 1774-1778.

Seulberger, H., Unger, C. M., and Risau, W. (1992). HT7, Neurothelin, Basigin, gp42 and OX-47--many names for one developmentally regulated immuno-globulin-like surface glycoprotein on blood-brain barrier endothelium, epithelial tissue barriers and neurons. Neurosci Lett *140*, 93-97.

Simionescu, M., Ghinea, N., Fixman, A., Lasser, M., Kukes, L., Simionescu, N., and Palade, G. E. (1988). The cerebral microvasculature of the rat: structure and luminal surface properties during early development. J Submicrosc Cytol *20*, 243-261.

Sims, D. E. (1986). The Pericyte - A review. Tissue and Cell *18*, 153-174.

Sixt, M., Engelhardt, B., Pausch, F., Hallmann, R., Wendler, O., and Sorokin, L. M. (2001). Endothelial cell laminin isoforms, laminins 8 and 10, play decisive roles in T cell recruitment across the blood-brain barrier in experimental autoimmune encephalomyelitis. J Cell Biol *153*, 933-946.

Sobel, R. A., Hinojoza, J. R., Maeda, A., and Chen, M. (1998). Endothelial cell integrin laminin receptor expression in multiple sclerosis lesions. Am J Pathol *153*, 405-415.

Sobel, R. A., Natale, J. M., and Schneeberger, E. E. (1987). The immunopathology of acute experimental allergic encephalomyelitis. IV. An ultrastructural immunocytochemical study of class II major histocompatiblity comples (Ia) expression. J Neuropathol Exp Neurol *46*, 239-249.

Sorokin, L., Girg, W., Gopfert, T., Hallmann, R., and Deutzmann, R. (1994). Expression of novel 400-kDa laminin chains by mouse and bovine endothelial cells. Eur J Biochem *223*, 603-610.

Stan, R. V., Ghitescu, L., Jacobson, B. S., and Palade, G. E. (1999). Isolation, cloning, and localization of rat PV-1, a novel endothelial caveolar protein. J Cell Biol *145*, 1189-1198.

Stanimirovic, D., and Satoh, K. (2000). Inflammatory mediators of cerebral endothelium: a role in ischemic brain inflammation. Brain Pathol *10*, 113-126.

Steffen, B. J., Breier, G., Butcher, E. C., Schulz, M., and Engelhardt, B. (1996). ICAM-1, VCAM-1, and MAdCAM-1 are expressed on choroid plexus epithelium but not endothelium and mediate binding of lymphocytes in vitro. Am J Pathol *148*, 1819-1838.

Steffen, B. J., Butcher, E. C., and Engelhardt, B. (1994). Evidence for involvement of ICAM-1 and VCAM-1 in lymphocyte interaction with endothelium in experimental autoimmune encephalomyelitis in the central nervous system in the SJL/J mouse. Am J Pathol *145*, 189-201.

Stevens, D. B., Chen, K., Seitz, R. S., Sercarz, E. E., and Bronstein, J. M. (1999). Oligodendrocyte-specific protein peptides induce experimental autoimmune encephalomyelitis in SJL/J mice. J Immunol *15*, 7501-7509.

Stewart, P. A. (2000). Development of the blood-brain barrier. In Morphogenesis of Endothelium, W. Risau, and G. M. Rubanyi, eds. (Amsterdam, Harwood Academic Publishers), pp. 109-122.

Stewart, P. A., and Wiley, M. J. (1981). Developing nervous tissue induces formation of blood-brain barrier characteristics in invading endothelial cells: A study using quail-chick transplantation chimeras. Dev Biol *84*, 183-192.

Tonra, J. R., Reiseter, B. S., Kolbeck, R., Nagashima, K., Robertson, R., Keyt, B., and Lindsay, R. M. (2001). Comparison of the timing of acute blood-brain barrier breakdown to rabbit immunoglobulin G in the cerebellum and spinal cord of mice with experimental autoimmune encephalomyelitis. J Comp Neurol *430*, 131-144.

Tran, E. H., Hoekstra, K., van Rooijen, N., Dijkstra, C. D., and Owens, T. (1998). Immune invasion of the central nervous system parenchyma and experimental allergic encephalomyelitis, but not leukocyte extravasation from blood, are prevented in macrophage-depleted mice. J Immunol *161*, 3767-3775.

Tsukada, N., Matsuda, M., Miyagi, K., and Yanagisawa, N. (1993). Cytotoxicity of T cells for cerebral endothelium in multiple sclerosis. J Neurol Sci *117*, 140-147.

Tsukita, S., Furuse, M., and Itoh, M. (1999). Structural and signalling molecules come together at tight junctions. Curr Opin Cell Biol *11*, 628-633.

Uhl, E., Pickelmann, S., Rohrich, F., Baethmann, A., and Schurer, L. (1999). Influence of platelet-activating factor on cerebral microcirculation in rats: part 2. Local application. Stroke *30*, 880-886.

Vajkoczy, P., Laschinger, M., and Engelhardt, B. (2001). Alpha4-integrin-VCAM-1 binding mediates G protein-independent capture of encephalitogenic T cell blasts to CNS white matter microvessels. J Clin Invest *108*, 557-565.

Vajkoczy, P., Ullrich, A., and Menger, M. D. (2000). Intravital fluorescence videomicroscopy to study tumor angiogenesis and microcirculation. Neoplasia (New York) *2*, 53-61.

Virtanen, I., Gullberg, D., Rissanen, J., Kivilaakso, E., Kiviluoto, T., Laitinen, L. A., Lehto, V. P., and Ekblom, P. (2000). Laminin alpha1-chain shows a restricted distribution in epithelial basement membranes of fetal and adult human tissues. Exp Cell Res *257*, 298-309.

Vleminckx, K., and Kemler, R. (1999). Cadherins and tissue formation: integrating adhesion and signaling. Bioessays *21*, 211-220.

von Andrian, U. H., and Engelhardt, B. (2003). Alpha4 integrins as therapeutic targets in autoimmune disease. N Engl J Med *348*, 68-72.

Wakai, S., and Hirokawa, N. (1978). Development of the blood-brain barrier to horseradish peroxidase in the chick embryo. Cell Tissue Res *195*, 195-203.

Wang, Y., Calder, V. L., Lightman, S. L., and Greenwood, J. (1995). Antigen presentation by rat brain and retinal endothelial cells. J Neuroimmunol *61*, 231-239.

Wilcox, C. E., Healey, D. G., Baker, D., Willoughby, D. A., and Turk, J. L. (1989). Presentation of myelin basic protein by normal guinea-pig brain endothelial cells and its relevance to experimental allergic encephalomyelitis. Immunology *67*, 435-440.

Wilcox, C. E., Ward, A. M., Evans, A., Baker, D., Rothlein, R., and Turk, J. L. (1990). Endothelial cell expression of the intercellular adhesion molecule-1 (ICAM-1) in the central nervous system of guinea pigs during acute and chronic relapsing experimental allergic encephalomyelitis. J Neuroimmunol *30*, 43-51.

Wolburg, H., and Lippoldt, A. (2002). Tight Junctions of the blood-brain barrier. development, composition and regulation. Vasc Pharmacol *28*, 323-337.

Wolburg, H., Neuhaus, J., Kniesel, U., Krauss, B., Schmid, E. M., Ocalan, M., Farrell, C., and Risau, W. (1994). Modulation of tight junction structure in blood-brain-barrier endothelial-cells - effects of tissue-culture, 2nd messengers and cocultured astrocytes. J Cell Sci *107*, 1347-1357.

Wolburg, H., Wolburg-Buchholz, K., Kraus, J., Rascher-Eggstein, G., Liebner, S., Hamm, S., Duffner, F., Grote, E.-H., Risau, W., and Engelhardt, B. (2003). Localization of claudin-3 in tight junctions of the blood-brain barrier is selectively lost during experimental autoimmune encephalomyelitis and human glioblastoma multiforme. Acta Neuropathol, online.

Wolburg, H., Wolburg-Buchholz, K., Liebner, S., and Engelhardt, B. (2001). OSP/claudin-11, claudin-1 and claudin-2 are present in tight junctions of choroid plexus epithelium of the mouse. Neurosci Lett *13*, 77-80.

Wolburg, K., Gerhardt, H., Schulz, M., Wolburg, H., and Engelhardt, B. (1999). Ultrastructural localization of adhesion molecules in the healthy and inflamed choroid plexus of the mouse. Cell Tiss Res *296*, 259-269.

Yednock, T. A., Cannon, C., Fritz, L. C., Sanchez Madrid, F., Steinman, L., and Karin, N. (1992). Prevention of experimental autoimmune encephalomyelitis by antibodies against alpha 4 beta 1 integrin. Nature *356*, 63-66.

Chapter A21

# IMMUNOMODULATION OF EAE: ALTERED PEPTIDE LIGANDS, TOLERANCE, AND TH1/TH2

Brendan Hilliard and Youhai H. Chen
*Department of Pathology and Laboratory Medicine, University of Pennsylvania, Philadelphia, PA 19104.*

**Abstract:**     Experimental autoimmune encephalomyelitis (EAE) is an animal model for human multiple sclerosis (MS). The disease is initiated by myelin-specific T cells, but mediated by cells of both the adaptive and innate immune system. Therefore, strategies targeting either lymphocytes or myeloid cells can be effective in modulating EAE. In this chapter, we will review some of the strategies that have been developed to treat or prevent EAE. Our emphasis will be on those strategies that are still in the developmental stage, but may have the potential to be used in the clinic in the future

Keywords:     EAE, tolerance, MS, deletion, anergy, suppression.

## Introduction

Studies using experimental models of multiple sclerosis (MS) have revealed important principles of autoimmune demyelination. The presence of self-reactive T cells directed against myelin antigens present in the central nervous system (CNS) is a defining characteristic of experimental autoimmune encephalomyelitis (EAE), the animal model that shares many of the clinical and pathological features of multiple sclerosis. [1-3]. In most experimental models, EAE is induced by priming specific T cells with immunogenic forms of myelin. Myelin antigens can be immunogenic if they are administered with adjuvants such as complete Freunds adjuvant (CFA). The adjuvants stimulate antigen presenting cells and result in the differentiation of autoreactive encephalitogenic T cells, which in most models of EAE are T helper (Th) 1 cells.

The importance of T cells in EAE [reviewed in [4]] was first established by the observations that 1) depleting CD4$^+$ helper T cells [5, 6] led to the suppression of the disease and 2) depleting CD4$^+$ T cells from activated lymph node cultures of myelin basic protein (MBP)-sensitized rats prevented the adoptive transfer of EAE [7, 8].

Helper T cells can be divided into different sub-populations, based on the types of cytokines they secrete [9]. Th1 cells are CD4$^+$ T cells that secrete a distinct set of cytokines including interferon (IFN)-γ, IL-2, and TNF-α. Th2 cells on the other hand secrete IL-4, IL-5 and IL-10. The development of these two classes of helper T cells appears to be mutually antagonistic because the cytokines of one type can downregulate the development of the other [10, 11]. For instance, IL-4 potentiates the development of Th2 responses while suppressing the development of IFN-γ secreting cells [12]. On the other hand, IL-12, which is produced by antigen presenting cells and NK cells and has a major role in the differentiation of IFN-γ secreting Th1 cells, suppresses Th2 cell development [13, 14]. The vast majority of encephalitogenic T cell clones secrete IFN-γ and IL-2 and, thus, are Th1 cells. Enhancing the development of Th1 cells exacerbates the disease [15]. For example, treatment of mice with IL-12 [16-18] or deficiency in IL-4 gene expression in mice [19] potentiates the development of EAE. Anti-IL-12 antibody treatment prevents EAE in mice while anti-IL-4 treatment increases the susceptibility of mice to EAE [15].

During the past decade, many advances have been made in the area of helper T cell activation. It is now clear that for a T cell to become activated, it must receive at least two signals. The first is an antigen specific signal that activates the T cell receptor by a peptide presented in the context of the major histocompatibility complex. The CD4 molecule is also involved in generating this first signal. The CD4 molecule is responsible for binding the major histocompatibility class II molecules on the antigen-presenting cell. The second signal comes from costimulatory receptors and their ligands. The most studied such ligand/receptor pairs are 1) the CD28 receptor and its ligands B7.1 (CD80) and B7.2 (CD86), and 2) the CD40 receptor and its ligand CD154 (CD40L). It has been reported that costimulation through B7.1 may lead to a Th1 bias whereas B7.2 costimulation may give rise to a Th2 type response [20].

In EAE, after the initial activation events, encephalitogenic T cells further undergo a differentiation program to become effector cells in the draining lymph nodes and spleen, and then migrate via the blood vasculature to the central nervous system. There they leave the blood stream by passing through the vascular endothelium forming the blood brain barrier and enter the CNS tissue. It is at this point that they reencounter antigen-presenting cells harboring specific myelin antigens and execute their effector functions.

EAE, therefore, is a consequence of concerted actions of myelin specific T cells, antigen presenting cells as well as non-specific effector cells such as macrophages and granulocytes [21]. Many antigen specific and non-specific strategies have been explored to treat or prevent EAE, some of which have already been incorporated into treatment regimens for MS patients. In this chapter, we will review some of these strategies. Our emphasis is on those that are still at the experimental stage, but may have the potential to be used in the clinic in the future.

# 1. Non-specific strategies

## 1.1 Macrophage-based strategies

Blood-derived macrophages and microglia, the resident macrophage-like cell of the CNS, are believed to be critically involved in the effector phase of EAE and MS. Macrophages phagocytosing myelin in active lesions can be easily detected in the CNS. Macrophages can be directly activated by myelin via CR3 [22]. EAE is suppressed in mice depleted of macrophages following administration of mannosylated liposome-encapsulated dichloromethylene diphosphonate ($Cl_2$ MDP) [23], which regulates the leukocyte influx to the CNS. In addition to directly phagocytosing myelin, macrophages may damage myelin by producing reactive oxygen species [22, 24]. Aminoquanidine, an inhibitor of nitric oxide synthase [25], suppresses EAE. Similarly, modulating arginine uptake and subsequent NO production by macrophages was effective in suppressing the disease [26].

## 1.2 Immunoglobulin injection.

Intravenous administration of immunoglobulins from healthy donors can have beneficial effects for patients with autoimmune diseases. It has been reported that normal human serum suppresses the development of EAE [27-31]. The suppression was associated with an inhibition of antigen specific Th1 type cytokine (IFN-$\gamma$, IL-2 and TNF-$\alpha$) production [29, 31] and T cell proliferation [28, 29]. This may be due to the induction of anergy in reactive T cells since the suppression of cytokine secretion could be reversed by IL-2 in vitro [29]. Two reports have shown that Igs have no suppressive effect on the adoptive transfer of EAE [27, 31]. However, Ig injection at the time of disease onset could be effective [31]. The protective effect may be due to the changes at the TCR-peptide/MHC interface. Activated T cells specific for MBP can bind IgG, presumably through the Fc receptor since binding could be inhibited by soluble Fc molecules [27].

# 1.3  Bone Marrow Replacement Therapy

Replacement  of the 'defective' autoimmune-susceptible immune system with a nondefective autoimmun resistant immmune  system represents a rational strategy for the treatment of autoimmune disease.  This necessitates the ablation  of  the immune system of susceptible mice (usually with lethal irradiation) followed  by transfer of allogeneic bone marrow from autoimmune-resistant  strains  autoimmune-susceptible recipient, with the concommitant immunodeficiency and risk  of infection shortly after the irradiation and graft versus host disease later.  EAE-susceptible animals with bone marrow from resistant animals were resistant to induction of EAE [32-38].

In only two reports susceptible mice remained susceptible to EAE after transfer of bone marrow from resistant strains [39, 40].
Unexpectedly in experiments in rats with arthritis, syngeneic bone marrow replacement therapy was as effective in suppressing disease as allogeneic bone marrow from resistant mice [41].  This surprising result provided a rationale for similar experiments in EAE which recapitulated the results in arthritis.  In chronic and relapsing models of EAE, treatment of mice with syngeneic bone marrow transfer at the acute stage of disease was also very effective with complete remission and only rare relapses.  Treatment at the chronic stage of disease was however not effective, presumably due to the recalcitrant nature of the preexisting demyelinated lesions in the CNS [42]. In experiments using high dose cyclophosphamide the time of treatment was important, treatment on the 6th day after the first of two immunizing injections one week apart was very effective [43].  Likewise treatment during acute disease was highly effective but treatment on the 9th day after the first immunization did not prevent EAE. The induction of tolerance was proposed as a mechanism in lower dose cyclophosphamide (30% of lethal) followed by sygeneic bone marrow although the evidence was far from conclusive [44, 45].  Even treatment with autologous or pseudoautologous bone marrow could reverse the disease course of EAE although relapses both spontaneous and induced did occur [38, 46, 47].  Treatment with syngeneic bone marrow cells though, had less relapses [36, 37].  This effect has been explained by the possible elimination of residual memory T cells after immune ablation by the 'graft versus-autoimmunity' reaction [48].  This

theory was supported by the finding that myeloablation resulted in the elimination of autoreactive T cells in the CNS [47]. These experiments suggest that bone marrow transfer in human MS patients has the potential to significantly suppress disease. The high risk factors involved in the procedure however, must be very seriously considered.

# 2. Antigen-specific strategies

Administration of tolerogenic forms of myelin antigens can inhibit the development of EAE in an antigen-specific manner. Because of the relative lack of side effects, this strategy is a very attractive alternative to the currently used therapies for MS. Tolerogenic antigens can be administered intravenously (i.v.), intranasally (i.n.), or orally. Although different mechanisms may be associated with different forms of tolerance induction, the following general mechanisms may operate in most systems [49].

a) Clonal deletion. Clonal deletion is the elimination of antigen specific cells by apoptosis.
b) Clonal anergy. Clonal anergy is the induction of functional hypo-responsiveness to antigens.
c) Active suppression. Active suppression is the induction of an antigen specific population of cells, which in turn antagonise or negatively regulate the antigen specific autoreactive cells. Immune deviation is the conversion of a Th1 type (interferon-$\gamma$ secretion, delayed type hypersensitivity) response to a Th2 type (IL-4 and IL-10 secretion) response. Since Th2 type response can antagonise the Th1 response, this mode of regulation may also be considered a component of active suppression.

# 2.1 Mucosal Tolerance

The existence of regulatory mechanisms in the mucosa can restrict the development of unwanted immune responses to common or harmless antigens encountered in the air or food. Exposure of the mucosal surfaces of the intestine or the naso-pharangial mucosa can induce tolerance to antigens. The induction of tolerance to myelin-specific antigens by oral [50-52] and nasal [53-57] route is a powerful and highly specific method to suppress EAE.

## 2.1.1 Oral Tolerance

The mucosae are the major ports of entry for foreign antigens. More than 100 kilograms of food antigens are processed each year by our gastrointestinal mucosa and an estimated 0.1 to 1% of these (100-1000 grams) are absorbed in un-degraded forms. To meet up with this challenge, more than 60% of our peripheral lymphoid tissue is deployed at the gut mucosa, where it secretes large quantities of IgA with various specificities [58, 59]. Yet, not all foreign antigens are infectious pathogens. In fact, most food antigens are beneficial nutrients that must be tolerated. Therefore, the gut immune system has to differentiate two types of antigens: beneficial dietary antigens and infectious pathogens. Failure to tolerate dietary antigens may lead to intestinal hypersensitivity as exemplified by food-sensitive enteropathies, and failure to expel infectious pathogens may contribute to infections [59, 60].

## 2.1.1.1 The unique properties of the gut mucosal immune system

Collectively, the gut-associated lymphoid tissue (GALT) makes up the largest lymphoid organ of the body. It consists of a large number of lymphoid follicles scattered throughout the entire intestinal tract. In many areas, the lymphoid follicles aggregate to form lymphoid nodules called Peyer's patches. The Peyer's patch is structurally similar to peripheral lymph node in that it contains separate zones for T and B cells. In the T cell zone of the Peyer's patch, i.e., the interfollicular region (IFR), high endothelial venules (HEL) are present, allowing extravasation of naive T cells to the mucosal lymphoid tissues. The B cell zone is located at the antiluminal side of the Peyer's patch and often exhibits characteristics of secondary follicles with activated B cells forming germinal centers.

However, the Peyer's patch differs from peripheral lymph node in one important aspect: it contains a subepithelial dome (SED) which is separated from the intestinal lumen by only one layer of epithelial cells. Some of the epithelial cells overlaying the Peyer's patch have a dendritic morphology and are called microfold (M) cells. The M cells are specialized epithelial cells which function to directly transfer antigen to underlying antigen-presenting cells (APC) [61, 62]. The APC present in the SED appears to be primarily dendritic cells; macrophages are rarely detected in the SED [63, 64]. The dendritic cell in the SED is the interdigital type but differs from those present in the interfollicular regions of the Peyer's patch or those in the

# Chapter A3

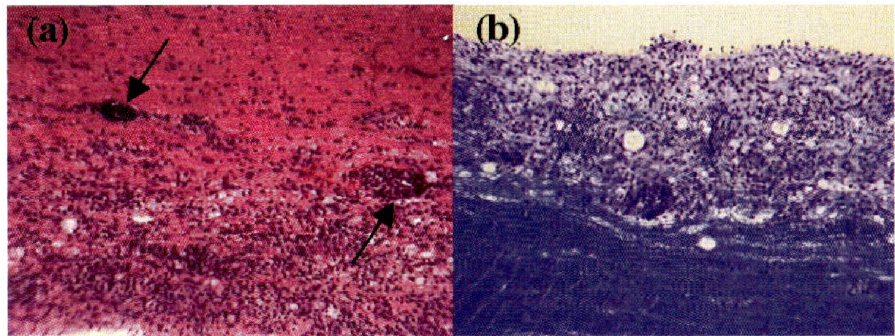

*Figure 3-1:* Serial longitudinal sections of spinal cord from a C57Bl/6 mouse with MOG induced EAE. (a) In the haematoxylin-eosin-stained section there is an intense mononuclear inflammatory infiltration of the peripheral white matter, with two perivascular cuffs at the margin of the affected area (arrows). (b) In the Luxol fast blue stained section the extent of the demyelination of this area of peripheral spinal cord white matter can be seen relative to the normally myelinated deeper tissue.

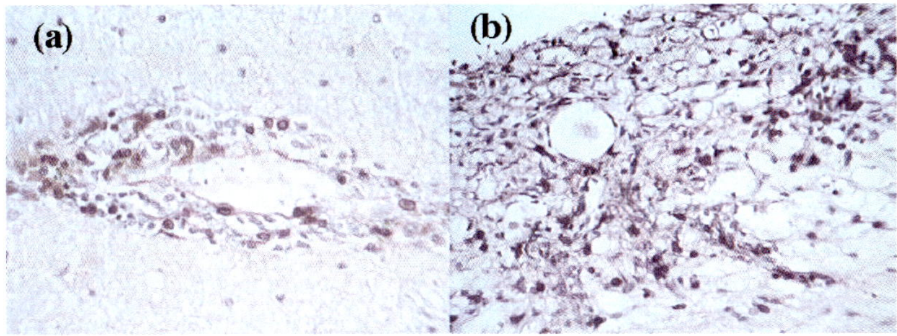

*Figure 3-2:* Immunohistochemical labelling of sections of spinal cord from a C57Bl/6 mouse with MOG induced EAE. These are formalin fixed, paraffin wax-embedded tissues that have been labelled for expression of CD3 using a cross-reactive polyclonal rabbit anti-human CD3 antiserum in an avidin-biotin immunohistochemical technique. (a) Shows T lymphocytes within a perivascular cuff in affected spinal cord white matter. (b) Shows an area of peripheral spinal cord white matter demyelination, throughout which are scattered individual T lymphocytes.

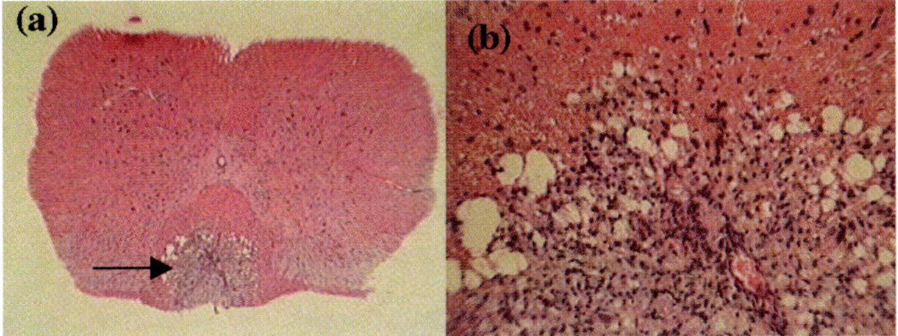

*Figure 3-3:* Sections of spinal cord from a C57Bl/6 mouse with MOG induced EAE-stained by haematoxylin-eosin. (a) There is an intense inflammatory and demyelinating lesion within the ventral white matter of the cord (arrowed). (b) Higher power view of the margin of this lesion with unaffected white matter, shows extensive macrophage-dominated inflammation.

# Chapter A3

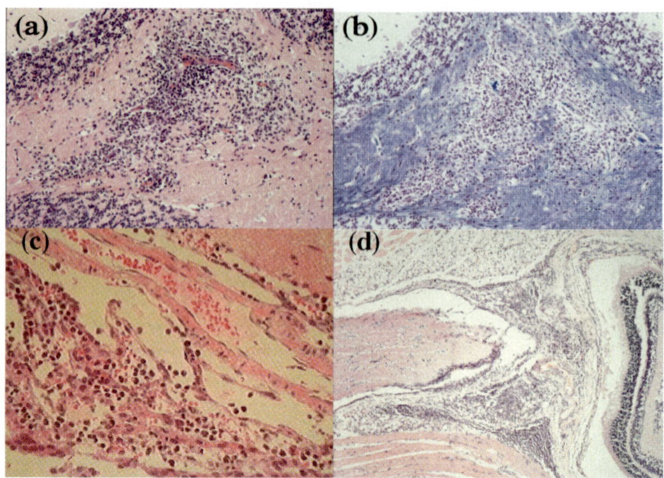

*Figure 3-4:* Tissues from C57Bl/6 mice with MOG induced EAE stained by haematoxylin and eosin (a, c and d) and Luxol fast blue (b). (a) High power view of cerebellar white matter showing a prominent area of perivascular cuffing with mononuclear inflammatory cells, with extension of the lesion into the parenchyma. (b) This serial section shows that demyelination is restricted to the area occupied by the inflammatory infiltrate, with uninflamed parenchyma showing normal staining reaction. (c) Section through the meninges, showing a mixed inflammatory infiltration throughout the meningeal layers. (d) Section through the retina, optic disc and optic nerve. There is an intense inflammatory infiltration surrounding the optic nerve (perineuritis). This comprises chiefly neutrophils and there are scattered neutrophils within the optic nerve tissue.

# Chapter A4

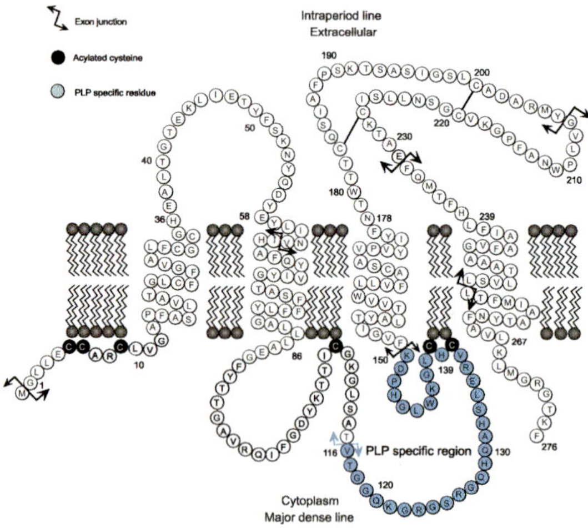


*Figure 2:* Proteolipid protein 1 structure. PLP1 is predicted to have four transmembrane domains, and both the amino and carboxy-termini in the cytoplamic compartment. A 35-amino acid long domain shaded in blue represents the PLP1-specific region of the protein. Cysteine residues that are predicted to be acylated are shaded in black.

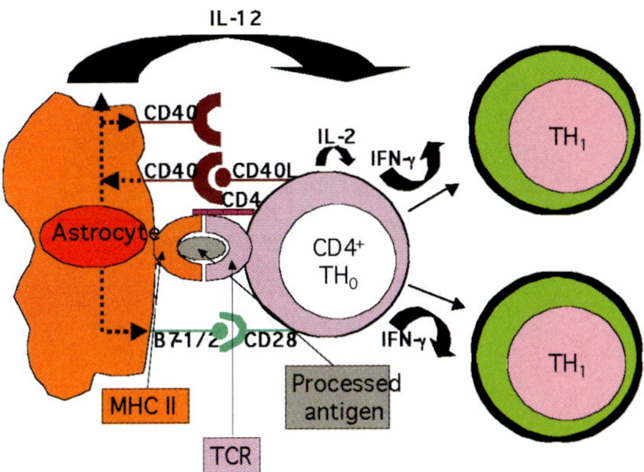



*Figure 1:* In order for antigen (Ag) presentation in the context of major histocompatibility complex (MHC) class II to occur, some absolute requirements have to be met. An MHC class II heterodimer on the cell surface of an antigen presenting cell (APC) contain and antigen binding groove, which can load a linear peptides that are typically 13-20 amino acids in length. A T cell can bind to Ag within the MHC class II binding groove with its T cell receptor. Only T cells that express the accessory surface molecule CD4 are capable of recognizing and binding linear peptide antigen (Ag) in the context of the MHC class II. Ag-activated T cells express CD40 ligand (CD40L), which engages to the costimulatory molecule CD40 on the cell surface of APC. Cross linking of CD40 and CD40L enhances expression of other costimulatory molecules on APC's, namely B7-1 (CD80), and B7-2 (CD86). Furthermore, engagement of CD40 leads to secretion of the cytokine IL-12, which induces the differentiation of naïve Th0 cells to effector CD4+ Th1 cells. These Th1 cells secrete increased amounts of proinflammatory cytokines like interferon (IFN)-g and IL-2. INF- g is required for induction of class II MHC molecules on nonprofessional antigen presenting cells (APC), including astrocytes. Some studies suggest that IFN-g also causes astrocytes to upregulate expression of B7 (B7-1 and B7-2) molecules, which are required for CD28 T cell costimulation. Thus, in cell-mediated immunity, cross-activation occurs between APC and T cell.

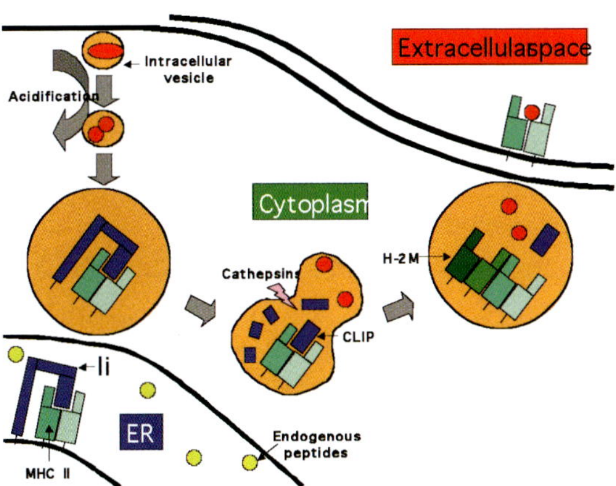



*Figure 2:* The endocytic compartment. Exogenous protein antigen (Ag) is taken up into an antigen presenting cell (APC) by endocytosis and broken down into peptide through acidification and proteolysis. The major histocompatibility complex (MHC) class II heterodimer is assembled in the endoplasmatic reticulum (ER) and then stabilized by the invariant chain (Ii). Within the ER, the Ii initially is a trimer that can bind three MHC class two heterodimers (not shown). The Ii blocks the MHC class II Ag binding groove, in order to prevent endogenous peptides in the ER from binding. Subsequently, the Ii targets the delivery of MHC class II dimers to the endosomal compartment, where it modulates proteolytic enzymes. Cathepsins are a group of cystein and aspartyl proteases that play a key role in the endosomal degradation of Ag and Ii. The last component of the Ii that remains bound to the MHC class II binding groove is named CLIP. The MHC class II like molecule (H-2M) catalyzes the release of CLIP and low stability peptides from the MHC class II Ag binding groove, and stabilizes empty MHC class II until high affinity peptide can be loaded. H-2M cannot bind Ag itself, and it is not expressed on the cell surface. Peptide loaded into the MHC class II binding groove can now be transported to the cell surface and presented to CD4+ T cells.

# Chapter A6

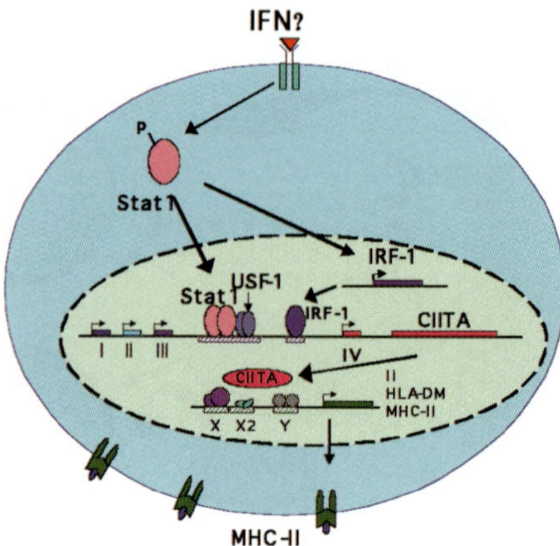



*Figure 3:* The major histocompatibility complex (MHC) class II transactivator (CIITA), a transcriptional co-activator, is the key molecule that directs interferon-gamma (IFNg)-inducible MHC class II expression in nonprofessional antigen presenting cells (APC) and constitutive MHC class II expression in professional APC. CIITA expression is controlled at the level of transcription by differential activation of multiple nonhomologous promoters. CIITA promoter, pIV, which contains a GAS site, E-box, and IRF element, directs IFNg-inducible CIITA expression in non-professional APC. It has been demonstrated that statins decrease IFNg inducible MHC class II surface expression through inhibition of CIITA pIV.

# Chapter A19

*Figure 1:* Expression of MMPs 2, 7 and 9 in perivascular inflammatory cuffs within the spinal cords of rats with experimental autoimmune encephalomyelitis.

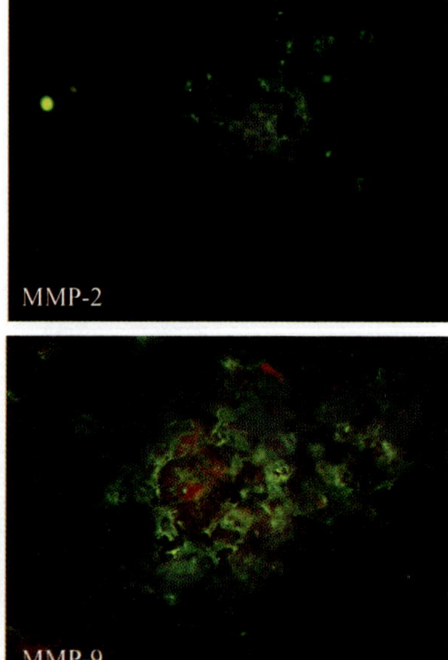

# Chapter A19

*Figure 1 (continued)*

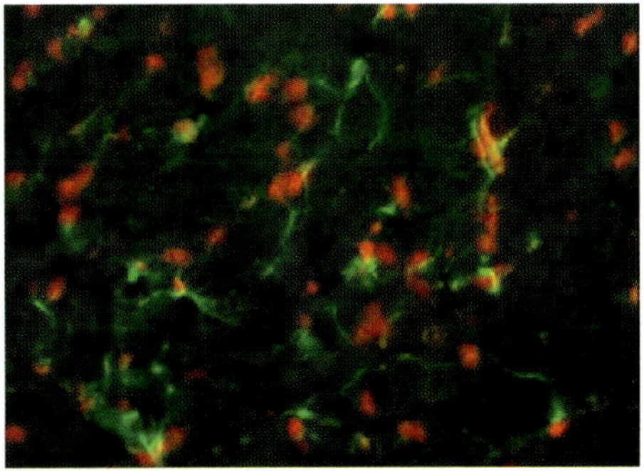

ADAM 17 expression in normal control rat brain parenchyma

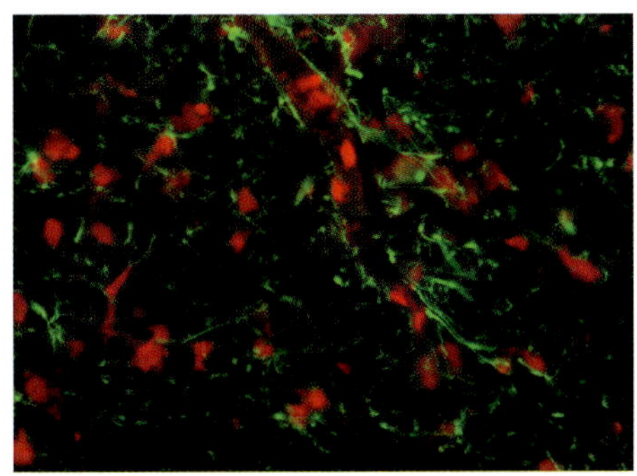

ADAM 17 (TACE) expression by perivascular astrocyctes in spinal cord of EAE rats

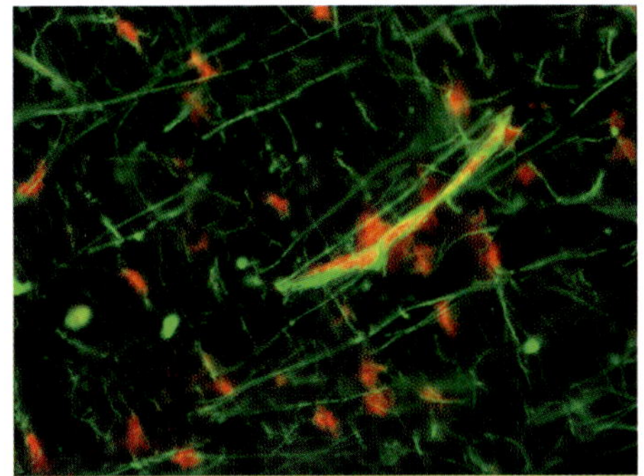

ADAM 17 expression by endothelial cells and astrocytes within spinal cord of EAE rats.

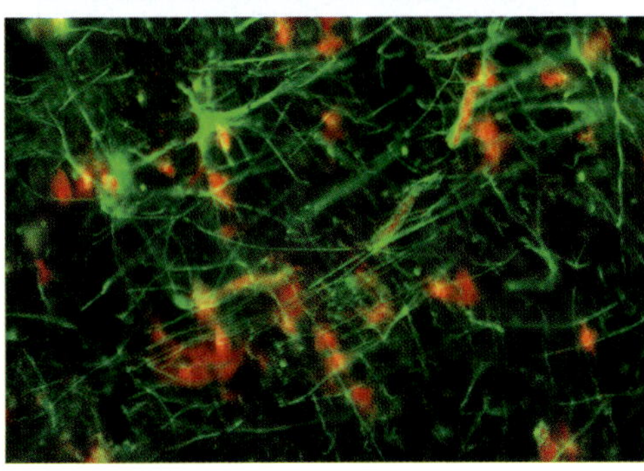

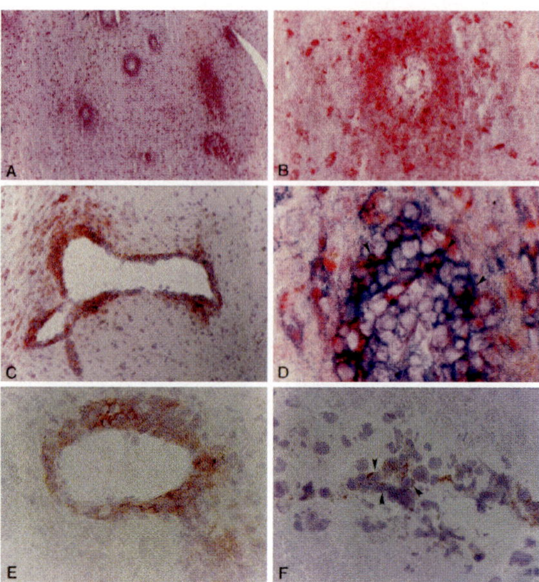

*Figure 2:* Expression of accessory molecules and cytokines in marmoset EAE brain. (a) Acid phosphatase (red) expressed by infiltrating macrophages and activated microglia (magnification = x40); (b) Detail of (a), showing extensive perivascular infiltration by macrophages (red)(x200); (c) CD40 (red) expressed on infiltrating mononuclear cells (x100); (d) Doublestaining showing macrophages with cytoplasmic acid phosphatase activity (red) and membrane CD40 expression (blue). Arrowheads indicate examples of macrophages expressing CD40 (x650); (e) CD40L (red) expressed on perivascular mononuclear cells (x200); (f) Scattered CD40L-expressing cells (red) within an infiltrate (x325). Figure reproduced with permission, from ref 17.

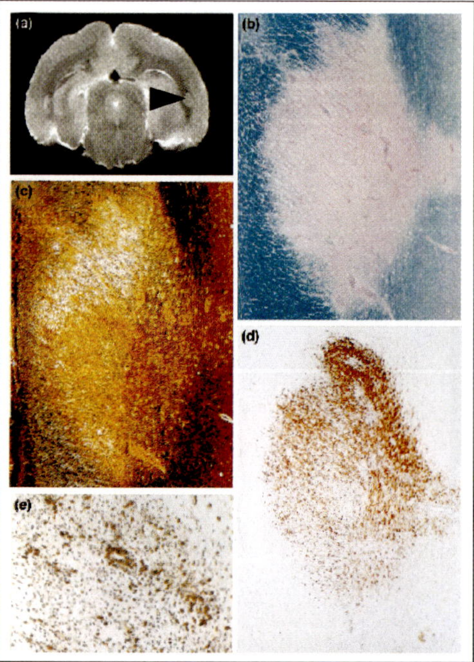

*Figure 4:* Histopathology of a magnetic resonance imaging (MRI)-detectable lesion in common marmosets. (a) coronal slice of a T2-weighted MRI recorded post mortem shows at least three MRI-detectable lesions (magnification = x2.5). The lesion indicated with the arrowhead was processed for histology. (b) Klüver-Barrera staining visualises myelin in blue, showing strong demyelination (x99). Within the same lesion axonal conservation was confirmed by Bielschowsky silver impregnation (x 99). The presence of numerous 27E10-positive macrophages (orange) (d) and CD3+ T-cells (brown) (e) classifies the lesion as late-active (x246). Figure reproduced with permission, from ref 14.

# Chapter B3

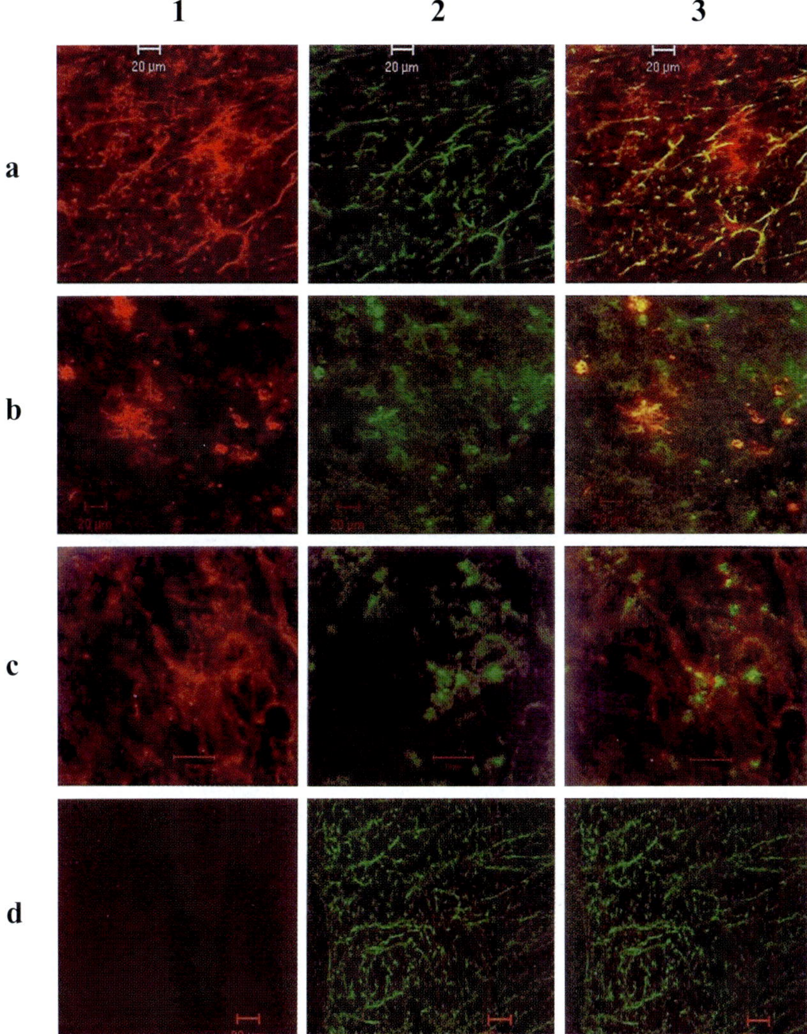

*Figure 2:* Immunohistologic analysis of spinal cord tissues from mice suffering from TMEV-induced demyelinating disease or EAE disease. Double immunostaining of frozen sections was done using both the BeAn antisera with a rhodamine conjugated secondary antibody (red) and different antibodies to cell specific markers with FITC-conjugated secondary antibodies (green). Column 1: viral antigen in (a) astrocytes, (b) microglia, (c) oligodendrocytes and (d) EAE. Column 2: (a and d) anti-GFAP for astrocytes, (b) MOMA-2 for macrophages, (c) anti-CNPase for oligodendrocytes. Column 3:colocalization of the two stains. Bar in row C measures 10 microns; all other bars measure 20 microns. (Reproduced with permission from J Neuroimmunol 118:256-267, 2001)

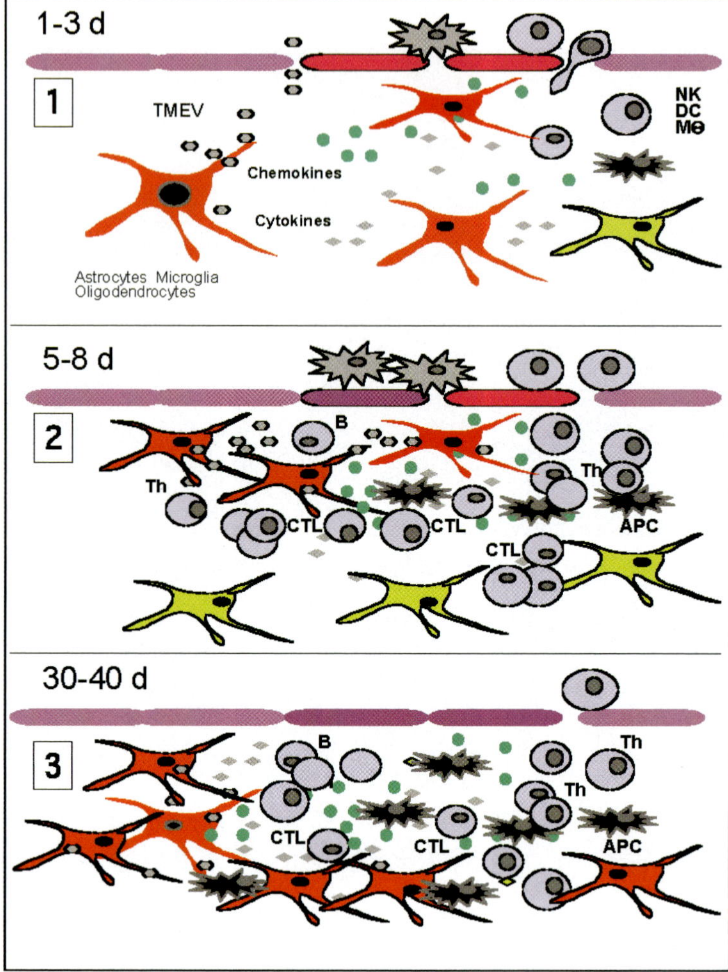



*Figure 1:* Schematic presentation of the steps involved in TMEV-induced demyelinating disease. (1) Early, (2) Acute and (3) Chronic responses in the CNS following viral infection. (1). At the Early stage, mainly chemokine, cytokine and adhesion molecule responses are directly induced in various glial cells upon viral infection. These molecules may further activate adjacent glial cells and facilitate cellular (NK, DC, macrophages, neutrophils) infiltration to the CNS. (2). At the Acute stage, various immune cells (B, Th and CTL) are infiltrated. These immune cells will be further stimulated/amplified by viral Ag-APCs (professional and non-professional). Virus will be neutralized and virus-infected cells will be eliminated by immune cells. (3). At the Chronic stage, viral persistence remains and continuous inflammatory responses are maintained in susceptible mice, whereas completely cleared viral persistence and the inflammatory response is resolved in resistant mice.

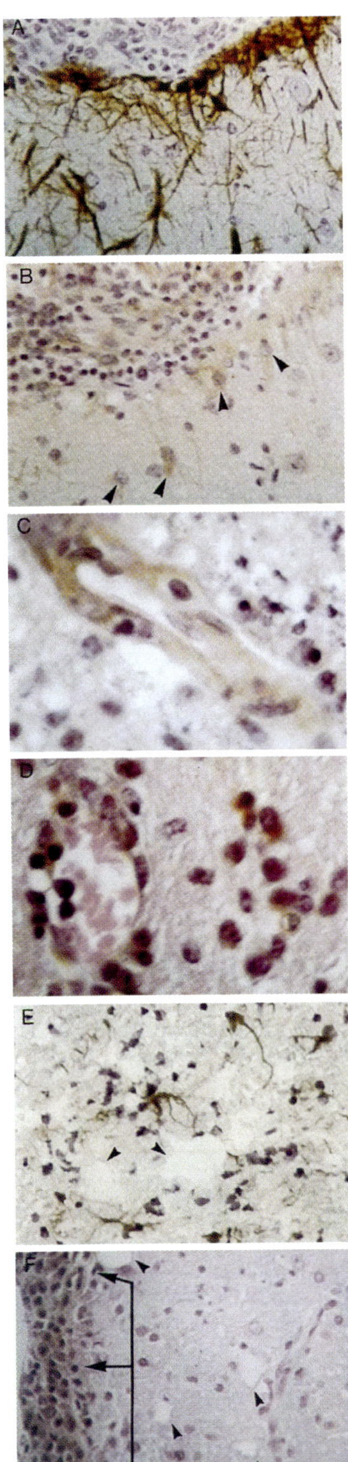

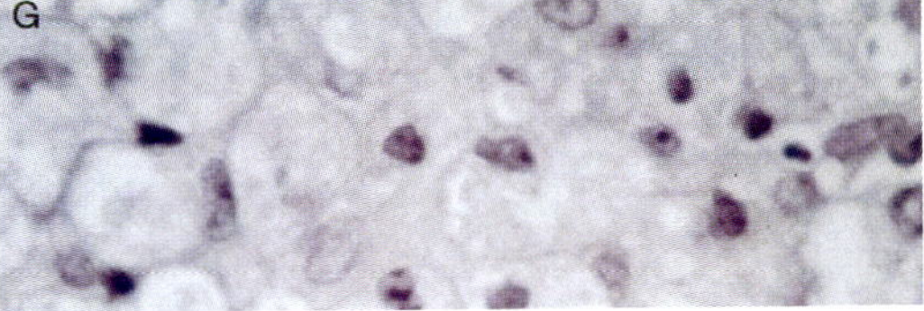

*Figure B8-1:* IHC localizations of GFAP (A) and iNOS protein in the CNS of SJL mice at 10 (A-C), 6 (D), 42 (E and F), and 66 (G) days p.i. (A) Encephalitis and poliomyelitis phase: an area of intense leptomeningeal inflammation accompanied by reactive astrogliosis in the subjacent brain parenchyma, which is highlighted by GFAP staining. (B) Same field from an immediately adjacent section exhibiting iNOS-like staining in reactive subpial astrocytes (arrowheads) as well as in scattered monocytes/macrophages in the overlying leptomeningeal infiltrate (top portion of photomicrograph). (C) Encephalitis and poliomyelitis phase: iNOS-like localization in hypertrophic endothelial cells of a blood vessel amidst an area of necrotizing encephalitis. (D) Encephalitis and poliomyelitis phase. iNOS-like staining in cells of the monocyte/macrophage lineage and in vascular endothelial cells from an area of spinal grey matter inflammation. (E) Early stage of spinal cord demyelination exhibiting a number of scattered iNOS-positive astrocytes in a background of incipient vacuolar changes (arrowheads) suggestive of early demyelination involving the lateral column. (F) Spinal cord demyelination phase: Focal, weak, and ill-defined iNOS-like immunoreactivity is detected among leptomeningeal infiltrates (the three arrows) which are encroaching upon the lateral funiculus of the spinal cord. Note vacuolar change in the lateral column (arrowheads). (G) Spinal cord demyelination phase: Lack of iNOS-like staining in large foamy (myelin-laden) macrophages in full-blown demyelinating lesions in the posterior columns of the thoracic cord. (Original magnifications: A, B, E, F  x400; C, D. G x1000)

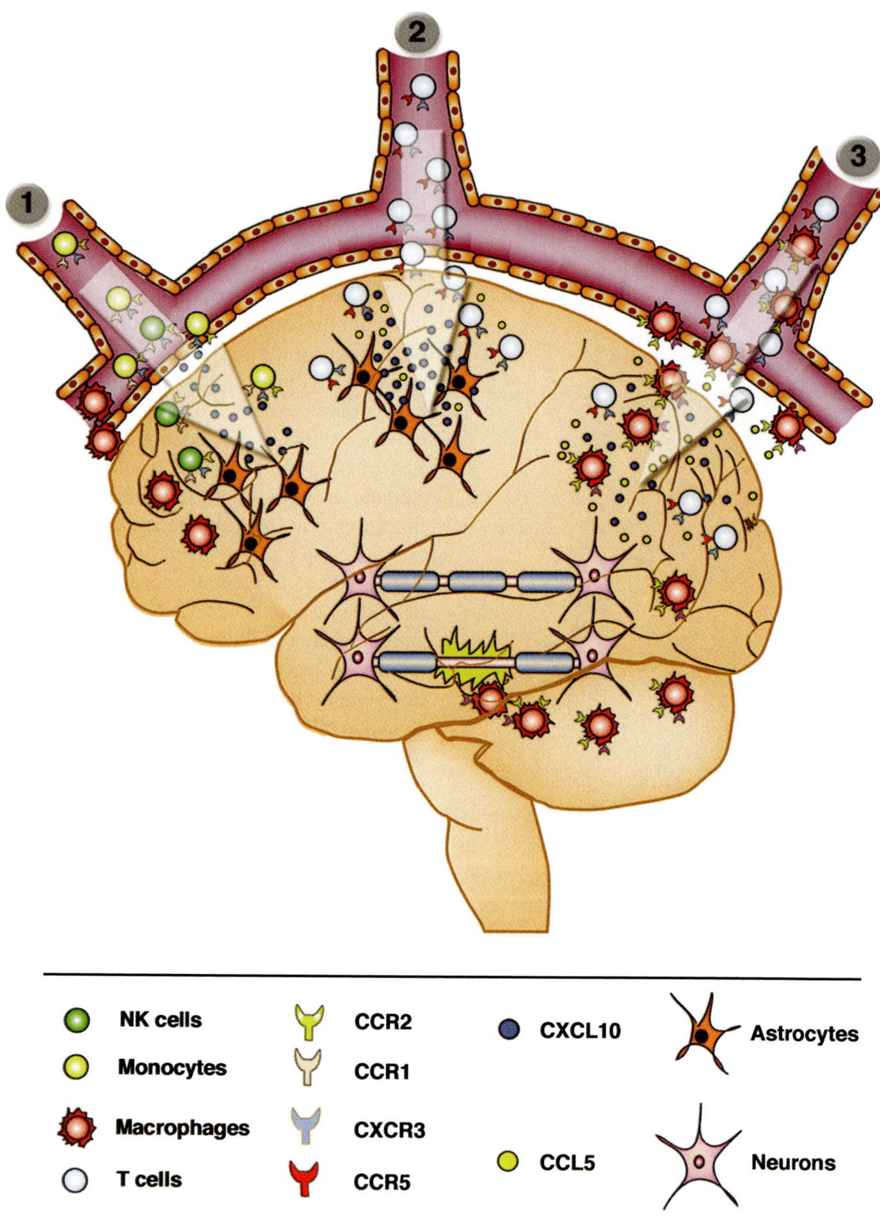



*Figure C9:* MHV infection of the CNS

paracortex of the lymph nodes: they express N418 and 2A1 but not NLDC-145 or M342 [63, 64]. Furthermore, Peyer's patch dendritic cells express 5-10 fold higher levels of MHC class II molecules than splenic dendritic cells [63, 64]. Unlike the T or B cell zones of the Peyer's patch, the SED contains both T and B cells [63-65].

The 'non-lymphoid' compartment of the mucosa is also different from that of most other organs: it contains relatively large number of lymphocytes. In fact, the 'non-lymphoid' compartment of the mucosa is often referred to as the 'diffused gut-associated lymphoid tissue' [59, 61, 62]. It consists of primarily the vili-associated epithelium and the underneath laminar propria. Inside the epithelium, a unique type of lymphocytes called intraepithelial lymphocytes are present; whereas in the laminar propria, T, B lymphocytes as well as plasma cells are abundant [58, 59]. Intraepithelial lymphocytes contain a sub-population of T cells that develop extra-thymically, most probably in intestinal epithelium. These cells express $\gamma\delta$ T cell receptors [66, 67]. Additionally, the epithelial cells of the small intestine express both class I and class II MHC molecules and can present antigen *in vitro* [58, 59].

For the development of mucosal immunity, a naive T cell must accomplish a minimum of four migration steps. First, it must extravasate at the interfollicular region of the Peyer's patch (or scattered lymphoid follicles of the intestine) through HEL. Second, it must become activated in IFR or SED and enter the B cell rich follicle of the Peyer's patch to activate specific B cells. Third, it must leave the Peyer's patch through lymph as memory T cells and re-enter systemic circulation. Finally, it must home to the laminar propria of the gut through specific "homing receptor" (e.g., $\alpha 4\beta 7$ integrin) [68-70]. Of note is that GALT is a member of the *common mucosal immune system* which serves not only the gastrointestinal tract, but also the bronchus, the lung and the urinary/genital tracts. Lymphocytes activated in the GALT will home not only to the gut but also to other mucosal tissues to effect immune function [69-71]. This is the cellular basis for oral vaccination against infections of non-gastrointestinal origins.

## 2.1.1.2 Mucosal immunization and tolerization

It is a long-recognized fact that oral administration of antigen can induce both immunity and tolerance. Oral immunization has been extensively exploited for vaccine development and is now being used for preventing a number of devastating diseases including polioviral and adenoviral infections [59, 72]. Oral tolerance on the other hand has only recently been exploited as a potential therapy for allergies, autoimmune diseases and transplant rejections [73, 74].

The potential application of oral vaccination and oral tolerization in clinical medicine has helped foster the growth of mucosal immunology during the past decades. Studies of the various cellular and molecular components of the mucosal immune system [64, 67, 72, 75-78] have enhanced tremendously our understanding the phenomena of oral immunization and tolerization. Characterization of TH1 and TH2 mucosal responses to various antigens revealed pivotal roles of cytokines in mucosal immunity [76, 79, 80]. Investigation of different pathways of tolerization led to the recognition of the importance of the antigen dosage in mucosal tolerance [74, 81, 82]. However, despite these progresses, an efficient manipulation of mucosal immune system, either for vaccination or for tolerance induction, remains to be a challenging task.

## 2.1.1.3 Deletion and non-deletion pathways of oral tolerance

Tolerance induced at the mucosa may be mediated by both deletion and non-deletion mechanisms [73, 74, 83, 84]. Although apoptotic cells can not be directly demonstrated in conventional animals (due to the extreme low frequency of antigen reactive cells), a role for deletion has been strongly suggested by the decrease in T cell frequency and by the lack of complete reversibility of the tolerance, especially when high dose antigens were used [85]. Using TcR transgenic animals [84] and conventional animals injected with transgenic T cells [86], we have demonstrated directly that mucosal exposure of antigen induced apoptosis and deletion of specific T cells. Because deletion eliminates specific cells from the system, this pathway of oral tolerance is usually not reversible and therefore the most effective.

However, deletion is not complete either in transgenic mice or conventional animals. In TcR transgenic mice, although more than 50% of transgenic T cells can be deleted in the Peyer's patch after five feedings with high dose ovalbumin (50-500 mg OVA per feeding), only about 20% decrease of transgenic cells was noted in other lymphoid organs [84]. In mice receiving an injection of TcR transgenic cells, deletion induced by high dose antigen may affect only about 50% of the transgenic cells [86] suggesting that non-deletional mechanisms also play important roles in mucosal tolerance. Non-deletion mechanisms of oral tolerance may involve clonal anergy of TH1 cells and immune deviation to TH2 cells. Clonal anergy is characterized by deficiency in interleukin (IL)-2 production and the reversibility of hypo-responsiveness by IL-2 [74, 81, 87]. By contrast, deviation is characterized by preferential activation of T cells producing TGF-β and TH2 cytokines. The deviated T cells may be not only non-pathogenic but also capable of downregulating TH1 responses through releasing their cytokines [73, 88].

It should be emphasized that deviation pathway leads to tolerance of TH1 but not TH2 cells and that the end result of immune deviation is partial or 'split' tolerance rather than full tolerance; this pathway has been shown to be a result of low dose antigen administration [81, 82, 88]. When high doses of antigens are used, both TH1 and TH2 cells are tolerized, and this may be mediated by deletion and/or anergy pathways [84, 89, 90].

## 2.1.1.4 Application of mucosal immunity and tolerance for immunotherapy.

Despite the success of polio oral vaccine, effective mucosal vaccines have not yet been developed for most infectious pathogens including human immunodeficiency virus (HIV) and various sexually transmitted microbes which invade the mucosa. Similarly, application of mucosal tolerance for treatment of autoimmune diseases and allergy has just begun to be explored; no clinical trials have yet been conducted successfully in humans, despite the bulk of evidence from animal studies that oral tolerization effectively ameliorates autoimmune diseases and allergies.

Thompson and Staines [91] and Nagler-Anderson et al. [92] initially described suppression of collagen induced arthritis by feeding type II collagen. Weiner's [93] and Whitacre's [74, 94] groups have studied suppression of experimental autoimmune encephalomyelitis by orally administered myelin antigens. Weiner's and others' laboratories have also investigated oral tolerance to suppress anti-viral immunity, allograft rejection (by feeding donor cells or MHC peptide) [95, 96] and autoimmune models of uveitis [83, 97, 98], diabetes [99] and adjuvant arthritis [100]. However, in none of these models, disease protection was complete especially when the antigens were given after the onset of the disease. Similarly, in many of these models, immune tolerance was partial, and preferential activation of certain arms of immunity was documented. Recently, Blanas et al. showed in a TcR transgenic model that oral antigen may exacerbate rather than suppress autoimmune diabetes [101]. Similarly, partial tolerance induced by intravenous antigen may delay the onset but exacerbate late stages of the disease in the marmoset model of autoimmune encephalomyelitis [102].

Using both intact TcR transgenic and adoptive transgenic mouse models, we showed that T cell activation preceded the development of mucosal tolerance [84, 86]. To determine whether blocking T cell activation during mucosal tolerization enhances the tolerance, we performed a series of experiments using B7-blocking agent CTLA4-Ig. This led to our unexpected finding that B7-blockade prevents the development of mucosal tolerance [103]. It is our belief that a thorough understanding of the cellular and molecular mechanisms of mucosal immunity and tolerance is essential for the success of mucosal antigen therapy.

### 2.1.2 Nasal Tolerance.

Intranasal (i.n.) administration of tolerogenic antigen can suppress responses not only against the tolerogen but also against other related antigens present at the same location. Thus, epitopes found in myelin basic protein can be used to regulate responses directed against other epitopes of the MBP or epitopes of the PLP, another component of the myelin. Bystander suppression of autoreactive T cells by regulatory T cells can be demonstrated following a single dose i.n. administration of autoantigen [53, 54]. The regulatory T cells appear to be short-lived since they could not regulate newly exported T cells from the thymus [54]. In rat, the bystander suppression induced by i.n. MBP was associated with the increased IL-4 and/or TGF-$\beta$ response and decreased IFN-$\gamma$ and TNF-$\alpha$ responses [56, 57]. Low doses of myelin antigen was sufficient to prevent the development of EAE whereas higher doses were necessary to suppress disease progression after active immunization. A role for regulatory CD8+ $\gamma\delta$ T cells in the airway has also been proposed [104, 105]. CD8+ $\gamma\delta$ T cells could suppress the adoptive transfer of diabetes to non-diabetic mice [106].

### 2.2 Intravenous Tolerance.

Intravenous administration of autoantigen has long been used to suppress EAE [107-109]. Tolerance was shown to be dependent on the presence of the encephalitogenic epitope in the tolerizing antigen. The timing of intravenous antigen can have a significant effect on the efficacy of the treatment. It is the most effective when the antigen is administered at the time of disease onset, or two weeks before active immunization, and the efficacy is reduced both before the time of the onset and after the establishment of the disease [110]. Tolerance induction by high dose i.v. antigen is mediated, at least in part, by the elimination of autoantigen specific T cells [111-113]. However, the induction of regulatory cells in

the spleen may also play a role [114].   In studies utilizing mice containing
the transgenic T cell receptor for the encephalitogenic epitope of MBP
(which spontaneously develop EAE), the disease course could be reversed
by repeated i.v. administration of the antigen.   This reversal was
accompanied by massive deletion of peripheral and thymic T cells and the
anergy of the remaining T cells in the periphery [115]. The lack of adequate
activating factors probably plays a role in the tolerance induction, and IL-12
can block the induction of i.v. tolerance [116].

Intravenous administration of antigen-coupled splenocytes also results
in specific suppression of EAE whose efficacy is dependent on the dose and
timing of the administration [117-119].   Anergy plays an important role in
the development of tolerance following this form of i.v. antigen
administration [120].

# 2.3 Altered Peptide Ligands

An altered peptide ligand (APL) is the MHC-binding antigenic peptide
modified by susstitution of one or more amino acids to change the MHC or
TCR binding characteristics.   It can induce tolerance to the original peptide
through several mechanisms including partial activation of specific T cells
[121, 122] and antagonism at the level of T cell activation [123].   A single
TCR can transmit different signals depending on the ligand it encounters
[121, 122, 124].   APLs have been used to regulate and modify the nature of
the T cell responses to antigens. The induction of anergy in both Th1 [125]
and Th2 [126] cells have been accomplished using altered peptide ligands.

A comparison of the response of Th1 and Th2 clones to the same
peptide/MHC complex revealed that different amino acid contact residues of
the peptide were important for TCR recognition and signaling by different
clones.  Altered peptide ligand may compete with native antigen peptide for
MHC binding to antagonise T cell activation [123].   In addition, APLs can
cross-react with native antigen specific T cell clones to induce qualitatively
and quantitatively different cytokine responses.   Altered peptide ligands,
therefore, can be used to regulate or alter immune responses.

There is ample evidence that APL can be used to suppress EAE.
Several different mechanisms may be responsible for the suppressive effect.
APL with high affinity for the MHC class II molecules can block peptide
MHC interaction [127, 128]. Regulatory cells were demonstrated following
administration of a proteolipid protein (PLP)-based APL.   This APL, when
co-injected with the original PLP peptide, could protect mice from clinical

EAE. However, the protected mice and mice immunized with APL alone showed substantial histological EAE [129, 130]. T cells isolated from the CNS of APL treated mice secreted TGF-$\beta$ and the suppression could be partially blocked by treatment with anti-TGF-$\beta$ antibody [130].

Superagonists, which are defined as altered peptide ligands with similar or lower MHC binding affinity, induce half-maximal effector responses at 100-fold lower concentrations than the original peptide [131]. An APL consisting of a single amino acid change replacing the lysine residue at position 4 of the MBP Ac1-11 with alanine (Ac1-11 [4Ala]) could bind to MHC II I-A$^u$ with a higher affinity and was a better stimulator in vitro for encephalitogenic T cell clones. However, it prevented EAE when administered in vivo [132] in (PL/JXSJL)F1 mice. Since a related but different APL with no protective effect could bind to I-A$^u$ with similar affinity, MHC blockade was ruled out as a mechanism for the EAE inhibition. Moreover, since the splenocytes from protected mice could not prevent mice from EAE, the generation of a regulatory cell population was likewise ruled out. Further experiments in this model using a superagonist amino acid substitutions at position 4 of the Ac 1-11 peptide Ac1-11 [4Tyr] demonstrated that stimulation with the weakest (original) antigen could expand the high affinity encephalitogenic T cells and cause disease. Ac1-11 [4Tyr] however deleted the high affinity T cells and stimulated a different population of T cells which had a lower affinity TCR [133]. Furthermore, responsive cells remaining after i.v. treatment with Ac1-11 [4Tyr] secreted Th2 type cytokines [134].

The modification of the antigenic peptides can have drawbacks, however. Immunization with an antagonistic APL effective in in vitro testing with a TCR transgenic T cell actually induced EAE in susceptible mice [135]. Recent human trials using APL have yielded unexpected results [136]. It is possible that tolerizing one population of encephalitogenic T cells could activate another. It is, therefore, necessary to carefully investigate the cellular and molecular mechanisms of APL action before it can be successfully applied in humans.


## Acknowledgments

We thank Ms. Nicole Stafford for excellent assistance in preparing this manuscript. This work was supported by grants from the National Institutes of Health (AI50059, NS40188, and NS40447).

# References


1. Glabinski, A.R., *et al.*, Murine experimental autoimmune encephalomyelitis: a model of immune-mediated inflammation and multiple sclerosis. *Methods Enzymol*, 1997. **288**: p. 182-90.

2. Wisniewski, H.M. and A.B. Keith, Chronic relapsing experimental allergic encephalomyelitis: an experimental model of multiple sclerosis. *Ann Neurol*, 1977. **1**(2): p. 144-8.

3. Goverman, J. and T. Brabb, Rodent models of experimental allergic encephalomyelitis applied to the study of multiple sclerosis. *Lab Anim Sci*, 1996. **46**(5): p. 482-92.

4. Zamvil, S.S. and L. Steinman, The T lymphocyte in experimental allergic encephalomyelitis. *Annu Rev Immunol*, 1990. **8**: p. 579-621.

5. Waldor, M.K., *et al.*, Reversal of experimental allergic encephalomyelitis with monoclonal antibody to a T-cell subset marker. *Science*, 1985. **227**(4685): p. 415-7.

6. Brostoff, S.W. and D.W. Mason, Experimental allergic encephalomyelitis: successful treatment in vivo with a monoclonal antibody that recognizes T helper cells. *J Immunol*, 1984. **133**(4): p. 1938-42.

7. Pettinelli, C.B. and D.E. McFarlin, Adoptive transfer of experimental allergic encephalomyelitis in SJL/J mice after in vitro activation of lymph node cells by myelin basic protein: requirement for Lyt 1+ 2- T lymphocytes. *J Immunol*, 1981. **127**(4): p. 1420-3.

8. Brostoff, S.W. and D.W. Mason, The role of lymphocyte subpopulations in the transfer of rat EAE. *J Neuroimmunol*, 1986. **10**(4): p. 331-40.

9. Mosmann, T.R. and R.L. Coffman, Heterogeneity of cytokine secretion patterns and functions of helper T cells. *Adv Immunol*, 1989. **46**: p. 111-47.

10. Street, N.E. and T.R. Mosmann, Functional diversity of T lymphocytes due to secretion of different cytokine patterns. *Faseb J*, 1991. **5**(2): p. 171-7.

11. Swain, S.L., *et al.*, IL-4 directs the development of Th2-like helper effectors. *J Immunol*, 1990. **145**(11): p. 3796-806.

12. Seder, R.A. and W.E. Paul, Acquisition of lymphokine-producing phenotype by CD4+ T cells. *Annu Rev Immunol*, 1994. **12**: p. 635-73.

13. Manetti, R., *et al.*, Natural killer cell stimulatory factor (interleukin 12 [IL-12]) induces T helper type 1 (Th1)-specific immune responses and inhibits the development of IL-4-producing Th cells. *J Exp Med*, 1993. **177**(4): p. 1199-204.

14. Seder, R.A., *et al.*, Interleukin 12 acts directly on CD4+ T cells to enhance priming for interferon gamma production and diminishes interleukin 4 inhibition of such priming. *Proc Natl Acad Sci U S A*, 1993. **90**(21): p. 10188-92.

15. Constantinescu, C.S., *et al.*, Modulation of susceptibility and resistance to an autoimmune model of multiple sclerosis in prototypically susceptible and resistant strains by neutralization of interleukin-12 and interleukin-4, respectively. *Clin Immunol*, 2001. **98**(1): p. 23-30.

16. Leonard, J.P., *et al.*, Regulation of the inflammatory response in animal models of multiple sclerosis by interleukin-12. *Crit Rev Immunol*, 1997. **17**(5-6): p. 545-53.

17. Constantinescu, C.S., *et al.*, IL-12 reverses the suppressive effect of the CD40 ligand blockade on experimental autoimmune encephalomyelitis (EAE). *J Neurol Sci*, 1999. **171**(1): p. 60-4.

18.     Constantinescu, C.S., *et al.*, Antibodies against IL-12 prevent superantigen-induced and spontaneous relapses of experimental autoimmune encephalomyelitis. *J Immunol*, 1998. **161**(9): p. 5097-104.

19.     Falcone, M., *et al.*, A critical role for IL-4 in regulating disease severity in experimental allergic encephalomyelitis as demonstrated in IL-4-deficient C57BL/6 mice and BALB/c mice. *J Immunol*, 1998. **160**(10): p. 4822-30.

20.     Kuchroo, V.K., *et al.*, B7-1 and B7-2 costimulatory molecules differentially activate the TH1/TH2 developmental pathways: application to autoimmune disease therapy. *Cell*, 1995. **80**: p. 707-18.

21.     Steinman, L., A few autoreactive cells in an autoimmune infiltrate control a vast population of nonspecific cells: a tale of smart bombs and the infantry. *Proc Natl Acad Sci U S A*, 1996. **93**(6): p. 2253-6.

22.     van der Laan, L.J., *et al.*, Macrophage phagocytosis of myelin in vitro determined by flow cytometry: phagocytosis is mediated by CR3 and induces production of tumor necrosis factor-alpha and nitric oxide. *J Neuroimmunol*, 1996. **70**(2): p. 145-52.

23.     Huitinga, I., *et al.*, Macrophages in T cell line-mediated, demyelinating, and chronic relapsing experimental autoimmune encephalomyelitis in Lewis rats. *Clin Exp Immunol*, 1995. **100**(2): p. 344-51.

24.     Ruuls, S.R., *et al.*, Reactive oxygen species are involved in the pathogenesis of experimental allergic encephalomyelitis in Lewis rats. *J Neuroimmunol*, 1995. **56**(2): p. 207-17.

25.     Zhao, W., *et al.*, Experimental allergic encephalomyelitis in the rat is inhibited by aminoguanidine, an inhibitor of nitric oxide synthase. *J Neuroimmunol*, 1996. **64**(2): p. 123-33.

26.     Martiney, J.A., *et al.*, Prevention and treatment of experimental autoimmune encephalomyelitis by CNI-1493, a macrophage-deactivating agent. *J Immunol*, 1998. **160**(11): p. 5588-95.

27.     Achiron, A., *et al.*, Suppression of experimental autoimmune encephalomyelitis by intravenously administered polyclonal immunoglobulins. *J Autoimmun*, 2000. **15**(3): p. 323-30.

28.     Pashov, A., *et al.*, Normal immunoglobulin G protects against experimental allergic encephalomyelitis by inducing transferable T cell unresponsiveness to myelin basic protein. *Eur J Immunol*, 1998. **28**(6): p. 1823-31.

29.     Pashov, A., *et al.*, A shift in encephalitogenic T cell cytokine pattern is associated with suppression of EAE by intravenous immunoglobulins (IVIg). *Mult Scler*, 1997. **3**(2): p. 153-6.

30.     Achiron, A., *et al.*, Intravenous gammaglobulin treatment in multiple sclerosis and experimental autoimmune encephalomyelitis: delineation of usage and mode of action. *J Neurol Neurosurg Psychiatry*, 1994. **57**(Suppl): p. 57-61.

31.     Achiron, A., *et al.*, Intravenous immunoglobulin treatment of experimental T cell-mediated autoimmune disease. Upregulation of T cell proliferation and downregulation of tumor necrosis factor alpha secretion. *J Clin Invest*, 1994. **93**(2): p. 600-5.

32.     Singer, D.E., M.J. Moore, and R.M. Williams, EAE in rat bone marrow chimeras: analysis of the cellular mechanism of BN resistance. *J Immunol*, 1981. **126**(4): p. 1553-7.

33.     Ben-Nun, A., H. Otmy, and I.R. Cohen, Genetic control of autoimmune encephalomyelitis and recognition of the critical nonapeptide moiety of myelin basic protein in guinea pigs are exerted through interaction of lymphocytes and macrophages. *Eur J Immunol*, 1981. **11**(4): p. 311-6.

34.  Pelfrey, C.M., F.J. Waxman, and C.C. Whitacre, Genetic resistance in experimental autoimmune encephalomyelitis. I. Analysis of the mechanism of LeR resistance using radiation chimeras. *Cell Immunol*, 1989. **122**(2): p. 504-16.

35.  Binder, T.A., *et al.*, Relative susceptibility of SJL/J and B10.S mice to experimental allergic encephalomyelitis (EAE) is determined by the ability of prethymic cells in bone marrow to develop into EAE effector T cells. *J Neuroimmunol*, 1993. **42**(1): p. 23-32.

36.  van Gelder, M. and D.W. van Bekkum, Treatment of relapsing experimental autoimmune encephalomyelitis in rats with allogeneic bone marrow transplantation from a resistant strain. *Bone Marrow Transplant*, 1995. **16**(3): p. 343-51.

37.  van Gelder, M., A.H. Mulder, and D.W. van Bekkum, Treatment of relapsing experimental autoimmune encephalomyelitis with largely MHC-matched allogeneic bone marrow transplantation. *Transplantation*, 1996. **62**(6): p. 810-8.

38.  van Gelder, M. and D.W. van Bekkum, Effective treatment of relapsing experimental autoimmune encephalomyelitis with pseudoautologous bone marrow transplantation. *Bone Marrow Transplant*, 1996. **18**(6): p. 1029-34.

39.  Korngold, R., *et al.*, Acute experimental allergic encephalomyelitis in radiation bone marrow chimeras between high and low susceptible strains of mice. *Immunogenetics*, 1986. **24**(5): p. 309-15.

40.  Lublin, F.D., *et al.*, Relapsing experimental allergic encephalomyelitis in radiation bone marrow chimeras between high and low susceptible strains of mice. *Clin Exp Immunol*, 1986. **66**(3): p. 491-6.

41.  van Bekkum, D.W., *et al.*, Regression of adjuvant-induced arthritis in rats following bone marrow transplantation. *Proc Natl Acad Sci U S A*, 1989. **86**(24): p. 10090-4.

42.  Burt, R.K., *et al.*, Effect of disease stage on clinical outcome after syngeneic bone marrow transplantation for relapsing experimental autoimmune encephalomyelitis. *Blood*, 1998. **91**(7): p. 2609-16.

43.  Karussis, D.M., *et al.*, Chronic-relapsing experimental autoimmune encephalomyelitis (CR-EAE): treatment and induction of tolerance, with high dose cyclophosphamide followed by syngeneic bone marrow transplantation. *J Neuroimmunol*, 1992. **39**(3): p. 201-10.

44.  Karussis, D.M., *et al.*, Prevention of experimental autoimmune encephalomyelitis and induction of tolerance with acute immunosuppression followed by syngeneic bone marrow transplantation. *J Immunol*, 1992. **148**(6): p. 1693-8.

45.  Karussis, D.M., *et al.*, Prevention and reversal of adoptively transferred, chronic relapsing experimental autoimmune encephalomyelitis with a single high dose cytoreductive treatment followed by syngeneic bone marrow transplantation. *J Clin Invest*, 1993. **92**(2): p. 765-72.

46.  van Gelder, M., E.P. Kinwel-Bohre, and D.W. van Bekkum, Treatment of experimental allergic encephalomyelitis in rats with total body irradiation and syngeneic BMT. *Bone Marrow Transplant*, 1993. **11**(3): p. 233-41.

47.  Burt, R.K., *et al.*, Syngeneic bone marrow transplantation eliminates V beta 8.2 T lymphocytes from the spinal cord of Lewis rats with experimental allergic encephalomyelitis. *J Neurosci Res*, 1995. **41**(4): p. 526-31.

48.  Marmont, A.M., Immunoablation followed or not by hematopoietic stem cells as an intense therapy for severe autoimmune diseases. New perspectives, new problems. *Haematologica*, 2001. **86**(4): p. 337-45.

49.  Miller, A., D.A. Hafler, and H.L. Weiner, Tolerance and suppressor mechanisms in experimental autoimmune encephalomyelitis: implications for immunotherapy of human autoimmune diseases. *Faseb J*, 1991. **5**(11): p. 2560-6.

50.  Whitacre, C.C., *et al.*, Oral tolerance in experimental autoimmune encephalomyelitis. *Ann N Y Acad Sci*, 1996. **778**: p. 217-27.

51.    Hafler, D.A. and H.L. Weiner, Antigen-specific immunosuppression: oral tolerance for the treatment of autoimmune disease. *Chem Immunol*, 1995. **60**: p. 126-49.

52.    Weiner, H.L., *et al.*, Antigen-driven peripheral immune tolerance. Suppression of organ-specific autoimmune diseases by oral administration of autoantigens. *Ann N Y Acad Sci*, 1991. **636**: p. 227-32.

53.    Metzler, B. and D.C. Wraith, Inhibition of experimental autoimmune encephalomyelitis by inhalation but not oral administration of the encephalitogenic peptide: influence of MHC binding affinity. *Int Immunol*, 1993. **5**(9): p. 1159-65.

54.    Metzler, B. and D.C. Wraith, Inhibition of T-cell responsiveness by nasal peptide administration: influence of the thymus and differential recovery of T-cell-dependent functions. *Immunology*, 1999. **97**(2): p. 257-63.

55.    Liu, J.Q., *et al.*, Inhibition of experimental autoimmune encephalomyelitis in Lewis rats by nasal administration of encephalitogenic MBP peptides: synergistic effects of MBP 68-86 and 87-99. *Int Immunol*, 1998. **10**(8): p. 1139-48.

56.    Shi, F.D., *et al.*, Nasal administration of multiple antigens suppresses experimental autoimmune myasthenia gravis, encephalomyelitis and neuritis. *J Neurol Sci*, 1998. **155**(1): p. 1-12.

57.    Bai, X.F., *et al.*, Nasal administration of myelin basic protein prevents relapsing experimental autoimmune encephalomyelitis in DA rats by activating regulatory cells expressing IL-4 and TGF-beta mRNA. *J Neuroimmunol*, 1997. **80**(1-2): p. 65-75.

58.    Brandtzaeg, P., *et al.*, *Chapter 1: Nature and properties of the human gastrointestinal immune system*, in *Immunology of the Gastrointestinal Tract*, K. Miller and S. Nicklin, Editors. 1987, CRC Press, Inc.: Boca Raton, Florida. p. 1-88.

59.    Brandtzaeg, P., Overview of the mucosal immune system. *Current Topics in Microbiology and Immunology*, 1989. **146**: p. 13-28.

60.    Mowat, A.M., The regulation of immune responses to dietary protein antigens. *Immunology Today*, 1987. **8**: p. 93-98.

61.    Neutra, M.R., A. Frey, and J.P. Kraehenbuhl, Epithelial M cells: gateways for mucosal infection and immunization. *Cell*, 1996. **86**(3): p. 345-8.

62.    Wolf, J.L. and W.A. Bye, The membranous epithelial (M) cell and the mucosal immune system. *Annual Review of Medicine*, 1984. **35**: p. 95-112.

63.    Kelsall, B.L. and W. Strober, Distinct populations of dendritic cells are present in the subepithelial dome and T cell regions of the murine Peyer's patch. *Journal of Experimental Medicine*, 1996. **183**: p. 237-47.

64.    Kelsall, B.L. and W. Strober, The role of dendritic cells in antigen processing in the Peyer's patch. *Ann. of N.Y. Acad. Sci.*, 1996. **778**: p. 47-54.

65.    Ermak, T.H. and R.L. Owen, Differential distribution of lymphocytes and accessory cells in mouse Peyer's patches. *Anatomical Record*, 1986. **215**: p. 144-152.

66.    Bandeira, A., *et al.*, Extrathymic origin of intestinal intraepithelial lymphocytes bearing T-cell antigen receptor γδ. *Proceedings of the National Academy of Sciences USA*, 1991. **88**: p. 43-47.

67.    Fujihashi, K., *et al.*, Immunoregulatory function of CD3$^+$, CD4$^-$, and CD8$^-$ T cells. γδ T cell receptor-positive T cells from nude mice abrogate oral tolerance. *Journal of Immunology*, 1989. **143**(11): p. 3415-3422.

68.    Berlin, C., *et al.*, Alpha 4 beta 7 integrin mediates lymphocyte binding to the mucosal vascular addressin MAdCAM-1. *Cell*, 1993. **74**(1): p. 185-5.

69.    Hamann, A., *et al.*, Role of alpha 4-integrins in lymphocyte homing to mucosal tissues in vivo. *Journal of Immunology*, 1994. **152**(7): p. 3282-93.

70.    Mackay, C.R., *et al.*, Phenotype, and migration properties of three major subsets of tissue homing T cells in sheep. *European Journal of Immunology*, 1996. **26**(10): p. 2433-9.

71.     Cepek, K.L., *et al.*, Adhesion between epithelial cells and T lymphocytes mediated by E-cadherin and the a4b7 integrin. *Nature*, 1994. **372**: p. 190-193.

72.     McGhee, J.R. and J. Mestecky, Oral immunization: a summary. *Current Topics in Microbiology and Immunology*, 1989. **146**: p. 233-237.

73.     Weiner, H.L., *et al.*, Oral tolerance: Immunologic mechanisms and treatment of murine and human organ specific autoimmune diseases by oral administration of autoantigens. *Annu. Rev. Immunol.*, 1994. **12**: p. 809.

74.     Whitacre, C.C., *et al.*, Oral tolerance in experimental autoimmune encephalomyelits. III. Evidence for clonal anergy. *Journal of Immunology*, 1991. **147**: p. 2155-2163.

75.     Fujihashi, K., *et al.*, Immunoregulatory functions for murine intraepithelial lymphocytes: $\gamma/\delta$ T cell receptor-positive (TCR$^+$) T cells abrogate oral tolerance while $\alpha/\beta$ TCR$^+\square$ T cells provide B cell help. $\square$*Journal of Experimental Medicine*, 1992. **175**(3): p. 695-707.

76.     Xu-Amano, J., *et al.*, Selective induction of Th$_2$ cells in murine Peyer's patches by oral immunization. *International Immunology*, 1992. **4**(4): p. 433-445.

77.     MacDonald, T.T. and J. Spencer, Ontogeny of the mucosal immune response. *Springer Seminars in Immunopathology*, 1990. **12**: p. 129-137.

78.     Li, Y., X.Y. Yio, and L. Mayer, Human intestinal epithelial cell-induced CD8+ T cell activation is mediated through CD8 and the activation of CD8-associated p56lck. *Journal of Experimental Medicine*, 1995. **182**(4): p. 1079-88.

79.     VanCott, J.L., *et al.*, Regulation of mucosal and systemic antibody responses by T helper cell subsets, macrophages, and derived cytokines following oral immunization with live recombinant Salmonella. *Journal of Immunology*, 1996. **156**(4): p. 1504-14.

80.     Daynes, R., *et al.*, Regulation of murine lymphokine production in vivo. III. The lymphoid tissue microenvironment exerts regulatory influences over T helper cell function. *Journal of Experimental Medicine*, 1990. **171**(April): p. 979-996.

81.     Friedman, A. and H.L. Weiner, Induction of anergy or active suppression following oral tolerance is determined by antigen dosage. *Proceedings of the National Academy of Science, USA*, 1994. **91**: p. 6688-6692.

82.     Gregerson, D.S., W.F. Obritsch, and L.A. Donoso, Oral tolerance in experimental autoimmune uveoretinitis: Distinct mechanisms of resistance are induced by low versus high dose feeding protocols. *Journal of Immunology*, 1993. **151**: p. 5751-5761.

83.     Rizzo, L.V., *et al.*, Interleukin-2 treatment potentiates induction of oral tolerance in a murine model of autoimmunity. *Journal of Clinical Investigation*, 1994. **94**(4): p. 1668-72.

84.     Chen, Y., *et al.*, Peripheral deletion of antigen-reactive T cells in oral tolerance. *Nature*, 1995. **376**: p. 177-180.

85.     Whitacre, C.C., *et al.*, Oral tolerance in experimental autoimmune encephalomyelitis. *Ann. of N.Y. Acad. Sci.*, 1996. **778**: p. 217-227.

86.     Chen, Y., J.-i. Inobe, and H.L. Weiner, Inductive events in oral tolerance in the TCR transgenic adoptive transfer model. *Cell. Immunol.*, 1997. **178**: p. 62-68.

87.     Van Houten, N. and S.F. Blake, Direct measurement of anergy of antigen-specific T cells following oral tolerance induction. *Journal of Immunology*, 1996. **157**(4): p. 1337-41.

88.     Chen, Y., *et al.*, Regulatory T cell clones induced by oral tolerance: suppression of autoimmune encephalomyelitis. *Science*, 1994. **265**: p. 1237-1240.

89.     Garside, P., *et al.*, T helper 2 cells are subject to high dose oral tolerance and are not essential for its induction. *Journal of Immunology*, 1995. **154**(11): p. 5649-55.

90.    Garside, P., *et al.*, Lymphocytes from orally tolerized mice display enhanced susceptibility to death by apoptosis when cultured in the absence of antigen in vitro. *American Journal of Pathology*, 1996. **149**(6): p. 1971-9.

91.    Thompson, H.S.G. and N.A. Staines, Gastric administration of type II collagen delays the onset and severity of collagen-induced arthritis in rats. *Clinical and Experimental Immunology*, 1986. **64**: p. 581-586.

92.    Nagler-Anderson, C., *et al.*, Suppression of type II collagen-induced arthritis by intragastric administration of soluble type II collagen. *Proceedings of the National Academy of Sciences USA*, 1986. **83**: p. 7443-7446.

93.    Higgins, P. and H.L. Weiner, Suppression of experimental autoimmune encephalomyelitis by oral administration of myelin basic protein and its fragments. *Journal of Immunology*, 1988. **140**: p. 440-445.

94.    Bitar, D.M. and C.C. Whitacre, Suppression of autoimmune encephalomyelitis by the oral administration of myelin basic protein. *Cellular Immunology*, 1988. **112**(2): p. 364-370.

95.    Sayegh, M.H., *et al.*, Induction of immunity and oral tolerance with polymorphic class II major histocompatability complex allopeptides in the rat. *Proceedings of the National Academy of Sciences USA*, 1992. **89**: p. 7762-7766.

96.    Sayegh, M.H., *et al.*, Down-regulation of the immune response to histocompatibility antigen and prevention of sensitization by skin allografts by orally administered alloantigen. *Transplantation*, 1992. **53**: p. 163-166.

97.    Nussenblatt, R.B., *et al.*, Inhibition of S-antigen induced experimental autoimmune uveoretinitis by oral induction of tolerance with S-antigen. *Journal of Immunology*, 1990. **144**(5): p. 1689-1695.

98.    Caspi, R.R., *et al.*, T cell lines mediating experimental autoimmune uveoretinitis (EAU) in the rat. *Journal of Immunology*, 1986. **136**: p. 928-33.

99.    Zhang, J.A., *et al.*, Suppression of diabetes in NOD mice by oral administration of porcine insulin. *Proc. Natl. Acad. Sci. USA*, 1991. **88**: p. 10252-56.

100.   Zhang, J.Z., *et al.*, Suppression of adjuvant arthritis in Lewis rats by oral administration of type II collagen. *Journal of Immunology*, 1990. **145**: p. 2489-2493.

101.   Blanas, E., *et al.*, Induction of autoimmune diabetes by oral administration of autoantigen. *Science*, 1996. **274**(5293): p. 1707-9.

102.   Genain, C.P., *et al.*, Late complications of immune deviation therapy in a nonhuman primate [see comments]. *Science*, 1996. **274**(5295): p. 2054-7.

103.   Samoilova, E.B., *et al.*, CTLA-4 is required for the induction of high dose oral tolerance. *International Immunology*, 1998. **10**(4): p. 491-8.

104.   Lahn, M., The role of gammadelta T cells in the airways. *J Mol Med*, 2000. **78**(8): p. 409-25.

105.   Pawankar, R., gammadelta T cells in allergic airway diseases. *Clin Exp Allergy*, 2000. **30**(3): p. 318-23.

106.   Harrison, L.C., *et al.*, Aerosol insulin induces regulatory CD8 gamma delta T cells that prevent murine insulin-dependent diabetes. *J Exp Med*, 1996. **184**(6): p. 2167-74.

107.   Levine, S., E.M. Hoenig, and M.W. Kies, Allergic encephalomyelitis: immunologically specific inhibition of cellular passive transfer by encephalitogenic basic proteins. *Clin Exp Immunol*, 1970. **6**(4): p. 503-17.

108.   Falk, G.A., M.W. Kies, and E.C. Alvord, Jr., Passive transfer of experimental allergic encephalomyelitis: mechanisms of suppression. *J Immunol*, 1969. **103**(6): p. 1248-53.

109.   Driscoll, B.F., M.W. Kies, and E.C. Alvord, Jr., Adoptive transfer of experimental allergic encephalomyelitis (EAE): prevention of successful transfer by treatment of donors with myelin basic protein. *J Immunol*, 1975. **114**(1 Pt 2): p. 291-2.

110.   Hilliard, B., E.S. Ventura, and A. Rostami, Effect of timing of intravenous administration of myelin basic protein on the induction of tolerance in experimental allergic encephalomyelitis. *Mult Scler*, 1999. **5**(1): p. 2-9.

111.   Racke, M.K., *et al.*, Intravenous antigen administration as a therapy for autoimmune demyelinating disease. *Ann Neurol*, 1996. **39**(1): p. 46-56.

112.   McFarland, H.I., *et al.*, Amelioration of autoimmune reactions by antigen-induced apoptosis of T cells. *Adv Exp Med Biol*, 1995. **383**: p. 157-66.

113.   Critchfield, J.M., *et al.*, T cell deletion in high antigen dose therapy of autoimmune encephalomyelitis. *Science*, 1994. **263**(5150): p. 1139-43.

114.   Hilliard, B.A., *et al.*, Mechanisms of suppression of experimental autoimmune encephalomyelitis by intravenous administration of myelin basic protein: role of regulatory spleen cells. *Exp Mol Pathol*, 2000. **68**(1): p. 29-37.

115.   Zhang, G.X., *et al.*, Reversal of spontaneous progressive autoimmune encephalomyelitis by myelin basic protein-induced clonal deletion. *Autoimmunity*, 1999. **31**(4): p. 219-27.

116.   Zhang, G.X., *et al.*, The role of IL-12 in the induction of intravenous tolerance in experimental autoimmune encephalomyelitis. *J Immunol*, 2002. **168**(5): p. 2501-7.

117.   Kennedy, K.J., *et al.*, Induction of antigen-specific tolerance for the treatment of ongoing, relapsing autoimmune encephalomyelitis: a comparison between oral and peripheral tolerance. *J Immunol*, 1997. **159**(2): p. 1036-44.

118.   Vandenbark, A.A., *et al.*, Prevention and treatment of relapsing autoimmune encephalomyelitis with myelin peptide-coupled splenocytes. *J Neurosci Res*, 1996. **45**(4): p. 430-8.

119.   Vandenbark, A.A., *et al.*, Myelin antigen-coupled splenocytes suppress experimental autoimmune encephalomyelitis in Lewis rats through a partially reversible anergy mechanism. *J Immunol*, 1995. **155**(12): p. 5861-7.

120.   Tan, L.J., *et al.*, Regulation of the effector stages of experimental autoimmune encephalomyelitis via neuroantigen-specific tolerance induction. III. A role for anergy/deletion. *Autoimmunity*, 1998. **27**(1): p. 13-28.

121.   Racioppi, L., *et al.*, Peptide-major histocompatibility complex class II complexes with mixed agonist/antagonist properties provide evidence for ligand-related differences in T cell receptor-dependent intracellular signaling. *J Exp Med*, 1993. **177**(4): p. 1047-60.

122.   Evavold, B.D. and P.M. Allen, Separation of IL-4 production from Th cell proliferation by an altered T cell receptor ligand. *Science*, 1991. **252**(5010): p. 1308-10.

123.   De Magistris, M.T., *et al.*, Antigen analog-major histocompatibility complexes act as antagonists of the T cell receptor. *Cell*, 1992. **68**(4): p. 625-34.

124.   Evavold, B.D., *et al.*, Complete dissection of the Hb(64-76) determinant using T helper 1, T helper 2 clones, and T cell hybridomas. *J Immunol*, 1992. **148**(2): p. 347-53.

125.   Sloan-Lancaster, J., B.D. Evavold, and P.M. Allen, Induction of T-cell anergy by altered T-cell-receptor ligand on live antigen-presenting cells. *Nature*, 1993. **363**(6425): p. 156-9.

126.   Sloan-Lancaster, J., B.D. Evavold, and P.M. Allen, Th2 cell clonal anergy as a consequence of partial activation. *J Exp Med*, 1994. **180**(4): p. 1195-205.

127.   Sakai, K., *et al.*, Prevention of experimental encephalomyelitis with peptides that block interaction of T cells with major histocompatibility complex proteins. *Proc Natl Acad Sci U S A*, 1989. **86**(23): p. 9470-4.

128.   Lamont, A.G., *et al.*, Inhibition of experimental autoimmune encephalomyelitis induction in SJL/J mice by using a peptide with high affinity for IAs molecules. *J Immunol*, 1990. **145**(6): p. 1687-93.

129.   Kuchroo, V.K., *et al.*, A single TCR antagonist peptide inhibits experimental allergic encephalomyelitis mediated by a diverse T cell repertoire. *J Immunol*, 1994. **153**(7): p. 3326-36.

130.   Santambrogio, L., M.B. Lees, and R.A. Sobel, Altered peptide ligand modulation of experimental allergic encephalomyelitis: immune responses within the CNS. *J Neuroimmunol*, 1998. **81**(1-2): p. 1-13.

131.   Dressel, A., *et al.*, Autoantigen recognition by human CD8 T cell clones: enhanced agonist response induced by altered peptide ligands. *J Immunol*, 1997. **159**(10): p. 4943-51.

132.   Smilek, D.E., *et al.*, A single amino acid change in a myelin basic protein peptide confers the capacity to prevent rather than induce experimental autoimmune encephalomyelitis. *Proc Natl Acad Sci U S A*, 1991. **88**(21): p. 9633-7.

133.   Anderton, S.M., *et al.*, Negative selection during the peripheral immune response to antigen. *J Exp Med*, 2001. **193**(1): p. 1-11.

134.   Pearson, C.I., W. van Ewijk, and H.O. McDevitt, Induction of apoptosis and T helper 2 (Th2) responses correlates with peptide affinity for the major histocompatibility complex in self-reactive T cell receptor transgenic mice. *J Exp Med*, 1997. **185**(4): p. 583-99.

135.   Anderton, S.M., *et al.*, Fine specificity of the myelin-reactive T cell repertoire: implications for TCR antagonism in autoimmunity. *J Immunol*, 1998. **161**(7): p. 3357-64.

136.   Pedotti, R., *et al.*, An unexpected version of horror autotoxicus: anaphylactic shock to a self-peptide. *Nat Immunol*, 2001. **2**(3): p. 216-22.

# THE ROLE OF COSTIMULATION IN EXPERIMENTAL AUTOIMMUNE ENCEPHALOMYELITIS

Michael K. Racke*, Robert B. Ratts, Rodney W. Stuart, Caishu Deng and Amy E. Lovett-Racke

*Department of Neurology and *The Center for Immunology, UT Southwestern Medical Center, Dallas, Texas, USA*

**Abstract:** Experimental allergic encephalomyelitis (EAE) is a T cell-mediated, autoimmune disorder characterized by central nervous system (CNS) inflammation and demyelination, features reminiscent of the human disease, multiple sclerosis (MS). In addition to the signal the encephalitogenic T cell receives through the T cell receptor (TCR), a second signal, termed costimulation, is required for complete T cell activation. The B7 family of cell surface molecules expressed on antigen presenting cells (APC) is capable of providing this second signal to T cells via two receptors, CD28 and CTLA-4. Our studies and those of others have shown that costimulation provided by B7 molecules to its ligand CD28 is important in the initiation of the autoimmune response in EAE. Further, it appears the costimulation provided by B7-1 is important in disease development, while B7-2 may play an important regulatory role. We and others have shown that B7/ CTLA-4 interaction plays a critical role in down-regulating the immune response in EAE. Other costimulatory pathways, such as CD40/CD40L and ICOS and its ligand, also play significant roles in the EAE model.

Previous work has shown that activated T cells and T cells of a memory phenotype are less dependent on costimulation than naive T cells. T cells reactive with myelin components that are involved in the pathogenesis of EAE and possibly MS would be expected to have been activated as part of the disease process. Building upon our prior work in the EAE model, we have tested the hypothesis that myelin-reactive T cells, which are relevant to the pathogenesis of CNS inflammatory demyelination, can be distinguished from naive myelin-reactive T cells by a lack of dependence upon costimulation for activation. It is hoped that therapies targeting costimulation may provide a means of conserving normal immune function whilst eliminating or suppressing autoreactive T cells, providing a more efficient means to treat autoimmune disease.

**Key words:** EAE, costimulation, second signal, CD28, CTLA-4, B7-1, B7-2, ICOS, memory T cells, multiple sclerosis

## Introduction

In the United States, it is estimated that 250,000-350,000 people have physician-diagnosed MS (Anderson et al., 1992). The majority of patients are diagnosed with the disease during their third and fourth decade of life, resulting in many patients having to suffer the effects of this disease for most of their adult life. The cause of MS remains elusive. However, because the disease is characterized by perivascular, mononuclear cell, inflammatory infiltrates and demyelination, features also characteristic of EAE, an autoimmune process is thought to be involved in disease pathogenesis (Martin et al., 1992; Raine, 1984; Arnason, 1983). Interestingly, both MS patients and normal controls have been shown to have T cells reactive to a number of myelin antigens, making it difficult to establish a relationship between myelin reactivity and the presence of disease. Epidemiological studies as well as studies examining identical twins suggest that both genetics and environment, particularly viral infections, may play a role in disease pathogenesis (Martin et al., 1992; Sadovnick and Ebers, 1993; Martyn, 1991). Three drugs have been approved by the FDA for use in the treatment of patients with relapsing and remitting MS (IFNB Multiple Sclerosis Study Group, 1995; Jacobs et al., 1996; Johnson et al., 1995). However, these agents are not a cure for the disease, so the need for the development of better treatment strategies in MS remains.

## Relevance of EAE to human MS

There are several animal models which have been used to study MS. In some of these models, disease is induced by viruses such as Theiler's virus (Miller and Karpus, 1994). Of the EAE models, the most commonly studied are those established in the Lewis rat and in several susceptible mouse strains. Our laboratory has used exclusively the murine model for a number of reasons. Many models of murine EAE result in a relapsing/remitting disease, similar to the early phase of most MS patients, while EAE in the Lewis rat is a monophasic illness. In chronic murine EAE, the pathology observed in the white matter is much more reminiscent of the pathology seen in the CNS of patients with MS, showing much more demyelination than the Lewis rat model. With the advent of transgenic and homologous recombination technology, it is becoming increasingly clear that many powerful molecular tools are becoming available to study the immune response in mice in pathological processes such as EAE.

## EAE as a model of organ-specific autoimmunity

The understanding of immunological mechanisms involved in autoimmune, inflammatory demyelination has advanced greatly through the investigation of EAE, an animal model of MS which can be induced either by immunization of susceptible animals with components of myelin or adoptive transfer of CD4+ T cells specific for myelin antigens (Martin et al., 1992; Zamvil and Steinman, 1990). It is the adoptive transfer experiments which have clearly established that EAE is a T cell-mediated autoimmune disease. Studies of various inbred mouse strains demonstrated that their ability to develop EAE following immunization with myelin varied significantly (Fritz and McFarlin, 1989). MHC class II background was initially thought to be the most important factor in conferring susceptibility to EAE, but studies have shown that other genetic loci are also involved in disease susceptibility (Fritz et al., 1985). Another variable in EAE studies has been the antigen used to induce disease. MBP and PLP are the major protein components of myelin and have been most studied as the disease-initiating antigen in EAE, although myelin oligodendrocyte glycoprotein (MOG) is being used increasingly in B6 mice (Martin et al., 1992; Zamvil and Steinman, 1990). Although EAE produced by immune responses against MBP or PLP is T cell mediated, it has also been shown that addition of antibodies to MOG, a minor glycoprotein on the outer surface of the oligodendrocyte, dramatically enhances demyelination in MBP-induced EAE (Linington et al., 1988).

**T cell costimulation**

The activation and differentiation of T cells require both antigen/MHC recognition and costimulatory signals. The signal conferred by the TCR determines the antigen-specificity of the response. The second signal, termed costimulation, is provided by accessory molecules on the antigen-presenting cell (APC) and appears to be necessary for T cell activation (see Figure 1). There are several receptor-ligand pairs which can provide costimulation. However , the signal provided by the B7 family of cell surface molecules with its receptors on T cells, CD28 and CTLA-4, appears to be the predominant one for T cell activation (June et al., 1994; Jenkins and Johnson, 1993; Bretscher, 1992).

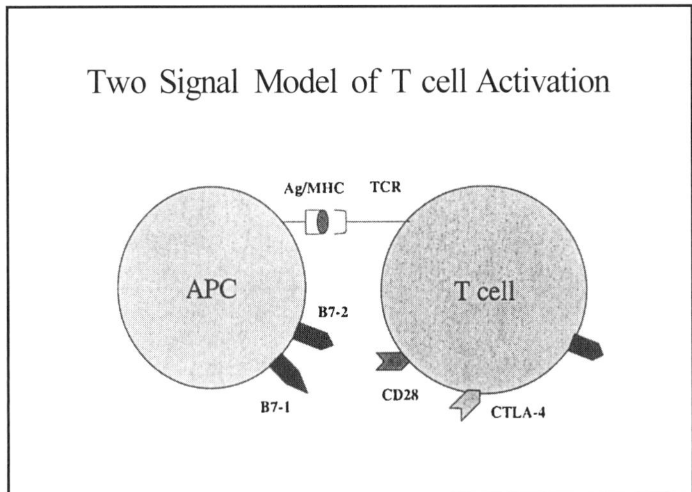


Figure 1. Trimolecular complex (TCR:Ag/MHC) forms signal one, while B7:CD28/CTLA-4 pathway composes important second signal.

At least two members of the B7 family of CD28 ligands have been defined, B7-1 (CD80) and B7-2 (CD86) (Freeman et al., 1993a,b,c; Hathcock et al., 1994; ). These molecules, although having modest homology, are each capable of providing costimulation to T cells for proliferation and IL-2 production. It is for this reason that a mouse genetically deficient for B7-1 was still immunocompetent (Freeman et al., 1993a). On the other hand, T cells from a mouse deficient in CD28 could not produce IL-2 after stimulation with the mitogenic lectin Con A, demonstrating that costimulation through this molecule is critical for IL-2 production (Green et al., 1994a).

B7-1 and B7-2 are expressed differentially on various APC and their kinetics of expression and binding also differ (Larsen et al., 1994; Freedman et al., 1991; Linsley et al., 1994). B7-2 is expressed by monocytes constitutively, while B7-1 can be induced on these APC with IFN-γ. B7 is expressed on B cell populations after an activation stimulus; B7-2 is expressed within 6 hours, while B7-1 expression peaks after 48 hours. The expression of B7-1 and B7-2 has also been shown to occur on T cells themselves (Sansom et al., 1993; Azuma et al., 1993). Interestingly, it appears that B7-2 expressed on T cells is unable to interact with CD28, yet is still able to interact with CTLA-4 (Greenfield et al., 1997). Work by our group demonstrated that T cells are the main cell type expressing B7 molecules in the CNS during various stages of EAE (Cross et al., 1999). Thus, as with CD28 and CTLA-4, B7 molecules themselves may be able to either enhance or inhibit immune responses.

Much has been learned about the role of costimulation in immune processes with a soluble fusion protein, CTLA-4Ig, which can prevent the interaction between B7 and CD28 or CTLA-4 (Gimmi et al., 1993). Administration of CTLA-4Ig prevented rat cardiac allograft rejection and pancreatic islet cell xenograft rejection in transplantation models in mice (Lin et al., 1993; Lenschow et al., 1992). In both instances, it appeared the mechanism of suppression involved the induction of antigen-specific tolerance. The costimulation provided by B7 also appears to be important for the development of encephalitogenic effectors and subsequent autoimmunity. Our laboratory has examined the role of B7:CD28/CTLA-4 interactions in the induction of EAE (Perrin et al., 1995a,b; Perrin et al., 1996a; Racke et al., 1995). In the adoptive transfer model of EAE, CTLA-4Ig was able to inhibit the proliferation and IL-2 production of MBP-specific lymph node cells during activation *in vitro*, diminishing the clinical disease observed upon subsequent transfer. Thus, B7-mediated costimulation was found to be an important factor in determining encephalitogenicity of myelin-specific T cells. It is important to note that once activated autoreactive cells had been injected into naive recipients, CTLA-4Ig intervention did not alter the course of disease. This would be consistent with the concept that activated T cells do not require costimulation to perform their effector function.

In active models of EAE, several intriguing observations have been made. Disease induced by immunization with MBP and pertussis toxin (PT) injection was inhibited by a single injection of CTLA-4Ig (Perrin et al., 1995b). On the other hand, in a model using a two immunization schedule without PT, clinical signs of EAE were enhanced following administration of multiple doses of CTLA-4Ig, yet was inhibited with a single injection 48 hours after the initial administration of encephalitogen (Racke et al., 1995). These paradoxical results suggested that B7 costimulation may also provide a regulatory role during an immune response.

Early studies indicated that CTLA-4 induced a positive costimulatory signal in conjunction with CD28 (Linsley et al., 1992). Evidence generated by our laboratory utilizing the EAE model suggested that the signaling through the CTLA-4 molecule actually mediated a negative regulatory function (Perrin et al., 1996b). In CD28-deficient mice, costimulation provided by B7+ APC did not transduce a positive signal (Green et al., 1994b). Furthermore, a mouse genetically deficient in CTLA-4 has a fatal phenotype secondary to a lymphoproliferative disorder (Tivol et al., 1995; Waterhouse et al., 1995). These studies all suggested that B7 mediated costimulation provided regulatory signals that limited the immune response.

We and others showed that B7-1 provides an important stimulus for the development of encephalitogenic T cells (Perrin et al., 1995a,b; Perrin et al., 1996a; Racke et al., 1995; Kuchroo et al., 1995; Miller et al., 1995). However, we were unable to demonstrate that B7-1 costimulation skewed development to the Th1 phenotype (secreting IFN-$\gamma$, IL-2, and LT) or that B7-2 costimulation preferentially resulted in Th2 cells (secreting IL-4-6), as described by Kuchroo et al. (1995). Interestingly, we did make similar observations that anti-B7-1 administration was protective while anti-B7-2 administration exacerbated EAE. Observations using mice deficient in either B7-1 or B7-2 suggests that both molecules can make significant contributions to the production of both IL-4 and IFN-$\gamma$. However, neither molecule plays an obligatory role in priming for the production of either effector cytokine (Schweitzer et al., 1997). Furthermore, in the non-obese diabetic mouse, the effect of administration of anti-B7 antibodies was reversed, with anti-B7-1 exacerbating disease and anti-B7-2 blocking the development of disease (Lenschow et al., 1995). However, as noted above, because the B7-2 expressed on T cells appears to be able to interact with CTLA-4, the addition of anti-B7-2 may result in the inhibition of a negative signal to the responding T cells, resulting in the enhanced disease observed in the EAE model. Thus, the role of the B7 family of costimulatory molecules in autoimmune diseases appears to be quite complicated.

Inhibition of the B7/CD28 pathway has been shown to have significant effects in a number of autoimmune models, including EAE. For example, NOD mice deficient in CD28 were totally resistant to the development of EAE (Akdis et al., 2000). However, T cells primed against proteolipid protein (PLP), which were also deficient in CD28, could proliferate and secrete IFN-$\gamma$ and TNF-$\alpha$ at wild type levels (Howard et al., 1999). Recent studies in the EAE model examining the role of CD28 and B7 molecules have utilized mice generated by homologous recombination which are deficient in these molecules. Initial studies examining susceptibility of B6 mice deficient in CD28 to EAE demonstrated that the T cells from these mice were impaired in their ability to enter the CNS (Perrin et al., 1999; Chitnis et al., 2001). These mice specifically developed a significant picture of meningitis, with clear impairment of T cells to directly enter the CNS parenchyma. Interestingly, studies carried out using B6 mice deficient in both B7-1 and B7-2 demonstrated that these mice were also resistant to the development of MOG-induced EAE  (Chang et al., 1999). This impairment was not due solely to impaired priming of MOG-reactive T cells, since wild type T cells transferred to B7-1 and B7-2 deficient mice were still resistant to the development of EAE. Again it appeared the major finding was that these cells could not invade the CNS parenchyma. These observations clearly suggest that CD28:B7 interaction may also play an important role for T cells to enter the CNS itself. In addition, it appears CD28 blockade is most effective in situations where the number of memory T cells for a particular auto-antigen is low, which may not be the case for all situations in autoimmunity.

**Costimulation requirements for naive versus memory T cells**
       The ability of the immune system to respond upon first contact with an antigen is critical for host defense against foreign pathogens. Understanding the requirements for expansion of T cells in the adaptive immune response for a particular antigen is not only important for host defense, but also important in understanding the pathogenesis of autoimmunity. Several functional differences between naive and memory T cells have been delineated. Naive T cells, which have not encountered their antigen since exiting the thymus, express high levels of CD45RA or RB and L-selectin and low levels of CD44. Memory T cells, which have encountered their antigen one or more times, have low expression of CD45RA/B (or high CD45RO in humans), low L-selectin, and high CD44 (Croft et al., 1994).
       It has been shown that naive T cells have distinct activation requirements in comparison to memory or activated T cells. For example, MHC-peptide complexes could stimulate primed TCR transgenic T cells, yet this same stimulus resulted in poor activation of those same TCR transgenic T cells when these cells were naive (Sagerstrom et al., 1993). Providing costimulation through CD28 resulted in activation and proliferation of the naive T cells. In addition, memory T cells have been shown to be considerably less dependent on accessory cell costimulation than naive T cells (Croft et al., 1994; Damle et al., 1992).  The activation state of myelin-reactive T cells has not been well characterized. Myelin-specific T cells with a mutation in the HPRT gene have been found at a higher frequency in MS patients than normal controls, suggesting that these cells have replicated *in vivo* and had the opportunity to mutate (Allegretta et al., 1990; Trotter et al., 1997).  In another study, the frequency of myelin-specific T cells expressing the IL-2 receptor, a marker of activation, was shown to be higher in MS patients (Zhang et al., 1994). Our own studies demonstrated that antigenic stimulation of murine MBP-specific lymph node cells made them much less

dependent on B7-mediated costimulation on subsequent antigenic stimulation *in vitro* (Racke et al., 1995). An important issue is whether costimulation-independent, myelin-reactive T cells in MS patients accumulate over time and whether this accumulation correlates with disease progression. This result will have important consequences with regard to the potential application of interference with costimulatory pathways as a treatment for human autoimmune diseases.

### Antigen determinant spreading in EAE and MS

The relapsing-remitting course of disease characteristic of both EAE and MS has suggested the possibility that T cell immunity to new myelin antigens may develop during chronic CNS inflammation and may play a role in subsequent relapses. This phenomenon, termed antigen-determinant or epitope spreading, occurs when T cell immunity has developed to antigens other than that antigen used to initiate the inflammatory response. Work performed in the laboratory of the late Dr. Dale McFarlin was the first to demonstrate that following the induction of EAE to one MBP epitope, T cell responses to another MBP peptide could be observed (McCarron et al., 1990; Lehmann et al., 1992). This process represented what is now known as intramolecular spreading. Later, work by Dr. Anne Cross demonstrated a response developed to PLP after EAE had been initiated by MBP, a phenomenon termed intermolecular epitope spreading (Cross et al., 1993; McRae et al., 1995). Recent studies have demonstrated that administration of Fab fragments that block B7-1 during remission in relapsing /remitting EAE were able to prevent epitope spreading and development of new relapses in EAE (Miller et al., 1995).

### Relevance to the therapy of multiple sclerosis

Our studies have addressed the question of whether memory T cells specific for an autoantigen such as MBP have different costimulatory requirements than naive cells specific for the same autoantigen. If we can demonstrate that memory T cells have reduced costimulatory requirements *in vivo*, this would have implications for the development of an immunotherapy targeting the B7:CD28/CTLA-4 pathway. Such an immunotherapy has several attractive features, the greatest being that it might be able to target autoreactive T cells without actually requiring the knowledge of the specific autoreactive antigens.

Our laboratory has performed several studies that have helped define the contribution of the B7:CD28/CTLA-4 pathway in the pathogenesis of EAE (Perrin et al., 1995a,b; Perrin et al., 1996a; Racke et al., 1995; Perrin et al., 1999; Cross et al., 1999; Ratts et al, 1999a,b). It is only much more recently that our laboratory has begun to engage in studies which attempt to characterize the immune response to myelin antigens in patients with MS (Lovett-Racke et al., 1997). The studies we have performed on the pivotal role of costimulation in the EAE model provided the intellectual basis for the experimental approaches we have used to study T cell activation requirements to myelin antigens in patients with MS.

### CD28 Costimulation Requirements in MS patients

Myelin-reactive T cells are present in both MS patients and controls (Pette et al., 1990; Martin et al., 1993), indicating that autoreactive T cells can exist in individuals without pathologic consequences. We wished to address whether there was a difference in the costimulatory requirements of MBP-reactive T cells in MS patients versus normal controls. The hypothesis tested was that MS patients would have myelin-reactive T cells that were less dependent on CD28-mediated costimulation because they had an activated or memory phenotype.

We decided to use an approach which would minimize the amount of *in vitro* manipulation, such as that which occurs with T cell cloning, because after repeated *in vitro* stimulation, T cells would have an activated or memory phenotype. Peripheral blood mononuclear cells (PBMC) from MS patients, stroke patients, or controls were plated at a density of 2.5 x $10^5$ cells/well in the presence of MBP and either αCD28 or control antibody. In addition, 48 wells without antigen and each antibody were plated. After 6 days in culture with MBP, $^3$H-methyl-thymidine was added and the cells were harvested on day 7. Positive wells were considered to be those wells with MBP and antibody (either αCD28 or control) that had counts greater than 2 SD above the mean of the control wells and also had a stimulation index greater than 2 (see Lovett-Racke et al., 1998 for details). Figure 2 shows the percentage of the MBP response inhibited by αCD28 in MS patients, stroke patients, and controls.

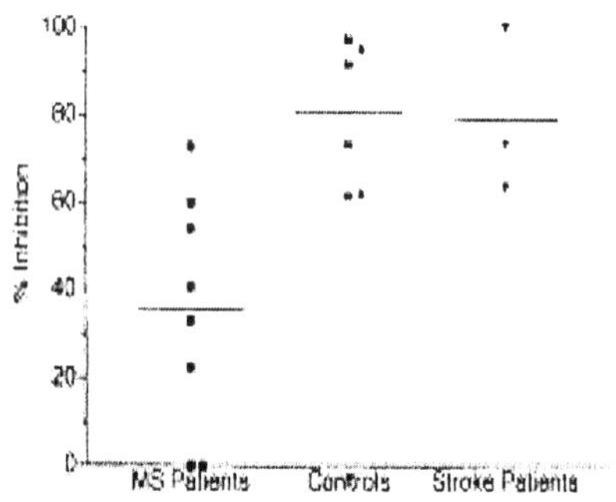


Figure 2. Blockade of CD28-mediated costimulation did not efficiently inhibit proliferation of MBP-reactive T cells in most MS patients. The percent inhibition of the number of positive wells in lymphocyte proliferation assays in the presence of anti-CD28 was determined for the patients and controls who had a response rate >3% with the control antibody. A two-tailed *t* test indicated a significant difference between the MS patients and controls. Reproduced from the Journal of Clinical Investigation, 1998, vol. 101, pp. 725-730 by copyright permission of the American Society for Clinical Investigation.

Approximately 80% of the MBP-specific proliferative response was inhibited in normal controls and stroke patients, while only 35% of the response was inhibited by αCD28 in MS patients (MS vs. control, p=0.0036; MS vs. stroke, p= 0.0317; control vs. stroke, NS). This data suggests that MBP-reactive T cells in MS patients are less dependent on costimulation, a characteristic of activated or memory T cells, than the MBP-reactive T cells in controls or stroke patients. The stroke patients represent an important control, because presumably after an infarct, macrophages that are recruited to the region of the infarcted tissue would subsequently have the opportunity to present MBP and other myelin antigens to the immune system. It was previously shown that B7-1 is expressed in the inflammatory lesions in the brains of MS patients, but not in cerebral infarcts, making it likely that priming to MBP would occur in MS, but not in stroke (Windhagen et al., 1995). As another control, we have examined the responses to tetanus toxoid (TT) in MS patients and normal controls. As one might expect, since almost everyone has had multiple immunizations to TT, patients and controls make an excellent response to TT that is not inhibited by the addition of αCD28 (Lovett-Racke et al., 1998). One might expect that costimulatory blockade should completely inhibit MBP-specific proliferation in healthy controls and conversely, have no effect on the proliferation of MBP-reactive T cells in MS patients. One explanation for the incomplete inhibition of MBP-reactive T cell proliferation in healthy controls following CD28 blockade may be due to cross-reactivity of MBP-reactive T cells with foreign antigens. Wucherpfennig and Strominger (1995) described MBP-reactive T cells that could proliferate to a variety of antigens derived from common viral pathogens, suggesting that these "MBP-specific" T cells may have been activated *in vivo* in controls in response to an infection and may have never encountered MBP, which is sequestered behind an intact blood-brain barrier. Limiting dilution analysis of MBP and TT-reactive T cells demonstrated that the frequency of MBP-reactive T cells able to respond when CD28 was blocked decreased 4-6 fold in controls, suggesting that these T cells are naive. In contrast, there was no significant difference in the frequency of MBP-reactive T cells with or without αCD28 in MS patients. There was also no difference in the frequency of TT-reactive cells for the controls or MS patients. This is consistent with the hypothesis that memory T cells in humans are less dependent on CD28-mediated costimulation to mount a proliferative response.

Our laboratory has a great deal of experience using CTLA-4Ig to block B7-mediated costimulation in the EAE model. Both CTLA-4Ig and αCD28 give similar responses in their ability to block an MBP-specific response in MS patients (Lovett-Racke et al., 1998). Just as important, CTLA-4Ig behaved similarly with regard to αCD28 in its inability to block the memory proliferative response to TT. Similar results blocking the proliferative response of MBP-specific T cells from normal controls using CTLA-4Ig have been demonstrated by others (Scholz et al., 1998). Thus, using both αCD28 and CTLA-4Ig has allowed us to accurately examine the costimulatory requirements of myelin-reactive T cells in MS patients. Recent work by Markovic-Plese et al. (2001) showed that these costimulation-independent T cells lack CD28 expression, yet can be stimulated to produce high amounts of IFN-γ and upregulate the IL-12Rβ2 chain. The authors hypothesized that these T cells would likely initiate immune responses in the CNS, because they would not require the expression of costimulatory molecules on professional APC.

**Costimulation Requirements in EAE**

In our prior studies examining costimulation in the EAE model, we have methodically examined the clinical outcome of blocking B7 molecules individually or together using either anti-B7-1 and anti-B7-2 antibodies or CTLA-4Ig (Perrin et al., 1995a,b; Perrin et al., 1996a; Racke et al., 1995; Perrin et al., 1999; Cross et al., 1999; Ratts et al, 1999a,b). Although we have identified certain regimens which appear to reduce clinical signs of EAE and reduce production of relevant inflammatory cytokines, we have never been able to completely eliminate the development of disease. This suggests the possibility that there may be a population of T cells that are B7-costimulation-independent and that these T cells will be able to mediate disease despite costimulatory blockade.

Recently, we performed experiments which demonstrated that blockade of the CD28 signal is actually responsible for reducing clinical disease following costimulatory blockade. MBP-specific LNC activated in the presence of $\alpha$CD28 Fab (10 $\mu$g/ml)  make less IL-2 and were less encephalitogenic when subsequently transferred into naive recipients (Perrin et al., 1999). Here again, although the encephalitogenic response was reduced, it was not eliminated, suggesting that there are encephalitogenic T cells which are not dependent on CD28-mediated costimulation. Another important observation was that blockade of CD28 during the first disease episode resulted in significant attenuation of subsequent disease course. However, this observation needs to be considered with recent findings using mice deficient in CD28 (Bachmeier et al., 1996). In an autoimmune myocarditis model, CD28 -/-  mice developed autoimmune heart disease, although it was less severe than that observed in heterozygous littermates. After primary immunization with myelin oligodendrocyte glycoprotein, CD28 -/- mice develop experimental autoimmune meningitis (EAM) instead of EAE (Perrin et al., 1999). EAM is characterized by neutrophil-dominated lesions within the leptomeninges and is reminiscent of aseptic meningitis. The different disease phenotypes seen in CD28-/- and CD28 +/+ mice demonstrate the importance of CD28 in T cell differentiation. In addition, breeding of NOD mice to CD28 -/- mice exacerbated autoimmunity, suggesting CD28 plays an important regulatory role in that disease model (Lenschow et al., 1996).


**Costimulation dependence of i.v. tolerance.**

We have also made some interesting observations with regard to the mechanisms of extrathymic tolerance secondary to high dose i.v. antigen therapy (Critchfield et al., 1994; Racke et al., 1996). The potential use of such a therapy in MS necessitates the need to further elucidate the mechanisms involved in tolerance induction. In our initial studies, we have learned that the V$\beta$8 transgenic mice, which do not express the V$\alpha$2 transgene, are useful in these studies (Ratts et al., 1999a,b). Unlike the V$\beta$8, V$\alpha$2 transgenic mice, these mice do not develop spontaneous EAE, however

100% of these animals develop EAE when immunized with either Ac1-11 or MBP in CFA, and do so without a requirement for pertussis toxin injection. These mice have a precursor frequency of MBP-reactive T cells ranging from 1/80,00-1/200,000 before priming with antigen, which is increased over naive B10.PL mice. LNC isolated from these mice without prior antigen priming can be shown to proliferate to MBP in a standard 4 day proliferation assay, thus we have used the proliferative response and cytokine secretion as a measure of tolerance induction.

Interactions between B7 molecules on antigen presenting cells and CTLA-4 on T cells have been shown to be important in establishing tolerance (Perez et al., 1997). We examined the kinetics of tolerance induction following i.v. administration of myelin basic protein (MBP) Ac1-11 in mice transgenic for a TCR Vβ8.2 gene derived from an encephalitogenic T cell clone specific for MBP Ac1-11 (Ratts e al., 1999a). Examination of the lymph node cell (LNC) response 10 days after antigen administration demonstrated an accentuation of i.v. tolerance induction with anti-CTLA-4 blockade. Anergy was induced in splenocytes by i.v. antigen administration as demonstrated by a decrease in MBP-specific proliferation and IL-2 production, and co-administration of anti-CTLA-4 potentiated this effect. In addition, i.v. antigen plus anti-CTLA-4 and CFA was not encephalitogenic. Interestingly, i.v. tolerance (a single injection) did not inhibit EAE, and anti-CTLA-4 administration did not alter this phenotype. These results suggest that while the majority of MBP-specific T cells are tolerized by i.v. antigen, and that this process is potentiated by anti-CTLA-4 administration, a population of T cells still remain that is quite effective in mediating EAE.


**Costimulation dependence of i.p. tolerance**.
Evidence suggests that costimulation provided by B7 molecules through CTLA-4 is important in establishing peripheral tolerance, whether by i.v. or i.p. antigen administration (Perez et al., 1997). In another study, we examined the kinetics of tolerance induction and T cell differentiation following i.p. administration of MBP Ac1-11 in mice transgenic for a TCR Vβ8.2 gene derived from an encephalitogenic T cell clone specific for MBP Ac1-11 (Ratts et al., 1999b). Examination of the lymph node cell (LNC) response after antigen administration demonstrated a dependence on CTLA-4 for i.p. tolerance induction. Examination of splenocyte responses suggested that i.p. antigen administration induced a Th2 response, which was potentiated by co-administration of anti-CTLA-4. Interestingly, i.p. tolerance was able to inhibit the induction of experimental autoimmune encephalomyelitis (EAE) and anti-CTLA-4 administration did not alter this phenotype, suggesting that CTLA-4 blockade did not block tolerance induction, but rather potentiated the Th2 phenotype of MBP-specific T cells, which likely was responsible for the inhibition of EAE. Thus, T cell differentiation and the dependence on CTLA-4 for tolerance induction following i.p. antigen administration differs between immune cells in the lymph node and spleen in the EAE model.

**Additional Costimulatory Pathways Regulating EAE**

While the B7:C28/CTLA-4 pathway has been the best studied in EAE, other pathways have also been investigated. For example, CD40L, a member of the TNF receptor family, is predominantly found on activated T cells, while its ligand CD40, can be found on B cells, dendritic cells, macrophages, astrocytes, and thymic epithelial cells. Professional APC constitutively express the costimulatory molecules CD40 and B7 and both are upregulated when the APC becomes activated (Howard et al., 1999b; Lefrancois et al., 2000; McCormick et al., 1999; van der Eertwegh et al., 1993). In CD4+ T cells, ligation of the TCR results in increased expression of CD40L and its expression is further increased by other costimulatory events. CD40 ligation leads to the upregulation of B7-1 and B7-2 on APC and thereby enhances the ability of APC to activate naïve T cells (Howard et al., 1999b). Thus, CD40L interaction with CD40 is important in T cell activation, particularly in its early stages. Studies conducted in SJL mice examining PLP139-151-induced EAE showed that anti-CD40L antibody could block progression of clinical disease, whether administered at the peak of disease or during disease remission. CNS inflammation was also reduced by anti-CD40L administration (Howard et al., 1999b). Thus, blockade of the CD40/CD40L pathway appears to be a promising method for inhibiting initial T cell activation and subsequent clonal expansion of autoreactive T cells.

Another recently described pathway is that of the inducible costimulator (ICOS) and its ligand B7-H2, which is expressed by B cells in the lymph node and also present in a number of nonlymphoid tissues. CD40 stimulation does not increase B7-H2 expression as it does with B7-1 and B7-2 (Yoshinaga et al., 1999). ICOS shares about 30-40% sequence homology with CD28 and CTLA-4 an dit maps to the same locus, suggesting that it likely arose via gene duplication. ICOS is not constitutively expressed by the T cell, but is upregulated after stimulation, reaching maximal expression in 48 hours (Zhang et al., 1997; Kouki et al., 2000; Barrat et al., 1999; Coyle et al., 2000; Mages et al., 2000). Interestingly, ICOS can be upregulated in the absence of CD28, however IL-2 production is less than if CD28 costimulation also exists (McAdam et al., 2000). ICOS expression does appear to be present constitutively on previously activated T cells, suggesting it may play a more important role in modifying the responses of memory and effector T cells.

Manipulating ICOS in the setting of diseases such as EAE has proven that complexity of ICOS signaling. ICOS blockade dramatically improved the resolution of clinical symptoms in EAE (Sporici et al., 2001), however susceptibility to EAE was greatly enhanced in ICOS-deficient mice (Chen et al., 2001). Hopefully future studies will determine whether targeting this costimulatory pathway will be beneficial in the treatment of human diseases such as MS.

## Conclusions

The studies described in this review demonstrate an important role for costimulation in both multiple sclerosis and its animal model EAE. Our current studies of MS patients, which are based on our prior work in the EAE model, will demonstrate whether costimulation-independent, myelin-reactive T cells develop over time with disease progression. More significantly, we will be able to determine whether these T cells have been expanded *in vivo* and will be able to follow the T cell repertoire to myelin antigens over time in MS patients. We feel these studies, initially developed in the EAE model, will provide important information relevant to the pathogenesis of MS and will have implications for the development of immunologically based therapies for this disease.

## ACKNOWLEDGMENTS

This work was supported by NIH grant RO1-NS-37513 (MKR). This work was also supported by grants from the National Multiple Sclerosis Society to MKR. ALR and MKR were also supported by the Yellow Rose Foundation. RWS was a fellow of the Doris Duke Charitable Foundation.

# REFERENCES

Akdis, CA, A Joss, AF Akdis, and B Kurt (2000). A molecular basis for T cell suppression by IL-10: CD28-associated IL-10 receptor inhibits CD28 tyrosine phosphorylation and phosphatidylinositol 3-kinase binding. FASEB J. 14: 1666-1668.

Allegretta , M, JA Nicklas, S Sriram,  and RJ Albertini (1990). T cells responsive to myelin basic protein in patients with multiple sclerosis. Science 247: 718-721.

Anderson, DW, JH Ellenberg, CM Leventhal, SC Reingold, M Rodriguez, and DH Silberberg (1992). Revised estimate of the prevalence of multiple sclerosis in the United States. Ann. Neurol. 31: 333-336.

Arnason, BG. (1983). Relevance of experimental allergic encephalomyelitis to multiple sclerosis. Neurol. Clin. 1:765-782.

Azuma, M, H Yssel, JH Phillips, H Spits, and LL Lanier (1993).  Functional expression of B7/BB1 on activated T lymphocytes. J. Exp. Med. 177: 845-850.

Bachmaier, C Pummerer, A Shahinian, J ionescu, N Neu, TW Mak, and JM Penninger (1996). Induction of autoimmunity in the absence of CD28 costimulation. J. Immunol. 157: 1752-1757.

Barrat, FJ, F Le Deist, M Benkerrou, P Bousso, J Feldmann, A Fischer, and G de Saint Basile (1999). Defective CTLA-4 cycling pathway in Chediak-Higashi syndrome: a possible mechanism for deregulation of T lymphocyte activation. Proc. Natl. Acad. Sci. USA 96: 8645-8650.

Bretscher, P (1992). The two-signal model of lymphocyte activation twenty-one years later. Immunol. Today 13: 74-76.

Chang TT, C Jabs, RA Sobel, VK Kuchroo, and AH Sharpe (1999). Studies in B7-deficient mice reveal a critical role for B7 costimulation in both induction and effector phases of experimental autoimmune encephalomyelitis. J. Exp. Med. 190: 733-740.

Chitnis T, N Najafian, KA Abdallah, V Dong, H Yagita, MH Sayegh, and SJ Khoury (2001). CD28-independent induction of experimental autoimmune encephalomyelitis. J. Clin. Invest. 107: 575-583.

Coyle AJ, Lehar S, C Lloyd, J Tian, T Delaney, S Manning, T Nguyen, T Burwell, H Schneider, JA Gonzalo, M Gosselin, LR Owen, CE Rudd, JC Gutierrez-Ramos (2000). The CD28-related molecule ICOS is required fo reffective T cell-dependent immune responses. Immunity 13: 95-105.

Critchfield, JM, MK Racke, JC Zuniga-Pflucker, B Cannella, CS Raine, J Goverman, and MJ Lenardo (1994). T cell deletion in high antigen dose therapy of autoimmune encephalomyelitis. Science 263:1139-1143.

Croft, M, LM Bradley, and SL Swain (1994). Naive versus memory CD4 T cell response to antigen: Memory cells are less dependent on accessory cell costimulation and can respond to many antigen-presenting cell types including resting B cells. J. Immunol. 152: 2675-2685.

Cross, AH, VK Tuohy, and CS Raine (1993). Development of reactivity to new myelin antigens during chronic relapsing autoimmune demyelination. Cell. Immunol. 146: 261-269.

Cross, A. H., J. A. Lyons, M. San, R. M. Keeling, G. Ku, and M. K. Racke. 1999. T cells are the main cell type expressing B7-1 and B7-2 in the central nervous system during acute, relapsing, and chronic experimental autoimmune encephalomyelitis. Eur. J. Immunol. 29: 1753-1761.

Damle, NK, K Klussman, PS Linsley, and A  Aruffo (1992). Differential costimulatory effects of adhesion molecules B7, ICAM-1, LFA-3, and VCAM-1 on resting and antigen-primed CD4+ T lymphocytes. J. Immunol. 148: 1985-1992.

Freedman, AS, GJ Freeman, K Rhynhart, and LM Nadler (1991). Selective induction of B7/BB-1 on interferon-g stimulated monocytes: A potential mechanism for amplification of T cell activation. Cell. Immunol. 137: 429-437.

Freeman, GJ, F Borriello, RJ  Hodes, H Reiser, KS Hathcock, G Laszlo, AJ McKnight, J Kim, L Du, DB Lombard, GS Gray, LM Nadler, and AH Sharpe (1993a). Uncovering of functional alternative CTLA-4 counter-receptor in B7-deficient mice. Science 262: 907-909.

Freeman, GJ, JG Gribben, VA Boussiotis, J Ng, VA Restivo, LA Lombard, GS Gray, and LM Nadler (1993b). Cloning of B7-2: A CTLA-4 counter-receptor that costimulates human T cell proliferation. Science 262: 909-911.

Freeman, GJ, F Borriello, RJ Hodes, H Reiser, JG Gribben, JW Ng, J Kim, JM Goldberg, K Hathcock, G Laszlo, LA Lombard, S Wang, GS Gray, LM Nadler, and AH Sharpe (1993c). Murine B7-2, an alternative CTLA4 counter-receptor that costimulates T cell proliferation and interleukin 2 production. J. Exp. Med. 178: 2185-2192.

Fritz, RB, MJ Skeen, CH Jen-Chou, M Garcia, and IK Egorov (1985). Major histocompatibility complex-linked control of the murine immune response to myelin basic protein. J. Immunol. 134: 2328-2332.

Fritz, RB and DE McFarlin (1989). Encephalitogenic epitopes of myelin basic protein. In: Sercarz, EE, ed. Antigenic determinants and immune response. Basel: Karger; pp.101-125.

Gimmi, CD, GJ Freeman, JG Gribben, G Gray, and LM Nadler (1993). Human T cell clonal anergy is induced by antigen presentation in the absence of B7 costimulation. Proc. Natl. Acad. Sci. USA 90: 6586-6590.

Green, JM, PJ Noel, AI Sperling, TL Walunas, GS Gray, JA Bluestone, and CB Thompson (1994a). Absence of B7-dependent responses in CD28-deficient mice. Immunity 1: 501-508.

Green, JM, PJ Noel, AI Sperling, TL Walunas, GS Gray, JA Bluestone, and CB Thompson (1994b). Absence of B7-dependent responses in CD28-deficient mice. Immunity 1: 501-508.

Greenfield, EA, E Howard, T Paradis, K Nguyen, F Benazzo, P McLean, P Hšllsberg, G Davis, DA Hafler, AH Sharpe, GJ Freeman, and VK Kuchroo (1997). B7.2 expressed by T cells does not induce CD28-mediated costimulatory activity but retains CTLA-4 binding: implications for induction of antitumor immunity to T cell tumors. J. Immunol. 158: 2025-2034.

Hathcock, KS, G Laszlo, C Pucillo, P Linsley, and RJ Hodes (1994). Comparative analysis of B7-1 and B7-2 costimulatory ligands: Expression and function. J. Exp. Med. 180: 631-640.

Howard, LM, AJ Miga, CL Vanderlugt, MC Dal Canto, JD Laman, RJ Noelle, and SD Miller (1999). Mechanisms of immunotherapeutic intervention by anti-CD40L (CD154) antibody in an animal model of multiple sclerosis. J. Clin. Invest. 103: 281-290.

The IFNB Multiple Sclerosis Study Group. Interferon beta-1b is effective in relapsing-remitting multiple sclerosis I: Clinical results of a multicenter, randomized, double-blind, placebo-controlled trial. Neurology 43:655-661.

Jacobs, LD, DL Cookfair, RA Rudick, and the Multiple Sclerosis Collaborative Research Group (1996). Intramuscular interferon beta-1a for disease progression in relapsing multiple sclerosis. Ann. Neurol. 39:285-294.

Jenkins, MK, and Johnson, JG (1993). Molecules involved in T-cell costimulation. Curr. Opin. Immunol. 5: 351-367.

Johnson, KP, BR Brooks, JA Cohen, and the Copolymer 1 MultipleSclerosis Study Group (1995). Copolymer 1 reduces relapse rate and improves disability in relapsing-remitting multiple sclerosis: Results of a phase III multicenter, double-blind, placebo-controlled trial. Neurology 45: 1268-1276.

June, CH, Bluestone, JA, Nadler, LM, and Thompson, CB (1994). The B7 and CD28 receptor families. Immunol. Today 15: 321-331.

Kouki, T, Y Sawai, CA Gardine, ME Fisfalen, ML Alegre, and LJ Degroot (2000). CTLA-4 polymorphism at position 49 in exon 1 reduces the inhibitory function of CTLA-4 an dcontributes to the pathogenesis of Graves' disease. J. Immunol. 165: 6606-6611.

Kuchroo, VK, MB Das, JA Brown, AM Ranger, SS Zamvil, RA Sobel, HL Weiner, N Nabavi, and LH Glimcher (1995). B7-1 and B7-2 costimulatory molecules activate differentially the Th1/Th2 developmental pathways: Application to autoimmune disease therapy. Cell: 80:707-718.

Larsen, CP, SC Ritchie, R Hendrix, PS Linsley, KS Hathcock, RJ Hodes, RP Lowry, and TC Pearson (1994). Regulation of immunostimulatory function and costimulatory molecule (B7-1 and B7-2) expression on murine dendritic cells. J. Immunol. 152: 5208-5219.

Lefrancois, L, JD ALtman, K Williams, and S Olson (2000). Soluble antigen and CD40 triggerring are sufficient to induce primary an dmemory cytotoxic T cells. J. Immunol. 164: 725-732.

Lehmann, PV, T Forsthuber, A Miller, and EE Sercarz (1992). Spreading of T cell autoimmunity to cryptic determinants of an autoantigen. Nature 358: 155-157.

Lenschow, DJ, Y Zheng, JR Thistlethwaite, A Montag, W Brady, MG Gibson, PS Linsley, and JA Bluestone (1992). Long-term survival of xenogeneic pancreatic islet grafts induced by CTLA-4Ig. Science 265: 1225-1227.

Lenschow, DJ, SC Ho, H Sattar, L Rhee, G Gray, N Nabavi, KC Herold, and JA Bluestone (1995). Differential effects of anti-B7-1 and anti-B7-2 monoclonal antibody treatment on the development of diabetes in the non-obese diabetic mouse. J. Exp. Med. 181: 1145-1155.

Lenschow, DJ, KC Herold, L Rhee, B Patel, A Koons, HY Qin, E Fuchs, B Singh, CB Thompson, and JA Bluestone (1996). CD28/B7 regulation of Th1 and Th2 subsets in the development of autoimmune diabetes. Immunity 5: 285-293.

Lin, H, SF Bolling, PS Linsley, RQ Wei, D Gordon, CB Thompson, and LA Turka (1993). Long-term acceptance of major histocompatibility mismatched cardiac allografts induced by CTLA4Ig plus donor-specific transfusion. J. Exp. Med. 178: 1801-1806.

Linington, C, M Bradl, H Lassman, C Brunner, and K Vass (1988). Augmentation of demyelination in rat acute allergic encephalomyelitis by circulating mouse monoclonal antibodies directed against a myelin/oligodendrocyte glycoprotein. Am. J. Pathol. 130: 443-454.

Linsley, PS, JL Greene, P Tan, J Bradshaw, JA Ledbetter, C Anasetti, and NK Damle (1992). Coexpression and functional cooperation of CTLA-4 and CD28 on activated T lymphocytes. J. Exp. Med. 176: 1595-1604.

Linsley, PS, JI Greene, W Brady, J Bajorath, JA Ledbetter, and R Peach (1994). Human B7-1 (CD80) and B7-2 (CD86) bind with similar avidities but distinct kinetics to CD28 and CTLA-4 receptors. Immunity 1: 793-801.

Liu, Y, and CA Janeway, Jr. (1992). Cells that present both specific ligand and costimulatory activity are the most efficient inducers of clonal expansion of normal CD4 T cells. Proc. Natl. Acad. Sci. USA 89: 3845-3849.

Lovett-Racke, AE, R Martin, HF McFarland, MK Racke, and U Utz (1997). Longitudinal study of myelin basic protein-specific T cell receptors during the course of multiple sclerosis. J. Neuroimmunol. 78:162-171.

Lovett-Racke, AE, Trotter, JL, Lauber, J, Perrin, PJ, June, CH, and Racke, MK (1998). Decreased costimulation dependence of myelin basic protein-reactive T cells on CD28-mediated costimulation in multiple sclerosis patients: a marker of activated/memory T cells. J. Clin. Invest. 101: 725-730.

Mages, HW, A Hutloff, C Heuck, K Buchner, H Himmelbauer, F Oliveri, RA Kroczek (2000). Molecular cloning and characterization of murine ICOS and identification of B7h as ICOS ligand. Eur. J. Immunol. 30: 1040-1047.

Markovic-Plese, S, I Cortese, KP Wandinger, HF McFarland, and R Martin (2001). CD4+CD28- costimulation-independent T cells in multiple sclerosis. J. Clin. Invest. 108: 1185-1194.

Martin, R, McFarland, HF, and McFarlin, DE (1992). Immunological aspects of demyelinating diseases. Annu. Rev. Immunol. 10:153-187.

Martin, R, R Voskuhl, M Flerlage, DE McFarlin, and HF McFarland (1993). Myelin basic protein-specific T-cell responses in identical twins discordant or concordant for multiple sclerosis. Ann. Neurol. 34: 524-535.

Martyn, CN (1991). The epidemiology of multiple sclerosis. In: Matthews, WB, ed. McAlpine's Multiple Sclerosis. 2nd ed. New York, NY: Churchill Livingstone; chapter 1.

McAdam, AJ, TT Chang, AE Lumelsky, EA Greenfield, VA Boussiotis, JS Duke-Cohan, T Chernova, N Malenkovich, C Jabs, VK Kuchroo, V Ling, M Collins, AH Sharpe, and GJ Freeman (2000). Mouse inducible costimulatory molecule (ICOS) expression is enhanced by CD28 costimulation and regulates differentiation of CD4+ T cells. J. Immunol. 165: 5035-5040.

McCarron, RM, RJ Fallis, and DE McFarlin (1990). Alterations in T cell antigen specificity and class II restriction during the course of chronic relapsing experimental allergic encephalomyelitis. J. Neuroimmunol. 29: 73-79.

McCormick, AL, L SAntos-Argumedo, MS Thomas, and AW Heath (1999). Cell surface expression of CD154 inhibits alloantibody responses: A mechanism for th eprevention of autoimmune repsonses against activated T cells? Cell. Immunol. 195: 157-161.

McRae, BL, CL Vanderlugt, MC Dal Canto, and SD Miller (1995). Functional evidence for epitope spreading in the relapsing pathology of EAE in the SJL/J Mouse. J. Exp. Med. 182: 75-85.

Miller, SD, and WJ Karpus (1994). The immunopathogenesis and regulation of T-cell mediated demyelinating diseases. Immunol. Today 15:356-361.

Miller, SD, CL Vanderlugt, DJ Lenschow, JG Pope, NJ Karandikar, MC Dal Canto, and JA Bluestone (1995). Blockade of CD28/B7-1 interaction prevents epitope spreading and clinical relapses of murine EAE. Immunity 3:739-745.

Perez, VL, L Van Parijs, A Biuckians, XX Zheng, TB Strom, and AK Abbas (1997). Induction of peripheral T cell tolerance in vivo requires CTLA-4 engagement. Immunity 6: 411-417.

Perrin, PJ, D Scott, CH June, and MK Racke (1995a). B7-mediated costimulation can either provoke or prevent clinical manifestations of experimental allergic encephalomyelitis. Immunol. Res. 14:189-199.

Perrin, PJ, D Scott, L Quigley, PS Albert, O Feder, GS Gray, R Abe, CH June, and MK Racke (1995b). The role of B7:CD28/CTLA4 in the induction of chronic relapsing experimental allergic encephalomyelitis. J. Immunol. 154:1481-1490.

Perrin, PJ, D Scott, TA Davis, GS Gray, MJ Doggett, R Abe, CH June, and MK Racke (1996a). Opposing effects of CTLA4-Ig and anti-CD80 (B7-1) plus anti-CD86 (B7-2) on experimental allergic encephalomyelitis. J. Neuroimmunol. 65: 31-39.

Perrin, PJ, JH Maldonado, TA Davis, CH June, and MK Racke (1996b). CTLA-4 blockade enhances clinical disease and cytokine production during experimental allergic encephalomyelitis. J. Immunol. 157:1333-1336.

Perrin, P. J., C. H. June, J. H. Maldonaldo, Robert Ratts, and M. K. Racke. 1999. Blockade of CD28 during in vitro activation of encephalitogenic T cells or after disease onset ameliorates experimental autoimmune encephalomyelitis J. Immunol. 163: 1704-1710.

Perrin, P. J., E. Lavi, S. A. Zekavat, C. A. Rumbley, and S. M. Phillips. (1999). Experimental autoimmune meningitis: A novel neurological disease in CD28-deficient mice demonstrate the importanceof CD28 in T cell differentiation. Clin. Immunol. 91: 41-49.

Pette, M, K Fujita, B Kitze, JN Whitaker, E Albert, L Kappos, and H Wekerle (1990). Myelin basic protein-specific T lymphocyte lines from MS patients and healthy individuals. Neurology 40: 1770-1776.

Racke, MK, DE Scott, L Quigley, GS Gray, R Abe, CH June, and PJ Perrin (1995). Distinct roles for B7-1 (CD80) and B7-2 (CD86) in the initiation of experimental allergic encephalomyelitis. J. Clin. Invest. 96: 2195-2203.

Racke, MK, JM Critchfield, L Quigley, B Cannella, CS Raine, HF McFarland, and MJ Lenardo (1996). Intravenous antigen administration as a therapy for autoimmune demyelinating disease. Ann. Neurol. 39: 46-56.

Raine, CS. (1984). Biology of disease. Analysis of autoimmune demyelination: its impact upon multiple sclerosis. Lab. Invest. 50:608-635.

Ratts, R. R., L. R. Arredondo, P. Bittner, P. J. Perrin, A. E. Lovett-Racke, and M. K. Racke. (1999a). The role of CTLA-4 in tolerance induction and T cell differentiation in experimental autoimmune encephalomyelitis: Intravenous antigen administration. Int. Immunol. (in press).

Ratts, R. R., L. R. Arredondo, P. Bittner, P. J. Perrin, A. E. Lovett-Racke, and M. K. Racke. (1999b). The role of CTLA-4 in tolerance induction and T cell differentiation in experimental autoimmune encephalomyelitis: Intraperitoneal antigen administration. Int. Immunol. (in press).

Sadovnick, AD, and GC Ebers (1993). Epidemiology of multiple sclerosis: a critical overview. Can. J. Neurol. Sci. 20:17-29.

Sagerstrom, CG, EM Kerr, JP Allison, and MM Davis (1993). Activation and differentiation requirements of primary T cells in vitro. Proc. Natl. Acad. Sci. USA 90: 8987-8991.

Sansom, DM, and ND Hall (1993). B7/BB1, the ligand for CD28, is expressed on repeatedly activated human T cells in vitro. Eur. J. Immunol. 23: 295-298.

Scholz, C, KT Patton, DE Anderson, GJ Freeman, and DA Hafler (1998). Expansion of autoreactive T cells in multiple sclerosis is independent of exogenous B7 costimulation. J. Immunol. 160:1532-1538.

Schweitzer, AN, F Borriello, RCK Wong, AK Abbas, and AH Sharpe (1997). Role of costimulators in T cell differentiation: Studies using antigen-presenting cells lacking expression of CD80 or CD86. J. Immunol. 158: 2713-2722.

Tivol, EA, F Borriello, AN Schweitzer, WP Lynch, JA Bluestone, and AH Sharpe (1995). Loss of CTLA-4 leads to massive lymphoproliferation and fatal multiorgan tissue destruction, revealing critical negative regulatory role of CTLA-4. Immunity 3: 541-547.

Trotter, JL, CA Damico, AH Cross, CM Pelfrey, RW Karr, X-T Fu, and HF McFarland (1997). HPRT mutant T-cell lines from multiple sclerosis patients recognize myelin proteolipid protein peptides. J. Neuroimmunol. 75: 95-103.

van den EErtwegh, AJ, RJ Noelle, M Roy, DM Shepherd, and A Aruffo (1993). In vivo CD40-gp39 interactions are essential for thymus-dependent humoral immunity.I. In vivo epxression of CD40 ligand, cytokines and antibody production delineates sites of cognate T-B cell interactions. J.Exp. Med. 178: 1555-1565.

Waterhouse, P, JM Penninger, E Timms, A Wakeham, A Shahinian, KP Lee, CB Thompson, H Griesser, and TW Mak (1995). Lymphoproliferative disorders with early lethality in mice deficient in Ctla-4. Science 170: 985-988.

Windhagen, A, J Newcombe, F Dangond, C Strand, MN Woodroofe, ML Cuzner, and DA Hafler (1995). Expression of costimulatory molecules B7-1 (CD-80), B7-2 (CD86), and interleukin 12 cytokine in multiple sclerosis lesions. J. Exp. Med. 182: 1985-1996.

Wucherpfennig, KW, and JL Strominger (1995). Molecular mimicry in T cell-mediated autoimmunity: viral peptides activate human T cell clones specific for myelin basic protein. Cell 80: 695-705.

Yoshinaga, SK, JS Whoriskey, S Khare, U Sarmiento, J Guo, T Horan, G Shih, M Zhang, MA Coccia, T Kohno, A Tafuri-Bladt, D Brankow, P Campbell, D Chang, L Chiu, T Dai, G Duncan, GS Elliott, A Hui, SM McCabe, S Scully, A Shahinian, CL Shaklee, G Van, and TW Mak (1999). T cell costimulation through B7RP-1 and ICOS. Nature 402: 827-832.

Zamvil SS, and Steinman L (1990).The lymphocyte in experimental allergic encephalomyelitis. Ann. Rev . Immunol. 8: 579-621.

Zhang, J, S Markovic-Plese, B Lacet, J Raus, HL Weiner, and DA Hafler (1994). Increased frequency of interleukin 2-responsive T cells specific for myelin basic protein and proteolipid protein in peripheral blood and cerebrospinal fluid of patients with multiple sclerosis. J. Exp. Med. 179: 973-984.

Zhang, Y, and JP Allison (1997). Interaction of CTLA-4 with AP50, a clathrin-coated pit adaptor protein. Proc. Natl. Acad. Sci. USA 94: 9273-9278.

Chapter A23

# EPITOPE SPREADING IN EAE

Andrea E. Edling and Vincent K. Tuohy
*Department of Immunology, Lerner Research Institute, Cleveland Clinic Foundation, Cleveland, OH, U.S.A.*

**Abstract:**     Intramolecular and intermolecular epitope spreading, the development of immune cells with new reactivities different from the original immunodominant response, plays an important role in the relapses and progression of EAE and has been implicated in the progression of MS. This chapter describes the process of epitope spreading in EAE, its impact on the development on tolerance, and the induction of peptide-specific tolerance and other immune intervention modalities that can prevent epitope spreading and its consequences.

**Key words:**     epitope spreading, EAE, MS, intramolecular, intermolecular, peptide-specific tolerance

## Introduction

Experimental allergic encephalomyelitis (EAE) is an inflammatory paralytic central nervous system (CNS) demyelinating autoimmune disease in rodents that serves as a widely used experimental model for studying multiple sclerosis (MS). In several mouse models of EAE a relapsing-remitting disease pattern of disease similar to that of MS develops (1). Studies have demonstrated that EAE is mediated by encephalitogenic $CD4^+$ proinflammatory T cells of the Th1 phenotype (2,3). Data from animal studies show that the autoimmune T cell repertoire is not fixed but changes as disease progresses. Many reports have indicated that the relapse and progression of EAE may be due to the acquisition of emerging T cell responses to new self-antigens during the course of disease. This acquired T cell neoreactivity is commonly referred to as epitope spreading or determinant spreading. Epitope spreading has been observed in both actively induced EAE and passively induced disease. Epitope spreading is typically demonstrated by proliferation, ELISPOT, or delayed type hypersensitivity (DTH) reactions to self epitopes not involved in initiation of disease. Epitope spreading during EAE has been shown to be a pathogenic process that spontaneously generates encephalitogenic T cells capable of inducing relapses and sustaining autoimmune demyelination. On the other hand the emergence of destructive new autoimmune T cell repertoires serves to provide more potential targets for developing antigen-specific immunotherapies capable of inhibiting or preventing the debilitating effects of chronic inflammatory self-recognition.

## EAE Disease Induction in Studying Epitope Spreading

EAE is a useful animal model system in which to study MS because both have many clinical, histopathologic, and immunopathologic similarities. EAE is generally induced in susceptible strains of rodents by immunization with CNS myelin proteins or their disease-inducing peptides, referred to as encephalitogenic determinants. The most common proteins used for inducing EAE include myelin basic protein (MBP;4,5), myelin proteolipid protein (PLP;6,7), and myelin oligodendrocyte glycoprotein (MOG;8). The adoptive transfer of MBP, PLP, or MOG specific CD4$^+$ T cells into naïve recipient mice can also induce acute and relapse-remitting EAE (9-11). Clinical disease symptoms emerge when activated CD4$^+$ T cells enter the central nervous system and recognize the encephalitogenic peptide epitopes presented in the context of major histocompatibility (MHC) class II molecules on CNS antigen presenting cells. Rodents that are susceptible to disease display an acute phase of disease followed at times by several relapsing episodes that are marked by flaccid limb paralysis.

In the SJL/J and (SWRxSJL)F$_1$ (SWXJ) mouse strains, antigen-induced EAE results in a relapsing-remitting form of disease. A moderate to severe initial acute disease followed by a recovery period characterizes the relapsing-remitting EAE (R-EAE) model. One or more relapses may occur over a period of weeks to months. This R-EAE mouse model system resembles the clinical features of relapsing-remitting MS patients that develop over many years. In both the SJL/J and SWXJ mouse strains the R-EAE animal model is often utilized by investigators to examine epitope spreading because disease can be induced by encephalitogenic epitopes of PLP, MBP and MOG that have been previously identified. In addition, a hierarchy of disease inducing epitopes has been demonstrated in both the SJL/J and SWXJ mouse models of EAE (12,13).

## Mouse Models for Examining Epitope Spreading

In SWXJ R-EAE the relapsing-remitting stage of disease often progresses to chronicity, thereby mimicking the clinical features of MS (14).

Immunization of SWXJ mice with either the p139-151, p104-117 or p178-191 peptides of PLP generates an acute disease followed by a relapsing-remitting course of EAE, with each new relapse progressively more severe than the previous (15;Figure 1A). Most SWXJ mice develop a chronic progressive stage of disease several months after EAE onset. Clinically, at this stage the mice exhibit the inability to initiate movement and show a significant loss in body weight. Severe demyelination and perivascular inflammatory foci develop in the CNS, particularly in the spinal cord. SJL/J mice also develop a relapsing-remitting EAE disease, similar to that seen in SWXJ mice and EAE can be induced in SJL/J mice with peptides from several myelin proteins including MBP (16), PLP (17) and MOG (18).

### **<u>Intramolecular Epitope Spreading during EAE Progression</u>**

Intramolecular epitope spreading in EAE was first described in studies by McCarron *et al.* (19) who showed that (SJLxPL)F$_1$ mice develop a new MHC class II-restricted response to MBP during relapse that is not observed during the initial inflammatory immune attack. In this study EAE was induced by adoptive transfer of I-A$^u$ restricted MBP1-37 specific T cells, whereas responses to the I-A$^s$ MBP89-169 determinant developed during disease progression.

Other studies by Lehmann and colleagues (20) expanded on these findings by analyzing T cell responses to MBP in the (SJLxB10.PL)F$_1$ mouse strain, which is highly susceptible to EAE. The investigators demonstrated that the specificity of the T cell repertoire in the (SJLxB10.PL)F$_1$ mice was restricted to the MBP1-11 immunodominant priming epitope during the inductive phase of EAE, but broadened to involve responses to the cryptic MBP epitopes p35-47, p81-100, and p121-140 as disease progressed. Most importantly immunization with whole MBP was not required for detection of epitope spreading which occurred even when mice were primed with the immunodominant MBP1-11 epitope. In other mouse strains such as SJL/J and B10.PL and in the Lewis rat EAE model, epitope spreading within the MBP protein has been detected (21, 22). Therefore, determinants of MBP that are cryptic after a primary immunization can become immunogenic during the disease course of EAE.

### **Intermolecular Epitope Spreading In EAE**

Intermolecular epitope spreading in EAE has been described in studies where immune responses specific for the priming CNS myelin autoantigen is associated with the detection of T cell responses specific for other CNS myelin proteins. Intermolecular spreading was first observed by Perry and colleagues (23) who analyzed lymph node T cell proliferative responses and DTH responses to PLP during the course of disease induced by immunization of (SJLxPL)F$_1$ with MBP. The investigators found *in vitro* T cell responses to PLP at 3 weeks post-immunization, and this response to PLP peaked during the first and second relapses at 5-7 weeks post-immunization.

In additional studies by Cross and colleagues (24), the authors adoptively transferred disease with MBP87-99 primed lymph node cells into naïve recipient mice and analyzed the lymph node and spleen cell reactivity to the encephalitogenic PLP p139-151, as well as to the priming MBP p87-99. Proliferative responses to the PLP 139-151 epitope were observed during the acute and chronic stages of EAE. The pathogenicity of intermolecular spreading was demonstrated by passive transfer of EAE into naïve SJL/J recipient mice with PLP139-151 stimulated spleen cells taken from SJL/J mice that had previously developed EAE by passive induction of EAE with MBP 87-99 stimulated cells. The results suggest that the autoimmune inflammatory CNS destruction mediated by T cells targeted against MBP leads to endogenous self priming to PLP.

Other studies by McRae *et al.* (12) demonstrated PLP intermolecular spreading during EAE induced in SJL/J mice by adoptive transfer of MBP84-104 primed T cells. Responses to the PLP139-151 spreading determinant were detected by proliferation and DTH during the emergence of the first disease relapse.

### The Epitope Spreading Cascade

In EAE primary autoreactivity associated with the clinical onset of disease regresses with time and often becomes undetectable during the course of disease. The emergence of secondary immune responses to autoreactive spreading epitopes is detected as disease progresses. Thus, the chronic progression of EAE involves a transition of autoreactivity from the primary initiating self-determinant to a defined cascade of secondary determinants that sustain the CNS inflammation (25).

In the SWXJ mouse strain, when EAE is induced by immunization with either PLP139-151, PLP178-191, or PLP104-117, the order in which new self-determinants are recognized during the course of disease follows a predictable and sequential pattern (13,15). By 12-16 weeks after immunization, responses to the priming determinants regress and are often undetectable by conventional proliferation assays despite the development of chronic progressive disease. This regression of primary autoreactivity occurs both in the periphery and CNS. SWXJ mice immunized with PLP139-151 develop a sequential pattern of neoreactivity that includes initially PLP249-273, MBP87-99 and PLP178-191 at 4, 8, and 12 weeks post-immunization, respectively (Figure 1A,B). After priming with the PLP104-117 determinant, responses to PLP139-151 occur early after EAE onset and are subsequently followed by responsiveness to MBP87-99.

Similarly, immunization PLP178-191 results in the development of responses to PLP139-151 and MBP87-99. This neoreactivity to spreading determinants has been detected up to 12 weeks post-immunization. In contrast to SWXJ mice, immunization of SJL/J mice with PLP139-151 results in a reverse spreading ordering with expansion of the PLP178-191 repertoire occurring before expansion of MBP84-104-specific T cells (12,26,27). The disease course in PLP139-151 or PLP178-191 induced R-EAE results in a moderate to severe acute stage followed by an incomplete remission. Primary relapse usually develops within 3 weeks after remission onset with a severity of disease not substantially greater than the acute phase. Responses to the relapsing PLP178-191 determinant develop after remission from onset of disease in PLP139-151 induced EAE. However, immune responses to the MBP84-104 spreading determinant begin to emerge during the second relapse. In MBP84-104 induced R-EAE in the SJL/J model, the disease is characterized differently. A mild acute stage of disease develops which is followed by complete remission. Approximately 30-40 days following the onset of remission, relapse occurs and an immune response to the immunodominant determinant PLP139-151 can be detected.

Epitope spreading has also been associated with disease progression in MS. Studies have focused on patients with isolated monosymptomatic demyelinating syndromes (IMDS) that have clinical disorders associated with progression to CDMS. In studies analyzing proliferative immune responses to PLP peptides, patients with monocentric monophasic IMDS responded to distinct PLP peptides (28). Responses to these defined PLP epitopes can be detected for more than 1 year (Figure 1C). However, the autoreactivity associated with the onset of symptoms in IMDS patients wanes over time and eventually becomes undetectable as disease progresses. Moreover, there is a shift in the responses to new PLP epitopes that appear to sustain the autoimmune inflammatory disease state. These results indicate that epitope spreading occurs during the development and progression of MS, similar to the epitope spreading cascade that develops during EAE progression.

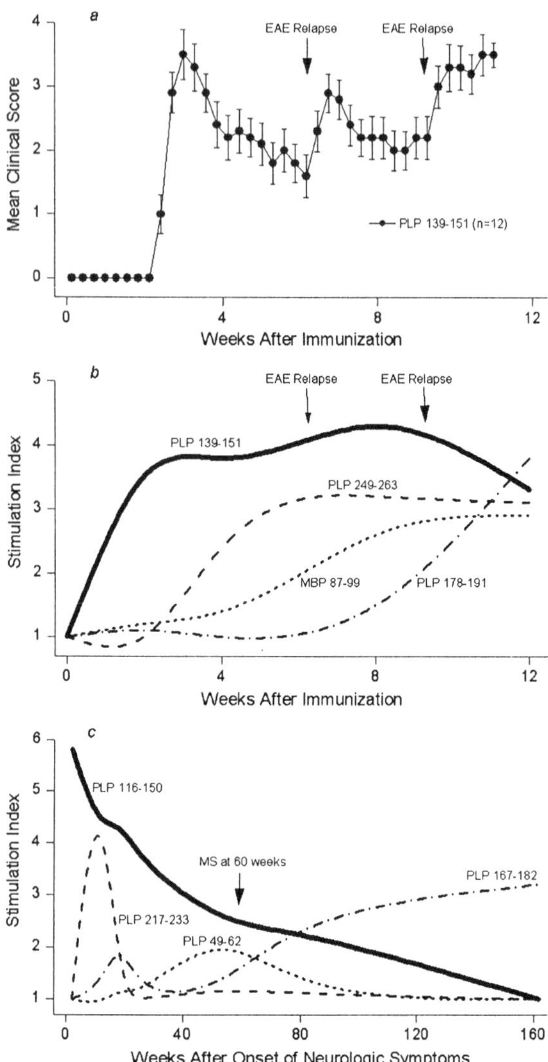



***Figure 1.*** *The Epitope Spreading Cascade Accompanies Progression Of EAE And MS.* (***a***) Immunization of SWXJ mice with PLP 139-151 results in a relapsing-remitting course of EAE that progresses to chronicity. Each relapse is followed by an incomplete recovery that results in a progressive accumulation of neurologic deficit. Similar clinical profiles occur following immunization with PLP 104-117 and PLP 178-191. Error bars show ±SEM. (***b***) A predictable sequential epitope spreading cascade accompanies progression to chronicity in EAE. Induction of EAE

by immunization of SWXJ mice with PLP 139-151 results in a temporally spaced cascade of neoautoreactivity involving PLP 139-151 → PLP 249-273 → MBP 87-99 → PLP 178-191. Similar sequential cascades occur during progression of EAE induced with the encephalitogenic PLP determinants p104-117 and p178-191. It should be noted that EAE progression continues long after responses to the priming p139-151 determinant have peaked. The data are represented as spline curve plots of mean splenocyte proliferative responses of at least three mice tested at each 0, 2, 4, 8, and 12 week time points. (*c*) The progression of MS is accompanied by the decline of primary autoreactivity associated with initial symptoms and by the concurrent emergence of epitope spreading. IMDS patient DL showed progression to CDMS 60 weeks after the initial onset of symptoms. Disease progression involved a sustained neo-autoreactivity to PLP 167-182 accompanied by a decline and eventual disappearance of responses to the PLP 116-150 determinant associated with disease onset. The data are represented as spline curve plots of mean proliferative responses of PBMC on each day tested. (Figure and legend from: Tuohy, V.K., Yu, M., Yin, L., Kawczak, J.A., Johnson, J.M. Mathisen, P.M., Weinstock-Guttman, B., and Kinkel R.P. 1998. The epitope spreading cascade during progression of experimental autoimmune encephalomyelitis and multiple sclerosis. Immunologic Rev. 164:93-100. Printed with permission from Blackwell Munksgaard, 1 Rosenørns Allè, P.O. Box 227, DK-1502, Copenhagen V, Denmark).

### Epitope Spreading to Self Determinants in Theiler's Murine Encephalomyelitis (TMEV)

TMEV is a murine pathogen in the picornavirus family that results in a persistent infection of CNS APCs. In SJL/J mice a persistent TMEV infection in the CNS produces a chronic progressive inflammatory demyelinating disease clinically similar to chronic progressive EAE, but not to relapsing-remitting disease (29). Demyelination in the TMEV model appears to be immune mediated and initiated by T cells specific for the viral epitopes. Recent studies have shown that epitope spreading of myelin self determinants contributes to the chronic progressive demyelination induced with TMEV.

Immune responses to TMEV epitopes develop at 5-7 days post-infection. However, T cell responses to myelin epitopes do not develop prior to disease onset, but are detected 30-35 days post-viral infection. CNS APCs persistently infected with TMEV present viral epitopes to virus-specific T cells thereby targeting T cell mediated myelin destruction in the CNS. During later stages in TMEV infection, T cell responses to a number of encephalitogenic myelin epitopes can be demonstrated (30,31). At 50-60 days post-infection T cell proliferative and DTH responses can be detected to the immunodominant epitope PLP139-151. T cell responses to several other myelin epitopes develop as the disease progresses. Four weeks after the detection of PLP139-151 responses, T cell responses to PLP56-70 and MOG92-106 appear followed 2 months later by responses to PLP178-191 and MBP84-104. These data demonstrate a pattern of epitope spreading to myelin proteins during the chronic stages of TMEV induced demyelination.

In the TMEV demyelination model epitope spreading may be viewed as a virus-specific T cell mediated inflammatory response which leads to the activation of resident CNS APCs that process and present endogenous self-epitopes to autoreactive T cells. Thus, Th1 autoreactivity is triggered by the inflammatory response to the viral infection and leads to a chronic autoimmune disease in the CNS by an epitope spreading process.

### Epitope Spreading and EAE Disease Progression

It is widely believed that newly acquired autoreactive T cell specificities develop during the course of EAE as a result of chronic tissue inflammation or damage and subsequent presentation of new self epitopes to T cells invading the target organ. Thus, the processing and presentation of newly acquired antigenic determinants lead to the activation of autoreactive T cells specific for spreading epitopes. One question that arises in autoimmunity is whether epitope spreading contributes to disease pathogenesis. This issue has been most difficult to address in human disease.

Evidence indicating a role for intramolecular epitope spreading in the pathogenesis of disease was observed in a PLP139-151-induced RR-EAE SJL/J disease model. T cells responding to the encephalitogenic PLP spreading epitope p178-191 were detected in spleens of PLP139-151-induced EAE mice. Importantly, the epitope spreading splenic PLP178-191 specific T cells demonstrated pathogenic potential by passively transferring disease into naïve SJL/J mice (12).

Vanderlugt *et al.* (32) demonstrated that T cells specific for endogenous myelin epitopes play a major role in the development of EAE relapses. In these studies splenic T cells isolated from SJL/J mice 7 weeks after induction of EAE with MBP84-104 specific T cells were able to transfer disease after *in vitro* activation with either MBP84-104 or the spreading determinant PLP139-151. Moreover, the autoreactive T cells specific for spreading epitopes were found in the CNS of mice before and during the primary disease relapse.

The pathogenicity of T cells responding to spreading determinants was also evaluated by adoptive transfer experiments in SWXJ mice (33). Seven weeks after priming with PLP139-151, spleen cells were activated with the PLP139-151 or MBP87-99 epitope *in vitro* and EAE was passively transferred into naïve SWXJ/J recipient mice. The EAE severity induced by T cells specific for the spreading MBP87-99 was similar to the severity induced by T cells specific for PLP139-151. Therefore, the neoautoreactive T cells were pathogenic and expressed a Th1 proinflammatory cytokine profile, secreting significant levels of IL-2 and IFN-γ.

More recent studies by Targoni et al. (34) demonstrate that throughout the course of EAE, a first stage of effector cells undergoes exhaustion in the CNS and periphery while a second stage of effector cells, developed by epitope spreading, emerges in the CNS. The investigators used enzyme-linked immunospot analysis (ELISPOT) to determine the frequency of the priming PLP139-151 or spreading PLP178-191 specific T cells in the CNS and periphery during disease progression. The frequency of T cells specific for the priming PLP139-151 epitope in the CNS and periphery was high during the onset of paralysis and then declined over time, completely disappearing 70 days after immunization. However, PLP178-191 specific T cells were detected in the CNS, prior to their detection in the periphery during the second episode of paralysis. Thus, the epitope spreading responses were shown to initiate in the targeted CNS tissue rather than in the periphery. Although the primary effector response declined to an undetectable level, mice continued to demonstrate disease progression. Presumably as a consequence of the active spreading repertoire these results confirm previous findings that epitope spreading plays a significant role in the progression of autoimmune disease.

### Peptide-Specific Tolerance in the Prevention of Epitope Spreading

Several studies indicate that induction of tolerance to spreading determinants inhibits EAE relapse and progression even when tolerance is induced after onset of disease. Vanderlugt *et al.* (32) showed that pretolerization with relapse associated epitopes 7 days before EAE induction did not have an effect on acute disease, but reduced the relapse rate in both PLP139-151 and MBP84-104 induced R-EAE. Tolerance induction with the relapsing-epitopes was also induced during the remission period following acute disease and resulted in a decrease in relapses.

Similar studies were conducted in the SWXJ mouse model. Mice were tolerized by intravenous injection of the MBP87-99 spreading determinant ten days after onset of EAE induced by immunization with PLP139-151. Mice tolerized to the spreading MBP 87-99 determinant showed consistent improvement in mean clinical score and delayed progression of clinical disease while mice tolerized to the non-spreading PLP104-117 determinant continued their progression to disability. The clinical improvement of MBP 87-99 tolerized mice was accompanied by a corresponding significant inhibition of MBP 87-99 DTH as measured by peptide-specific ear swelling. These studies demonstrate that epitope spreading is not simply an epiphenomenon but rather plays a pathogenic role in the development of relapse and chronic progression of autoimmune disease (33).

### Additional Therapeutic Strategies Targeting Epitope Spreading in EAE

It has been shown that progression of EAE may be significantly inhibited when Th2-like T cells specific for priming determinants are adoptively transferred after onset of disease (35,36). Yin *et al.* (37) examined the efficacy of preemptive targeting of the epitope spreading cascade by adoptive transfer of regulatory T cells after onset of disease. A panel of high IL-10 producing Th2/Tr1-like T cell lines were generated by transfecting primed lymph node cells with a transgene construct designed to produce mouse IL-10 regulated by a mouse IL-2 promoter. The investigators found that after induction of EAE with PLP 139-151, adoptive transfer of MBP 87-99 specific Th2/Tr1-like T cells resulted in a marked and prolonged inhibition of disease progression accompanied by a significant decrease in spinal cord demyelination. The therapeutic effect involved a Tr1 immune deviation of the response to the MBP 87-99 spreading determinant and a bystander inhibition of the recall IFN-γ response to the priming PLP 139-151 immunogen. These results indicate that the clinical course of autoimmune demyelinating disease may be dramatically altered by preemptively creating an immune deviated epitope spreading cascade. However, such deviation may produce cascades involving distinctly different targeted self-antigens as has been shown in the response of NOD mice to beta cell antigens in insulin dependent diabetes mellitus (38).

Another target of therapy for inhibiting epitope spreading in EAE is blockade of B7-1 mediated co-stimulation (26,32). The activation of T cells requires two signals: an antigen-specific signal through the T cell receptor and MHC complex and co-stimulatory signals through the CD28 molecule expressed on T cells, and B7-1 (CD80) expressed on APCs (39,40). B7-1 is upregulated during R-EAE disease progression. When anti-B7-1 F(ab) fragment therapy is administered in SJL/J mice, endogenous self priming and epitope spreading are blocked and clinical relapses and CNS pathology are reduced. The clinical and pathological effects appear to be long lasting, even when treatment is administered late in the disease stage. The mechanisms involved in the inhibition of disease may include anergy induction of T cells specific for relapse associated epitopes. Thus, the co-stimulation antagonist may have a therapeutic application in MS as a result of its ability to prevent endogenous self priming and epitope spreading.

Finally, it is quite possible that the therapeutic effects of IFN-β in MS may substantially be due to inhibition of IL-12 dependent epitope spreading (41). McRae *et al.* (42,43) have shown that IFN-β abrogates Th1 differentiation *in vitro* by inhibiting T cell dependent CD40-induced production of IL-12 from monocyte derived DC, a requirement for priming Th0 → Th1 cells (44,45). We have found that the therapeutic effect of treating SWXJ mice with IFN-β after EAE onset is accompanied by a dramatic inhibition of IL-12 production in *ex vivo* splenocyte responsiveness to the PLP 139-151 immunogen (46). Moreover, this IFN-β induced inhibition of IL-12 is accompanied by a marked failure to develop neoautoreactivity to determinants involved in the epitope spreading cascade. These results support the view that IFN-β inhibits IL-12 dependent endogenous self-priming *in vivo* resulting in aborted epitope spreading and subsequent inhibition of disease progression.

In summary, epitope spreading is a pathogenic process that leads to relapse and chronic progression of autoimmune disease. During development of autoimmune disease a predictable sequential pattern of epitope spreading emerges. This ordered epitope spreading cascade may provide a basis for targeted peptide-specific immunotherapy after onset of clinical disease in MS.

## ACKNOWLEDGMENTS

This work was supported by grants NS-36054, NS-37476 and AI-51837 from the National Institute of Health, USA and grants RG-3070 and FA 1445-A-1 from the National Multiple Sclerosis Society, USA

# REFERENCES

1.    Martin, R., H.F. McFarland, and D.E. McFarlin. 1992. Immunological aspects of demyelinating disease. *Annu. Rev. Immunol.* 10:153-187.

2.    Pettinelli, B.B. and D.E. McFarlin. 1981. Adoptive transfer of EAE in SJL/J mice after in vitro activation of lymph node cells by myelin basic protein: requirement for Lyt1$^+$2$^-$ T lymphocytes. *J. Immunol.* 127:1420-1423.

3.    Sriram, S., and C.A. Roberts. 1986. Treatment of established chronic relapsing EAE with anti-L3T4 antibodies. *J. Immunol.* 136:4464-4469.

4.    Fritz, R.B., C.-H.J. Chou, and D.E. McFarlin. 1983. Relapsing murine experimental allergic encephalomyelitis induced by myelin basic protein. *J. Immunol.* 130:1024-1026.

5.    Chou, C.-H.J., R. Shapira, and R.B. Fritz. 1983. Encephalogenic activity of the small form of mouse myelin basic protein in the SJL/J mouse. *J. Immunol.* 130:2183-2186.

6.    Trotter, J.L., H.B. Clark, K.G. Collins, C.L. Wegeschiede, and J.D. Scarpellini. 1987. Myelin proteolipid protein induces demyelination disease in mice. *J. Neurol. Sci.* 79:173-188.

7.    Tuohy, V.K., R.A. Sobel, and M.B. Lees. 1988. Myelin proteolipid protein induces EAE. Variation of disease expression in different strains of mice. *J. Immunol.* 140:1868-1873.

8.     Amor, S., N. Groome, C. Linington, M.M. Morris, K. Dornmair, M.V. Gardinier, J.M. Matthieu, and D. Baker. 1994. Identification of epitopes of myelin oligodendrocyte glycoprotein for the induction of experimental allergic encephalomyelitis in SJL and Biozzi AB/H mice. *J. Immunol.* 153:4349-4356.

9.     Yamamura, T., T. Namikawa, M. Endoh, T. Kunishita, and T Tabira. 1986. Passive transfer of EAE induced by proteolipid protein. *J. Neurol. Sci.* 76:269-275.

10.    Van der Veen, R.C., J.L. Trotter, H.B. Clark, and J.A. Kapp. 1989. The adoptive transfer of chronic relapsing EAE with lymph node cells sensitized to myelin proteolipid protein. *J. Neuroimmunol.* 21:183-191.

11.    McRae, B.L., M.K. Kennedy, L.J. Tan, M.C. Dal Canto, and S.D. Miller. 1992. Induction of active and adoptive chronic relapsing EAE using an encephalitogenic epitope of proteolipid protein. *J. Neuroimmunol.* 38:229-240.

12.    McRae, B.L., C.L. Vanderlugt, M.C. Dal Canto, and S.D. Miller. 1995. Functional evidence for epitope spreading in the relapsing pathology of experimental autoimmune encephalomyelitis. *J. Exp. Med.* 182:75-85.

13.    Yu, M., J.M. Johnson, and V.K. Tuohy. 1996. A predictable sequential determinant spreading cascade invariably accompanies progression of experimental autoimmune encephalomyelitis: A basis for peptide-specific therapy after onset of clinical disease. *J. Exp. Med.* 183:1777-1788.

14.    Yu, M., A. Nishiyama, B.D. Trapp, and V.K. Tuohy. 1996. Interferon-beta inhibits progression of relapsing-remitting experimental autoimmune encephalomyelitis. *J. Neuroimmunol.* 64:91-100.

15.    Tuohy, V.K., M. Yu, L.Yin, J.A. Kawczak, J.M. Johnson, P.M. Mathisen, B. Weinstock-Guttman, and R.P. Kinkel. 1998. The epitope spreading cascade during progression of experimental autoimmune encephalomyelitis and multiple sclerosis. *Immunol. Rev.* 164:93-100.

16.    Kono, D.H., J.L. Urban, S.J. Horvath, D.G. Ando, R.A. Saavedra, and L. Hood. 1988. Two minor determinants of myelin basic protein induce experimental allergic encephalomyelitis in SJL/J mice. *J. Exp. Med.* 168:213-227.
17.    Tuohy, V.K., Z. Lu, R.A. Sobel, R.A. Laursen, and M.B. Lees. 1989. Identification of an encephalitogenic determinant of myelin proteolipid protein for SJL mice. *J. Immunol.* 142:1523-1526.

18.    Amor, S., N. Groome, C. Linington, M.M. Morris, K. Dornmair, M.V. Gardinier, J.M. Matthieu, and D. Baker. 1994. Identification of epitopes of myelin oligodendrocyte glycoprotein for the induction of experimental allergic encephalomyelitis in SJL and Biozzi AB/H mice. *J. Immunol.* 153:4349-4356.

19.   McCarron, R.M., R.J. Fallis, and D.E. McFarlin. 1990. Alterations in T cell antigen specificity and class II restriction during the course of chronic relapsing experimental allergic encephalomyelitis. *J. Neuroimmunol.* 29:73-79.

20.   Lehmann, P.V., T. Forsthuber, A. Miller, and E.E. Sercarz. 1992. Spreading of T-cell autoimmunity to cryptic determiants of an autoantigen. *Nature.* 358:155-157.

21.   Bhardwaj, V., V. Kumar, I.S. Grewal, T. Dao, P.V. Lehmann, H.M. Geysen, and E.E. Sercarz. 1994. T cell determinant structure of myelin basic protein in B10.PL, SJL/J, and their F1s. *J. Immunol.* 152:3711-3719.

22.   Mor, F., and I.R. Cohen. 1993. Shifts in the epitopes of myelin basic protein recognized by Lewis rat T cell before, during, and after the induction of experimental autoimmune encephalomyelitis. *J. Clin. Invest.* 92:2199-2206.

23.   Perry, L.L., E.Barzaga-Gilbert, and J.L. Trotter. 1991. T cell sesitization to proteolipid protein in myelin basic protein-induced relapsing experimental allergic encephalomyelitis. *J. Neuroimmunol.* 33:7-15.

24.   Cross, A.H., V.K. Tuohy, and C.S. Raine. 1993. Development of reactivity to new myelin antigens during chronic relapsing autoimmune demyelination. *Cell. Immunol.* 146:261-269.

25.   Tuohy, V.K., M. Yu, L. Yin, J.A. Kawczak, and R.P. Kinkel. 1999. Spontaneous regression of primary autoreactivity during chronic progression of experimental autoimmune encephalomyelitis and multiple sclerosis. *J. Exp. Med.* 189:1033-1042.

26.   Miller, S.D., C.L. Vanderlugt, D.J. Lenschow, J.G. Pope, N.J. Karandikar, M.C. Dal Canto, and J.A. Bluestone. 1995. Blockade of Cd28/B7-1 interaction prevents epitope spreading and clinical relapses of murine EAE. *Immunity.* 3:739-745.

27.   Vanderlugt, C.L., W.S. Begolka, K.L. Neville, Y. Katz-Levy, L.M. Howard, T.N. Eager, J.A. Bluestone, and S.D. Miller. 1998. The functional significance of epitope spreading and its regulation by co-stimulatory molecules. *Immunol. Rev.* 164:63-72.

28.   Tuohy, V.K., M. Yu, B.Weinstock-Guttman, and R.P. Kinkel. 1997. Diversity and plasticity of self-recognition during the development of multiple sclerosis. *J. Clin. Invest.* 99:1682-1690.

29.   Lipton, H.L. 1975. Theiler's virus infection in mice: An unusual biphasic disease process leading to demyelination. *Infect. Immun.* 11:1147-1155.

30.   Miller, S.D., C.L. Vanderlugt, W.S. Begolka, W. Pao, R.L. Yauch, K.L. Neville, Y. Katz-Levy, A. Carrizosa, and B.S. Kim. 1997. Persistent infection with

Theiler's virus leads to CNS autoimmunity via epitope spreading. *Nature Med.* 3:1133-1136.

31.   Miller, S.D. and T.N. Eagar. 2001. Functional role of epitope spreading in the chronic pathogenesis of autoimmune and virus-induced demyelinating diseases. *Advances in Exp. Med. and Biol.* 490:99-107.

32.   Vanderlugt C.L., K.L. Neville, K.M. Nikcevich, T.N. Eager, J.A. Bluestone, and S.D. Miller. 2000. Pathologic role and temporal appearance of newly emerging autoepitopes in relapsing experimental autoimmune encephalomyelitis. *J. Immunol.* 164:670-678.

33.   Yu, M., J.M. Johnson, and V.K. Tuohy. 1996. Generation of autonomously pathogenic neo-reactive Th1 cells during the development of the determinant spreading cascade in murine autoimmune encephalomyelitis. *J. Neurosci. Res.* 45:463-470.

34.   Targoni, O.S., J. Baus, H.H. Hofstetter, M.D. Hesse, A.Y. Karulin, B. O. Boehm, T.G. Forsthuber, and P.V. Lehmann. 2001. Frequencies of neuroantigen-specific T cell in the central nervous system  versus the immune periphery during the course of experimental allergic encephalomyelitis. *J. Immunol.* 166:4757-4764.

35.   Kuchroo, V.K., M.P. Das, J.A. Brown, A.M. Ranger, S.S. Zamvil, R.A. Sobel, H.L. Weiner, N. Nabavi, and L.H. Glimcher. 1995. B7-1 and B7-2 costimulatory molecules activate differentially the Th1/Th2 developmental pathways: application to autoimmune therapy. *Cell* 80:707-718

36.   Mathisen, P.M., M. Yu, J.M. Johnson, J.A. Drazba, and V.K. Tuohy. 1997. Treatment of experimental autoimmune encephalomyelitis with genetically modified memory T cells. *J. Exp. Med.* 186:159-164.

37.   Yin, L., A.E. Edling, J. A. Kawczak, P.M. Mathisen, T. Nanvati, J.M. Johnson, and V.K. Tuohy. 2001. Pre-emptive targeting of the epitope spreading cacade with genetically modified regulatory t cells during autoimmune demyelinating disease. *J.immunol.* 167:6105-6112.

38.   Tian, J., Lehmann, P.V., and Kaufman, D.L. 1997. Determinant spreading of T helper cell 2 (Th2) responses to pancreatic islet autoantigens. *J. Exp. Med.* 186:2039-2043.

39.   Jenkins MK. and J.G. Johnson. 1993. Molecules involved in T-cell costimulation. *Current Opinion in Immunology. 5:361-367.*

40.   Lenschow DJ., T.A. Walunas, and J.A. Bluestone. 1996. CD28/B7 system of T cell costimulation. *Annual Review of Immunology. 14:233-258.*

41.   Karp, C.L., Biron, C.A., Irani, D.N. 2000. Interferon β in multiple sclerosis: is IL-12 suppression the key? *Immunol. Today* 21, 24-28.

42. McRae, B.L., Semnani, R.T., Hayes, M.P., van Seventer, G.A. 1998. Type I IFNs inhibit human dendritic cell IL-12 production and Th1 cell development. *J. Immunol.* 160, 4298-4304.

43. McRae, B.L., Beilfuss, B.A., van Seventer, G.A. 2000. IFN-β differentially regulates CD40-induced cytokine secretion by human dendritic cells. *J. Immunol.* 164, 23-28.

44. Abbas, A.K., Murphy, K.M., Sher, A. 1996. Functional diversity of helper T lymphocytes. *Nature (Lond.)* 383, 787-793.

45. Trinchieri, G. 1998. Immunobiology of interleukin-12. *Immunologic Res.* 17, 269-278.

46. Tuohy, V.K., Yu, M., Yin, L., Mathisen, P.M., Johnson, J.M.,Kawczak, J.A. 2000. Modulation of the IL-10/IL-12 cytokine circuit by interferon-beta inhibits the development of epitope spreading and disease progression in murine autoimmune encephalomyelitis.*J. Neuroimmunol.* 111:55-63.

Chapter  A24

# APOPTOTIC CELL DEATH IN EXPERIMENTAL AUTOIMMUNE ENCEPHALOMYELITIS
*Apoptosis of effector cells as a safe mechanism in the termination of an autoimmune inflammatory attack*

Andrew Chan and Ralf Gold
*Department of Neurology, Clinical Research Group for Multiple Sclerosis and Neuroimmunology, Julius-Maximilians University, D-97080 Würzburg, Germany*

Abstract:     Particularly in the vulnerable CNS with a low capacity for regeneration specialized mechanisms must be active for the fast and gentle elimination of dysregulated autoaggressive immune cells.  In EAE, local apoptosis of autoimmune T-cells has been identified as a safe means for the removal of these unwanted cells. T-cell apoptosis *in situ* followed by phagocytic clearance of apoptotic remnants by glia assures a minimum of detrimental bystander damage to the local parenchyma and down-regulates the local inflammatory reaction. The pharmacological augmentation of local apoptosis of inflammatory effector cells might gain therapeutic importance also in human neuroimmunological diseases such as multiple sclerosis.

Key words:   T-cell, phagocytosis, Experimental autoimmune neuritis, astrocyte, microglia, glucocorticosteroids, multiple sclerosis

## 1.     INTRODUCTION

Multicellular organisms face the task of safely disposing unwanted cells under physiological and pathological conditions. The active and tightly regulated cell death program during apoptosis is considered to play a major role in the physiological control of cell turnover while at the other end dysregulation of apoptosis has been shown to contribute to the pathogenesis of different autoimmune diseases.

During the last decade the role of apoptosis also in neuroimmunological diseases has been more clearly defined. In EAE and experimental autoimmune neuritis (EAN), especially apoptosis of the pathogenic effector cells has been thoroughly investigated. This review will briefly introduce definitions and detection methods for apoptosis. We will focus on apoptotic cell death of effector T-cells in EAE, proposed mechanisms of T-cell apoptosis, functional consequences and finally its potential therapeutic implications.

## 1.1    Apoptosis: definition and detection

Apoptosis is a specialized, morphologically and biochemically distinct form of eukaryotic cell death. Introduced in 1972 by Kerr and colleagues, the term apoptosis served to describe a set of morphological features commonly observed during cell death in various tissues and cell types which were different to those observed during necrotic cell death (1). The stereotypic series of cellular changes comprise blebbing of the plasma membrane and condensation of chromatin at the periphery of the nucleus, followed by disintegration of the cell into multiple membrane-enclosed vesicles. In contrast to the lytic processes observed during necrosis, cell death by apoptosis is not associated with secondary inflammatory tissue responses, e.g. towards potentially harmful intracellular contents of the dying cells (2). Thus, apoptosis is considered to provide a safe and gentle cell death mechanism. *In vivo*, apoptosis also appears to be a fast and efficacious mechanism for the elimination of unwanted cells, with completion of the cell death program within 4-5 hrs at least in certain tissues and under defined conditions (3).

As the traditional definitions of apoptosis were mainly based on morphological criteria, electron microscopy served as the gold standard in the detection of apoptosis. Meanwhile, the better understanding of underlying biochemical processes during apoptosis has led to the introduction of an array of detection methods at the molecular level (4,5). For example, intravital internucleosomal DNA-fragmentation which generates fragments of 180 bp and multiples thereof can  reliably be detected using TUNEL or *in situ* nick translation techniques. The identification of a set of aspartate-directed cysteine proteases (Caspases) whose activation underly many of the observed morphological changes during apoptosis has further broadened the spectrum of detection methods. Thus, either the activated caspases or their proteolytic cleavage products (e.g. cytokeratin 18, Poly(ADP-Ribose)-Polymerase PARP, APP, actin cleavage products) can be identified. Also specific alterations of the cell membrane (e.g. loss of phospholipid asymmetry detected with annexin-V) or mitochondrial changes (e.g. cytochrome c release, "permeability transition") serve as the basis for commercially available detection methods for apoptosis. However, many of these detection methods have to be interpreted cautiously (4). For example DNA-fragmentation does not only occur during apoptosis, but also during necrosis. Since different intracellular pathways lead to typical apoptotic features, a combination of different molecular detection methods is feasible. Thus, typical morphological features of apoptosis can also occur in the absence of oligonucleosomal DNA-fragmentation as well as in enucleated cells (6,7).

## 1.2      T-cell apoptosis in EAE and CNS inflammation

T-cells autoreactive to CNS-antigens such as MBP can also occur in healthy individuals (8,9). Albeit more often in MS-patients, surges of increased frequencies of circulating myelin-reactive T-cells can also be observed in healthy subjects, possibly driven by cross-reactive environmental antigens (10). These activated autoreactive T-cells are capable of entering the CNS-parenchyma and thus have the potential to induce a local immune response when they encounter their specific antigen in the context of appropriate restriction molecules (11,12). Therefore, specialized anatomic barriers such as the blood brain barrier or the absence of lymphatic drainage are not sufficient to prevent immune-mediated damage in the CNS, arguing for other mechanisms in the physiological control of autoreactive inflammatory cells *in situ*. Local apoptosis of pathogenic T-cells has been identified as a major immunological defense mechanism in the "immune-privileged" CNS, leading to an inhibition of inflammation or, once it has encroached, to rapid and non-destructive elimination of the inflammatory infiltrate (review in (13,14).

First reports on cell death by apoptosis in inflammatory brain lesions of Lewis rat EAE were from Pender and colleagues (15). Using morphological criteria the majority of the dying cells appeared to be lymphocytes and oligodendrocytes. Apoptosis of $\alpha/\beta$-T-cells was subsequently confirmed by the same group using pre-embedding immunolabelling techniques (16). By combined T-cell immunohistochemistry, molecular labeling techniques and ultrastructural criteria, Schmied and coworkers analyzed T-cell apoptosis in the spinal cord quantitatively during the time course of different rat EAE models (17). Of all apoptotic cells, 64% were identified as T-lymphocytes, mostly expressing the $\alpha/\beta$-T cell receptor. While another 9% of the apoptotic cells were classified as oligodendrocytes, apoptosis of macrophages was only rarely observed. However, a considerable proportion of apoptotic cells could not be identified immunocytochemically, due to the advanced degeneration of the cells.

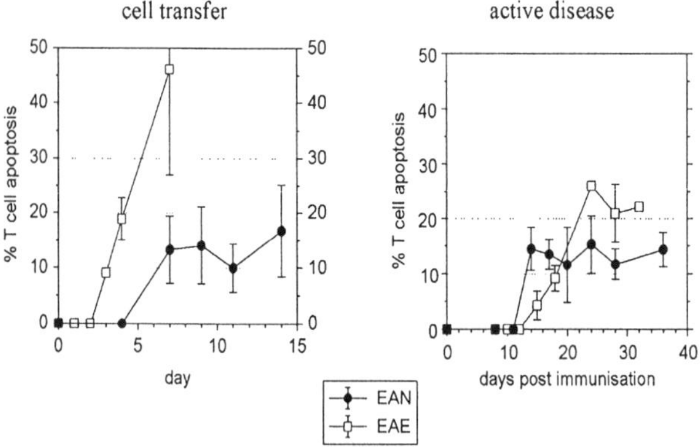


*Figure 1.* Time course of T-cell apoptosis *in situ.* Apoptosis during the natural course of experimental autoimmune encephalomyelitis (EAE) or neuritis (EAN), passively induced by cell transfer or by active immunization with MBP (EAE) or peripheral nerve myelin (EAN). Ordinate: percentage of apoptotic T-cells (mean +/- SD). Modified after (18).

During Lewis rat cell transfer (AT)-EAE using MBP-specific T-cells the degree of T-cell apoptosis was minimal at early stages on day 4, but peaked at the time of recovery from disease at day 7 with apoptosis rates of up to 49% (figure 1). Also in the active disease model prevalence of T-cell apoptosis *in situ* was highest when animals had recovered from the disease.

Apoptosis of inflammatory cells *in situ* also occurs in chronic relapsing EAE models, especially during clinical relapses (15,19,20,21). In the CNS of adoptively transferred chronic relapsing EAE in SJL/J mice, apoptosis of CD4+ T-cells and microglia/brain macrophages could readily be observed, while oligodendrocytes and astrocytes did not exhibit TUNEL positivity (19).

T-cell apoptosis *in situ* has not only been identified in EAE but also in coronavirus-mediated encephalomyelitis (22). Apoptosis of T-lymphocytes also occurs in inflammatory human brain lesions, most prominently in acute disseminated leukoencephalomyelitis (ADEM) (23), an acute monophasic disease which closely mimics pathological changes of acute EAE. T-cell apoptosis can also be observed in active MS lesions albeit to a lesser extent, possibly due to the chronicity of the disease (24). The similar degree of T-cell apoptosis in the acutely autoimmune inflamed rodent (EAE) and human

CNS (ADEM) (25) with approximately 30% of all invading T-cells undergoing apoptosis highlights the importance of the local apoptotic destruction of inflammatory T-cells in the termination of CNS-inflammation.

In contrast to the high T-cell apoptosis rate observed in the CNS, the PNS and other peripheral tissues appear to have a lower capacity to induce T-cell apoptosis locally. T-cell apoptosis could also be observed in sciatic nerves of Lewis rats with active or AT-EAN (26). Whereas the time course was similar to EAE, the extent of T-cell apoptosis was markedly different: highest levels were also found during recovery, with typical T-cell rates of approximately 10% (figure 1). However, in spite of this lower apoptosis rate, also in the PNS local T-cell apoptosis can be regarded as an efficient mechanism of cell elimination. Thus, given a time frame of 4-5 hrs for the completion of apoptosis, one could assume that up to 50% of an inflammatory infiltrate would be eliminated within 24 hrs.

Negligible degrees of T-cell apoptosis were  found in T-cell mediated autoimmune diseases of muscular tissue. T-cell apoptosis was less than 0.5% in CD8-T-cell dominated experimental autoimmune myositis and biopsies of different human myopathies of presumed autoimmune etiology (27,28). Thus, T-cell inflammation in muscle is not cleared by apoptosis *in situ*, which could contribute to the non-self-limiting nature of these diseases. Even in HIV-associated polymyositis and -neuropathy there is virtually no relevant T-cell apoptosis *in situ,* in spite of the pathophysiological relevance of T-cell apoptosis in HIV-infection (29).

Also T-cell infiltrates in the skin do not appear to be eliminated by local apoptosis. In Lewis rat EAE, there was negligible T-cell apoptosis in the dermal tissue adjacent to the sensitization site, despite a heavy T-cell infiltration (17). Furthermore, in skin biopsies of patients with dermatomyositis and lupus erythematosus, T-cell apoptosis was virtually absent (30)(Chan, Gold, unpublished observations).

Given the obvious disparity in T-cell apoptosis between the "immunoprivileged" CNS and other non-immunologically protected sites, tissue-specific local mechanisms must be active in the CNS, that lead to a very efficient apoptotic clearance of pathogenic T-cells.

## 1.3     Possible mechanisms of T-cell apoptosis in the CNS

An important question in the elucidation of possible mechanisms of T-cell apoptosis concerns the specificity of the dying cells. Thus, apoptosis-inducing mechanisms such as activation induced cell death via triggering of the T-cell receptor would lead to selective cell death only of autoantigen-specific T-cells. In contrast, apoptosis also of non-antigen specific "bystander" T-cells, secondarily recruited into the lesion, would suggest non-selective mechanisms of apoptosis induction.

To elucidate this further, Pender and colleagues investigated T-cell apoptosis in Lewis rat EAE passively induced with a MBP-specific T-cell clone using the Vβ8.2+ T-cell receptor, which is the predominant T-cell receptor element in MBP-induced Lewis rat EAE (31,32). Apoptosis and T-cell receptor usage was then analyzed in lymphocytes isolated and enriched from spinal cord. The frequency of Vβ8.2+ T-cells was about sevenfold higher in the apoptotic cell population than in the non-apoptotic T-cells. Moreover, using MBP- and ovalbumin-specific T-cells, it appeared that MBP-specific T-cells were readily eliminated from the CNS by apoptosis, while T-cells specific for ovalbumin, an antigen not present in the Lewis rat, survived in the CNS and recirculated to peripheral lymphoid organs. However, these studies could not exclude that also ovalbumin specific T-cells underwent apoptosis in the CNS. Also, altered physicochemical properties of collapsed apoptotic cells could have precluded their quantitative recovery from spinal cord during gradient centrifugation. Moreover, other MBP-reactive T-cell receptor elements besides Vβ8.2, proven to be encephalitogenic were not analyzed in these studies (33,34).

Using T-cells stably expressing specific genomic markers, Lassmann and colleagues could demonstrate that both MBP-specific and ovalbumin-specific T-cells undergo apoptosis in the CNS of EAE-animals (35). In addition, the occurence of T-cell apoptosis during EAE in bone marrow chimeras with different MHC-haplotypes of the resident glial cells and the passively transferred T-cells indicates that T-cell apoptosis is not dependent on antigen-specific mechanisms (35). Thus, both encephalitogenic, antigen-specific T-cells as well as secondarily recruited bystander T-lymphocytes appear to undergo apoptosis *in situ*.

The molecular mechanisms leading to T-cell apoptosis in the CNS *in situ* are as yet incompletely understood (14). *In vivo*, the TNF-receptor-1 (TNFR1)-mediated death pathway appears to play a major role in the induction of T-cell apoptosis in the CNS (36). Thus, in EAE of TNFR1-or TNF/lymphotoxin-deficient mice, local T-cell apoptosis was decreased. TNF-α has been demonstrated to be a potent inducer of T-cell apoptosis (37). Following this line, administration of anti-TNF-α antibody decreases local T-cell apoptosis in EAE of Lewis rats undergoing high-dose antigen-therapy with MBP, where the release of high amounts of TNF-α can be observed *in situ* (38).

T-cell apoptosis in the CNS may also be mediated via the Fas/Fas Ligand (CD95/CD95L)-pathway (review in (14,39)). Thus, one proposed mechanism is the ligation of T-cell CD95 by CD95L expressed by host cells resident in or secondarily recruited into the CNS (40). In addition, intrathecal infusion of FasL suppresses Lewis rat EAE and augments apoptosis of inflammatory cells *in situ* (41).

Since T-cell apoptosis appears to be tissue specific, resident glial cells might sensitize T-cells towards cell death directly or through soluble factors. Antigen presentation to MBP-specific T-cells by rat microglia results in T-cell apoptosis *in vitro*, which can be prevented by exogenous IL-2 (42).

Also astrocytes, which act as non-professional antigen presenting cells (APC), render T-cells susceptible to apoptosis induced by glucocorticosteroids *in vitro* (43). Glucocorticosteroids markedly increased T-cell apoptosis when added to T-cell astrocyte co-cultures during late T-cell activation stages whereas there was no effect when thymus cells were used as antigen presenters. These results could argue for a scenario, where an infiltrating T-cell is primed for an apoptotic stimulus by a resident non-professional APC, and subsequently undergoes apoptosis under hormonal influences of the microenvironment or systemic changes. However, T-cell apoptosis during recovery from EAE also occurs in the CNS of adrenalectomized animals, arguing for additional mechanisms (44).

Among others, further proposed mechanisms of local CNS T-cell apoptosis comprise the action of reactive oxygen intermediates and nitric oxide, or cell death due to deprivation of growth factors such as IL-2 (45,46). Yet, the precise role of the different putative mechanisms as well as other to date unknown CNS-specific factors in the induction of T-cell apoptosis remain to be elucidated.

## 1.4    Phagocytosis of apoptotic cells in the central nervous system

The high degree of apoptosis of inflammatory cells in the autoimmune inflamed human and rodent CNS raises the need for an efficient and tightly controlled removal mechanism for these unwanted cells. A key event in the resolution of an inflammatory infiltrate is the nonphlogistic and thus safe phagocytic clearance of dying, yet intact leukocytes undergoing apoptosis (review in (47,48). Apoptosis labels unwanted cells with signals that direct the recognition, uptake and subsequent degradation by tissue-specific phagocytes, thus preventing the spilling of potentially harmful contents and inhibiting secondary immune responses directed towards the dying cells. In addition to the mere clearing of cell remnants, the ingestion of apoptotic cells also actively elicits phagocyte responses that modulate immune reactions and inflammation.

In Lewis rat EAE, phagocytosis of apoptotic lymphocytes by macrophages/microglia, oligodendrocytes and astrocytes *in situ* has first been described by electron microscopy (49). Using an *in vitro* phagocytosis model, we have further investigated the phagocytosis of apoptotic T-cells by different glial cell elements and its functional consequences (50). Lewis rat microglia efficiently phagocytose specifically apoptotic, encephalitogenic MBP-specific T-cells. This process is differentially regulated by Th1-/Th2-type cytokines (51). The phagocytosis of apoptotic T-cells by Lewis rat microglia is more efficient than by other glial cell elements such as astrocytes. Moreover, phagocytosis of apoptotic T-cells leads to a profound downregulation of microglial immune functions with a suppression of pro-inflammatory cytokines and microglial T-cell activation, thus silencing the microglial phagocyte in the inflammatory context (52). Figure 2 gives a hypothetical model of the phagocytosis of apoptotic T-cells in the CNS.

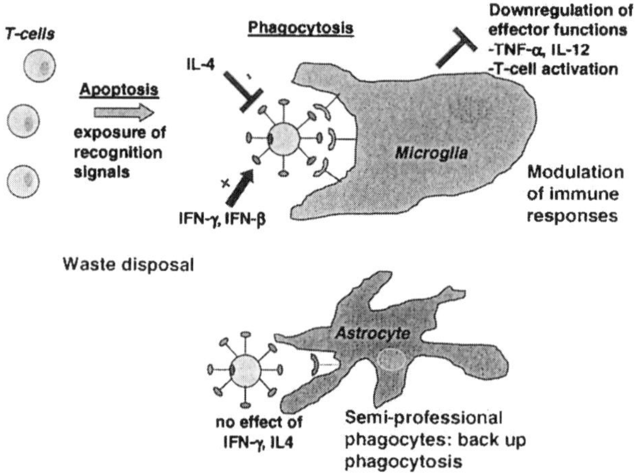


*Figure 2.* T-cells undergoing apoptosis exhibit specific "eat me" recognition signals for phagocytes. Phagocytosis by microglia is differentially regulated by cytokines ("waste disposal"). The uptake of apoptotic cells leads to a down-regulation of different microglial immune-functions ("modulation of immune responses"). Astrocytes have a lower capacity to phagocytose apoptotic cells and may provide a "backup" mechanism. Modified from (50).

*In vitro*, rat microglia/brain macrophages have also been demonstrated to phagocytose apoptotic neurons via lectin-, integrin-, and phosphatidylserine-dependent mechanisms (53).

## 1.5      Induction of T-cell apoptosis by immunotherapy

Current therapeutic approaches in presumably immune-mediated, demyelinating diseases of the CNS such as MS aim at the rapid termination of the inflammatory process, thereby hastening clinical recovery and potentially also preventing subsequent demyelinative and axonal tissue damage (54). Glucocorticosteroids (GS) are potent anti-inflammatory drugs, whose therapeutic efficacy in neuroimmunological diseases has especially been established in immune neuropathies, MS, lupus erythematosus and cerebral vasculitis. Intravenously administered high-dose GS serve as the current mainstay in the therapy of acute MS-relapses(55). GS can mediate their pleiotropic anti-inflammatory effects via different mechanisms, e.g. modulation of cell activation, cytokine expression, secretion of inflammatory mediators, leucocyte migration and the reduction of local edema (56). These effects are mediated via the cytosolic GS-receptor (GSR) and can be blocked by the GSR antagonist RU 468 at low GS concentrations. At higher concentrations, GS appear to induce cell death directly, possibly via non-genomic, physicochemically mediated effects on the plasma membrane (56).

In Lewis rat EAE, iv. methylprednisolone (MP) therapy administered at the clinical disease maximum increased T-cell apoptosis in the spinal cord *in situ* and decreased T-cell infiltration (57). This effect was clearly dose-dependent with a dosage of at least 10 mg/kg body weight MP required to achieve an increase of T-cell apoptosis and a marginal decrease in T-cell infiltration. Higher dosages of 50 mg/kg body weight MP were superior in both respects, and were also effective in mild EAE, in contrast to the lower dosage of MP. The strong apoptosis-promoting effect of GS has also been demonstrated on peripheral blood leukocytes (PBL) of MS-patients (58). After intravenous high-dose corticosteroid treatment, apoptosis of PBLs was markedly augmented in different MS-subgroups, predominantly affecting CD4-positive T cells.

Also in the peripheral nervous system, high-dose GS augment T-cell apoptosis (59). In Lewis rat EAN, a 4-5-fold increase of T-cell apoptosis could be observed in the sciatic nerve after therapeutic administration of GS (10 mg/kg body weight prednisolone). Similar results could be observed in Lewis rat experimental autoimmune myositis (EAM), a model for human idiopathic myosites of presumed autoimmune-inflammatory etiology (27).

Here, a clear increase of apoptotic endomysial T-cells could be demonstrated, with up to 50% of these apoptotic T-cells being CD8-positive. This indicated that glucocorticosteroids also induce CD8-T-cell apoptosis, even in organs where normally T-cell apoptosis does not occur.


The type 1 interferon (IFN) IFN-$\beta$, is one of the current mainstays in the immunomodulatory therapy of MS (60). IFN-$\beta$ appears to affect particularly inflammatory aspects of MS, with a reduction of the relapse rate and positive effects on inflammatory changes observed in MRI scans. In the experimental models, IFN-$\beta$ has been shown to inhibit progression of relapsing-remitting EAE (61), while early discontinuation of IFN-$\beta$ led to exacerbation of active EAE (62). While the exact mechanism of action of IFN-$\beta$ is still unknown, numerous putative mechanisms have been proposed (review in (63)). Whether the therapeutic benefit of IFN-$\beta$ is also based on the induction of apoptosis in inflammatory cells is still controversially discussed (64,65). In Lewis rat AT-EAE, no major increase of T-cell apoptosis *in situ* could be observed after treatment with IFN-$\beta$ (66). However, IFN-$\beta$ clearly increases the phagocytosis of apoptotic, encephalitogenic, MBP-specific T-cells by microglia *in vitro* (Chan, A., Seguin, R., submitted). Thus, theoretically IFN-$\beta$ could help to engulf and eliminate T-cells driven into apoptosis by other therapeutically used agents, e.g. corticosteroids.

Antigen-specific therapy with CNS or PNS-autoantigens has been shown to be effective in EAE and EAN (67,68). The underlying mechanism appears to involve T-cell receptor reengagement at an appropriate stage of the cell cycle which leads to T-cell apoptosis (69). In Lewis rat cell transfer (AT)-EAE, high dose antigen (guinea pig MBP) therapy not only led to apoptotic cell death of invading T-cells *in situ* but also of resident oligodendrocytes (38). This was accompanied by a profound modulation of the cytokine network with a  rapid induction of the Th1-type cytokines TNF-$\alpha$ and IFN-$\gamma$ as well as inducible nitic oxide synthase. Neutralization of TNF-$\alpha$ *in vivo* led to a decrease in T-cell and oligodendrocyte apoptosis but did not change the beneficial clinical effect of high-dose antigen therapy. This argues for a central role of TNF-$\alpha$ as a mediator of local apoptosis which may have different functional effects. Similar results could also be observed in antigen-specific therapy of Lewis rat AT-EAN using P2-protein (70,71).


## Conclusions

Apoptosis of inflammatory effector cells constitutes an effective protective mechanism in the autoimmune-inflamed  CNS. Thus, the selective induction of apoptosis in pathogenic cells is an attractive therapeutic target in neuroimmunological diseases. However, apoptosis of resident CNS cells represents also a major mechanism in the pathogenesis of different neurodegenerative disorders, where selective blockade of apoptotic pathways may be benificial. Therefore, the identification of the molecular, tissue- and cell-specific factors that lead to local apoptosis remain to be elucidated in order to take advantage of this phylogenetically old and important defense mechanism.

## Acknowledgements

The authors thank Dr. J. Schmidt, Dr. A. Weishaupt and Dr. U.K. Zettl for their experimental contributions. We thank Dr. H.P. Hartung, Dr. H. Lassmann and Dr. K.V. Toyka for stimulating discussions. The experimental work of the authors is supported by the Deutsche Forschungsgemeinschaft, Hertie-Stiftung, Deutsche Multiple Sklerose Gesellschaft and Wilhelm Sander Stiftung.

**References**

1. Kerr JFR, Wyllie AH, Currie AR. Apoptosis: a basic biological phenomenon with wide-ranging implications in tissue kinetics. Br J Cancer 1972;26:239-257
2. Cohen JJ, Duke RC, Fadok VA, Sellins KS. Apoptosis and programmed cell death in immunity. Annu Rev Immunol 1992;10:267-293
3. Bursch W, Paffe S, Putz B, Barthel G, Schulte-Hermann R. Determination of the length of the histological stages of apoptosis in normal liver and in altered hepatic foci of rats. Carcinogenesis 1990;11:847-853
4. Stadelmann C, Lassmann H. Detection of apoptosis in tissue sections. Cell Tissue Res 2000;301:19-31
5. Kaufmann SH, Hengartner MO. Programmed cell death: alive and well in the new millennium. Trends Cell Biol 2001;11:526-534
6. Oberhammer F, Wilson JW, Dive C et al. Apoptotic death in epithelial cells: cleavage of DNA to 300 and/or 50 kb fragments prior to or in the absence of internucleosomal fragmentation. EMBO J 1993;12:3679-3684
7. Schulze-Osthoff K, Walczak H, Dröge W, Krammer PH. Cell nucleus and DNA fragmentation are not required for apoptosis. J Cell Biol 1994;127, No.1:15-20
8. Martin R, Jaraquemada D, Flerlage M et al. Fine specificity and HLA-restriction of myelin basic protein- specific T cell lines from multiple sclerosis patients and healthy individuals. J Immunol 1990;145:540-548
9. Pette M, Fujita K, Wilkinson D et al. Myelin autoreactivity in multiple sclerosis: recognition of myelin basic protein in the context of HLA-DR2 products by T lymphocytes of multiple-sclerosis patients and healthy donors. Proc Natl Acad Sci U S A 1990;87:7968-7972
10. Pender MP, Csurhes PA, Greer JM et al. Surges of increased T cell reactivity to an encephalitogenic region of myelin proteoplipid protein occur more often in patients with multiple sclerosis than in healthy subjects. J Immunol 2000;165:5322-5331

11.  Wekerle H, Linington C, Lassmann H, Meyermann R. Cellular immune reactivity within the CNS. Trends Neurosci 1986;9:271-277

12.  Hickey WF, Hsu BL, Kimura H. T-lymphocyte entry into the central nervous system. J Neurosci Res 1991;28:254-260

13.  Gold R, Hartung HP, Lassmann H. T-cell apoptosis in autoimmune diseases: termination of inflammation in the nervous system and other sites with specialized immune-defense mechanisms. Trends Neurosci 1997;20:399-404

14.  Pender MP, Rist MJ. Apoptosis of inflammatory cells in immune control of the nervous system: role of glia. Glia 2001;36:137-144

15.  Pender MP, Nguyen KB, McCombe PA, Kerr JFR. Apoptosis in the nervous system in experimental allergic encephalomyelitis. J Neurol Sci 1991;104:81-87

16.  Pender MP, McCombe PA, Yoong G, Nguyen KB. Apoptosis of $\alpha\beta$ T lymphocytes in the nervous system in experimental autoimmune encephalomyelitis: its possible implications for recovery and acquired tolerance. J Autoimmun 1992;5:401-410

17.  Schmied M, Breitschopf H, Gold Ret al. Apoptosis of T Lymphocytes in experimental autoimmune encephalomyelitis. Am J Pathol 1993;143:446-452

18.  Gold R, Hartung HP, Lassmann H. T cell apoptosis in autoimmune diseases: termination of inflammation in the nervous system and other sites with specialized immune-defense mechanisms. Trends Neurosci 1997;20:399-404

19.  Bonetti B, Pohl J, Gao YL, Raine CS. Cell death during autoimmune demyelination: Effector but not target cells are eliminated by apoptosis. J Immunol 1997;159:5733-5741

20.  Malipiero U, Frei K, Spanaus KS et al. Myelin oligodendrocyte glycoprotein-induced autoimmune encephalomyelitis is chronic/relapsing in perforin knockout mice, but monophasic in Fas- and Fas ligand-deficient *lpr* and *gld* mice. European J Immunol 1997;27:3151-3160

21.  Suvannavejh GC, Dal Canto MC, Matis LA, Miller SD. Fas-mediated apoptosis in clinical remissions of relapsing experimental autoimmune encephalomyelitis. J Clin Invest 2000;105:223-231

22.  Barac-Latas V, Wege H, Lassmann H. Apoptosis of T-lymphocytes in Corona Virus induced encephalomyelitis. Reg Immunol 1995;6:355-357

23.  Bauer J, Stadelmann C, Bancher C, Jellinger K, Lassmann H. Apoptosis of T lymphocytes in acute disseminated encephalomyelitis. Acta Neuropathol 1999;97:543-546

24.  Ozawa K, Suchanek G, Breitschopf H et al. Patterns of oligodendroglia pathology in multiple sclerosis. Brain 1994;117:1311-1322

25.  Bauer J, Rauschka H, Lassmann. Inflammation in the nervous system: the human perspective. Glia 2001;36:235-243

26.  Zettl UK, Gold R, Toyka KV, Hartung H-P. In situ demonstration of T cell activation and elimination in the peripheral nervous system during experimental autoimmune neuritis in the Lewis rat. Acta Neuropathol Berl 1996;91:360-367

27.  Schneider C, Matsumoto Y, Kohyama K, Toyka KV, Hartung HP, Gold R. Experimental autoimmune myositis in the lewis rat: lack of spontaneous T-cell apoptosis and therapeutic response to glucocorticosteroid application. J Neuroimmunol 2000;107:83-87

28.  Schneider C, Gold R, Dalakas MC et al. MHC Class I-mediated cytotoxicity does not induce apoptosis in muscle fibers nor in inflammatory T cells: studies in patients with polymyositis, dermatomyositis and inclusion body myositis. J Neuropathol Exp Neurol 1996;55:1205-1209

29.  Schneider C, Dalakas MC, Toyka KV, Said G, Hartung HP, Gold R. T-cell apoptosis in inflammatory neuromuscular disorders associated with human immunodeficiency virus infection. Arch Neurol 1999;56:79-83

30.  Adler S, Zettl H, Brück W, Zimmermann R, Zettl UK. The role of apoptotic cell death and bcl-2 protein expression in dermatomyositis. Eur J Dermatol 1997;7:413-416

31. Tabi Z, McCombe PA, Pender MP. Apoptotic elimination of V beta 8.2+ cells from the central nervous system during recovery from experimental autoimmune encephalomyelitis induced by the passive transfer of V beta 8.2+ encephalitogenic T cells. European J Immunol 1994;24:2609-2617

32. Tabi Z, McCombe PA, Pender MP. Antigen-specific down-regulation of myelin basic protein-reactive T cells during spontaneous recovery from experimental autoimmune encephalomyelitis: further evidence of apoptotic deletion of autoreactive T cells in the central nervous system. Int Immunol 1995;7:967-973

33. Gold R, Giegerich G, Hartung H-P, Toyka KV. T-cell receptor (TCR) usage in Lewis rat experimental autoimmune encephalomyelitis: TCR β-chain-variable-region Vβ8.2-positive T cells are not essential for induction and course of disease. Proc Natl Acad Sci USA 1995;92:5850-5854

34. Chan A, Gold R, Giegerich G et al. Usage of Vβ3.3 T-cell receptor by myelin basic protein-specific encephalitogenic T-cell lines in the lewis rat. J Neurosci Res 1999;58:214-225

35. Bauer J, Bradl M, Hickley WF et al. T-cell apoptosis in inflammatory brain lesions: destruction of T cells does not depend on antigen recognition. Am J Pathol 1998;153:715-724

36. Bachmann R, Eugster HP, Frei K, Fontana A, Lassmann H. Impairment of TNF-receptor-1 signaling but not fas signaling diminishes T-cell apoptosis in myelin oligodendrocyte glycoprotein peptide-induced chronic demyelinating autoimmune encephalomyelitis in mice. Am J Pathol 1999;154:1417-1422

37. Zheng L, Fisher G, Miller RE, Peschon J, Lynch DH, Lenardo MJ. Induction of apoptosis in mature T cells by tumor necrosis factor. Nature 1995;377:348-

38. Weishaupt A, Jander S, Brück W et al. Molecular mechanisms of high-dose antigen therapy in experimental autoimmune encephalomyelitis: rapid induction of Th1-type cytokines and inducible nitric oxide synthase. J Immunol 2000;

39. Zipp F, Krammer PH, Weller M. Immune (dys)regulation in multiple sclerosis: role of the CD95/CD95 ligand system. Immunol Today 1999;20:550-554

40. Sabelko-Downes KA, Cross AH, Russell JH. Dual role for Fas ligand in the initiation of and recovery from experimental allergic encephalomyelitis. J Exp Med 1999;189:1195-1205

41. Zhu B, Luo L, Chen Y, Paty DW, Cynader MS. Intrathecal Fas Ligand infusion strengthens immunoprivilege of central nervous system and suppresses experimental autoimmune encephalomyelitis. J Immunol 2002;169:1561-1569

42. Ford AL, Foulcher E, Lemckert FA, Sedgwick JD. Microglia induce CD4 T lymphocyte final effector function and death. J Exp Med 1996;184:1737-1745

43. Gold R, Schmied M, Tontsch U et al. Antigen presentation by astrocytes primes rat T lymphocytes for apoptotic cell death A model for T-cell apoptosis in vivo. Brain 1996;119:651-659

44. Smith T, Schmied M, Hewson AK, Lassmann H, Cuzner ML. Apoptosis of T cells and macrophages in the central nervous system of intact and adrenalectomized Lewis rats during experimental allergic encephalomyelitis. J Autoimmun 1996;9:167-174

45. Zettl UK, Mix E, Zielasek J, Stangel M, Hartung H-P, Gold R. Apoptosis of myelin-reactive T cells induced by reactive oxygen and nitrogen intermediates in vitro. Cell Immunol 1997;178:1-8

46. Issazadeh S, Abdallah K, Chitnis T et al. Role of passive T-cell death in chronic experimental autoimmune encephalomyelitis. J Clin Invest 2000;105:1109-1116

47. Savill J, Fadok V. Corpse clearance defines the meaning of cell death. Nature 2000;407:784-788

48. Fadok VA, Bratton DL, Henson PM. Phagocyte receptors for apoptotic cells: recognition, uptake, and consequences. J Clin Invest 2001;108:957-962

49. Nguyen KB, Pender MP. Phagocytosis of apoptotic lymphocytes by oligodendrocytes in experimental autoimmune encephalomyelitis. Acta Neuropathol 1998;95:40-46
50. Magnus T, Chan A, Savill J, Toyka KV, Gold R. Phagocytic removal of apoptotic, inflammatory lymphocytes in the central nervous system by microglia and its functional implications. J Neuroimmunol 2002;130:1-9
51. Chan A, Magnus T, Gold R. Phagocytosis of apoptotic inflammatory cells by microglia and modulation by different cytokines: mechanism for removal of apoptotic cells in the inflamed nervous system. Glia 2001;33:87-95
52. Magnus T, Chan A, Grauer O, Toyka KV, Gold R. Microglial phagocytosis of apoptotic inflammatory T cells leads to down-regulation of microglial immune activation. J Immunol 2001;167:5004-5010
53. Witting A, Muller P, Herrmann A, Kettenmann H, Nolte C. Phagocytic clearance of apoptotic neurons by Microglia/Brain macrophages In vitro: involvement of lectin-, integrin-, and phosphatidylserine-mediated recognition. J Neurochem 2000;75:1060-1070
54. Trapp BD, Peterson J, Ransohoff RM, Rudick R, Mörk S, Bö L. Axonal transection in the lesions of multiple sclerosis. New Engl J Med 1998;338:278-285
55. Beck RW, Cleary PA, Anderson MMJ et al. A randomized, controlled trial of corticosteroids in the treatment of acute optic neuritis. The Optic Neuritis Study Group. New Engl J Medicine 1992;326:581-588
56. Gold R, Buttgereit F, Toyka KV. Mechanism of action of glucocorticosteroid hormones: possible implications for therapy of neuroimmunological disorders. J Neuroimmunol 2001;117:1-8
57. Schmidt J, Gold R, Zettl UK, Hartung HP, Toyka KV. T-cell apoptosis in situ in experimental autoimmune encephalomyelitis following methylprednisolone pulse therapy. Brain 2000;123:1431-1441
58. Leussink VI, Jung S, Merschdorf U, Toyka KV, Gold R. High-dose methylprednisolone therapy in multiple sclerosis induces apoptosis in peripheral blood leukocytes. Arch Neurol 2001;58:91-97
59. Zettl UK, Gold R, Toyka KV, Hartung H-P. Intravenous glucocorticosteroid treatment augments apoptosis of inflammatory T cells in experimental autoimmune neuritis (EAN) of the lewis rat. J Neuropathol Exp Neurol 1995;54:540-547
60. Rolak LA. Multiple sclerosis treatment 2001. Neurol Clin 2001;19:107-118
61. Yu M, Nishiyama A, Trapp BD, Tuohy VK. Interferon-β inhibits progression of relapsing-remitting experimental autoimmune encephalomyelitis. J Neuroimmunol 1996;64:91-100
62. van der Meide PH, De Labie MCDC, Ruuls SRet al. Discontinuation of treatment with IFN-beta leads to exacerbation of experimental autoimmune encephalomyelitis in Lewis rats. Rapid reversal of the antiproliferative activity of IFN-beta and excessive expansion of autoreactive T cells as disease promoting mechanisms. J Neuroimmunol 1998;84:14-23
63. Rieckmann P. The effects of interferon-β on cytokines and immune responses. In: Interferon therapy of Multiple Sclerosis. Reder AT (ed). Marcel Dekker Inc., New York, 1997;161-191
64. Kaser A, Deisenhammer F, Berger T, Tilg H. Interferon-β1b augments activation-induced T-cell death in multiple sclerosis patients. Lancet 1999;353:1413-1414
65. Zipp F, Beyer M, Gelderblom H, Wernet D, Zschenderlein R, Weller M. No induction of apoptosis by IFN-beta in human antigen-specific T cells. Neurology 2000;54:485-487
66. Schmidt J, Sturzebecher S, Toyka KV, Gold R. Interferon-beta treatment of experimental autoimmune encephalomyelitis leads to rapid nonapoptotic termination of T cell infiltration. J Neurosci Res 2001;65:59-67
67. Critchfield JM, Racke MK, Zuniga Pflucker JC et al. T cell deletion in high antigen dose therapy of autoimmune encephalomyelitis. Science 1994;263:1139-1143

68. Weishaupt A, Gold R, Gaupp S, Giegerich G, Hartung H-P, Toyka KV. Antigen therapy eliminates T cell inflammation by apoptosis: Effective treatment of experimental autoimmune neuritis with recombinant myelin protein P2. Proc Natl Acad Sci USA 1997;94:1338-1343

69. Liblau R, Tisch R, Bercovici N, McDevitt HO. Systemic antigen in the treatment of T-cell-mediated autoimmune diseases. Immunol Today 1997;12:599-604

70. Weishaupt A, Gold R, Hartung T et al. Role of TNF-alpha in high-dose antigen therapy in experimental autoimmune neuritis: inhibition of TNF-alpha by neutralizing antibodies reduces T-cell apoptosis and prevents liver necrosis. J Neuropathol Exp Neurol 2000;59:368-376

71. Weishaupt A, Bruck W, Hartung T, Toyka KV, Gold R. Schwann cell apoptosis in experimental autoimmune neuritis of the Lewis rat and the functional role of tumor necrosis factor-alpha. Neurosci Lett 2001;306:77-80.

Chapter A25

# ENVIRONMENTAL INFLUENCES IN EXPERIMENTAL AUTOIMMUNE ENCEPHALOMYELITIS


Cris S Constantinescu
*Division of Clinical Neurology, University of Nottingham, UK*



**Abstract:** Environmental factors, in particular infectious agents, are thought to have a major influence on the development and course of MS. Some of these influences are also reflected in the animal model, EAE. In this chapter, the role of infectious agents in the development and course of autoimmunity in EAE is discussed. Other environmental agents including trauma, solar radiation exposure, temperature, stress, toxins, are discussed in terms of their relevance to MS and EAE.

**Key words:** EAE, MS, infection, autoimmunity, trauma, solar radiation, UV light, temperature, stress, toxins


## Introduction

Like several other diseases with an autoimmune mechanism and polygenic susceptibility, including type I diabetes mellitus and rheumatoid arthritis, the most commonly accepted working model of MS development is that it is an autoimmune disease triggered in genetically susceptible individuals by some environmental factor. MS thus arises as a result of an interplay of varying degrees of environmental and genetic factors. Such an interplay is important not only in determining the susceptibility of an individual to the disease but also possibly other aspects of the disease such as severity or course.

Experimental autoimmune encephalomyelitis (EAE) is in many respects the prototypical organ specific, T cell mediated autoimmune disease having provided many clues to the immune mechanisms of MS. As a useful animal model, EAE ideally must reflect some of these environmental influences that play a role in MS. Moreover, EAE being generally induced in strains of animals, which are very well characterized immunogenetically, the genetic contribution remains equal across the strain or substrain, and the environmental conditions can be varied to provide clues to their contribution to the susceptibility and course of EAE.

**Infectious agents**

Of the many environmental factors implicated in MS, infectious agents have been most consistently investigated [1]. The possibility that MS is caused by an infectious agent has been entertained since the first detailed clinico-pathological description of MS, and is supported by several lines of evidence. The possibility that, in addition, infectious agents, directly or through their immunological consequences, may modify disease activity is also being increasingly substantiated by studies.

Many infectious agents have been implicated in MS, including spirochetes, bacteria, and viruses (table)

**Table 1.**

Some infectious agents linked to MS

Adenovirus
Borrellia burgdorferi
Canine distemper virus
Chlamydia pneumoniae
Coronavirus
EBV
Enterovirus
HSV 1, 2, 6, 8
Herpes zoster
HTLV-1 and 2
Measles
Mumps
Papovavirus
Rubella
Simian virus 5
Theiler's encephalomyelitis virus

Some of these infectious agents are linked to MS because of evidence of an increased titer of antibodies against them in the CSF of MS patients, for example measles and herpes zoster, or in their blood, for example EBV. Many, however, are linked to MS because of the existence of spontaneous or experimentally induced demyelination in animal models. These aspects are discussed in detail in the section of this book dedicated to viral models of demyelination.

However, like in MS, and underscoring the utility of EAE as a model, infectious agents can influence aspects of EAE.

Self-reactive lymphocytes recognizing neuroantigens exist not only in MS patients but also in normal subjects. Moreover, the precursor frequency of these cells does not appear to be higher in MS patients. However, the neuroantigen-reactive T cells from MS patients are in higher state of activation than those of normal subjects [2]. Activated T cells are more capable of crossing the blood brain barrier and there, where they encounter their cognate antigen, they can cause the inflammatory demyelination damage. In relapsing MS, as in relapsing models of EAE, the reactivation of these cells may be a major event in the pathogenesis of relapses.

The role of infectious agents in these situations has been investigated, as it is recognized that infectious events (or immunizations) can trigger MS relapses or the first manifestations of the disease.

Several mechanisms may operate whereby autoreactive T cells can be activated and their action in EAE is discussed below.

*Superantigen activation*

Superantigens (SAG) are viral or bacterial products capable of activating a large number of T cells bearing the same T cell receptor (TCR)V chain [3]. SAG include several bacterial enterotoxins such as staphylococcal enterotoxin A (SEA) and B (SEB) and have been implicated in several human diseases. SAB bind the Vβ region of the TCR outside the groove normally occupied by the antigenic peptide, and also the MHC class II molecule on the surface of antigen presenting cells. In EAE, and possibly in MS there may be a bias in the Vβ TCR usage in particular in the early stages of the disease, before significant epitope spreading has occurred. SAG activation of neuroantigen-reactive T cells bearing the corresponding TCR may lead to a relapse in EAE. This is indeed the case. SAG SEB, which predominantly binds Vβ TCR in mice has been demonstrated to induce relapses in myelin basic protein (MBP)-induced EAE in PL/J and PL/J x SJL/J F1 mice, whose MBP-reactive T cells predominantly use TCR Vβ8 [4, 5]. Moreover, we confirmed this ability of SEB to induce relapses [6]. In addition, we demonstrated that SEA (which does not bind Vβ8 in mice) also induces significant relapses. Therefore, although it was originally proposed that Vβ8 TCR confers specific tendency for autoimmunity [7], the ability of SAG to reactivate autoimmunity in EAE is not entirely dependent on Vβ8 TCR.

What is the mechanism by which SAG induce relapses? It is known that SAG preferentially activate Th1 cells [8, 9]. It is also known that superantigens induce a variety of proinflammatory cytokines and other inflammatory mediators. We demonstrated that SAG induce IL-12, the major factor in the development of Th1-type immune responses. Indeed antibodies against IL-12 p40 blocked SAG-induced relapses in PL/J x SJL/J F1 mice. Moreover, IL-12 administration mimicked the effects of SAG, inducing itself relapses [6]. This result was also demonstrated in MBP-induced EAE in Lewis rats [10].

A similar response has been found with TNF. TNF administration induced relapses [11] while administration of neutralizing TNF antibodies partially suppressed relapses [4].

Several anti-inflammatory cytokines including TGF-$\beta$, IFN-$\tau$, and IL-10 were capable of suppressing SAG-induced relapses of EAE.

It appears most likely that the mechanism of activation of potentially pathogenic T cells in EAE by superantigens, in addition to activation of a large number of T cells using the same TCR, involves tipping the balance of pro- and anti-inflammatory cytokines toward a more inflammatory environment.

Interestingly, SAG were not only shown to reactivate autoimmunity by inducing relapses in EAE, but also to modify susceptibility to EAE. Interleukin-6 (IL-6) is a proinflammatory cytokine also known to be induced by SAG. Besides IL-12 (p40) it is a proinflammatory cytokine whose deficiency is consistently known to render mice completely resistant to EAE [12, 13]. Administration of SEB to IL-6 knockout mice has overcome their resistance to EAE [14].

These results are highly relevant for MS. It is known that relapses or the initial attack are often preceded by infections. It is also known that SAG can activate human neuroantigen-reactive T cells [15]. This may represent an important mechanism whereby infectious agents can (re)activate MS. Knowledge of this possibility also has implications for the treatment of MS. For example, glatiramer acetate, currently successfully used in the treatment of relapsing remitting MS, promiscuously binds MHC class II inside the peptide binding groove but not requiring antigen processing [16]. Although initially developed as a mimic of MBP, it suppresses not only MBP-induced EAE but also PLP-induced EAE [17]. This promiscuous MHC binding makes it a potentially very useful immunomodulatory treatment, and it was shown to suppress autoimmune uveitis as well [18]. However, while it suppresses antigen-specific T cell proliferative responses in vitro to nominal antigen, it does not suppress SAG-induced responses, suggesting that its blockade of the peptide-binding site in the MHC class II groove, does not affect SAG binding. We have recently confirmed this observation in human SEB-stimulated T cells (Constantinescu C, Robins A et al, unpublished observations).

*Molecular mimicry*

Another possible activation mechanism for autoreactive T cells with potential to cause or reactivate autoimmunity is molecular mimicry.  The concept implies the existence of a structural similarity and an immunological cross-reactivity between self antigens of the host and an infectious agents such as bacteria, viruses, or yeasts, which elicits an immune response, after which T or B cells developed against these agents respond against cross-reactive self-determinants.  A most convincing and biologically relevant example of molecular mimicry in human autoimmunity and neuroimmunology is the Guillain-Barre syndrome (GBS) [19].  This is another autoimmune demyelinating disease of the nervous system involving the peripheral nerves.  There is convincing evidence for a bacterial etiology in up to 50% of patients.  Although typically monophasic, the presentation can be different (for example the various proportions of axonal, myelin, or cranial nerve involvement) and the distribution across age groups in different part of the world can also be different. *Campylobacter jejuni* is a common gastrointestinal pathogen that causes diarrhea. It also, less frequently causes GBS.  However, up to two thirds of patients with GBS have serological evidence of a recent *C jejuni* infection.  Of these serogroup Penner 19 is overrepresented, while, on the other hand, rarely associated with gastroenteritis. Different serogroups express different lipopolysaccharides (LPS) closely resembling human cell surface glycolipids.  Infection with *C jejuni* can thus lead to production of anti-ganglioside antibodies, which cross-react with human glycolipids.  The most common such cross-reactive self-antigen is ganglioside GM1.  Anti-GM1 antibodies are found in approximately 70% of patients with C. jejuni-associated GBS and in about 30% of patients with GBS but without serological evidence of recent *C jejuni* infection.

Other anti-ganglioside antibodies are associated with specific clinical forms of GBS, for example anti-GQ1b with the Miller Fischer variant with prominent cranial nerve involvement and anti-GalNac-GD1a antibodies with distal motor variant.  These antibodies can also be triggered by *C jejuni* as well as other infections.

Other infection can result in GBS, notably *Mycoplasma pneumoniae*. Molecular mimicry may play an important role here as well. It has been shown that *M pneumoniae* triggers anti-GalC antibodies, which, in high titre are associated with a demyelinating GBS.  It is important to note here that these glycolipids are also present in abundance in the CNS myelin therefore may represent potential autoantigens in MS and EAE in addition to GBS. Moreover, EAE has been induced by immunization with GalC.

Although a significant component of neuropathology in GBS is antibody and complement mediated, a role for a T cell participation in molecular mimicry involving glycolipids, traditionally considered T-independent antigens, is likely to exist, and this mechanism likely to be relevant to T cell mediated autoimmunity such as occurs in EAE and MS. It has been shown that GM1 antibody production by B cells of GBS patients is T cell dependent [20]. Moreover, self-glycolipids reactive T cells have bee found both in normal individuals and in patients with MS [21]. In addition, EAE is more severe when the immunogenic brain homogenate contains the glycolipid fraction than when this fraction has been removed [22]. Therefore, a T cell and B cell cross reactive immune response between infectious agent components such as LPS and myelin glycolipids is a conceivable mechanism whereby molecular mimicry generates and activates self-reactive lymphocytes in CNS inflammatory demyelination.


Another important example of molecular mimicry with great relevance to EAE and MS is the hepatitis B virus (HBV). Its polymerase has been found to have sequence homology and antigenic cross-reactivity with an encephalitogenic epitope on MBP. Moreover, immunization of animals with HBV leads to a histologically evident EAE-like inflammatory demyelination [23]. It is thus conceivable that T cells against antigenic determinants of the HBV polymerase recognize cross-reactive MBP epitopes and become self-reactive. Wucherpfennig et al screened viral sequences for potential homology with the peptide predicted to bind MHC class II groove of the commonest MS-associated haplotype, HLA-DR2 (DRB1*1501), and demonstrated that some of these peptides stimulated myelin antigen (MBP)-reactive T cells from patients with MS [24].

*Bystander activation*

Bystander activation is the mechanism in which T cells are activated by exposure to self-antigen in the situations in which an infection has created an inflammatory environment favorable for the development of Th1 type T cells, or in an evironment which allows the expression of an antigen normally not exposed to the immune system, for example a sequestered antigen. The latter situation is closely linked with the phenomenon of epitope spreading, which is known to occur in EAE and MS, whereby previously unexposed (cryptic) epitopes become exposed during the tissue damage accompanying the initial event [25]. The initial event may be the initial inflammatory demyelinating attack, but also potentially an event with another pathogenesis such as injury, ischemia, or indeed, infection. The latter can be exemplified by the fact that infection with viruses that cause experimental demyelination such as coronavirus or TMEV eventually leads to spreading of the epitope to self-determinants including the known neuroantigens of EAE such as MBP [26].

Another example of bystander activation with a sequestered antigen is that of αB-crystallin. This is a protein belonging to the heat shock protein family. It has been shown that αB-crystallin induces marked T cell responses in individuals with MS, more prominenent than responses to other neuroantigens such as MBP or PLP [27]. αB-crystallin is normally sequestered, not exposed to the immune system, not being on the surface of oligodendrocytes, but the response against it may develop later in the course of the disease via epitope spreading [28]. Moreover, EBV has been shown to induce expression of B-crystallin in B cells, which then present it in an HLA-DR-restricted fashion [29]. Since serological evidence of EBV exposure is present in nearly 100% of people with MS [30], it is conceivable that a mechanism of activation of neuroantigen (in this case αB-crystallin)-reactive T cells is induction by this virus in the context of a non-sequestered environment.

The creation of an inflammatory environment for bystander activation of autoreactive T cells by infectious agents has been convincingly demonstrated (and it may in great part explained why in some EAE induction models there is need for adjuvants (see chapter on adjuvants in EAE). Mice transgenic for a TCR specific for MBP do not develop EAE when kept in a germ free environment; however, when placed in a germ-containing environment they develop spontaneous EAE [31]. It is known that bacterial products other than SAG are involved in this (re)activation of autoimmune disease. LPS of gram-negative bacteria is an example of such infectious product. It was shown to be involved in the reactivation of experimental arthritis by bacterial overgrowth in the bowel [32]. In EAE, CpG-rich bacterial DNA, which has been shown to have multiple immunostimulating activities largely centered on induction of proinflammatory cytokines, modifies resistance to disease of prototypically resistant mice, rendering them susceptible [33].

A common feature of these bacterial products is their ability to induce proinflammatory cytokines and in particular, IL-12, a key cytokine for the development of Th1 responses, required for EAE. Also adjuvants such as the *Mycobaterium tuberculosis*-based complete Freund's adjuvant (CFA) all induce IL-12 [34, 35].

A plausible working hypothesis, therefore, is that the final common pathway of activation of autoimmunity in EAE (and in MS) by infectious agents is through their effects on proinflammatory cytokines and in particular IL-12 and/or the related IL-23, with which IL-12 shares a subunit. In terms of how this applies to MS, it is likely that not all infections, but those infections preferentially inducing a Th-1-type immune response may trigger or reactivate autoimmunity. In support of this is the fact that the viral sequences found to have homology with HLA-DR2 binding motif capable of activating CNS-reactive T cells from MS patients belong to relatively common human viruses such as EBV [24]. A substantial proportion of people without MS have evidence of EBV exposure. Thus, a large number of HLA-DR2+ individuals would be expected to develop MS if that were the unique requirement for self-reactive T cell activation. The ability to build a response in which these activated T cells have a Th1 phenotype, which may depend on a combination of host and pathogen factors may represent one of the additional requirements necessary for induction of autoimmunity.

An infectious agent capable of skewing the immune response toward Th2 may conceivably confer the opposite mechanism, protection against Th1 autoimmunity in EAE. It has been shown that coexistent Th2 responses dampen the mangnitude of a Th1 response. This may reflect the finding that the prevalence of asthma and allergic conditions is decreased in MS [36]. Some infections may also have a protective role against autoimmunity. This has been used to explain the increasing rates of autoimmune diseases including MS, paralleled by decreasing rates of infectious diseases in the developed countries [37].

*Other possible mechanisms*

Conceivably, other mechanisms may contribute to the pathology of EAE. Cytokines potentially induced by infectious agents may have directs effects on the CNS including oligodendrocyte apoptosis or permeabilization of the blood brain barrier.

## Trauma

Whether head trauma can trigger the onset of MS or attacks of established MS remains controversial [38, 39]. However, it is known that head trauma can cause a breakdown of the blood brain barrier, thus potentially facilitating the entry of autoreactive T cells in the CNS. Recent experiments in EAE have shown that focal brain injury following the immunization can lead to more severe neuropathology, mediated in part by chemokines [40].

## Solar radiation

A well-described feature of the epidemiology of MS is that its prevalence increases with increasing latitude. This holds true for both hemispheres [41]. Although in part this feature can be described on the basis of genetic factors taking into consideration various population migration patterns [42], it strongly suggests that an environmental factor linked to latitude plays a role in the susceptibility to MS. Of all potential factors investigated, the most consistently associated with MS has been the decreased level of exposure to solar radiation [43]. In addition, at the same latitude, there is an inverse altitude gradient, in that the prevalence of MS in mountainous areas is significantly lower than in neighboring regions in industrial valleys [44]. This further supports the concept that sunlight exposure has a protective effect against MS. Sunlight contains ultraviolet and visible light. Both have the potential to affect MS and EAE.

### Ultraviolet light

UV light from sunlight has numerous biological activities including many effects on the immune system. It is therefore used as a therapeutic agent in a number of diseases. It induces apoptosis in several cell types [45]. It also modifies properties of lymphocytes potentially changing the autoreactive cells and rendering them tolerant. Photopheresis, a procedure in which circulating blood cells are exposed extracorporeally to UV light after administration of a photosensitizing agent, and then reintroduced in the circulation, is based on this principle. Although established in cutaneous T cell lymphoma and graft versus host disease, photopheresis has had conflicting results in MS. In an uncontrolled report, two patients with relapsing remitting disease benefited [46], while in a placebo-controlled study in progressive MS it was not beneficial [47]. In EAE, a recent study shows that photopheresis decreases the number of relapses in a relapsing model in rats.

A major role of UV light is in the supply of the Vitamin D. Diet, in particular fish and vitamin D-enriched milk provide only a part of a human's daily requirement of Vitamin D. The majority of this requirement is provided by the Vitamin D synthesis in the skin. Pre-vitamin D3 is synthesized in the skin from 7-dehydrocholesterol in a reaction catalyzed by UV light. It then isomerizes to form Vitamin D3, which is then converted to 25-hydroxy-Vitamin D3in the liver and then to the active 1,25 Dihydroxy-Vitamin D3 (calcitriol) in the kidney.

Goldberg, noting a high prevalence of MS in areas with low solar radiation exposure suggested that this might be related to low vitamin D3, and that this vitamin is important in myelin biosynthesis [43]. There is no evidence for a role of vitamin D3 in myelination, but there is increasing evidence for its multiple roles in regulation of immune functions. It has been shown that vitamin D3 and its biologically active derivatives suppress the major Th1 inducing cytokine, IL-12 [48]. Thus, its deficiency may lead to an enhanced Th1 immune response. The epidemiological evidence suggesting that MS patients may have a vitamin D3 deficiency was confirmed in studies in which 25-hydroxy-vitamin D3 was actually measured in MS patients. These showed a very significant proportion of vitamin D3 deficiency or insufficiency. Moreover, there was evidence of significantly reduced bone mass in women with MS compared to age-matched controls [49].

In EAE, Cantorna et al [50] showed that oral administration of calcitriol prevented the onset of the disease, and its administration at the first clinical signs of EAE suppressed further progression. In further studies [51] the same group showed that calcitriol also is capable of reversing the clinical signs of disease in mice that already had clinically severe disease.

The mechanisms of action of calcitriol in EAE are not entirely elucidated. However, it is likely that its acts on the balance between pro- and anti-inflammatory (Th1/Th2) cytokines in favor of the latter. Cantorna et al [51] presented evidence that calcitriol induces anti-inflammatory cytokines TGF-$\beta$1 and IL-4. On the other hand, another study demonstrating suppression of EAE by vitamin B12 showed that the major mechanism was inhibition of IL-12 [52].

Vitamin D3 is not only effective in EAE. It was also successfully used in other experimental models of autoimmune disease including collagen-induced arthritis (where it was effective in both preventing and suppressing disease), [53] and insulin dependent diabetes mellitus in NOD mice [54]. It also has a beneficial effect in ameliorating the human inflammatory, presumed autoimmune disease, psoriasis [55].

*Visible light*

A fundamental component of sunlight is, of course, visible light. There is evidence that visible light is also important in MS and EAE. Daylight is also closely linked with the biological clock. One possibility explaining the latitude gradient for MS is that day length may be related to MS prevalence. There is anecdotal evidence of relapses being more frequent in the spring and autumn. The biological clock that maintains daily and seasonal rhythmicity may have an influence on the susceptibility to and manifestations of, autoimmune disease including EAE and MS.

In mammals the central clock is in the suprachiasmatic nucleus (SCN), which sends signals to the pineal gland to produce its hormone, melatonin, in a cyclical fashion. Melatonin feeds back onto the SCN and can modulate the clock cycle [56]. Melatonin can also modulate the body temperature cycle. The body temperature is high during the day, when melatonin levels are low [57].

Melatonin has a number of immune effects, in particular immunostimulatory activities [58]. It enhances production of IFN-$\gamma$ and appears to favor a Th1 bias of T cells. Interestingly, this action has been shown to be due to the ability of melatonin to induce the major cytokine driving Th1 responses, IL-12 [59]. In a study of diurnal rhythmicity of cytokine production in humans, the analysis of the ratio of IFN-$\gamma$ to IL-10 indicates that the strongest Th1 bias coincided to the highest levels of melatonin [60]. Removing production of melatonin by pinealectomy can suppress experimental autoimmune disease including collagen-induced arthritis. This has subsequently been demonstrated in EAE as well [61].

The peak production of melatonin is at night and the nadir is during daytime. Visible light suppresses production of melatonin. Thus conceivably increased daylight exposure time may be associated with lower susceptibility to Th1 mediated autoimmunity, in part possible due to decreased melatonin-enhanced Th1 cytokine production. This may explain why the geographic distribution of MS is related to the degree of sunlight exposure.

In experimental models of autoimmune diseases, this has been confirmed: constant light ameliorated experimental arthritis. In addition, exposure of animals with EAE to constant visible light significantly suppressed disease activity [62].

Melatonin can act directly upon cells of the immune system including T and B cells which have been shown to possess melatonin receptors [57]. We have used a melatonin receptor antagonist, Luzindole, injected at nighttime in the dark during peak melatonin production, and significantly suppressed EAE disease activity [63].

These results concur in suggesting that sunlight' visible light spectrum via suppression of melatonin, and UV light spectrum via provision of vitamin D3, and their effects on the immune system may contribute to the susceptibility to and manifestations of EAE and MS.

An additional and opposing aspect of melatonin needs to be considered. By modulation the body temperature clock, it lowers the body temperature. This has led to relief of several temperature-dependent symptoms in demyelinating disease. It is known that increases in body temperature can worsen symptoms of MS, usually in a transient fashion (Uhthoff's phenomenon). Conversely, lowering the body temperature improves symptoms of MS, possibly by improving conduction through demyelinated nerve fibers. This has been shown in human and experimental demyelination. Thus melatonin, which lowers the body temperature, can lead to short term symptomatic relief of MS symptoms. In the long term, however, its immunostimulating, Th1 biasing effects may make this hormone detrimental, and the above evidence of the positive effects of its suppression on EAE supports this.

Although speculative, another aspect of sunlight exposure may be protective against inflammatory demyelination: its ability to stimulate melanin production. Even within the same latitude, dark skinned individuals are at lower risk of developing MS than light skinned individuals. In EAE, with notable exceptions, susceptible strains are albino, while resistant ones are dark. This may be related to the protective effect of melanin, for example its known ability to scavenge and suppress free radicals.


**Stress**

Stress can be physical (trauma) or psychological (emotional). This part of the discussion refers to the latter. The role of stress in MS remains controversial. It is very difficult to assess and quantify stress. Many differences in the threshold, degree of perception, and cultural responses to stress between individuals are likely. Stress is so much a part of everyday life that envisioning a stress free environment is difficult in humans. Because of the ubiquity of stress, prospective studies in MS linking disease activity to stressful events are also difficult, and retrospective studies pose problems with controls. As with major physical trauma, major stressful events are most easily remembered, which may bias retrospective studies. However, a study in MS has suggested that stress may have a protective effect. Investigating the frequency of relapses and new onset of MS in Israel during the period of general stress to the whole population of Israel, of the Gulf War, Nisipeanu and Korczyn [64] have shown it to be significantly lower than expected. On the other hand, stress is frequently blamed, usually by patients themselves, for the onset of MS or for an exacerbation in the MS symptoms.

In solving this apparent discrepancy, here again animal models can be very helpful. Although emotional stress in humans is obviously a very complex phenomenon, its simplified direct biological correlate is a strong sympathetic response. The immunological phenomena that ensue are mediated by the presence of both beta2 and alpha2 adrenergic receptors on cells of the immune system. It is thought that Th1 cells possess beta2 receptors and thus are more amenable to immunomodulation via these receptors, while Th2 cells do not. Therefore, creating conditions representing the biological correlate of a stress free environment in an experimental animal can be accomplished by sympathectomy. Chemical sympathectomy can be achieved by using 6-hydroxydopamine hydrobromide, which becomes oxidized in the sympathetic nerve terminals and destroys them resulting in long term sympathectomy in the neonate and temporary axotomy in the adult. The short-term results of either procedure are similar, and consistent with a significant enhancement of immune functions.

This can be reflected in EAE as well. Animals having undergone permanent or temporary sympathectomy have been shown to exhibit more severe CNS inflammation [65, 66]. In adoptive transfer experiments between sympathectomized and non-sympathectomized mice, it has been shown that donor T cells from sympathectomized animals induce more severe EAE, and when the recipient is sympathectomized, the EAE is also more severe, suggesting that both T cells and APC from sympathectomized animals contribute to the increased severity of EAE [67].

In addition to sympathectomy, experimental situations inducing stress in the animal have been tested for its effect on EAE. These have included maternal deprivation, a swimming test, thermal stress etc. These investigations have consistently shown a protective effect of stress on EAE.

Taken together, these results suggest not only that EAE reflects reliably the relationship between MS and stress, but also that sympathectomy as is reliable as a surrogate for creation of "stress-free" conditions while adrenergic stimulation mimics stress.
This is further confirmed by the fact that beta adrenergic agonists such as terbutaline or isoproterenol mimic stress by inducing similar, suppressive changes in the immune system. These effects were also seen on EAE [68].

After the initial sympathetic stimulation and immune suppression by stress, there appears to be an overshoot of immune stimulation as stress declines or the stressful situation has passed. It has been shown that after exposure of medical students to examination stress with the corresponding immune suppressive effect, there was a hyper-responsiveness in immune functions 4-6 weeks after the examination. It is also possible that chronic stress induces different changes than acute or one-time stressful conditions. This may reconcile the discrepancies between the studies showing a protective effect of stress on EAE and MS and situations in which worsening or onset of MS were attributed to stress.

### Diet

Besides the importance of vitamin D3 (with the contribution of diet and sunlight) as discussed above, there are several other associations of diet and MS, which, despite much investigation, remain controversial. The most consistent aspects linking diet and MS are the positive association with the consumption of meats (in particular those preserved by curing or smoking) and the prevalence of MS, and the inverse relationship with the use of fish and vegetable oils rich in polyunsaturated fatty acids. The protective effect of omega-6 and omega-3 fatty acids has been investigated in double blind trials. They appear to decrease the severity of relapses. They reduce production of proinflammatory cytokines and eicosanoids [69]. Similarly, omega-6 fatty acids have a beneficial effect in EAE. This effect appears to be mediated by enhancing TGF-beta1 and prostaglandin (PG) E2 production [70].

In addition to vitamin D3, retinoids such as vitamin A appear to be important in skewing the immune response toward a Th2, anti-inflammatory type. In experimental animals, a diet poor in vitamin A resulted in an enhanced Th1-type responses and administration of retinoids led to suppression of EAE. The mechanism has been shown to be via suppression of IL-12 and enhancement of IL-4 production. [71-73].

## Toxins

Several toxins have been implicated in MS but none have been consistently and definitively demonstrated to have role in MS. The more notable ones have been occupational exposure to Zinc and organic solvents.

With regard to EAE, these have not been studied extensively. However, several toxins have been shown to induce experimental demyelination, and, although not necessarily involved in human demyelinating disease, provide important insights into the pathology or demyelination, the biology of myelin producing cells and the glial and neural response to demyelination. They can also answer important questions regarding mechanisms of remyelination and the use of novel mechanisms to induce or enhance it.

Several toxin-induced demyelinating models exist.

### *Cuprizone-induced demyelination*

The drug Cuprizone (biscyclohexanone, oxaldihydrazone) represents an excellent model fro demyelination [74-78]. When the drug is added to the diet of weanling mice for several weeks or longer, a near-total, primary demyelination of the superior cerebellar peduncle develops. The demyelination is caused by oligodendrocyte degeneration, but there is relative preservation of axons. There is a continuum between myelin degeneration via formation of vacuoles and death of oligodendrocytes. The prevalence of these pathological processes is in part dependent of the concentration of the toxin, suggesting that the demyelination via vacuolation does not represent a specific myelinopathy but rather an dose-dependent oligodendrocyte toxicity. The myelin debris is removed by microglia/macrophages and astrocytes. The relatively preserved axons become invested in glial processes. There is a remarkable glial response to demyelination, characterized by the occurrence of immature glial cells, both of astrocytic and oligodendroglial nature. There is also survival of some of the mature oligodendrocytes.

A great advantage of this model is that it allows the detailed study of the kinetics and mechanisms of remyelination in the CNS.

With the re-institution of a normal, cuprizone-free diet, remyelination begins within a week, and continues until nearly all residual axons are remyelinated. The pattern resembles that of myelination as part of the normal CNS development, with spiral wrapping around axons. Cellular sources of remyelination largely represent residual surviving adult oligodendrocytes and new cells differentiated from immature precursor forms.   The possibility of perineuronal satellite cells contributing to remyelination has also been proposed.  Like in EAE and MS, remyelination is incomplete: the maximal thickness reached by new myelin sheets is ½ that of normal sheets, and the internodal distance is shortened.

Importantly, recurrent demyelination can be demonstrated in this system. This can be accomplished by repeated exposure to cuprizone.  This recurrent demyelination appears to be more protracted and the remyelination potential is more reduced.  This is attributed to reduced numbers of surviving oligodendrocytes. The intensity of the glial reaction to the injury is lower compared to the initial reaction, and this may also in part explain the less efficient remyelination.  In addition, this model of re-exposure to cuprizone shows that remyelinated axons are not more susceptible to demyelination than axons that have not previously been demyelinated.  This finding, if it can be extrapolated to other mechanisms of demyelination, has important implications for EAE and MS.

*Ethidium bromide-induced demyelination.*
Another model of toxin-induced demyelination is the ethidium bromide exposure model [79]. This is accomplished by the intracisternal injection of ethidium bromide in rats.   The toxin induces status spongiosus and oligodendrocyte degeneration.  Many axons are demyelinated within 6 days of injection.    The   myelin   debris   is   phagocytosed   largely   by microglia/macrophages. Such cells also infiltrate between myelin lamellae. There can be complete oligodendrocyte degeneration in areas of myelin loss.

There is remyelination in this model as well. Remyelinated axons are thinner.   In addition, Schwann cells may contribute to remyelination, a feature recognised in some forms of MS with predominanlty myelopathic features in Japan.

There are additional models of toxin-induced demyelination including diphtheria toxin and lysolecithin induced demyelination [80]

**Effects of temperature**

The effects of temperature on clinical and experimental demyelination have been investigated since the report by Uhthoff of worsening symptoms of optic neuritis associated with an increased in body temperature caused by physical exertion [81]. It is now recognized that any other mechanisms of elevating the body temperature (even within the normal range, even by as little as 0.5C) can cause an increased perception of symptoms of demyelination (Uhthoff phenomenon), and that such symptoms are not limited to visual disturbances [82].

Conversely, a decrease in body temperature can improve symptoms of demyelination. This has been shown in humans with optic neuritis given ice cream or iced water [83, 84]. Although Uhthoff's phenomenon is not exclusively limited to primary demyelinating conditions, having been reported in giant cell arteritis or brain tumors, the ice test seems to allow distinguishing optic neuritis from other clinically similar conditions, and has been suggested as an auxiliary differential diagnostic tests.

These phenomena have been studies in experimental models of demyelination as well.
**Lower temperatures increased conduction while high temperatures impair it . [85-87]**


**Hyperbaric oxygen**

Although hyperbaric oxygen has long been advocated by some authors as a treatment for multiple sclerosis, no convincing evidence from controlled trials has been obtain to support its use. There are anecdotal accounts of symptomatic improvement with hyparbaric oxygen treatment.

Reports in EAE are also somewhat controversial. Theoretically the high pressure provided during hyperbaric treatments can affect the blood brain barrier and increase its permeability, thus potentially facilitating invasion of CNS by myelin-reactive cells. However, an opposite effect on the blood brain barrier has been proposed.

Dysbarism, as occurs in scuba diving accidents, may result in permeabilization of the blood brain barrier and autopsies of fatal cases have shown demyelination in the spinal cord similar to that seen in MS and EAE.

Few studies have investigated the effects of hyperbaric oxygen treatment in EAE. The effects may be dual, and may depend on the timing of the treatment relative to the immunization.

An old study revealed immunosuppressive effects of hyperbaric oxygen, with an associated positive effect on EAE in guinea pigs [88]. These effects depended on the duration of treatment and gas pressure. They, were, however, sustained only briefly after discontinuation of treatment. The immunosuppression was demonstrated by inhibition of cellular reactivity to myelin basic protein and tuberculin.

More recent studies using magnetic resonance (MRI) imaging showed that hyperbaric oxygen treatment does not reduce the disruption of the blood brain barrier or the cerebral edema seen in EAE [89].  On the contrary, it was shown to increase the blood brain barrier breakdown.  When given at the time of immunization as a preventive treatment, it caused amelioration in the clinical course of EAE, but when given 11 days after immunization in attempt to modify EAE, it had no effect [89].

### Conclusion

In conclusion, the various environmental influences on MS can be investigated in the animal model, EAE. The mechanisms by which infectious agents can have an impact on MS and autoimmunity in general can be carefully dissected in the EAE model.  Other factors such as UV and visible light exposure, stress, temperature, trauma, and their relevance to MS can be explored.

# ACKNOWLEDGMENTS

Dr. Constantinescu's research has been supported in part by the NIH, University of Nottingham, The Stanhope Trust and the Multiple Sclerosis Society of the United Kingdom and Northern Ireland.

### References

1.  Rose, J.W. and J.E. Greenlee, Multiple sclerosis: the role of pathogens in demyelination, in Multiple sclerosis. Immunology, Pathology, and Pathophysiology, R.M. Herndon, Editor. 2003, Demos: New York. p. 113-133.

2.  Zhang, J., S. Markovic-Plese, B. Lacet, J.C.M. Raus, H.L. Weiner, and D.A. Hafler, Increased frequency of interleukin-2-responsive T cell specific for myelin basic protein and proteolipid protein in peripheral blood and cerebrospinal fluid of patients with multiple sclerosis. J Exp Med, 1994. **179**: p. 973-84.

3.  Kotzin, B.L., D.Y.M. Leung, J. Kappler, and P. Marrack, Superantigens and their potential role in human disease. Adv Immunol, 1993. **54**: p. 99-166.

4.  Brocke, S., A. Gaur, C. Piercy, A. Gautam, K. Gijbels, C.G. Fathman, and L. Steinman, Induction of relapsing paralysis in experimental autoimmune encephalomyelitis by bacterial superantigens. Nature, 1993. **365**: p. 642-644.

5.  Schiffenbauer, J., H.M. Johnson, E.J. Butfiloski, L. Wegrzyn, and J.M. Soos, Staphylococcal enterotoxins can reactivate experimental allergic encephalomyelitis. Proc Natl Acad Sci USA, 1993. **90**: p. 8543-6.

6. Constantinescu, C.S., M. Wysocka, B. Hilliard, E.S. Ventura, E. Lavi, G. Trinchieri, and A. Rostami, Antibodies against IL-12 prevent superantigen-induced and spontaneous relapses of experimental autoimmune encephalomyelitis. J Immunol, 1998. **161**: p. 5097-5104.

7. Heber-Katz, E., The autoimmune T cell receptor: epitopes, idiotypes and malatopes. Clin Immunol Immunopathol, 1990. **55**: p. 1-8.

8. Nagelkerken, L., K.J. Gollob, M. Tielemans, and R.L. Coffman, Role of transfroming growth factor b in the preferential induction of T helper cells type 1 by staphylococcal enterotoxin B. Eur J Immunol, 1993. **23**: p. 2301-6.

9. Miethke, T., C. Wahl, K. Heeg, B. Echtenacher, P.H. Krammer, and H. Wagner, T cell-mediated lethal shock triggered in mice by the superantigen staphylococcal enterotoxin B: critical role of tumor necrosis factor. J Exp Med, 1992. **175**: p. 191-8.

10. Smith, T., A.K. Hewson, C.I. Kingsley, J.P. Leonard, and M.L. Cuzner, Interleukin-12 induces relapse in experimental allergic encephalomyelitis in the Lewis rats. Am J Pathol, 1997. **150**: p. 1909-17.

11. Crisi, G.M., L. Santambrogio, G.M. Hochwald, S.R. Smith, J.A. Carlino, and G.J. Thorbecke, Staphylococcus enterotoxin B and tumor-necrosis factor-α induced relapses of experimental allergic encephalomyelitis: protection by transforming growth factor-b and interleukin-10. Eur J Immunol, 1995. **25**: p. 3035-40.

12. Eugster, H.P., K. Frei, M. Kopf, H. Lassmann, and A. Fontana, IL-6-deficient mice resist myelin oligodendrocyte glycoprotein-induced autoimmune encephalomyelitis. Eur J Immunol, 1998. **28**: p. 2178-87.

13. Samoilova, E.B., J.L. Horton, B. Hilliard, T.S. Liu, and Y. Chen, IL-6-deficient mice are resistant to experimental autoimmune encephalomyelitis: roles of IL-6 in the activation and differentiation of autoreactive T cells. J Immunol, 1998. **161**: p. 6480-6.

14. Eugster, H.P., K. Frei, F. Winkler, U. Koedel, W. Pfister, H. Lassmann, and A. Fontana, Superantigen overcomes resistance of IL-6-deficient mice towards MOG-induced EAE by a TNFR1 controlled pathway. Eur J Immunol, 2001. **31**: p. 2302-12.

15. Burns, J., K. Littlefield, J. Gill, and J.L. Trotter, Bacterial toxin superantigens activate human T lymphocytes reactive with human autoantigens. Ann Neurol, 1992. **32**: p. 352-7.

16. Fridkis-Hareli, M., D. Teitelbaum, E. Gurevich, I. Pecht, C. Brautbar, O.J. Kwon, T. Brenner, R. Arnon, and M. Sela, Direct binding of myelin basic protein and synthetic coploymer 1 to class II major histocompatibility complex molecules on living antigen presenting cells-specificity and promiscuity. Proc Natl Acad Sci USA, 1994. **91**: p. 4872-6.

17. Teitelbaum, D., M. Fridkis-Hareli, R. Arnon, and M. Sela, Copolymer 1

inhibits chronic relapsing experimental allergic encephalomyelitis induced by proteolipid protein (PLP) peptides in mice and interferes with PLP-specific T cell responses. J Neuroimmunol, 1996. **64**: p. 209-17.

18.     Zhang, M., C.-C. Chan, B. Vistica, V. Hung, B. Wiggert, and I. Gery, Copolymer 1 inhibits experimental autoimmune uveitis. J Neuroimmunol, 2000. **103**: p. 189-94.

19.     Moran, A.P. and M.M. Prendergast, Molecular mimicry in campylobacter jejuni and helicobacter pylori lipopolysaccharides: contribution of gastrointestinal infections to autoimmunity. J Autoimmun, 2001. **16**: p. 241-56.

20.     Heidenreich, F., L. Leifeld, and T. Jovin, T-cell dependent activity of ganglioside GM1-specific B cells in Guillain-Barre syndrome and multifocal motor neuropathy. J Neuroimmunol, 1994. **49**: p. 97-108.

21.     Shamshiev, A., A. Donda, I. Carena, L. Mori, L. Kappos, and G. De Libero, Self glycolipids as T-cell autoantigens. Eur J Immunol, 1999. **29**: p. 1667-75.

22.     Schwerer, B., H. Lassmann, K. Kitz, H. Bernheimer, and H.M. Wisniewski, Fractionation of spinal cord tissue affects its activity to induce chrinic relapsing experimental encephalomyelitis. Acta Neuropathol, 1981. **7(suppl)**: p. 165-8.

23.     Fujinami, R.S. and M.B. Oldstone, Aminoacid homology between the encephalitogenic site of myelin basic protein and virus: mechanism for autoimmunity. Science, 1985. **230**: p. 1043-5.

24.     Wucherpfennig, K. and J.L. Strominger, Molecular mimicry in T cell-mediated autoimmunity: viral peptides activate human T cell clones specific for myelin basic protein. Cell, 1995. **80**: p. 695-705.

25.     Lehmann, P.V., T. Forsthuber, A. Miller, and E. Sercarz, Spreading of T cell autoimmunity to cryptic determinants of an autoantigen. Nature, 1992. **358**: p. 155-7.

26.     Miller, S.D., C.L. Vanderlugt, W.S. Begolka, W. Pao, R.L. Yauch, K.L. Nevile, Y. Katz-Levy, A. carrizosa, and B.S. Kim, persistent infection with Theiler's virus leads to autoimmunity via epitope spreading. Nature Med, 1997. **3**: p. 1133-6.

27.     van Noort, J.M., A.C. van Sechel, J.J. Bajramovic, M. el Ougamiri, C.H. Polman, H. Lassmann, and R. Ravid, The small heat-shock protein alpha B-crystallin as candidate autoantigen in multiple sclerosis. Nature, 1993. **375**: p. 798-801.

28.     Steinman, L., Multiple sclerosis. Presenting an odd autoantigen. Nature, 1995. **375**: p. 739-40.

29.     van Sechel, A.C., J.J. Bajramovic, M.J. van Stipdonk, C. Persoon-Deen, S.B. Geutskens, and J.M. van Noort, EBV-induced expression and HLA-DR-restricted presentation by human B cells of alpha B-crystallin, a

candidate autoantigen in multiple sclerosis. J Immunol, 1999. **162**: p. 129-35.

30. Bray, P.F., L.C. Bloomer, V.C. Salmon, M.H. Bagley, and P.D. Larsen, Epstein-Barr virus infection and antibody synthesis in patients with multiple sclerosis. Arch Neurol, 1983. **40**: p. 406-8.

31. Goverman, J., A. Woods, L. Larson, L.P. Weiner, L. Hood, and D.M. Zaller, Transgenic mice that express a myelin basic protein-specific T cell receptor develop spontaneous autoimmunity. Cell, 1993. **72**: p. 551-60.

32. Lichtman, S.N., J. Wang, R.B. sartor, C. Zhang, D. Bender, F.G. Dalldorf, and J.H. Schwab, Reactivation of arthritis induced by small bowel bacterial overgrowth in rats: role of cytokines, bacteria, and bacterial polymers. Infect Immun, 1995. **63**: p. 2295-301.

33. Segal, D.M., J.T. Chang, and E.M. Shevach, CpG oligonucleotides are potent adjuvants for the activation of autoreactive encephalitogenic T cells in vivo. J Immunol, 2000. **164**: p. 5683-8.

34. Ladel, C.H., G. Szalay, D. Riedel, and S.H. Kaufmann, Interleukin-12 secretion by Mycobacterium tuberculosis infected macrophages. Infect Immun, 1997. **65**: p. 1936-8.

35. Mahon, B.P., M.S. Ryan, F. Griffin, and K.H. Mills, Interleukin-12 is produced by macrophages in response to live or killed bordetella pertussis and enhances the efficacy of an acellular pertussis vaccine by promoting induction of Th1 cells. Infect Immun, 1996. **64**(5295-5301).

36. Oro, A.S., T.J. Guarnio, R. Driver, L. Steinman, and D.T. Umetsu, Regulation of disease susceptibility: decreased prevalence of IgE-mediated allergic disease in patients with multiple sclerosis. J Allergy Clin Immunol, 1996. **97**: p. 1402-8.

37. Bach, J.F., The effect of infections on susceptibility to autoimmune and allergic diseases. New Engl J Med, 2002. **347**: p. 911-920.

38. Cook, S.D., Trauma does not precipitate multiple sclerosis. Arch Neurol, 2000. **57**: p. 1077-8.

39. Poser, C.M., Trauma to the central nervous system may result in formation or enlargement of multiple sclerosis plaques. Arch Neurol, 2000. **57**: p. 1074-6.

40. Sun, D., T. M., T.A. Newman, K. Krivacic, M. Phillips, A. Chernosky, P. Gill, T. Wei, K.J. Griswold, R.M. Ransohoff, and R.O. Weller, Role of chemokines, neuronal projections, and the blood brain barrier in the enhancement of cerebral EAE following focal brain damage. J Neuropathol Exp Neurol, 2000. **59**: p. 1031-43.

41. Kurtzke, J. and M.T. Wallin, Epidemiology, in Multiple Sclerosis: Diagnosis, Medical Management and Rehabilitation, J.S. Burks and K.P. Johnson, Editors. 2000, Demos: New York. p. 49-71.

42. Arnason, B.G.W. and W.J. Wojcik, Multiple sclerosis: an update. Progress

in Neurology, 2000. **2**: p. 3-22.

43.     Goldberg, P., Multiple sclerosis: vitamin D and calcium as environmental determinants of prevalence ( a viewpoint). Part 1: Sunlight, dietary factors and epidemiology. Int J Environment Studies, 1974. **6**: p. 19-27.

44.     Kurtzke, J.F., On the fine structure of the distribution of multiple sclerosis. Acta Neurol Scand, 1967. **43**: p. 257.

45.     Kulms, D. and T. Schwarz, Independent contribution of three different pathways to ultraviolet-B-induced apoptosis. Biochem Pharmacol, 2002. **64**: p. 837-41.

46.     Poehlau, D., M. Rieks, T. Postert, R. Westerhausen, S. Busch, K. Hoffmann, P. Altmeyer, and H. Przuntek, Photopheresis--a possible treatment of multiple sclerosis?: report of two cases. J Clin Apheresis, 1997. **12**: p. 154-5.

47.     Rostami, A.M., R.A. Sater, S.J. Bird, S. Galetta, R.E. Farber, M. Kamoun, D.H. Silberberg, R.I. Grossman, and D. Pfohl, A double-blind, placebo-controlled trial of extracoporeal photopheresis in chronic progressive multiople sclerosis. Mult Scler, 1999. **5**: p. 198-203.

48.     D'Ambrosio, D., M. Cippitelli, M.G. Cocciolo, D. Mazzeo, P. Di Lucia, R. Lang, F. Sinigaglia, and P. Panina-Bordignon, Inhibition of IL-12 production by 1,25-dihydroxyvitamin D3. Involvement of NF-kappaB downregulation in transcriptional repression of the p40 gene. J Clin Invest. **101**: p. 252-62.

49.     Nieves, J., F. Cosman, J. Herbert, V. Shen, and R. Lindsay, High prevalence of vitamin D deficiency and reduced bone mass in multiple sclerosis. Neurology, 1994. **44**: p. 1687-92.

50.     Cantorna, M.T., C.E. Hayes, and H.F. DeLuca, 1,25-dihydroxyvitamin D3 reversibly blocks the progression of relapsing encephalomyelitis, a model of multiple sclerosis. Proc Natl Acad Sci USA, 1996. **93**: p. 7861-4.

51.     Cantorna, M.T., W.D. Woodward, C.E. Hayes, and H.F. DeLuca, 1,25-dihydroxyvitamin D3 is a positive regulator for the anti-encephalitogenic cytokines TGF$\beta$1 and IL-4. J Immunol, 1998. **160**: p. 5314-9.

52.     Mattner, F., S. Smiroldo, F. Galbiati, M. Muller, P. Di Lucia, P.L. Poliani, G. Martino, P. Panina-Bordignon, and L. Adorini, Inhibition of Th1 development and treatment of chronic-relapsing experimental allergic encephalomyelitis by a non-hypercalcemic analogue of 1,25-dihydroxyvitamin D(3). Eur J Immunol, 2000. **30**: p. 498-508.

53.     Cantorna, M.T., C.E. Hayes, and H.F. DeLuca, 1,25-dihydroxycholecalciferol inhibits the progression of arthritis in murine models of human arthritis. J Nutr, 1998. **128**: p. 68-72.

54.     Mathieu, C., J. Laureys, H. Sobis, M. Vandeputte, M. Waer, and R. Bouillon, 1,25-Dihydroxyvitamin D3 prevents insulitis in NOD mice. Diabetes, 1992. **41**: p. 1491-5.

55.    Perez, A., R. Raab, T.C. Chen, A. Turner, and M.F. Holick, Safety and efficacy of oral calcitriol (1,25-dihydroxyvitamin D3) for the treatment of psoriasis. Br J Dermatol, 1996. **134**: p. 1070-8.

56.    Dunlop, J.C., Molecular basis for the circadian clock. Cell, 1999. **96**: p. 271-90.

57.    Brzezinski, A., Melatonin in humans. N Engl J Med, 1997. **336**: p. 186-95.

58.    Maestroni, G.J., The immunotherapeutic potential of melatonin. Expert Opin Investig Drugs, 2001. **10**: p. 467-76.

59.    Garcia-Maurino, S., D. Pozo, A. Carrillo-Vico, J.R. Calvo, and J.M. Guerrero, Melatonin activates Th1 lymphocytes by increasing IL-12 production. Life Sci, 1999. **65**: p. 2143-50.

60.    Petrovsky, N. and L.C. Harrison, Diurnal rhythmicity of human cytokine production: a dynamic disequilibrium in T helper cell type 1/T helper type 2 balance? J Immunol, 1997. **158**: p. 5163-8.

61.    Sandyk, R., Influence of the pineal gland on the expression of experimental allergic encephalomyelitis: possible relationship to the acquisition of multiple sclerosis. Int J Neurosci, 1997. **90**: p. 129-33.

62.    Schneider, Suppression of experimental allergic encephalomyelitis by microwaves from fluorescent lamps., in The Epidemiology of Multiple Sclerosis, M. Alter and J.F. Kurtzke, Editors. 1968, CC Thomas: Springfield. p. 144-171.

63.    Constantinescu, C.S., Luzindole, a melatonin receptor antagonist, suppresses experimental autoimune encephalomyelitis. Pathobiology, 1997. **65**: p. 190-4.

64.    Nisipeanu, P. and A.D. Korczyn, Psychological stress as risk factor for exacerbations in multiple sclerosis. Neurology, 1993. **43**: p. 1311-2.

65.    Chelmica-Schorr, E., M. Checinski, and B.G.W. Arnason, Chemical symapthectomy augments the severity of experimental allergic encephalomyelitis. J Neuroimmunol, 1988. **17**: p. 347-50.

66.    Leonard, J.P., F.J. MacKenzie, H.A. Patel, and M.L. Cuzner, Hypothalamic noradrenergic pathways exert an influence on neuroendocrine and clinical status in experimental autoimmune encephalomyelitis. Brain Behav Immun, 1991. **5**: p. 328-38.

67.    Chelmica-Schorr, E., M.N. Kwasniewski, and R.I. Wollmann, Sympathectomy augments adoptively transferred experimental allergic encephalomyelitis. J Neuroimmunol, 1992. **37**: p. 9-103.

68.    Wiegmann, K., S. Muthyala, D.H. Kim, B.G. Arnasaon, and E. Chelmica-Schorr, Beta-adrenergic agonists suppress chronic/relapsing experimental allergic encephalomyelitis (CREAE) in Lewis rats. J Neuroimmunol, 1995. **56**: p. 201-6.

69.    Gallai, V., P. Sarchielli, A. Trquattrini, M. Franceschinii, A. Floridi, C. Firenze, A. Alberti, D. Di Benedetto, and E. Straglioto, Cytokine secretion

and eicosanoid production in the peripheral blood mononuclear cells of MS patients undergoing dietary supplementation with n-3 polyunsaturated fatty acids. J Neuroimmunol, 1995. **56**: p. 143-53.

70.    Harbige, L.S., L. Layward, M.M. Morris-Downes, D.C. Dumonde, and S. Amor, The protective effects of omega-6 fatty acids in experimental autoimmune encephalomyelitis (EAE) in relation to transforming growth factor-beta 1 (TGF-beta-1) up-regulation and increased prostaglandin E2 (PGE2) preoduction. Clin Exp Immunol, 2000. **122**: p. 445-52.

71.    Vladutiu, A. and N. Cringulescu, Suppression of experimental allergic encephalomyelitis by vitamin A. Experientia, 1968. **24**: p. 718-9.

72.    Massacesi, L., A.L. Abbamondi, C. Giorgi, F. Sarlo, F. Lolli, and L. Amaducci, Suppression of experimental allergic encephalomyelitis by retnoic acid. J Neurol Sci, 1987. **80**: p. 55-64.

73.    Racke, M.K., D. Burnett, S.H. Pak, P.S. Albert, B. Cannella, C.S. Raine, D.E. McFarlin, and D.E. Scott, Retinoid treatment of experimental allergic encephalomyelitis. IL-4 production correlates with improved disease course. J Immunol, 1995. **154**: p. 450-8.

74.    Love, S., Cuprizone neurotoxicity in the rat: morphological observations. J Neurol Sci, 1988. **84**: p. 223-37.

75.    Blakemore, W.F., The response of oligodendrocytes to chemical injury. Acta Neurol Scand, 1984. **100 (Suppl)**: p. 3-8.

76.    Ludwin, S.K., Central nervous system demyelination and remyelination in the mouse: an untrastructural study of cuprizone toxicity. Lab Invest, 1978. **39**: p. 597-612.

77.    Hiremath, M.M., Y. Saito, G.W. Knapp, J.P. Ting, and G.K. Matsushima, Microglial/macrophage accumulation during cuprizone-induced demyelination in C57BL/6 mice. J Neuroimmunol, 1998. **92**: p. 38-49.

78.    McMahon, E.J., K. Suzuki, and G.K. Matsushima, Peripheral macrophage recruitment in cuprizone-induced CNS demyelination despite an intact blood-brain barrier. J Neuroimmunol, 130. **130**(32-45).

79.    Yajima, K. and K. Suzuki, Demyelination and remyelination in the rat central nervous system folowing ethidium bromide injection. Lab Invest, 1979. **41**: p. 385-92.

80.    Blakemore, W.F., Observations on remyelination in the rabbit spinal cord follwing demyelination induced by lysolecithin. Neuropathol Appl Neurobiol, 1978. **4**: p. 47-59.

81.    Uhthoff, W., Untersuchungen uber bei der multiplen Herdsklerose vorkommenden Augenstorungen. Arch Psychit Nervenkr, 1890. **21**(55-116-303-410).

82.    Guthrie, T.C. and D.A. Nelson, Influence of temperature changes on multiple sclerosis: critical review of mechanisms and research potential. J Neurol Sci, 1995. **129**: p. 1-8.

83.    Scherokman, B.J., J.B. Selhorst, E.A. Waybright, B. Jabbari, G.E. Bryan, and C.G. Maitlamd, Improved optic nerve conduction with ingestion of ice water. Ann Neurol, 1985. **17**: p. 418-9.

84.    Gottlob, I. and W. Heider, Effect of hypothermia on pattern VEP and pattern ERG in optic neuritis. Clin Vis Sci, 1988. **2**: p. 277-82.

85.    Pender, M.P. and T.A. Sears, The pathophysiology od acute experimenatl allergic encephalomyelitis in the rabbit. Brain, 1984. **107**(699-726).

86.    Stanley, G.P. and M.P. Pender, The pathophysiology of chronic relapsing experimental allergic encephalomyelitis in the Lewis rat. Brain, 1991. **114**: p. 1827-53.

87.    Heininger, K., W. Fierz, B. Schafer, H.P. Hartung, P. Wehling, and K.V. Toyka, Electrophysiological investigations in adoptively transferred experimental autoimmune encephalomyelitis in the Lewis rat. Brain, 1989. **112**: p. 537-53.

88.    Warren, J., M.R. Sacksteder, and C.A. Thuning, Oxygen immunosuppression: modification of experimental allergic encephalomyelitis in rodents. J Immunol, 1978. **121**: p. 315-2.

89.    Namer, I.J., J. Steibel, P. Poulet, Y. Mauss, J.P. Armspach, B. Eclancher, and J. Chambron, Hyperbaric oxygen treatment in acute experimental allergic encephalomyelitis. Contribution of magnetic resonance imaging study. Neuroimage, 1994. **1**: p. 308-12.

Chapter A26

# HORMONAL AND GENDER INFLUENCES ON EXPERIMENTAL AUTOIMMUNE ENCEPHALOMYELITIS

Christopher Gilmore, Cris S Constantinescu and Caroline C Whitacre*
*Division of Clinical Neurology, University of Nottingham, UK and *Department of Molecular Virology, Immunology and Medical Genetics, Ohio State University, Columbus, OH, USA*

**Abstract:**     This chapter highlights the important interactions between the CNS, the endocrine system and the immune system that influence the course of EAE. In particular, we focus on the bi-directional relationship between the hypothalamic-pituitary-adrenal (HPA) axis and EAE, and the influences of gender and sex hormones on EAE.

**Key words:**     EAE, glucocorticoids, hypothalamic-pituitary-adrenal axis, sex hormones

Experimental autoimmune encephalomyelitis (EAE) is studied as a model for the human demyelinating disease, multiple sclerosis (MS). Nearly all of the therapeutic strategies which have been tried in MS have first been tested in the EAE model system.   EAE can be induced in all common laboratory animals by the injection of central nervous system (CNS) myelin proteins and adjuvant or by the transfer of disease-causing activated T lymphocytes. EAE has been studied most intensively using the following animal species/strain and antigen immunization regimens: Lewis rats immunized with myelin basic protein (MBP), DA rats immunized with myelin oligodendrocyte glycoprotein (MOG), SJL (H-2$^s$) mice immunized with myelin proteolipid protein (PLP)$_{139-151}$ peptide, B10.PL (H-2$^u$) mice immunized with MBP, and C57BL/6 (H-2$^b$) mice immunized with MOG$_{35-55}$ peptide.   In the aforementioned models, EAE is characterized by clinical neurologic signs of ascending hindlimb paralysis and CNS histopathologic changes consisting of demyelination and inflammatory cell infiltration into the brain and spinal cord. EAE is mediated by CD4+ MHC Class II-restricted Th1 lymphocytes, and IL-2, interferon gamma (INF-γ) and tumor necrosis factor alpha (TNF-α) have been shown to be important contributors to disease pathogenesis. Because EAE is a Th1-mediated disease, the disease course can be dramatically affected by immune modulatory drugs, hormones, cytokine ablation, and other therapies directed at T cells and their products. In this chapter, we review the influences of steroid and sex hormones on the disease course and pathogenesis of EAE.


## 1.  Glucocorticoids

MacPhee et al. observed markedly elevated serum corticosteroid levels in Lewis rats with MBP-induced EAE, with peak levels occurring at the time of maximal clinical disease[1]. The rats started to recover while glucocorticoid levels were still maximal and, as the clinical signs resolved, levels decreased to normal. Corticosteroid levels increased after the appearance of CNS inflammatory changes, but before the onset of paralysis, suggesting endogenous steroid production is not triggered by the stress of the paralytic attack. It is proposed that macrophage production of pro-inflammatory cytokines, such as IL-1β and IL-6, contributes to activation of the HPA axis in EAE[2,3]. Blockade of IL-1 receptors in Lewis rats manifesting MBP-induced EAE reduces corticosteroid levels[2]. IL-1 may stimulate the HPA axis by acting on the anterior pituitary, stimulating adrenocorticotrophic hormone (ACTH) release, or the hypothalamus, stimulating corticotrophin releasing factor (CRF) release[3,4].


### 1.1 Glucocorticoids modulate the severity of EAE

It has been recognized for over fifty years that ACTH and cortisol can prevent the development of EAE[5]. The importance of endogenous glucocorticoid production in mediating EAE recovery is highlighted by studies on adrenalectomized animals, exogenous glucocorticoid administration, and the pharmacological blockade of the HPA axis[1,6,7]. MacPhee et al. demonstrated that animals adrenalectomized before EAE induction, and treated with exogenous steroids to maintain normal resting corticosterone levels, rapidly developed clinical disease and died[1]. When steroid replacement was adjusted to mimic the levels produced in EAE the disease followed a non-fatal course, while higher steroid levels suppressed clinical disease completely.

Pharmacological blockade of the HPA axis also worsens EAE[6,8]. Reder et al. observed that treatment with RU 486 (a glucocorticoid receptor antagonist), from the time of immunization with guinea pig spinal cord, amplified the clinical and histological severity of EAE in Lewis rats, while dexamethasone therapy blocked clinical disease[6]. Similarly, Bolton et al. showed RU486 worsened chronic relapsing EAE in Biozzi ABH mice, while dexamethasone suppressed disease[8].

Stefferl et al. observed the corticosteroid response to MOG-induced chronic relapsing EAE in DA rats, characterized by an initial episode of disability, followed by a clinical remission and then a relapse, with chronic progressive disability and death[9]. In the initial phase corticosteroid levels rose markedly and then dropped as the animals entered remission. However, relapse was not accompanied by a second rise in corticosteroid levels, but rather a further decrease as disease progressed. The authors suggest the poor HPA axis response during relapse is due to a combination of reduced production of pro-inflammatory mediators and HPA hypo-responsiveness to these mediators. During relapse, the HPA axis was hypo-responsive to IL-1β (but not to disease-unrelated psychological stress) and IL-1β levels in the spinal cord were reduced in comparison with the initial phase of the disease. Supplementing corticosterone, to maintain steroid levels at those seen during the initial phase of disease, delayed relapse. Sternberg et al. reported that Lewis rats exhibit a deficit in HPA axis responsiveness to inflammatory and other stress mediators which may contribute to a high susceptibility to autoimmune disease[10]. In comparison to the Lewis rat, the PVG rat, which is not susceptible to EAE, mounts a more vigorous response to stress[11]. Adrenalectomy allows EAE induction in some, but not all, resistant rat strains[7,11].

## 1.2 The influence of restraint stress on EAE

The importance of endogenous glucocorticoids in mediating EAE recovery is supported by work on restraint stress (RST). Levine et al. first demonstrated that stress induced by 3 hours of daily restraint (i.e. by taping down all four limbs), initiated before immunization with spinal cord homogenate, decreased both the incidence and the clinical and histological severity of EAE in the Lewis rat[12]. Dowdell et al. extended these findings to male B10.PL mice which exhibit a relapsing-remitting disease course following immunization with guinea pig MBP[13]. 12 hours of RST dailystarted before disease induction and administered during the active night cycle, reduced the incidence, severity and relapse frequency of EAE. Furthermore, they demonstrated that the HPA axis plays a greater role than the sympathetic nervous system in mediating the protective effects of EAE. Blocking glucocorticoid action with RU 486 and aminoglutethimide (a glucocorticoid synthesis inhibitor) partially reversed the RST-induced suppression of EAE, while nadolol (a β2-adrenergic antagonist) had no effect. The duration of RST is important. While 9 hours of RST/day suppressed EAE in Lewis rats, 1 hour/day enhanced disease[14].

### 1.3 Abrupt steroid withdrawal can worsen EAE

Reder et al. demonstrated that abrupt withdrawal of glucocorticoids can exacerbate EAE, using Lewis rats immunized with guinea pig spinal cord[6]. Untreated rats developed EAE 9 to 11 days after immunization and then completely recovered over 7 to 12 days. Animals treated with dexamethasone for 3 weeks from the time of immunization showed no clinical evidence of EAE. However, when the dexamethasone was stopped suddenly (or rapidly reduced over 4 days), rather than being reduced gradually, severe clinical and histological disease was provoked. Interestingly, disease was also provoked when untreated rats, which had recovered spontaneously from EAE, received a 5 day "pulse" of dexamethasone. Clearly this work has potential implications on steroid treatment regimens for Multiple Sclerosis.

### 1.4 Mechanisms of action of glucocorticoids in EAE

The mechanism of action of glucocorticoids in the amelioration of EAE is unknown. *In vivo*, glucocorticoids are likely to interact with the immune system at a number of levels (for reviews see[15-17]), limiting the generation and trafficking of encephalitogenic cells to the CNS, and by inhibiting the activity of these cells. In EAE encephalitogenic T cells primarily produce Th1 cytokines (IL-2, IFN-$\gamma$, TNF), while Th2 cytokines (IL-4, IL-10) help disease recovery. There is *in vitro* evidence that glucocorticoids mediate a Th1 to Th2 shift in cytokine production[16]. *In vivo*, a reduction in both Th1 and Th2 cytokines, rather than a Th2 shift, was observed when RST reduced disease severity in relapsing EAE[13,18].

Steroids may reduce the proliferative capacity of lymphocytes or induce apoptosis of autoreactive T cells[16]. RST and corticosteroid therapy have been shown to reduce T cell numbers in the lymph nodes and spleen of mice with EAE[13]. Smith et al. described T-cell infiltration and apoptosis in spinal cord lesions of Lewis rats with MBP-induced EAE, observing reduced T-cell apoptosis in adrenalectomized rats[19]. In addition, treatment of EAE that is equivalent to the standard glucocorticoid treatment of relapses in Multiple Sclerosis induces T cell apoptosis[20]. Glucocorticoids may also inhibit lymphocyte trafficking to the CNS, by altering endothelial cell expression of adhesion molecules[15]. Recent *in vitro* work has suggested that decreased synthesis of inducible nitric oxide synthase (iNOS) is responsible for mediating dexamethasone-induced protection of rat microglial cells from inflammatory injury[21].

## 1.5 Neonatal glucocorticoid therapy increases EAE susceptibility in adult rats

Glucocorticoids play an important role in the development of the immune system[16]. Bakker et al. demonstrated that treating Wistar rats with dexamethasone for the first three days of life resulted in an increase in incidence and severity of EAE when rats were immunized with MBP at 8 weeks of age[22]. This was associated with a reduction in corticosteroid response to endotoxin. Neonatal dexamethasone therapy was also shown to reduce endotoxin-induced TNF-$\alpha$ and IL-1$\beta$ production by macrophages in adult rats, suggesting that a diminished macrophage response contributes to the HPA hypo-reactivity that predisposes to EAE.

## 2. Gender influences on EAE

Gender-related differences in susceptibility and severity of EAE are well recognised. Many models of EAE parallel multiple sclerosis, where females are more susceptible to disease than males[23,24] and pregnancy is associated with reduced disease activity[25,26]. Increased female susceptibility to EAE has been observed in both active and adoptive models[24,27]. In adoptive EAE gender differences influence both the ability of the donor to generate encephalitogenic T cells and the ability of the recipient to manifest disease[24]. It should be noted that important differences exist between different EAE models, depending on the species, strain and method of disease induction. For example in the Lewis rat males may be more susceptible to disease and less amenable to immunomodulation than females[28]. This may be related to lower production of corticosteroids and higher numbers of immunologically competent cells[28] Seven different mouse strains were evaluated for gender differences in susceptibility to actively induced EAE[29]. In the SJL, females exhibited greater severity regardless of immunizing antigen (PLP or MOG peptides) as did the ASW, both strains on an H-2$^s$ genetic background. However, male B10.PL and PL/J mice, both H-2$^u$, showed more severe disease than females. No sex differences were seen in the C57BL/6, NZW or NOD strains[29].

Sex hormones and sex-linked genetic factors are likely to influence gender differences in susceptibility to autoimmune disease[30]. Sex hormones may act directly on androgen or estrogen receptors expressed by immune cells, influencing antigen presentation, cytokine production, homing of immune cells or lymphocyte activation (for review see[31]). Sex hormones may contribute to the female preponderance for autoimmune disease by affecting the T cell balance between Th1 and Th2 cytokine production. Several groups suggest the increased severity of EAE in female mice is associated with increased levels of Th1 cytokines, while males produce higher levels of Th2 cytokines and develop less severe or frequent disease[23,31,32].

## 2.1 Gender differences in ability to generate encephalitogenic T cells

Using an adoptive transfer model of EAE, Kim et al. demonstrated that lymphocytes derived from male SJL mice following immunization with MBP are less encephalitogenic, with female-derived cells transferring more severe disease[23]. Male-derived cells only induced disease when transferred at higher doses. Similarly, Bebo et al. demonstrated T cells derived from PLP-immunized female SJL mice transferred more severe disease than T cells from male mice[33]. Kim et al. observed that, following immunization with MBP, draining lymph nodes derived from males contained fewer lymphocytes compared to females[23]. Furthermore, the proliferative response of these lymphocytes to MBP and non-MBP stimulation was reduced. Bebo et al., in contrast, found no differences in proliferation between sexes when spleen cells from $PLP_{139-151}$-immunized mice were stimulated with neuro-antigen *in vitro*[33].

Kim et al. suggested that reduced IL-12 during antigen-specific stimulation is central to the decreased encephalitogenicity of MBP-specific T lymphocytes derived from males[23]. When draining lymph node cells from immunized mice were stimulated *in vitro* with MBP, cells from males produced lower levels of IL-12 and IFN-γ. When stimulated with MBP in the presence of exogenous IL-12, lymphocytes from male mice induced more severe disease.

The B10.PL strain provides an interesting comparison to the results observed in the SJL. In the B10.PL immunized with MBP, males exhibit more severe disease than females during the acute disease period[29]. In this strain, male lymph node cells demonstrate a greater proliferative response to MBP than female cells. Moreover, upon stimulation with MBP *in vitro*, male lymphoid cells showed higher levels of IFNγ and TNFα and lower levels of IL-10 than female lymphoid cells.

Several studies have examined the role of androgens in regulating T cell cytokine secretion. Bebo et al. suggest that the presence of androgen in the lymphoid microenvironment during EAE induction alters the development of encephalitogenic T cells, causing a shift towards Th2 cytokine production and contributing to gender differences in EAE susceptibility[33]. Spleen cells, removed from female $PLP_{139-151}$-immunized SJL mice and stimulated *in vitro* with neuro-antigen, secreted increased IFN-γ while male T cells produced increased IL-10. When T cell lines from immunized females were stimulated *in vitro* in the presence of exogenous androgens they secreted less IFN-γ and more IL-10 than untreated cell lines and transferred less severe disease. Interestingly, the Th2 shift induced by androgen therapy was maintained during subsequent stimulations in the absence of androgen. Other work has also reported a bias in autoantigen-specific T cell responses toward the Th2 phenotype in males with EAE. Increased IL-4 and reduced IFN-γ production has been implicated in the mechanism of reduced encephalitogenicity of T cells derived from SJL male mice following immunization with MBP or PLP$_{139-151}$[27,34].

## 2.2 Gender differences in the effector phase of EAE

When MBP-specific T cells are transferred adoptively to female and male SJL mice, female recipients show more severe clinical and histological disease[24,32]. Dalal et al. suggested the protective effect of testosterone on EAE is mediated by enhancing IL-10 production by encephalitogenic T cells[32]. MBP-specific T cells derived from spleens of male mice during the effector phase of EAE produced higher levels of IL-10 than female-derived cells. IFN-γ and IL-4 production was not different. Treating females with dihydrotestosterone, initiated before injection of MBP-specific T cells, enhanced IL-10 production and resulted in less severe disease.

## 2.3 Role of sex hormones in determining gender differences in EAE

### 2.3. 1. Androgens

The effect of exogenous androgens on enhancing Th2 cytokine production has already been discussed. Hormonal manipulation by castration of male animals provides further evidence of the effect of androgens on disease course. Bebo et al. immunized SJL mice with PLP$_{139-151}$, inducing a monophasic disease in male mice but a relapsing disease in females[31]. However, when males were castrated before immunization, the majority developed relapsing disease. This was associated with increased activated IFN-γ-producing T cells in the spinal cord at relapse. The authors suggest androgens may modify leucocyte migration to the CNS[31].

### 2.3.2. Estrogens

Pregnancy levels of estrogen are protective against a number of autoimmune disorders including EAE, as will be discussed[30]. It has been proposed that estrus or diestrus levels of 17β-estradiol (E2, the primary hormone of the ovulatory cycle) enhance Th1 immunity and susceptibility to cell-mediated autoimmune diseases, whereas the high levels seen in pregnancy shift the response toward Th2 immunity and protection from cell-mediated pathology[35]. Indeed, *in vitro* studies on human T cell lines have shown a biphasic dose effect of E2, with low doses increasing TNF-α production, and doses approximating those of pregnancy decreasing TNF-α production and increasing IL-10 production[36].

Bebo et al. challenged this hypothesis, demonstrating low dose estrogens can provide protection against EAE in SJL mice immunized with PLP$_{139-151}$[37]. Estrogen therapy with either E2 (at estrus and diestrus levels) or estriol (E3, at sub-pregnancy levels) reduced the clinical and histological severity of active EAE, with both genders being equally sensitive to estrogen. Furthermore, low dose estrogens (E2 or E3) reduced the capacity of lymphocytes from female mice to transfer EAE, with recipients of splenocytes from estrogen-treated animals showing a delay in onset and reduction in severity of disease. Low dose E2 therapy initiated after the appearance of clinical disease had no effect on disease severity suggesting differentiated effector cells are less sensitive to estrogens.

The protective effects of estrogens are also highlighted by the increased incidence and severity of EAE seen in ovariectomized female mice[29,38,39]. Gender differences in EAE susceptibility appear to be primarily due to the protective effect of testosterone[40]. It should be noted that low-dose E2 therapy has produced disparate results in different mouse strains. For example, Jansson et al. observed no disease amelioration when castrated B10.RIII female mice received E2 at ovulatory cycle doses[38].

The mechanisms by which estrogens inhibit EAE are poorly understood. While high dose estrogen therapy mediates a shift towards Th2 cytokine production[41,42], low dose estrogens can be protective in the absence of such a shift[37,43]. *In vitro*, estrogens alter the function of various immunocompetent cells and estrogen receptors are expressed on a variety of immune cells, including T cells and macrophages[30,44,45]. Recent evidence suggests that the intracellular estrogen receptor alpha (ER$\alpha$) mediates estrogen-induced disease protection in EAE[42]. ER$\alpha$-knockout mice were not protected from disease by E3 treatment, while control mice showed reduced disease severity with E3[42]. *In vitro*, estrogen can regulate the expression of endothelial cell adhesion molecules and some groups have suggested that estrogens regulate lymphocyte trafficking to the CNS[39,46]. For example, Matejuk et al. observed that E2-mediated suppression of EAE was accompanied by a dramatic reduction in CNS cellularity and chemokine synthesis[39]. A recent DNA microarray study in BV8S2 transgenic mice immunized with MBP suggests that E2 therapy regulates the expression of a number of genes for several cytokines, chemokines and adhesion molecules[47].

Ito et al. suggested that down-regulation of TNF-$\alpha$ is an important pathway by which estrogens ameliorate EAE[43]. Low dose E2 therapy (five times less than pregnancy levels), initiated before immunization with MOG$_{33-55}$, suppressed the clinical and histological manifestations of EAE in C57BL/6 female mice. E2 suppressed EAE in IL-4 knockout, IL-10 knockout, and IFN-$\gamma$ knockout mice to a similar extent, suggesting E2 regulation of EAE occurs independently of inducing changes in these cytokines. The frequency of TNF-$\alpha$ producing cells in the CNS and spleens of estrogen-treated animals was decreased profoundly. Reduced TNF-$\alpha$ production has also been observed in lymphocytes from $MOG_{33-55}$-immunized C57BL/6 male mice treated with E3[42].

Sex steroids may also act indirectly via the HPA axis. Suppression of MBP-induced EAE with restraint stress is more pronounced in the female Lewis rat than in the male[18]. The female Lewis rat exhibits higher levels of basal and stress-induced corticosteroid production than males[28]. The presence of an estrogen-response element in the promoter region of the CRH gene further suggests that estrogen-mediated regulation of glucocorticoid production is a mechanism by which estrogens protect mice from EAE[30].

## 3. Pregnancy

A protective effect of pregnancy on EAE has been reported in acute models of EAE in rats, rabbits and guinea pigs [25,26]. Recently, Langer-Gould et al examined the effect of pregnancy on a relapsing model of $PLP_{139-151}$-induced EAE in SJL mice[50]. Late pregnancy improved disability and protected against relapses in mice with pre-existing EAE, without an associated improvement in CNS inflammatory lesions. They also demonstrated that mice immunized during mid and late pregnancy had a reduced incidence of EAE. We have confirmed these observations and shown that mice immunized during the post-partum period exhibit increased severity of clinical signs and increased EAE relapses. A shift towards humoral (Th2) immunity, which is thought to be beneficial for maintenance of the foetal allograft, has been demonstrated during pregnancy[17]. This shift may explain observations that Th1-mediated diseases, such as multiple sclerosis and rheumatoid arthritis, remit during pregnancy while Th2-mediated conditions, such as systemic lupus erythematatosus, are exacerbated[17].

### 3.1. Mediators of beneficial effects of pregnancy

Potential candidate hormones responsible for the amelioration of EAE during pregnancy include E3, E2 and progesterone, all of which increase during late pregnancy and can induce Th2 shifts in cytokine secretion *in vitro*[36,48].

### 3.1.1. Estrogens

Estrogen at pregnancy levels hasbeen shown to decrease the severity of EAE in mice[38]. E3 has profound effects on EAE. Kim et al. demonstrated that E3 decreases the ability of encephalitogenic T cells to induce EAE in non-pregnant SJL/J female mice[41]. E3 therapy at late pregnancy levels, commenced before the adoptive transfer of MBP-specific T cells, reduced the clinical and histological severity of EAE. This was accompanied by a Th2 shift, as evidenced by an increase in MBP-specific antibodies of the IgG1 isotype and increased IL-10 production by MBP-specific T cells obtained from E3 treated animals. E3 therapy also reduced the severity of EAE after disease induction, when commenced at the onset of clinical signs. Again this was associated with increased IL-10 production during the MBP-specific response. The concept of an E3-induced Th2 shift is also supported by the work of Liu et al. who treated C57BL/6 male mice with pregnancy levels of E3 before immunization with $MOG_{33\text{-}55}$, protecting them from disease[42]. They observed reductions in the pro-inflammatory cytokines TNF-α, IFN-γ and IL-2, and increased levels of the Th2 cytokine IL-5.

E2 has also been shown to have beneficial effects at pregnancy levels. Jansson et al. studied chronic relapsing EAE in female B10.RIII mice induced with $MBP_{89\text{-}101}$[38]. Ovariectomy led to earlier disease onset. High levels of E2 delayed onset, but not incidence or severity, in both castrated and normal female mice. In castrated females pregnancy levels of E3 delayed disease onset for longer than pregnancy levels of E2, suggesting E3 might be more potent. In contrast Bebo et al. detected no difference in the incidence or severity of $PLP_{139\text{-}151}$-induced EAE between E2 and E3 treated female mice[37].

### 3.1. 2. Progestogens

*In vitro*, progesterone induces a Th1 to Th2 shift in cytokine production[48]. Despite this, progesterone has not been shown to ameliorate disease activity in EAE[41]. Implanting progesterone pellets into non-pregnant SJL/J female mice before the adoptive transfer of MBP-specific T cells had no effect on EAE[41]. However, progesterone has been shown to be neuroprotective and thus in the long term it may also have a beneficial effect on neuronal loss or dysfunction in inflammatory demyelination[49].

### 3.2 Evidence for another serum factor that suppresses EAE in pregnancy

Not all studies implicate a Th2 shift in ameliorating EAE during pregnancy. Langer-Gould et al. suggest the presence of a transient circulating serum factor in late pregnancy that suppresses T cell activation[50]. T cells derivedfrom mice immunized with neuroantigen during late pregnancy, and stimulated *in vitro* with $PLP_{139\text{-}151}$, showed no differences in cytokine secretion profiles or proliferation compared with virgin controls. However, when this stimulation occurred in the presence of sera taken from mice in late pregnancy, proliferation and IL-2 production weredecreased compared with stimulation in the presence of normal mouse sera. Similarly, late pregnancy sera diminished the proliferative response and IL-2 production of auto-reactive cells from virgin controls.

### 4. Other hormones in EAE

### 4.1 Prolactin

Riskind et al. demonstrated that female Lewis rats immunized with guinea pig spinal cord homogenate experienced a rise in prolactin levels before the onset of clinical signs[51]. Prolactin suppression with bromocriptine therapy reduces disease severity in these animals, either when started before immunization, or at the onset of disease[51,52]. Dijkstra et al. also used bromocriptine therapy in a chronic relapsing EAE model. Starting therapy at the onset of the first attack reduced the duration of the second attack[52].

## 4.2. Thyroid Hormones

Thyroid hormone (T4) has potential application in promoting re-myelination in EAE by recruiting CNS progenitor cells and channelling them into the oligodendrocyte lineage. T4 is required for normal oligodendrocyte maturation and myelination[53]. Calza et al. induced EAE in Lewis rats using guinea pig spinal cord tissue[54]. A large number of proliferating cells and undifferentiated precursors were present in the spinal cords of the animals. T4 treatment, commenced at disease onset, reduced the number of cells expressing proliferation markers and markers for undifferentiated precursors, and up-regulated expression of markers for oligodendrocyte progenitors. This effect of T4 on differentiation of progenitor cells towards the oligodendrocyte lineage was found only in EAE rats and not in controls.

## 4.3. Growth Hormone-Releasing Hormone (GHRH)

Ikushima et al. recently suggested a role for GHRH in the development of EAE, demonstrating C57BL/6J mice deficient of the GHRH gene were resistant to MOG-induced EAE[55].

## 5. Conclusion

Interactions between hormones and the immune system are clearly important ina wide variety of immune responses. Sex and steroid hormones exert profound effects on the course and disease progression of EAE. A predominant mechanistic theme of both hormone systems is a Th1 to Th2 shift in cytokine production, but these hormones are likely to exert their effects by multiple mechanisms. Identification of these mechanisms will lead to a greater understanding of autoimmune disease of the CNS.

ACKNOWLEDGMENTS

Dr. Constantinescu's research has been supported in part by the NIH, University of Nottingham, The Stanhope Trust and the Multiple Sclerosis Society of the United Kingdom and Northern Ireland. Dr. Whitacre's research has been supported by the NIH and the National Multiple Sclerosis Society of the United States.

# REFERENCES

1.     MacPhee, I. A., Antoni, F. A. & Mason, D. W. Spontaneous recovery of rats from experimental allergic encephalomyelitis is dependent on regulation of the immune system by endogenous adrenal corticosteroids. *J Exp Med* **169**, 431-45 (1989).

2.     del Rey, A., Klusman, I. & Besedovsky, H. O. Cytokines mediate protective stimulation of glucocorticoid output during autoimmunity: involvement of IL-1. *Am J Physiol* **275**, R1146-51 (1998).

3.     Besedovsky, H. O. & del Rey, A. Immune-neuro-endocrine interactions: facts and hypotheses. *Endocr Rev* **17**, 64-102 (1996).

4.     Bernton, E. W., Beach, J. E., Holaday, J. W., Smallridge, R. C. & Fein, H. G. Release of multiple hormones by a direct action of interleukin-1 on pituitary cells. *Science* **238**, 519-21 (1987).

5.     Moyer, A., Jervis, G., Black, J., Koprowski, H. & Cox, H. Action of adrenocorticotrophic hormone in experimental allergic encephalomyelitis of the guinea pig. *Proc Soc Exp Biol Med* **75**, 387 (1950).

6.     Reder, A. T., Thapar, M. & Jensen, M. A. A reduction in serum glucocorticoids provokes experimental allergic encephalomyelitis: implications for treatment of inflammatory brain disease. *Neurology* **44**, 2289-94 (1994).

7.     Levine, S., Wenk, E. J., Muldoon, T. N. & Cohen, S. G. Enhancement of experimental allergic encephalomyelitis by adrenalectomy. *Proc Soc Exp Biol Med* **111**, 383-5 (1962).

8.     Bolton, C., O'Neill, J. K., Allen, S. J. & Baker, D. Regulation of chronic relapsing experimental allergic encephalomyelitis by endogenous and exogenous glucocorticoids. *Int Arch Allergy Immunol* **114**, 74-80 (1997).

9.     Stefferl, A. et al. Disease progression in chronic relapsing experimental allergic encephalomyelitis is associated with reduced inflammation-driven production of corticosterone. *Endocrinology* **142**, 3616-24 (2001).

10.    Sternberg, E. M. et al. Inflammatory mediator-induced hypothalamic-pituitary-adrenal axis activation is defective in streptococcal cell wall arthritis-susceptible Lewis rats. *Proc Natl Acad Sci U S A* **86**, 2374-8 (1989).

11.    Mason, D., MacPhee, I. & Antoni, F. The role of the neuroendocrine system in determining genetic susceptibility to experimental allergic encephalomyelitis in the rat. *Immunology* **70**, 1-5 (1990).

12.    Levine, S., Strebel, R., Wenk, E. J. & Harman, P. J. Suppression of experimental allergic encephalomyelitis by stress. *Proc Soc Exp Biol Med* **109**, 294-8 (1962).

13.    Dowdell, K. C., Gienapp, I. E., Stuckman, S., Wardrop, R. M. & Whitacre, C. C. Neuroendocrine modulation of chronic relapsing experimental autoimmune encephalomyelitis: a critical role for the hypothalamic-pituitary-adrenal axis. *J Neuroimmunol* **100**, 243-51 (1999).

14.    Griffin, A. C., Lo, W. D., Wolny, A. C. & Whitacre, C. C. Suppression of experimental autoimmune encephalomyelitis by restraint stress: sex differences. *J Neuroimmunol* **44**, 103-16 (1993).

15.    Padgett, D. A. & Glaser, R. How stress influences the immune response. *Trends Immunol* **24**, 444-8 (2003).

16.    Ashwell, J. D., Lu, F. W. & Vacchio, M. S. Glucocorticoids in T cell development and function*. *Annu Rev Immunol* **18**, 309-45 (2000).

17.    Elenkov, I. J. & Chrousos, G. P. Stress hormones, proinflammatory and antiinflammatory cytokines, and autoimmunity. *Ann N Y Acad Sci* **966**, 290-303 (2002).

18.    Whitacre, C. C., Dowdell, K. & Griffin, A. C. Neuroendocrine influences on experimental autoimmune encephalomyelitis. *Ann N Y Acad Sci* **840**, 705-16 (1998).

19.    Smith, T., Schmied, M., Hewson, A. K., Lassmann, H. & Cuzner, M. L. Apoptosis of T cells and macrophages in the central nervous system of intact and

adrenalectomized Lewis rats during experimental allergic encephalomyelitis. *J Autoimmun* **9**, 167-74 (1996).

20.  Schmidt, J. et al. T-cell apoptosis in situ in experimental autoimmune encephalomyelitis following methylprednisolone pulse therapy. *Brain* **123** ( **Pt 7**), 1431-41 (2000).

21.  Golde, S., Coles, A., Lindquist, J. A. & Compston, A. Decreased iNOS synthesis mediates dexamethasone-induced protection of neurons from inflammatory injury in vitro. *Eur J Neurosci* **18**, 2527-37 (2003).

22.  Bakker, J. M. et al. Neonatal dexamethasone treatment increases susceptibility to experimental autoimmune disease in adult rats. *J Immunol* **165**, 5932-7 (2000).

23.  Kim, S. & Voskuhl, R. R. Decreased IL-12 production underlies the decreased ability of male lymph node cells to induce experimental autoimmune encephalomyelitis. *J Immunol* **162**, 5561-8 (1999).

24.  Voskuhl, R. R., Pitchekian-Halabi, H., MacKenzie-Graham, A., McFarland, H. F. & Raine, C. S. Gender differences in autoimmune demyelination in the mouse: implications for multiple sclerosis. *Ann Neurol* **39**, 724-33 (1996).

25.  Mertin, L. A. & Rumjanek, V. M. Pregnancy and the susceptibility of Lewis rats to experimental allergic encephalomyelitis. *J Neurol Sci* **68**, 15-24 (1985).

26.  Evron, S., Brenner, T. & Abramsky, O. Suppressive effect of pregnancy on the development of experimental allergic encephalomyelitis in rabbits. *Am J Reprod Immunol* **5**, 109-13 (1984).

27.  Bebo, B. F., Jr., Vandenbark, A. A. & Offner, H. Male SJL mice do not relapse after induction of EAE with PLP 139-151. *J Neurosci Res* **45**, 680-9 (1996).

28.  Griffin, A. C. & Whitacre, C. C. Sex and strain differences in the circadian rhythm fluctuation of endocrine and immune function in the rat: implications for rodent models of autoimmune disease. *J Neuroimmunol* **35**, 53-64 (1991).

29.  Papenfuss, T.L., et al. Sex differences in experimental autoimmune encephalomyelitis in multiple murine strains. *J. Neuroimmunol.*, in press, 2004.

30.  Whitacre, C. C. Sex differences in autoimmune disease. *Nat Immunol* **2**, 777-80 (2001).

31.  Bebo, B. F., Jr. et al. Gonadal hormones influence the immune response to PLP 139-151 and the clinical course of relapsing experimental autoimmune encephalomyelitis. *J Neuroimmunol* **84**, 122-30 (1998).

32.  Dalal, M., Kim, S. & Voskuhl, R. R. Testosterone therapy ameliorates experimental autoimmune encephalomyelitis and induces a T helper 2 bias in the autoantigen-specific T lymphocyte response. *J Immunol* **159**, 3-6 (1997).

33.  Bebo, B. F., Jr., Schuster, J. C., Vandenbark, A. A. & Offner, H. Androgens alter the cytokine profile and reduce encephalitogenicity of myelin-reactive T cells. *J Immunol* **162**, 35-40 (1999).

34.  Cua, D. J., Hinton, D. R. & Stohlman, S. A. Self-antigen-induced Th2 responses in experimental allergic encephalomyelitis (EAE)-resistant mice. Th2-mediated suppression of autoimmune disease. *J Immunol* **155**, 4052-9 (1995).

35.  Whitacre, C. C., Reingold, S. C. & O'Looney, P. A. A gender gap in autoimmunity. *Science* **283**, 1277-8 (1999).

36.  Gilmore, W., Weiner, L. P. & Correale, J. Effect of estradiol on cytokine secretion by proteolipid protein-specific T cell clones isolated from multiple sclerosis patients and normal control subjects. *J Immunol* **158**, 446-51 (1997).

37.  Bebo, B. F., Jr. et al. Low-dose estrogen therapy ameliorates experimental autoimmune encephalomyelitis in two different inbred mouse strains. *J Immunol* **166**, 2080-9 (2001).

38.  Jansson, L., Olsson, T. & Holmdahl, R. Estrogen induces a potent suppression of experimental autoimmune encephalomyelitis and collagen-induced arthritis in mice. *J Neuroimmunol* **53**, 203-7 (1994).

39.     Matejuk, A. et al. 17 beta-estradiol inhibits cytokine, chemokine, and chemokine receptor mRNA expression in the central nervous system of female mice with experimental autoimmune encephalomyelitis. *J Neurosci Res* **65**, 529-42 (2001).

40.     Voskuhl, R. R. & Palaszynski, K. Sex hormones in experimental autoimmune encephalomyelitis: implications for multiple sclerosis. *Neuroscientist* **7**, 258-70 (2001).

41.     Kim, S., Liva, S. M., Dalal, M. A., Verity, M. A. & Voskuhl, R. R. Estriol ameliorates autoimmune demyelinating disease: implications for multiple sclerosis. *Neurology* **52**, 1230-8 (1999).

42.     Liu, H. B. et al. Estrogen receptor alpha mediates estrogen's immune protection in autoimmune disease. *J Immunol* **171**, 6936-40 (2003).

43.     Ito, A. et al. Estrogen treatment down-regulates TNF-alpha production and reduces the severity of experimental autoimmune encephalomyelitis in cytokine knockout mice. *J Immunol* **167**, 542-52 (2001).

44.     Polanczyk, M. et al. The protective effect of 17beta-estradiol on experimental autoimmune encephalomyelitis is mediated through estrogen receptor-alpha. *Am J Pathol* **163**, 1599-605 (2003).

45.     Matejuk, A., Dwyer, J., Hopke, C., Vandenbark, A. A. & Offner, H. 17Beta-estradiol treatment profoundly down-regulates gene expression in spinal cord tissue in mice protected from experimental autoimmune encephalomyelitis. *Arch Immunol Ther Exp (Warsz)* **51**, 185-93 (2003).

46.     Cid, M. C. et al. Estradiol enhances leukocyte binding to tumor necrosis factor (TNF)-stimulated endothelial cells via an increase in TNF-induced adhesion molecules E-selectin, intercellular adhesion molecule type 1, and vascular cell adhesion molecule type 1. *J Clin Invest* **93**, 17-25 (1994).

47.     Matejuk, A., Dwyer, J., Zamora, A., Vandenbark, A. A. & Offner, H. Evaluation of the effects of 17beta-estradiol (17beta-e2) on gene expression in experimental autoimmune encephalomyelitis using DNA microarray. *Endocrinology* **143**, 313-9 (2002).

48.     Miyaura, H. & Iwata, M. Direct and indirect inhibition of Th1 development by progesterone and glucocorticoids. *J Immunol* **168**, 1087-94 (2002).

49.     Labombarda, F. et al. Cellular basis for progesterone neuroprotection in the injured spinal cord. *J Neurotrauma* **19**, 343-55 (2002).

50.     Langer-Gould, A., Garren, H., Slansky, A., Ruiz, P. J. & Steinman, L. Late pregnancy suppresses relapses in experimental autoimmune encephalomyelitis: evidence for a suppressive pregnancy-related serum factor. *J Immunol* **169**, 1084-91 (2002).

51.     Riskind, P. N., Massacesi, L., Doolittle, T. H. & Hauser, S. L. The role of prolactin in autoimmune demyelination: suppression of experimental allergic encephalomyelitis by bromocriptine. *Ann Neurol* **29**, 542-7 (1991).

52.     Dijkstra, C. D. et al. The therapeutic effect of bromocriptine on acute and chronic experimental allergic encephalomyelitis. *Ann Neurol* **31**, 450-1 (1992).

53.     Rodriguez-Pena, A. Oligodendrocyte development and thyroid hormone. *J Neurobiol* **40**, 497-512 (1999).

54.     Calza, L., Fernandez, M., Giuliani, A., Aloe, L. & Giardino, L. Thyroid hormone activates oligodendrocyte precursors and increases a myelin-forming protein and NGF content in the spinal cord during EAE. *Proc Natl Acad Sci U S A* **99**, 3258-63 (2002).

55.     Ikushima, H., Kanaoka, M. & Kojima, S. Cutting edge: Requirement for growth hormone-releasing hormone in the development of experimental autoimmune encephalomyelitis. *J Immunol* **171**, 2769-72 (2003).

Chapter A27

# EXPERIMENTAL AUTOIMMUNE ENCEPHALOMYELITIS IN PRIMATES

Paul A. Smith, Sandra Amor, and Bert A. 't Hart
*Biomedical Primate Research Center, Rijswijk, Netherlands*

**Abstract:**  Due to their genetic and immunological proximity to humans, the common marmoset, *Callithrix jacchus*, provides an important bridge between rodent-based research and the human disease. Experimental autoimmune encephalomyelitis (EAE) induced in the outbred common marmoset provides a highly reproducible model of multiple sclerosis (MS) phenotypically characterised by neurological deficits, inflammatory lesions of the cerebral white matter and spinal cord, primary demyelination and axonal destruction. A fused placental blood circulation *in utero*, unique amongst monkeys, results in an immune tolerance between chimeric twins which allows adoptive transfer experiments assessing the relative contributions of T- and B-cells to pathogenesis. Overall, the ability to perform serial blood sampling, quantitative magnetic resonance imaging (MRI) and histopathological techniques, which are comparable to the human MS situation, means that the common marmoset EAE model provides an excellent test system for investigating the relationship between neurological deficits and pathological processes and the evaluation of new therapeutics.

**Key words:**  common marmoset, experimental autoimmune encephalomyelitis (EAE), magnetic resonance imaging (MRI), myelin oligodendrocyte glycoprotein (MOG), non-human primate, pre-clinical model

## Introduction

In this chapter we review a new non-human primate experimental model of multiple sclerosis (MS) that bridges the experimental data between inbred rodent strains and the disease in humans. The animal model experimental autoimmune encephalomyelitis (EAE) shares many of the clinical and pathological phenotypes of MS. To date, EAE induced in susceptible inbred rodent strains has highlighted many possible therapeutic candidate molecules although few have translated to clinical efficacy. Clearly the extrapolation of experimental data from genetically homogenous laboratory rodents to the heterologous human population has been hampered by the wide microbiological, genetic and immunological diversity.[1]

Until recently, reproducible non-human primate EAE models were restricted to two macaque species, namely rhesus (*M. mulatta*) and the cynomolgus monkey (*M. fascicularis*). *M. mulatta* appeared to be the most susceptible of the *Macaca* species to disease induced with whole brain homogenate or myelin basic protein. EAE in the rhesus macaque usually followed an acute course with large haemorrhagic/necrotic lesions observed in most animals, these data suggested that the acute pathological event induced a severe inflammatory necrosis rather than selective demyelination.[2] Thus, EAE in the rhesus monkey most resembled the acute fulminant forms of MS, such as post-infectious leuko-encephalomyelitis, rather than the more common chronic MS phenotypes.


# 1.        EAE IN THE COMMON MARMOSET


## 1.1        The common marmoset

The common marmoset (*Callithrix jacchus*) is a small sized monkey originating from the Atlantic rainforest in north and north-east Brazil (Figure 1). Common marmosets are born as non-identical twins however due to a fused placental blood circulation *in utero* they share a natural bone marrow chimerism, a unique feature among monkeys. Chimeric twins have, in effect, haemopoetic cell systems that are educated in the same thymic environment resulting in a permanent state of mutual allospecific tolerance.

**Figure 1:**  *The common marmoset (Picture* *was a kind gift from Dr. Susanne Rensing, German Primate Centre)*

The evolutionary divergence of 65 million years between mice and humans has resulted in several known discrepancies in both the innate and adaptive immunity.[1] In contrast, humans and common marmosets share a high degree of similarity at the level of cytokine mediated T- and B-cell responses,[3] cell surface markers,[4-6] T-cell receptor $V\beta$ repertoire,[7] co-stimulatory molecules,[8] major histocompatibility complex (MHC) class II genes,[9,10] immunoglobulin $V_H$ repertoire,[11] and myelin proteins.[12,13]

## 1.2    Myelin-induced EAE

Outbred common marmosets immunised with human brain-derived myelin from MS patients, emulsified within Freund's complete adjuvant, were characterised by a clinically and pathologically heterogeneous disease pattern. Despite the outbred population EAE incidence was 100%, with the vast majority of monkeys characterised by a variable number of neurological episodes followed by a chronic-progressive phenotype.[14,15] Whilst chronic-progression to hind-limb paralysis (score 3) resulted in a humane end-point, the time duration between disease onset and non-remitting paralysis varied considerably.

| Multiple Sclerosis | | Experimental Autoimmune Encephalomyelitis | | |
| --- | --- | --- | --- | --- |
| | | Marmoset | Rhesus | Rodent |
| *Epidemiology* | | | | |
| Population | Outbred | Outbred | Outbred | Inbred |
| Prevalence/ incidence | 1:1000 | 100% | MOG 100% MBP 60% | 100% in susceptible strains |
| Cause | Unknown | Immunisation | Immunisation | Immunisation |
| Clinical Course | | Relapsing-remitting chronic-progressive | Acute, fulminant | Monophasic, relapsing remitting or chronic progressive |
| MHC association | HLA-DR2 | *Caja-DRB*W1201* | *Manu-DPB1*01* | Strain-dependent |
| *Pathology* | | | | |
| Demyelination | Present | Present | Probable | Present |
| Inflammation | Focal | Mainly focal | Mainly focal | Focal and diffuse |
| Remyelination | Present | Present | Present | Strain/model dependent |
| Axonal pathology | Extensive in acute MS, variable in chronic MS | Present, limited | Strong in all lesions | Strain/model dependent |
| MRI analysis | Yes | Yes | Yes | Only in rats |
| *Immunology* | | | | |
| Induction | Not testable | Myelin, MBP/PLP, MOG, MOG14-36 | Myelin, MBP, MOG, MOG 35-56 | Myelin proteins or peptides. |
| Adoptive Transfer | Not testable | Yes, chimeric twins | Yes, autologous | Yes |
| B-cell/antibodies | Plasma cells, Ig/C | Plasma cells, Ig/C, anti-MOG ↑ demyelination | Not investigated | Anti-MOG ↑ demyelination |
| Pre-clinical testing | Limited possibilities | Possible, low quantity of reagents needed | Suboptimal model | Reagents may not be species cross-reactive. |

**Table 1: Comparison of multiple sclerosis with non-human primate EAE models**

Isolated lymphoid organs from necropsied monkeys with chronic neurological progression demonstrated more prominent mononuclear cell (MNC) *in vitro* proliferative responses to the quantitatively minor myelin protein myelin oligodendrocyte glycoprotein (MOG) compared to major myelin proteins, such as proteolipid protein (PLP) and myelin basic protein (MBP).[16] It is pertinent to mention here that all experiments were performed with recombinant protein encompassing the Ig-like extracellular domain of human MOG (amino acids 1-125; rhMOG$^{Igd}$). Myelin-induced EAE was characterised by active lesions within the brain and spinal cord and immunohistochemical analysis revealed the presence of many macrophages containing myelin degredation products and the expression of activation markers 27E10, MRP14 and CD40. Also, activated T-cells positive for the CD40 ligand protein CD154 were found at these sites, although at lower numbers. Since CD40-CD154 ligation provides one of the important co-stimulatory signals needed for Th1-cell stimulation by antigen presenting cells (APC), it was not surprising that within these cell clusters a variety of immunomodulating molecules were expressed (Figure 2).[17-19]

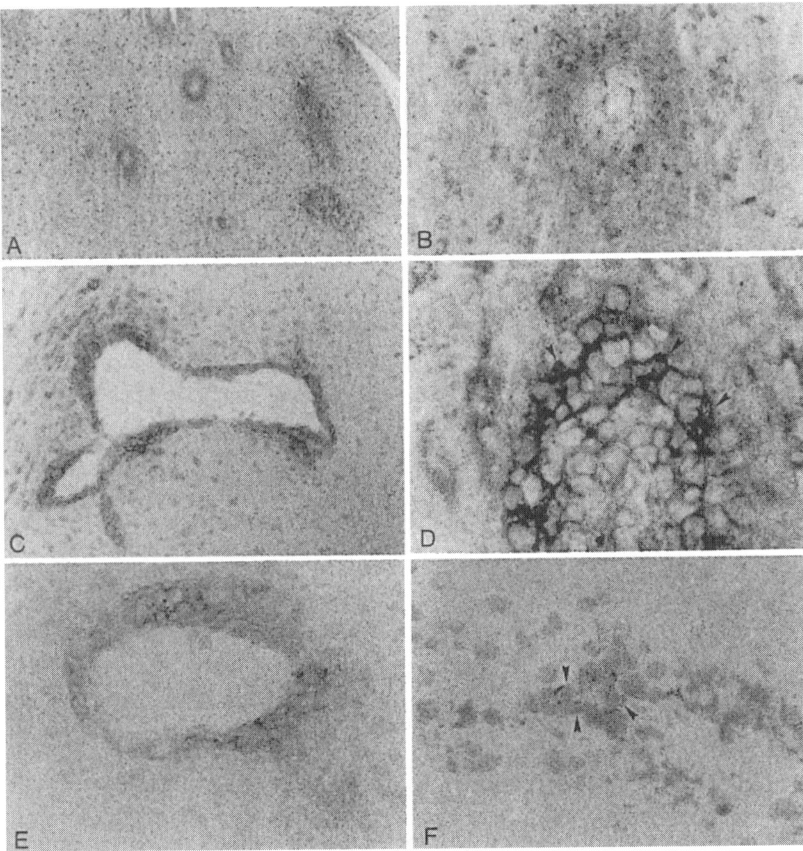

*Figure 2:* *Expression of accessory molecules and cytokines in marmoset EAE brain. (a) Acid phosphatase (red) expressed by infiltrating macrophages and activated microglia (magnification = x40); (b) Detail of (a), showing extensive perivascular infiltration by macrophages (red)(x200); (c) CD40 (red) expressed on infiltrating mononuclear cells (x100); (d) Doublestaining showing macrophages with cytoplasmic acid phosphatase activity (red) and membrane CD40 expression (blue). Arrowheads indicate examples of macrophages expressing CD40 (x650); (e) CD40L (red) expressed on perivascular mononuclear cells (x200); (f) Scattered CD40L-expressing cells (red) within an infiltrate (x325). Figure reproduced with permission, from ref 17.*

The prevalence of MOG-specific antibodies in areas of myelin disruption suggested a causal relationship between anti-MOG humoral responses and lesion formation.[20] Furthermore, the induction of demyelination and clinical signs in the common marmoset were dependent on the presence of anti-MOG antibodies.[20,21] Fine specificity of anti-MOG antibody responses were characterised by testing necropsy sera against 23-mer peptides encompassing the MOG extracellular Ig-like domain. Robust antibody responses to conformational epitopes within the Ig-like domain of MOG, but minor antibody reactivity to MOG peptides was present following myelin-induced EAE. Occasional responses to recombinant human MBP, PLP (purified from human brain), or $\alpha$B-crystallin were also observed.

## 1.3     MOG-induced EAE

To further elucidate the contribution of MOG-specific autoreactivity during EAE common marmosets were immunised with rhMOG[Igd]. Similar to myelin-induced EAE, rhMOG[Igd] immunisation resulted in complete susceptibility to chronic-progressive EAE (Figure 3)

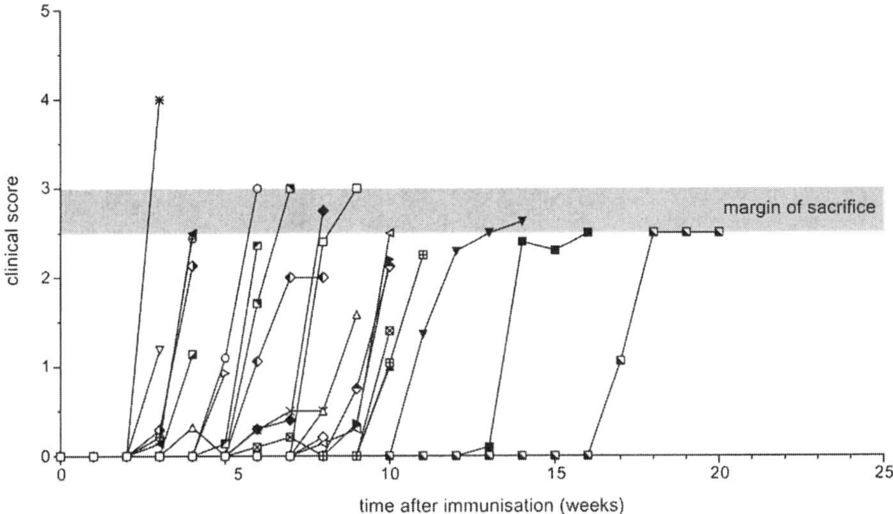


*Figure 3:* Clinical heterogeneity in rhMOG-immunised marmosets. Twenty-six marmosets (2-5 years old) were immunised with 100 mg rhMOG in CFA. The clinical scores represent: 0.5=apathy, loss of appetite; 1.0=lethargy, anorexia, flaccid tail; 2.0=ataxia, sensory loss,, blindness; 2.5=incomplete paralysis one-(hemi-) or two-sided (paraparesis); 3.0=complete paralysis hind part of the body (hemi-/paraplegia); 4.0=complete paralysis of whole body (quadriplegia). Figure reproduced with permission, from ref 2.

This 100% susceptibility within the outbred population was found to map to the monomorphic class II allele *Caja-DRB*W1201*.[9,16] This allele was found to be present in all monkeys from four different outbred populations and the generation of T-cell lines from unrelated animals demonstrated diverse patterns of T-cell reactivities, however the epitope pMOG$_{24-36}$ was common to all. Subsequent MHC isotype blocking and cross-presentation studies revealed that the *Caja-DRB*W1201* was the restriction element for the activation of CD4$^+$ pMOG$_{14-36}$-specific T-cells.[16] This conclusion was further supported by the observation that B-cell lines from the evolutionary ancestor of *Callithrix jacchus*, *Callithrix penicillata*, could present the pMOG$_{14-36}$ peptide in a recognisable fashion, as the *Caja-DRB*W12* locus in this species also contains the *–DRB*W1201* allele. B-cell lines from the more distantly related cotton-top tamarin (*Saguinus oedipus*), which lacks the *Caja-DRB*W1201* homologue, failed to stimulate peptide specific T-cell lines. Importantly, the marmoset *Caja*-DR is equivalent to the human HLA-DR, which has been shown to exert a strong genetic influence on MS susceptibility.[22]

Adoptive transfer of CD4$^+$ pMOG$_{14\text{-}36}$ T-cell lines into naïve MHC compatible recipients induced early pathological signs of EAE. [23] Interestingly, pMOG$_{14\text{-}36}$ induced EAE was characterised by only mild clinical disease, inflammatory oedema and perivascular infiltrates of MNC but an absence of demyelination. Whilst serum antibodies from pMOG$_{14\text{-}36}$ immunised marmosets recognised linear epitopes of MOG they were unable to bind to the protein expressed within a MOG-transfected cell line (unpublished data). The onset of clinical disease following rhMOG$^{Igd}$ immunisation correlated with broadening of the T-cell repertoire to numerous MOG epitopes. No such association could be demonstrated between the presence of autoantibodies and disease onset (unpublished data).

## 2    MAGNETIC RESONANCE IMAGING

As lesions of both MS and EAE are localised to the CNS *in vivo* characterisation requires the use of non-invasive imaging techniques. To date, the diagnostic imaging modality of choice is magnetic resonance (MR) imaging (MRI).[24,25] The most frequently used imaging parameters are T1-weighting (T1W) with gadolinium-diethylenetriamine-penta-acetic acid (Gd-DTPA) contrast enhancement and T2-weighting (T2W). Inflammatory active demyelinating brain lesions can be visualised following intravenously injected Gd-DTPA extravasation through the disrupted ('leaky') blood-brain barrier (BBB), giving rise to a focal increase of the T1W NMR signal intensity. T2W images reveal alterations in structural alteration of CNS tissues with all lesion stages, from early inflammation to late chronic stages, characterised as hyperintense regions. Quantification of T2W MRI positive lesions equalled those found by pathological analysis and thus T2W can be utilised to discern spatial distribution and total lesion load.[26] Lesions with contrast-induced regional signal enhancement correlated reasonably well with disease activity both in MS[27-30] and EAE.[31] Essentially the same MRI techniques were used for visualisation and initial characterization of brain white matter lesions in the marmoset EAE model.[14,32]

By utilising NMR-spectroscopy (MRS) markers for inflammation, such as lactate, myelin degradation (choline, inositol), or axonal degradation (N-acetyl-aspartate) can be measured in individual lesions. Through a combination of MRI and MRS analysis a reasonable impression of the neuropathological aspect of a lesion can be formed, although very little is still known about the MRI correlates to distinct pathophysiological processes that give rise to lesion formation.

## 3.    PATHOLOGY

Pathological examination of common marmosets with EAE revealed the presence of inflammatory lesions with predilection sites in the cerebral white matter and spinal cord. In contrast, the cerebellum and medulla oblongata remained relatively spared with only occasional perivascular lesions present.[14] Large confluent plaques containing lymphocytes and macrophages were present within the cerebrum. The maturation stage of lesions were assessed using previously published criteria of MS lesions.[33] Based upon the presence of inflammatory T-cells, myelin degradation products in phagocytic macrophages, as well as astrocytic scar formation, all animals presented active as well as inactive lesions. Furthermore, immunoglobulin deposition was present in both active and inactive lesions and this co-localised with complement factor C9 in demyelinating lesions.[14,34] These data suggested that Ig- and complement-dependent mechanisms played an active role in the destruction of myelin sheaths, as described in MS.[35,36] Remyelinated lesions, or shadow plaques due to the presence of very thin LFB-positive myelin sheaths, were observed in inactive, early and late actively demyelinating lesions.

## 3.1    Axonal damage

In rhMOG-immunised marmosets a substantial lesion load can occasionally be observed within the brain of animals without evidence of neurological dysfunction. A similar discrepancy between clinical and pathological features has also been reported in MS patients.[37] Recent attention has highlighted the significant contribution of axonal pathology to the neurological dysfunction in MS patients, even at early stages of the disease.[38,39] Either staining by conventional silver staining or monoclonal antibodies directed against two markers of early axonal pathology in MS, namely human β-amyloid precursor protein (β-APP) and unphosphorylated neurofiliaments (SMI-32) revealed axonal suffering in demyelinated lesions.[14,40] Early active lesions showed β-APP accumilation and staining with SMI-32. These markers of early axonal pathology (suffering) were absent in late active lesions. Finally, silver staining of old chronic lesions revealed marked axonal destruction.

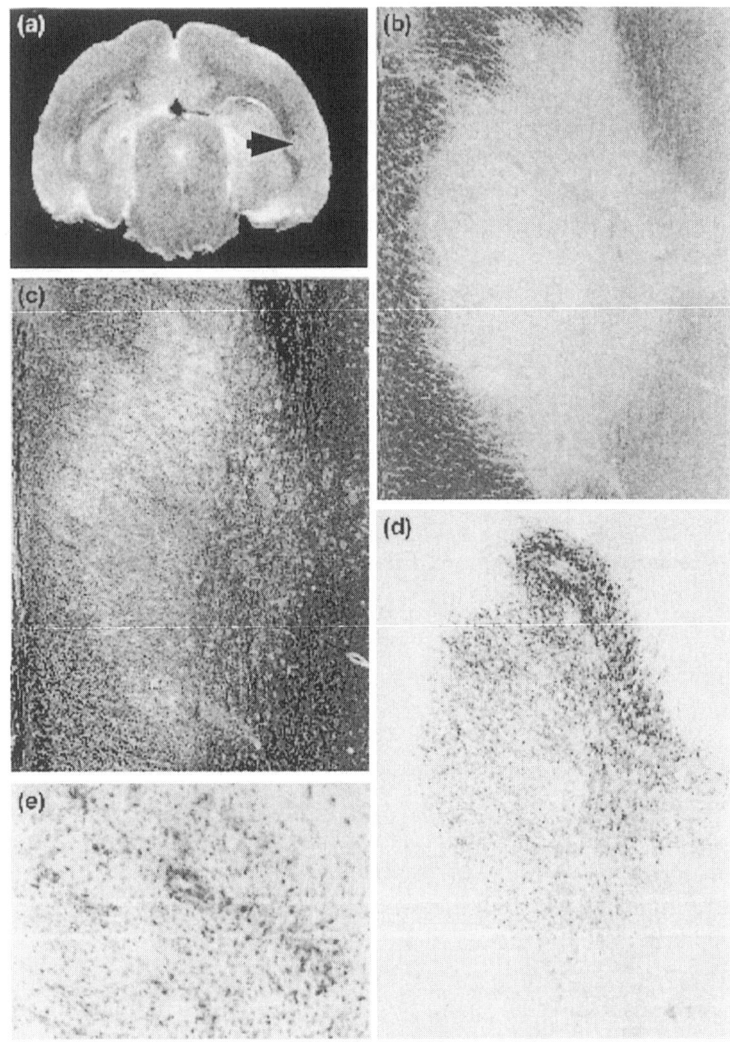

*Figure 4:* Histopathology of a magnetic resonance imaging (MRI)-detectable lesion in common marmosets. (a) coronal slice of a T2-weighted MRI recorded post mortem shows at least three MRI-detectable lesions (magnification = x2.5). The lesion indicated with the arrowhead was processed for histology. (b) Klüver-Barrera staining visualises myelin in blue, showing strong demyelination (x99). Within the same lesion axonal conservation was confirmed by Bielschowsky silver impregnation (x 99). The presence of numerous 27E10-positive macrophages (orange) (d) and $CD3^+$ T-cells (brown) (e) classifies the lesion as late-active (x246). Figure reproduced with permission, from ref 14.

## 3.2    Pre-clinical model.

As outlined early, the disappointing translation of therapeutic candidate molecules from rodent EAE models to MS have highlighted the necessity to establish model systems with greater genetic and immunological proximity to humans. The common marmoset EAE model facilitates the investigation into the relationship between neurological deficits and pathological processes. The ability to perform serial blood sampling, quantitative MRI and histopathological techniques, which are comparable to the human MS situation, means that the autoimmune common marmoset EAE model provides an excellent test system for new therapeutics.

Previously we have demonstrated the therapeutic potential of a chimeric anti-CD40 antibody (5D12) therapy[41] as well as the protective effect of a fully human anti-IL12p40 antibody (CNTO1275) which neutralised interleukin (IL)-12/23.[18] More recently, these antibodies were shown to suppress the activity and enlargement of existing brain lesions (unpublished data).

## 4.    CONCLUDING REMARKS

Due to their genetic and immunological proximity to humans, non-human primates provide an important bridge between rodent-based research and the human disease. The two most investigated models are EAE in rhesus monkeys and common marmosets. As myelin or myelin antigen-induced EAE in the rhesus monkey resembles acute fulminant forms of MS its pre-clinical value remains limited. However, EAE induced in the common marmoset provides a reproducible model of chronic with several immunological advantages (Table 1). Their small size and relative ease of breeding and handling reduce the amount of therapeutic agent needed. Marmoset MHC typing and TCR characterisation permit epitope specificity to be described in detail. Widespread primary demyelination, the hallmark of MS, occurs in the marmoset CNS. Finally, the immune tolerance among chimeric twins allows adoptive transfer experiments assessing relative contributions of T- and B-cells to pathogenesis.

**ACKNOWLEDGMENTS**

This work was supported in part by the Dutch Foundation for Support of MS Research, The Netherlands Organization for Scientific Research (NOW), the Dutch Cancer Foundation (NKB) and the MS Society of the UK and Northern Ireland

# REFERENCE LIST

1.      Mestas J. and Hughes CC. 2004. Of Mice and Not Men: Differences between Mouse and Human Immunology. *J Immunol* **172**(5): 2731-2738.

2.      Brok HP, Bauer J, Jonker M, Blezer E, Amor S, Bontrop RE, Laman JD, 't Hart BA. 2001. Non-human primate models of multiple sclerosis. *Immunol Rev.* **183**:173-185.

3.      Quint DJ, Buckham SP, Bolton EJ, Solari R, Champion BR, Zanders ED. 1990. Immunoregulation in the common marmoset, Calithrix jaccus: functional properties of T and B lymphocytes and their response to human interleukins 2 and 4. *Immunology.* **69**(4):616-621.

4.      Neubert R, Foerster M, Nogueira AC, Helge H. 1996. Cross-reactivity of antihuman monoclonal antibodies with cell surface receptors in the common marmoset. *Life Sci.* **58**(4):317-324.

5.      Foerster M, Delgado I, Abraham K, Gerstmayr S, Neubert R. 1997. Comparative study on age-dependent development of surface receptors on peripheral blood lymphocytes in children and young nonhuman primates (marmosets). *Life Sci.* **60**(10):773-785.

6.      Brok HP, Hornby RJ, Griffiths GD, Scott LA, 't Hart BA. 2001. An extensive monoclonal antibody panel for the phenotyping of leukocyte subsets in the common marmoset and the cotton-top tamarin. *Cytometry.* **45**(4):294-303.

7.      Uccelli A, Oksenberg J, Jeong M, Genain C, Rombos T, Jaeger E, Giunti D, Lanchbury J, Hauser S. 1997. Characterization of the TCRB chain repertoire in the New World monkey Callithrix jacchus. *J Immunol.* **158**(3):1201-1207.

8.      Villinger F, Bostik P, Mayne A, King CL, Genain CP, Weiss WR, Ansari AA. 2001. Cloning, sequencing, and homology analysis of nonhuman primate Fas/Fas-ligand and co-stimulatory molecules. *Immunogenetics.* **53**(4):315-328.

9.      Antunes SG, de Groot NG, Brok H, Doxiadis G, Menezes AA, Otting N, Bontrop RE. 1998. The common marmoset: a new world primate species with limited Mhc class II variability. *Proc Natl Acad Sci U S A.* **95**(20):11745-11750.

10.     Bontrop RE, Otting N, de Groot NG, Doxiadis GG. 1999. Major histocompatibility complex class II polymorphisms in primates. *Immunol Rev.* **167**:339-350.

11.     von Budingen HC, Tanuma N, Villoslada P, Ouallet JC, Hauser SL, Genain CP. 2001. Immune responses against the myelin/oligodendrocyte glycoprotein in experimental autoimmune demyelination. *J Clin Immunol.* **21**(3):155-170.

12.     Della Gaspera B, Delarasse C, Lu , Rodriquez D, Lachapelle F, Genain C, Dautigny A, Pham-Dinh D. (2001) Alternative splicing of the MOG gene across species. *J. Neuroimmunol.* **118**: A259.

13.     Mesleh MF, Belmar N, Lu CW, Krishnan VV, Maxwell RS, Genain CP, Cosman M. 2002. Marmoset fine B cell and T cell epitope specificities mapped onto a homology model of the extracellular domain of human myelin oligodendrocyte glycoprotein. *Neurobiol Dis.* **9**(2):160-172.

14.     't Hart BA, Bauer J, Muller HJ, Melchers B, Nicolay K, Brok H, Bontrop RE, Lassmann H, Massacesi L. 1998. Histopathological characterization of magnetic resonance imaging-detectable brain white matter lesions in a primate model of multiple sclerosis: a correlative study in the experimental autoimmune encephalomyelitis model in common marmosets (Callithrix jacchus). *Am J Pathol.* **153**(2):649-663

15.     't Hart BA, van Meurs M, Brok HP, Massacesi L, Bauer J, Boon L, Bontrop RE, Laman JD. 2000. A new primate model for multiple sclerosis in the common marmoset. *Immunol Today.* **21**(6):290-297.

16.     Brok HP, Uccelli A, Kerlero De Rosbo N, Bontrop RE, Roccatagliata L, de Groot NG, Capello E, Laman JD, Nicolay K, Mancardi GL, Ben-Nun A, 't Hart BA. 2000. Myelin/oligodendrocyte glycoprotein-induced autoimmune encephalomyelitis in common marmosets: the encephalitogenic T cell epitope pMOG24-36 is presented by a monomorphic MHC class II molecule. *J Immunol.* **165**(2):1093-1101

17.     Laman JD, van Meurs M, Schellekens MM, de Boer M, Melchers B, Massacesi L, Lassmann H, Claassen E, 't Hart BA. 1998. Expression of accessory molecules and cytokines in acute EAE in marmoset monkeys (*Callithrix jacchus*). *J Neuroimmunol* **86**(1): 30-45.

18.     Brok HP, van Meurs M, Blezer E, Schantz A, Peritt D, Treacy G, Laman JD, Bauer J, 't Hart BA. 2002. Prevention of Experimental Autoimmune Encephalomyelitis in Common Marmosets Using an Anti-IL-12p40 Monoclonal Antibody. *J Immunol* **169**(11): 6554-6563.

19.     Laman JD, 't Hart BA, Brok H, Meurs M, Schellekens MM, Kasran A, Boon L, Bauer J, Boer M, Ceuppens J. 2002. Protection of marmoset monkeys against EAE by treatment with a murine antibody blocking CD40 (mu5D12). *Eur J Immunol* **32**(8): 2218-2228.

20.   Genain CP, Nguyen MH, Letvin NL, Pearl R, Davis RL, Adelman M, Lees MB, Linington C, Hauser SL. 1995. Antibody facilitation of multiple sclerosis-like lesions in a nonhuman primate. *J Clin Invest.* **96**(6):2966-2974.

21.   McFarland H, Lobito AA, Johnson MM, Nyswaner JT, Frank JA, Palardy GR, Tresser N, Genain CP, Mueller JP, Matis LA, Lenardo MJ. 1999. Determinant spreading associated with demyelination in a nonhuman primate model of multiple sclerosis. *J Immunol.* **162**(4):2384-90.

22.   Dyment DA, Ebers GC, Sadovnick AD. 2004. Genetics of multiple sclerosis. *Lancet Neurol* **3**(2): 104-110.

23.   Villoslada P, Abel K, Heald N, Goertsches R, Hauser SL, Genain CP. 2001. Frequency, heterogeneity and encephalitogenicity of T cells specific for myelin oligodendrocyte glycoprotein in naive outbred primates. *Eur J Immunol.* **31**(10):2942-2950.

24.   Thorpe JW, Barker GJ, MacManus DG, Moseley IF, Tofts PS, Miller DH. 1994. Detection of multiple sclerosis by magnetic resonance imaging. *Lancet* **344**(8931):1235.

25.   Young IR, Hall AS, Pallis CA, Legg NJ, Bydder GM, Steiner RE. 1981. Nuclear magnetic resonance imaging of the brain in multiple sclerosis. *Lancet.* **2**(8255):1063-1066.

26.   Harris JO, Frank JA, Patronas N, McFarlin DE, McFarland HF. 1991. Serial gadolinium-enhanced magnetic resonance imaging scans in patients with early, relapsing-remitting multiple sclerosis: implications for clinical trials and natural history. *Ann Neurol.* **29**(5):548-55.

27.   Thompson AJ, Kermode AG, Wicks D, MacManus DG, Kendall BE, Kingsley DP, McDonald WI. 1991. Major differences in the dynamics of primary and secondary progressive multiple sclerosis. *Ann Neurol.* **29**(1):53-62.

28.   Barkhof F, Scheltens P, Frequin ST, Nauta JJ, Tas MW, Valk J, Hommes OR. 1992. Relapsing-remitting multiple sclerosis: sequential enhanced MR imaging vs clinical findings in determining disease activity. *AJR Am J Roentgenol.* **159**(5):1041-1047.

29.   van Walderveen MA, Barkhof F, Hommes OR, Polman CH, Tobi H, Frequin ST, Valk J. 1995. Correlating MRI and clinical disease activity in multiple sclerosis: relevance of hypointense lesions on short-TR/short-TE (T1-weighted) spin-echo images. *Neurology.* **45**(9):1684-1690.

30.   Hawkins CP, Munro PM, MacKenzie F, Kesselring J, Tofts PS, du Boulay EP, Landon DN, McDonald WI. 1990. Duration and selectivity of

blood-brain barrier breakdown in chronic relapsing experimental allergic encephalomyelitis studied by gadolinium-DTPA and protein markers. *Brain.* **113** (Pt 2):365-378.

31.    Morrissey SP, Stodal H, Zettl U, Simonis C, Jung S, Kiefer R, Lassmann H, Hartung HP, Haase A, Toyka KV. 1996. In vivo MRI and its histological correlates in acute adoptive transfer experimental allergic encephalomyelitis. Quantification of inflammation and oedema. *Brain.* **119** (Pt 1):239-248.

32.    't Hart BA, Vogels J, Bauer J, Brok HP, Blezer E. 2004. Non-invasive measurement of brain damage in a primate model of multiple sclerosis. *Trends Mol. Med.* **10**(2): 85-91.

33.    Bruck W, Porada P, Poser S, Rieckmann P, Hanefeld F, Kretzschmar HA, Lassmann H. 1995. Monocyte/macrophage differentiation in early multiple sclerosis lesions. *Ann Neurol.* **38**(5):788-796.

34.    van Beek, van Meurs et al. (2004). Decay-accelerating factor (CD55) is expressed by neurons in response to chronic but not acute autoimmune CNS inflammation associated with complement activation. *Submitted for publication.*

35.    Lucchinetti CF, Bruck W, Rodriguez M, Lassmann H. 1996. Distinct patterns of multiple sclerosis pathology indicates heterogeneity on pathogenesis. *Brain Pathol.* **6**(3):259-274.

36.    Storch MK, Stefferl A, Brehm U, Weissert R, Wallstrom E, Kerschensteiner M, Olsson T, Linington C, Lassmann H. 1998. Autoimmunity to myelin oligodendrocyte glycoprotein in rats mimics the spectrum of multiple sclerosis pathology. *Brain Pathol.* **8**(4):681-694.

37.    Filippi M, Rocca MA, Rovaris M. 2002. Clinical trials and clinical practice in multiple sclerosis: conventional and emerging magnetic resonance imaging technologies. *Curr Neurol Neurosci Rep.* **2**(3):267-276.

38.    Ferguson B, Matyszak MK, Esiri MM, Perry VH. 1997. Axonal damage in acute multiple sclerosis lesions. *Brain.* **120** (Pt 3):393-399.

39.    Trapp BD, Ransohoff R, Rudick R. 1999. Axonal pathology in multiple sclerosis: relationship to neurologic disability. *Curr Opin Neurol.* **12**(3):295-302.

40.    Mancardi G, Hart B, Roccatagliata L, Brok H, Giunti D, Bontrop R, Massacesi L, Capello E, Uccelli A. 2001. Demyelination and axonal damage in a non-human primate model of multiple sclerosis. *J Neurol Sci* **184**(1): 41-49.

41.     Boon L, Brok HP, Bauer J, Ortiz-Buijsse A, Schellekens MM, Ramdien-Murli S, Blezer E, van Meurs M, Ceuppens J, de Boer M, 't Hart BA, Laman JD. 2001. Prevention of experimental autoimmune encephalomyelitis in the common marmoset (Callithrix jacchus) using a chimeric antagonist monoclonal antibody against human CD40 is associated with altered B cell responses. *J Immunol.* **167**(5):2942-2949.

Part B

THEILER'S MURINE ENCEPHALOMYELITIS
VIRUS (TMEV) – INDUCED DEMYELINATION

Chapter B1

# HISTOPATHOLOGY IN THE THEILER'S VIRUS MODEL OF DEMYELINATION


Ure, D. R. and Rodriguez, M.
*Deparments of Neurology and Immunology, Mayo Medical and Graduate School, Rochester, MN*



Abstract:    The Theiler's virus model shows all the major features of MS pathology - demyelination, inflammation, axonal injury, and remyelination. Extensive qualitative and quantitative characterization of the pathology has provided important insights into disease mechanisms, particularly how they relate to neurological function.

Key words:    Theiler's virus, histopathology, demyelination, multiple sclerosis


Among the viral models of MS, one of the best characterized is inflammatory demyelination in the mouse CNS induced by Theiler's murine encephalomyelitis virus (TMEV). A primary reason why the TMEV model is relevant to MS is the similarities in pathology between the two diseases. Both are characterized by inflammation, demyelination, axonal injury, and remyelination to varying degrees. Each of these pathological features varies widely and somewhat independently in the TMEV model, largely as a result of mouse strain. This chapter will focus primarily on the spinal cord pathology in the SJL mouse strain following infection with the Daniel's strain of TMEV. SJL mice develop inflammatory demyelination in the spinal cord and brainstem starting at 3 weeks postinfection (pi). It persists for the life of the mouse, accompanied by extensive axonal injury, minimal remyelination, and progressive accumulation of neurological deficits – a model of chronic, progressive MS. This disease differs in important ways from that induced by the closely related BeAn strain of virus, which includes infected cell types, pathology, and disease course. Dal Canto, Lipton, and colleagues provided many of the early descriptions of TMEV pathology. In the past decade our laboratory has extended these observations, but more importantly we have analyzed each of the major pathological features quantitatively and related them to motor dysfunction.

## INFLAMMATORY DEMYELINATION

Inflammation is the earliest pathologic event in spinal cords from infected mice, only slightly preceding demyelination. Inflammation comprises the infiltration of blood-derived cells, particularly T and B lymphocytes and monocytes, and activation of resident microglia and astrocytes. Monocyte-derived macrophages and activated microglia account for over half of the inflammatory cells in spinal cords, as determined by fluorescence activated cell sorting (FACS) of isolated, inflammatory cells. They are nearly indistinguishable from each other morphologically and phenotypically and are widely distributed throughout the white matter where virus persists. T and B lymphocytes, including plasma cells, each account for approximately a quarter of the isolated cells. Within the T cell population, CD4 cells outnumber CD8 cells approximately 3:1. B cells and CD4 cells reside predominantly around blood vessels, forming perivascular cuffs, whereas CD8 cells distribute widely throughout the parenchyma. Meningeal inflammation also is common. The relative abundance of each cell type changes little over the course of disease.

The onset of demyelination coincides with the inflammatory response. Electron and light microscopy of plastic-embedded sections show several characteristics of demyelination in common among all mouse strains. Demyelination and inflammation co-localize to areas of virus expression, as shown by in situ hybridization and immunohistochemistry for TMEV. Demyelination is not obviously centered on blood vessels in chronic disease. Normally myelinated axons often are seen adjacent to perivascular cuffs, suggesting that the perivascular cells do not ubiquitously injure myelin. Demyelinated lesions develop only rostral to the second lumbar level, concentrating in the thoracic cord and secondarily in the cervical cord. Ventral and lateral columns are consistently lesioned, whereas the dorsal column is more variably affected. The ventral-most aspect of the dorsal column is the only area where lesions rarely develop. Demyelination usually begins at the cord periphery, but during advanced disease may extend the entire width of the white matter.

Demyelination first appears as myelin splitting, vesiculation, or thickening of the myelin lamellae associated with axon collapse. Swelling and vesiculation of inner glial loops, the most distal oligodendrocyte processes, also are seen early and support the hypothesis that demyelination represents a dying-back oligodendrogliopathy (1,2). Myelin figures are abundant within phagocytes during early demyelination. Myelin "stripping" by macrophages has been reported (3), but it is not clear whether macrophages initiate attack on completely normal myelin or alternatively that macrophages are responding to molecular changes that target the myelin for degradation. Often early stage lesions have a patchy appearance, in which degenerating myelin, phagocytes, and other inflammatory cells are interspersed among normal axon-myelin profiles. In other lesions myelin degeneration occurs uniformly in a region, which results in demarcation from normal white matter. In some mouse strains the lesions develop a highly vacuolated, spongiform appearance. Early stage lesions are found at any time postinfection, suggesting that the pathogenic mechanisms responsible for demyelination are ongoing throughout disease.

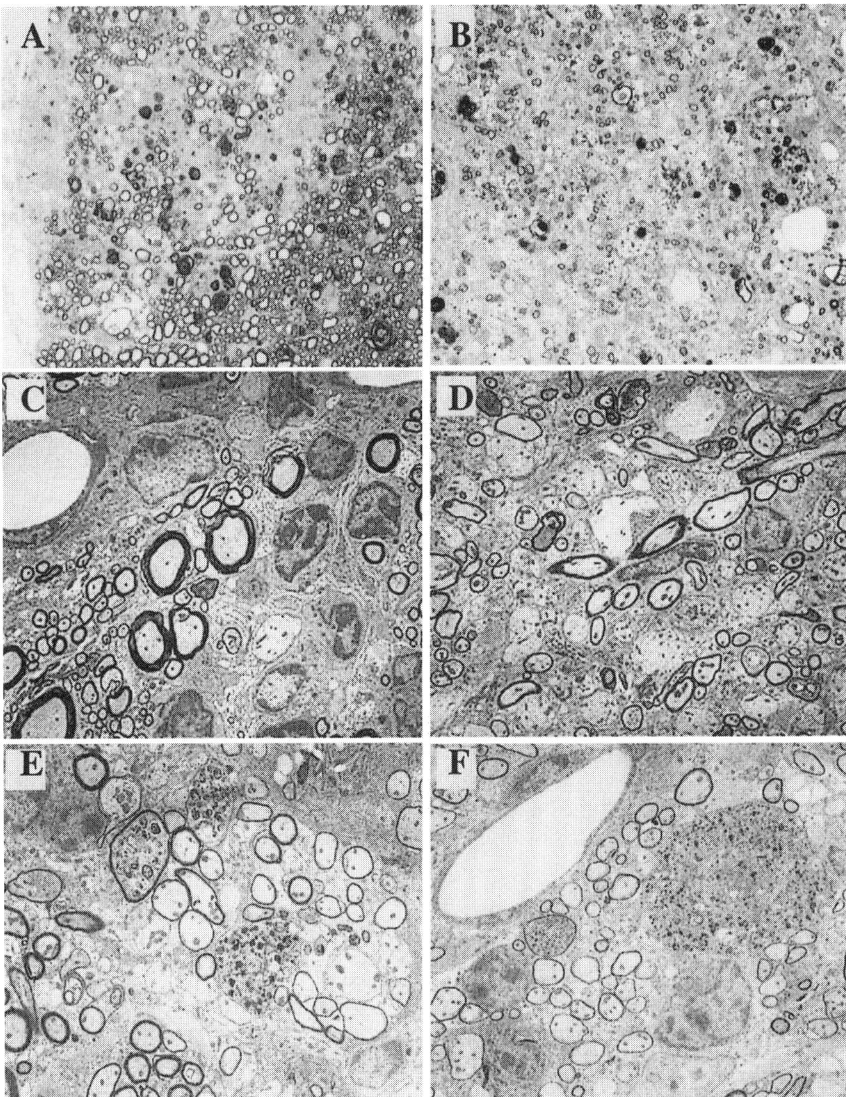

*Figure 1.* SJL spinal cord lesions. (A) Early stage lesion showing patchy demyelination ; numerous phagocytic cells with myelin debris. (B) Late stage lesion showing inflammati nearly complete demyelination, and many vacuolated cells. (C) Perivascular inflammat with partial preservation of myelin. (D) Chronic demyelination with minimal remyelinati (E) and (F) Axonal injury indicated by large accumulations of abnormal organelles demyelinated, partially remyelinated, and sometimes swollen axons.

Morphometric data indicate that the proportion of white matter with inflammatory demyelination in SJL mice reaches a plateau by 3 months pi, but the cord continues to atrophy beyond this time (4). Late stage, or chronic lesions take on a "cleaned out" appearance in which the center of the lesion is devoid of myelin figures, being replaced by cells filled with lipid vacuoles. The vacuolated cells are more common in SJL mice than many other strains. Some likely are foamy macrophages and microglia containing end-products of myelin degradation, but others likely are oligodendrocytes (5,6). Inflammatory cells and glia persist in chronic lesions, and axons are either completely demyelinated or remyelinated to varying degrees.

Demyelination is quantified in our lab morphometrically by camera lucida tracings of lesions on a digital tablet (7). Ten cross-sections spanning the second cervical to second lumbar level of each spinal cord are analyzed, and lesion areas are expressed as a percentage of total white matter area (lesion load). Beyond 3 months pi lesions occupy 5-10 % of the total white matter. This percentage seems low, but in fact the proportion of axons that course through lesions at least at some point in the cord is much higher – at least 30%. This value represents the fraction of white matter area that is lesioned in the single cross-section with the most expansive lesions from each cord (8). Lesion tracing also reveals that most lesions extend several millimeters rostro-caudally. This does not necessarily imply that axons completely demyelinated over this distance, as some remyelination occurs. However, the probability is high that conduction in every axon in the lesion is affected at least somewhere along its length.

Myelin degeneration probably occurs in many ways. It may reflect a direct effect of toxic factors such as proteases, cytokines, complement, or free radicals upon the myelin sheath, with sparing of oligodendrocyte cell bodies. A second possibility is secondary dying-back of myelin following a primary insult to the oligodendrocyte cell body. Oligodendrocyte cell bodies show degenerative pathology in lesions, consisting of swelling, vacuolation, and the presence of degenerating, electron-dense organelles. Reductions in proteolipid protein mRNA expression occur well in advance of any morphological change and probably contribute to myelin disruption (9,10). Some demyelination also must be associated with Wallerian degeneration, as axonal loss occurs in chronically diseased mice.

One mechanism of oligodendrocyte death and myelin disruption is direct, virus-induced cytolysis or apoptosis, although its extent in immunocompetent mice is not clear. The Daniel's strain of TMEV kills oligodendrocytes in culture (5,11). It also infects oligodendrocytes in vivo, including cell bodies, inner and outer glial loops, and Schmidt-Lanterman incisures, and colocalizes to degenerating oligodendrocytes (2,12-14). A direct role for virus in causing demyelination is supported by infection of athymic, nude mice, which lack most T lymphocytes. Infected and degenerating oligodendrocytes and demyelination are present in these mice, although the overall extent of demyelination is relatively small (15,16). In T and B lymphocyte-deficient RAG -/- (recombinase activating gene) mice we have not observed demyelination after direct inoculation of the brain or spinal cord, although 3-week survival of these mice may be insufficient to realize the full effects of oligodendrocyte infection. Virus persistence and neuronal necrosis in the gray matter in nude and RAG-/- mice highlight an important role of the immune system in protecting neurons from infection.

The immune system plays a prominent role in demyelination insofar as many experimental manipulations such as drug or antibody treatment, adoptive cell transfer, or genetic knockout influences pathology and disease course. One nice example is the adoptive transfer of splenocytes into immunodeficient SCID mice, which normally succumb to encephalitis within 3 weeks of infection (17). Transfer of too few cells does not protect against death, and too many cells leads to nearly complete virus clearance and little pathology. However, transfer of intermediate numbers of cells results in virus persistence and demyelination, an independent effect of both CD4 and CD8 lymphocytes. The tenuous balance of TMEV immunity and immunopathology illustrated by this study is the reason why the contributions of specific immune mechanisms are difficult to discern. Manipulations designed to reduce immunopathology also impact antiviral immunity and alter viral pathogenesis. For example, cyclophosphamide depletes T lymphocytes and reduces the extent of demyelination when administered at an early stage of disease (e.g. 1-wk pi) but kills many mice both early and late (e.g. 6-months pi), presumably due to viral recrudescence (18-20). Similarly, antibody-mediated depletion of CD4 cells early in disease and genetic knockout of CD4 also cause high viral titers and mortality. Depletion of CD4 cells late in disease reduces clinical symptoms but does not affect demyelination (20-23).

In mice that normally do not develop demyelination, elimination of Class I MHC activity through genetic knockout of β2microglobulin (β2m-/-) makes them susceptible to demyelination. Interestingly, the converted mice develop CD4-mediated DTH responses and demyelination equivalent to SJL mice but no obvious disease symptoms (8,24-26). On the SJL background β2microglobulin knockout results in more demyelination than in wildtype animals but no greater incidence of disease (27). CD8-/- mice on both resistant and susceptible backgrounds also exhibit increased demyelination compared to wildtype but little or no deficit (23). Virus titers are elevated in β2m-/- and CD8-/- strains (23,27,28). Therefore, both the Class II MHC-CD4 and Class I MHC–CD8 pathways are important in virus control and in turn CNS pathology.

Other treatments that exacerbate pathology in some strains of mice include cobra venom factor, anti-mu antibodies, anti-interferon-gamma (IFNγ) antibodies, and knockout of Vβ loci of the T cell receptor (29,30). These results are consistent with protective, antiviral effects of complement, B lymphocytes, IFNγ, and a diverse T cell repertoire. Other cytokines that reduce demyelination and virus expression early in disease include interleukin-6, tumor necrosis factor-alpha (TNFα), and transforming growth factor-beta2 (TGFβ2) (31-33). In contrast, cyclosporin A and macrophage-depleting agents reduce demyelination, consistent with a deleterious role for interleukin-2 release and macrophage infiltration (34,35). Thus, a wide variety of cells and molecules regulate demyelination in the TMEV model.

## AXONAL INJURY

Ultrastructural studies suggest that axons mostly are preserved in MS lesions as compared to other neurological diseases, but axonal injury does occur variably and is more common than previously believed. Likewise, most axons present in TMEV lesions look normal ultrastructurally, although demyelinated axons contain more mitochondria than normally myelinated axons (36). The abundance of mitochondria in lesions implies a mechanism to focally regulate ATP supply and meet the energetic demands of demyelinated axons. Some axons, however, show clear signs of dysfunction or degeneration, including overcrowding of axoplasm with abnormal organelles, darkened axoplasm, and irregular axonal outlines. Degenerating axons are found in all states of myelination – completely demyelinated, remyelinated, or normally myelinated (36). Completely demyelinated axons may be most susceptible to injury by inflammatory cells or secreted factors such as nitric oxide (37,38). Degenerating, remyelinated or normally myelinated axons probably represent Wallerian degeneration following axonal transection in lesions elsewhere in the white matter. Neurofilament immunohistochemistry and Bielschowsky staining also show axonal disruption.

An EM morphometric analysis of 48,827 axons showed that large-diameter axons are preferentially lost in chronically diseased mice (36). A light microscopy procedure that quantified surviving, normally myelinated axons in parts of the white matter that were free of lesions (e.g. lumbar) verified the dropout of large axons, and secondarily medium axons, (7). Axon loss was present at 3 months pi and increased at 7 months pi despite a constant level of demyelination. The loss translated into atrophy of the white matter columns in the cervical and thoracic cord up to 25% compared to uninfected mice (4). Extrapolation of the axon loss from sample images to the entire cord estimated a 23% reduction in axon number and 37% loss in axon area overall (39).

Another method that documents axonal injury is retrograde neuronal labeling with the fluorescent tracer, Fluoro-Gold. Administration of Fluoro-Gold into the lower thoracic cord results in its accumulation within a few days in cell bodies of all descending neurons in the brain that have intact and functional projections to the site of Fluoro-Gold application. In 9-month-infected mice the number of neurons that accumulate tracer is reduced by 60 – 93% compared to uninfected mice, depending on the neuronal population studied (vestibulospinal, reticulospinal, rubrospinal). Vestibulospinal neurons show the largest reduction, consistent with their projection through ventral columns where lesions are most common. Thus, most descending neurons are transected in chronically diseased mice, or their retrograde uptake and/or trafficking mechanisms are completely disrupted somewhere along the axon length (39).

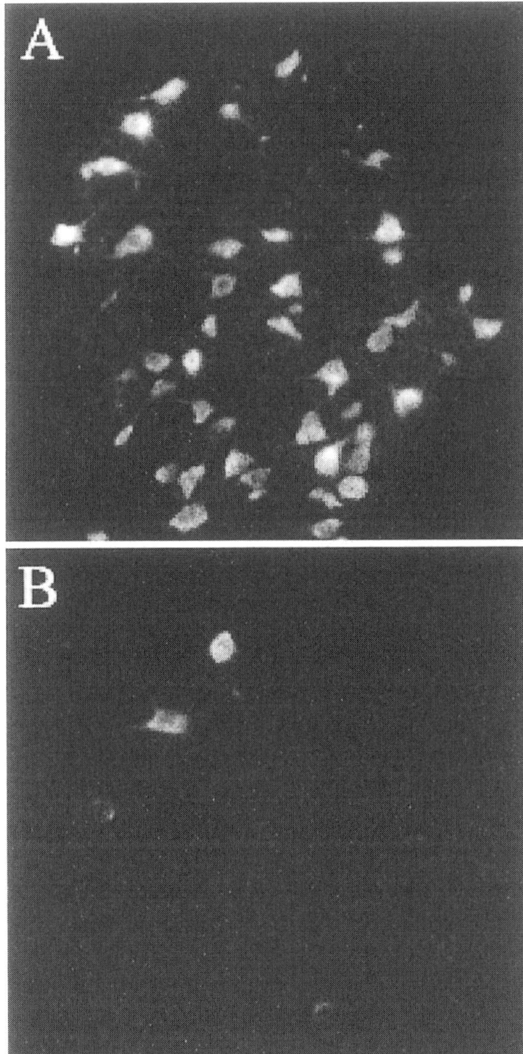

*Figure 2.* Axonal injury shown by retrograde neuronal labeling of vestibulospinal neurons. (A) Uninfected mouse with many labeled cell bodies in the vestibular nucleus. (B) Chronically infected mouse with few labeled neurons due to axon injury in the spinal cord.

The extent of axonal pathology is the major predictor of neurological deficits in the TMEV model. In most SJL mice neurological deficits become obvious not until beyond the time at which lesion load reaches a plateau. Most informative is the finding that motor dysfunction measured by rotarod assay in chronically diseased mice correlates linearly with all measures of axonal injury - spinal cord atrophy (4), dropout of axons as measured by light and electron microscopy (4,36), reduction in retrograde labeling, and remarkably even the number of degenerating axons in electron micrographs from single lesions (36). The importance of axon integrity in maintaining normal function also is exemplified in β2m-/- mice, which develop demyelination but no neurological deficits (26). β2m-/- mice develop similar lesion loads, lesion expanses, and lesion distributions as SJL mice, but show more retrograde labeling indicative of axonal preservation (8). This finding has several important implications. First, it indicates that demyelination and axonal loss can occur independently and at different rates. Secondly, it emphasizes that the quality or extent of CNS pathology can not be inferred in all cases from the clinical presentation of disease or lack thereof. Like β2m-/- mice without obvious deficits, people also develop subclinical demyelination (40). Thirdly, the data suggest that CD8 T cells may specifically injure axons, which is supported by the lack of deficits in CD8-/- mice (23) and by axon transection by CD8 cells in a culture model (41). These findings are relevant to MS because CD8 cells are the major type of lymphocyte in many MS lesions and their presence correlates with axonal pathology (42).

## REMYELINATION

Remyelination improves impulse conduction (43). It also may protect axons by shielding them from inflammatory insults or by providing trophic support, although this hypothesis has not been thoroughly tested. Remyelination in MS patients occurs sporadically and its regulation is poorly understood. As in MS, remyelination in the TMEV model can be mediated by oligodendrocytes or by Schwann cells that migrate into the injured CNS (44). Remyelination by oligodendrocytes, which is most common in the TMEV model, appears as thin, compacted myelin (45) and usually begins at the edge of lesions adjacent to normal myelin. Schwann cell myelin is thicker than oligodendrocyte myelin, and the Schwann cell body juxtaposes to the axon-myelin unit. Schwann cell remyelination usually begins near the periphery of the cord. Prior to remyelination, lesions show nearly complete demyelination and largely are void of degenerative myelin figures (chronic lesion), implying that catabolism of myelin debris or associated events are necessary for remyelination to proceed. This chronology also supports the conclusion that the thin myelin sheaths indeed represent remyelination rather than partial demyelination. A reduction in inflammation also is prognostic of repair, as the best remyelination occurs in areas with the least inflammation. Astrocytic processes often associate with remyelinated axons.

The extent and timing of spontaneous remyelination varies widely among different mouse strains. SJL mice develop little remyelination even up to 1 year pi. Lymphocytes are partly responsible for the poor repair, especially CD4 cells, as their depletion by antibodies or genetic knockout induces remyelination (20). Depleting lymphocytes for the goal of promoting remyelination is problematic, however, because it also elevates virus expression and kills many mice. In MS, treatment with lymphocyte-depleting antibodies reduces inflammation but either has no effect on clinical disease or increases disability and worsens MRI parameters of neuronal pathology

(46,47). Thus, lymphocytes appear to have both protective and destructive properties in demyelinating diseases. Macrophages and microglia also likely have multiple roles.

Remyelination is inducible in SJL mice not only by immunosuppression but also by treatment with certain antibodies (48,49). The first demonstration of this effect was the treatment of diseased mice with serum or polyclonal antibody generated against spinal cord homogenate (50,51). Many remyelination-promoting antibodies of murine and human origin now have been identified and share certain characteristics. All recognize oligodendrocytes at an intermediate developmental stage and all are polyreactive. Some of the antibodies are polyclonal, for

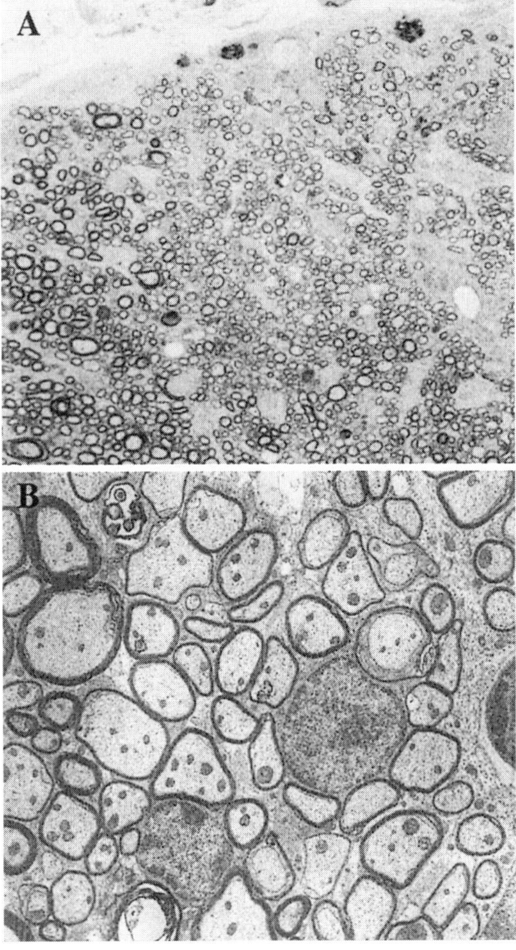

*Figure 3.* Remyelination induced with oligodendrocyte- reactive antibodies in SJL mice. (A) Light microscopy shows nearly complete lesion repair. (B) Electron microscopy shows oligodendrocytes and thin myelin sheaths around most axons.

example recognizing myelin basic protein or copolymer-1 (52,53), whereas others are monoclonal. Two mechanisms are proposed for their activity. Firstly, binding to oligodendrocytes or their progenitors might trigger proliferation or differentiation directly. In support of this hypothesis we have observed calcium fluxes and changes in phosphorylation in oligodendrocytes upon antibody binding (54). Secondly, the antibodies might modulate immune mechanisms sufficiently to release the block on remyelination (55). Induced remyelination in SJL mice has not yet shown clinical benefit; combining remyelination with other immunosuppressive treatments may be necessary. In support, PLJ-CD4-/- mice show late remyelination and improvement in motor function (56).

In conclusion, investigation of the pathology of TMEV disease has yielded invaluable clues about disease mechanisms and potential, reparative treatments. Prominent examples include á role of CD8 T cells in neuronal injury, and the potential for antibodies to promote remyelination. The Theiler's model provides a rare opportunity for rigorous quantitation of pathology in a demyelinating disease.

## REFERENCES


1. **Ludwin SK, Johnson ES.** Evidence for a "dying-back" gliopathy in demyelinating disease. Ann Neurol. 1981;9:301-5.

2. **Rodriguez M.** Virus-induced demyelination in mice: "dying back" of oligodendrocytes. Mayo Clin Proc. 1985;60:433-38.

3. **Dal Canto MC, Lipton HL.** Primary demyelination in Theiler's virus infection. An ultrastructural study. Lab Invest. 1975;33:626-37.

4. **McGavern DB, Murray PD, Rivera-Quinones C, Schmelzer JD, Low PA, Rodriguez M.** Axonal loss results in spinal cord atrophy, electrophysiological abnormalities and neurological deficits following demyelination in a chronic inflammatory model of multiple sclerosis. Brain. 2000;123 Pt 3:519-31.

5. **Rodriguez M, Siegel LM, Hovanec-Burns D, Bologa L, Graves MC.** Theiler's virus-associated antigens on the surfaces of cultured glial cells. Virology. 1988;166:463-74.

6. **Wong K, Armstrong RC, Gyure KA, Morrison AL, Rodriguez D, Matalon R et al.** Foamy cells with oligodendroglial phenotype in childhood ataxia with diffuse central nervous system hypomyelination syndrome. Acta Neuropathol. 2000; 100:635-46.

7. **McGavern DB, Murray PD, Rodriguez M.** Quantitation of spinal cord demyelination, remyelination, atrophy, and axonal loss in a model of progressive neurologic injury. J Neurosci Res. 1999;58:492-504.

8. **Ure DR, Rodriguez M.** Preservation of neurologic function during inflammatory demyelination correlates with axon sparing in a mouse model of multiple sclerosis. Neuroscience. 2002;111:399-411.

9. **Ozden S, Aubert C, Gonzalez-Dunia D, Brahic M.** In situ analysis of proteolipid protein gene transcripts during persistent Theiler's virus infection. J Histochem Cytochem. 1991;39:1305-9.

10. **Rodriguez M, Prayoonwiwat N, Howe C, Sanborn K.** Proteolipid protein gene expression in demyelination and remyelination of the central nervous system: a model for multiple sclerosis. J Neuropathol Exp Neurol. 1994;53:136-43.

11. **Graves MC, Bologa L, Siegel L, Londe H.** Theiler's virus in brain cell cultures: lysis of neurons and oligodendrocytes and persistence in astrocytes and macrophages. J Neurosci Res. 1986;15:491-501.

12. **Rodriguez M, Leibowitz JL, Lampert PW.** Persistent infection of oligodendrocytes in Theiler's virus-induced encephalomyelitis. Ann Neurol. 1983;13:426-33.

13. **Aubert C, Chamorro M, Brahic M.** Identification of Theiler's virus infected cells in the central nervous system of the mouse during demyelinating disease. Microb Pathog. 1987;3:319-26.

14. **Tsunoda I, Kurtz CI, Fujinami RS.** Apoptosis in acute and chronic central nervous system disease induced by Theiler's murine encephalomyelitis virus. Virology. 1997;228:388-93.

15. **Roos RP, Wollmann R.** DA strain of Theiler's murine encephalomyelitis virus induces demyelination in nude mice. Ann Neurol. 1984;15:494-99.

16. **Rosenthal A, Fujinami RS, Lampert PW.** Mechanism of Theiler's virus-induced demyelination in nude mice. Lab Invest. 1986;54:515-22.

17. **Rodriguez M, Pavelko KD, Njenga MK, Logan WC, Wettstein PJ.** The balance between persistent virus infection and immune cells determines demyelination. J Immunol. 1996;157:5699-709.

18. **Lipton HL, Dal Canto MC.** Theiler's virus-induced demyelination: prevention by immunosuppression. Science. 1976;192:62-64.

19. **Roos RP, Firestone S, Wollmann R, Variakojis D, Arnason BG.** The effect of short-term and chronic immunosuppression on Theiler's virus demyelination. J Neuroimmunol. 1982;2:223-34.

20. **Rodriguez M, Lindsley MD.** Immunosuppression promotes CNS remyelination in chronic virus-induced demyelinating disease. Neurology. 1992;42:348-57.

21. **Welsh CJ, Tonks P, Nash AA, Blakemore WF.** The effect of L3T4 T cell depletion on the pathogenesis of Theiler's murine encephalomyelitis virus infection in CBA mice. J Gen Virol. 1987;68:1659-67.

22. **Rodriguez M, Sriram S.** Successful therapy of Theiler's virus-induced demyelination (DA strain) with monoclonal anti-Lyt-2 antibody. J Immunol. 1988;140:2950-2955.

23. **Murray PD, Pavelko KD, Leibowitz J, Lin X, Rodriguez M.** CD4(+) and CD8(+) T cells make discrete contributions to demyelination and neurologic disease in a viral model of multiple sclerosis. J Virol. 1998;72:7320-7329.

24. **Fiette L, Aubert C, Brahic M, Rossi CP.** Theiler's virus infection of beta 2-microglobulin-deficient mice. J Virol. 1993;67:589-92.

25. **Pullen LC, Miller SD, Dal Canto MC, Kim BS.** Class I-deficient resistant mice intracerebrally inoculated with Theiler's virus show an increased T cell response to viral antigens and susceptibility to demyelination. Eur J Immunol. 1993;23:2287-93.

26. **Rivera-Quinones C, McGavern D, Schmelzer JD, Hunter SF, Low PA, Rodriguez M.** Absence of neurological deficits following extensive demyelination in a class I-deficient murine model of multiple sclerosis. Nat Med. 1998;4:187-93.

27. **Begolka WS, Haynes LM, Olson JK, Padilla J, Neville KL, Dal Canto M et al.** CD8-deficient SJL mice display enhanced susceptibility to Theiler's virus infection and increased demyelinating pathology. J Neurovirol. 2001;7:409-20.

28. **Aubagnac S, Brahic M, Bureau JF.** Viral load increases in SJL/J mice persistently infected by Theiler's virus after inactivation of the beta(2)m gene. J Virol. 2001;75:7723-26.

29. **Rodriguez M, Kenny JJ, Thiemann RL, Woloschak GE.** Theiler's virus-induced demyelination in mice immunosuppressed with anti-IgM and in mice expressing the xid gene. Microb Pathog. 1990;8:23-35.

30. **Rodriguez M, Pavelko K, Coffman RL.** Gamma interferon is critical for resistance to Theiler's virus-induced demyelination. J Virol. 1995;69 :7286-90.

31. **Paya CV, Leibson PJ, Patick AK, Rodriguez M.** Inhibition of Theiler's virus-induced demyelination in vivo by tumor necrosis factor alpha. Int Immunol. 1990;2:909-13.

32. **Rodriguez M, Pavelko KD, McKinney CW, Leibowitz JL.** Recombinant human IL-6 suppresses demyelination in a viral model of multiple sclerosis. J Immunol. 1994;153:3811-21.

33. **Drescher KM, Murray PD, Lin X, Carlino JA, Rodriguez M.** TGF-beta 2 reduces demyelination, virus antigen expression, and macrophage recruitment in a viral model of multiple sclerosis. J Immunol. 2000;164:3207-13.

34. **Rodriguez M, Quddus J.** Effect of cyclosporin A, silica quartz dust, and protease inhibitors on virus-induced demyelination. J Neuroimmunol. 1986;13:159-74.

35. **Rossi CP, Delcroix M, Huitinga I, McAllister A, van Rooijen N, Claassen et al.** Role of macrophages during Theiler's virus infection. J Virol. 1997;71:3336-40.

36. **Sathornsumetee S, McGavern DB, Ure DR, Rodriguez M.** Quantitative ultrastructural analysis of a single spinal cord demyelinated lesion predicts total lesion load, axonal loss, and neurological dysfunction in a murine model of multiple sclerosis. Am J Pathol. 2000;157:1365-76.

37. **Redford EJ, Kapoor R, Smith KJ.** Nitric oxide donors reversibly block axonal conduction: demyelinated axons are especially susceptible. Brain. 1997;120:2149-57.

38. **Smith KJ, Kapoor R, Hall SM, Davies M.** Electrically active axons degenerate when exposed to nitric oxide. Ann Neurol. 2001;49:470-476.

39. **Ure D, Rodriguez M.** Extensive injury of descending neurons demonstrated by retrograde labeling in a virus-induced murine model of chronic inflammatory demyelination. J Neuropathol Exp Neurol. 2000;59:664-78.

40. **Engell T.** A clinical patho-anatomical study of clinically silent multiple sclerosis. Acta Neurol Scand. 1989;79 :428-30.

41. **Medana I, Martinic MA, Wekerle H, Neumann H.** Transection of major histocompatibility complex class I-induced neurites by cytotoxic T lymphocytes. Am J Pathol. 2001;159:809-15.

42. **Neumann H, Medana IM, Bauer J, Lassmann H.** Cytotoxic T lymphocytes in autoimmune and degenerative CNS diseases. Trends Neurosci. 2002;25:313-19.

43. **Smith EJ, Blakemore WF, McDonald WI.** Central remyelination restores secure conduction. Nature. 1979;280:395-96.

44. **Dal Canto MC, Barbano RL.** Remyelination during remission in Theiler's virus infection. Am J Pathol. 1984;116:30-45.

45. **Ludwin SK, Maitland M.** Long-term remyelination fails to reconstitute normal thickness of central myelin sheaths. J Neurol Sci. 1984;64:193-98.

46. **Coles AJ, Wing MG, Molyneux P, Paolillo A, Davie CM et al.** Monoclonal antibody treatment exposes three mechanisms underlying the clinical course of multiple sclerosis. Ann Neurol. 1999;46:296-304.

47. **Paolillo A, Coles AJ, Molyneux PD, Gawne-Cain M, MacManus D, Barker GJ et al.** Quantitative MRI in patients with secondary progressive MS treated with monoclonal antibody Campath 1H. Neurology. 1999;53:751-57.

48. **Warrington AE, Asakura K, Bieber AJ, Ciric B, Van K, V et al.** Human monoclonal antibodies reactive to oligodendrocytes promote remyelination in a model of multiple sclerosis. Proc Natl Acad Sci U S A. 2000;97:6820-6825.

49. **Bieber AJ, Warrington A, Pease LR, Rodriguez M.** Humoral autoimmunity as a mediator of CNS repair. Trends Neurosci. 2001;24:S39-S44.

50. **Rodriguez M, Lennon VA, Benveniste EN, Merrill JE.** Remyelination by oligodendrocytes stimulated by antiserum to spinal cord. J Neuropathol Exp Neurol. 1987;46:84-95.

51. **Rodriguez M, Lennon VA.** Immunoglobulins promote remyelination in the central nervous system. Ann Neurol. 1990;27:12-17.

52. **Rodriguez M, Miller DJ, Lennon VA.** Immunoglobulins reactive with myelin basic protein promote CNS remyelination. Neurology. 1996;46:538-45.

53. **Ure DR, Rodriguez M.** Polyreactive antibodies to glatiramer acetate promote myelin repair in murine model of demyelinating disease. FASEB Journal. 2002;16:1260-1262.

54. **Paz Soldan MM, Warrington AE, Bieber AJ, Ciric B, van Keulen V, Pease LR et al.** Remyelination promoting antibodies activate distinct Ca++ influx pathways in astrocytes and oligodendrocytes: Relationships to the mechanism of myelin repair. Cell Mol Neurosci. 2002;in press.

55. **Miller DJ, Bright JJ, Sriram S, Rodriguez M.** Successful treatment of established relapsing experimental autoimmune encephalomyelitis in mice with a monoclonal natural autoantibody. J Neuroimmunol. 1997;75:204-9.

56. **Murray PD, McGavern DB, Sathornsumetee S, Rodriguez M.** Spontaneous remyelination following extensive demyelination is associated with improved neurological function in a viral model of multiple sclerosis. Brain. 2001;124:1403-16.

Chapter B2

# TMEV AND NEUROANTIGENS: MYELIN GENES AND PROTEINS, MOLECULAR MIMICRY, EPITOPE SPREADING, AND AUTOANTIBODY-MEDIATED REMYELINATION

Ikuo Tsunoda, and Robert S. Fujinami
*Department of Neurology, University of Utah School of Medicine, 30 North 1900 East, Salt Lake City, Utah 84132*

Abstract:     The Theiler's murine encephalomyelitis (TMEV) model has been used to study the interactions of virus, myelin and anti-neuroantigen autoimmunity. TMEV and myelin can interrelate during virus entry and persistence. On virus entry, TMEV might use peripheral myelin $P_0$ protein as a virus receptor. For persistence, TMEV seems to require myelin functional proteins or structural myelin itself. Here, myelin and oligodendrocyte loss and downregulation of myelin genes would lead to demyelination, but might limit virus spread in the central nervous system. Unlike experimental allergic encephalomyelitis (EAE), a pathogenic role of anti-myelin autoimmunity is unclear in TMEV infection. Anti-myelin autoantibodies have been detected in TMEV infection. Among them, only anti-galactocerebroside (GC) antibody is shown to be myelinotoxic, and has molecular mimicry with TMEV. Myelin-specific T cells play no role in initiation or progression of demyelination in the first two to three months after TMEV infection. However, cellular autoimmunity against several myelin antigens (epitope spreading) can be detected during the late chronic stage. Using the TMEV model, epitope spreading and autoantibody-mediated remyelination have been investigated by recombinant TMEV and anti-neuroantigen (natural) antibodies, respectively

Key words:     Animal models, Autoimmune diseases of the nervous system, Demyelinating diseases, Galactosylceramides, Myelin basic proteins, myelin proteolipid protein, Multiple sclerosis, Picornaviridae infections

# 1.     INTRODUCTION

## 1.1     Overview

The Theiler's murine encephalomyelitis virus (TMEV) infection model of multiple sclerosis (MS) has provided excellent paradigms to study the interactions between the virus, myelin and immune system (1-5). In this chapter, we will first review the associations between TMEV and myelin genes and proteins. We will next discuss possible roles of anti-myelin immunity including molecular mimicry and epitope spreading in the pathogenesis of demyelination. Finally, we will introduce two unique model systems using TMEV infection (i) to study the pathogenic role(s) of molecular mimicry and (ii) to explore potential therapeutic agents, remyelination-promoting autoantibodies, in demyelinating diseases. An understanding of the diverse interactions between viruses, myelin and host immune responses to neuroantigens may shed light on the etiology and pathogenesis of MS.

## 1.2     Theiler's murine encephalomyelitis virus

In 1934, TMEV was isolated by Max Theiler, who was awarded a Nobel Prize for developing the first effective vaccine against yellow fever (6,7). TMEV belongs to the genus *Cardiovirus*, family *Picornaviridae* (2). Picornaviruses are small, nonenveloped negative stranded RNA viruses. Several picornaviruses, such as poliovirus and coxsackievirus, can cause neurological disease in humans (8). However, TMEV infects mice and rats *in vivo* and causes a demyelinating disease only in mice. Thus, TMEV has not been identified as a causative agent in any human disease; its association with dermatomyositis (9) and Vilyuisk encephalomyelitis (10,11) are thus far inconclusive.

The Theiler's original (TO) subgroup, which includes Daniels (DA), BeAn 8386, and WW strains, produces a biphasic disease in susceptible mice inoculated intracerebrally [reviewed in (2)]. During the first week (acute phase), the mice develop an acute polioencephalomyelitis where mice occasionally show clinical signs such as weight loss. Virus antigen is expressed primarily in neurons in the central nervous system (CNS) (12). Subsequently, about one month post infection (p.i.) (chronic phase), the mice show signs of spastic paralysis as a result of demyelination within the white matter of the spinal cord. Areas of demyelination are associated with perivascular infiltration of mononuclear cells (MNC). The disease progresses without evidence of remission, either clinically or pathologically, with the exception of CD-1 mice infected with the WW strain of TMEV

(13).    During the chronic phase, viral antigens are expressed within oligodendrocytes (14,15), astrocytes and macrophage/microglia but not in neurons [reviewed in (4)].

TMEV-induced demyelinating disease is one of a few animal models for the progressive forms of MS (16).  Although there are many animal models for MS, their clinical courses are either acute monophasic or relapsing-remitting (RR).  These animal models are only useful as experimental models for RR-MS.  Therefore, the TMEV model is a unique animal model to study the pathogenesis for the other three types of MS: primary progressive (PP)-, secondary progressive (SP)-, and progressive relapsing (PR)-MS (17).

The precise mechanisms of demyelination in TMEV infection are not clear.  There are two major hypotheses: (i) direct virus infection of myelin forming cells, i.e. oligodendrocytes, and (ii) immune-mediated mechanisms. Infection with live TMEV is necessary to induce demyelination, since neither ultraviolet-inactivated TMEV nor cDNAs encoding individual TMEV capsid proteins alone cause demyelination (18).  Virtually all immune cells, including T cells, B cells and macrophages, have been shown to play important roles [reviewed in (1-4)].  In contrast to experimental allergic encephalomyelitis (EAE) (3) and other virus-induced demyelinating diseases (19), adoptive transfer of uninfected immune cells from TMEV-infected mice does not induce demyelination in the recipient mice (20).  In addition, acquired immunity is neither necessary, nor sufficient in some TMEV model systems; TMEV can persistently infect the CNS and induce demyelination in: major histocompatibility complex (MHC) class I-deficient mice (21,22), MHC class II-deficient mice (23), CD4-deficient mice (24), or CD8-deficient mice (24).  TMEV also causes demyelination in the spinal cord of moth-eaten mice, lacking protein tyrosine phosphatase SHP-1, as early as five days p.i., when no acquired immune responses are normally generated (25).

## 2.    MYELIN GENES, VIRUS ENTRY, AND VIRUS PERSISTENCE

### 2.1    Peripheral nerve myelin protein for TMEV entry

During a natural infection, TMEV spreads from mouse to mouse by the oral-fecal route (26,27).  In order to cause CNS infection by this route, the virus must get from the alimentary tract into the CNS, either hematogenously or intra-axonally.  Since the strains of the TO subgroup of TMEV could be isolated from feces up to 154 days p.i. (28), invasion into

**Table 1. Myelin gene, virus entry and virus persistence**

| Virus Entry | Virus Persistence | |
|---|---|---|
| | TMEV | Host |
| CNS | ⬇ | ⬇ |
| Receptor unknown | | |
| PNS | requires | downregulates |
| $P_0$ protein as a virus receptor? | MBP and PLP ⬅ ➡ functionally and/or | MBP and PLP mRNA actively or passively and/or |
| | requires myelin itself ⬅ ➡ as a structure | loses oligodendrocyte and myelin (demyelination) |
| Evidence: *In vitro* assays | Evidence: No virus persistence in shiverer and rumpshaker mice | Evidence: *In situ* hybridization, TUNEL, myelin staining, EM |

**Abbreviations**: CNS, central nervous system; EM, electron microscopy; MBP, myelin basic protein; PLP, myelin proteolipid protein; PNS, peripheral nervous system; TMEV, Theiler's murine encephalomyelitis virus; TUNEL, terminal deoxynucleotidyl trasnferase-mediated dUTP-biotin nick-end labeling.

the CNS via the neural route from a persistent infection in the intestine may be one pathway, as originally suggested by Theiler and Gard (27). Myelin is a multiple lamellar membrane that sheathes axons and facilitates the saltatory conduction of nerve impulses. $P_0$ protein is a member of the immunoglobulin (Ig) gene superfamily, and constitutes 50% of the total myelin proteins present in the peripheral nervous system (PNS). Using virus overlay and virus binding assays, Libbey *et al* (29) identified $P_0$ protein as a potential receptor for TMEV (Table 1). They suggested that the use of the $P_0$ protein in Schwann cells as a receptor may be one mechanism by which TMEV spreads from the gastrointestinal tract to the CNS *in vivo*. In the CNS, TMEV receptors on neurons during the acute phase and on glial cells and macrophages during the chronic phase have not been identified.

## 2.2 Myelin protein requirement for TMEV persistence

Myelin basic protein (MBP) and myelin proteolipid protein (PLP) are the major component of CNS myelin; many other proteins and glycoproteins are present to a lesser extent. Bihl *et al* (30) demonstrated that myelin gene mutation could affect the persistence of TMEV. Susceptibility/resistance to chronic TMEV infection has been mapped to several loci including one tentatively located in the telomeric region of chromosome 18, close to the myelin basic protein locus (*Mbp* locus) (31,32). To determine if the *Mbp* gene influences viral persistence, Bihl *et al* (30) inoculated DA virus into C3H mice bearing the shiverer (*shi*) mutation, a 20-kb deletion in the gene. Whereas TMEV-infected wild-type C3H +/+ mice had intermediate levels of virus persistence on days 20 and 45 p.i., an increase in viral persistence was detected in heterozygous C3H +/*shi* mice, and no virus persistence was detected in homozygous C3H *shi/shi* mice. The increased susceptibility of the heterozygous shiverer mice was also seen in (C57BL/6 x C3H +/*shi*) F1 mice (1), although C57BL/6 mice are resistant to TMEV persistence. Furthermore, no viral RNA was detected 20 days after DA virus infection of C3H/101H mice homozygous for a point mutation in the gene coding for PLP (*PLP1*, previously *PLP*), the rumpshaker (*rsh/rsh*) mutation (no data was available in control C3H/101H (+/+) mice, though) (30). In these studies, infection of oligodendrocytes, the presence of demyelination and anti-TMEV immune responses were not tested. The number of oligodendrocytes has been reported to be normal, or even slightly increased in uninfected shiverer and rumpshaker mutant mice, although the myelin is grossly deficient. Therefore, these results suggest that TMEV persistence is linked to myelin and functional alteration, for example, may change the virus's ability to replicate in, or traffic through, the myelin sheath. Although the *MBP* gene or some other *MBP*-linked locus has been suggested to influence susceptibility to MS, other studies have not demonstrated linkage between MS and the *MBP* gene (33,34).

## 2.3    Myelin gene downregulation in TMEV infection

The genes encoding myelin proteins, such as MBP and PLP, are expressed by oligodendrocytes in the CNS, and their transcripts are markers for activity of maintenance or reproduction of myelin. Yamada *et al* (35) demonstrated the relationship between MBP and PLP mRNAs, viral RNA, and demyelination within the spinal cord during TMEV infection. At four weeks p.i., decreased levels of MBP and PLP mRNAs were associated with the presence of viral RNA and demyelination. By 8 or 12 weeks p.i., however, extensive demyelination was associated with a significant decrease of MBP and PLP mRNAs but not with the presence of viral RNA. These results indicate that a viral infection of white matter in the early phase of TMEV infection leads to demyelination, but that, later in the course of disease, some other mechanism, such as an immunopathologic process, drives extensive demyelination. Subsequent reports by other groups confirmed the decrease in PLP mRNA expression, not only in infected- but also in uninfected-oligodendrocytes in the spinal cord (36-38).

The decrease of MBP and PLP mRNAs in the lesions indicates either a loss of oligodendrocytes or suppression of the myelin metabolism of the oligodendrocyte without cell lysis. The former is supported by the finding of apoptosis of oligodendrocytes in demyelinating lesions in TMEV infection (39). The latter possibility is supported by the fact that some viruses are known to alter the "luxury" or differentiated function of a cell, without disturbance of its vital function, i.e., ability to survive (40). Using electron microscopy, Rodriguez (15) demonstrated degenerative changes in the most distal processes of oligodendrocytes in TMEV infection. Although this 'dying-back oligodendrogliopathy' has not been directly linked to suppression of myelin metabolism, similar ultrastructural pathology has also been described in MS (41). Also, in the polyomavirus-induced human demyelinating disease, progressive multifocal leukoencephalopathy (PML), JC virus has been reported to cause demyelination not only by the lytic infection of oligodendrocytes, but also via the suppression of myelin gene expression [reviewed in (42)].

Although suppression of MBP and PLP mRNAs as well as demyelination results in neurological functional deficit, this might benefit the host by suppressing viral replication in the CNS. As we discussed in the previous section, persistent viral infection seems to require either functionally or structurally intact myelin. Therefore, suppression of myelin function and loss of myelin (demyelination) might serve a protective role by suppressing virus persistence in TMEV infection (Table 1). Demyelinating diseases might represent the oligodendrocytes' futile attempt to suppress viral replication, which ultimately results in potentially irreversible myelin damage. On the other hand, incomplete suppression of the oligodendrocyte functions might actually benefit the virus, since the heterozygous shiverer (+/*shi*) mutation resulted in an increase in viral persistence in the CNS.


## 3.        HUMORAL AUTOIMMUNE RESPONSES TO MYELIN

### 3.1        Anti-myelin antibodies in TMEV infection

Damage to CNS tissue liberates CNS antigens into the cerebrospinal fluid (CSF) compartment. MBP can be detected in the CSF of patients with active myelin destruction caused by MS or other processes, such as CNS infarction. Rauch *et al* (43) found MBP in the CSF of 14 of 19 mice infected with DA virus 12 weeks p.i. They also detected serum anti-MBP antibody in 10 of 14 TMEV-infected mice. Welsh *et al* (44) also demonstrated that antibody responses to myelin developed in CBA mice by day 28 after BeAn virus infection and reached a peak between day 42 to 56 p.i. Anti-myelin antibody production was prevented by depletion of CD4$^+$ T cells. However, there was no correlation between the titer of anti-myelin antibodies and the severity of clinical disease.

Using Western blot analysis, Cash *et al* (45) studied the specificity of Ig secreted by B cells isolated from the spinal cord of SJL/J mice infected with DA virus. They found antibodies directed against two unidentified nonviral proteins with molecular weights of 26 and 28kDa; these were present only in the white matter of infected mice and were not found in purified myelin.

Rodriguez *et al* (46) showed that, like in MS, increased amounts of Ig in the CSF and oligoclonal IgG bands were detected in TMEV-infected mice on day 21 p.i. However, IgG in the serum and CSF was primarily directed at virus antigen rather than at normal myelin components, although rare examples of myelin sheaths positive for IgG were found by immunoelectron microscopy in TMEV-infected mice. On the other hand, Roos *et al* (47) demonstrated that the heterogeneity of anti-TMEV antibody in the CSF was less restricted than oligoclonal IgG bands seen in MS or subacute sclerosing panencephalitis (SSPE). Rodriguez *et al* (46) also showed that complement depletion on day 15 p.i. resulted in exacerbation of demyelination on day 21 p.i. Although the authors suggested no pathogenic role of humoral immunity from this study, the treatment can hamper antibody-mediated clearance of TMEV and day 21 might be too early to assess effector mechanisms in TMEV infection, where overt demyelination is generally seen later.

A direct pathogenic role for anti-MBP antibody in demyelination is unlikely, since anti-MBP antibody has been demonstrated not to be myelinotoxic either *in vivo* or *in vitro* (48). This is in contrast to antibodies against galactocerebroside (GC) or myelin oligodendrocyte glycoprotein (MOG), whose adoptive transfer to mice with EAE exacerbates demyelination (49). Functionally, Barbano and Dal Canto (20) demonstrated that sera from mice infected with TMEV for two to three months did not injure myelinating cultures *in vitro*. This study suggests that there is no myelinotoxic humoral factors in sera from TMEV-infected mice.

Anti-myelin antibody might indirectly affect myelin antigen presentation to class II-restricted T cell by receptor-mediated antigen uptake in TMEV infection (50). Myelin-specific surface Ig can serve to capture myelin antigens on the surface of B cells (51). The B cells could then internalize and present myelin antigen to specific T cells. In addition, professional antigen presenting cells (APCs) could exploit soluble anti-myelin antibody for antigen capture by Fc receptor mediated uptake of myelin-containing immune complexes. This would lead to augmentation of myelin-specific T cell responses, and also might contribute to determinant spreading (see a later section) in TMEV infection, as suggested for other autoimmune diseases (52). The potential beneficial role of anti-neuroantigen antibodies will be discussed in the last section of this chapter.

## 3.2 Molecular mimicry between TMEV and galactocerebroside

Molecular mimicry between host protein and microorganism has been known for many years. A cross-reacting immune response initially generated against a virus that also reacts with myelin components could lead to demyelination. However, Rauch *et al* (43) found that anti-MBP antibody from TMEV infected mice did not cross-react with TMEV antigens. Lack of cross reaction between MBP and capsid proteins VP1, VP2, and VP3 was also supported by a competition radioimmunoassay with sera from rabbit subcutaneously injected with TMEV capsid proteins (53).

On the other hand, Fujinami *et al* (54) raised a monoclonal antibody (mAb) H8 that reacts with TMEV and GC from spleen cells of BALB/c mice infected with DA virus intracerebrally. GC is a major lipid component of myelin. The mAb H8 reacts with viral capsid protein VP1 and efficiently neutralizes TMEV *in vitro*. Intraneural injection of mAb H8 into the sciatic nerve induced Schwann cell destruction and demyelination (55). When injected intravenously into mice with EAE, mAb H8 increased demyelinating lesions within the spinal cord by ten-fold (56). To determine whether antibodies similar to mAb H8 exist during viral infection, mice were infected with TMEV and sera were collected at various times after infection. Antibodies to GC were detectable ten days p.i. These results suggest that antibody(s) of the H8 type is generated and could contribute to demyelination *in vivo* in TMEV infection.

Interestingly, Fujinami *et al* (57) also demonstrated that another TMEV-neutralizing antibody (H7) could react with VP1 but not with GC. Passive administration of mAb H7 resulted in the clearance of virus and survival of athymic mice. Both H7 and H8 antibodies can neutralize TMEV, yet only H8 is myelinotoxic (Table 2). Further investigation is needed to clarify which immunologic factors, such as T helper cells, favors generation of each type of anti-VP1 antibody *in vivo*. The presence of "myelinotoxic activity" has been demonstrated in the CSF and sera in demyelinating diseases; only GC and MOG are identified as target molecules of demyelinating antibodies of which passive transfer exacerbates, but not initiates, demyelination in the CNS *in vivo* (58). Anti-GC antibody has also been detected in some MS patients (59).

**Table 2. Destructive or beneficial autoimmunity in TMEV infection**

| Effector Mechanism | P.I. Days | Destructive | Beneficial |
|---|---|---|---|
| Molecular mimicry | 10 days | Antibody to TMEV cross reacts with GC and exacerbates demyelination | Virus neutralization |
| Epitope spreading | 2-3 months | MBP-, PLP-, MOG-specific T cells induce DTH responses | CNS regeneration |
| Natural autoantibody | N.E. | Autoantibody binds to CNS antigens | Remyelination |

**Abbreviations:** CNS, central nervous system; DTH, delayed-type hypersensitivity; GC, galactocerebroside; MBP, myelin basic protein; MOG, myelin oligodendrocyte glycoprotein; N.E., not examined in TMEV infection, but detected in SCH-immunized mice; P.I., TMEV post infection; PLP, myelin proteolipid protein; SCH, spinal cord homogenate; TMEV, Theiler's murine encephalomyelitis virus.

# 4. CELLULAR AUTOIMMUNE RESPONSES TO MYELIN

## 4.1 Myelin-specific T cell responses play no role in demyelination in the first two to three months after TMEV infection

Epitope spreading (determinant spreading) is defined as the diversification of epitope specificity from the initial dominant epitope-specific immune response to subdominant epitopes [reviewed in (60)]. Epitope spreading can play an active role in the relapse or progression of autoimmune diseases, as was first suggested in EAE by Lehmann *et al* (61). After initial immune-mediated tissue destruction, tissue debris is taken up by macrophages, dendritic cells or B cells, and presented to tissue-specific T cells. Spreading of the specificity of an immune response can be from one epitope to another on the same molecule (intramolecular epitope spreading, for example, from $PLP_{139-151}$ to $PLP_{178-191}$) or from an epitope on one molecule to one on a different molecule (intermolecular epitope spreading, for example, from $PLP_{139-151}$ to $MOG_{92-106}$) (60).

Although tissue destruction occurs less than one week after TMEV infection, epitope spreading is not observed until two to three months after TMEV infection [reviewed in (62)]. Barbano and Dal Canto (20) demonstrated that spleen cells from TMEV-infected mice did not injure myelinating cultures and did not proliferate in response to MBP or spinal cord homogenate (SCH) *in vitro*. In addition, these spleen cells did not produce disease *in vivo* by adoptive transfer into naïve mice, regardless of whether *in vitro* stimulation was with MBP or SCH. Similarly, Miller *et al* (63,64) demonstrated a lack of delayed-type hypersensitivity (DTH) responses and T cell proliferation against whole MBP and PLP, their encephalitogenic peptides $MBP_{91-104}$ and $PLP_{139-151}$, and SCH in TMEV-infected mice 23, 60, 69, 90 and 127 days p.i.

Comparative studies using the EAE model induced by neuroantigens also demonstrated no pathogenic role of myelin specific T cells during the early chronic phase of TMEV infection. EAE is induced by subcutaneous injection of myelin antigens emulsified with complete Freund's adjuvant (CFA), while emulsification with incomplete Freund's adjuvant (IFA) can induce anergy to myelin antigens, rendering resistance to subsequent EAE induction. Following MBP or SCH emulsified in IFA injection, Lang *et al* (65) injected mice with DA virus or SCH/CFA for EAE induction. They found that pretreatment with myelin antigens/IFA suppressed EAE induction, but not TMEV-induced demyelination at 45 days p.i. Another protocol for neuroantigens-specific tolerance induction that can suppress

EAE also failed to alter the development of TMEV-induced demyelinating disease on day 65 p.i. (64). Similarly, using a treatment regimen reported to suppress EAE, Drescher *et al* (66) demonstrated that oral tolerization with myelin did not alter TMEV-induced demyelination at day 45 p.i., supporting a non-autoimmune mechanism of TMEV-induced demyelination. Although TMEV-infected mice develop demyelinating disease one month after infection and disease progresses continuously from then on, myelin-specific cellular immunity contributes to neither initiation nor progression of demyelinating disease at least for the first two to three months after TMEV infection.

## 4.2     Epitope spreading during the late chronic phase of TMEV infection

Although TMEV-infected mice develop extensive inflammatory demyelination with progressive disability by two to three months after infection, neurological disease seems to progress for the rest of the life of infected mice. While most studies on TMEV-induced demyelination have used mice infected with TMEV for one to two months p.i., the Miller group has examined the role of epitope spreading in initiating anti-myelin auto-immune responses in SJL/J mice during the late chronic phase of disease, more than two to three months after BeAn strain of TMEV infection (60). As reviewed in the previous section, before three months p.i., TMEV-infected mice develop CD4$^+$ T cells specific for TMEV, but not for neuroantigens. However, the Miller group demonstrated PLP$_{139-151}$-specific T cell proliferation in TMEV-infected mice 87 days, but not 33 days, after infection, and DTH response to PLP$_{139-151}$ was first demonstrable at day 52 (67). Peripheral T-cell responses to myelin epitopes develop during progressive disease in a hierarchical order, beginning with the dominant PLP$_{139-151}$ peptide, but only after the onset of demyelination. As disease progresses, responses to various other less-dominant encephalitogenic myelin epitopes, such as PLP$_{178-191,}$ PLP$_{56-70}$ and MOG$_{92-106}$ develop. Although there are some inconsistencies as to the onset date of the anti-neuroantigen specific immune response between the reports by Miller's group, overt anti-myelin immune responses seem to be detectable around three months after TMEV infection, which argues against these responses contributing to myelin damage before this time.

Autoreactivity could be initiated, in part, by the processing of myelin debris and subsequent presentation of myelin epitopes by infiltrating macrophages and resident perivascular cells/microglia in the CNS. Macrophages and microglia harvested from the CNS during the early chronic phase, 40-42 days after TMEV infection, presented viral peptides, but not myelin peptide (68). However, by 90 days post-infection, CNS APCs presented both viral and myelin epitopes to Th1 and Th2 cell lines and hybridomas specific for TMEV or PLP peptides in the absence of exogenous antigen *in vitro*, indicating these APCs could endogenously process and present both viral and myelin antigens (68,69). Importantly, anti-CD86 (B7-2) antibody predominantly inhibited $PLP_{139-151}$-specific T cell proliferation mediated by endogenous antigen presentation by CNS APCs from TMEV-infected mice (72% inhibition), compared with anti-CD80 (B7-1) (44% inhibition). This is in contrast to CD80 predominance in SJL/J mice with RR-EAE. Therefore, the costimulatory dependence of epitope spreading differs between TMEV infection and RR-EAE. *In vivo* costimulatory blockade of CD28 (ligand for CD80 and 86) on days 0, 2, 4, 6 and 9 after TMEV infection inhibited anti-TMEV immune responses and increased virus persistence, accelerating epitope spreading (70). This CD28 blocking would have been more intriguing, if it had been administered in TMEV-infected mice during the early chronic phase, possibly during the priming period of anti-PLP responses, where the treatment could inhibit epitope spreading without suppression of anti-TMEV immunity.

Most recently, Miller's group demonstrated that the treatment of TMEV-infected mice with an MP4 fusion protein, consisting of PLP and MBP, mildly attenuated disease progression (40-110 days p.i.) and resulted in decreased inflammatory cell infiltration in the CNS (71). Paradoxically, $PLP_{139-151}$-specific splenic T cell proliferative and interferon (IFN)-γ responses were enhanced in the MP4-treated mice, while $PLP_{139-151}$-specific DTH responses were downregulated. In addition, after 120-140 days p.i., MP4-treated mice showed disease severity comparable to control animals. These results indicated that MP4-treatement in TMEV infection did neither induce tolerance via deletion or anergy, nor immune deviation from Th1 to Th2 responses. In contrast, the MP4 treatment completely inhibited both the acute attack and relapse of PLP- and MBP-induced EAE with significant suppression of PLP- and MBP-specific lymphoproliferative responses (72). Administration of MP4 also markedly ameliorates the course of established EAE. These results suggest that the immunological parameters of epitope spreading differ between TMEV infection and EAE.

Another group also demonstrated myelin specific-DTH responses during the late chronic phase (36-48 weeks p.i.), but not during the early chronic phase (4 to 12 weeks p.i.), indicating autoimmune responses do not play a major role in the initiation of demyelination in TMEV infection (73). There was no correlation between clinical score and the strength of the DTH response to myelin autoantigens in mice tested at 36-48 weeks. Therefore, the authors concluded that anti-myelin autoimmunity might be an additional factor that contributes to lesion progression during the late chronic stage, although autoimmunity to myelin might represent only an epiphenomena, being induced only after damage to myelin produced by other mechanisms.

It is unknown why epitope spreading can be detected only during the late chronic stage of TMEV infection. In addition to virally induced tissue injury and immunomodulation, mouse strain and aging might contribute to epitope spreading in TMEV-infected SJL/J mice. Naïve SJL/J mice have a high frequency of $PLP_{139-151}$ specific T cells, and it increases with age (74,75). Moreover, aged SJL/J mice tend to have immunological abnormalities: a high incidence of reticulum cell sarcoma and paraproteinemia with enlargement of lymph nodes and splenomegaly (76).

A pathological role for epitope spreading is controversial even in EAE (74) and difficult to verify. Epitope spreading might simply reflect release of autoantigens secondary to myelin destruction, and may not have pathogenic significance. However, Dal Canto *et al* (77) showed that lymph node cells harvested from mice 70 days after TMEV infection produced demyelination in organotypic cultures after stimulation with $PLP_{139-152}$ but not after stimulation with whole MBP. The authors suggested that antigen spreading to PLP, but not to MBP, might play an important role in demyelination. It would be intriguing to test whether adoptive transfer of the PLP peptide-stimulated lymphocytes from TMEV-infected mice could cause demyelination in naïve mice *in vivo*, while possible contamination with live transferred virus in cultures and the effect of whole PLP or MBP peptide stimulation of lymph node cells should also be addressed in future experiments. On the other hand, autoimmunity in the CNS, including MBP-specific T cells, has been demonstrated to have a protective or beneficial nature under certain circumstances (78). Therefore, one cannot rule out the possibility that anti-myelin autoimmunity might play a neuroprotective role during the late chronic phase of TMEV infection, a time when severe axonal injury and loss has been demonstrated [reviewed in (5)].

## 5.        TMEV INFECTION AS MODELS TO STUDY MOLECULAR MIMICRY AND REMYELINATION

### 5.1        Recombinant TMEV as a molecular mimicry model

Using TMEV, Olson *et al* (79) established a virus-induced molecular mimicry model of MS. They engineered a nonpathogenic TMEV (ΔClaI-BeAn) encoding $PLP_{139-151}$. Infection with the $PLP_{139-151}$-encoding TMEV (PLP139-BeAn) led to rapid-onset of paralytic disease within ten days. As early as 14 days p.i., inflammatory demyelinating lesions were detected in the spinal cord, and PLP$_{139\text{-}151}$-specific T cell proliferation and DTH response were detected in the spleen. Similar early-onset disease with anti-PLP responses was also observed in mice infected with a TMEV encoding altered peptide ligand (APL) of PLP$_{139\text{-}151}$ with an amino acid substitution at the T cell receptor contact residue (H147A PLP139-BeAn). PLP$_{139\text{-}151}$-specific T cells, but not TMEV-specific T cells, from PLP139-BeAn-infected mice transferred demyelinating disease to naïve recipients, and induction of PLP$_{139\text{-}151}$-specific tolerance before infection prevented clinical disease (80). The authors suggest that both PLP139-BeAn and H147A PLP139-BeAn viruses induced PLP specific T cells, leading to early onset of demyelinating disease. No information is available about pathology and viral replication one week after PLP139-BeAn virus infection. It is also unknown whether this model is similar to TMEV infection or PLP$_{139\text{-}151}$-induced EAE clinically and histologically.

In this study, TMEV expressing an encephalitogenic PLP epitope as well as an APL of PLP could induce early-onset disease in the absence of a stimulus associated with CFA. This is in contrast to the infection of mice with a vaccinia virus construct expressing the entire coding region of PLP, which failed to induce CNS disease without a second challenge by CFA (81). Thus, TMEV might provide sufficient signals to induce autoreactive T cells in some circumstances. Importantly, the APL of PLP that is encoded in H147A PLP139-BeAn is not encephalitogenic even when administered with CFA in mice (82). This may mimic the situation of a recent clinical trial in MS, in which administration of an APL of MBP into MS patients resulted in induction of MBP-specific T cell proliferation and exacerbation of clinical signs (83,84).

## 5.2 Anti-neuroantigen antibodies as a therapeutic agent in TMEV infection

In contrast to the classical view of the anti-myelin immune responses playing a pathogenic role in CNS demyelinating disease, autoimmunity against myelin might play a beneficial role by promoting CNS regeneration: axonal regeneration and myelin regeneration (remyelination). In the EAE model, injection of a myelin component with IFA ameliorates clinical and pathological demyelinating disease. This therapeutic effect could be due to: 1) suppression of Th1 cell-mediated autoimmunity and 2) production of anti-myelin antibody that plays a direct role in remyelination. This latter role is in agreement with *in vitro* observations of antibody-mediated oligodendrocyte stimulation. Based on the initial observations by Traugott *et al* (85) of remyelination in EAE, Lang *et al* (86) conducted similar experiments with mice infected with the DA strain of TMEV. At three months p.i., mice treated with subcutaneous injections of MBP and GC emulsified in IFA showed substantial CNS remyelination compared with control treated animals. In the following paper, however, they demonstrated that pretreatment of mice with SCH/IFA or MBP/IFA did not suppress clinical and histological disease in TMEV infection (65).

The discovery that certain autoantibodies can be beneficial in an animal model of CNS demyelination suggests that the presumed pathogenic role of antibodies in MS may need to be reevaluated. Using the TMEV model, Rodriguez and colleagues have shown a beneficial role of anti-neuroantigen antibodies due to their promotion of remyelination [reviewed in (87)]. Spontaneous remyelination is impaired in SJL/J mice infected with the DA strain of TMEV. Rodriguez and Lennon (88) observed that passive transfer of either antiserum or purified IgG from uninfected syngeneic animals immunized with SCH promoted remyelination by oligodendrocytes (CNS-type), but not by Schwann cells (PNS-type), in TMEV-infected mice. While astrocytes were common within areas of remyelination, inflammatory cells were less numerous in remyelinating areas than in demyelinating lesions. *In vivo* autoradiography with [$^3$H] thymidine demonstrated that the majority of radiolabeled cells were MNCs, while remyelinated areas contained GC$^-$-proliferating cells that might be immature oligodendrocytes, morphologically (89). However, subsequent experiments demonstrated that a similar extent of oligodendrocyte proliferation was also detected in mice treated with control sera, suggesting proliferation of oligodendrocytes is not sufficient for new myelin synthesis (90). Anti-SCH serum also stimulated proliferation of primary glial cell cultures *in vitro* (91). The effect of the treatment on anti-TMEV antibody production and virus persistence seemed to be inconclusive. A mild increase of anti-TMEV antibody titer was detected by an enzyme-linked immunosorbent assay (ELISA), but not by a plaque reduction assay; an increase of viral persistence was reported by immunohistochemistry against viral antigen, but not by virus plaque assays of CNS homogenates (92). Rodriguez *et al* (93) also demonstrated that treatment with antiserum or purified IgG directed against MBP promoted remyelination after TMEV infection without altering virus persistence or serum TMEV-specific antibody responses.

The Rodriguez group also generated a remyelination-promoting IgM mAb, SCH94.03, from the spleen cells of mice injected with SCH. SCH94.03 showed reactivity toward a red blood cell antigen, spectrin and several other protein antigens, astrocytes and oligodendrocytes, but not with MBP (94,95). SCH94.03 injection suppressed T cell infiltration and increased TMEV persistence in the CNS, while levels of TMEV-specific IgG were similar to controls (96). SCH94.03 neither reacted with any TMEV proteins as determined by Western blot, nor neutralized TMEV *in vitro* (97). Treatment with SCH94.03 also reduced relapse rate, demyelination and meningitis in SJL/J mice with EAE induced by adoptive transfer of MBP$_{91\text{-}103}$-specific T cells (98). These results are consistent with the hypothesized immunomodulatory function of autoantibodies, although neither immune responses to MBP nor extent of remyelination was explored in this EAE model. Interestingly, however, SCH94.03 increased the rate of oligodendrocyte remyelination in a nonimmune, toxin-induced model of lysolecithin-induced demyelination (99).

Other IgM mAbs, SCH79.08, O1, O4, A2B5, and HNK-1, have also been shown to promote remyelination in TMEV models. SCH79.08 was established from mice immunized with SCH, reacts with MBP and shows intracellular, but not cell surface, staining of oligodendrocytes (100). O1, O4, A2B5, and HNK-1 antibodies recognize differentiation stage-specific surface antigens on oligodendrocytes. We do not know, however, whether these antibodies against oligodendrocyte precursors actually recognized the cells *in vivo*. Immunoperoxidase staining for markers expressed on presumed glial progenitor cells (O4, A2B5, and G$_{D3}$) was not successful in the adult mouse spinal cord in TMEV infection, even though the reagents worked in the neonatal rat optic nerve system *in vitro* (90).

Several antibodies join a list of immunoglobulins that enhance remyelination in the Theiler's virus model (101). The list includes pooled human intravenous IgG (IVIg), polyclonal human IgM, and two human IgM mAbs, sHIgM22 and sHIgM46, both of which were isolated from Waldenstrom's macroglobulinemia. Neither IVIg nor human IgM bound to human oligodendrocytes *in vitro*, while the two IgM mAbs bound to oligodendrocytes. Polyclonal human IgM and sHIgM22, but neither polyclonal IgG, anti-SCH IgG, nor IVIg, accelerated the rate of remyelination following lysolecithin-induced demyelination in SJL/J mice (99,102). Therefore, the authors suggested that IgG and IgM antibodies might function to enhance remyelination through different mechanisms.

The precise mechanisms of remyelination by the autoantibodies are unknown. It is unclear whether there is one single common mechanism which explains promotion of remyelination in all remyelination-promoting sera, IgG, and IgM antibodies. Because many of the remyelination-promoting antibodies bind to oligodendrocytes and/or myelin, a direct effect on the recognized cells has been hypothesized. However, since the remyelination-promoting autoantibodies have varying specificities, it is unlikely that each of the antibodies functions directly through a common antigen or receptor. The authors also suggested opsonization of myelin debris due to IgM antibody by phagocytes might enhance clearance of cellular debris from the damaged area, allowing the normal process of CNS repair to progress. Since Fc receptors on phagocytes do not generally bind to the Fc portion of IgM, IgM likely interacts with C3b via its Cμ1 domain, thereby allowing antibody-antigen complexes containing IgM to indirectly mediate phagocytosis. Although by this mechanism, C3b, once fixed, can promote uptake via complement receptors on macrophages, the roles of complement and Fc receptor have not been tested in this model. Immunomodulation by autoantibody is also hypothesized. However, no data are available as to how autoantibody treatment might influence cellular immune responses to virus or myelin antigen.

Several other studies need to be performed to clarify the mechanism of remyelination, such as immunological and histological time course studies over a four-week antibody treatment period and *in vivo* trafficking of injected antibody. In addition, we do not know why autoantibodies stimulate CNS-type remyelination by oligodendrocytes, but not PNS-type remyelination by Schwann cells, despite the fact that the PNS also expresses antigens, including MBP, recognized by the autoantibodies (95). It should be noted that most remyelination-promoting studies used mice four to six months after TMEV infection, in contrast to the other TMEV research groups who use mice one to two months after TMEV infection. As discussed earlier, the mechanisms of demyelination might be different at these two time points.

Interestingly, among the autoantibodies, several (SCH94.03, SCH79.08, O1, O4, HNK-1), but not all (A2B5, sHIgM22), monoclonal antibodies have been shown to have characteristics of natural autoantibodies (100,101,103). Natural autoantibodies, usually IgMs whose physiological significance is unknown, are present in the serum of normal unimmunized animal, and represent a substantial fraction of the total Ig repertoire (104). Natural autoantibodies show polyreactivity toward multiple self and nonself antigens, and are typically encoded by unmutated germ line Ig genes (103). It has been shown that some viruses may act as polyclonal B cell activators, may release antigens not ordinarily recognized by the host immune system, or may augment the response to many antigens (105,106). Any of these mechanisms may explain the induction of autoantibodies or an increase of the levels of natural autoantibodies.

It is attractive to hypothesize that the presence or absence of remyelination-promoting antibody might alter the disease course in demyelinating diseases. Here, a lack of the autoantibodies would prevent remyelination in DA virus-induced progressive disease in SJL/J mice. It may be intriguing to test whether the autoantibody is involved in the pathogenesis of CD-1 mice infected with the WW strain of TMEV, who develop a relapsing remitting disease course, characterized by extensive PNS- and CNS-type remyelination (13).

Clinically, comparison of the levels of autoantibodies in patients with RR-MS versus PP- and SP-MS might be important. From a patient with MS, a clone of a possible natural antibody was established (107). This oligo-dendrocyte–reactive IgM monoclonal antibody, DS1F8, is polyreactive and its antigen-binding domains are encoded by germ line genes. In addition, Matsiota *et al* (108) found that MS patients often had elevated CSF antibody levels against many autoantigens, but not against MBP, suggesting a local expansion of B cells producing natural autoantibodies. Whether these antibodies have any effect on demyelination is unknown.

## 6.    CONCLUSION

In this chapter, we have reviewed the interactions of TMEV, myelin and anti-myelin immune responses. On virus entry, TMEV might use peripheral myelin protein $P_0$ as a virus receptor, and require myelin genes functionally or myelin itself structurally for virus persistence. In turn, the host down-regulates myelin gene expression and/or loses oligodendrocytes and myelin. While this leads to demyelination, it might also be beneficial to the host since it prevents virus spread in the CNS. Anti-TMEV humoral immune responses can help virus clearance, while some anti-viral antibodies might

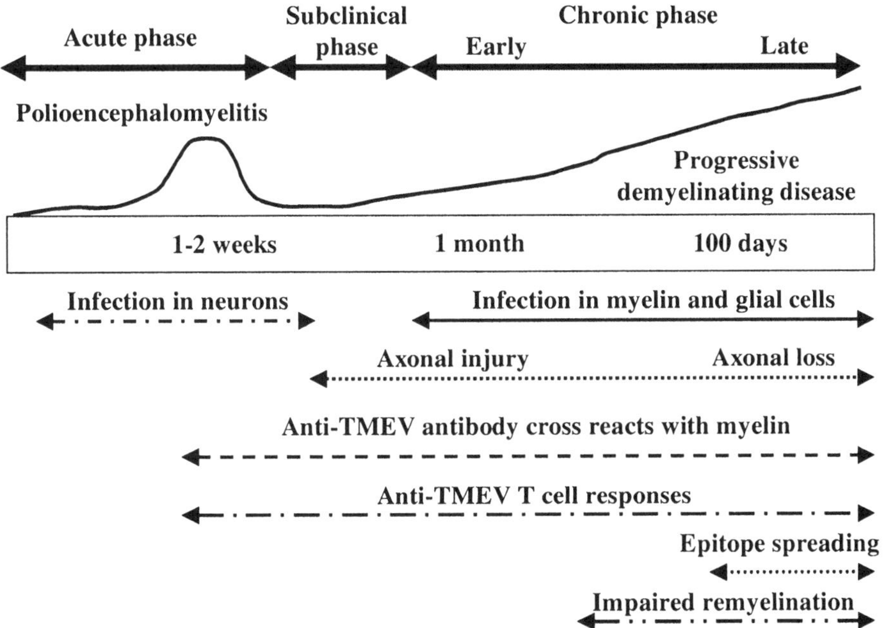


*Figure 1.* Myelin infection, anti-myelin responses, and other effector mechanisms during TMEV disease course.

cross-react with host myelin molecules, including GC, leading to exacerbation of demyelination. During the early chronic phase, cellular immune responses to TMEV, but not to myelin, were detected. However, epitope spreading seems to develop during the late chronic phase, and T cell responses to encephalitogenic epitopes of PLP, MBP and MOG, can be detected. Impaired remyelination could also contribute to disease progression. Here, the presence or absence of remyelination-promoting antibody might be important in disease remission. Of all the above interactions between virus, myelin, and autoimmune responses, no single mechanism may be solely responsible for demyelination. However, all of the interactions potentially contribute to progression of demyelinating disease in which a variety of different pathogenic mechanisms, such as oligodendrocyte infection and apoptosis, axonal injury and anti-TMEV immune responses, seem to act in parallel or sequence to cause myelin destruction (Figure 1).

## ACKNOWLEDGEMENTS

The authors would like to thank Melina V. Jones, PhD, Jane E. Libbey, MS and Lisa K. Peterson, BS for many helpful discussions, and Nancy K. Burgess, BS, Isaac Z. M. Igenge and Kenneth B. Scott for their technical assistance. We are grateful to Ms. Kathleen Borick for preparation of the manuscript. This was supported by the NIH grant NS34497.

## REFERENCES

1. Monteyne P, Bureau J-F, Brahic M. The infection of mouse by Theiler's virus: from genetics to immunology. Immunol Rev 1997;159:163-76.
2. Roos RP. "Pathogenesis of Theiler's murine encephalomyelitis virus-induced disease." In *Molecular Biology of Picornaviruses*, B. L. Semler, E. Wimmer, eds. Washington, D.C.: ASM Press, pp. 427-435, 2002.
3. Tsunoda I, Fujinami RS. Two models for multiple sclerosis: Experimental allergic encephalomyelitis and Theiler's murine encephalomyelitis virus. J Neuropathol Exp Neurol 1996;55:673-86.
4. Tsunoda I, Fujinami RS. "Theiler's murine encephalomyelitis virus." In *Persistent Viral Infections*, R. Ahmed, I. Chen, eds. Chichester: John Wiley & Sons Ltd, pp. 517-526, 1999.
5. Tsunoda I, Fujinami RS. Inside-out versus outside-in models for virus induced demyelination: Axonal damage trigger demyelination. Springer Semin Immunopathol 2002;24:105-25.
6. Theiler M. Spontaneous encephalomyelitis of mice--A new virus disease. Science 1934;80:122.
7. Bendiner E. Max Theiler: Yellow jack and the jackpot. Hosp Pract 1988;23:211-44.

8. Jubelt B, Miller JR. "Viral infections." In *Merritt's Neurology*, 10th Edition, L. P. Rowland, ed. Philadelphia: Lippincott Williams & Wilkins, pp. 134-162, 2000.

9. Rosenberg NL, Rotbart HA, Abzug MJ, Ringel SP, Levin MJ. Evidence for a novel picornavirus in human dermatomyositis. Ann Neurol 1989;26:204-9.

10. Lipton HL, Friedmann A, Sethi P, Crowther JR. Characterization of Vilyuisk virus as a picornavirus. J Med Virol 1983;12:195-203.

11. Goldfarb LG, Gajdusek DC. Viliuisk encephalomyelitis in the Iakut people of Siberia. Brain 1992;115:961-78.

12. Tsunoda I, Iwasaki Y, Terunuma H, Sako K, Ohara Y. A comparative study of acute and chronic diseases induced by two subgroups of Theiler's murine encephalomyelitis virus. Acta Neuropathol (Berl) 1996;91:595-602.

13. Dal Canto MC, Barbano RL. Remyelination during remission in Theiler's virus infection. Am J Pathol 1984;116:30-45

14. Knobler RL, Rodriguez M, Lampert PW, Oldstone MBA. Virologic models of chronic relapsing demyelinating disease. Acta Neuropathol Suppl (Berl) 1983;9:31-7.

15. Rodriguez M. Virus-induced demyelination in mice: "dying back" of oligodendrocytes. Mayo Clin Proc 1985;60:433-8.

16. Tsunoda I, Kuang L-Q, Theil DJ, Fujinami RS. Antibody association with a novel model for primary progressive multiple sclerosis: Induction of relapsing-remitting and progressive forms of EAE in *H2$^s$* mouse strains. Brain Pathol 2000;10:402-18.

17. Lublin FD, Reingold SC. Defining the clinical course of multiple sclerosis: results of an international survey. National Multiple Sclerosis Society (USA) Advisory Committee on Clinical Trials of New Agents in Multiple Sclerosis. Neurology 1996;46:907-11.

18. Tolley ND, Tsunoda I, Fujinami RS. DNA vaccination against Theiler's murine encephalomyelitis virus leads to alterations in demyelinating disease. J Virol 1999;73:993-1000.

19. Watanabe R, Wege H, ter Meulen V. Adoptive transfer of EAE-like lesions from rats with coronavirus-induced demyelinating encephalomyelitis. Nature 1983;305:150-3.

20. Barbano RL, Dal Canto MC. Serum and cells from Theiler's virus-infected mice fail to injure myelinating cultures or to produce *in vivo* transfer of disease. The pathogenesis of Theiler's virus-induced demyelination appears to differ from that of EAE. J Neurol Sci 1984;66:283-93.

21. Fiette L, Aubert C, Brahic M, Pena-Rossi C. Theiler's virus infection of $\beta_2$-microglobulin-deficient mice. J Virol 1993;67:589-92.

22. Pullen LC, Miller SD, Dal Canto MC, Kim BS. Class I-deficient resistant mice intracerebrally inoculated with Theiler's virus show an increased T cell response to viral antigens and susceptibility to demyelination. Eur J Immunol 1993;23:2287-93.

23. Njenga MK, Pavelko KD, Baisch J, Lin X, David C, Leibowitz J, Rodriguez M. Theiler's virus persistence and demyelination in major histocompatibility complex class II-deficient mice. J Virol 1996;70:1729-37.

24. Murray PD, Pavelko KD, Leibowitz J, Lin X, Rodriguez M. CD4$^+$ and CD8$^+$ T cells make discrete contributions to demyelination and neurologic disease in a viral model of multiple sclerosis. J Virol 1998;72:7320-9.

25. Massa PT, Ropka SL, Saha S, Fecenko KL, Beuler KL. Critical role for protein tyrosine phosphatase SHP-1 in controlling infection of central nervous system glia and demyelination by Theiler's murine encephalomyelitis virus. J Virol 2002;76:8335-46.

26. Theiler M. Spontaneous encephalomyelitis of mice, a new virus disease. J Exp Med 1937;65:705-19.

27. Theiler M, Gard S. Encephalomyelitis of mice. III. Epidemiology. J Exp Med 1940;72:79-90.

28. Brownstein D, Bhatt P, Ardito R, Paturzo F, Johnson E. Duration and patterns of transmission of Theiler's mouse encephalomyelitis virus infection. Lab Anim Sci 1989;39:299-301.

29. Libbey JE, McCright IJ, Tsunoda I, Wada Y, Fujinami RS. Peripheral nerve protein, P0, as a potential receptor for Theiler's murine encephalomyelitis virus. J NeuroVirol 2001;7:97-104.

30. Bihl F, Pena-Rossi C, Guénet J-L, Brahic M, Bureau J-F. The shiverer mutation affects the persistence of Theiler's virus in the central nervous system. J Virol 1997;71:5025-30.

31. Bureau J-F, Montagutelli X, Bihl F, Lefebvre S, Guénet J-L, Brahic M. Mapping loci influencing the persistence of Theiler's virus in the murine central nervous system. Nat Genet 1993;5:87-91.

32. Aubagnac S, Behrens L, Bihl BF, Brahic M, Bureau J-F, Vigneau S. "Genetic control of Theiler's virus infection." In *Genes and Viruses in Multiple Sclerosis*, O. R. Hommes, M. Clanet, H. Wekerle, eds. Amsterdam, Netherlands: Elsevier Science B.V., pp. 57-59, 2001.

33. Rose J, Gerken S, Lynch S, Pisani P, Varvil T, Otterud B, Leppert M. Genetic susceptibility in familial multiple sclerosis not linked to the myelin basic protein gene. Lancet 1993;341:1179-81.

34. Graham CA, Kirk CW, Nevin NC, Droogan AG, Hawkins SA, McMillan SA, McNeill TA. Lack of association between myelin basic protein gene microsatellite and multiple sclerosis. Lancet 1993;341:1596.

35. Yamada M, Zurbriggen A, Fujinami RS. The relationship between viral RNA, myelin-specific mRNAs, and demyelination in central nervous system disease during Theiler's virus infection. Am J Pathol 1990;137:1467-79.

36. Ozden S, Aubert C, Gonzalez-Dunia D, Brahic M. *In situ* analysis of proteolipid protein gene transcripts during persistent Theiler's virus infection. J Histochem Cytochem 1991;39:1305-9.

37. Ozden S, Aubert C, Bureau J-F, Gonzalez-Dunia D, Brahic M. Analysis of proteolipid protein and P0 transcripts in mice infected with Theiler's virus. Microb Pathog 1993;14:123-31.

38. Rodriguez M, Prayoonwiwat N, Howe C, Sanborn K. Proteolipid protein gene expression in demyelination and remyelination of the central nervous system: a model for multiple sclerosis. J Neuropathol Exp Neurol 1994;53:136-43.

39. Tsunoda I, Kurtz CIB, Fujinami RS. Apoptosis in acute and chronic central nervous system disease induced by Theiler's murine encephalomyelitis virus. Virology 1997;228:388-93.

40. Oldstone MBA. "Virus can alter cell function without causing cell pathology: Disordered function leads to imbalance of homeostasis and disease." In *Concepts in Viral Pathogenesis*, A. L. Notkins, M. B. A. Oldstone, eds. New York: Springer-Verlag, pp. 269-276, 1984.

41. Rodriguez M, Scheithauer BW, Forbes G, Kelly PJ. Oligodendrocyte injury is an early event in lesions of multiple sclerosis. Mayo Clin Proc 1993;68:627-36.

42. Gordon J, Gallia GL, Valle LD, Amini S, Khalili K. Human polyomavirus JCV and expression of myelin genes. J NeuroVirol 2000;6 Suppl 2:S92-S97.

43. Rauch HC, Montgomery IN, Hinman CL, Harb W, Benjamins JA. Chronic Theiler's virus infection in mice: Appearance of myelin basic protein in the cerebrospinal fluid and serum antibody directed against MBP. J Neuroimmunol 1987;14:35-48.

44. Welsh CJR, Tonks P, Nash AA, Blakemore WF. The effect of L3T4 T cell depletion on the pathogenesis of Theiler's murine encephalomyelitis virus infection in CBA mice. J Gen Virol 1987;68:1659-67.

45. Cash E, Bandeira A, Chirinian S, Brahic M. Characterization of B lymphocytes present in the demyelinating lesions induced by Theiler's virus. J Immunol 1989;143:984-8.

46. Rodriguez M, Lucchinetti CF, Clark RJ, Yakash TL, Markowitz H, Lennon VA. Immunoglobulins and complement in demyelination induced in mice by Theiler's virus. J Immunol 1988;140:800-6.

47. Roos RP, Nalefski EA, Nitayaphan S, Variakojis R, Singh KK. An isoelectric focusing overlay study of the humoral immune response in Theiler's virus demyelinating disease. J Neuroimmunol 1987;13:305-14.

48. Seil FJ. "Effects of humoral factors on myelin in organotypic cultures." In *Multiple Sclerosis: Current Status of Research and Treatment*, R. M. Herndon, F. J. Seil, eds. New York: Demos Publications, pp. 33-50, 1994.

49. Linington C, Bradl M, Lassmann H, Brunner C, Vass K. Augmentation of demyelination in rat acute allergic encephalomyelitis by circulating mouse monoclonal antibodies directed against a myelin/oligodendrocyte glycoprotein. Am J Pathol 1988;130:443-54.

50. Lanzavecchia A. Receptor-mediated antigen uptake and its effect on antigen presentation to class II-restricted T lymphocytes. Annu Rev Immunol 1990;8:773-93.

51. Roth R, Gee RJ, Mamula MJ. B lymphocytes as autoantigen-presenting cells in the amplification of autoimmunity. Ann N Y Acad Sci 1997;815:88-104.

52. Mori T, Ono K, Hakozaki M, Kasukawa R, Kochi H. Autoantibodies of sera from patients with primary biliary cirrhosis recognize the $\alpha$ subunit of the decarboxylase component of human branched-chain 2-oxo acid dehydrogenase complex. J Hepatol 2001;34:799-804.

53. Rubio N, Cuesta A. Lack of cross-reaction between myelin basic proteins and putative demyelinating virus envelope proteins. Mol Immunol 1989;26:663-8.

54. Fujinami RS, Zurbriggen A, Powell HC. Monoclonal antibody defines determinant between Theiler's virus and lipid-like structures. J Neuroimmunol 1988;20:25-32.

55. Fujinami RS. Immune responses against myelin basic protein and/or galactocerebroside cross-react with viruses: implications for demyelinating disease. Curr Top Microbiol Immunol 1989;145:93-100.

56. Yamada M, Zurbriggen A, Fujinami RS. Monoclonal antibody to Theiler's murine encephalomyelitis virus defines a determinant on myelin and oligodendrocytes, and augments demyelination in experimental allergic encephalomyelitis. J Exp Med 1990;171:1893-907.

57. Fujinami RS, Rosenthal A, Lampert PW, Zurbriggen A, Yamada M. Survival of athymic (*nu/nu*) mice after Theiler's murine encephalomyelitis virus infection by passive administration of neutralizing monoclonal antibody. J Virol 1989;63:2081-7.

58. Lassmann H, Brunner C, Bradl M, Linington C. Experimental allergic encephalomyelitis: The balance between encephalitogenic T lymphocytes and demyelinating antibodies determines size and structure of demyelinated lesions. Acta Neuropathol (Berl) 1988;75:566-76.

59. Arnon R, Crisp E, Kelley R, Ellison GW, Myers LW, Tourtellotte WW. Anti-ganglioside antibodies in multiple sclerosis. J Neurol Sci 1980;46:179-86.

60. Vanderlugt CL, Miller SD. Epitope spreading in immune-mediated diseases: Implications for immunotherapy. Nat Rev Immunol 2002;2:85-95.

61. Lehmann PV, Forsthuber T, Miller A, Sercarz EE. Spreading of T-cell autoimmunity to cryptic determinants of an autoantigen. Nature 1992;358:155-7.

62. Miller SD, Olson JK, Croxford JL. Multiple pathways to induction of virus-induced autoimmune demyelination: lessons from Theiler's virus infection. J Autoimmun 2001;16:219-27.

63. Miller SD, Clatch RJ, Pevear DC, Trotter JL, Lipton HL. Class II-restricted T cell responses in Theiler's murine encephalomyelitis virus (TMEV)-induced demyelinating disease. I. Cross- specificity among TMEV substrains and related picornaviruses, but not myelin proteins. J Immunol 1987;138:3776-84.

64. Miller SD, Gerety SJ, Kennedy MK, Peterson JD, Trotter JL, Tuohy VK, Waltenbaugh C, Dal Canto MC, Lipton HL. Class II-restricted T cell responses in Theiler's murine encephalomyelitis virus (TMEV)-induced demyelinating disease. III. Failure of neuroantigen-specific immune tolerance to affect the clinical course of demyelination. J Neuroimmunol 1990;26:9-23.

65. Lang W, Wiley C, Lampert P. Theiler's virus encephalomyelitis is unaffected by treatment with myelin components. J Neuroimmunol 1985;9:109-13.

66. Drescher KM, Rivera-Quinones C, Lucchinetti CF, Rodriguez M. Failure of treatment with Linomide or oral myelin tolerization to ameliorate demyelination in a viral model of multiple sclerosis. J Neuroimmunol 1998;88:111-9.

67. Miller SD, Vanderlugt CL, Begolka WS, Pao W, Yauch RL, Neville KL, Katz-Levy Y, Carrizosa A, Kim BS. Persistent infection with Theiler's virus leads to CNS autoimmunity via epitope spreading. Nat Med 1997;3:1133-6.

68. Katz-Levy Y, Neville KL, Girvin AM, Vanderlugt CL, Pope JG, Tan LJ, Miller SD. Endogenous presentation of self myelin epitopes by CNS-resident APCs in Theiler's virus-infected mice. J Clin Invest 1999;104:599-610.

69. Katz-Levy Y, Neville KL, Padilla J, Rahbe S, Begolka WS, Girvin AM, Olson JK, Vanderlugt CL, Miller SD. Temporal development of autoreactive Th1 responses and endogenous presentation of self myelin epitopes by central nervous system-resident APCs in Theiler's virus-infected mice. J Immunol 2000;165:5304-14.

70. Neville KL, Dal Canto MC, Bluestone JA, Miller SD. CD28 costimulatory blockade exacerbates disease severity and accelerates epitope spreading in a virus-induced autoimmune disease. J Virol 2000;74:8349-57.

71. Neville KL, Padilla J, Miller SD. Myelin-specific tolerance attenuates the progression of a virus-induced demyelinating disease: implications for the treatment of MS. J Neuroimmunol 2002;123:18-29.

72. Elliott EA, McFarland HI, Nye SH, Cofiell R, Wilson TM, Wilkins JA, Squinto SP, Matis LA, Mueller JP. Treatment of experimental encephalomyelitis with a novel chimeric fusion protein of myelin basic protein and proteolipid protein. J Clin Invest 1996;98:1602-12.

73. Borrow P, Welsh CJR, Tonks P, Dean D, Blakemore WF, Nash AA. Investigation of the role of delayed-type-hypersensitivity responses to myelin in the pathogenesis of Theiler's virus-induced demyelinating disease. Immunology 1998;93:478-84.

74. Takacs K, Chandler P, Altmann DM. Relapsing and remitting experimental allergic encephalomyelitis: a focused response to the encephalitogenic peptide rather than epitope spread. Eur J Immunol 1997;27:2927-34.

75. Anderson AC, Nicholson LB, Legge KL, Turchin V, Zaghouani H, Kuchroo VK. High frequency of autoreactive myelin proteolipid protein-specific T cells in the periphery of naive mice: Mechanisms of selection of the self-reactive repertoire. J Exp Med 2000;191:761-70.

76. Wanebo HJ, Gallmeier WM, Boyse EA, Old LJ. Paraproteinemia and reticulum cell sarcoma in an inbred mouse strain. Science 1966;154:901-3.

77. Dal Canto MC, Calenoff MA, Miller SD, Vanderlugt CL. Lymphocytes from mice chronically infected with Theiler's murine encephalomyelitis virus produce demyelination of organotypic cultures after stimulation with the major encephalitogenic epitope of

myelin proteolipid protein. Epitope spreading in TMEV infection has functional activity. J Neuroimmunol 2000;104:79-84.

78. Cohen IR, Schwartz M. Autoimmune maintenance and neuroprotection of the central nervous system. J Neuroimmunol 1999;100:111-4.

79. Olson JK, Croxford JL, Calenoff MA, Dal Canto MC, Miller SD. A virus-induced molecular mimicry model of multiple sclerosis. J Clin Invest 2001;108:311-8.

80. Olson JK, Eagar TN, Miller SD. Functional activation of myelin-specific T cells by virus-induced molecular mimicry. J Immunol 2002;169:2719-26.

81. Theil DJ, Tsunoda I, Rodriguez F, Whitton JL, Fujinami RS. Viruses can silently prime for and trigger central nervous system autoimmune disease. J NeuroVirol 2001;7:220-7.

82. Kuchroo VK, Greer JM, Kaul D, Ishioka G, Franco A, Sette A, Sobel RA, Lees MB. A single TCR antagonist peptide inhibits experimental allergic encephalomyelitis mediated by a diverse T cell repertoire. J Immunol 1994;153:3326-36.

83. Bielekova B, Goodwin B, Richert N, Cortese I, Kondo T, Afshar G, Gran B, Eaton J, Antel J, Frank JA, McFarland HF, Martin R. Encephalitogenic potential of the myelin basic protein peptide (amino acids 83-99) in multiple sclerosis: Results of a phase II clinical trial with an altered peptide ligand. Nat Med 2000;6:1167-75.

84. Genain CP, Zamvil SS. Specific immunotherapy: One size does not fit all. Nat Med 2000;6:1098-100.

85. Traugott U, Stone SH, Raine CS. Chronic relapsing experimental autoimmune encephalomyelitis. Treatment with combinations of myelin components promotes clinical and structural recovery. J Neurol Sci 1982;65-73.

86. Lang W, Rodriguez M, Lennon VA, Lampert PW. Demyelination and remyelination in murine viral encephalomyelitis. Ann N Y Acad Sci 1984;436: 98-102.

87. Miller DJ, Rodriguez M. Spontaneous and induced remyelination in multiple sclerosis and the Theiler's virus model of central nervous system demyelination. Microsc Res Tech 1995;32:230-45.

88. Rodriguez M, Lennon VA. Immunoglobulins promote remyelination in the central nervous system. Ann Neurol 1990;27:12-7.

89. Rodriguez M, Pierce ML, Thiemann RL. Immunoglobulins stimulate central nervous system remyelination: Electron microsco microscopic and morphometric analysis of proliferating cells. Lab Invest 1991;64:358-70.

90. Prayoonwiwat N, Rodriguez M. The potential for oligodendrocyte proliferation during demyelinating disease. J Neuropathol Exp Neurol 1993;52:55-63.

91. Rodriguez M, Lennon VA, Benveniste EN, Merrill JE. Remyelination by oligodendro-cytes stimulated by antiserum to spinal cord. J Neuropathol Exp Neurol 1987;46:84-95.

92. Patick AK, Thiemann RL, O'Brien PC, Rodriguez M. Persistence of Theiler's virus infection following promotion of central nervous system remyelination. J Neuropathol Exp Neurol 1991;50:523-37.

93. Rodriguez M, Miller DJ, Lennon VA. Immunoglobulins reactive with myelin basic protein promote CNS remyelination. Neurology 1996;46:538-45.

94. Miller DJ, Rodriguez M. A monoclonal autoantibody that promotes central nervous system remyelination in a model of multiple sclerosis is a natural autoantibody encoded by germline immunoglobulin genes. J Immunol 1995;154:2460-9.

95. Miller DJ, Njenga MK, Parisi JE, Rodriguez M. Multi-organ reactivity of a monoclonal natural autoantibody that promotes remyelination in a mouse model of multiple sclerosis. J Histochem Cytochem 1996;44:1005-11.

96. Miller DJ, Njenga MK, Murray PD, Leibowitz J, Rodriguez M. A monoclonal natural autoantibody that promotes remyelination suppresses central nervous system inflam-

mation and increases virus expression after Theiler's virus-induced demyelination. Int Immunol 1996;8:131-41.

97.  Miller DJ, Sanborn KS, Katzmann JA, Rodriguez M.  Monoclonal autoantibodies promote central nervous system repair in an animal model of multiple sclerosis. J Neurosci 1994;14:6230-8.

98.  Miller DJ, Bright JJ, Sriram S, Rodriguez M.  Successful treatment of established relapsing experimental autoimmune encephalomyelitis in mice with a monoclonal natural autoantibody. J Neuroimmunol 1997;75:204-9.

99.  Pavelko KD, van Engelen BGM, Rodriguez M.  Acceleration in the rate of CNS remyelination in lysolecithin-induced demyelination. J Neurosci 1998;18:2498-505.

100.  Asakura K, Miller DJ, Pease LR, Rodriguez M.  Targeting of IgMκ antibodies to oligodendrocytes promotes CNS remyelination. J Neurosci 1998;18:7700-8.

101.  Warrington AE, Asakura K, Bieber AJ, Ciric B, Van Keulen V, Kaveri SV, Kyle RA, Pease LR, Rodriguez M.  Human monoclonal antibodies reactive to oligodendrocytes promote remyelination in a model of multiple sclerosis. Proc Natl Acad Sci U S A 2000;97:6820-5.

102.  Bieber AJ, Warrington A, Asakura K, Ciric B, Kaveri SV, Pease LR, Rodriguez M. Human antibodies accelerate the rate of remyelination following lysolecithin-induced demyelination in mice. Glia 2002;37:241-9.

103.  Asakura K, Miller DJ, Pogulis RJ, Pease LR, Rodriguez M.  Oligodendrocyte-reactive O1, O4, and HNK-1 monoclonal antibodies are encoded by germline immunoglobulin genes. Brain Res Mol Brain Res 1995;34:283-93.

104.  Avrameas S.  Natural autoantibodies: from 'horror autotoxicus' to 'gnothi seauton'. Immunol Today 1991;12:154-9.

105.  Fujinami RS, Oldstone MBA, Wroblewska Z, Frankel ME, Koprowski H.  Molecular mimicry in virus infection: Crossreaction of measles virus phosphoprotein or of herpes simplex virus protein with human intermediate filaments. Proc Natl Acad Sci USA 1983;80:2346-50.

106.  Fujinami RS, Oldstone MBA.  "Cross-reaction of measles virus phosphoprotein with a human intermediate filament: molecular mimicry during virus infection." In *Nonsegmented Negative Strand Viruses*, Academic Press, Inc., pp. 427-434, 1984.

107.  Kirschning E, Jensen K, Dübel S, Rutter G, Hohenberg H, Will H. Primary structure of the antigen-binding domains of a human oligodendrocyte-reactive IgM monoclonal antibody derived from a patient with multiple sclerosis. J Neuroimmunol 1999;99:122-30.

108.  Matsiota P, Blancher A, Doyon B, Guilbert B, Clanet M, Kouvelas ED, Avrameas S. Comparative study of natural autoantibodies in the serum and cerebrospinal fluid of normal individuals and patients with multiple sclerosis and other neurological diseases. Ann Inst Pasteur Immunol 1988;139:99-108.

Chapter B3

# THE ROLE OF ASTROCYTES, OLIGODENDROCYTES, MICROGLIA AND ENDOTHELIAL CELLS IN TMEV INFECTION

MC Dal Canto and CL VanderLugt
*Department of Pathology, Northwestern University Medical School, Chicago, IL*

**Abstract**:    Chronic TMEV infection in susceptible strains leads to an inflammatory demyelinating disease modeling MS. TMEV antigen can be found abundantly in microglia and infiltrating macrophage, but viral growth is restricted in these cells. Viral persistence has been seen in astrocytes, microglia, and oligodendrocytes. However, infiltrating macrophage and, to some extent, resident microglia play an important role in antigen presentation, inflammation and epitope spreading.

**Key words**:    Theiler's virus infection; demyelination; astrocytes; microglial cells; oligodendrocytes; viral persistence; epitope spreading

## 1.    INTRODUCTION

### 1.1    TMEV infection

*Theiler's murine encephalomyelitis virus* (TMEV) is an enteric pathogen of mice belonging to the *Picornaviridae* family (1). Two groups of TMEVs include high neurovirulent strains such as GDVII and low neurovirulence TMEVs such as BeAn and DA. Intracerebral inoculation with GDVII results in a rapidly fatal encephalitis in which grey matter neurons are infected and lysed. In contrast, BeAn and DA strains experimentally injected into susceptible strains of mice result in an initial acute phase of grey matter involvement followed by a chronic phase of viral persistence, inflammation and demyelination in the white matter of the spinal cord (2), TMEV-induced demyelinating disease (TMEV-IDD).

The extent of acute phase grey matter involvement and the proposed mechanisms of the chronic phase demyelination depend largely on the strain of virus and the strain of mice infected. The initial acute grey matter involvement following inoculation of the DA strain into SJL mice is characterized by microglial proliferation and neuronal necrosis. Surviving mice go on to develop TMEV-IDD. In contrast, intracerebral inoculation of BeAn into SJL mice results in a very limited and clinically silent early acute phase grey matter disease. Chronic demyelination in the BeAn/SJL model appears to be due to the Th1 inflammatory response rather than direct lysis of virally infected oligodendrocytes.

Histopathologically, TMEV-IDD resembles the human demyelinating disease multiple sclerosis (MS).

## 1.2 TMEV-IDD as a model of MS

There is evidence for a viral etiology (epidemiology) of multiple sclerosis (3, 4), making TMEV-IDD an important model of MS. Although no one virus has been shown to be consistently associated with MS, early infection may trigger events, through molecular mimicry (5) or epitope spreading (6), that eventually result in disease. Like TMEV-IDD, MS is an immune-mediated demyelinating disease characterized by perivascular CD4+ T cell and mononuclear cell infiltration, with subsequent primary demyelination of axonal tracts, leading to progressive paralysis (7). MS is generally considered to have an autoimmune component, however a direct cause-effect relationship between myelin reactivity and disease has not been established. Interestingly, although TMEV-IDD is due initially to a persistent CNS viral infection, autoimmune anti-myelin Th1 responses are seen during the chronic phase of disease.

Three interesting areas of investigation in TMEV-IDD include identification of the CNS cells harbouring the persistent infection, the antigen-presenting cells (CNS APCs) responsible for presenting viral antigen and inducing the Th1 inflammatory response leading to demyelination, and the CNS cells involved in epitope spreading, from viral epitopes to myelin targets, in this chronic inflammatory disease.

## 2. THE ROLE OF CNS GLIA IN TMEV-IDD

### 2.1 Persistent infection

Demyelination and the resulting clinical disease symptoms are not the consequence of viral lysis of oligodentrocytes, but are immune mediated, due to the inflammatory response in the CNS (8, 9). Viral persistence is however required for chronic disease, since mouse strains which clear the virus do not go on to chronic demyelination that characterizes TMEV-IDD. Therefore it is important to determine which cells harbor the persistent virus.

Initial observations using immuno-histochemistry demonstrated that viral antigens could be found in abundance in the spinal cord of TMEV infected mice in macrophage/microglial cells and in astrocytes (10, 11). In contrast to the abundance of viral antigen, infections virus has been consistently shown to be low ($<10^4$ PFU). Viral persistence is routinely demonstrated by plaque assays of the spinal cord and brain tissue of infected mice. Historically, macrophage/microglia have been thought to be the major reservour of infectious virus. Macrophage depletion with mannosylated liposomes abrogated TMEV persistence (12). However, it had also been demonstrated that viral replication was blocked in these cells (11, 12).

Recently, Lipton et. al. analysed the copy number of TMEV genomes, plus- to minus-strand ratios, and full-length. Interestingly, large numbers of viral genomes were detected (13). To some extent, an abundance of viral antigen and viral genomes combined with very low amounts of infectious virus can be explained by neutralization of infectious virus via the immune response and its restricted growth in macrophages (13, 14), blocked after viral RNA replication.

Although macrophage/microglia have historically been thought to be the predominant cell type supporting TMEV replication, other cell types may be important for persistence. Both oligodendrocytes and astrocytes have been suggested to harbor TMEV *in vivo*. Oligodendrocytes isolated from infected SJL mice contained amounts of viral genome equal on a per cell basis to that of macrophages (13) and an oligodendrocyte line proved to be significantly more productive in terms of infectious virus (PFU) per cell than macrophages. However, most early immuno-histochemical studies showed little if any viral antigen in oligodendrocytes during TMEV-IDD.

Although astrocytes have been considered by some to be an unlikely source of viral persistence, early immunohistochemical studies demonstrated viral antigen in these cells during TMEV-IDD (10), and more recent work in our lab suggests rethinking this historic view (15). *De novo* viral protein synthesis was seen in astrocytes and, to a lesser extent, in oligodentrocytes and microglia, when these cells (obtained from neonatal mice) were infected *in vitro* in the presence of $^{35}$S-methionine. Additionally, when a GFP-BeAn virus was used to infect microglia, oligodendrocytes and astrocytes *in vitro*, similar results were seen. *De novo* protein synthesis (GFP is translated as part of the viral leader) was clearly seen in astrocytes. Plaque assays and RNAse protection assays confirmed the findings that astrocytes from neonatal mice were productively infected *in vitro* to a greater extent than oligodendrocytes. Minimal cytopathic effects resulted from this infection. The most remarkable finding was that the ratio of viral replication in astrocytes versus microglial cells was 100 to 4 as quantitated by Image Quant analysis in a Phosphor Imager measuring negative RNA strand production in these cells. The production of negative strand RNA is the most sensitive method to measure true viral replication (Fig 1).

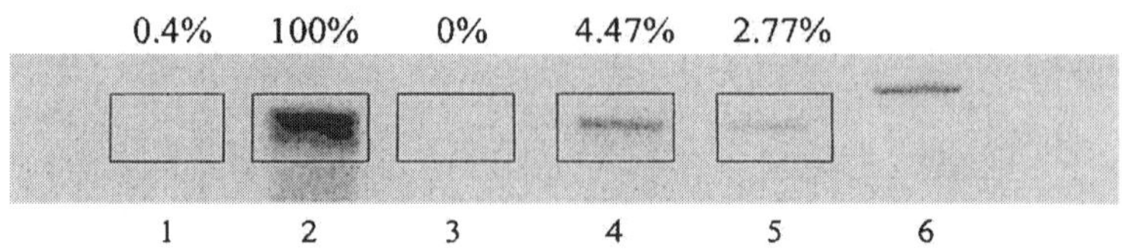


Fig 1. Rnase protection assay for the detection of TMEV replication. A positive RNA transcript, containing 208 specific TMEV nucleotides, was protected by the replicative, negative strand of TMEV in infected astrocytes and microglial cells. (1) Uninfected astrocytes, (2) infected astrocytes, (3) uninfected microglia, (4) infected microglia, (5) microglia infected in the presence of INF-gamma and (6) 258 base transcript probe. The amount of RNA protected was quan titated by Image Quant analysis in a Phosphor Imager. The values represent amount of the RNA protected in microglia as a percentage of the amount protected in astrocytes, which was assigned a value of 100. (Reproduced with permission from J Neuroimmunol 118:256-267, 2001)

While providing new data regarding the unexpected capacity of astrocytes to become infected by TMEV, these studies also confirm that viral transcription is restricted in microglia, as earlier studies had showed. Confocal microscopy of immuno-histochemical stains of the brain and spinal cord from mice with TMEV-IDD confirmed the *in vitro* results as staining with anti-TMEV antibody co-localized with GFAP staining. In fact, most astrocytes stained positive for viral antigen (Fig 2) (15).

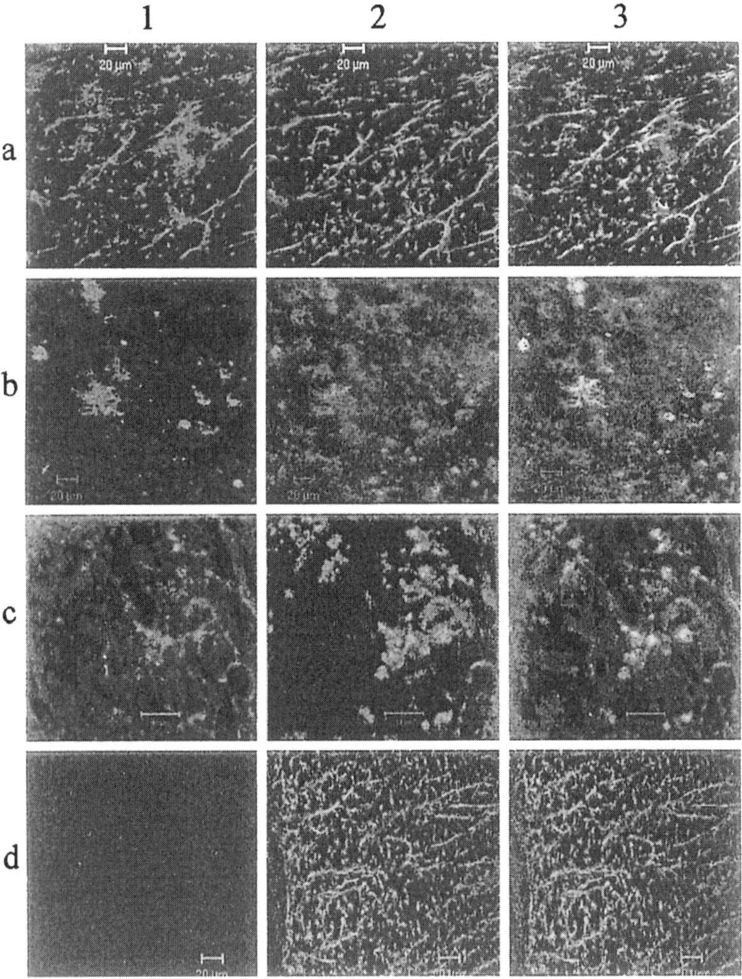

Fig 2. Immunohistologic analysis of spinal cord tissues from mice suffering from TMEV-induced demyelinating disease or EAE disease. Double immunostaining of frozen sections was done using both the BeAn antisera with a rhodamine conjugated secondary antibody (red) and different antibodies to cell specific markers with FITC-conjugated secondary antibodies (green). Column 1: viral antigen  in (a) astrocytes, (b) microglia, (c) oligodendrocytes and (d) EAE. Column 2: (a and d) anti-GFAP for astrocytes, (b) MOMA-2 for macrophages, (c) anti-CNPase for oligodendrocytes. Column 3:colocalization of the two stains. Bar in row C measures 10 microns; all other bars measure 20 microns. (Reproduced with permission from J Neuroimmunol 118:256-267, 2001)

The relative resistance of oligodendrocytes to TMEV infection is of interest and supports earlier studies that showed different responses to infection by mature oligodendrocytes when compared with an immortalized, developmentally arrested oligodendroglial cell line (16). Those studies demonstrated that infected mature oligodendrocytes, similarly to astrocytes, revealed very little, if any, cytopathic effect. In addition, when synthesis of mRNA for myelin proteins, such as MBP and MOG was investigated, no suppression could be demonstrated. This contrasted with the demonstration of both cytopathic effects and suppression of MBP and MOG mRNA synthesis in an immortalized line of oligodendrocytes which was arrested at an early developmental level. These studies suggest that adult, differentiated oligodendrocytes are quite resistant to TMEV cytopathic effects, both structurally and functionally, although a low level infection is permitted. On the other hand, developing oligodendrocytes are more susceptible to TMEV-induced injury, a conclusion that correlates with previous studies demonstrating direct oligodendroglial cytopathic effects in neonatal mice (17).

The importance of oligodendroglial resistance to TMEV infection is also underscored by the ability of oligodendrocytes to be stimulated into a remyelinating mode after demyelination produced by different strains of TMEV. Without stimulation, very little remyelination can be seen in SJL/J mice, in which demyelination is generally quite severe. In strains of mice in which the demyelinating process is more subdue, or when using certain TMEV strains, spontaneous remyelination can occur, however, most of the remyelinating activity is carried out by invading Schwann cells, and only a minority of the repair process is spontaneously carried out by oligodendrocytes (reviewed in 18). Recently, however, Rodriguez and his colleagues have demonstrated that oligodendroglial cells may be induced to produce abundant remyelination when appropriately stimulated.

Their initial studies utilized anti-spinal cord antiserum to stimulate remyelination in TMEV-infected mice (19). Later, IgM monoclonal antibodies were prepared, and these showed the same capability of enhancing oligodendroglia-mediated remyelination (20, 21). These studies suggested that binding of antibodies to the surface of oligodendrocytes initiated a series of events eventually leading to remyelination. Subsequently, these investigators demonstrated that an IgM antibody present in human serum was also capable to enhance remyelination in the CNS of demyelinated animals (22). These results were confirmed by using a recombinant human IgM antibody generated by DNA-mediated gene transfer (23). *In vitro* studies showed binding of this antibody to oligodendrocytes resulting in enhanced Ca entry into the cells. Studies *in vivo*, utilizing chronically TMEV-infected mice, clearly showed enhanced remyelination. These studies strongly suggest that IgM antibody binding to the oligodendroglial surface initiates a signaling cascade, that eventually results in enhanced remyelination. These studies open the possibility of using antibody therapy in MS patients to stimulate myelin repair.

Summary: During ongoing TMEV-IDD macrophage/microglia phagocytose viral particles but do not allow viral replication, blocked after transcription. Oligodendrocytes may be infected, but the bulk of active infectious virus replication is found in astrocytes. Oligodendrocytes, even after infection are resistant to cytolysis and they may produce substantial remyelination when stimulated with antibody binding to their surface.

## 2.2 Antigen presentation and inflammation

Although viral persistence is necessary for disease, the immune response itself, not direct viral lysis, appears to cause the demyelination characteristic of TMEV-IDD. Therefore, another important area of investigation includes the identification of the cells in the CNS important in initiating and sustaining the immune response resulting in CNS inflammation and demyelination.

The brain has been considered an immune-privileged site due to the blood-brain-barrier (BBB). The BBB includes cerebrovascular endothelial cells (CVEs) and astrocytes. CVEs differ from endothelial cells in the capillaries of other organs by their continuous tight junctions and lack of transendothelial pathways. The foot processes from astrocyes cover almost the entire surface of the CVE basement membrane. Together these cells provide a continuous cellular barrier between the blood and the interstitial fluid limiting the entry of immune cells such as T lymphocytes and antibodies.

Activation of T cells requires two signals. The antigen-specific signal through the major histocompatibility complex (MHC) class II plus peptide and costimulatory signals, most notably through the B7 molecules on APCs. Efficient antigen presenting cells must take in antigen, process it into peptides which then are bound in the groove of the MHC class II molecule on its surface. Additionally, costimulatory molecules like CD40, B7-1 and B7-2 must be expressed in sufficient quantities for effective activation of CD4+ T cells. CNS cells express low to no levels of (MHC) class II, costimulatory and adhesion molecules and are therefore, under normal circumstances, not capable of activating T lymphocytes.

During an inflammatory immune response, however, the BBB is disrupted and activated T cells along with infiltrating peripheral monocytes/macrophage begin to traffic into the CNS. How this initiation of CNS inflammation occurs and which resident CNS cells serve as antigen presenting cells (APCs), for initiation and during ongoing inflammation, has been the subject of much investigation.

When treated *in vitro* with IFN-γ, SJL astrocytes (isolated from neonatal mice) upregulated adhesion molecules ICAM-1 and VCAM-1, in addition to MHC class II, invariant chain, H2-M, and costimulatory molecules CD40 and B7-1. These IFN-γ-treated astrocytes were capable of processing PLP139-151 peptide from intact proteolipid protein (PLP) and induced B7-1 dependent activation of Th1 lines and lymph node T cells (24). Soos *et al* also demonstrated that astrocyte clones are capable of processing and presenting protein antigen to activate T cells (25).

SJL astrocytes isolated from neonatal mice were compared with cloned SJL CVEs. *In vitro* studies demonstrated that although neither astrocytes nor CVEs constitutively express MHC class II, both cells up-regulated expression when incubated with IFN-γ alone. TNF-α inhibits IFN-γ-induced MHC class II up-regulation on CVEs, but not on astrocytes. Limited expression of costimulatory and adhesion molecules were found on CVEs as compared to astrocytes. B7-1, ICAM-1 and VCAM are found on astrocytes and upregulated by IFN-γ. CD40, another important costimulatory molecule, was expressed on neither CVEs or astrocytes. B7-1 and VCAM was very low on CVEs but VCAM was upregulated strongly by treatment with IFN-γ and TNF-α together. Not surprisingly, astrocytes, but not CVEs, elicit MHC class II-restricted T cell responses (26). These data suggest that astrocytes, in the inflammatory milieu of TMEV-IDD, can function as CNS APCs. CVEs, on the other hand, are less likely to participate. The question of whether CVEs could present antigen at the BBB to myelin-specific T cells remains unanswered.

Microglia are more efficient than astrocytes at processing and presenting antigen to T cells (27). Microglia are bone marrow-derived, macrophage-like cells of the CNS which migrate into CNS during development and are considered to be resident macrophage populations. Microglia are phagocytic and can process and present antigen, including the dominant myelin epitope in the SJL mouse, PLP139-151. When infected *in vitro* with TMEV, microglia upregulate MHC class II/I, cytokines including the "anti-viral" IFN-α/β, co-stimulatory molecules including B7-1/2 and CD40, and adhesion molecules like ICAM-1 to a similar extent as IFN-γ-treated microglia. Infected microglia can also process and present endogenous viral antigens to a virus specific T cell line (28). In MS, microglia were shown to phagocytoze myelin and express MHC class II plus costimulatory molecules (29).

Summary: Once CNS inflammation has been initiated, peripheral macrophage certainly play a major role in the inflammatory demyelination characteristic of MS and TMEV-IDD. Microglia, the resident macrophage cell of the CNS, are fully capable of participating in the demyelinating inflammation. They are likely infected, although abortively, and are able to process and present viral antigens to T cells. Astrocytes may also participate in this process since they are not only infected, but are also capable APCs when in an inflammatory environment. CVEs, however, are unlikely to take part in antigen processing and presentation during TMEV-IDD, although their proposed role in the initial migration of antigen-specific T cells across the BBB remains a possibility.

## 2.3 Epitope spreading

TMEV-IDD is initiated by virus infection and chronic disease progression requires viral persistence. In addition to infiltrating macrophage, CNS resident microglia and astrocytes are not only infected, but are also capable of processing and presenting viral antigens to virus-specific T cells, thereby triggering CNS inflammation. Interestingly, although virus-specific T cell responses are detectable throughout TMEV-IDD, T cell responses against self myelin antigens arise at around day 50 post infection (6). This autoimmune component may have a role in the chronic nature of TMEV-IDD.

The mechanisms underlying the initiation and progression of MS are not well understood, however, like TMEV-IDD, viral infections may be important in induction while disease progression appears to be autoimmune. Several mechanisms in which virally induced disease may trigger autoimmunity have been hypothesized, including molecular mimicry, in which autoreactive T cells cross-reactive with a viral epitope are directly activated, non-specific T cell activation by a viral superantigen, and epitope spreading.

Epitope spreading is a process in which tissue debris (myelin in this case) is phagocytosed, processed and presented in an inflammatory environment. As a result of this presentation of self epitopes, self-reactive T cells are activated. Epitope spreading plays a major role in the progression of myelin destruction in the SJL mouse during relapsing experimental autoimmune encephalomyelitis (R-EAE), another MS model (30). For more information on epitope spreading in EAE see section A, chapter 23.

In TMEV-IDD, initiation of myelin damage is associated with the activation of mononuclear inflammatory cells by proinflammatory cytokines from TMEV-specific CD4+ T helper 1 (Th1) cells responding to viral epitopes presented by CNS resident APCs harbouring persistent virus. T cell proliferation to major encephalitogenic epitopes on PLP and myelin basic protein (MBP) (epitope spreading) can be detected by day 50 and delayed-type hypersensitivity (DTH) by day 90 PI, only after animals exhibit histological demyelination and clinical disease (31). The site of T cell activation during epitope spreading is not clear, however, CNS APCs may be important in this process.

In fact, endogenous presentation of viral and myelin epitopes by plastic-adherent CNS-resident mononuclear cells (CNS APCs) isolated from the spinal cords of mice with ongoing TMEV-IDD was clearly demonstrated at 90 days PI (32). In these assays, myelin-specific T cell lines were induced to proliferate by CNS APCs without the addition of antigen (endogenous presentation of antigen already present within and on the APCs). This antigen presentation was B7- dependent and MHC class II restricted. No endogenous presentation of viral epitopes was seen in APCs isolated from EAE (a non-viral model of MS) and no endogenous presentation of myelin epitopes could be detected using plastic-adherent CNS-resident mononuclear cells harvested from TMEV-IDD mice prior to onset of demyelination and clinical signs. These data suggest that immune mediated virally-induced inflammatory demyelination leads to presentation of myelin debris by CNS APCs. Autoreactive T cells are then activated and participate in the inflammatory response. For more information on autoreactive T cell activation in TMEV see section B, chapter 3 and 6.

The functional significance of epitope spreading was recently explored utilizing organotypic spinal cord explants from neonatal SJL/J mice. T cells were isolated from mice 50 days after TMEV infection, at the time when sensitization had been shown to be mainly limited to the major encephalitogenic PLP epitope. These cells, after stimulation with PLP, but not other myelin proteins were capable of demyelinating the myelinated cultures. These results strongly suggest that sensitization to myelin proteins in chronic TMEV infection is functionally significant (33).

Summary: Although viral infection of peripheral macrophages, CNS resident microglia, and astrocytes initiates the demyelination of TMEV-IDD, ongoing presentation of both viral and myelin antigen by these CNS APCs certainly perpetuates the inflammatory demyelination seen in this model of MS.


1. **Theiler M.** Spontaneous encephalomyelitis of mice, a new virus disease. J Exp Med.1937;65:705-719.
2. **Lipton HL.** Theiler's virus infection in mice: An unusual biphasic disease process leading to demyelination. Infect Immunol. 1975;11:1147-1155.
3. **Waksman BH.** Multiple sclerosis: More genes versus environment. Nature. 1995;377:105-106.
4. **Kurtzke JF.** Epidemiologic evidence for multiple sclerosis as an infection. Clin Microbiol Rev. 1993;6:382-427.
5. **Miller SD, Olson JK, Croxford JL.** Multiple pathways to induction of virus-induced autoimmune demyelination: lesssons from Theiler's virus infection. J Autoimmun. 2001;16:219-227.
6. **Miller SD, VanderLugt CL, Smith Begolka W, Pao W, Yauch RL, Neville KL, et al.** Persistent infection with Theiler's virus leads to CNS autoimmunity via epitope spreading. Nature Med. 1997;3:1133-1136.
7. **Wekerle H.** Immunopathogenesis of multiple sclerosis. Acta Neurologica. 1991;13:197-204.
8. **Dal Canto MC, Lipton HL.** Primary demyelination in Theiler's virus infection. An ultrastructural study. Lab Invest. 1975;33:626-637.
9. **Clatch RJ, Melvold, RW, Miller SD, Lipton HL.** Theiler's murine encephalomyelitis virus (TMEV)-induced demyelinating disease in mice is influenced by the H-2D region: correlation with TMEV-specific delayed-type hypersensitivity. J Immunol. 1985;135:1408-1414.
10. **Dal Canto MC, Lipton HL.** Ultrastructural immunohistochemical localization of virus in acute and chronic demyelinating Theiler's virus infection. Am J Pathol. 1982;106:20-29.
11. **Lipton HL, Twaddle G, Jelachich ML.** The predominant virus antigen burden is present in macrophages in Theiler's murine encephalomyelitis virus-induced demyelinating disease. J Virol. 1995;69:2525-2533.
12. **Pena Rossi C, Delcroix M, Huitinga I, McAllister A, van Rooijen N, Claassen E, et al.** Role of macrophages during Theiler's virus infection. J Virol. 1997;71:3336-3340.

13. **Trottier M, Kallio P, Wang W, Lipton HL.** High numbers of viral RNA copies in the central nervous system of mice during persistent infection with Theiler's virus. J Virol. 2001;75:7420-7428.

14. **Clatch RJ, Miller SD, Metzner R, Dal Canto MC, Lipton HL.** Monocytes/macrophages isolated from the mouse central nervous system contain infectious Theiler's murine encephalomyelitis virus (TMEV). Virology 1990;176:244-254.

15. **Zheng L, Calenoff MA, Dal Canto MC.** Astrocytes, not microglia, are the main cells responsible for viral persistence in Theiler's murine encephalomyelitis virus infection leading to demyelination. J Neuroimmunol. 2001;118:256-267.

16. **Qi Y, Dal Canto MC.** Effect of Theiler's murine encephalomyelitis virus and cytokines on cultured oligodendrocytes and astrocytes. J Neurosci Res. 1996;45:364-374.

17. **Penney JB, Wolinski JS.** Neuronal and oligodendroglial infection by the WW strain of Theiler's virus. Lab Invest 1979;40:324-330.

18. **Dal Canto MC.** Experimental models of virus-induced demyelination. In: Handbook of Multiple Sclerosis, SD Cook (Ed), Marcel Dekker, Inc, New York-Basel 2001, pp 67-113.

19. **Rodriguez M, Lennon VA, Benveniste EN, Merrill JE.** Remyelination by oligodendrocytes stimulated by antiserum to spinal cord. J Neuropath Exp Neurol 1987;46:84-95.

20. **Asakura K, Miller DJ, Murray P, Bansai R, Pfeiffer SE, Rodriguez M.** Monoclonal autoantibody SCH94.03, which promotes central nervous system remyelination, recognizes an antigen on the surface of oligodendrocytes. J Neurosci Res. 1996;43:272-281.

21. **Asakura K, Miller DJ, Pease LR, Rodriguez M.** Targeting of IgM kappa antibodies to oligodendrocytes promotes CNS remyelination. J Neurosc. 1998;18:7700-7708.

22. **Warrington AE, Asakura K, Bieber AJ, Ciric B, Van Keulen V, Kaveri SV, et al.** Human monoclonal antibodies reactive to oligodendrocytes promote remyelination in a model of multiple sclerosis. Proc Natl Acad Sci USA 2000;97:6820-6825.

23. **Mitsunaga Y, Ciric B, Van Keulen V, Warrington AE, Paz Soldan M, Bieber AJ, et al.** Direct evidence that a human antibody derived from patient serum can promote myelin repair in a mouse model of chronic-progressive demyelinating disease. FASEB Journal 2002;16:1325-1327.

24. **Tan L, Gordon KB, Mueller JP, Matis LA, Miller SD.** Presentation of proteolipid protein epitopes and B7-1-dependent activation of encephalitogenic T cells by IFN-gamma-activated SJL astrocytes. J of Immunol. 1998;160:4271

25. **Soos JM, Morrow J, Ashley TA, Szente BE, Bikoff EK, Zamvil SS.** Astrocytes express elements of the class II endocytic pathway and process central nervous system autoantigen for presentation to encephalitogenic T cells. J of Immunol. 1998;161:595-5966.

26. **Girvin AM, Gordon KB, Welsh CJ, Clipstone NA, Miller SD.** Differential abilities of central nervous system resident endothelial cells and astrocytes to serve as inducible antigen-presenting cells. Blood. 2002;99:3692-3702.

27. **Aloisi F, Ria F, Penna G, Adorini L.** Microglia are more efficient than astrocytes in antigen processing and in Th1 but not Th2 cell activation. J Immunol. 1998;160:4671-4680.

28. **Olson JK, Girvin AM, Miller SD.** Direct activation of innate and antigen-presenting functions of microglia following infection with Theiler's virus. J of Virol. 2001;75:9780-9789.

29. **Aloisi F, Ria F, Adorini L.** Regulation of T-cell resonses by CNS antigen-presenting cells: different roles for microglia and astrocytes. Immunol Today 2000;21:141.

30. **Mc Rae BL, VanderLugt CL, Dal Canto MC, Miller SD.** Functional evidence for epitope spreading in the relapsing pathology of EAE in the SJL mouse. J Exp Med. 1995;182:75-83.

31. **Katz-Levy Y, Neville KL, Padilla J, Rahbe S, Begolka WS, Girvin AM, et al.** Temporal development of autoreactive Th1 responses and endogenous presentation of self myelin epitopes by central nervous system-resident APCs in Theiler's virus-infected mice. J Immunol. 2000;165:5304-5314.

32. **Katz-Levy K, Neville KL, Girvin AM, Vanderlugt DL, Pope JG,Tan LJ, et al.** Endogenous presentation of self myelin epitopes by CNS-resident APCs in Theiler's virus-infected mice. J Clin Invest. 1999;104:599-610.

33. **Dal Canto MC, Calenoff MA, Miller SD, Vanderlugt CL.** Lymphocytes from mice infected with Theiler's murine encephalomyelitis virus produce demyelination of organotytpic cultures after stimulation with the major encephalitogenic epitope of myelin proteolipid protein. Epitope spreading in TMEV infection has functional activity. J Neuroimmunol. 2000;104:79-84.

# Chapter B4

# IMMUNOGENETICS, RESISTANCE, AND SUSCEPTIBILITY TO THEILER'S VIRUS INFECTION

Roger W. Melvold and Stephen D. Miller
*Department of Microbiology and Immunology, University of North Dakota School of Medicine and Health Sciences, Grand Forks, ND and Department of Microbiology-Immunology and the Interdepartmental Immunobiology Center, Northwestern University Medical School, Chicago, IL*

Abstract:    The genetic basis for differential susceptibility to Theiler's virus-induced demyelinating disease is complex and often confusing, similar to the situation seen in analyzing the genetic risk for human multiple sclerosis. Numerous genetic loci have been identified as influencing the likelihood of disease development, most prominently among them the class I *H2-D* locus which has the single most important effect of any of the loci studied thus far. Genetic analysis is complicated, however, by the fact that susceptibility to the disease is not inherited as a consistently dominant or recessive trait, and that the relevant loci can differ from one mouse strain to another. The system has proved amenable to the utilization of numerous transgenic and knock-out gene systems to analyze the disease process.

Key words:    Theiler's murine encephalomyelitis virus (TMEV), demyelinating disease, multiple sclerosis, major histocompatibility complex (MHC), H2, H2-D, genetics

Multiple sclerosis (MS) has a suspected infectious etiology based on epidemiologic evidence that viral infections may play a role in initiation and/or exacerbation of the disease (1, 2), but also has a clear genetic basis (3-9). Similarly, there is clearly a genetic component to the risk of developing demyelinating disease in mice infected with Theiler's murine encephalomyelitis virus (TMEV) (10-18). TMEV is a single-stranded RNA virus belonging to the *Picornaviridae* family (19). TMEV-induced demyelinating disease (TMEV-IDD) is a natural chronic-progressive CNS demyelinating disease with similarities to primary-progressive MS (20, 21). CNS pathology in TMEV-IDD is initiated by viral capsid protein-specific $CD4^+$ T cells targeting virus persisting in CNS microglia, but autoreactive myelin-specific $CD4^+$ Th1 responses arise via epitope spreading and contribute to chronic pathology (22, 23). Myelin destruction is mediated largely by the production of the pro-inflammatory cytokines, IFN-$\gamma$ and TNF-$\alpha$ (24). Both HLA and numerous non-HLA loci have been implicated in contributing to the risk of human MS, although in many cases the statistical associations are low as might be anticipated for a genetically complex disease in an outbred population (25, 26). Similarly, multiple genes are involved in determining whether the anti-viral immune responses in TMEV-infected mice will eventually lead to the destructive inflammatory responses associated with demyelination. In some cases, effects on human and murine demyelinating diseases have been attributed to homologous loci in the two species (27).

There are clear differences among various inbred strains of mice in susceptibility to the development of TMEV-IDD disease (Table 1), in terms of incidence, severity and time of onset. Some strains such as SJL/J, DBA/2J and SWR/J are highly susceptible and consistently develop severe demyelinating disease within a month of infection. Other strains such as C57BL/6 and C57BL/10 are highly resistant, with only an infrequent individuals developing disease after infection. Strains such as CBA and C3H (both $H2^k$) develop relatively mild forms of the disease. BALB/c mice ($H2^d$) can be either resistant or susceptible, depending upon the substrain (28, 29). However, even the susceptible BALB/c substrains tend to develop only moderately severe disease (2), and the severity of disease within some strains can also vary considerably (29). C57BR mice ($H2^k$) are generally considered resistant (30), but do begin to develop signs of disease after a lengthy interval, 100+ days, following infection (Miller, personal communication).

*Table 1.* Susceptible and Resistant Inbred and Congenic Mouse Strains

| H2 type | Susceptible Strains | Resistant Strains |
|---|---|---|
| $H2^b$ | 129Sv | C57L, C57BL/6, C57BL/10 |
| $H2^d$ | BALB/c [1], DBA/2, NZW [2], B10.D2 [3] | BALB/c [1] |
| $H2^f$ | B10.M [3] | |
| $H2^k$ | AKR, C3H, CBA, B10.BR [3] | C57BR, B10.K [3] |
| $H2^q$ | SWR, DBA/1, FVB | |
| $H2^s$ | SJL | |
| $H2^r$ | B10.RIII [3] | |
| $H2^u$ | PL/J | |
| $H2^p$ | P/J | |
| $H2^z$ | NZB [2] | |

1    BALB/cByJ and BALB/cCum substrains are resistant.    BALB/cJ and BALB/cAnNCr substrains are susceptible (28).

2    Melvold (unpublished observations)

3    Reference 31

Resistance and susceptibility are not consistently dominant or recessive traits, but vary with the particular hybrid generated. Hybrids between different combinations of resistant and susceptible strains (Table 2) are sometimes susceptible, sometimes resistant, and sometimes intermediate. The outcome, therefore, is dependent upon the sum of the allelic interactions between the multiple loci that may vary between the susceptible and resistant parental strains. In comparisons among B10 congenics, where the non-*H2* genotypes are held constant, resistance is inherited as the dominant trait and susceptibility as the recessive trait (31). However, when both *H2* and non-*H2* genotypes are varied, this is not always true and the interactions between relevant loci may be complex. For example, hybrids between resistant C57BL/6 and susceptible DBA/2 mice are resistant, while those between resistant BALB/cByJ and susceptible DBA/2 mice are susceptible (32-34). Likewise, when susceptible SWR mice are crossed with either the resistant C57BL/6 or C57L strains (both carrying $H2\text{-}D^b$), the two types of resulting hybrid progeny have markedly different susceptibilities, despite the very close similarities of the two types of resistant parents (35). Comparable results have been found in EAE, another animal model for multiple sclerosis (36) where some crosses between resistant and susceptible strains produced resistant hybrid progeny, while other crosses produced susceptible progeny.

Human multiple sclerosis occurs more frequently in women than in men. In the TMEV model, differences in susceptibility between the two sexes have not been frequently noted, possibly because many studies employed only one sex or the other. In some strain combinations, however, such differences have been reported. Among backcross and F2 progeny between the susceptible C57L and susceptible SJL strains, females were more frequently affected than males, by ratios ranging from 2:1 to 7:1. Similarly, among SWXL recombinant-inbred mice, generated between the susceptible SWR and resistant C57L strains, females were affected approximately 3 times as frequently as males (37).

*Table 2.* Hybrids Between Susceptible and Resistant Strains

| Resistant Strain | Susceptible Strain | H2 genotype | Status | Reference |
|---|---|---|---|---|
| BALB/cByJ | SJL | $H2^b/H2^s$ | Intermediate (60% affected) | 38, 39 |
| C57BL/6 | SJL | $H2^b/H2^s$ | Resistant (5-15% affected) | 40 |
| C57L | SJL | $H2^b/H2^s$ | Susceptible (100% affected) | 41 |
| BALB/cByJ | DBA/2 | $H2^d/H2^d$ | Susceptible (85% affected) | 34 |
| C57BL/6 | DBA/2 | $H2^b/H2^d$ | Resistant (0-10% affected) | 32, 33 |
| C57BL/10 | DBA/2 | $H2^b/H2^d$ | Resistant | 42 |
| BALB/cByJ | BALB/cAnNCr | $H2^d/H2^d$ | Resistant (30% affected) | 28 |
| BALB/cByJ | SWR | $H2^d/H2^q$ | Susceptible (94% affected) | 35 |
| C57BL/6 | SWR | $H2^b/H2^q$ | Intermediate (44% affected) | 35 |
| C57L | SWR | $H2^b/H2^q$ | Susceptible | 35 |
| BALB/cByJ | PL/J | $H2^d/H2^u$ | Susceptible (78% affected) | Melvold (personal comm.) |
| C57BL/6 | PL/J | $H2^b/H2^u$ | Resistant (25% affected) | Melvold (personal comm.) |

Immune response differences exist between strains in terms of immunodominant TMEV epitopes. For example, resistant C57BL/6 mice show stronger T proliferative responses to TMEV capsid proteins VP1 and VP3, while susceptible SJL mice proliferative more strongly to VP2 (43). In DTH assays, the two strains react equally to VP1 and VP3, but SJL shows a slightly stronger reactivity to VP2. No pattern of epitope specificity has been found, however, that is consistent across a variety of resistant and susceptible strains or H2 haplotypes.

Several specific loci have been shown to influence the development of demyelinating disease following TMEV infection. In addition, associations between disease and specific DNA polymorphisms have identified several chromosomal segments as containing loci influencing disease. The loci and chromosomal segments affecting TMEV persistence or actual susceptibility to TMEV-induced demyelination are enumerated in Table 3. Lipton and Melvold (40) initially identified the involvement of both *H2* and non-*H2* genes. Numerous additional investigations (44-47) have demonstrated that major histocompatibility complex (MHC) genes are frequently of major effect in the differential susceptibility of inbred mouse strains to TMEV-IDD. Subsequent studies in several laboratories, utilizing different resistant-susceptible strain combinations, mapped the primary *H2* contribution to the class I *H2-D* locus (42, 48-50). β2-microglobulin deficient mice that are unable to express class I MHC molecules, and thus lack CD8$^+$ T cells, displayed an increased susceptibility (51-54). The importance of the MHC in the TMEV model is similar to human MS where statistical association of particular HLA genes with disease has been frequently reported (3, 7, 9, 25, 26).

Naturally existing polymorphisms at *H2-D* are usually important, with *H2-D$^b$* usually associated with resistance, and *H2-D$^{s,q,p,u,z}$* alleles associated with susceptibility. The *H2-D$^d$* and *H2-D$^k$* alleles are found in both susceptible and resistant strains, illustrating the joint effects of both *H2* and non-*H2* genomes. Utilization of strains carrying mutated or transgenic *H2-D* alleles confirmed the importance of this locus. Mutations of either the α1 or α2 domain of the *H2-D$^d$* allele (49) or the *H2-D$^b$* allele (42, 55) in resistant mice resulted in a change to the susceptible phenotype. Transgenic insertion of *H2-D* alleles associated with resistance into normally susceptible mice resulted in resistance (56-57). The importance of this *H2* class I locus suggests that H2-D restricted CD8$^+$ T cells have an important role in determining whether or not disease ensues after TMEV infection. CD8$^+$ cytotoxic T cells generated in TMEV-infected animals are directed preferentially against H2-D-restricted capsid proteins (58-61), although their precise role in pathogenesis or protection is yet unclear. At one week post-infection, the CNS of resistant mice contained TMEV-specific CTLs that were only H2-D restricted, although both H2-D and H2-K restricted TMEV-specific CTLs were present in the spleen (58). Susceptible strains had no TMEV-specific CTLs in the CNS during this same time period (58).

The importance of MHC genes in differential susceptibility to TMEV-induced demyelination (40, 42, 44-50), as well as their statistical association with the risk of multiple sclerosis in human populations (3-9) has led to experiments in which human HLA genes are inserted into mice to assess their effect on the development of demyelinating disease. This approach has led to insights in many disease models (62). Rodriguez and David (63) inserted the human HLA-B27 gene into susceptible B10.D2-*H2$^{dm1}$* mice, and found that mice carrying the transgene exhibited less demyelination and lower expression of TMEV antigen than did non-transgenic controls. In later experiments, mice lacking β2m and intrinsic class II gene expression were transfected with human DQ6 or DQ8 genes (64). While both types of transgenic mice exhibited similar levels of demyelination by 120 days post-infection, those with DQ8 developed demyelinated lesions more rapidly than did those with DQ6 (64).

Non-*H2* loci are also important in determining resistance or susceptibility (Table 3), as is readily apparent from comparing resistant BALB/cByJ and susceptible DBA/2J mice, both carrying $H2^d$. Utilizing various strain combinations, effects on susceptibility have been mapped to *Theiler's murine encephalomyelitis virus demyelination-1 (Tmevd-1)* on chromosome 6 (38, 39, 47), *Tmevd-2* on chromosome 3 (65), *Tmevd-3* and *Tmevd-4* on chromosome 14 (66) and *Tmevd-5* on chromosome 11 (67). Several loci affecting TMEV persistence (e.g., *Tmevp-1*, *-2*, *-3* and *–pg1*) have also been identified (68-73), several of them on chromosome 10. Loci (*Tmevd-6, -7, -8, -9*) have recently been identified on chromosomes 1, 5 and 15 (74) that affect the severity of clinical signs. Differences between resistant B6 and susceptible SJL mice that affect the persistence of TMEV have been mapped to chromosomes 10 and 18, near the *Ifg (interferon-γ)* and *Mbp (myelin basic protein)* loci, (69, 70). A number of additional unnamed loci have been identified as well (see Table 3).

Rodriguez *et al* (75, 76) assessed the effects of deletions (50 - 70%) of T cell receptor Vβ genes, finding that intermediately susceptible B10.RIII mice (70% of Vβ genes deleted) had increased demyelination and viral loads. However, no other effects of deletion were found in normally resistant or susceptible animals. Bahk *et al* (77) found associations between susceptibility and particular VβCβ polymorphisms. The involvement of chromosomal regions near or including T cell receptor genes is paralleled in humans by observations that some Vα and Vβ alleles are not randomly represented in MS patients (78-81). Even such closely related strains as C57BL/6, C57BL/10 and C57L (all $H2^b$) have differences in non-*H2* genes affecting susceptibility evident in hybrid progeny produced by crossing these three strains with susceptible strains (35). For example, (D2 x B10.D2) F1 hybrids are more susceptible than (D2 x B6.C-H-2$^d$)F1 hybrids (Melvold, personal communication).

*Table 3.* Specific Loci/Chromosomal Segments Affecting Susceptibility

| Locus | Chromosome | Effect | References |
| --- | --- | --- | --- |
| **MHC** | | | |
| *H2* (entire complex) | 17 | Affects susceptibility | 40, 42, 44-47 |
| *H2-D* | 17 | Affects susceptibility | 42, 44, 48-50 |
| | | | |
| **Non-MHC** | | | |
| *Tmevd-1* | 6 (near TCRβ) | Affects susceptibility | 38 |
| *Tmevd-2* | 3 (near *Car-2*) | Affects susceptibility | 65 |
| *Tmevd-3* | 14 | Affects susceptibility | 66 |
| *Tmevd-4* | 14 | Affects susceptibility | 66 |
| *Tmevd-5* | 11 | Affects susceptibility | 67 |
| *Tmevd-6* | 1 | Affects severity of symptoms | 74 |
| *Tmevd-7* | 5 | Affects severity of symptoms | 74 |
| *Tmevd-8* | 15 | Affects severity of symptoms | 74 |
| *Tmevd-9* | 1 | Affects severity of symptoms | 74 |
| *Tmevp-1* | unknown | Affects TMEV persistence | 68 |
| *Tmevp-2* | 10 | Affects TMEV persistence | 69-71 |
| *Tmevp-3* | 10 | Affects TMEV persistence | 71-73 |
| *Tmevpg1* | 10 | Affects TMEV persistence | 73 |
| Tcrβ | 6 | Deletions/polymorphisms affect susceptibility | 75-77 |
| *Eae3* | 3 (near *Tmevd-2*) | Affects susceptibility | 34 |
| *shiverer* | 18 | Affects susceptibility | 82 |
| Unnamed | 18 | Affects TMEV persistence | 70 |
| Unnamed | multiple loci | Expressed in non-hematopoietic tissues | 72 |
| Unnamed | ?? | Differs among SWR, C57BL/6 and C57 strains | 35 |
| UDP-galactose transporter | 11 | Required for entry of TMEV into mammalian cells | 83 |

Comparisons between susceptible DBA/2J and resistant BALB/cByJ mice have implicated the same region of chromosome 3 as *Tmevd-2* and suggest the possibility that *Tmevd-2* may be identical (or very near) to *Eae3*, a gene governing susceptibility to EAE (experimental autoimmune encephalomyelitis), an autoimmune model for MS (34). Interestingly, a region of human chromosome 5 that is homologous to the region of mouse chromosome 15 containing *Eae2* has been implicated in risk of MS (27).

Additional comparisons employing different combinations of susceptible and resistant strains indicate that the relevant differential loci vary from one combination to another. For example, the differences between susceptible DBA/2 and resistant C57BL/6 mice map to both the *H2-D* and *Tmevd-2* loci (65) while differences between resistant BALB/c and susceptible SJL mice are attributable jointly to *H2-D* and *Tmevd-1* (38, 39, 47). Interestingly, the *H2* complex is not always involved. Comparisons between resistant B6 and susceptible SWR mice indicate that susceptibility is due to a single non-*H2* locus other than *Tmevd-1* and *Tmevd-2*, while differences between C57L and SWR are strongly affected by *H2*, as well as non-*H2*, genes (35). Thus, the genetic basis for susceptibility or resistance to TMEV-IDD disease is a complex, multigenic one.

The genetically determined susceptible/resistant status of mice is not, however, immutable. Several resistant strains become susceptible if treated with various immunopotentiating agents prior to infection with TMEV. Normally resistant B6D2F1 hybrids become susceptible if treated with either low dose (2 - 20 mg/kg body weight) cyclophosphamide or low dose irradiation (50 - 300 rads) prior to infection (32, 33). C57BL10 and BALB/cByJ mice become susceptible if given low dose (200 - 300 rads) irradiation shortly prior to infection (28, 84-86). Thus exposure to such agents can reverse the intrinsic genetic "programming" of resistant animals, apparently by killing or inhibiting TMEV-specific CD8[+] and/or other regulatory T cells that can inhibit the inflammatory demyelination (85, 86). It should be noted that comparable treatments of normally susceptible animals do not have substantive effects on the severity of disease (28, 32, 33). However, Pullen *et al* (87) demonstrated that normally resistant C57BL/6 mice became susceptible if given LPS prior to infection, possibly due to alterations the innate immune response during infection. The difference between the *H2*-identical susceptible BALB/cAnNCr and resistant BALB/cByJ substrains is consistent with a single unmapped locus (28), and appears to be due to the lack of a CD4[+] T cell subset required for activation of CD8[+] T cells capable of preventing the development of inflammatory demyelination (88). These results are similar to those reported in EAE by Lando *et al* (36, 89). The differences between various immunopotentiating agents in their ability to convert specific resistant strains suggests that the genetic basis for resistance may differ among those resistant strains. Thus the risk for an animal to develop demyelinating disease, whether induced by autoimmunity or by infection, involves at least three elements: (a) exposure to infection by TMEV, (b) genotype and (c) exposure to agents which can modulate the effects of genotype.

Utilization of mice bearing deleted or inactivated molecules suspected to be involved in the disease process has provided additional insight into the mechanisms regulating TMEV-IDD (Table 4). Deletions affecting expression of either MHC class I or class II molecules are associated with an increase in susceptibility. Loss of MHC expression in resistant animals , due to an absence of $\beta$2-microglobulin, inhibits CD8$^+$ T cell function and is associated with development of TMEV-induced demyelination (51-53). In strains that are already susceptible, this leads to exacerbation of clinical signs (54). Similarly, loss of MHC class II molecules leads to early phase neurological disease, although the pathogenesis may be different. The loss of class II molecules inhibits CD4$^+$ T cell function in generating both cellular and antibody responses and may eliminate much of the animal's ability to resist the infection altogether, leading to a viral encephalitis (90, 91). The loss of IFN-$\alpha$ and IFN-$\beta$ receptors from normal resistant FVB mice led to a rapid infection and death due to loss of neural tissue after TMEV infection. Those who carried deletions of the IFN-$\gamma$ receptor developed a late phase disease with demyelination (92).

*Table 4.* Loci Involved in Functions Pertinent to TMEV-IDD

| Locus | Function | Reference |
| --- | --- | --- |
| Beta-2-microglobulin | Loss results in non-expression of MHC I molecules | 51-54 |
| MHC class II | Deletion of MHC II genes is associated with susceptibility | 90, 91 |
| CD4 | Loss leads to TMEV persistence and demyelination | 93 |
| Interferon-$\alpha$, -$\beta$, -$\gamma$ receptors | Deletion causes loss of cytokines associated with viral infection and inflammation | 92 |
| L-selectin | Reduced number of CD8$^+$ T cells in CNS | 94 |
| Perforin | Involved in killing of virally infected cells | 95, 96 |

Normally resistant C57BL/6 mice bearing knock-out mutations of the L-selectin gene, involved in targeting of activated T cells to sites of infection, exhibited diminished CNS infiltration of CD8$^+$ T cells following TMEV infection (94).   This diminished infiltration, however, did not appear to inhibit CNS clearance of TMEV.   Loss of the perforin gene, utilized by cytotoxic CD8$^+$ T cells to destroy TMEV-infected cells, led to increased viral load and demyelination in otherwise resistant animals (95, 96).

In summary, the genetic influence on the development of TMEV-induced demyelination is an exceedingly complex one similar to the genetic loci influencing susceptibility to MS.   Not only are numerous loci, both MHC and non-MHC involved, but the precise combinations of such genes that are present in a particular animal have considerable influence on the outcome. A gene that may promote demyelination in one genetic setting, may promote resistance in another.   A major role is clearly evident for loci of the *H2* complex, particularly the *H2-D* class I locus.   However, the precise mechanisms by which these genes, and their products and the other genes with which they interact, ultimately produce the resistant or susceptible phenotypes still requires considerable investigation.

## References


1.   Olson, J.K., Croxford, J.L. and Miller, S.D.   2001.   Virus-induced autoimmunity: Potential role of viruses in initiation, perpetuation, and progression of T cell-mediated autoimmune diseases. Viral Immunol. 14:227-250.
2.   Kurtzke, J.F.  1993.  Epidemiologic evidence for multiple sclerosis as an infection.  Clin. Microbiol. Rev. 6:382-427.
3.   Compston, A and Sawyer,S.  2002.  Genetic analysis of multiple sclerosis.  Curr. Neurol. Neurosci. Rep. 2:259-266.
4.   Kantarci, O.H., de Andrade, M. and Weinshenker, B.G.  2002.  Identifying disease modifying genes in multiple sclerosis. J. Neuroimmunol. 123:144-159.
5.   Sandovick, S.  2002.  The genetics of multiple sclerosis.  Clin. Neurol. Neurosurg. 104:199-202.
6.   Fukazawa, T., Sasaki, H., Kikuchi, S., Hamada, T. and Tashiro, K.  2000.  Genetics of multiple sclerosis.  Biomed. Pharmacother. 54:103-106.
5.   Oksenberg, J.R. and Barcellos, L.F.:  2000.  The complex genetic aetiology of multiple sclerosis.
8.   Compston, A.  2000.  The genetics of multiple sclerosis.  Clin. Chem. Lab. Med. 38:133-135.
9.   Willer, C.J. and Ebers, G.C.  2000.  Susceptibility to multiple sclerosis: interplay between genes and environment.  Curr. Opin. Neurol. 13:241-147.
10.  Lipton, H.L. and M.C. Dal Canto.  1979.  Susceptibility of inbred mice to encephalomyelitis virus.  Infect. Immun. 26 : 369-374.
11.  Lipton, H.L., Miller, S.D., Melvold, R.W. and Fujinami, R.S.  1986.  "Theiler's murine encephalomyelitis virus (TMEV) infection in mice as a model for multiple sclerosis." In *Concepts in Viral Pathogenesis II*, A.L. Notkins and M.B.A. Oldstone, eds., pp. 248 - 254, Springer-Verlag, New York.

12. Rodriguez, M., Pease, L.R. and C.S. David. 1986. Immune-mediated injury of virus-infected oligodendrocytes. A model of multiple sclerosis. Immunology Today 7:359-366.

13. Blankenhorn, E.P. and Stranford, S.A. 1992. Genetic factors in demyelinating disease: genes that control demyelination due to experimental allergic encephalomyelitis (EAE) and Theiler's murine encephalomyelitis virus. Regional Immunol. 4 : 331-343.

14. Miller, S.D., Karpus, W.J., Pope, J.G., Dal Canto, M.C. and Melvold, R.W. 1994. "Theiler's murine encephalomyelitis virus-induced demyelinating disease." In: *Animal Models for Autoimmune Diseases,* I.R. Cohen, ed., pp. 23-38, Academic Press, New York.

15. Oleszak, E.L., Kuzmak, J., Good, R.A. and Platsoucas, C.D. 1995. Immunology of Theiler's murine encephalomyelitis virus infection. Immunol. Res. 14 : 13-33.

16. Dal Canto, M.C., Kim, B.S., Miller, S.D. and Melvold, R.W. 1996. Theiler's Murine Encephalomyelitis Virus (TMEV)-induced demyelination: a model for human multiple sclerosis. Methods 10:453-461

17. Monteyne, P., Bureau, J.F. and Brahic, M. 1997. The infection of mouse by Theiler's virus: from genetics to immunology. Immunol. Rev. 159:163-176.

18. Brahic, M. and Bureau, J.F. 1998. Genetics of susceptibility to Theiler's virus infection. BioEssays 20:627-633.

19. Pevear, D.C., Calenoff, M., Rozhon, E. and Lipton, H.L. 1987. Analysis of the complete nucleotide sequence of the picornavirus Theiler's virus indicates that it is closely related to cardioviruses. J. Virol. 61:1507-1516.

20. Lipton, H.L. 1975. Theiler's virus infection in mice: An unusual biphasic disease process leading to demyelination. Infect. Immun. 11:1147-1155.

21. Miller, S.D. and Gerety, S.J. 1990. Immunologic aspects of Theiler's murine encephalomyelitis virus (TMEV)-induced demyelinating disease. Semin. Virol. 1:263-272.

22. Karpus, W.J., Pope, J.G. Peterson, J.D. Dal Canto, M.C. and Miller, S.D. 1995. Inhibition of Theiler's virus-mediated demyelination by peripheral immune tolerance induction. J. Immunol. 155:947-957.

23. Miller, S.D., Vanderlugt C.L., Begolka W.S., Pao W., Yauch, R.L., Neville, K.L., Katz-Levy, Y., Carrizosa, A. and Kim, B.S. 1997. Persistent infection with Theiler's virus leads to CNS autoimmunity via epitope spreading. Nature Med. 3:1133-1136.

24. Begolka, W.S., Vanderlugt, C.L., Rahbe, S.M., and Miller, S.D. 1998. Differential expression of inflammatory cytokines parallels progression of CNS pathology in two clinically distinct models of multiple sclerosis. J. Immunol. 161:4437-4446.

25. Haines, J.L., Ter-Minassian, M., Bazyk A. et al (26 additional authors). 1996. A complete genomic screening for multiple sclerosis underscores a role for the major histocompatibility complex. Nature Genetics 13 : 469-471.

26. Liblau, R. and Gautam, A.M. 2000. HLA, molecular mimicry and multiple sclerosis. Rev. Immunogenet. 2:95-104.

27. Kuokkanen, S., Sundvall, M., Terwilliger, J.D., Tienari, P.J., Wikstrom, J., Holmdahl, R. Petterson, U. and Pelronen, L. 1996. A putative vulnerability locus to multiple sclerosis maps to 5p14 - p12 in a region syntenic to the murine locus Eae2. Nature Genetics 13:477-480.

28. Nicholson, S., Peterson, J.D., Miller, S.D., Dal Canto, M.C. and Melvold, R.W. 1994. BALB/c substrain variation in susceptibility to Theiler's murine encephalomyelitis virus (TMEV)-induced demyelinating disease. J. Neuroimmunol. 52:19-24.

29. Simas, J.P. and J.K. Fazakerley. 1996. The course of disease and persistence of virus in the central nervous system varies between individual CBA mice infected with the BeAn strain of Theiler's murine encephalomyelitis virus. J. Gen. Virol. 77:2701-2711.

30. Clatch, R.J., Lipton, H.L. and Miller, S.D. 1987. Class II-restricted T cell responses in Theiler's murine encephalomyelitis virus (TMEV)-induced demyelinating disease. II. Survey of host immune responses and central nervous system virus titers in inbred mouse strains. Microbial Pathogenesis 3:327-337.

31. Patick, A.K., Pease, L.R., David, C.S. and Rodriguez, M. 1990. Major histocompatibility complex-conferred resistance to Theiler's virus-induced demyelinating disease is inherited as a dominant trait in B10 congenic mice. J. Virol. 64:5570-5576

32. Olsberg, C.A., A. Pelka, T.M. Creighton, S.D. Miller, C.R. Waltenbaugh, M.C. Dal Canto,H.L. Lipton and R.W. Melvold. 1993. Induction of Theiler's murine encephalomyelitis virus (TMEV)-induced demyelinating disease in genetically resistant mice. Regional Immunol. 5 : 1-10.25.

33. Pelka, A.E., Olsberg, C.A., Miller, S.D., Waltenbaugh, C.R., Dal Canto, M.C. and Melvold, R.W. 1993. Effects of irradiation on resistance to Theiler's murine encephalomyelitis virus (TMEV)-induced demyelinating disease. Cell. Immunol. 152:440-455.

34. Teuscher, C., Rhein, D.M., Livingstone, K.D., Paynter, R.A., Doerge, R.W., Nicholson, S.M. and Melvold, R.W. 1997. Evidence that *Tmevd2* and *eae3* may represent either a common locus or members of a gene complex controlling susceptibility to immunologically mediated demyelination in mice. J. Immunol. 159:4930-4934.

35. Nicholson S.M., Jokinen, D.M., Dal Canto, M.C., Kim, B.S. and Melvold, R.W. 1995. Genetic analysis of susceptibility of the SWR strain to TMEV-induced demyelinating disease. J. Neuroimmunol. 59:19-28.

36. Lando, Z., Teitelbaum, D. and Arnon, R. 1980. Induction of experimental allergic encephalomyelitis in genetically resistant strains of mice. Nature 287:551-552.

37. Kappel, C.A., Melvold, R.W. and Kim, B.S. 1990. Influence of sex on susceptibility in the Theiler's murine encephalomyelitis virus model for multiple sclerosis. J. Neuroimmunol. 29:15-19.

38. Melvold, R.W., Jokinen, D.M. Knobler, R.L. and Lipton, H.L. 1987. Variations in genetic control of susceptibility to Theiler's murine encephalomyelitis virus (TMEV)-induced demyelinating disease. I. Differences between susceptible SJL/J and resistant BALB/c strains map near the T cell beta chain constant gene on chromosome 6. J. Immunol. 138:1429-1433.

39. Knobler, R.L., Lublin, M.D., Linthicum, D.S., Cohn, M, Melvold, R., Lipton, H.L., Taylor, B.S. and Beaner, W.G. 1988. Genetic regulation of susceptibility and severity of demyelination. Ann. N.Y. Acad. Sci. 540:735-737.

40. Lipton, H.L. and Melvold .W. 1984. Genetic analysis of susceptibility to Theiler's virus-induced demyelinating disease in mice. J. Immunol. 132 : 1821-1825.

41. Dal Canto, M.C., Melvold, R.W. and Kim, B.S. 1995. A hybrid between a resistant and a susceptible strain of mouse alters the pattern of Theiler's murine encephalomyelitis virus-induced white matter disease and favors oligodendrocyte-mediated remyelination. Mult. Scler. 1:95-103

42. Melvold, R.W., Jokinen, D.M,. Miller, S.D., Dal Canto, M.C. and Lipton, H.L. 1987. "H-2 genes in TMEV-induced demyelination, a model for multiple sclerosis." In *Major Histocompatibility Genes and Their Role in Immune Function*, C. David, ed., pp. 735-745, Plenum, New York.

43. Miller, S.D. , Clatch, R.J. and Lipton, H.L.  1988.  Fine specificity of T-cell-mediated immune responses of susceptible and resistant strains in Theiler's murine encephalomyelitis virus-induced demyelinating disease.

44. Rodriguez, M., David, C.S. and Pease, L.R. 1987. "The contribution of the MHC gene products to demyelination by Theiler's virus." In *Major Histocompatibility Genes and Their Role in Immune Function.* C. David, ed., pp. 747-756, Plenum, New York, .

45. Rodriguez, M. and David,.C.S.  1985.  Demyelination induced by Theiler's virus: influence of the H-2 haplotype.  J. Immunol. 135:2145-2148.

46. Clatch, R.J., Melvold, R.W., Dal Canto, M.C., Miller, S.D. and Lipton,H.L.  1987. The Theiler's murine encephalomyelitis  virus (TMEV) model for multiple sclerosis shows a strong influence of the murine equivalents of HLA-A, B and C.  J. Neuroimmunol. 15:121-135.

47. Kappel, C.A., Dal Canto, M.C., Melvold, R.W. and Kim, B.S.  1991.  Hierarchy of effects of the major histocompatibility complex and T cell receptor β-chain genes in susceptibility to Theiler's murine encephalomyelitis virus-induced demyelinating disease. J. Immunol. 147 4322-4326.

48. Clatch, R.J., Melvold, R.W., Miller, S.D. and Lipton, H.L.  1985.  Theiler's murine encephalomyelitis virus (TMEV)-induced demyelinating disease in mice is influenced by the H-2D region: correlation with TMEV-specific delayed type hypersensitivity.  J. Immunol. 135:1408-1414.

49. Rodriguez, M., Leibowitz, J. and  David,.C.S.  1986.  Susceptibility to Theiler's virus-induced demyelination.  Mapping of the gene within the H-2D region.  J. Exp. Med. 163:620-631.

50. Azoulay-Cayla, A., Syan, S., Brahic, M. and Bureau, J.F.  2001.  Roles of the H-2D(b) and H-K(b) genes in resistance to persistent Theiler's murine encephalomyelitis virus infection of the central nervous system.  J. Gen. Virol. 82:1043-1047.

51. Rodriguez, M., A.J. Dunkel, R.L. Thiemann, J. Leibowitz, M. Zijlstra and R. Jaenisch. 1993.  Abrogation of resistance to Theiler's virus-induced demyelination in *H-2$^b$* mice deficient in beta 2-microglobulin. J. Immunol. 151 : 266-278.

52. Fiette, L., Aubert, C., Brahic, M. and Pena Rossi, C. 1993.  Theiler's virus infection of β2-microglobulin-deficient mice. J. Virol. 67:589-592.

53. Pullen, L.C., Miller, S.D., Dal Canto, M.C. and Kim, B.S.  1993.  Class-I deficient resistant mice intracerebrally inoculated with Theiler's virus show an increased T cell response to viral antigens and susceptibility to demyelination.  Eur. J. Immunol. 23:2287-2293.

54. Begolka, W.S., Haynes, L.M., Olson, J.K., Padilla, J., Neville, K.L., Dal Canto, M., Palma, J., Kim, B.S. and Miller, S.D.  2001.  CD8-deficient SJL mice display enhanced susceptibility to Theiler's virus infection and increased demyelinating pathology.  J. Neurovirol. 7:409-420.

55. Lipton, H.L., Melvold, R., Miller, S.D. and Dal Canto, M.C.  1995.  Mutation of a major histocompatibility class I locus, H-2D, leads to an increased virus burden and disease susceptibility in Theiler's virus-induced demyelinating disease.  J. Neurovirol. 1:138-144.

56. Azoulay, A., Brahic, M. and Bureau, J.F.  1994.  FVB mice transgenic for the H-2D$^b$ gene become resistant to persistent infection by Theiler's virus. J. Virol. 68:4049-4052.

57. Rodriguez, M. and David, C.S.  1995.  H-2D$^d$ transgene suppresses Theiler's virus-induced demyelination in susceptible strains of mice. J. Neurovirol. 1:111-117.

58. Lin, X., Pease, L.R. and Rodriguez, M. 1997. Differential generation of class I H-2D-versus H-2K-restricted cytotoxicity against a demyelinating virus following central nervous system infection. Eur. J. Immunol. 27:963-970.

59. Borson, N.D., Paul, C., Lin, X., Nevala,, W.K. Strausbaugh, M.A., Rodriguez, M. and Wettstein, P.J. 1997. Brain-infiltrating cytolytic T lymphocytes specific for Theiler's virus recognize H-2D$^b$ molecules complexed with a viral VP2 peptide lacking a consensus anchor residue. J. Virol. 71:5244 - 5250.

60. Dethlefs, S., Escriou, N., Brahic, M., Van der Werf, S. and Larsson-Sciard, E. 1997. Theiler's virus and Mengo virus induce cross-reactive cytotoxic T lymphocytes restricted to the same immunodeterminant VP2 epitope in C57BL/6 mice. J. Virol. 71:5361-5365.

61. Azoulay-Cayla, A., Dethlefs, S., Perarnau, B., Larsson-Sciard, E.L., Lemonnier, F.A., Brahic, M. and Bureau, J.F. 2000. H-2D(b-/-) mice are susceptible to persistent infection by Theiler's virus. J. Virol. 74:5470-5476

62. Taneja, V. and David, C.S. 1999. HLA class II transgenic mice as models of human diseases. Imm. Rev. 169:67-69.

63. Rodriguez, M., Pedrinaci, S. and David, C. 1993. Human class I major histocompatibility complex transgene prevents virus-induced demyelination in susceptible mutant B10.D2dm1 mice. Ann. Neurol. 33:208-212

64. Pavelko, K.D., Drescher, K.M., McGavern, D.B., David, C.S. and Rodriguez, M. 2000. HLA-DQ polymorphism influences progression of demyelination and neurologic deficits in a viral model of multiple sclerosis. Mol. Cell. Neurosci. 15:495-509.

65. Melvold. R.W., Jokinen, D.M., Miller, S.D., Dal Canto, M.C. and Lipton, H.L. 1990. Identification of a locus on mouse chromosome 3 involved in differential susceptibility to Theiler's murine encephalomyelitis virus (TMEV)-induced demyelinating disease. J. Virol. 64:686-690.

66. Bureau, J.F., Drescher, K.M., Pease, L.R., Vikoren, T., Delcroix, M., Zoecklein, L., Brahic, M. and Rodriguez, M. 1998. Chromosome 14 contains determinants that regulate susceptibility to Theiler's virus-induced demyelination in the mouse. Genetics 148: 1941-1949.

67. Aubagnac, S., Brahic, M. and Bureau, J.F. 1999. Viral load and a locus on chromosome 11 affect the late clinical disease caused by Theiler's virus. J. Virol. 73:7965-7971.

68. Aubagnac, S, Brahic, M. and Bureau, J.F. 2001. Viral load increases in SJL/J mice persistently infected by Theiler's virus after inactivation of the beta(2)m gene. J. Virol. 75:7723-7726.

69. Bureau, J.F., Montagutelli, X., Lefebre, S., Guenet, J-L., Pla, M. and Brahic, M. 1992. The interaction of two groups of murine genes determines the persistence of Theiler's virus in the central nervous system. J. Virol. 66:4698-4704

70. Bureau, J.F., Montagutelli, X., Bihl, F., Lefebvre, S., Guenet, J.L. and Brahic, M. 1993. Mapping loci influencing the persistence of Theiler's virus in the murine central nervous system. Nat. Genet. 5:87-91.

71. Bihl, F., Brahic, M. and Bureau, J.F. 1999. Two loci, *Tmevp2* and *Tmevp3*, located on the telomeric region of chromosome 10, control the persistence of Theiler's virus in the central nervous system of mice. Genetics 152:385-392.

72. Aubagnac, S., Brahic, M. and Bureau, J.F. 2002. Bone marrow chimeras reveal non-H-2 hematopoietic control of susceptibility to Theiler's virus persistent infection. J. Virol. 76:5807-5812.

73. Vigneau, S., Levillayer, F., Crespeau, H., Cattolico, L., Caudron, B., Bihl, F., Robert, C., Brahic, M., Weissenbach, J. and Bureau, J.F. 2001. Homology between a 173-kb region

from mouse chromosome 10, telomeric to the *Ifng* locus, and human chromosome 12q15. Genomics 78:206-213

74. Butterfield, R.J., Roper, R.J., Rhein, D.M., Melvold, R.W., Haynes, L., Ma, R.Z., Doerge, R.W. and Teuscher, C. Sex-specific QTL govern susceptibility to Theiler's murine encephalomyelitis virus-induced demyelination (TMEVD). Genetics (in press).

75. Rodriguez, M., Patick, A.K. Pease, L.R. and David, C.S. 1992. Role of T cell receptor V beta genes in Theiler's virus-induced demyelination of mice. J. Immunol. 148 921-927.

76. Rodriguez, M., Nabozny, G.H., Thiemann, R.L. and David, C.S. 1994. Influence of deletion of T cell receptor V beta genes on the Theiler's virus model of multiple sclerosis. Autoimmunity 19:221-230.

77. Bahk, Y.Y., Kappel, C.A., Rasmussen G. and Kim B.S. 1997. Association between susceptibility to Theiler's virus-induced demyelination and T-cell receptor Jbetal-Cbetal polymorphism rather that Vbeta deletion. J. Virol. 71:4181-4185.

78. Beall, S.S, Concannon, P., Charmley, P., McFarland, H.F., Gatti, R.A., Hood, L.E., McFarlin, D.E. and Biddison, W.E. 1989. The germline repertoire of T cell receptor β-chain genes in patients with chronic progressive multiple sclerosis. J. Neuroimmunol. 21:59-66.

79. Oksenberg, J.R., M. Sherritt, A.B. Begovich, H.A. Erlich, C.C. Bernard, L.L. Cavalli-Sforza and L. Steinman. 1989. T-cell receptor Vα and Cα alleles associated with multiple sclerosis and myasthenia gravis. Proc. Natl. Acad. Sci. USA 86:988-992.

80. Seboun, E., Robinson, M.A., Doolittle, T.H., Clulla, T.A., Kindt, T.J. and Hauser, S.L. 1989. A susceptibility locus for multiple sclerosis is linked to the T cell receptor β chain complex. Cell 57:1095-1100.

81. Hashimoto, L.L., Mak, T.W. and Ebers, G.C. 1992. T-cell alpha chain polymorphisms in multiple sclerosis. J. Neuroimmunol. 40:41-48.

82. Bihl, F., Pena-Rossi, C., Guenet, J.L., Brahic, M. and Bureau, J.F. 1997. The shiverer mutation affects the persistence of Theiler's virus in the central nervous system. J. Virol. 71:5025-5030

83. Hertzler, S., Kallio, P. and Lipton, H.L. 2001. UDP-galactose transporter is required for Theiler's virus entry into mammalian cells. Virology 286:336-344.

84. Rodriguez, M., Patick, A.K. and Pease, L.R. 1990. Abrogation of resistance to Theiler's virus-induced demyelination in C57BL mice by total body irradiation. J. Neuroimmunol. 26:189-199.

85. Nicholson, S.M., Dal Canto, M.C., Miller, S.D. and Melvold, R.W. 1996. Adoptively transferred CD8[+] T lymphocytes provide protection against TMEV-induced demyelinating disease in BALB/c mice. J. Immunol. 156:1276-1283.

86. Haynes, L.M., Vanderlugt, C.L., Dal Canto, M.C., Melvold, R.W. and Miller S.D. 2000. CD8[+] T cells from Theiler's virus-resistant BALB/cByJ mice downregulate pathogenic virus-specific CD4[+] T cells. J.Neuroimmunol. 106:43-52.

87. Pullen, L.C., Park, S.H., Miller, S.D., Dal Canto, M.C. and Kim, B.S. 1995. Treatment with bacterial LPS renders genetically resistant C57BL/6 mice susceptible to Theiler's virus-induced demyelinating disease. J. Immunol. 155:4497-1503.

88. Karls, K.A., Denton, P.W. and Melvold, R.W. 2002. Susceptibility to Theiler's murine encephalomyelitis virus-induced demyelinating disease in BALB/cAnNCr mice is related to absence of a CD4[+] T-cell subset. Mult. Scler. 8:469-474.

89. Lando, Z., Teitelbaum, D. and Arnon, R. 1979. Effect of cyclophosphamide on suppressor cell activity in mice unresponsive to EAE. J. Immunol. 123:2156-2160.

90.  Fiette, L., Aubert, C., Brahic, M. and Pena Rossi, C.  1996.  Infection of class II-deficient mice by the DA strain of Theiler's virus. J. Virol. 70:4811-4815.

91.  Njenga, M.K., Pavelko, K.D., Baisch, J., Lin, S., David, C., Leibowitz, J. and Rodriguez, M.  1996.  Theiler's virus persistence and demyelination in major histocompatibility complex class II-deficient mice. J. Virol: 70:1729-1737

92.  Fiette, L., Aubert, C., Muller, U., Aguet, M., Brahic, M. and Bureau, J.F.  1995.  Theiler's virus infection of 129Sv mice that lack the interferon alpha/beta or interferon gamma receptors. J. Exp. Med. 181:2069-2076.

93.  Rodriguez, M., Roos, R.P., McGavern, D., Zoecklein, L., Pavelko, K., Sang, H. and Lin, X.  2000.  The CD4-mediated immune response is critical in determining the outcome of infection using Theiler's viruses with VP1 capsid protein point mutations.  Virol. 275:9-19.

94.  Zhang, X., Brewer, L., Walcheck, B., Johnson, A., Pease, L.R. and Njenga, M.K.  2001.  Theiler's virus-infected L-selectin-deficient mice have decreased infiltration of CD8(+) T lymphocytes in central nervous system but clear the virus.  J. Neuroimmunol. 116:178-187

95.  Palma, J.P., Lee, H.G., Mohindru, M., Kang, B.S., Dal Canto, M., Miller, S.D. and Kim, B.S.  2001.  Enhanced susceptibility to Theiler's virus-induced demyelinating disease in perforin-deficient mice. J. Neuroimmunol. 116:125-135

96.  Rossi, C., McAllister, A., Tanguy, M., Kagi, D. and Brahic, M.  1998.  Theiler's virus infection of perforin-deficient mice. J. Virol. 72:4515-4519

Chapter B5

# THE ROLE OF T CELLS AND THE INNATE IMMUNE SYSTEM IN THE PATHOGENESIS OF THEILER'S VIRUS DEMYELIATING DISEASE


Julie K. Olson and Stephen D. Miller
*Department of Microbiology and Immunology, Northwestern University Feinberg School of Medicine, Chicago, IL*



**Abstract**:  The immune response to TMEV involves activation of both the innate and adaptive immune responses.  The innate immune response is the initial response by the host to a viral infection.  The adaptive immune response follows the initial response and is mediated by $CD8^+$ and $CD4^+$ T cell and antibody responses specific to the viral antigens.  CNS pathogenesis in TMEV-IDD is associated with long-term virus persistence in CNS-resident microglia and macrophages which leads to the activation of a virus-specific $CD4^+$ Th1 response.  Myelin destruction is initiated by the bystander effector functions of CNS-resident mononuclear cells which are activated both directly by the innate immune response to persistent TMEV infection and indirectly by pro-inflammatory cytokines released by virus-specific Th1 cells.  Initial myelin destruction leads to the release of myelin antigens and the subsequent activation of myelin epitope-specific autoreactive $CD4^+$ Th1 cells via epitope spreading which are largely responsible for chronic disease progression.



Key words:  Theiler's murine encephalomyelitis virus (TMEV), demyelinating disease, multiple sclerosis, epitope spreading, molecular mimicry

# 1. Introduction

Theiler's murine encephalomyelitis virus' (TMEV) infection triggers a two-step immune response directed towards the virus. The initial response is the innate immune response that occurs immediately following virus infection which is geared to control the early stages of virus replication. The second response is the adaptive or virus-specific immune response that arises with a few days and which is necessary to clear the virus infection from mouse strains which are resistant to induction of TMEV-induced demyelinating disease (TMEV-IDD). In susceptible mouse strains, however, virus is not cleared from the mouse resulting in a persistent infection of central nervous system (CNS) mononuclear cells including resident microglia and infiltrating peripheral macrophages. CNS viral persistence leads to a cascade of events in which pro-inflammatory cytokines released from virus-specific CD4$^+$ T cells cause initial macrophage-mediated myelin tissue damage, release of endogenous myelin antigens, and eventual activation of myelin epitope-specific autoreactive CD4$^+$ T cells via the process of *epitope spreading*. Thus, adaptive immune responses to both viral and self epitopes appear to be involved in the chronic-progressive pathogenesis of TMEV-IDD.

This chapter will begin with discussion of the distinct roles of TMEV-specific CD4$^+$ and CD8$^+$ T cell responses in both virus clearance and the initiation of CNS pathology. Secondly, the role of myelin-specific autoreactive CD4$^+$ T cells in chronic progressive disease will be discussed. Finally, the innate immune response to TMEV will be discussed in relation to the activation of persistently infected CNS mononuclear cells and their role as effector cells in axonal demyelination and as antigen presenting cells in CNS activation of both virus-specific and autoreactive T cells.

## 2. Role of virus-specific T cell responses in disease resistance/ susceptibility

Susceptibility to TMEV-induced demyelinating disease has been associated with the CD8$^+$ T cell response by the finding that resistance/susceptibility is linked to the H2-D locus of MHC class I gene (1-4). A potent CD8$^+$ T cell response has been shown to mediate viral clearance in genetically resistant mice and a weak anti-TMEV CTL response and inability to clear the infection is associated with CNS viral persistence and subsequent demyelinating disease in susceptible mouse strains. Virus persistence in the CNS is associated with the later development of the myelin-specific autoimmune responses leading to the chronic phase of disease in these mice. Mutations in the H-2D MHC class I locus of resistant mice leads to viral persistence and development of demyelinating lesions in these mice (5). CD8[+] T cell depletion in the resistant strains confers virus persistence and demyelination while CD8[+] T cell depletion in susceptible strains does not alter demyelination (6-9). TMEV infection of β2-microglobulin (β2M)-deficient mice, which lack expression of MHC class I expression and thus lack functional CD8[+] T cells, further demonstrates the critical role of CD8[+] T cells in disease resistance. Infection of β2M-deficient mice on the resistant C57BL/6 background resulted in the development of pathologic signs of demyelinating disease, while infection of β2M-deficient mice on the susceptible SJL background resulted in enhanced demyelination, earlier onset of disease and increased levels of persistence virus in the CNS (10,11). Further, adoptive transfer of CD8[+] T cells from TMEV-infected resistant mice into susceptible mice protected them from demyelinating disease following TMEV infection (12). Recent studies have identified the H-2D restricted epitopes from the TMEV capsid proteins that are recognized by CD8[+] T cells in TMEV infected resistant mice (13). Likewise, functional CD8[+] T cells specific for viral epitopes have been isolated from the CNS of infected susceptible SJL mice in which the epitopes are H-2K restricted (14). Therefore, the TMEV-specific CD8[+] T cells are activated following infection in both susceptible and resistant mice strains, however, the susceptible mice do not have the H-2D restricted response that has been linked to efficient viral clearance from the CNS.

In contrast to virus-specific CD8[+] T cell responses which are associated with disease resistance, susceptibility to TMEV-IDD is correlated with a strong CD4[+] T cell response to viral antigens. TMEV-IDD is a CD4[+] T cell dependent disease as demonstrated by numerous observations. Depletion of CD4[+] T cells in SJL/J mice infected with TMEV dramatically decreased disease incidence and delayed onset of demyelinating disease while a similar depletion of CD8[+] T cells did not lead to a decrease in demyelinating disease (6-8). The MHC class II-restricted CD4[+] T cell response can be recognized in infected mice by high level delayed type hypersensitivity (DTH) responses to viral antigens beginning at 7-10 days post-infection and continuing throughout the late chronic phase of disease (1,15). Mice resistant to TMEV-IDD develop a low level DTH response to viral antigens, but develop similar virus-specific T cell proliferative responses and serum antibody levels as susceptible mice (16). The immunodominant CD4[+] T cell determinant in the susceptible SJL/J mice has been mapped to the TMEV VP2 capsid protein, specifically VP2 70-86 (17). Systemic transfer of VP2

70-86 specific CD4$^+$ Th1 cells into SJL/J mice infected with a suboptimal dose of TMEV increased the incidence and severity of demyelinating disease (6). Thus, VP2 74-86 specific CD4$^+$ T cells have functional immunopathologic potential in virus infected mice. Minor CD4$^+$ T cell epitopes have also been identified on other TMEV capsid proteins, VP1 (VP1 233-250) and VP3 (VP3 24-37) (18,19) and may also contribute to the immune-mediated demyelination.

TMEV-IDD is initiated by TMEV-specific CD4$^+$ T cells which respond to persistent virus in the CNS (Figure 1). TMEV-specific CD4$^+$ T cell

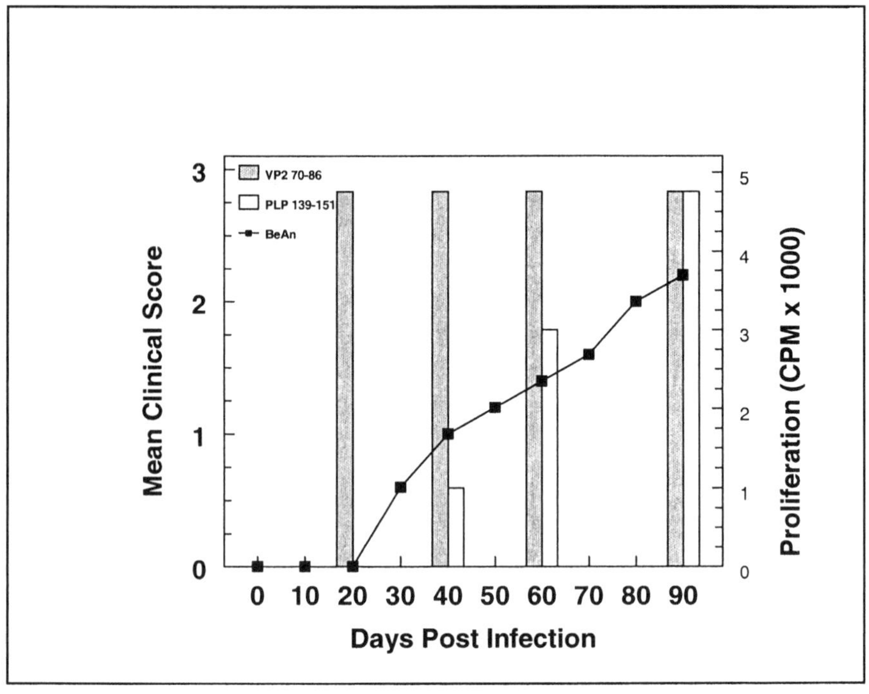


**Figure 1. Temporal development of TMEV-IDD clinical disease and CD4$^+$ T cell responses.** SJL/J mice infected with BeAn strain of TMEV develop a chronic progressive demyelinating disease associated with clinical signs (line graph). CD4$^+$ T cell responses to viral antigen (VP2 70-86) and myelin antigen (PLP139-151) were assayed from infected mice at various times following infection by splenic proliferation assays (bar graph).

responses are detected within 4-5 days following infection and persist for at least six months, while T cell responses to myelin basic protein (MBP) and proteolipid protein (PLP) are not detected until 50-60 days post infection which is after disease onset which typically occurs 30-40 days post-infection (20,21). Proof of the critical pathologic role for TMEV-specific CD4$^+$ T cells in *initiation* of demyelination is demonstrated by the observation that tolerance induced by intact TMEV virions coupled to syngeneic splenocytes prior to or shortly after virus infection, which specifically anergizes virus-specific Th1 responses (22,23), results in a dramatic reduction in the incidence and severity of demyelinating lesions and clinical disease in SJL/J mice subsequently infected with TMEV (24). Tolerance specifically reduced the virus-specific CD4$^+$ Th1 response in the mice and reduced the number of CD4$^+$ T cells infiltrating the CNS (24). Thus, the virus-specific CD4$^+$ T cells appear to play contrasting roles by contributing to virus clearance by serving as helper cells for virus-specific CD8$^+$ CTLs and B cells, but also by contributing to the initiation of demyelinating disease.

## 3. Role of myelin-specific T cell responses in TMEV-IDD

Neuroantigen-specific T cells do not play a significant role in disease initiation. As described previously, although responses to TMEV epitopes arise within 4-5 days post-infection, MBP and PLP responses are not detected prior to disease onset (30-40 days PI) (20,25) and induction of peripheral tolerance to mouse spinal cord homogenate (MSCH - a heterogeneous mixture of multiple neuroantigens) 7-14 days prior to TMEV infection fails to affect the initiation of TMEV-IDD and the accompanying virus-specific T cell and antibody responses (26). However, this tolerogenic regimen is extremely effective in preventing clinical and histologic signs of MSCH-induced relapsing EAE (R-EAE) and the accompanying neuroantigen-specific DTH responses (26,27). In addition, there appears to be no cross-reactivity between CD4$^+$ T cells specific for the immunodominant TMEV epitopes and myelin proteins indicating that the wildtype BeAn strain of TMEV contains no epitopes which *mimic* the immunodominant myelin epitopes in the SJL mouse (28).

Although myelin-specific responses appear to not be involved in disease initiation, they play an important role in the chronic stages of the disease process. Approximately 45-55 days post-infection (2-3 weeks after the onset of clinical disease), when myelin damage reaches a critical threshold, CD4$^+$ T cell responses to the immunodominant myelin PLP139-151 epitope appear (28), and as disease progresses T cell responses to multiple additional encephalitogenic myelin epitopes (*e.g.*, PLP56-70, MOG92-106, PLP178-191, and MBP84-104) arise in an ordered progression (21,28). Following i.c. infection with the wildtype BeAn strain of TMEV, anti-myelin epitope responses arise via epitope spreading as indicated by the late appearance of anti-myelin T cells and lack of cross-reactivity between T cells specific for TMEV epitopes and myelin proteins and peptides and vice versa (28). Thus, epitope spreading appears to play a role in progression of TMEV-IDD similar to the roles we have described that intermolecular and intramolecular epitope spreading play in mediating relapses in two different models of myelin peptide-induced relapsing EAE (R-EAE) in the SJL mouse (29,30).

Relevant to the pathologic role of these autoreactive T cells, recent evidence has shown that PLP139-151-specific T cells can be found in the CNS of mice with chronic TMEV-IDD at 65 days post-infection(31) and that these autoreactive T cells can demyelinate *in vitro* organotypic CNS cultures (32).   More importantly, we have reported that induction of peripheral tolerance to MP4 (a recombinant fusion protein comprising the 21.5 kD isoform of human MBP and a recombinant variant of human PLP) at 45 days post-infection attenuated disease progression and decreased demyelination and inflammatory cell infiltration in the CNS (31).   In addition, antigen presenting cells isolated from the CNS of infected mice present myelin epitopes to myelin-specific CD4$^+$ T cells *in vitro* (33).   Thus, autoreactive CD4$^+$ T cell responses to myelin epitopes develop only after clinical signs of demyelinating disease are evident, however these responses play a major pathologic role in mediating chronic progression of the disease (Figure 1). Collectively, these results are consistent with a model (Figure 2) wherein virus-specific CD4$^+$ Th1 cells responding to the persistent virus infection of CNS APCs secrete pro-inflammatory cytokines that recruit and activate mononuclear inflammatory cells that initiate myelin damage in the CNS (34).  The damaged myelin is then processed and presented by CNS APCs and infiltrating macrophages to infiltrating autoreactive CD4$^+$ T cells which then secrete pro-inflammatory cytokines to perpetuate the chronic pathologic process.  The epitope spreading hypothesis is also supported by observations that CNS APCs isolated from TMEV-infected mice 60-90 days post-infection endogenously present both virus and multiple myelin antigens to specific CD4$^+$ T cell lines, whereas CNS APCs isolated at 40 days post-infection endogenously present virus, but not myelin antigens (33).   Thus, epitope spreading is the primary mechanism for initiation of the autoimmune response in TMEV-IDD (28).

## 4. TMEV-IDD Model of Molecular Mimicry

As described above, myelin-specific autoreactive CD4$^+$ T cell responses in TMEV-IDD are initiated by epitope spreading; however, autoimmune

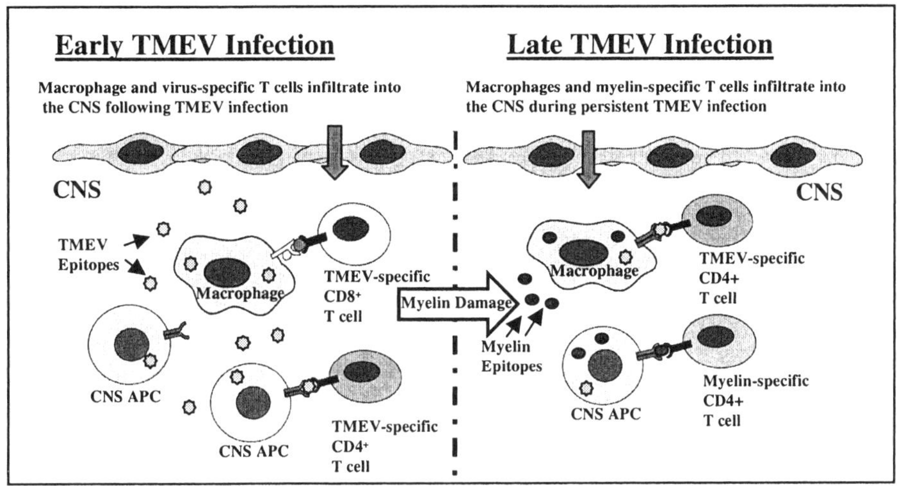


**Figure 2. Model of Initiation and Progression of TMEV-Induced Demyelinating Disease.** Infection of susceptible SJL/J mice leads to an inefficient CD8[+] CTL response resulting in virus persistence in the CNS. TMEV-specific CD4[+] T cells then infiltrate the CNS and release pro-inflammatory cytokines, promoting infiltration and activation of peripheral macrophages and CNS-resident APCs (microglia and astrocytes) which cause bystander myelin damage. Via the process of epitope spreading, myelin debris is then phagocytized and processed by microglia/macrophages and endogenous myelin epitopes are presented to myelin-specific autoreactive T cells which play the predominant role in chronic immunopathology..

responses can also be initiated by molecular mimicry during infection. Molecular mimicry is the activation of cross-reactive T cells during a virus infection which recognize both a viral epitope and a self epitope due to sequence homology. We have recently provided conclusive evidence that CD4[+] T cell-mediated molecular mimicry can be initiated following virus infection. A nonpathogenic mutant of TMEV was used to construct a model to test the ability of a virus to encode a self sequence and initiate autoimmunity by molecular mimicry (35). Infection of SJL/J mice with a recombinant TMEV containing the immunodominant PLP139-151 epitope resulted in a rapid-onset, severe demyelinating disease mediated specifically by PLP139-151 specific CD4[+] Th1 cells (35). The pathologic role of autoreactive T cells in mediating the early-onset demyelinating disease was demonstrated by the findings that tolerance induced with PLP139-151-coupled splenocytes prior to infection protected SJL mice from disease development and serial transfer of PLP139-151 specific CD4[+] T cells from infected mice to a naïve recipient resulted in induction of demyelinating disease (36). More relevant to proof of autoimmunity via molecular mimicry, infection of mice with recombinant TMEV containing a mimic PLP139-151 sequence derived from the protease IV gene of *Haemophilus influenzae* resulted in an early onset demyelination concomitant with induction of CD4$^+$ T cell responses cross-reactive with PLP139-151 (35). Therefore, although infection with wildtype TMEV does not result in induction of autoimmunity via molecular mimicry, the virus provides the structure and specificity to facilitate study of a molecular mimicry model of virus-induced demyelinating disease.


## 5. Innate Immune Response Induced by TMEV Infection

The innate immune response occurs immediately following infection in response to pathogen-associated molecular patterns (PAMPs) integral to the pathogen, such as double-stranded RNA (dsRNA) derived from virus infection, lipopolysaccharide (LPS) from gram negative bacteria and peptidoglycan from gram positive bacteria. These molecular patterns are recognized by Toll like receptors (TLRs) which are a class of innate immune receptors identified in humans and mice. At least ten TLRs have been described in humans (37) and all require MyD88 in their signaling that leads to the activation of NF$_k$B resulting in expression of multiple innate immune cytokines, including type I interferons (IFN-$\alpha$ and IFN-$\beta$), IL-1, IL-12, TNF-$\alpha$, *etc.* In virus infections, type I interferons are produced in response to dsRNA (38) which is commonly found in the replication cycles of multiple viruses, including TMEV, but which is not found in mammalian cells (39). The cellular response to dsRNA is dependent on the dsRNA-activated protein kinase R (PKR), a dormant enzyme directly activated by binding of dsRNA which leads, via the activation of NF$\kappa$B, to the production of type I IFNs (40). In turn, type I IFN signaling takes place through an IFN receptor complex consisting of two alpha chains (type I IFN receptor) complexed with Jak1 and Tyk2. These kinases phosphorylate Stat1 and Stat2 respectively eventually leading to upregulation of MHC class I and II and many other downstream effects (41). A recent study has also reported that TLR3 binds dsRNA suggesting a TLR pathway for virus recognition by the innate immune response (42).

Relevant to the chronic-progressive pathogenesis of TMEV-IDD, we have shown that naïve microglia could be persistently infected with TMEV *in vitro* (43). Significantly, these infected microglia were activated to upregulate expression of cytokines involved in innate immunity (TNF-$\alpha$, IL-6, IL-18 and, most importantly, type I interferons) along with upregulation of MHC class I and II, IL-12, various co-stimulatory molecules (B7-1, B7-2, CD40 and ICAM-1) and chemokines (MCP-1 and MIP-1$\alpha$). Most significantly, TMEV-infected microglia were able to efficiently process and present both endogenous virus epitopes and exogenous myelin epitopes to inflammatory CD4$^+$ Th1 cells whereas microglia exposed to UV-inactivated TMEV remained inefficient APCs. Thus, productive TMEV infection of microglia activates these cells to initiate an innate immune response which may lead to the activation of naïve and memory virus- and myelin-specific adaptive immune responses within the CNS and, in addition, causes them to elaborate pro-inflammatory molecules (e.g. iNOS and TNF-$\alpha$) involved in myelin destruction. Collectively, these results suggest that astrocytes, activated microglia, and CNS-infiltrating macrophages (Figure 2), exposed to the pro-inflammatory cytokine environment encountered during immune-mediated TMEV demyelination and/or upon direct infection with TMEV have the potential of stimulating both activated (involved in initiation of disease) and naive (possibly involved in epitope spreading) virus/myelin-specific T cells which gain access to the CNS.

## 6. Conclusions

The immune response following TMEV infection of genetically susceptible mouse strains leading to chronic immune-mediated demyelination is a complex and multi-step process involving activation of innate immune signals within persistently infected CNS-resident APCs, CD4$^+$ and CD8$^+$ anti-viral T cells, and autoreactive myelin-specific CD4$^+$ T cells. Myelin destruction is initiated by virus-specific CD4$^+$ T cells responding to persistent virus infection resulting in the release of pro-inflammatory chemokines and cytokines which activate CNS-resident mononuclear to cause bystander myelin destruction. Subsequent release of myelin antigens into the inflammatory environment leads to epitope spreading and the activation of myelin-specific CD4$^+$ T cells that continue to perpetuate the chronic demyelinating process.

## 7. References

1.   Clatch R.J., Melvold R.W., Miller S.D. and Lipton H.L. 1985. Theiler's murine encephalomyelitis virus (TMEV)-induced demyelinating disease in mice is influenced by the H-2D region: correlation with TMEV-specific delayed-type hypersensitivity. J. Immunol. 135:1408-1414.

2.   Melvold R.W., Jokinen D.M., Miller S.D., Dal Canto M.C., Lipton H.L. 1987. H-2 genes in TMEV-induced demyelination, a model for multiple sclerosis. In H-2 Antigens. C. S. David, ed. Plenum Publishing Corporation, New York, pp 735-745.

3.   Rodriguez M., Leibowitz J. and David C.S.  1986.  Susceptibility to Theiler's virus-induced demyelination. Mapping of the gene within the H-2D region.  J. Exp. Med. 163:620-631.

4.   Azoulay-Cayla A., Syan S., Brahic M. and Bureau J.F. 2001.  Roles of the H-2D(b) and H-K(b) genes in resistance to persistent Theiler's murine encephalomyelitis virus infection of the central nervous system. J. Gen. Virol. 82:1043-1047.

5.   Lipton H.L., Melvold R., Miller S.D., Dal Canto M.C. and Jensen K.  1995.  Mutation of a major histocompatibility class I locus, H-2D, leads to an increased virus burden and disease susceptibility in Theiler's virus-induced demyelinating disease.  J. Neurovirol. 1:138-144.

6.   Gerety S.J., Rundell M.K., Dal Canto M.C. and Miller S.D.  1994.  Class II-restricted T cell responses in Theiler's murine encephalomyelitis virus (TMEV)-induced demyelinating disease. VI. Potentiation of demyelination with and characterization of an immunopathologic CD4$^+$ T cell line specific for an immunodominant VP2 epitope.  J. Immunol. 152:919-929.

7.   Borrow P., Tonks P., Welsh C.J.R. and Nash A.A.  1992.  The role of CD8$^+$ T cells in the acute and chronic phases of Theiler's murine encephalomyelitis virus-induced disease in mice. J. Gen. Virol. 73:1861-1865.

8.   Welsh C.J., Tonks P., Nash A.A. and Blakemore W.F.  1987.  The effect of L3T4 T cell depletion on the pathogenesis of Theiler's murine encephalomyelitis virus infection in CBA mice. J. Gen. Virol. 68:1659-1667.

9.   Murray P.D., Pavelko K.D., Leibowitz J., Lin X. and Rodriguez M.  1998.  CD4$^+$ and CD8$^+$ T cells make discrete contributions to demyelination and neurologic disease in a viral model of multiple sclerosis. J. Virol. 72:7320-7329.

10.  Pullen L.C., Miller S.D., Dal Canto M.C. and Kim B.S.  1993.  Class I-deficient resistant mice intracerebrally inoculated with Theiler's virus show an increased T cell response to viral antigens and susceptibility to demyelination. Eur. J. Immunol. 23:2287-2293.

11.  Begolka W.S., Haynes L.M., Olson J.K., Padilla J., Neville K.L., Canto M.D., Palma J., Kim B.S. and Miller S.D.  2001.  CD8-deficient SJL mice display enhanced susceptibility to Theiler's virus infection and increased demyelinating pathology.  J. Neurovirol. 7:409-420.

12.  Nicholson S.M., Dal Canto M.C., Miller S.D. and Melvold R.W.  1996.  Adoptively transferred CD8$^+$ lymphocytes provide protection against TMEV-induced demyelinating disease in BALB/c mice. J. Immunol. 156:1276-1283.

13.  Lyman M.A., Lee H.G., Kang B.S., Kang H.K. and Kim B.S.  2002.  Capsid-specific cytotoxic T lymphocytes recognize three distinct H-2D(b)-restricted regions of the BeAn strain of Theiler's virus and exhibit different cytokine profiles. J. Virol. 76:3125-3134.

14.  Kang B.S., Lyman M.A. and Kim B.S.  2002.  Differences in avidity and epitope recognition of CD8(+) T cells infiltrating the central nervous systems of SJL/J mice infected with BeAn and DA strains of Theiler's murine encephalomyelitis virus. J. Virol. 76:11780-11784.

15. Clatch R.J., Lipton H.L. and Miller S.D. 1986. Characterization of Theiler's murine encephalomyelitis virus (TMEV)-specific delayed-type hypersensitivity responses in TMEV-induced demyelinating disease: correlation with clinical signs. J. Immunol. 136:920-927.

16. Clatch R.J., Lipton H.L. and Miller S.D. 1987. Class II-restricted T cell responses in Theiler's murine encephalomyelitis virus (TMEV)-induced demyelinating disease. II. Survey of host immune responses and central nervous system virus titers in inbred mouse strains. Microb. Pathogen. 3:327-337.

17. Gerety S.J., Clatch R.J., Lipton H.L., Goswami R.G., Rundell M.K. and Miller S.D. 1991. Class II-restricted T cell responses in Theiler's murine encephalomyelitis virus-induced demyelinating disease. IV. Identification of an immunodominant T cell determinant on the N- terminal end of the VP2 capsid protein in susceptible SJL/J mice. J. Immunol. 146:2401-2408.

18. Yauch R.L., Kerekes K., Saujani K. and Kim B.S. 1995. Identification of a major T cell epitope within VP3(24-37) of Theiler's virus in demyelination-susceptible SJL/J mice. J. Virol. 69:7315-7318.

19. Yauch R.L. Kim B.S. 1994. A predominant viral epitope recognized by T cells from the periphery and demyelinating lesions of SJL/J mice infected with Theiler's virus is located within VP1(233-244). J. Immunol. 153:4508-4519.

20. Miller S.D., Clatch R.J., Pevear D.C., Trotter J.L. and Lipton H.L. 1987. Class II-restricted T cell responses in Theiler's murine encephalomyelitis virus (TMEV)-induced demyelinating disease. I. Cross-specificity among TMEV substrains and related picornaviruses, but not myelin proteins. J. Immunol. 138:3776-3784.

21. Katz-Levy Y., Neville K.L., Padilla J., Rahbe S.M., Begolka W.S., Girvin A.M., Olson J.K., Vanderlugt C.L. and Miller S.D. 2000. Temporal development of autoreactive Th1 responses and endogenous antigen presentation of self myelin epitopes by CNS-resident APCs in Theiler's virus-infected mice. J. Immunol. 165:5304-5314.

22. Peterson J.D., Karpus W.J., Clatch R.J. and Miller S.D. 1993. Split tolerance of Th1 and Th2 cells in tolerance to Theiler's murine encephalomyelitis virus. Eur. J. Immunol. 23:46-55.

23. Karpus W.J., Peterson J.D. and Miller S.D. 1994. Anergy *in vivo*: Down-regulation of antigen-specific CD4[+] Th1 but not Th2 cytokine responses. Int. Immunol. 6:721-730.

24. Karpus W.J., Pope J.G., Peterson J.D., Dal Canto M.C. and Miller S.D. 1995. Inhibition of Theiler's virus-mediated demyelination by peripheral immune tolerance induction. J. Immunol. 155:947-957.

25. Barbano R.L. Dal Canto M.C. 1984. Serum and cells from Theiler's virus-infected mice fail to injure myelinating cultures or to produce in vivo transfer of disease. The pathogenesis of Theiler's virus-induced demyelination appears to differ from that of EAE. J. Neurol. Sci. 66:283-293.

26. Miller S.D., Gerety S.J., Kennedy M.K., Peterson J.D., Trotter J.L., Tuohy V.K., Waltenbaugh C., Dal Canto M.C. and Lipton H.L. 1990. Class II-restricted T cell

responses in Theiler's murine encephalomyelitis virus (TMEV)-induced demyelinating disease. III. Failure of neuroantigen-specific immune tolerance to affect the clinical course of demyelination. J. Neuroimmunol. 26:9-23.

27. Kennedy M.K., Dal Canto M.C., Trotter J.L. and Miller S.D. 1988. Specific immune regulation of chronic-relapsing experimental allergic encephalomyelitis in mice. J. Immunol. 141:2986-2993.

28. Miller S.D., Vanderlugt C.L., Begolka W.S., Pao W., Yauch R.L., Neville K.L., Katz-Levy Y., Carrizosa A. and Kim B.S. 1997. Persistent infection with Theiler's virus leads to CNS autoimmunity via epitope spreading. Nature Med. 3:1133-1136.

29. McRae B.L., Vanderlugt C.L., Dal Canto M.C. and Miller S.D. 1995. Functional evidence for epitope spreading in the relapsing pathology of experimental autoimmune encephalomyelitis. J. Exp. Med. 182:75-85.

30. Miller S.D., McRae B.L., Vanderlugt C.L., Nikcevich K.M., Pope J.G., Pope L. and Karpus W.J. 1995. Evolution of the T cell repertoire during the course of experimental autoimmune encephalomyelitis. Immunol. Rev. 144:225-244.

31. Neville K.L., Padilla J. and Miller S.D. 2002. Myelin-specific tolerance attenuates the progression of a virus-induced demyelinating disease: Implications for the treatment of MS. J. Neuroimmunol. 123:18-29.

32. Dal Canto M.C., Calenoff M.A., Miller S.D. and Vanderlugt C.L. 2000. Lymphocytes from mice chronically infected with Theiler's murine encephalomyelitis virus produce demyelination of organotypic cultures after stimulation with the major encephalitogenic epitope of myelin proteolipid protein. Epitope spreading in TMEV infection has functional activity. J. Neuroimmunol. 104:79-84.

33. Katz-Levy Y., Neville K.L., Girvin A.M., Vanderlugt C.L., Pope J.G., Tan L.J. and Miller S.D. 1999. Endogenous presentation of self myelin epitopes by CNS-resident APCs in Theiler's virus-infected mice. J. Clin. Invest. 104:599-610.

34. Miller S.D., Vanderlugt C.L., Begolka W.S., Pao W., Neville K.L., Yauch R.L. and Kim B.S. 1997. Epitope spreading leads to myelin-specific autoimmune responses in SJL mice chronically infected with Theiler's virus. J. Neurovirol. 3 Suppl 1:S62-S65.

35. Olson J.K., Croxford J.L., Calenoff M., Dal Canto M.C. and Miller S.D. 2001. A virus-induced molecular mimicry model of multiple sclerosis. J. Clin. Invest. 108:311-318.

36. Olson J.K., Eagar T.N. and Miller S.D. 2002. Functional activation of myelin-specific T cells by virus-induced molecular mimicry. J. Immunol. 169:2719-2726.

37. Chuang T. Ulevitch R.J. 2001. Identification of hTLR10: a novel human Toll-like receptor preferentially expressed in immune cells. Biochim. Biophys. Acta 1518:157-161.

38. Hoffmann J.A., Kafatos F.C., Janeway C.A. and Ezekowitz R.A. 1999. Phylogenetic perspectives in innate immunity. Science 284:1313-1318

39. Medzhitov R. Janeway C.A., Jr. 1997. Innate immunity: impact on the adaptive immune response. Curr. Opin. Immunol. 9:4-9.

40. Zamanian-Daryoush M., Mogensen T.H., DiDonato J.A. and Williams B.R. 2000. NF-kappaB activation by double-stranded-RNA-activated protein kinase (PKR) is mediated

through NF-kappaB-inducing kinase and IkappaB kinase.  Mol. Cell Biol.  20:1278-1290.

41. Sen G.C. Lengyel P.  1992.  The interferon system. A bird's eye view of its biochemistry. J. Biol. Chem.  267:5017-5020.

42. Alexopoulou L., Holt A.C., Medzhitov R. and Flavell R.A.  2001.  Recognition of double-stranded RNA and activation of NF-kappaB by Toll- like receptor 3.  Nature 413:732-738.

43. Olson J.K., Girvin A.M. and Miller S.D.  2001.  Direct activation of innate and antigen presenting functions of microglia following infection with Theiler's virus.  J. Virol. 75:9780-9789.

# Chapter B6

# CYTOKINES, CHEMOKINES AND ADHESION MOLECULES IN TMEV-IDD

Byung S. Kim[1], Alyson C. Fuller[1], and Chang-Sung Koh[2]

[1]*Department of Microbiology-Immunology, the Finberg Medical School of Northwestern University, Chicago, Il 60611 and [2]Department of Medical Technology, Shinshu University School of Allied Medical Sciences, Matsumoto, Japan*

**Abstract:**     Cytokines, chemokines and adhesion molecules play pivotal roles in the initiation and progression of TMEV-induced demyelinating disease. Furthermore, the same cytokines, chemokines and adhesion molecules are also important in induction of protective immune responses that are necessary to eliminate persistent viral infection. Thus, the type, level and timing of the initial responses of cytokines, chemokines and adhesion molecules upon viral infection, as well as consequent immune responses, will likely determine the pathogenic or protective outcome. The previous trials to intervene the development and progression of disease have been focused on the inflammatory Th1 responses. However, recent studies have indicated that these molecules induced directly by viral infection are important in clearing viral persistence and shaping adaptive immune responses. Therefore, investigations to block the initiation of the inflammatory responses, as the early events of cellular infiltration to CNS and cytokines/chemokines involved will be very valuable.

**Key words:**     Chemokines, Cytokines, TMEV, Astrocyte, Microglia, Oligodendrocyte

Intracerebral infection of susceptible strains of mice with Theiler's murine encephalomyelitis virus (TMEV) leads to a chronic immune-mediated demyelinating disease of the central nervous system (CNS) that is very similar to human multiple sclerosis (MS). There are two subgroups of TMEV; the first group, which includes the GDVII and FA viruses, causes a rapid fatal encephalitis. The BeAn and DA viruses belong to the second subgroup, and cause a biphasic neurological disease in susceptible mice such as SJL/J (1). However, the clinical and histopathological symptoms of BeAn and DA differ in the acute phase of disease. Infection with the DA strain leads to an acute polioencephalitis in the gray matter causing flaccid limb paralysis and neuron degeneration followed by chronic viral persistence in the white matter of the spinal cord leading to demyelination. The BeAn strain, on the other hand, is characterized by a clinically undetectable acute phase followed by a very similar chronic phase and demyelination in the white matter. Thus, the TMEV strain used in the experiments may affect the experimental outcome and consequent interpretations.

Demyelination in the white matter leading to clinical symptoms is associated primarily with cell-mediated immune responses, in particular inflammatory Th1 responses to both viral and autoantigens (2, 3). However, the role of cytotoxic T cells in pathogenesis of demyelination is not yet clear, although this cell population appears to be important for clearing viral persistence resulting in the protection from TMEV-induced demyelinating disease. The major $CD4^+$ and $CD8^+$ T cell epitopes of TMEV in resistant and susceptible mice have recently been determined; therefore studying the anti-viral immune response in the CNS is now feasible (2, 4, 5). Early studies with cytokines have mainly been focused on defining the Th types. Analysis of cytokines, chemokines and adhesion molecules involved in TMEV-induced demyelinating disease is quite complex since these molecules are produced by not only various activated immune cells but also many different CNS resident glial cells directly upon viral infection.

## 1.    CYTOKINES

Due to the inflammatory nature of the immune response to TMEV, the cytokines produced by resident CNS cells and infiltrating immune cells are likely to play a key role in the pathogenesis of TMEV. TMEV-induced demyelinating disease (TMEV-IDD) appears to be mediated by $CD4^+$ T cells that gain access to the CNS after viral infection. The T cells that are found within demyelinating lesions of the CNS preferentially produce IFN-$\gamma$ (5, 6), therefore the initial establishment of a T helper type 1 (Th1) response may be critical for the pathogenesis of TMEV. While the pathogenic role of $CD4^+$ T cells is well accepted, the role of $CD8^+$ T cells is controversial. However, it is conceivable that virus specific $CD8^+$ T cells are critical for viral clearance, as well as pathogenesis via IFN-$\gamma$ production (5). A fine line of balance of various cytokines produced by glial cells, antigen presenting cells (APC), natural killer (NK) cells, as well as helper T cells and $CD8^+$ T cells, may contribute to protection or pathogenesis. Due to the inflammatory nature of immune responses in the CNS following infection with TMEV, the majority of the previous studies have focused on the cytokines associated with Th1 and/or Th2 responses.

## 1.1      Cytokines in the CNS following viral infection

Many studies have examined the production of various cytokines throughout the course of TMEV-IDD. Susceptible SJL/J mice demonstrate an early production of the inflammatory cytokines, Interferon (IFN)-γ, tumor necrosis factor (TNF)-α, and also the anti-inflammatory cytokines IL-10 and IL-4 within the CNS of infected mice during the preclinical phase of disease following viral infection, and these levels continue to rise into the chronic phase of disease (7, 8). Attempts have been made to determine the order of cytokine production in the CNS following viral infection and the results somewhat vary, i.e. Th1 associated cytokine production precedes Th2 associated cytokines in some studies (7, 9) and virtually the same time by others (8). Both resistant C57BL/6 and susceptible SJL/J mice show an early production of IFN-γ, IL-1, IL-6, IL-12 and TNF-α (10, 11). However, SJL/J mice show greater levels of these cytokines, particularly tumor growth factor (TGF)-β (11). It has been postulated that this increase in TGF-β may be responsible for the insufficient anti-viral immune response in SJL/J mice. Cytokine secretion during the acute phase is sustained during the chronic phase of disease in SJL/J mice. C57BL/6 mice, on the other hand, show very little cytokine secretion in the CNS at this late time point (10, 11). These studies indicate the importance of individual cytokines on the development and progression of disease. However, the cellular origin of the cytokines in these studies has not been elucidated and may reflect the level of many different inflammatory responses, in addition to the Th responses.

## 1.2      Cell types involved in cytokine production

The earlier studies have heavily emphasized on cytokine production by T cells. Due to the difficulties in isolating T cell population from the CNS, peripheral T cells specific for viral epitopes have frequently been analyzed. These early studies indicated that virus-specific T cells predominantly produce IFN-γ, a potent proinflammatory cytokine. Recent advancements in cytokine detection with small numbers of antigen-specific T cells utilizing ELISPOT and intracellular cytokine staining accompanied by flow cytometric analysis, have permitted to assess cytokine production by various infiltrating T cell populations. Again, the majority of virus-reactive T cells, including Th and CTL produce proinflammatory cytokines, like IFN-γ and TNF-α, although some subpopulations produce anti-inflammatory cytokines such as IL-4/IL-5 and/or IL-10 ((5, 6), Lyman and Kim, unpublished data).

Previous studies have indicated that the main reservoir of viral replication is microglia/macrophages in the CNS and depletion of macrophages protect mice from TMEV-IDD. Our initial studies indicate that the majority (>50%)

of microglias in the CNS of TMEV-infected SJL/J mice produce TNF-α, compared to a minor population (<5%) of macrophages (Mohindru and Kim, unpublished data). Infection of other glial cells (e.g. astrocytes, oligodendrocytes) within the CNS is also crucial for viral persistence. However, isolation and maintenance of these glial cells from infected adult mice are rather difficult.

Consequently, astrocytes, oligodendrocytes and microglia derived from neonatal mice have been utilized to investigate whether virus infection of these cell types directly induces the production of chemokines, cytokines and other surface and secreted molecules. Early studies using cultured murine astrocytes illustrated TNF-α production following TMEV infection (12). Recent studies with isolated glial cells, astrocytes in particular, indicate that the expression of various proinflammatory cytokine genes (e.g. IL-12, IL-6, TNFα, IL-1, IFNβ) is directly induced by TMEV infection via the NFκB pathway (Palma and Kim, unpublished). The cytokine induction in astrocyte cultures following TMEV infection has been further extended by using pathogenic and low-pathogenic viruses (13). This study shows that key proinflammatory cytokines, such as IL-12 and IL-1, are vigorously induced by the pathogenic virus, whereas much reduced levels by low-pathogenic virus. In addition, TMEV infection similarly induces the upregulation of several pro-inflammatory cytokines and molecules in isolated microglia cultures. Among these molecules are TNF-α, IL-6, IL-18, type I IFNs, and IL-12 (14).

In addition to the glial cells, it is possible that TMEV infection may also induce proinflammatory cytokines in professional antigen presenting cells, which critically affect the initial commitment of Th1/Th2 differentiation (13). Infection of isolated dendritic cells and macrophages with TMEV results in preferential upregulation of IL-12 production over IL-10, suggesting the importance of cytokines directly induced by TMEV infection in subsequent inflammatory immune responses.

## 1.3     Effects of manipulation of cytokine levels

In order to study the role of cytokines in the progression of demyelinating disease, the effects of these cytokines on the development of disease have been extensively investigated following the neutralization or overexpression of the key cytokines. Since TMEV-induced demyelinating disease is considered as an immune-mediated inflammatory disease, most of the early studies have focused on either inhibiting inflammatory cytokines or enhancing anti-inflammatory cytokine production (15).

IL-12 secreted by antigen presenting cells is critical in the Th1 commitment by naïve Th cells. By neutralizing IL-12 using a monoclonal antibody (mAb), the immune response to TMEV and the disease progression have been studied in the absence of an effective Th1 response, which has been determined to be critical for the pathogenesis of TMEV-IDD. Two separate studies have yielded very different results. The first study showed that the elimination of IL-12 leads to suppression of disease, decrease IFN-γ and TNF-α production with a concomitant increase in the Th2 type cytokines IL-4 and IL-10 (16). The second report indicated that anti-IL-12 mAb treatment had no effect on TMEV-mediated pathology (17). The use of different strain of virus (BeAn and DA respectively), as well as the differences in the timing and dose of mAb treatments may have affected the experimental outcome.

IFN-γ is produced by many different cell types, including NK/NKT, Th as well as cytotoxic T cells and affects the immune system in many different ways. This cytokine is critically important for upregulation of proinflammatory cytokines and chemokines, activation of macrophages and NK cells, as well as molecules involved in cellular trafficking. The production of IFN-γ is drastically upregulated in the CNS following virus infection; therefore it is possible that the elimination of this cytokine may lead to protection from TMEV-IDD. Surprisingly, depletion of IFN-γ by the use of neutralizing mAbs or IFN-γ receptor-deficient mice, lead to an exacerbation of clinical symptoms rather than a protection from disease (18-20). However, intracerebral administration of recombinant IFN-γ also lead to exacerbation of disease (18). A likely explanation is that IFN-γ is necessary to activate microglias/macrophages, CTL and NK cells, that are critical for protective immunity. In the absence of IFN-γ, the initial anti-viral immune response may not be able to effectively clear virus from the CNS hence leading to enhanced disease. On the other hand, an excess amount of IFN-γ may lead to a highly inflammatory environment, which ultimately recruits and promotes more pathogenic Th1 cell responses in the CNS to inflict neurological damage.

TNF-α is another important pro-inflammatory cytokine produced in significant quantities early after viral infection by both lymphocytes and glial cells in the CNS. Both susceptible and resistant mice produce TNF-α, but susceptible mice produce higher levels during both the acute and chronic phases of disease (10, 11, 21). Administration of neutralizing anti-TNF-α mAb to susceptible mice at the time of disease onset suppresses the development of both clinical and histopathological signs of disease (21). In contrast, administration of anti-TNF-α prior to and during the course of infection does not suppress disease (22) and injection of recombinant TNF-α at those time points ameliorates disease. Thus, treatment regimens that differ in timing of administration appear to be critical for the outcome of the cytokine effects. Taken together, TNF-α may also have dual functions:

protective role in the initial stages of viral infection, perhaps by activating innate immunity and pathogenic effects after establishment of viral persistence by promoting inflammatory responses.


## 2.        CHEMOKINES

Chemokines are low molecular weight chemotactic proteins that bind and signal through G protein coupled receptors. In many inflammatory diseases including MS, chemokines play an important role in mediating the infiltration of various inflammatory cells into sites of tissue damage. Much work has been done studying the role of chemokine/chemokine receptor expression in other animal models of MS including experimental autoimmune encephalomyelitis (EAE) and mouse hepatitis virus (MHV)-induced demyelinating disease. However, very little work determining the importance of these molecules in TMEV-IDD has been reported. Viral infection is known to induce various proinflammatory chemokines in many different cell types, which are detected in the inflammation sites induced by different infectious agents. In addition, activation of glial cells by certain chemokines or cytokines also results in further induction of the same or additional chemokines. Since the transmigration of inflammatory cells into the CNS is dependent on chemokines, these molecules are likely involved in the pathogenesis of TMEV-IDD.

Studies using the BeAn strain of TMEV showed early expression of monocyte chemoattractant protein (MCP)-1 (or CCL2), macrophage inflammatory protein (MIP)-1$\alpha$ (CCL3), IFN-$\gamma$ inducible protein of 10 KDa (IP-10 or CXCL10), regulated on activation, normal T-cell expressed and secreted (RANTES, CCL5), and MIP-1$\beta$ (CCL4) mRNAs in the CNS but not the periphery after infection with TMEV (23). As early as six days post infection with the DA strain of TMEV, mRNAs for IP-10, RANTES, and MCP-1 were detected in the brains of infected mice (24). There was a resurgence of the same chemokines during the chronic phase of disease and the chemokine gene expression appears to correspond well with viral persistence rather than the level of adaptive T cell responses (25). These chemokines are also found after infection with different TMEV strains, including GDVII and DA displaying different pathogenicity (26).

The above studies of chemokine gene expression *in vivo* strongly suggest the importance of these molecules for inflammatory responses in the CNS during TMEV infection. However, the cell types responsible for chemokine production as well as the sequence of the events are still elusive. Recently, the direct activation levels of chemokine gene expression after viral infection have been examined in various CNS cell cultures (27). Upon infection with

TMEV, the expression of RANTES, IP-10, MCP-1, MIP-1β, MIP-1α, and MIP-2 genes were upregulated in astrocytes, microglia, as well as oligo-dendrocytes, although the levels were somewhat different depending on the cell types. In whole brain cultures the pattern of chemokine expression is identical to that of astrocytes cultures (27), suggesting the importance of chemokine contribution by infected astrocytes in the CNS. Similar to viral infection, the proinflammatory cytokines, such as IFN-γ and TNF-α, were also able to upregulate the expression of chemokines by astrocytes. These results indicate that the identical chemokines are induced secondarily by stimulation with cytokines that are initially produced by virus-infected resident glial cells, infiltrating antigen producing cells, natural killer cells, as well as T cells in the CNS.

Studies assessing the role of chemokines in the pathogenesis of and/or protection from TMEV-IDD have not yet been reported. If the pathogenic mechanisms involved in TMEV-IDD are similar to the MHV-induced demyelinating disease (28, 29), inhibition of RANTES and/or IP-10 that are directly and primarily produced in astrocytes after TMEV infection (27), may decrease the pathogenesis of demyelinating disease. Such studies will provide important information regarding the role and mechanisms of various chemokines in the initiation and maintenance of inflammatory responses in the CNS leading to TMEV-induced chronic demyelinating disease. These studies will also present opportunity to compare to those involved in either autoimmune and/or other virally induced demyelinating disease models.

# 3.     ADHESION MOLECULES

Adhesion molecules are important in the T cell-APC interaction, cellular trafficking, and transmigration of inflammatory cells into the CNS. While other adhesion molecules (e.g. VCAM-1) may be involved in the patho-genesis of TMEV-IDD, the role of intracellular adhesion molecule (ICAM)-1 (CD54) has been primarily studied. ICAM-1, expressed on APC is a mem-ber of the immuno-globulin superfamily and interacts with lymphocyte function associated antigen (LFA)-1 (or CD11a), expressed on T cells, granulocytes, monocytes and macrophages. Because the interaction between these molecules is critical for T cell activation as well as extravasation of lymphocytes into the CNS, attempts to inhibit this interaction were made by administrating neutralizing antibodies. Blocking the interaction leads to decreased incidence and severity of TMEV-IDD accompanied by decreased numbers of infiltrating CD4[+], CD8[+] and MHC class II[+] cells in the CNS resulting in inhibition of inflammatory demyelination. In addition, this treatment also decreased the levels of virus-specific DTH, T cell proliferation, and the production of IFN-γ and TNF-α with an increase in the production of IL-10 (30). In contrast, similar administration of neutralizing mAb against LFA-1 and/or ICAM-1 in another report showed no effect on the onset/severity or the extent of demyelination or inflammation (31). The first study administered the antibody prior to and during infection while the second study only began antibody treatment after virus infection during the acute phase of disease (7 d post-infection). ICAM-1 may be necessary for the migration of T cells through the CNS endothelium, therefore blocking ICAM-1 prior to infection may prevent this migration and hence the subsequent inflammation and demyelination. Thus, it is conceivable that all the necessary inflammatory T cell populations have already been infiltrated into the CNS at 7 days post-infection. Local amplification of T cells may occur in the CNS (32) and continuous cellular infiltration may not be necessary to sustain chronic inflammatory demyelinating disease in the presence of viral persistence. In addition, the ICAM-1/LFA-1 interaction functions as an important costimulatory signal for T cell activation and clonal expansion. Without this signal, T cell response to TMEV may be compromised or deviated to the non-pathogenic Th2 response.

The role of CD28/B7 (CD80/86) costimulatory molecules in the development of TMEV-IDD has also been investigated. Treatment with mAbs to B7-1 (CD80), but not B7-2 (CD86), resulted in significant inhibition of the Th1 response to viral epitopes and the consequent development of disease (33). However, blocking the costimulation at the time of TMEV infection enhances disease severity (34), suggesting that such a blockade may also inhibit the development of protective immune responses permitting unchecked viral replication. Thus, utilization of immuno-suppressive regimes may be difficult to balance between promoting the protective immune responses and inhibiting pathogenic inflammatory responses. The timing of the treatments appears to be critically important in determining the clinical outcome.


## 4.    ROLE IN PATHOGENESIS OF DISEASE

The pathogenic mechanisms involved in virally induced demyelinating disease appear to be complex. Both the protection and pathogenesis are likely dependent on the levels of various chemokines, cytokines as well as adhesion molecules upregulated following viral infection. In general, the innate chemokine and cytokines responses induced upon viral infection are important in protecting the host by providing anti-viral functions. However, in the presence of viral persistence in susceptible mice, such innate responses may continuously stimulate pathogenic T cell responses leading to

CNS damage (2). For example, susceptible SJL/L mice are known to be deficient in NK/NKT cell compartment(s). Thus, the early viral clearance may be compromised and allow establishment of viral persistence in the CNS, despite the vigorous innate chemokine and cytokine responses after TMEV infection. Under the circumstances, chemokines and cytokines as well as adhesion molecules upregulated in various glial cells by viral infection will facilitate influx of various inflammatory cells, including virus-specific T cells (Figure 1). In addition, these chemokines and cytokines may also further stimulate adjacent glial cells to produce similar chemokines and cytokines, resulting in amplification of inflammatory responses in the CNS. Furthermore, these chemokines/cytokines as well as viral infections will further activate professional and non-professional antigen presenting cells (13, 35) and may reduce the threshold of T cell activation, leading to pathogenic Th1 cell responses specific for viral antigens as well as neuronal autoantigens (3). For example, proinflammatory cytokines such IFN-$\gamma$ and TNF-$\alpha$ produced by activated Th1 cells stimulate astrocytes to enable effective presentation of viral epitopes to T cells for further T cell activation and expansion (36). Also, such cytokines further activate resident microglias and macrophages that are likely involved in tissue damage of CNS. In addition, certain cytokines such as TNF-$\alpha$ may also be directly cytotoxic to oligodendrocytes responsible for myelin production (37).

However, the same chemokines and cytokines as well as upregulated adhesion molecules may also be necessary to contain viral replication. For example, strong infiltrating Th1 as well as CTL responses that are effective for viral clearance in the CNS are necessary to be protective to TMEV-IDD. These virus-specific immune responses are also heavily dependent on the initial innate chemokine/cytokine responses. Therefore, investigations of these initial chemokine/cytokine responses in resistant and susceptible mice will be valuable in understanding the pathogenic/protective mechanisms involved in TMEV-IDD. The previous attempts to ameliorate this virally induced demyelinating disease by simply skewing against the Th1 response have met inconsistent outcomes. Often times, attempts to suppress inflammatory responses may have lead to viral expansion and persistence, resulting in exacerbation of chronic disease. On the other hand, this disease is immune-mediated and inflammatory immune responses need to be inhibited in order to prevent/treat the disease. Thus, fast and complete elimination of viral persistence before sustaining immune damage of the CNS tissues would be ideal for preventing virally induced immune-mediated diseases. Alternatively, selective upregulation of strong anti-viral immunity, while inhibiting inflammatory Th responses, may be essential for treating virally induced demyelinating diseases.

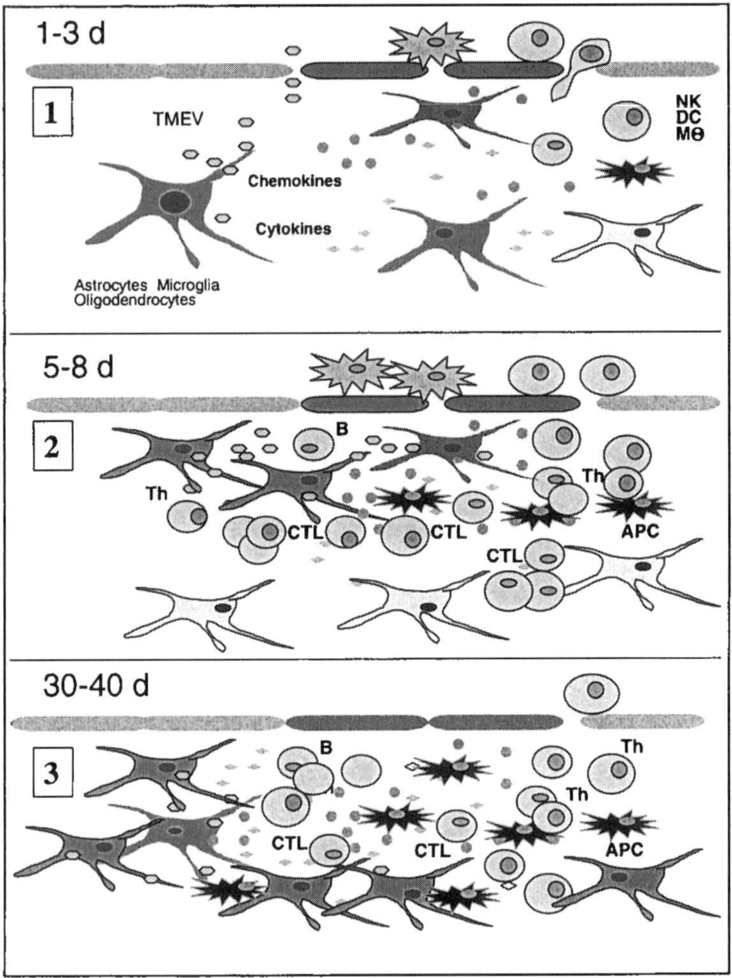



Figure 1: Schematic presentation of the steps involved in TMEV-induced demyelinating disease. (1) Early, (2) Acute and (3) Chronic responses in the CNS following viral infection. (1). At the Early stage, mainly chemokine, cytokine and adhesion molecule responses are directly induced in various glial cells upon viral infection. These molecules may further activate adjacent glial cells and facilitate cellular (NK, DC, macrophages, neutrophils) infiltration to the CNS. (2). At the Acute stage, various immune cells (B, Th and CTL) are infiltrated. These immune cells will be further stimulated/amplified by viral Ag-APCs (professional and non-professional). Virus will be neutralized and virus-infected cells will be eliminated by immune cells. (3). At the Chronic stage, viral persistence remains and continuous inflammatory responses are maintained in susceptible mice, whereas completely cleared viral persistence and the inflammatory response is resolved in resistant mice.

1.    Dal Canto, M. C., Kim, B.S., Miller, S.D., Melvold, R.W. 1996. Theiler's murine encephalomyelitis virus (TMEV)-induced demyelination: a model for human multiple clerosis. *Methods 10:453.*
2.    Kim, B. S., M. A. Lyman, B. S. Kang, H. K. Kang, H. G. Lee, M. Mohindru, and J. P. Palma. 2001. Pathogenesis of virus-induced immune-mediated demyelination. *Immunol Res 24:121.*
3.    Miller, S. D., Vanderlugt, C. L., Begolka, W. S., Pao, W., Yauch, R. L., Neville, K. L., Katz-Levy, Y., Carrizosa, A., Kim, B. S. 1997. Persistent infection with Theiler's virus leads to CNS autoimmunity via epitope spreading. *Nat. Med. 3:1133.*
4.    Lyman, M. A., H. G. Lee, B. S. Kang, H. K. Kang, and B. S. Kim. 2002. Capsid-specific cytotoxic T lymphocytes recognize three distinct H-2D(b)-restricted regions of the BeAn strain of Theiler's virus and exhibit different cytokine profiles. *J Virol 76:3125.*
5.    Kang, B. S., M. A. Lyman, and B. S. Kim. 2002. The majority of infiltrating CD8+ T cells in the central nervous system of susceptible SJL/J mice infected with Theiler's virus are virus specific and fully functional. *J. Virol 76:6577.*
6.    Yauch, R. L., and B. S. Kim. 1994. A predominant viral epitope recognized by T cells from the periphery and demyelinating lesions of SJL/J mice infected with Theiler's virus is located within VP1(233-244). *J. Immunol. 153:4508.*
7.    Yauch, R. Y., J. P. Palma, H. Yahikozawa, C.-S. Koh, and B. S. Kim. 1998. Role of individual T cell epitopes of Theiler's virus in the pathogenesis of demyelination correlates with the ability to induce a Th1 response. *J. Virol. 72:6169.*
8.    Begolka, W. S., C. L. Vanderlugt, S. M. Rahbe, and S. D. Miller. 1998. Differential expression of inflammatory cytokines parallels progression of central nervous system pathology in two clinically distinct models of multiple sclerosis. *J Immunol 161:4437.*
9.    Sato, S., S. L. Reiner, M. A. Jensen, and R. P. Roos. 1997. Central nervous system cytokine mRNA expression following Theiler's murine encephalomyelitis virus infection. *J Neuroimmunol 76:213.*
10.   Palma, J. P., S. H. Park, and B. S. Kim. 1996. Treatment with lipopolysaccharide enhances the pathogenicity of a low-pathogenic variant of Theiler's murine encephalomyelitis virus. *Journal of Neuroscience Research 45:776.*
11.   Chang, J. R., E. Zaczynska, C. D. Katsetos, C. D. Platsoucas, and E. L. Oleszak. 2000. Differential expression of TGF-beta, IL-2, and other cytokines in the CNS of Theiler's murine encephalomyelitis virus-infected susceptible and resistant strains of mice. *Virology 278:346.*
12.   Sierra, A., and N. Rubio. 1993. Theiler's murine encephalomyelitis virus induces tumour necrosis factor-alpha in murine astrocyte cell cultures. *Immunology 78:399.*
13.   Palma, J. P., R. L. Yauch, H. K. Kang, H. G. Lee, and B. S. Kim. 2002. Preferential induction of IL-10 in APC correlates with a switch from Th1 to Th2 response following infection with a low pathogenic variant of Theiler's virus. *J Immunol 168:4221.*
14.   Olson, J. K., A. M. Girvin, and S. D. Miller. 2001. Direct activation of innate and antigen-presenting functions of microglia following infection with Theiler's virus. *J Virol 75:9780.*

15. Kim, B. S., J. P. Palma, A. Inoue, and C. S. Koh. 2000. Pathogenic immunity in Theiler's virus-induced demyelinating disease: a viral model for multiple sclerosis. *Arch. Immunol. Ther. Exp. 48:373.*

16. Inoue, A., C. S. Koh, M. Yamazaki, H. Yahikozawa, M. Ichikawa, H. Yagita, and B. S. Kim. 1998. Suppressive effect on Theiler's murine encephalomyelitis virus-induced demyelinating disease by the administration of anti-IL-12 antibody. *J Immunol 161:5586.*

17. Bright, J. J., M. Rodriguez, and S. Sriram. 1999. Differential influence of interleukin-12 in the pathogenesis of autoimmune and virus-induced central nervous system demyelination. *J Virol 73:1637.*

18. Pullen, L. C., S. D. Miller, M. C. Dal Canto, P. H. Van der Meide, and B. S. Kim. 1994. Alteration in the level of interferon-gamma results in acceleration of Theiler's virus-induced demyelinating disease. *Journal of Neuroimmunology 55:143.*

19. Fiette, L., C. Aubert, U. Muller, S. Huang, M. Aguet, M. Brahic, and J. F. Bureau. 1995. Theiler's virus infection of 129Sv mice that lack the interferon alpha/beta or interferon gamma receptors. *Journal of Experimental Medicine 181:2069.*

20. Rodriguez, M., K. Pavelko, and R. L. Coffman. 1995. Gamma interferon is critical for resistance to Theiler's virus-induced demyelination. *Journal of Virology 69:7286.*

21. Inoue, A., C. S. Koh, H. Yahikozawa, N. Yanagisawa, H. Yagita, Y. Ishihara, and B. S. Kim. 1996. The level of tumor necrosis factor-alpha producing cells in the spinal cord correlates with the degree of Theiler's murine encephalomyelitis virus-induced demyelinating disease. *International Immunology 8:1001.*

22. Paya, C. V., P. J. Leibson, A. K. Patick, and M. Rodriguez. 1990. Inhibition of Theiler's virus-induced demyelination in vivo by tumor necrosis factor alpha. *Int Immunol 2:909.*

23. Hoffman, L. M., B. T. Fife, W. S. Begolka, S. D. Miller, and W. J. Karpus. 1999. Central nervous system chemokine expression during Theiler's virus-induced demyelinating disease. *J Neurovirol 5:635.*

24. Murray, P. D., K. Krivacic, A. Chernosky, T. Wei, R. M. Ransohoff, and M. Rodriguez. 2000. Biphasic and regionally-restricted chemokine expression in the central nervous system in the Theiler's virus model of multiple sclerosis [In Process Citation]. *J Neurovirol 6 Suppl 1:S44.*

25. Ransohoff, R. M., T. Wei, K. D. Pavelko, J. C. Lee, P. D. Murray, and M. Rodriguez. 2002. Chemokine expression in the central nervous system of mice with a viral disease resembling multiple sclerosis: roles of CD4+ and CD8+ T cells and viral persistence. *J Virol 76:2217.*

26. Theil, D. J., I. Tsunoda, J. E. Libbey, T. J. Derfuss, and R. S. Fujinami. 2000. Alterations in cytokine but not chemokine mRNA expression during three distinct Theiler's virus infections. *J Neuroimmunol 104:22.*

27. Palma, J. P., and B. S. Kim. 2001. Induction of selected chemokines in glial cells infected with Theiler's virus. *J Neuroimmunol 117:166.*

28. Lane, T. E., M. T. Liu, B. P. Chen, V. C. Asensio, R. M. Samawi, A. D. Paoletti, I. L. Campbell, S. L. Kunkel, H. S. Fox, and M. J. Buchmeier. 2000. A central role for CD4(+) T cells and RANTES in virus-induced central nervous system inflammation and demyelination. *J Virol 74:1415.*

29.     Liu, M. T., H. S. Keirstead, and T. E. Lane. 2001. Neutralization of the chemokine CXCL10 reduces inflammatory cell invasion and demyelination and improves neurological function in a viral model of multiple sclerosis. *J Immunol 167:4091.*

30.     Inoue, A., C. S. Koh, M. Yamazaki, M. Ichikawa, M. Isobe, Y. Ishihara, H. Yagita, and B. S. Kim. 1997. Anti-adhesion molecule therapy in Theiler's murine encephalomyelitis virus-induced demyelinating disease. *Int Immunol 9:1837.*

31.     Rose, J. W., C. T. Welsh, K. E. Hill, M. K. Houtchens, R. S. Fujinami, and J. J. Townsend. 1999. Contrasting effects of anti-adhesion molecule therapy in experimental allergic encephalomyelitis and Theiler's murine encephalomyelitis. *J Neuroimmunol 97:110.*

32.     Kang, J. A., M. Mohindru, B. S. Kang, S. H. Park, and B. S. Kim. 2000. Clonal expansion of infiltrating T cells in the spinal cords of SJL/J mice infected with Theiler's virus. *J Immunol 165:583.*

33.     Inoue, A., C. S. Koh, M. Yamazaki, and H. Yagita. 1999. Effect of Anti-B7-1 and Anti-B7-2 mAb on Theiler's Murine Encephalomyelitis Virus-Induced Demyelinating Disease. *J Immunol 163:6180.*

34.     Neville, K. L., M. C. Dal Canto, J. A. Bluestone, and S. D. Miller. 2000. CD28 costimulatory blockade exacerbates disease severity and accelerates epitope spreading in a virus-induced autoimmune disease. *J Virol 74:8349.*

35.     Girvin, A. M., K. B. Gordon, C. J. Welsh, N. A. Clipstone, and S. D. Miller. 2002. Differential abilities of central nervous system resident endothelial cells and astrocytes to serve as inducible antigen-presenting cells. *Blood 99:3692.*

36.     Palma, J. P., R. L. Yauch, S. Lang, and B. S. Kim. 1999. Potential role of CD4+ T cell-mediated apoptosis of activated astrocytes in Theiler's virus-induced demyelination. *J Immunol 162:6543.*

37.     Qi, Y., and M. C. Dal Canto. 1996. Effect of Theiler's murine encephalomyelitis virus and cytokines on cultured oligodendrocytes and astrocytes. *J Neurosci Res 45:364.*

Chapter B7

# MOLECULAR DETERMINANTS OF TMEV PATHOGENESIS

Stephen T. Guest and Raymond P. Roos
*Department of Neurology, Committee on Virology, University of Chicago Medical Center.*

**Abstract**:
Strains of Theiler's murine encephalomyelitis virus (TMEV) can be divided into two subgroups based on their distinct biological activities. Members of the GDVII subgroup are highly neurovirulent and cause an acute, fatal encephalomyelitis. In contrast, members of the TO subgroup cause a mild encephalomyelitis early after infection followed by a late disease that is characterized by virus persistence, a low level of virus gene expression, and inflammatory demyelination.
Chimeric viruses constructed between members of the two subgroups have shown that the viral determinants for these distinct disease phenotypes are polygenic, and that interactions between different determinants on the viral genome are critical in determining disease phenotype. A point mutation in the internal ribosome entry site (IRES) of the 5'untranslated region of the GDVII subgroup strain can result in a virus with reduced neurovirulence that is able to persist. These results demonstrate that determinants for virus persistence are not unique to the TO subgroup strains, and identify the viral IRES as playing a critical role in determining TMEV pathogenesis. We have identified an 18kD protein known as L* that is present in TO subgroup strains but not GDVII subgroup stains, and appears to play a key role in virus persistence and demyelination. Disruption of L* expression results in a virus that grows well in cell culture but is unable to persist in the central nervous system or induce demyelination. Further investigations of TMEV-induced disease should provide insight into mechanisms that underlie virus persistence and virus-induced demyelination.

**Key words**:
Theiler's murine encephalomyelitis virus, persistence, demyelination, L*, internal ribosome entry site

# INTRODUCTION

Theiler's murine encephalomyelitis virus (TMEV) refers to a group of closely related viruses in the *Picornaviridae* family, genus *cardiovirus*. Strains of TMEV can be divided into two subgroups based on their distinct biological activities (Table 1). Members of the GDVII subgroup, which include GDVII and FA strains, are highly neurovirulent in weanling mice and display a large plaque phenotype when grown on cultured cells. Intracranial (IC) inoculation with as little as one plaque-forming unit (pfu) causes a severe, acute encephalomyelitis that is associated with a progressive flaccid paralysis of the hind limbs that leads to death within days to a couple weeks post-inoculation (PI). Members of the GDVII subgroup replicate in neurons and astrocytes of the gray matter in the central nervous system (CNS) and are not observed to induce demyelination. Members of the second TMEV subgroup, Theiler's original (TO) subgroup, include DA, BeAn, WW, Yale and TO strains. These strains, which display a small plaque phenotype when grown in cultured cells, fail to kill a weanling mouse despite IC inoculations of $>10^6$ pfu. During the first two weeks PI, TO subgroup strains replicate in neurons and astrocytes of the gray matter of the CNS, and induce a subclinical encephalomyelitis. By 3 weeks PI the encephalomyelitis resolves, at which time virus is primarily found in glia and microglial cells of the spinal cord white matter and is associated with foci of inflammatory demyelination. The white matter lesions peak at about 6 weeks, but continue throughout the life of the mouse. The inflammatory demyelination observed is similar to that seen in multiple sclerosis. Virus persists in the CNS with a restricted expression of viral genes, i.e., genome is present with relatively little viral protein or infectious virus production.

*Table B7.* Phenotypic Differences Between TMEV Subgroups

| Phenotype | TO Subgroup Strains (DA, BeAn, TO, WW, Yale, etc.) | GDVII Subgroup Strains (GDVII, FA) |
|---|---|---|
| Neurovirulence (50% lethal dose) | **–** <br> ($>10^6$ PFU) | **++** <br> (1 PFU) |
| Early Disease (Encephalomyelitis) | **+** | **+** |
| Late Disease (restricted virus expression, inflammation and white matter demyelination) | **+** | **–** |

The complete genome of the GDVII strain as well as two TO subgroup strains, DA and BeAn, have been sequenced. Sequence analysis demonstrated 90% identity at the nt level and 95% identity at the amino acid (aa) level among the different strains of TMEV from the two subgroups [2, 3]. While this degree of similarity is relatively high, the differences that are present are located throughout the viral genome, suggesting that determinants for the very distinct disease phenotypes caused by the GDVII vs. TO subgroup strains are likely to be scattered over the TMEV genome.

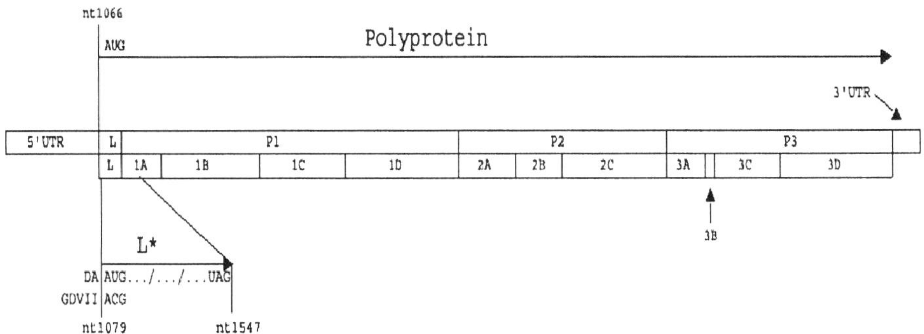


*Figure B7.* The TMEV genome of approximately 8,100 nucleotides. The polyprotein coding region is flanked upstream by a 5'UTR and downstream by a 3'UTR. The coding region for the polyprotein begins with the coding region for Leader (L) protein and is followed by genomic regions P1, P2 and P3. P1 is divided into 1A, 1B, 1C and 1D coding regions, which synthesize, respectively, the capsid proteins VP4, VP2, VP3 and VP1. P2 (which includes 2A, 2B and 2C) and P3 (which includes 3A, 3B, 3C and 3D) encode various nonstructural proteins. The initiating AUG codon for L* is located at nt 1079 for strain DA while the GDVII strain has an ACG in this position.

## RECOMBINANT AND MUTANT VIRUS STUDIES

The generation of full-length cDNA clones derived from TO and GDVII subgroup strains of TMEV has made it possible to rapidly engineer chimeric cDNAs and recombinant viruses between the two subgroups. Studies utilizing these recombinant viruses have been aimed at determining the regions of the viral genome that are necessary for the subgroup-specific phenotypes, such as GDVII subgroup neurovirulence and TO subgroup persistence, demyelination, and restricted virus gene expression.

The recombinant virus studies have demonstrated that phenotypes such as virus neurovirulence, persistence and demyelinating activity are polygenic. Replacement of the capsid coding region of the TO subgroup strain DA with the capsid coding region from GDVII strain results in a recombinant virus that is almost as neurovirulent as wild type (wt) GDVII virus [4]; replacement of the 5' UTR along with the capsid coding region of DA with the corresponding regions of GDVII results in a virus with neurovirulence that is equal to that of the parental GDVII virus. This result demonstrates that the GDVII capsid coding region plays a key role in determining neurovirulence, and suggests that the DA capsid plays a role in virus persistence and demyelination (although later studies described below indicate that the latter suggestion needs to be modified). Surprisingly, replacement of DA regions coding for the carboxyl half of VP2, all of VP3, all of VP1, and the amino end of 2A with the corresponding regions of GDVII resulted in a virus (GD1B-2A/DA virus) that persisted and induced demyelination in mice, suggesting that a key determinant for the persistent, demyelinating phenotype may lie within the 5' half of DA VP2 [5] (but again, this suggestion was modified by later studies).

Another interpretation of the results from studies of GD1B-2A/DA virus is that GDVII strain does indeed contain determinants for persistence and demyelination, but that GDVII does not persist or demyelinate due to its extreme neurovirulence. In fact, investigations showed that virus persistence and demyelination can occur in a series of recombinant viruses in which different parts of the DA genome that span the length of the genome from the 5' terminus to the 3'UTR were replaced with the corresponding segments of the GDVII genome, suggesting that there is no single determinant for virus persistence and demyelination on the DA genome that cannot be substituted for by the corresponding region of the GDVII genome, i.e., GDVII contains determinants for virus persistence and demyelination [6].

The issue as to whether GDVII strain contains determinants for persistence and demyelination that are not apparent because of the virus' extreme neurovirulence was also examined by testing whether attenuation of a GDVII strain to a certain level will result in a virus that persists and demyelinates. It is clear that GDVII strain can be sufficiently attenuated so that it fails to efficiently grow in the CNS, and therefore will not persist or demyelinate; an example of one such virus is the leaderless GDVII virus, which is completely attenuated for growth in the CNS and therefore unable to persist [7]. We recently investigated this topic by examining GDVII strains with mutations in the IRES that alter the binding of the cellular factor polypyrimidine tract binding protein. These mutant viruses were attenuated, and at least one, GD91 [8], persisted in the CNS but failed to induce demyelination (Viktorova et al., manuscript in preparation). These results demonstrate that: TMEV persistence does not require the presence of the capsid of TO subgroup strains; attenuated GDVII strains (with one nucleotide change, in the 5'UTR, compared to GDVII wt virus) are capable of persistence; persistence is not sufficient for demyelination. The lack of demyelinating activity by the attenuated GDVII virus may result from the absence of L* protein (see later) and/or the presence of GDVII rather than DA capsid proteins. Our studies demonstrate the importance of IRES-binding proteins in determining disease phenotype, a theme that recurs in discussions of the varying levels of L* vs. the polyprotein synthesis depending on different neural cell types.

Interactions between parts of the viral genome are not unexpectedly important in the virus' biological activities. Further studies with GD1B-2A/DA virus showed that introduction of a single, additional mutation in the DA backbone at aa residue 141 of VP2 from Lys to Asn (where Asn is the amino acid present in GDVII at this position, and corresponds to the position of Lys in DA), resulted in a virus (R5 virus) with increased neurovirulence that neither persisted nor induced demyelination [9]. These results demonstrated that interactions between parts of the genome can have dramatic effects on disease phenotype. In this case, interactions of a particular aa residue (141) in VP2 (that is normally present in GDVII) along with the capsid segment containing half of GDVII VP2, all of VP3 and all of VP1 presumably interfered with the activity of determinants for the persistent demyelinating phenotype of the TO subgroup strain.

The interactions between aa141 and the capsid can be better understood by taking advantage of the available three-dimensional crystal structure of TMEV [10, 11, 12]. In these structures, aa141 lies at the tip of the EF loop in the VP2 puff on the rim of the canyon region of the viral capsid coat. It is this canyon region that is thought to be the site of interactions between the virion and its still unknown cellular receptor(s). Therefore, it has been proposed that the importance of aa141 lies in its ability to alter interactions with cellular receptors present on specific cell types [13]. The presence of Asn in this position along with part of the GDVII capsid may prevent interactions of the virion with cellular receptors, specifically on cells in the white matter of the spinal cord. Support for this hypothesis has been provided by studies showing that R5 virus fails to spread from the gray matter to the white matter of the spinal cord of nude mice, whereas DA strain spreads efficiently [13]. The absence of a competent immune system in these mice indicates that the inability of this recombinant virus to spread is independent of the host immune response, providing further support that the Asn interferes with interactions between the virion and specific cellular receptors present on cells in the white matter of the spinal cord.

Although much has been learned from recombinant virus studies there are clear shortcomings to this approach. At times, the substitution of a genome segment of one strain into the genome of a strain from the other subgroup can disrupt key intratypic interactions important for disease phenotype or even viability of the virus.   For example, when two segments of the GDVII genome were substituted separately into the DA genome neither of the two resultant recombinant viruses displayed the phenotype of a recombinant virus which contained both of the GDVII genome segments [4]. In addition, some intertypic recombinant viruses have a defect in a critical function of the virus, such as assembly, leading to poor virus growth *in vitro*, and therefore preventing animal studies [14].   For these reasons, studies using recombinant viruses as a way of clarifying subgroup-specific disease phenotypes have been generally replaced by investigations involving site-specific mutation of particular nucleotides or aa residues.

# THE ROLE OF L* IN TO SUBGROUP STRAIN DISEASE

## L* in DA Strain-Infected Cells

We recently identified a coding region of TO subgroup strains that appears critical in virus persistence and demyelinating activity.  This region codes for an 18 kD protein that we have called L*.  L* was initially identified following *in vitro* translation of the DA RNA genome, but not of GDVII [15].   The L* protein is not an artifact of the *in vitro* translation system since L* is synthesized in DA, but not GDVII infected cells.  L* synthesis starts at 4 hours PI, with an expression profile that closely mimics that of the viral polyprotein; the protein is stable throughout the course of the virus infection [16, 17].

L* is synthesized from an initiating AUG located 13nt downstream from the polyprotein's AUG of TO subgroup strains [18], out of frame with the polyprotein's reading frame [17].  Sequence analysis of different TMEV strains has shown that the initiating AUG for the L* reading frame is present in the RNA genome of all TO subgroup strains, while the genome of GDVII subgroup strains contains an ACG at this site in the genome; this result explains the apparent absence of L* synthesis following translation of GDVII RNA.

Comparison of the nucleotide sequence of the L* reading frame among different strains of the TO subgroup has shown that the predicted amino acid sequence of the L* reading frame is more highly conserved than the aa sequence of the overlapping polyprotein reading frame.  This high degree of conservation of the L* reading frame suggests that L* has a function that is important for the virus.  Additionally, the presence of an L* AUG in TO, but not GDVII, subgroup strains suggests that it plays a role in a TO subgroup-specific phenotype, such as persistence, demyelination, and restricted virus gene expression.  These suggestions were found to be valid, as described below.

## L* and TO Subgroup Strain Persistence and Demyelination

L* has been shown to play a critical role in determining virus persistence and the demyelinating phenotype of TO subgroup strains.  Work from our laboratory has shown that viruses containing mutations that prevent L* protein expression, whether by mutating the L* initiating AUG to ACG (DAL*-1 virus) or by creating stop codons in the L* reading frame (DAL*stop virus), grow at wt levels in cultured cells but are unable to persist and induce demyelination in the CNS [17, 19, Viktorova et al. unpublished].  DAL*-1 virus induces an acute encephalomyelitis similar to that produced by wt virus and grows to a similar titer at 1 week PI; however, at 45 days PI, there is no evidence of the DAL*-1 viral genome and scant evidence of inflammation or demyelination in the CNS.  Both DAL*-1 and DAL*stop viruses are cleared following infection, with little evidence of demyelination, indicating that the actual L* coding region rather than the additional translational start site is key to the mutant phenotype.

The importance of L* in TO subgroup persistence and demyelination was initially called into question by Michiels and colleagues [20].  These investigators were working with a *different* full-length clone of DA, pTMDA, than the one used by our group.  Surprisingly, Michiels and colleagues found that substitution of an ACG mutation for the L* initiating AUG in pTMDA resulted in a virus (which we will call TMDAL*-1 virus) that persisted and demyelinated almost as well as the wt TMDA virus. However, a more recent study from this group found that: TMDAL*-1 virus does in fact produce small amounts of the L* protein, presumably through initiation at the ACG; the small amount of L* produced by TMDAL*-1 virus is presumably able to facilitate virus persistence and demyelination, however, mutations which introduce a stop codon in the L* reading frame of pTMDA (and produce *no* L*) result in viruses that fail to persist and demyelinate [21].  Together, these studies support our thesis that L* plays a key role in determining TO subgroup persistence and demyelination.

Although L* is important for TO subgroup persistence and demyelination, the recombinant virus studies at first glance suggest that L* may not be essential for these activities.  For example, some GDVII/DA recombinant viruses that replace the DA L* AUG with the GDVII ACG are able to persist and demyelinate. However, Michiels and colleagues demonstrated that a small amount of L* is synthesized from the ACG of wt GDVII, suggesting that L* may be synthesized in these recombinant viruses and that L* may indeed be essential for virus persistence and demyelination.

The precise mechanism(s) by which L* allows TO subgroup strains to persist and induce demyelination is unclear.  There is evidence that L* perturbs the immune response, and therefore may play a role in allowing the virus to escape clearance by the host's immune system [22].  DA wt virus infection of certain mouse strains (e.g., C57Bl/6 mice) that are resistant to the persistent, demyelinating infection induced by TO subgroup strains generate an anti-virus cytolytic T-lymphocyte (CTL) response 7 DPI (and this response clears the virus), while mouse strains (e.g., SJL mice) susceptible to TO-induced disease do not generate a virus-specific CTL (and the absence of this response fosters virus persistence) [23, 24].  In contrast to this latter situation, inoculation of SJL mice with DAL*-1 virus results in a virus-specific CTL response, virus clearance and an absence of the demyelinating disease [22].  Our recent data involving wt and DAL*-1 virus infections of CD4-/- and CD8-/- mice as well as adoptive transfer studies suggest that L* inhibits a CD4+ T cell –dependent activity that is capable of clearing DA virus from susceptible mice [Lin, X, Ma, X, Rodriguez, M, and Roos, R.P., submitted]

## The Role of L* in Microglial Infection

L* may play a role in TMEV infection of microglia, a key CNS cell in which the virus persists.  *In vitro* experiments have shown that members of the GDVII subgroup are growth restricted in cultured macrophage cell lines, whereas the DA strain grows to high titer in these cells [25].  Additional studies showed that a recombinant GDVII-type virus engineered to contain the L* coding region from the DA strain gained the ability to productively infect cultured macrophages.  In an analogous way, a DA strain that lacks L*, as is the case with DAL*-1 virus, lost the ability to productively infect cultured macrophages [26].  These results suggest that the L* protein is necessary for efficient infection of macrophages by TMEV.  In addition, work from our laboratory has shown that the GDVII strain and DAL*-1 virus are significantly more efficient at inducing apoptosis of cultured macrophages than DA, suggesting that the L* protein has an antiapoptotic effect in macrophages [19].  In summary, L* allows the virus to efficiently replicate in macrophage cells, a major site for virus persistence, and the virus persistence in this cell type is enhanced by inhibition of virus-induced apoptotic cell death.

# Cell-Type Specific Regulation of L* Synthesis

The critical role of the L* protein in the TO subgroup strain-induced persistent CNS infection and demyelinating disease highlights the importance of regulating translation of L* in addition to the polyprotein. We have hypothesized that the presence of cell-specific IRES-binding proteins regulate whether ribosomes initiate translation at the L* rather than the polyprotein start site, leading to different ratios of L* to polyprotein synthesis in different cell types. Neural cells in which ribosomes preferentially synthesize L* over the polyprotein (with an associated decrease in synthesis of the polyprotein) might have the following important properties key to TO subgroup strain disease: translation from the L* reading frame may lead to a restricted expression of the viral polyprotein; a combination of increased L* production and decreased polyprotein production may help the virus avoid recognition and clearance by the immune system; increased L* production may prevent virus-induced apoptosis in cells that support virus persistence. Our studies found that the ratio of L* synthesis to VP2 synthesis varied in rabbit reticulocyte lysates in comparison to BHK-21 cells, and also in different tissue culture cell types [27]. In addition, recent investigations have shown that the presence of different IRES-binding proteins in different cell types regulates the efficiency of translation of the viral polyprotein of TMEV and has a profound effect on disease phenotype [8].

# CONCLUSION

TMEV infection provides an excellent model system for the study of virus persistence and virus-induced demyelinating disease. The identification of the critical role of the L* protein, capsid proteins, and the viral 5'UTR in determining the demyelinating disease caused by the TO subgroup strains represent important steps towards a full understanding of TMEV-induced disease. Continued investigations of the L* protein and its function should provide additional insight into the mechanism by which TO subgroup strains of TMEV persist and induce demyelination in the host. Identification of factors involved in regulating translation of the L* reading frame in various cell types *in vivo* will also enhance our understanding of TMEV-induced disease. Lastly, identification of the host cell receptor(s) and an understanding of the interaction of receptor with the virus capsid proteins will further answer questions about virus tropism and spread in the host CNS.

## ACKNOWLEDGMENTS

Research described in this chapter was the result of work supported by NIH grant #5RO1NS37958-04 and a grant from the National Multiple Sclerosis Society.

## REFERENCES

1. Bandyopadhyay, P.K., Wang, C., & Lipton, H.L. 1992. Cap-independent translation by the 5' untranslated region of Theiler's murine encephalomyelitis virus. J. Virol. **66**: 6249-6256.

2. Ohara, Y., Stein, S., Fu, J.L., Stillman, L., Klaman, L., & Roos, R.P. 1988. Molecular cloning and sequence determination of DA strain of Theiler's murine encephalomyelitis viruses. Virology. **164**(1): 245-255.

3. Pevear, D.C., Borkowski, J., Calenoff, M., Rozhon, E., & Lipton, H.L. 1987. Analysis of the complete nucleotide sequence of the picornavirus Theiler's murine encephalomyelitis virus (TMEV) indicates that it is closely related to the cardioviruses. J. Virol. **61**: 1507-1516.

4. Zhang, L., Senkowski, A., Shim, B., & Roos, R.P. 1993. Chimeric cDNA studies of Theiler's murine encephalomyelitis virus neurovirulence. J. Virol. **67**(7): 4404-4408.

5. Fu, J.L., Stein, S., Rosenstein, L., Bodwell, T., Routbort, M., Semler, B.L., & Roos, R.P. 1990. Neurovirulence determinants of genetically engineered Theiler viruses. Proc. Natl. Acad. Sci. USA. **87**(11): 4125-4129.

6. Fu, J., Rodriguez, M., & Roos, R.P. 1990. Strains from both Theiler's virus subgroups encode a determinant for demyelination. J. Virol. **64**(12): 6345-6348.

7. Calenoff, M.A., Badshah, C.S., Dal Canto, M.C., Lipton, H.L., & Rundell, M.K. 1995. The leader polypeptide of Theiler's virus is essential for neurovirulence but not for virus growth in BHK cells. J Virol. **69**(9): 5544-5549.

8. Pilipenko, E., Viktorova, E., Guest, S., Agol, V., & Roos, R.P. 2001. Cell-specific proteins regulate viral RNA translation and virus-induced disease. EMBO J. **20**(23): 6899-6908.

9. Jarousse, N., Grant, R.A., Hogle, J.M., Zhang, L., Senkowski, A., Roos, R.P., Michiels, T., Brahic, M., & McAllister, A. 1994. A single amino acid change determines persistence of a chimeric Theiler's virus. J. Virol. **68**(5): 3364-3368.

10. Luo, M., Toth, K.S., Zhou, L., Pritchard, A., & Lipton, H.L. 1996. The structure of a highly virulent Theiler's murine encephalomyelitis virus (GDVII) and implications for determinants of viral persistence. Virology. **220**(1): 246-250.

11. Luo, M., He, C., Toth, K.S., Zhang, C.X., & Lipton, H.L. 1992. Three-dimensional structure of Theiler murine encephalomyelitis virus (BeAn strain). Proc Natl Acad Sci U S A. **89**(6): 2409-2413.

12. Grant, R.A., Filman, D.J., Fujinami, R.S., Icenogle, J.P., & Hogle, J.M. 1992. Three-dimensional structure of Theiler's virus. Proc. Natl. Acad. Sci. USA. **89**: 2061-2065.

13. Jarousse, N., Fiette, L., Grant, R.A., Hogle, J.M., McAllister, A., Michiels, T., Aubert, C., Tangy, F., Brahic, M., & Pena Rossi, C. 1994. Chimeric Theiler's virus with altered tropism for the central nervous system. J. Virol. **68**(5): 2781-2786.

14.  Pritchard, A.E., Jensen, K., & Lipton, H.L. 1993. Assembly of Theiler's virus recombinants used in mapping determinants of neurovirulence. J. Virol. **67**: 3901-3907.

15.  Roos, R.P., Kong, W.-P., & Semler, B.L. 1989. Polyprotein processing of Theiler's murine encephalomyelitis viruses. J. Virol. **63**: 5344-5353.

16.  Obuchi, M., Odagiri, T., Asakura, K., & Ohara, Y. 2001. Association of L* protein of Theiler's murine encephalomyelitis virus with microtubules in infected cells. Virology. **289**(1): 95-102.

17.  Chen, H.H., Kong, W.P., Zhang, L., Ward, P.L., & Roos, R.P. 1995. A picornaviral protein synthesized out of frame with the polyprotein plays a key role in a virus-induced immune-mediated demyelinating disease. Nat Med. **1**(9): 927-931.

18.  Kong, W.P. & Roos, R.P. 1991. Alternative translation initiation site in the DA strain of Theiler's murine encephalomyelitis virus. J. Virol. **65**(6): 3395-3399.

19.  Ghadge, G.D., Ma, L., Sato, S., Kim, J., & Roos, R.P. 1998. A protein critical for a Theiler's virus-induced immune system-mediated demyelinating disease has a cell type-specific antiapoptotic effect and a key role in virus persistence. J. Virol. **72**(11): 8605-8612.

20.  Van Eyll, O. & Michiels, T. 2000. Influence of the Theiler's virus L* protein on macrophage infection, viral persistence, and neurovirulence. J. Virol. **74**: 9071-9077.

21.  Van Eyll, O. & Michiels, T. 2002. Non-AUG-Initiated Internal Translation of the L* Protein of Theiler's Virus and Importance of This Protein for Viral Persistence. J Virol. **76**(21): 10665-10673.

22.  Lin, X., Roos, R.P., Pease, L.R., Wettstein, P., & Rodriguez, M. 1999. A Theiler's virus alternatively initiated protein inhibits the generation of H-2K-restricted virus-specific cytotoxicity. J. Immunol. **162**(1): 17-24.

23.  Lin, X., Thiemann, N.R., Pease, L.R., & Rodriguez, M. 1995. VP1 and VP2 capsid proteins of Theiler's virus are targets of H-2D-restricted cytotoxic lymphocytes in the central nervous system of B10 mice. Virology. **214**(1): 91-99.

24.  Dethlefs, S., Brahic, M., & Larsson-Sciard, E.L. 1997. An early, abundant cytotoxic T-lymphocyte response against Theiler's virus is critical for preventing viral persistence. J. Virol. **71**(11): 8875-8878.

25.  Ohara, Y., Himeda, T., Asakura, K., & Sawada, M. 2002. Distinct cell death mechanisms by Theiler's murine encephalomyelitis virus (TMEV) infection in microglia and macrophage. Neurosci Lett. **327**(1): 41-44.

26.  Obuchi, M., Yamamoto J., Odagiri, T., Uddin, M.N., Iizuka, H., Ohara, Y. 2000. L* protein of Theiler's murine encephalomyelitis virus is required for virus growth in a murine macrophage-like cell line. J. Virol. **74**(10): 4898-4901.

27.  Yamasaki, K., Weihl, C.C., & Roos, R.P. 1999. Alternative translation initiation of Theiler's murine encephalomyelitis virus. J. Virol. **73**(10): 8519-8526.

# Chapter B8

# NITRIC OXIDE IN TMEV

Emilia L. Oleszak[1], Jacek Kuzmak[1], Arun Varadhachary[2] and
Christos D. Katsetos[3]
[1]*Fels Institute for Cancer Research and Molecular Biology and Department of Anatomy and
Cell Biology,* [2]*Department of Microbiology and Immunology, Temple University School of
Medicine;* [3]*Drexel University, College of Medicine and St. Christopher's Hospital for
Children, Philadelphia PA.*

**Abstract:**     We and others have previously investigated the role of inducible nitric oxide
synthase (iNOS) on early acute and late chronic demyelinating disease induced
by Theiler's Murine Encephalomyelitis Virus (TMEV). Infection of
susceptible SJL mice with this virus serves as an excellent model of virus-
induced demyelinating disease, such as multiple sclerosis (MS). iNOS
transcripts and protein were detected in brains and spinal cords of TMEV-
infected SJL mice during early acute disease, which resembles
polioencephalomyelitis. Similar level of expression of iNOS has been found in
resistant B6 mice, which develop only early acute disease. Weak iNOS
staining was detected in reactive astrocytes and in leptomeningeal infiltrates in
TMEV-infected SJL mice at 42 days post infection (p.i.), corresponding to
early phase of chronic demyelinating disease, but not at 66 and 180 days p.i.
corresponding to advanced and terminal stages of the disease, respectively.
Results from other laboratories demonstrated that, blocking of NO by
treatment of TMEV-infected SJL mice with amino guanidine (AG), a specific
inhibitor of NO resulted in delay of late chronic demyelinating disease.
However this protective effect of NO inhibitor depended on the temporal
phase of the disease, type of cells expressing iNOS and the time of
administration of AG. The results from our laboratory suggests that NO
expressed during early acute disease is beneficial to the host through induction
of apoptosis of infiltrating T cells and resolution of encephalitis, but its role in
myelin/oligodendrocytes damage during late chronic demyelinating disease is
not clear and it may depend on availability of superoxide and  formation of
peroxynitrite.

**Key words:**     nitric oxide, TMEV, multiple sclerosis

# 1.    INTRODUCTION

Theiler's virus (TMEV) induces a biphasic disease in susceptible strains of mice (such as SJL). Early acute disease (polioencephalomyelitis) is characterized by replication of the virus in gray matter(1-4). This phase of disease is associated with neuronophagia and inflammatory infiltrates in the cerebral gray matter and anterior horn cells of spinal cord(4, 5). Within two to three weeks the virus is partially cleared and approximately 35 days p.i. susceptible mice develop late chronic demyelinating disease. The virus persists at very low level in microglia/macrophages, astrocytes and oligodendrocytes(1, 5-9). Heavy inflammatory infiltrates comprised of $CD4^+$ and $CD8^+$ T cells, macrophages, and few B cells are present, exclusively in the spinal cord(7, 9, 10). Demyelinating lesions are associated with perivascular and parenchymal inflammatory infiltrates. TMEV-induced late chronic demyelinating disease is an excellent animal model for human MS(1, 2, 7, 9, 11). Resistant strains of mice (such as C57BL/6) develop only early acute disease, clear the virus completely and do not develop demyelinating disease. MS is the most common inflammatory demyelinating disease in humans. It has been suggested that T cells specific for a myelin component such as MBP (Myelin Basic protein, PLP (proteolipid) and/or MOG (myelin oligodendrocytes protein), and activated microglia/macrophages participate in myelin damage(12-15). Epidemiological evidence suggests that MS could be triggered by virus (es) acquired before puberty.

The mechanism(s) of the pathogenesis of inflammatory demyelinating diseases is not well understood. Although oligodendrocytes/myelin may be damaged by a direct attack of cytotoxic T cells, other cells, including $CD4^+$ T cells, activated macrophages, and microglia may contribute to myelin destruction by the production of cytokines, such as IL-1, IFN-γ, and TNF-α,.and reactive oxygen and reactive nitrogen species(14, 16). Nitric Oxide (NO) is a short lived, highly reactive molecule with free radical properties. It is synthesized by nitric oxide synthase (NOS) by converting of L-arginine to L-citrulline, reviewed in(17, 18). There are three isoforms of NOS; NOS-type I and NOS-type III are calcium dependent and are constitutively expressed. This results in the production of physiological levels of NO which acts as second messenger in the NO signaling pathway in the neuronal and cardiovascular system. The NOS-type II (iNOS, inducible NOS) is calcium independent and is produced in large amounts during innate and adaptive immune responses(19-24). NO produced during an innate immune response clearly contributes to the killing of intracellular microorganisms and parasites(25-27). Less is known about a role of NO in the generation and regulation of an adaptive immune response, including inflammation.

It has been suggested that NO may exhibit cytotoxic, regulatory, and immunosuppressive properties, including induction of apoptosis and autoimmune disease, reviewed in (28-30). Such a wide array of biological actions of NO suggests that iNOS may have different functions in different cells. The immunoregulatory action of NO is likely mediated through its effect on cytokines and other inflammatory molecules by regulation of the transcription factor, NF-activating protein-1 (AP-1) by targeting the Th1/Th2 balance(29, 31-33). On the other hand, NO produced by microglia/macrophages could be a potent neurotoxin and mediates TNF-α neurotoxicity toward oligodendrocytes(16, 34-36). The toxic effect of NO is likely attributed to the nitrosylation of target iron-sulfur proteins, including enzymes involved in DNA replication and repair and in blocking mitochondrial respiration(37, 38). The uncoupling of the electron transfer chain in mitochondria will result in production of oxygen free radicals. Reaction of NO with a superoxide will lead to formation of extremely toxic peroxynitrite (ONOO⁻)(39-41). Peroxidations of membranes and swollen oligodendrocyte cell bodies have been described in the CNS of patients with MS(42). Nitrosylation of tyrosine in proteins has been demonstrated in a number of neurodegenerative and inflammatory conditions of the CNS(43-45).

In the inflammatory conditions of CNS, such as viral encephalitis and EAE (experimental allergic encephalomyelitis), iNOS is expressed in macrophages, microglia and/or astrocytes, and possibly neurons but not oligodendrocytes(41, 43, 45-51). The role of iNOS has been examined in a number of encephalitis's induced by viruses, such as corona, flavi, rhabdo, Borna, rabies, herpes, Japanese encephalitis, Venezuelan encephalitis and several others, revealing that NO may contribute to the pathogenesis of the disease(52-60). Alternatively, NO may have a protective role in a viral infection.

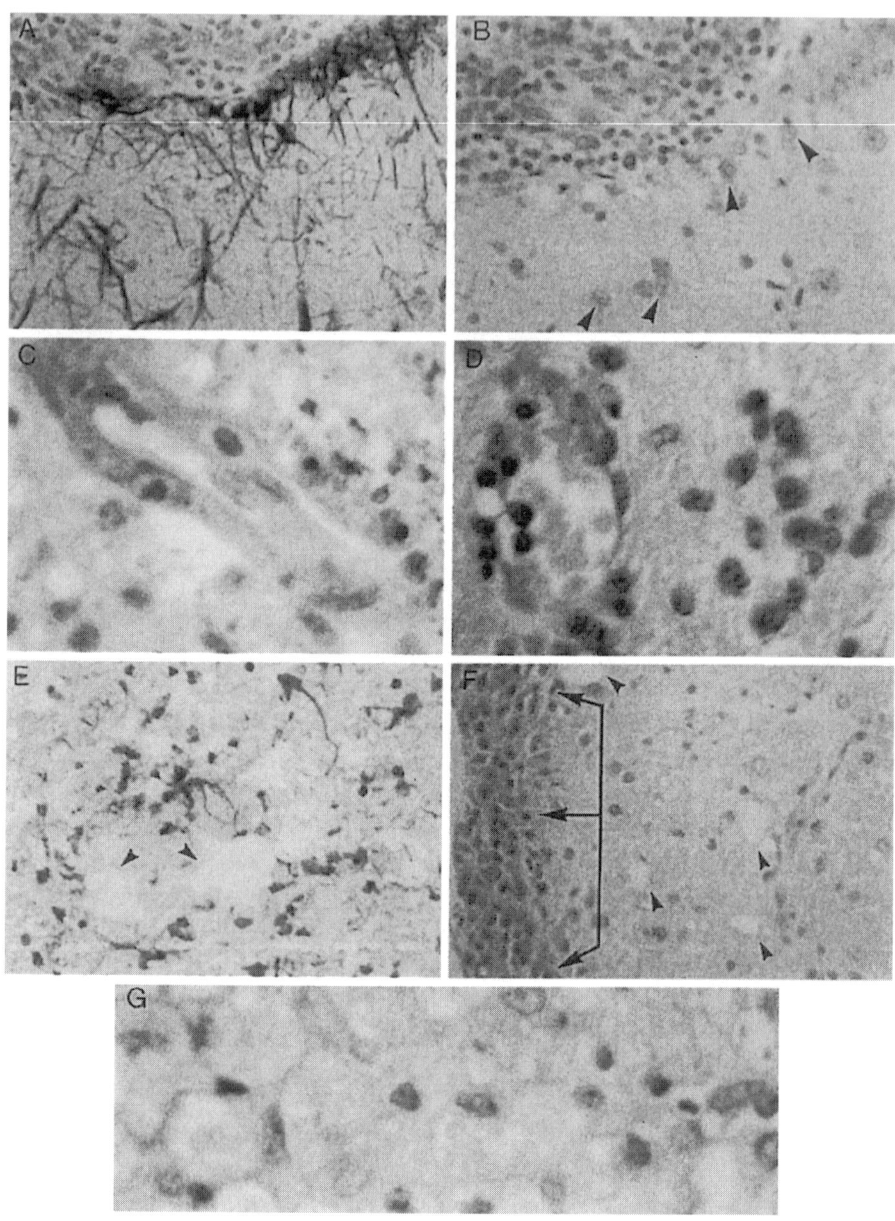

*Figure B8-1.*

IHC localizations of GFAP (A) and iNOS protein in the CNS of SJL mice at 10 (A-C), 6 (D), 42 (E and F), and 66 (G) days p.i. (A) Encephalitis and poliomyelitis phase: an area of intense leptomeningeal inflammation accompanied by reactive astrogliosis in the subjacent brain parenchyma, which is highlighted by GFAP staining. (B) Same field from an immediately adjacent section exhibiting iNOS-like staining in reactive subpial astrocytes (arrowheads) as well as in scattered monocytes/macrophages in the overlying leptomeningeal infiltrate (top portion of photomicrograph). (C) Encephalitis and poliomyelitis phase: iNOS-like localization in hypertrophic endothelial cells of a blood vessel amidst an area of necrotizing encephalitis. (D) Encephalitis and poliomyelitis phase. iNOS-like staining in cells of the monocyte/macrophage lineage and in vascular endothelial cells from an area of spinal grey matter inflammation. (E) Early stage of spinal cord demyelination exhibiting a number of scattered iNOS-positive astrocytes in a background of incipient vacuolar changes (arrowheads) suggestive of early demyelination involving the lateral column. (F) Spinal cord demyelination phase: Focal, weak, and ill-defined iNOS-like immunoreactivity is detected among leptomeningeal infiltrates (the three arrows) which are encroaching upon the lateral funiculus of the spinal cord. Note vacuolar change in the lateral column (arrowheads). (G) Spinal cord demyelination phase: Lack of iNOS-like staining in large foamy (myelin-laden) macrophages in full-blown demyelinating lesions in the posterior columns of the thoracic cord. (Original magnifications: A, B, E, F x400; C, D. G x1000)

## 1.1    iNOS in TMEV infection.

We have examined the expression of iNOS transcripts and protein in the CNS of TMEV-infected mice during early acute and late chronic demyelinating disease(61). Both susceptible (SJL) and resistant (B6) strain of mice were infected i.c. with the DA strain of TMEV and infected mice were sacrificed at 0, 3, 6, 10 days p.i. (SJL and B6), corresponding to early acute disease and at 39, 42, 66–67 and 180 days p.i. corresponding to late chronic demyelinating disease (SJL only). Mock-infected mice were injected i.c. with medium alone and they were sacrificed at 6 and 39 days p.i. iNOS transcripts were determined by RT-PCR, followed by Southern blotting, as described(61); the presence of iNOS protein was determined by immunohistochemical staining using affinity-purified polyclonal rabbit anti-iNOS antibody (61). These antibodies are iNOS specific and do not react with NOS type I and NOS type II.

Although both strains of TMEV-infected mice develop polioencephalomyelitis (gray matter inflammatory disease), confirmed by histological analysis (Fig.1), none of the animals showed overt neurological signs. We did not detected any transcripts for iNOS in brain and spinal cord of mock-infected animals examined at day 6 and 39 p.i. and very little iNOS transcripts were present at day 3 p.i. In contrast, high levels of expression of iNOS transcripts were detected in the CNS of TMEV-infected SJL and B6 mice at 6 and 10 days p.i. In general, higher levels of iNOS expression were detected in brains of SJL and B6 mice than in the spinal cords. Immunohistochemical staining of CNS tissue enabled the identification of cells which expressed iNOS during early acute disease. At 6 and 10 days p.i. iNOS was expressed in strains of TMEV-infected mice in areas of intense inflammation in reactive astrocytes (Figs. 1A & B), monocytes/macrophages and hypertrophic endothelial cells (Figs. 1C, 1D). Few perivascular monocytes-like cells in leptomeninges were also iNOS positive, while lymphocytes were negative(61). Normal endothelial cells and smooth muscles cells were consistently negative for iNOS. To confirm that reactive astrocytes were the major cell type expressing iNOS, we stained adjacent sections with antibodies to GFAP (glial fibrillary acidic protein) or with antibodies to iNOS. Cells which were positive for iNOS were also positive for GFAP. The profile of iNOS staining in the spinal cord was similar to that described for encephalitis. We have not detected any iNOS-like immunoreactivity in normal interfascicular fibrous astrocytes, protoplasmic astrocytes, Bergmann's glia or oligodendrocytes. Rod-shaped microglia-like cells were also consistently negative. iNOS protein was essentially undetectable in the brain and spinal cord of naive mice.

We have examined expression of iNOS in late chronic demyelinating disease(61). iNOS transcripts were detected in spinal cords of TMEV-infected SJL mice at 39 days p.i., which corresponds to the beginning of inflammatory demyelinating disease (61). Less iNOS transcripts were detected in brains of these mice. Immunohistochemical staining confirmed expression of iNOS protein in fibrillary astrocytes in areas of inflammation (Fig.1E) and in leptomeningeal infiltrate (albeit weak) at 42 days p.i. (Fig.1F). iNOS protein was not detected in the CNS of TMEV-infected B6 mice at 42 and 67 days p.i. and, as expected, there were no overt pathological changes in the brains and spinal cords of these mice. In contrast, extensive inflammatory disease, demyelination and gliosis were demonstrated in TMEV-infected SJL mice at day 39, 67 and 180 p.i. However, we did not detect any iNOS transcripts nor iNOS protein at day 67 p.i. (advanced phase of demyelinating disease) or at 180 days p.i. (terminal phase of disease) associated with hind leg paralysis and incontinence). Foamy-myelin-laden macrophages were always negative (Fig.1G).

## 1.1.1    DISCUSSION

We have demonstrated that during early acute disease induced in TMEV-infected SJL and B6 mice iNOS is expressed in the brain and spinal cord of infected animals, predominantly in hypertrophic astrocytes, monocytes/macrophages and endothelial cells in the area of inflammation (61). During late chronic demyelinating disease, iNOS has been detected in fibrous astrocytes and meningeal infiltrates of spinal cord but only during the early stages (39 days p.i.) of late chronic demyelinating disease (SJL mice) and not at 66days p.i. or at 180 days p.i. (terminal stage of the disease). We have also demonstrated that during early acute disease there is high level of apoptosis of T cells undergoing activation induced cell death, while there is paucity of apoptosis of T cells infiltrating the CNS of TMEV-infected SJL mice during late chronic demyelinating disease (unpublished observations). We suggest that NO produced by reactive astrocytes and macrophages/microglia during early acute disease triggers apoptosis of infiltrating T cells. Conversely, lack of expression of iNOS in the CNS during late chronic demyelinating disease (days 66 and 180 p.i.) may relate to the very low level of apoptosis of these T cells. Induction of apoptosis of MBP-specific T cells by NO producing astrocytes (functioning as antigen presenting cells) has been described(62). Molina-Holgado(63) recently reported that in brain astrocytes infected in vitro with TMEV there is high expression of iNOS and that IL-4 and IL-10 down regulate expression of iNOS by modulation of NF-κB activity. We have demonstrated high level of expression of proinflammatory cytokines in the CNS of TMEV-infected mice during early acute and late chronic demyelinating disease (64). However, IL-4 transcripts have been detected primarily during late chronic demyelinating disease. It is therefore possible to suggest that IL-4 down regulates expression of NO in astrocytes in SJL mice during late chronic demyelinating disease, which may result in inhibition of apoptosis. According to this postulate, expression of iNOS during early acute disease would be beneficial to the host, because NO could be responsible for eliminating infiltrating T cells through down regulation of Bcl-2 and induction of AICD, leading to restoration of homeostasis in the CNS. Thus, NO may function as an anti-inflammatory or immunosuppressive molecule during early acute disease. Immunosuppressive effect of macrophage-derived NO on T cells has also been reported (25, 28, 29). However, lack of iNOS expression (down regulated perhaps by IL-4/NF-κB) during late chronic demyelinating disease in TMEV-infected mice may be detrimental to the host by blocking apoptosis of infiltrating T cells. On the other hand, down regulation of NO production in the CNS of TMEV-infected mice may represent an effort of the host to curb formation of peroxinitrite, an extremely toxic molecule, which may lead to peroxidation of membranes and formation of nitrotyrosine. Whether nitrotyrosine is present in the CNS of TMEV-infected mice is not known. NO is a weak oxidant and acts rather like an antioxidant and only in reaction with superoxide it forms peroxynitrite, very toxic molecule. We and others demonstrated the expression of iNOS and nitrotyrosine in reactive astrocytes and monocytes/macrophages in Multiple Sclerosis lesions(43, 45). In general the expression of iNOS and nitrotyrosine seems to depend on the extent of lesion activity (acute vs. chronic).

The unexpected protective or destructive role of NO in induction and progression of EAE is also well known, reviewed in(65).

The effect of specific iNOS inhibitor, amino guanidine (AG) on mice infected with DA strain of TMEV has been examined(66). Treatment of

TMEV-infected mice starting at day 28 p.i. till day 49 p.i. significantly decreased inflammation, demyelination and axonal necrosis. This treatment was associated with decreased apoptosis of CNS infiltrating cells, presumably monocytes or lymphocytes. Early treatment (starting at day 7 p.i.) was not effective(66). When TMEV-infected mice were treated starting at day 14 p.i., only demyelination and axonal necrosis were decreased while the level of inflammation was not changed(66). These results suggest that inflammation and demyelination and axonal damage may be differentially down regulated by treatment of TMEV-infected mice with NO inhibitors at different times of infection. The outcome of treatment may depend on the level of activation of macrophages/microglia and astrocytes, which produce NO but also on availability of superoxide. The role of iNOS in TMEV induced demyelinating disease has also been examined using BeAn strain of the virus (instead of DA strain routinely employed in our laboratory). The BeAn strain of TMEV replicates with different kinetics in the CNS of SJL mice(67). For example, mice infected with BeAn show clinical signs of disease such as extensor spasm and hind leg of paralysis at 50 days p.i., whereas, mice infected with DA strain of TMEV do not show these signs earlier than 180 days p.i. Treatment of SJL mice infected with BeAn strain of TMEV with AG during the effector phase of the disease (initiating treatment at 15 days p.i.) resulted in significant delay of inflammatory demyelinating disease (67). Interestingly, iNOS was detected at low levels in monocytes/macrophages (MOMA-positive) in leptomeninges and perivascular spaces, but not in astrocytes, in TMEV infected mice between 0 and 15 days p.i.(67). The maximum expression of iNOS was observed at day 60 p.i. In contrast, iNOS expression was high in the CNS of SJL mice infected with DA strain of TMEV between 6 and 10 days p.i., but was not detectable at 66 and 180 days p.i. Treatment of TMEV (BeAn)-infected mice with AG in the induction phase (starting at day 1 p.i.) did not alter the course of the disease(67). The authors suggested that protective effect of AG on inflammatory demyelinating disease observed in BeAn infected mice when treatment had begun 15 days p.i. was associated with suppression of activation and proliferation of inflammatory macrophages and lymphocytes(67). The results of these experiments suggest that the effect of NO on inflammatory demyelinating disease induced by these two strains of TMEV may depend on the kinetics of disease, time of appearance of inflammatory infiltrates, type of cells which express iNOS and time of administration of the inhibitor.

## ACKNOWLEDGEMENTS

The support of the National Multiple Sclerosis Society (grant RG2942) and the N. Eleanor Dana Charitable Trust is greatly acknowledged. We thank Mr. Brad E. Hoffman for assistance in editing of the chapter.

## REFERENCES

1. **Oleszak EL, Kuzmak J, Good RA, Platsoucas. CD.** Immunobiology of TMEV infection. Immunology of Theiler's murine encephalomyelitis virus infection. *Immunologic Res.* 1995;1(13-33.).
2. **Dal Canto M, Lipton. HL.** Multiple sclerosis. Animal model: Theiler's virus infection in mice. Am. *J. Pathol.* 1977;4(497-500.).
3. **Levy M, Aubert C, Brahic. M.** Theiler(s virus replication in brain macrophages cultured in vitro. *J. Virol.* 1992;3(3188-3193.).
4. **Lipton HL.** Characterization of the TO strains of Theiler's mouse encephalomyelitis viuses. *Infect. Immun.* 1978;1(869-872.).
5. **Lipton HL.** Theiler's virus infection in mice: an unusual biphasic disease process leading to demyelination. *Infect. Immunol.* 1975;0(1147-1155.).
6. **Lindsley MD, Rodriguez. M.** Characterization of the inflammatory response in the central nervous system of mice susceptible or resistant to demyelination by Theiler(s virus. *J. Immunol.* 1989;6(2677-2682.).
7. **Tsunoda I, Fujinami RS.** Two models for multiple sclerosis: experimental allergic encephalomyelitis and Theiler's murine encephalomyelitis virus. *J Neuropathol Exp Neurol.* 1996;55(6):673-86.
8. **Brahic M, Stroop WG, Baringer. JR.** Theiler's virus persists in glial cells during demyelinating disease. *Cell.* 1981;1(123-128.).
9. **Rodriguez M, Oleszak E, Leibowitz J.** Theiler's murine encepha-lomyelitis: a model of demyelination and persistence of virus. *CRC Crit. Rev. Immunol.* 1987;7(325-365.).
10. **Theiler M.** Spontaneous encephalomyelitis of mice: a new virus disease. *J. Exp. Med.* 1937;3(705-719.).
11. **Kim BS, Palna JP.** Immune mechanisms of Theiler's virus-induced demyelination. *Exp Mol Med.* 1999;31(3):115-21.
12. **Martin R, McFarland H.** *Immunology of Multiple Sclerosis and experimental allergic encephalomyelitis.* 1st ed London: Chapman & Hall Medical; 1997. (Raine CS, McFarland HF, Tourtellotte WW, eds. Multiple sclerosis : clinical and pathogenic basis).
13. **Brosnan CF, Raine CS.** Mechanisms of immune injury in multiple sclerosis. *Brain Pathol.* 1996;6(3):243-57.
14. **Wekerle H.** *Immunology of Multiple Sclerosis.* 3rd ed London: Churchill Livingstone; 1998. (Compston A, ed. McAlpine's multiple sclerosis).
15. **Prineas J, McDonald W.** *Demyelinating Disease.* 6th ed London, New York: Oxford University Press; 1997. (Greenfield JG, Lantos PL, Graham DI, eds. Greenfield's Neuropathology).
16. **Selmaj K, Raine CS, Farooq M, Norton WT, Brosnan CF.** Cytokine cytotoxicity against oligodendrocytes. Apoptosis induced by lymphotoxin. *J Immunol.* 1991;147(5):1522-9.
17. **Marlena MA.** Nitric oxide synthase structure and mechanism. *J. Biol. Chem.* 1993;11(12231-12234.).
18. **Nathan C, Xie. QW.** Regulation of biosynthesis of nitric oxide. *J. Biol. Chem.* 1994;11(13725-13728.).

19.**Bredt DS.** Endogenous nitric oxide synthesis: biological functions and pathophysiology. *Free Radic Res.* 1999;31(6):577-96.

20.**Knowles RG, Moncada S.** Nitric oxide synthases in mammals. *Biochem J.* 1994;298 ( Pt 2):249-58.

21.**Marletta MA, Yocr PS, Lyengar R, Leaf CD, Wishnok. JS.** Macrophage oxidation of L-arginine to nitrite and nitrate: nitric oxide is an intermediate. *Biochemistry.* 1988;1(8706-8711.).

22.**Nussler AK, Billiar. TR.** Inflammation, immunoregulation, and inducible nitric oxide synthase. *J. Leukocyte Biol.* 1993;2(171-178.).

23.**Bogdan C, Rollinghoff M, Diefenbach A.** The role of nitric oxide in innate immunity. *Immunol Rev.* 2000;173:17-26.

24.**Di Rosa F, Barnaba V.** Persisting viruses and chronic inflammation: understanding their relation to autoimmunity. *Immunol Rev.* 1998;164:17-27.

25.**Gazzinelli RT, Oswald IP, Hieny S, James SL, Sher A.** The microbicidal activity of interferon-gamma-treated macrophages against Trypanosoma cruzi involves an L-arginine-dependent, nitrogen oxide-mediated mechanism inhibitable by interleukin-10 and transforming growth factor-beta. *Eur J Immunol.* 1992;22(10):2501-6.

26.**Chan J, Tanaka K, Carroll D, Flynn J, Bloom BR.** Effects of nitric oxide synthase inhibitors on murine infection with Mycobacterium tuberculosis. *Infect Immun.* 1995;63(2):736-40.

27.**Xie Q, Kawakami K, Kudeken N, Zhang T, Qureshi MH, Saito A.** Different susceptibility of three clinically isolated strains of Cryptococcus neoformans to the fungicidal effects of reactive nitrogen and oxygen intermediates: possible relationships with virulence. *Microbiol Immunol.* 1997;41(9):725-31.

28.**Shi FD, Flodstrom M, Kim SH, et al.** Control of the autoimmune response by type 2 nitric oxide synthase. *J Immunol.* 2001;167(5):3000-6.

29.**Kolb H, Kolb-Bachofen. V.** Nitric oxide; A pathogenic factor in autoimmunity. *Immunol. Today.* 1992;1(157-160.).

30.**Kolb H, Kolb-Bachofen V.** Nitric oxide in autoimmune disease: cytotoxic or regulatory mediator? *Immunol Today.* 1998;19(12):556-61.

31.**Moulian N, Truffault F, Gaudry-Talarmain YM, Serraf A, Berrih-Aknin S.** In vivo and in vitro apoptosis of human thymocytes are associated with nitrotyrosine formation. *Blood.* 2001;97(11):3521-30.

32.**Marks-Konczalik J, Chu SC, Moss J.** Cytokine-mediated transcriptional induction of the human inducible nitric oxide synthase gene requires both activator protein 1 and nuclear factor kappaB-binding sites. *J Biol Chem.* 1998;273(35):22201-8.

33.**Nathan C, Xie QW.** Nitric oxide synthases: roles, tolls, and controls. *Cell.* 1994;78(6):915-8.

34.**Dawson VL, Dawson TM, London ED, Bredt DS, Snyder. SH.** Nitric oxide mediates glutamate neurotoxicity in primary cultures. *Proc. Natl. Acad. Sci. USA.* 1991;4(6368-6371.).

35.**Merrill JE, Ignarro LJ, Sherman MP, Melinek J, Lane. TE.** Microglial cell cytotoxicity of oligodentrocytes is mediated through nitric oxide. *J. Immunol.* 1993;6(2132-2141.).

36.**Mitrovic B, Ignarro LJ, Montestruque S, Smoll A, Merrill. JE.** Nitric oxide as a potential pathological mechanism in demyelination: its differential effects on primary glial cells in vitro. *Neuroscience.* 1994;3(575-585.).

37.**Wink DA, Kasprzak KS, Maragos CM, et al.** DNA deaminating ability and genotoxicity of nitric oxide and its progenitors. *Science.* 1991;11(1001-1003.).

38. **Stuehr DJ, Nathan CF.** Nitric oxide. A macrophage product responsible for cytostasis and respiratory inhibition in tumor target cells. *J Exp Med.* 1989;169(5):1543-55.

39. **Beckman JS, Crow. JP.** Pathological implications of nitric oxide, superoxide and peroxynitrite formation. *Biochem. Soc. Trans.* 1993;1(330-334.).

40. **Lipton SA, Choi YB, Pan ZH, et al.** A redox-based mechanism for the neuroprotective and neurodestructive effects of nitric oxide and related nitrosocompounds. *Nature.* 1993;15(626-631.).

41. **Zielasek J, Tausch M, Toyka KV, Hartung. HP.** Production of nitrite by neonatal rat microglial cells/brain macrophages. *Cell. Immunol.* 1992;6(111-120.).

42. **Hunter MI, Nlemadim BC, Davidson DL.** Lipid peroxidation products and antioxidant proteins in plasma and cerebrospinal fluid from multiple sclerosis patients. *Neurochem Res.* 1985;10(12):1645-52.

43. **Liu JS, Zhao ML, Brosnan CF, Lee SC.** Expression of inducible nitric oxide synthase and nitrotyrosine in multiple sclerosis lesions. *Am J Pathol.* 2001;158(6):2057-66.

44. **Liu B, Gao HM, Wang JY, Jeohn GH, Cooper CL, Hong JS.** Role of nitric oxide in inflammation-mediated neurodegeneration. *Ann N Y Acad Sci.* 2002;962:318-31.

45. **Oleszak EL, Zaczynska E, Bhattacharjee M, Butunoi C, Legido A, Katsetos CD.** Inducible nitric oxide synthase and nitrotyrosine are found in monocytes/macrophages and/or astrocytes in acute, but not in chronic, multiple sclerosis. *Clin Diagn Lab Immunol.* 1998;5(4):438-45.

46. **Bagasra O, Michaels FH, Zheng YM, et al.** Activation of the inducible form of nitric oxide synthase in the brains of patients with multiple sclerosis. *Proc. Natl. Acad. Sci. USA.* 1995;4(12041-12045.). '

47. **Bö L, Dawson TM, Wesselingh S, et al.** Induction of nitric oxide synthase in demyelinating regions of multiple sclerosis brains. *Ann. Neurol.* 1994;2(778-786.).

48. **Galea E, Reis DJ, Feinstein. DL.** Cloning and expression of inducible nitric oxide synthase from rat astrocytes. *J. Neurosci. Res.* 1994;2(406-414).

49. **Hewett SJ, Corbett JA, McDaniel ML, Choi. DW.** IFN-? and IL-1ß induce nitric oxide formation from primary mouse astrocytes. *Neurosci. Lett.* 1993;7(229-232.).

50. **Minc-Golomb D, Tsarfaty I, Schwartz. JP.** Expression of inducible nitric oxide synthase by neurones following exposure to endotoxin and cytokine. Br. *J. Pharmacol.* 1994;5(720-722.).

51. **Van Dam A-M, Bauer J, Man-A-Hing WKH, Marquette C, Tilders FJH, Berkenbosch. F.** Appearance of inducible nitric oxide synthase in the rat central nervous system after rabies virus infection and during experimental allergic encephalomyelitis but not after peripheral administration of endotoxin. *J. Neurosci. Res.* 1995;2(251-260.).

52. **Reiss CS, Komatsu T.** Does nitric oxide play a critical role in viral infections? *J Virol.* 1998;72(6):4547-51.

53. **Wu GF, Pewe L, Perlman S.** Coronavirus-induced demyelination occurs in the absence of inducible nitric oxide synthase. *J Virol.* 2000;74(16):7683-6.

54. **Lane TE, Paoletti AD, Buchmeier MJ.** Disassociation between the in vitro and in vivo effects of nitric oxide on a neurotropic murine coronavirus. *J Virol.* 1997;71(3):2202-10.

55. **Bartholdy C, Nansen A, Christensen JE, Marker O, Thomsen AR.** Inducible nitric-oxide synthase plays a minimal role in lymphocytic choriomeningitis virus-induced, T cell-mediated protective immunity and immunopathology. *J Gen Virol.* 1999;80 ( Pt 11):2997-3005.

56. **Schoneboom BA, Catlin KM, Marty AM, Grieder FB.** Inflammation is a component of neurodegeneration in response to Venezuelan equine encephalitis virus infection in mice. *J Neuroimmunol.* 2000;109(2):132-46.

57. **Saxena SK, Mathur A, Srivastava RC.** Induction of nitric oxide synthase during Japanese encephalitis virus infection: evidence of protective role. *Arch Biochem Biophys.* 2001;391(1):1-7.
58. **Kreil TR, Eibl MM.** Nitric oxide and viral infection: NO antiviral activity against a flavivirus in vitro, and evidence for contribution to pathogenesis in experimental infection in vivo. *Virology.* 1996;219(1):304-6.
59. **Koprowski H, Zhen YM, Heber-Katz E, et al.** In vivo expression of inducible nitric oxide synthase in experimentally induced neurologic diseases. *Proc. Natl. Acad. Sci.* 1993;4(3024-3027.).
60. **Torre D, Pugliese A, Speranza F.** Role of nitric oxide in HIV-1 infection: friend or foe? *Lancet Infect Dis.* 2002;2(5):273-80.
61. **Oleszak EL, Katsetos CD, Kuzmak J, Varadhachary A.** Inducible nitric oxide synthase in Theiler's murine encephalomyelitis virus infection. *J Virol.* 1997;71(4):3228-35.
62. **Xiao BG, Xu LY, Yang JS, Huang YM, Link H.** An alternative pathway of nitric oxide production by rat astrocytes requires specific antigen and T cell contact. *Neurosci Lett.* 2000;283(1):53-6.
63. **Molina-Holgado E, Arevalo-Martin A, Castrillo A, Bosca L, Vela JM, Guaza C.** Interleukin-4 and interleukin-10 modulate nuclear factor kappaB activity and nitric oxide synthase-2 expression in Theiler's virus-infected brain astrocytes. *J Neurochem.* 2002;81(6):1242-52.
64. **Chang JR, Zaczynska E, Katsetos CD, Platsoucas CD, Oleszak EL.** Differential expression of TGF-beta, IL-2, and other cytokines in the CNS of Theiler's murine encephalomyelitis virus-infected susceptible and resistant strains of mice. *Virology.* 2000;278(2):346-60.
65. **Willenborg DO, Staykova MA, Cowden WB.** Our shifting understanding of the role of nitric oxide in autoimmune encephalomyelitis: a review. *J Neuroimmunol.* 1999;100(1-2):21-35.
66. **Rose JW, Hill KE, Wada Y, et al.** Nitric oxide synthase inhibitor, aminoguanidine, reduces inflammation and demyelination produced by Theiler's virus infection. *J Neuroimmunol.* 1998;81(1-2):82-9.
67. **Iwahashi T, Inoue A, Koh CS, Shin TK, Kim BS.** Expression and potential role of inducible nitric oxide synthase in the central nervous system of Theiler's murine encephalomyelitis virus-induced demyelinating disease. *Cell Immunol.* 1999;194(2):186-93.

Chapter B9

# THEILER'S MURINE ENCEPHALOMYELITIS VIRUS (TMEV)-INDUCED DEMYELINATION
*Apoptosis in TMEV infection*

[1]Mary Lou Jelachich, and [2]Howard L. Lipton
*[1]Northwestern University Medical School, Department of Neurology, Chicago, IL 60611,*
*[2]Northwestern University, Department of Biochemistry, Molecular Biology, and Cell Biology,*
*Evanston, IL 60208, [1,2] Evanston-Northwestern Healthcare Research Institute MS Research,*
*Evanston, IL 60201*



**Abstract**:     Apoptosis, one aspect of programmed cell death, is initiated by cell death receptor-ligand interactions and in response to cellular stress. Environmental stressors, such as increased temperature, and intracellular stressors, such as viral infection, result in the activation of caspase cascades leading to cell death.     Intracerebral inoculation of susceptible mouse strains with the picronavirus TMEV results in the chronic infection of the central nervous system white matter and concomitant demyelination. This review examines the evidence for TMEV-induced apoptosis in vitro and in vivo.



**Keywords**:     TMEV, apoptosis, demyelination, picornavirus


## DEFINITION OF APOPTOSIS

Apoptosis is defined as the initiation of a cascade of cellular processes leading to cell death. Hallmarks of apoptosis include membrane inversion and exposure of phosphatidylserine residues, blebbing, fragmentation of the nucleus, chromatin condensation, and DNA degradation. These features distinguish apoptosis from necrosis, which results from injury and is characterized by cell swelling, no nuclear condensation, and cell rupture (1-3). There are two other recently described phenomena of programmed cell death (PCD): apoptosis-like PCD, which presents some features of apoptosis but lacks densely packed chromatin; and necrosis-like PCD, which is typified by the absence of chromatin condensation but utilizes a signaling pathway (4).

Cellular degradation occurs through a cascade of cysteine proteases that cleave at aspartic acid residues in the P1 position (caspases) and target a restricted set of proteins necessary for the cell's integrity (2). Caspases have been classified as "initiator" or "apical" (caspases 2, 8, 9, 10, 11), which begin the apoptotic cascade (5-8) and "effector" or "executioner"caspases (caspase 3, 6, 7), which cleave vital cellular proteins (9).

## Pathways to Cell Death

There are two well-recognized pathways of apoptosis induction (10, 11), an intrinsic pathway that depends on the cell's ability to recognize cytoplasmic and/or nuclear damage, and an extrinsic pathway that relies on signaling through death receptors expressed on the cell surface (Figure 1).

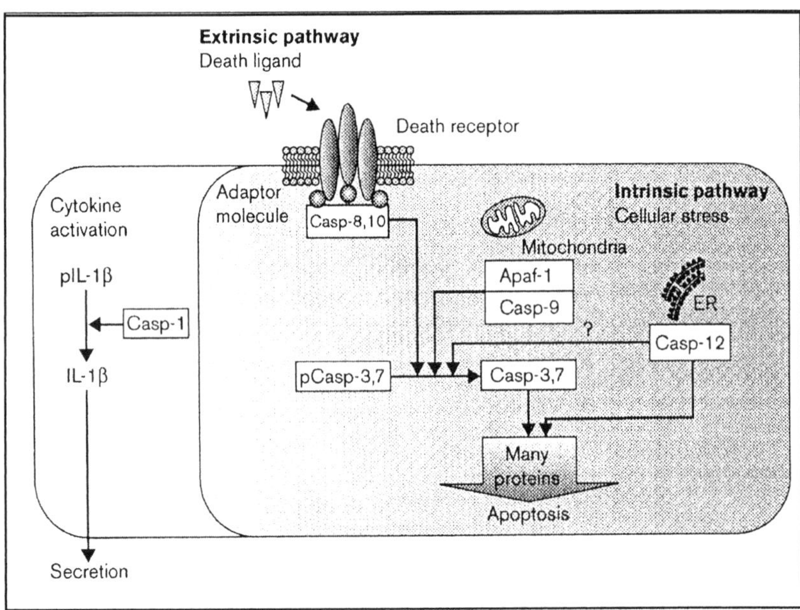


*Figure 1. Schematic representation of the caspase pathways. The extrinsic pathway, which is initiated by trimerization of death receptors by their ligands, leads to activation of initiator caspases 8 and 10, and results in the cleavage of the executioner caspases 3 and 7. The intrinsic pathway, which is activated by cellular stress, leads to mitochondrial membrane permeability, cytochrome c release, the formation of an Apaf-1-caspase 9 complex with subsequent cleavage of caspase 9 and activation of the executioner caspases. "Reprinted from Current Opinion in Structural Biology, 10, Marcus G. Grutter, Caspases: key players in programmed cell death, 649-655. Copyright (2000), with permission from Elsevier".*

## Intrinsic Pathway

The intrinsic apoptosis pathway is activated when the cell detects intracellular metabolic disturbances and forms a complex called the apoptosome, which consists of cytochrome c, Apaf-1 (apoptosis-activating factor-1), and caspase 9. Such alterations can be induced chemically by compounds that affect DNA integrity (topoisomerase, staurosporine), or oxidative stress (12-15) or environmentally by heat, radiation, nutrient deprivation (16) or virus infection (4, 17). In any case, it appears that mitochondrial permeability and subsequent release of cytochrome c are essential for the formation of the apoptosome, activation of caspase 9 and ensuing cell death through caspase 3 activation (18). Recent evidence by Lassus et al. (8) using adenoviral oncogene E1A-transformed human fibroblasts and several cancer cell lines suggest that caspase 2 may be responsible for mitochondrial membrane permeability that initiates the formation of the apoptosome. However, O'Reilly et al. (19) have reported that caspase 2 is not required for thymocyte or neuronal apoptosis. Whether caspase 2 is an early integral component of the intrinsic apoptotic pathway or a cell specific component in the apoptotic process remains to be determined.

The basic debate in this field concerns the role of mitochondria in apoptosis, i.e. do mitochondria and bcl-2 proteins control apoptosis by activating the caspases or do caspases trigger apoptosis which results in mitochondrial rupture and amplification of the caspase cascade (20).

## Extrinsic Pathway

The extrinsic pathway is dependent on signaling through the cell surface death receptors by their ligands, tumor necrosis factor (TNF)-α/TNF-receptor, Fas/FasL, and TRAIL/ TRAIL-receptors (2, 9). TNF-α and Fas have been studied extensively, while characterization of the more recently defined TRAIL and its receptors lag behind. Ligand binding and receptor oligomerization leads to the formation of the death-inducing signal complex (DISC) and recruitment of the Fas-adapter protein, FADD, through death domains (DDs). FADD, in turn, recruits caspase 8 and 10 through death effector domains (DEDs) located at the other end of the molecule from DDs. Activation of caspase 8 by binding to the DISC promotes cleavage of downstream caspases 3, 6, 7 (9). These smaller effector/executioner caspases degrade various cellular components, leading to the cell's demise.

Truncation of the Bcl-2 interacting protein, Bid, by caspase 8 leads to Bcl-2 inhibition, mitochondrial permeability, and amplification through apoptosome formation (9).

# TMEV-INDUCED APOPTOSIS

## Cytopathic Effect vs. Apoptosis

While the relationship between virus-induced cytopathic effect (CPE) and virus-induced apoptosis is unclear, the effect of many virus infections on cells is now described as apoptotic.  Both DNA and RNA viruses have devised strategies to inhibit or curtail apoptosis, some viruses have both anti- and pro-apoptotic strategies, and still others depend on apoptosis for their dissemination (for reviews see (4, 21)).  Apoptotic response to infection by the picornaviruses TMEV and poliovirus (small RNA viruses) appears to depend on the cell's permissiveness for virus infection, i.e., cells that restrict virus replication undergo apoptosis, while CPE predominates in permissive cells (22-24).  Agol et al. (24) have described a switch from a pro-apoptotic program early in poliovirus replication in HeLa cells to CPE during the later stages of a productive virus infection.   When poliovirus replication is restricted, the switch to CPE does not occur.   In TMEV, morphologic changes and DNA laddering indicative of apoptosis could only be discerned in cells that restricted virus replication (22, 23)

## TMEV-Induced Apoptosis in Cultured Cells

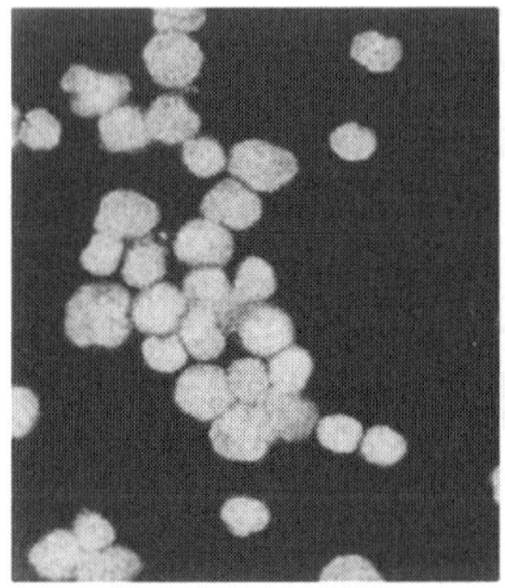
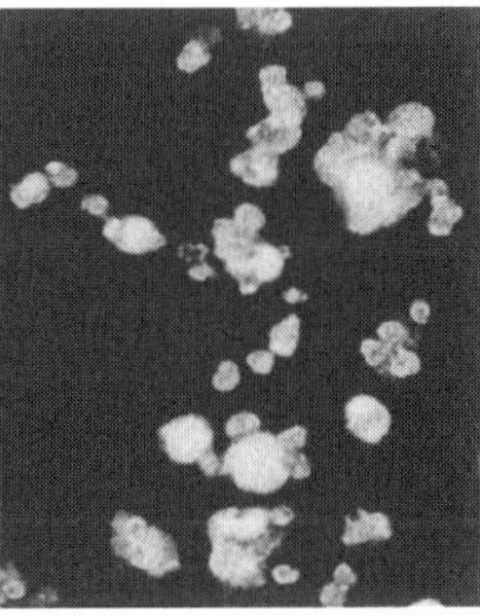

The activation/differentiation state of the macrophage determines whether this cell is susceptible to TMEV infection.   While precursor macrophage cell lines do not support virus replication, the differentiated cells become infected and undergo apoptosis (Figure 2) (25, 26). Because macrophages appear to bear the largest virus-antigen burden *in vivo* (27)

*Figure 2.     Apoptosis and TMEV antigen expression in undifferentiated (left) or differentiated (right) M1 cells.   Cells were infected with TMEV-BeAn virus (MOI = 10) and examined 20 h later for virus antigen (green) by staining with polyclonal rabbit anti-TMEV followed by a FITC-labeled secondary antibody; nuclei (pink) were visualized by DAPI and PI staining.*

and are necessary for persistence (28), macrophage cell lines are the best-studied restrictive cells. Analyses of several macrophage cell lines for their ability to support TMEV replication (22, 26, 29) revealed ~100-fold lower virus replication and titers than in permissive BHK-21 cells, but abundant virus antigen and extensive apoptosis.

In a recent analysis of TMEV replication in bone-marrow-derived macrophages from SJL/J and interferon-$\alpha/\beta$ (IFN-$\alpha/\beta$) receptor-negative mice (30), maximal numbers (30-50%) of TMEV-positive cells were detected 27-54 h after infection, but after 13 days in culture, viable cells showed no evidence of virus antigen. At the low MOI used in these cultures, the TMEV-infected cells produced sufficient IFN-$\alpha/\beta$ to protect the remaining uninfected cells from virus infection and apoptosis. We recently found that mouse peritoneal macrophages undergo apoptosis as a result of TMEV infection at a high MOI (20-50), although about 50% the TUNEL+ cells did not contain virus antigen (manuscript in preparation). The fact that only one-half of the cells in both bone marrow derived- and peritoneal macrophage cultures are infected suggests that in these cultures cells may be at different stages of macrophage differentiation/activation. However, macrophage subsets have not been identified with available markers, and the effect of cell cycle on susceptibility to TMEV infection has not been explored in primary cells. Alternatively, the amount of IFN-$\alpha/\beta$ in cultures may selectively protect some cell subsets.

TMEV infection has been studied in microglia, the resident central nervous system (CNS) macrophages (29, 31) (32). In one study (31), the TMEV-infected microglial cell line Ra2 showed no evidence of apoptosis as assessed by DNA degradation and TUNEL, whereas both were evident in the macrophage cell line P388D1. In both cell lines, virus titers were very low. Although virus-induced apoptosis or virus replication was not directly measured in another study of primary microglia, TMEV-infected cells showed up-regulated expression of surface markers and cytokines associated with antigen presentation and activation, and were persistently infected for 1-2 weeks as determined by flow cytometric detection of virus antigen (32). The data suggest that microglia are persistently infected without evidence of cell death and may be the reservoir for continual virus infection and concurrent inflammation.

A study by Zheng et al. (33) of primary CNS cultures for TMEV infection of glial cells and apoptosis indicated that both astrocytes and oligodendrocytes supported virus replication (titers $8 \times 10^7$ and $1 \times 10^7$ pfu/ml, respectively), whereas microglia showed much lower titers ($3 \times 10^4$ pfu/ml). Oligodendrocytes exhibited incomplete CPE and microglia showed DNA laddering indicative of apoptosis. Since astrocytes supported virus replication without evidence of apoptosis, the authors suggested that astrocytes may be the primary cell type responsible for TMEV persistence in the CNS of susceptible mice. By contrast, Olson et al. (32) reported that TMEV antigen persisted in microglia for 2 weeks without evidence of CPE or apoptosis. Thus, there is still controversy about the primary cell type infected and undergoing apoptosis, possibly due to differences in isolation procedures from neonatal brains or virus concentrations and strains used for infection.

Anderson and colleagues (34) have recently characterized TMEV-infected neonatal cerebellar explant cultures, which produced virus within 24 h after infection and maintained a constant yield of ~$10^4$ pfu/culture for 72 h. Lactic dehydrogenase release at 48 h but not at 72 h indicated initial necrosis and recovery, respectively, whereas apoptosis increased steadily until 72 h after infection. Microscopic examination of infected cultures at 72 h revealed apoptotic nuclei and condensed axons separated from the surrounding myelin, but no myelin destruction. The investigators concluded that TMEV can directly induce apoptosis in CNS-derived neuronal granule cells without contribution of an anti-viral immune response. Since oligodendrocytes could not be identified in the explants, infection and apoptosis of cells that actually produce the myelin sheath could not be determined. Moreover, this type of study addresses only the acute phase of TMEV infection, not virus persistence and apoptosis in the chronic phase of disease.

### TMEV-Induced Apoptosis in Animals

Intracerebral inoculation of TMEV in susceptible SJL mice results in a biphasic CNS disease, an initial polio-like disease primarily affecting neurons, and a chronic phase in which the virus persists in glial cells not neurons and leads to an inflammatory, demyelinating pathology.

TUNEL analysis of TMEV-infected spinal cord sections revealed numerous apoptotic cells in acute disease (1-2 weeks) in mice infected with DA virus (35-37) and in the chronic disease (> 30 days) in mice infected with BeAn or DA virus (36, 38, 39). Although Tsunoda et al. (37) reported the frequency of apoptotic cells as "seldom" in the DA-induced chronic disease (>30 days after infection),

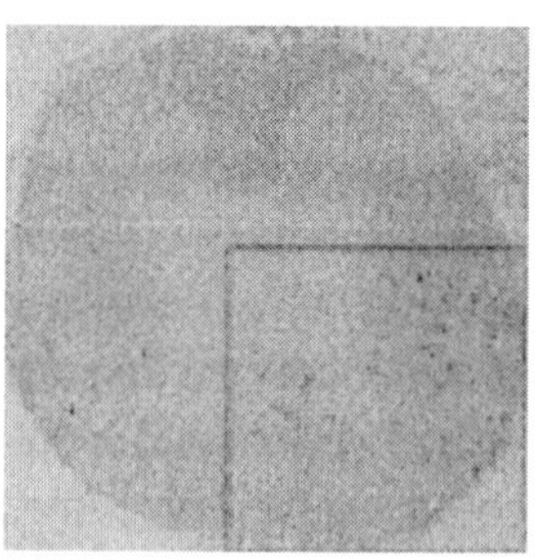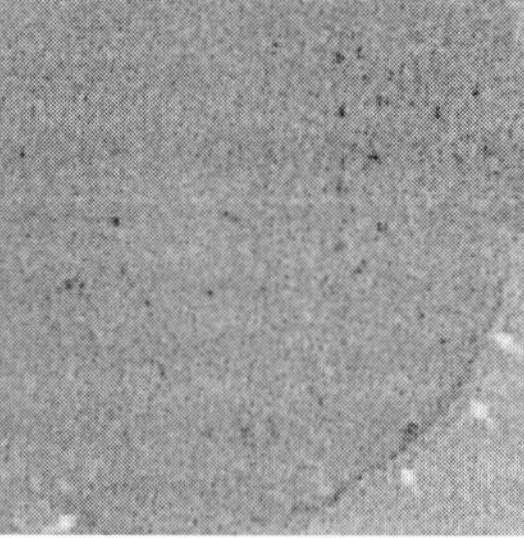

*Figure 3. TMEV BeAn-infected SJL/J mouse spinal cord 69 days after infection. Apoptotic cells are labeled by TUNEL staining (50X) (left). The boxed area is in the left panel is shown at higher magnification on the right (200X). TUNEL-positive cells are prominent in areas of demyelination.*

some apoptotic cells could be identified as oligodendrocytes and macrophages, but not astrocytes. In contrast, Palma et al. (35) identified mainly astrocytes as TUNEL-positive 120 days after BeAn infection of SJL mice. Very few CD4+ cells were apoptotic, and other apoptotic cells were not identified. At this late stage of infection, the number of hypertrophic fibrillary astrocytes was increased indicating an active astrogliosis. In other studies of spinal cord segments examined at earlier times after infection, TUNEL-positive cells were readily observed, although the cell phenotypes were not defined (36, 39). Examination of apoptosis during the course of chronic TMEV-induced demyelinating disease revealed an increase in mean number of apoptotic cells as the disease progressed (Figure 3) (38). Approximately 75% of the apoptotic cells were identified as T cells, 10% as macrophages, and 1% or less as astrocytes; about 14% of TUNEL-positive cells were left unidentified, but were postulated to be B cells and/or oligodendrocytes. These data differ from those of Palma et al. (35), who identified apoptotic astrocytes and not T cells, but agree with Zheng et al. (33), who reported apoptosis in microglia and not astrocytes. In summary, studies of TMEV-induced apoptosis show that either few or many cells undergo apoptosis in vivo, the majority of apoptotic cells are either astrocytes or T cells, after infection microglia either die by apoptosis or survive, and, apoptosis either does or does not play a role in pathogenesis. Clearly, more definitive studies are required to determine the role of apoptosis in TMEV-induced pathogenesis.

# PICORNAVIRUS ANTI-APOPTOTIC MECHANISMS

Infection of susceptible mice with the low-neurovirulence TMEV strains DA and BeAn results in a persistent virus infection and chronic demyelinating disease, while high-neurovirulence strain GDVII infection is fatal in 7-10 days with as little as 1 pfu. The DA and BeAn strains translate an 18 kb protein, L* from an alternative reading frame in the leader sequence that is not present in the high-neurovirulence strains (40). Ghadge et al. (41) reported that GDVII virus infection induced apoptosis much more efficiently than DA in a macrophage cell line, P388D1, and that DA virus with a mutation to eliminate L* translation resembled GDVII in its ability to quickly shut-off host protein synthesis and induce apoptosis. Thus, L*'s function was thought to be anti-apoptotic. However, the virus yield was unchanged in that study; if L* acts to delay the onset of apoptosis, virus replication should proceed for longer periods and virus titers should increase, whereas in the absence of L*, apoptosis should be faster and virus titers lower.

In studies of several other investigators using a different macrophage cell line (J774-1), recombinant DA viruses translating L* replicated in cells, whereas DA virus constructed to express the GDVII leader or mutated DA L* sequences did not (42, 43). Those authors suggested that L* is required for virus growth in macrophages, but apoptosis kinetics were not assessed with these recombinant viruses. The definitive role of L* in apoptosis induction awaits further clarification.

An interesting viral anti-apoptotic mechanism is the reduction of tumor necrosis factor (TNF)-α receptor by poliovirus non-structural proteins (44). In that study retroviral vectors with non-capsid genes were analyzed for their ability to inhibit induction of apoptosis by TNF-α and Fas. Interestingly, while protein 2A, which turns off host protein synthesis, sensitized cells to TNF-α-induced apoptosis, proteins 2B and 3A, with unclear functions, suppressed it. Analysis of the mechanism by which protein 3A suppresses apoptosis revealed that brefeldin A, like poliovirus 3A, reduced the level of TNF-α receptor (44). Those experiments implicate Golgi transport inhibition as the mechanism of TNF-α removal from the cell surface. Whether TMEV proteins have a similar function remains to be determined.


## ROLE OF APOPTOSIS
## IN TMEV-INDUCED DEMYELINATING DISEASE

That apoptotic cells exist in the spinal cord of mice with TMEV-induced demyelinating disease (TMEV-IDD) has been well-documented, although no consensus exists about the apoptotic cell phenotype. It is also known that perforin-deficient mice are unable to clear the virus and die within 2 weeks (45), indicating that the cytolytic machinery of the CD8+ T cells is required for control of the initial virus infection. Finally, virus antigen is known to persist for the lifetime of the infected animal. What is not clear is how apoptosis might contribute to the pathogenesis of the disease. One hypothesis is that macrophage phagocytosis of infected, apoptotic cells account for the continuing presence of virus antigen in macrophages in the spinal cord, so that apoptosis might play only a passive role in the persistence. Apoptosis of infected macrophages/microglia might also serve to restrict virus infection by suppressing virus production keeping virus titers low and allowing protection of the surrounding cellular environment by IFN-α/β. Apoptotic macrophages/microglia might also release IL-1, an inflammatory cytokine, which may propagate an inflammatory response (see Figure 1).

The presence of apoptotic T cells in TMEV-IDD suggests that activation-induced cell death (AICD) is taking place, since there is no evidence that T cells become infected with TMEV. AICD in TMEV-susceptible mouse strains has not been examined. If AICD is defective, it is possible that activated T cells remain in the spinal cord too long, resulting in an overabundance of cytokines and maintenance of an inflammatory response. Alternatively, if AICD occurs too quickly, the stimulating antigen may not be efficiently eliminated.

Induction of AICD in T cells involves Fas/FasL and TRAIL/TRAIL-R interactions (46), molecules that are increased in MS patients (47). Recently, the anti-apoptotic protein Bcl-$X_L$ was found to be up-regulated in multiple sclerosis patients (48), raising the possibility of a dysregulated balance between pro- and anti-apoptotic proteins in this disease and suggesting a promising area of research.

While oligodendrocyte- and astrocytes-induced apoptosis by direct BeAn virus infection in vivo remains controversial, virus infection of derived cell lines in vitro does result in apoptosis. Besides the elimination the myelin-producing oligodendrocytes, phagocytosis of these cells and myelin debris by macrophages might result in efficient antigen presentation of myelin peptides capable of stimulating antigen-specific T cells and propagating an additional inflammatory response. This expansion would result in further CNS damage and exacerbation of the clinical symptoms in the mice.

Further studies of TMEV-IDD are needed to address the role of L* in persistence and pathogenesis, the contribution of apoptotic astrocytes, macrophages and oligodendrocytes to pathogenesis, and the role of apoptotic dysregulation in the disease process. Understanding the mechanism of apoptosis control in this model system would offer insights and direction for MS research.

REFERENCES

1. **Sartorius U, Schmitz I, Krammer PH.** Molecular mechanisms of death-receptor-mediated apoptosis. *Chembiochem.* 2000; 2:20-29.

2. **Hengartner MO.** The biochemistry of apoptosis. *Nature.* 2000; 407:770-776.

3. **Grutter MG.** Caspases: key players in programmed cell death. *Curr Opin Struct Biol.* 2000; 10:649-655.

4. **Hay S, Kannourakis G.** A time to kill: viral manipulation of the cell death program. *J Gen Virol.* 2002; 83:1547-1564.

5.     **Sprick MR, Rieser E, Stahl H, Grosse-Wilde A, Weigand MA, Walczak H.** Caspase-10 is recruited to and activated at the native TRAIL and CD95 death-inducing signalling complexes in a FADD-dependent manner but can not functionally substitute caspase-8. *EMBO J.* 2002; 21:4520-4530.

6.     **Werner AB, de Vries E, Tait SW, Bontjer I, Borst J.** TRAIL receptor and CD95 signal to mitochondria via FADD, caspase-8/10, Bid, and Bax but differentially regulate events downstream from truncated Bid. *J Biol Chem.* 2002; 277:40760-40767.

7.     **Shikama Y, Shen L, Yonetani M, Miyauchi J, Miyashita T, Yamada M.** Death effector domain-only polypeptides of caspase-8 and -10 specifically inhibit death receptor-induced cell death. *Biochem Biophys Res Commun.* 2002; 291:484-493.

8.     **Lassus P, Opitz-Araya X, Lazebnik Y.** Requirement for caspase-2 in stress-induced apoptosis before mitochondrial permeabilization. *Science.* 2002; 297:1352-1354.

9.     **Budd RC.** Death receptors couple to both cell proliferation and apoptosis. *J Clin Invest.* 2002; 109:437-441.

10.    **Roy S, Nicholson DW.** Cross-talk in cell death signalling. *J Exp Med.* 2000; 192:F21-F25.

11.    **Sun XM, MacFarlane M, Zhuang J, Wolf BB, Green DR, Cohen GM.** Distinct caspase cascades are initiated in receptor-mediated and chemical-induced apoptosis. *J Biol Chem.* 1999; 274:5053-5060.

12.    **Wen J, You KR, Lee SY, Song CH, Kim DG.** Oxidative stress-mediated apoptosis. The anticancer effect of the sesquiterpene lactone parthenolide. *J Biol Chem.* 2002;277:38954-38964.

13.    **Pervaiz S, Clement MV.** Hydrogen peroxide-induced apoptosis: oxidative or reductive stress? *Methods Enzymol.* 2002; 352:150-159.

14.    **Curtin JF, Donovan M, Cotter TG.** Regulation and measurement of oxidative stress in apoptosis. *J Immunol Methods.* 2002; 265:49-72.

15.    **Akao M, Ohler A, O'Rourke B, Marban E.** Mitochondrial ATP-sensitive potassium channels inhibit apoptosis induced by oxidative stress in cardiac cells. *Circ Res.* 2001; 88:1267-1275.

16.    **Zanghi JA, Fussenegger M, Bailey JE.** Serum protects protein-free competent Chinese hamster ovary cells against apoptosis induced by nutrient deprivation in batch culture. *Biotechnol Bioeng.* 1999; 64:108-119.

17.    **Xu XN, Screaton GR, McMichael AJ.** Virus infections: escape, resistance, and counterattack. *Immunity.* 2001; 15:867-870.

18.    **Kumar S, Vaux DL.** Apoptosis: A cinderella caspase takes center stage. *Science.* 2002; 297:1290-1291.

19.    **O'Reilly LA, Ekert P, Harvey N, et al.** Caspase-2 is not required for thymocyte or neuronal apoptosis even though cleavage of caspase-2 is dependent on both Apaf-1 and caspase-9. *Cell Death Differ.* 2002; 9:832-841.

20.    **Finkel E.** The mitochondrion: is it central to apoptosis? *Science.* 2001; 292:624-626.

21.    **Roulston A, Marcellus RC, Branton PE.** Viruses and apoptosis. *Annu. Rev. Microbiol.* 1999; 53:577-628.

22.    **Jelachich ML, Lipton HL.** Restricted Theiler's murine encephalomyelitis virus infection in murine macrophages induces apoptosis. *J Gen Virol.* 1999; 80:1701-1705.

23.    **Jelachich ML, Lipton HL.** Theiler's murine encephalomyelitis virus kills restrictive but not permissive cells by apoptosis. *J Virol.* 1996; 70:6856-61.

24.    **Agol VI, Belov GA, Bienz K, et al.** Two types of death of poliovirus-infected cells: caspase involvement in the apoptosis but not cytopathic effect. *Virology.* 1998; 252:343-353.

25.    **Shaw-Jackson C, Michiels T.** Infection of macrophages by Theiler's murine encephalomyelitis virus is highly dependent on their activation or differentiation state. *J Virol.* 1997; 71:8864-8867.

26.    **Jelachich ML, Bramlage C, Lipton HL.** Differentiation of M1 myeloid precursor cells into macrophages results in binding and infection by Theiler's murine encephalomyelitis virus and apoptosis. *J Virol.* 1999; 73:3227-3235.

27.    **Lipton HL, Twaddle G, Jelachich ML.** The predominant virus antigen burden is present in macrophages in Theiler's murine encephalomyelitis virus-induced demyelinating disease. *J Virol.* 1995; 69:2525-2533.

28.    **Rossi CP, Delcroix M, Huitinga I, et al.** Role of macrophages during Theiler's virus infection. *J of Virol.* 1997; 71:3336-3340.

29.    **Obuchi M, Yamamoto J, Uddin N, Odagiri T, Iizuka H, Ohara Y.** Theiler's murine encephalomyelitis virus (TMEV) subgroup strain-specific infection in neural and non-neural cell lines. *Microbiol Immunol.* 1999; 43:885-892.

30.    **Martinat C, Mena I, Brahic M.** Theiler's virus infection of primary cultures of bone marrow-derived monocytes/macrophages. *J Virol.* 2002; 76:12823-33.

31.    **Ohara Y, Himeda T, Asakura K, Sawada M.** Distinct cell death mechanisms by Theiler's murine encephalomyelitis virus (TMEV) infection in microglia and macrophage. *Neurosci Lett.* 2002; 327:41-44.

32.    **Olson JK, Girvin AM, Miller SD.** Direct activation of innate and antigen-presenting functions of microglia following infection with Theiler's virus. *J Virol.* 2001; 75:9780-9.

33.    **Zheng L, Calenoff MA, Dal Canto MC.** Astrocytes, not microglia, are the main cells responsible for viral persistence in Theiler's murine encephalomyelitis virus infection leading to demyelination. *J Neuroimmunol.* 2001; 118:256-67.

34.    **Anderson R, Harting E, Frey MS, Leibowitz JL, Miranda RC.** Theiler's murine encephalomyelitis virus induces rapid necrosis and delayed apoptosis in myelinated mouse cerebellar explant cultures. *Brain Res.* 2000; 868:259-67.

35.    **Palma JP, Yauch RL, Lang S, Kim BS.** Potential role of CD4+ T cell-mediated apoptosis of activated astrocytes in Theiler's virus-induced demyelination. *J Immunol.* 1999; 162:6543-6551.

36.    **Rose JW, Hill KE, Wada Y, et al.** Nitric oxide synthase inhibitor, aminoguanidine, reduces inflammation and demyelination produced by Theiler's virus infection. *J Neuroimmunol.* 1998; 81:82-9.

37.    **Tsunoda I, Kurtz CI, Fujinami RS.** Apoptosis in acute and chronic central nervous system disease induced by Theiler's murine encephalomyelitis virus. *Virology.* 1997; 228:388-93.

38.    **Schlitt BP, Felrice M, Jelachich ML, Lipton HL.** Apoptotic cells, including macrophages, are prominent in Theiler's virus-induced inflammatory demyelinating lesions. *J Virol.* 2003: In Press.

39.    **Drescher KM, Murray PD, Lin X, Carlino JA, Rodriguez M.** TGF-beta 2 reduces demyelination, virus antigen expression, and macrophage recruitment in a viral model of multiple sclerosis. *J Immunol.* 2000; 164:3207-3213.

40.    **Kong WP, Roos RP.** Alternative translation initiation site in the DA strain of Theiler's murine encephalomyelitis virus. *J Virol.* 1991; 65:3395-9.

41.    **Ghadge GD, Ma L, Sato S, Kim J, Roos RP.** A protein critical for a Theiler's virus-induced immune system-mediated demyelinating disease has a cell type-specific antiapoptotic effect and a key role in virus persistence. *J Virol.* 1998; 72:8605-8612.

42.    **Takata H, Obuchi M, Yamamoto J, et al.** L* protein of the DA strain of Theiler's murine encephalomyelitis virus is important for virus growth in a murine macrophage-like cell line. *J Virol.* 1998; 72:4950-4955.

43.    **Obuchi M, Yamamoto J, Odagiri T, Uddin MN, Iizuka H, Ohara Y.** L* protein of Theiler's murine encephalomyelitis virus is required for virus growth in a murine macrophage-like cell line. *J Virol.* 2000; 74:4898-4901.

44.    **Neznanov N, Kondratova A, Chumakov KM, et al.** Poliovirus protein 3A inhibits tumor necrosis factor (TNF)-induced apoptosis by eliminating the TNF receptor from the cell surface. *J Virol.* 2001; 75:10409-20.

45.    **Rossi CP, McAllister A, Tanguy M, Kagi D, Brahic M.** Theiler's virus infection of perforin-deficient mice. *J Virol.* 1998; 72:4515-4519.

46.    **Martínez-Lorenzo MJ, Alava MA, Gamen S, et al.** Involvement of APO2 ligand/TRAIL in activation-induced death of Jurkat and human peripheral blood T cells. *Euro J of Immunol.* 1998; 28:2714-2725.

47.    **Zipp F.** Apoptosis in multiple sclerosis. *Cell Tissue Res.* 2000; 301:163-171.

48.    **Waiczies S, Weber A, Lunemann JD, Aktas O, Zschenderlein R, Zipp F.** Elevated Bcl-$X_L$ levels correlate with T cell survival in multiple sclerosis. *J Neuroimmunol.* 2002;126(1-2):213-220.

Part C

CORONA VIRUS-INDUCED DEMYELINATION

Barnett, E. M., M. D. Cassell, et al. (1993). "Two neurotropic viruses, herpes simplex virus type 1 and mouse hepatitis virus, spread along different neural pathways from the main olfactory bulb." Neuroscience **57**: 1007-1025.

Barthold, S. W. (1988). "Olfactory neural pathway in mouse hepatitis virus nasoencephalitis." Acta. Neuropathol. **76**: 502-506.

Barthold, S. W. and A. L. Smith (1983). "Mouse hepatitis virus S in weanling Swiss mice following intranasal inoculation." Lab. Anim. Sci. **33**: 355-360.

Buchmeier, M. J., H. A. Lewicki, et al. (1984). "Murine hepatitis virus-4 (strain JHM) - induced neurologic disease is modulated in vivo by monoclonal antibody." Virology **132**: 261-270.

Cheever, F. S., J. B. Daniels, et al. (1949). "A murine virus (JHM) causing disseminated encephalomyelitis with extensive destruction of myelin." J. Exp. Med. **90**: 181-194.

Fleury, H. J. A., R. D. Sheppard, et al. (1980). "Further ultrastractrual observations of morphogenesis and myelin pathology in JHM virus encephalomyelitis." Neuropath. Appl. Neurobiol. **6**: 165-179.

Herndon, R. M., D. E. Griffin, et al. (1975). "Mouse hepatitis virus induced recurrent demyelination." Arch. Neurol. **32**: 32-35.

Herndon, R. M., D. L. Price, et al. (1977). "Regeneration of oligodendroglia during recovery from demyelinating disease." Science **195**: 693-694.

Hirano, N., T. Murakami, et al. (1981). "Comparison of mouse hepatitis virus strains for pathogenicity in weanling mice infected by various routes." Arch. Virol. **70**: 69-73.

Knobler, R. L., M. V. Haspel, et al. (1981). "Mouse hepatitis virus type 4 (JHM strain)- induced fatal central nervous system disease, part 1 (genetic control and the murine neurone as the susceptible site for disease)." J. Exp. Med. **153**: 832-843.

Knobler, R. L., P. W. Lampert, et al. (1982). "Virus persistence and recurring demyelination produced by a temperature sensitive mutant of MHV-4." Nature **298**: 279-280.

Kristensson, K., K. V. Holmes, et al. (1986). "Increased levels of myelin basic protein transcript gene in virus-induced demyelination." Nature **322**: 544-547.

Lampert, P. W., J. K. Sims, et al. (1973). "Mechanism of demyelination in JHM virus induced encephalomyelitis, electron microscopic studies." Acta. Neuropathol **24**: 76-85.

Lavi, E., S. P. Fishman, et al. (1988). "Limbic encephalitis following inhalation of murine coronavirus MHV-A59." Lab. Invest. **58**: 31-36.

Lavi, E., D. H. Gilden, et al. (1984). "Persistence of MHV-A59 RNA in a slow virus demyelinating infection in mice as detected by in situ hybridization." J. Virol. **51**: 563-566.

Lavi, E., D. H. Gilden, et al. (1986). "The organ tropism of mouse hepatitis virus A59 is dependent on dose and route of inoculation." Lab. Anim. Sci. **36**: 130-135.

Lavi, E., D. H. Gilden, et al. (1984). "Experimental demyelination produced by the A59 strain of mouse hepatitis virus." Neurology **34**: 597-603.

Lavi, E., A. Suzumura, et al. (1989). "Induction of MHC class I antigens on glial cells is dependent on persistent mouse hepatitis virus infection." J. Neuroimmunol. **22**: 107-111.

Massa, P. A., R. Dorries, et al. (1986). "Viral particles induce Ia antigen expression on astrocytes." Nature **320**: 543-546.

Perlman, S., G. Jacobsen, et al. (1990). "Identification of the spinal cord as a major site of persistence during chronic infection with a murine coronavirus." Virology 175: 418-426.

Robb, J. A., C. W. Bond, et al. (1979). "Pathogenic murine coronaviruses. III. Biological and biochemical characterization of temperature sensitive mutants of JHMV." Virology 94: 385-399.

Sorensen, O., R. Durge, et al. (1982). "In vivo and in vitro models of demyelinating disease: endogenous factors influencing demyelinating disease caused by mouse hepatitis virus in rats and mice." Infect. Immun. 37: 1248-1260.

Stohlman, S. A. and L. P. Weiner (1981). "Chronic central nervous system demyelination in mice after JHM virus infection." Neurology 31: 38-44.

Sun, N., D. Grzybicki, et al. (1995). "Activation of astrocytes in the spinal cord of mice chronically infected with a neurotropic coronavirus." Virology 213: 482-493.

Sun, N. and S. Perlman (1995). "Spread of a neurotropic coronavirus to spinal cord white matter via neurons and astrocytes." J. Virol. 69: 633-641.

Suzumura, A., E. Lavi, et al. (1988). "Induction of glial cell MHC antigen expression in neurotropic coronavirus infection: characterization of the H-2 inducing soluble factor elaborated by infected brain cells." J. Immunol. 140: 2068-2072.

Suzumura, A., E. Lavi, et al. (1986). "Coronavirus infection induces H-2 antigen expression on oligodendrocytes and astrocytes." Science 232: 991-993.

Virelizier, J. L., A. D. Dayan, et al. (1975). "Neuropathological effects of persistent infection of mice by mouse hepatitis virus." Infect. Immn. 12: 1127-1140.

Wang, F. I., S. A. Stohlman, et al. (1990). "Demyelination induced by murine hepatitis virus JHM strain (MHV-4) is immunologically mediated." J. Neuroimmunol. 30: 31-41.

Watanabe, R., H. Wege, et al. (1984). "Adoptive transfer of EAE-like lesion from rats with coronavirus-induced demyelinating encephalomyelitis." Nature 305: 150.

Wege, H., S. Siddell, et al. (1982). "The biology and pathogenesis of coronaviruses." Adv. Virol. Immunol. 99: 165-200.

Weiner, L. P. (1973). "Pathogenesis of demyelination induced by a mouse hepatitis virus (JHM virus)." Arch. Neurol. 28: 298-303.

Chapter  C2

# THE ROLE OF ASTROCYTES, MICROGLIA, AND ENDOTHELIAL CELLS IN CORONAVIRUS-INDUCED DEMYELINATION:

*Induction of cytokines and other signaling mechanisms*


Yun Li and Ehud Lavi[*]

*Division of neuropathology, Department of Pathology and Laboratory  Medicine, University of Pennsylvania, School of Medicine, Philadelphia, PA 19104.  *Present address: Department of Pathology and Laboratory Medicine, Weill Medical College of Cornell University, New York, NY 10021*



Abstract:    Infection of neurotropic strains of coronaviruses in susceptible animals results in acute encephalomyelitis followed by a chronic demyelinating disease, similar to multiple sclerosis (MS).   Although the mechanism of chronic coronavirus-induced demyelination is not entirely clear, studies show that it is mostly immune-mediated. Astrocytes, microglia and endothelial cells play an important role in normal functions as well as in immunologic and pathologic processes in the central nervous system (CNS).   The interaction between coronaviruses and these cells induces various inflammatory mediators including cytokines, chemokines, MHC and NO, which acting in concert are involved in the pathogenesis of demyelination.


Key words:    astrocytes, microglia, endothelial cells, coronavirus, demyelination

## 1. INTRODUCTION

Astrocytes, microglia and endothelial cells play an important role in the pathogenesis of coronavirus, mouse hepatitis virus (MHV)-induced disease in mice in several ways.  All three cell-types are targets to various degrees of MHV infection during acute and chronic disease.  Secondly, astrocytes and microglia are also targets of apoptotic cell death, presumably as a direct consequence of infection.  In addition, astrocytes and microglia serve as reactive proliferating cells, as normally seen in viral encephalitis and other types of CNS injury.  Finally all three cell-types are considered part of the CNS innate immune system, and as such are responsible for proinflammatory immune reactions. These reactions involve secretory functions of a variety of pro-inflammatory molecules including cytokines, chemokines, upregulation of major histocompatibility complex (MHC) and NO molecules and a variety of downstream signal transduction and related molecules. CNS disease, especially demyelination could be a result of a combination of viral cytopathic effect on cells and the secondary pathologic effect of toxic inflammatory mediators, thus fulfilling the concept of autoimmunity. Indirect proof for this concept comes from studies in which immuno-suppression ameliorated chronic MHV-induced demyelination. In the following paragraphs we will briefly review the normal functions of astrocytes, microglia and endothelial cells and evaluate how these functions are affected in relationship to the pathogenesis of virus-induced CNS disease.

## 1.1. Astrocytes

Historically two groups of glial cells were identified in the CNS: the macroglia, including astrocytes, oligodendrocytes, and ependymal cells, and microglia. More recently microglial cells have been identified as bone marrow derived and not of neuroectodermal origin, and therefore cannot be considered as true glial cells.

Astrocytes, the main CNS glial cells have a number of important physiological properties related to CNS homeostasis. Astrocytes have a dynamic role in regulating neuronal function (Sofroniew et al. 1999; Haydon 2000) by the release of neurotrophic factors, guidance of neuronal development, contributing to the metabolism of neurotransmitters, and regulating extracellular pH and $K^+$ levels. Astrocytes influence the formation and maintenance of the blood-brain barrier (BBB), a structure that serves to limit entry of blood-borne elements into the CNS. Astrocytic foot processes are in close apposition to the abluminal surface of the microvascular endothelium of the BBB, and soluble factors secreted by astrocytes appear to be involved in the maintenance of the BBB. Thus, astrocytes contribute to both the structural and functional integrity of the BBB (Wolburg-H 1995).

Astrocytes also serve as one of the immune effecter cells in the CNS. Astrocytes play a significant role during infectious and autoimmune diseases of the CNS. As part of the BBB, astrocytes are in close proximity to endothelial cells and secrete cytokines, chemokines and adhesion molecules (Lee et al. 2000). Cultured astrocytes express a wide range of molecules with neurotropic properties such as nervous growth factor (NGF), glial-derived growth factor (GDGF), and ciliary neurotropic factor (CNTF). In response to activation by interferons astrocytes express class I and class II MHC antigens. The expression of the costimulatory molecules B7 and

CD40 in astrocytes is controversial (Aloisi et al. 1998; Soos et al. 1999; Nguyen and Benveniste 2000).

The ability of astrocytes to function as bona fide antigen presenting cells (APCs) is also controversial. Early studies documented the ability of class II MHC-positive astrocytes to function as APCs in vitro (Fierz et al. 1985; Fontana et al. 1986; Takiguchi and Frelinger 1986). These studies clearly indicate that astrocytes exposed to IFN-γ have the capacity to express molecules such as class II MHC, invariant chain, H2-M, B7-1, and B7-2, enabling them to efficiently process and present self antigens and activate both naive and memory T cells. However, there is also a large body of evidence indicating that astrocytes function as nonprofessional APCs by promoting mainly Th2 responses and/or apoptosis of T cells, which may be important for recovery from Th1-mediated CNS inflammation (Meinl et al. 1994; Weber et al. 1994).

## 1.2. Microglia

Microglia constitutes 12% of the cells in the CNS, which are the principal immune cells in the CNS and play a significant role in the host defense against invading microorganisms. The CNS endogenous microglia share many properties with macrophages, having developed from a common precursor cells (Nelson et al. 2002). Features common to microglia and systemic macrophages include the expression of innate immune receptors and the ability to phagocytose pathogens, cells or cellular debris (Ling and Wong 1993; Williams et al. 1994).

The functions of microglia in the CNS include phagocytosis, antigen presentation, production and release of cytokines, eicosanoids, complement components, and excitatory amino acids such as glutamate, oxidative radicals, and nitric oxide (Gehrmann et al. 1995). Three states of microglia have been identified based on developmental and pathophysiologic studies: (1) the resting, ramified microglia present in normal CNS; (2) the activated non-phagocytic microglia found in areas involved in CNS inflammation (3) the reactive, phagocytic microglia observed in areas of trauma and infection. Microglia can express class I and II MHC antigens, Fc receptors (I-III), complement receptors (CR1, CR2, CR4), β2-integrins, intercellular adhesion molecule-1 (ICAM-1), and costimulatory molecules B7-1 and B7-2 (Minagar et al. 2002). Numerous studies have confirmed the role of microglia as important APCs within the CNS (Aloisi et al. 2000; Becher et al. 2000), while the role of   astrocytes remains controversial.

## 1.3. Endothelial cells

Traditionally, the CNS has been regarded as an immunologically privileged site. The CNS lacks lymphatic vessels and is sealed by the BBB.

The cellular basis for the BBB is at the levels of both the CNS microvascular endothelial cells and the choroid plexus epithelial cells. The BBB is an important mechanism for protecting the brain from fluctuations in plasma composition and from circulating agents such as neurotransmitters and xenobiotics capable of disturbing neural function (Abbott and Romero 1996). The barrier also plays an important role in the homeostatic regulation of the brain microenvironment necessary for the stable and coordinated activity of neurons (Abbott 2002). The brain endothelium has lower levels of endocytosis/transcytosis than peripheral capillaries, but has a number of specific transport and enzyme systems, which regulate molecular traffic across the endothelial cells. The brain endothelium also contains specific enzymatic systems such as monoamine oxidase that support the protective and detoxifying roles of the BBB. Cerebral microvascular endothelial cells are also considered as potential APCs because of their large cumulative surface and their unique anatomical location between circulating T cells and the extra-vascular sites of antigen exposure. Thus BBB covers a number of static and dynamic properties that enable the endothelium to protect and regulate the brain microenvironment (Abbott and Romero 1996).

The BBB phenotype develops under the influence of associated brain cells, especially astrocytes. *In vitro* cell culture models have provided a great deal of information about the key role of astrocytes in the induction of the BBB phenotype in brain endothelium (Bauer and Bauer 2000). The chemical nature of the glial-produced inductive signal(s) is currently unclear. TGF-β (Tran et al. 1999), GDNF (Utsumi et al. 2000), bFGF (Sobue et al. 1999), IL-6 and hydrocortisone (Hoheisel et al. 1998) have been shown to modulate endothelial differentiation and induction. In addition to a role in barrier induction and maintenance, astrocytes may play active roles in modulating BBB permeability by the release of several humoral agents including glutamate, aspartate, taurine, ATP, ET-1, NO, TNFα, MIP2 and IL-1β, although the regulation of this release is not well understood (Abbott 2002). Interestingly, it has recently been shown that endothelial cells induce the differentiation of astrocyte precursor cells into astrocytes in vitro, which is mediated by endothelial cell production of leukemia inhibitory factor (LIF) (Mi et al. 2001). Thus endothelium and astrocytes are involved in two-way induction. It is clear that endothelial cells are involved in both long- and short-term chemical communication with neighboring cells, with astrocytes being of particular importance.

Maintenance of the adult BBB appears to depend on continuing exchange of inductive signals between glia and endothelium, and disturbance of this induction may be instrumental in several neuropathologies involving BBB dysfunction, such as tumors and MS.

## 2. Animal models of coronavirus induced demyelination

Coronaviruses constitute a large group of positive-stranded RNA viruses that are associated with a wide variety of respiratory, gastrointestinal, and neurological diseases in animals and humans. However there is no conclusive evidence that any human neurological diseases, such as MS, may occur as a result of human coronavirus infection.

MS is an inflammatory demyelinating disease of the CNS that is characterized by mononuclear cell infiltration into the CNS, myelin degradation and oligodendrocyte loss. It is the most common neurological disease affecting young adults (Ewing and Bernard 1998). The etiology and pathogenesis of MS have yet to be elucidated but are probably multifactorial, involving both genetic and environmental factors (Ewing and Bernard 1998). Several viruses have been associated with demyelinating processes including human coronaviruses (HCoV) and murine coronavirus (MHV). Intracerebral infection of susceptible mice with neuroadapted strains of mouse hepatitis virus (MHV) results in an acute encephalomyelitis followed by a chronic demyelinating disease similar to the pathology of the human MS (Lavi et al. 1984; Lavi et al. 1986; Houtman and Fleming 1996; Lane and Buchmeier 1997), serving as models for this diseases. The two strains of MHV that have been used in the majority of studies of MHV infection in the CNS are MHV-A59 and JHM (MHV-4).

The mechanism of MHV induced demyelination is still not well understood. Although it was initially believed to result from direct viral lysis of oligodendrocytes, chronic demyelination was demonstrated to be largely immune mediated in more recent reports. The immune response to MHV infection is critical in host defense as well as the development of demyelination (Houtman and Fleming 1996; Wu et al. 2000). However, the precise role of individual components of the immune system in this process is not known and need intense investigation.

## 3. The role of astrocytes, microglia, and endothelial cells in MHV-induced CNS demyelination

The fact that the nude mice were unable to clear virus and still develop demyelination suggests that conventionally educated T cells are not an essential component for demyelination. That means that either the $\gamma\delta$ subset of T cells, natural killer cells, or microglia, astrocytes, or even endothelial cells, may participate and/or be responsible for the demyelinating process.

### 3.1 The innate immune system of the CNS

To protect the host from the invasion of foreign organisms and pathogenic insults, the immune system has evolved into two parts: one is responsible for immediate, relatively generic, action against external agents, known as the "the innate immune system". The other component of the immune system responds specifically to an external threat and requires plasticity and memory, which is called the "adaptive or acquired immune system". These two systems are not separate, but are functionally intertwined and the actions of one have a profound effect on the other (Medzhitov and Janeway 1997).

B and T lymphocytes are the cellular members of the adaptive immune system. These cells are generated in primary lymphoid tissues (bone marrow, thymus) and re-circulated in secondary lymphoid structures (lymphnode, spleen). The innate immune system on the other hand uses mainly phagocytic cells (including monocytes/macrophages, and polymorphonuclear phagocytes) as a first line of defense against foreign structures. Astrocytes and microglia are the principal innate immune cells in the CNS concomitant with inflammatory brain disease and play a significant role in the host defense against invading microorganisms.

## 3.2 The immune response within the CNS upon viral infection

Although the CNS has been characterized as an immune privileged site, it is also the site of blood-borne inflammation, either in response to exogenous antigens (infection) or as a result of disrupted peripheral tolerance to self antigens (autoimmnunity). The microenvironment of the CNS, shaped by its structural and cellular components, participates in the physiological immune process that occurs within the CNS and is at least in part responsible for the extent of immune response within the CNS upon viral infections. Glial and microglial cells play a pivotal role in both acute and chronic phases of coronavirus infection, which could lead to the production of a variety of inflammatory molecules in glial cells including cytokines, chemokines, MHC and nitric oxide (NO) that have been associated with the pathogenesis of demyelination.

## 3.4. MHC class I and II expression

The MHC class I and II antigens play an important role in demyelination following infection by MHV of resident cells of the CNS. Infection of the CNS with MHV-A59, a neurotropic murine coronavirus, induces class I MHC antigen expression on oligodendrocytes and astrocytes (Suzumura et al. 1986; Suzumura et al. 1988; Lavi et al. 1989), cells that do not normally express detectable MHC surface antigens. The role of MHC class I in demyelination remains complex because of reports indicating that demyelination can occur in MHV-infected mice that lack either stable expression of the MHC class I molecule or functional CD8+ T cells (Gombold et al. 1995). Because MHC class I antigens play a key role in interaction between cytotoxic T cells and target cells, the induction of MHC class I antigens could potentially allow glial cells to interact with or become a target for immunocytes.

The MHC class II molecules play a critical role in induction of immune responses through presentation of processed antigens to CD4$^+$ T-helper cells. Although class II MHC molecules are essential for lymphocyte development, antigen presentation, and T-cell activation, inappropriate class II expression has been implicated in several autoimmune diseases, including rheumatoid arthritis, inflammatory bowel disease, and MS (Grusby and Glimcher 1995). Class II MHC molecules are normally expressed on professional APCs, such as B cells, macrophages, dendritic cells, and thymic epithelium, expression on other cell types, including astrocytes, can be induced and/or regulated by cytokines, neurotransmitters, and neuropeptides. Astrocytes were the first CNS cell type shown to express class II MHC molecules upon IFN-γ stimulation in vitro (Wong et al. 1984). The cytokine TNF-α, while having no influence alone on class II MHC expression, enhances IFN-γ-induced class II MHC expression on astrocytes (Panek et al. 1994). Infection of astrocyte cultures derived from MHV-susceptible Lewis rats with MHV-JHM resulted in induction of MHC class II, whereas no such induction was observed in astrocytes from disease-resistant Brown Norway rats (Massa et al. 1986; Massa et al. 1986). A number of soluble mediators have been shown to inhibit class II MHC expression in astrocytes, which include TGF-β, IL-1β, IFN-β, IL-4, glutamate, vasoactive intestinal peptide, norepinephrine, and nitric oxide (Rohn et al. 1996). The expression of MHC II on phagocytic microglia implies the capacity of microglia as APCs. Phagocytic microglia is often regarded as potentially cytotoxic in both acute and chronic CNS disorders. MHC class II expression by phagocytic microglia also corresponds to the elevated expression of potentially cytotoxic substances in the CNS, among which nitric oxide and TNF-α are notably involved in the death of oligodendrocytes and demyelination (Benveniste 1997; Eugster et al. 1999). A study on the demyelinating *twitcher* mouse has shown that the number of reactive microglia and the concomitant demyelination is significantly reduced when the *twitcher* mouse carries an MHC II null background (Matsushima et al. 1994). Therefore, MHC II expression by activated microglia in chronic neurodegenerative disorders may indicate that the activated microglia become cytotoxic (Zhang et al. 2001).

Expression of MHC class II has been shown to be important in demyelinating disease in some animal models, such as TMEV and EAE (Borrow and Nash 1992). However, mice deficient in MHC class II expression are capable of having demyelination following infection with MHV-JHM (Houtman and Fleming 1996). Therefore, the expression of MHC class I or II may play some role in the MHV-induced demyelination, but neither one is absolutely necessary for this process.

## 3.5. Cytokines and chemokines production

Astrocytes and microglia produce a variety of cytokines and chemokines upon stimulation with a variety of factors including viral infection. Although expression of cytokines and chemokines is controlled and balanced in the normal, healthy, CNS, aberrant expression occurs in CNS diseases such as AD, MS, HAD, Parkinson's disease, and in brain injury/trauma. In many of these diseases, astrocytes and microglia are responsible for some of the production of cytokines and chemokines, whereas blood-borne inflammatory cells are responsible for the rest. Both *in vitro* and *in vivo* studies have documented the ability of astrocytes to produce interleukin-1, -6, and -10 (IL-1, -6, and -10); interferon (IFN)-α, and -β; colony-stimulating factors GM-CSF, M-CSF, and G-CSF; TNF-α; TGF-β; and chemokines including RANTES, IL-8, monocyte chemoattractant protein-1 (MCP-1), and IFN-γ-inducible protein-10 (IP-10) (Dong and Benveniste 2001). Cytokines that have been demonstrated to be produced by microglia are IL-1, 2, 3, 4, 6, 8, 10, IFN-γ, TNF-α, TGF-β, and CSF. Cytokine/chemokine communications between microglia and astrocytes are involved in the balance of protective and destructive actions by these cells. Both cell types may play a dual role, amplifying the effects of inflammation and mediating cellular damage as well as protecting the CNS. Thus, interactions between T lymphocytes, microglia, and astrocytes play a major role in the pathogenesis of nerological diseases such as MS.

Chemokines are small chemotactic cytokines that modulate leukocyte recruitment and activation during inflammation. Chemokines have been implicated in a variety of normal CNS functions (i.e., neuronal progenitor migration, axon guidance and adhesion, oligodendrocyte proliferation, and intercellular communication) although more and more evidence supports their role in CNS disease and injury such as Alzheimer's disease and HAD. In particular, chemokines are induced in the demyelinating disease MS, and numerous animal models for demyelination including experimental autoimmune encephalomyelitis (EAE), Theiler's and MHV-induced demyelinating disease, experimental autoimmune neuritis, and twitcher, a murine model of globoid cell leukodystrophy (Glabinski and Ransohoff 1999; Hoffman et al. 1999; Kieseier et al. 2000).

Many cells in the body can secrete chemokines, including astrocytes and microglia. MHV infection of the CNS results in an orchestrated expression of chemokine genes including IFN inducible protein of 10 kDa/ CXCL10, monokine induced by IFN-γ/CXCL9, CCL2, CRG-2, MCP-1, MCP-3, RANTES, Mig, macrophage inflammatory protein-1β/CCL4, CCL5, and CCL7 (Lane et al. 1998). Neutralization of IP-10 or Mig at the time of infection with MHV results in increased mortality, higher viral loads and markedly decreased T cell infiltration into the CNS (Liu et al. 2000; Liu et al. 2001). In contrast, mice treated with RANTES antiserum have delayed viral clearance, decreased T cell infiltration, and significantly less demyelination than do untreated mice, but show no change in mortality (Lane et al. 2000). Also certain chemokine receptors including CCR1, CCR2 and CCR5, have been shown to regulate the gateway for inflammatory cell entry into the CNS. CCR1-deficient mice have reduced CNS inflammatory responses in the MOG-induced EAE model (Rottman et al. 2000), and CCR5-deficient mice have negligible inflammation after disseminated *C. neoformans* infection of the CNS (Huffnagle et al. 1999). Therefore, chemokines have an important role in host defense as well as demyelination induced by viral infection by attracting T lymphocytes and macrophages into the CNS.

Cytokines are a large and diverse group of polypeptides ranging in size from 8 to 26 kDa, which major function is activation of the immune system and inflammatory responses. In most acute encephalitis cytokines such as IL-1, TNF-α, IL-6 and IFN-γ are detected in the CNS of infected mice. Depletion of IFN-γ leads to decreased virus clearance and greater mortality. However, neutralization of TNF-α did not appear to affect either T cell recruitment or virus clearance. Except for astrocytes and microglia, edothelial cells are another possible cellular source of cytokine including IL-1, IL-6 and TNF-α, as well as some cytokines receptors IL-1R1 and IL-1R2 .

In our laboratory we analyzed the ability of MHVs with different neurotropic phenotypes to modulate the expression of cytokines in astrocytes and microglia in primary cultures from newborn C57BL/6 mice. The results show that infections of cultures with neurotropic viruses MHV-A59 and JHM up-regulate IL-1β, IL-6, IL-12p40, IL-15 and TNF-α mRNA in both astrocytes and microglia, significantly more than infection with the non-neurotropic virus MHV-2. Similar up-regulation was observed in mouse brains during acute encephalitis and in mouse spinal cords during the chronic inflammatory demyelinating disease but a different profile was seen in the liver during acute hepatitis. There are close interactions between the five upregulated cytokines. Studies of MHV-A59 infections in cultures derived from cytokine knockout mice indicate that IL-12 and TNF-α are the upstream molecules that cause the subsequent up-regulation of the other three cytokines. Furthermore we found that infection with neurotropic MHVs induce up-regulation of a set of chemokines in astrocytes and microglia, including MIP-1, MIP-3, MCP-2, MCP-5, MIG. This pattern of cytokine/chemokine response in astrocytes and microglia is similar to the Th1 response of lymphocytes and appears to be associated with the neurotropic properties of the virus strain. These studies allow us to study the unique contribution of CNS local immune cells to the process of MHV disease and thus provide new insights into the pathogenesis of MHV-induced neurological diseases as well as the general role of brain immune cells and cytokines in the process of immune-mediated diseases of the CNS (Li et al. 2004).

Several reports studied the role of cytokines in the chronic demyelinating process. The cytokines IL-1β, TNF-α, IL-6 are expressed by astrocytes which are located near the sites of demyelination in chronically infected spinal cords (Perlman 1998). A wide range of cytokines have been detected in MS lesions, and IL-1α and β, IL-6, TNF-α, and IFN-γ may participate in the CNS demyelination process (Benveniste 1998). Cytokines such as IFN-γ, TGF-β, IL-1, IL-4, and IL-10 may regulate class II MHC expression (Han et al. 1999; O'Keefe et al. 1999). Cytokines induce increased expression of chemokines, stimulate adhesion molecules on endothelial cells, and allow leukocyte entry into the CNS. Thus cytokines are involved in the initial phase as well as the development stages of demyelination.

Studies using knockout mice have shown that neither perforin, Fas-FasL, TNF-α, IL-10 or IFN-γ are required for demyelination (Haring and Perlman 2001). Our own studies in knockout mice indicate that deficiency of IL-6 and IL-12p40 does not decrease the level of demyelination induced by MHV-A59 infection compared to wide type C57BL/6 mice. Thus, none of the single cytokines plays a crucial role in demyelination, however, we still did not rule out that a combination of cytokines may be required for demyelination. Chemokines may play a role in demyelination. RANTES is a proinflammatory chemokine that act as a chemoattractant for a variety of lymphocytic and myeloid cell types including monocytes and granulocytes (Lane et al. 2000). Inhibition of RANTES has been shown to reduce MHV induced demyelination (Lane et al. 1999). An alternative hypothesis is that MHV could potentially damage oligodendrocytes by disrupting the function of astrocytes, which in turn affect oligodendrocytes (Sun and Perlman 1995).

## 3.6. Regulation of Th1/Th2 development

Activation of CD4[+] T-helper (Th) cells within the CNS plays an important role in regulating immune responses, inflammation and ultimately repair, during a variety of CNS diseases. The types of Th cells are distinguished by the profiles of cytokines that they produce. Th1 cells produce the cytokines IL-2, IL-12, IFN-γ, LT-α, and tumor necrosis factor (TNF), leading to macrophage activation, inflammation, and tissue damage (Rengarajan et al. 2000). Th1 cells have been implicated in the pathogenesis of CNS autoimmune diseases such as MS and EAE (Owens et al. 2001).  In contrast, Th2 cells produce the cytokines IL-3, IL-4, IL-5, IL-6, IL-7, IL-10, IL-11, and IL-13, that mediate humoral immune responses and inhibit macrophage functions (Rengarajan et al. 2000).  Within the CNS, Th2 type cytokines play a role in down-regulating Th1 responses and macrophage activation.  The presence of IL-12 and IFN-γ facilitates Th1 development; while IL-4 promotes Th2 differentiation.

Cytokines are the most prominent factors determining the development and polarization of Th cells.  However these cytokines are not produced by T cells but by APCs such as DCs, B lymphocytes and macrophages. Astrocytes and microglia are also the source of cytokines within the CNS that can influences Th1 versus Th2 responses (Aloisi et al. 1997; Stalder et al. 1997).  According to our own studies IL-12, the crucial cytokine facilitating Th1 development, can be produced by astrocytes/microglia upon MHV infection in vitro. Regulation of class II MHC, B7, and CD40 expression are also involved in the T cells polarization.

## 3.7. NOS activation and NO production

Nitric oxide (NO) is a pleiotropic molecule with important functions in vascular regulation, neuronal function, and immunological processes.  Of the three isoforms of nitric oxide synthase (NOS), NOS2 is recognized as an important inflammatory mediator, with both protective and immunopathological capabilities (Nathan and Xie 1994).  Macrophage production of NO by NOS2 is part of the effector phase of the immune response to numerous pathogens, including viruses (MacMicking et al. 1997; Reiss and Komatsu 1998).  NOS2-generated NO may contribute to the pathology of demyelination by exerting cytotoxic effects on oligodendrocytes (Merrill et al. 1993) and by regulating proinflammatory factors such as cytokine and chemokine secretion (Remick and Villarete 1996; Brenner et al. 1997; Merrill and Murphy 1997).

Both astrocytes and microglia express NOS-2 and release NO upon virus stimulation in vitro.  Moreover, these cells are the likely source of NOS-2 expression during CNS inflammation *in vivo* (Mitrovic et al. 1996).  During the acute phase of MHV infection of mice, expression of NOS2 by macrophages is up-regulated, whereas NOS2 synthesis is confined to astrocytes during the chronic demyelinating disease (Sun et al. 1995; Grzybicki et al. 1997).

The role of NO and NOS2 in MHV-induced demyelination is still controversial.  Inducible NOS2 transiently contributes to MHV-induced demyelination.  Inhibition of NOS2/NO by aminoguanidine (AG), a selective inhibitor of NOS2 activity, slows the progression of MHV-induced demyelination by controlling inflammation and modulating chemokine expression in the CNS (Lane et al. 1999). More recent studies show that NOS2 function is not required for demyelination in mice infected with MHV-JHM (Wu et al. 2000). Although NO may have a transient role in the early steps of demyelination, it is not necessary for disease to develop. Infection of NOS2 knockout (-/-) and NOS2(+/+) mice with MHV resulted in similar kinetics of viral clearance from the brain and comparable levels of demyelination. MHV-infected NOS2(-/-) mice displayed a marked decrease in mortality as compared to infected NOS2(+/+) mice, which correlated with a significant decrease in the number of apoptotic cells present in the brain. These studies indicate that NOS2-generated NO contributes to apoptosis of neurons but not demyelination following MHV infection (Chen and Lane 2002).

Of note, MHV-induced demyelination differs from that observed in other model systems since inhibition of NO production by AG ameliorates demyelination in rodents with adoptive EAE and mice infected with Theiler's murine encephalomyelitis virus (Cross et al. 1994; Rose et al. 1998).

## 3.8. The disruption of the BBB

The BBB has been assumed to be an effective barrier for blood cells. However, during viral infections or autoimmune diseases of the CNS, T lymphocytes are activated and migrate into the CNS and initiate the cellular events leading to inflammation and demyelination within the CNS white matter. The endothelial BBB has been considered the obvious place of entry of circulating lymphocytes into the CNS. The vascular endothelium is the first element that leukocytes encounter. Immune-specific interactions between lymphocytes and cerebral endothelium have been excluded as an initial mechanism. Adhesion interaction between T cells and endothelia must precede transmigration (Hart and Fabry 1995). Certain adhesion molecules seem to be critically involved in this process for example intercellular adhesion molecule 1(ICAM-1) and vascular cell adhesion molecule 1(VCAM-1) are up-regulated on the endothelial cells. Dissociated cultures of cerebral endothelial cells prepared from surgical resections of adult CNS tissue express an array of adhesion molecules whose expression is modulated by encounter with T cells or the cytokines they produce (Calabresi et al. 1997). Chemical mediators, particularly chemokines, regulate the synthesis, surface expression, and avidity of adhesion molecules. The most important adhesion molecule pairs are the selectins (E, L and P), the immunoglobulins ICAM-1 and VCAM-1, and the beta 2 and beta 1 integrins (e.g., LFA-1 and VLA-4), which play a role in a number of pathological processes (Cotran and Mayadas-Norton 1998). Various inflammatory mediators, primarily pro-inflammatory cytokines including

TNF-$\alpha$ and IL-1$\beta$, activate endothelial cells. Endothelial cells can also be induced to express MHC class II and B7 in vitro (Prat et al. 2000). Furthermore, they inhibit antigen presentation by other APC in co-cultures. The basis for this inhibition is still unknown.

The BBB may have a protective role against spreading of MHV-A59 into the CNS, by specific restriction of viral entry into endothelial cells of cerebral origin (Godfraind et al. 1997).

## 4. CONCLUSIONS

The effector mechanisms of coronavirus-induced demyelination are complex and largely redundant. The brain resident cells respond to infection by coronavirus by producing various inflammatory mediators including cytokines, chemokines, MHC and NO, which could act in concert to orchestrate an inflammatory pathology in the CNS. This argues for an indirect immunologic mechanism by which coronavirus infection could be implicated in a pathogenesis of the CNS diseases. However, one can conclude from all these data that no single molecule is absolutely required for coronavirus-induced demyelination, neither cytokines, MHC class I or II, nor NOS2. Furthermore, no specific immune cell population is essential for coronavirus-induced demyelination, neither CD4 T cells, CD8 T cells, nor hematogenous macrophages (Gombold et al. 1995; Sutherland et al. 1997). Therefore, coronavirus-induced demyelination may be a result of multiple mechanisms, which still need further investigation.

## References

Abbott, N. J. (2002). "Astrocyte-endothelial interactions and blood-brain barrier permeability." J Anat **200**(6): 629-38.
Abbott, N. J. and I. A. Romero (1996). "Transporting therapeutics across the blood-brain barrier." Mol Med Today **2**(3): 106-13.
Aloisi, F., G. Penna, et al. (1997). "IL-12 production by central nervous system microglia is inhibited by astrocytes." J Immunol **159**(4): 1604-12.
Aloisi, F., F. Ria, et al. (2000). "Regulation of T-cell responses by CNS antigen-presenting cells: different roles for microglia and astrocytes." Immunol Today **21**(3): 141-7.

Aloisi, F., F. Ria, et al. (1998). "Microglia are more efficient than astrocytes in antigen processing and in Th1 but not Th2 cell activation." <u>J Immunol</u> **160**(10): 4671-80.

Bauer, H. C. and H. Bauer (2000). "Neural induction of the blood-brain barrier: still an enigma." <u>Cell Mol Neurobiol</u> **20**(1): 13-28.

Becher, B., A. Prat, et al. (2000). "Brain-immune connection: immuno-regulatory properties of CNS-resident cells." <u>Glia</u> **29**(4): 293-304.

Benveniste, E. N. (1997). "Role of macrophages/microglia in multiple sclerosis and experimental allergic encephalomyelitis." <u>J Mol Med</u> **75**(3): 165-73.

Benveniste, E. N. (1998). "Cytokine actions in the central nervous system." <u>Cytokine Growth Factor Rev</u> **9**(3-4): 259-75.

Borrow, P. and A. A. Nash (1992). "Susceptibility to Theiler's virus-induced demyelinating disease correlates with astrocyte class II induction and antigen presentation." <u>Immunology</u> **76**(1): 133-9.

Brenner, T., S. Brocke, et al. (1997). "Inhibition of nitric oxide synthase for treatment of experimental autoimmune encephalomyelitis." <u>J Immunol</u> **158**(6): 2940-6.

Calabresi, P. A., L. R. Tranquill, et al. (1997). "Increases in soluble VCAM-1 correlate with a decrease in MRI lesions in multiple sclerosis treated with interferon beta-1b." <u>Ann Neurol</u> **41**(5): 669-74.

Chen, B. P. and T. E. Lane (2002). "Lack of nitric oxide synthase type 2 (NOS2) results in reduced neuronal apoptosis and mortality following mouse hepatitis virus infection of the central nervous system." <u>J Neurovirol</u> **8**(1): 58-63.

Cotran, R. S. and T. Mayadas-Norton (1998). "Endothelial adhesion molecules in health and disease." <u>Pathol Biol (Paris)</u> **46**(3): 164-70.

Cross, A. H., T. P. Misko, et al. (1994). "Aminoguanidine, an inhibitor of inducible nitric oxide synthase, ameliorates experimental autoimmune encephalomyelitis in SJL mice." <u>J Clin Invest</u> **93**(6): 2684-90.

Dong, Y. and E. N. Benveniste (2001). "Immune function of astrocytes." <u>Glia</u> **36**(2): 180-90.

Eugster, H. P., K. Frei, et al. (1999). "Severity of symptoms and demyelination in MOG-induced EAE depends on TNFR1." <u>Eur J Immunol</u> **29**(2): 626-32.

Ewing, C. and C. C. Bernard (1998). "Insights into the aetiology and pathogenesis of multiple sclerosis." <u>Immunol Cell Biol</u> **76**(1): 47-54.

Fierz, W., B. Endler, et al. (1985). "Astrocytes as antigen-presenting cells. I. Induction of Ia antigen expression on astrocytes by T cells via immune interferon and its effect on antigen presentation." <u>J Immunol</u> **134**(6): 3785-93.

Fontana, A., P. Erb, et al. (1986). "Astrocytes as antigen-presenting cells. Part II: Unlike H-2K-dependent cytotoxic T cells, H-2Ia-restricted T cells are only stimulated in the presence of interferon-gamma." <u>J Neuroimmunol</u> **12**(1): 15-28.

Gehrmann, J., Y. Matsumoto, et al. (1995). "Microglia: intrinsic
        immuneffector cell of the brain." Brain Res Brain Res Rev **20**(3):
        269-87.
Glabinski, A. R. and R. M. Ransohoff (1999). "Chemokines and chemokine
        receptors in CNS pathology." J Neurovirol **5**(1): 3-12.
Godfraind, C., N. Havaux, et al. (1997). "Role of virus receptor-bearing
        endothelial cells of the blood-brain barrier in preventing the spread
        of mouse hepatitis virus-A59 into the central nervous system." J
        Neurovirol **3**(6): 428-34.
Gombold, J. L., R. M. Sutherland, et al. (1995). "Mouse hepatitis virus A59-
        induced demyelination can occur in the absence of CD8+ T cells."
        Microb Pathog **18**(3): 211-21.
Grusby, M. J. and L. H. Glimcher (1995). "Immune responses in MHC class
        II-deficient mice." Annu Rev Immunol **13**: 417-35.
Grzybicki, D. M., K. B. Kwack, et al. (1997). "Nitric oxide synthase type II
        expression by different cell types in MHV-JHM encephalitis
        suggests distinct roles for nitric oxide in acute versus persistent virus
        infection." J Neuroimmunol **73**(1-2): 15-27.
Han, Y., Z. H. Zhou, et al. (1999). "TNF-alpha suppresses IFN-gamma-
        induced MHC class II expression in HT1080 cells by destabilizing
        class II trans-activator mRNA." J Immunol **163**(3): 1435-40.
Haring, J. and S. Perlman (2001). "Mouse hepatitis virus." Curr Opin
        Microbiol **4**(4): 462-6.
Hart, M. N. and Z. Fabry (1995). "CNS antigen presentation." Trends
        Neurosci **18**(11): 475-81.
Haydon, P. G. (2000). "Neuroglial networks: neurons and glia talk to each
        other." Curr Biol **10**(19): R712-4.
Hoffman, L. M., B. T. Fife, et al. (1999). "Central nervous system
        chemokine expression during Theiler's virus-induced demyelinating
        disease." J Neurovirol **5**(6): 635-42.
Hoheisel, D., T. Nitz, et al. (1998). "Hydrocortisone reinforces the blood-
        brain barrier properties in a serum free cell culture system."
        Biochem Biophys Res Commun **244**(1): 312-6.
Houtman, J. J. and J. O. Fleming (1996). "Dissociation of demyelination and
        viral clearance in congenitally immunodeficient mice infected with
        murine coronavirus JHM." J Neurovirol **2**(2): 101-10.
Houtman, J. J. and J. O. Fleming (1996). "Pathogenesis of mouse hepatitis
        virus-induced demyelination." J Neurovirol **2**(6): 361-76.
Huffnagle, G. B., L. K. McNeil, et al. (1999). "Cutting edge: Role of C-C
        chemokine receptor 5 in organ-specific and innate immunity to
        Cryptococcus neoformans." J Immunol **163**(9): 4642-6.
Kieseier, B. C., K. Krivacic, et al. (2000). "Sequential expression of
        chemokines in experimental autoimmune neuritis." J Neuroimmunol
        **110**(1-2): 121-9.
Lane, T. E., V. C. Asensio, et al. (1998). "Dynamic regulation of alpha- and
        beta-chemokine expression in the central nervous system during

mouse hepatitis virus-induced demyelinating disease." J Immunol
**160**(2): 970-8.

Lane, T. E. and M. J. Buchmeier (1997). "Murine coronavirus infection: a
paradigm for virus-induced demyelinating disease." Trends
Microbiol **5**(1): 9-14.

Lane, T. E., H. S. Fox, et al. (1999). "Inhibition of nitric oxide synthase-2
reduces the severity of mouse hepatitis virus-induced demyelination:
implications for NOS2/NO regulation of chemokine expression and
inflammation." J Neurovirol **5**(1): 48-54.

Lane, T. E., M. T. Liu, et al. (2000). "A central role for CD4(+) T cells and
RANTES in virus-induced central nervous system inflammation and
demyelination." J Virol **74**(3): 1415-24.

Lavi, E., D. H. Gilden, et al. (1986). "The organ tropism of mouse hepatitis
virus A59 in mice is dependent on dose and route of inoculation."
Lab Anim Sci **36**(2): 130-5.

Lavi, E., D. H. Gilden, et al. (1984). "Experimental demyelination produced
by the A59 strain of mouse hepatitis virus." Neurology **34**(5): 597-
603.

Lavi, E., A. Suzumura, et al. (1989). "Induction of MHC class I antigens on
glial cells is dependent on persistent mouse hepatitis virus
infection." J Neuroimmunol **22**(2): 107-11.

Lee, S. J., T. Zhou, et al. (2000). "Differential regulation and function of Fas
expression on glial cells." J Immunol **164**(3): 1277-85.

Li, Y., L. Fu, et al. (2004). "Coronavirus neurovirulence correlates with the
ability of the virus to induce proinflammatory cytokine signals from
astrocytes and microglia." J Virol **78**(7): 3398-406.

Ling, E. A. and W. C. Wong (1993). "The origin and nature of ramified and
amoeboid microglia: a historical review and current concepts." Glia
**7**(1): 9-18.

Liu, M. T., D. Armstrong, et al. (2001). "Expression of Mig (monokine
induced by interferon-gamma) is important in T lymphocyte
recruitment and host defense following viral infection of the central
nervous system." J Immunol **166**(3): 1790-5.

Liu, M. T., B. P. Chen, et al. (2000). "The T cell chemoattractant IFN-
inducible protein 10 is essential in host defense against viral-induced
neurologic disease." J Immunol **165**(5): 2327-30.

MacMicking, J., Q. W. Xie, et al. (1997). "Nitric oxide and macrophage
function." Annu Rev Immunol **15**: 323-50.

Massa, P. A., R. Dorries, et al. (1986). "Viral particles induce Ia antigen
expression on astrocytes." Nature **320**: 543-546.

Massa, P. T., H. Wege, et al. (1986). "Analysis of murine hepatitis virus
(JHM strain) tropism toward Lewis rat glial cells in vitro. Type I
astrocytes and brain macrophages (microglia) as primary glial cell
targets." Lab. Invest. **55**: 318-327.

Matsushima, G. K., M. Taniike, et al. (1994). "Absence of MHC class II
molecules reduces CNS demyelination, microglial/macrophage

infiltration, and twitching in murine globoid cell leukodystrophy."
     Cell **78**(4): 645-56.
Medzhitov, R. and C. A. Janeway, Jr. (1997). "Innate immunity: impact on
     the adaptive immune response." Curr Opin Immunol **9**(1): 4-9.
Meinl, E., F. Aloisi, et al. (1994). "Multiple sclerosis. Immunomodulatory
     effects of human astrocytes on T cells." Brain **117** ( **Pt 6**): 1323-32.
Merrill, J. E., L. J. Ignarro, et al. (1993). "Microglial cell cytotoxicity of
     oligodendrocytes is mediated through nitric oxide." J Immunol
     **151**(4): 2132-41.
Merrill, J. E. and S. P. Murphy (1997). "Inflammatory events at the blood
     brain barrier: regulation of adhesion molecules, cytokines, and
     chemokines by reactive nitrogen and oxygen species." Brain Behav
     Immun **11**(4): 245-63.
Mi, H., H. Haeberle, et al. (2001). "Induction of astrocyte differentiation by
     endothelial cells." J Neurosci **21**(5): 1538-47.
Minagar, A., P. Shapshak, et al. (2002). "The role of macrophage/microglia
     and astrocytes in the pathogenesis of three neurologic disorders:
     HIV-associated dementia, Alzheimer disease, and multiple
     sclerosis." J Neurol Sci **202**(1-2): 13-23.
Mitrovic, B., J. Parkinson, et al. (1996). "An in Vitro Model of
     Oligodendrocyte Destruction by Nitric Oxide and Its Relevance to
     Multiple Sclerosis." Methods **10**(3): 501-13.
Nathan, C. and Q. W. Xie (1994). "Nitric oxide synthases: roles, tolls, and
     controls." Cell **78**(6): 915-8.
Nelson, P. T., L. A. Soma, et al. (2002). "Microglia in diseases of the central
     nervous system." Ann Med **34**(7-8): 491-500.
Nguyen, V. T. and E. N. Benveniste (2000). "Involvement of STAT-1 and
     ets family members in interferon-gamma induction of CD40
     transcription in microglia/macrophages." J Biol Chem **275**(31):
     23674-84.
O'Keefe, G. M., V. T. Nguyen, et al. (1999). "Class II transactivator and
     class II MHC gene expression in microglia: modulation by the
     cytokines TGF-beta, IL-4, IL-13 and IL-10." Eur J Immunol **29**(4):
     1275-85.
Owens, G. P., M. P. Burgoon, et al. (2001). "The immunoglobulin G heavy
     chain repertoire in multiple sclerosis plaques is distinct from the
     heavy chain repertoire in peripheral blood lymphocytes." Clin
     Immunol **98**(2): 258-63.
Panek, R. B., Y. J. Lee, et al. (1994). "Characterization of astrocyte nuclear
     proteins involved in IFN-gamma- and TNF-alpha-mediated class II
     MHC gene expression." J Immunol **153**(10): 4555-64.
Perlman, S. (1998). "Pathogenesis of coronavirus-induced infections.
     Review of pathological and immunological aspects." Adv Exp Med
     Biol **440**: 503-13.

Prat, A., K. Biernacki, et al. (2000). "B7 expression and antigen presentation by human brain endothelial cells: requirement for proinflammatory cytokines." J Neuropathol Exp Neurol **59**(2): 129-36.

Reiss, C. S. and T. Komatsu (1998). "Does nitric oxide play a critical role in viral infections?" J Virol **72**(6): 4547-51.

Remick, D. G. and L. Villarete (1996). "Regulation of cytokine gene expression by reactive oxygen and reactive nitrogen intermediates." J Leukoc Biol **59**(4): 471-5.

Rengarajan, J., S. J. Szabo, et al. (2000). "Transcriptional regulation of Th1/Th2 polarization." Immunol Today **21**(10): 479-83.

Rohn, W. M., Y. J. Lee, et al. (1996). "Regulation of class II MHC expression." Crit Rev Immunol **16**(3): 311-30.

Rose, J. W., K. E. Hill, et al. (1998). "Nitric oxide synthase inhibitor, aminoguanidine, reduces inflammation and demyelination produced by Theiler's virus infection." J Neuroimmunol **81**(1-2): 82-9.

Rottman, J. B., A. J. Slavin, et al. (2000). "Leukocyte recruitment during onset of experimental allergic encephalomyelitis is CCR1 dependent." Eur J Immunol **30**(8): 2372-7.

Sobue, K., N. Yamamoto, et al. (1999). "Induction of blood-brain barrier properties in immortalized bovine brain endothelial cells by astrocytic factors." Neurosci Res **35**(2): 155-64.

Sofroniew, M. V., T. G. Bush, et al. (1999). "Genetically-targeted and conditionally-regulated ablation of astroglial cells in the central, enteric and peripheral nervous systems in adult transgenic mice." Brain Res **835**(1): 91-5.

Soos, J. M., T. A. Ashley, et al. (1999). "Differential expression of B7 co-stimulatory molecules by astrocytes correlates with T cell activation and cytokine production." Int Immunol **11**(7): 1169-79.

Stalder, A. K., A. Pagenstecher, et al. (1997). "Lipopolysaccharide-induced IL-12 expression in the central nervous system and cultured astrocytes and microglia." J Immunol **159**(3): 1344-51.

Sun, N., D. Grzybicki, et al. (1995). "Activation of astrocytes in the spinal cord of mice chronically infected with a neurotropic coronavirus." Virology **213**(2): 482-93.

Sun, N. and S. Perlman (1995). "Spread of a neurotropic coronavirus to spinal cord white matter via neurons and astrocytes." J Virol **69**(2): 633-41.

Sutherland, R. M., M. M. Chua, et al. (1997). "CD4+ and CD8+ T cells are not major effectors of mouse hepatitis virus A59-induced demyelinating disease." J Neurovirol **3**(3): 225-8.

Suzumura, A., E. Lavi, et al. (1988). "Induction of glial cell MHC antigen expression in neurotropic coronavirus infections. Characterization of the H-2-inducing soluble factor elaborated by infected brain cells." J Immunol **140**(6): 2068-72.

Suzumura, A., E. Lavi, et al. (1986). "Coronavirus infection induces H-2
     antigen expression on oligodendrocytes and astrocytes." <u>Science</u>
     **232**(4753): 991-3.
Takiguchi, M. and J. A. Frelinger (1986). "Induction of antigen presentation
     ability in purified cultures of astroglia by interferon-gamma." <u>J Mol</u>
     <u>Cell Immunol</u> **2**(5): 269-80.
Tran, N. D., J. Correale, et al. (1999). "Transforming growth factor-beta
     mediates astrocyte-specific regulation of brain endothelial
     anticoagulant factors." <u>Stroke</u> **30**(8): 1671-8.
Utsumi, H., H. Chiba, et al. (2000). "Expression of GFRalpha-1, receptor for
     GDNF, in rat brain capillary during postnatal development of the
     BBB." <u>Am J Physiol Cell Physiol</u> **279**(2): C361-8.
Weber, F., E. Meinl, et al. (1994). "Human astrocytes are only partially
     competent antigen presenting cells. Possible implications for lesion
     development in multiple sclerosis." <u>Brain</u> **117 ( Pt 1)**: 59-69.
Williams, K., E. Ulvestad, et al. (1994). "Immune regulatory and effector
     properties of human adult microglia studies in vitro and in situ." <u>Adv</u>
     <u>Neuroimmunol</u> **4**(3): 273-81.
Wolburg H, R. W. (1995). Formation of the blood-brain barrier. <u>Neuroglia</u>.
     K. H. Ransom BR New York, Oxford University Press**:** 763-776.
Wong, G. H., P. F. Bartlett, et al. (1984). "Inducible expression of H-2 and
     Ia antigens on brain cells." <u>Nature</u> **310**(5979): 688-91.
Wu, G. F., L. Pewe, et al. (2000). "Coronavirus-induced demyelination
     occurs in the absence of inducible nitric oxide synthase." <u>J Virol</u>
     **74**(16): 7683-6.
Zhang, S. C., B. D. Goetz, et al. (2001). "Reactive microglia in
     dysmyelination and demyelination." <u>Glia</u> **34**(2): 101-9.

Chapter C3

# AXONS AND NEURONS IN CORONAVIRUS-INDUCED DEMYELINATION


Ajai A. Dandekar and Stanley Perlman
*Departments of Pediatrics and Microbiology and Interdisciplinary Program in Immunology,
Univerisity of Iowa, Iowa City, Iowa, USA*



Abstract:     Infection of mice with the coronavirus mouse hepatitis virus induces primary
demyelination in susceptible strains of rodents. Although demyelination is the
primary pathological process detected in the central nervous system of infected
mice, axonal dysfunction and damage also occur concomitantly with
demyelination. This process is T cell mediated, with either CD4 or CD8 T
cells sufficient for MHV-induced axonal damage. A striking feature is that
axonal damage occurs early in the disease process, at nearly the same time as
demyelination is first observed. Axonal damage in MHV-infected mice has
many similarities with the parallel process in humans with multiple sclerosis.

Key words:    Coronavirus, demyelination, axonal damage, T cells


## 1.      INTRODUCTION

The human disease multiple sclerosis (MS) is characterized by focal
demyelinating lesions throughout the white matter of the CNS (1, 2). In
multiple sclerosis, immune-mediated damage to oligodendrocytes and/or the
myelin sheaths accounts for this pathology. The process of demyelination
results in electrical conduction deficits and other alterations in axonal
physiology, and previously had been assumed to account for the clinical
signs and symptoms of MS.

As described elsewhere in this volume, several older studies identified
axonal damage as part of the disease process in MS (3). A series of recent
reports confirmed that permanent damage to neurons and their axons also
occurs within these demyelinating lesions (4-6). In one study (7), N-acetyl

aspartate, a neurotransmitter, was diminished throughout the CNS in patients with progressive MS, while those patients with relapsing-remitting MS exhibited reduced levels of N-acetyl aspartate only in areas of demyelination. This reduction in N-acetyl aspartate correlated with axonal loss detected by either magnetic resonance imaging or electron microscopy of involved tissue.

Other studies of patients with MS have examined the characteristics of axonal damage, using immunohistochemistry with antibodies to nonphosphoneurofilament H or amyloid precursor protein to visualize damaged axons (5, 6). The preponderance of axonal damage occurred in demyelinating lesions, with the remainder of damage found adjacent to these lesions.

Axonal damage in MS may account for a significant amount of the clinical signs and symptoms seen in the progressive phase of disease. Progressive MS is poorly correlated with the size or number of demyelinating lesion seen by MRI, suggesting that axonal pathology, not demyelination, may be the primary cause of the irreversible deficits that are observed (8). One interpretation is that demyelination and remyelination may explain the relapsing-remitting phase, while progressive disease results from irreversible axonal pathology (9).

Axonal damage has been reported in several animal models of MS, including in rodents with experimental autoimmune encephalomyelitis (EAE) and in mice with demyelination induced by Theiler's encephalomyelitis virus or mouse hepatitis virus, strain JHM (MHV) (10-13). Although it had long been believed that MHV-associated demyelination occurs primarily as a result of virus-induced destruction of oligodendrocytes, accumulating data from multiple studies indicate that CD4 and CD8 T cells are essential in the pathological process (14, 15) (see also chapter C4 of this volume). In this review, we summarize data showing that this T cell-mediated demyelination occurs concomitantly with axonal damage in MHV-infected mice.

## 2.    TEMPOROSPATIAL PROFILE OF AXONAL DAMAGE IN MHV INFECTION

As described elsewhere in this volume, several models of MHV-induced demyelination are studied in different laboratories. In one model, splenocytes are transferred from MHV immune mice to syngeneic immunodeficient mice [mice with severe combined immunodeficiency or with genetic disruption of recombination activation gene 1 ($RAG1^{-/-}$)] infected with the neuroattenuated variant of MHV, 2.2-V-1 (16). Both SCID and RAG1$^{-/-}$ mice lack B and T lymphocytes (17) and as such are unable to mount an adaptive immune response to MHV. While immunocompetent C57Bl/6 (B6) mice develop demyelination ten to twelve days after intracranial inoculation with MHV, their RAG1$^{-/-}$ or SCID counterparts develop, instead, a fatal acute encephalitis at 14-18 days post infection (p.i.). Clinical and histological evidence of demyelination, with accompanying macrophage/microglia infiltration, can be detected within seven to ten days of transfer.

Most recent work has been performed using MHV-infected RAG1$^{-/-}$ mice as recipients. In this adoptive transfer model, either CD4 or CD8 T cells can mediate demyelination and disease. T cells are necessary for the demyelinating process since transfer of splenocytes depleted of CD4 and CD8 T cells results in no demyelination. RAG1$^{-/-}$ recipients of CD4 T cell-enriched splenocytes (CD8 T cell-depleted) or CD8 T cell-enriched splenocytes (CD4 T cell-depleted) also develop demyelination, although with different kinetics than recipients of undepleted splenocytes. CD4 T cell enrichment results in a rapid course of disease, with mortality by day 7 p.i., while CD8 T cell enrichment results in a protracted course of disease (as compared to wild type), with mortality at day 14-16 p.i. or later. Strikingly, up to 50% demyelination is observed in recipients of CD8 T cell-enriched populations by 15 days p.i. This model system was used to determine the kinetics of axonal damage and relationship to demyelination in MHV-infected mice.

## 2.1 The relationship between demyelination and axonal damage in MHV-induced disease

Spinal cords were harvested from MHV-infected RAG1-/- mice at 7 days after adoptive transfer (10 days p.i.) of MHV immune splenocytes. We examined zinc formalin-fixed, paraffin-embedded sections for demyelination and axonal damage. Areas of myelin damage were determined using the chemical stain luxol fast blue (LFB) with quantification as previously described (18). We further assessed the distribution of macrophages/microglia using mAb to the macrophage-specific protein F4/80 (Serotec, Oxford, England) and viral antigen (using the MHV nucleocapsid-specific mAb 5B11.2, provided by Dr. M. Buchmeier, The Scripps Research Institute, La Jolla, CA) by immunohistochemistry. Additionally we stained for axonal damage using an antibody specific for nonphosphoneurofilament H (mAb SMI-32 (Sternberger Monoclonals, Lutherville, MD). This protein is largely found in damaged axons, but is also expressed in the cell body and proximal processes of a fraction of unaffected neurons.

In areas of demyelination, there was abundant staining with SMI-32, indicating axonal damage. Examination of the staining pattern revealed continuous axonal staining, suggestive of intact, demyelinated axons, discontinuous staining patterns consistent with Wallerian degeneration of the axon, and terminal ovoids indicative of axonal transection (Figure C3-1). We detected a large infiltrate of macrophages/microglia in areas of demyelination, consistent with a role for these cells as the terminal effectors of MHV-induced demyelination. It seems likely that these cells also partly mediate axonal damage. These areas of demyelination had relatively little viral antigen staining, suggesting that myelin and axonal damage occurred during the process of viral clearance by infiltrating lymphocytes and macrophages/microglia.

Next we analyzed areas of white matter adjacent to demyelinating lesions. These periplaque regions exhibited abundant staining for virus antigen, with a modest infiltration of macrophages as compared to areas of demyelination. This most likely represents an early infiltrate of macrophages into virus-infected white matter. This infiltration of macrophages was accompanied by roughly half the level of staining for nonphosphoneurofilament H as detected in areas of frank demyelination.

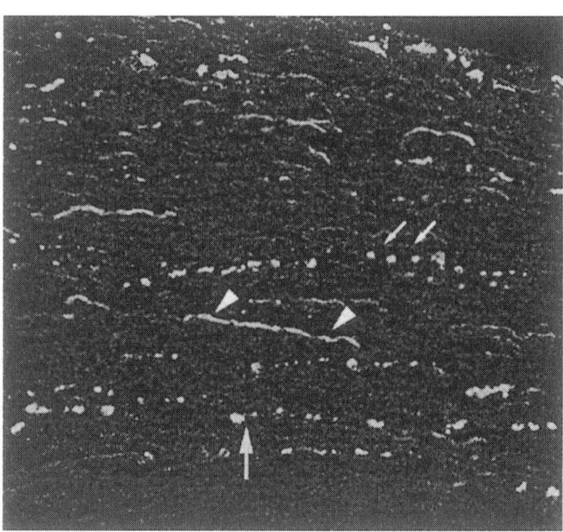

*Figure C3-1.* MHV-induced axonal damage. Midsagittal spinal cord sections from MHV-infected mice were stained with mAb SMI-32. An area of demyelination within the white matter is shown. Three different types of axonal pathology are indicated. Arrowheads: a demyelinated axon. Small arrows: a degenerating axon. Large arrow: a terminal ovoid, consistent with axonal transection.

Areas of normal appearing white matter distant from demyelinating lesions, as assessed by staining with LFB, were generally devoid of macrophages and viral antigen, although scattered virus-infected cells were occasionally detected in these regions of the spinal cord. These areas had only limited staining with mAb SMI-32. This staining may have resulted from direct viral damage to the axon, since viral antigen has been detected throughout axons (19). Alternatively, and we believe more likely, axonal damage was occurring distal to a demyelinating lesion.

Two methods were used to quantify the amount of axonal damage in the spinal cords of MHV-infected mice. In one method, midsagittal sections of whole spinal cords were stained with mAb SMI-32, analyzed by confocal microscopy, photographed and digitalized. The numbers of pixels fluorescing above background were counted (Figure C3-2A). This method quantifies total damage throughout the spinal cord, but because the majority of the spinal cord does not exhibit demyelination, tended to blunt differences between samples. To address this issue, the number of SMI-32 positive blebs in areas of demyelination, areas adjacent to demyelination, or normal appearing white matter was counted in a blinded fashion (Figure C3-2B). Quantification of the amount of axonal damage (Figure C3-2B) revealed that there were roughly twice as many SMI-32 positive axons in areas of demyelination, on average, than in adjacent areas, while there was minimal damage in distant normal-appearing areas of white matter.

Finally, we investigated the kinetics of axonal damage in relationship to the appearance of demyelination. Spinal cords of mice at 4.5 days p.i. were

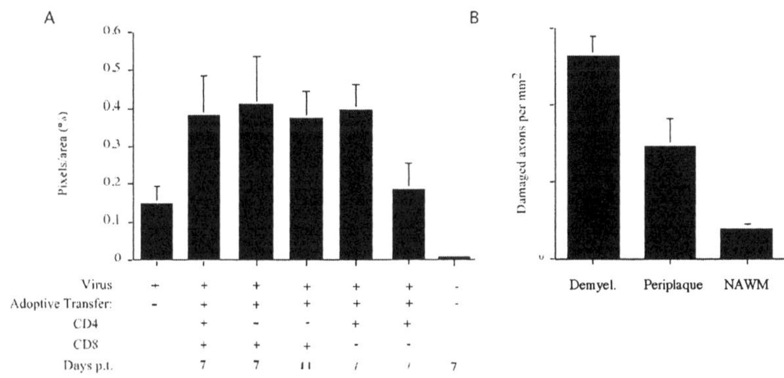

*Figure C3-2.* Quantification of axonal damage in MHV-infected mice. A, Whole spinal cords (midsagittal sections) were stained with SMI-32, photographed and digitalized, and the number of positive pixels was counted and expressed as a percentage of total spinal cord area.

B, the number of terminal ovoids in areas of demyelination, areas adjacent to demyelination ("periplaque" areas) or areas of normal appearing white matter (NAWM) were counted.

analyzed using the stains described above. This is the first time point at which demyelination is detectable. Surprisingly, axonal damage, as measured by SMI-32 positivity, was found in these mice. This occurred only in areas of early macrophage infiltration and viral antigen, suggesting that demyelination and axonal damage occur concomitantly, and are mediated at least in part by the same effector agents.

## 2.2      Contribution of CD4 and CD8 cells to axonal damage

As previously published (16), there are substantial differences in clinical disease and in demyelination between recipients of CD4 T cell- and CD8 T cell-enriched splenic populations. We reasoned that these differences in phenotype could be related to differences in the amount of axonal damage mediated by these two cell types. Therefore, the adoptive transfer system was used to elucidate the contribution of CD4 and CD8 T cells to the pathogenesis of axonal damage. MHV-infected RAG1$^{-/-}$ mice received splenocytes enriched for either CD4 or CD8 T cells three days p.i. We quantified the amount of axonal damage by measuring the number of pixels above background in spinal cords stained with mAb SMI-32 and analyzed by confocal microscopy.

Mice receiving splenocytes depleted of both CD4 and CD8 T cells did not develop demyelination and also did not develop axonal damage (Figure C3-2A). Mice that received splenocytes treated with complement only exhibited 0.4% of mAb SMI-32 immunoreactivity across the spinal cord. Surprisingly, no difference in total mAb SMI-32 immunoreactivity could be detected between the spinal cords of recipients of CD4 T cell- or CD8 T cell-enriched splenocytes. The amount of axonal damage was not different statistically from that observed in recipients of undepleted populations of cells. In addition, the amount of axonal damage was not well   correlated with the extent of demyelination. These results show that axonal damage did not explain the differences in clinical disease observed between the recipients of undepleted and T cell subset-enriched splenocytes.

## 3.       FUTURE DIRECTIONS

Axonal damage appears to be important in the pathological process in all demyelinating diseases, including MS, EAE, and viral models of demyelination, including MHV-infected mice (5, 10-13). In the model of MHV-induced demyelination described herein, clinical disease and demyelination occur rapidly and reproducibly after adoptive transfer of immune T cells, making it ideal for the dissection of the processes leading to axonal damage in demyelinating disease.

Although CD4 and CD8 T cells both contribute to demyelination and axonal damage, the mechanisms by which these processes are mediated remains to be determined. CD8 T cell-mediated demyelination is interferon-$\gamma$ mediated, whereas CD4 T cell-mediated demyelination is increased in the absence of this cytokine (20, 21). Demyelination itself is associated with changes in the cytoskeletal structure of the associated axon (22). In mice with genetic disruption of the proteolipid protein gene, axonal dysfunction is observed in the presence of normally compacted myelin (23, 24). In some models, demyelinated axons appear to be able to compensate for the loss of myelin by upregulating ion channels necessary to maintain conduction (25, 26). Other demyelinated axons fail to do so, leading to severe conduction defects and axonal dysfunction. It is likely that T cells and macrophages/microglia, by creating a pro-inflammatory milieu, contribute to the pathogenesis of axonal injury.

Important issues to be resolved include identifying the effector molecules and cells that cause axonal damage and determining if axonal injury is a necessary consequence of demyelination. Once the mechanisms of axonal dysfunction and damage are better understood, it may be possible to design therapy to minimize axonal disease in MHV-infected mice, and ultimately, in humans with MS.

# REFERENCES


1.  **Noseworthy JH**. Progress in determining the causes and treatment of multiple sclerosis. *Nature*. 1999;399:A40-A47.
2.  **Hemmer B, Archelos JJ, Hartung HP**. New concepts in the immunopathogenesis of multiple sclerosis. *Nat Rev Neurosci*. 2002;3:291-301.
3.  **Lassmann H**. The pathology of multiple sclerosis and its evolution. *Philos Trans R Soc Lond B Biol Sci*. 1999;354:1635-40.
4.  **De Stefano N, Matthews PM, Narayanan S, Francis GS, Antel JP, Arnold DL**. Axonal dysfunction and disability in a relapse of multiple sclerosis: longitudinal study of a patient. *Neurology*. 1997;49:1138-41.
5.  **Trapp B, Peterson J, Ransohoff R, Rudick R, Monk S, Bo L**. Axonal transection in the lesions of multiple sclerosis. *New Engl. J. Med.* 1998;338:278-285.
6.  **Ferguson B, Matyszak M, Esiri M, Perry V**. Axonal damage in acute multiple sclerosis lesions. *Brain*. 1997;120:393-399.

7.    **Matthews PM, De Stefano N, Narayanan S, et al**. Putting magnetic resonance spectroscopy studies in context: axonal damage and disability in multiple sclerosis. *Sem. Neurol.* 1998;18:327-36.

8.    **Brex PA, Ciccarelli O, O'Riordan JI, Sailer M, Thompson AJ, Miller DH**. A longitudinal study of abnormalities on MRI and disability from multiple sclerosis. *N Engl J Med.* 2002;346:158-64.

9.    **Waxman SG**. Demyelinating diseases--new pathological insights, new therapeutic targets. *N Engl J Med.* 1998;338:323-5.

10.   **Kornek B, Storch MK, Weissert R, et al**. Multiple sclerosis and chronic autoimmune encephalomyelitis: a comparative quantitative study of axonal injury in active, inactive, and remyelinated lesions. *Am J Pathol.* 2000;157(1):267-76.

11.   **Raine CS, Cross AH**. Axonal dystrophy as a consequence of long-term demyelination. *Lab. Invest.* 1989;60:714-725.

12.   **Sathornsumetee S, McGavern DB, Ure DR, Rodriguez M**. Quantitative ultrastructural analysis of a single spinal cord demyelinated lesion predicts total lesion load, axonal loss, and neurological dysfunction in a murine model of multiple sclerosis. *Am J Pathol.* 2000;157(4):1365-76.

13.   **Dandekar AA, Wu G, Pewe LL, Perlman S**. Axonal damage is T cell mediated and occurs concomitantly with demyelination in mice infected with a neurotropic cornavirus. *J. Virol.* 2001;75:6115-6120.

14.   **Stohlman SA, Bergmann CC, Perlman S**. Mouse hepatitis virus. In: Ahmed R, Chen I, eds. *Persistent Viral Infections*. New York: John Wiley & Sons, Ltd.; 1998:537-557.

15.   **Haring J, Perlman S**. Mouse hepatitis virus. *Curr Opin Microbiol.* 2001;4:462-6.

16.   **Wu GF, Dandekar AA, Pewe L, Perlman S**. CD4 and CD8 T cells have redundant but not identical roles in virus-induced demyelination. *J. Immunol.* 2000;165:2278-2286.

17.   **Mombaerts P, Iacomini J, Johnson RS, Herrup K, Tonegawa S, Papaloannou VE**. RAG-1-deficient mice have no mature B and T lymphocytes. *Cell.* 1992;68:869-877.

18.   **Xue S, Sun N, van Rooijen N, Perlman S**. Depletion of blood-borne macrophages does not reduce demyelination in mice infected with a neurotropic coronavirus. *J. Virol.* 1999;73:6327-6334.

19.   **Pasick J, Kalicharran K, Dales S**. Distribution and trafficking of JHM coronavirus structural proteins and virions in primary neurons and the OBL-21 neuronal cell line. *J. Virol.* 1994;68:2915-2928.

20.   **Pewe L, Haring J, Perlman S**. CD4 T-cell-mediated demyelination is increased in the absence of gamma interferon in mice infected with mouse hepatitis virus. *J Virol.* 2002;76:7329-33.

21.   **Pewe LL, Perlman S**. CD8 T cell-mediated demyelination is IFN-$\gamma$ dependent in mice infected with a neurotropic coronavirus. *J. Immunol.* 2002;168:1547-1551.

22.   **Kirkpatrick LL, Brady ST**. Modulation of the axonal microtubule cytoskeleton by myelinating Schwann cells. *J Neurosci.* 1994;14:7440-50.

23.   **Boison D, Stoffel W**. Disruption of the compacted myelin sheath of axons of the central nervous system in proteolipid protein-deficient mice. *Proc Natl Acad Sci U S A.* 1994;91:11709-13.

24. **Griffiths I, Klugmann M, Anderson T, et al**. Axonal swellings and degeneration in mice lacking the major proteolipid of myelin. *Science*. 1998;280:1610-1613.

25. **Rivera-Quinones C, McGavern D, Schmelzer J, Hunter S, Low P, Rodriguez M**. Absence of neurological deficits following extensive demyelination in a class I-deficient murine model of multiple sclerosis. *Nature Med*. 1998;4:187-193.

26. **Felts PA, Baker TA, Smith KJ**. Conduction in segmentally demyelinated mammalian central axons. *J Neurosci*. 1997;17:7267-77.

Chapter C4

# THE ROLE OF T CELLS IN CORONA-VIRUS-INDUCED DEMYELINATION

Cornelia C Bergmann[1], Stephen A Stohlman[1], and Stanley Perlman[2]
*Keck School of Medicine University of Southern California, Los Angeles, CA 90033[1] and University of Iowa, Iowa City, Iowa 52242[2]*

**Abstract:**     Mice infected with neurotropic strains of coronavirus develop acute encephalomyelitis and eliminate infectious virus. However, control of acute infection is incomplete resulting in persistence of viral RNA in the central nervous system (CNS) associated with ongoing primary demyelination. A high prevalence of virus specific CD8 and CD4 T cells within the CNS correlates with *ex vivo* cytolytic activity and IFN-γ secretion, which are both required for virus reduction during the acute infection. Although most infected cell types are susceptible to perforin mediated clearance, IFN-γ is required for controlling infection of oligodendrocytes. Furthermore, by enhancing class I expression and inducing class II expression within resident CNS cells IFN-γ optimizes T cell receptor dependent functions. In addition to its direct anti viral activity, these multifactorial effects make IFN-γ more essential than perforin for viral control. CD4 T cells enhance CD8 T cell expansion, survival and effectiveness. Although both CD8 and CD4 T cells are retained within the CNS during persistence, they cannot control viral recrudescence in the absence of humoral immunity. Demyelination can be mediated by either CD8 or CD4 T cells; however, although a variety of effector molecules have been excluded, a dominant common denominator remains elusive. Thus concerted efforts to control infection coincide with a variety of potential mechanisms causing chronic demyelinating disease.

**Key words:**     CNS, demyelination, IFN-γ, perforin, T cells, mouse hepatitis virus

## INTRODUCTION

Neurotropic strains of mouse hepatitis virus (MHV) produce an acute demyelinating encephalomyelitis in rodents (reviewed in 1-3).

Following infection with the JHM strain of MHV (JHMV) survivors have no detectable infectious virus by 2 weeks p.i., but viral antigen (Ag) and RNA persist within the CNS, predominantly in spinal cords up to 2 years p.i. Although persistently infected mice exhibit few clinical abnormalities, histological examination shows ongoing primary CNS demyelination consistent with the pathological changes characterizing the human demyelinating disease multiple sclerosis. Experimental CNS infection by MHV thus provides an excellent model to study the role of inflammatory cells during ongoing demyelination associated with persisting virus.

The majority of research focusing on the role of T cells in MHV pathogenesis has used the JHMV 2.2v-1 monoclonal antibody (Ab) neutralization escape mutant (1-3). This variant has several advantages in studying virus-induced demyelination when compared to parental JHMV and other variants: i) Most mice survive the infection, with demyelination observed in all survivors. ii) The cellular tropism of most JHMV strains for microglia, astrocytes and oligodendroglia is retained; neurons are only rarely infected. iii) Hepatitis, which is transient following infection with the MHV-A59 strain and may confound analysis of CNS specific immune responses, is rare following i.c. infection. Much of the data discussed within this chapter pertain to this variant.

Several independent studies demonstrate that both CD4 T cells and CD8 T cells are required to control acute virus replication within the CNS, thereby protecting from an otherwise lethal infection (1-3). However, T cell-mediated protection against JHMV is generally provided at the cost of immune-mediated demyelination (1-7). Demyelination also occurs in mice with T cell deficiencies following infection with either JHMV (5) or MHV-A59 (8). Nevertheless, mice defective in B cells or anti-virus antibody secretion all develop severe demyelination (9,10), negating a prominent role of humoral immune responses. Irrespective of the controversies regarding mediators and perpetuators of demyelination, the final effector cells are believed to be activated macrophages/microglia (1,2,6,11). This chapter focuses on mechanisms of viral clearance and how activated T cells may contribute to immune pathology in JHMV-infected mice with an emphasis on the complexity of interactions involved.

## CONCERTED ANTI-VIRAL EFFORTS OF T CELLS

Although neutrophils, macrophages, NK cells and B cells constitute early infiltrates, T cells, most prominently the CD8 subset, provide the most critical anti-viral functions (3). A protective role for T cells in acute JHMV pathogenesis was demonstrated by the failure of both SCID and nude mice, genetically deficient in T cells, to clear JHMV resulting in lethality within 12-16 days (5,6). Selective depletion of either CD4 or CD8 T cells both diminished viral clearance, implicating a contribution of both T cell subsets (1-3). In addition, adoptive transfer of either virus specific CD8 T cells or CD4 T cells protected mice from acute encephalomyelitis (1-3). However, whereas CD8 T cells afforded protection by clearing infectious virus, distinct CD4 T cell donor populations varied in their ability to mediate direct anti viral function (1). The precise role that T cells have in viral clearance and MHV-induced demyelination is an area of active investigation, as discussed below.

## T Cell Priming, CNS Recruitment, and Effector Function

Mapping of T cell epitopes has greatly facilitated phenotypic and functional analysis of both CD8 and CD4 T cell expansion and recruitment into the CNS (3; see Chapter C6). Virus specific CD8 T cells are detectable in the CNS by day 6 p.i. and accumulate to high frequencies within the CNS by day 8 p.i. as demonstrated using MHC class I/peptide tetramers, intracellular IFN-$\gamma$ staining, as well as ELISPOT technology (3,12-16). During acute infection, up to 40% of CD8 T cells in the CNS respond to a single immunodominant epitope in both BALB/c and C57BL/6 mice (3,12-16). By contrast, the frequency of virus specific CD8 T cells in the periphery is ~10-20 fold lower (3,12,13). Detection of virus specific CD8 T cells in cervical lymph nodes (CLN) and spleen 2 days prior to the CNS supports priming, expansion, and differentiation in the periphery, rather than the CNS (17). The high prevalence of virus specific T cells in the CNS correlates with virus specific *ex vivo* cytolysis by CNS derived cells, a property not detected in unfractionated spleen or CLN cells (3). Although these observations suggested acquisition of effector function within the CNS, virus specific cytolytic effectors can be demonstrated following CD8 T cell enrichment from splenocytes at day 7 p.i. (17). The rapid accumulation of armed effector CTL results in a potent regional response which efficiently reduces viral titers, but does not prevent persistence. Furthermore, despite protection and

enhanced viral clearance achieved by transfer of JHMV specific CD8 T cells, virus could still be detected in oligodendrocytes (18), a major target of replication for the JHMV 2.2v-1 variant (1-3,18-20).

# Distinct Anti-Viral Functions of IFN-γ and Perforin

The suggestion that oligodendrocytes are more resistant to CD8 T cell mediated anti-viral function compared to other glial cell types prompted investigation of T cell effector functions utilized for protection. Anti-microbial T cell mediators include perforin, granzymes, and Fas-FasL, as well as soluble agents such as IFN-γ, TNF and selected chemokines (21;22). A unique aspect of JHMV pathogenesis is that the immunological mechanisms clearing infectious virus are cell type specific (19,20). Studies in genetically deficient mice revealed that replication in astrocytes and microglia is controlled via perforin mediated cytolysis, but not Fas/FasL interactions (19,23). Furthermore, abrogation of TNF-α by mAb treatment did not affect viral clearance (24). By contrast, IFN-γ controls replication in oligodendrocytes (20). Both cytolytic and IFN-γ effector mechanisms peak between days 8-10 p.i., coincident with maximal T cell infiltration (13,14,25). The inability to achieve sterile immunity, despite concerted action of these anti-viral mediators, suggested downregulatory mechanisms or viral evasion from T cell function. Although CTL escape variants play a prominent role in disease progression of weanling mice protected by neutralizing Ab, they appear to play no role in immune competent adult mice (3; see Chapter C6). However, the number of CNS T cells gradually decrease after day 10 p.i., despite the continued presence of Ag positive cells and viral RNA (13-15). Furthermore, Ag specific cytolytic activity at a single cell level rapidly declines by day 14 p.i. and remains absent thereafter (13). Loss of cytolytic activity is independent of either demyelination (14) or viral recrudescence in Ab deficient mice (9,12). The apparent inability to enhance T cell recruitment/function even during virus recrudescence (9,12) suggests restricted access or Ag driven apoptosis; however, there is no evidence to support increased apoptosis by high Ag levels *in vivo* (26). Whether CD8 T cells are downregulated by release of inhibitory factors within the local CNS environment or by the presence of suppressor cells is unknown.

Contrasting the loss of cytolysis, mRNA and ELISPOT analysis suggests that IFN-γ secretion in the CNS appears to continue after clearance of infectious virus (13,25). Furthermore, unlike the permanent loss of perforin mediated cytolysis, IFN-γ mRNA levels increase during JHMV reactivation in the CNS of B cell deficient μMT mice (26). The contradictory finding of reduced numbers of IFN-γ secreting cells in μMT mice (12) may reside in potential apoptosis of μMT derived CD8 T cells upon *in vitro* stimulation, as apoptosis is not increased *in vivo* (26). Irrespective of the extent of IFN-γ secretion, the balance between loss of cytolytic activity but continued cytokine secretion appears to reflect an attempt to control infection while reducing CNS immunopathology.

Overall the results suggest that IFN-γ is more vital than perforin in the hierarchy of immune effectors controlling JHMV infection and mortality. The nonredundant roles for both IFN-γ and perforin are more clearly evident by pathogenesis studies in mice deficient for both perforin and IFN-γ (27). These mice exhibit ongoing viral replication in all glial cell types coincident with a high mortality rate by day 14 p.i. Adoptive transfers demonstrate that wt memory CD8 T cells were most efficient in reducing virus, while perforin deficient T cells were slightly less efficient. IFN-γ deficient T cells had no affect on reducing viral titers. The inefficiency of IFN-γ deficient donor CD8 T cells to provide protection coincided with diminished class I upregulation and an absence of class II expression on resident microglia. IFN-γ mediated upregulation of class I Ag processing expression (28) is especially crucial within the CNS due to the paucity of MHC expression in the healthy, quiescent CNS (29,30). As both IFN-γ and TNF-α release as well as perforin mediated killing are strictly dependent on MHC–TCR contact (31), CD8 T cells initially recruited by chemokines (32) are likely to be less effective in exerting effector function *in vivo* in an IFN-γ deficient environment. In addition to its direct anti-viral role, IFN-γ thus enhances both CD8 and CD4 T cell function by upregulating MHC class I and class II presentation. The correlation between Ag recognition and elicitation of effector function has clear consequences for the efficiency of distinct effector functions *in vivo*. Whereas release of IFN-γ can be triggered by a subset of antigen presenting cells and act on a distal, non MHC expressing cell types, perforin mediated cytolysis can only be exerted on cells presenting viral Ag. The multifactorial effects of IFN-γ within the CNS thus shed light on the higher efficacy of IFN-γ compared to perforin in anti JHMV effector mechanisms.

## Auxilliary and Direct Anti-Viral Functions of CD4 T Cells

The precise role of CD4 T cells as direct anti-viral mediators are less well understood, especially their contribution to anti-viral IFN-γ secretion. The majority of CD4 T cells accumulating within the CNS are virus specific (33), similar to CD8 T cells. Depletion of CD4 T cells prevents JHMV clearance from the CNS (1-3). Although transferred JHMV specific CD4 T cells protect from acute disease, the mechanisms of protection may be unrelated to IFN-γ mediated control (1). CD4 T cells also provide crucial accessory function by enhancing expansion of virus specific CD8 T cells and maintaining CD8 T cell viability within the infected CNS (12,34). Although depletion of CD4 T cells does not prevent CD8 T cells from entering the CNS parenchyma, it compromises their viability, as indicated by increased apoptosis (34); the mechanisms involved remain obscure. Predominant localization of CD4 T cells in the perivascular and subarachnoid spaces during JHMV infection is consistent with recent data demonstrating that Fas/FasL interactions, a mechanism of CD4 T cell mediated cytolysis, does not contribute to JHMV clearance, encephalomyelitis or demyelination (23,34).

# T CELL FUNCTION DURING PERSISTENCE

A hallmark of JHMV persistence and ongoing demyelination is significant CNS retention of virus specific CD8 and CD4 T cells (12-16). Persisting T cells appear tightly linked to persisting virus as a source for chronic activation, as shown by the disappearance of T cells from the CNS of mice infected with a JHMV variant that does not persist (15). However, the percentages of virus specific cells within the CD8 T cell compartment remain remarkably similar throughout infection, negating selective enrichment or loss of cells with virus specificity (13-16). Despite an apparently dormant T cell state, several observations support an active turnover process: i) Persisting CD8 T cells exhibit restricted fine-specificity (35); ii) A reversed pattern of immunodominance emerges comparing acute and persistently infected CB6F1 (H-2$^{dxb}$) mice (13); and iii) CD69 expression is sustained during persistence, indicative of chronic CD8 T cell activation (13). Sustained CD69 expression has been observed on persisting, yet activated CD8 T cells in the CNS, lung and liver, following other viral infections (36-38).

Despite an apparent role for continued Ag presentation in T cell survival within the CNS, there is also evidence to support the contrary. CD8 T cells are readily maintained in the CNS following challenge of influenza virus immune mice with a heterologous neurotropic influenza virus (36). Unlike the JHMV model, there was no evidence for persistent infection and these cells retain *ex vivo* cytolytic capacity. The phenotype of these cells is reminiscent of peripheral memory CD8 T cells found in nonlymphoid organs (39), suggesting that T cell retention within the CNS may in part reflect peripheral memory T cells. However, in the JHMV model ultimate disappearance of all T cells in the absence of JHMV RNA in both brains and spinal cords suggests persisting RNA supports active T cell maintenance by ongoing chemokine release and/or continued Ag presentation. The source and renewal of such persisting T cells is under investigation.

The sufficiency of T cells for protection during MHV persistence has been challenged by the discovery of viral recrudescence in mice deficient in Ab (8,9,26). The necessity of Ab in controlling persistent infection is supported by the accumulation of virus specific Ab secreting cells within the CNS of wt mice (40). T cells may nevertheless contribute to control during the chronic phase, perhaps by secreting IFN-$\gamma$ and supporting B cell differentiation and/or survival.


# T CELLS AS CORRELATES OF DEMYELINATION

T cells control viral replication in several models of viral CNS infection, but are also key mediators of the pathological changes associated with encephalitis and/or demyelination (1-7,41,42). During JHMV infection both CD4 or CD8 T cells alone are capable of inducing demyelination (7). Unlike TMEV induced demyelination, there is little evidence for involvement of an autoimmune CD4 T cell component or epitope spreading. A prominent role for CD4 T cells is however indicated by clinical improvement as well as remyelination of JHMV infected mice treated with anti-IP-10 Ab (32). IP-10 treatment specifically reduced CD4, not CD8 T cells within the CNS, suggesting CD4 T cell dependent pathology in persistently infected mice. Secretion of RANTES leading to increased macrophage recruitment may be one mechanism (32). Mice deficient in perforin, IFN-$\gamma$, TNF-$\alpha$, NOS2 and IL-10 all develop demyelination similar to that in wt mice (19,20,24,43,44), suggesting none of these effector molecules alone dominate the demyelinating process. Recent findings demonstrating that CD4 and CD8 T cells contribute to demyelination via different mechanisms (45,46) further complicate understanding the network of effector molecules involved. Infection of mice deficient in recombination activation gene 1 activity (RAG$^{-/-}$) or SCID mice results in no demyelination. However, adoptive transfer of splenocytes from mice immunized with JHMV resulted in demyelination within 7 days post transfer (5-7). Splenocytes enriched for either CD4 or CD8 T cells mediated demyelination, although no demyelination was observed if both subsets were deleted. Further, reconstitution of infected RAG$^{-/-}$ mice using T cell subsets selectively deficient in T cell effector functions indicate that IFN-$\gamma$, but not perforin or TNF-$\alpha$ is crucial for CD8 T cell mediated demyelination (45).

By contrast, the absence of IFN-γ severely enhances demyelination mediated by CD4 T cells (46). Surprisingly, increased demyelination was not necessarily correlated with the amount of activated macrophages/microglia within the white matter. Furthermore, although virus replication was diminished to some extent after adoptive transfer of CD4 or CD8 T cells from wt or immunodeficient donors, only transfer of populations from immunocompetent B6 mice effected virus clearance (7). Contrasting a detrimental role of IFN-γ in demyelination by T cell transfers into infected H-2$^b$ RAG$^{-/-}$ mice, adoptive transfers of memory CD8 T cells into H-2$^d$ SCID mice suggest that both IFN-γ and perforin together propagate the demyelination process, but that CD8 T cell IFN-γ secretion alone does not increase pathology (Parra & Stohlman, unpublished). Finally, demyelination following JHMV infection can also be caused by non virus-specific bystander CD8 T cells, albeit only following activation by cognate Ag (47). Bystander CD4 T cells do not cause demyelination (Haring & Perlman, unpublished). In a recent study demyelination was shown to be γδ T cell-mediated in nude mice (48).

The complex linkages between IFN-γ, MHC class I and II expression, and chemokines have made it extremely challenging to delineate distinct pathways leading to demyelination. For example, impaired MHC and costimulatory molecule expression on resident CNS cells in the absence of IFN-γ are likely to ameliorate perforin and TNF mediated cell damage, but at the cost of increased virus load. Furthermore, macrophages/glia activation may be reduced, in turn affecting chemokine secretion (32). Together the current evidence suggests that JHMV induced demyelination is influenced by the balance between virus replication predominantly in oligodendroglia, distinct immune effector functions mediated by both CD4 and CD8 T cells and their effects on glia/macrophage activation.

## CONCLUSIONS

T cell mediated immune responses within the CNS can be both beneficial by clearing infectious agents, and detrimental by destroying tissue, triggering autoimmunity, and activating resident cells. The CNS responds rapidly to coronavirus infection by chemokine mediated recruitment of specific as well as bystander T cells. Recognition of cognate Ag presented by MHC is crucial in triggering Ag-specific T cell function both during virus and auto-Ag induced inflammation. However, prolonged T cell stimulation by foreign as well as self Ag promotes immune pathology, such as demyelination. The ubiquitous upregulation of class I during inflammation allows both IFN-γ and perforin mediated anti-viral functions. Whereas perforin secretion is directed, only affecting cytolysis of Ag presenting cells, IFN-γ can act on neighboring and distal cells not expressing MHC. Soluble mediators, i.e. IFN-γ and RANTES, may have detrimental effects by perpetuating chemokine secretion by astrocytes and recruiting activated macrophages. Efforts to link sustained immune activation to ongoing chemokine secretion and Ag recognition during viral persistence will provide clues for candidate factors promoting macrophage/microglia dependent demyelination and CNS pathogenesis.


# REFERENCES


1.	Stohlman S.A., Bergmann C.C., Perlman S. "Mouse hepatitis virus." In *Persistent Viral Infections*. R. Ahmed and I. Chen, eds. New York, NY: John Wiley & Sons Ltd, 1999.
2.	Perlman S.R., Lane T.E., and Buchmeier M.J. "Coronaviruses: Hepatitis, hepatitis, and central nervous system disease." In *Effects of Microbes on the Immune System*. Vol. I. M.W. Cunningham and R.S. Fujinami, ed. Philadelphia, PA: Lippincott Williams & Wilkins, 1999.
3.	Marten N.W., Stohlman S.A., Bergmann C.C. MHV infection of the CNS: Mechanisms of immune mediated control. Viral Immunol. 2001; 14:1.
4.	Houtman J.J., Fleming J.O. Pathogenesis of mouse hepatitis virus-induced demyelination. J. Neurovirol. 1996; 2:361.
5.	Houtman, J.J., Fleming J.O. Dissociation of demyelination and viral clearance in congenitally immunodeficient mice infected with murine coronavirus JHM. J. NeuroVirol. 1996; 2: 101.
6.	Wu G.F., Perlman S. Macrophage infiltration, but not apoptosis, is correlated with immune-mediated demyelination following murine infection with a neurotropic coronavirus. J. Virol. 1999; 73:8771.
7.	Wu G.F., Dandekar A.A., Pewe L., Perlman S. CD4 and CD8 T cells have redundant but not identical roles in virus-induced demyelination. J. Immunol. 2000; 165:2278.
8.	Sutherland R.M., Chua M.-M., Lavi E., Weiss S.R., Paterson Y. CD4$^+$ and CD8$^+$ T cells are not major effectors of mouse hepatitis virus A59-induced demyelinating disease. J. NeuroVirol. 1997; 3:225.
9.	Ramakrishna C., Stohlman S.A., Atkinson R., Schlomchik M.J., Bergmann C.C. Mechanisms of central nervous system persistence: critical role of antibody and B cells J. Immunol. 2002; 168: 1204.
10.	Matthews A.E., Weiss S.R., Shlomchik M.J., Hannum L.G., Gombold J.L. Paterson Y. Antibody is required for clearance of infectious hepatitis virus A59 from the central nervous system, but not the liver. J. Immunol. 2001; 167:5254-5263.
11.	Xue S., Sun N., van Rooijen N., Perlman S. Depletion of blood-borne macrophages does not reduce demyelination in mice infected with a neurotropic coronavirus. J. Virol. 1999 73:6327.

12. Bergmann C. C., Ramakrishna C., Kornacki M. Stohlman, S.A. Impaired T cell immunity in B cell deficient mice following central nervous system infection. J. Immunol. 2001; 167:1575.

13. Bergmann C.C., Altman J.D., Hinton D., Stohlman S.A. Inverted imunodominance and impaired cytolytic function of CD8+ T cells during viral persistence in the CNS. J. Immunol. 1999; 163: 3379.

14. Marten N.W., Stohlman S.A., Atkinson R., Hinton D.A., Bergmann C.C. Contributions of CD8$^+$ T cells and viral spread to demyelinating disease. J. Immunol. 2000; 164:4080.

15. Marten N.W., Stohlman S.A., Bergmann C.C. Role of viral persistence in retaining CD8$^+$ Tcells within the CNS. J. Virol. 2000; 74:7903.

*16.* Pewe L., Heard S.B., Bergmann C., Dailey M.O., Perlman S. Selection of T cell escape mutants in mice infected with a neurotropic coronavirus: Quantitative estimate of TCR Diversity using MHC/peptide tetramers. J. Immunol. 1999; 163:6106-6113.

17. Marten N.W., Stohlman S.A., Zhou J., Bergmann C.C. Kinetics of virus specific CD8 T cell expansion and trafficking following central nervous system infection. J. Virol. *Submitted*

18. Stohlman S.A., Bergmann C., van der Veen R., Hinton D. Mouse hepatitis virus - specific cytotoxic T lymphocytes protect from lethal infection without eliminating virus from oligodendroglia. J. Virol. 1995; 69:684.

19. Lin M.T., Stohlman S.A., Hinton D.R. Mouse hepatitis virus is cleared from the central nervous systems of mice lacking perforin-mediated cytolysis. J. Virol. 1997; 71:383.

20. Parra B., Hinton D.R., Marten N.W., Bergmann C.C., Lin M.T., Yang C.S., Stohlman S.A. IFN-γ is required for viral clearance from central nervous system oligodendroglia. J. Immunol. 1999. 162: 1641.

21. Wodarz D., Christensen J.P., Thomsen A.R. The importance of lytic and nonlytic immune responses in viral infections. Trends in Immunol. 2000; 23:194.

22. Harty J. T., Tvinnereim A. R., White D.W. CD8$^+$ T cell effector mechanisms in resistance to infection. Ann. Rev. Immunol. 2000; 18:275.

23. Parra B., Lin M.T., Stohlman S.A., Bergmann C.C., Atkinson R., Hinton D.A. Fas-Fas ligand interactions do not contribute to the pathogenesis of mouse hepatitis virus in the presence of perforin mediated cytolysis. J. Virol. 2000; 74:2447.

24. Stohlman S.A., Hinton D.R., Cua D., Dimacali E., Sensintaffar J., Tahara S., Hofman F. Yao Q. Tumor necrosis factor expression during mouse hepatitis virus induced demyelination. J. Virol. 1995; 69:5898.

25. Parra B., Hinton D.R., Lin M.T., Cua D.J., Stohlman S.A. Kinetics of cytokine mRNA expression in the central nervous system following lethal and nonlethal coronavirus-induced acute encephalomyelitis. Virology 1997; 233:260.

26. Lin M.T., Hinton D.R., Marten N.W., Bergmann C.C., Stohlman S.A. Antibody prevents virus reactivation within the central nervous system. J. Immunol. 1999; 162:7358.

27. Bergmann C.C., Parra B., Hinton D.R., Chandran R., Morrison M., Stohlman S.A. Perforin mediated effector function within the CNS requires IFN-γ mediated MHC upregulation. J. Immunol. *Submitted.*

28. Früh, K., Yang Y. Antigen presentation by MHC class I and its regulation by interferon γ. Cur. Opin. Immunol. 1999; 11:76.

29. Hickey W.F. Basic principles of immunological surveillance of the normal central nervous system. Glia 2001; 36:118.

30. Sedgwick J.D., Hickey W.F. "Antigen presentation in the central nervous system." In *Immunology of the Central Nervous System.* R. W. Keane and W. F. Hickey, eds. Oxford, UK: University Press, 1997.

31. Slifka M.K., Rodriguez F., Whitton J.L. Rapid on/off cycling of cytokine production by virus-specific CD8$^+$ T cells. Nature 1999; 401:76.

32. Glass W.G., Chen B.P., Liu M.T., Lane T.E. Mouse hepatitis virus infection of the central nervous system: chemokine-mediated regulation of host defense and disease. Viral Immunol. 2002; 2:261-272.
33. Haring J.S., Pewe L.L., Perlman S. High-magnitude, virus-specific CD4 T-cell response in the central nervous system of coronavirus-infected mice. J. Virol. 2001; 75:3043.
34. Stohlman S., Bergmann C., Cua D., Lin M., Ho S., Hinton D. CTL effector function within the central nervous system requires CD4+ T cells. J. Immunol. 1998; 160:2896.
35. Marten N.W., Stohlman S.A., Smith Begolka W., Miller S.D., Dimicali M., Stohl S., Goverman J., Bergmann C.C. Selection of CD8+ T cells with highly focused specificity during viral persistence in the central nervous system. J. Immunol. 1999; 162:3905.
36. Hawke S., Stevenson P.G., Freeman S., Bangham C.R.M.. Long-term persistence of activated cytotoxic T lymphocytes after viral infection of the central nervous system. J. Exp. Med. 1998; 187:1575
37. Ziegler S.F., Ramsdell R., Alderson M.R. The activation antigen CD69. Stem Cells 1994; 12:456.
38. Hogan R.J., Cauley L.S., Ely K.H., Cookenham T., Roberts A.D., Brennan J.W., Monard S., Woodland D.L. Long-term maintenance of virus-specific effector memory CD8 T cells in the lung airways depends on proliferation. 2002; 169: 4976.
39. Masopust, D., V. Yezys, A.L. Marzo, and L. Lefrancois. 2001. Preferential localization of effector memory cells in nonlymphoid tissue. Science 291:2413.
40. Tschen, S.-I, C.C. Bergmann, C. Ramakrishna, S. Morales, R. Atkinson and S.A. Stohlman. 2002. Recruitment kinetics and composition of antibody secreting cells within the CNS following viral encephalomyelitis. J. Immunol. 168: 2922.
41. Dorries R. The role of T cell mediated mechanisms in virus infections of the nervous system. Curr. Topics Microbiol. Immunol. 2001; 253:219.
42. Binder G.K., Griffin D.E. Interferon-gamma-mediated site-specific clearance of alphavirus from CNS neurons. Science 2001; 293:303.
43. Wu G. F., Pewe L., Perlman S. Coronavirus-induced demyelination occurs in the absence of inducible nitric oxide synthase. J. Virol. 2000; 74:7683.
44. Lin M.T., Hinton D. R., Para B., Stohlman S.A, van der Veen R.C. The role of IL-10 in mouse hepatitis virus-induced encephalomyelitis. Virology 1998; 245:270
45. Pewe L., Perlman S. Cutting Edge: CD8 T cell-mediated demyelination is IFN-$\gamma$ dependent in mice infected with a neurotropic coronavirus. J. Immunol. 2002; 168:1547.
46. Pewe L., Haring J., Perlman S. CD4 T-cell-mediated demyelination is increased in the absence of gamma interferon in mice infected with mouse hepatitis virus. J. Virol. 2002; 76:7329.
47. Haring J.S., Pewe L.L., Perlman S. Bystander CD8 T cell-mediated demyelination after viral infection of the central nervous system. J. Immunol. 2002; 169:1550.
48. Dandekar, A. , Perlman, S. Virus-induced demyelination in nude mice is mediated by $\gamma\delta$ T cells. Amer. J. Pathol. 2002; 161:1255.

Chapter C5

# THE ROLE OF HUMORAL IMMUNITY IN MOUSE HEPATITIS VIRUS INDUCED DEMYELINATION

C. Ramakrishna[1], S. Tschen[2], C. C. Bergmann[2], S. A. Stohlman[1,3]
*Departments of Neurology[1], Pathology[2] and Molecular Microbiology[3], Keck School of Medicine, University of Southern California, Los Angeles, CA*

**Abstract:** Pathogenesis induced by mouse hepatitis virus (MHV) infection of rodents is characterized by acute viral encephalomyelitis and demyelination which progresses to a persistent CNS infection associated with ongoing myelin loss, pathologically similar to multiple sclerosis (MS). Although humoral immunity appears redundant for the control of acute virus replication, it is vital in maintaining virus at levels detectable only by RNA analysis. T cell mediated control of acute infection cannot be sustained in antibody (Ab) deficient mice, resulting in virus reactivation. The protective role of Ab during persistence is strongly supported by detection of Ab in the cerebrospinal fluid of MHV infected rodents and maintenance of virus specific Ab secreting cells (ASC) in the CNS long after virus clearance. Ab mediated neutralization constitutes the major mechanism of protection, although fusion inhibition also plays a minor role. Delayed accumulation of ASC, concomitant with a decline in T cell function, assures control of residual virus while minimizing T cell mediated pathology. Although there is little evidence for a detrimental role of Ab in demyelination, an association between Ab mediated protection and remyelination is unclear.

**Key words:** CNS, demyelination, mouse hepatitis virus, B cells, Antibody.

## HUMORAL IMMUNITY AND MULTIPLE SCLEROSIS

MS is a chronic demyelinating disease of the CNS characterized by inflammation and white matter destruction, but of unknown etiology (1). The chief pathological features are focal areas of myelin loss associated with an inflammatory response consisting of B cells, T cells and macrophages (2). The relative contribution of each of these individual immune components remains poorly understood. A variety of cellular and humoral abnormalities have been observed in MS patients (2). The seminal findings of elevated Ig in CSF (3) as well as humoral involvement in demyelination during experimental allergic encephalomyelitis (EAE) (4) were the first hints implicating humoral components in the demyelinating disease. Increased B cell abundance as well as detection of myelin specific Ab in acute lesions with active ongoing demyelination compared to older, inactive lesions have supported their possible involvement in the disease pathogenesis (5,6). Although T cell studies have predominated this field of research based on the detection of activated T cells in MS plaques and involvement of T cells in EAE, current research in both MS and MOG protein induced EAE has re-ignited interest in the potential role of humoral immunity in the pathogenesis of demyelinating disease (2,7).

One hallmark of MS is that cerebral spinal fluid (CSF) in ~90% of patients is characterized by the presence of oligoclonal immunoglobulin (Ig) bands which exhibit restricted isoelectric focusing points (3,7). The CSF Ig response is sustained by resident B cells not represented in peripheral compartments (8). The IgG sequences feature extensive somatic mutations suggesting active antigen (Ag) driven B cell selection and clonal expansion (8). Oligoclonal bands are also encountered routinely following CNS infections with measles virus, Herpes simplex virus-1 and mumps virus; however, the majority of the Ig are specific for the causal agent (9). Moreover, detailed studies into the clonality of B cells in both CSF and MS plaques has revealed no single dominant Ag target. These Ab express varied Ag specificities from myelin components, oligodendrocyte protein, viruses, cell nuclei, endothelial cells, fatty acids, gangliosides, and axolemma (7,10). Auto Ab could potentially contribute to demyelination by various mechanisms including Ab dependent cell mediated cytotoxicity (ADCC), stimulation of Fc receptors expressed on NK cells or macrophages/microglia resulting in release of inflammatory molecules, opsonization of myelin resulting in phagocytosis by macrophages, or complement activation. Direct evidence of Ab mediated mechanisms in MS lesions was revealed with the detection of IgG deposition on the borders of the actively demyelinating plaques along with the presence of activated complement fragments and complexes (11). Ab eluted from MS plaques preferentially bind to CNS self Ag (7), although this reactivity to a variety of CNS Ag could also be due to liberation of Ag as a consequence of inflammatory tissue destruction. It thus appears that a concerted action of humoral and cellular factors could participate in the induction and/or maintenance of an inflammatory response specific for CNS Ag.

## MHV PATHOGENESIS

Infection of the CNS with MHV produces an acute encephalomyelitis associated with a focal loss of myelin. Virus replicates in various CNS cell types including astrocytes, oligodendroglia and microglia / macrophages. Distinct mechanisms of CD8 T cell mediated control are required to clear virus from the CNS as described elsewhere in this book (12,13). As virus replication is controlled, focal myelin loss however, is increased, suggesting a temporal lag between the maximal number of virus infected cells and the loss of myelin. The inability of cell mediated immunity to completely eliminate virus results in a non-productive persistent infection confined to the CNS, predominantly in spinal cord oligodendroglia. Persistent infection is associated with ongoing myelin loss and repair. These traits make MHV infection of the CNS along with Theiler's murine encephalitis virus infection and EAE the three primary murine models to study mechanisms underlying both encephalomyelitis and demyelination (12,13).

## HUMORAL IMMUNITY AND ACUTE INFECTION

Analysis of viral pathogenesis in mice depleted of T cell subsets, distinct T cell functions or humoral components demonstrated that acute infection by MHV-JHM strain (JHMV) is controlled by cellular immune responses (12,13). Humoral responses are redundant during acute infection as shown by initial virus clearance in the CNS of B cell deficient mice with similar kinetics to wt mice (14,15). Unlike the almost complete clearance of JHMV which expresses the immunodominant spike protein derived S510 epitope, infection with the MHV-A59 strain, which lacks this CTL epitope revealed minimal clearance in the B cell deficient mice, thus emphasizing the dominant role of cellular immunity (16). Similarly, mice deficient in T cells were unable to clear virus (12,13). Consistent with a redundant early role of humoral immunity, examination of serum anti-viral Ab titers showed that IgM and IgG are detectable only after the majority of the infectious virus has already been cleared from the CNS (12,17). Moreover, examination of the two biological activities of the Ab response directed at the spike protein, i.e., neutralization and fusion inhibition (17), determined that while neutralizing Ab was initially detectable by day 10 p.i., when most of infectious virus had already been cleared from the CNS, fusion inhibiting Ab was not detected until much later (day 21 p.i.). Thus the delayed appearance of serum Ab relative to virus clearance further supported a minor, if any role of Ab early in the adaptive host response to acute infection.

Transfer of monoclonal Ab specific for different virus structural proteins including spike, nucleocapsid and matrix proteins prior to or concomitant with JHMV infection nevertheless resulted in protection from virus induced mortality, but was not always associated with reductions in virus replication (12,18-20). The mechanisms of protection by pre-established Ab may merely reside in neutralization of the bolus of infectious virus prior to establishing CNS infection. Alternatively, Ab may modify tropism by prevention of infection of critical cells specifically neurons, thereby increasing survival rates (18,19). In other models, suckling mice weaned on immunized dams or transgenic mice which express neutralizing IgA in the milk are protected from acute JHMV induced encephalomyelitis; however, the role of Ab in these models is unclear (21,22). A variable percentage of protected weanling mice exhibit a delayed onset of acute encephalitis associated with demyelination. However, delayed onset with progressive disease is associated with CTL escape variants rather than Ab escape mutants or the loss of Ab mediated protective mechanisms (21).

Irrespective of the apparent redundancy of Ab in controlling acute infection, flow cytometric analysis of CNS inflammatory cells from infected wt mice revealed that mature B cells (CD19$^+$/sIg$^+$) are rapidly recruited into the CNS along with T cells (12,17). Furthermore, following the peak of virus replication, a low but constant percentage of B cells (5-7%) is retained in the CNS during viral persistence (12,17). A potential Ab independent, innate anti-viral effector function of B cells *in vivo* was initially suggested by the observation that B cells from naïve mice exhibit the ability to lyse virus infected target cells *in vitro*. Lysis is mediated through an interaction between viral receptor expressed at high levels on B cells and the viral spike protein expressed on the surface of infected cells (23). This unique cytolysis is inhibited by Ab specific for the viral spike protein, which *in vivo* is detected only after CNS virus replication has been initially controlled. Hence the absence of Ab early during infection could allow B cells to interact with virus infected cells *in vivo*, thereby contributing to viral control. However, convincing evidence for this mechanism was negated by studies in transgenic mice containing B cells unable to secrete anti-viral Ab (15). These mice cleared virus with similar kinetics as B cell deficient mice, excluding an effect of innate B cell function *in vivo*.

B cells are recruited to and activated in secondary lymphoid organs following infection (24,25). It has been shown that Ag within the CNS drains to cervical lymph nodes (CLN), which are the major sites of B cell activation following intrathecal administration of Ag (25). However, virus specific Ab secreting cells (ASC) appear in the spleen prior to detection in the CLN (17). While there was no major increase in virus specific IgM ASC, the frequencies of IgG ASC increases almost four fold in the CLN compared to spleen suggesting that the majority of initial activation occurs in the spleen, followed by transient CLN accumulation. No virus specific ASC were observed in blood, consistent with previous data suggesting that ASC do not migrate by this route (25). It is assumed that activated B cells/plasmablasts traffic to the target organ containing Ag where they differentiate to plasma cells and secrete Ab (25). Recruitment kinetics of anti-viral ASC into the CNS revealed that low frequencies detected prior to maximal viral replication increased only slightly as virus replication was controlled (17). Thus, based on: 1) the inability of B cells to express detectable anti-viral activity, 2) the paucity of plasma cells in the peripheral lymphoid compartments at any time post infection and 3) the appearance of serum Ab and virus specific ASC within the CNS following viral clearance, humoral immunity appears to have play a redundant role during acute MHV infection of the CNS.


# HUMORAL IMMUNITY AND VIRAL PERSISTENCE


Humoral immunity protects the CNS from infection by prevention of virus dissemination from peripheral sites. However, if CNS infection is already well established prior to induction of Ab responses, Ab can help further reduce infectious virus (26) or maintain viral persistence at non-detrimental levels (17). The beneficial effect of an early Ab response was demonstrated in the rat model of JHMV infection. Resistant Norway rats exhibit a more rapid and robust neutralizing Ab response, determined by detection of serum and CSF Ab and ASC compared to susceptible Lewis rats (27). Incomplete virus clearance in B cell deficient mice at days 12-14 p.i. support that neutralizing Ab, which first appears by day 10 p.i., indeed may act in concert with T cells during the later stages of acute infection to completely eradicate infectious virus from CNS, particularly as resident CD8 T cells lose their CTL activity (15,28). This prompted detailed analysis of the emergence of ASC relative to the decline in T cell activity within the CNS. Virus specific ASC were barely detectable in the peripheral compartments following virus clearance in mice (17). Furthermore, despite early recruitment of ASC into the CNS, only a minority was virus specific (17). However, the limited recruitment of virus specific ASC into the CNS towards the end of the acute phase, at days 10-12 p.i., was followed by a dramatic increase, which continued up to day 21 p.i. While there was a slight decrease in virus specific IgM ASC, frequencies of IgG ASC in the CNS remained constant. The majority of the ASC at this time point were of the IgG2b isotype indicating massive expansion of this subtype specific Ab, with fewer numbers of IgG2a and IgG1 ASC. Preferential expansion of ASC during this period of infection, concomitant with down regulation of CTL activity and inflammation, implies a major role for ASC in CNS during both the terminal phase of acute infection and during chronic infection.

Virus specific ASC are retained in the CNS, although there is a decline following the peak of expansion. The relatively long-term presence of ASC accompanied by the decline of other inflammatory cells, including T cells and macrophages, in addition to reactivation of virus in B cell deficient mice enforce the role of Ab in controlling viral persistence in the CNS. Following virus clearance from the CNS in wt mice, viral Ag can be detected beyond day 35 p.i. and viral RNA for up to a year (13). Virus specific ASC were found at higher frequencies in the CNS compared to bone marrow or any other peripheral compartment even at 90 days p.i. (Tschen & Stohlman, unpublished). This is in contrast to detection of plasma cells predominantly in the bone marrow following resolution of peripheral infections (24). The persistence of plasma cells within the CNS appears to be a common trait of many viral CNS infections (26,29). While the majority of Ag is cleared from the brain during persistence, viral Ag and RNA are detectable for longer periods of time in the spinal cord (13). Spinal cords also exhibited increased ASC by histology compared to brain, thereby signifying a causal relationship between prolonged retention of ASC and viral Ag / RNA (16). A variety of cellular and soluble factors could also contribute to the survival of these cells in the CNS: cytokines including TNF-$\alpha$, interleukins 1, 2, 4, 6 and 10, nerve growth factor, or stimulation with CD40L on T cells and Fc receptors on microglia / macrophages. The possible presence of chemokines such as SDF-$\alpha$, MIP-3$\beta$ or BLC could also contribute towards long-term retention of plasma cells in the brain. Although trapping of B cells cannot be excluded, these data suggest that the CNS provides a nourishing microenvironment that facilitates retention and maintenance of ASC.

Neutralizing Ab is maintained at relatively high levels in the serum beyond 90 days p.i. (Tschen, unpublished). Given the short half life of Ab, maintenance of Ab secretion may be attributed to one or a combination of factors: (i) re-exposure to Ag due to persistent low grade chronic viral infection; (ii) structural homologies of viral and CNS Ag generating cross reactive responses; (iii) persistent Ag presentation by dendritic cells; (iv) presence of long lived plasma cells in either bone marrow or the CNS. Work is ongoing to find a causal relationship between long-lived plasma cells and possible sustaining factors, viral control, and demyelination.

Virus recrudescence in the CNS of Ab deficient mice strongly implies a more crucial protective function of persisting ASC versus T cells during chronic infection (14-17). Virus reactivation is accompanied by sustained clinical disease and high mortality rates in contrast to recovery of wt mice that had cleared infectious virus (14-16). A similar picture was observed in transgenic mice, which contained B cells that could not secrete anti-viral Ab (15,16). Following initial virus clearance, virus reactivated in the CNS of the transgenic mice by day 14 p.i. albeit at a slower, but statistically insignificant rate compared to B cell deficient mice, thus ruling out a prominent role of innate B cell effector function (15,16). In addition to multiple Ag positive cells compared to wt mice, histological examination suggested differential Ag tropism in the transgenic mice containing B cells as compared to B cell deficient mice during reactivation (15). While Ag was detected in all CNS cell types in B cell deficient mice, similar to acute infection, the transgenic mice containing B cells harbored Ag predominantly in oligodendrocytes. Preferential elimination of virus from astrocytes and microglia, suggests either B cell lytic activity during persistence or increased class II mediated stimulation of virus specific CD4 T cells. Irrespective of potential B cell mediated lysis in transgenic mice, this mechanism is irrelevant in wt mice, which have neutralizing Ab in the CNS at this time (15). Oligodendroglial tropism in these mice is reminiscent of the resistance of these cells to perforin-mediated cytolysis (30). However, it appears that following initial virus clearance, the CNS environment preempts further T cell mediated antiviral effects. There is no evidence for either increased CD8 T cell recruitment or re-expression of CTL activity in response to reactivation in Ab deficient mice (15,28). Such an altered CNS environment, likely resulting from T cell mediated or virus induced pathology, emphasizes the necessity for alternative, less destructive modes of protection during recovery, such as provided by Ab.

These observations are reinforced by protection of B cell deficient mice treated with polyclonal neutralizing Ab at time points simulating the appearance of serum neutralizing Ab in wt mice following i.c. inoculation (14-16). Ab transfer reduced virus to almost undetectable levels in recrudescing recipients at 21 days p.i., compared to high titers in the CNS of control mice. These results in B cell deficient mice, in conjunction with the burst of virus specific ASC in wt mice between days 12 and 21 p.i., clearly suggested that control of viral persistence in glial cells is solely dependent on anti-viral Ab. Further investigation of the mechanisms involved in Ab mediated protection revealed that only neutralizing Ab and to a limited extent, fusion inhibiting Ab could reduce virus reactivation (31). Monoclonal Ab specific for matrix, nucleocapsid protein or even spike protein with no neutralizing activities were unable to suppress virus reactivation (31). Suppression of virus reactivation appears to require the constant presence of Ab, as virus reemerged following discontinuation of Ab administration resulting in increased mortality rates. The transient nature of virus control is comparable to other infectious models wherein immunocompromised mice required continuous infusion of Ab to control virus replication (26,32).

Other humoral components do not appear to play a major role in virus clearance. Analysis of MHV infection in mice devoid of complement (C3) revealed no role for complement in virus clearance (16,18). Furthermore, complete clearance of virus in FcR deficient mice (16) and the ability of neutralizing F(ab)2 fragments to afford protection (33) indicate a minor role for ADCC mechanisms in virus clearance, although each mechanism may suffice to make individual contributions.


## ANTIBODY AND DEMYELINATION

Despite the presence of oligoclonal Ig bands in CSF of MS patients, there is little evidence for the involvement of Ab responses during most experimental virus induced demyelinating diseases. Robust demyelination is induced by both JHMV and MHV-A59 strains in B cell deficient mice as well as transgenic mice containing B cells, but deficient in virus specific Ab (14-16). These data support multiple studies implicating T cells and macrophages/microglia as the prominent players in the initiation of demyelination during MHV infection (12,13). However demyelination in Ab deficient mice does not exclude other humoral mechanisms, including complement and FcR, as contributors to the demyelinating process. Furthermore, infection of nude mice demonstrated the presence of demyelinating plaques albeit to a lesser degree than wt mice, supporting a role for humoral immune responses in the demyelinating process (34). Infected SCID and RAG-/- mice, deficient in both B and T cells, by contrast, do not develop demyelination. Adoptive transfer of splenocytes from immunized nude mice into infected SCID mice however failed to induce demyelination (34), leaving a pathogenic role for B cells controversial. Nevertheless, a pathogenic role for humoral immunity was also suggested by the detection of IgG, complement (C9) deposits and plasma cells in areas of actively demyelinating plaques in Lewis rats (35). While an Ab-dependent role in initiation of demyelination is not established, the presence of ASC with viral as well as yet unidentified specificities in the CNS does not exclude a possible involvement in the demyelinating process. Following Ag clearance in wt mice, demyelination is reduced compared to increasing levels of demyelination in B cell deficient mice (14,15). Whether this is due to higher viral load or absence of B cells or Ab is unclear. Although the effect of Ab on remyelination has not been studied in this model, Ab mediated control of viral recrudescence and thereby prevention of further virus induced cell death, results in less severe demyelination in wt mice as compared to infected B cell deficient mice. Furthermore, preliminary analysis of B cell deficient mice protected by the transfer of Ab, suggests an ameliorating, rather than detrimental effect on demyelination, similar to the TMEV model (31,36). These results support an overall beneficial role of Ab in MHV induced CNS pathogenesis, predominantly by enhancing elimination of infectious virus and controlling viral persistence. Further studies are required to assess potential effects of Ab and other components of humoral immunity on the dynamic processes of myelin loss and repair.

## CONCLUSIONS

In summary, despite unequivocal evidence that acute CNS infection is controlled by T cells with minimal B cell involvement, B cells and Ab regulates viral persistence. Virus reactivation in the CNS in mice devoid of antiviral Ab results in increased demyelination. A protective role of B cells is supported by accumulation of virus specific ASC following T cell mediated clearance. The prominent mechanism of Ab mediated protection appears to reside in its neutralizing activity. Furthermore, humoral immune mediated control does not appear to initiate or enhance the severity of demyelination. Thus, an evolving model of biphasic immune control of viral infection of the CNS has developed with acute clearance and initiation of demyelination being controlled by T cells while viral persistence and possibly repair is regulated by humoral immunity.

## REFERENCES:

1. Correale J., de Los Milagros Bassani Molinas M. Oligoclonal bands and antibody responses in Mulitple sclerosis. J Neurol. 2002. 249: 375-89.
2. Connor O., Kevin C., Bar-Or A., Hafler D. A. The Neuroimmunology of Multiple Sclerosis: Possible roles of T and B lymphocytes in Immunopathogenesis. J Clin Immunol. 2001. 21: 81-92.
3. Kabat E. A., Glusman M., Knaub V. Quantitative estimation of the albumin and gamma globulin in normal and pathologic cerebrospinal fluid by immunochemical methods. Am J Med. 1948. 4: 653-62.

4. Bornstein M. B., Appel S. H. The application of tissue culture to the study of experimental allergic encephalomyelitis. I. Patterns of demyelination. J Neuropathol Exp Neurol. 1961. 20: 141-7.

5. Esiri M. M., Immunoglobulin-containing cells in multiple sclerosis plaques. Lancet. 1977. ii: 478-80.

6. Genain C. P., Cannella B., Hauser S. L., Raine C. S. Identification of autoantibodies associated with myelin damage in multiple sclerosis. Nat. Med. 1999. 5: 170-5.

7. Cross A. H., Trotter J. L., Lyons J-A. B cells and antibodies in CNS demyelinating disease. J NeuroImm. 2001. 112: 1-14.

8. Colombo M., Dono M., Gazzola P., Roncella S., Valetto A., Chiorazi N., Mancardi G. L., Ferrarini M. Accumulation of clonally related B lymphocytes in the cerebrospinal fluid of multiple sclerosis patients. 2001. J Immunol. 164: 2782-89.

9. Gilden D. H., Devlin M. E., Burgoon M. P., Owens G. P. The search for virus in multiple sclerosis brain. Mult Scler. 1996. 2: 179-83.

10. Archelos J. J., Storch M. K., Hartung H-P. The role of B cells and autoantibodies in Multiple sclerosis. Annals Neurol 2000. 47: 694-706.

11. Lumsden C. E. The immunopathogenesis of the multiple sclerosis plaque. Brain Res. 1971. 28: 365-90.

12. Stohlman, S. A., Bergmann C. C., Perlman S. "Mouse hepatitis virus." In *Persistent Viral Infections* R. Ahmed, I. Chen. ed. New York, NY: Wiley Publishers. 1999.

13. Marten N. W., Stohlman S. A., Bergmann C. C. MHV infection of the CNS: Mechanisms of immune-mediated control. Viral Immunol. 2001. 14: 1-18.

14. Lin M. T., Hinton D. R., Marten N. W., Bergmann C. C., Stohlman S. A. Antibody prevents virus reactivation within the central nervous system. J Immunol. 1999. 162: 7358-68.

15. Ramakrishna, C., Stohlman S. A.,. Atkinson R., Shlomchik M. J.,. Bergmann C. C. Mechanisms of central nervous system viral persistence: the critical role of antibody and B cells. J Immunol. 2002. 168: 1204-11.

16. Matthews A. E., Weiss S. R., Shlomchik M. J., Hannum L. G., Gombold J. L., Paterson Y. Antibody is required for clearance of infectious murine hepatitis virus A59 from the central nervous system, but not the liver. J Immunol. 2001. 167: 5254-63.

17. Tschen, S., Bergmann C. C., Ramakrishna C., Morales S., Atkinson R., Stohlman S. A. 2002. Recruitment kinetics and composition of antibody secreting cells within the central nervous system following viral encephalomyelitis. J. Immunol. 168: 2922-29.

18. Fleming J. O., Shubin R. A., Sussman M. A., Casteel N., Stohlman S. A. Monoclonal antibodies to the matrix (E1) glycoprotein of mouse hepatitis virus protect mice from encephalitis. Virology. 1989. 168: 162-7.

19. Buchmeier M. J., Lewicki H. A., Talbot P. J., Knobler R. L. Murine hepatitis virus-4 (strain JHM)-induced neurologic disease is modulated in vivo by monoclonal antibody. Virology. 1984. 132: 261-70.

20. Nakanaga K., Yamanouchi K., Fujiwara K. Protective effect of monoclonal antibodies on lethal mouse hepatitis virus infection in mice. J Virol. 1986. 59: 168-71.

21. Perlman S., Schelper R., Bolger E., Ries D. Late onset, symptomatic, demyelinating encephalomyelitis in mice infected with MHV-JHM in the presence of maternal antibody. Microb Pathog. 1987. 2: 185-94.

22. Kolb A. J., Pewe L., Webster J., Perlman S., Whitelaw C. B., Siddel S. G. Virus-neutralizing monoclonal antibody expressed in milk of transgenic mice provides full protection against virus induced encephalitis. J. Virol. 2001. 75:2803-9.

23. Morales S., Parra B., Ramakrishna C., Blau D. M., Stohlman S. A. B cell mediated lysis of cells infected with the neurotropic JHM strain of mouse hepatitis virus. Virol. 2001. 286: 160-7.

24. Slifka M. K., Ahmed R. Long lived plasma cells: a mechanism for maintaining persistent antibody production. Curr Opin Immunol. 1998. 10: 252-8.

25. Knopf P.M., Harling-Berg C. J., Cserr H. F., Basu B., Sirulnick E. J., Nolan S. C., Park J. T., Keir G., Thompson E. T., Hickey W. F. Antigen-Dependent intrathceal antibody synthesis in the normal rat brain: Tissue Entry and local retention of Antigen specific B cells. J Immunol. 1998. 161: 692-701.
26. Griffin D., Levine B., Tyor W., Ubol S., Despres P. The role of antibody in recovery from alphavirus encephalitis. Immunol Rev. 1997. 159: 155-61.
27. Schwender S., Imrich H., Dorries R. The pathogenic role of virus-specific antibody secreting cells in the central nervous system of rats with different susceptibility to coronavirus-induced demyelinating encephalitis. Immunol. 1991. 74: 533-8.
28. Bergmann C. C., Ramakrishna C., Kornacki M.,, Stohlman S. A.  Impaired T cell immunity in B cell-deficient mice following viral central nervous system infection.  J. Immunol. 2001. 167:1575-83.
29. Smith-Norowitz T. A., Sobel R. A., Mokhatarian F. B cells and antibodies in the pathogenesis of myelin injury in Semliki Forest Virus encephalomyelitis. Cell Immunol. 2000. 200: 27-35.
30. Lin M. T., Stohlman S. A., Hinton D. R. Mouse hepatitis virus is cleared from the central nervous systems of mice lacking perforin-mediated cytolysis. J Virol. 1997. 71:383-91.
31. Ramakrishna C., Bergmann C. C., Atkinson R., Stohlman S. A. Control of central nervous system persistence by neutralizing antibody. J. Virol. Submitted.
32. Furrer, E., T. Bilzer, L. Stitz, and O. Planz. Neutralizing antibodies in persistent Borna disease virus infection: prophylactic effect of gp94-specific monoclonal antibodies in preventing encephalitis. J Virol. 2001. 75: 943-51.
33. Lamarre A., Talbot P. J. Protection from lethal Coronavirus infection by immunoglobulin fragments. J Immunol. 1995. 154: 3975-84.
34. Houtman J. J., Fleming J. O. Dissociation of demyelination and viral clearance in congenitally immunodeficient mice infected with murine coronavirus JHM. J Neurovirol. 1996. 2: 101-10.
35. Zimprich F., Winter J., Wege H., Lassmann H. Coronavirus induced primary demyelination: indications for the involvement of a humoral immune response. Neuropath Applied Neurobiol. 1991. 17: 469-84.
36. Drescher K. M., Pease L. R., Rodriguez M. Antiviral immune responses modulate the nature of central nervous system (CNS) disease in a murine model of multiple sclerosis. Immunol Rev. 1997. 159:177-93.

Chapter C6

# THE ROLE OF T CELL EPITOPES IN CORONAVIRUS INFECTION

Taeg S. Kim* and Stanley Perlman*,(
*Affiliation: *Interdisciplinary Program in Immunology, (Departments of Microbiology and Pediatrics, University of Iowa, Iowa City, IA .*

Abstract:     Multiple MHV-specific CD4 and CD8 T cell epitopes have been identified in C57Bl/6 and BALB/c mice. In particular, at least two CD8 T cell epitopes are recognized in C57Bl/6 mice. In one model of MHV persistence, mutations are detected in the immunodominant CD8 T cell epitope recognized in this strain. These mutations contribute to virus persistence and to the development of more severe clinical disease.

Key words: Coronavirus, CTL escape, demyelination

## 1.        INTRODUCTION

In most viral infections, host defense mechanisms are efficient and result in rapid clearance of the virus. Stable memory populations of T and B cells develop and protect the host from subsequent infections with the same agent. Viruses, in turn, have evolved mechanisms to evade the host immune response and are able to establish long-term persistent infections. Much has been learned over the past several about the strategies that viruses use to evade the immune response [reviewed in (1)]. This review will be focused on one strategy used by viruses to persist, i.e., the selection of cytotoxic CD8 T cell (CTL) escape variants. The first part of this article will discuss aspects of the T cell immune response to MHV, with a focus on CD4 and CD8 T cell epitopes recognized in rodents. Second, CTL escape in the context of mice infected with the neurotropic JHM strain of mouse hepatitis virus (MHV) will be described in detail.

## 2.  CELL-MEDIATED IMMUNITY DURING MHV INFECTION OF RODENTS

The importance of CD4 and CD8 T cells in virus clearance in rodents infected with MHV is well documented, as described elsewhere in this volume (Chapter C4). In several studies, the CD8 and CD4 T cell epitopes recognized by MHV-specific lymphocytes have been identified. C57Bl/6 (B6) and BALB/c mice are used in most studies of MHV-induced pathogenesis and, as a consequence, most of the epitopes are H-2$^b$- or H-2$^d$-restricted. A summary of the epitopes identified thus far is shown in Table 1. These epitopes were identified using either splenocytes after *in vitro* stimulation or, alternatively, lymphocytes harvested from the CNS of infected mice. In the latter case, cells were analyzed in direct *ex vivo* cytotoxicity assays. In BALB/c mice, a CD8 T cell epitope encompassing residues 318-326 of the nucleocapsid (N) protein (epitope N318), was immunodominant (2) whereas, in B6 mice, two CD8 T cell epitopes encompassing residues 510-518 and 598-605 of the surface (S) protein (epitopes S510 and S598) were recognized (3, 4). Both are located within a hypervariable region of the protein that can tolerate deletion or mutation without loss of viability (5). CD4 T cell epitopes have also been identified in both strains of mice. In BALB/c mice, CD4 T cell epitopes were identified within the N protein whereas in C57BL/6 mice such epotopes were located within the S and transmembrane (M) proteins (6-8). These CD4 T cell epitopes are present outside the hypervariable region of the S protein.

**Table 1**. Summary of epitopes recognized by CD4 and CD8 T cells in MHV-infected rodents

| Cell type | Host | Protein | Residues |
|---|---|---|---|
| CD8 T cell[2] | BALB/c mice | N | 318-326 |
| CD8 T cell[3,4] | B6 mice | S | 510-518 |
| CD8 T cell[4] | B6 mice | S | 598-605 |
| CD4 T cell[9] | Lewis rat | N | 361-458 |
| CD4 T cell[7] | B6 mice | M | 128-147 |
| CD4 T cell[8,10] | B6 mice | S | 329-343 |
| CD4 T cell[10] | BALB/c mice | S | 329-343 |
| CD4 T cell[6] | BALB/c mice | N | 266-279 |
| CD4 T cell[8] | B6 mice | S | 358-372 |
|  |  | S | 408-422 |

## 3.　　CTL ESCAPE IN MHV-INFECTED MICE

### 3.1　Overview of selection of CTL escape variants in viral encephalomyelitis

Infection with virulent variants of the JHM strain of MHV cause an acute fatal encephalitis in all susceptible rodents. Disease is largely confined to the central nervous system, with variable, albeit minor, infection of the liver. A persistent infection was noted in several studies when either mice were infected with attenuated strains of virus or when infection with virulent virus was modified by treatment with anti-viral antibodies or T cells (11). In one model, clinical disease manifested by hindlimb paralysis and histological evidence of demyelination developed several weeks after virus inoculation (12). In this model, suckling B6 mice (10 to 14 days old) were inoculated intranasally with a virulent variant of MHV. If nursed by naïve dams, they uniformly developed an acute encephalitis. However, when nursed by dams previously immunized

**Table 2**. Summary of mutations in epitope S510 identified in B6 mice with hindlimb paralysis.

| Position | 1 | 2 | 3 | 4 | 5 | 6 | 7 | 8 | 9 |
|---|---|---|---|---|---|---|---|---|---|
| Contact | | *M | M | T | M | T | M/T | | M |
| WT epitope | **C** | **S** | **L** | **W** | **N** | **G** | **P** | **H** | **L** |
| Variant epitopes | | F(2)$^\Delta$ | R(2) | R(7) | H(1) | V(2) | H(2) | | |
| | | Y(1) | F(3) | G(1) | D(2) | R(1) | L(6) | | |
| | | P(2) | P(2) | C(1) | S(12) | E(1) | | | |
| | | | | | T(1) | | | | |
| *(untested)* | | | | | K(1) | | | R(1) | F(1) |

The immunogenicity of the mutated epitopes was tested in cytotoxicity assays. The amino acid changes listed above resulted in a greater than 20 –fold decrement in ability to sensitize cells for CTL lysis , except for the L to P change noted at position 2 (4 fold decrement) (15). Some mutations were not analyzed in these assays *(untested)*. *MHC (M) or TCR (T) contact residues based on published data (16). $^\Delta$ Numbers in parenthesis represent the number of animals from which the indicated mutations were detected.

to MHV, they were protected from the acute encephalitis. A variable fraction (40-90%) developed clinical disease and a demyelinating encephalomyelitis 3-8 weeks after inoculation. Infectious virus could be isolated from all mice with clinical disease but not from those that remained asymptomatic. Subsequent analysis of viral RNA harvested from the infected CNS revealed the presence of mutations in the immunodominant CD8 T cell epitope (epitope S510) in all samples of RNA and viruses isolated from symptomatic mice (Table 2) (13). These mutations abrogated recognition by epitope S510-specific CD8 T cells in direct *in vivo* cytotoxicity assays using CNS-derived cells. These mutations were detected at early times and mutations were not detected in epitope S598 or in the regions flanking the epitopes (14). Mutations in epitope S510 were not detected in most mice that remained asymptomatic or in persistently infected mice with severe combined immunodeficiency (SCID) (14). These results suggest that CTL escape variants were selected by immune pressure and consequently contributed to the establishment of virus persistence and to the development of a demyelinating encephalomyelitis.

## 3.2 Requirements for selection of CTL escape variants in the CNS

1. Immunodominance. It is conceivable that a CTL response directed against a single immunodominant epitope favored the emergence of virus mutants abrogating CTL recognition. In B6 mice infected with MHV, epitope S510 was immunodominant, with up to 50% of the CD8 T cell response in the CNS directed at this epitope, as determined by either direct *ex vivo* intracellular interferon-γ assays or by staining with MHC class I/peptide tetramers (17, 18). The CD8 T cells were fully activated as measured by cytokine secretion, cytotoxic function and expression of activation markers, at least during the acute infection. This strong immune response to the immunodominant epitope is likely to be a major factor facilitating the selection of CTL escape variants. In support of this, mutations in the subdominant CD8 T cell epitope (epitope S598), an epitope with substantially less functional avidity have never been detected. This was true even when suckling mice were infected with virus mutated in epitope S510 so that epitope S598 became the dominant target for the CD8 T cell response (19). In other viral infections, epitopes with strong functional avidity were preferentially mutated at early times after inoculation (20). One apparent exception to this observation is that mutations were never identified in the immunodominant CD4 T cell epitope recognized in B6 mice (epitope

M133-147, Table 1), but mutations were detected in a subdominant CD4 T cell epitope (epitope S328-347, Table 1). However, the biological significance of this mutation is not known since CD4 T cell recognition to the mutated epitope was not affected by this mutation (8).

2. Hypervariable region. It is imperative that the CTL epitope mutations do not inactivate the virus. The two CTL epitopes (epitopes S510 and S598) recognized in B6 mice are located within a region of the protein that is highly permissive to both deletions and amino acid changes without the loss of infectivity (5). Presence of epitopes within this hypervariable region may provide an optimal genetic background for CTL escape, while keeping virus viable. Epitope S598 is also located in this hypervariable region, but mutations in this epitope were not detected. At present, it is not known whether the lack of mutations in epitope S598 reflect an effect on virus viability. It is now possible to address this issue, using recently described methods of reverse genetics (21). Alternatively, there may be insufficient immune pressure to select CTL escape mutations because the epitope has a low functional avidity (4) or because it is expressed relatively poorly on the cell surface. Mutations in epitopes with high functional avidity were detected at early times during the course of other persistent viral infections whereas mutations in epitopes with low functional avidity were not detected until disease had progressed (20, 22, 23).

3. Infection of suckling B6 mice. CTL escape variants were selected only in B6 mice (H-2$^b$ background) inoculated at the suckling stage (10-14 days old) with virulent MHV-JHM and protected from acute encephalitis by nursing with dams previously immunized to MHV. In other models of MHV persistence, immunocompetent adult mice were inoculated with attenuated strains of MHV (11). Virus was cleared under these conditions, but viral RNA and ongoing demyelination were readily detected. CTL escape mutants were never detected in these mice. The host strain is also important in the selection of CTL escape mutants. Suckling BALB/c mice never develop a late onset demyelinating encephalomyelitis. The immunodominant CD8 T cell epitope recognized in BALB/c mice is located in a protein (nucleocapsid [N] protein) that is highly conserved among coronaviruses (2, 24). Consistent with the lack of clinical disease, mutations are never detected in this epitope (N318-326).

One explanation for the observation that CTL escape only occurred in suckling B6 mice relates to the role of anti-MHV antibody in pathogenesis. Vertical transmission of maternal neutralizing antibody plays a crucial role not only in providing protection from acute encephalitis but also for establishing a milieu for selection of CTL escape variants. Levels of neutralizing antibodies need to be sufficient to control the acute infection but insufficient to achieve sterile immunity. Our recent data showed that, as maternal antibody waned, production of anti-MHV antibody by infected suckling B6 mice was substantially lower than by suckling BALB/c mice. Most striking was the lack of anti-MHV antibody producing cells in the CNS of B6 mice infected at the suckling stage (A. Dandekar and SP, unpublished data). Since anti-MHV antibodies provide a crucial role in preventing recrudescence of infectious virus in infected mice (25), lack of locally produced antibody may allow virus to continue to replicate and therefore predispose to the selection of CTL escape mutants. Interestingly, suckling BALB/c mice, which never show evidence of CTL escape, mount a robust humoral immune response in the CNS. Of note, Brown Norway, but not Lewis rats, mount a strong anti-MHV antibody response in the CNS. MHV causes a clinically silent demyelinating encephalomyelitis in Brown Norway rats, whereas Lewis rats develop symptomatic clinical disease, with elevated levels of infectious virus present. These results also suggest an important role for CNS-derived anti-MHV antibody in controlling the infection (26).

4. Monospecific polyclonal epitope S510-specific CTL response. As discussed above, CTL immune pressure focused on an immunodominant CTL epitope may favor selection of CTL escape variants by virtue of the significant loss of immune recognition when the epitope is mutated. However, this phenomenon may be less likely if CD8 T cells specific to the epitope are polyclonal (27). To determine the diversity of the epitope S510-specific T cell response, TCR Vβ element usage and the complexity of the complementary determining region 3 (CDR3) were determined. Epitope S510-specific CD8 T cells were obtained from the CNS of acutely and chronically infected mice by staining with peptide S510-loaded MHC class I tetramers and FACS analysis. A large number of different Vβ elements were detected within the epitope S510-specific population. Furthermore, analysis of CDR3 regions from this population revealed the presence of 400-600 different clonotypes per infected mouse, with many sequences common in all animals (17). This polyclonal response was not protective against the CTL escape, however, since it was functionally monospecific, with a strong focus on the center of the MHC class I/peptide complex (15).

5. Anatomical location of persistent virus. Since the CNS is a site of immune privilege, CTL escape mutants, once selected, may be able to replicate more efficiently in an environment with a less than optimal anti-MHV immune response. This may contribute to high level virus persistence, and eventually, to the development of clinical disease.

# 4. BIOLOGICAL SIGNIFICANCE OF CTL ESCAPE VARIANTS IN MHV-INFECTED MICE

Accumulating data suggest that evasion of the immune response by selection of CTL escape mutants occurs in several persistent infections, including humans infected with HIV, hepatitis B virus or hepatitis C virus and primates infected with SIV or hepatitis C virus (20, 28-30). A causal relationship between the presence of CTL escape variants and clinical outcomes is difficult to establish in infected humans or primates. Thus, the maternal antibody protection model described above is useful for determining the role of these variants in virus pathogenesis.

Pewe et al. (19) showed that escape from the CTL response was a major factor in disease progression in MHV-infected mice. Infection of suckling mice with CTL escape mutant virus resulted in higher mortality and morbidity than observed in mice infected with wild type virus. Furthermore, sequencing of infectious virus isolated from chronically infected mice revealed a lack of heterogeneity among isolates from a single animal. These data indicated that viral variants were selected during early infection and, once selected, they were fixed permanently, possibly due to continuous positive selection by CTL, ultimately leading to clinical disease.

# 5. CONCLUDING REMARKS

CTL escape has been demonstrated in several persistent infections. MHV-infected mice provide an excellent model for determining the role of these phenomena in viral pathogenesis. The results thus far show that CTL escape variants have a selective advantage and enhance clinical disease and demyelination. Although a specific virus has not been identified as an etiological agent in patients with MS, the development of CTL escape mutants provides a plausible mechanism for how a persistent, virus infection controlled by the immune system could progress to cause more significant disease. Further efforts will be directed at determining the mechanism of CTL escape and to determine why CTL escape has been identified in only a fraction of persistent viral infections.

# REFERENCES.

1. **Tortorella D, Gewurz B, Furman M, Schust D, Ploegh H-, 2000.** Viral subversion of the immune system. *Ann. Rev. Immunol.* 2000;18:861-926.

2. **Bergmann C, McMillan M, Stohlman SA.** Characterization of the Ld-restricted cytotoxic T-lymphocyte epitope in the mouse hepatitis virus nucleocapsid protein. *J. Virol.* 1993;67:7041-49.

3. **Bergmann CC, Yao Q, Lin M, Stohlman SA.** The JHM strain of mouse hepatitis virus induces a spike protein-specific D^b-restricted CTL response. *J. Gen. Virol.* 1996;77:315-325.

4. **Castro RF, Perlman S.** CD8+ T cell epitopes within the surface glycoprotein of a neurotropic coronavirus and correlation with pathogenicity. *J. Virol.* 1995;69:8127-8131.

5. **Parker SE, Gallagher TM, Buchmeier MJ.** Sequence analysis reveals extensive polymorphism and evidence of deletions within the E2 glycoprotein gene of several strains of murine hepatitis virus. *Virology.* 1989;173:664-673.

6. **Van der Veen RC.** Immunogenicity of JHM virus proteins: Characterization of a CD4+ T cell epitope on nucleocapsid protein which induces different T-helper cell subsets. *Virology.* 1996;225:339-346.

7. **Xue S, Jaszewski A, Perlman S.** Identification of a CD4$^+$ T cell epitope within the M protein of a neurotropic coronavirus. *Virology.* 1995;208:173-179.

8. **Xue S, Perlman S.** Antigen specificity of CD4 T cell response in the central nervous system of mice infected with mouse hepatitis virus. *Virology.* 1997;238:68-78.

9. **Wege H, Schliephake A, Korner H, Flory E, Wege H.** An immunodominant CD4+ T cell site on the nucleocapsid protein of murine coronavirus contributes to protection against encephalomyelitis. *J. Gen. Virol.* 1993;74:1287-1294.

10. **Heemskerk M, Schoemaker H, Spaan W, Boog C.** Predominance of MHC class II-restricted CD4$^+$ cytotoxic T cells against mouse hepatitis virus A59. *Immunology.* 1995;84:521-527.

11. **Stohlman SA, Bergmann CC, Perlman S.** Mouse hepatitis virus. In: Ahmed R, Chen I, eds. *Persistent Viral Infections.* New York: John Wiley & Sons, Ltd.; 1998:537-557.

12. **Perlman S, Schelper R, Bolger E, Ries D.** Late onset, symptomatic, demyelinating encephalomyelitis in mice infected with MHV-JHM in the presence of maternal antibody. *Microb. Pathog.* 1987;2:185-194.

13. **Pewe L, Wu G, Barnett EM, Castro R, Perlman S.** Cytotoxic T cell-resistant variants are selected in a virus-induced demyelinating disease. *Immunity.* 1996;5:253-262.

14. **Pewe L, Xue S, Perlman S.** Cytotoxic T cell-resistant variants arise at early times after infection in C57Bl/6 but not in SCID mice infected with a neurotropic coronavirus. *J. Virol.* 1997;71:7640-7647.

15. **Pewe L, Perlman S.** Immune response to the immunodominant epitope of mouse hepatitis virus is polyclonal, but functionally monospecific in C57Bl/6 mice. *Virology.* 1999;255:106-116.

16. **Hudrisier D, Mazarguil H, Oldstone MBA, Gairin JE.** Relative implication of peptide residues in binding to major histocompatibility complex class I H-2D^b: Application to the design of high-affinity, allele-specific peptides. *Mol. Immunol.* 1995;32:895-907.

17. **Pewe L, Heard SB, Bergmann CC, Dailey MO, Perlman S.** Selection of CTL escape mutants in mice infected with a neurotropic coronavirus: Quantitative estimate of TCR diversity in the infected CNS. *J. Immunol.* 1999;163:6106-6113.

18. **Bergmann CC, Altman JD, Hinton D, Stohlman SA.** Inverted immunodominance and impaired cytolytic function of CD8+ T cells during viral persistence in the central nervous system. *J. Immunol.* 1999;163:3379-3387.

19. **Pewe L, Xue S, Perlman S.** Infection with cytotoxic T-lymphocyte escape mutants results in increased mortality and growth retardation in mice infected with a neurotropic coronavirus. *J. Virol.* 1998;72:5912-5918.

20. **O'Connor DH, Allen TM, Watkins DI.** Cytotoxic T-lymphocyte escape monitoring in simian immunodeficiency virus vaccine challenge studies. *DNA Cell Biol.* 2002;21:659-64.
21. **Masters PS.** Reverse genetics of the largest RNA viruses. *Adv Virus Res.* 1999;53:245-64.
22. **Vogel TU, Friedrich TC, O'Connor DH, et al.** Escape in one of two cytotoxic T-lymphocyte epitopes bound by a high- frequency major histocompatibility complex class I molecule, Mamu-A*02: a paradigm for virus evolution and persistence? *J Virol.* 2002;76:11623-36.
23. **Mothe BR, Horton H, Carter DK, et al.** Dominance of CD8 responses specific for epitopes bound by a single major histocompatibility complex class I molecule during the acute phase of viral infection. *J Virol.* 2002;76:875-84.
24. **Parker MM, Masters PS.** Sequence comparison of the N genes of five strains of the coronavirus mouse hepatitis virus suggests a three domain structure for the nucleocapsid protein. *Virology.* 1990;179:463-8.
25. **Lin MT, Hinton DR, Marten NW, Bergmann CC, Stohlman SA.** Antibody prevents virus reactivation within the central nervous system. *J. Immunol.* 1999;162:7358-68.
26. **Schwender S, Imrich H, Dorries R.** The pathogenic role of virus-specific antibody-secreting cells in the central nervous system of rats with different susceptibility to coronavirus-induced demyelinating encephalitis. *Immunology.* 1991;74:533-8.
27. **Ishikawa T, Kono D, Chung J, et al.** Polyclonality and multispecificity of the CTL response to a single viral epitope. *J. Immunol.* 1998;161:5843-5850.
28. **Erickson S, de Sauvage F, Kikly K, et al.** Decreased sensitivity to TNF but normal T-cell development in TNF-receptor-2-deficient mice. *Nature.* 1994;372:560-563.
29. **Borrow P, Shaw GM.** Cytotoxic T-lymphocytes escape viral variants: how important are they in viral evasion of immune clearance in vivo? *Immunol/ Rev.* 1998;164:37-51.
30. **Perlman S, Wu G.** Selection of and evasion from cytotoxic T cell responses in the central nervous system. In: Buchmeier M, Campbell I, eds. *Neurovirology: Viruses and the Brain.* San Diego, CA: Academic Press; 2001:219-242.

Chapter C7

# CORONAVIRUSES AND NEUROANTIGENS:
*myelin proteins, myelin genes*

Pierre J. Talbot, Annie Boucher, Pierre Duquette [1], and Edith Gruslin
*INRS-Institut Armand-Frappier, Institut national de la recherche scientifique, Université du Québec, Laval, Québec, Canada H7V 1B, and [1] Pavillon Notre-Dame, Centre hospitalier de l'Université de Montréal, Université de Montréal, Montréal, Québec, Canada H2L 4K8*

**Abstract:** Multiple sclerosis (MS) is an autoimmune disease in which autoreactive T cells specific to central nervous system (CNS) myelin antigens are activated. Although disease etiology remains unknown, coronaviruses are suspected to be involved in MS pathology. Molecular mimicry, the recognition of two antigens by a single immune cell, could be the mechanism explaining the link between a viral infection and MS through activation of myelin-reactive T cells by a virus infection in a genetically predisposed individual. Evidence supporting this hypothesis in humans has been accumulated in our laboratory. Human coronavirus (HCoV) – myelin cross-reactive T-cell lines (TCL) were predominantly found in MS patients compared to patients with other neurological or inflammatory diseases, or healthy controls. Moreover, virus-myelin T cell cross-reactivity was confirmed at the clonal level. Molecular mimicry between infectious pathogens such as the ubiquitous human respiratory coronaviruses could, in genetically susceptible individuals, play a role leading to the development of MS. Together with other possible mechanisms such as bystander effects, epitope spreading or even superantigenic activities, this pathogen-associated immune induction could play a role in maintaining and broadening the autoimmune response associated with MS pathology.

**Key words:** autoimmunity, T cells, central nervous system, virus, molecular mimicry

## 1. INTRODUCTION

Multiple sclerosis (MS) is an autoimmune disease of the central nervous system (CNS) characterized by inflammation and myelin destruction. Even though MS was described more than 130 years ago, its exact etiology remains unknown. A genetic predisposition is suggested and HLA genes are suspected to be involved (1, 2). Environmental factors are also associated with the triggering of this pathology. Indeed, epidemiological studies with identical twins demonstrate a low MS concordance between siblings and specific patterns of worldwide disease distribution point out the role for environmental factors, such as microbial agents (3, 4).

Amongst environmental factors potentially related to MS is a long list of more than 20 viruses. These viruses have been associated with MS by different experimental approaches: isolation of virus particles from tissues of MS patients (5), antibodies specific to virus found in central nervous system (CNS) of MS patients (6, 7) or demonstration of virus gene expression in the CNS of MS patients (8-14).

## 2.        NEUROANTIGENS

Tolerance is described as a mechanism established to create and maintain a state of non-response to autoantigens. In the CNS, some proteins were found to activate the immune system: myelin basic protein (MBP), proteolipid protein (PLP), myelin associated glycoprotein (MAG), myelin oligodendrocyte glycoprotein (MOG), myelin oligodendrocyte basic protein (MOBP), CNPase (2',3'- cyclic nucleotidyl 3'-phosphodiesterase), transaldolase, S100$\beta$ and $\alpha$-B-crystallin (15, 16). All of these proteins were found at one time or another to activate autoreactive T lymphocytes in MS patients and therefore are suspected in maintaining or amplifying autoimmune reactions observed in the disease (17-20). Those neuroantigens, previously cryptic, might have been liberated in the CNS milieu following damages to the myelin sheath, and made available for antigen presentation.

How tolerance is broken in MS is still not understood. The suggested potential mechanisms are related to the initiation or propagation of the autoimmunity observed in MS. Superantigens could lead to a potent non-specific activation of a given T-cell receptor (TCR) V$\beta$ family. Non-specific bystander effects of inflammation could also hold a key role. Moreover, determinant (epitope) spreading would broaden the immune response towards the recognition of new autoantigens. Finally, molecular mimicry between self-proteins and microbes represents another possible mechanism that could be involved in breaking tolerance and could explain the association of several infectious pathogens with MS etiology and exacerbations.

## 3.  MS AND CORONAVIRUSES

Coronaviruses are enveloped virus with a diameter of approximately 120 nm and that exhibit characteristic crown-shaped projections on their surface (21). Their genome consists of a very long positive-stranded RNA. Coronaviruses are responsible for 10 to 35% of common cold in humans (22) and were associated with more serious diseases such as respiratory distress in newborns (23), severe acute respiratory syndrome (SARS), and severe diarrhea (24).

The association between coronavirus and MS is strengthened by observations in mice. Indeed, murine coronavirus infection of susceptible mice leads to an inflammatory demyelination. This virus-induced demyelination is similar to MS in several aspects, providing a widely studied model for this neurological autoimmune disease (23, 24).

In humans, it was demonstrated using different experimental approaches that coronaviruses are neuroinvasive, i.e. that they can reach the CNS. First, virus was isolated from the brains of MS patients (27). Second, titers of antibodies specific to human coronaviruses were described to be higher in cerebrospinal fluids of MS patients than controls (28). Third, coronavirus RNA was detected in the brain of MS patients by RT-PCR and by *in situ* hybridization (11, 12, 14).

Furthermore, coronavirus neurotropism, the capacity to infect CNS cells, has also been demonstrated in humans. An acute infection by human coronaviruses was demonstrated in primary cultures of human microglia and astrocytes (29) as well as oligodendrocytes (Talbot *et al.,* unpublished data). In addition, a persistent infection by coronavirus was demonstrated in cell lines from nervous system (30, 31).

These observations demonstrated that coronavirus could reach the CNS and infect neural cells. Although it cannot be concluded from these observations that coronaviruses do cause MS, it suggests an association between CNS viral infection and this neurologic disease. In addition, epidemiological studies have shown that MS relapses were often preceded with respiratory tract infections (30). Interestingly, coronavirus reinfections are in fact possible (22).

## 4.  T CELL CROSS-REACTIVITY TO CORONAVIRUS AND MYELIN PROTEINS

Several mechanisms have been proposed to explain how a viral infection could lead to an autoimmune disease. Amongst these mechanisms is molecular mimicry. According to this model, a non-self agent first activates an immune cell. Then, the cell recognizes a self-element that shares antigenic conformation with the pathogen and directs a response towards it. Individuals who are genetically predisposed to respond to this antigenic determinant of a pathogen having a similar conformation to a determinant on a self-antigen could develop an autoimmune response following infection. Shared sequences or similar determinant conformations between coronavirus and myelin proteins such as myelin basic protein (MBP) and proteolipid protein (PLP) have been identified (33-35). Molecular mimicry provides a unifying mechanism that could explain both the genetic and environmental aspects in the triggering of MS.

The molecular mimicry hypothesis would indeed explain observation of T cell lines (TCL) cross-reactive to both myelin antigen (Ag) and human coronavirus (HCoV) we have reported in MS patients (36). In this study, peripheral blood lymphocytes from MS patients and healthy controls were used to select long-term MBP- and HCoV-reactive T cell lines. A summary of these results is shown in Table 1. MBP T-cell lines were found as frequently in MS patients than in healthy controls. All the participants tested seropositive for coronavirus, as expected (22). Interestingly, TCL from ten of sixteen MS patients and two of fourteen control subjects showed cross-reactivity between myelin and viral antigens, for a proportion of such lines of 29% in MS patients and only 1.3% in healthy donors. Such cross-reactivity is thus highly significantly observed in MS patients. Therefore, in genetically predisposed individuals, T cells could be primed and activated following a systemic infection; the activated lymphocyte would enter the CNS and could establish a response towards myelin antigens. Myelin-reactive T lymphocytes could also have a role in broadening the immune response and increasing tissue damage. Coronavirus-myelin cross-reactive lymphocytes could thus initiate, maintain and/or amplify autoimmunity.

*Table 1.* Overall number of coronavirus-myelin cross-reactive TCL produced from MS patients and healthy donors. Cross-reactive TCL are obtained in a highly significant proportion (p<0.0001) in MS patients versus controls, with 29% out of TCL generated from the peripheral blood compared to only 1.3% (Talbot, 1996).

| Donors | n | Lines (CD4+) | Cross-reactive T-cell lines | | |
|---|---|---|---|---|---|
| | | | Lines | n | % |
| MS | 16 | 134 | 39 | 10 | 29 |
| Healthy Controls | 14 | 155 | 2 | 2 | 1.3 |

Although these findings regarding T-cell cross-reactivity between human coronavirus and myelin antigen are consistent with the molecular mimicry hypothesis, experiments at the clonal level were needed to prove that a single T cell was indeed activated by both viral and myelin antigens.

## 5. CLONAL T-CELL CROSS-REACTIVITY

Studies conducted at the molecular level reinforce and confirm the molecular mimicry hypothesis in MS (37). It was essential to examine coronavirus-myelin cross-reactivities at the T cell clonal level since TCL are made of a heterogeneous cell population: more than one TCR could have been involved in the measured cross-reactive response. Similar to the previous study involving TCL, long-term primary T cell clones (TCC) were also derived from the peripheral blood of MS patients. Selecting antigens used in this study were both known human coronavirus serotypes (229E and OC43) as well as two CNS proteins, MBP and PLP. Briefly, TCL were first selected with either one of the selecting antigen and positive cells were then cloned by limiting dilution; tritiated thymidine incorporation assays determined antigenic-specitifies (38). Table 2 shows that a total of 145 monospecific TCC were obtained from twenty-two patients, out of thirty-two patients studied. Many TCC were positive to coronavirus antigens, with about 80% of the TCC cultures selected with either HCoV-229E or HCoV-OC43.

Interestingly, ten cross-reactive long-term primary TCC were also cultured from six out of thirty-two MS patients (Table 2). While some patients led to very poor TCC production, other patients could lead to more then one cross-reactive TCC. These TCC were selected with myelin antigens (PLP and MBP, both encephalitogenic in genetically predisposed rodents) and HCoV. Half of these patients were HLA-DRB1*1501, a genetic susceptibility associated with MS (1). TCC produced were CD4+. HLA-typing and antigenic specificities of all ten cross-reactive TCC are described in Table 3. The selecting antigen is identified for each clone. Interestingly, two TCC selected with viral antigens recognized both myelin antigens. The TCR Vβ chains were also identified for six of those TCC and hypervariable regions were also determined through sequencing analysis (37).

Even if the method used to obtain the TCC is reliable, clonality was double-checked by diversity PCR. TCR were sequenced and only one TCR β chain was identified by PCR on several bacterial colonies issued from the same transformation. Assuming that the TCC obtained bear only one TCR, this study shows that a single TCR can recognize two different antigens, one from a human coronavirus, and the other from a myelin antigen that is targeted by T lymphocytes in MS. Such a cross-reactive phenomenon could be an elegant framework to explain the origin of an autoimmune disease as MS.

Thus, the T cell cross-reactivity between HCoV and myelin proteins in MS patients was confirmed at the single cell level. Combined with our previous results of highly significant coronavirus-myelin T-cell cross-reactivities in MS patients compared to healthy controls (36), it appears that molecular mimicry could represent a signicant pathogenic mechanism associated with MS pathogenesis.

*Table 2.* Overall number of both monospecific and coronavirus-myelin cross-reactive TCC produced from the peripheral blood of thirty-two MS patients. Selecting antigens used for that experiment were both known serotypes of human coronavirus, HCoV-229E and HCoV-OC43, as well as two CNS proteins, MBP and PLP. Long-term primary human TCC were generated by limiting dilution of antigen-specific TCL. (37, 38)

| TCC | Donors (n) | Number of TTC Produced with Viral Antigens | | Number of TCC Produced with Myelin Antigens | | Total |
|---|---|---|---|---|---|---|
| | | HCoV-229E | HCoV-OC43 | MBP | PLP | |
| Monospecific | 22/32 | 80 | 34 | 19 | 12 | 145 |
| Cross-reactive | 6/32 | 4 | 2 | 2 | 2 | 10 |

*Table 3.* Antigenic specificity patterns of ten coronavirus-myelin cross-reactive TCC obtained from six MS patients. The HLA-typing for each patient from which those TCC were produced is shown. The selecting antigen is identified in bold.

| TCC | HLA-DR | Antigenic Specificity | |
|---|---|---|---|
| P7-a | 11,17 | **HCoV-OC43** | MBP |
| P8-a | 15,17 | **HCoV-229E** | MBP |
| P12-a | 13 | **PLP** | HCoV-229E |
| P12-b | 13 | **PLP** | HCoV-229E |
| P12-c | 13 | **MBP** | HCoV-229E |
| P13-a | 15 | **MBP** | HCoV-229E |
| P13-b | 15 | **HCoV-OC43** | MBP and PLP |
| P22-a | 13,15 | **HCoV-229E** | MBP |
| P22-b | 13,15 | **HCoV-229E** | MBP and PLP |
| P24-a | 7,17 | **HCoV-229E** | MBP |

# 6.      MS SPECIFICITY OF T-CELL CROSS-REACTIVITY

In order to verify whether coronavirus-myelin T-cell cross-reactivity is more significantly associated with MS, TCL were established from the peripheral blood of patients having neurological or inflammatory disease and from healthy controls (Talbot *et al.,* unpublished data). In this study, TCL cross-reactive for both myelin (MBP, PLP) and human coronavirus antigens were found in four out of eleven MS patients, in a proportion of 3.4% of all TCL derived from the peripheral blood. Only one out of ten patients having neurological or inflammatory disease showed such cross-reactive TCL, for a proportion of 0.5% of the overall TCL produced from the peripheral blood. The latter proportion (0.5%) of cross-reactive TCL was obtained from healthy donors, where a cross-reactive TCL toward myelin and coronaviral Ag was found in only one out of twelve donors. These results are shown in Table 4. Cross-reactive TCL were found more often in MS patients. It is too early to say that such antigenic pattern is clearly related to MS, however, it remains of great interest to explore those antigens regarding molecular mimicry and neurodegeneresence.

Thus, cross-reactive TCL for myelin proteins and HCoV seem to be more abundant in the peripheral blood MS patients than in healthy patients. Moreover, the cross-reactivity observed appears to be found in MS, an autoimmune neurological disease, but not in other inflammatory or neurological diseases.

## 7. CONCLUSIONS

The etiology of MS, an autoimmune neurologic disease, is unknown. Coronaviruses are suspected to be involved in MS pathology. Molecular mimicry, the recognition of two antigens by one single immune cell, could be the mechanism explaining the link between a viral infection and MS. Evidence of molecular mimicry between coronavirus and myelin antigens have been described in humans by our laboratory. Cross-reactive TCL were found predominantly in MS patients compared to patients with other neurological or inflammatory diseases, or healthy controls. Moreover, the T cell cross-reactivity was confirmed at the clonal level. However, it is important to keep in mind that the role of molecular mimicry does not exclude the participation of other immune mechanisms such as bystander effects, determinant spreading or even superantigens in the triggering or development of this autoimmune disease. All those mechanisms could play a role in maintaining or broadening the autoimmune response. The association of infection by a ubiquitous virus such as the human coronavirus with the development of multiple sclerosis in genetically susceptible individuals most likely represents the result of an aberrant induction of immune responses towards myelin proteins, possibly coupled to a persistent virus infection of the central nervous system that may activate glial cells (39), induce neuronal loss (40) and be associated with local immunopathology (41).

*Table 4.* To assess whether coronavirus-myelin cross-reactive TCL were only found in MS or could also be detected in other diseases, 13 MS patients, 10 patients suffering either of a neurological or an inflammatory disease, as well as 12 healthy controls were studied in comparison. For each group, TCL were produced and selected with MBP, HCoV-229E or HCoV-OC43, then were tested with a proliferation assay against both CNS proteins and both human coronavirus serotypes. The coronavirus-myelin cross-reactive TCL are shown in bold; those are also compared to the overall monospecific TCL number produced during the experiment to determine the percentage of cross-reactive TCL generated per studied group of donors. N.D. indicates that the test was not performed and results are not determined (selecting antigen indicated at the top of each column and antigen used for proliferation assay indicated on the left).

| | Multiple Sclerosis | | | Neurological and Inflammatory Diseases | | | Healthy Controls | | |
|---|---|---|---|---|---|---|---|---|---|
| | MBP | HCoV-229E | HCoV-OC43 | MBP | HCoV-229E | HCoV-OC43 | MBP | HCoV-229E | HCoV-OC43 |
| MBP | N.D. | **3** | **1** | N.D. | **1** | 0 | N.D. | 0 | 0 |
| PLP | 5 | 0 | 0 | 2 | 0 | 0 | 2 | N.D. | N.D. |
| HCoV-229E | **1** | N.D. | 0 | 0 | N.D. | 5 | N.D. | N.D. | 1 |
| HCoV-OC43 | **2** | 2 | N.D. | 0 | 1 | N.D. | **1** | 2 | N.D. |
| Overall Monospecifc TCL produced | 123 | 50 | 53 | 55 | 90 | 37 | 71 | 84 | 49 |
| Myelin-Virus Cross-reactive TCL/Total | 7/226 (3.1%) | | | 1/182 (0.5%) | | | 1/204 (0.5%) | | |

# ACKNOWLEDGEMENTS

The work from our laboratory that is herein cited was performed with grant support from the Institute of Infection and Immunity of the Canadian Institutes for Health Research, or the Multiple Sclerosis Society of Canada. The latter also provided a studentship to Annie Boucher. Scholarship support from the Fonds de la recherche en santé du Québec, and more recently a Canada Research Chair in Neuroimmunovirology to Pierre J. Talbot, are gratefully acknowledged.

Expert technical help from Francine Lambert was instrumental in a large part of the cited studies. Mélanie Tremblay participated in the initial stages of the study on MS specificity of TCL cross-reactivity.

# REFERENCES


1 - Ebers, G.C., Kukay, K., Bulman, D.E., Sadovnick, A.D., Rice, G., Anderson, C., Armstrong, H., Cousin, K., Bell, R.B., Hader, W., Paty, D.W., Hashimoto, S, Oger, J., Duquette P., Warren, S., Gray, T., O'Connor, P., Nath, A., Auty, A., Metz, L., Francis, G., Paulseth, J.E., Murray, T.J., Pryse-Phillips, W., Risch, N., et al. (1996) A full genome search in multiple sclerosis. Nat Genet 13:472-476.

2 - Cristen, J., Willer , A., Ebers, G.C. (2000) Susceptibility to multiple sclerosis: interplay between genes and environment. Curr Op Neurol 13: 241-247.

3 - Kurtzke, J.F. (1993) Epidemiologic evidence for multiple sclerosis as an infection. Clin Microb Rev 6:382-427.

4 - Sadvonick, A.D., Dyment, D., Ebers, G.C. (1997) Genetic epidemiology of multiple sclerosis. Epidemiol Rev 19:99-106.

5 - Johnson, R.C. (1985) Viral aspects of multiple sclerosis. In: Handbook of clinical neurology: demyelinating diseases. Koetsier JC Eds, Elsevier 3:319-336.

6 - Salmi, A., Reunanen, M., Ilonen, J. (1981) Possible viral etiology of multiple sclerosis. In: Internationl Congress Series Neurology. Katsuki, S., Tsybaki, T., Toyokura, Y. Eds, Excerpta Medica, 416-431.

7 - Bray, P.F., Luka, J., Bray, P.F., Culp, K.W., Sclight, J.P. (1992) Antibodies against Epstein-Barr nuclear antigen (EBNA) in multiple sclerosis CSF and two pentapeptide sequences identities between EBNA and myelin basic protein. Neurology, 42:1798-1804.

8 - Haase, A.T., Ventura, P., Gibbs, C.J., Toutelotte, W.W. (1981) Measles virus nucleotide sequences: detection by hybridization in situ. Science 212:672-674.

9 - Cosby, S.L., McQuaid, S. Taylor, M.J., Bailey, M., Rima, B.K., Martin, S.J., Allen, I.V. (1989) Examination of 8 cases of multiple sclerosis and 56 neurological and non-neurological controls for genomic sequences of measles virus, canine distemper virus, simian virus-5 and rubella virus. J Gen Virol 70:2027-2036.

10 - Reddy, E.P., Sanberg-Wohlheim, M., Mettus, R.V., Ray, P.E., De Freitas, E., Koprowski, H. (1989) Amplification and molecular cloning of HTLV-1 sequences from DNA of multiple sclerosis patients. Science 243:529-533.

11 - Murray, R.S., Brown, B., Brian, D., Cabirac, G.F. (1992) Detection of coronavirus RNA and antigen in multiple sclerosis brain. Ann Neurol 31:525-533.

12 - Stewart, J.N., Mounir, S., Talbot, P.J. (1992) Human coronaviruses gene expression in the brains of multiple sclerosis patients. Virology 191:502-505.

13 - Challoner, P.B., Smith, K.T., Parker, J.D., MacLeod, D.L., Coulter, S.N. Rose, T.M., Schultz, E.R., Bennett, J.L., Garber, R.L., Chang, M., Schad, P.A., Stewart, P.M., Nowinski, R.C., Brown, J.P., Burner, G.C. (1995) Plaque-associated expression of human herpesvirus 6 in multiple sclerosis. Proc Natl Acad Sci USA 92:7440-7444.

14 - Arbour, N., Day, R., Newcombe, J., Talbot, P.J. (2000) Neuroinvasion by human respiratory coronaviruses. J Virol 74:8913-8971.

15 - Schmidt, S., 1998. Candidate autoantigens in multiple sclerosis. Multiple Sclerosis 5: 147-160.

16 - Holz, A., Bielekova, B., Martin, R., Oldstone, M.B.(2000) Myelin-associated oligodendrocyte basic protein: identification of an ecephalitogenic epitope and association with multiple sclerosis. J Immunol 164:1103-1109.

17 - Zhang, J., Markovic-Plese, S., Lacet, B., Raus, J., Weiner, H.L., Hafler, D.A. (1994) Increased frequency of interleukin 2-responsive T cells specific for myelin basic protein and proteolipid protein in the peripheral blood and cerebrospinal fluid of patients with multiple sclerosis. J Exp Med 179:973-984.

18 - Van Noort, J.M. (1995) The small heat shock protein α–B-crystallin as candidate autoantigen in multiple sclerosis. Nature 375:798-801.

19 - Schmidt, S., Linington, C., Zipp, F., Sotgiu, S., de Wall Malefyt, R., Wekerle, H., Holfeld, R. (1997) Multiple sclerosis: comparaison of the human T-cell response to S100 beta and myelin basic protein reveals parallels to rat experimental autoimmune panencephalitis. Brain 120:1437-1445.

20 - Colombo, E., Banki, K., Tatum, A.H., Daucher, J., Ferrante, P., Myrrau, R.S., Phillips, P.E., Perl, A., (1997) Comparative analysis of antibody and cell-mediated autoimmunity to transaldolase and myelin basic protein in patients with multiple sclerosis. J Clin Invest 99:1238-1250.

21 - Lai, M.M.C., Cavanagh, D. (1997) The molecular biology of coronaviruses. Adv Vir Res, 48:1-100.

22 - Myint, S.H. (1994) Human coronaviruses – a brief review. Rev Med Virol 4:35-46.

23 – Sizun, J., Soupre, D., Legrand, M.C., Giroux, J.D., Rubio, S., Cauvin, J.M., Chastel, C., Alix, D., de Parscau, L. (1995) Neonatal nosocomial respiratory infection with coronavirus: a prospective study in a neonatal intensive care unit. Acta Paed 84:617-20.

24 – Resta, S., Luby, J.P., Rosenfeld, C.R., Siegel, J.D. (1985) Isolation and propagation of a human enteric coronavirus. Science 229:978-981.

25 - Wege, H. (1995) Immunopathological aspects of coronavirus infections. Springer Semin Immunopathol 17:133-148.

26 - Lane, T.E. , Buchmeier, M.J. (1997) Murine coronavirus infection: a paradigm for virus-induced demyelinating diseases. Trends Microbiol 5:9-14.

27 - Burks, J.S., DeVald, B.L., Jankovsky, L.D., Gerdes, J.C. (1980) Two coronaviruses isolated from central nervous system tissue of two multiple sclerosis patients. Science 209:933-934.

28 - Salmi, A., Ziola, B., Hovi, T., Reunanen, M. (1982) Antibodies to coronaviruses OC43 and 229E in multiple sclerosis patients. Neurology 32:292-295.

29 - Bonavia, A., Arbour, N., Yong, W.V., Talbot, P.J. (1997) Infection of primary cultures of human neural cells by human coronaviruses 229E and OC43. J Virol 71:800-806.

30 - Arbour, N., Côté, G., Lachance, C., Tardieu, M., Cashman, N.R., Talbot, P.J. (1999) Acute and persistent infection of human neural cell lines by human coronaviruses OC43. J Virol 73:3338-3350.

31 - Arbour, N., Ekandé, S., Côté, G., Lachance, C., Chagnon, F., Cahsman, N.R., Talbot, P.J. (1999) Persistent infection of human oligodendrocytic and neuronal cell lines by human coronaviruses 229E. J Virol 73:3326-3337.

32 - Marrie, R.A., Wolfson, C., Sturkenboom, M.C., Gout, O., Heinzlef, O., Roullet, E., Abenhaim, L. (2000) Multiple sclerosis and antecedent infections: a case-control study. Neurology 54:2307-2310.

33 - Jahnke, U, Fisher, E.H., Alvord, E.C. (1985) Sequence homology between certain viral proteins and proteins related to encephalomyelitis and neuritis. Science 229:282-284.

34 - Shaw, S.Y., Laursen, R.A., Lees, M.B. (1986) Analogous amino acid sequences in myelin proteolipid and viral proteins. FEBS Lett 207:266-270.

35 - Jouvenne, P., Mounir, S., Stewart, J.N., Richardson, C.D., Talbot, P.J. (1992) Sequence analysis of human coronavirus 229E messenger RNAs 4 and 5-evidence for polymorphism and homology with myelin basic protein. Virus Res 22:125-141.

36 - Talbot, P.J., Paquette, J.S., Ciurli, C., Antel, J.P., Ouellet, F. (1996) Myelin basic protein and human coronavirus 229E cross-reactive T cells in multiple sclerosis. Ann Neurol 39:233-240.

37 - Boucher, A., Duquette, P., Denis, F., Talbot, P.J. (2003) Long-term coronavirus-myelin cross-reactive T-cell clones derived from multiple sclerosis patients. Manuscript n preparation.

38 - Boucher, A., Duquette, P., Denis, F., Talbot, P.J. (2003) Generation of antigen specific long-term T cell clones. Manuscript in preparation.

39 - Edwards, J., Denis, F., Talbot, P.J. (2000) Activation of glial cells by human coronavirus OC43. J Neuroimmunol 108:73-81.

40 - Jacomy, H., Talbot, P.J. (2003) Vacuolating encephalitis in mice infcted by human coronavirus OC43. Virology. In press.

41 – Talbot, P.J.., Arnold, D., Antel, J.P. (2001) Virus-induced autoimmune reactions in the nervous system. Curr Top Microbiol Immunol 253:247-271.

Chapter C8

# CORONAVIRUS-INDUCED DEMYELINATION AND SPONTANEOUS REMYELINATION
*Growth factor expression and function*

Regina C. Armstrong[1,2], Jeffrey M. Redwine[2,3], and Donna J. Messersmith[1,4]
*[1]Department of Anatomy, Physiology, and Genetics, and [2]Program in Neuroscience, Uniformed Services University of the Health Sciences, 4301 Jones Bridge Rd., Bethesda, MD 20814-4799; rarmstrong@usuhs.mil*
*[3]Current affiliation: Neurome Inc., La Jolla, CA 92037*
*[4]Current affiliation: National Center for Biotechnology Information, Bethesda, MD 20894*

Abstract:    MHV-A59 coronavirus infection produces a transient episode of demyelination that is followed by spontaneous remyelination. This paradigm provides a complex lesion environment to examine cellular and molecular mechanisms involved in successful CNS remyelination. Our work in this model has focused on the roles of platelet-derived growth factor and fibroblast growth factor 2 in regulating oligodendrocyte progenitor responses required for remyelination.

Key words:    platelet-derived growth factor, fibroblast growth factor, estrous cycle, gender, demyelinating disease, remyelination, coronavirus, cuprizone

## 1.    INTRODUCTION

Insufficient remyelination results in prolonged neurological impairment in demyelinating disease states, such as multiple sclerosis. A critical determinant of remyelination is regulation of oligodendrocyte lineage responses. Surviving and/or newly generated oligodendrocyte lineage cells must be recruited to appropriate sites within demyelinated tissues and induced to differentiate and form myelin. Each of these oligodendrocyte lineage cell responses appears to be regulated by signals within the lesion environment, such as growth factors, cytokines, and cell-cell interactions.

The pathology of multiple sclerosis (MS) lesions is heterogeneous between patients, with at least four fundamentally different patterns of demyelination (12). Therefore, analysis of experimental models of demyelination with distinct mechanisms of pathogenesis is warranted. In addition, different experimental models have advantages for examining specific aspects within the course of demyelinating diseases. The mouse model of murine hepatitis strain A59 (MHV-A59) coronavirus infection serves as a relevant model for analyzing the cellular and molecular components involved in spontaneous remyelination. The complexity of MHV-A59 lesions includes infiltration of CD8+ and CD4+ T cells, B lymphocytes producing immunoglobulins, macrophages, and reactive glial cells (15,21). These lesion components are variably exhibited among categories of MS lesions.

The potential function of molecules that can promote remyelination in MS lesions, such as growth factors, is ideally analyzed in the context of a complex lesion environment due to contributing effects of cytokines, chemokines, infiltrating lymphocytes, and reactive cells. However, this complex lesion milieu can also make it difficult to delineate effects that are specific to the remyelination process. For this purpose, analysis of oligodendrocyte lineage responses is facilitated by comparison with a simpler lesion model, such as ingestion of cuprizone (14). Growth factor effects common to experimental lesions of diverse pathogenesis, such as MHV-A59 and cuprizone models, are most likely to be applicable more generally to demyelinating diseases.

This chapter will review recent findings of the expression and function of specific growth factors in MHV-A59 and cuprizone models of spontaneous remyelination. In addition, the complexity of the MHV-A59 model will be exemplified by discussion of the modulation of the disease course in correlation with gender and estrous cycle status. This modulation of the MHV-A59 disease course is also relevant to modulation of MS disease activity.

## 2.       DISEASE SEVERITY IN THE MHV-A59 MODEL

Intracranial infection of female C57Bl/6 mice at 28 days of age with 1000 plaque forming units (PFU) of MHV-A59 virus produces a characteristic progression of demyelinating disease. Demyelination begins within the first week post-injection (wpi), with more extensive areas of myelin degeneration by 2 wpi (1,9). During this progression of demyelination, mice exhibit loss of motor function, and virus is present in cells of the white matter (9). In subsequent weeks, clearance of virus and myelin debris occurs, remyelination is initiated, and recovery of motor function proceeds (9,21). However, among mice similarly infected with MHV-A59, there is variability in the proportion of mice that exhibit distinct results from asymptomatic, mild paresis, severe paralysis, to mortality. Early studies reported mortality rates that ranged from 11% (1) to 70% (11).

To take advantage of knockout mice for the analysis of growth factors and cellular components involved in spontaneous remyelination in this model, we needed to optimize the reproducibility of the MHV-A59 infection outcome. This objective led us to explore more quantitative methods to monitor disease progression and identify determinants of variation in disease severity, including mortality.

Intracranial MHV-A59 injection of 1000 PFU at 28 days of age has been used to maximize demyelination and minimize mortality compared to infection of younger mice (1,2,21). However, in addition to an age factor, a correlation was also observed between weight and mortality. Female 28-day old mice that eventually died during the acute infection weighed $12 \pm 0.34$ g (mean $\pm$ s.e.; n = 25) while survivors weighed 13.5 g $\pm 0.22$ (mean $\pm$ s.e.; n = 51) ($p < 0.0005$, unpaired $t$-test). Accordingly, subsequent experiments used only mice weighing 13-15 g at 28 days of age.

Variability of infection outcome still occurred among female mice within these weight and age restrictions. Measures were already in place to minimize experimental variability from animal husbandry and experimental treatment. Additional tests were conducted to quantify the neurological effects of MHV-A59 lesions and determine whether the infection outcomes were clearly different, which would suggest an additional factor contributing to infection outcome. A "hang time" test of limb motor function and grasping was chosen to assess the result of MHV-A59 infection because the lesion predilection sites include the spinal cord white matter adjacent to the ventral root exit zones throughout all spinal cord levels (8,10,21).

To perform the hang time test, mice were placed onto a wire cage top, and once the mice had gripped the bars the top was turned upside-down and held horizontally above a table (similar to ref. 23). The length of time that each mouse held onto the bars was recorded, up to a maximum of 60 seconds. Control mice easily grasped the cage top bars with paws, often curled tails around the bars, and moved around to explore while upside-down (Figure 1A). In contrast, mice with impaired limb motor function could not grasp cage top bars with paws, rarely curled tails around bars, and dropped off prior to the 60 seconds maximum time limit (Figure 1B). Inability to hold on was associated with demyelination, as evident in previous reports of the histopathology of MHV-A59 infected mice (1,21).

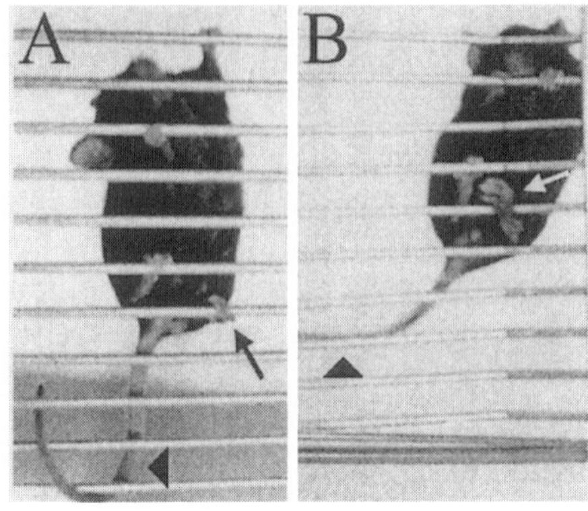


*Figure 1.* **Control (A) and MHV-A59 infected (B) mice shown hanging from a cage top during hang time tests.** The control mouse can easily grasp the bars with paws (black arrow) and tail (black arrowhead). The MHV-A59 infected mouse has hind limb and tail impairment so that the paws cannot normally grasp the bars (white arrow) and the tail is limp (black arrowhead).

The time course of individual mouse hang time performance over 4 wpi showed that some mice had only mild motor dysfunction while others had severe motor dysfunction (Figure 2). Operational definitions for mild and severe groups can be made several ways. One method is to sum hang times over the disease course, through 4 wpi when many mice have fully recovered. Clustering of subgroups of mice is evident as either a low summed hang time (severe motor dysfunction) or a high summed hang time (mild motor dysfunction). However, to allow mice to be used prior to the recovery phase, a practical set of criteria was needed by 2 wpi. After testing the hang time of mice at 7, 10, and 14 days post-infection (dpi), a mouse that had 2 or 3 hang time values of > 30 seconds <u>or</u> recovered to a 60 second hang time by 14 dpi was categorized as only mildly affected. A mouse that had 2 or 3 hang times ≤ 30 seconds <u>and</u> did not recover to 60 seconds by 14 dpi was categorized as severely affected.

In many models of experimental demyelination involving the spinal cord, disease severity is commonly quantified by a clinical scoring system. An adapted clinical rating has been applied in the MHV-A59 model (21). The clinical score scale is a measure of impairment based upon observed paresis or paralysis of limbs, so that 0 = no observable paresis/paralysis, 1-5 = paresis/paralysis in 1-5 limbs, respectively (such as limp tail, abnormal limb contracture or extension, dragging of limb, abnormal lateral movement of limb), and 6 = morbidity. The clinical scores of mildly affected mice over time are significantly less than the clinical scores of severely affected mice (Figure 2). However, the limited unit range of the clinical score scale does not detect individual variation as well as the hang time test in the MHV-A59 model.

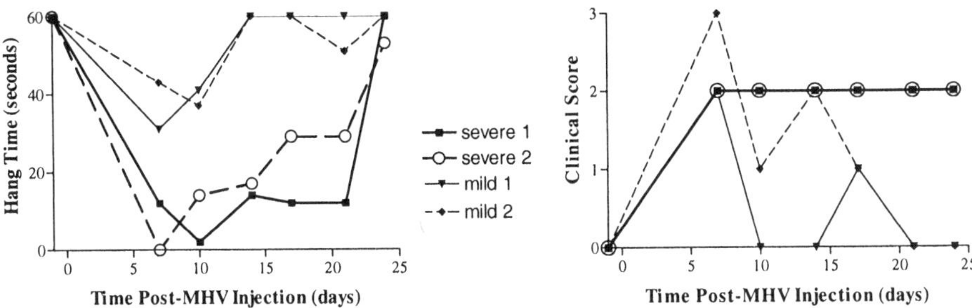

*Figure 2.* **Examples of hang time and clinical score results**. Individual records are shown for 2 mildly affected mice and 2 severely affected mice as representative examples of the variability in disease severity. The same mice are shown in both graphs. The hang times shown on each day are the averages of 3 trials.

## 3. MHV-A59 DISEASE SEVERITY RELATIVE TO GENDER AND ESTROUS CYCLE

Sex hormones have a role in disease susceptibility and severity in MS, which has been replicated in experimental autoimmune encephalomyelitis (EAE) in SJL mice (25). While females are more susceptible to MS than males (26), a worse survival curve is associated with male gender in primary progressive MS (6). In females with relapsing-remitting MS, worsening disease activity has been reported to vary with estrous cycle stage, but results attempting to correlate further to hormone levels have produced conflicting results (4,19,24,29).

To identify an underlying factor in the variability of disease severity in our MHV-A59 model, estrous cycle stage was monitored using vaginal smears (7)(Figure 3). Severely affected female mice were more likely to be in diestrus or metestrus during the early effector stage of the pro-inflammatory response (4 dpi). Estradiol and progesterone are relatively low at these stages of the estrous cycle. Female mice in proestrus or estrus (at 4 dpi), when estradiol and progesterone levels are elevated, more frequently experienced a milder outcome after MHV-A59 infection. Interestingly, EAE was ameliorated in SJL mice treated with estradiol at a sustained high dose, regardless of the dose of progesterone used in combination (25). However, the dose of estradiol needed to significantly decrease EAE severity was above levels that occur naturally during the estrous cycle (25). Thus, the MHV-A59 pathogenesis may be very sensitive to variation of hormone levels during the estrous cycle.

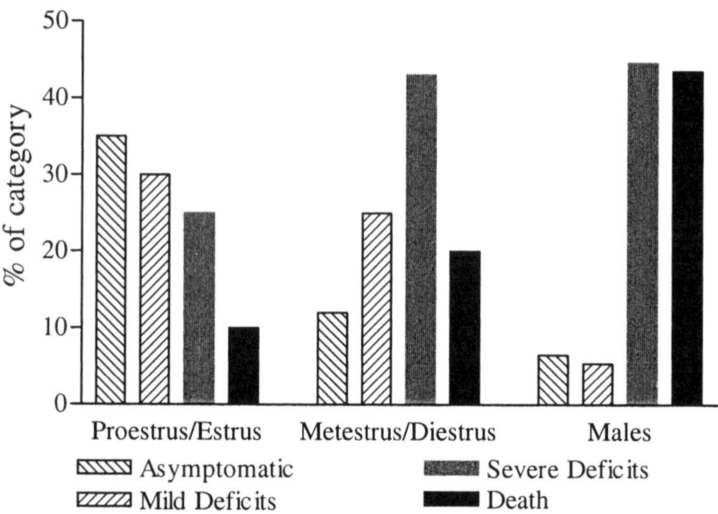


*Figure 3.* **Outcome of MHV-A59 infection in female mice, at different stages of the estrous cycle, and male mice**. The outcome of infection in each mouse was based upon hang time score using the classification described in the text. Each value represents mice combined from at least 3 separate infection experiments. Female mice, Proestrus/Estrus $n = 20$, Metestrus/Diestrus, $n = 129$; combined from 6 separate sets of infections; sampling difference reflects the proportional difference in the duration of the stages within the 4 day mouse menstrual cycle. Male mice, $n = 92$; combined from 3 separate sets of infections.

Male mice in this MHV-A59 model clearly had a more severe disease course and higher rate of mortality than female mice. A Th1 cytokine response, specifically interferon gamma, is required for clearing of the MHV-A59 virus from oligodendrocytes and resolution of the disease as a model of transient demyelination (18). Male hormonal conditions favor a bias towards Th2 cytokine production rather than a Th1 response (28), which may correlate with a lack of interferon gamma production in lesions of males and explain the increased rates of mortality compared to females.

## 4. GROWTH FACTOR EXPRESSION FOLLOWING DEMYELINATION

Discussion of growth factor pathways in MHV-A59 lesions will focus on oligodendrocyte lineage responses during remyelination. Given the variability of MHV-A59 disease severity noted above, studies of growth factors in this model utilize only the mice that are classified as "severe" to provide a reproducible status. Also, when possible, findings in the MHV-A59 model will be compared with the cuprizone model to establish common mechanisms.

Experimental models with different pathogenesis indicate that, when remyelination occurs, oligodendroglial repopulation of demyelinated lesions may result from a common mechanism: local proliferation and recruitment of oligodendrocyte progenitor cells. *In vitro* studies predict that specific growth factors regulate the oligodendrocyte progenitor responses needed for myelination during development and for remyelination (*i.e.* proliferation, migration, differentiation, and survival). Among the growth factors known to regulate the oligodendrocyte lineage, platelet-derived growth factor (AA homodimer, PDGF) and fibroblast growth factor 2 (FGF2) warrant high priority for analysis.

*In vitro* studies have documented the ability of PDGF and FGF2 to regulate responses of oligodendrocyte lineage cells isolated from normal neonatal and adult CNS. During development, neonatal oligodendrocyte progenitors proliferate in response to PDGF or FGF2, and grow as a self-renewing cell line in the combination of PDGF and FGF2 (5,16,22). PDGF and FGF2 can also enhance the proliferation of oligodendrocyte progenitor cells isolated from normal adult rodent CNS (13,27). FGF2 treatment of more mature stages of the oligodendrocyte lineage inhibits terminal differentiation (3).

*In vivo* studies have supported the hypothesis that growth factors may regulate oligodendrocyte progenitor responses involved in myelin repair. At the onset of remyelination (4 wpi) in spinal cords of MHV-A59 lesioned mice, increased expression of PDGF-A and FGF2 ligands, as well as the corresponding receptors, indicated the potential for each to act upon oligodendrocyte lineage cells involved in remyelination. Immunostaining for PDGF-A ligand demonstrated increased expression in reactive astrocytes in remyelinating lesions (21). *In situ* hybridization to detect PDGF-A mRNA revealed a 2.9 fold increase in the number of labeled cells in sections from MHV-A59 infected mice vs. PBS-injected control mice (21). Similar *in situ* hybridization of FGF2 demonstrated markedly increased signal in reactive astrocytes and microglia localized within and near white matter lesions (17). Over the course of MHV-A59 disease progression, FGF2 mRNA transcript abundance peaked at 4 wpi as greater than 5-fold over non-lesioned levels (17). At a corresponding stage in cuprizone demyelination, a similar increase was observed for FGF2 (2) and PDGF-A mRNA (unpublished observation). Interestingly, in both models the increased signal for FGF2 is clearly localized to the lesion while the increase in cells with detectable PDGF-A is much more widely distributed in the tissue. These distributions may indicate differences in the influence of these growth factors on the oligodendrocyte lineage cell response.

In addition to demonstrating the increased expression of PDGF-A and FGF2 mRNA, expression of corresponding receptors on oligodendrocyte lineage cells has been examined *in vivo*. In lesioned white matter undergoing remyelination after MHV-A59 infection, the number of cells expressing FGF receptors increased as much as 5-fold (17). Oligodendrocyte progenitors expressed receptors for both FGF2 and PDGF (specifically PDGF$\alpha$R, the alpha form of PDGF receptor) (7,20,21). Oligodendrocytes also expressed FGF receptors (19,20).

In both the MHV-A59 model and the cuprizone model, the oligodendrocyte progenitor proliferative response corresponds with expression of PDGF$\alpha$R (2,21). During early remyelination, there is a dramatic increase in the number of cells that are expressing PDGF$\alpha$R mRNA transcripts and have incorporated bromodeoxyuridine (BrdU; indicative of DNA synthesis) during a 4-hour terminal *in vivo* pulse. In contrast, mature oligodendrocytes did not exhibit substantial proliferation. Taken together, these studies support the prediction that PDGF and FGF2 may play a role in the successful generation of oligodendrocyte lineage cells and subsequent remyelination observed in this mouse model; the ligands and receptors are expressed at an appropriate stage of the disease process and with the appropriate cellular localization, and association with a proliferation indicator.

A specific role of PDGF and FGF2 in regulating proliferation during remyelination has been demonstrated by culturing oligodendrocyte lineage cells from spinal cords of MHV-A59 infected mice. Approximately 20% of the white matter is demyelinated throughout the rostrocaudal levels of the spinal cord as a result of MHV-A59 infection (8). This extent of tissue involvement is sufficient for isolation of a population of oligodendrocyte progenitor cells that is significantly more proliferative than the population derived from normal adult mouse spinal cord (1). PDGF and FGF2, produced by reactive astrocytes and microglia within the cultures, are significant mitogens regulating this proliferative response (8). Interestingly, in cultures derived from different time points after MHV-A59 infection, the oligodendrocyte progenitor response varies with the disease stage; proliferation is more robust during the demyelination phase while differentiation is more extensive during remyelination (8).

Knockout and transgenic mice can test the function of growth factors throughout the progression of demyelination and remyelination, as has been done for FGF2 knockout mice (2). In both MHV-A59 and cuprizone models, oligodendroglial repopulation of lesions was enhanced in the absence of FGF2. During early remyelination, oligodendrocyte lineage cell proliferation and survival were not altered by the FGF2 genotype – possibly because other growth factors sufficiently supported these lineage responses (see ref. 2).

Thus, during remyelination, absence of FGF2 may promote oligodendroglial regeneration by enhancing oligodendrocyte progenitor maturation. This *in vivo* distinction among potential roles of FGF2 in differentiation versus proliferation demonstrates that transgenic and knockout mouse models are important for testing growth factor roles in the context of the lesion environment throughout the disease progression. The expression and activity of growth factors, and other regulatory molecules, change throughout the disease course. In addition, the oligodendrocyte lineage population changes with progressive stages of the disease course. Therefore, the effect of a given growth factor on the oligodendrocyte lineage population may vary *in vivo* during the disease course.

## 5. CONCLUSION

Future therapeutic developments may make it possible to intervene to promote remyelination for MS patients in which the disease progression can be arrested. Effective treatments to promote repair of demyelinated lesions will need to be tailored to the specific pathology of the patient and adapted for the disease stage. Animal models are clearly an important component for identifying and testing the potential roles of growth factors as candidate interventions to promote remyelination.

## ACKNOWLEDGEMENTS

We gratefully acknowledge the support of this work by the National Institutes of Health and the National Multiple Sclerosis Foundation.

## REFERENCES

1. Armstrong R, Friedrich VL, Jr., Holmes KV, Dubois-Dalcq M. In vitro analysis of the oligodendrocyte lineage in mice during demyelination and remyelination. J Cell Biol. 1990;111:1183-1195.
2. Armstrong RC, Le TQ, Frost EE et al. Absence of fibroblast growth factor 2 promotes oligodendroglial repopulation of demyelinated white matter. J Neurosci. 2002;22:8574-8585
3. Bansal R, Pfeiffer SE. Regulation of oligodendrocyte differentiation by fibroblast growth factors. Adv Exp Med Biol. 1997;429:69-77

4.  Bansil S, Lee HJ, Jindal S et al. Correlation between sex hormones and magnetic resonance imaging lesions in multiple sclerosis. Acta Neurol Scand. 1999;99:91-94

5.  Bogler O, Wren D, Barnett SC et al. Cooperation between two growth factors promotes extended self-renewal and inhibits differentiation of oligodendrocyte-type-2 astrocyte (O-2A) progenitor cells. Proc Natl Acad Sci U S A. 1990;87:6368-6372

6.  Cottrell DA, Kremenchutzky M, Rice GP et al. The natural history of multiple sclerosis: a geographically based study. 5. The clinical features and natural history of primary progressive multiple sclerosis. Brain. 1999;122 ( Pt 4):625-639

7.  Freeman ME. The neuroendocrine control of the ovarian cycle in the rat. In: Knobil E, Neill JD, eds. The Physiology of Reproduction. 2nd ed. New York, NY: Raven Press, 1994:613-658

8.  Frost EE, Nielsen JA, Le TQ, Armstrong RC. PDGF and FGF2 regulate oligodendrocyte progenitor responses to demyelination. Journal of Neurobiology. 2003; 54:457-472.

9.  Jordan CA, Friedrich VL, Jr., Godfraind C et al. Expression of viral and myelin gene transcripts in a murine CNS demyelinating disease caused by a coronavirus. Glia. 1989;2:318-329

10. Jordan CA, Friedrich VL, Jr., de Ferra F et al. Differential exon expression in myelin basic protein transcripts during central nervous system (CNS) remyelination. Cell Mol Neurobiol. 1990;10:3-18

11. Kristensson K, Holmes KV, Duchala CS et al. Increased levels of myelin basic protein transcripts in virus-induced demyelination. Nature. 1986;322:544-547

12. Lucchinetti C, Bruck W, Parisi J et al. Heterogeneity of multiple sclerosis lesions: implications for the pathogenesis of demyelination. Ann Neurol. 2000;47:707-717

13. Mason JL, Goldman JE. A2B5+ and O4+ Cycling progenitors in the adult forebrain white matter respond differentially to PDGF-AA, FGF-2, and IGF-1. Mol Cell Neurosci. 2002;20:30-42

14. Matsushima GK, Morell P. The neurotoxicant, cuprizone, as a model to study demyelination and remyelination in the central nervous system. Brain Pathol. 2001;11:107-116.

15. Matthews AE, Weiss SR, Paterson Y. Murine hepatitis virus--a model for virus-induced CNS demyelination. J Neurovirol. 2002;8:76-85

16. McKinnon RD, Matsui T, Dubois-Dalcq M, Aaronson SA. FGF modulates the PDGF-driven pathway of oligodendrocyte development. Neuron. 1990;5:603-614.

17. Messersmith DJ, Murtie JC, Le TQ et al. Fibroblast growth factor 2 (FGF2) and FGF receptor expression in an experimental demyelinating disease with extensive remyelination. J Neurosci Res. 2000;62:241-256.

18. Parra B, Hinton DR, Marten NW et al. IFN-gamma is required for viral clearance from central nervous system oligodendroglia. J Immunol. 1999;162:1641-1647
19. Pozzilli C, Falaschi P, Mainero C et al. MRI in multiple sclerosis during the menstrual cycle: relationship with sex hormone patterns. Neurology. 1999;53:622-624
20. Redwine JM, Blinder KL, Armstrong RC. In situ expression of fibroblast growth factor receptors by oligodendrocyte progenitors and oligodendrocytes in adult mouse central nervous system. J Neurosci Res. 1997;50:229-237.
21. Redwine JM, Armstrong RC. In vivo proliferation of oligodendrocyte progenitors expressing PDGFalphaR during early remyelination. J Neurobiol. 1998;37:413-428.
22. Richardson WD, Pringle N, Mosley MJ et al. A role for platelet-derived growth factor in normal gliogenesis in the central nervous system. Cell. 1988;53:309-319
23. Sango K, McDonald MP, Crawley JN et al. Mice lacking both subunits of lysosomal beta-hexosaminidase display gangliosidosis and mucopolysaccharidosis. Nat Genet. 1996;14:348-352
24. Smith R, Studd JW. A pilot study of the effect upon multiple sclerosis of the menopause, hormone replacement therapy and the menstrual cycle. J R Soc Med. 1992;85:612-613
25. Voskuhl RR, Palaszynski K. Sex hormones in experimental autoimmune encephalomyelitis: implications for multiple sclerosis. Neuroscientist. 2001;7:258-270
26. Whitacre CC, Reingold SC, O'Looney PA. A gender gap in autoimmunity. Science. 1999;283:1277-1278
27. Wolswijk G, Noble M. Cooperation between PDGF and FGF converts slowly dividing O-2A adult progenitor cells to rapidly dividing cells with charac-teristics of O-2A perinatal progenitor cells. J Cell Biol. 1992;118:889-900.
28. Zaffaroni M, Ghezzi A. The prognostic value of age, gender, pregnancy and endocrine factors in multiple sclerosis. Neurol Sci. 2000;21:S857-860
29. Zorgdrager A, De Keyser J. Menstrually related worsening of symptoms in multiple sclerosis. J Neurol Sci. 1997;149:95-97

Chapter C9

# CHEMOKINES IN CORONAVIRUS-INDUCED DEMYELINATION

Matthew J. Trifilo[*], Michael T. Liu, William G. Glass, and Thomas E. Lane[*1]
*Department of Molecular Biology and Biochemistry, University of California, Irvine, CA 92697-3900

Abstract: Inflammation within the central nervous system (CNS) is critical in the development of the neuropathology associated with the human demyelinating disease multiple sclerosis (MS). Recent studies have identified a family of soluble proinflammatory molecules called chemokines that are able to direct leukocyte infiltration into the CNS in response to infection or injury. Identification of chemokines within and around demyelinating lesions in MS patients indicate a potential role for these molecules in contributing to the pathogenesis of MS. To address this issue, we have used mouse hepatitis virus (MHV) infection of the CNS to understand the dynamic interaction of chemokine expression as it relates to inflammation and neuropathology. Our results indicate that chemokine expression within the CNS results in persistent recruitment of both T lymphocytes and macrophages and results in subsequent myelin destruction. Herein, we demonstrate the complexity of the chemokine response to MHV infection of the CNS and the delicate balance that exists between host defense and development of disease.

Key words: Chemokines, T lymphocytes, Macrophages, Demyelination

## 1.    INTRODUCTION

Mouse hepatitis virus (MHV) is a single strand, positive sense RNA virus that is a member of the Coronaviradae family. As with many viruses, the disease induced by MHV infection depends upon a variety of different factors such as host and viral genetics as well as the dose and route of inoculation. For example, intracranial (i.c.) infection of susceptible strains of mice such as C57BL/6 (H-2$^b$ background) with the neuroattenuated variant MHV strain J2.2v-1 results in infection of glial cells and oligodendrocytes resulting in an acute encephalomyelitis accompanied by white matter destruction. Type I interferons are expressed early following infection and this correlates with recruitment of neutrophils and natural killer (NK) cells into the CNS. Although not completely understood, it is thought that these components of the innate immune response control viral spread prior to the entry of virus-specific T cells as well as participate in amplifying chemokine expression within the CNS that results in the induction of the adaptive immune response. CD8$^+$ T cells are the primary effector component of the adaptive immune response and are able to eliminate infectious virus from the host through secretion of IFN-$\gamma$ (oligodendrocytes) and perforin-mediated cytolysis (astrocytes and microglia) (1,2). CD4$^+$ T cells can also directly participate in reducing viral burden presumably through the release of antiviral cytokines such as IFN-$\gamma$. However, CD4$^+$ T cells also play an essential role in host defense by providing support for CD8$^+$ T cells (3,4). Although both CD4$^+$ and CD8$^+$ T cells are potent anti-viral effector cells within the CNS, sterile immunity is not achieved and viral RNA and protein persist primarily within white matter tracts in which astrocytes and oligodendrocytes are the predominant cellular reservoirs for virus.

Viral persistence in white matter tracts results in a chronic demyelinating disease in which foci of demyelination are associated with areas of viral RNA/antigen. Clinically, mice develop loss of tail tone and a partial to complete hind-limb paralysis. Recent reports have indicated that MHV-induced demyelination is complex and involves immunopathologic responses against viral antigens expressed in infected tissues. MHV infection of immunosuppressed or immunodeficient mice results in high titers of virus within the CNS and increased mortality but no detectable demyelination (5). Further, adoptive transfer of MHV-immune splenocytes restores demyelination to the infected recipients suggesting a role for immune cells in driving demyelination (5). We have recently demonstrated that CD4$^{-/-}$ mice displayed a significant reduction in the severity of demyelination as compared to CD8$^{-/-}$ and immunocompetent wildtype mice suggesting an important role for CD4$^+$ T cells in amplifying the severity of white matter destruction (6). Additional support for T cells in contributing to demyelination comes from a recent report by Wu *et al.* (3) that demonstrated that both CD4$^+$ and CD8$^+$ T cells are important in mediating myelin destruction following adoptive transfer into MHV-infected RAG1$^{-/-}$ mice. Although B cells and/or antibodies have been shown to play an important role in controlling the recrudescence of virus during the chronic stage of disease, they are not important in MHV-induced demyelination (7). We and others have found that macrophages/microglia are also important in contributing to demyelination (3,6,8). Current evidence suggests that demyelination in MHV-infected mice is not the result of epitope spreading and induction of an immune response against neuroantigens as has recently been reported to occur during Theiler's virus-induced demyelination (9). It is likely that T cells present within the CNS of MHV-infected mice are specific for viral antigens and these cells drive demyelination by producing and/or influencing the production of cytokines/chemokines that amplify and support inflammation. Therefore, collective evidence points to a role for inflammatory T cells in contributing to macrophage/microglia infiltration and activation that results in myelin destruction. Collectively, these studies illustrate the complexity of the MHV system as it relates to the immune response and demyelination.

Given the importance of both T cells and macrophages in amplifying the severity of myelin destruction in persistently infected mice, it is clear that maintenance of a chronic inflammatory response is important with regards to the neuropathology of this model system. Numerous cytokines including TNF-$\alpha$, IL-1$\beta$, and IL-6 are produced by activated astrocytes within the CNS of persistently infected mice (10). Working in concert or individually, these molecules may contribute to disease and myelin destruction by attracting T cells and macrophages into the CNS. However, the soluble signals that regulate CNS invasion by inflammatory cells likely consist of additional factors. With this in mind, studies from our laboratory have focused on chemokines (*chemo*tactic cyto*kines*) and the role these molecules have in MHV-induced demyelination. This chapter will focus on recent studies evaluating the functional significance of chemokine expression as it relates to disease development following MHV infection of the CNS.

## 1.1    Chemokines and Chemokine Receptors

Chemokines are a family of small molecular weight, secreted proteins that have been shown to target the migration of specific populations of leukocytes (Reviewed in 11). The chemokines are currently divided into four subfamilies based upon the arrangement of conserved cysteine residues in the amino-terminus of the protein (structural criteria) and what leukocytes are targeted, e.g. monocytes, macrophages, granulocytes, neutrophils, and T lymphocytes (functional criteria)(11-13). A variety of lymphoid and nonlymphoid tissues can produce chemokines, often in connection with the initiation or progression of inflammation (11). However, recent evidence also identifies a role for chemokines in the regulation of normal leukocyte migration into, out of, and within lymphoid tissues and organs (14,15).

Much attention has focused on two distinctive groups of chemokines, the CXC and CC chemokines, because of their appearance in numerous disease models, including MS (11,16,17). The CXC family can be further subdivided into two groups based upon the presence of a glutamic acid-leucine-arginine (ELR) motif on the amino terminus. The appearance of the ELR motif encodes for chemotaxis of neutrophils, whereas its absence dictates the recruitment of lymphocytes and monocytes (11,13,18,19). In general, the CC chemokines share chemoattractant properties with non-ELR CXC chemokines, attracting monocytes, T lymphocytes, basophils, and eosinophils and have little or no effect on neutrophils (11,13).

As mentioned above, chemokines can be produced in the presence or absence of inflammation. However, much attention has focused on the beneficial and detrimental role of chemokines in leukocyte recruitment in response to infection (11,13, 20-28). A wide variety of cell types can produce chemokines in response to pathogen, including monocytes, macrophages, lymphocytes, neutrophils, endothelial cells, fibroblasts, and astrocytes. Although many of these cells can produce a diverse array of chemokines, specific chemokine production appears to be linked to the type and strength of the stimuli present. Chemokines are able to induce leukocyte migration into tissue by promoting the adherence and extravasation of leukocytes across the vascular endothelium. Once inside the inflamed tissue, leukocytes are able to migrate across chemokine gradients towards the highest localized concentration. These gradients are established through the binding of the highly basic chemokines to the acidic extracellular matrix on the luminal side of the endothelium. This interaction may also allow for reduced chemokine diffusion and help establish a fixed gradient towards the site of infection (29).

Although in vitro studies have shown extensive redundancy in the chemotactic properties of chemokines, in vivo studies utilizing knockout mice and neutralizing-antibodies have suggested that in response to viral infection of the CNS, chemokines are able to selectively attract distinct leukocyte populations (20-24). Importantly, accumulating evidence indicates that this non-redundant recruitment may influence a variety of CNS diseases by initiating events culminating in neuroinflammation (20,30,31). In addition to their prominent role in inflammation, chemokines may also directly contribute to disease progression by exerting a cytotoxic effect upon host tissue (32). Furthermore, there is increasing evidence that chemokines are able to influence the activation and differentiation (Th1 vs. Th2) of T cells (22,33).

Chemokines act through binding specific receptors present on the surface of numerous cell types. The majority of chemokine receptors belong to the seven-transmembrane G protein-coupled receptor superfamily. Unlike the specificity observed between most ligand receptor pairs, chemokines exhibit a high level of promiscuity, capable of productively binding to multiple chemokine receptors and chemokine receptors can functionally bind multiple chemokines (34-37). Following chemokine binding to its cognate receptor, signaling events are initiated that result in various cellular processes including increases in intracellular calcium, production of cytokines and chemokines, adhesion to the endothelial matrix, and chemotaxis (38). Interestingly, numerous CXC and CC chemokines can also bind to a receptor identical to the Duffy blood group antigen that is present on the surface of erythrocytes (39-41). Binding of chemokines to these receptors does not result in intracellular signaling events, instead it appears to reduce the level of circulating chemokines that may decrease unwanted interactions between chemokines and leukocytes within the blood, maximizing the ability of circulating leukocytes to migrate to highly concentrated chemokine levels at the site of infection (40,42,43).

Leukocytes can limit chemokine signaling by effectively down-regulating chemokine receptor expression or by uncoupling the downstream signaling events from the chemokine receptor, allowing for immediate control of chemokine signaling. Additionally, leukocytes exposed to high concentrations of an individual chemokine can become desensitized via phosphorylation of chemokine receptors limiting leukocyte movement at the site of infection (34,44).

Chemokine receptors can be identified on nearly all lymphocyte populations. However, the expression pattern of certain receptors appears to correlate with the activation state of the cell. For this reason, specific chemokine receptors have been utilized to determine the activation state of specific sub-populations of T lymphocytes. For example, expression of the chemokine receptor CCR7, which is necessary for migration into secondary lymphoid tissues, has been localized to the surface of naïve T cells and a subpopulation of memory cells (45). In contrast, T cells expressing the chemokine receptors CXCR3, CCR1 and/or CCR5 selectively migrate into inflamed tertiary tissues, indicating an important role for these chemokine receptors and their respective ligands in promoting an inflammatory response (20,24,45-47).

## 1.2    Chemokines and Multiple Sclerosis

Expression of chemokines has been associated with demyelinating plaque lesions present in MS patients (48,49). Hvas and colleagues (48) demonstrated expression of the CC chemokine CCL5 by infiltrating T-lymphocytes surrounding MS plaque lesions. Elevated levels of the non-

ELR CXC chemokines CXCL10, CXCL9 as well as CCL5 were found in the CSF of MS patients during periods of clinical attack (49). In addition, astrocyte expression of CXCL10 has been reported in plaque lesions present in MS patients (49). Recent work has demonstrated a direct correlation between clinical progression in the severity of MS with leukocyte infiltration into the CNS suggesting that production of CXCL10, CXCL9, and CCL5 may contribute to the pathogenesis of MS by recruiting T cells and macrophages into the CNS (49). Further, MS patients display increased expression of CXCR3 and CCR5, the cellular receptors for CXCL10, CXCL9, and CCL5, respectively, within the CNS as compared to control patients (49). T cells expressed both CXCR3 and CCR5 whereas macrophages/microglia were found to express CCR5 (49). Further support for chemokines and chemokine receptors in contributing to recruitment of activated leukocytes comes from a recent study by Baranzini et al. (50) that demonstrated increased expression of CCR1 (signaling receptor for CCL5 and CCL3) and CCR5 within the brains of MS patients. Collectively, these data emphasize the need for a better understanding the functional significance of chemokine and chemokine receptor expression as it relates to demyelination in MS patients.

## 1.3    MHV Infection and Chemokine Gene Expression

MHV infection of the CNS results in a well-orchestrated expression of chemokine genes which appears to be dictated, in part, by viral burden (51, Table I). CXCL10 mRNA transcripts are detected within 1 day following infection and co-localize with areas of MHV replication. At this time, the cellular source of CXCL10 transcripts are primarily ependymal cells lining the lateral ventricle (51). By day 6 p.i., virus has spread throughout the brain parenchyma and a robust inflammatory response, characterized primarily by $CD4^+$ and $CD8^+$ T cells and macrophages, is established within the brain. Chemokines expressed at this time include CXCL9, CXCL10, CCL2, CCL3, CCL5 and CCL7 (51). *In situ* double-labeling revealed that astrocytes are the predominant cellular source of mRNA transcripts for CXCL9 and CXCL10 during the acute stage of disease (21,51). In addition, MHV-infection of mouse cultured astrocytes results in the rapid induction of numerous chemokine genes including CXCL1, CXCL10, CCL2, CCL4, and

**Table I. Summary of chemokine expression with clinical and pathological features following MHV infection of the CNS**

| | Days Postinfection | | | |
|---|---|---|---|---|
| | **3** | **7** | **12** | **35** |
| **Clinical signs** | None | Limp tail, ruffled fur,hunched | Limp tail, ruffled fur, hunched, awkward gait | Limp tail, ruffled fur, hunched, awkward gait |
| **Chemokines[a]** | CXCL10 CCL9 CCL5 CCL2 CCL3 CXCL11 | CXCL10 CXCL9 CCL5 CCL2 CCL3 CXCL11 | CXCL10 CCL5 | CCL5 CXCL10 |
| **Viral Titer[c]** | +++ | ++ | BD[b] | BD |
| **Viral RNA** | + | ++++ | +++ | ++ |
| **Demyelination[d]** | - | + | ++ | +++ |

[a]Brains and spinal cords were removed at indicated time points, and total RNA extracted and chemokine mRNA transcripts were evaluated by RPA. Chemokine mRNA transcripts are listed in order of abundance.

[b]BD= Below Detection.

[c]Viral RNA detectable by in situ hybridization.

[d]Demyelination was determined by Luxol fast blue staining of paraffin-embeded brain and spinal cord sections.

CCL7 (6). UV-inactivated MHV was sufficient to trigger low level chemokine gene expression indicating that viral replication is not necessary to induce expression of these genes (51). *In vivo* analysis of the chemokine receptor mRNA profile indicates increased expression of CCR1, CCR5, and CXCR3 during the acute stage of disease (20,24). Confocal microscopy revealed that macrophages (determined by F4/80 antigen expression) express CCR5 (24). In addition, studies have shown that both CD4$^+$ and CD8$^+$ T cells infiltrating into the brain express CXCR3 (20).

## 1.4    Functional Significance of Chemokine Expression during MHV-Induced Demyelination

CXCL10 and CCL5 are the predominant chemokines expressed in the CNS of persistently infected mice undergoing demyelination (51) (Table I). Moreover, CXCR3 and CCR5 expression (receptors for CXCL10 and CCL5, respectively) is also detected at this time (20,24). These findings have parallels in human neurodegenerative diseases. Glial cell expression of CCL5 and CXCL10 as well as CXCR3-and CCR5-positive mononuclear cells have been reported within demyelinating lesions present in patients with MS (49,52). Moreover, increased levels of CXCL9, CXCL10 and CCL5 are present within the cerebral spinal fluid of MS patients during periods of clinical attack and their presence correlates with increased numbers of inflammatory cells (49). Therefore, the overlap in chemokine and chemokine receptor expression profiles within the demyelinating CNS of MHV-infected mice and MS patients provides an opportunity to assess the functional significance of these molecules as it relates to demyelination in the MHV model system.

## 1.5 CCL5 and Disease

Intracranial infection of $CD4^{-/-}$ and $CD8^{-/-}$ mice with MHV resulted in increased mortality and delayed clearance of virus from the CNS demonstrating an important role for T lymphocytes in host defense against MHV-induced CNS disease (6). Interestingly, infected $CD4^{-/-}$ mice displayed significantly less severe inflammation and demyelination as compared to $CD8^{-/-}$ and wildtype C57BL/6 mice. FACS analysis of the cellular infiltrate present within the CNS of infected mice revealed that $CD4^{-/-}$ mice contain fewer activated macrophages and significantly lower levels of CCL5 mRNA transcripts and protein when compared to $CD8^{-/-}$ and C57BL/6 mice. These data suggested that $CD4^{+}$ T cells are the predominant source of CCL5 following MHV infection of the CNS, although it is also possible that $CD4^{+}$ T cells influence expression of CCL5 by other cell populations through the release of cytokines and/or chemokines. Additional cellular sources such as $CD8^{+}$ T cells, macrophage and glial cells must be considered due to the fact that CCL5 mRNA transcripts and protein are detected, albeit at lower levels, within the CNS of $CD4^{-/-}$ mice. In light of the fact that CCL5 exerts a potent chemotactic effect on both T cells and macrophages, these data suggest that the reduction in macrophage infiltration and the severity of demyelination into the CNS of $CD4^{-/-}$ mice is causally related to the reduced CCL5 levels observed.

To provide a direct test of CCL5 importance in contributing to MHV-induced CNS inflammation and demyelination, MHV-infected C57BL/6 mice were treated with anti-CCL5 antisera immediately following MHV infection and the severity of disease evaluated. Treatment lead to a disease in C57BL/6 mice similar to the phenotype observed in $CD4^{-/-}$ mice with mice displaying a significant reduction in the severity of demyelination. Decreased macrophage infiltration in anti-CCL5 treated mice correlated with the reduced severity of demyelination supporting the observations with

MHV-infected $CD4^{-/-}$ mice (6). These observations reinforce the functional significance of CCL5 expression during virus-induced CNS disease indicating that this chemokine has an important role in the recruitment of macrophages into the CNS following MHV infection. Moreover, administration of neutralizing anti-CCL5 antibodies to MHV-infected mice in which disease e.g. demyelination and paralysis is already established resulted in a marked improvement in neurologic disease that correlated with reduced T cell infiltration and a reduction in the severity of myelin destruction (Glass and Lane, unpublished observations). Collectively, these data highlight the important role of CCL5 signaling in both host defense and disease pathogenesis by regulating leukocyte accumulation within the CNS of MHV-infected mice.

Having established that CCL5, a major signaling ligand of CCR5, is important in contributing to demyelination in MHV-infected mice by attracting macrophages into the CNS, studies were performed to further characterize the contributions of CCR5 to neuroinflammation and demyelination in MHV-infected mice. Analysis of CCR5 expression within the CNS of MHV-infected mice revealed that the majority of cells expressing this receptor were macrophage/microglia (24). MHV infection of $CCR5^{-/-}$ mice did not result in an increase in either morbidity or mortality as compared to infected $CCR5^{+/+}$ mice. In addition, clearance of virus from the CNS of infected $CCR5^{-/-}$ mice was not impaired and this correlated with equivalent levels of infiltrating T cells into the CNS when compared to MHV-infected $CCR5^{+/+}$ mice. These data indicate that the expression of this receptor does not contribute to host defense. However, demyelination was reduced in MHV-infected $CCR5^{-/-}$ mice and this correlated with reduced levels of infiltrating macrophages as compared to $CCR5^{+/+}$ mice. These data suggest that CCL5 signaling through CCR5 is an important mechanism whereby macrophages are able to enter the CNS and contribute to demyelination. Additional support for the importance of CCR5 signaling in leukocyte migration comes from recent studies from our laboratory examining the outcome of MHV infection of CCL3 knock-out mice. In addition to CCL5, CCL3 is able to bind and signal through CCR5 and trigger T cell and macrophage activation and trafficking. Intracranial infection of $CCL3^{-/-}$ mice resulted in diminished T cell and macrophage accumulation within the CNS that correlated with a significant reduction in demyelination as compared to MHV-infected wild-type mice (22). Therefore, these data demonstrate that CCR5 signaling by either CCL3 or CCL5 enhances both T cell and macrophage trafficking and accumulation within the CNS of MHV-infected mice and this correlates with the severity of myelin destruction.

Signaling through CCR5 appears to selectively regulate the trafficking of subsets of T cells into the CNS of MHV-infected mice. Adoptive transfer of virus-specific CD4$^+$ T cells derived from MHV-immunized CCR5$^{+/+}$ mice and expanded to the immunodominant epitope present in the matrix protein spanning residues 133-147 into MHV-infected RAG1$^{-/-}$ mice resulted in T cell accumulation within the CNS and increased CCL5 expression that correlated with macrophage infiltration and demyelination (53). In contrast, adoptive transfer of virus-specific CD4$^+$ T cells obtained from immunized CCR5$^{-/-}$ mice resulted in only limited numbers of T cells present within the CNS and diminished CCL5 expression that correlated with a reduction in macrophage infiltration and demyelination. Examination of chemokine receptor expression on virus-specific CD4$^+$ T cells derived from CCR5$^{+/+}$ and CCR5$^{-/-}$ mice indicated a marked decrease in mRNA transcripts for several receptors. Notably, CXCR3, which is the signaling receptor for CXCL10, was significantly reduced. These data imply that CCR5 signaling may regulate expression of additional chemokine receptors which enhance T cell trafficking into target tissues. In contrast, CCR5 signaling is not required for trafficking of virus-specific CD8$^+$ T cells into the CNS (Glass and Lane, unpublished observations). However, CCR5 does appear to contribute to effector functions of CD8$^+$ T cells as CD8$^+$ T cells lacking CCR5 exhibit increased production of IFN-$\gamma$ and increased cytolytic activity as compared to wildtype mice. These data highlight the importance of chemokines and chemokine receptors with regards to the trafficking and activation of specific subsets of T cells as it relates to host defense and disease progression in the MHV model system.

## 1.6     CXCL10 and Disease

As mentioned above, the pattern of chemokine expression as well as the appearance of the respective chemokine receptors on the surface of activated T cells suggested that these molecules may control the migration of lymphocytes following viral infection of the CNS. Although many chemokines are expressed at specific times during the course of MHV infection, one of the most prominent chemokines expressed is CXCL10 (51). Analysis of the functional contributions of CXCL10 during the acute stage of disease indicated a direct role in activated CD4$^+$ and CD8$^+$ T cell chemotaxis into the CNS. The correlation between T cell infiltration and demyelination following MHV infection as well as the appearance of CXCL10 during the chronic stage of MHV disease suggested that this chemokine may also contribute to T cell mediated myelin destruction. To address the functional contributions of CXCL10 to MHV induced demyelination, mice were treated with anti-CXCL10 neutralizing antibodies 12 days following i.c. infection. Strikingly, anti-CXCL10 treatment not only stopped further progression of myelin destruction, but decreased the severity of clinical symptoms as compared to mice treated with control antibody (31). Histological analysis revealed that reduced clinical symptoms correlated with a dramatic reduction in the level of demyelination within anti-CXCL10-treated mice. Furthermore, this decrease in behavioral deficits correlated with the presence of increased levels of remyelination within the CNS as assessed by electron microscopy. Analysis of T cell infiltration following anti-CXCL10 treatment indicated a dramatic reduction in CD4$^+$ T cells while CD8$^+$ T cell levels remained similar to control. The preferential effect on CD4$^+$ T cells was in contrast to effects observed following anti-CXCL10 treatment during the acute stage of disease, where both CD4$^+$ and CD8$^+$ T cell migration was compromised (20). However, analysis of CXCR3 expression on infiltrating lymphocytes during both the acute and chronic stage of MHV infection indicated that while both CD4$^+$ and CD8$^+$ T cells express abundant CXCR3 expression during the acute disease, CD4$^+$ T cells preferentially express increased levels of the CXCL10 receptor during chronic disease, suggesting that receptor expression is responsible for selective CD4$^+$ T cell migration (31). Regardless, these results indicate that CD4$^+$ T cell infiltration can contribute not only to myelin destruction, but also inhibit remyelination following MHV infection. These results indicate that one mechanism by which CXCL10 can contribute to demyelination is by recruiting activated CD4$^+$ T cells into the CNS that can participate in disease pathogenesis.

## 1.7 Conclusion

MHV infection of the CNS provides a consistent, reliable model in which to study not only host response to viral infection but also to determine the contributions of chemokines to neuroinflammation and demyelination. Although the studies mentioned above indicate a complex web of chemokines and chemokine receptor expression, analysis of individual chemokines utilizing knockout mice and neutralizing antibodies has indicated specific and selective roles for these molecules within the CNS. As depicted in Figure 1, chemokines are expressed within the CNS throughout MHV infection, and have now been shown to participate in the recruitment and/or activation of cells based on their temporal expression. Indeed, early expression of CXCL10 contributes to the development of the innate response that can both control viral replication and further amplify chemokine production aiding in the establishment of the adaptive immune response (Trifilo and Lane, unpublished observations) (Figure 1, **1**). By day 7 p.i., robust expression of numerous CXC and CC chemokines (see Table I) are detected within the CNS. Highlighted is CXCL10 and CCL5 expression that are critical for the infiltration of T cells (CD4$^+$ and CD8$^+$) and activated

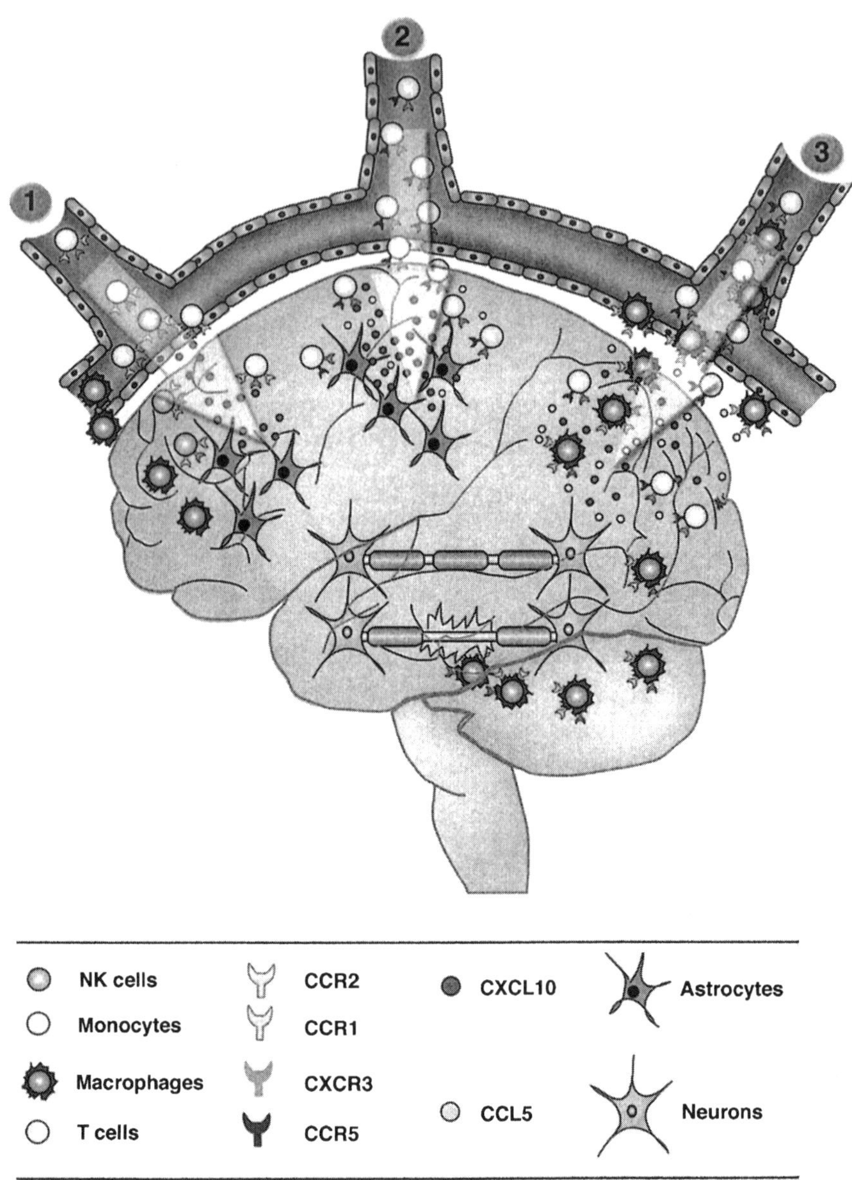



*Figure C9-1.* MHV infection of the CNS

macrophages (20,24) (Figure 1, **2**). T cell infiltration mediates clearance of replicating virus from the CNS by 12 days p.i. through the release of cytokines (ie. IFN-γ) and cell mediated cytolysis (CTL). Although replicating virus can no longer be detected, viral protein and RNA persist after 12 days p.i., resulting in chronic expression of both CXCL10 and CCL5 (Figure 1, **3**). These chemokines induce the chronic recruitment of activated T cells and macrophages that subsequently participate in demyelination. This model reveals the protective and destructive properties chemokines posses based on both the level and time of expression.

Collectively, our data indicate that chemokines can play specific and selective roles in T lymphocyte and macrophage recruitment within the CNS (CXCL10 and CCL5) as well as contribute to the activation of virus-specific T cells (CCL3). More importantly, although numerous chemokines are detected within the CNS following MHV infection, it is clear that these molecules function in a nonredundant manner, thus meriting further studies on chemokines with regards to the role of these molecules in viral-induced CNS disease. Finally, the data clearly demonstrate that chemokines and their receptors may represent viable targets in modulating the severity of human nueroinflammatory diseases including MS.

## ACKNOWLEDGEMENTS

This work was supported by National Institutes of Health grants 37336 and 41249 and the National Multiple Sclerosis Society Research Grant RG 30393A1/T to T.E.L. W.G.G. and M.T.L. were supported by National Institutes of Health training grants A107319-12 and T32NS07444, respectively.

## REFERENCES

1.  Lin MT, Hinton DR, Stohlman SA. Mouse hepatitis virus is cleared from the central nervous system in mice lacking perforin-mediated cytolysis. *J Virol* 1997 **71**:383-391.
2.  Parra B, Hinton DR, Marten NW, Bergmann CC, Ln MT, Yang CS, Stohlman SA. IFN-gamma is required for viral clearance from central nervous system oligodendroglia. *J Immunol* 1999 **162**:1641-1647.
3.  Wu GF, Dandekar AA, Pewe L, Perlman S. CD4 and CD8 T cells have redundant but not identical roles in virus-induced demyelination. *J Immunol* 2000 **165**:2278-2286.
4.  Stohlman SA, Bergmann CC, Lin MT, Cua DJ, Hinton DR. CTL effector function within the central nervous system requires CD4+ T cells. *J Immunol* 1998 **160**:2896-2904.
5.  Wang F, Stohlman SA, Fleming JO. Demyelination induced by murine hepatitis virus JHM (MHV-4) is immunologically mediated. *J Neuroimmunol* 1990 **30**:31-41.

6.   Lane TE, Liu MT, Chen BP, Asensio VC, Samawi RM, Paoletti AD, Campbell IL, Kunkel SL, Fox HS, Buchmeier MJ. A central role for CD4$^+$ T cells and RANTES in virus-induced central nervous inflammation and demyelination. *J Virol* 2000 **74**:1415-1424.

7.   Lin MT, Hinton DR, Marten NW, Bergmann CC, Stohlman SA. Antibody prevents virus reactivation within the central nervous system. *J Immunol* 1999 **162**:7358-7368.

8.   Wu GF, Perlman S. Macrophage infiltration, but not apoptosis, is correlated with immune-mediated demyelination following murine infection with a neurotropic coronavirus. *J Virol* 1999 **73**:8771-8780.

9.   Miller SD, Vanderlugt CL, Begolka WS, Pao W, Yaunch RL, Neville KL, Katz-Levy Y, Carrizosa A, Kim BS. Persistent infection with Theiler's virus leads to CNS autoimmunity via epitope spreading. *Nat Med* 1997 **3**:1133-1136.

10.  Sun N, Grzybicki D, Castro R, Murphy S, Perlman S. Activation of astrocytes in the spinal cord of mice chronically infected with a neurotropic coronavirus. *Virology* 1995 **213**:482-493.

11.  Luster AD. Chemokines-chemotactic cytokines that mediate inflammation. *New Engl J Med* 1998 **338**:436-445.

12.  Clark-Lewis I, Kim K, Rajarathan K, Gong J, Dewald B, Moser B, Baggiolini M, Sykes B. Structure activity relationships of chemokines. *J Leuk Biol* 1995 **57**:703-711.

13.  Taub DD, Oppenheim JJ. Chemokines, inflammation and the immune system. *Ther Immunol* 1994 **1**:229-246.

14.  Cyster JG. Chemokines and cell migration in secondary lymphoid organs. *Science* 1999 **286**:2098-2102.

15.  Cyster JG. Leukocyte migration: scent of the T zone. *Cur Biol* 2000 **10**:R30-33.

16.  Huang D, Han Y, Rani MR, Glabinski A, Trebst C, Sorensen T, Tani M, Wang J, Chien P, O'Bryan S, Bielecki B, Zhou ZL, Majumder S, Ransohoff RM. Chemokines and chemokine receptors in inflammation of the nervous system: manifold role and exquisite regulation. *Immunol Rev* 2000 **177**:52-67.

17.  Glabinski AR, Ransohoff RM. Chemokines and chemokine receptors in CNS pathology. *J Neurovirol* 1999 **5**:3-12.

18.  Farber JM. Mig and IP-10: CXC chemokines that target lymphocytes. *J Leukoc Biol* 1997 **61**:246-257.

19.  Piali L, Weber C, LaRosa G, Mackay CR, Springer TA, Clark-Lewis I, Moser B. The chemokine receptor CXCR3 mediates rapid and shear-resistant adhesion-induction of effector T lymphocytes by the chemokines IP-10 and Mig. *Eur J Immunol* 1998 **28**:961-972.

20.  Liu MT, Chen BP, Oertel P, Buchmeier MJ, Armstrong D, Hamilton TA, Lane TE. Cutting Edge: the T cell chemoattractant IFN-inducible protein 10 is essential in host defense against viral-induced neurologic disease. *J Immunol* 2000 **165**:2327-2330.

21.  Liu MT, Armstrong D, Hamilton TA, Lane TE. Expression of Mig (monokine induced by interferon-g) is important in T lymphocyte recruitment and host defense following viral infection of the central nervous system. *J Immunol* 2001 **166**:1790-1795.

22.  Trifilo MJ, Bergmann CC, Kuziel WA, Lane TE. The CC chemokine ligand 3 (CCL3) regulates CD8$^+$ T cell effector function and migration following viral infection. *J Virol* 2003 in press.

23. Chen BP, Kuziel WA, Lane TE. Lack of CCR2 results in increased mortality and impaired leukocte activation and trafficking following infection of the central nervous system with a neurotropic coronavirus. *J Immunol* 2001 **167**:4585-4592.

24. Glass WG, Liu MT, Kuziel WA, Lane TE. Reduced macrophage infiltration and demyelination in mice lacking the chemokine receptor CCR5 following infection with a neurotropic coronavirus. *Virology* 2001 **288**:8-17.

25. Murray PD, Krivacic K, Chernosky A, Wei T, Ransohoff RM, Rodriquez M. Biphasic and regionally-restricted chemokine expression in the central nervous system in the Theiler's virus model of multiple sclerosis. *J Neurovirol* 2000 **6**:S44-52.

26. Ransohoff RM, Wei T, Pavelko KD, Lee JC, Murray PD, Rodriguez, M. Chemokine expression in the central nervous system of mice with a viral disease resembling multiple sclerosis: roles of CD4+ and CD8+ T cells and viral persistence. *J Virol* 2002 **76**:2217-2224.

27. Olszewski MA, Huffnagle GB, McDonald RA, Lindell DM, Moore BB, Cook DN, Toews GB. The role of macrophage inflammatory protein-1 alpha/CCL3 in regulation of T cell-mediated immunity to Cryptococcus neoformans infection. *J Immunol* 2000 **165**:6429-6436.

28. Huffnagle GB, Strieter RM, Standiford TJ, McDonald RA, Burdick MD, Kunkel SL, Toews GB. The role of monocytes chemotactic protein-1 (MCP-1) in the recruitment of monocytes and CD4+ T cells during a pulmonary Cryptococcus neoformans infection. *J Immunol* 1995 **155**:4790-4997.

29. Tanaka Y, Adams DH, Hubscher S, Hirano H, Siebenlist U, Shaw S. T-cell adhesion induced by proteoglcan-imobilized cytokine MIP-1 beta. *Nature* 1993 **361**:79-82.

30. Karpus WJ. Chemokines and central nervous system disorders. *J Neurovirol* 2001 **7**:493-500.

31. Liu MT, Keirstead HS, Lane TE. Neutralization of the chemokine CXCL10 reduces inflammatory cell invasion and demyelination and improves neurological function in a viral model of multiple sclerosis. *J Immunol* 2001 **167**:4091-4097.

32. Furie MB, Randolph GJ. Chemokines and tissue injury. *Am J Pathol* 1995 **146**:1287-1301.

33. Karpus WJ, Lukacs NW, Kennedy KJ, Smith WS, Hurst SD, Barrett TA. Differential CC chemokine-induced enhancement of T helper cell cycle cytokine production. *J Immunol* 1997 **158**:4129-4136.

34. Murphy P. The molecular biology of leukocyte chemoattractant receptors. *Ann Rev Immunol* 1994 **12**:593-633.

35. Kelvin D, Michiel D, Johnston J, Lloyd A, Sprenger H, Oppenheim J, Wang JM. Chemokines and serpentines: the molecular biology of chemokine receptors. *J Leukocyte Biol* 1993 **54**:604-612.

36. Horuk R. Molecular properties of the chemokine receptor family. *Trends Pharmacol Sci* 1994 **15**:159-165.

37. Horuk R. The interleukin-8-receptor family: from chemokines to malaria. *Immunol Today* 1994 **15**:169-174.

38. Ward SG, Bacon K, Westwick J. Chemokines and T lymphocytes: more than an attraction. *Immunity* 1998 **9**:1-11.

39. Horuk R, Chitnis CE, Darbonne WC, Colby TJ, Rybicki A, Hadley TJ, Miller H. A receptor for the malarial parasite Plasmodium vivax: the erythrocyte chemokine receptor. *Science* 1993 **261**:1182-1184.

40. Chaudhuri A, Polyakove J, Zbrzezna V, Williams K, Gulati S, Pogo A. Cloning of glycoprotein D cDNA, which encodes the major subunit of the Duffy blood group system and the receptor for Plasmodium vivax malaria parasite. *Proc Nat Acad Sci USA* 1993 **90**:10793-10797.

41. Chaudhuri A, Zbrzezna V, Polyakova J, Pogo A, Hesselgesser J, Horuk R. Expression of the Duffy antigen in K562 cells: evidence that it is the human chemokine erythrocyte receptor. *J Biol Chem* 1994 **269**:7835-7838.

42. Neote K, Darbonne W, Ogez J, Horuk R, Schall T. Identification of a promiscuous inflammatory peptide receptor on the surface of red blood cells. *J Biol Chem* 1993 **268**:12247-12249.

43. Horuk R, Colby T, Darbonne W, Schall T, Neote K. The human erythrocyte inflammatory peptide (chemokine) receptor. Biochemical characterization, solubilization, and development of a binding assay for the soluble receptor. *Biochemistry* 1993 **32**:5733-5738.

44. Ben-Baruch A, Michiel DF, Oppenheim JJ. Signals and receptors involved in recruitment of inflammatory cells. *J Biol Chem* 1995 **270**:11703-11706.

45. Sallusto F, Mackay CR, Lanzavecchia A. The role of chemokine receptors in primary, effector, and memory immune responses. *Annu Rev Immunol* 2000 **18**:593-620.

46. Hamann A, Klugewitz K, Austrup F, Jablonski-Westrich D. Activation induces rapid and profound alterations in the trafficking of T cells. *Eur J Immunol* 2000 **30**:3207-3218.

47. Potsch C, Vohringer D, Pircher H. Distinct migration patterns of naïve and effector CD8 T cells in the spleen: correlation with CCR7 receptor expression and chemokine reactivity. *Eur J Immunol* 1999 **29**:3562-3570.

48. Hvas J, Bernard CC. Molecular detection and quantitation of the chemokine RANTES mRNA in neurological brain. *APMIS* 1998 **106**:598-604.

49. Sorensen TL, Tani M, Jensen J, Pierce V, Lucchinetti C, Folcik VA, Qin S, Rottman J, Sellebjerg F, Strieter RM, Frederiksen JL, Ransohoff RM. Expression of specific chemokines and chemokine receptors in the central nervous system of multiple sclerosis patients. *J Clin Invest* 1999 **103**:807-815.

50. Baranzini SE, Elfstrom C, Chang SY, Butunoi C, Murray R, Higuchi R, Oksenberg JR. Transcriptional analysis of multiple sclerosis brain lesions reveals a complex pattern of cytokine expression. *J Immunol* 2000 **165**: 6576-6582.

51. Lane TE, Asensio VC, Yu N, Paoletti AD, Campbell IL, Buchmeier MJ. Dynamic regulation of alpha and beta chemokine expression in the central nervous system during mouse hepatitis virus-induced demyelinating disease. *J Immunol* 1998 **160**:970-978.

52. Balashov KE, Rottman JB, Weiner HL, Hancock WW. CCR5(+) and CCR3(+) T cells are increased in multiple sclerosis and their ligands MIP-1-alpha and IP-10 are expressed in demyelinating brain lesions. *Proc Natl Acad Sci USA* 1999 **96**:6873-6878.

53. Glass WG, Lane TE. Functional expression of chemokine receptor CCR5 on CD4(+) T cells during virus-induced central nervous system disease. *J Virol* 2003 **77**: 191-198.

Chapter C10

# CORONAVIRUS RECEPTORS

Fumihiro Taguchi
*National Institute of Neuroscience, NCNP, 4-1-1 Ogawahigashi, Kodaira, Tokyo 187-8502, Japan*

**Abstract:** The major receptor for murine coronavirus, mouse hepatitis virus (MHV), is identified as a protein, cell-adhesion molecule 1 in the carcinoembryonic antigen family (CEACAM1), which is classified in the immunoglobulin superfamily. There are four CEACAM1 isoforms, with either four or two ectodomains, resulting from an alternative splicing mechanism. CEACAM1 is expressed on the epithelium and in endothelial cells of a variety of tissues and hemopoietic cells, and functions as a homophilic and heterophilic adhesion molecule. It is used as a receptor for some bacteria as well. The N terminal domain participates in mediating homophilic adhesion. This domain is also responsible for binding to the MHV spike (S) protein; the CC' face protruding in this domain interacts with an N terminal region of the S protein composed of 330 amino acids (called S1N330). The binding of CEACAM1 with MHV S protein induces S protein conformational changes and converts fusion-negative S protein to a fusion-positive form. The allelic forms of CEACAM1 found among mouse strains are thought to be an important determinant for mouse susceptibility to MHV.

**Key words:** CEACAM1, cell adhesion molecule, carcinoembryonic antigen, mouse hepatitis virus

## 1. INTRODUCTION

The Coronavirus family includes a number of different viruses that infect a variety of animal species, causing numerous diseases, mainly in organs of the enteric, respiratory and central nervous systems. They are classified into three distinct groups in terms of serological cross-reactivity and sequence homology. Group 1 consists of porcine transmissible gastroenteritis virus, human coronavirus (HCoV) 229E, feline infectious peritonitis virus and so on. Group II includes mouse hepatitis virus (MHV), HCoV-OC43, bovine coronavirus and some others. Group III is comprised of avian coronaviruses; infectious bronchitis virus and turkey coronavirus. All of these viruses infect animals in a highly species-specific fashion, although some of them can experimentally infect animals different from their natural hosts. The receptor protein for group I viruses has been revealed to be an aminopeptidase N, while the receptor for MHV in group II is a protein classified in the immunoglobulin (Ig) superfamily. The receptors for other viruses in group II, as well as for viruses in group III, have not yet been identified. In this chapter, I describe the receptor for MHV as well as the interaction of MHV receptor and the virus spike (S) protein. The receptors of other coronaviruses can be found in a review article (1).

## 2. RECEPTORS FOR MHV

2.1 Discovery of MHV receptor proteins

MHV infects mice, but few other species. The major target organs are the liver, intestine and central nervous system. This host-species specificity or organ tropism of MHV has been thought to be determined mainly by the cellular receptor for MHV. A series of studies carried out by Kay Holmes and her colleagues, which began with analysis of differential susceptibility to MHV infection among mouse strains, has led to the finding of a major MHV receptor molecule. Boyle et al. found that the plasma membranes isolated from MHV-susceptible BALB/c mouse hepatocytes or enterocytes contained a 110 to 120-kDa protein that binds to MHV particles, but those derived from MHV-resistant SJL mice lacked such a protein (2). This finding suggested that the difference in MHV susceptibility among mouse strains is determined by this protein, presumably the MHV receptor. By using monoclonal antibody (MAb) CC-1 specific to 110-120 kDa protein from BALB/c, they purified a protein of ca. 110 kDa and determined the amino acid sequence in its N terminal region, from which the 110-kDa protein was postulated to be a glycoprotein classified in the carcinoembryonic antigen (CEA) family (3, 4). Finally, they isolated a gene encoding this protein, which was revealed to be cell adhesion molecule 1 in the CEA family of the Ig superfamily [formerly called biliary glycoprotein1 (Bgp1) and currently termed CEACAM1] (5). MHV non-permissive BHK cells transfected with this gene were converted to MHV-susceptible cells, indicating that this molecule is the receptor for MHV. It was also found that MHV-resistant SJL mice express a homologous protein (6, 7).

Two other species of glycoprotein, Bgp2 (8) and pregnancy-specific glycoprotein (9), both of which belong to CEA family members, were thereafter found to serve as the MHV receptor in mouse species. However, none of these are as highly efficient as CEACAM1 in terms of receptor functionality or receptor utility by MHV strains. Human CEA glycoprotein works as an MHV receptor as well (10).

## 2.2. Structure and functional regions of major MHV receptor proteins

CEACAM1 is a member of the Ig superfamily and its prototypical 120-kDa glycoprotein consists of four ectodomains (in the order of N, A1, B and A2 from the N terminus), a transmembrane region (TM) and a cytoplasmic tail (Cy) (Fig. 1, 11). The N domain is similar to an Ig-variable domain, and the three other domains resemble a C2 Ig-constant domain. Four different isoforms of CEACAM1 are known to exist, and have been produced by alternative splicing (Fig. 1). Two of the isoforms have 4 ectodomains and the other two have 2 domains, consisting of an N terminal and A2 domains, one of which has either a short or long Cy. The two-domain protein is 48 to 58 kDa in size. CEACAM1 has two allelic forms, CEACAM1$^a$ and CEACAM1$^b$ (Fig. 1). The former is expressed in most laboratory mouse strains, while the latter, insofar as is currently known, is expressed only in MHV-resistant SJL mice (12). In wild mice, however, both of those forms are widely distributed (13). The major structural differences between CEACAM1$^a$ and CEACAM1$^b$ lie in the N domain, which differs in 29 of its 108 amino acids (6, 7). CEACAM1$^a$ is 10- to 100-fold higher than CEACAM1$^b$ in terms of receptor function (14, 15). There is no apparent difference in virus-binding activity as examined by a neutralization test between the 4-ectodomain isoform and the 2-domain CEACAM1$^a$ (16). However, mice deleting the 4-domain CEACAM1$^a$ and expressing the 2-domain isoform alone are more resistant to MHV than those expressing both of the 4- and 2-domain isoforms (17). Thus, there could be a difference between them in terms of MHV receptor function in mice. On the contrary, CEACAM1$^b$ isoform containing 4 domains neutralizes MHV-A59 strain more efficiently than does the isoform containing 2 domains (N and A2 domains), while both of these isoforms showed similar neutralization activity to MHV-JHM strain (16), suggesting a virus-strain specificity in the interaction with CEACAM1. The N domain is responsible for receptor function (18). Since the CEACAM1 splice variant deleting the A1 and B domains is functional, then it is evident that these domains are not necessary for receptor function. The CEACAM1 isoform containing the N and second A1 domains is also

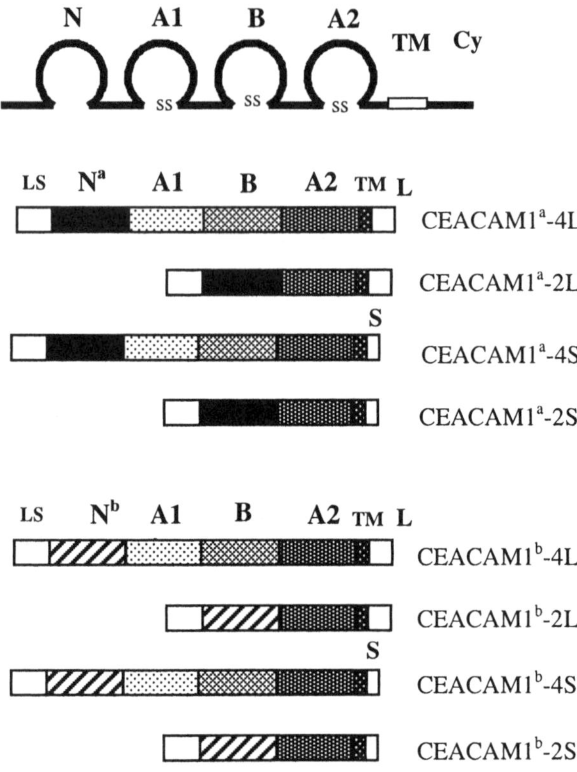

Fig. 1. Schematic structure of a major MHV receptor CECACM1: Four distinct isoforms of CECAM1 exist in each CEACAM1[a] and CECAM1[b] allelic forms. There are four or two ectodomains resembling Ig-V or -C domain. Also, long (L) or short (S) cytoplasmic tail (Cy) is known. LS; leader sequence, TM; transmembrane domain.

functional, indicating that the fourth A2 domain is not absolutely critical. Although CEACAM1 consisting of N domain alone bound MHV, it did not work as a functional receptor when expressed on CEACAM1-negative cells (18). However, the chimeric CEACAM1 having the N domain linked to the mouse poliovirus receptor homolog deleting N domain that has a binding specificity to poliovirus served as a functional receptor for MHV (19). As a result of these reports, it is believed that the N domain alone is sufficient for receptor function; however, when expressed on the cell surface, CEACAM1 containing the N domain alone was buried among the various molecules expressed on the cell surface because of its shortness and hence it failed to bind to viruses (19). By using a soluble form of

CEACAM1, we have recently found that CEACAM1 with N domain alone converted MHV S protein from a fusion-negative to fusion-positive form (20). Collectively, the N domain is sufficient for the MHV receptor function. Detailed analysis using N domain deletion mutants of CEACAM1 showed that a stretch composed of 19 amino acids (aa34 to 52) in the N domain is particularly important for receptor function (21). The stretch is located in the CC' loop in the N domain composed by β-strands and is supposed to protrude from the N domain (22, 23). The difference in receptor function observed between CEACAM1[a] and CEACAM1[b] results from a 6-amino-acid difference in the above-described 19 amino acids in the N domain (14).

## 2.3. Distribution and biological function of CEACAM1

CEACAM1 is reported to be distributed in various cells in different organs, not only in the target organs of MHV, but also in those in which MHV infection has not been detected. A high level of CEACAM1 expression is reported on epithelium and endothelial cells of a variety of tissues, and in hemopoietic cells, such as monocytes, macrophages, granulocytes and their precursors, the B cells, activated T cells and thymic stromal cells (11, 24). It is also demonstrated on both apical membranes of epithelial cells as well as on sites of cell-cell contact (e.g., between hepatocytes, stratified epithelia, junctional epithelium that forms a transition zone between gingival epithelium and teeth, and between pericytes and endothelial cells of blood vessel walls). Furthermore, during early mouse embryonic development, CEACAM1 is abundantly expressed in endodermal and mesenchymal derivatives (25), but is not detected by immune histochemistry in any type of cells in the brain (26). However, MHV infection in the brain was blocked by anti-CEACAM1 MAb CC-1 (27). Also, a CEACAM1 isoform with 2 domains was detected by RT-PCR in the brain (28). These findings suggest that the CEACAM1 molecule is expressed, albeit in very small amounts, in some cell populations of the brain.

The major biological function of CEACAM1 is cell adhesion. It serves as both a homophilic and heterophilic adhesion molecule (11). Homophilic adhesion, confirmed by in vitro studies of rodent and human CEACAM1, is thought to be important in the embryonic organization of the intestinal epithelium and liver hepatocytes , in placental trophoblasts, during muscle and tooth development and vascularization of the central nervous system (24). CEACAM1 also plays an important role in neutrophil activation and adhesion during inflammatory responses (29), lymphoregulation and immunosurveillance (30), angiogenesis (31), and the negative regulation of cell proliferation (11, 24). Heterophilic adhesion of CEACAM1 to other

CEACAM family members has been shown (32). Also, heterophilic adhesion to *Opa* proteins of *Neisseria gonorrhoeae*, *Neisseria meningitis* and *Haemophilus influenzae* mediates their infections (33, 34), indicating that CEACAM1 is a receptor for those bacteria. This also facilitates bacterial colonization of the gut and bacterial phagocytosis by neutrophils and is involved in the initial tethering of granulocytes to E-selectin on the endothelium prior to their transendothelial migration during inflammatory responses. It was recently shown that homophilic adhesion of CEACAM1 involves N-terminal domain interactions. The GFCC'C" face of the N domain, which includes the MHV binding site CC' region, is responsible for homophilic interaction (22).

## 3. INTERACTION OF CEACAM1 AND MHV S PROTEIN

The MHV S protein comprising a petal-like projection on virion surface is the ligand for the CEACAM1 molecule. The projection is composed of two or three molecules of the S1-S2 heterodimer derived from S protein. S protein is a type I glycoprotein. It is synthesized and co-translationally glycosylated as a 150-kDa, protein, becomes as 180-200 kDa protein after modification of glycans and cleaved by a host-derived proteinase into two subunits, N terminal S1 and C terminal S2 (35). S1 comprises an outer knob-like structure of the spike, and S2 consists of the stem-like part beneath the knob (36). S1 and S2 units are associated by non-covalent linkage, and they can be easily dissociated from each other by denaturing reagents or even during a purification process. Alpha-helices constructed by the heptad repeats in the S2 play an important role for oligomerization of S1-S2 heterodimers, though there is another determinant in the S1 for oligomerization (37). Following its synthesis, the S protein is incorporated into the envelope of viral particles after interaction with viral integral membrane protein in the internal compartments from endoplasmic reticulum to the Golgi apparatus. The S protein is also transported to the plasma membrane.

The N terminal region in the S1 composed of 330 amino acids (S1N330) is responsible for binding to CEACAM1 (38). Among S proteins of a variety of MHV strains, there are three conserved regions in S1N330 (S1N330-1, -II and -III) composed of 10 or more identical amino-acid stretches (39). Site-directed mutagenesis analysis suggested that two of these regions, S1N330-1 and –II, located far from one another, are involved in receptor binding (39). Studies using MHV variants containing mutations in S1N330-I confirmed the significance of this region in receptor binding (40). S1N330-III was recently suggested to be responsible for virus entry into the cell in combination with a region in the S2 (41). Denaturing of S1N330 abolished the receptor-binding activity, indicating that the tertiary structure composed of different regions in the S1N330 or/and the dimerization of S1 which takes place within S1N330 (37) is important.

The interaction of MHV S protein and CEACAM1 leads to the S protein functional conversion from a fusion-negative to a fusion-positive form (42). Recently, it was also shown that this functional activation is accompanied with conformational changes in the S protein; receptor-bound S protein has a fraction resistant to proteinase digestion, while receptor-unbound S protein is susceptible (43). These functional and structural changes of the S protein greatly resemble those of the envelope protein of retroviruses that take place after they bind to their receptors (44, 45), suggesting that MHV enters cells in a fashion similar to that of retroviruses.

## 4. MHV RECEPTOR AND ITS IMPLICATIONS FOR MOUSE SUSCEPTIBILITY TO MHV

A number of investigators have reported that BALB/c, C57BL and most other mouse strains, later revealed to have a *Ceacam1ᵃ (1ᵃ)* gene, are susceptible, while SJL mice with a *Ceacam1ᵇ (1ᵇ)* gene are resistant (46, 47). Genetic analyses indicated that a single dominant gene located on chromosome 7 is responsible for susceptibility to MHV (47). The *Ceacam1* could be a gene determining susceptibility, since 1) the 1ᵃ makes mice susceptible and 2) *Ceacam1* is also mapped on chromosome 7 (48). The expression of either 1ᵃ or 1ᵇ in CEACAM1-negative cells converted them to MHV susceptible, suggesting that allelic differences in receptor proteins were not sufficient to explain the differences in mouse susceptibility to MHV (6, 7). In detailed studies, however, cells transiently expressing 1ᵃ were 10 to 100 times more sensitive to MHV than were cells expressing 1ᵇ (14, 15), indicating a small, but significant difference between 1ᵃ and 1ᵇ. This also suggested that the MHV receptor expressed in SJL is still functional. If the receptor allele controls MHV susceptibility, then SJL should be relatively, but not completely, resistant to MHV. SJL mice are, in fact, resistant to MHV when challenged with a low dose of virus, but susceptible when inoculated with a high dose of virus (49, 50). These observations suggest that *Ceacam1* is a gene controlling MHV susceptibility. Of 120 mice of (BALB/c x SJL) F2 and backcrossed mice to SJL, all mice with *1ᵃ/1ᵃ* and *1ᵃ/1ᵇ* genotypes were susceptible, while all mice with *1ᵇ/1ᵇ* genotype were resistant after infection with a low dose of virus (12). This implies the MHV receptor gene and MHV-susceptibility gene are identical, and if not, they are located within 0.86 cM on chromosome 7. To finally examine whether the receptor gene is identical to the gene that controls MHV susceptibility, gene replacement is a useful strategy. The MHV susceptibility of BALB/c mice in which $1^a/1^a$ is replaced by $1^b/1^b$ and SJL mice in which $1^b/1^b$ is replaced with $1^a/1^a$ will conclusively establish whether the MHV receptor gene is the gene which controls the MHV susceptibility of mice.

# 5. REFERENCES


1) Holmes, K. Coronavirus receptors in "The coronaviridae" ed. by Siddell S. pp55-71 1995 Plenum press.

2) Boyle J.F., Weismiller DG., Holmes KV. Genetic resistance to mouse hepatitis virus correlates with absence of virus-binding activity on target tissues. J. Virol. 1987;61: 185-89.

3) Williams R., Jiang GS, Snyder SW, Frana MF, Holmes KV. Purification of the 110-kilodalton glycoprotein receptor for mouse hepatitis virus (MHV)-A59 from mouse liver and identification of a nonfunctional, homologous protein in MHV-resistant SJL/J mice. J Virol. 1990; 64: 3817-23

4) Williams RK, Jiang GS, Holmes KV. Receptor for mouse hepatitis virus is a member of the carcinoembryonic antigen family of glycoproteins. Proc Natl Acad Sci U S A.1991; 88: 5533-36.

5) Dveksler GS, Pensiero MN, Cardellichio CB, Williams RK, Jiang GS, Holmes KV. et al. Cloning of the mouse hepatitis virus (MHV) receptor: expression in human and hamster cell lines confers susceptibility to MHV. J Virol. 1991; 65: 6881-91.

6) Dveksler GS, Dieffenbach CW, Cardellichio CB, McCuaig K, Pensiero MN, Jiang GS. et al. Several members of the mouse carcinoembryonic antigen-related glycoprotein family are functional receptors for the coronavirus mouse hepatitis virus-A59. J Virol. 1993; 67: 1-8.

7) Yokomori K Lai MM. The receptor for mouse hepatitis virus in the resistant mouse strain SJL is functional: implications for the requirement of a second factor for viral infection. J Virol. 1992;66: 6931-38.

8) Nedellec P, Dveksler GS, Daniels E, Turbide C, Chow B, Basile AA, et al. Bgp2, a new member of the carcinoembryonic antigen-related gene family, encodes an alternative receptor for mouse hepatitis viruses. J Virol.1994; 68: 4525-37.

9) Chen DS, Asanaka M, Yokomori K, Wang F, Hwang SB, Li HP, et al.. A pregnancy-specific glycoprotein is expressed in the brain and serves as a receptor for mouse hepatitis virus. Proc Natl Acad Sci U S A. 1995; 92; 12095-99.

10) Chen DS, Asanaka M, Chen FS, Shively JE, Lai MM. Human carcinoembryonic antigen and biliary glycoprotein can serve as mouse hepatitis virus receptors. J Virol. 1997; 71: 1688-91.

11) Beauchemin N, Draber P, Dvekssler G, Gold P, Gray-Owen,S, Grunert F. et al. Redefined nomenclature for members of the carcinoembryo-nic antigen family. Exp Cell Res. 1999; 252: 243-49.

12) Ohtsuka ., Taguchi F. Mouse susceptibility to mouse hepatitis virus infection is linked to viral receptor genotype. J Virol. 1997; 71: 8860-63.

13) Ohtsuka N, Tsuchiya K, Honda E, Taguchi F. A study on mouse hepatitis virus receptor genotype in the wild mouse. Adv. Exp. Med. Biol. 2001; 494: 237-40

14) Rao PV, Kumari S, Gallagher TM. Identification of a contiguous 6-residue determinant in the MHV receptor that controls the level of virion binding to cells. Virology.1997; 229: 336-48.

15) Ohtsuka N, Yamada YK, Taguchi F. Difference in virus-binding activity of two distinct receptor proteins for mouse hepatitis virus. J Gen Virol. 1996; 77: 1683-92.

16) Zelus BD, Wessner DR, Williams RK, Pensiero MN, Phibbs FT, de Souza M, et al. Purified, soluble recombinant mouse hepatitis virus receptor, Bgp1(b), and Bgp2 murine coronavirus receptors differ in mouse hepatitis virus binding and neutralizing activities. J Virol. 1998; 72: 7237-44.

17) Blau D., Turbide C, Tremblay M, Olson M, Letourneau S, Michaliszyn E, et al. Targeted disruption of the Ceacam1 (MHVR) gene leads to reduced susceptibility of mice to mouse hepatitis virus infection. J Virol. 2001; 75: 8173-86.

18) Dveksler GS, Pensiero MN, Dieffenbach CW, Cardellichio CB, Basile AA, Elia P.E, et al. Mouse hepatitis virus strain A59 and blocking antireceptor monoclonal antibody bind to the N-terminal domain of cellular receptor. Proc Natl Acad Sci U S A. 1993; 90: 1716-20.

19) Dveksler GS, Basile AA, Cardellichio CB, Holmes KV. Mouse hepatitis virus receptor activities of an MHVR/mph chimera and MHVR mutants lacking N-linked glycosylation of the N-terminal domain. J. Virol. 1995; 69: 543-46

20) Miura H, Taguchi F. (unpublished observation)

21) Wessner DR, Shick PC, Lu JH, Cardellichio CB, Gagneten SE, Beauchemin N, et al. Mutational analysis of the virus and monoclonal antibody binding sites in MHVR, the cellular receptor of the murine coronavirus mouse hepatitis virus strain A59. J Virol. 1998; 72: 1941-48.

22) Watt SM, Teixeira AM, Zou GQ, Doyonnas R, Zhang Y, Grunert F,et al. Homophilic adhesion of human CEACAM1 involves N-terminal domain interactions: structural analysis of the binding site. Blood. 2002; 98: 1469-79

23) Tan K, Zelus BD, Meijers R, Liu J, Berfelson J, Duke N. et al. Crystal structure of murine sCEACAM1a [1,4]: a coronavirus receptor in the CEA family. EMBO J. 2002; 21: 2076-86

24) Obrink B. CEA adhesion molecules: multifunctional proteins with signal-regulatory properties. Curr Opin Cell Biol.1997; 9: 616-26.

25) Daniels ., Letourneau S, Turbide C, Kuprina N., Rudinskaya T., Yazova AC, et al. Biliary glycoprotein 1 expression during embryogenesis: correlation with events of epithelial differentiation, mesenchymal-epithelial interactions, absorption, and myogenesis. Dev Dyn. 1996; 206: 272-90.

26) Godfraind C, Langreth SG, Cardellichio CB, Knobler R, Coutelier JP, Dubois-Dalcq M, et al. Tissue and cellular distribution of an adhesion molecule in the carcinoembryonicantigen family that serves as a receptor for mouse hepatitis virus. Lab Invest. 1995; 73: 615-27

27) Smith A, Cardellichio CB, Winograd DF, de Souza MS, Barthold SW, Holmes KV. Monoclonal antibody to the receptor for murine coronavirus MHV-A59 inhibits viral replication in vivo. J Infect. Dis. 1991;163: 879-82

28) Yokomori K, Lai MM. Mouse hepatitis virus utilizes two carcinoembryonic antigens as alternative receptors. J Virol. 1992; 66: 6194-99.

29) Skubitz KM, Campbell KD, Skubitz APN, CD66a, CD66b, CD66c and CD66d each independently stimulate neutrophils. J. Leukoc. Biol. 1996; 60: 106-17

30) Morale V, Christ A, Watt SM, Kim HS Johnson KW, Utku N. et al. Regulation of human intestinal intraepithelial lymphocyte cytolytic function by billiary glycoprotein (CD66a). J. Immunol. 1999; 163: 1363-70

31) Ergun S, Kilik N, Ziegeler G, Hansen A, Nollau P, Gotze J, et al. CEA-related cell adhesion molecule 1: a potent angiogenic factor and a major effector of vascular endothelial growth factor. Mol Cell. 2000; 5: 311-20.

32) Stocks SC, Kerr MA, Haslett C, Dransfield I. CD66-dependent neutrophil activation: a possible mechanism for vascular selectin-mediated regulation of neutrophil adhesion. J. Leukoc. Biol 1995; 58: 40-8

33) Virji M, Watt S., Barker ., Makepeace K, Doyonnas R. The N-domain of the human CD66a adhesion molecule is a target for Opa proteins of Neisseria meningitidis and Neisseria gonorrhoeae. Mol Microbiol. 1996; 22: 929-39.

34) Virji M, Evans D, Griffith J, Hill D., Serino L, Hadfield A. et al. Carcinoembryonic antigens are targeted by diverse strains of typable and non-typable Haemophilus influenzae. Mol. Microbiol. 2000;36: 784-95

35) Sturman LS, Ricard CS, Holmes KV. Proteolytic cleavage of the E2 glycoprotein of murine coronavirus: Activation of cell fusing activity of virions by trypsin and separation of two different 90K cleavage fragments. J. Virol. 1985; 56: 904-11

36) De Groot RJ, W. Luytjes MC, Horzinek BAM, .Van der Zeijst S, SpaanWJM, Lenstra JA. Evidence for a coiled-coil structure in the spike of coronaviruses. J. Mol. Biol. 1987;196: 963-66

37) :Lewicki D, Gallagher T Quaternary structure of coronavirus spikes in complex with CEACAM cellular receptors. J. Boil. Chem. 2002; 277:19727-34

38) Kubo H, Yamada YK, Taguchi F. Localization of neutralizing epitopes and the receptor-binding site within the amino-terminal 330 amino acids of the murine coronavirus spike protein. J Virol. 1994; 68: 5403-10.

39) Suzuki H, Taguchi F. Analysis of the receptor-binding site of murine coronavirus spike protein. J Virol. 1996; 70: 2632-36.

40) Saeki K, Ohtsuka N, Taguchi F. Identification of spike protein residues of murine coronavirus responsible for receptor-binding activity by use of soluble receptor-resistant mutants. J Virol. 1987; 71: 9024-31.

41) Matsuyama S, Taguchi F. Communication between S1N330 and a region in S2 of murine coronavirus spike protein is important for virus entry into cells expressing CEACAM1b receptor. Virology 2002; 295: 160-71

42) Taguchi F, Matsuyama S. Soluble receptor potentiates receptor-independent infection by murine coronavirus. J. Virol. 2002; 76: 950-58

43) Matsuyama S, Taguchi F. Receptor-induced conformational changes of murine coronavirus spike protein. J. Virol. 2002;76: inp ress.

44) Damico R, Bates P. Soluble receptor –induced retraviral infection of receptor-deficient cells. J. Virol. 2000; 74: 6469-75.

45) Chen DC, Kim PS. HIV entry and its inhibition. Cell, 1998;29: 681-84

46) Stohlman SA, Frelinger JA. Resistance to fatal central nervous system disease by mouse hepatitis virus, strain JHM. 1. Genetic analysis. Immunogenet. 1978; 6: 277-81

47) Smith MS, Click RE, Plagemann GW. Control of mouse hepatitis virus replication in macrophages by a recessive gene on chromosome 7. J. Immunol. 1984; 134: 428-32

48) Mouse genome database

49) Knobler RL, Haspel MV, Oldstone MAB. Mouse hepatitis virus type 4 (JHM strain)-induced fatal central nervous system disease. 1. Genetic control and the murine neuron as the susceptible site of disease. J. Exp. Med. 1981; 153: 832-43

50) Taguchi F. (unpublished observation)

Chapter C11

# APOPTOSIS IN MHV-INDUCED DEMYELINATION


Talya Schwartz[§] and Ehud Lavi*
*Department of Pathology and Laboratory Medicine, University of Pennsylvania, School of Medicine, Philadelphia, PA.. Present address: [§]National Institute of Dental and Craniofacial Research, National Institutes of Health, Bethesda MD 20892, *Department of Pathology and Laboratory Medicine, Weill Medical College of Cornell University, New York, NY 10021*



Abstract:    Mouse hepatitis virus (MHV) induced encephalitis, hepatitis and chronic demyelinating disease involves inflammatory and parenchymal cell death. TUNEL assays and electron microscopy reveal that brain parenchymal cells such as neurons, astrocytes, microglia/macrophages and oligodendrocytes are undergoing apoptosis during acute and chronic infection with a neurotropic strain MHV-A59. Although apoptosis is a normal phenomenon in helping the turnover of inflammatory cells, apoptosis of brain parenchymal cells may have significant implications for the pathogenesis of viral-induced demyelination.

Key words:    apoptosis, astrocytes, microglia, macrophages, coronavirus, demyelination


## INTRODUCTION

Coronavirus mouse hepatitis virus (MHV) strain A59 causes severe acute hepatitis, focal meningoencephalitis, and chronic demyelinating disease. It serves as an experimental model for multiple sclerosis (MS). A closely related strain (MHV-2) causes acute hepatitis and meningitis without encephalitis or demyelination. The pathogenesis of MHV-induced demyelination in mice is not completely understood and a potential mechanism of apoptosis was suggested. Previous studies showed apoptotic T cells, astrocytes, and oligodendrocytes in demyelinating areas following infection with JHM, another neurotropic-demyelinating strain of MHV (Barac-Latas *et al* 1997).

However, the distribution of demyelination has been suggested to correlate better with macrophage infiltration than with the apoptotic cells (Wu and Perlman 1999). In the chronic demyelinating disease induced by Theiler's virus, apoptotic astrocytes were found in demyelinating lesions (Palma *et al* 1999). In experimental allergic encephalitis (EAE), the disease was much milder in mice lacking Fas-Fas ligand molecules, suggesting that apoptosis plays a role in the disease (Waldner *et al* 1997).


## APOPTOSIS IN MHV-INDUCED DISEASE

To study apoptosis in MHV-A59 disease we used the experimental model system of MHV-A59 and MHV-2 infection in 4-week-old male C57BL/6 and B6MRL-Fas-lpr (lpr) mice (Schwartz, Fu et al. 2001; Schwartz, Fu et al. 2002). Mice were injected intracerebrally (I.C.) with $2.5 \times 10^3$, 25 and 5 PFU of MHV-A59, 25 and 5 PFU of MHV-2. B6MRL-Fas-lpr mice were injected I.C. with 25 and 5 PFU of MHV-A59. Mock-infected mice were injected I.C. with L2 cell lysate. Mice were sacrificed on days 1,3,5,7,9,11,and 30 (2-3 mice per time point per virus, during the acute stage and 52 mice on day 30). Brain, spinal cord and liver were removed and placed in 10% normal buffered formalin and embedded in paraffin. Five μm thick sections were stained with LFB and H&E. Detection of in situ DNA fragmentation was done with fluorescein *in situ* cell death detection kit (Boehringer Mannheim, Indianapolis, Indiana) as specified by the manufacturer. Double labeling for TUNEL with viral antigen, and TUNEL with specific markers for astrocytes, oligodendrocytes, and macrophages was performed by immunohistochemistry. The analysis was performed by the avidin-biotin-phosphatase based technique (Biomeda, Foster City, CA) using fast-red or vector blue (Vector laboratories, Burlingame, CA) as a staining substrate and a 1:200 dilution of rabbit anti-MHV-A59 polyclonal antibody, 1:100 dilution of monoclonal glial fibrillary acidic protein (Lee *et al.* 1984), 1:500 carbonic anhydrase II (Cammer *et al* 1985) or 1:50 rat anti mouse F4/80 antigen (Serotec Inc., Raleigh, NC), respectively.

TUNEL staining was detected in the brains and livers of mice infected with MHV-A59 and MHV-2. Extensive liver apoptosis was observed in both MHV-2 and MHV-A59 infections. Co-localization of A59 viral antigen and TUNEL staining was detected in hepatocytes. Apoptosis was found in the brain parenchyma of MHV-A59 infected mice and meningial apoptosis was found in both infections. The kinetics, intensity and pattern of apoptosis correlated with the inflammatory events. Mock-infected mice were apoptosis-negative in all tissues. No apoptosis was identified in the spinal cord during the acute stage.

TUNEL staining was observed exclusively in the spinal cords with demyelinating lesions of 15 mice infected with MHV-A59. No apoptosis was detected in the spinal cords of 3 MHV-A59 infected mice without demyelination, 28 MHV-2 infected mice and 10 control mice. TUNEL staining was negative in the brains and the livers of all infected mice.

To assess the role of the apoptotic cascade in the demyelination process, lpr mice were infected with MHV-A59 virus. Two of 5 mice in both groups developed demyelination, however the extent of demyelination in the spinal cord was significantly less in the lpr mice compared to wt mice (10% and 34% demyelinating quadrants respectively).

Recently mice were infected with MHV-A59 and simultaneously implanted with subcutaneous ALZET osmotic pumps that were filled with a pan-caspase inhibitor Z-Val-Ala-D1-Asp-fluromethylketone (Z-VAD-fmk). This treatment has been previously shown to be effective in the treatment of apoptosis in EAE and also reduce the amount of demyelination. The treatment was given continuously at a dose of 1 mg per day, at a delivery rate of 1microliter per hour, for 30 days, beginning 1 day prior to the infection with the virus. While encephalitis was not different between the treated and untreated mice, demyelination was significantly reduced in the treated mice. In 4 mice receiving A59 and anti-pan-caspase treatment demyelination was found in only 3/76 spinal cord quadrants (3.94%), whereas 6 untreated A59-infected mice in the same experiment had 18/120 demyelinating quadrants of spinal cord (15%), and mice receiving only treatment without infection had no demyelination. Although these studies are preliminary due to small number of animals, they may support an important role for apoptosis in autoimmune demyelination as in EAE and Theiler's virus induced demyelination.

Double labeling for TUNEL-positive nuclei and specific markers for astrocytes, oligodendrocytes and macrophages was observed. Quantification studies demonstrated 3-5% oligodendrocytes, 1-2% astrocytes and 70% macrophages, double stained.


## DISCUSSION


Apoptosis was observed in mice following infection with MHV-A59 and MHV-2. The kinetics, intensity and pattern of the apoptotic staining correlated well with the distribution of inflammation. However, apoptosis was found in both inflammatory cells (macrophages and possibly lymphocytes), and parenchymal cells such as hepatocytes, oligodendrocytes, neurons and astrocytes. The involvement of parenchymal cells in CNS apoptosis and the location and temporal relationship between apoptosis and pathology (either encephalitis or demyelination), may indicate that apoptosis is indeed a significant factor in the pathogenesis of the disease and not a merely epiphenomenon. This conclusion is also supported by the reduced demyelination in animals lacking the Fas-lpr pro-apoptotic molecule and mice receiving prolonged pan-caspase inhibitor treatment. Since these treatments affect both inflammatory cells and parenchymal cells the result may be the balance between two opposing trends. A recent report documented the ability of oligodendeocytes to undergo apoptosis in culture upon infection with MHV-A59 (Liu, Cai et al. 2003). This in vitro model greatly supports the hypothesis that death of oligodendrocytes and demyelination due to MHV are caused at least in part by apoptosis. A previous study did not find any difference in the extent of demyelination between wt and lpr mice 13 days following JHM infection (Parra *et al* 2000). However, the different results may be due to the differences in strains of the virus or time points examined.

In conclusion, apoptosis may play an important role in both acute and chronic MHV disease. The relationship between apoptosis, inflammation and tissue damage is yet to be defined, but can possibly be defined through further dissection of the apoptotic cascade.


## ACKNOWLEDGMENTS

This work was supported by a grant from the National Multiple Sclerosis Society (RG 2615-B-2). We thank Elsa Aglow for histology expertise.


## REFERENCES

Barac-Latas V., G. Suchanken, H. Breitschop, A. Stuehler, H. Wege, H. Lassmann. Patterns of oligodendrocytes pathology in coronavirus induced sub-acute demyelinating encephalomyelitis in the Lewis rat.1997. Glia 19:1-12.

Cammer W., Sacchi R., Sapirstein V. Immunohistochemical locaiozation of carbonic anhydrase in the spinal cords of normal nd mutant (shiverer) adult mice with comparison among fixation methods.1985 J Histochem Cytochem. 33(1):45-54.

Lee VM, Page CD., Wu HL., Schlaepfer WW. Monoclonal antibodies to gel-excised glial filament protein and their reactivities with other intermediate filament proteins.1984. J Neurochem. 42(1):25-32.

Liu, Y., Y. Cai, et al. (2003). "Induction of caspase-dependent apoptosis in cultured rat oligodendrocytes by murine coronavirus is mediated during cell entry and does not require virus replication." J. Virol. **77**: 11952-11963.

Palma JP, Yauch RL., Lang S., Kim BS. Potential role of CD4+ T cell mediated apoptosis of activated astrocytes in Theiler's virus induced demyelination. 1999. J. Immunol 162:6543-6551.

Parra B., Lin MT, Stohlman SA, Bergmann C., Atkinson R., Hinton D. Contribution of Fas-Fas ligand interaction to the pathogenesis of mouse hepatitis virus in the central nervous system. 2000. J Virol. 74(5):2447-2450.

Schwartz, T., L. Fu, et al. (2001). "Programmed cell death in MHV-induced demyelination." Adv Exp Med Biol **494**: 163-7.

Schwartz, T., L. Fu, et al. (2002). "Differential induction of apoptosis in demyelinating and nondemyelinating infection by mouse hepatitis virus." J Neurovirol **8**(5): 392-9.

Sungwhan AN., Chen CJ, Yu X., Leibowitz JL., Makino S. Induction of apoptosis in murine coronavirus infected cultured cells and demonstration of E protein as an apoptosis inducer. 1999. J Virol. 73(9):7853-7859.

Waldner H., R.A. Sobel, E. Howard, V. K. Kuchroo. Fas- and FasL deficient mice are resistant to induction of autoimmune encephalomyelitis. 1997 J. Immunol. 159:3100-3103.

Wu GF, Perlman S. Macrophages infiltration, but not apoptosis, is correlated with immune mediated demyelination following murine infection with a neurotropic coronavirus. 1999. J Virol 73(10):8771-8780.

Chapter C12

# THE ROLE OF METALLOPROTEINASES IN CORONA VIRUS INFECTION

Norman W. Marten and Jiehao Zhou
*Keck School of Medicine, University of Southern California, Los Angeles, CA 90033*

**Abstract:**    Infection with neurotropic strains of mouse hepatitis virus (MHV) results in rapid leukocyte infiltration into the central nervous system (CNS). The inflammatory response controls virus replication but fails to mediate sterile clearance. The persistence of viral RNA and inflammatory cells within the CNS is associated with the development of ongoing demyelination. Matrix metalloproteinases (MMPs) are a family of proteases involved in degradation of the extracellular matrix (ECM). During inflammatory responses MMPs are thought to play a significant role in breaking down the basement membrane surrounding blood vessels as well as parenchymal ECM thereby facilitating leukocyte infiltration. MMPs have also been associated with activation of chemokines and perhaps more significantly the degradation of myelin proteins and generation of autoantigens. Recent examination of MMP expression during MHV infection suggests that MMP-3, -9 and -12 are involved in the inflammatory response. The proinflammatory effects of these MMPs are likely tempered by induction of tissue inhibiter of metalloproteinase-1 expression.

**Key words:**    Central nervous system (CNS), demyelination, matrix metalloproteinase (MMP), mouse hepatitis virus, tissue inhibiter of matrix metalloproteinase (TIMP).

## INTRODUCTION

Mouse hepatitis virus (MHV) can cause respiratory, enteric, hepatic or neurologic infections depending on the viral strain, age of the host and route of infection (33). MHV infection of the central nervous system (CNS) as a model of demyelinating disease has been studied extensively using strains derived from either the neurotropic JHM strain (MHV-JHM; MHV-4 serotype) or the MHV-A59 strain, which is both hepatotropic and neurotropic (33,43). As described by others, and ourselves acute MHV infection of the CNS induces a potent inflammatory response involving neutrophils, macrophages, NK, B and T cells (33,43,52). In order to reach the site of infection, however, these cells must overcome physical barriers represented by the blood brain barrier as well as the dense extracellular matrix (ECM) of the CNS parenchyma. The ability of leukocytes to extravasate from blood vessels and cross the parenchymal tissues is presumably dependent on the activity of extracellular proteases. One family of proteases, which are often associated with inflammatory responses are the matrix metalloproteinases (MMPs). The descriptions of MMP regulation and function during normal physiological as well as pathological conditions have been described in numerous publications (32,23,41), thus only a brief summary is provided herein.


## BACKGROUND


MMPs comprise a large group of endoproteinases that, in conjunction with other proteases, mediate degradation during remodeling of the extracellular matrix (ECM). Twenty four different MMPs have been identified and classified as collagenases, gelatinases, stromelysins, matrilysin and membrane type MMPs (MT-MMPs) based on their substrate specificities and protein domain structures (24,34). These proteases are produced by a wide range of cell types, including endothelial, inflammatory and stromal cells (18,23,32). Regulation of MMP activity has been classically defined as occurring at three distinct levels. Primary regulation of MMP activity occurs at the transcriptional level (5,12,22,30,40). MMP gene expression can be induced by a wide range of signals including acute phase cytokines such as IL-1$\alpha$ and -$\beta$, IL-6 and TNF-$\alpha$, and chemokines such as MCP-1, MIP1$\beta$, and RANTES (6,12,22). A second level of regulation for MMP activity is post-translational, as MMPs require proteolytic cleavage of their proenzyme form to become activated (4,34). Following translation, most proMMPs are released into the extracellular space. By contrast, a subgroup of membrane associated MMPs, the MT-MMPs, may play a role in the processing of secreted proMMPs to their active form (20). The final level of control for MMP activity resides in their regulation by a small group of specific inhibitors, the tissue inhibitors of metalloproteinases (TIMPs) (7). TIMPs function primarily as specific inhibiters, binding to MMPs in a 1:1 stoicheometry. Interestingly, there is evidence that TIMPs may also act as chaperone proteins during the processing of MMPs from proenzyme to active form, suggesting that they can promote as well as inhibit MMP enzymatic activity, at least for MMP-2 (42,44).

MMPs and their inhibitors are involved in remodeling of the ECM during normal physiological conditions. However they have also been associated with diverse pathological processes from rheumatoid arthritis to tumor invasion and metastasis (16,25). A number of MMPs and TIMPs have also been associated with the CNS inflammatory/demyelinating disease multiple sclerosis (MS) in humans (2,32) as well as its rodent model experimental autoimmune encephalitis (EAE) (10,38). In the case of MS, these include MMP-1, -2, -3 -7 and -9 and TIMP-1 (2,11,32). These MMPs and TIMPs have been detected by immunohistochemistry in both inflammatory infiltrates and activated glia within acute MS lesions as well as perivascular infiltrates. Following EAE induction in mice, MMP-3, -9, -12 and -14 as well as TIMP-1 gene expression were induced within the CNS (38). MMP-1 and -10 were also induced although to a lesser degree than the aforementioned genes (38). The induction of MMP-7 mRNA expression was not detected but the presence of MMP-7 protein has also been reported during EAE in rats (10). The consistent association of MMP and TIMP expression with inflammatory demyelinating disease suggests that MMPs play either a direct or indirect role in promoting CNS pathology (2,27,32). However, as discussed below, identifying specific MMPs and their exact role in promoting inflammatory/demyelinating has proven to be difficult.


# MMP AND TIMP EXPRESSION DURING CORONAVIRUS INFECTION

*In vitro* infection of human astrocytic and microglial cell lines with human coronavirus induced up-regulation of MMP-2 and -9 protein suggesting that virus infection can up-regulate MMP expression within the CNS (15). The mechanism by which MMPs were up-regulated by infection where not known, however, virus infection of the glial cell lines induced expression of IL-6, TNF-$\alpha$ and MCP-1, which are all inducers of MMP expression (15). Transfer of non-infectious supernatants from infected cultures to uninfected cells was sufficient to induce expression of MMP-2 and -9. MMP expression, including MMP-2 and -9, is also elevated in naïve transgenic mice, which express IL-6 and TNF-$\alpha$ within the CNS (38). These observations suggest that viral infections induce expression of chemokines/cytokines, which in turn induce MMP expression in neighboring cells (15,38).

Recent examination of MMP gene expression within the brains of mice infected with lethal and attenuated MHV-JHM variants showed that MMP-3 and -12 mRNA were rapidly up regulated during acute infections (51, 53). By contrast, infection did not alter transcription of the genes encoding MMP-2, -9, -11 or -14, which were constitutively expressed in the CNS of naïve BALB/c mice (51). These data are summarized in Table 1. Kinetic analysis of MMP expression indicated that MMP-3 and MMP-12 expression peaked by 6 days post infection (p.i.) in accord with viral load, however, expression was significantly higher in lethally infected mice (51,52). Comparisons of MMP/TIMP expression in infected irradiated and control mice revealed relatively little loss in MMP or TIMP expression in immunosuppressed mice (51). These results indicated that CNS resident cells were responsible for the majority of MMP-3, and -12 gene expression detected within the brain and that this gene induction was relatively independent of the presence of inflammatory cell infiltration. Expression of MMPs constitutively expressed in naïve mice (MMP-2, -9, -11, -14) were unaffected by immunosuppression.

*Table 1: MMP and TIMP expression within the CNS during MHV infection[a].*

| Virus induced expression | Constituative expression | Not Detected |
| --- | --- | --- |
| MMP-3 | MMP-2 | MMP-1 |
| MMP-9[b] | MMP-11 | MMP-7 |
| MMP-12 | MMP-14 | |
| TIMP-1 | TIMP-2 | |
| | TIMP-3 | |

[a] RNA was prepared from brains of MHV-JHM infected mice. Relative gene expression was determined by RNase protection assay.
[b] Analysis of MMP-9 expression by RNase protection assay did not reveal an increase in MMP-9 expression in RNA prepared from total brain. However, analysis of MMP expression from CD45[hi] inflammatory cells isolated from the CNS indicated that MMP-9 expression was increased during MHV infection.

Four TIMPs have been characterized and all are capable of inhibiting a range of MMPs (7,23). Basal expression of TIMP-1 is low but readily induced in response to a multitude of inflammatory signals (5,7). By comparison, TIMP-2, -3 and -4 are expressed constitutively and alterations in their expression are less frequently associated with inflammatory responses (26,31,38). Examination of naïve and MHV-JHM infected brains revealed that TIMP-2 and -3 were unaffected by the infection (51) (Table 1). By contrast, MHV infection induced rapid expression of TIMP-1 (51) (Table 1). TIMP-1 expression peaked by 6 p.i. in conjunction with MMP-3 and -12 expression (51,52). TIMP-1 expression was higher in lethally infected mice compared to sublethally infected mice suggesting that expression is dependent on viral load and/or virulence. TIMP-1 expression has been reported to be associated with both tissue resident cells as well as most inflammatory cells (7). Expression levels of TIMP-1 were higher in irradiated, immuncompromised mice compared to control animals, indicating that most TIMP-1 is produced by CNS resident cells (51).

Analysis of whole brain samples for MMP RNA and protein failed to detect induction of MMP-9 expression during MHV infection (51), which was in sharp contrast to its prominence in MS. However, subsequent analysis of individual cell populations sorted from infected brains indicated that MMP-9 gene expression is up regulated several fold among infiltrating inflammatory cells (Zhou, unpublished data) (Table 1). Similarly, a strong induction in TIMP-1 expression within the inflammatory cell populations (Zhou, unpublished data) suggests that T cells are, at least in part, a source of both MMP-9 and TIMP-1 as both of these genes are co-expressed by T cells (35). Furthermore, analysis of cell extracts prepared from infiltrating cells revealed that MHV infection also induced a sharp rise in the level of intracellular MMP-9 protein (51, 53). The increase in intracellular MMP-9 protein was attributed to the presence of polymorphonuclear cell infiltrates in response to infection (51, 53). Neutrophil synthesis of MMP-9 differs from other cell types in two key ways. First, in contrast to mononuclear cells, which release MMP-9 directly into the extracellular space, neutrophils store MMP-9 proenzyme in granules for release during degranulation (35), therefore permitting neutrophils to rapidly release large quantities of MMP-9 in the absence of *de novo* synthesis. Second, expression of MMP-9 in neutrophils is not linked to co-expression of TIMP-1 as is the case for T cells (35). Thus, MMP-9 from neutrophils is regulated at the level of degranulation as opposed to gene transcription and co-expression of TIMPs. These data suggest MMP-9 is likely released from both mononuclear and polymorphonuclear infiltrates during JHMV infection. Thus, with the exception that MMP-7 was not detected during MHV infection (51), the overall expression patterns for both MMPs (-3, -9 and -12) and TIMP-1 during MHV infection, EAE and MS are extremely similar.

## ROLE OF MMPS IN DEMYELINATION

Linkage between MMP/TIMP expression during demyelinating disease has been well chronicled (32,38,51). By contrast, determining a definitive role for MMPs during demyelinating disease has proven to be far more elusive. During inflammatory processes within the CNS, MMPs most likely play two key roles: 1) Breaking down the basal lamina surrounding the blood brain barrier, thereby allowing inflammatory cells to enter the CNS (35) ; and 2) breaking down connective tissues linking tightly packed CNS resident cells, thereby allowing inflammatory cells to migrate along chemokine gradients towards the sites of infection/injury (18,50). Other key mechanisms in the genesis of inflammatory demyelination have been associated with MMPs. These include activation of cytokines and chemokines, such as the conversion of TNF-α from pro to active form (8) and increasing the potency of the chemokine IL-8 (48). Furthermore, activated MMP-3 can convert pro-MMP-9 into it active form (13). Thus MMP-3 produced by CNS resident cells may be involved in activating pro-MMP-9 synthesized by inflammatory cells. Activation of these proinflammatory proteins by MMPs increases the potential for immune mediated damage to the CNS.

A number of MMPs including -3, -9, and -12, which are common to many demyelinating diseases, are capable of breaking down proteins of the myelin sheath (8,9,39). The consequences of this are two fold. First, induction of an inflammatory response to a foreign antigen within the CNS may non-specifically initiate destruction of the myelin sheath through the release of MMPs. Second, generation of potential autoreactive self-antigens by MMP mediated breakdown of myelin proteins may initiate specific responses against oligodendroglia resulting in additional demyelination (36,39).

Of the MMPs associated with MHV infection of the CNS, MMP-9 and its specific inhibitor, TIMP-1, are of particular interest for several reasons. MMP-9 and TIMP-1 are produced by both activated inflammatory as well as CNS resident cells (19,35,47). MMP-9 is specific for type IV collagens, which make up the basement membrane surrounding the blood brain barrier (49). This suggests that MMP-9 contributes to the ability of activated inflammatory cells to cross this barrier. Support for the role of MMP-9 in T cell trafficking comes from MS patients undergoing IFN-β treatment, one of the few therapeutics shown to have a positive effect. Although no single conclusive mechanism for the anti-inflammatory activity of IFN-β has been demonstrated, MMP-9 gene expression is down regulated by IFN-β (30,45,46). Conversely, TIMP-1, which inhibits MMP-9, is up regulated by IFN-β (37). This suggests that one of the mechanisms by which IFN-β works is through inhibition of MMP-9 mediated inflammatory cell trafficking. These data are supported by treatment of EAE mice with MMP inhibitors, which have generally prevented disease induction or reduced clinical symptoms when treatment was initiated after active disease was apparent (17,21,28).


## CONCLUSIONS


MMP and TIMP expression are closely associated with the human demyelinating disease MS as well as experimental animal models of demyelinating disease. MMP expression within the CNS is most likely up regulated by the presence of inflammatory cytokines. However, as stated previously, confirmation of a specific role for MMPs in the inflammatory process is lacking. The use of knockout mice to determine the role of specific MMPs in inflammatory responses have generated mixed results, particularly in regard to blood brain barrier breakdown and inflammatory cell infiltration across endothelial cell barriers (1,3,14,29). This may in part be explained by the large number of MMPs with overlapping specificities as well as the presence of other proteases. Thus, although MMP expression is clearly associated with inflammatory responses during demyelinating disease, the role of individual MMPs as active participants or simply as markers remains to be determined.

# REFERENCES

1.	Allport, J.R., Lim, Y.C., Shipley, J.M., Senior, R.M., Shapiro, S.D., Matsuyoshi, N., Vestweber, D., and Luscinskas, F.W. Neutrophils from MMP-9- or neutrophil elastase-deficient mice show no defect in transendothelial migration under flow in vitro. J. Leukoc. Biol. 2002; 71:821.

2.	Anthony, D. C., Ferguson, B., Matyzak, M.K., Miller, K.M., Esiri, M.M., and Perry, V.H. Differential matrix metalloproteinase expression in cases of multiple sclerosis and stroke. Neuropathol. Appl. Neurobiol. 1997; 23:406.

3.	Asahi, M., Wang, X., Mori, T., Sumii, T., Jung, J.C., Moskowitz, M.A., Fini, M.E., and Lo, E.H. Effects of matrix metalloproteinase-9 gene knock-out on the proteolysis of blood-brain barrier and white matter components after cerebral ischemia. J. Neurosci. 2001; 21:7724.

4.	Birkedal-Hansen, H., Moore, W. G., Bodden, M.K., Windsor, L.J., Birkedal-Hansen B., DeCarlo, A., and Engler, J.A. Matrix Metalloproteinases: a review. Crit. Rev. Oral. Biol. Med. 1993; 4:197.

5.	Bond, M., Fabunmi, R.P., Baker, A.H., and Newby, A.C. Synergistic upregulation of Metalloproteinase-9 by growth factors and inflammatory cytokines: an absolute requirement for transcription factor NF-kB. FEBS Lett. 1998; 435:29.

6.	Bond, M., Baker, A.H., and Newby, A.C. Nuclear factor kappaB activity is essential for matrix metalloproteinase-1 and -3 upregulation in rabbit dermal fibroblasts. Biochem. Biophys. Res. Commun. 1999; 264:561.

7.	Brew, K., Dinakarpandian, D., and Nagase, H. Tissue inhibitors of metalloproteinases: evolution, structure and function. Biochim. Biophys. Acta. 2000; 1477:267.

8.	Chandler, S., Cossins, J., Lury, J., and Wells, G. Macrophage metalloelastase degrades matrix and myelin proteins and processes a tumour necrosis factor-alpha fusion protein. Biochem. Biophys. Res. Commun. 1996; 228:421.

9.	Chandler S., Coates, R., Gearing, A., Lury, J., Wells, G., and Bone, E. Matrix metalloproteinases degrade myelin basic protein. Neurosci. Lett. 1995; 201:223.

10.  Clements J.M., Cossins, J.A., Wells, G.M., Corkill, D.J., Helfrich, K., Wood, L.M., Pigott, R., Stabler, G., Ward, G.A., Gearing, A.J., and Miller, K.M. Matrix metalloproteinase expression during experimental autoimmune encephalomyelitis and effects of a combined matrix metalloproteinase and tumour necrosis factor-alpha inhibitor. J. Neuroimmunol. 1997; 74:85.

11.  Cossins, J.A., Clements, J.M., Ford, J., Miller, K.M., Pigott, R., Vos, W., Van der Valk, P., and De Groot, C.J. Enhanced expression of MMP-7 and MMP-9 in demyelinating multiple sclerosis lesions. Acta. Neuropathol. 1997; 94:590.

12.  Cross, A.K., and Woodroofe, M.N. Chemokine modulation of matrix metalloproteinase and TIMP production in adult rat brain microglia and a human microglial cell line in vitro. Glia 1999; 28:183.

13.  Cuzner, M.L. and Opdenakker, G. Plasminogen activators and matrix metalloproteases, mediators of extracellular proteolysis in inflammatory demyelination of the central nervous system. J. Neuroimmunol. 1999; 94:1.

14.  Dubois, B., Masure, S., Hurtenbach, U., Paemen, L., Heremans, H., van den Oord, J., Sciot, R., Meinhardt, T., Hammerling, G., Opdenakker, G., and Arnold, B. Resistance of young gelatinase B-deficient mice to experimental autoimmune encephalomyelitis and necrotizing tail lesions. J. Clin. Invest. 1999; 104:1507.

15.  Edwards, J.A., Denis, F., and Talbot, P.J. Activation of glial cells by human coronavirus OC43 infection. J. Neuroimmunol. 2000; 108:73.

16.  Giannelli G., Falk-Marzillier, J., Schiraldi, O., Stetler-Stevenson, W.G., and Quaranta, V. Induction of cell migration by matrix metalloprotease-2 cleavage of laminin-5. Science. 1997; 277:225.

17.  Gijbels, K., Galardy, R.E., and Steinman, L. Reversal of experimental autoimmune encephalomyelitis with a hydroxamate inhibitor of matrix metalloproteases. J. Clin. Invest. 1994; 94:2177.

18.  Goetzl, E.J., Banda, M.J., and Leppert, D. Matrix metalloproteinases in immunity. J. Immunol. 1996; 156:1.

19.  Gottschall, P.E. and Yu, X. Cytokines regulate gelatinase A and B (matrix metalloproteinase 2 and 9) activity in cultured rat astrocytes. J. Neurochem. 1995; 64:1513.

20.  Hartung, H.P., and Kieseier, B.C. The role of matrix metalloproteinases in autoimmune damage to the central and peripheral nervous system. J. Neuroimmunol. 2000; 107:140.

21.  Hewson, A.K., Smith, T., Leonard, J.P., and Cuzner, M.L. Suppression of experimental allergic encephalomyelitis in the Lewis rat by the matrix metalloproteinase inhibitor Ro31-9790. Inflamm. Res. 1995; 44:345.

22.  Horton, W.E., Jr., Udo, I., Precht, P., Balakir, R., and Hasty, K. Cytokine inducible matrix metalloproteinase expression in immortalized rat chondrocytes is independent of nitric oxide stimulation. In Vitro Cell. Dev. Biol. Anim. 1998; 34:378.

23.  Johansson, N., Ahonen, M., and Kahari, V.M. Matrix metalloproteinases in tumor invasion. Cell. Mol. Life. Sci. 2000; 57:5.

24.  Kleiner, D.E. Jr and Stetler-Stevenson, W.G. Structural biochemistry and activation of matrix metalloproteases. Curr. Opin. Cell. Biol. 1993; 5:891.

25.  Konttinen, Y.T., Ainola, M., Valleala, H., Ma, J., Ida, H., Mandelin, J., Kinne, R.W., Santavirta, S., Sorsa, T., Lopez-Otin, C., and Takagi, M. Analysis of 16 different matrix metalloproteinases (MMP-1 to MMP-20) in the synovial membrane: different profiles in trauma and rheumatoid arthritis. Ann. Rheum. Dis. 1999; 58:691.

26.  Leco, K.J., Hayden, L.J., Sharma, R.R., Rocheleau, H., Greenberg, A.H., and Edwards, D.R. Differential regulation of TIMP-1 and TIMP-2 mRNA expression in normal and Ha-ras-transformed murine fibroblasts. Gene. 1992; 117:209.

27.	Lichtinghagen, R., Seifert, T., Kracke, A., Marckmann, S., Wurster, U., and Heidenreich, F. Expression of matrix metalloproteinase-9 and its inhibitors in mononuclear blood cells of patients with multiple sclerosis. J. Neuroimmunol. 1999; 99:19.

28.	Liedtke, W., Cannella, B., Mazzaccaro, R.J., Clements, J.M., Miller, K.M., Wucherpfennig, K.W., Gearing, A.J., and Raine, C.S. Effective treatment of models of multiple sclerosis by matrix metalloproteinase inhibitors. Ann. Neurol. 1998; 44:35.

29.	Liu, Z., Shipley, J.M., Vu, T.H., Zhou, X., Diaz, L.A., Werb, Z., and Senior, R.M. Gelatinase B-deficient mice are resistant to experimental bullous pemphigoid. J. Exp. Med. 1998; 188:475.

30.	Ma, Z., Qin, H., and Benveniste, E.N. Transcriptional suppression of matrix metalloproteinase-9 gene expression by IFN-gamma and IFN-beta: critical role of STAT-1alpha. J. Immunol. 2001; 167:5150.

31.	Madtes, D.K., Elston, A.L., Kaback, L.A., and Clark, J.G. Selective induction of tissue inhibitor of metalloproteinase-1 in bleomycin-induced pulmonary fibrosis. Am. J. Respir. Cell. Mol. Biol. 2001; 24:599.

32.	Maeda, A. and Sobel, R.A. Matrix metalloproteinases in the normal human central nervous system, microglial nodules, and multiple sclerosis lesions. J. Neuropathol. Exp. Neurol. 1996; 55:300.

33.	Marten, N.W., Stohlman, S.A., and Bergmann, C.C. MHV infection of the CNS: mechanisms of immune-mediated control. Viral. Immunol. 2001; 14:1.

34.	Nagase, H. and Woessner, J. F., Jr. Matrix metalloproteinases. J. Biol. Chem. 1999; 274:21491.

35.	Opdenakker, G., Van den Steen, P. E., Dubois, B., Nelissen, I., Van Coillie, E., Masure, S., Proost, P., and Van Damme, J. Gelatinase B functions as regulator and effector in leukocyte biology. J. Leukoc. Biol. 2001; 69:851.

36.	Opdenakker, G., Van den Steen, P.E., and Van Damme, J. Gelatinase B: a tuner and amplifier of immune functions. Trends Immunol. 2001; 22:571.

37.	Ozenci, V., Kouwenhoven, M., Teleshova, N., Pashenkov, M., Fredrikson, S., and Link, H. Multiple sclerosis: pro- and anti-inflammatory cytokines and metalloproteinases are affected differentially by treatment with IFN-beta. J Neuroimmunol. 2000; 108:236.

38.	Pagenstecher, A., Stalder, A.K., Kincaid, C.L., Shapiro, S.D., and Campbell, I.L. Differential expression of matrix metalloproteinase and tissue inhibitor of matrix metalloproteinase genes in the mouse central nervous system in normal and inflammatory states. Am. J. Pathol. 1998; 152:729.

39.	Proost P., Van Damme, J., and Opdenakker, G. Leukocyte gelatinase B cleavage releases encephalitogens from human myelin basic protein. Biochem. Biophys. Res. Commun. 1993; 192:1175.

40.	Reis, C. and Petrides, P.E. cytokine regulation of Matrix Metalloproteinase activity and its regulatory dysfunction in disease. Biol. Chem. Hoppe-Seyler. 1995; 376:345.

41.	Rudolph-Owen, L.A., Hulboy, D.L., Wilson, C.L., Mudgett, J., and Matrisian, L.M. Coordinate expression of matrix metalloproteinase family members in the uterus of normal, matrilysin-deficient, and stromelysin-1-deficient mice. Endocrinology. 1997;138:4902.

42.	Sato, H., Takino, T., Kinoshita, T., Imai, K., Okada, Y., Stetler Stevenson, W.G., and Seiki, M. Cell surface binding and activation of gelatinase A induced by expression of membrane-type-1-matrix metalloproteinase (MT1-MMP). FEBS Lett. 1996; 385: 238.

43. Stohlman, S.A., Bergmann, C.C., and Perlman, S. Mouse hepatitis virus. In *Persistent Viral Infections*, Vol. 26. p. 537. R. A. a. I. Chen, ed. John Wiley & Sons, 1999.

44. Strongin, A.Y., Collier, I., Bannikov, G., Marmer, B.L., Grant, G.A., and Goldberg, G.I. Mechanism of cell surface activation of 72-kDa type IV collagenase. Isolation of the activated form of the membrane metalloprotease. J. Biol. Chem. 1995; 270:5331.

45. Stuve, O., Dooley, N.P., Uhm, J.H., Antel, J.P., Francis, G.S., Williams, G., and Yong, V.W. Interferon beta-1b decreases the migration of T lymphocytes in vitro: effects on matrix metalloproteinase-9. Ann Neurol. 1996; 40:853.

46. Trojano, M., Avolio, C., Liuzzi, G.M., Ruggieri, M., Defazio, G., Liguori, M., Santacroce, M.P., Paolicelli, D., Giuliani, F., Riccio, P., and Livrea, P. Changes of serum sICAM-1 and MMP-9 induced by rIFNbeta-1b treatment in relapsing-remitting MS. Neurology. 1999; 53:1402.

47. Uhm, J.H., Dooley, N.P., Oh, L.Y., and Yong, V.W. Oligodendrocytes utilize a matrix metalloproteinase, MMP-9, to extend processes along an astrocyte extracellular matrix. Glia. 1998; 22:53.

48. Van den Steen, P.E., Proost, P., Wuyts, A., Van Damme, J., and Opdenakker, G. Neutrophil gelatinase B potentiates interleukin-8 tenfold by aminoterminal processing, whereas it degrades CTAP-III, PF-4, and GRO-alpha and leaves RANTES and MCP-2 intact. Blood. 2000; 96:2673.

49. Vissers, M.C., Winterbourn, C.C. Gelatinase contributes to the degradation of glomerular basement membrane collagen by human neutrophils. Coll. Relat. Res. 1988; 8:113.

50. Yong, V.W., Krekoski, C.A., Forsyth, P.A., Bell, R., and Edwards, D.R. Matrix metalloproteinases and diseases of the CNS. Trends Neurosci. 1998; 21:75.

51. Zhou, J., Stohlman, S.A., Atkinson, R., Hinton, D.R., and Marten, N.W. Matrix metalloproteinase expression correlates with virulence following neurotropic mouse hepatitis virus infection. J. Virol. 2002; 76:7374.

52. Zhou, J., Stohlman, S.A., Hinton, D.R., and Marten, N.W. Neutrophils modulate inflammation during viral induced encephalitis. In press.

53. Zhou, J., Stohlman, S.A., Marten, N.W., Hinton, D.R. Regulation of matrix metalloproteinase (MMP) and tissue inhibitor of matrix metalloproteinase (TIMP) genes during JHMV infection of the central nervous system. Adv. Exp. Med. Biol. 2001; 494:329.

Chapter C13

# MOLECULAR DETERMINANTS OF CORONAVIRUS MHV- INDUCED DEMYELINATION

Li Fu and Ehud Lavi*

*Division of Neuropathology, Department of Pathology & Laboratory Medicine, University of Pennsylvania school of medicine. Philadelphia, PA 19104-6100. *Present address: Department of Pathology and Laboratory Medicine, Weill Medical College of Cornell University, New York, NY 10021*

Abstract:     Mouse hepatitis virus (MHV) is a member of the coronavirus family of the *nidovirales* order. MHV is an enveloped virus with single-stranded, positive genomic RNA of about 31kb. Infection of susceptible strains of mice with the MHV-JHM and A59 strains results in acute encephalomyelitis and chronic demyelinating disease with features similar to the human demyelination disease multiple sclerosis (MS). Because the mechanism of demyelination in MS is not completely understood, various experimental models, including MHV infection in mice, have been used to study the pathogenesis of inflammatory autoimmune demyelination. The spike (S) glycoprotein of MHV has been implicated as the most critical genomic determinant of MHV pathogenesis and demyelination. However, other genes and proteins are likely to contribute to MHV pathogenesis as well.

Key words:     Mouse hepatitis virus (MHV), Coronaviruses, Nidoviruses, Molecular determinants, Demyelination, Multiple Sclerosis (MS).

## THE MHV GENOME

The MHV genome is approximately 31 kb of single-stranded RNA of positive polarity. The genomic organization consists of 7 functional genes, 5 of which encode structural proteins (Fig 1). Genes 2b, 3, 5b, 6 and 7 encode the structural proteins hemagglutinin-esterase (HE) in some strains of MHV, spike (S), membrane (M), envelope (E) and nucleocapsid (N) [1]. The remaining genes encode non-structural proteins. The nucleotide sequences have been determined for 3 strains of MHV: A59, JHM and MHV-2. The structural protein genes are clustered at the 3′ of the genome, whereas the 5′ third of the genome consists of a single gene encoding a large polyprotein comprising of viral polymerase, proteases and associated nonstructural proteins. The nucleotide sequence at the 5′ end contains a 65 to 85 nucleotide-long leader RNA, followed by 200 to 500 non-translated nucleotides. The first 60% of the length of the genome, roughly 20 Kb from the 5′ end, consists of two overlapping open reading frames, ORF 1a and 1b, that encode the viral RNA- dependent RNA polymerase, proteases and helicase. At the overlapping region between ORF 1a and 1b there is a 7-nucleotide slippery sequence and a pseudoknot structure, essential for ribosomal frame shifting. Several of the non-structural proteins are not essential for viral replication *in vitro* but may play a role in viral pathogenesis. An important aspect of MHV biology is the high frequency of RNA-RNA recombination between different, but closely related MHVs, when mixed in the same culture. RNA recombination is likely an important mechanism of viral evolution and may contribute to viral pathogenesis.

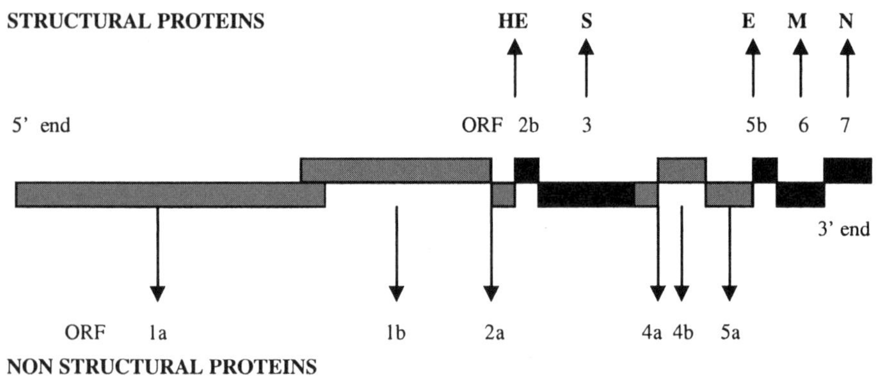


**Figure 1 .** Schematic diagram of the MHV genomic organization.


# STRUCTURAL-FUNCTIONAL RELATIONSHIP OF MHV

The virions of MHV contain a nucleocapsid protein (N), which binds to the viral RNA to form a helically symmetric nucleocapsid. Surrounding the nucleocapsid is a membrane envelope, which contains 3 or 4 major glycoproteins: the spike (S), the membrane (M), the small envelope protein (E), and in some strains such as MHV-JHM, the hemagglutinin-esterase protein (HE) [1-4].

Upon infection of susceptible cells the MHV genomic RNA serves as a template for the negative strand RNA synthesis, which, in turn, is used for the synthesis of genomic and subgenomic mRNAs [2]. Each mRNA has a leader RNA of about 70 nucleotides at the 5' end [3]. The exchange of leader RNA is a common phenomenon during coronavirus replication and mRNA 1 can be synthesized during MHV infection with or without a 9-nucleotide sequence (UUUAUAAAC) located immediately downstream of the leader RNA of the 5' terminus of MHV [4]. Studies of naturally occurring and artificially constructed defective interfering (DI) RNAs have shown that both 5'- and 3'-terminal portions of the MHV genome are required for making the DIs functional [5, 6]. More specifically, the 5' cis-acting elements essential for MHV-JHM and A59 DI RNA replication have been mapped to 5' 474 nucleotides of which at least 446 nucleotides are required [6]. At the 3' end of the MHV genome 417 to 463 nucleotides is the minimal number of nucleotides required for replication [5]. Using interspecies-targeted RNA recombination, MHV-JHM gene 4 was genetically inactivated. Interestingly, the virus lacking gene 4 replicated in tissue culture cells with similar kinetics as the wild type virus. Thus ORF 4 is not essential for viral growth in tissue culture or in the CNS of the infected host [7]. Likewise, the deletion of the 2a/HE gene cluster did not affect viral growth in vitro; deletion of the 4/5a gene cluster merely decreased viral replication rate by 10 fold [8].

The function of the M protein has been recognized as the predominant constituent of virion architecture. However, the function of HE protein has yet to be established other than the hemagglutinating and acetyl-esterase activities [9]. The HE protein forms disulfide-linked homodimers and strains of viruses that express this protein show a second, shorter fringe of peplomers on their virions. The small envelope protein E, has been shown to be translated both in vivo and in vitro. The E and M proteins are both necessary and sufficient for the formation and extracellular release of particles appearing structurally identical to MHV virions, thus suggesting that the E and M proteins are important for virion assembly [10, 11].

The N protein of MHV is a 50-60 kDa phosphoproteins with overall high basic amino acid content and 3 structural domains [12]. The functions of the N protein involve interactions between N and viral RNAs [13, 14]. The internal gene I, a large open reading frame embedded entirely within the 5' half of N gene, has been proved not essential for the replication of MHV either in tissue culture or in its natural host [15]. The middle domain of N is responsible for RNA binding. N binds certain regions of coronavirus RNA more efficiently than nonviral RNA and the binding specificity for encapsidation and packaging in the context of virus-infected cells may be influenced by interactions with the membrane protein (M) or possibly other viral proteins [16].

## THE SPIKE GLYCOPROTEIN (S)

Forming the spikes on the virion surface, the spike protein (S) of coronavirus MHV has a variety of important biological activities. The S protein is a transmembrane glycoprotein of about 180 kDa with 4139 nucleotides in length, encoding two posttranslational subunits: S1, predicted to form the globular head, and S2, predicted to form the stalk of the glycoprotein. The MHV S protein plays a critical role in viral pathogenesis, including functions in both viral entry and spread within the host [17, 18]. The S1 portion may independently bind to cellular receptors and the S2 portion is responsible for fusion of viral and cellular membranes [19]. The S protein is also the target for both humoral and cellular immune responses [20-22]. The S protein contains determinants of neurovirulence and demyelination [23-26].

Studies to determine molecular determinates of MHV pathogenesis have been limited by the large size of the MHV genome. For many years the size of the MHV genome presented an obstacle for the construction of a full-length cDNA clone, which would be amenable for genetic manipulations. The construction of a plasmid vector, pMH54, containing the entire 3' of the MHV-A59 genome downstream from the hemagglutinin esterase gene and two marker restriction sites *avrII* and *sbfI*, marked a new stage in pathogenesis studies. The availability of this construct allowed the introduction of different S genes into the background of MHV-A59 and the introduction of site directed mutatation into S or other genes [27, 28]. Using targeted RNA recombination between the trandfected and manipulated pMH54 plasmid and an infectious virus, various properties of the S protein were studied in the past few years.

Replacing the S gene of MHV-A59 with the S gene of the highly neurovirulent MHV-4, increased the virulence of the new recombinant virus substantially [23, 29] Alterations in both S1 and S2 resulted in reduced viral replication in the brain, associated with decrease in neurovirulence [25]. Mutations and deletions in the hypervariable region (HVR) suggested that the HVR was important for neurovirulence [25, 30-32]. In addition, the H716D amino acid substitution correlated with a delayed fusion phenotype, but without loss of hepatotropism [33, 34]. Q159L amino acid substitution affected hepatotropism and demyelination [35, 36] and L1114R (within the S2 region) has been speculated to play a role in the conformation of spike and in the dynamic interaction between S1 and S2 [37, 38]. The S gene was also found to contain determinants of hepatotropism [39].

Infection of MHV-A59 in 4-week-old C57BL/6 mice produces a biphasic disease with acute hepatitis and meningoencephalitis followed by chronic CNS demyelination [40-42]. Thus infection of mice with MHV-A59 has been used as an animal model for the human demyelinating disease multiple sclerosis (MS) [43-46]. MHV-2, previously recognized as mainly hepatotropic virus with only weakly neurotropic properties [47-50], was shown to produce severe hepatitis and meningitis, without encephalitis and demyelination [26, 51]. The weak neurotropic properties of MHV-2 made this strain ideal for a comparative study with the closely related, neurotropic, demyelinating strain MHV-A59 [52]. We first sequenced the entire genome of MHV-2 and compared it with that of MHV-A59 [52]. We then studied the role of the S gene in demyelination by replacing the S gene of MHV-A59 with that of MHV-2, using targeted RNA recombination. The new recombinant viruses (Penn98-1, 2) were more virulent than A59, but unable to induce demyelination in C57BL/6 mice [26]. Thus we hypothesized that demyelination determinants map to the S gene of MHV-A59.

To further explore which specific regions of the S gene are directly responsible for the demyelinating phenotype, we studied a set of random recombinant viruses (ML-7, ML-8, ML-10, and ML-11) that were produced by co-infection of cultures with MHV-2 and LA-7, a ts mutant of MHV-A59 [53]. All of these recombinant viruses were unable to induce demyelination *in vivo*. *In vitro* study ascertained that this group of recombinant viruses was able to replicate to the same titers as their parental viruses in L2 cells tissue cultures. Morphologically, the viruses ML-7, ML-8 and ML-10 produced large plaques and syncytia (similar to MHV-A59), whereas ML-11 produced small plaques as MHV-2. Sequencing of the S genes of ML-7, ML-8 and ML-10 revealed that they contained three common mutations, I375M, L652I, and T1087N [54]. The I375M mutation is located downstream from the receptor-binding site, the L652I mutation is located in the Hypervariable (HVR) region of the S gene and the T1087N mutation is located between the two-heptad repeat regions. Since these viruses were derived from a recombination between LA-7 and MHV-2, we also sequenced the S gene of LA-7 and we confirmed that the S gene of LA-7 contained the exact same three mutations implicating LA-7 as the source of these mutations. In addition, the sequence results of the S gene of ML-11 confirmed that the S gene of ML-11 was identical and was derived from to the S gene of MHV-2 [51].

We then investigated the role of the three amino acid substitutions in MHV-A59-induced demyelination. We first produced viruses with all 3 amino acid substitutions by targeted recombination. The virulence of Penn2K-1 and 2 was dramatically reduced (LD50>50,000pfu) although Penn2K-1 and 2 exhibit similar kinetics of replication as their parental virus

MHV-A59 in L2 cell cultures. Pathogenesis studies of Penn2K-1 and 2 showed only mild encephalitis and hepatitis when high doses were used for infection (50,000pfu I.C). No demyelination was found with any dose of virus used for infection. Thus we hypothesized that one of the 3 amino acid substitutions or a combination of two or three of these mutations marked genomic determinants of CNS demyelination.

Using site directed mutagenesis and targeted recombination; we then made a series of recombinant viruses containing each one of the 3 points mutation and each of the combined 2 point mutations. In vitro studies showed that all of these recombinant viruses exhibited similar kinetics of replication in L2 cell culture. The virulence of all of these recombinant viruses was reduced compared to wtR-A59. The recombinant viruses with the T1087N mutation (Penn99-1 and 2) were the most virulent (LD50=20,000pfu) among these recombinant viruses. The virulence of all other recombinant viruses, including the I375M point mutation (Penn01-1 and 2), the L652I point mutation (Penn01-3 and 4) and I375M-L652I two point mutation (Penn2K3 and 4), I375M, and T1087N two point mutation (Penn01-5 and 6), L652I, and T1087N two point mutation (Penn01-7 and 8), were all significantly attenuated (LD50>50,000pfu). The reduced virulence correlated well with the reduced in severity of hepatitis and encephalitis. Interestingly, demyelination was not influenced by the reduction of virulence because most of the recombinant viruses with one or two combined amino acid mutations were significantly less virulent but demyelination-positive. It further confirmed that the induction of demyelination is a separate pathologic property that does not depend on the intensity of prior encephalitis or hepatitis during the acute phase [26]. Penn2K-3 and 4, the combination of mutations in S1 (I375-M, L652I) did not produce demyelination in C57BL/6 mice even when mice were infected with the highest dose (50,000pfu, I.C.). We then searched for additional mutations around amino acid 375 and eventually found that N345S was a single point mutation that abolished demyelination (Fu et al, manuscript in preparation). Thus we identified a region in the S1 gene downstream from the receptor-binding domain that controls demyelination.

## SUMMARY.

The S gene controls a variety of biologic functions many of them are important for pathogenesis. Our laboratory identified a region around amino acid 345 of the S gene, outside and downstream from the receptor-binding domain that appears to control demyelination. Further studies are underway to define the functions of this region in virus-host interactions. Crystallization studies will determine whether the three dimensional folding of this S1 region of the molecule may enable a spatial interaction between this site and the other two mutation sites which may affect viral host interaction and subsequently pathogenesis. More detailed mutagenesis, coupled with functional studies such as studies of receptor binding, S1-S2 separation, and fusion properties, will also be helpful in elucidating the mechanism of MHV neuropathogenesis.

# REFERENCES


1. Yu, X., et al., *Mouse hepatitis virus gene 5b protein is a new virion envelope protein.* Virology, 1994. **202**(2): p. 1018-23.
2. Sawicki, S.G. and D.L. Sawicki, *Coronavirus transcription: subgenomic mouse hepatitis virus replicates intermediates function in RNA synthesis.* J. Virol., 1990. **64**: p. 1051-1056.
3. Lai, M.M., et al., *Studies on the mechanism of RNA synthesis of a murine coronavirus.* Adv Exp Med Biol, 1984. **173**: p. 187-200.
4. Zhang, X. and M.M. Lai, *A 5'-proximal RNA sequence of murine coronavirus as a potential initiation site for genomic-length mRNA transcription.* J Virol, 1996. **70**(2): p. 705-11.
5. Liu, Y.J., Lai, M. M.C, *The 3' untranslated region of coronavirus RNA is required for subgenomic mRNA transcription from a defective interfering RNA.* J. Virol., 1993. **70**: p. 7236-7240.
6. Luytjes, W., H. Gerritsma, and W.J. Spaan, *Replication of synthetic defective interfering RNAs derived from coronavirus mouse hepatitis virus-A59.* Virology, 1996. **216**(1): p. 174-83.
7. Ontiveros, E., et al., *Inactivation of expression of gene 4 of mouse hepatitis virus strain JHM does not affect virulence in the murine CNS.* Virology, 2001. **289**(2): p. 230-8.
8. de Haan, C.A., et al., *The group-specific murine coronavirus genes are not essential, but their deletion, by reverse genetics, is attenuating in the natural host.* Virology, 2002. **296**(1): p. 177-89.
9. Brian, D.A., Hogue, B. G., and Kienzle, T. E., *The coronavirus hemagglutinin esterase glycoprotein.* The coronaviridae, 1995: p. 191-217.
10. Risco, C., Anton, I. M., Enjuanes, L., and Carrascosa, J. L., *The Transmissible Gastroenteritis Coronavirus Contains a Spherical Core Shell Consisting of M and N Proteins.* J. Virol., 1996. **70**: p. 4773-4777.
11. Lai M.M, a.D.C., *The molecular biology of coronaviruses.* Adv. Virol. Res, 1997. **48**: p. 1-100.
12. Parker, M.M. and P.S. Masters, *Sequence comparison of the N genes of five strains of the coronavirus mouse hepatitis virus suggests a three domain structure for the nucleocapsid protein.* Virology, 1990. **179**(1): p. 463-8.

13.   Denison, M.R., et al., *The putative helicase of the coronavirus mouse hepatitis virus is processed from the replicase gene polyprotein and localizes in complexes that are active in viral RNA synthesis.* J Virol, 1999. **73**(8): p. 6862-71.

14.   van der Meer, Y., et al., *Localization of mouse hepatitis virus nonstructural proteins and RNA synthesis indicates a role for late endosomes in viral replication.* J Virol, 1999. **73**(9): p. 7641-57.

15.   Fischer, F., et al., *The internal open reading frame within the nucleocapsid gene of mouse hepatitis virus encodes a structural protein that is not essential for viral replication.* J Virol, 1997. **71**(2): p. 996-1003.

16.   Sturman, L.S., Holmes, K. V., and Behnke, J., *Isolation of Coronavirus Envelope Glycoproteins and Interaction with the Viral Nucleocapsid.* J. Virol., 1980. **33**: p. 449-462.

17.   Williams, R.K., G.S. Jiang, and K.V. Holmes, *Receptor for mouse hepatitis virus is a member of the carcinoembryonic antigen family of glycoproteins.* Proc Natl Acad Sci U S A, 1991. **88**(13): p. 5533-6.

18.   Boyle, J.F., Weismiller, D. G., and Holmes, K. V., *Genetic resistance to mouse hepatitis virus correlates with the absence of virus binding activity on target tissue.* J. Virol., 1987. **61**: p. 185-189.

19.   Gallagher, T.M., Buchmeier, M., *Coronavirus Spike Proteins in Viral Entry and Pathogenesis.* Virology, 2001. **279**: p. 371-374.

20.   Fleming, J.O., et al., *Antigenic relationships of murine coronaviruses: analysis using monoclonal antibodies to JHM (MHV-4) virus.* Virology, 1983. **131**(2): p. 296-307.

21.   Bergmann, C.C., et al., *The JHM strain of mouse hepatitis virus induces a spike protein-specific Db-restricted cytotoxic T cell response.* J Gen Virol, 1996. **77** ( **Pt 2** ): p. 315-25.

22.   Castro, R.F., et al., *Coronavirus-induced demyelination occurs in the presence of virus-specific cytotoxic T cells.* Virology, 1994. **200**(2): p. 733-43.

23.   Phillips, J.J., et al., *Pathogenesis of chimeric MHV4/MHV-A59 recombinant viruses: the murine coronavirus spike protein is a major determinant of neurovirulence.* J Virol, 1999. **73**(9): p. 7752-60.

24.   Phillips, J.J. and S.R. Weiss, *MHV neuropathogenesis: the study of chimeric S genes and mutations in the hypervariable region.* Adv Exp Med Biol, 2001. **494**: p. 115-9.

25.   Phillips, J.J., et al., *Multiple regions of the murine coronavirus spike glycoprotein influence neurovirulence.* J Neurovirol, 2001. **7**(5): p. 421-31.

26.   Das Sarma, J., et al., *Demyelination determinants map to the spike glycoprotein gene of coronavirus mouse hepatitis virus.* J Virol, 2000. **74**(19): p. 9206-13.

27.     Fischer, F., et al., *Analysis of a recombinant mouse hepatitis virus expressing a foreign gene reveals a novel aspect of coronavirus transcription.* J Virol, 1997. **71**(7): p. 5148-60.

28.     Kuo, L., et al., *Retargeting of coronavirus by substitution of the spike glycoprotein ectodomain: crossing the host cell species barrier.* J Virol, 2000. **74**(3): p. 1393-406.

29.     Phillips, J.J., et al., *Murine coronavirus spike glycoprotein mediates degree of viral spread, inflammation, and virus-induced immunopathology in the central nervous system.* Virology, 2002. **301**(1): p. 109-20.

30.     Dalziel, R.G., et al., *Site-specific alteration of murine hepatitis virus type 4 peplomer glycoprotein E2 results in reduced neurovirulence.* J Virol, 1986. **59**(2): p. 463-71.

31.     Fleming, J.O., et al., *Pathogenic characteristics of neutralization-resistant variants of JHM coronavirus (MHV-4).* Adv Exp Med Biol, 1987. **218**: p. 333-42.

32.     Gallagher, T.M., S.E. Parker, and M.J. Buchmeier, *Neutralization-resistant variants of a neurotropic coronavirus are generated by deletions within the amino-terminal half of the spike glycoprotein.* J Virol, 1990. **64**(2): p. 731-41.

33.     Hingley, S.T., et al., *MHV-A59 fusion mutants are attenuated and display altered hepatotropism.* Virology, 1994. **200**(1): p. 1-10.

34.     Gombold, J.L., S.T. Hingley, and S.R. Weiss, *Identification of peplomer cleavage site mutations arising during persistence of MHV-A59.* Adv Exp Med Biol, 1993. **342**: p. 157-63.

35.     Leparc-Goffart, I., et al., *Altered pathogenesis of a mutant of the murine coronavirus MHV-A59 is associated with a Q159L amino acid substitution in the spike protein.* Virology, 1997. **239**(1): p. 1-10.

36.     Leparc-Goffart, I., et al., *Targeted recombination within the spike gene of murine coronavirus mouse hepatitis virus-A59: Q159 is a determinant of hepatotropism.* J Virol, 1998. **72**(12): p. 9628-36.

37.     Saeki, K., N. Ohtsuka, and F. Taguchi, *Identification of spike protein residues of murine coronavirus responsible for receptor-binding activity by use of soluble receptor-resistant mutants.* J Virol, 1997. **71**(12): p. 9024-31.

38.     Wang, F.I., J.O. Fleming, and M.M. Lai, *Sequence analysis of the spike protein gene of murine coronavirus variants: study of genetic sites affecting neuropathogenicity.* Virology, 1992. **186**(2): p. 742-9.

39.     Navas S., Seo, SH, Chua MM, Sarma JD, Lavi E, Hingley ST, Weiss SR, *The spike protein of murine coronavirus determines the ability of the virus to replicate in the liver and cause hepatitis.* Journal of virology, 2001. **75**: p. 2452-2457.

40.     Lavi, E., et al., *Experimental demyelination produced by the A59 strain of mouse hepatitis virus.* Neurology, 1984. **34**(5): p. 597-603.

41.    Lavi, E., et al., *The organ tropism of mouse hepatitis virus A59 in mice is dependent on dose and route of inoculation.* Lab Anim Sci, 1986. **36**(2): p. 130-5.

42.    Lavi, E., et al., *Coronavirus mouse hepatitis virus (MHV)-A59 causes a persistent, productive infection in primary glial cell cultures.* Microb Pathog, 1987. **3**(2): p. 79-86.

43.    Knobler, R.L., et al., *Selective localization of wild type and mutant mouse hepatitis virus (JHM strain) antigens in CNS tissue by fluorescence, light and electron microscopy.* J Neuroimmunol, 1981. **1**(1): p. 81-92.

44.    Lavi, E., et al., *MHV-A59 pathogenesis in mice.* Adv Exp Med Biol, 1984. **173**: p. 237-45.

45.    Perlman, S., et al., *Identification of the spinal cord as a major site of persistence during chronic infection with a murine coronavirus.* Virology, 1990. **175**(2): p. 418-26.

46.    Houtman, J.J. and J.O. Fleming, *Pathogenesis of mouse hepatitis virus-induced demyelination.* J Neurovirol, 1996. **2**(6): p. 361-76.

47.    Hirano, N., et al., *Replication and plaque formation of mouse hepatitis virus (MHV-2) in mouse cell line DBT culture.* Arch Gesamte Virusforsch, 1974. **44**(3): p. 298-302.

48.    Hirano, N., K. Ono, and N. Goto, *Establishment of persistently infected DBT cell cultures with mouse hepatitis virus strains, MHV-2 and MHV-3.* Nippon Juigaku Zasshi, 1984. **46**(5): p. 757-60.

49.    Yokomori, K., L.R. Banner, and M.M. Lai, *Heterogeneity of gene expression of the hemagglutinin-esterase (HE) protein of murine coronaviruses.* Virology, 1991. **183**(2): p. 647-57.

50.    Yamada, Y.K., et al., *Acquired fusion activity of a murine coronavirus MHV-2 variant with mutations in the proteolytic cleavage site and the signal sequence of the S protein.* Virology, 1997. **227**(1): p. 215-9.

51.    Das Sarma, J., et al., *Sequence analysis of the S gene of recombinant MHV-2/A59 coronaviruses reveals three candidate mutations associated with demyelination and hepatitis.* J Neurovirol, 2001. **7**(5): p. 432-6.

52.    Das Sarma, J., et al., *Mouse hepatitis virus type-2 infection in mice: an experimental model system of acute meningitis and hepatitis.* Experimental & Molecular Pathology, 2001. **71**(1): p. 1-12.

53.    Keck, J.G., et al., *RNA recombination of murine coronavirus: recombination between fusion-positive mouse hepatitis virus A59 and fusion-negative mouse hepatitis virus 2.* J. Virol., 1988. **62**: p. 1989-1998.

54.    Das Sarma, J., et al., *Sequence analysis of the S gene of recombinant MHV-2/A59 coronaviruses reveals three candidate mutations associated with demyelination and hepatitis.* Journal of Neurovirology, 2001. **7**(5): p. 432-6.

Part D

# DEMYELINATION INDUCED BY OTHER VIRUSES

Chapter D1

# SEMLIKI FOREST VIRUS INDUCED DEMYELINATION

John K. Fazakerley

*Centre for Infectious Diseases, College of Medicine and Veterinary Medicine, University of Edinburgh, Summerhall, Edinburgh, EH9 1QH, UK.*

Abstract:     SFV infection of the laboratory mouse provides an excellent model to study age-related virus virulence, virus neuroinvasion and the pathogenesis of acute encephalitis and demyelination. The A7(74) strain is avirulent in weaned mice. Following intraperitoneal infection this virus is neuroinvasive and crosses the blood-brain barrier to establish focal infections in neurons and oligodendrocytes. Virus replication is for the most part restricted and non-destructive in these cells and in the absence of immune responses virus persists. In immunocompetent animals focal lesions of inflammatory demyelination develop throughout the CNS including the olfactory nerves. $CD8^+$ T-lymphocytes are required for the generation of these lesions. Animals have abnormal visual evoked responses and may develop limb paralysis.

Key words:    Semliki Forest virus, demyelination.

## 1.      VIRUS STRAINS

SFV is an alphavirus of the *Togaviridae*. Naturally the virus is found in Africa where it has been shown to infect horses and man. The virus is spread by mosquitoes. In man infection results in a mild febrile illness (Mathiot et al, 1990). SFV has been studied for many years in the laboratory mouse as an experimental model system to understand viral and immunological events in viral encephalitis and virus induced CNS demyelination (Atkins et al, 1985; Fazakerley and Buchmeier, 1993; Fazakerley et al, 1997; Fazakerley and Walker 2003). The virus is relatively simple, it has an approximately 12 kb genome of single-stranded positive sense RNA. This has been cloned (Garoff et al, 1980; Takkinen et al, 1986) and as with many positive stranded RNA viruses, studies of site directed mutants are proving to be a powerful approach to mapping areas of the genome responsible for individual phenotypic characteristics (Glasgow et al, 1994; Santagati et al, 1995; Santagati et al, 1998). The genome has a replicase gene which undergoes sequential post-translational processing to produce a complex of four polypeptides (nsP1-4) which mediate RNA replication. The viral structural proteins (the capsid C, and 3 glycoproteins, E1, E2 and E3) are encoded within a separate transcriptional unit controlled by a subgenomic promoter. A major determinant of virulence resides in the nsP3 protein, a component of the viral replicase (Tuittila et al, 2000). There are a number of strains of the virus derived from independent isolates by passage through mice or cell culture. The most commonly studied strains are L10, prototype and A7 and A7(74) which are respectively, virulent and avirulent in adult mice. As with all virus infections the route of inoculation, the dose of virus, the age of the mice and the genetics of both virus and host affect the outcome of infection. All SFV strains are virulent in neonatal and suckling mice by all routes of inoculation. A number of viruses generated by mutagenesis have also been studied, from the point of view of demyelinating disease, the most notable of these is M9 (Barrett et al, 1980).


## 2.     AGE-RELATED VIRULENCE AND RESTRICTED REPLICATION

In neonatal mice all strains of this virus spread rapidly around the brain and infection is rapidly fatal. As the mice age some strains are less able to spread in the CNS and there is a sharp age-related virulence (Fleming, 1977; Oliver et al, 1997; Oliver and Fazakerley, 1998). In the mature CNS the A7(74) strain has a restricted replication and remains confined to perivascular foci (Pathak and Webb, 1978; Fazakerley et al, 1993; Oliver et al, 1997). This age-related virulence is not a function of the maturity of specific immune responses as this strain of the virus remains confined to small, predominantly perivascular foci in athymic *nu/nu* mice or mice with severe combined immunodeficiency (Fazakerley et al, 1993; Fazakerley and Webb, 1987b; Amor et al, 1996). Interestingly, widespread CNS infection can be established in adult mice by treatment with gold compounds prior to infection. It is unclear how this works but gold in common with other heavy metals is thought to have an affect on neuronal differentiation (Scallan and Fazakerley, 1999). Events in age-related virulence are well described with intranasal infection. Following intranasal infection SFV can track along the olfactory nerve to enter the CNS in the olfactory bulb (Kaluza et al, 1987;

Oliver and Fazakerley, 1998). For the A7(74) strain of SFV, spread of infection along olfactory and other CNS pathways is dependant upon neuronal maturity (Oliver et al, 1997; Oliver and Fazakerley, 1998). The virus can only replicate efficiently in immature developing neurons. Following intranasal infection, in the immature brain of weanling mice virus can be observed to sequentially infect the olfactory nerve layer, periglomerular cells, mitral cells and granule cells of the olfactory bulb and then to exit into secondary olfactory areas such as the olfactory tubercle and the anterior olfactory nucleus (Oliver and Fazakerley, 1998). In the immature brain this wave of infection is followed by a wave of apoptotic cell death (Allsopp and Fazakerley, 2000; Fazakerley and Allsopp, 2001). Interestingly and in complete contrast, infection of mature neurons appears to be non-destructive and can be persistent (Fazakerley and Webb, 1987b; Amor et al, 1996; Donnelly et al, 1997). This phenomenon of restricted virus replication in mature CNS cells may be highly relevant to demyelinating disease. It is also observed with measles virus in subacute sclerosing panencephalitis and in animal models (Schneider-Schaulies et al, 1995). The inability of viruses to replicate fully or efficiently in mature CNS cells coupled with a relative resistance of these highly specialized and irreplaceable cells to initiate programmed cell death upon infection may be crucial in establishing CNS infection (Allsopp and Fazakerley, 2000; Fazakerley 2001). This may only be eliminated by immune mediated destruction and consequent demyelination (Fazakerley and Walker, 2003).

## 3.    ENTRY INTO AND SPREAD WITHIN THE BRAIN

SFV has the advantage of being neuroinvasive as well as neurotropic, allowing study of virus entry into the CNS and the functioning of the blood-brain barrier. In weaned (> 3 weeks of age) mice following intraperitoneal (ip) inoculation, SFV replicates in muscles including skeletal, smooth and cardiac muscle (Pusztai et al, 1971; Amor et al, 1996). A high titer plasma viraemia is detectable by 24 hours. This usually falls after 48 hours and virus is generally undetectable in the blood by 4 days (Pusztai et al, 1971; Fazakerley et al, 1993). All strains which have been studied (L10, A7, A7(74)) are neuroinvasive. Original electronmicroscopic studies of infected mouse brains did not observe virus replication in cerebral vascular endothelial cells and the authors suggested virus was transported without replicating in them (Pathak and Webb, 1974). However, an in vitro study in immature mouse endothelial cells showed good virus replication (Dropulic and Masters, 1989). Virus replication was also subsequently observed in suckling mouse primary cerebral vascular endothelial cell cultures (Soilu-Hanninen et al., 1994). In the same study, viral RNA and proteins were detected in brain endothelial cells by in situ hybridization and immunostaining between days 1 and 7 post-infection with the highest frequency of positive cells on days 3 and 4. Other in situ hybridization and immunostaining studies have not reported positive cerebral vascular endothelial cells (Fazakerley et al, 1993; Gates et al, 1984; Balluz et al, 1993).

Once across this barrier the virus infects adjacent cells resulting in striking perivascular foci of infection (Fazakerley et al, 1993). It is likely that other strains also enter the CNS predominantly by this route. Infection of the brain via the olfactory nerves can also occur (Kaluza et al, 1987; Oliver and Fazakerley, 1998). Whether this is of relevance following ip infection is not clear. All strains of the virus appear to infect both neurons and oligodendrocytes but not astrocytes and only rarely meningeal, ependymal or choroid plexus cells (Balluz et al, 1993; Pathak and Webb, 1983).

In weaned mice, SFV infection of the CNS occurs by 24h. Thereafter titers rise, rapidly with the virulent strains (eg L10) until death at 3 or 4 days, or more slowly with the avirulent strains (eg A7(74)) to reach a peak around 3 days after which titers begin to fall (Pusztai et al, 1971; Fazakerley et al, 1993). Avirulent A7(74) virus cannot be detected in the brain by plaque assay after day 8, by the more sensitive $ICLD_{50}$ after day 11 and by in situ hybridization after day 14 (Jagelman et al, 1978; Suckling et al, 1978; Fazakerley et al, 1993). RT-PCR studies following M9 infection indicate that viral RNA can be detected in the brain for even longer periods, >90 days (Donnelly et al, 1997). Following A7(74) infection antibody producing plasma cells in the brain and intrathecal anti-viral antibody are detectable for months suggesting persistence of viral antigen (Parsons and Webb, 1984; Parsons and Webb, 1989). In the absence of specific immune responses the A7(74) virus can persist in CNS cells, apparently without damage and essentially for the life of the mouse (Fazakerley et al, 1983; Amor et al, 1996).


## 4.      BLOOD-BRAIN BARRIER CHANGES

With avirulent strains, early infection of the adult CNS detectable by 24 h is followed a few days later by a biphasic mononuclear cell pleocytosis (Parsons and Webb, 1982b). Measurement of albumin and IgG in the cerebrospinal fluid demonstrates a disturbance of the blood-brain barrier between days 4 and 10 (Parsons and Webb, 1982a; SoiluHanninen et al,

1994). Disruption of the blood-brain barrier at day 5 has also been observed by the detection of fibrinogen in the brain parenchyma adjacent to cerebral blood vessels (SoiluHanninen et al, 1994). The cause of the blood brain barrier disruption is unclear; it could be a direct viral effect, for example destruction of endothelial cells as is observed in culture (SoiluHanninen et al, 1994), a direct result of the passage of inflammatory cells across the barrier or it could be mediated by cytokines, for example interferon-gamma. The leaky barrier does corresponds to the time of increasing inflammatory cell infiltration and reduction of brain virus titers. Pleocytosis is apparent from 4 days and subsides by 9 days but cell counts in the cerebrospinal fluid do not return to normal for weeks and immunoglobulins are synthesized locally by invading B-lymphocytes (Parsons and Webb, 1984; Parsons and Webb, 1989).

## 5.    ANTIBODY RESPONSES

SFV A7(74) infection of BALB/c mice results in a rapid and neutralizing serum IgM antibody response followed rapidly by an IgG2a response and a much slower IgG1 response (Fazakerley et al, 1993). SCID mice which have no serum anti-viral antibodies have both a persistent viraemia and a persistent CNS infection (Amor et al, 1996), whereas *nu/nu* mice control the viraemia as seen in immunocompetent mice but *nu/nu* mice have a persistent CNS infection. The *nu/nu* mice produce serum anti-viral IgM but not IgG (Suckling et al, 1982) suggesting that this may be controlling the viraemia but unable to eliminate the CNS infection. That this is the case is confirmed by resolution of persistent viraemia but not CNS infection by transfer of day 7 *nu/nu* mouse serum to infected SCID mice (Amor et al, 1996). In contrast, transfer of high titer IgG sera from immunocompetent BALB/c mice to SCID mice eliminates both the viraemia and CNS virus, at least that detectable by infectivity assay (Amor et al, 1996). Whether this is complete eradication of all viral sequences is not clear since it has not been checked by techniques such as RT-PCR. Antibodies are highly effective in protecting mice from challenge with a lethal dose of SFV. A number of B-cell epitopes have been mapped on the viral glycoproteins (Boere et al, 1984; Snijders et al, 1991) and even a single monoclonal antibody directed to the E2 envelope glycoprotein has been shown to protect mice against a virulent infection (Boere et al, 1983). Combinations of linear B-cell epitopes and B and T cell epitopes have been tested as potential vaccine candidates (Snijders et al, 1991; Snijders et al, 1992a).

## 6.      INFLAMMATORY RESPONSES

Mice infected with SFV have good delayed-type hypersensitivity responses to the virus (Kraaijeveld et al, 1979) and T-cell epitopes for some haplotypes have been mapped (Snijders et al, 1991; Snijders et al, 1992b; Snijders et al, 1992a). Expression of cellular adhesion molecules on cerebral vascular endothelial and infiltrating mononuclear cells, cytokine expression and the phenotype of infiltrating mononuclear cells during SFV A7(74) infection have been studied in two systems BALB/c and Biozzi ABH mice (SoiluHanninen et al, 1994; SoiluHanninen et al, 1997; Morris et al, 1997; Parsons and Webb, 1989; Smith et al., 2000). It is possible that these two mouse strains differ in the kinetics of events but the following presents a synthesis. By immunostaining, interleukin (IL) $-1\alpha$ and $1\beta$, IL-10, and TGF-$\beta$ are detectable at 3 days; GMCSF, IL-2 and TNF$\alpha$ at 7 days and IFN-$\gamma$ and IL-6 at 10 days. Increased expression of ICAM-I on cerebral vascular endothelial cells and adjacent cells occurs from day 3, with a peak somewhere between days 5 and 10. Expression of VCAM-I is maximal around 7 to 10 days and is limited to the blood vessels. Major histocompatibility type-I and $-$II expression are detectable on endothelial and other blood vessel associated cells by 7 days. Immunostaining studies indicate that CD8[+] cells, LFA-1[+] cells, Mac-1[+] cells and VLA-4[+] cells are all first observed in the infected brain at 3 days and that CD8[+] cells dominate the early phase of infection while CD4[+] T-lymphocytes are detectable from 7 days to at least 35 days. Numbers of B220[+] B-lymphocytes increase dramatically from 7 days to a peak at 21 days. The parenchymal MHC-I[+] cells form a network surrounding the foci of infection in which the level of MHC-I staining decreases with distance from the infected cells. MHC-II[+] cells, probably pericytes can be observed adjacent to blood vessels in areas of infection and MHC-II[+] astrocytes and microglial cells are present in the areas of infection and inflammation (Morris et al, 1997).


## 7.      NEUROPATHOLOGY

Semliki Forest virus (SFV) infection of adult mice results in a demyelinating meningoencephalomyelitis. An intense inflammatory response characterized by perivascular cuffing with invading cells spreading into the parenchyma is apparent histologically from 3 days (Parsons and Webb, 1982b). Areas of inflammatory infiltrates correspond to the areas of infection (Subak-Sharpe et al, 1993; Morris et al, 1997). Perivascular cuffs are a prominent feature of the neuropathology and are maximal between 7 and 10 days. Demyelination is first apparent by luxol fast blue staining of paraffin sections around 14 days post-infection but small focal lesions are apparent by electron microscopy at 10 days (Suckling et al, 1978; Kelly et al, 1982). In the optic nerve there are lesions of demyelination and changes in visually evoked responses and axonal transport (Tremain and Ikeda, 1983). These are features which also occur in MS. Adoptive transfer studies in SCID and *nu/nu* mice and studies in mice immunosuppressed with total body irradiation, cyclophosphamide, cyclosporin or cycloleucine all demonstrate that T-cell responses are pathogenic and are required to generate the lesions of demyelination (Fazakerley et al, 1983; Fazakerley and Webb, 1987c; Fazakerley and Webb, 1987b; Fazakerley and Webb, 1987a; Amor and Webb, 1987; Amor et al, 1996). Treatment of infected mice with anti-VLA-4 antibodies dramatically reduces CNS cellular infiltrates and demyelination (Smith et al, 2000). Depletion of CD8[+] cells with a monoclonal antibody greatly reduces demyelination whereas depletion of CD4[+] cells reduces the extent of the inflammation but not that of demyelination (Subak-Sharpe et al, 1993). Given the requirement for CD8[+] cells and that the virus infects oligodendrocytes, the most likely mechanism of demyelination would seem to be CD8[+] T-cell destruction of virally infected oligodendrocytes, though this remains to be proven following for example, adoptive transfer of SFV-specific cytotoxic T-cell clones or lines into *nu/nu* or SCID mice. Mice infected with SFV A7(74) generate CNS antigen specific autoantibodies, are more sensitive to the induction of experimental allergic encephalomyelitis and have T-cells which cross react between viral and CNS epitopes (Amor and Webb, 1988); Mokhtarian and Swoveland, 1987; Mokhtarian et al, 1996). At present the role of autoimmune responses in the generation, exacerbation or perpetuation of SFV A7(74) induced demyelinating lesions remains unclear.


## ACKNOWLEDGEMENTS

I am grateful to the UK Multiple Sclerosis Society which has over many years supported the study of Semliki Forest virus induced demyelinating disease.


## REFERENCES

Allsopp TE, Fazakerley JK (2000). Altruistic cell suicide and the specialized case of the virus-infected nervous system. *Trends Neurosci* 23:284-290.
Amor S, Scallan MF, Morris MM, Dyson H, Fazakerley JK (1996). Role of immune responses in protection and pathogenesis during Semliki Forest virus encephalitis. *J Gen Virol* 77:281-291.

Amor S, Webb HE (1987). The effect of cycloleucine on SFV A7(74) infection in mice. *Br J Exp Pathol* 68:225-235.

Amor S, Webb HE (1988). CNS pathogenesis following a dual viral infection with Semliki Forest (alphavirus) and Langat (flavivirus). *Br J Exp Pathol* 69:197-208.

Atkins GJ, Sheahan BJ, Dimmock NJ (1985). Semliki Forest virus infection of mice: a model for genetic and molecular analysis of viral pathogenicity. *J Gen Virol* 66:395-408.

Balluz IM, Glasgow GM, Killen HM, Mabruk MJ, Sheahan BJ, Atkins GJ (1993). Virulent and avirulent strains of Semliki Forest virus show similar cell tropism for the murine central nervous system but differ in the severity and rate of induction of cytolytic damage. *Neuropathol Appl Neurobiol* 19:233-239.

Barrett PN, Sheahan BJ, Atkins GJ (1980). Isolation and preliminary characterization of Semliki Forest virus mutants with altered virulence. *J Gen Virol* 49:141-147.

Boere WA, Benaissa-Trouw BJ, Harmsen M, Kraaijeveld CA, Snippe H (1983). Neutralizing and non-neutralizing monoclonal antibodies to the E2 glycoprotein of Semliki Forest virus can protect mice from lethal encephalitis. *J Gen Virol* 64:1405-1408.

Boere WA, Harmsen T, Vinje J, Benaissa-Trouw BJ, Kraaijeveld CA, Snippe H (1984). Identification of distinct antigenic determinants on Semliki Forest virus by using monoclonal antibodies with different antiviral activities. *J Virol* 52:575-582.

Donnelly SM, Sheahan BJ, Atkins GJ (1997). Long-term effects of Semliki Forest virus infection in the mouse central nervous system. *Neuropathol Appl Neurobiol* 23:235-241.

Fazakerley JK (2001). Neurovirology and developmental neurobiology. *Adv Vir Res* 56:73-124.

Fazakerley JK, Allsopp TE (2001). Programmed cell death in virus infections of the nervous system. *Curr Top Microbiol Immunol* 253:95-119.

Fazakerley JK, Amor S, Nash AA (1997). Animal model systems of MS. 255-273.

Fazakerley JK, Amor S, Webb HE (1983). Reconstitution of Semliki forest virus infected mice, induces immune mediated pathological changes in the CNS. *Clin Exp Immunol* 52:115-120.

Fazakerley JK, Buchmeier MJ (1993). Pathogenesis of virus-induced demyelination. *Adv Virus Res* 42:249-324.

Fazakerley JK, Pathak S, Scallan M, Amor S, Dyson H (1993). Replication of the A7(74) strain of Semliki Forest virus is restricted in neurons. *Virology* 195:627-637.

Fazakerley JK, Walker R (2003). Virus demyelination. *J Neurovirol* (in press).

Fazakerley JK, Webb HE (1987a). Cyclosporine enhances virally induced T-cell-mediated demyelination. The effect of cyclosporine on a demyelinating virus infection. *J Neurol Sci* 78:35-50.

Fazakerley JK, Webb HE (1987b). Semliki Forest virus-induced, immune-mediated demyelination: adoptive transfer studies and viral persistence in nude mice. *J Gen Virol* 68:377-385.

Fazakerley JK, Webb HE (1987c). Semliki Forest virus induced, immune mediated demyelination: the effect of irradiation. *Br J Exp Pathol* 68:101-113.

Fleming P (1977). Age-dependent and strain-related differences of virulence of Semliki Forest virus in mice. *J Gen Virol* 37:93-105.

Garoff H, Frischauf A, Simons K, Lehrach H, Delius H (1980). Nucleotide sequence of cDNA coding for Semliki Forest virus membrane glycoproteins. *Nature* 288:236-241.

Gates MC, Sheahan BJ, Atkins GJ (1984). The pathogenicity of the M9 mutant of Semliki Forest virus in immune- compromised mice. *J Gen Virol* 65:73-80.

Glasgow GM, Killen HM, Liljestrom P, Sheahan BJ, Atkins GJ (1994). A single amino acid change in the E2 spike protein of a virulent strain of Semliki Forest virus attenuates pathogenicity. *J Gen Virol* 75:663-668.

Jagelman S, Suckling AJ, Webb HE, Bowen ETW (1978). The pathogenesis of avirulent Semliki Forest virus infections in athymic nude mice. *J Gen Virol* 41:599-607.

Kaluza G, Lell G, Reinacher M, Stitz L, Willems WR (1987). Neurogenic spread of Semliki Forest virus in mice. *Arch Virol* 93 :97-110.

Kelly WR, Blakemore WF, Jagelman S, Webb HE (1982). Demyelination induced in mice by avirulent Semliki Forest virus. II. An ultrastructural study of focal demyelination in the brain. *Neuropathol Appl Neurobiol* 8:43-53.

Kraaijeveld CA, Harmsen M, Khader Boutahar-Trouw B (1979). Delayed-type hypersensitivity against Semliki Forest virus. *Infect Immun* 23:217-221.

Mathiot CC, Grimaud G, Garry P, Bouquety JC, Mada A, Daguisy AM, Georges AJ (1990). An outbreak of human Semliki Forest virus infections in Central African Republic. *Am J Trop Med Hyg* 42:386-393.

Mokhtarian F, Shi Y, Grob D (1996). Epitopes of Semliki Forest virus which evoke cross-reactive immune- response with peptides of the central-nervous-system. *Faseb Journal* 10:2092-2092.

Mokhtarian F, Swoveland P (1987). Predisposition to EAE induction in resistant mice by prior infection with Semliki Forest virus. *J Immunol* 138:3264-3268.

Morris MM, Dyson H, Baker D, Harbige LS, Fazakerley JK, Amor S (1997). Characterization of the cellular and cytokine response in the central nervous system following Semliki Forest virus infection. *J Neuroimmunol* 74:185-197.

Oliver KR, Fazakerley JK (1998). Transneuronal spread of Semliki Forest virus in the developing mouse olfactory system is determined by neuronal maturity. *Neurosci* 82:867-877.

Oliver KR, Scallan MF, Dyson H, Fazakerley JK (1997). Susceptibility to a neurotropic virus and its changing distribution in the developing brain is a function of CNS maturity. *J Neurovirol* 3:38-48.

Parsons LM, Webb HE (1982a). Blood brain barrier disturbance and immunoglobulin G levels in the cerebrospinal fluid of the mouse following peripheral infection with the demyelinating strain of Semliki Forest virus. *J Neurol Sci* 57:307-318.

Parsons LM, Webb HE (1982b). Virus titers and persistently raised white cell counts in cerebrospinal-fluid in mice after peripheral infection with demyelinating Semliki Forest virus. *Neuropathol Appl Neurobiol* 8:395-401.

Parsons LM, Webb HE (1984). Specific immunoglobulin G in serum and cerebrospinal fluid of mice infected with the demyelinating strain of Semliki Forest virus. *Microbios Letters* 25: 135-140.

Parsons LM, Webb HE (1989). Identification of immunoglobulin-containing cells in the central nervous system of the mouse following infection with the demyelinating strain of Semliki Forest virus. *Br J Exp Pathol* 70:247-255.

Pathak S, Webb HE (1974). Possible mechanisms for the transport of Semliki Forest virus into and within mouse brain: An electron microscopic study. *J Neurol Sci* 23:175-184.

Pathak S, Webb HE (1978). An electron-microscopic study of avirulent and virulent Semliki Forest virus in the brains of different ages of mice. *J Neurol Sci* 39:199-211.

Pathak S, Webb HE (1983). Semliki Forest virus multiplication in oligodendrocytes in mouse-brain with reference to demyelination. *J Physiol (London)* 339:17-17.

Pusztai R, Gould E, Smith H (1971). Infection pattern in mice of an avirulent and virulent strain of Semliki Forest virus. *Br J Exp Pathol* 52:669-677.

Santagati MG, Maatta JA, Itaranta PV, Salmi AA, Hinkkanen AE (1995). The Semliki Forest virus E2 gene as a virulence determinant. *J Gen Virol* 75:47-52.

Santagati MG, Maatta JA, Roytta M, Salmi AA, Hinkkanen AE (1998). The significance of the 3'-nontranslated region and E2 amino acid mutations in the virulence of Semliki Forest virus in mice. *Virology* 243:66-77.

Scallan MF, Fazakerley JK (1999). Aurothiolates enhance the replication of Semliki Forest virus in the CNS and the exocrine pancreas. *J Neurovirol* 5:392-400.

Schneider-Schaulies S, Schneider-Schaulies J, Dunster LM, ter M, V (1995). Measles virus gene expression in neural cells. *Curr Top Microbiol Immunol* 191:101-116.

Smith JP, Morris-Downes M, Brennan FR, Wallace GJ, Amor S (2000). A role for alpha4-integrin in the pathology following Semliki Forest virus infection. *J Neuroimmunol* 106:60-68.

Snijders A, Benaissa-Trouw BJ, Oosterlaken TA, Puijk WC, Posthumus WP, Meloen RH, Boere WA, Oosting JD, Kraaijeveld CA, Snippe H (1991). Identification of linear epitopes on Semliki Forest virus E2 membrane protein and their effectiveness as a synthetic peptide vaccine. *J Gen Virol* 72:557-565.

Snijders A, Benaissatrouw BJ, Snippe H, Kraaijeveld CA (1992a). Immunogenicity and vaccine efficacy of synthetic peptides containing Semliki Forest virus B-cell and T-cell epitopes. *J Gen Virol* 73:2267-2272.

Snijders A, Benaissatrouw BJ, Visservernooy HJ, Fernandez I, Snippe H, Kraaijeveld CA (1992b). A delayed-type hypersensitivity-inducing T-cell epitope of Semliki Forest virus mediates effective T-helper activity for antibody- production. *Immunol* 77:322-329.

SoiluHanninen M, Eralinna JP, Hukkanen V, Roytta M, Salmi AA, Salonen R (1994). Semliki-Forest virus infects mouse-brain endothelial-cells and causes blood-brain-barrier damage. *J Virol* 68:6291-6298.

SoiluHanninen M, Roytta M, Salmi AA, Salonen R (1997). Semliki Forest virus infection leads to increased expression of adhesion molecules on splenic T-cells and on brain vascular endothelium. *J Neurovirol* 3:350-360.

Subak-Sharpe I, Dyson H, Fazakerley JK (1993). In vivo depletion of CD8+ T cells prevents lesions of demyelination in Semliki Forest virus infection. *J Virol* 67:7629-7633.

Suckling AJ, Jagelman S, Webb HE (1982). Immunoglobulin synthesis in nude (nu/nu), nu/+ and reconstituted nu/nu mice infected with a demyelinating strain of Semliki Forest virus. *Clin Exp Immunol* 47:283-288.

Suckling AJ, Pathak S, Jagelman S, Webb HE (1978). Virus associated demyelination: a model using avirulent Semliki Forest virus infection of mice. *J Neurol Sci* 36:147-154.

Takkinen K, Kalkkinen N, Soderlund H (1986). Complete nucleotide-sequence of the nonstructural protein genes of Semliki Forest virus. *J Cell Biochem* 291-291.

Tremain KE, Ikeda H (1983). Physiological deficits in the visual-system of mice infected with Semliki Forest virus and their correlation with those seen in patients with demyelinating disease. *Brain* 106:879-895.

Tuittila MT, Santagati MG, Roytta M, Maatta JA, Hinkkanen AE (2000). Replicase complex genes of Semliki Forest virus confer lethal neurovirulence. *J Virol* 74:4579-4589.

Chapter D2

# THE PATHOGENESIS OF CANINE DISTEMPER VIRUS INDUCED DEMYELINATION
## A biphasic process


Wolfgang Baumgärtner[1] and Susanne Alldinger[2]
*[1]Institut für Pathologie der Tierärztlichen Hochschule Hannover, Germany, [2]Institut für Veterinär-Pathologie der Justus-Liebig-Universität Giessen, Germany*



Abstract:     Canine distemper is induced by canine distemper virus (CDV), a morbillivirus of the Paramyxoviridae family closely related to measles virus. CDV infection frequently results in the affection of the central nervous system (CNS). Besides a polioencephalitis, distemper leukoencephalitis (DL) is the main sequela of CDV CNS infection. Demyelination in DL appears to be a biphasic process. Initiation of demyelination is ascribed to a direct action of the virus as there is a prominent intralesional expression of viral proteins and mRNA. Early lesions are accompanied by the presence of few CD8+ lymphocytes and single CD4+ cells and an up-regulation of the major histocompatibility complex class II (MHCII). In this phase, tumour necrosis factor-α (TNF-α), CD44, a hyaluronate receptor, and matrix-metalloproteinases (MMPs) as well as their inhibitors (TIMPs) are up-regulated. Plaque progression seems to be an immunopathological process. The strong reduction or absence of viral protein and mRNA expression is associated with a vigorous immune response. CD8+ lymphocytes dominate within the lesions  while CD4+ lymphocytes and B cells are mainly found perivascularly. Simultaneously, there is a strong up-regulation of MHCII and the pro-inflammatory cytokines interleukin (IL) -6, IL-8, IL-12, and TNF-α. The pro-inflammatory cytokines Il-1ß, IL-2, and interferon-γ are lacking. The anti-inflammatory cytokines IL-10 and transforming growth factor-ß are not influenced by the CDV infection. CD44 , TIMPs and single MMPs are strongly diminished.

Summarized, in CDV induced demyelinating leukoencephalitis, plaque genesis is a biphasic process with various factors associated with lesion initiation or progression. Understanding the mechanisms of demyelination in these two phases might allow  insights into the pathogenesis of demyelinating diseases of man.

Key words:     canine distemper, brain, demyelination, immunopathology, extracellular matrix, dog

# 1.    INTRODUCTION

## 1.1    Canine distemper virus

Canine distemper (CD) was decribed for the first time in Europe in 1761 (Mitscherlich, 1938) and the virus etiology was confirmed by Carré in 1905 (Fankhauser, 1982). The natural host spectrum of CDV comprises all families of the order Carnivora (Deem et al., 2000). In humans, the participation of CDV in the pathogenesis of Paget's disease of bone (Gordon et al., 1991, Mee et al., 1998, Mee et al., 1999, Helfrich et al., 2000, Ooi et al., 2000) and multiple sclerosis (MS; Cook et al., 1978 and 1987, Vollmer and Waxman, 1991, Hodge and Wolfso, 1997, Hernan et al., 2001) has been speculated but lacks final verification.
Canine distemper virus belongs to the genus morbillivirus of the Paramyxoviridae family and is closely related to measles virus (Pringle, 1999). CDV is an enveloped negative stranded virus (Kingsbury, 1990). It contains 6 structural proteins, the nucleocapsid (NP), phospho (P), and the large (L) protein (core proteins) and the membrane (M), hemagglutinin (H), and fusion (F) protein (surface proteins; Hall, 1980, Örvell, 1980, Diallo, 1990). The NP proteins catalyze transcription and replication in the host cell (Kingsbury, 1990). The M protein is involved in virus maturation as it connects the surface glycoproteins with the nucleocapsid (Kingsbury, 1990). The F protein manages the fusion of the virus with the host cell or the infected cell to the neighbour cell (Diallo, 1990). The H protein mediates the attachment of the virus to the host cell by neuraminic acid binding receptors (Hall et al., 1980; Kingsbury, 1990) and determines viral tropism and cytopathogenicity (von Messling et al., 2001). The function of the non-structural proteins V and C is not clear. Recently, another non-structural protein has been detected that seems to allow virus replication in different species (Wang et al., 1998). CD9, a tetraspan transmembrane protein is associated with CDV induced cell-cell fusion but not virus-cell fusion (Schmid et al., 2000). Signalling lymphocyte activation molecules (SLAMs; CD150) act as cellular receptors for CDV and other morbilliviruses (Tatsuo et al., 2001, Tatsuo and Yanagi, 2002). There is only one serotype but there are several cocirculating genotypes of the virus (Haas et al., 1997). CDV strains display different properties. Some strains are associated with polioencephalitis (Snyder Hill) while others induce demyelinating leukoencephalitis (R252, A75-17; Summers and Appel, 1994).

## 1.2    Forms of canine distemper encephalitis

Canine distemper encephalitis can be classified in different subtypes according to  morphological changes and  brain areas affected. The type of neuropathological changes depend on the virus strain, age and immune status of the affected animal (Krakowka and Koestner, 1976, Summers et al., 1984a, Pearce-Kelling et al., 1990, Raw et al., 1992). **Polioencephalitis** is a rare manifestation of CDV infection and is predominantly observed in the cortex and the nuclei of the brain stem, neurons and protoplasmic astrocytes being the mainly affected cell populations (Baumgärtner et al., 1999). Distemper polioencephalitis includes the subgroups "old dog encephalitis" (Lincoln et al., 1973, Imagawa et al., 1980), inclusion body encephalitis (Nesseler et al., 1997, and 1999), and post-vaccinal encephalitis (Hartley, 1974, Bestetti et al., 1978).

In contrast, distemper **leukoencephalitis** (DL) is a common finding and has been known for a long time. "Swelling of myelinated nerve fibres" was described by Cerletti in his work "about various forms of encephalitis and myelitis in dogs with canine distemper" as early as 1912. Scherer (1944), a German neuropathologist was among the first to compare DL white matter lesions to acute multiple sclerosis and he is the one who gave the impulse to use canine distemper as a model to study the pathogenesis of human demyelinating diseases (Koestner, 1975, DalCanto and Rabinowitz, 1982). Lesions in DL are mainly found in the cerebellum and less frequently in the cerebral white matter and the spinal cord (Baumgärtner et al., 1999). Demyelination is consistently observed  in fibre tracts in close proximity to the ventricles like the rostral medullary velum, cerebellar peduncles, and optic tracts (Summers and Appel, 1994). Studies tracking the route of CDV invasion and the mode of distribution of the virus in the brain showed ependymal infection and spread of the virus to the subependymal white matter indicating CNS infection along the cerebrospinal fluid (CSF) pathways (Vandevelde et al., 1985a). This finding matches the observation of Higgins et al. (1982a) who found a productive infection of the choroid plexus of the fourth ventricle and ependymal cells. In addition, hematogenous spread of CDV in the CNS is mediated by the association of the virus with lymphocytes (Rochborn, 1958, Summers et al., 1979) and plasma (Krakowka et al., 1989) or via infection of endothelial cells and subsequent spread to pericytes and astrocytic processes (Axtehelm, Baumgärtner, 1989). DL lesions are categorized in acute, subacute non-inflammatory, subacute inflammatory, chronic, and sclerotic (Krakowka et al., 1985, Alldinger et al., 1993, Gaedke et al., 1999). Acute lesions are characterized by focal vacuolation and mild astro- and microgliosis.

Subacute non-inflammatory plaques display primary demyelination, astrogliosis with formation of gemistocytes and multinucleated giant cells, and microgliosis occasionally accompanied by malacia and scavenging gitter cells. Axonal injury is expressed by spheroid formation. Subacute inflammatory and chronic lesions are infiltrated by mononuclear inflammatory cells accumulated in the parenchyma orin the perivascular space. In sclerotic plaques demyelinated areas are replaced by astrocytic scar tissue (Baumgärtner et al., 1989, Alldinger et al., 1993, Wünschmann et al., 1999, Gaedke et al., 1999).

Plaque genesis was intensely studied in naturally and experimentally CDV infected dogs (McCullough et al., 1974a and c, Raine, 1976, Summers et al., 1979, Higgins et al., 1982b, Summers and Appel, 1987). In experimentally infected dogs, primary changes such as vacuolation and spongious disaggregation of the subependymal and subpial white matter are light microscopically observed 24 days post infection (d.p.i.). The vacuols grow in number and size to form a demyelinated plaque. Transmission electron microscopically the vacuolations turn out to be an edema of the myelin sheath (Summers et al., 1987). According to Wisniewski et al. (1972) and Higgins et al. (1982a) the separation starts at the intraperiod line. The edema of the myelin sheath compresses axons and spheroids are formed. This change is named complex necrosis by Higgins et al. (1982b). The plaque periphery constitutes a growth zone, where myelin edema can be observed (Summers et al., 1987). Loss of compact myelin starts 27 d.p.i. in experimentally infected dogs. Astrocytes separate intact myelin segmentally from the axon (Higgins et al., 1982b). Summers, (1987) however, observed, that loss of compact myelin begins by spiral wrapping of cytoplasmic processes of macrophages followed by concentric stripping of myelin lamellae away from the axon. Myelin fragments are found as extracellular myelin droplets (Summers et al., 1987) or in the cytoplasm of macrophages/microglia (Wisniewski et al., 1972, Raine, 1976, Summers et al., 1987), astrocytes (Raine, 1976, Summers et al., 1987) or in single oligodendrocytes (Summers et al., 1987). Remyelination is feature of older demyelinating lesions in the presence of intact axons (Wisniewski et al., 1972), McCullough et al., 1974b, Higgins et al., 1982b, Summers et al., 1987).

## 2.    DEMYELINATING DISTEMPER ENCEPHALITIS

### 2.1    Cell tropism

The pathogenesis of CDV induced demyelination is still largely unknown. In experimental studies, CDV appears in the CNS 9 d.p.i. (Appel, 1969) and is found in astrocytes, neurons, microglia, ependymal cells, leptomeningeal cells, and choroid plexus epithelial cells (Appel, 1969, Vandevelde, 1980, Alldinger et al., 1993a; Wünschmann et al., 1999). Infection of **oligodendrocytes** has been the focus of numerous studies in the last years with contradictory results. Using light and transmission electron microscopy, a direct oligodendrocyte was observed by several investigators (McCullough et al. 1974b, Raine 1976, Blakemore et al. 1989). However, other studies did not detect oligodendrocyte infections in vivo (Vandevelde et al., 1983) and several in vitro investigations did not succeed in confirming an oligodendrocyte infection in primary brain cell cultures infected with different CDV strains (Zurbriggen and Vandevelde, 1983, Vandevelde et al., 1985b, Zurbriggen et al., 1986, 1987a and b). Employing molecular techniques, the question on the role of oligodendrocytes in CDV infection was resumed. Zurbriggen et al. (1993) demonstrated a restricted infection of oligodendrocytes. Furthermore, Graber et al. (1995) and Zurbriggen et al. (1998) found a down-regulation of the expression of a variety of oligodendroglial genes after restricted CDV infection. The same group investigated the participation of apoptosis or necrosis in cell death induced by CDV in vivo and in vitro and found, that neither of them played a significant role (Schobesberger et al., 1999). Recently, it has been shown, that oligodendrocytes are present in chronic lesions, and thus demyelination precedes oligodendrocyte loss in CDV induced encephalitis (Schobesberger et al., 2002).

**Astrocytes** are the main cell population infected by CDV. In non-inflammatory lesions in the cerebellar white matter of dogs with acute distemper, doublelabelling techniques unveiled, that 64% of astrocytes within a lesion were infected and this cell population provides 95% of all infected cells (Mutinelli et al., 1988). Infected astrocytes form gemistocytes, syncytia (Summers and Appel, 1985) and have a glassy cytoplasm (Vandevelde et al., 1983). They lose their cell processes and glial fibrillary acidic protein (GFAP) positive fibrils are newly arranged being concentrated around the nucleus (Zurbriggen et al., 1986). Among other CNS cell populations astrocytes display cytoplasmic and intranuclear inclusion bodies. The paradoxical intranuclear accumulation of the NP protein of an RNA virus seems to be a sign of cellular stress (Oglesbee and Krakowka, 1993).

## 2.2    Viral persistence

In spite of a strong virus-specific intrathecal immune response, early lesions without overt inflammation can be observed in the vicinity of chronic lesions with prominent invasion of inflammatory cells (Alldinger et al., 1993). Clearence of viral antigen is associated with the presence of inflammation and intrathecal virusspecific antibodies. However, these chronic lesions, devoid of viral antigen coexist with acute plaques containing a high viral load (Bollo et al., 1986). A detailed immunohistological investigation upon the distribution of CDV proteins and their epitopes in different lesion types showed, that in acute and subacute lesions without inflammation, expression of the M, H, and F proteins is only slightly diminished compared to the core proteins. However, plaques with severe inflammation are either devoid of viral antigen or exhibit NP- and P-specific immunoreactivity restricted to the periphery with simultaneous reduction or loss of surface proteins. The intracellular distribution of the proteins varies. Exclusively the NP and P proteins are found in the cell nucleus and the cell processes in addition to the cytoplasm (Alldinger et al.,1993). It can be assumed, that the nucleocapsid itself is infectious as shown for measles virus (Rozenblatt et al., 1979). The presence of the core proteins in cell processes is suspicious of a possible spread of the virus along this pathway in order to escape immune surveillance (Alldinger et al., 1993) as has been shown in cell culture studies (Zurbriggen et al., 1995J Virol ). The hypothesis of the intralesional persistence of viral mRNA in the absence of proteins was disproved by molecular studies (Müller et al., 1995, Gaedke et al., 1999). In the grey matter, however, more CDV RNA than protein is found suggestive of  impaired translation of CDV and a possible mode of viral persistence (Müller et al. 1995). Studies on distemper inclusion body polioencephalitis support the hypothesis of restricted grey matter infection as a mechanism of viral persistence (Nesseler et al., 1999). According to in vitro investigations, persistence of CDV in the brain appears to be due to non-cytolytic selective spread of the virus with very limited budding (Vandevelde and Zurbriggen, 1995; Zurbriggen et al., 1995). An analysis of the nucleotide- and deduced amino-acid sequences of the NP and M genes yielded results supporting the notion, that the NP and the M gene harbour determinants of viral persistence (Stettler et al., 1997).

## 2.3    Immunopathology

CDV-associated immunosuppression (Krakowka et al., 1975a) is an important manifestation of canine distemper. It is caused by the lytic effect of the virus on lymphocytes and macrophages during the caute phase of the disease and can be found months after viral clearence (Krakowka et al., 1975a and 1980). The recovery of the lymphatic tissue, coincides with the invasion of the CNS by inflammatory cells. Immunophenotyping of the cellular immune response in DL revealed, that only few inflammatory cells are found in acute and subacute non-inflammatory  lesions. In this phase, CD8+ cells dominate and might be involved in viral clearence or contribute as antibody-independent cytotoxic T cells to early lesion development (Wünschmann et al., 1999). This hypothesis is supported by the finding, that occurrence of T cells in early demyelinating lesions  correlates with sites of viral replication and coincides with the demonstration of an early immune response against the nucleocapsid protein of CDV. T cell migration to the CNS is suspected to be triggered by activated microglia secreting chemokines. Hence, a marked IL-8 activity is present in the CSF of dogs with acute lesions. Resting T cells of the early stages of the disease might facilitate the later development of the intrathecal immune response and associated immunopathological complications (Tipold et al., 1999). Subacute inflammatory and chronic lesions are massively infiltrated by inflammatory cells. CD8+ cells may function as cytotoxic effectors. CD4+ cells are suspected of initiating a delayed-type hypersensitivity reaction in these older lesions (Wünschmann et al., 1999). Immunpathological mechanisms involved in CD plaque pathogenesis are emphasized by up-regulation of the major histocompatibility complex class II (MHCII) in DL brains. Following CDV infection, there is a mild up-regulation of this molecule in acute increasing to severe in chronic lesions. MHCII is present on microglia, endothelial, meningeal, choroid plexus epithelial, ependymal, and intravascular cells. Virtually all cells forming perivascular infiltrates are MHCII-positive. Although MHCII is up-regulated throughout the white matter an accentuated expression is found in CDV-antigen positive acute and subacute plaques. In chronic plaques MHCII expression is especially prominent in areas with reduced or absent viral antigen expression indicating that non-viral antigens may play an important role as triggering molecules in the chronic phase of demyelination (Alldinger et al., 1996).

The term "bystander demyelination" was used to describe the myelin damaging effect of proteolytic enzymes released by stimulated macrophages (Cammer et al. 1978) to explain myelin destruction in the absence of obvious oligodendrocyte infection. Other macrophage mediators, including reactive oxygen species (ROS), seemed to be involved in DL demyelination (Bürge et al., 1989; Griot et al., 1989a and b and 1990).

The humoral immune response, i.e. the intracerebral presence of B cells and their importance for intrathecal antibody production during chronic demyelinating distemper leukoencephalomyelitis is well documented (Vandevelde et al., 1982, 1986, Alldinger et al., 1996). B cells especially producing immunoglobulins of the IgG subtype are found, at variance within chronic lesions and in perivascular infiltrates (Vandevelde et al., 1982, Alldinger et al., 1996). Besides a prominent CDV-specific humoral immune response in serum and cerebrospinal fluid (Vandevelde et al., 1982, 1986, Rima et al., 1991) anti-myelin specific antibodies are found (Krakowka et al., 1973, Vandevelde et al., 1986). Consequently, a mechanism of demyelination based on complement-dependent antibody mediated humoral cytotoxicity is discussed (Vandevelde et al., 1982). However, the pathological significance of anti-myelin antibodies remained doubtful as high titres of these antibodies are also detected in dogs with resolving lesions (Vandevelde et al., 1986). Furthermore, antibody-dependent T-cell mediated cytotoxicity was suspected as many intralesional and perivascular cells are T lymphocytes and there is a simultaneous occurrence of perivascular B and CD4+ cells (Vandevelde et al., 1982, Alldinger et al., 1996, Wünschmann et al., 1999). The simultaneous decline of intrathecal antibody production and clinical improvement were suggestive of a deleterious role of the humoral immune response (Vandevelde et al., 1982). Cell to cell communication and interaction between immune cells and resident or activated brain cells were investigated by studying cytokine expressions in DL. Accordingly, demyelination is frequently associated with interleukin (IL)-1, -12, tumour necrosis factor (TNF) and transforming growth factor (TGF) mRNA expression in the blood of dogs with natural CD. IL-6 transcripts are only found in animals with early CNS lesions. Simultaneous occurrence of pro- and anti-inflammatory cytokines in whole blood preparation from most of the dogs with CD, favored the hypothesis of a complex most likely disease stage dependent orchestrated cytokine expression (Gröne et al., 1998). In the CSF of dogs naturally infected with CDV, TNF and IL-6, associated with disease exacerbation and IL-10 and TGF, indicative of remission, are often observed simultaneously and can not be assigned to specific disease stage. IL-10 remains the most frequently detected cytokine indicating that the stage of inactivity prevails (Frisk et al., 1999). An immunohistochemical in situ study revealed, that IL-1 is expressed to varying degrees in all types of lesions and is most often present in CD3 (T lymphocytes) and BS-I (microglia/brain macrophages) positive cells in early plaques and in cells comprising perivascular cuffs found in chronic lesions. IL-6 expression is present in all lesions and follows a similar distribution pattern as IL-1. IL-12 is observed only in individual perivascular cells. TNF-α is mainly associated with astrocytes and might be involved in the pathogenesis of early demyelination in DL (Gröne et al., 2000). According to an in situ semi-quantitative RT-PCR study, there is a prominent up-regulation of the pro-inflammatory cytokines IL-6, IL-8, IL-12, and TNF-α in early distemper CNS lesions. Other pro-inflammatory cytokines such as IL-1, IL-2, and interferon (IFN)-γ are not detected. Conversely, IL-10 and TGF-ß, though detectable in most distemper cases, experienced no significant up-regulation following CDV infection and lesion development. The lack of increased expression of anti-inflammatory cytokines might represent an early derailment of the immune response in demyelinating distemper leukoencephalitis (Markus et al., 2002).

## 2.4    Extracellular matrix

In view of the importance of astrocytes as major source of extracellular matrix (ECM) proteins (Montgomery, 1994) and their significance for maintaining structure and relationships in the brain, the consequences of a CDV infection of this cell population seemed to be an important question to be addressed. Asher et al. (1991) demonstrated, that a hyaluronate-based ECM must exist in the canine CNS as glial hyaluronate binding protein (GHAP) is demonstrated immunoelectron microscopically in the space between myelin sheaths and astrocytic processes. The importance of the ECM is stressed by the observation, that In acute and subacute demyelinating encephalitis, a plaque-associated CD44-up-regulation is found that parallels astrocyte activation. CD44 plays a dual role for the progression of the demyelination process. Ligation of this receptor might induce chemokines and cytokines and hence initiate and perpetuate the inflammatory process. In chronic lesions, CD44 expression declines together with the number of GFAP-positive astrocytes. In addition, in this plaque type, CD44 is expressed on the cell membrane of perivascular mononuclear cells, possibly marking memory cells (Alldinger et al., 2000). Matrix-metalloproteinases (MMPs) , zinc dependent enzymes, that cleave molecules of the extracellular matrix and their inhibitors (TIMPs) are most prominently up-regulated in acute and subacute non-inflammatory lesions and are expressed mostly by astrocytes and microglia. In this phase, they might contribute to lesion initiation. In older lesions, MMP and TIMP expressions decline apart from MMP-11, -12, and –13 favoring the hypothesis of the

TIMP-MMP imbalance as the motor for lesion progression (Miao et al., 2002).


# 3.     CONCLUSIONS

Demyelinating distemper leukoencephalitis represents a biphasic disease. Lesion initiation is most likely directly virus-induced and is supported by the actions of CD8+ cells, inflammatory cytokines including TNF-$\alpha$ as well as specific MMPs. The second or immunopathological phase of demyelination is associated with a massive infiltration of inflammatory cells, antigen presentation, a preponderance of pro-inflammatory cytokines and a TIMP-MMP imbalance in the absence of  virus. Obviously, CDV does not lead to myelin destruction on a direct pathway, but interferes with sensitive control mechanisms between brain cells, immune cells myelin, and the extracellular matrix.


# ACKNOWLEDGEMENTS

The authors' projects were supported by grants of the Deutsche Forschungsgemeinschaft and the Hertie-Stiftung.
The authors want to thank Annette Artelt for skilful technical assistence and Ute Zeller for photographic support.


# REFERENCES

1.    **Alldinger S, Fonfara S, Kremmer E, Baumgärtner W**. Up-regulation of the hyaluronate receptor CD44 in canine distemper demyelinated plaques. Acta Neuropathol 2000;99:138-146. [PMID: 10672320]
2.    **Alldinger S, Baumgärtner W, Örvell C**. Restricted expression of viral surface proteins in canine distemper encephalitis. Acta Neuropathol. 1993;85:635-645. [PMID: 7687812]
3.    **Alldinger S, Wünschmann A, Baumgärtner W, Voß C, Kremmer E**. Up-regulation of major histocompatibility complex class II antigen expression in the central nervous system of dogs with spontaneous canine distemper virus encephalitis. Acta Neuropathol. 1996; 92: 273-280. [PMID: 8870829]
4.    **Appel MJG**. Pathogenesis of canine distemper. Am J Vet Res. 1969;30:1167-1182

5. **Asher R, Perides G, Vanderhaeghen J-J, Bignami A.** Extracellular matrix of central nervous system white matter: demonstration of an hyaluronate-protein complex. J Neurosci Res. 1991;28:410-421. [PMID: 1713274]

6. **Axthelm MK, Krakowka S.** Canine distemper virus: the early blood-brain barrier lesion. Acta Neuropathol (Berl). 1987;75:27-33. [PMID: 3434212]

8. **Baumgärtner W, Örvell C, Reinacher M.** Naturally occurring canine distemper virus encephalitis: distribution and expression of viral polypeptides in nervous tissues. Acta Neuropathol. 1989;78:504-512. [PMID: 2683562]

9. **Bestetti G, Fatzer R, Fankhauser R.** Encephalitis following vaccination against distemper and infectious hepatitis in the dog. Acta Neuropathol (Berl). 1978;43:69-75. [PMID: 676689]

10. **Blakemore WF, Summers BA, Appel MJG.** Evidence of oligodendrocyte infection and degeneration in canine distemper encephalomyelitis. Acta Neuropathol. 1989;77:550-553. [PMID: 2718748]

11. **Bollo E, Zurbriggen A, Vandevelde M, Fankhauser R.** Canine distemper virus clearence in chronic inflammatory demyelination. Acta Neuropathol (Berl). 1986;72:69-73. [PMID: 3825508]

12. **Bürge T, Griot C, Vandevelde M, Peterhans E.** Antiviral antibodies stimulate production of reactive oxygen species in cultured canine brain cells infected with canine distemper virus. J Virol. 1989;63:2790-2797. [PMID: 2724413]

13. **Cammer W, Bloom BR, Norton WT, Gordon S.** Degradation of basic protein in myelin by neutral proteases secreted by stimulated macrophages: a possible mechanism of inflammatory demyelination. Proc Natl Acad Sci USA. 1978;75:1554-1558. [PMID: 148651]

14. **Cerletti U.** Über verschiedene Encephalitis- und Myelitisformen bei an Staupe erkrankten Hunden. Zur Kenntnis der sogenannten progressiven Paralyse des Hundes. Z f d ges Neurol und Psychiatr Orig. 1912;9:521-563.

16. **Cook SD, Dowling PC, Russell WC.** Multiple sclerosis and canine distemper. Lancet. 1978;1:605-606. [PMID: 76140]

17. **Dal Canto MC, Rabinowitz SG.** Experimental models of virus-induced demyelination of the central nervous system. Ann Neurol. 1982;11:109-127. [PMID: 6280582]

18. **Deem SL, Spelman LH, Yates RA, Montali RJ.** Canine distemper in terrestrial carnivores: a review. J ZooWildl Med. 2000;31:441-451. [PMID: 11428391]

19. **Diallo A.** Morbillivirus group: genome organisation and proteins. Vet Microbiol. 1990;23:155-163. [PMID: 2205968]

20. **Fankhauser R.** Hundestaupe-Geschichte einer Krankheit. Schweiz Arch Tierheilk. 1982;124:245-256. [PMID: 7048524]

21. **Frisk AL, Baumgärtner W, Gröne A.** Dominating interleukin-10 mRNA expression induction in cerebrospinal fluid cells of dogs with natural canine distemper virus induced demyelinating and non-demyelinating CNS lesions. J Neuroimmunol . 1999;97: 102-109. [PMID: 10408963]

22. **Gaedke K, Zurbriggen A, Baumgärtner W.** Lack of correlation between virus nucleoprotein and mRNA expression and the inflammatory response in demyelinating distemper encephalitis indicates a biphasic disease process. Eur J Vet Pathol. 1999;5:9-20.

23. **Glaus T, Griot C, Richard A, Althaus U, Herrschkowitz N, Vandevelde M** . Ultrastructural and biochemical findings in brain cell cultures infected with canine distemper virus. Acta Neuropathol. 1990;80:59-67. [PMID: 2360417]

24. **Gordon MT, Anderson DC, Sharpe PT**. Canine distemper virus localised in bone cells of patients with Paget`s disease. Bone. 1991;12:195-201. [PMID: 1910961]

25. **Gordon MT, Mee AP, Anderson DC, Sharpe PT**. Canine distemper virus transcripts sequenced from pagetic bone. Bone Miner. 1992;19:159-174. [PMID: 1345324]

26. **Graber HU, Müller CF, Vandevelde M, Zurbriggen A**. Restricted infection with canine distemper virus leads to down-regulation of myelin gene transcription in cultured oligodendrocytes. Acta Neuropathol. 1995;90:312-318. [PMID: 8525806]

27. **Griot C, Vandevelde M, Peterhans E, Stocker R.** Selective degeneration of oligodendrocytes mediated by reactive oxygen species. Free Rad Res Comm. 1990;14:181-193. [PMID: 1965721]

28. **Griot C, Bürge T, Vandevelde M, Peterhans E.** Antibody-induced generation of reactive oxygen radicals by brain macrophages in canine distemper encephalitis: a mechanism for bystander demyelination. Acta Neuropathol . 1989a;78:396-403. [PMID: 2782050]

30. **Gröne A, Alldinger S, Baumgärtner W.** Interleukin-1β, -6, -12 and tumor necrosis factor-α expression in brains of dogs with canine distemper virus infection. J Neuroimmunol. 2000; 110:20-30. [PMID: 11024531]

31. **Gröne A, Frisk AL, Baumgärtner W.** Cytokine mRNA expression in whole blood samples from dogs with natural canine distemper infection. Vet Immunol Immunopathol. 1998;65:11-27. [PMID: 9802573]

32. **Hall WW, Lamb RA, Choppin PW**. Polypeptides of canine distemper virus: synthesis in infected cells and relatedness to the polypeptides of other morbilliviruses. Virol. 1980;100:433-339. [PMID: 7352374]

33. **Hartley WJ**. A post-vaccinal inclusion body encephalitis in dogs. Vet Pathol. 1974;11:301-312. [PMID: 4156617]

34. **Helfrich MH, Hobson RP, Grabowski PS, Zurbriggen A, Cosby SL, Dickson GR, Fraser WD, Ooi CG, Selby PL, Crisp AJ, Wallace RG, Kahn S, Ralston SH**. A negative search for a paramyxoviral etiology of Paget's disease of bone: molecular, immunological, and ultrastructural studies in UK patients. J Bone Miner Res. 2000:15:2315-2329. [PMID: 11127197]

35. **Hernan MA, Zhang SM, Lipworth L, Olek MJ, Ascherio A**. Multiple sclerosis and age at infection with common viruses. Epidemiology. 2001;12:301-306. [PMID: 11337603]

36. **Higgins RJ, Child G, Vandevelde M**. Chronic relapsing demyelinating encephalomyelitis associated with persistent spontaneous canine distemper virus infection. Acta Neuropathol. 1989;77:441-444. [PMID: 2711831]

37.  **Higgins RJ, Krakowka SG, Metzler AE, Koestner A**. Experimental canine distemper encephalomyelitis in neonatal gnotobiotic dogs. A sequential ultrastructural study. Acta Neuropathol (Berl). 182a;57:287-295. [PMID: 7136507]

38.  **Higgins RJ, Krakowka SG, Metzler AE, Koestner A**. Primary demyelination in experimental canine distemper virus induced encephalomyelitis in gnotobiotic dogs. Sequential immunologic and morphologic findings. Acta Neuropathol (Berl). 1982b;58:1-8. [PMID: 7136512]

39.  **Hodge MJ, Wolfson C**. Canine distemper virus and multiple sclerosis. Neurol . 1997;49:62-69. [PMID: 9270694]

40.  **Imagawa DT, Howard EB, van Pelt LF, Ryan CP, Bui HD, Shapshak P**. Isolation of canine distemper virus from dogs with chronic neurological diseases. Proc Soc Exp Biol Med. 1980;164:355-362. [PMID: 7403107]

42.  **Koestner A**. Animal models of human diseases: distemper associated demyelinating encephalomyelitis. Am J Pathol . 1975;78:361-364. [PMID: 1090185]

43.  **Krakowka S, McCullough B, Koestner A, Olsen R**. Myelin-specific autoantibodies associated with central nervous system demyelination in canine distemper virus infection. Infect Immun. 1973;8:819-827. [PMID: 4584054]

44.  **Krakowka S**. Canine distemper virus infectivity of various blood fractions for central nervous system vasculature. J Neuroimmunol. 1989;21:75-80. [PMID: 2908882]

45.  **Krakowka S, Axthelm MK, Johnson GC**. Canine distemper virus. In: Olsen RG, Krakowka S, Blakeslee JR, eds. Comparative pathology of viral diseases. Vol. 2. CRC Press Inc.; 1985:137-164.

46.  **Krakowka S, Koestner A**. Age-related susceptibility to infection with canine distemper virus in gnotobiotic dogs. J Infect Dis. 1976;134:629-632. [PMID : 1003016]

47.  **Krakowka S, Olson R, Confer A. Koestner A, McCullough B**. Serological response to canine distemper viral antigens in gnotobiotic dogs infected with canine distemper virus. J Infect Dis. 1975b;132: 384-392. [PMID: 1185008]

48.  **Krakowka S, Cockerell G, Koestner A**. Effects of canine distemper virus infection on lymphoid functions in vitro and in vivo. Infect Immun. 1975a;11:1069-1078. [PMID: 1091560]

49.  **Krakowka S, Higgins RJ, Koestner A**. Canine distemper virus: review of structural and functional modulation in lymphoid tissue. Am J Vet Res. 1980;41:284-292. [PMID: 6989302]

50.  **Lincoln SD, Gorham JR, Davis WC, Ott RL**. Studies of old dog encephalitis. II. Electron microscopic and immunohistologic findings. Vet Pathol. 1973;10:124-129. [PMID: 4588526]

51.  **Lincoln SD, Gorham JR, Ott RL, Hegreberg GA**. Etiologic studies of old dog encephalitis. I. Demonstration of canine distemper viral antigen in the brain of two cases. Vet Pathol. 1971;8:1-8. [PMID: 4945659]

52.  **Markus S, Failing K, Baumgärtner W**. Increased expression of pro-inflammatory cytokines and lack of up-regulation of anti-inflammatory cytokines in early distemper CNS lesions. J Neuroimmunol. 2002;125:30-41. [PMID: 11960638]

53.  **McCullough B, Krakowka S, Koestner A, Shadduck J**. Demyelinating activity of canine distemper virus isolates in gnotobiotic dogs. J Infect Dis. 1974c;130:343-350. [PMID : 4443615]

54.  **McCullough B, Krakowka S, Koestner A**. Experimental canine distemper virus-induced lymphoid depletion. Am J Pathol. 1974a;74:155-170. [PMID: 4809312]

55.  **McCullough B, Krakowka S, Koestner A**. Experimental canine distemper virus-induced demyelination. Lab Invest. 1974b;31:216-222. [PMID: 4413368]

56.  **Mee AP**. Paramyxoviruses and Paget's disease: the affirmative view. Bone. 1999;24: 19-21. [PMID: 10321921]

57.  **Mee AP, Dixon JA, Hoyland JA, Davies M, Selby PM, Mawer EB**. Detection of canine distemper virus in 100% of Paget's disease sample by in situ-reverse transcriptase-polymerase chain reaction. Bone. 1998;23: 171-175. [PMID: 9701477]

59.  **Miao Q, Baumgärtner W, Failing K, Alldinger S**. Up-regulation of matrix-metalloproteinases and their inhibitors in demyelinating canine distemper encephalitis. Acta Neuropathol. 2002;104:560.

60.  **Mitscherlich E**. Über Züchtungsversuche des Virus der Hundestaupe. Dtsch Tierärztl Wschr. 1938;46: 497-502.

61.  **Montgomery DL**. Astrocytes: form, functions, and roles in disease. Vet Pathol. 1994;31:145-167. [PMID: 8203078]

62.  **Müller CF, Fatzer RS, Beck K, Vandevelde M. Zurbriggen A**. Studies on canine distemper virus persistence in the central nervous system. Acta Neuropathol. 1995;89: 438-445. [PMID: 7618441]

63.  **Mutinelli F, Vandevelde M, Griot C, Richard A**. Astrocytic infection in canine distemper virus-induced demyelination. Acta Neuropathol. 1988;77:333-335. [PMID: 2922996]

64.  **Nesseler A, Baumgärtner W, Zurbriggen A, Örvell C**. Restricted virus protein translation in canine distemper virus inclusion body polioencephalitis. Vet Microbiol. 1999;69:23-28. [PMID: 10515265]

66.  **Oglesbee M, Krakowka S**. Cellular stress response induces selective intranuclear trafficking and accumulation of morbillivirus major core protein. Lab Invest. 1993;68:109-117. [PMID: 8423670]

67.  **Ooi CG, Walsh CA, Gallagher JA, Fraser WD**. Absence of measles virus and canine distemper virus transcripts in long-term bone marrow cultures from patients with Paget's disease of bone. Bone. 2000;27:417-421. [PMID: 10962354]

68.  **Örvell C**. Structural poypeptides of canine distemper virus. Arch Virol. 1980;66: 193-206. [PMID: 6893798]

69.  **Pearce-Kelling, S., Mitchell, W.J., Summers, B.A., Appel, M.J.G**. (1990). Growth of canine distemper virus in cultured astrocytes: relationship to in vivo persistence and disease. Microb Pathogen 8: 71-82. [PMID: 2333034]

70. **Pearce-Kelling S, Mitchell WJ, Summers BA, Appel, MJG**. Virulent and attenuated canine distemper virus infects multiple dog brain cell types in vitro. Glia. 1991;4:408-416. [PMID: 1834561]

72. **Pringle CR**. Virus taxonomy at the XIth International Congress of Virology, Sydney, Australia. Arch Virol. 1999;144: 2065-2070. [PMID: 10550679]

73. **Raine CS**. On the development of CNS lesions in natural canine distemper encephalomyelitis. J Neurol Sci. 1976;30:13-28. [PMID: 978220]

75. **Rima BK, Duffy N, Mitchell WJ, Summers BA, Appel MJG**. Correlation between humoral immune responses and presence of virus in the CNS in dogs experimentally infected with canine distemper virus. Arch Virol. 1991;121:1-8. [PMID: 1759903]

77. **Rozenblatt S, Koch T, Pinhasi O, Bratosin S**. Infective substructures of measles virus from acutely and persistently infected cells. J Virol. 1979;32:329-333. [PMID: 120450]

78. **Scherer HJ**. Die akute Multiple Sklerose des Hundes. In: Vergleichende Pathologie des Nervensystems der Säugetiere. Georg Thieme Verlag Leipzig,;1944:282-300.

79. **Schmid E, Zurbriggen A, Gassen U, Rima B, ter Meulen V, Schneider-Schaulies J**. Antibodies to CD9, a tetraspan transmembrane protein, inhibits canine distemper virus-induced cell-cell fusion but not virus-cell fusion. J Virol. 2000;74:7554-7561. [PMID: 10906209]

80. **Schobesberger M, Zurbriggen A, Doherr MG, Weissenböck H, Vandevelde M, Lassmann H, Griot C.** Demyelination precedes oligodendrocyte loss in canine distemper virus-induced encephalitis. Acta Neuropathol (Berl). 2002;103:11-19. [PMID: 11837742]

81. **Schobesberger M, Zurbriggen A, Summerfield A, Vandevelde M, Griot C**. Oligodendroglial degeneration in distemper: apoptosis or necrosis? Acta Neuropathol . 1999;97:279-287. [PMID: 10090676]

82. **Stettler M, Beck K, Wagner A, Vandevelde M, Zurbriggen A.** Determinants of persistence in canine distemper viruses. Vet Microbiol. 1997;57:83-93. [PMID: 9231983]

83. **Summers BA, Appel MJG**. Demyelination in canine distemper encephalomyelitis: an ultrastructural analysis. J Neurocytol. 1987;16:871-881. [PMID: 3450794]

84. **Summers BA, Appel MJG**. Syncytia formation: an aid in the diagnosis of canine distemper encephalomyelitis. J Comp Pathol. 1985;95:425-435. [PMID: 4031136]

85. **Summers BA, Greisen HA, Appel MJG**. Early events in canine distemper demyelinating encephalomyelitis. Acta Neuropathol (Berl). 1979;46:1-10. [PMID: 452851]

86. **Summers BA, Appel MJG** (1994). Aspects of canine distemper virus and measles virus encephalomyelitis. Neuropathol Appl Neurobiol. 1994;20:525-534. [PMID: 7898614]

87. **Summers BA, Greisen HA, Appel MJG**. Canine distemper encephalomyelitis: variation with virus strain. J Comp Pathol. 1984;94:65-75. [PMID: 6699231]

88.  **Tatsuo H, Ono N, Yanagi Y**. Morbilliviruses use signaling lymphocyte activation molecules (CD150) as cellular receptors. J Virol. 2001;75:5842-5850. [PMID: 11390585]

89.  **Tatsuo H, Yanagi Y**. The morbillivirus receptor SLAM (CD150). Microbiol Immunol. 2002;46:135-142. [PMID: 12008921]

90.  **Tipold A, Moore P, Zurbriggen A, Bürgener I, Barben G, Vandevelde M**. Early T cell response in the central nervous system in canine distemper virus infection. Acta Neuropathol (Berl). 1999;97:45-56. [PMID: 9930894]

91.  **Vandevelde M, Bichsel P, Cerruti-Sola S, Steck A, Kristensen F, Higgins RJ**. Glial proteins in canine distemper virus-induced demyelination. Acta Neuropathol (Berl). 1983;59:269-276. [PMID: 6191513]

92.  **Vandevelde M, Kristensen B, Braund KG, Greene CE, Swango LJ, Hoerlein BF**. Chronic canine distemper virus encephalitis in mature dogs. Vet Pathol. 1980;17:17-29. [PMID: 7352359]

94.  **Vandevelde M, Zurbriggen A, Higgins RJ, Palmer D**. Spread and distribution of viral antigen in nervous canine distemper. Acta Neuropathol. 1985a;67:211-218. [PMID: 4050335]

95.  **Vandevelde M, Zurbriggen A**. The neurobiology of canine distemper virus infection. Vet Microbiol. 1995;44:271-280. [PMID: 8588322]

96.  **Vandevelde M, Zurbriggen A, Dumas M, Palmer D**. Canine distemper virus does not infect oligodenrocytes in vitro. J Neurol Sci. 1985b;69: 133-137. [PMID: 3897461]

97.  **Vandevelde M, Kristensen F, Kristensen B, Steck AJ, Kihm U**. Immunological and pathological findings in demyelinating encephalitis associated with canine distemper virus infection. Acta Neuropathol (Berl). 1982;56:1-8. [PMID: 6175160]

98.  **Vandevelde M, Zurbriggen A, Steck A, Bichsel P**. Studies on the intrathecal humoral immune response in canine distemper encephalitis. J Neuroimmunol. 1986;11:41-51. [PMID: 3944250]

100. **von Messling V, Zimmer G, Herrler G, Haas L, Cattaneo R**. The hemagglutinin of canine distemper virus determines tropism and cytopathogenicity. J Virol. 2001;75:6418-6427. [PMID: 11413309]

101. **Wang LF, Michalski WP, Yu M, Pritchard LI, Crameri G, Shiell B, Eaton BT**. A novel P/V/C gene in a new member of the paramyxoviridae family, which causes lethal infection in humans, horses, and other animals. J Virol. 1998;72:1482-1490. [PMID: 9445051]

102. **Wisniewski H, Raine CS, Kay WJ**. Observations on viral demyelinating encephalomyelitis. Canine distemper. Lab Invest. 1972;26:589-599. [PMID: 5021319]

103. **Wünschmann A, Alldinger S, Kremmer E, Baumgärtner W**. Identification of CD4+ and CD8+ T cell subsets and B cells in the brain of dogs with spontaneous acute, subacute-, and chronic-demyelinating distemper encephalitis. Vet Immunol Immunopathol. 1999;67:101-116. [PMID: 10077417]

104. **Lamb RA, Kolakofsky D**. Paramyxoviridae: The viruses and their replication. In: Knipe DM, Howley PM, eds. Virology. 4th ed. Philadelphia: Lippincott Williams&Wilkins; 2001;1305-1443.

105. **Zurbriggen A, Graber HU, Vandevelde M**. Selective spread and reduced virus release leads to canine distemper virus persistence in the nervous system. Vet Microbiol. 1995;44:281-288. [PMID: 8588323]

106. **Zurbriggen A, Graber HU, Wagner A, Vandevelde M**. Canine distemper virus persistence in the nervous system is associated with noncytolytic selective virus spread. J Virol. 1995;69:1678-1686. [PMID: 7853504]

107. **Zurbriggen A, Vandevelde M**. Canine distemper virus-induced glial cell changes in vitro. Acta Neuropathol (Berl). 1983;62:51-58. [PMID: 6659878]

108. **Zurbriggen A, Müller C, Graber HU, Vandevelde M**. Canine distemper virus and myelin mRNAs in oligodendrocytes. Schweiz Arch Neurol Psychiatr. 1994;145: 17. [PMID: 7533928]

109. **Zurbriggen A, Schmid I., Graber HU, Vandevelde M**. Oligodendroglial pathology in canine distemper. Acta Neuropathol. 1998;95:71-77. [PMID: 9452824]

110. **Zurbriggen A, Vandevelde M, Bollo E**. Demyelinating, non-demyelinating and attenuated canine distemper virus strains induce oligodendroglial cytolysis in vitro. J Neurol Sci. 1987a;79: 33-41. [PMID: 2440997]

111. **Zurbriggen A, Vandevelde M, Dumas M**. Secondary degeneration of oligo-dendrocytes in canine distemper virus infection in vitro. Lab Invest. 1986;54:424-431. [PMID: 2421102]

112. **Zurbriggen A, Vandevelde M, Dumas M, Griot C, Bollo E**. Oligodendroglial pathology in canine distemper virus infection in vitro. Acta Neuropathol (Berl). 1987b;74: 366-373. [PMID: 3687388]

113. **Zurbriggen A, Yamawaki M, Vandevelde M**. Restricted canine distemper virus infection of oligodendrocytes. Lab Invest. 1993;68:277-284. [PMID: 8450647]

114. **Haas L, Liermann H, Harder TC, Barrett T, Löchelt M, von Messling V, Baumgärtner W, Greiser-Wilke I**. Analysis of the H gene, the centzral untranslated region and the proximal coding part of the F gene of wild-type and vaccine canine distemper viruses. Vet Microbiol. 1999;69:15-18. [PMID: 10515263]

# INDEX